Fundamentals of Structured COBOL Programming

Fundamentals of Structured COBOL Programming

Fourth Edition

Carl Feingold
West Los Angeles College

wcb

Wm. C. Brown Co. Publishers
Dubuque, Iowa

Book Team

Robert B. Stern Senior Editor
Nicholas Murray Associate Developmental Editor
Anthony L. Saizon Designer
David A. Welsh Production Editor
Mavis M. Oeth Permissions Editor

---

wcb group

Wm. C. Brown Chairman of the Board
Mark C. Falb Executive Vice President

wcb

Wm. C. Brown Company Publishers, College Division

Lawrence E. Cremer President
David Wm. Smith Vice President, Marketing
E. F. Jogerst Vice President, Cost Analyst
David A. Corona Assistant Vice President,
Production Development and Design
James L. Romig Executive Editor
Marcia H. Stout Marketing Manager
Janis M. Machala Director of Marketing Research
Marilyn A. Phelps Manager of Design
William A. Moss Production Editorial Manager
Mary M. Heller Visual Research Manager

Cover photo by Herbert Jackson

Library of Congress Catalog Card Number: 82-071272

ISBN 0-697-08173-7

2-08173-01

Printed in the United States of America

Brief Contents

Contents

3

Identification and Environment Divisions 58

4

Data Division 70

5

Structured Programming 130

6

Procedure Division: Input/Output Operations 181

7

Procedure Division: Arithmetic Operations 234

8

Procedure Division: Sequence Control Statements 275

9

Procedure Division: Data Manipulation Statements 335

10

Table Handling 375

11

Sort-Merge Feature 479

15
Declaratives and Linkage Sections 689

16
COBOL Programming Techniques 727

Preface

COBOL has emerged as one of the leading programming languages in use today. Support from its users, computer manufacturers, and governmental agencies assure that the programming language will meet the needs of the user, not only today, but in the future. The wide and popular acceptance of COBOL as a standard business programming language has made it essential for institutions of learning to offer COBOL programming courses.

The relative ease of COBOL programming with its self-documenting features makes it an ideal introductory programming language for students, whether they be programmers, business students, accountants, or others interested in learning computer programming.

The text was written in accordance with the American National Standard (ANSI) Programming Language 1974. These are the guidelines, which are universally accepted in the industry for COBOL programming. These standards are responsible for specifying both the form and interpretation of programs expressed in COBOL. The purpose of these guidelines is to promote a high degree of machine independence of such programs in order to permit their use in a variety of data processing systems.

The purpose of this text is to provide the general reader with a broad and comprehensive coverage of all the latest features of American National Standard (ANSI) COBOL, which has been universally adopted by all computer manufacturers. The latest concepts of structured programming are used throughout the text. Prior programming knowledge is not required because the introductory chapters provide sufficient background material for the beginning student and serve as a review for continuing students and programmers.

The text, designed so that the user can write programs early in the course, provides an overall view of the functions and uses of all divisions of COBOL. Numerous illustrations and thorough explanations accompany each segment of the COBOL language, thereby making the text useful as both a learning tool and a reference manual. Most of the illustrations are complete within themselves, freeing the user from reading through pages of narrative for an explanation of the illustration. Each segment of the COBOL language is completely explained and illustrated so that the user does not have to "wade" through numerous examples to find the basic formats, uses, functions, or rules for applying each of the COBOL elements.

Being extremely comprehensive, the text provides the user with the introductory concepts of programming through the basic components of COBOL programming and advanced COBOL programming concepts. It is a step-by-step problem-oriented approach to COBOL programming covering both basic and advanced COBOL topics. The text contains a wealth of illustrated examples that help the reader progress from problem definition to solution.

New to the Fourth Edition

This fourth edition represents a major revision; some chapters have been combined while others have been subdivided and new chapters have been added. All the salient features of the previous edition have been retained.

All chapters have been carefully and extensively revised with expanded explanations, numerous examples and illustrations, and programming problems. Every COBOL entry is described and illustrated in great detail. The text is meant to be all-inclusive, with little or no need for reference to computer manufacturer's manuals.

Some of the major additions to the fourth edition are:

1. The extensive use of many new HIPO (Hierarchy plus Input, Processing and Output) charts, the latest innovation in structured COBOL programming.
2. Illustrative programs for study have been added following the summaries for most chapters, including HIPO charts and flowcharts. Documentation is complete for the first programs (following chapters 4 and 5). Starting with chapter 6, only selected IPO charts and flowcharts are shown. By this point, students should understand what would be done in the charts not shown. Note that no IPO charts are shown for the illustrative programs accompanying chapter 12, or for the first two in chapter 13, since the nature of the programs is clear enough without them.
3. The readability has been further simplified so that students with little knowledge of data processing can understand the many facets of COBOL programming.

4. Further COBOL coverage including character string manipulation, file processing, subroutines, etc., add to the features that were included in the previous edition. This will enable the instructors to select from the many features contained in the text to suit their individual course needs.
5. ANSI 1974 COBOL standards are used throughout the text. This will provide the user with the latest universally accepted standards in COBOL programming.
6. Additional appendices provide the user with invaluable reference material that will greatly aid in the debugging process.
7. The chapter objectives and summaries should greatly simplify the comprehension of the material covered in each chapter.

Some of the specific revisions from the previous edition are:

1. The Basic Computer Components chapter of the previous edition has been eliminated. This will allow the user to advance to the programming phase more quickly.
2. The large Procedure Division chapter of the previous edition has been subdivided into four more manageable chapters. This should aid in the comprehension of the important Procedure Division material.
3. The Sort Feature chapter now precedes the Report Writer Feature chapter. This is a more logical approach as the data is usually sorted prior to entering the Report Writer.
4. The COBOL DIFFERENCES-ANS 1968 COBOL and ANS 1974 COBOL has been retained as there are still many installations using ANS 1968 COBOL and many that are in the process of changing or using both compilers, so it is important for the user to be familiar with the differences between the two.

Here is a brief summary of the contents of the chapters:

Chapter 1 prepares the reader for the programming phase including the tools necessary to represent logical solutions to data processing problems in the form of flowcharts and/or HIPO charts.

Chapter 2 describes the format for writing COBOL source programs and the precise rules for the use of COBOL statements so that the number of diagnostic errors can be kept at a minimum. The reader is now ready to write a complete simple COBOL program for listing data.

Chapter 3 describes the functions and formats of the Identification Division and the Environment Division.

Chapter 4 describes the characteristics of the information to be processed by the object program.

Chapter 5 prepares the reader for writing programs using the latest techniques of structured programming.

Chapters 6 through 9 describe the format and various units of data used in the Procedure Division including the functions of the input/output verbs, the arithmetic operations of the computer and the arithmetic verbs, the flow of control through a computer and the various verbs that may alter the flow control, and the data manipulation verbs that may alter the format of the data.

Chapter 10 describes the table handling feature and how it may be applied to data processing problems involving the use of arrays and tables.

Chapter 11 describes the sort and merge features and how they may be applied to data processing problems.

Chapter 12 describes the operation of the report writer feature. This feature provides the facility for producing reports by specifying the physical appearance of a report rather than requiring specifications of detailed procedures necessary to produce that report.

Chapter 13 describes the methods for the operation and processing of direct access files organized in different manners, including the creation, accessing, updating, and the addition of records to these files.

Chapter 14 describes additional and optional features that are available and that can be used to simplify the coding process.

Chapter 15 explains the use of declaratives and the Linkage Section in providing for the continued processing of files in case of errors and interprogram communication.

Chapter 16 describes the various programming techniques that can be used to increase the efficiency of COBOL programs.

In the appendices the important features of Debugging COBOL programs, Job-Control language, differences between ANS 1968 and ANS 1974 COBOL and useful reference material will be found.

Pedagogy

Extensive material for review and assignments follows each chapter:

Questions for Review

Matching Questions

Exercises

Problems

Illustrative programs with HIPO charts and flowcharts follow most chapters.

Appendix D contains ten programming problems, including input data, from which the instructor can select term project assignments.

The text is aimed at any course in COBOL programming, both at the introductory and advanced levels. It can be used as a one-semester course covering as many chapters as necessary for course requirements or as a two-semester course where chapters 1 through 9 can be assigned as the basic course with chapters 10 through 16 serving as an advanced COBOL course.

Whether a one-semester or two-semester approach is used, the problem programs in the appendix should be assigned, as this will provide the student with the overall knowledge from problem definition to solution.

Supplementary Material

The text is accompanied by an instructor's manual and a workbook.

Instructor's Manual

The instructor's manual includes a brief commentary on each chapter, stating the intent of each chapter, and offering useful teaching suggestions and a summary of important points. It also provides answers to all of the end-of-chapter material as well as programs for the ten problems in appendix D.

Workbook

The workbook contains additional exercises and assignments relating to each chapter. Many programs are worked through, with emphasis on the use of HIPO charts and pseudocode.

The text is intended to provide a comprehensive coverage of ANS 1974 COBOL programming language. It should provide the basic instruction material necessary to write simple as well as advanced structured COBOL programs.

Acknowledgments

The following information is reprinted from *COBOL Edition 1965,* published by the Conference on Data Systems Languages (CODASYL), and printed by the U.S. Government Printing Office.

"Any organization interested in reproducing the COBOL report and specifications in whole or in part, using ideas taken from this report as the basis for an instruction manual or for any other purpose is free to do so. However, all such organizations are requested to reproduce this section as part of the introduction to the document. Those using a short passage, as in a book review, are requested to mention "COBOL" in acknowledgment of the source, but need not quote this entire section.

"COBOL is an industry language and is not the property of any company or group of companies, or of any organization or group of organizations.

"No warranty, expressed or implied, is made by any contributor or by the COBOL Committee as to the accuracy and functioning of the programming system and language. Moreover, no responsibility is assumed by any contributor, or by the committee, in connection therewith.

"Procedures have been established for the maintenance of COBOL. Inquiries concerning the procedures for proposing changes should be directed to the Executive Committee of the Conference on Data Systems Languages.

"The authors and copyright holders of the copyrighted material used herein

> FLOW-MATIC (Trademark of Sperry Rand Corporation), Programming for the Univac (R) I and II, Data Automation Systems copyrighted 1958, 1959, by Sperry Rand Corporation; IBM Commercial Translator Form No. F28–8013, copyrighted 1959 by IBM; FACT, DSI 27A52602760, copyrighted 1960 by Minneapolis-Honeywell

have specifically authorized the use of this material in whole or in part, in the COBOL specifications. Such authorization extends to the reproduction and use of COBOL specifications in programming manuals of similar publications."

I am indebted to the IBM Corporation for allowing the use of numerous illustrations and material in appendixes B and F that have made the text more meaningful. Special thanks also to the Burroughs Corporation for permitting the use of other problems and illustrations. Some of the programming problems have been adapted for COBOL from other texts of mine—*Introduction to Assembler Language Programming,* and *RPG II Programming*—with the permission of the publisher, Wm. C. Brown Company Publishers.

I would like to thank the following people who reviewed the manuscript and made many constructive suggestions for this fourth edition: Charles W. Butler, University of Arkansas at Fayetteville; Lawrence H. Gindler, Southwest Texas State University; Seth A. Hock, Columbus Technical Institute; John B. Lane, Edinboro State College; John M. Lloyd, Montgomery College; and Eileen Wrigley, Community College of Allegheny County.

For her special contribution to many improvements in the coordination of text and illustrations, thanks are due to Sr. Mary Kenneth Keller of Clarke College. Sr. Mary Kenneth also prepared most of the IPO charts shown with the illustrative programs. Others were prepared by Teresa Nickeson of the WCB Information Services Department. Rus Caughron, also with WCB, was instrumental in the final arrangement of the illustrations.

My last special thanks are to my wife, Sylvia, who served as chief typist, confidant, and proofreader. Without her encouragement, the book would never have been written.

An Overview Of COBOL Programming

Chapter Outline

Introduction
Introduction to Programming
- Analysis
- Design
  - HIPO Charts
  - Flowcharts
    - System Flowcharts
    - Program Flowcharts
  - Program Development
  - Testing
  - Implementation
  - Acceptance by Management/User

Introduction to COBOL
General Description of COBOL
- Popularity of COBOL
- COBOL As a Common Language
- Advantages of COBOL
- Disadvantages of COBOL
- Objectives of COBOL

The Compiler
- Checking the Program

Divisions of COBOL
- Identification Division
- Environment Division
- Data Division
- Procedure Division

Writing the First COBOL Program
- Coding Form
- An Address-Listing Program for Study

Summary

Chapter Objectives

The learning objectives of this chapter are:

1. To prepare you for the programming phase of COBOL in the proper analysis of the problem, writing the program to solve the particular problem, operating the program, and planning the program.
2. To acquaint you with the tools necessary to help you find logical solutions to complex data processing problems in the form of flowcharts and HIPO charts.
3. To learn the general characteristics of COBOL, its advantages and disadvantages, popularity, and objectives.
4. To study the compiler and its importance in the translation process.
5. To explain the principal functions of the Identification, Environment, Data, and Procedure divisions.
6. To prepare you to write your first COBOL program.

Introduction

A computer program is much more than a set of detailed instructions. It is the outcome of the programmer's applied knowledge of the problem and the operation of the computer system as well. The programmer receives a problem statement that contains a job description, a description of the different types of data involved, and layout specifications for input and output of the data. After a complete analysis of the problem statement, the completed problem statement must be translated into flowcharts and/or HIPO (hierarchy plus input, process, output) charts. The flowcharts and HIPO charts are translated into a computer program, which instructs the computer on how the data will be received, how the data will be processed, and what output should be developed.

Not so many years ago, when computer programming was a new "art," a certain mystique surrounded any individual who could actually tell one of the great beasts how to do something. The programmer could, and did, fill the program with mysterious names or coding tricks and jealously stood guard over the subsequent use of the program. All this went a long way toward increasing the programmer's job security (Who else could make sense of the thing?), but may have laid the groundwork for many of the inefficient habits still found in numerous programming groups. The programs themselves were relatively small and often short-lived.

Today, most frequently used programs are large and complicated, and probably the original authors (or author) are no longer responsible for the program or are gone from the area because of promotions and job changes, etc. A large portion of a programmer's time is spent making changes to programs that he or she did not write.

Introduction to Programming

Running an automatic machine is such an easy job that it is easy to overlook the many hours that have to be spent in designing, testing, and adjusting all of the automatic features until the operation becomes effortless. The same meticulous care and preparation is necessary before the computer can function automatically. It must be instructed on how to perform every task. This job of writing detailed instructions for the computer and making them work correctly is called *programming.*

A data-processing problem is born from the needs of company management; for example, a report about monthly sales, a payroll plan for employees, or an inventory control system. This results in a precise problem statement, which has been developed by the system analyst. It should contain a job description, special processing information, a description of the data desired, and the layout specifications for input and output of the data. The information contained in the problem statement will be further translated into the necessary tools that enable the programmer to both solve and program the problem. In the absence of such information, the programmer must develop the tools necessary for solving the problem.

A computer program is a set of instructions arranged in a proper sequence to cause the computer to perform a particular process. The steps required to develop computer programs include the following:

1. Analysis of the problem.
2. Design of the solution.
3. Development of programs.
4. Program testing.
5. Implementation of the application.
6. Acceptance by management or the user.

Analysis

A problem must be thoroughly analyzed before any attempt is made at a solution. This requires that boundary conditions be established so that the solution neither exceeds the objectives of management nor becomes too narrow to encompass all the necessary steps. The *output* needs should be clearly stated. The necessary *inputs* to produce the desired results should be carefully studied and described.

Design

To illustrate design techniques, a simple problem has been selected, the *Stock Status Report* (figure 1.1). This is, in fact, only part of a much larger application involving an inventory system. Some applications may be such that a team of programmers are needed to complete the system in reasonable time. Because of this, certain techniques have been devised to help coordinate team efforts. However,

Job Description

Stock status report

Code a COBOL program that will produce a report listing all the active items in a company's inventory. Print asterisks on the detail line for each item that needs replenishing, that is, the quantity on hand is equal to or less than the order point quantity. Accumulate the total value of all active items in stock.

File information

Input file
File: Inventory master
File organization: Sequential
Record length: 128
Label: ITMSTR
Access method: Sequential

Output file
Print positions: 120

Calculations

For each active item in inventory:

1. Find the value of the quantity on hand by multiplying the unit cost times the new quantity on hand.
2. Add the on hand value of each item to an accumulator.
3. Compare the new quantity on hand to the order point. Print asterisks if the new quantity on hand is equal to or less than the order point.
4. Count the active items in stock. Inactive records have a "1" in position 72 of the record.

Stock status report run

Inventory master → System/XX → Stock status report

Input Layout

Inventory master record

CODE	Record identification code—'AA'
ITEMNO	Item number
UCOST	Unit cost (2 decimal positions)
UPRICE	Unit price (2 decimal positions)
DESCR	Description
ORDPT	Order point (0 decimal positions)
DELETE	Activity code—blank for active, '1' for inactive
ONORD	On order (0 decimal positions)
NOHAND	New on hand (0 decimal positions)
OOHAND	Old on hand (0 decimal positions)

Record length is 128

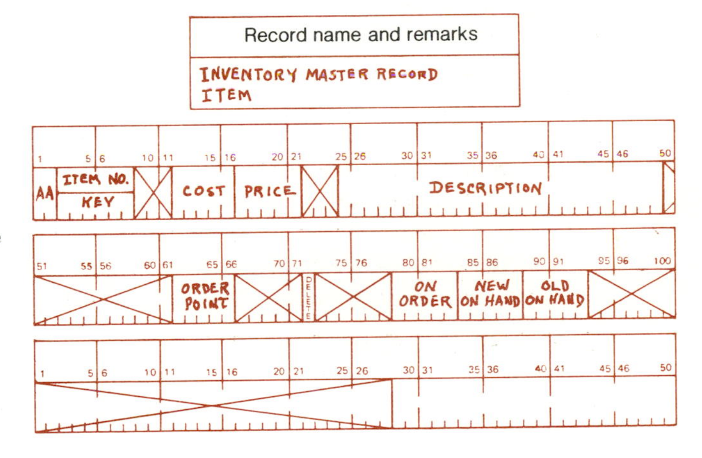

Output Layout

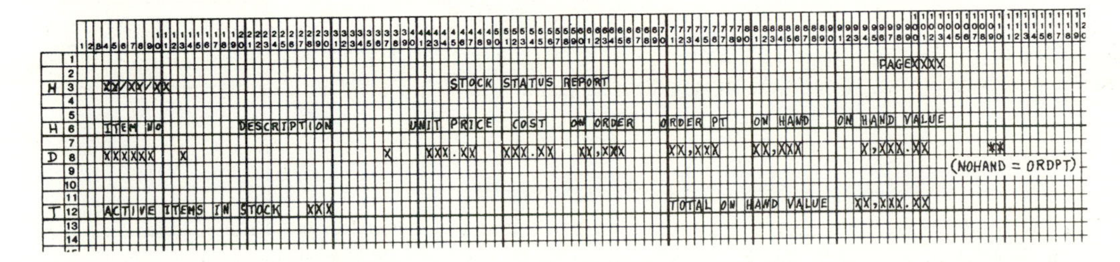

Heading lines on all pages
Double space detail lines
Top margin 5 lines
Page body 55 lines
Bottom margin 6 lines
Footing area starts at line 50

Figure 1.1 Stock status report.

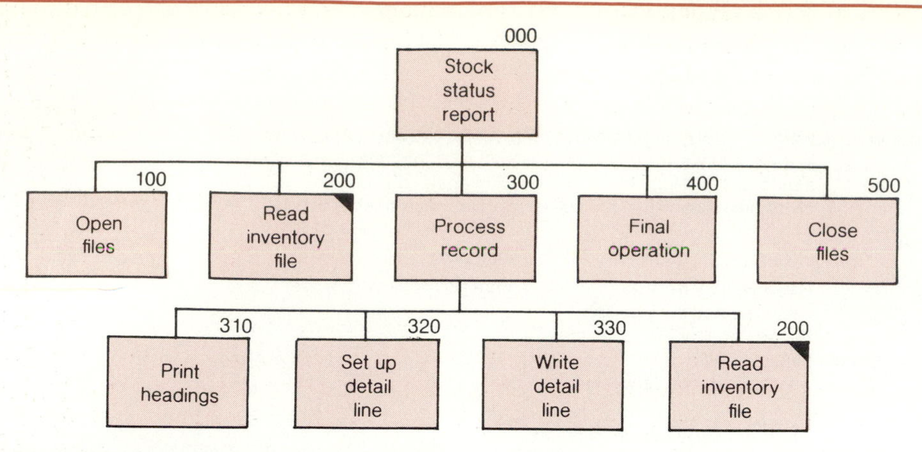

Figure 1.2 Hierarchical chart for stock status report.

regardless of size, it is profitable to design programs with care. The tools presented here are a means to that end.

HIPO Charts

Figure 1.2 shows a *hierarchical chart* for the Stock Status Report. This chart has been prepared as part of the design phase. It shows the program, which produces the report, as broken down into modules that perform particular functions in the program. The hierarchical structure shows the relation of the modules to each other. The first level, numbers 100, 200, 300, 400, and 500 constitute the main line of the program. The next level shows that module 300 is further divided into submodules. Note that a module that is repeated at another level such as the module marked 200 is indicated with a triangle in the upper right-hand corner.

This hierarchical chart is a part of a design technique known as HIPO. HIPO is an acronym standing for "*H*ierarchy plus *I*nput, *P*rocess, *O*utput." For each module shown on the hierarchical chart,

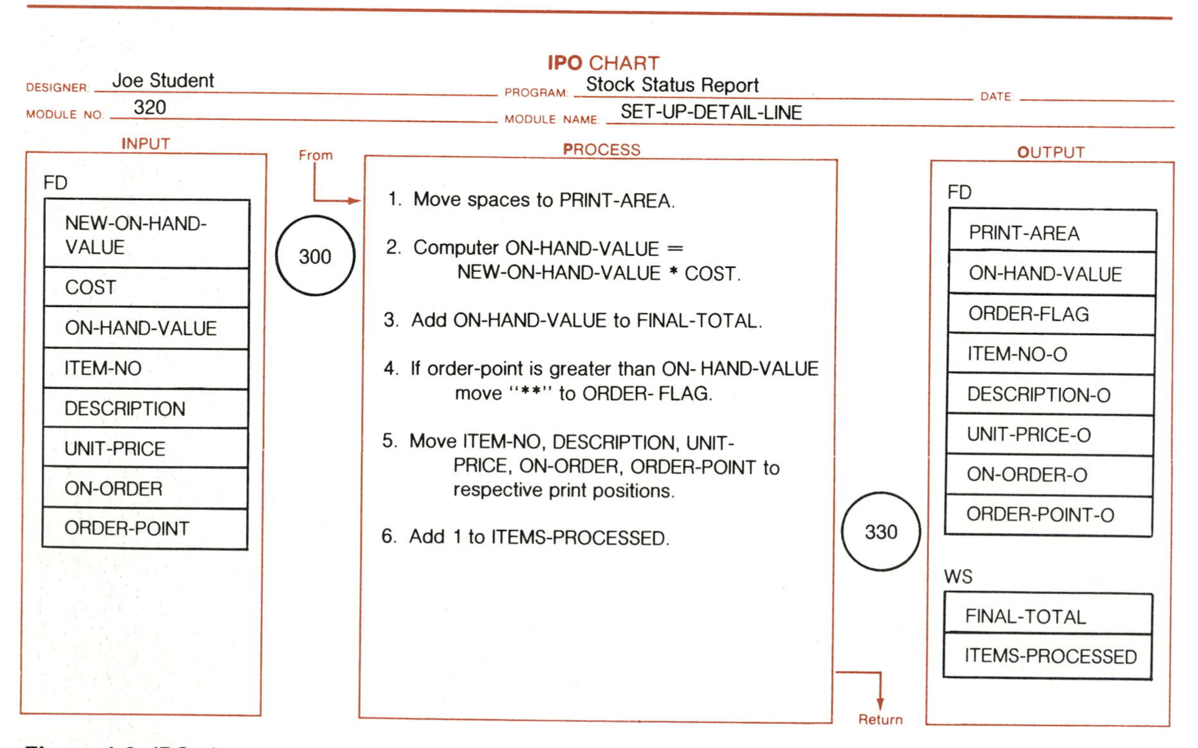

Figure 1.3 IPO chart for stock status report.

another chart called IPO (Input, Process, Output) is made. Figure 1.3 is an example of this for module 320. Note that as the name of this chart suggests, the inputs to the module are listed along with their source; then the processing of these inputs is given; and finally, the outputs from the processing are shown. The function of the module in the example was to set up a detail line for printing. Some of the information was obtained from the inventory file which was stored on a disk. The rest of the information was computed from the information in the file. An examination of the input layout shows what was stored on the disk. The output layout gives the format of the detail line and what information must be moved to it.

Flowcharts

The flowchart is a graphic representation of the flow of information through a system in which the information is converted from the source document to the final reports. Because most data processing applications involve a large number of alternatives, decisions, exceptions, etc., it would be impractical to state these possibilities verbally. The value of a flowchart is that it can show graphically, at a glance, the organized procedures and data flows so that their apparent interrelationships are readily understood by the reader. Such relationships would be difficult to abstract from a detailed narrative text. Meaningful symbols are used in place of narrative statements. The flowchart is the "roadmap" by which the data travels through the entire system.

System Flowcharts

There are two types of flowcharts widely used in data processing operations: the system flowchart, representing the flow of data through all parts of a system, and the program flowchart, wherein the emphasis is on the computer decisions and processes. A system flowchart is normally used to illustrate the overall objectives to data and nondata processing personnel. The flowchart provides a picture indicating what is to be accomplished. Emphasis is on the documents and the work stations they must pass through. The flowchart can also be applied where source media is converted to a final report or stored in files.

Many symbols depicting documents and operations are used throughout the system flowchart. The symbols are designed so that they are meaningful without too much further comment or text. Card symbols are used to indicate when the input or output may be a card. Document symbols are used to represent the printed reports (see figure 1.4).

The system flowchart is usually prepared on one sheet of paper so as to facilitate the presentation of the overall picture of the system to administrative personnel and executives. It indicates the job to be done without detailing the steps involved.

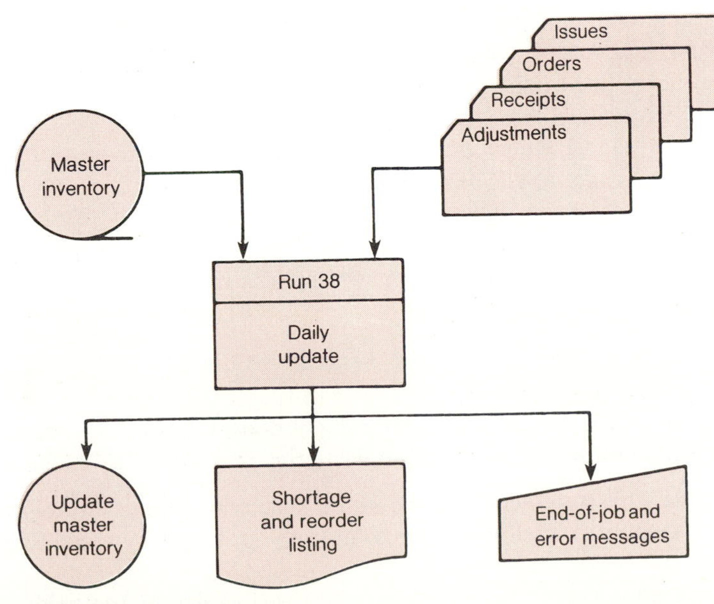

This flowchart demonstrates the use of symbols in the presentation of a system. Master inventory is an input magnetic tape; four different types of cards serve as input to this particular run 38-daily update. The resulting output will be an update master inventory magnetic tape, a shortage and reorder listing in the form of printed output, and end-of-job and error messages on the typewriter console. The detailed steps of processing are not mentioned; they will be described on the program flowchart for this particular job.

Figure 1.4 System flowchart.

Program Flowcharts

A program flowchart is a graphic representation of the procedures by which data is to be processed. The chart provides a picture of the problem solution, the program logic used for coding, and the processing sequences. Specifically, it is a diagram of the operations and decisions to be made and the sequence in which they are to be performed by the machine. The major functions and sequences are shown, and if any detail is required, a *block diagram* is prepared. (See figure 1.5 B, C, D, and E.).

After the HIPO charts have been prepared for the Stock Status Report, the programmer may wish to show the logic of the program in the form of a program flowchart. Figure 1.5 A shows the main line program for the Stock Status Report. In the main line program there are several modules containing the COBOL verb PERFORM. Each of the modules may have a separate flowchart, but it is especially important that the PERFORMed modules have flowcharts. Figure 1.5 B, C, D, and E shows how the logic of some of these procedures have been described in detail. Even though you may not fully understand all that is implied by these charts, at this time, it is still worthwhile to study this design system. It will be helpful to you as you begin to design programs of your own.

The program flowchart shows the relationship of one part of the program to another, and can be used to experiment or verify the accuracy of different approaches to coding the application. Where large segments of the program are indicated, a single processing symbol may be used and the detail for the segment shown in a separate block diagram, which would be used for the machine coding. Once the flowchart has been proven sound and the procedures developed, it may be used for coding the program.

A program flowchart should provide:

1. A pictorial diagram of the problem solution to act as a roadmap of the program.
2. A symbolic representation of the program logic used for coding, desk checking, and debugging while testing all aspects of the program.
3. Verification that all possible conditions have been considered and taken care of.
4. Documentation of the program, necessary to give an unquestionable historical reference record.
5. Aid in the development of programming and coding.

These are the important features and phases of program flowcharting:

1. It provides the programmer with a means of visualizing the entire program during its development. The sequence, the arithmetic and logical operations, the input and outputs of the system, and the relationship of one part of the program to another are all indicated.
2. The system flowchart will provide the various inputs and outputs, the general objective of the program, and the general nature of the operation. A program flowchart will be prepared for each run and will serve as means of experimenting with the program specifically, in order to achieve the most efficient program.
3. Starting with symbols representing the major functions, the programmer must develop the overall logic by depicting blocks for input and output, identification decisions, etc.
4. After the overall logic has been developed by the programmer, he/she will extract the larger segments of the program and break them down into smaller, detailed block diagrams.
5. After the flowchart has been proven sound, the coding for the program will commence.
6. Upon completion of the coding, the program will be documented for further modification, which will always occur after the testing, installation, and operational stages.
7. Final documentation should involve the overall main logic, system flowcharts, program flowcharts, and the detailed block diagrams. The general system flowcharts help in the understanding of the more detailed program flowcharts.

Program Development

When the program design has been completed, the actual programming is done. Observe that the design has provided the programmer with all the information needed to write the program. This includes:

1. *Precise Statement of the Problem.* This statement must be exact, specifying what the program is to accomplish. For example, "To compute social security tax, multiply gross pay by social security rate to arrive at the FICA tax, etc."
2. *List of Inputs.* All sample copies of inputs to be used together with the size of the fields, type (alphabetic or numeric), control fields, etc., should be included.

Figure 1.5 Stock status report flowcharts.

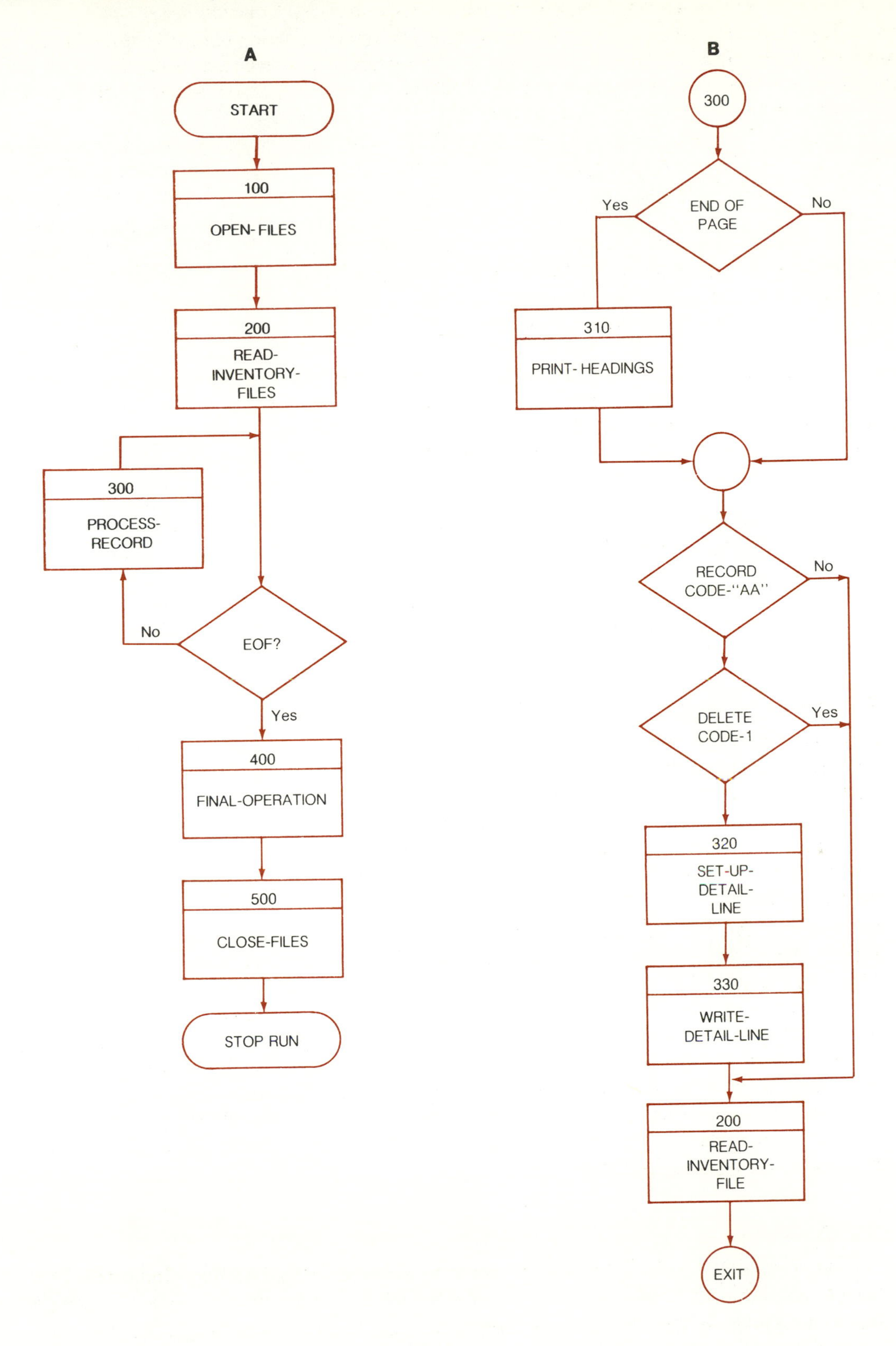

Figure 1.5 *(continued)*

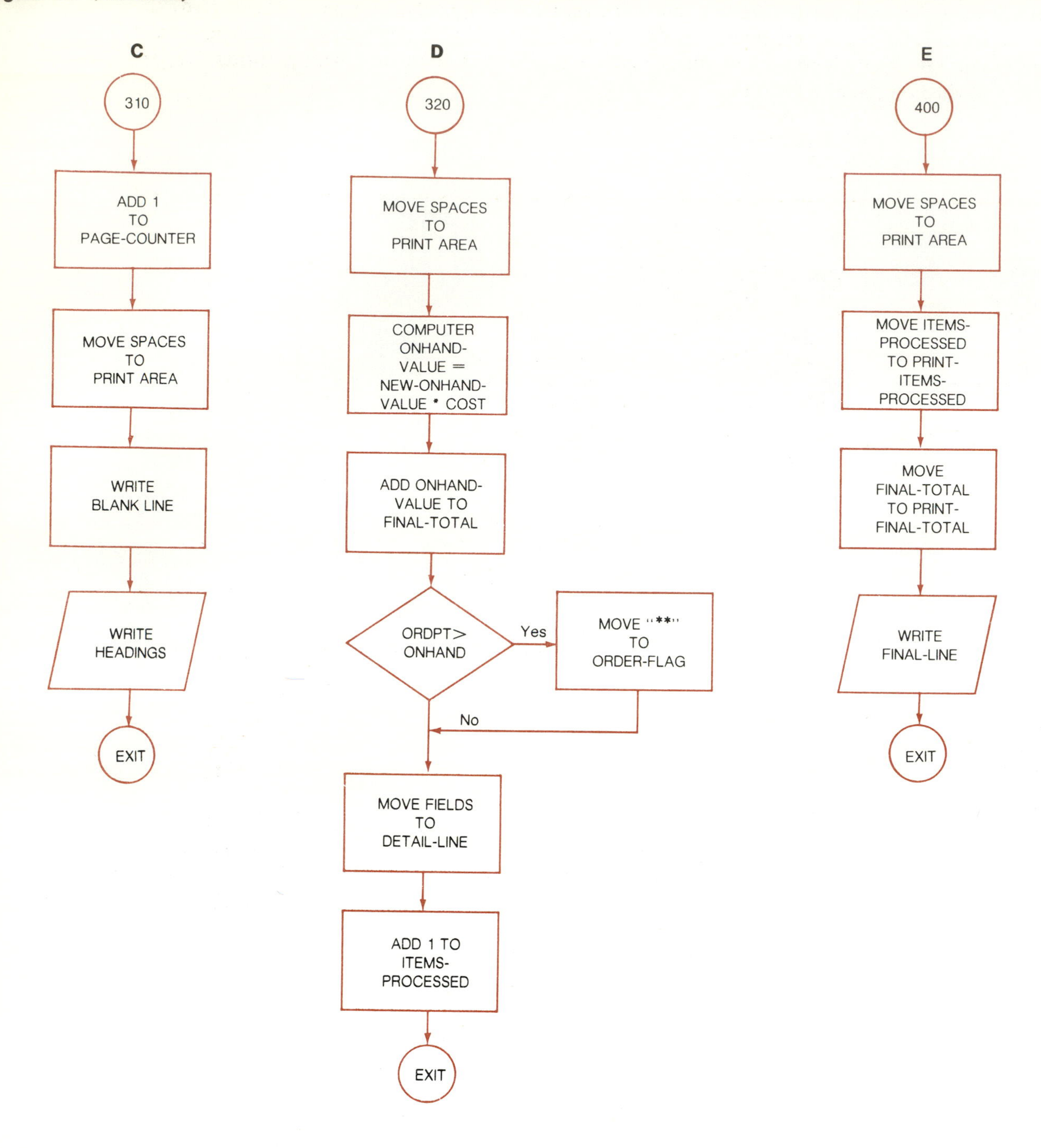

3. *Outputs Desired.* Samples of all outputs should be included with all headings indicated. The number of copies desired, type and size of paper to be used, tape density (if used)—are some of the items that should be included in this section.
4. *Flowcharts.* All necessary system flowcharts, HIPO charts, and program flowcharts should be included. A system flowchart represents the flow of data through all parts of a system, the HIPO charts show the relationships and functions of each module, and the program flowchart shows the logic used in the process.

The programmer must have a knowledge of a computer language, in this case COBOL, in order to write the proper instructions for the computer. The actual writing of the program is simplified by the care exerted in designing the program. Moreover, good design will assure the proper functioning of the program and save many hours of rewriting. Figure 1.6 shows the steps from writing through an acceptable production run of a program.

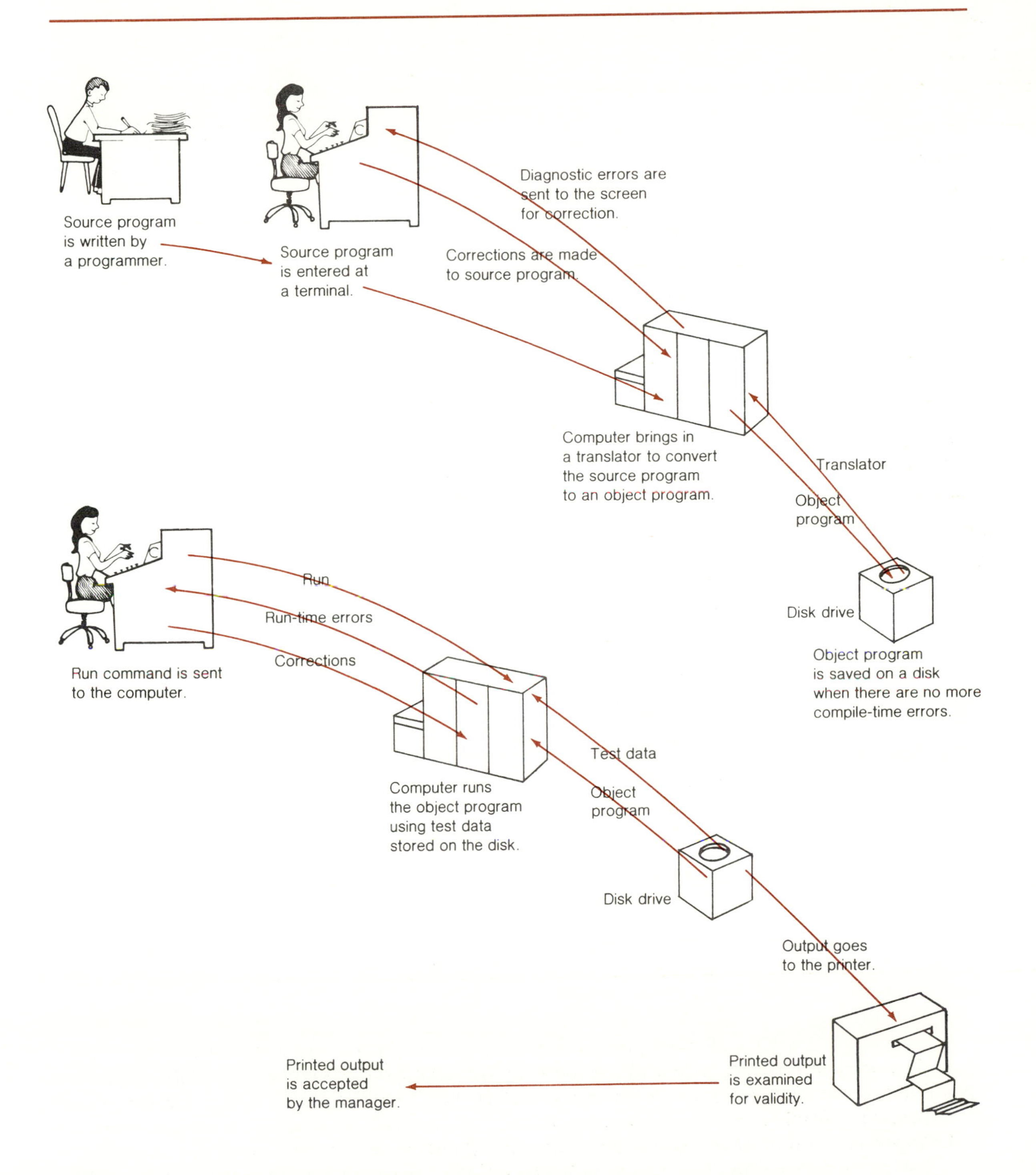

Figure 1.6 Program translation, testing and production run.

Testing
Programs must be tested by entering them into the computer system to be translated from COBOL to the language of the machine. COBOL is the souce language, and the machine language becomes the object code.

Just as a typist may make mistakes, the person entering either a program or data may make transcription mistakes at the keyboard when communicating with the computer. Such mistakes can be corrected easily by proofreading a listing of the program after it has been entered. This should be the first step in eliminating errors. For the beginning programmer there is another level of mistake which involves the syntax of the computer language. The rules for writing instructions in COBOL must be learned and practiced with care. With such practice, programmers soon become competent in using COBOL and eliminating syntax errors.

A more subtle and less obvious type of error involves the logic of the program and the algorithms that are used as a solution to a specific problem. When these are in error the program may "run," that is, produce output, but the report may not be valid. Considerable checking must be done to make sure that the program does indeed, and in all cases, produce the correct result. A failure to do this leads to the "horror stories" about so-called computer error.

Implementation
When the program has been thoroughly tested, then production runs can be made. This means that the program is ready for use in a system. Even at this stage there should be some monitoring of the output. When others are using a program that they did not write, problems can occur which were not anticipated by the programmer.

Acceptance By Management/User
The final affirmation of the program should be made by management and those who use it. The programmer should be grateful if errors are found even at this late date, and be willing to correct them. This will add to the integrity of the programmer, and should not be considered as negative criticism. At this stage the final documentation should be added to the design phase documents. These include copies of the program, production runs, and reports generated by the program.

Introduction to COBOL

COBOL is defined as a *CO*mmon *B*usiness *O*riented *L*anguage. As such it is the result of the efforts of computer users in both industry and government to establish a language for programming business data processing applications. The committee was formed in 1959 (at the insistence of the Department of Defense) for the express purpose of producing a common business language that could be processed on the various computers without any reprogramming. The federal government, one of the largest users of data processing equipment, was being faced with the enormous task and expense of reprogramming each time a different type of computer was installed. Some questions to be answered were:

Could a language be created for existing computers that could also be utilized on future computing systems?

Could a language be developed that would fit the rapidly changing and expanding requirements of management?

With the need to produce a large number of computer programs in a short period of time, could a language be developed that would permit existing programming staffs to be augmented with relatively inexperienced programmers?

As data processing installations grew in size and complexity, it became apparent that a new programming tool was necessary, one in which the source language was the language of the businessperson. None of the existing compilers could be used since they were mathematical in nature and not geared for business applications. However, experience gained in the creation of the algebraic compilers pointed the way for the creation of the more complex data processing compilers.

The initial specifications for COBOL were presented in a report by the *CO*nference on *DA*ta *SY*stems *L*anguage (CODASYL) in April of 1960. The group consisted of computer professionals representing the United States Government, manufacturers of computer equipment, universities, and users.

This group, confronted with the difficulty of program exchange among users of computer equipment, was inspired to meet the challenge presented by the situation. At the first meeting the conference agreed upon the development of a common language for the programming of commercial problems that was capable of continuous change and development. The proposed language would be problem oriented, machine independent, and would use a syntax closely resembling English-like statements, thus avoiding the use of special symbols as much as possible. Here are some typical COBOL statements:

```
SUBTRACT DEDUCTIONS FROM GROSS GIVING NET-AMOUNT.

MULTIPLY UNITS BY LIST-PRICE GIVING BILLING-AMT.

IF ON-HAND IS LESS THAN MINIMUM-BALANCE
    PERFORM REORDER-ROUTINE
ELSE
    NEXT SENTENCE.
```

The combined effort of the group was utilized to produce a business-oriented language that would permit a single expression of a program to be compiled on any computer then operative or contemplated in the future. This would reduce the reprogramming costs and provide a means for interchange of computer programs among the users.

COBOL is especially efficient in the processing of business problems. *Business data processing is characterized by the processing of many files, used repeatedly, requiring relatively few calculations and many output reports.* A payroll application is a good example of a business application owing to the fact that with a limited amount of input (time cards and personnel records, for example), many output reports are produced, such as updated personnel files, payroll checks, payroll registers, deduction reports, and various other internal and external reports. Business problems involve relatively little algebraic or logical processing; instead, they usually manipulate large files of similar records in a relatively simple manner. This means that COBOL emphasizes the description and handling of data items and input/output records.

The first special specifications for COBOL were written in 1960 and improved, refined, and standardized by subsequent meetings of the CODASYL committee. In 1970, a standard COBOL was approved by the American National Standards Institute (ANSI), an industry-wide association of computer users and manufacturers. This standard was called American National Standards (ANS) COBOL.

There are two basic versions of COBOL in use today. One is based on the standards developed by the American National Standards Institute in 1968 known as ANSI 1968 COBOL. The other is a more recent revised set of standards released in 1974 known as the ANSI 1974 COBOL. This text conforms to the ANSI 1974 COBOL standards. Since most COBOL programs written today conform to either the 1968 or 1974 standards, the programmer should be familiar with the differences between the two versions. (The COBOL differences between ANSI 1968 and ANSI 1974 COBOL appear in the appendix.)

At the present time, the ANSI committee is working on a revised version of the 1974 COBOL, which will be known as ANSI 1980 COBOL.

General Description of COBOL

The *CO*mmon *B*usiness *O*riented *L*anguage (COBOL) is a near-English programming language designed primarily for programming business applications on computers.

When you think of "Common Business Oriented Language" as COBOL's official title, unfortunately, each one of the words of the title can be taken in various ways, so they require a little explaining. The title is intended to convey the basic purpose of COBOL: to be one language for all computers—a standard language for programming business problems.

The words "business oriented" also require a little explaining. They have three implications:

First, that COBOL is business procedure oriented, rather than machine oriented because a COBOL program consists of descriptions of (1) the procedures according to which data files are to be processed, (2) what the contents of the data file are, and (3) what input/output devices the data files are assigned to.

Second, that COBOL is particularly applicable to business data processing problems, as opposed to scientific problems because tasks like preparing reports and updating files fall into the business category, to which COBOL is oriented. Tasks that involve trigonometric functions (sines, cosines, etc.) or

Boolean algebra (logical ands, logical ors, etc.) are examples of tasks that fall into the scientific category. For the most part, business data processing involves moving data around—putting data in and out, rearranging it, changing its appearance, comparing items of data, locating desired items, etc. Arithmetic operations are involved, but they are generally limited to adding, subtracting, multiplying, dividing, and occasionally exponentiating (raising a number to a power or finding a root of a number). COBOL is designed for these kinds of operations. Jobs that involve complicated mathematics are harder to program in COBOL. As a general rule, such tasks are in the scientific category.

Third, that COBOL is a language that business people can understand after a short introduction to the language, whether or not they have had any programming experience.

COBOL is described as near-English because its free form enables the programmer to write in such a way that the final result can be read easily, with the general flow of logic understood by persons not closely allied with the details of the problem as the programmer.

From a computer user's standpoint, COBOL offers several advantages:

COMPATIBILITY—COBOL makes it possible for the first time to use the same program on different computers with a minimum of change. Reprogramming can be reduced to making minor modifications in the COBOL source program, and re-compiling for the new computer.

STANDARDIZATION—The standardization of a computer programming language overcomes the communication barrier which exists among programming language systems which are oriented to a single computer or a single family of computers.

COMMUNICATION—Easier communication between decision-making management, the systems analyst, the programmer, the coding technician, and the operator is established.

AUTOMATIC UNIFORM DOCUMENTATION—Easily understandable English documentation, provided automatically by the compiler, facilitates program analysis and thus simplifies any future modifications in the program.

COMPLETELY DEBUGGED PROGRAMS—Programs produced by the COBOL compiler are free from clerical errors.

CORRECTIONS AT ENGLISH LEVEL—Corrections and modifications in program logic may be made at the English level.

EASE OF TRAINING—New programming personnel can be trained to write productive programs with COBOL in substantially less time than it takes to train them in machine coding.

FASTER AND MORE ACCURATE PROGRAMMING—The English language notation expressed by the user and the computer-acceptable language produced by the COBOL computer ensure greater programming accuracy and a reduction in programming time.

REDUCTION IN PROGRAMMING COSTS—The ability to program a problem faster reduces the cost of programming. Also, reprogramming costs are greatly reduced since a program run on one system may be easily modified to run on another without being entirely recoded.

COBOL has emerged as the leading processing language in the business world and is enjoying wide acceptance in the data processing market.

Popularity of COBOL

COBOL is a high-level computer language that is procedure-oriented and relatively machine independent. COBOL was designed with the programmer in mind in that it frees him/her from the many machine-oriented instructions of other languages and allows him/her to concentrate on the logical aspect of a program. The program is written in an English-like syntax that looks and reads like ordinary business English. Organization of the language is simple in comparison to machine-oriented language—much easier to teach to new programmers, thus reducing training time.

The following are some of the reasons advanced for the popularity of COBOL today.

1. COBOL has been continuously standardized by repeated meetings of the CODASYL committee to improve the language and to guarantee its responsiveness to the data processing needs of the community.

2. COBOL has been designed to meet the needs of users today and in the future at decreasing costs to all concerned.
3. COBOL is the only language translator supported by the users, including the federal government.
4. COBOL users will be skeptical of any new equipment without COBOL capabilities; therefore, it is incumbent upon the computer manufacturers to participate wholeheartedly in all technological progress in COBOL.
5. COBOL is the major data processing language available today. It is included in more software packages of computer manufacturers than any other language.
6. COBOL has proven that it is machine-independent, in that it can be processed through various computer configurations with the minimum of program change.
7. Although COBOL was primarily designed for commercial users, it has evolved as a highly sophisticated language in other areas of data processing.
8. COBOL is not plagued by computer obsolescence since it is constantly being revised to accommodate the newer computers.
9. COBOL has a self-documentary feature in that the English language statements are easily understood by managers and nonprogramming personnel.

After a decade of dedicated effort by a small group of data processing professionals, COBOL has emerged as the leading language in the data processing community. The continued voluntary efforts at standardization and technological improvements in the COBOL language will guarantee its responsiveness to the needs of information by management, and its ability to survive the ever-changing data processing field.

In addition to its popularity as a batch processing programming language, COBOL is gaining popularity as an interactive programming language where data is entered directly into the computer from remote locations through terminals. COBOL has also been added to many software packages for microcomputers.

COBOL As a Common Language

A great deal of controversy arises when considering whether COBOL is truly a common language—that is, a programming language that can be compiled on any configuration of any computer. COBOL programs are written for computers of a certain minimum storage capacity, so certain small computers are thus excluded from the use of COBOL.

Can a programmer write a more efficient program in COBOL if he/she is familiar with the hardware? The answer is, of course, yes. He/she can take advantage of many programming approaches offered by the different computer manufacturers to reduce programming time and the number of storage positions required.

COBOL is not completely common as yet, but it is rapidly approaching this objective. It offers more commonality than any other processor presently in use. It is hoped that the continuing meetings of the CODASYL committee will make COBOL an even more useful tool in the future.

The efficiency of COBOL has steadily increased, to the point whereby a COBOL program is more efficient than that of a new programmer who codes a program on a one-for-one basis in some assembly language for a particular machine. However, the COBOL program is not as efficient as an object program produced by an experienced programmer in the symbolic language of the individual computer.

Advantages of COBOL

1. The principal advantage of the COBOL system is its advancement of communication. The ability to use English-like statements solves language difficulties that have often existed between the experienced programmer and decision-making management.
2. The program is written in the English language, thus removing the programmer from the individual machine or symbolic language instructions required in the program. Although a knowledge of the individual instructions (symbolic and machine) are not required in COBOL programming, that knowledge is very useful in the writing of an efficient program if the programmer possesses some knowledge of the hardware and coding of the particular computer.

3. Pretested modules of input and output are included in the COBOL processor, which relieve the programmer of the tedious task of writing input and output specifications and testing them.
4. The programmer is writing in a language that is familiar to him or her, which reduces the documentation required since the chance for clerical error is diminished. Generally, the quality and the quantity of documentation provided by the COBOL compiler is far superior to that of other language processors. The printed output resulting from the compilation provides a highly desirable simplification of person-to-person and person-to-machine communication problems—a welcome improvement.
5. While COBOL is not completely machine-independent, a program written for one type of machine can be easily converted for use on another with minimum modification. The standardization of a COBOL program provides this benefit.
6. Because of the separate divisions in COBOL, a large program can be broken down into various segments, and each programmer may write one division. The format definition can be made available to all programmers engaged in the problem.
7. Nonprogrammers and managers can read the COBOL program in English, which provides them with the opportunity of judging the logic of the program.
8. During the compilation phase, the COBOL language processor generates a list of diagnostics. A diagnostic is a statement provided by the compiler that indicates all errors (except the logic errors) in a source program. Because diagnostics effect the measurement of compiling efficiency, these as well as compiling speed become important conditions for measuring the superior attributes of COBOL. This advantage derived from the attribute of COBOL can materially reduce the "debugging" time.

Disadvantages of COBOL

Most of the disadvantages result from the failure of personnel to fully understand the language and its use—as, for example:

1. The expectation that a single COBOL program will provide a permanent solution without ever reprogramming.
2. Assuming that the programmer need be taught only the COBOL language without any knowledge of the hardware or the operation of the computer.
3. COBOL will not generate a sophisticated program similar to one written in the actual language of the particular computer.
4. COBOL processors will operate only with a computer having a certain storage capacity. The newer COBOL compilers have drastically reduced the storage requirements. With the introduction of ever-larger storage units, this problem has been greatly reduced.

Objectives of COBOL

1. To provide standardized elements in entry format that can be used on all computers regardless of make or model—a single common language that can be used by all.
2. To provide a source program that is easy to understand because it is written in the English language. Nonprogrammers can understand the logic of the program as well as the programmers can.
3. To provide a language that is oriented primarily toward commercial applications. Thus the opportunity is provided for business people to participate in the programming.

Although COBOL is oriented toward the problem rather than toward the particular machine, there are major differences in computers that have to be allowed for and adjusted to within the framework of the common language. These adjustments are usually minor, and the programmer with a COBOL knowledge can learn these on the job.

Because it uses English-language descriptions of application requirements, COBOL is especially designed for those who can best define their data processing needs. With a minimum of training and only a basic familiarity with the computing system as prerequisites, accountants, systems and procedures analysts, and many other members of operating management can use COBOL and the computer effectively.

The Compiler

Computers can't understand English, so what is the point of writing programs in English? Programs written in English serve to communicate data processing procedures to people, in addition to serving as source programs to a computer.

COBOL is similar to the English language in the use of words, sentences and paragraphs. The programmer can use English words and conventional arithmetic symbols to direct and control the operations of a computer.

```
ADD QUANTITY TO ON-HAND.

MULTIPLY GROSS-EARN BY SS-RATE GIVING SS-TAX.

IF Y-T-D-EARN IS LESS THAN SS-LIMIT PERFORM SS-PROC.
```

Each of the above sentences is understandable by the computer, but they must first be translated into the particular machine language of the computer before the program can be executed. During the compilation stage, a special system program known as a compiler is first entered into the computer.

The COBOL system consists of two basic elements: the *source program,* which is a set of rules or instructions that carry out the logic of the particular data processing application; and the *compiler,* the intermediate routine that converts the English-like statements of COBOL into computer-acceptable instructions. Since the COBOL language is directed primarily at those unfamiliar with machine coding, terms common to business applications rather than to computing systems are used in the language. The compilation process is shown in figure 1.7.

Obviously, neither the computer nor the method of operating it is an end in itself. Rather, the purpose of a data processing system is to achieve, in the most efficient and economical way possible, solutions to the various applications that occur in the normal functioning of any business, educational, or governmental installation. With the advent of larger, more complex data processing systems, the burden of the programmer could conceivably increase to the point where problem solution becomes subordinate to the intricate methods of computer operation and direction. To preclude this possibility, innovations are constantly being made in an area that has come to be known as *software.*

Precoded software programs, which are in large measure often considered extensions of hardware capabilities, free the programmer from exacting machine considerations and allow him/her to devote more time to the logic of the problem. These routines, which may vary from simple input-output and diagnostic routines to the more sophisticated routines that effect mass conversions of data being fed to the computer, are provided as special software packages with each data processing system. Certainly, the software packages effect a great saving in coding time and in problem preparation by performing many jobs normally undertaken by the programmer. Furthermore, because many sources of programming errors are removed, costly machine time is conserved.

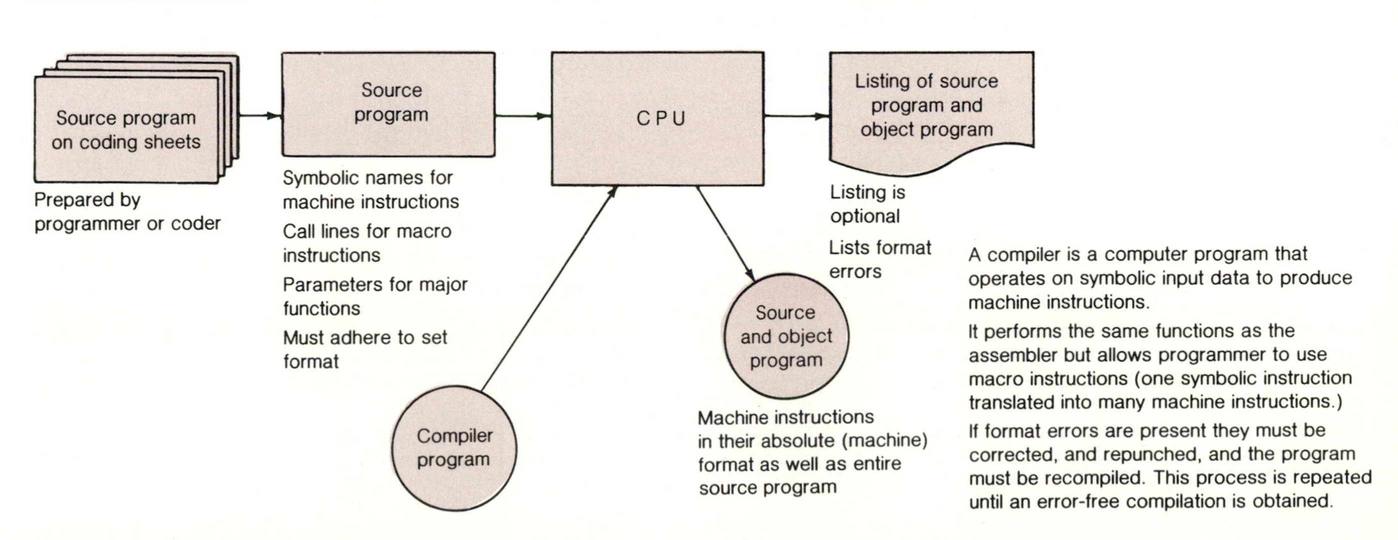

Figure 1.7 The compilation process.

The compiler, the most intricate of the software routines, is a master routine which takes a program, one written in some elementary form of problem statement (e.g., COBOL English-like statements), and translates it into instructions acceptable to the computer. The translated program is then fed back to the computer to be processed. The initial program that the compiler translates is called the *source program.* The machine-coded program produced from the translation is referred to as the *object program.*

The source program (COBOL symbolic program) is read into the computer and translated into a usable set of machine instructions (see figure 1.7). Thus the combination of COBOL reserved words and symbols are transformed into a machine language program (object program). This object program will be used to process the data at execution time to provide the desired outputs. The machine language program produced may be used at once or may be stored on an external medium where it may be called in when needed. This object program may be used repeatedly to process data without any further compiling.

Checking the Program

The programmer's job does not end with coding. The computer program goes through many steps before the desired results are produced, and the programmer is involved with most of them.

After the successful coding and compilation of the source program has been completed, the next phase in the computer checkout procedure is to check the resultant machine-language program with the data. Few programs are written that work correctly the first time they are tried with actual data.

The two major types of errors that cause difficulty in programs are the *logical errors* and the *clerical errors. Logical errors result from poor analysis of the problem.* For example, the failure to anticipate the possibility of an employee having a negative amount of pay due to excessive payroll deductions will result in an error. The omission of a test for this condition could result in a paycheck being written for this negative amount. Errors of this type can result from a lack of understanding by the programmer of all the possibilites of this data-processing application. A more comprehensive analysis of the problem can eliminate many errors like this.

Clerical errors occur during the coding stage. A programmer in error may assign two different symbolic names for the same item or may use the wrong operation code symbol or format, or may omit one or more necessary entries to correctly define fields.

Divisions of COBOL

COBOL is an acronym for the phrase *CO*mmon *B*usiness *O*riented *L*anguage. The COBOL system, which includes a compiler in addition to the COBOL language, is used to state all the facets of a business-oriented problem and to convert the statements into a form usable by a computer.

A business-oriented data processing problem can be broken down into four distinct groups of logically related information. The first is the identification of the problem, such as, is it an inventory control problem, a personnel accounting problem, a payroll problem, or an inventory problem? Also included in this group is information such as the assignment to solve the problem, when, and where. The second group of logically related information is the data processing environment in which the problem is to be solved. That is, what computer will be used to compile the program and run the job? What peripheral equipment is necessary to run the job? What other programs are necessary? The information in this group is also useful when a program has been written on one computer environment and it is desired to run it in another computer environment. The third group of logically related information is that which consists of the data to be processed and the processed data that is desired. Each file, both input and output, is described in terms of unique data items. In addition to this, the organization of the files must be described. These three groups of logically related information, the identification of the problem, the information related to the data processing environment, and the description of the data, all can be considered the problem statement. The fourth group of logically related information can be considered as the procedure(s) by which the data is to be processed to solve the problem. In this group of information, the programmer states in a step-by-step manner exactly what is to be done to the data to produce new or additional data.

The COBOL language is structured to accommodate the four groups of logically related information in four named divisions. These divisions are the Identification Division, the Environment Divi-

IDENTIFICATION DIVISION	identifies the program to computer.

Provides all of the necessary documentation for the program such as:
- The program name and number.
- The programmer's name.
- The system or application to which the program belongs.
- The security restrictions on the use of the program.
- A brief description of the processing performed and the output produced.
- The dates on which the program was written and compiled.

ENVIRONMENT DIVISION	describes the computer to be used and the hardware features to be used in the program.

CONFIGURATION SECTION.

SOURCE-COMPUTER. What computer will be used for compilation?

OBJECT-COMPUTER. What computer will be used for running the compiled object program?

SPECIAL-NAMES. What names have you assigned to the sense (alteration) switches and the channels of the paper tape loop on the printer?

INPUT-OUTPUT SECTION.

FILE-CONTROL. What name and hardware device have you assigned to each file used by the object program?

DATA DIVISION	defines the characteristics of the data to be used including the files, record layouts, and storage areas.

FILE SECTION. For each file named in the FILE-CONTROL paragraph above:
- The file name.
- The record name.
- The layout of the record—the name, location, size, and format of each field.

WORKING-STORAGE SECTION. The size, format, and content of every counter, storage area, or constant value used by the program.

PROCEDURE DIVISION	consists of a series of statements directing the processing of the data according to the program logic as expressed in the detailed program flowchart or HIPO chart.

The individual processing steps are written as COBOL-language statements. This division is divided into programmer-created paragraphs, each containing all of the procedure statements that constitute one particular routine.

Figure 1.8 The four divisions of a COBOL program.

sion, the Data Division, and the Procedure Division. Every COBOL program contains these four divisions in the same aforementioned order. The Identification Division is used to identify by name the source program (that which the programmer writes) and the outputs of a compilation. Other information that can be included in the Identification Division is the name of the programmer, the date written, and such other information relevant to the purpose of the program. The Environment Division is that part of the program in which the computer(s) to be used for compiling and running the program is described. In the Environment Division, names may be assigned to peripheral equipment and the features of the files directly related to the hardware may be described. In the Data Division, the files of data the program processes or creates and the unique individual records of these files are described. Data is written according to a standard job format rather than an equipment-oriented format. In the Procedure Division, a step-by-step logical process is written to instruct the computer to process the input data.

Each of the four divisions must be placed in its proper sequence, begin with a division header, and abide by all the format rules for a particular division. The particular functions of each division are summarized in figure 1.8.

Identification Division

The Identification Division contains the information necessary to identify the source program that is written and compiled. A unique data-name is assigned to the source program.

The intended use of this division is to supply information to the reader. The name of the program must be given in the first paragraph. The programmer may also include information as to when the program was written, by whom, any security information relative to the program and any other information desired.

Environment Division

Although COBOL is to a large degree machine-independent, there are some aspects of programming that depend upon the particular computer to be used and the associated input/output devices. The Environment Division is one division that is machine-dependent, since it contains the necessary information about the equipment that will be used to compile and execute the source program. This division provides a standard method of expressing those aspects of a data-processing problem that depend upon the physical characteristics of a specific computer. In this division, the hardware features of the compiling as well as the executing computer are specified.

Each data file to be used in the program must be assigned to an input or output device. If special input or output techniques are to be used in the program, they have to be specified in this division. Any special-names assigned to hardware devices must be stipulated here.

To transfer a COBOL program from one computer to another, the Environment Division would have to be modified or even replaced to make the source computer compatible with the new computer.

Data Division

The Data Division describes the data that the object program is to accept as input, to manipulate, to create, or to produce as output. The division describes the formats and the detailed characteristics of the input and output data to be processed by the object program. Data to be processed falls into three categories:

1. That which is contained in files and enters or leaves the internal memory of the computer from a specified area or areas.
2. That which is developed internally and placed into intermediate or working storage, or placed into specific format for output reporting.
3. Constants that are defined by the programmer.

The programmer attaches unique names to the files, and records within the files and the items within the records. All files that are named in the Environment Division must be described therein.

In addition to the file and record descriptions of the data, work areas and constants to be used in the program must be described in the Working-Storage Section of this division.

Procedure Division

The Procedure Division contains the procedures required to solve a given problem. This division specifies the actions expected of the object program to process the data to achieve the desired results. The sequential order of the processing steps and also any alternate paths of actions where necessitated by decisions encountered during the processing are specified in this division.

This division is usually written from the program flowchart and/or HIPO charts. The names of the data described in the Data Divison are used to write the statements, employing program verbs to direct the computer to perform some action. The main types of action that may be specified are input and output, arithmetic, data transmission, and sequence control. Some examples of each appear in figure 1.9; these and other verbs are explained later in the text, with formats for each.

Writing the First COBOL Program

In the study of any programming language, it is important to write programs as soon as possible. COBOL is based on English; it uses English words and certain syntax rules derived from English. However,

Input and Output	*Arithmetic*	*Data Transmission*	*Sequence Control*
OPEN	ADD	MOVE	GO TO
READ	SUBTRACT	INSPECT	PERFORM
WRITE	MULTIPLY		STOP
CLOSE	DIVIDE		
ACCEPT	COMPUTE		
DISPLAY			

Figure 1.9 Examples of program verbs.

because it is a computer language, it is more precise than English. The programmer must, therefore, learn the rules that govern COBOL and follow them exactly. These rules are detailed later, beginning in the next chapter. Since COBOL source programs are written in an English-like language using paragraphs, sentences, statements, verbs, etc., it is possible to write a program using only the minimal requirements in each division. This should give one a general picture of how a COBOL program is put together and how it is compiled and executed.

Each of the divisions that constitute a COBOL source program (Identification, Environment, Data, and Procedure) is written in its stated order. Each division is placed in its logical sequence, each has its necessary logical function in the program, and each uses information developed in the divisions preceding it.

Coding Form

The COBOL coding form provides a standard method for writing COBOL source programs and aids the programmer in writing his/her source program for the subsequent entry into the computer. Only the minimum necessary entries to write a "file-free" program will be illustrated at this time.

The program sheet, despite its necessary restrictions, is relatively free-form. The programmer should note, however, that the rules for using it are precise and must be followed exactly.

The general format for the COBOL coding form is as follows:

Positions	**Description**	
1–6	Represents the sequence number area. The sequence number has no effect in the source program and need not be written. The sequence number provides a check on the order of the entries.	
7	Is the continuation position. Explained in a later chapter.	
8–11 12–72	*represent Area A represent Area B	These positions are used for writing the COBOL source program.
73–80	are used to identify the program.	

*Area A, positions 8 through 11, is reserved for the beginning of division headers, section-names, paragraph-names, level indicators, and certain level numbers. All other entries start at Area B, positions 12 through 72.

Figure 1.10 shows the special form from which a COBOL source program is written. Positions 1–6 of the coding form are used for sequencing and positions 8–72 are used to write COBOL statements. Position 7 is used as a continuation indicator.

Some entries in a COBOL program must begin in Area A (positions 8–11) and others must begin in Area B (positions 12–72).

Each sentence must be terminated by a period and at least one space (more spaces may be used).

COBOL CODING FORM

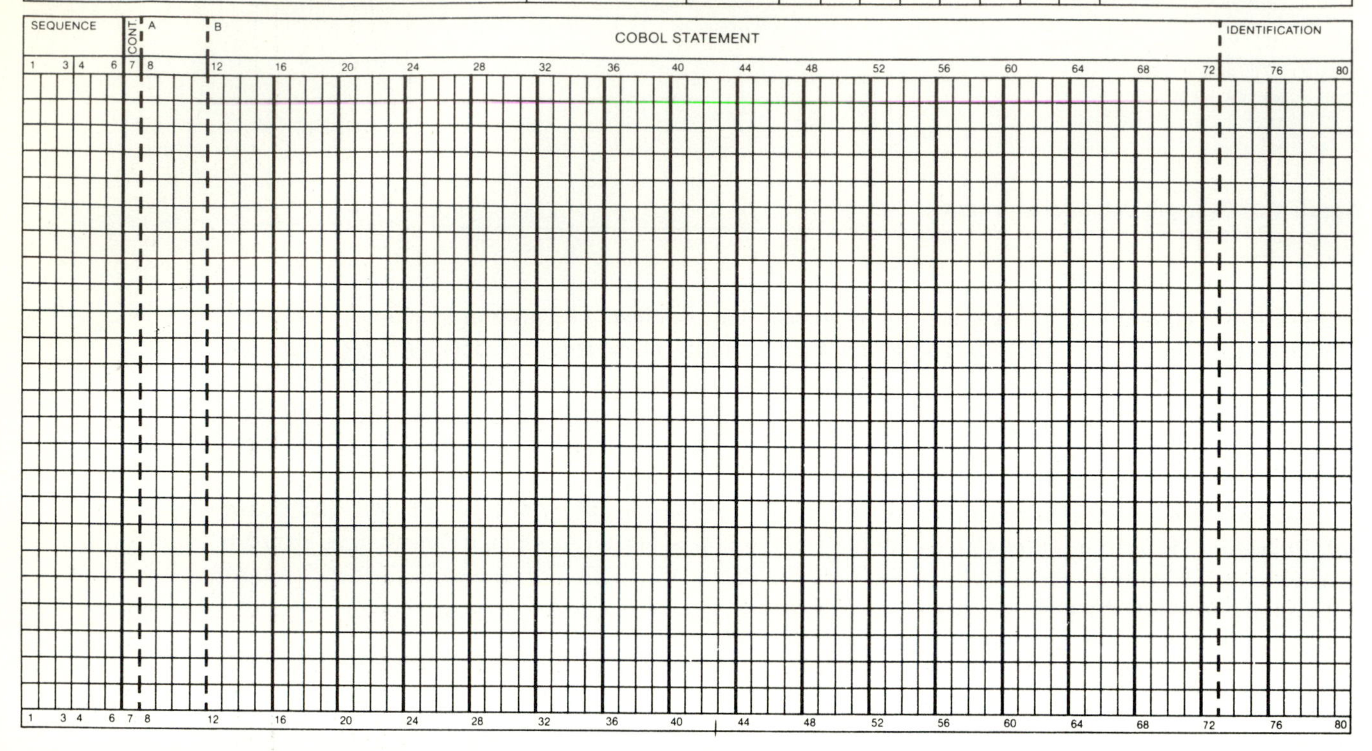

SYSTEM	PUNCHING INSTRUCTIONS	PAGE OF
PROGRAM	GRAPHIC	CARD FORM NUMBER
PROGRAMMER / DATE	PUNCH	

SEQUENCE | CONT. | A | B | COBOL STATEMENT | IDENTIFICATION

1 3 4 6 7 8 12 16 20 24 28 32 36 40 44 48 52 56 60 64 68 72 76 80

Figure 1.10 COBOL coding form.

To illustrate how a COBOL program is written, we will write a simplifed procedure whereby we will use a microcomputer as a terminal to access a central computer via a communication network. The Apple computer is used as the microcomputer and the central computer is the Honeywell series 60 level 66.

The initial steps involve identifying ourselves to the computer through the use of identification codes and passwords. The source program is then input and output is requested, such as

```
      TT01 18  01/28/82 20:05:00
106 36247
STATION TT01 CONNECTED ON NPS LINE 3

laccd  his timesharing on  01/28/82 at 20.071  channel 6244 ets

user id --         
password--         
#############          These statements identify us to the system.
iiiiiiiiiii        
hhhhhhhhhhhh       
qylcpbjvvxcd       

*
*
```

The COBOL source program statements are entered here (see the following for listing of original source program input and the COBOL source program formatted).

```
010$$M,T(:,8,16)
020$$MONITOR,JOUT
30$:IDENT:WEST,STEVE
040$:OPTION:CBL74
050$:CBL74
60$$SELECT(WEST/COBOL1)
070$:EXECUTE
080$:SYSOUT:P1,ORG
090$:DATA:C1
100$$SELECT(WEST/C1DATA)
110$:ENDJOB
```

These are job control statements that inform the system to compile and execute the source COBOL program.

```
*
SNUMB 8016W
8016W EXECUTING     @ 20.083
8016W -01    EXECUTING     @ 20.085
8016W -02    EXECUTING     @ 20.086
8016W OUTPUT WAITING     ID=SM. @ 20.088
normal termination

JOUT INVOKED.
function?
```

The aforementioned statements are issued by the system, stating that the program was compiled and executed and that the output is ready.

After proper acceptance by the central system, the computer will proceed to compile and execute our program and provide us with a listing of the source program as well as the output asked for. This program will list data as it is read into the computer system.

The following source program, together with its output, is intended only to show you at this time how a COBOL-compiled program appears.

This program can be useful to retain as it can be used to list a source program for desk checking prior to submittal to a computer for compilation and execution. This can save valuable time in debugging.

```
000100$$M,T(:,7,8,12,16,20,24,28,32)}
```

These are format codes for the program.

```
000110::IDENTIFICATION DIVISION.
000120::PROGRAM-ID.
000130:::READ-AND-WRITE.
000140:*PURPOSE.
000150:*:THIS PROGRAM WILL LIST DATA USING THE COBOL LANGUAGE.
000160::ENVIRONMENT DIVISION.
000170::CONFIGURATION SECTION.
000180::SOURCE-COMPUTER. LEVEL-66-ASCII.
000190::OBJECT-COMPUTER. LEVEL-66-ASCII.
000200::INPUT-OUTPUT SECTION.
000210::FILE-CONTROL.
000220:::SELECT FILE-IN ASSIGN TO C1-CARD-READER.
000230:::SELECT FILE-OUT ASSIGN TO P1-PRINTER.
000240::DATA DIVISION.
000250::FILE SECTION.
000260::FD:FILE-IN
000270:::CODE-SET IS GBCD
000280:::LABEL RECORDS ARE OMITTED
000290:::DATA RECORD IS READ-IN.
000300::01:READ-IN::::PIC X(80).
000310::FD:FILE-OUT
000320:::CODE-SET IS GBCD
000330:::LABEL RECORDS ARE OMITTED
000340:::DATA RECORD IS OUT-PUT.
000350::01:OUT-PUT.
000352:::05 CARRIAGE-CHAR:PIC X.
000354:::05:OUT-PUT-RECORD:PIC X(132).
000360::WORKING-STORAGE SECTION.
000370::01:FLAGS.
000380:::05:MORE-DATA-FLAG:PIC XXX VALUE "YES".
000390::::88 MORE-DATA:        VALUE "YES".
000400::::88 NO-MORE-DATA:        VALUE "NO".
000410::PROCEDURE DIVISION.
000420::MAIN-ROUTINE.
```

```
000430:::OPEN:INPUT    FILE-IN
000440::::OUTPUT  FILE-OUT.
000450:::READ FILE-IN
000452::::AT END MOVE "NO" TO MORE-DATA-FLAG.
000470:::PERFORM PROCESS-ROUTINE
000475::::UNTIL NO-MORE-DATA.
000480:::CLOSE:FILE:IN
000490::::FILE-OUT.
000500:::STOP RUN.
000510::PROCESS-ROUTINE.
000515:::MOVE SPACE TO CARRIAGE-CHAR.
000520:::MOVE READ-IN TO OUT-PUT-RECORD.
000530:::WRITE OUT-PUT
000535::::AFTER ADVANCING 1 LINE.
000540:::READ FILE-IN
000545::::AT END MOVE "NO" TO MORE-DATA-FLAG.
```

The aforementioned is the original COBOL source program input with line numbers. The line numbers are stripped as per formatted program below.

```
7015W 01    02-01-82    11.071     HISI SERIES 60 LEVEL 66 COBOL CB4.1DS-2A
                                 READ-A    PAGE       1

      ISN XSN    C     TEXT

        1          IDENTIFICATION DIVISION.
        2          PROGRAM-ID.
        3              READ-AND-WRITE.
        4         *PURPOSE.
        5         *    THIS PROGRAM WILL LIST DATA USING THE COBOL LANGUAGE.
        6          ENVIRONMENT DIVISION.
        7          CONFIGURATION SECTION.
        8          SOURCE-COMPUTER. LEVEL-66-ASCII.
        9          OBJECT-COMPUTER. LEVEL-66-ASCII.
       10          INPUT-OUTPUT SECTION.
       11          FILE-CONTROL.
       12              SELECT FILE-IN ASSIGN TO C1-CARD-READER.
       13              SELECT FILE-OUT ASSIGN TO P1-PRINTER.
       14          DATA DIVISION.
       15          FILE SECTION.
       16          FD  FILE-IN
       17              CODE-SET IS GBCD
       18              LABEL RECORDS ARE OMITTED
       19              DATA RECORD IS READ-IN.
       20          01  READ-IN               PIC X(80).
       21          FD  FILE-OUT
       22              CODE-SET IS GBCD
       23              LABEL RECORDS ARE OMITTED
       24              DATA RECORD IS OUT-PUT.
       25          01  OUT-PUT.
       26              05 CARRIAGE-CHAR     PIC X.
       27              05  OUT-PUT-RECORD   PIC X(132).
       28          WORKING-STORAGE SECTION.
       29          01  FLAGS.
       30              05  MORE-DATA-FLAG   PIC XXX VALUE "YES".
       31                  88 MORE-DATA             VALUE "YES".
       32                  88 NO-MORE-DATA          VALUE "NO".
       33          PROCEDURE DIVISION.
       34          MAIN-ROUTINE.
       35              OPEN     INPUT   FILE-IN
       36                       OUTPUT  FILE-OUT.
       37              READ FILE-IN
       38                       AT END MOVE "NO" TO MORE-DATA-FLAG.
       39              PERFORM PROCESS-ROUTINE
       40                       UNTIL NO-MORE-DATA.

7015W 01    02-01-82    11.071     HISI SERIES 60 LEVEL 66 COBOL CB4.1DS-2A
                                 READ-A    PAGE       2

      ISN XSN    C     TEXT

       41              CLOSE    FILE-IN
       42                       FILE-OUT.
       43              STOP RUN.
```

```
44         PROCESS-ROUTINE.
45             MOVE SPACE TO CARRIAGE-CHAR.
46             MOVE READ-IN TO OUT-PUT-RECORD.
47             WRITE OUT-PUT
48                     AFTER ADVANCING 1 LINE.
49             READ FILE-IN
50                     AT END MOVE "NO" TO MORE-DATA-FLAG.

THERE WERE 50 SOURCE INPUT LINES.
THERE WERE NO DIAGNOSTICS.
```

The aforementioned is the formatted COBOL source program compiled.

```
THIS IS A LISTING IN COBOL.
THIS IS AN OUTPUT LINE, TOO.
THIS IS ALSO AN OUTPUT LINE.
```

The aforementioned is the output from the program.

```
*
**cost:  $    0.46 to date:   $   551.11= 55%
**on at 13.178 - off at 13.241 on   02/01/82
```

The aforementioned shows the accounting distribution indicating the cost of the job and cost to date as well as time required to run the job.

An Address-Listing Program for Study

For further study, here is a program that uses a file. A program is to be written to list all records of a customer file. The customer file is in 80-position records containing the name and address of each customer. The customer listing is to contain the above information. The remarks that follow are identified with letters corresponding to the lettered sections of the program.

A. Identifies the program to the computer and provides other information necessary for the documentation of the program.

B. Specifies that the program is to be compiled and executed on an IBM-370 computer. The input-output devices to be used are as follows:

   A model 2540 card reader is to be used for input data.

   A model 1403 printer is to output the data in printed format.

C. One input file is to be used. The file contains 80-position records with the name and address of each customer.

D. The output file will contain 133 characters (one carriage control character and 132 print positions). There is a margin of 10 positions on the left with five blank spaces between each item in the output printed line.

E. These flags will be used to signal when the last record has been processed.

F. The MAIN-ROUTINE paragraph contains the overall logic of the program.

   The two files will be opened, which prepares them for processing and specifies which will be used for input and which will be used for output.

   The first READ statement reads the first record of the file. The AT END phrase specifies the operation to be performed after the last record has been processed. (Even though this is highly unlikely with the first record, the READ format requires an AT END phrase.) The program will then branch to the PROCESS-ROUTINE paragraph where the input data will be moved to the output areas ready for printing.

   The WRITE statement will print each line double space as specified.

   The READ statement will read all subsequent records.

   The PROCESS-ROUTINE paragraph will be repeated for each record until the last record has been processed and then the program will return to the next statement after PERFORM in the MAIN-ROUTINE paragraph.

In the MAIN-ROUTINE paragraph, the CLOSE statement, will terminate the processing of the files.

The STOP RUN statement will terminate the processing of the run and return the system to the operating system.

```
    IDENTIFICATION DIVISION.
    PROGRAM-ID.
        ADDRESS-LISTING.
    AUTHOR.
(A)     C FEINGOLD.
    DATE-WRITTEN.
        JAN 12 1982.
    DATE-COMPILED.
        JAN 12 1982.
   *PURPOSE.
        THIS PROGRAM LISTS ALL OF OUR CUSTOMERS.

    ENVIRONMENT DIVISION.
    CONFIGURATION SECTION.
    SOURCE-COMPUTER.                          IBM-370.
(B) OBJECT-COMPUTER.                          IBM-370.
    INPUT-OUTPUT SECTION.
        SELECT CUSTOMER-FILE-IN               ASSIGN TO UR-254OR-S-INFILE.
        SELECT CUSTOMER-REP-OUT               ASSIGN TO UR-1403-S-OUTFILE.

    DATA DIVISION.
    FILE SECTION.
    FD  CUSTOMER-FILE-IN
        LABEL RECORDS ARE OMITTED
        RECORD CONTAINS 80 CHARACTERS
        DATA RECORD IS CUSTOMER-RECORD-IN.
(C) 01  CUSTOMER-RECORD-IN.
        05  NAME-IN.
            10  INIT-IN                       PICTURE X(4).
            10  LAST-NAME-IN                  PICTURE X(12).
        05  HOME-ADDR-IN                      PICTURE X(40).
        05  FILLER                            PICTURE X(24).

    FD  CUSTOMER-REP-OUT
        LABEL RECORDS ARE OMITTED
        RECORD CONTAINS 133 CHARACTERS
        DATA RECORD IS CUSTOMER-RECORD-OUT.
    01  CUSTOMER-RECORD-OUT.
        05  FILLER                            PICTURE X(10).
(D)     05  NAME-OUT.
            10  LAST-NAME-OUT                 PICTURE X(12).
            10  FILLER                        PICTURE X(5).
            10  INIT-OUT                      PICTURE X(4).
            10  FILLER                        PICTURE X(5).
        05  HOME-ADDR-OUT                     PICTURE X(40).
        05  FILLER                            PICTURE X(57).

    WORKING-STORAGE SECTION.
    01  FLAGS.
(E)     05  MORE-DATA-FLAG                    PICTURE XXX      VALUE 'YES'.
            88  MORE-DATA                                      VALUE 'YES'.
            88  NO-MORE-DATA                                   VALUE 'NO'.

    PROCEDURE DIVISION.
    MAIN-ROUTINE.
        OPEN    INPUT     CUSTOMER-FILE-IN
                OUTPUT    CUSTOMER-REP-OUT.
        READ CUSTOMER-FILE-IN
            AT END MOVE 'NO' TO MORE-DATA-FLAG.
        PERFORM PROCESS-ROUTINE
            UNTIL NO-MORE-DATA.
        CLOSE   CUSTOMER-FILE-IN
                CUSTOMER-REP-OUT.
```

```
        STOP RUN.

    PROCESS-ROUTINE.
        MOVE SPACES TO CUSTOMER-RECORD-OUT.
        MOVE INIT-IN TO INIT-OUT.
        MOVE LAST-NAME-IN TO LAST-NAME-OUT.
        MOVE HOME-ADDR-IN TO HOME-ADDR-OUT.
        WRITE CUSTOMER-RECORD-OUT
            AFTER ADVANCING 2 LINES.
        READ CUSTOMER-FILE-IN
            AT END MOVE 'NO' TO MORE-DATA-FLAG.
```

Summary

A computer program is the outcome of the programmer's applied knowledge of the problem and the operation of the computer system. The programmer must analyze the problem, prepare a program to solve the particular problem and operate the program.

After a complete analysis of the problem statement, flowcharts and/or HIPO charts are written. (HIPO is an acronym for *H*ierarchy plus *I*nput, *P*rocess, and *O*utput.) The flowcharts and HIPO charts will be subsequently translated into a computer program which instructs the computer on how the data will be received and processed and what output should be developed.

Programming entails writing detailed instructions for the computer and making them work correctly. A computer program is a set of instructions arranged in proper sequence to cause the computer to perform a particular process. The steps required to develop computer programs are:

1. *Analysis* A problem must be thoroughly analyzed before any attempt at programming is made.
2. *Design* Certain techniques are used to design a program, such as HIPO charts and flowcharts.

   HIPO charts provide a means of precisely defining user requirements by expanding the essential lines of communication between the user and the data processing department. The flowchart is a graphic representation of the flow of information through a system in which the information is converted from source document to final output. A system flowchart represents the flow of data through all parts of a system without detailing the steps involved, whereas the program flowchart emphasizes the computer decisions and processes. The program flowchart is used to write the source program.

   Flowcharts depict the organization and logic of programs, whereas HIPO charts show the functions.
3. *Program development* When the program design has been completed, the actual programming is done. This includes a precise statement of the problem, a list of inputs, outputs desired and flowcharts.
4. *Testing* Programs must be tested by entering them into the computer system to be translated from the COBOL source language into the language of the particular computer.
5. *Implementation* When the program has been thoroughly tested, production runs can be made.
6. *Acceptance by management/user* The final affirmation of the program should be made by management and those who use the reports.

The COBOL language is a near-English programming language used throughout the world for programming business data processing applications.

The reasons for the popularity of COBOL are: it has been continuously standardized by repeated meetings of committees (the current version is ANSI COBOL 1974); it is designed to meet the needs of the users today as well as in the future; it is the only language translator supported by its users; it is included in more software packages than any other language; it is not plagued by computer obsolescence; and it has many self-documenting features that are easily understood by nonprogrammers.

COBOL is not truly a common language as certain small computers are excluded from the use of COBOL.

The principal advantages of COBOL are: it uses English-like statements; it uses pretested input/output modules; COBOL is not machine-dependent; because of its separate divisions, large COBOL programs may be broken down into segments; it provides a quick means of program implementation; it reduces the cost of computer conversion; it reduces costs for training programmers and guarantees a method of standard documentation; nonprogrammers can read COBOL programs; and simple diagnostic messages make it easy to debug COBOL programs.

The principal disadvantages of COBOL result from the failure of personnel to fully understand the language such as; expecting a single COBOL program to provide a permanent solution to a problem without ever reprogramming; assuming that the programmer need only be taught COBOL language without any knowledge of the hardware; COBOL will not generate a sophisticated program as one written in the actual language of the particular computer; and COBOL will operate only with computers with certain storage capacities.

The main objectives of COBOL are: to provide standardized elements in entry format that may be used in any computer; and to provide a source program that is easily understood by programmers and nonprogrammers and is oriented towards business applications.

The COBOL system consists of two basic elements; the *source program,* which is a set of instructions that carry out the logic of the particular data processing application, and the *compiler,* the intermediate routine that will convert the English-like statements of COBOL into computer-acceptable statements. The machine-coded program produced from the translation is referred to as the *object program,* which is then fed back to the computer to be processed with data.

The programmer's job does not end with the coding. The programmer must check the program for accuracy. The two major types of errors that cause difficulty in programs are logical errors, which result from poor analysis of the problem, and clerical errors, which occur during the coding stage, and which result from the incorrect use of the COBOL language.

There are four divisions in a COBOL program that must be coded in the following sequence: the Identification Division, which identifies the program to the computer; the Environment Division, which describes the computer to be used in the program and the hardware features to be used; the Data Division, which describes the characteristics of the data to be used; and the Procedure Division, which consists of a series of statements directing the processing of the data according to the program logic.

The COBOL coding form provides a method for writing COBOL source programs. Some entries in a COBOL program must begin in Area A (positions 8–11) and others must begin in Area B (positions 12–72). Position 7 is used as a continuation indicator.

The first program entails the use of a microcomputer acting as a terminal to access a central computer. It is a simple program that reads records and lists them in the same format. This program may be used by the programmer in the subsequent coding of COBOL programs to list programs and provide a desk check of the souce program before the compilation stage.

Questions for Review

1. What should a problem statement contain?
2. What is a computer program?
3. Briefly describe the necessary steps to develop computer programs.
4. How are HIPO charts and flowcharts used in the design phase of computer program development.
5. Why was the federal government interested in developing a common business language?
6. When did the first CODASYL committee meet and what were their major objectives?
7. What is meant by COBOL?
8. List the major reasons for the popularity of COBOL.
9. What is a common language? Is COBOL a common language?
10. What are the main advantages of COBOL as a programming language?
11. Briefly describe the main characteristics of the COBOL programming language.
12. What are the two basic elements of the COBOL system and their main purposes?
13. Briefly explain the important role of software in COBOL programming.
14. What is a compiler?
15. What is a source program? an object program?
16. What are the two major types of errors that cause difficulty in programs?
17. What are the four elements that must be provided for in a source program?
18. What are the main functions of each of the COBOL divisions?

Matching Questions

Match each item with its proper description.

_____ 1. Password
_____ 2. Compiler
_____ 3. Computer program
_____ 4. Clerical errors
_____ 5. Flowchart
_____ 6. Object program
_____ 7. Coding form
_____ 8. HIPO
_____ 9. Source program
_____10. Logical errors

A. Initial subprogram written by programmer that is to be translated.
B. Program broken down into modules which perform particular functions in the program.
C. Standard method for writing COBOL source programs.
D. Result from poor analysis of the problem.
E. Intermediate routine that converts English-like statements of COBOL into computer-acceptable instructions.
F. Result from erroneous use of symbolic names, operation codes, etc.
G. A set of instructions arranged in a proper sequence to cause the computer to perform a particular process.
H. Identify oneself to computer.
I. Graphic representation of the flow of information through a system in which the information is converted from the source document to the final result.
J. Machine-coded program produced from the translation.

Exercises

1. Identify the following steps in the compilation of a COBOL source program.

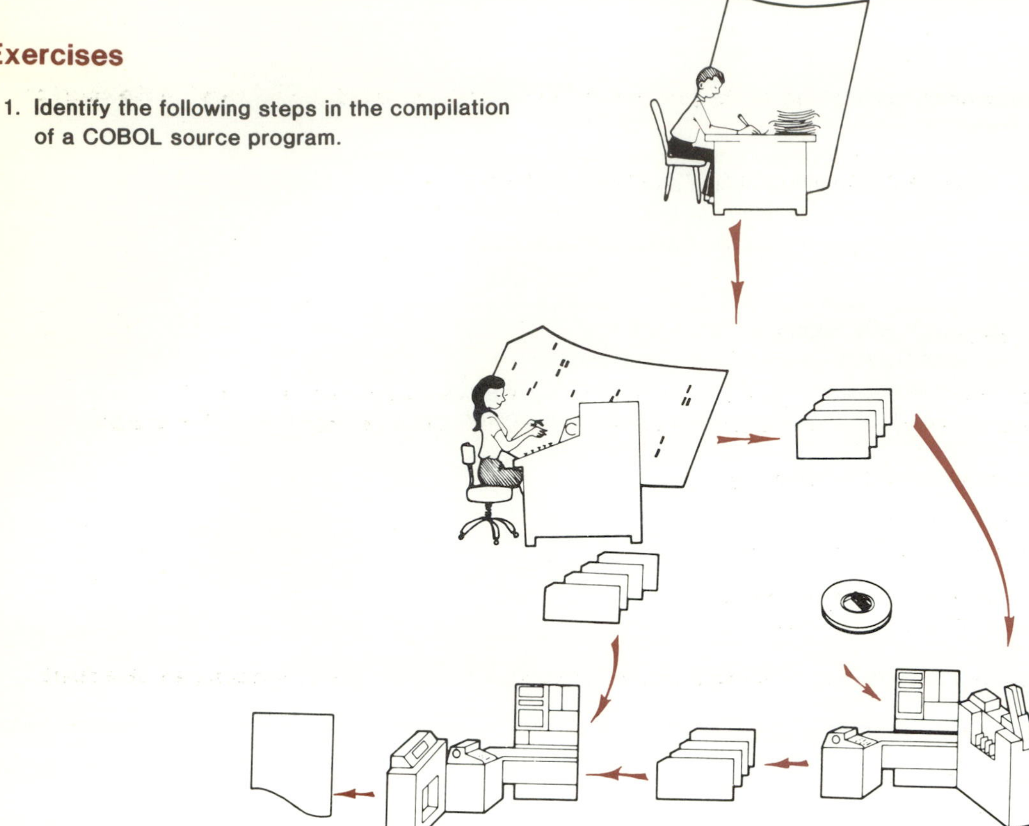

2. Number the following in their proper sequence of appearance in a COBOL source program.
   _____ a. DATA DIVISION.
   _____ b. ENVIRONMENT DIVISION.
   _____ c. PROCEDURE DIVISION.
   _____ d. IDENTIFICATION DIVISION.
3. Match each item with its description.

_____	1. Procedure Division	A.	Describes the computer to be used and the hardware features to be used in the program.
_____	2. Identification Division	B.	Describes the characteristics of the information to be processed.
_____	3. Data Division	C.	Identifies the program to the computer.
_____	4. Environment Division	D.	Specifies the logical steps necessary to process the data.

4. There are two areas defined on a COBOL coding form called area A and area B. Area A includes positions _____ - _____ and area B includes positions _____ - _____ .
5. The division of a COBOL program used to identify the program is the _____ Division.
6. You would look in the Environment Division of a COBOL program to identify the
   a. input/output required in processing.
   b. computer(s) on which the program is to be compiled and executed.
   Which statement(s) is correct?
7. If you wanted to know what type data is recorded on the records of a particular file, you would look in the _____ Division.
8. To identify input/output devices required by a COBOL program, you would look in the _____ Division.
9. The Data Division of a COBOL program
   a. identifies the program.
   b. describes records to be used in a program.
   Which statement(s) is correct?
10. The Procedure Division of a COBOL program contains instructions
    a. for solving a data-processing problem.
    b. specifying mathematical calculations.
    c. to direct input and output operations.
    d. all of the above.
    Which statement(s) is correct?

Problems

1. **In registering for classes, prepare a flowchart of the procedures and decisions necessary to enroll in the correct courses.**

2. **Given a file of records of students containing the following information:**

   Student ID Number
   Names
   Sex
   Age
   Class code: Freshman, sophomore, junior, senior.
   Grade point average.
   a. Prepare a program flowchart which will list the freshman female students between the ages of 18–20.
   b. Prepare a flowchart that will list all 21-year-old junior students with a grade point average of B (3.0 or better).

3. **Prepare a system flowchart showing the following:**

Inputs	Magnetic tape—Inventory file. Punched cards—Transaction cards.
Outputs	Magnetic tape—Updated Inventory file. Printer report—Transaction register.

4. **In an inventory file containing quantity, class of stock, stock number and amount, prepare a program flowchart that will**

   a. print all items of the file.
   b. accumulate all the amounts in class stock 34.
   c. count the number of items in the file.

5. **Prepare a HIPO chart to calculate SDI tax in a payroll procedure. Some of the information in the payroll record include the following:**

   a. Social Security number.
   b. Employee name.
   c. Accumulated earnings—previous week.
   d. Current weekly earnings.
   Required:
   a. Read all records.
   b. Check to see if accumulated earnings exceed $9,000 this week or last week.
   c. Calculate SDI tax for
      (1) employees who have not exceeded SDI limit this week or last week.
      (2) employees who have reached SDI limit this week.
      (3) employees who have reached SDI limit last week.

Writing COBOL Programs

Chapter Outline

Chapter Objectives

The learning objectives of this chapter are:

1. To describe the reference format which is a standard method for writing COBOL source programs.
2. To explain the precise rules for using the COBOL program form so that the number of diagnostic errors will be kept at a minimum.
3. To describe the program structure of COBOL as to divisions, sections, paragraphs, and entries so that the programmer is familiar with the overall picture of a COBOL program and the function of each unit within the program.
4. To describe the various elements of a COBOL program such as names, words, constants, punctuation characters, statements, sentences, and the role each element plays in a COBOL program.
5. To write a simple COBOL program using files.

Introduction

COBOL structure rules are an important part of the COBOL programming effort. The programmer should become familiar with these structure rules before any attempt is made to write COBOL programs. Many diagnostic errors are generated because these rules have been violated. So strict adherence to these rules can save valuable time in the debugging phase of COBOL programming.

Reference Format

The reference format, which provides a standard method for describing COBOL source programs, is defined in terms of character positions on an input/output device. The COBOL compiler accepts source programs written in reference format and produces an output listing of the source program input in reference format. The form for the reference is provided by the COBOL program sheet.

The rules for spacing given in the discussion of the reference format takes precedence over all other rules for spacing.

COBOL Program Sheet Format

The source program is written by the programmer on a COBOL Program Sheet Coding Form (figure 2.1). The program sheet provides the programmer with a standard method of writing COBOL source programs. Despite the necessary restrictions, the program is written in rather free form. However, there are precise rules for using this form. Unless these rules are followed, especially in respect to spacing, many diagnostic errors will be generated unnecessarily.

The program sheet is designed so that it can be readily entered into the computer. Each line of the form requires a separate entry. The form also provides for all positions of an eighty-position record. Unnumbered boxes will not be entered.

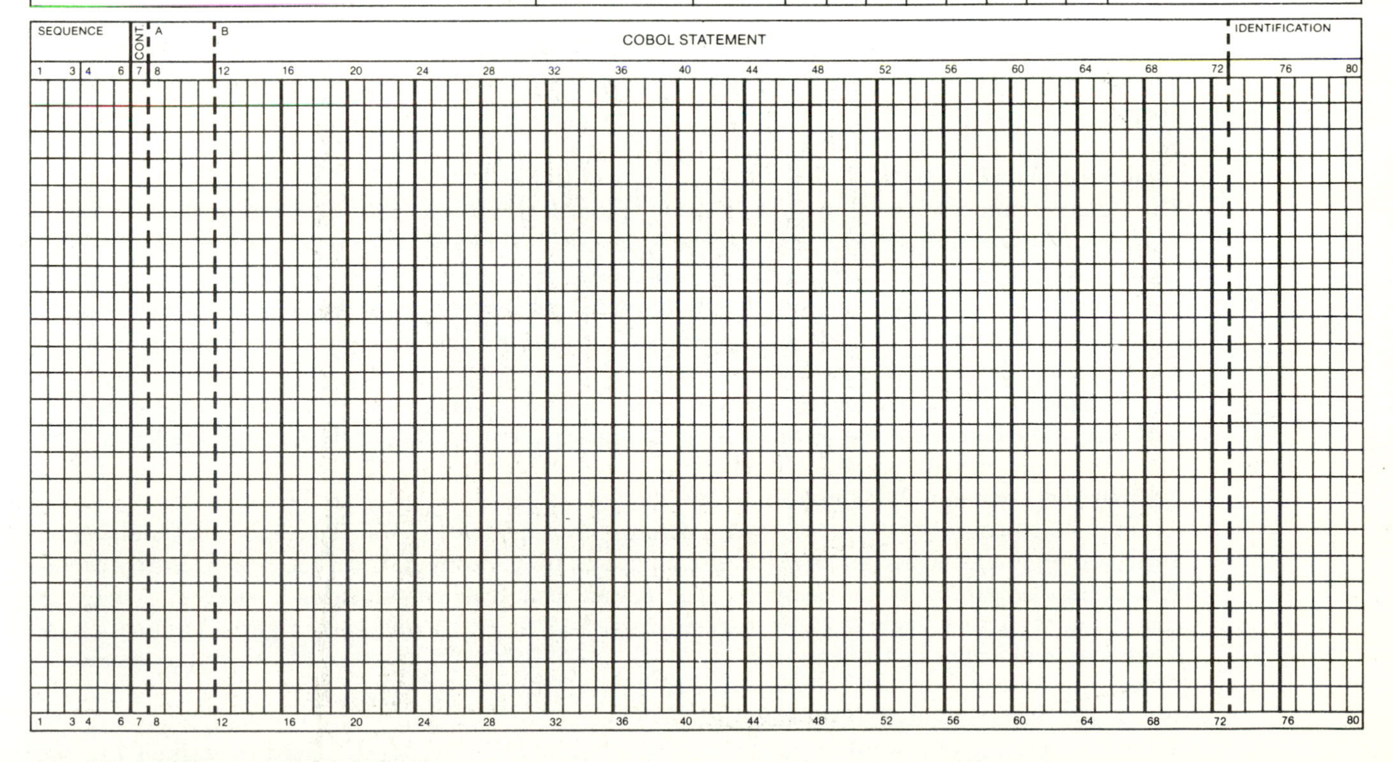
COBOL CODING FORM

SYSTEM		PUNCHING INSTRUCTIONS	PAGE OF
PROGRAM		GRAPHIC	CARD FORM NUMBER
PROGRAMMER	DATE	PUNCH	

SEQUENCE	CONT.	A	B	COBOL STATEMENT	IDENTIFICATION
1 3 4 6	7	8	12	16 20 24 28 32 36 40 44 48 52 56 60 64 68 72	76 80

Figure 2.1 COBOL program sheet.

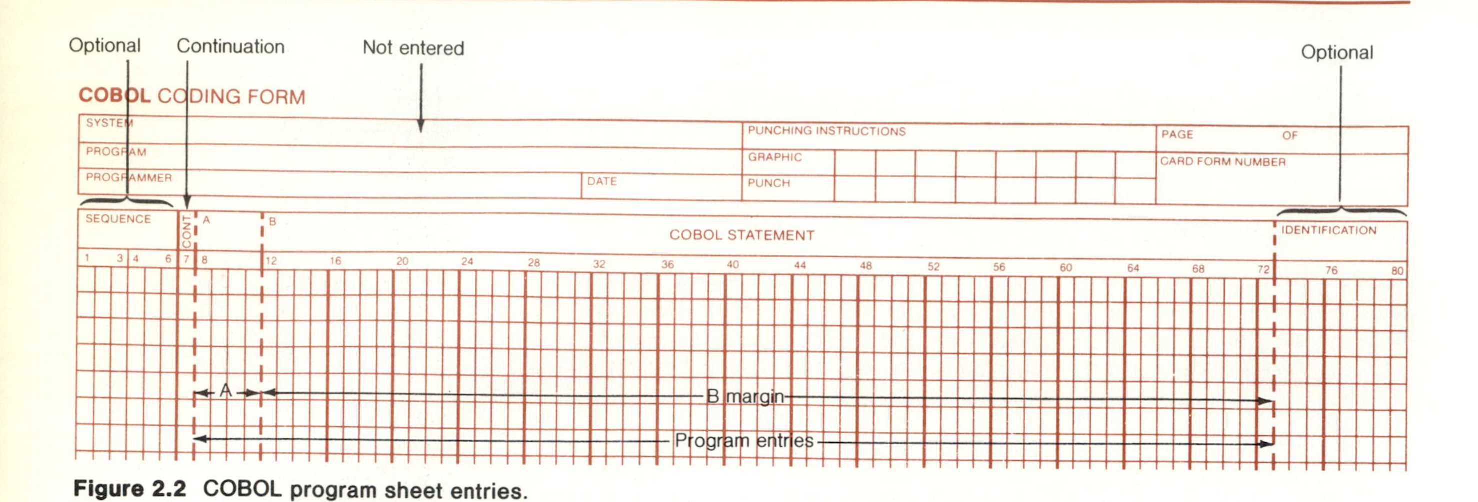

Figure 2.2 COBOL program sheet entries.

Care should be taken in the sequence of the instructions as the COBOL compiler will execute the program exactly as written. Entries from the program sheet serve as the initial input medium to the COBOL compiler. The compiler accepts the source program as written in the prescribed program sheet reference format and produces an output listing in the same format. Ample room in sequence numbers should be left for possible insertion of additional or "patch" instructions.

All characters should be written distinctly so that there aren't any questions as to the nature of the instructions. A data entry operator will enter them exactly as written. Punctuation symbols, especially periods, can cause numerous errors when incorrectly used. And characters such as Z and 2, zero and O (letter), should be clearly indicated in the punching instructions of the form so that there will be no doubt as to what characters were intended.

Reference Format Representation

The sequence number area occupies six character positions (1–6). The continuation indicator area is the seventh character position. Area A occupies character positions 8, 9, 10, and 11. Area B occupies 61 character positions; it begins immediately in character position 12 and terminates in character position 72. The areas are shown in figure 2.2.

Sequence Number (1–6)

The sequence number consisting of six digits is used to identify numerically each record to the COBOL compiler. The use of these sequence numbers is optional and has no effect on the object program. But it is a good practice to use these sequence numbers, as they will provide a control on the sequence of the lines if a correction or if an insertion is to be made into the program. The single-character insertion permits one or more lines of coding to be inserted between existing lines. Position 6 normally contains a zero except where insertions are made.

If sequence numbers are used, they must be in ascending sequence. The compiler will check the sequence and indicate any sequence errors. No sequence check is made if the columns are left blank.

Continuation Indicator (7) Hyphen (-)

A coded statement may not extend beyond position 72 of the coding form. To continue a statement on a succeeding line, it is not necessary to use all spaces up to position 72 on the first line. Any excess spaces are disregarded by the compiler. If a statement must be continued on a succeeding line, the continuing word must begin at the B margin (position 12 or any column to the right of position 12). A continuation indicator is not necessary.

Words and Numeric Literals Continuation

When the splitting of a word or numeric literal is necessary, a hyphen is placed in position 7 of the continuation line to indicate that the first nonblank character follows the last nonblank character of the

COBOL CODING FORM

```
SYSTEM                                   PUNCHING INSTRUCTIONS                  PAGE        OF
PROGRAM                                  GRAPHIC  N  A  N  A  N  A              CARD FORM NUMBER
PROGRAMMER                       DATE    PUNCH    Ø  O  1  I  2  Z
                         N- NUMERIC          A- ALPHABETIC

SEQUENCE  CONT A   B              COBOL STATEMENT                              IDENTIFICATION
1  3 4  6  7   8   12

ØØ3Ø2Ø     PROCESS-ROUTINE.
ØØ3Ø3Ø         READ FILE-IN AT END PERFORM CLOSE-PROC.       MOVE INVENTORY-
ØØ3Ø4Ø-        MASTER TO NEW-MASTER.
          *    THE CONTINUED COBOL WORD IS INVENTORY-MASTER
ØØ3Ø6Ø         ADD TOTAL-AMT-1, TOTAL-AMT-2, GIVING NEW-TOTAL.     MULTIPLY 741
ØØ3Ø7Ø-        675.3 BY NEW GIVING NEW-AMT.
          *    THE CONTINUED COBOL NUMERIC LITERAL IS 741675.3
```

The proper coding procedure for the above would be to start the MOVE INVENTORY statement on a new line, thus avoiding the continuation of the COBOL word, and at the same time adhering to the coding principal of having each new statement start on a new line. The continuation of a numeric literal should be avoided and never used unless absolutely necessary. The MULTIPLY statement should have been written on a new line.

Figure 2.3 Continued COBOL word and numeric literals—example.

continued line without an intervening space. If the hyphen is omitted, the compiler will assume that the word or numeric literal was complete on the previous line and will insert an automatic blank character after it. The examples in figure 2.3 show continuation of COBOL words and numeric literals on succeeding lines. This type of coding should be avoided; it is confusing to the reader of the program and may lead to errors in programming.

Nonnumeric Literal Continuation

Unlike the word or numeric literal, the nonnumeric literal, when continued, must be carried out to position 72, since spaces are considered part of the literal. To split a nonnumeric literal, the last character must be written in position 72, a hyphen placed in position 7 of the continuation (next) line, and a quotation mark placed in the B margin (position 12 or anywhere to the right of position 12) and the literal continued. A final quotation mark at the end of the literal terminates the entry (see figure 2.4).

It is strongly recommended that statements not be continued on succeeding lines, if possible, as this may be confusing to the reader of the program.

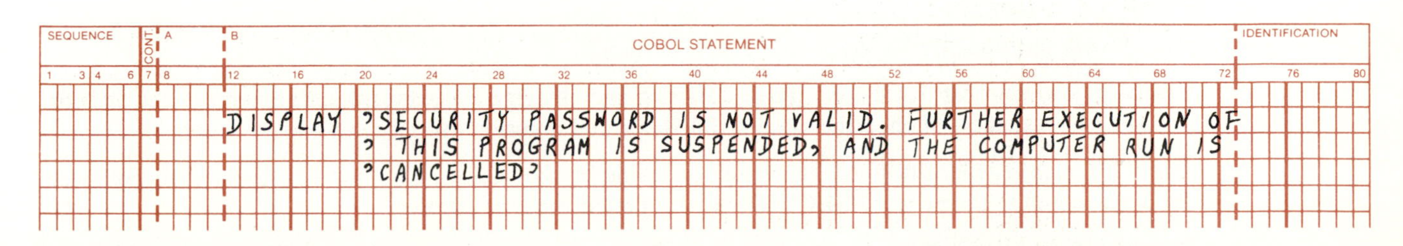

```
SEQUENCE  CONT A   B              COBOL STATEMENT                              IDENTIFICATION
1  3 4  6  7   8   12

               DISPLAY 'SECURITY PASSWORD IS NOT VALID. FURTHER EXECUTION OF
                       ' THIS PROGRAM IS SUSPENDED, AND THE COMPUTER RUN IS
                       'CANCELLED'
```

Figure 2.4 Continuation of a nonnumeric literal.

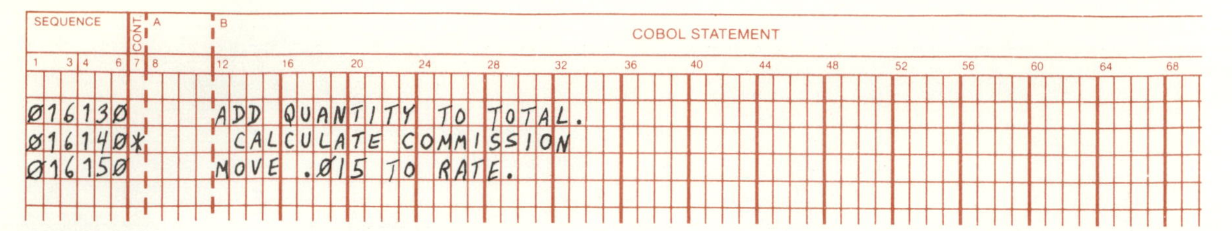

Figure 2.5 Comment line—example.

Comment Line

An asterisk (*) in position 7 indicates that the entire line is a *comment line.* A comment line is printed on the compiler listing as documentation only. A comment line can appear as any line in the source program. Starting in position 8 of the comment line, any combination of characters from the COBOL character set may be used (figure 2.5).

How Elements Are Written in Entries (Positions 8–72)

The text of a COBOL program is made up of statements. Each statement is a program entry. Each element in an entry must be written in full on one line. An exception to the rule is a nonnumeric literal which may be written in two or more lines if necessary.

Spacing between Elements

Each element must be separated from the next element by at least one space. (*Note:* If an element ends in position 72, it is treated as if it were followed by a space.) There are exceptions to the rule; certain symbols must be written directly next to other elements with no spaces left between.

The following symbols must not be followed by a space.

(Left parenthesis.

" Beginning quotation mark. (However, a space may be the first character of the literal enclosed by quotation marks.)

The following symbols must not be preceded by a space. (ANSI 1974 COBOL relaxed the punctuation rules with regard to spaces. Spaces may now optionally precede the comma, period, or semicolon and may optionally precede or follow a left parenthesis.)

. Period used to end an entry.

, ; Comma or semicolon used to punctuate an entry.

) Right parenthesis.

" Ending quotation mark (However, a space may be the last character of the literal enclosed by quotation marks.)

Rules for Program Entries

Some program entries in a COBOL program must begin in area A and others must begin in area B. An entry beginning in area A should begin in the first character position in that area.

Entries that begin in area A. The following entries are required to be in area A (character positions 8–11); they customarily belong in character position 8.

Division headers must be the first line in a division and on a line by themselves. The name of the division followed by a space and then the word DIVISION and a period must be entered.

Section headers generally must be written on a line by themselves. The name of the section followed by a space and then the word SECTION and a period must be entered.

Paragraph headers need not be written on a line by themselves, but may optionally be written on a separate line. The paragraph name is followed by a period and a space. Succeeding statements may be

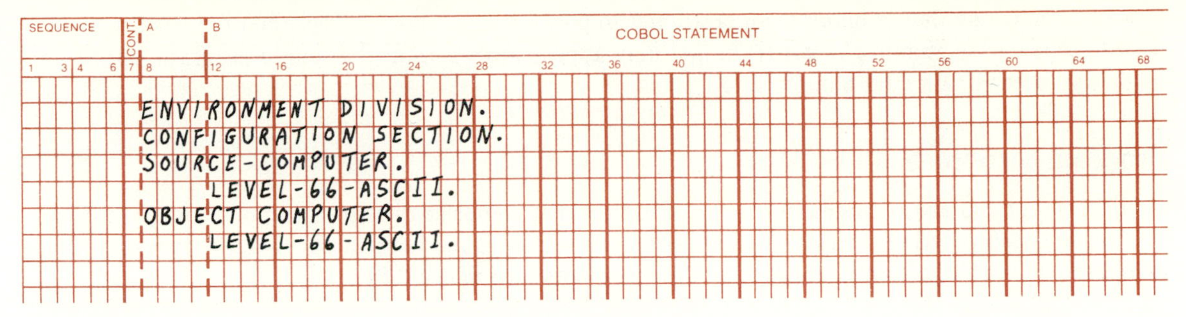

Figure 2.6 Division, section, paragraph headers—examples.

written on the same line starting in area B. Examples of division, section, and paragraph headers appear in figure 2.6.

File descriptions (found in the Data Division) start with a two-letter reserved word called "level indicator," such as FD; the level indicator must be written in area A, but the rest of the file description must be written in area B (see figure 2.7).

Two special headers in the Procedure Division—DECLARATIVES and END DECLARATIVES must be written on a line by themselves.

Level numbers (01–49, 66, 77, 88) found in the item description entries in the Data Division (as distinct from the level indicators mentioned above) may begin in area A if so desired. Usually 01 and 77 level numbers, which are required to be written in area A, are written in area A with other level numbers indented in area B to improve readability.

An entry that is required to start in area A must begin on a new line. Also, if an entry is required to appear on a line by itself, the next entry must begin on a new line.

If an entry is too long to be completed on one line and must be continued on another line, the continuation of the entry is written in area B of the next line (see continuation indicator rules).

Good practice dictates the writing of short entries, leaving the remainder of a line blank. Individual statements on separate lines aid in the debugging of the COBOL program and increase the readability of the program. Insertions and corrections of programs are simplified by using short entry statements on single lines.

Identification Code (73–80)

These columns are used for the names of the program for identification purposes. Any character, including blanks, in the COBOL character set may be used. These codes are particularly useful in keeping programs separate from each other. A group of COBOL program entries could be checked in positions 73–80 to assure that there are no other program entries in the group. This code has no effect on the object program or compilation, and may be left blank.

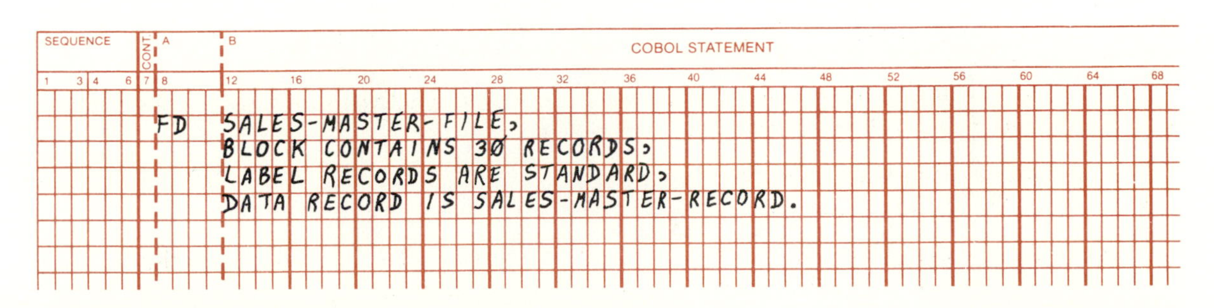

Figure 2.7 File description entry—sample entries.

Blank Lines

A blank line is one that is blank from position 7 to position 72 inclusive. These blank lines are usually used to separate segments of a program and may appear anywhere in the source program, except immediately preceding a continuation line.

Program Structure

COBOL programs are arranged in a series of entries that comprise divisions, sections, and paragraphs. A division is composed of a series of sections, while a section is made up of paragraphs. Paragraphs are composed of a series of sentences containing statements (see figure 2.8).

Divisions

Four divisions are required in every COBOL program. They are Identification, Environment, Data, and Procedure. These divisions must always be written in the foregoing sequence. A fixed name header consisting of the division name followed by a space, the word DIVISION, and a period must appear on a line by itself.

Sections

All divisions do not necessarily contain sections. The Environment and Data Divisions always contain sections with fixed names. Sections are never found in the Identification Division, while the Procedure Division sections are optional and are created by the programmer if needed.

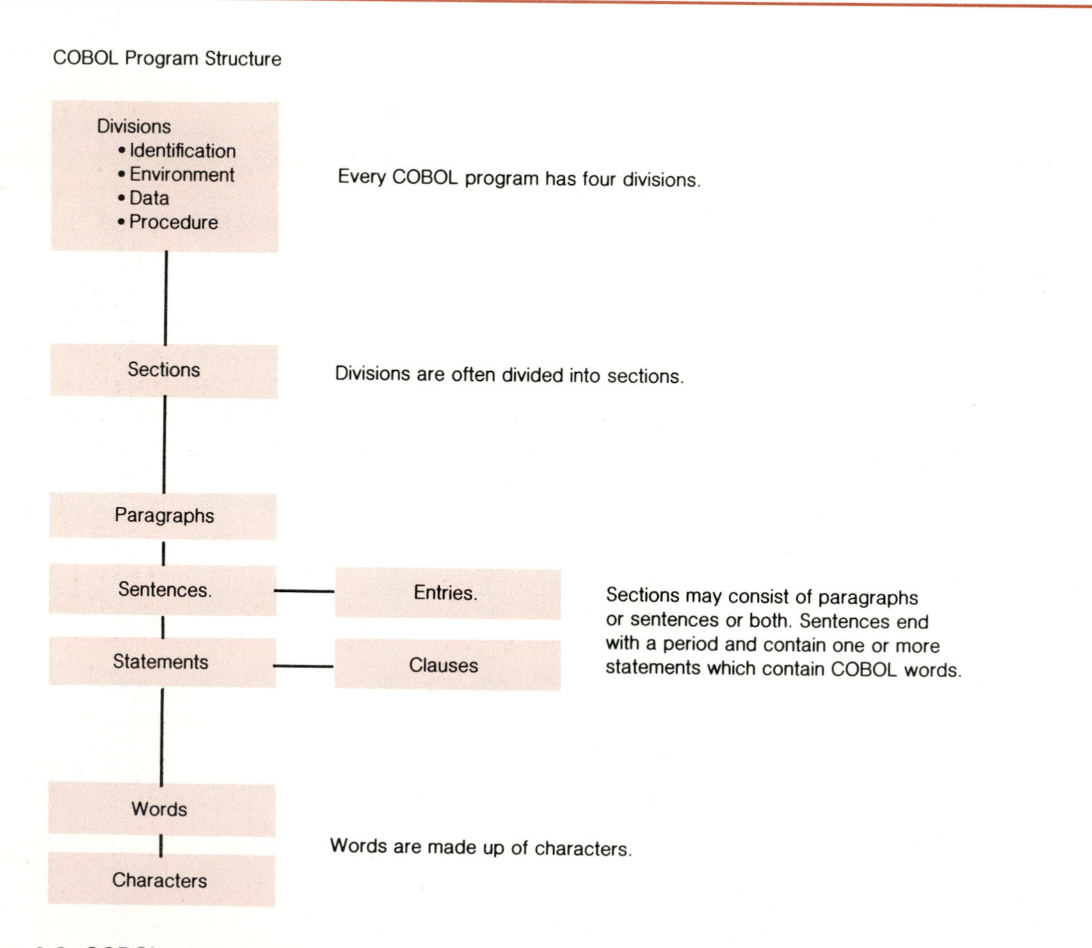

Figure 2.8 COBOL program structure.

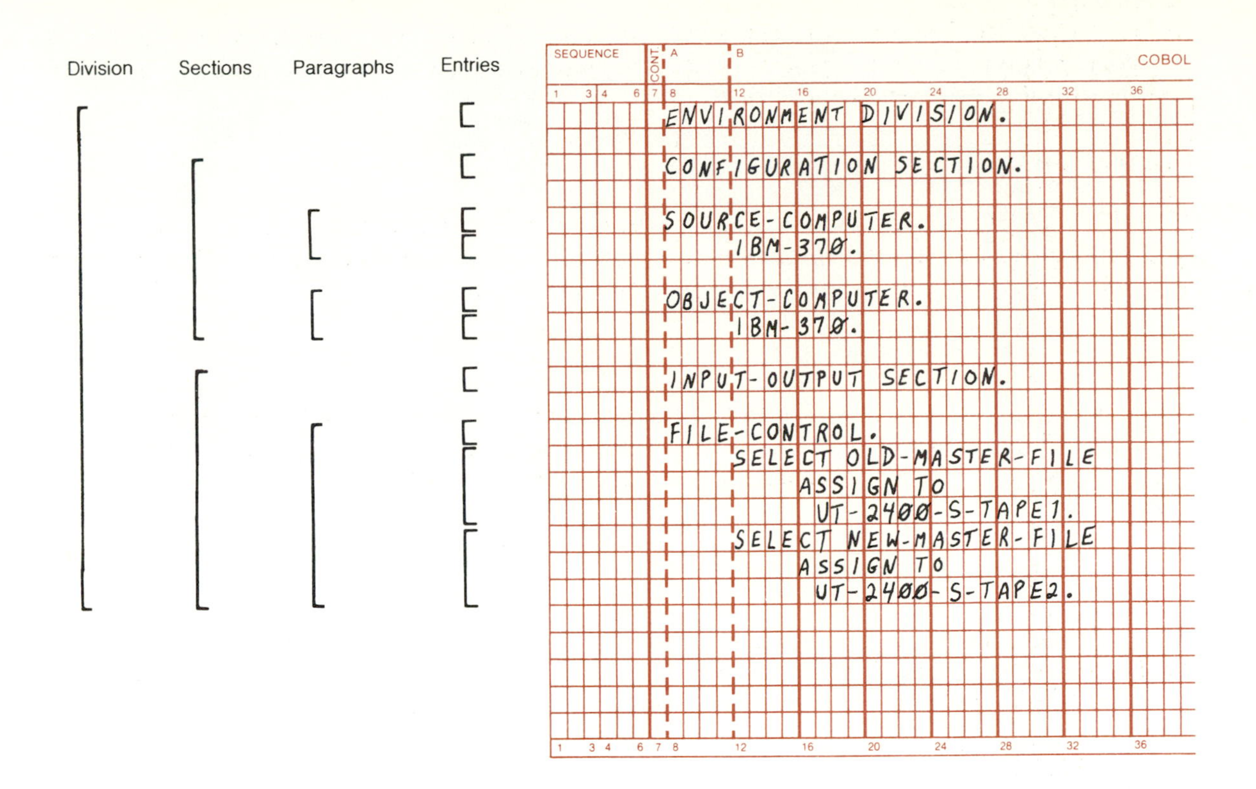

Division
Sections
Paragraphs
Entries
SEQUENCE
CONT.
A
B
COBOL
ENVIRONMENT DIVISION.
CONFIGURATION SECTION.
SOURCE-COMPUTER.
IBM-370.
OBJECT-COMPUTER.
IBM-370.
INPUT-OUTPUT SECTION.
FILE-CONTROL.
SELECT OLD-MASTER-FILE
ASSIGN TO
UT-2400-S-TAPE1.
SELECT NEW-MASTER-FILE
ASSIGN TO
UT-2400-S-TAPE2.

Figure 2.9 COBOL program structure—example.

The beginning of each section is preceded by the name of the section followed by a space, the word SECTION, and a period. The header must appear on a line by itself unless otherwise noted. (See rules for margin entries.)

A section consists of paragraphs in the Environment Division and Procedure Division and entries in the Data Division. The section ends immediately before the next section or at the end of the division.

Paragraphs

A paragraph is a segment or module performing a function. All divisions except the Data Division contain paragraphs. In the Identification and Environment Divisions, the paragraphs all have fixed names. The paragraph names in the Procedure Division are supplied by the programmer.

Each paragraph is identified by a paragraph name followed by a period and a space. Paragraph headers *do not* contain the word PARAGRAPH. The paragraph header need not appear on a line by itself; however, it must be the first entry, and can be followed on the same line by a series of entries. (A paragraph header entry consists of either a reserved word or a data-name and a period.)

A paragraph ends immediately before the next paragraph-name or section-name or at the end of the division.

Entries

Entries (sentences) consist of a series of statements terminated by a period and a space. These statements must follow precise format rules as to sequence.

In order to write a COBOL program, the programmer should become familiar with the basic components of COBOL programming. There are many terms, rules, entry formats, and program structures to be learned before any attempt at COBOL programming is made. Figure 2.9 shows examples of the four structural elements discussed.

Digits 0 through 9

Letters A through Z

Special characters:

	Blank or space
+	Plus sign
−	Minus sign or hyphen
*	Check protection symbol, asterisk
/	Slash
=	Equal sign
>	Inequality sign (greater than)
<	Inequality sign (less than)
$	Dollar sign
,	Comma
.	Period or decimal point
'	Quotation mark
(	Left parenthesis
)	Right parenthesis
;	Semicolon

The following characters are used for words:

0 through 9
A through Z
- (Hyphen)

The following characters are used for punctuation:

'	Quotation mark
(	Left parenthesis
)	Right parenthesis
,	Comma
.	Period
;	Semicolon

The following characters are used in arithmetic expressions:

+	Addition
−	Subtraction
*	Multiplication
/	Division
**	Exponentiation

The following characters are used in relation tests:

>	Greater than
<	Less than
=	Equal to

Figure 2.10 COBOL character set.

Terms

Source Computer. This computer is used to compile the source program and is usually the same computer that is used for the object computer.

Object Computer. This is the computer upon which the machine language program will be processed.

Character Set. The most basic and indivisible unit of the language is the character.

The complete set of COBOL characters consists of fifty-one characters. These are the characters (alphabetic, numeric, and special characters or symbols) the manufacturer has included in the COBOL programming package (see figure 2.10).

Names

Names are a means of establishing words to identify certain data within a program. A symbolic name is attached to an item that is being used in the program. All reference to the item will be through that name, although the value may change many times throughout the execution of the program. The name must be unique or identified with the particular group of which it is a part.

Rules for the Assignment of Names

1. Names may range from one to thirty characters in length.
2. No spaces (blanks) may appear within a name.
3. Names may be formed from the alphabet, numerals, and the hyphen. No special characters may appear in a name except the hyphen.
4. Although the hyphen may appear in a name, no name may begin or end with a hyphen.
5. The paragraph-name may consist entirely of numerals, but all other names must have at least one alphabetic character.
6. Names which are identical must be qualified with a higher level name (see qualification rules).

Types of Names

Data-Names

A grouping of contiguous characters treated as data is called a *data item.* Data-names are words that are assigned by the programmer to identify data items in the COBOL program. For example, TRAN-AMOUNT could be the data-name of an item containing the amount of a transaction. Here are some other examples of data names:

```
DISBURSEMENTS
    QUANTITY
    RECORD-DATE
    DOCUMENT-NUMBER
    CLASS-STOCK
    STOCK-NUMBER
    UNIT-PRICE
    AMOUNT
    UNIT
    DESCRIPTION
```

Data-names are devised and used by the programmer as needed in a program. The programmer must name and define each data item in a program so that the item can be accessed during the execution of the program. For example, if a program is written to process incoming accounts, one of the data items in the record would probably be the account number. The programmer assigns a name such as ACCOUNT-NUMBER to this data item and uses the name ACCOUNT-NUMBER each time the item is referred to.

If the same data-name is assigned to more than one data item then the data item must be made unique when referred to in the Procedure Division. A nonunique name must be followed by the necessary words that make it unique (see Qualification of Data-Names). All data items in the Data Division must be identified by a unique name or qualified data-names.

Paragraph-Names

Paragraph-names are data-names that are attached to various segments of the Procedure Division. These names are used for reference by the program in a decision-making operation. The basic concept of computer programming is the ability of the program to leave the sequential order to another part of the program for further processing.

Section-Names

Section-names are data-names that are attached to various sections of the Procedure Division. Figure 2.11 shows examples of paragraph-names and section-names.

Section-names in the Environment and Data Divisions are fixed while section-names in the Procedure Division are created by the programmer as needed in the program. The Procedure Division section-name must abide by all rules for creation of data-names.

Qualification of Data-Names

Every data-name that defines an element in a COBOL source program must be unique, either because no other name has the identical spelling and hyphenation, or because the name exists within a hierarchy of names such that references to the name can be made unique by mentioning one or more of the higher levels of the hierarchy. The higher levels are called *qualifiers* and the process that specifies uniqueness is called *qualification.*

Qualification is accomplished by placing a data-name or paragraph-name, and one or more phrases, each composed of the qualifier preceded by IN or OF (IN and OF are logically equivalent). Thus, if an item in MASTER-RECORD is called STOCK-NUMBER, and an item in DETAIL-RECORD is also called STOCK-NUMBER, then the data-name must appear in qualified form, since the data-name STOCK-NUMBER is not unique. All references to this name must appear as STOCK-NUMBER OF MASTER-RECORD or STOCK-NUMBER IN DETAIL-RECORD, whichever name is intended.

```
PROCEDURE DIVISION.
BEGIN.
    DISPLAY ' START-EXAMPLE ' UPON CONSOLE.
    SORT SORTFILE.
        DESCENDING KEY KEY1, KEY2,
        ASCENDING KEY KEY3,
        DESCENDING KEY KEY4
        INPUT PROCEDURE INP-PROC
        OUTPUT PROCEDURE OUT-PROC.
    STOP RUN.

INP-PROC SECTION.
INPUT-ROUTINE.
    OPEN     INPUT     INFILE.
    READ INFILE
        AT END MOVE 'NO' TO MORE-INPUT-FLAG.
    PERFORM PROCESS-INPUT-ROUTINE
        UNTIL NO-MORE-INPUT.
    CLOSE INFILE.
    DISPLAY ' END OF INPUT ' UPON CONSOLE.

PROCESS-INPUT-ROUTINE.
    MOVE INREC TO SRT-REC.
    RELEASE SRT-REC.
    READ INFILE
        AT END MOVE 'NO' TO MORE-INPUT-FLAG.

OUT-PROC SECTION.
OUTPUT-ROUTINE.
    OPEN     OUTPUT     OUTFILE.
    RETURN SORTFILE
        AT END MOVE 'NO' TO MORE-OUTPUT-FLAG.
    PERFORM PROCESS-OUTPUT-ROUTINE
        UNTIL NO-MORE-OUTPUT.
    CLOSE     OUTFILE.
    DISPLAY ' END OF OUTPUT ' UPON CONSOLE.

PROCESS-OUTPUT-ROUTINE.
    MOVE SRT-REC TO OUTREC.
    WRITE OUTREC.
    RETURN SORTFILE
        AT END MOVE 'NO' TO MORE-OUTPUT-FLAG.
```

Section-names

Paragraph-names

Paragraph-names

Figure 2.11 **Paragraph-Names and Section-Names—examples.**

The following lines show examples of such identical data-names and four statements that qualify them.

```
01  MASTER-RECORD.
    05  ID-INFO
        10  STOCK-NUMBER
            .
            .
            .
01  DETAIL-RECORD.
    05  ID-INFO
        10  STOCK-NUMBER
            .
            .
            .
MOVE STOCK-NUMBER OF ID-INFO IN MASTER-RECORD . . .
MOVE STOCK-NUMBER IN ID-INFO OF DETAIL-RECORD . . .
MOVE STOCK-NUMBER OF MASTER-RECORD . . .
MOVE STOCK-NUMBER OF DETAIL-RECORD . . .
```

The aforementioned example illustrates a number of points.

1. Either connector (IN or OF) may be used.
2. The order in which the qualifiers appear must proceed from lower level to higher level.
3. It is not necessary to include all intermediate levels of qualification unless a data-name could not be uniquely determined if such intermediate levels were omitted.

Enough qualification must be mentioned to make the name unique; however, it may not be necessary to mention all levels of the hierarchy. Within the Data Division, all data-names used for qualification must be associated with a level indicator or level number. Therefore, two identical data-names must not appear as entries subordinate to a group item unless they are capable of being made unique through qualification. In the Procedure Division, two identical paragraph-names must not appear in the same section.

The rules for qualification are as follows:

1. Each qualifier must be of a successive higher level and within the same hierarchy as the name it qualifies.
2. The same name must not appear at two levels in a hierarchy.
3. If a data-name is assigned to more than one data item in a source program, the data-name must be qualified each time it is referred to in the Environment, Data, and Procedure divisions.
4. A paragraph-name must not be duplicated within a section. When a paragraph-name is qualified by a section-name, the word SECTION must not appear. A paragraph-name need not be qualified when referred to from within the same section.
5. A data-name can be qualified even though it does not need qualification.

Words

A COBOL word is a character-string (a character-string is a sequence of contiguous characters) chosen from the character set that form a user-defined word or a reserved word. A word is followed by a space or by a right parenthesis, comma, or semicolon.

User-defined words are COBOL words that must be supplied by the user to satisfy the format of a statement. The rules for the formation of user-defined words are the same as those for the formation of names.

Reserved Words

A reserved word is a COBOL word that is one of a specified list of words that may be used in a COBOL source program, but must not appear in the program as user-defined words. Reserved words can be used only as indicated in the general formats.

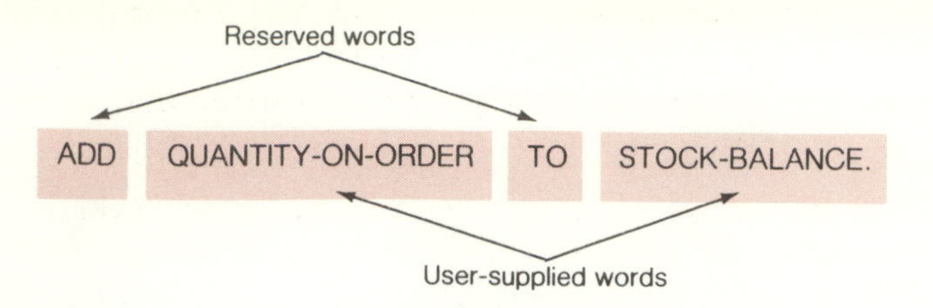

Figure 2.12 Reserved words and user-supplied words—example.

Reserved words have preassigned meanings and must not be altered, misspelled, or changed in any manner from the specific purpose of the word. Each of these COBOL reserved words has a special meaning to the compiler; hence, it should not be used out of context. COBOL reserved words may appear in nonnumeric literals (enclosed in quotation marks). When appearing in this form, they lose their meanings as reserved words; therefore they violate no syntactical rules. Figure 2.12 shows both user-supplied and reserved words.

A list of reserved words is available for all computers and must be checked before attempting to program, since there are slight differences in the lists for different computers.

Interpretation of Words Used in COBOL Statements

The foregoing are used in describing the format of the COBOL statements and are not used in the actual programming of the statements. The notation to be used in the remainder of the text to show the general forms of COBOL statements is shown in figure 2.13.

Reserved Words. Reserved words are printed entirely in upper-case (capital) letters.

Key Words. A key word is a reserved word whose presence is required when the format in which the word appears is used in a source program. The use of the reserved word is essential to the meaning and the structure of the COBOL statement. Within each format such words are upper case and underlined. The three types of key words are:

1. Verbs such as ADD and READ.
2. Required words which appear in statement and entry format.
3. Words which have a specific functional meaning such as NEGATIVE, SECTION, etc.

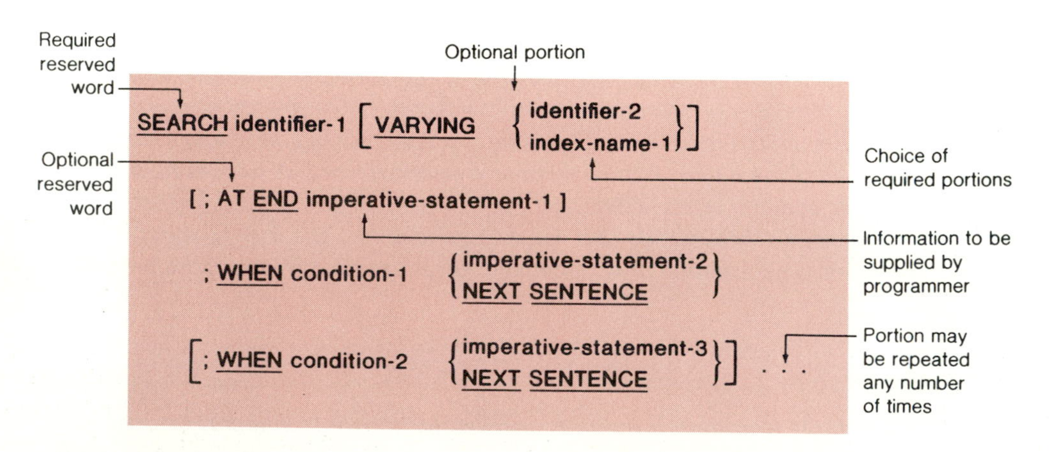

Figure 2.13 COBOL words in a general format of a COBOL statement.

Optional Words. Reserved words become optional words in the format in which they appear if they are not underlined. These words appear at the user's option to improve the readability. The presence or absence of an optional word does not affect the compilation of the program. However, misspelling of an optional word or its replacement by another word is prohibited.

Lower Case Letter Words. Words printed in lower-case format represent information that must be supplied by the programmer.

Bracketed Words [ ]. Words appearing in brackets must be included or omitted depending on the requirements of the program.

Braced Words { }. Braces enclosing words indicate that at least one of the words enclosed must be included.

Ellipsis . . . . Ellipsis points immediately following the format indicate that the words may appear any number of times.

Constants

A constant is an actual value of data that remains unchanged during the execution of the program. The value for the constant is supplied by the programmer at the time the program is loaded into storage.

There are two types of constants used in COBOL programming: literals and figurative constants.

Literals

Literals are character-strings wherein the value is determined by the ordered set of characters of which the literal is a part. A literal may be *named* or *unnamed.* A *named literal* has a data-name assigned to it with a fixed value stipulated in the Working-Storage Section of the Data Division. It is a method of giving a data-name to a constant value.

An *unnamed literal* has the actual value specified at the time it is being used in the Procedure Division and does not require any separate definition in the Data Division. The two examples that follow show both named and unnamed numeric laterals.

Example 1

```
DATA DIVISION.
    .
    .
    .
77  FICA-RATE PICTURE V999 VALUE .067.
PROCEDURE DIVISION.
    .
    .
    .
MULTIPLY GROSS BY FICA-RATE GIVING FICA-TAX.
```

Example 2

```
MULTIPLY GROSS BY .067 GIVING FICA-TAX.
```

In example 1, we are using a named numeric literal which must be described in the Working-Storage Section of the Data Division before it can be used in the Procedure Division.

In example 2, we are using an unnamed numeric literal which need not be described anywhere as it is used exactly as written.

One of the principal advantages of using a named numeric literal is that it can be easily changed. For example, if the rate of FICA-RATE changes, as it often does in a payroll application, all that need be changed is the description in the Working-Storage-Section. If an unnamed numeric literal was used, it would have to be changed in every procedural statement that it appeared in, which may be rather cumbersome.

Every literal belongs to one of two types, numeric or nonnumeric.

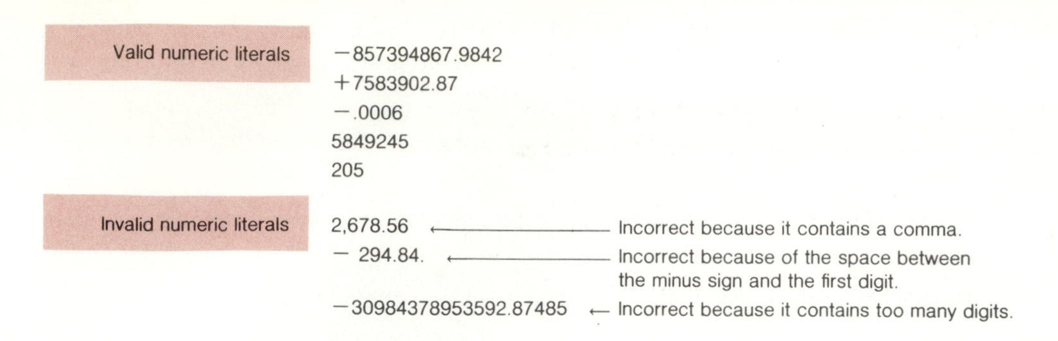

Valid numeric literals
−857394867.9842
+7583902.87
−.0006
5849245
205
Invalid numeric literals
2,678.56
Incorrect because it contains a comma.
− 294.84.
Incorrect because of the space between the minus sign and the first digit.
−30984378953592.87485
Incorrect because it contains too many digits.

Figure 2.14 Valid and invalid numeric literals—examples.

Numeric Literal

A numeric literal is a character-string whose characters are selected from the digits "0" through "9", the plus or minus sign, and the decimal point. The value of the literal is implicit in the characters themselves. Thus 842 is both the literal as well as the value.

Rules for Numeric Literals

1. A numeric literal may contain from one to eighteen digits.
2. It may contain a sign only in the leftmost character position. If no sign is indicated, the compiler will assume the value to be positive. No space is permitted between the sign and the literal.
3. It may contain a decimal point anywhere in the literal except as the rightmost character. Integers may be written without decimal points. The decimal point is treated as an assumed decimal point.
4. It may contain only one sign and/or one decimal point.
5. It must not be enclosed in single quotation marks.

Figure 2.14 shows some examples of valid and invalid numeric literals. Their use is shown in the following lines:

```
ADD 150.50 TO AMOUNT.
MOVE −8967 TO BALANCE.
MULTIPLY GROSS BY .01 GIVING SDI-TAX.
```

Nonnumeric Literal

A nonnumeric literal is a character-string enclosed by single quotation marks. (Some computers use double quotation marks for the same purpose.) The string of characters may consist of any character in the computer's character set including nonCOBOL characters. Some examples follow:

```
'9%'
'JANUARY 1982'
'NOT IN FILE'
'THIS IS A COBOL PROGRAM'
'1234567'
```

Rules for Nonnumeric Literals

1. A nonnumeric literal must be enclosed in single (or double) quotation marks.
2. It can be used only for display purposes. The literal must not be used for computation. Only numeric literals may be used in computation.
3. It may contain from 1 to 120 characters.
4. It may contain any character in the character set of the particular computer, including blanks, special characters (except quote marks), and reserved words. The following example shows a nonnumeric literal correctly used:

```
77  MESSAGE-4     PICTURE A(12)
          VALUE 'OUT OF STOCK'.
```

Name	Uses
HIGH-VALUE(S)	are assigned the highest value in the computer collating sequence.
LOW-VALUE(S)	are assigned the lowest value in the computer collating sequence.
ZERO ZEROS ZEROES	are assigned the value 0.
SPACE SPACES	are assigned to one or more spaces (blanks).
QUOTE QUOTES	are assigned to one or more quotation mark characters (apostrophes).

ALL "any literal" will represent a continuous sequence of "any literal."
The singular and plural forms of figurative constants are interchangeable.

Figure 2.15 Figurative constants.

(*Note:* Signs and/or decimal points are not included in the size count of a numeric literal but are counted in the size of a nonnumeric literal. Quotation marks are not considered part of a nonnumeric literal and therefore are not included in the size count. A figurative constant may be used in place of a literal wherever a literal appears in the format.)

Figurative Constants

A figurative constant is a compiler-generated value referenced by the use of certain reserved words. These words are frequently used in programming so that the programmer is relieved of the responsibility of assigning names for commonly used constants. Figure 2.15 shows a list of these constants. A MOVE ZEROS TO WORK statement will fill the entire WORK area with zeros. Similarly, MOVE SPACES TO OUT will blank out the entire area OUT. ZERO, ZEROS, ZEROES may be used interchangeably when singular or plural forms are desired. Usage of singular or plural forms does not affect the execution of the statement. They are only used to improve the readability of the statement.

The figurative constant must *not* be enclosed in quotation marks. It may be used whenever a literal appears in a format, except whenever the literal is restricted to numeric characters, then the only figurative constant permitted is ZERO (ZEROS, ZEROES).

Punctuation Characters

Punctuation characters are important in the successful execution of a COBOL program. Unless the correct symbols are used, many diagnostic errors can be generated during the compilation phase. Figure 2.16 shows the symbols and their meanings as used to punctuate entries.

Symbol	Name	Meaning
.	PERIOD	is used to terminate entries.
,	COMMA	is used to separate operands, clauses in a series of entries.
;	SEMICOLON	is used to separate clauses and statements.
'	QUOTATION MARK	is used to enclose nonnumeric literals.
( )	PARENTHESIS	is used to enclose subscripts.

Figure 2.16 Punctuation symbols and meanings.

Separators

A separator is a string of one or more punctuation characters. The rules for forming separators are as follows:

1. The punctuation character *space* is a separator. Whenever a space is used as a separator in a program, more than one space may be used.
2. The punctuation characters *comma, semicolon,* and *period,* when immediately followed by a space, are separators. These separators may appear in a COBOL source program only where explicitly permitted.
3. The punctuation characters *right parenthesis* and *left parenthesis* are separators. Parenthesis may appear only in balanced pairs of left and right parenthesis.
4. The punctuation character *quotation mark* is a separator. An opening quotation mark must be immediately preceded by a space or a left parenthesis; a closing quotation mark must be immediately followed by one of the following separators: space; comma; semicolon; or right parenthesis.

   Quotation marks may appear only in balanced pairs bounding nonnumeric literals, except when the literal is continued (see continuation rules).
5. The separator *space* may, as an option, immediately precede all separators except
   a. as specified by format rules, or
   b. as the separator quotation mark. In this case, a preceding space is considered as part of the nonnumeric literal and not as a separator.
6. The separator *space* may, as an option, immediately follow any separator except the opening quotation mark. In this case, a following space is considered as part of the nonnumeric literal and not as a separator.

The punctuation characters comma and semicolon are shown in some formats. Where shown in the format, they are optional and may be included or omitted in the coding. In the source program, these two punctuation characters are interchangeable and either one may be used when one of them is shown in the format. Neither character may appear immediately preceding the first clause of an entry or paragraph.

If desired, a comma or semicolon may be used between statements in the Procedure Division. They are not required for correct COBOL execution of a program, but may be included to improve the readability of the program.

Paragraphs within the Identification and Procedure Divisions, and the entries within the Environment and Data Divisions must be terminated by the separator period.

Statements and Sentences

A statement is a syntactically valid combination of words and symbols in the Procedure Division used to express a thought in COBOL. The statement combines COBOL-reserved words together with programmer-defined operands.

A COBOL statement may be either a simple or a compound expression. A *simple* statement would specify one action while a *compound* statement, usually joined by a logical operator (and, or), will specify more than one form of action.

Imperative Statements

There are three types of statements: imperative; conditional; and compiler-directing. An *imperative statement indicates a specific unconditional action to be taken by an object program.* An imperative statement is any statement that is neither a conditional nor a compiler-directing statement. An imperative statement may consist of a sequence of imperative statements, each possibly separated from the next by separators. Some examples of imperative statements follow:

```
PERFORM READ-RECORD.
SUBTRACT 1 FROM HOUR GIVING LAST-HOUR
MOVE ZEROS TO WORKAREA-1.
WRITE PRINT-LINE FROM OUT-LINE
    AFTER ADVANCING 2 LINES.
```

Conditional Statements

A conditional statement specifies that the truth value of a condition is to be determined and that the subsequent action of the object program is dependent on the truth value. An example of a compound conditional statement appears in the following lines:

```
IF SCORE > 84
    AND SCORE < 93
    MOVE 'B' TO GRADE
ELSE
    PERFORM C-ROUTINE.
```

An *imperative* statement directs the program to perform a particular operation under *all* conditions, while a *conditional* statement specifies the operation to be performed only if the condition is *satisfied (true) or not (false).*

Compiler-Directing Statements

A compiler-directing statement causes the compiler to take specific action during compilation time. For example, a record description may be used in several programs. When this is the case, the segment should be stored in what is called a source statement library on a disk. A programmer may direct the compiler to COPY the record description into a program when needed. For example, suppose the library entry named WORKREC consists of the following Data Division record

```
01  WORKREC.
    05  AAA.
        10  HOURS-WORKED     PICTURE 99.
        10  SICK-LEAVE       PICTURE 99.
        10  HOLIDAY-PAY      PICTURE 99.
```

The user may copy this record into his source program with the first of the following two statements:

```
01  WORK-RECORD    COPY    WORKREC.
01  NEXT-RECORD.
```

After the record has been copied, the compiled source program will appear as if it had been written as:

```
01  WORK-RECORD
    05  AAA.
        10  HOURS-WORKED     PICTURE 99.
        10  SICK-LEAVE       PICTURE 99.
        10  HOLIDAY-PAY      PICTURE 99.
01  NEXT-RECORD
```

A sentence is a sequence of one or more statements, the last of which is terminated by a period and a space. A sentence may be either imperative, conditional, or compiler-directing.

Writing a COBOL Program Using Files

To illustrate how a COBOL program is written using files, a simplified procedure will be created that will list all records that have been read by a card reader and output on a line printer.

Identification Division

A name must be assigned to the program and, optionally, other information that will serve to document the program.

The Identification Division will appear as follows:

```
IDENTIFICATION DIVISION.
PROGRAM-ID. LISTING.
*PURPOSE. THIS PROGRAM WILL LIST ALL CARDS THAT ARE BEING READ.
```

PROGRAM-ID informs the compiler that the unique name LISTING was chosen for the program.

The COMMENTS statement serves to state the purpose of the program. Other information, such as the name of the programmer, the date of the program, etc., may also be added.

Environment Division

The purpose of the Environment Division is to describe the hardware configuration of the compiling computer (source computer) and the computer on which the object program is run (object computer). It also describes the relationship between the files and input/output media.

The Environment Division will appear as follows:

```
ENVIRONMENT DIVISION.
CONFIGURATION SECTION.
SOURCE-COMPUTER. XEROX-530.
OBJECT-COMPUTER. XEROX-530.
INPUT-OUTPUT SECTION,
FILE-CONTROL.
    SELECT FILE-IN ASSIGN TO READER.
    SELECT FILE-OUT ASSIGN TO PRINTER.
```

The Configuration Section specifies that the computer to be used for the compilation and the computer to be used for the execution of the program will be one and the same—XEROX 530.

Next, the files to be used in the program must be identified and assigned to specific input/output devices. This is done in the Input-Output Section. The Input-Output Section gives the necessary information to control the transmission of data between the external media and the object program. The entry is written in the File-Control paragraph, which associates the files with the external media. The Select clause is used to name each file in the user's COBOL program. File-In and File-Out are the file names supplied by the programmer. Every file name must have a File Description entry in the Data Division.

The Assign clause permits the files to be assigned to particular hardware devices. Thus File-In, the input file, will be associated with the card reader (*reader* is the implementor name for the card reader). File-Out, the output file, will be associated with the line printer (*printer* is the implementor name for the line printer).

Data Division

The Data Division describes data that the object program accepts as input in order to manipulate, create, or produce as output. Data falls into two categories:

External data, which is contained in files, and enters or leaves the internal memory of the computer from specified areas.

Data developed or stored internally. This type of data is placed into storage within the computer's internal memory.

The Data Division is divided into two sections: File Section, and Working-Storage Section. (These sections will be discussed in greater detail further along in the text).

The File Section defines the data stored on external files; all such data must be described here before it can be processed by the COBOL program. Each file is defined by a file description entry and one or more record description entries. Record description entries are written individually following the file description entry.

The Working-Storage Section describes records and noncontiguous data items that are not part of external data files, but instead are developed and processed internally.

The Data Division will appear as follows:

```
DATA DIVISION.
FILE SECTION.
FD  FILE-IN
    LABEL RECORDS OMITTED.
01  CARD-IN                       PICTURE X(80).
FD  FILE-OUT
    LABEL RECORDS OMITTED.
01  RECORD-OUT.
    05  CARRIAGE-CONTROL          PICTURE X.
    05  PRINT-OUT                 PICTURE X(132).
WORKING-STORAGE SECTION.
01  FLAGS.
    05  MORE-DATA                 PICTURE XXX       VALUE 'YES'.
```

The file description entry (FD) begins at Area A and is followed by at least one space. The name of the file is entered in Area B. The remainder of the file description entry is made up of independent clauses describing the file. The level indicator FD introduces File-In itself and informs the compiler that each entry written within File-In will be referred to as Card-In. Level number 01 identifies the record, Card-In. The level number must begin in Area A and the name of the record must begin in Area B. The Label Records clause is required in all FD entries. Since no labels are used in this program, the OMITTED option is used.

The second FD entry describes the output file, File-Out. The same rules apply as the first FD entry with the exception that Record-Out has a subordinate level. The concept of levels is a basic attribute of COBOL. The highest level is FD, the next highest level is 01. Level numbers 02–49 may subdivide the record, and the subdivisions themselves can be further subdivided if need be. *The smaller the subdivision, the larger the level number must be.*

Since the first print position is reserved for the carriage control character, no printing will take place in the first position. We wish to single space our listing, so we will leave our carriage control position blank. Later in the text, we will learn more about variable spacing. Level number 05 will begin in Area B with one or more intervening spaces followed by the data-name, Carriage-Control, and the Picture clause, which indicates a length of one character. The 05 Print-Out indicates the actual print area of 132 positions for one line of printing.

The Working-Storage Section entries set up a flag to check for an end-of-file condition in the Procedure Division. The initial setting of the flag is 'YES,' which will be tested later in the program and eventually changed to 'NO' when the last record has been processed.

Procedure Division

The Procedure Division contains the COBOL statements that give the computer step-by-step instructions for handling the data to be processed by the program. Procedures are logically successive instructions that process the data. Execution of the object program begins with the first statement in the procedures and continues in logical sequence. The end of the Procedure Division and the physical end of the COBOL source program is the physical position in the source program after which no procedures appear.

The Procedure Division will appear as follows:

```
PROCEDURE DIVISION.
MAIN-ROUTINE.
    OPEN    INPUT    FILE-IN,
            OUTPUT   FILE-OUT.
    READ FILE-IN
        AT END MOVE 'NO' TO MORE-DATA.
    PERFORM PROCESS-ROUTINE
        UNTIL MORE-DATA = 'NO'.
    CLOSE   FILE-IN
            FILE-OUT.
    STOP RUN.

PROCESS-ROUTINE.
    MOVE SPACES TO RECORD-OUT.
    MOVE CARD-IN TO PRINT-OUT.
    WRITE RECORD-OUT
        AFTER ADVANCING 1 LINE.
    READ FILE-IN
        AT END MOVE 'NO' TO MORE-DATA.
```

Each paragraph in the Procedure Division must have a paragraph-name. The Procedure Division in our example is divided into two paragraphs. The first paragraph—MAIN-ROUTINE—specifies the overall logic of the program. The second paragraph—PROCESS-ROUTINE—specifies the processing steps that will be performed for each record. After the last record has been processed, the control of the program will return to the MAIN-ROUTINE.

The Open statement does the following: makes the records contained in File-In and File-Out available for processing; establishes a line of communication with each file; checks to make sure that each is available for use; brings the first record of File-In into special areas of internal storage known as buffers; and does other housekeeping chores. Each file must be defined as Input or Output or both (I-O). Each file named in the Open statement must have been described in the Data Division. An input file must be opened to enable the Read statement to obtain records for processing. The files can now be accessed.

Note: The names of the files are the same as those mentioned in the Environment and Data divisions.

The READ FILE-IN AT END MOVE 'NO' TO MORE-DATA statement makes the first record of the file, FILE-IN, available for processing. The AT END phrase is necessary in the sentence to check for an end-of-file condition. If the end-of-file condition has been reached, the message 'NO' will be moved to the MORE-DATA field to signal the necessary procedures to be taken when the end of file has been reached. Still, it is highly unlikely that an end-of-file condition will be reached with the first record. (The phrase AT END is required in this format.)

The PERFORM PROCESS-ROUTINE UNTIL MORE-DATA = 'NO' statement will control the continual reading and processing of records. The PROCESS-ROUTINE paragraph will be performed over and over again until an end-of-file condition is reached, and then control of the program will return to the next statement after the PERFORM in the MAIN-ROUTINE.

The PROCESS-ROUTINE paragraph contains the necessary statements for the reading and printing of the records. The statements in the PROCESS-ROUTINE paragraph will be processed next as follows:

> The statement MOVE SPACES TO RECORD-OUT will clear out the print area by moving blanks to the record in the output file. SPACES is a COBOL-reserved word (explained in detail in the Figurative Constants section) that will be moved to the designated area.
>
> The MOVE CARD-IN TO PRINT-OUT statement moves the necessary data from the input file to the output file ready for printing. The Move statement does not mean an actual physical movement of data. Instead it means that the data items from Card-In are copied into Print-Out. Data items within Card-In are not destroyed when a Move statement is executed, but are still available for further processing if necessary.
>
> The WRITE RECORD-OUT AFTER ADVANCING 1 LINE statement causes the record to be recorded on an output device specified for the file in the Environment Division; its format would be determined by the Data Division description of the file. The AFTER ADVANCING option indicates that the report is to be single-spaced. Notice that the format requires the name of a *record*, not that of a file. The file must have been opened before the record can be written into it.

The READ statement with its appropriate AT END phrase is used to process all subsequent records. The end of file will be signaled as previously explained and control will return to MAIN-ROUTINE.

After the return to the MAIN-ROUTINE after the last record has been processed, the program must be terminated. The following statements are used to terminate the program:

> The CLOSE statement terminates the processing of the files and releases the storage areas used as buffers for those files. It is not necessary to stipulate whether the files were used for input or output or both. Any file that was previously opened must be closed before the run is stopped.
>
> The STOP RUN statement terminates the execution of the object program. Ending COBOL procedures are initiated, and control of the computer is returned to the operating system.

Summary

The reference format provides a standard method of describing COBOL source programs. This format is defined in terms of character positions on an input/output device. The rules for spacing in the reference format take precedence over all other rules for spacing.

The COBOL program sheet format provides the programmer with a standard method of writing COBOL source programs. The sequence number (positions 1–6) is optional but provides a control on the sequence of lines. The continuation indicator, a hyphen in position 7, is used to continue split words, numeric literals, or nonnumeric literals in succeeding lines.

An asterisk (*) in position 7 indicates that the entire line is a comment. A comment line is printed on the compilation listing as documentation only.

Program statements are written in positions 8–72. Each element of a statement must be separated from the next by at least one space. There are, however, certain symbols, such as left parenthesis and beginning quotation mark, that must not be followed by a space. Whereas symbols such as periods, commas, semicolons, right parenthesis, ending quotation mark, must not be preceded by a space. (ANSI 1974 COBOL permits spaces optionally to precede the comma, period, or semicolon and optionally to precede or follow a left parenthesis.)

Certain program entries, such as division headers, section headers, paragraph headers, file description entries, two special headers in the Procedure Division (DECLARATIVES and END DECLARATIVES), and level numbers 01 and 77, must begin in Area A (positions 8–11), while all other entries must begin in Area B (positions 12–72).

The identification code (positions 73–80) is used for the name of the program for identification purposes only and has no effect on the object program or compilation, and may be left blank.

Blank lines, blank in positions 7–72, are usually used to separate segments of a program and may appear anywhere in the program.

COBOL programs are arranged in a series of entries that comprise divisions, sections, and paragraphs. A division is composed of a series of sections, while a section is made up of paragraphs. Paragraphs are composed of a series of sentences containing statements. Four divisions are required in every COBOL program in the following sequence: Identification, Environment, Data, and Procedure divisions.

All divisions do not necessarily contain sections. The Environment and Data divisions always contain sections with fixed names, but sections are never found in the Identification Division, while the Procedure Division sections are optional and created by the programmer if required in the program.

All divisions except the Data Division contain paragraphs. A paragraph is a segment or module performing a function.

Entries (sentences) consist of a series of statements terminated by a period.

There are certain terms used in COBOL programming, such as SOURCE-COMPUTER, the computer used to compile the program, and the OBJECT-COMPUTER, the computer used to execute the program.

The COBOL character set consists of fifty-one characters.

Names are a means of establishing words to identify certain data within a program. All references to an item will be through the name. The assignment of names must abide by certain rules such as, names may range from one to thirty characters, no spaces may appear within a name, names may be formed from the alphabet, numerals and hyphen (the hyphen must be imbedded within the name). There must be at least one alphabetic character in a name, except the paragraph name, which may consist entirely of numerals. The name must be unique or otherwise qualified. The types of names are: data-names that are assigned by the programmer, which are used to identify data items in a COBOL source program; paragraph-names, which are data-names attached to various segments of the Procedure Division; and section-names, which are data-names assigned to various sections of the Procedure Division.

Qualification is a method of specifying uniqueness for data-names by mentioning one or more higher levels of the hierarchy for the name.

A COBOL word is a character-string (a character-string is a sequence of contiguous characters) chosen from the character set that form a user-defined word or reserved word. User-defined words are COBOL words that must be supplied by the user to satisfy the format of a statement. A reserved word is a COBOL word that is one of a specified list of words that may be used in a COBOL source program, but must not appear as user-defined words. Reserved words have preassigned meanings and must not be altered, misspelled, or changed in any manner from the specific purpose of the word. COBOL-reserved words appear in upper case in formats.

A keyword is a reserved word whose presence is required when the format in which the word appears is used in a source program. Within each format such words appear in upper case and are underlined.

Optional words are reserved words that may or may not be used in the format in which they appear. They appear in upper case but are not underlined in the format. Lower-case words represent information that must be supplied by the programmer. Bracketed words may be included or omitted depending upon the requirements of the program. Braced words indicate that at least one of the words in the braces must be included in the format. Ellipsis indicate that words may appear any number of times.

A constant is an actual value of data that remains unchanged during the execution of the program. There are two types of constants used in COBOL programs: literals and figurative constants. Literals are character-strings wherein the value is determined by the ordered set of characters of which the literal is a part. A literal may be named, where a data-name with a fixed value is stipulated in the Working-Storage Section of the Data Division, or unnamed, where the actual value is specified at the time it is used in the Procedure Division.

Every literal belongs to one of two types. A numeric literal is a character string whose characters are selected from the digits 0–9, the plus or minus sign, and the decimal point. The value of the literal is implicit in the characters themselves and is limited to eighteen digits exclusive of the sign and/or decimal point. The literal must not be enclosed in quotation marks.

Nonnumeric literals are character-strings enclosed in quotation marks, which may consist of any character in the computer's character set including nonCOBOL characters.

A figurative constant is a compiler-generated value referenced by the use of certain reserved words. These words are frequently used in programming to relieve the programmer of the responsibility of assigning data-names for commonly used constants, such as zeros and spaces.

Punctuation characters are used as separators of data elements in a COBOL program. The punctuation characters include: space, comma, semicolon, period, right and left parenthesis, and quotation marks. Where space is indicated, more than one space may be used. Comma, semicolon, and period, where immediately followed by a space, are considered separators. Right and left parenthesis and a closing quotation mark must be immediately followed by one of the separators. Quotation marks may appear only in balanced pairs except when a nonnumeric literal is to be continued on a succeeding line. The separator space may precede or follow any separator with certain exceptions such as, space must not precede a closing quotation mark or follow an opening quotation mark. The punctuation characters comma and semicolon, where shown in some formats, are interchangeable and optional to improve readability. Paragraphs within the Identification and Procedure divisions, and entries within the Environment and Data Divisions must be terminated by the separator period.

A statement is a syntactical valid combination of words and symbols written in the Procedure Division to express a thought. A statement may be simple, specifying one action, or compound, usually employing a logical operator, and specifying more than one action.

There are three types of statements: an imperative statement directs the program to perform a particular action under all conditions; a conditional statement specifies the operations to be performed only if the condition is satisfied (true) or not (false); and a compiler-directing statement causes the compiler to perform some specifications during compilation time.

A sentence is a sequence of one or more statements, the last of which is terminated by a period and space. A sentence may be imperative, conditional, or compiler-directing.

Questions for Review

1. Explain the purpose of the COBOL program coding sheet.
2. What is the importance of the sequence number on a programming coding form?
3. What is the purpose of the continuation indicator? What are some of the principal uses?
4. What is a comment line and how is it written on a coding form?
5. How are elements entered in COBOL entries on coding forms?
6. What are the rules for program entries?
7. Explain the major program structure of COBOL.
8. What should the programmer familiarize him/herself with before attempting to write a program in COBOL?
9. Define the following terms: source-computer, object-computer, and character set.
10. Explain the different types of names used in COBOL programming and their expressed purpose.
11. What are the rules for assigning names?
12. What is meant by qualification and what are the rules for qualification of data-names?
13. What is a reserved word and how is it used?
14. What are the rules for the use of words?
15. What is a constant?
16. What is a literal?
17. What is a named literal? an unnamed literal?
18. What are the rules for the use of numeric literals?
19. What are the rules for the use of nonnumeric literals?
20. What is a figurative constant? What is its main advantage to a programmer? Give an example of a figurative constant.
21. Why is the correct use of punctuation characters so important to the successful execution of a COBOL program?
22. What is a separator and what are the rules for forming separators?
23. What is a statement? What are the different forms of statements?
24. What is a sentence?

Matching Questions

A. *Match each item with its proper description.*

_____ 1. Names	A. Series of statements terminated by a period and a space.
_____ 2. Paragraph	B. A series of paragraphs.
_____ 3. Blank lines	C. Most basic and indivisible unit of the language.
_____ 4. Reference format	D. A character string chosen from the character set.
_____ 5. Qualification	E. Means of establishing words to identify certain data within a program.
_____ 6. Entries	F. Grouping of continguous characters treated as a unit.
_____ 7. Data item	G. Standard method for describing the COBOL source program.
_____ 8. Section	H. Used to separate segments of a program.
_____ 9. Character	I. Process that specifies uniqueness.
_____10. Word	J. Segment of module performing a function.

B. *Match each item with its proper description.*

_____ 1. Literals	A. String of one or more punctuation characters.
_____ 2. Reserved words	B. Character string bound by quotation marks.
_____ 3. Imperative statement	C. A specific unconditional action to be taken by an object program.
_____ 4. Constant	D. Preassigned meaning.
_____ 5. Sentence	E. Sequence of one or more statements, the last of which is terminated by a period and a space.
_____ 6. Statement	F. Unchanging value during execution of a program.
_____ 7. Key word	G. Compiler-generated value referenced by the use of certain reserved words.
_____ 8. Separator	H. Required in COBOL entry.
_____ 9. Figurative constant	I. Value is determined by order set of characters of which it is a part.
_____10. Nonnumeric literal	J. Syntactically valid combination of words and symbols in the Procedure Division used to express a thought in COBOL.

Exercises

Multiple Choice: Indicate the best *answer (questions 1-25).*

1. Which is the correct way to write "minus one-half" as a numeric literal?
   a. −½
   b. −.05
   c. −.5
   d. None of the above.
2. Which of the following literals are valid nonnumeric literals?
   a. 'JANUARY, 1982'
   b. 'NOT IN FILE'
   c. '50'
   d. All of the above.
3. According to the rules for using the COBOL program sheet, it is all right for a line of the form to be
   a. filled in completely.
   b. left partly blank.
   c. left entirely blank.
   d. all of the above.
4. An entry may be continued on the next line or lines of a program sheet
   a. whether or not it could be written on one line.
   b. only if it is too long to fit on one line.
   c. only if it is started on margin A.
   d. only if it is started on margin B.
5. When an entry provides a choice of required portions, they are stacked with a pair of braces like this:

$$\left\{\begin{matrix}\text{POSITIVE}\\\text{NEGATIVE}\\\text{ZERO}\end{matrix}\right\}$$

   In this case, the programmer

   a. must write one, and only one of the words.
   b. must write one and may write all three.
   c. may write one, or none, of the words.
   d. may write one, since it is a programmer choice.
6. When an entry provides a choice of optional portions, they are stacked within a pair of brackets like this:

$$\left[\begin{matrix}\text{UNIT}\\\text{UNITS}\end{matrix}\right]$$

   In this example, the programmer

   a. may omit UNIT only, as an option, in the entry.
   b. may write UNIT or UNITS, or both in the entry.
   c. must write either UNIT or UNITS in the entry.
   d. may write UNIT or UNITS, or neither in the entry.
7. COBOL entries are written in positions 8 through 72. The program entry positions of the program sheet are divided into two margins. Which of the following entries is *not* required to begin in margin A?
   a. section headers
   b. paragraph headers
   c. level numbers
   d. level indicators
8. A division header (such as Identification Division) must begin in the A margin, and
   a. no part of it may be written in the B margin.
   b. most of it will be written in the B margin.
   c. writing in the B margin is optional.
   d. continue into the next A margin below.
9. Records are usually subdivided into smaller items. Some or all of these items may be further subdivided into smaller items.
   a. The maximum number of subdivisions is 77.
   b. The level numbers can go up to 88.
   c. The level number can be as large as desired.
   d. The level number may not be larger than 49.

10. When two data items (having the same name) are referenced, _______________ must be qualified?
    a. either one
    b. only one
    c. both
    d. neither one
11. The figurative constant ZERO may be used as a
    a. numeric literal.
    b. nonnumeric literal.
    c. either a numeric or a nonnumeric literal.
    d. None of the above.
12. The figurative constant SPACE may be used as a
    a. numeric literal.
    b. nonnumeric literal.
    c. either a numeric or a nonnumeric literal.
    d. None of the above.
13. The Procedure Division is required to contain
    a. no paragraphs at all.
    b. at least one paragraph.
    c. just one paragraph.
    d. any even number of paragraphs.
14. Must a paragraph have a name if it is the only paragraph in the Procedure Division?
    a. it's up to the programmer
    b. no
    c. yes
    d. it's optional
15. If a paragraph name is used, it
    a. must be composed of reserved words.
    b. can be duplicated in the same section.
    c. is programmer-supplied.
    d. None of the above.
16. What is the difference between a *sentence* and a *statement?* Pick the *incorrect* choice.
    a. A sentence is an entry; therefore, it must be terminated by a period.
    b. A statement specifies an action to be taken and is found within a sentence.
    c. Each sentence contains at least one statement, but it may contain more than one.
    d. The smallest unit of expression in the Procedure Division is not the statement.
17. If a numeric literal is unsigned, we can assume
    a. it really has no value.
    b. it can be positive or negative.
    c. its value is unknown.
    d. that it is positive.
18. Which of the following language elements is composed by the programmer?
    a. reserved words
    b. data-names
    c. optional words
    d. punctuation characters
19. The header for one of the four divisions always starts in what way?
    a. The word division followed by a division name followed by a space and a period.
    b. The word division followed by division name followed by a period.
    c. Division name followed by the word division followed by a space and a period.
    d. Division name followed by the word division and a period.
20. Sections are not found in which division?
    a. Procedure
    b. Identification
    c. Environment
    d. Data
21. All of the divisions contain paragraphs except which one?
    a. Procedure
    b. Identification
    c. Environment
    d. Data

22. In what division is the names of all paragraphs *not* fixed?
    a. Environment
    b. Procedure
    c. Data
    d. Identification
23. The period, which is a language element, when utilized in entries is
    a. sometimes used.
    b. always used.
    c. optionally used.
    d. never used.
24. A period (.) is followed by a space.
    a. always
    b. sometimes
    c. never
    d. at times
25. Which of the following is an example of a nonnumeric literal?
    a. '33166'
    b. PAGE 144 MISSING
    c. 1506798
    d. 12-3E2
26. For each name below, decide what rule, if any, is being violated.
    a. 5
    b. SYSTEM/370
    c. OVERFLOW
    d. ECONOMIC-ORDER-QUANTITY-COMPUTATION
    e. ENTRY-PROCESS
    f. HEADING LINE
    g. F.I.C.A.
27. Indicate the incorrect numeric literals in the following list.
    a. −8573956894183456.98
    b. −.00015
    c. 2,900.56
    d. −192.85
    e. .5
    f. SIX
28. Rewrite the following entry correctly.
    COMPUTE GROSS,ROUNDED = (HOURS*RATE) + OTPAY.
29. Correct the following data-names.
    a. SPACE
    b. JOB 1
    c. 12345*
    d. LEVEL-1
    e. −10T4
    f. TAX/RATE
30. Match each item with its proper description.

_____ 1. Comment line	A. Provides programmer with a standard method of writing COBOL source programs.
_____ 2. B margin	B. Series of statements terminated by a period and a space.
_____ 3. Special-names	C. Hyphen in position 7.
_____ 4. Names	D. Positions 8 through 11.
_____ 5. Program sheet	E. Words to identify certain data within a program.
_____ 6. Continuation indicator	F. Positions 12 through 72.
_____ 7. A margin	G. Mnemonic names assigned to various components in the Environment Division.
_____ 8. Entries	H. An asterisk in position 7.

31.

SEQUENCE	CONT	A	B COBOL STATEMENT	IDENTIFICATION
			MULTIPLY AMOUNT OF CLIENT-PURCHASES BY TRADE-DISCOUNT, GIVING	

Which of the lines below shows a correct way of completing this entry?

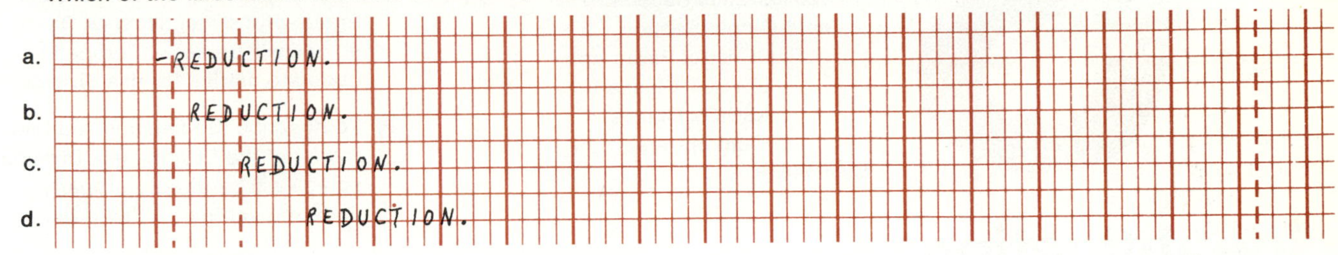

32. Number the following in an ascending to descending order.
 _____ a. Data-name.
 _____ b. Division headers.
 _____ c. Paragraph-name.
 _____ d. Section-name.

Problems

1. Identify the purposes and uses of numbered items.

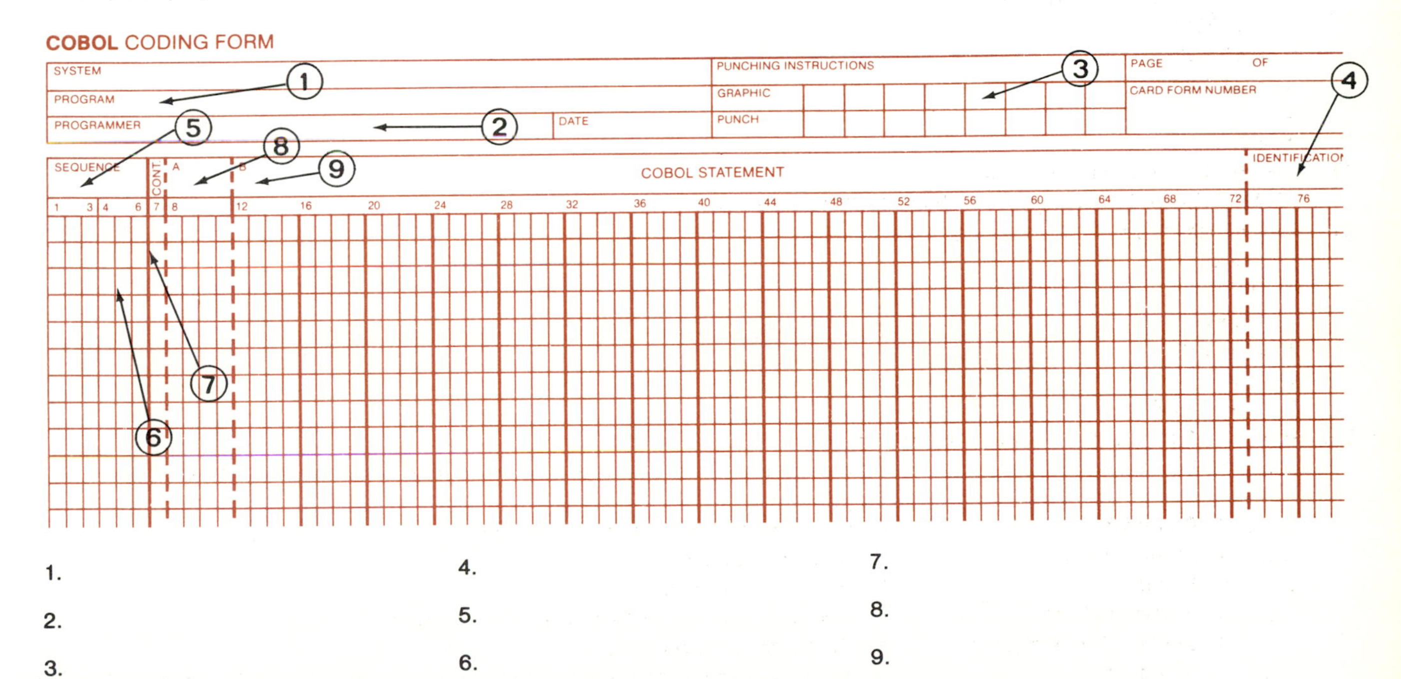

1.
2.
3.
4.
5.
6.
7.
8.
9.

2. Write a COBOL program to list 80-column cards on the printer, double spaced. Be sure to code all four divisions.

Identification and Environment Divisions

Chapter Outline

Chapter Objectives

The learning objectives of this chapter are:

1. To describe the function and the format of the Identification Division so you will be able to properly code the entries for this division.
2. To describe the function and the format of the Environment Division so you will be able to properly code the entries for this division.

Identification Division

The Identification Division is the first and simplest of all four divisions to code, and must be included in every COBOL program.

Function. The function of this division is to identify both the source program and the resultant output listing. In addition, the programmer may include any other information that is considered vital to the program, such as the name of the programmer, the date the program was written, the date of the compilation of the source program, and such other information as desired.

In the format, paragraph headers identify the type of information contained in the paragraph. The name of the program must be given in the first paragraph, which is the PROGRAM-ID paragraph. The other paragraphs are optional and may be included in this division at the programmer's choice.

Structure

The following is the general format of the paragraphs in the Identification Division; it defines the purpose and order of presentation in the source program. (The general format is shown in figure 3.1.)

1. The Identification Division must begin with the reserved words IDENTIFICATION DIVISION followed by a period on a line by itself.
2. The *PROGRAM-ID paragraph* supplies the name by which a program is identified. The general format is PROGRAM-ID. program-name. The program-name must conform to the rules for the formation of user-defined words.
   a. The PROGRAM-ID paragraph must contain the name of the program and must be present in every program.
   b. The program-name identifies the source listing and all listings pertaining to a particular program.
3. *Optional paragraphs* (shown in brackets in figure 3.1).
   a. The *AUTHOR* paragraph is used to supply the name or otherwise identify the author of the program. The general format is AUTHOR. comment-entry.
   b. The *INSTALLATION* paragraph is used to supply the name or otherwise identify the installation at which the source program was written. The general format is INSTALLATION. comment-entry.
   c. The *DATE-WRITTEN* paragraph is used to supply the date in which the program was written. The general format is DATE-WRITTEN. comment-entry.
   d. The *DATE-COMPILED* paragraph provides the compilation date in the Identification Division source program listing. The general format is DATE-COMPILED. comment-entry. The paragraph-name DATE-COMPILED causes the current date to be inserted during program compilation. If a DATE-COMPILED paragraph is present, it is replaced during compilation with a paragraph in the form of DATE-COMPILED. current-date.

```
IDENTIFICATION DIVISION.

PROGRAM-ID.  program-name.

[AUTHOR.  [ comment-entry ]  ...]

[INSTALLATION.  [ comment-entry ]  ...]

[DATE-WRITTEN.  [ comment-entry ]  ...]

[DATE-COMPILED.  [ comment-entry ]  ...]

[SECURITY.  [ comment-entry ]  ...]
```

Figure 3.1 Identification Division—format

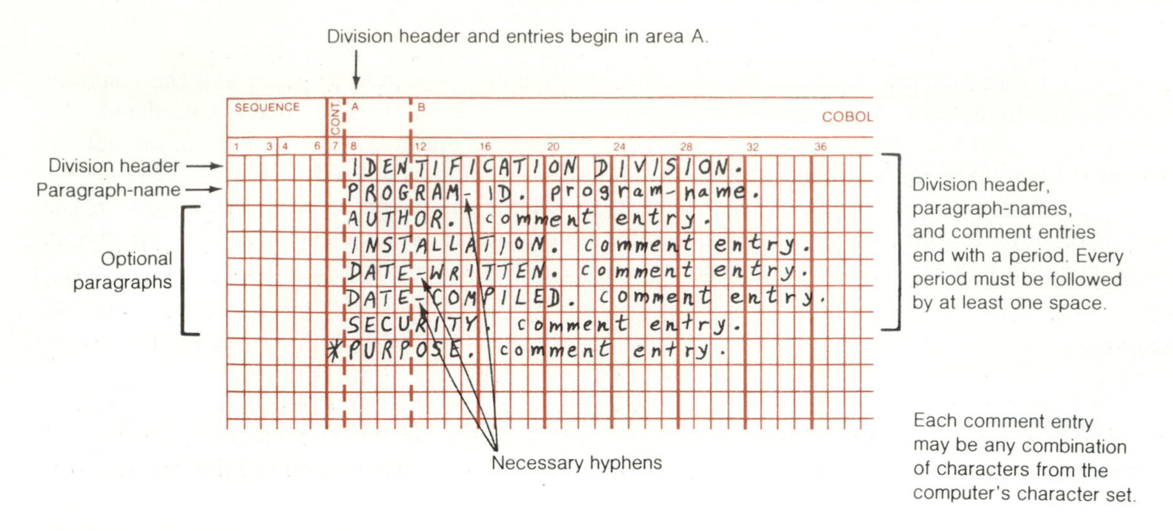
Division header and entries begin in area A.
SEQUENCE
CONT.
A
B
COBOL
Division header
Paragraph-name
Optional paragraphs
IDENTIFICATION DIVISION.
PROGRAM-ID. program-name.
AUTHOR. comment entry.
INSTALLATION. comment entry.
DATE-WRITTEN. comment entry.
DATE-COMPILED. comment entry.
SECURITY. comment entry.
*PURPOSE. comment entry.
Necessary hyphens
Division header, paragraph-names, and comment entries end with a period. Every period must be followed by at least one space.
Each comment entry may be any combination of characters from the computer's character set.

Figure 3.2 Guide for coding Identification Division entries.

This paragraph can be particularly useful in debugging programs, as it will supply the most recent compilation listing in the DATE-COMPILED paragraph, thus avoiding the necessity of looking through many compilation listings to find the most current.

e. The *SECURITY* paragraph is used to supply the level of security attached to the program by the installation or the programmer. The general format is SECURITY. comment-entry.

The comment-entry in the aforementioned paragraphs may be any combination of characters from the computer's character set. The continuation of the comment-entry by the use of the hyphen in the indicator area is not permitted; however, the comment-entry may be continued on one or more lines. The comment-entry has no effect on the operation of a COBOL program.

Each program must contain the first two paragraphs listed. The other paragraphs are *optional* and may be included to meet the needs of the programmer. The optional paragraphs, if used, must be pre-

```
       IDENTIFICATION DIVISION.
       PROGRAM-ID.
           EXPENSES.
       AUTHOR.
           CHARLES BROWN.
       INSTALLATION.
           DYNAMIC DATA DEVICES, INC.
       DATE-WRITTEN.
           NOVEMBER 9, 1982.
       DATE-COMPILED
           NOVEMBER 10, 1982.
       SECURITY.
           COMPANY-CONFIDENTIAL; AVAILABLE TO
           AUTHORIZED PERSONNEL ONLY.
      *PURPOSE.
      *    PRODUCES A WEEKLY LISTING OF ALL
      *    OPERATING EXPENSES, BY DEPARTMENT.
```

Figure 3.3 Identification Division entries—examples.

sented in the order shown in the format with the proper paragraph-names. The programmer has complete freedom as to what is to be coded in these optional paragraphs. A guide for coding the Identification Division appears in figure 3.2.

A comment line (*[asterisk] in the continuation indicator position 7) may be used to stipulate the purpose of the program. The example in figure 3.3 includes such a comment line.

Environment Division

All aspects of a data processing problem that depend upon the physical characteristics of a specific computer are expressed in the Environment Division. This division is the one division of COBOL that is machine-dependent; here the programmer must become familiar with the characteristics and special name of the machine upon which the particular source program is to be run. Any changes in a computer requires many changes in this division. A link is provided between the logical concept of the files and the physical aspects of the devices upon which the files will be processed and stored.

Function. The function of the Environment Division is to specify the configuration of the computer that will be used to compile the source program, as well as the configuration of the computer that will execute the object program. In addition, all input and output files will be assigned to individual hardware devices.

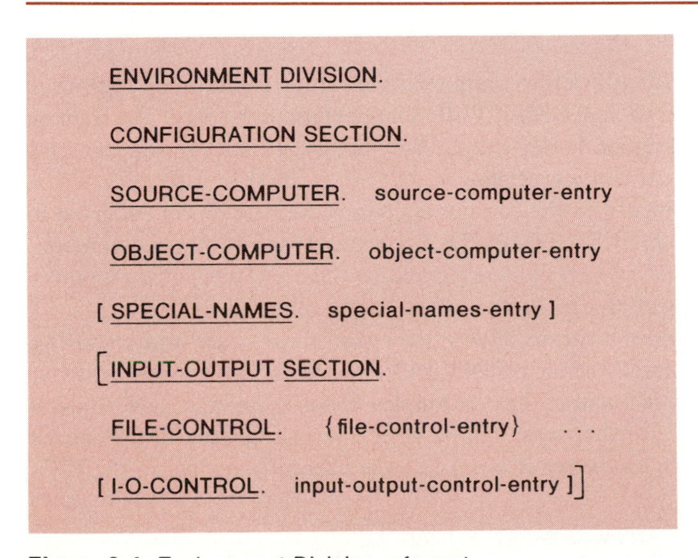

Figure 3.4 Environment Division—format.

The Environment Division must be included in every COBOL source program. This division provides a standard method of expressing those aspects of a data processing problem that depends upon the physical characteristics of a specific computer. In this division, the compiling computer and the executing computer are specified. In addition, the information relating to Input-Output control, special hardware or operating system characteristics, and control techniques can also be presented. Figure 3.4 shows the general structure of the Environment Division.

The Environment Division must follow the Identification Division and must begin with the reserved words ENVIRONMENT DIVISION followed by a period on a line by itself.

Two sections make up the Environment Division: the Configuration Section and the Input-Output Section.

Configuration Section

The Configuration Section provides program documentation for the hardware characteristics of the computer used for compilation and of the computer used to execute the object program. Provisions are

CONFIGURATION SECTION.

SOURCE-COMPUTER. source-computer-entry

OBJECT-COMPUTER. object-computer-entry

[SPECIAL-NAMES. special-names-entry]

Figure 3.5 Configuration Section—format.

included in this section for relating specific hardware and operating systems to user-specified mnemonic names.

Note: The Configuration Section is optional for many computers. When it is used, the information in the SOURCE-COMPUTER and OBJECT-COMPUTER paragraphs is used for documentation purposes only. It should be noted that these entries are treated as comments and are written in accordance with the computer manufacturer's specifications.

The Configuration Section specifies the overall characteristics of the computer involved in the compilation and the execution of a COBOL program. The section is divided into three paragraphs: SOURCE-COMPUTER, OBJECT-COMPUTER, and the SPECIAL-NAMES paragraphs. Figure 3.5 shows the general format for the Configuration Section.

1. The *SOURCE-COMPUTER* paragraph identifies the computer upon which the source program is to be compiled. The general format is SOURCE-COMPUTER. source-computer-entry. The source-computer-entry is a fixed system-name assigned to the computer by the individual installation, and provides a means for identifying equipment configurations.
2. The *OBJECT-COMPUTER* paragraph identifies the computer upon which the object program is to be executed. The general format is OBJECT-COMPUTER. object-computer-entry (memory-size). The object-computer-entry is a fixed system-name assigned to the computer. The memory-size phrase is used for program documentation and has no effect on the object program.
3. The *SPECIAL-NAMES* paragraph provides a means of relating specific hardware and operating features to user-specified names. The general format is SPECIAL-NAMES. special-names entry. The special-names-entry is a user specified name. Two examples follow: the first line equates TRANSACTION-FILE to TRANFILE, so that both can be used interchangeably in the source program; the second line specifies that the character R is used as the currency symbol in the PICTURE clause.

```
SPECIAL NAMES. TRANSACTION-FILE IS TRANFILE.
SPECIAL NAMES. CURRENCY SIGN IS 'R'.
```

In the next example, the function name CO1 has been assigned to channel 1 (first printing line of a form) so that the form can be skipped to the top of a page whenever necessary.

```
SPECIAL NAMES. CO1 IS SKIP-TO-1.
```

The SOURCE-COMPUTER and the OBJECT-COMPUTER paragraphs are required in all COBOL source programs and are used for documentation purposes only (see the two examples just given), while the SPECIAL-NAMES paragraph is optional and included only when the user-specified mnemonic names are used for specific hardware and operating system features. The format requirements for coding a typical Configuration Section are shown in figure 3.6.

Input-Output Section

The Input-Output Section must be included in any COBOL source program if there are any input and output files required. As most programs involve the processing of files, this section is required in most programs.

The Input-Output Section is concerned with the definition of the input and output devices as well as the most efficient method of handling data between the devices and the object program.

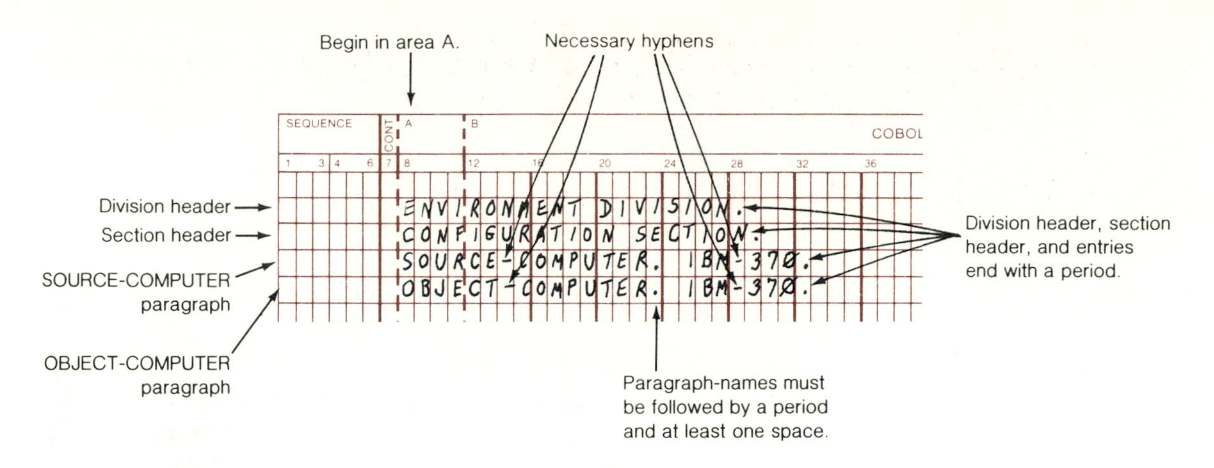

Figure 3.6 Format requirements for Configuration Section.

The Input-Output Section provides the information needed to control the transmission and handling of data between the external devices and the object program.

This section is divided into two paragraphs: the File-Control paragraph, which names each file used in the program and identifies the media on which each file is located; and the I-O-Control paragraph, which specifies any special Input/Output control techniques to be used in the object program. The individual clauses that make up these paragraphs may appear in any sequence within their respective sentences or paragraphs, but must begin at the B margin. Figure 3.7 shows the general format for this section.

File-Control Paragraph

The File-Control paragraph names and associates files with external media. The names of files given in the file description entries in the Data Division are assigned to input/output devices. There is a relationship between the file entries in the three divisions. The Data Division entries specify the characteristics and the structure of the data within these files. The Procedure Division will specify the READ and WRITE entries for these files. The input/output device that will be used to read or write will be determined by the Environment Division entry, which names the input/output devices assigned to the particular file.

SELECT Clause

The SELECT clause is used to name files within a COBOL source program. The SELECT entry must begin with the word SELECT followed by the file-name, and must be given for each file referred to by the COBOL source program. A separate SELECT clause is required for each file name in the Data Division.

File-Name. This is the unique name of the file assigned in the file description entry in the Data Division of the source program. This name will also be used in entries in the Procedure Division.

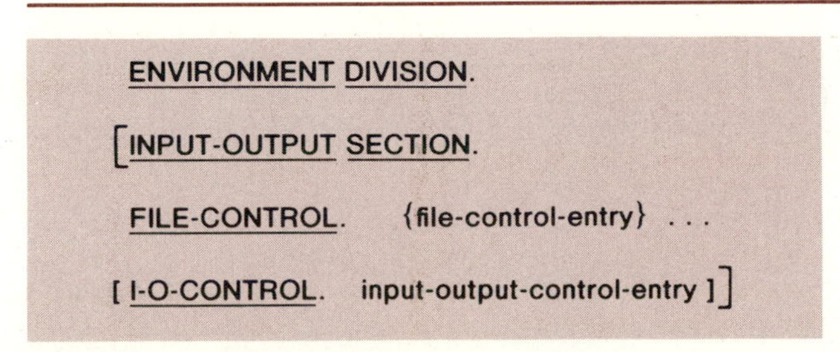

Figure 3.7 Input-Output Section—format.

SELECT [OPTIONAL] file-name

ASSIGN TO implementor-name [, implementor-name-2] . . .

Figure 3.8 SELECT and ASSIGN clauses—format.

Each file described in the Data Division must be named once, and only once, as file-name in the File-Control paragraph. Each file specified in the file control entry must have a file description entry in the Data Division.

Optional. This is a key word that may be specified only for input files accessed sequentially, yet may not be present each time the object program is executed. If this option is used and the file is not present at object time, the first READ statement causes the control to be passed to the imperative statement following the key words AT END.

ASSIGN Clause

The ASSIGN clause specifies the association of the file referenced by file-name to a storage device. Figure 3.8 shows the general format for both SELECT and ASSIGN clauses. The operating system associates the files or devices allocated to the job. The implementor-name specifies the external storage device. Figure 3.9 shows a specific file name and implementor name.

I-O-Control Paragraph

The I-O-Control paragraph defines special control techniques to be used in the object program. It may specify certain conditions in an object program, such as which checkpoints to establish, which storage areas to be shared by different files, the location of files on multiple-file reels, and the optimization techniques.

If special techniques or conditions need to be defined in the program, the I-O-Control paragraph is used; otherwise, the entire paragraph and its associated clauses may be omitted.

The features of the I-O-Control paragraph and other features of the Environment Division are discussed later in the text. An example of a complete Environment Division for a particular computer system and application, including the I-O-Control paragraph, appears in figure 3.10.

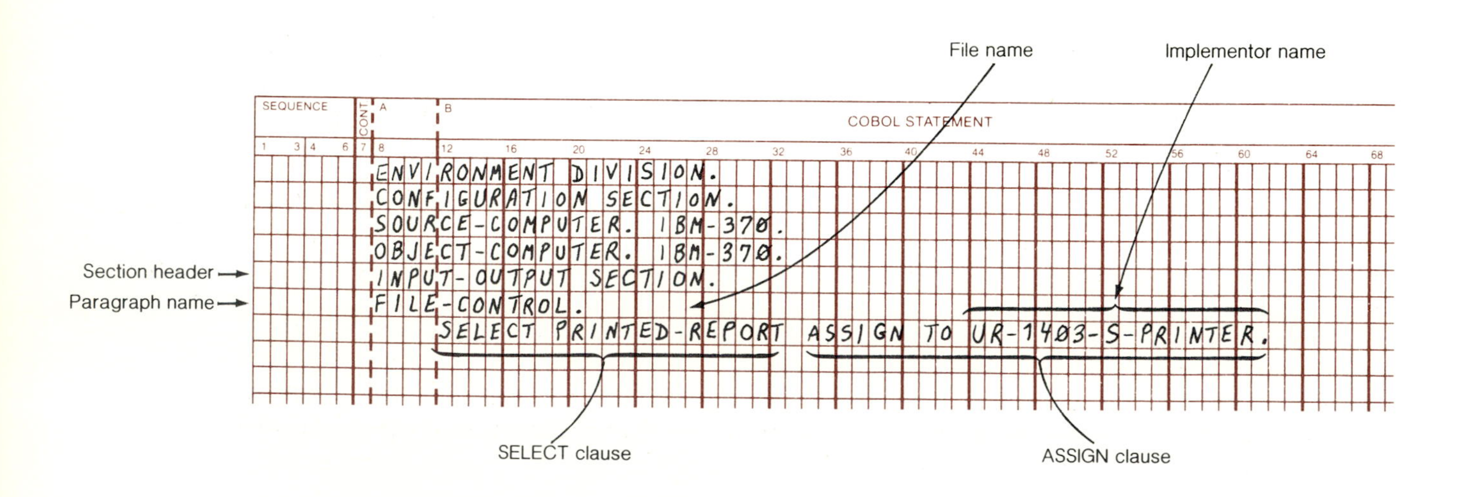

Figure 3.9 Format requirements for Input-Output Section entries.

```
ENVIRONMENT DIVISION.
CONFIGURATION SECTION.
SOURCE-COMPUTER.                NCR-CENTURY-200.
OBJECT-COMPUTER.                NCR-CENTURY-200
                                MEMORY SIZE 32000 CHARACTERS.
SPECIAL-NAMES.    SWITCH-5 ON STATUS IS PROCESS-FLIGHTFILE,
                  OVERHAUL-FILE IS OVFILE.

INPUT-OUTPUT SECTION.
FILE-CONTROL.
*     FILE SPECIFICATIONS WORKSHEETS FOR ALL FILES
      SELECT TRANFILE           ASSIGN TO NCR-TYPE-41.
      SELECT FLIGHTFILE         ASSIGN TO NCR-TYPE-00.
      SELECT AIRFILE            ASSIGN TO NCR-TYPE-41.
      SELECT NEWAIRFILE         ASSIGN TO NCR-TYPE-41.
      SELECT OVERHAUL-FILE      ASSIGN TO NCR-TYPE-25.

I-O-CONTROL.
          RERUN EVERY END OF REEL OF NEWAIRFILE.
```

The SELECT entries in the Input-Output Section make the following hardware assignments: TRANFILE is assigned to a magnetic tape unit; FLIGHTFILE is assigned to the card reader; AIRFILE is assigned to a magnetic tape unit; NEWAIRFILE is assigned to a magnetic tape unit and has a rescue dump taken at the end of every reel; OVERHAUL-FILE is assigned to the printer.

Figure 3.10 Environment Division entries—examples.

Summary

The Identification Division and the Environment Division must be included in every COBOL source program. The Identification Division identifies both the source program and the resultant output listing, while the Environment Division provides a standard method of expressing those aspects of a data processing problem that depend upon the physical characteristics of a specific computer.

The Identification Division is the first and simplest of all four divisions to code. Besides the identification of the source program and resultant output listing, the programmer may include any other information that is considered vital to the program, such as the name of the programmer, the date the program was written, and such other information as desired.

The only required entries in the Identification Division are the name of the division, which must appear on a line by itself, and the PROGRAM-ID paragraph, which identifies the program.

The other optional paragraphs in the Identification Division may be included to meet the needs of the programmer. They include AUTHOR paragraph, which supplies the name of the programmer; the INSTALLATION paragraph, which identifies the installation where the source program is to be compiled and run; the DATE-WRITTEN paragraph, which is used to supply the date in which the program was written; the DATE-COMPILED paragraph, which causes the current date to be inserted during program compilation; and the SECURITY paragraph, which is used to supply the level of security attached to the program by the programmer.

A comment line may be used in the Identification Division to stipulate the purpose and what the program is to accomplish.

In the Environment Division, the compiling and executing computers, the information relating to Input-Output Control, special hardware and operating characteristics and control techniques are specified.

The Environment Division follows the Identification Division and is divided into two sections: the Configuration Section and the Input-Output Section.

The Configuration Section provides program documentation for the hardware characteristics of the computer used for the compilation of the source program and of the computer used for the execution of the object program. Provisions are also included for relating specific hardware and operating features to user-specified mnemonic names.

The Configuration Section is divided into three paragraphs: the SOURCE-COMPUTER paragraph, which specifies the computer upon which the source program will be compiled; the OBJECT-

COMPUTER paragraph, which specifies the computer upon which the object program is to be executed; and the SPECIAL-NAMES paragraph, which provides a means of relating specific hardware and operating features to user-specified names.

The SOURCE-COMPUTER and OBJECT-COMPUTER paragraphs are required in all COBOL source programs and are used for documentation purposes only, while the SPECIAL-NAMES paragraph is optional and included only when the user-specified mnemonic names are used for specific hardware and operating features.

The Input-Output Section is concerned with the definition of the input and output devices as well as the most efficient method of handling data between the devices and the object program. This section is divided into two paragraphs: the File-Control paragraph, which names each file used in the program and identifies the media on which each file is located; and the I-O-CONTROL paragraph, which specifies any special input/output techniques to be used in the object program.

The SELECT clause is used to name files within a COBOL source program. Each file described in the Data Division must be named in the File-Control paragraph.

The ASSIGN clause specifies the association of the file referenced by file-name to a storage device.

The I-O-Control paragraph defines special control techniques to be used by the object program.

Questions for Review

1. What is the function of the Identification Division?
2. What are the required entries and the purpose of each in the Identification Division?
3. Briefly describe the optional paragraphs and their purpose in the Identification Division.
4. How may the purpose of the program be stated in the Identification Division?
5. What is the importance of the Environment Division and what is its main purpose?
6. What are the sections of the Environment Division and what is the main function of each section?
7. What are the main functions of the Source-Computer, Object-Computer, and Special-Names paragraphs?
8. What are the main functions of the SELECT and ASSIGN clauses?
9. What is the purpose of the I-O-Control paragraph and what is it customarily used for?

Matching Questions

Match each item with its proper description.

______ 1. Input-Output Section	A. Used to name files within a COBOL source program.
______ 2. Program-ID	B. Identifies source listing and all listings pertaining to a particular program.
______ 3. SELECT Clause	C. A means of relating specific hardware and operating features to user-specified names.
______ 4. Environment Division	D. Provides program documentation for the hardware characteristics of computer uses for compilation and execution of programs.
______ 5. I-O-Control Paragraph	E. Comment line used to stipulate what the program is to accomplish.
______ 6. Program-name	F. Supplies name by which a program is identified.
______ 7. Configuration Section	G. Provides the information needed to control the transmission and handling of data between the external devices and the object program.
______ 8. ASSIGN Clause	H. Provides a standard method of expressing those aspects of a data processing problem that depends upon the physical characteristics of a computer.
______ 9. Purpose	I. Specifies the association of the file references by file-name to a storage device.
______10. Special-Names Paragraph	J. Defines control techniques to be used in the object program.

Exercises

Multiple Choice: Indicate the best *answer (questions 1-13).*

1. The File-Control paragraph of the Environment Division is important because in it every input or output file is assigned to an input or output
   a. file.
   b. data file.
   c. device.
   d. data cell.
2. Each entry in the File-Control paragraph of the Environment Division begins with the word SELECT, followed immediately by
   a. file number.
   b. file name.
   c. device number.
   d. device name.
3. The Program-ID paragraph in the Identification Division is required in every program. All of the other paragraphs are
   a. ignored.
   b. optional.
   c. required.
   d. None of the above.
4. Which section belongs to the Environment Division?
   a. Configuration.
   b. File-Control.
   c. I-O-Control.
   d. Special-Names.

5-7. When a/an ______________ section is written, the ______________ paragraph must be included, but the __________ paragraph may be omitted if no special techniques or conditions are defined.
   a. Configuration.
   b. File-Control.
   c. I-O-Control.
   d. Input-Output.

8. Which of the following is optional in a COBOL program?
   a. Identification Division.
   b. Program-ID
   c. Configuration Section.
   d. Environment Division.
   e. None of the above.
9. The following entries are required in the Identification Division.
   a. Division header.
   b. Program-ID paragraph.
   c. Program name entry.
   d. All of the above.
10. Since the format of each optional paragraph in the Identification Division is enclosed in a separate set of brackets, we can conclude that
    a. we may not choose to write more than one optional paragraph.
    b. we must write one or more of the optional paragraphs.
    c. we must write any number, or none, of the optional paragraphs.
    d. None of the above.
11. Of the following entries, which one is correctly written?
    a. IDENTIFICATION DIVISION.
       PROGRAM-ID.SALES-ANALYSIS.
    b. IDENTIFICATION DIVISION
       PROGRAM-ID. CARD-TO-TAPE.
    c. IDENTIFICATION DIVISION.
       PROGRAM-ID. PAYROLL-MASTER
    d. None of the above.

12. Which of the following PROGRAM-ID names is written correctly?
    a. DATA
    b. 'INVENTORY-CONTROL-REPORT'
    c. INVENTORY-MASTER
    d. PAYROLL*
13. In the following list of Identification Division names, which one is incorrect?
    a. AUTHOR
    b. TITLE
    c. SECURITY
    d. INSTALLATION
14. Match each item with its proper paragraph name.

_____ 1. Program-ID	A. Non-Military
_____ 2. Author	B. May 17 1982
_____ 3. Date-Written	C. J. Morse
_____ 4. Date-Compiled	D. Payroll04
_____ 5. Installation	E. District Office
_____ 6. Security	

15. In the Environment Division, indicate with a check mark which of the following must be written at the A margin.

_____ SOURCE-COMPUTER.	_____ SELECT Clause.
_____ CONFIGURATION SECTION.	_____ FILE-CONTROL.
_____ SPECIAL-NAMES.	_____ ASSIGN Clause.

16. Match each item with its proper device class.

_____ Utility	A. Magnetic disks and data cells.
_____ Unit Record	B. Magnetic tape and magnetic drum.
_____ Direct Access	C. Card Readers, Card Punches and Printers.

17. Determine which of the following is written correctly. Give reasons for your answers.
    a. IDENTIFICATION DIVISION.
       PROGRAM-ID.SALES-REPORT.
    b. IDENTIFICATION DIVISION.
       PROGRAM-ID. DISK-TO-DISK.
    c. IDENTIFICATION DIVISION.
       PROGRAM-ID.-PERSONNEL-MASTER.
18.

```
ENVIRONMENT DIVISION.
CONFIGURATION SECTION.
SOURCE-COMPUTER.     IBM-370.
OBJECT-COMPUTER.     IBM-370.
SPECIAL-NAMES.   CO4 IS TO-COMMENT.
INPUT-OUTPUT SECTION.
FILE-CONTROL.
     SELECT INCOME-FILE
          ASSIGN TO UR-2540R-INFILE.
```

In the Environment Division segment just given,
    a. the SELECT statement should begin in the A margin.
    b. the Special-Names paragraph should be another division.
    c. the File Section should replace the Input-Output Section.
    d. none of the above.
Select the best answer.

Problems

1. **Write the Identification Division, using all required and optional entries for the following:**

   The Acme Manufacturing Company is initiating an inventory control system. You are the programmer assigned to write the program. The program is restricted to production control personnel and is to be run at the Boston Center.

2. **Write the necessary entries in the Identification and Environment Divisions for the following:**

   The name of the program is EXAMPLE-2.
   The program is to be run at the CDP branch.
   Include all optional paragraphs that you consider necessary in the Identification Division. Be sure to state the purpose of the program. In addition, code the necessary entries in the Environment Division for both the Configuration and Input-Output Sections that will
   a. be compiled and executed on your computer, and
   b. create a file called CUSTOMER-FILE on the printer.

3. **Write the Environment Division for the following systems flowchart. The program will be compiled and executed on an IBM 370 model 155 computer with the following hardware assignments:**

SYS005	1403	Printer
SYS009	2540R	Card Reader
SYS006	2540P	Card Punch
SYS011	2400	Magnetic Tape
SYS012	2400	Magnetic Tape

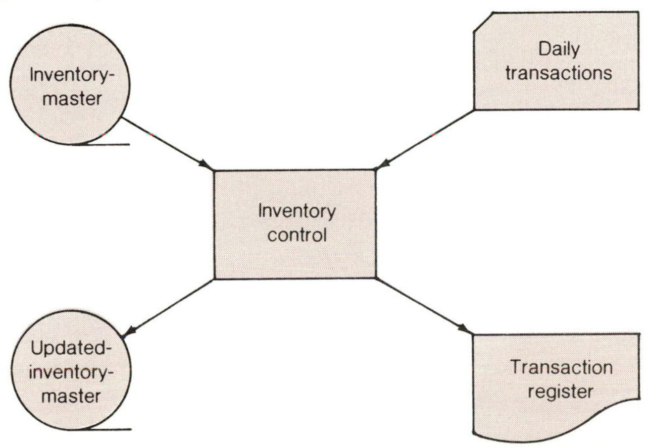

4. Write the Environment Division for the following systems flowchart using the same hardware assignments as problem 3.

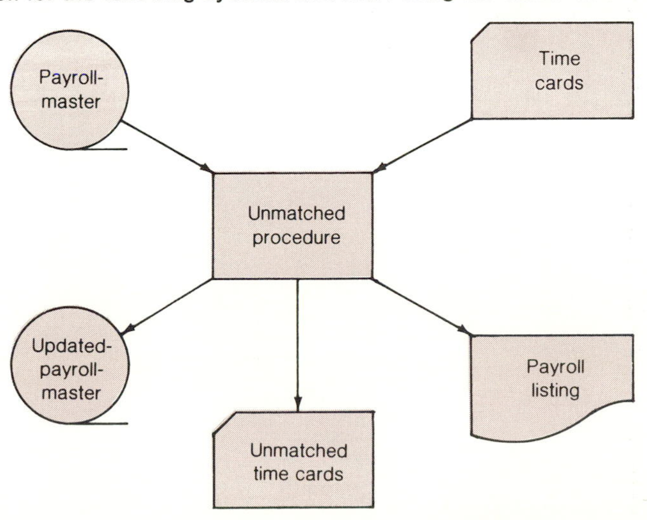

Data Division

Chapter Outline

Chapter Objectives

The learning objectives of this chapter are:

1. To describe the characteristics of the information processed by the object program.
2. To explain the various terms used in the handling of data.
3. To describe the characteristics of physical and logical records.
4. To describe the manner in which the data is organized in both records and files so that you will be able to properly code these entries in the Data Division.
5. To explain the functions and uses of the various clauses so that they can be properly used in your program.
6. To explain the use of the Working-Storage Section in the defining of constants and work areas.

Introduction

The Data Division is one of the most complex of the four divisions of COBOL. Certainly it demands much more of the programmer's time than the Identification and Environment divisions. The complexity of the coding depends on the complexity of the data itself; and the programmer's job is easier if he or she is intimately familiar with the layouts of the records and files that he or she is trying to describe in COBOL.

The Data Division describes the characteristics of the information to be processed by the object program. The separation of divisions provides the programmer with flexibility, as the Procedure Division is interwoven with the Data Division. The programmer can give his/her complete time to describing files and records without being concerned with the procedures that will process the data. That is, the task of describing data has been logically separated from the task of processing data; this separation is an important feature of COBOL. In practice, this means that each record will be described once, and the same record description may be used in every program that processes that record. It also means that all programmers will use the same names to refer to data items, and it means that we are justified in studying about the Data Division without worrying about the Procedure Division at the same time. The only relation that we must keep in mind is the Working-Storage Section, which will contain work areas and constants that will become necessary as we code the Procedure Division.

The manner in which data is organized and stored has a major effect upon the efficiency of the object program. To make data as computer-independent as possible, the characteristics of the data are described in relation to a standard data format rather than to an equipment-oriented format. This standard data format is oriented toward general data processing applications and uses the decimal system to represent numbers, and the remaining characters in the COBOL character set to represent nonnumerical data items.

Function: To describe data files and records in the files as well as items in working-storage. The structure of each record is usually shown, with items described in the order in which they appear in the record, and with a breakdown of smaller items within the larger items.

The Data Division describes the data that the object program is to accept as input, to manipulate, to create, or to produce as output. Data to be processed falls into three categories:

1. The data in the files that are entering or leaving the internal storage areas of the computer.
2. The data in the work areas of the computer that have been developed internally by the program.
3. Constant data that is to be used by the program.

Study figure 4.1. Identify these three categories as shown there.

The structure of each record within a file is usually shown with the items described in the sequence in which they appear in the record.

The Data Division begins with the header DATA DIVISION at the A margin followed by a period on a line by itself. *Each of the sections within the Data Division has a fixed name.* The sections are followed by the word SECTION and a period, and are on a line by themselves. These sections consist of entries rather than paragraphs. Each entry must contain:

1. A level indicator or level number.
2. A data-name or other name (FILLER).
3. A series of clauses defining the data that may be separated by commas. The clauses may be written in any sequence by the programmer (except the REDEFINES clause). Each entry must be terminated by a period and a space.

Figure 4.1 illustrates these entries.

Units of Data Terminology

Several terms make up typical Data Division entries. Terms like "file," "record," and "independent item" need further explanations. Each of these terms has a precise meaning and these definitions must be fully understood before any attempt is made to code the Data Division.

Data in COBOL source programs are referred to by various names. They are:

Item. An item is considered a field and is a storage area used to contain a particular type of information. The COBOL programmer reserves areas in which data will be stored while it is being

```
       DATA DIVISION.

       FILE SECTION.

       FD  PURCHASING-FILE.
           RECORD CONTAINS 8Ø CHARACTERS
           LABEL RECORDS ARE OMITTED
           DATA RECORD IS PURCHASING-RECORD.

       Ø1  PURCHASING-RECORD.
           Ø5  COMMODITY.
                 1Ø  NUMBER                PICTURE 9(12).
                 1Ø  DESCRIPTION           PICTURE X(3Ø).
           Ø5  PURCHASE.
                 1Ø  NUMBER                PICTURE 9(8).
                 1Ø  DATE                  PICTURE 9(6).
           Ø5  UNITS-PURCHASED             PICTURE 9(6).
           Ø5  UNIT-COST                   PICTURE 9(4)V99.
           Ø5  TOTAL-COST                  PICTURE 9(6)V99.
           Ø5  CARD-CODE                   PICTURE X(4).

       FD  PURCHASE-REPORT-FILE
           RECORD CONTAINS 133 CHARACTERS
           LABEL RECORDS ARE OMITTED
           DATA RECORD IS PURCHASE-REPORT-LINE.

       Ø1  PURCHASE-REPORT-LINE.
           Ø5  FILLER                      PICTURE X(1Ø).
           Ø5  COMMODITY-NUMBER            PICTURE 9(5)B99B9(5).
           Ø5  FILLER                      PICTURE X(6).
           Ø5  COMMODITY-NAME              PICTURE X(3Ø).
           Ø5  FILLER                      PICTURE X(6).
           Ø5  PURCHASE-DATE               PICTURE 99B99B99.
           Ø5  FILLER                      PICTURE X(4).
           Ø5  QUANTITY                    PICTURE Z,ZZZ,ZZ9.
           Ø5  FILLER                      PICTURE X(6).
           Ø5  COST-PER-UNIT               PICTURE $$,$$$.99.
           Ø5  FILLER                      PICTURE X(4).
           Ø5  PURCHASE-COST               PICTURE $$,$$$,$$$.99.
           Ø5  FILLER                      PICTURE X(14).

       WORKING-STORAGE SECTION.

       77  OLD-NUMBER                      PICTURE 9(12).
       77  QUANTITY-TOTAL                  PICTURE 9(7)           VALUE ZERO.
       77  PURCHASE-COST-TOTAL             PICTURE 9(7)V99        VALUE ZERO.
```

Figure 4.1 Data Division entries—sample.

processed. The data itself will change with each record that is read or outputted except for constant data.

An item can fall into three possible categories:

1. *Elementary item.* This is the smallest unit available that is not divided into smaller units. For example, in figure 4.1, 10 NUMBER and 10 DESCRIPTION are elementary items.
2. *Group item.* This is a larger item that is composed of a named sequence of one or more elementary items. A referral to a group item applies to the entire area of elementary items. In figure 4.1, 05 COMMODITY and 05 PURCHASE are group items.
3. *Independent item.* An independent item is an elementary item appearing in the Working-Storage Section that is not a record or part of a record. These areas are usually used as work areas or to contain constant data. Examples of independent items appear in figure 4.2.

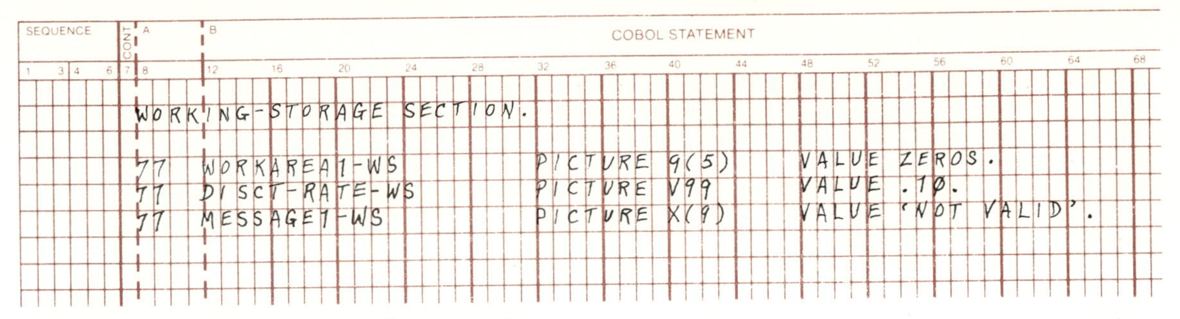

Figure 4.2 Independent items entries—sample.

Logical Record and File Concept

The approach taken in defining file information is to distinguish between the physical aspects of the file and the conceptual characteristics of the data contained in the file. Figure 4.3 illustrates some of the physical aspects of files. For example, we can see that files may be stored on a disk or in a card file, and that printed reports also constitute files.

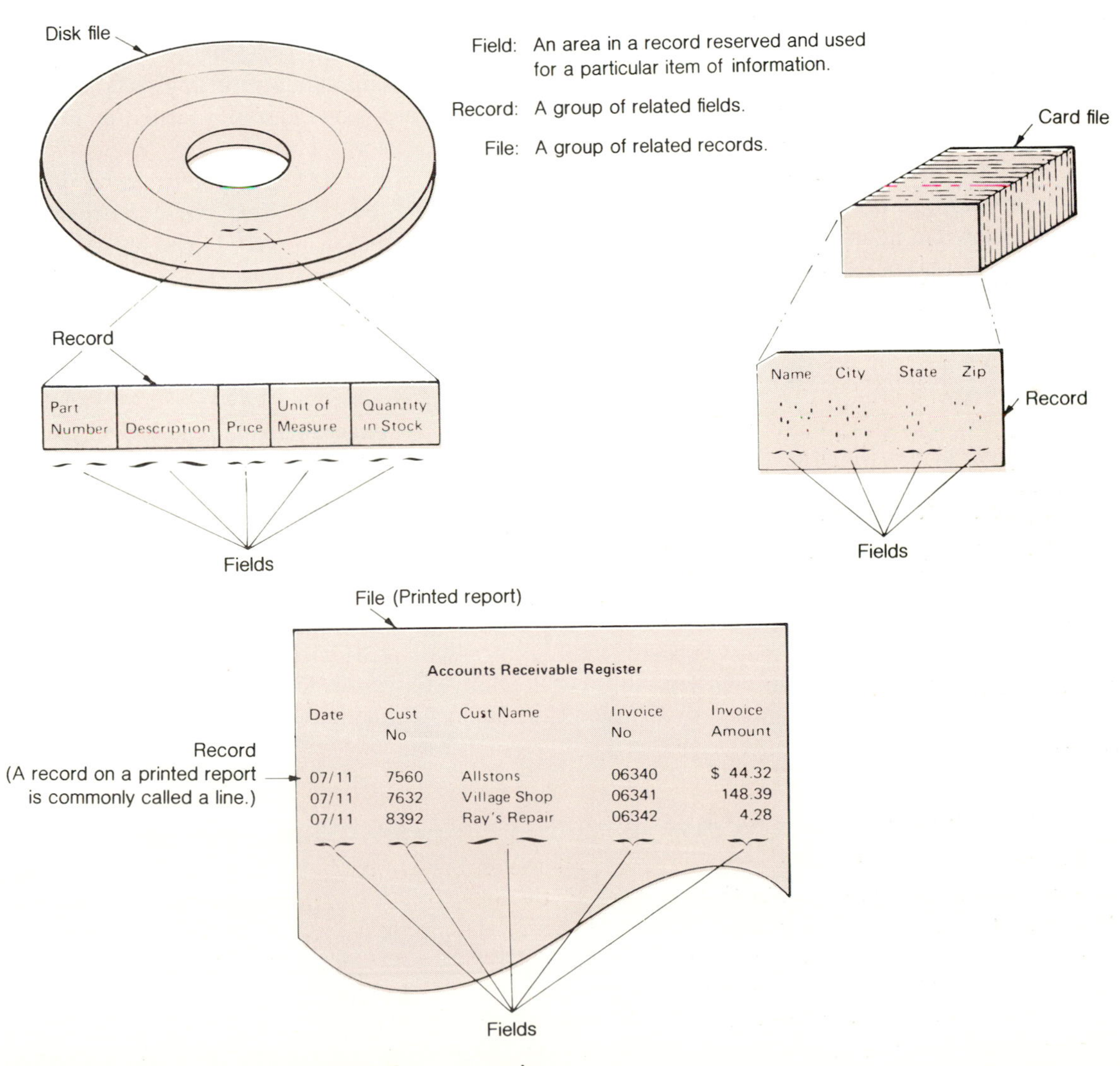

Figure 4.3 Fields (items), records, files—examples.

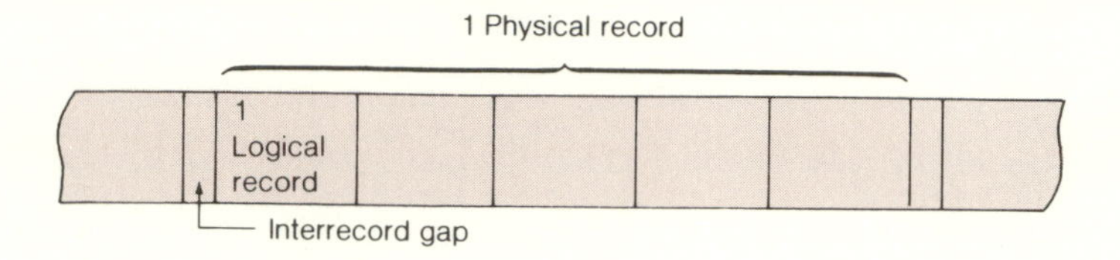

Figure 4.4 Physical record—example.

Physical Aspects of a File

The physical aspects of a file describe the data as it appears in the input or output devices and includes such features as:

1. The grouping of logical records within the physical limitations of the file medium and
2. The means by which the file can be identified.

Conceptual Characteristics of a File

The conceptual characteristics of a file are the explicit definitions of each logical entity within the file itself. In a COBOL program, the input and output statements refer to one logical record.

It is important to distinguish between a physical record and a logical record. *A COBOL logical record is a group of related information, uniquely identifiable, and treated as a unit.*

Data Record. The data record is usually considered to be the group item comprising several related items. It also is referred to as the "logical record." A logical record is one unit of information in a file of like units; for example, an item of inventory or an employee's record.

A physical record is a physical unit of information whose size and record mode is convenient to a particular input or output device for the storage of data. It may be made up of one or more logical records as shown in figure 4.4, or the logical record may itself be a physical record. The logical record is normally the unit of each program that processes input and output operations.

File. A file is composed of a series of related data records. The data records may have the same or varying lengths.

Block. A block is referred to as the "physical record" consisting of a series of logical records. A *physical record* is a group of characters or records which is treated as an entity when moved into and out of main storage. When data is stored on magnetic tape or direct-access devices, the logical records are grouped in blocks. Each read or write operation may transfer an entire block of data to or from main storage at one time and to or from an input/output device. Each logical record within the block is then processed separately.

COBOL source language statements provide a means of describing the relationship between physical and logical records. Once this relationship is established, only logical records are made available to the program.

Label Records. Label records are normally used for files that are stored on magnetic tape or direct-access devices. The record usually contains information relative to the file. Card files do not contain any label records. Label records are generally written at the beginning and end of a file, as shown in figure 4.5.

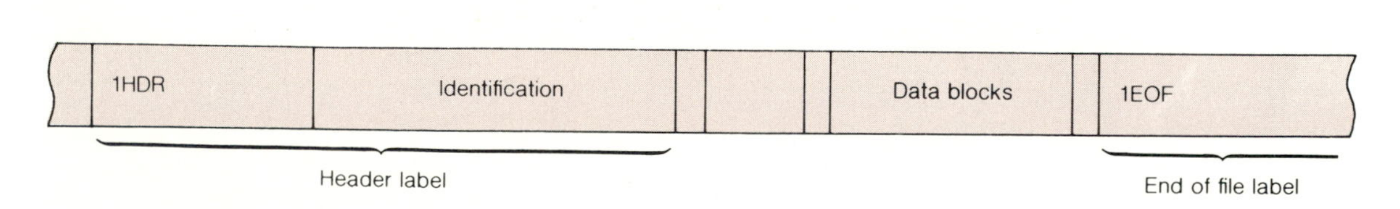

Figure 4.5 Label records—example.

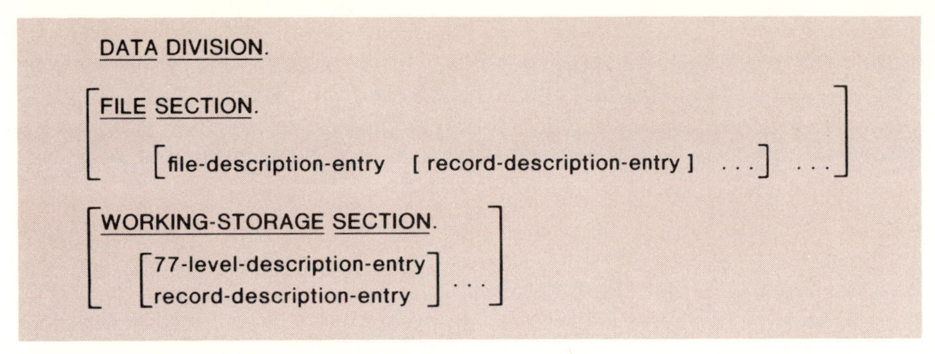

Figure 4.6 Structure of Data Division.

Organization

Data Description

In the discussion of data description, a distinction must first be made between the record's external description and its internal content.

External description refers to the physical aspect of a file, such as the way the file appears on an external medium. For example, the number of logical records per physical record describe the grouping of records in the file. The physical aspects of a file are specified in the file description entries.

A COBOL record usually consists of groups of related information that are treated as an entity. The explicit description of the contents of each record defines its internal characteristics. For example, the type of data to be contained within each field of a logical record is an internal characteristic. This type of information about each field of a particular record is grouped into a record description entry.

To repeat, the function of the Data Division is to describe files and records within the files, as well as items in working storage. The structure of each record is usually shown with items described in the order in which they appear in the record and with a breakdown of smaller items within larger items.

Only the two most frequently used sections, the File Section and the Working-Storage Section, will be discussed at this time. Formats for these two sections appear in figure 4.6. Other, less common sections are discussed later in the text.

The File Section defines the structure of the data files. Each file is defined by a file description entry and one or more record description entries. Record description entries are written immediately

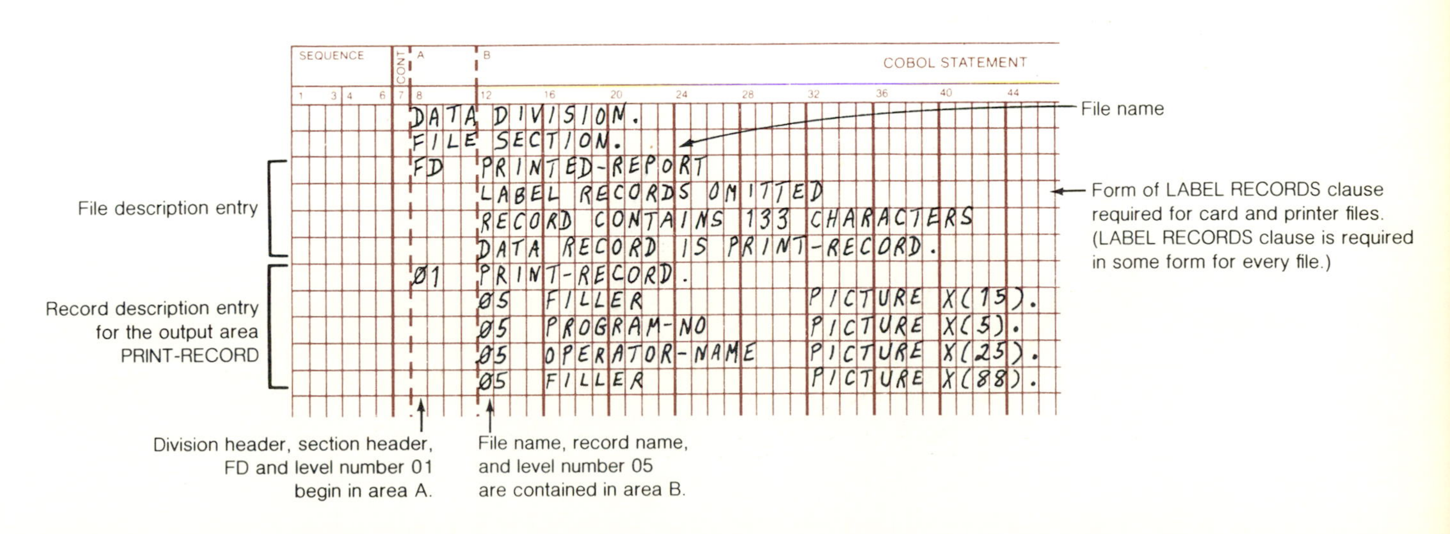

Figure 4.7 Format requirements for file description and record description entries in the file section.

following the file description entry. Figure 4.7 shows an example of a file description entry and an associated record description.

The Working-Storage Section describes records and non-contiguous data items that are not part of external data files, but rather are developed and processed internally. It also describes data items whose values are assigned in the source program and do not change during the execution of the object program. It is also used to set up the necessary work areas needed during the program execution.

File Section

Every program that processes input or output files is required to have a File Section. Since most programs employ files in the processing, a File Section appears in most programs. The File Section describes the characteristics of the file, and the record descriptions contained in those files.

For every file named in the SELECT clause in the Environment Division, a file description entry must appear in the Data Division.

File Description Entry

In a COBOL program, the file description (FD) entry represents the highest level of organization in the File Section. Figure 4.8 gives the general format for a file description entry. The File Section header is followed by a file description entry consisting of a level indicator (FD), a file-name, and a series of independent clauses. The FD clauses specify the size of the logical and physical records, the presence or absence of label records, and the value of label items, and the names of the data records of which the file is composed. The entry itself is terminated by a period. Figure 4.9 shows another example of a file description entry (see also figure 4.7). *NOTE:* It is important to remember that there should be *no* periods after any of the clauses in the file description (FD) entry except the last clause used. Unnecessary periods will generate many diagnostic errors in the source program listing.

Level Indicator

The file description entry always begins with the level indicator "FD," which is a reserved word. The indicator must be written at Area A, and is the only entry in the file description entry that begins in Area A.

```
DATA DIVISION.

[ FILE SECTION. ]

[FD  file-name

  [; BLOCK CONTAINS  [ integer-1 TO ]  integer-2  { RECORDS    } ]
                                                  { CHARACTERS }

  [; RECORD CONTAINS  [ integer-3 TO ]  integer-4 CHARACTERS ]

  [; LABEL  { RECORD IS   }  { STANDARD } ]
            { RECORDS ARE }  { OMITTED  }

   ; VALUE OF implementor-name-1 is  { data-name 1 }
                                     { literal-1   }

       [, implementor-name-2 IS  { data-name-2 } ] ... ]
                                 { literal-2   }

  [; DATA  { RECORD IS   }  data-name-3  [ , data-name-4 ]  ... ]]
           { RECORDS ARE }
```

Figure 4.8 File description entry—format.

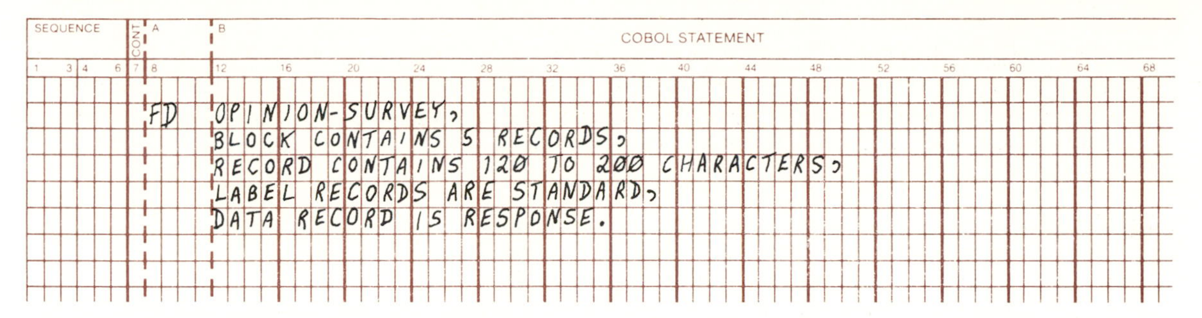


```
FD  OPINION-SURVEY,
    BLOCK CONTAINS 5 RECORDS,
    RECORD CONTAINS 120 TO 200 CHARACTERS,
    LABEL RECORDS ARE STANDARD,
    DATA RECORD IS RESPONSE.
```

Figure 4.9 File description entry—example.

File-Name

The file-name always follows the level indicator. The name is supplied by the programmer and must be the same as stipulated in the SELECT clause in the Environment Division. The name must begin in Area B.

The file-name should be meaningful to the contents of the file and state whether the file is an input or output file such as CUSTOMER-FILE-IN, indicating that this is a customer file to be used as input, or PRINTER-OUTPUT, which would designate a file to be used as output in a printed format. The use of such names will be helpful in documenting the program as well as assisting in the debugging of the source program.

In the FD entry, the programmer must specify whether the file contains records used to label the file in addition to the records of data; this is specified in the LABEL RECORDS clause and is required in every FD entry, whether or not the file contains any labels. All other clauses may be used in the FD entry at any time. Some clauses may be required under certain conditions to explain the nature and form of input or output, while other clauses serve only as documentation to assist the reader of the source program.

The order of clauses is not important to the program.

BLOCK CONTAINS Clause

The BLOCK CONTAINS clause specifies the size of a physical record. The size of a physical record may be stated in terms of RECORDS in a block or the number of CHARACTERS in a block. When the number of CHARACTERS per block is given, the clause specifies the *largest* number of characters that the longest block in storage will occupy in integer-2. If both integer-1 and integer-2 are shown, they refer to the minimum and maximum size of the physical record respectively. The format for this clause and an example appear in figure 4.10.

The BLOCK CONTAINS clause states the number of logical records or characters per physical record. The clause may be omitted when there is only one complete logical record in a physical record or when the hardware device assigned to the file has one and only one physical record.

If the size of the physical record is stated in terms of RECORDS, the RECORDS phrase must be used, otherwise the compiler would assume the phrase CHARACTERS.

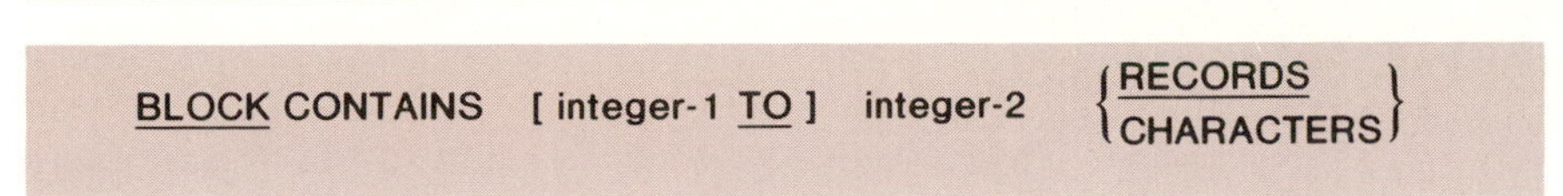

BLOCK CONTAINS [integer-1 TO] integer-2 {RECORDS | CHARACTERS}

BLOCK CONTAINS 20 RECORDS

Figure 4.10 BLOCK CONTAINS Clause format and example.

RECORD CONTAINS [integer-1 TO] integer-2 CHARACTERS

RECORD CONTAINS 80 CHARACTERS

Figure 4.11 RECORD CONTAINS Clause format and example.

RECORD CONTAINS Clause

The RECORD CONTAINS clause specifies the size of logical records contained in the file. If the record does not range in size, this clause will specify how many characters will appear in the longest record. If the record has a range of record sizes, integer-1 will indicate the size of the smallest record, and integer-2 will indicate the size of the largest record. The format for this clause and an example appear in figure 4.11.

Regardless of whether or not this clause is included, the record lengths are determined by the compiler from the record description entries. Since the size of each record is completely defined within the record description entry, this clause is never required. However, it can prove useful for both documentation and debugging purposes. Since the compiler determines the record length of each record from the record description entry, any variance in size between the record length determined by the compiler and the size specified in the RECORD CONTAINS clause will generate a diagnostic error. The use of this optional clause will assure that all items of the record have been properly defined in the record description entry, and will also provide the reader of the program with the record size without the necessity of counting the characters in the PICTURE clause of the record description entry.

LABEL RECORDS Clause

The LABEL RECORDS clause specifies whether labels are present and, if present, identifies the labels. Usually, magnetic tape files are labeled at the beginning to identify the file and the tape unit, with another label at the end of the file to provide a control to signal the end or indicate if there are more tapes in the file.

The LABEL RECORDS clause specifies the presence of standard or nonstandard labels in a file, or the absence of labels. This clause is required to appear in every file description entry. This clause may indicate that the label records are omitted or standard. The format for this clause and an example appear in figure 4.12.

Omitted. This option is used where there are no explicit labels for the file or where the existing file labels are nonstandard. This option may also be specified with nonstandard labels that the user wishes not to be processed by a label declarative.

Standard. The Standard option is used for labels that exist for a file and have the standard label format for the particular computer.

VALUE OF Clause

The VALUE OF clause specifies the description of an item in the label records associated with the file

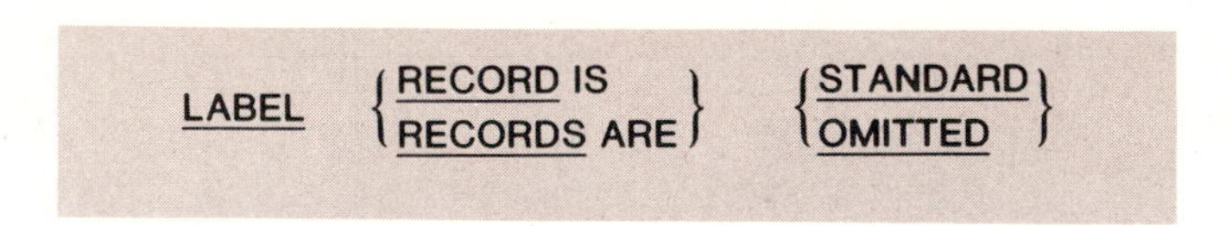

LABEL RECORDS ARE OMITTED

Figure 4.12 LABEL RECORDS Clause format and example.

```
VALUE OF implementor-name-1 IS {data-name-1}
                                {literal-1  }

    [, implementor-name-2 IS {data-name-2}] ...
                             {literal-2  }
```

VALUE OF IDENT IS DESC

Figure 4.13 VALUE OF Clause format and example.

and serves only as documentation. To specify the required value of identifying data items in the label record for the file, the programmer must use the VALUE OF clause.

For an input file, the appropriate label routine checks to see if the value of implementor-name-1 is equal to the value of literal-1, or of data-name-1, whichever has been specified. For an output file, at the appropriate time, the value of implementor-name-1 is made equal to the value of literal-1, or of a data-name-1, whichever has been specified.

A figurative constant may be substituted in the format wherever a literal is specified. The format and an example appear in figure 4.13.

DATA RECORDS Clause

The DATA RECORDS clause serves only as documentation and tells the compiler what the name or names of each of the records in a file are. The name of each record is the data-name supplied by the programmer. The presence of more than one data-name indicates that the file has more than one data record. Two or more record descriptions may occupy the same storage area for a given file. These records need not have the same description, and may be of differing sizes, differing formats, etc. When records of differing sizes are defined, the size of each record written is equal to the length of the largest record defined. The order in which these records are written is not significant.

Conceptually, all data records within a file share the same area. This is in no way altered by the presence of more than one type of record within a file.

Below the file description entry, each record-name must also appear in the level 01 entry in the record description clauses. This DATA RECORDS clause is never required in the file description entry. The format and an example appear in figure 4.14.

Record Description Entry

The format of the File Section requires that each file description must be followed by one or more record descriptions—one record description for each type of record in the file. A record description consists of a set of data description entries, which describe the characteristics of each item in a record. Every entry must be described in the same order as that in which it appears in the record and must indicate whether

```
DATA {RECORD IS   }  data-name-1  [, data-name-2] ...
     {RECORDS ARE }
```

DATA RECORDS ARE EXPENSE-DETAIL, DEPARTMENT-TOTAL.

Figure 4.14 DATA RECORD Clause format and example.

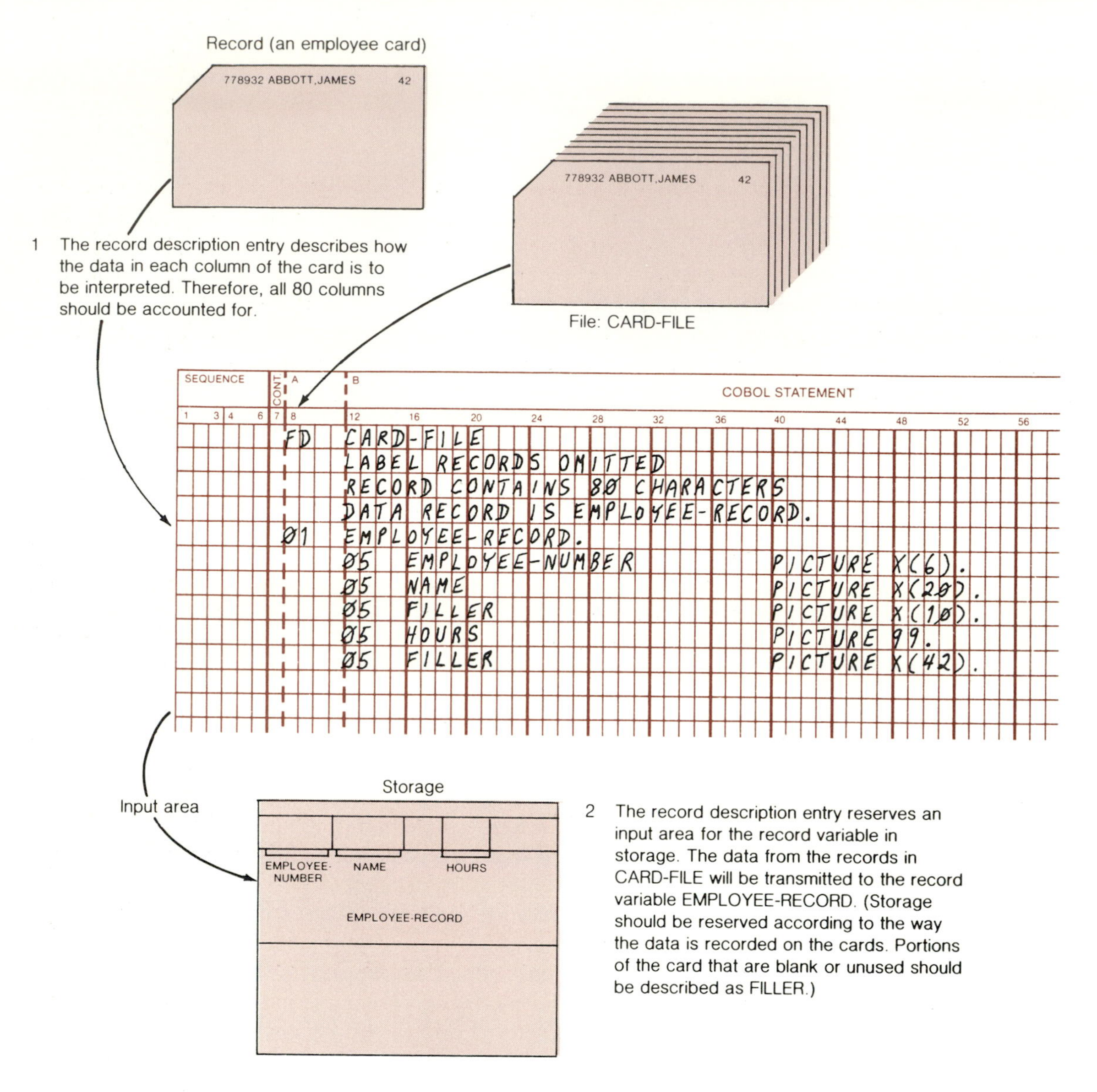

Figure 4.15 Record description entry and its relation to the external and internal file.

the items are related to each other. Figure 4.15 shows the relationship between a record punched on a card and its description in the program. The storage area reserved by the record description as shown conforms to the record description. Each record description entry consists of a level number, a data-name or FILLER, and a series of clauses followed by a space. The entry must be terminated by a period.

Some records may be divided into smaller units as follows:

1. Each entry must be given a level number beginning with a 01 for the record, with succeeding entries given higher-level numbers.
2. In subdividing an entry, the level numbers need not be consecutive. Level numbers 01–49 may be used for entries in the File Section.

Elementary Items are not further subdivided. Elementary items may be part of a group, or may be an independent item (not part of a group).

Group Items consist of all items under it until a level number equal to or less than the group number is reached.

Indentation. Item descriptions are usually indented to show the reader the relationship of the items within the group. Indenting is not required. If one entry at a given level is indented, then all similar entries should be indented for consistency.

Level Numbers

Concepts of Levels
A level concept is inherent in the structure of a logical record. This concept arises from the need to specify subdivisions of a record for the purpose of data reference. Once a subdivision has been specified, it may be further subdivided to permit more detailed data referral.

A system of level numbers is used to indicate the organization of elementary items and group items. Since records are the most inclusive data item, level numbers for records start at 01. Less inclusive data items are assigned higher (not necessarily successive) level numbers not greater than 49. Special level numbers 66, 77, and 88 are exceptions to this rule. Separate entries are written in the source program for each level number used.

Assignment of Level Numbers
The system of level numbers shows the organization of elementary and group items.

Level numbers are the first items of a record description entry.

1. Level numbers 01 and 77 must begin at the A margin (A margin and Area A are synonymous) followed by data-names and associated clauses beginning at the B margin (B margin and Area B are synonymous). All other level numbers may begin at the A or B margin, with the data-names and associated clauses beginning at the B margin.
2. At least one space must separate a level number from its data-name.
3. Separate entries are written for each level number.
4. A single-digit level number may be written as a space followed by a digit or as a zero followed by a digit.

01 Level number indicates that the item is a record. Since records are the most inclusive data items, the level number for a record must be 1 or 01. A record is usually composed of related elementary items, but may be an elementary item itself.

Level numbers 02–49 are used for subdivisions of group-related record items (not necessarily in consecutive order). Figure 4.16 shows the breakdown of a record into group and elementary items. Notice that the level numbers are not consecutive, which permits further subdivision or the addition of new group items as desired later.

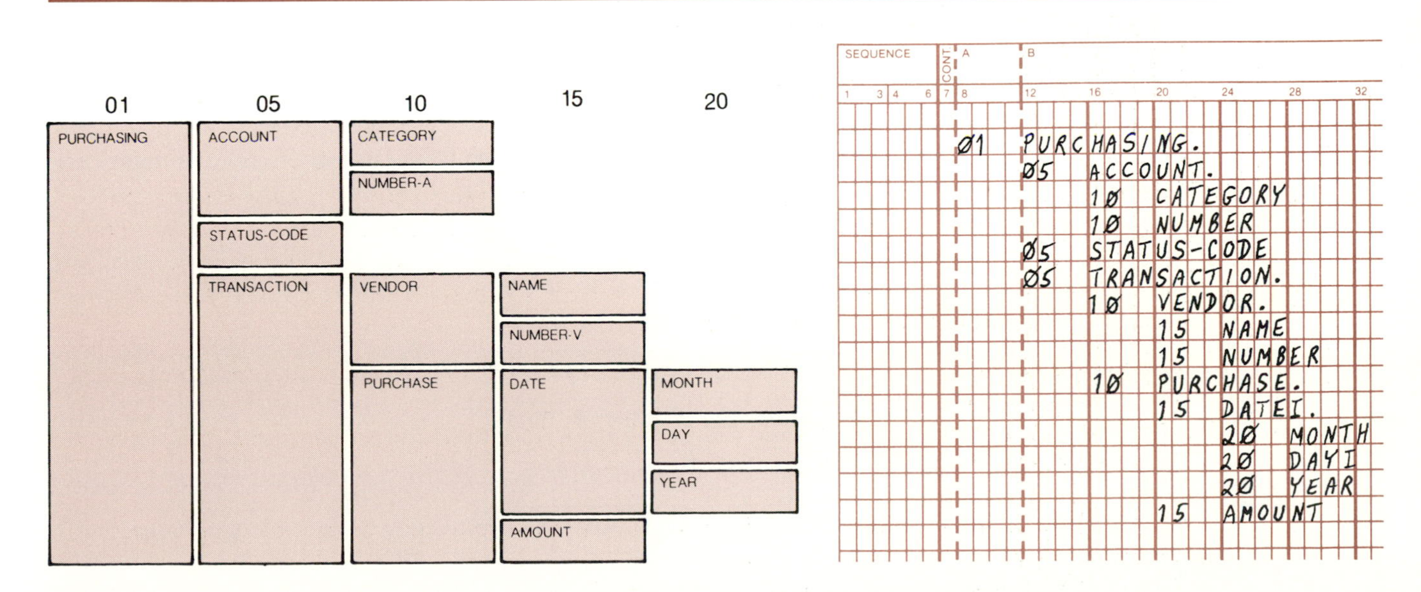

Figure 4.16 Level numbers—examples.

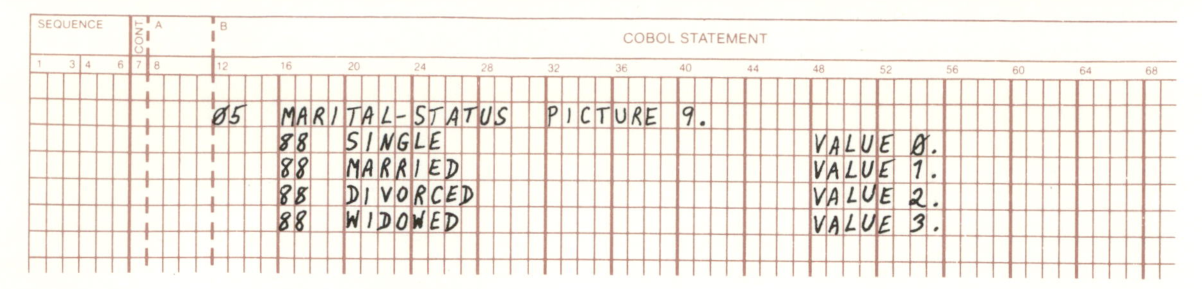

```
       05  MARITAL-STATUS  PICTURE 9.
           88  SINGLE                          VALUE 0.
           88  MARRIED                         VALUE 1.
           88  DIVORCED                        VALUE 2.
           88  WIDOWED                         VALUE 3.
```

Figure 4.17 Level number 88 entries—example.

There are several special level numbers used within data items where there is no real concept of level. These level numbers are:

66 Level number is used for names of elementary items or groups described by a RENAMES clause for the purpose of regrouping data items (see RENAMES clause section for an example of the function of the clause).

77 Level number is used to identify an independent elementary item in the Working-Storage Section. The item is not part of any record and is not related to any other item. Level 77 is usually used for noncontiguous items to define a work area or to store constant data. See figures 4.1 and 4.2 for examples.

88 Level number designates a CONDITION-NAME entry and is used to assign values to particular items during execution time. A name is furnished to values that the preceding item assumes. It does not reserve any storage area. Level 88 is associated only with elementary items. (See CONDITION-NAME clause section for examples of the use of level-88 entries.) Some examples appear in figure 4.17.

Condition-Names

Condition-names are assigned to an item that may have various values. The data item itself is called a *condition variable* and may assume a specific value, set of values, or range of values. Condition-names are often used in the Procedure Division as a test condition (usually with an IF statement) to specify certain conditions for branching to another part of the program.

The condition-name is the name of the *value* of an item, not the *name* of an item. An item description entry is required to define the item itself. Level-88 entries must follow immediately after the item description entry for the item with which they are associated. A condition-name is useful only if you know which item it is associated with. In the series of entries shown in figure 4.18, you should know that MALFUNCTION is a condition-name associated with TYPE-OF-CALL. Condition names must *follow* the elementary item with which they are associated. In figure 4.18, two condition-names are associated with TYPE-OF-CALL.

In condition-name condition, a condition-variable is tested to determine whether or not its value is equal to one of the values associated with a condition-name in the Data Division. Condition-names are

```
       01  SERVICE-HISTORY.
           05  MACHINE-NUMBER      PICTURE 9(8).
           05  TYPE-OF-CALL        PICTURE 9.
               88  PREVENTIVE-MAINTENANCE      VALUE 7.
               88  MALFUNCTION                 VALUE 4.
           05  DOWN-TIME           PICTURE 999V9.
```

Figure 4.18 Condition-name—example.

used in IF sentences; used properly, they make the sentences much more meaningful to the reader. Having defined a condition-name as shown here:

```
10  PRIORITY-CODE, PICTURE X.
    88  HIGHEST-PRIORITY, VALUE 'G'.
```

The programmer can write a sentence such as:

```
IF HIGHEST-PRIORITY, PERFORM FILL-ORDER-AT-ONCE.
```

Without the condition name, the programmer would have to write:

```
IF PRIORITY-CODE = 'G', PERFORM FILL-ORDER-AT-ONCE.
```

If a condition-name is associated with a range of values or ranges of values (that is, the VALUES ARE clause contains at least one 'literal THRU literal' phrase), then the condition variable is tested on whether or not its value falls into this range, including the end values. (Discussion and examples of the ranges of values will be found later in the text).

Data-Name or FILLER

Data-Name. Each item in the record description entry must contain either a data-name or the reserved word FILLER immediately following the level number beginning at the B margin. The data-name permits the programmer to refer to items individually in procedural statements.

A data-name should be meaningful. For instance, the assignment of the name should be descriptive, indicating the type of data it contains, and should also include some reference to the area of the Data Division in which it appears, such as: STOCK-NUMBER-IN would indicate that the data item appears in an input record; EMPLOYEE-NUMBER-WS would indicate that the data item appears in the Working-Storage Section; or QUANTITY-OUT would indicate that the item appears in an output record. This suggestion can be of great value to the programmer in debugging the source program by eliminating the need to search the entire program for data-name entries.

The data-name must be unique (not a reserved word) or must be properly qualified if not unique. The data-name can be made unique by either spelling the data-name differently from any other data-name used in the program, or through qualification of a nonunique name (see Qualification of Names section).

The data-name refers to the name of the storage area that contains the data, not to a particular value; the item referred to may assume numerous values during the execution of the program. For example, the storage area referred to as SCORE in figure 4.19 may assume any integer value up to 100.

In addition to the rules mentioned earlier in the Qualification of Names section, the following rules apply to data-names in the Data Division.

1. The highest possible qualifier would be the name of the file; thus it is possible for two records to have the same name.
2. The highest possible qualifier in the Working-Storage Section would be the record-name; thus all record-names in this section must be unique. The data-names for all independent items (level 77) must be unique since they can never be qualified.

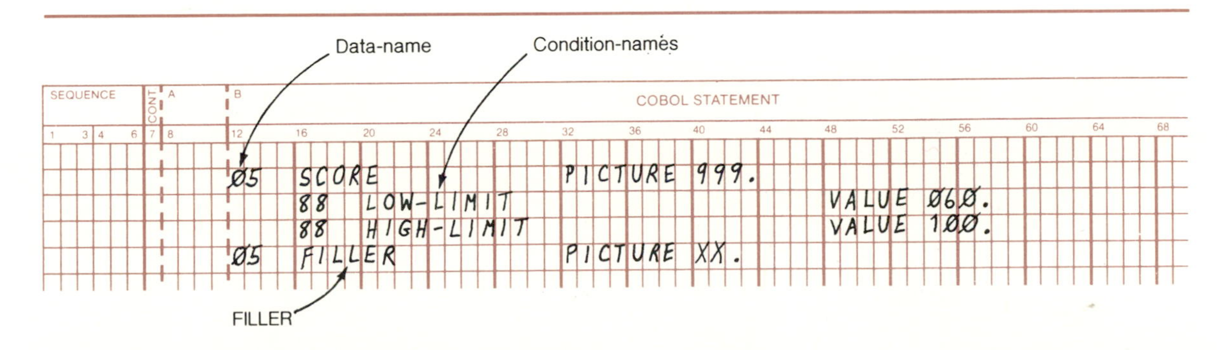

Figure 4.19 Data-names, condition-names, and FILLER— examples.

```
level number   { data-name }
               { FILLER    }
    [ REDEFINES Clause ]
    [ BLANK WHEN ZERO Clause ]
    [ JUSTIFIED Clause ]
    [ OCCURS Clause ]
    [ PICTURE Clause ]
    [ SYNCHRONIZED Clause ]
    [ USAGE Clause ]
    [ VALUE Clause ].
```

Figure 4.20 Record description format.

FILLER. The reserved word FILLER (see fig. 4.19) may be used in place of a data-name. The name cannot be referenced by any procedural statements. Its primary use is in the description of items that will not be referred to because the information contained is not necessary for the processing of the program. FILLER *does not always represent a blank area.*

Each record description entry may consist of one or more independent clauses that provide information about the data item. The most commonly used independent clauses are the USAGE, PICTURE, and VALUE clauses. The general format appears in figure 4.20.

USAGE Clause

In COBOL programming, data can be processed in different formats. Thus, one can take advantage of the flexibility of the data representation. It is not the purpose of this text to discuss the reasons for having different codes, nor to explain which codes are best for which purposes—these are "system design" concepts, not COBOL considerations.

The USAGE clause describes the form in which the data is stored in the computer's main storage. The possible USAGE clause formats appear in figure 4.21.

Rules for USAGE clauses are:

1. The clause may be written at any level.

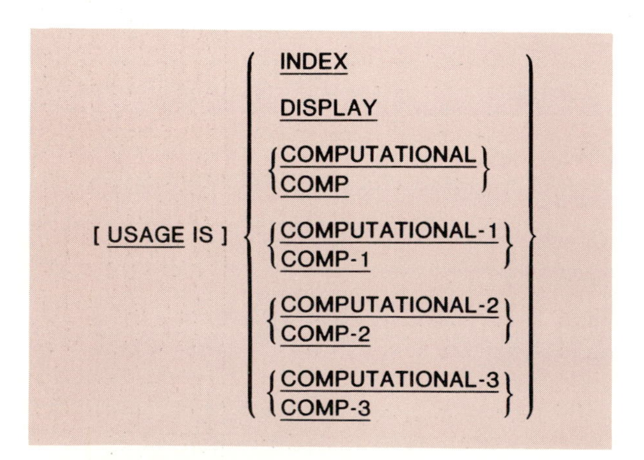

Figure 4.21 USAGE Clause formats.

2. If the clause is written at a group level, it applies to each elementary item within the group:

```
05 PAYMENT, COMPUTATIONAL-3.
   10 AMOUNT-DUE      PICTURE S9(6)V99.
   10 AMOUNT-PAID     PICTURE S9(6)V99.
```

Both elementary items will be COMPUTATIONAL-3.

3. The usage of an elementary item must not contradict the explicit usage of the group item of which the item is a part:

```
01 STOCK-TRANSFER.
   05 STOCK-NUMBER      PICTURE X(7).
   05 DESCRIPTION       PICTURE X(15).
   05 UNITS-OF-STOCK    PICTURE S9(8)     COMPUTATIONAL.
```

4. The usage of an elementary item is assumed DISPLAY if some other usage clause is not specified at the group or elementary level.

DISPLAY. This option specifies that one character is stored in each byte of the item. This corresponds to the form in which the information is represented for initial input or for final printed output. If the item is used to store numeric data, an operational sign may be included.

COMPUTATIONAL CLAUSE. All items in a computational clause represent values to be used in arithmetic operations, and must be numeric. If a group USAGE clause is used, it is only the elementary items that have that usage, since the group item cannot be used in computation.

Most computers require that the SYNCHRONIZED clause be added to all definitions of COMPUTATIONAL clauses that require alignment or the level number 01 descriptions that contain such items. The SYNCHRONIZED clause assures proper alignment of an elementary item on computer memory boundaries. The intent of the clause is to efficiently align data items on integral storage boundaries.

COMPUTATIONAL. This option is specified for binary data items. One binary digit is stored in each bit of the item except the leftmost, which will contain the operational sign. Such items have the decimal equivalent consisting of the decimal digits 0 through 9 plus the operational sign.

The amount of storage to be occupied by a binary item depends on the number of digits in its PICTURE clause.

Digits in PICTURE clause	Storage Occupied
1–4	2 bytes (halfword)
5–9	4 bytes (fullword)
10–18	8 bytes (2 fullwords not necessarily a doubleword)

The PICTURE clause of an item having COMPUTATIONAL usage may contain only 9s, the operational sign character S, the implied decimal point V, and one or more Ps. An operational sign character S must appear in COMPUTATIONAL usage items.

Items are aligned at the nearest halfword, fullword, or doubleword boundary.

COMPUTATIONAL-1. This option specifies that the item is stored in short precision internal floating-point format. Such items are four bytes in length and aligned on the next fullword boundary.

The data code is internal floating point, short (fullword) format.

COMPUTATIONAL-2. This option specifies that the item is stored in long precision internal floating-point format. Such items are eight bytes in length and are aligned on the next double-word boundary.

The data code is internal floating-point long (doubleword) format.

Both COMPUTATIONAL-1 and COMPUTATIONAL-2 options have special formats designed for floating-point arithmetic operations. Part of the item may be stored in binary form, and part may be stored in hexadecimal format.

PICTURE clauses are treated as comments with internal floating-point items.

The uses of COMPUTATIONAL-1 and COMPUTATIONAL-2 will be discussed later in the text, as these two clauses are not often used in decimal processing.

Item	Value	Description	Internal Representation*	
External decimal	−1234	DISPLAY PICTURE 9999	Z1 | Z2 | Z3 | F4 (Byte)	Note that internally, the D4, which represents −4, is the same bit configuration as the EBCDIC character M.
		DISPLAY PICTURE S9999	Z1 | Z2 | Z3 | D4 (Byte)	
Binary	−1234	COMPUTATIONAL PICTURE S9999	1111 | 1011 | 0010 | 1110 (S; Byte)	Note that, internally, negative binary numbers appear in two's complement form.
Internal decimal	+1234	COMPUTATIONAL-3 PICTURE 9999	01 | 23 | 4F (Byte)	
		COMPUTATIONAL-3 PICTURE S9999	01 | 23 | 4C (Byte)	

*Codes used in this column are as follows:

Z = zone, equivalent to hexadecimal F, bit configuration 1111

Hexadecimal numbers are their equivalent meanings are:
F= non-printing plus sign (treated as an absolute value)
C= internal equivalent of plus sign, bit configuration 1100
D= internal equivalent of minus sign, bit configuration 1101

S = sign position of a numeric field; internally,
1 in this position means the number is negative
0 in this position means the number is positive

Figure 4.22 Internal representation of numeric items.

COMPUTATIONAL-3. This option is specified for an item that is stored in packed decimal format (two digits per byte) with the low-order four bits of the rightmost byte containing the operational sign.

The PICTURE clause of COMPUTATIONAL-3 usage may contain only 9s, the operational sign S, the assumed decimal point V, and one or more Ps.

The data code is internal decimal (packed decimal format). For a description of internal decimal, see figures 4.22 and 4.23. Specifying the usage as COMPUTATIONAL-3 can increase the effectiveness

If the usage is:	Then the data code is:	Which means that:
Display	External decimal—also called BCD (binary-coded decimal), or EBCDIC (extended binary coded decimal interchange code)	One character is stored in each byte of the item; if the item is used to store a number, the rightmost byte may contain an operational sign in addition to a decimal digit.
Computational	Binary	One binary digit is stored in each bit of the item, except the left-most bit, in which the operational sign is stored.
Computational-3	Internal decimal—also called packed decimal	Two decimal digits are stored in each byte of the item, except the rightmost byte, in which one digit and the operational sign are stored.

Figure 4.23 USAGE Clause meanings.

of a program, if the data is to be used repeatedly in computations. Packed decimal format is more efficient for the storing of data; it uses less storage space and saves processing steps. The following example shows an easy and efficient way to set up three accumulators in the Working-Storage Section. The entries define a record named TOTALS, made up of MINOR-TOTAL (five digits), MAJOR-TOTAL (seven digits), and FINAL-TOTAL (nine digits). Each elementary item is in packed decimal form and contains a sign.

```
01  TOTALS, COMPUTATIONAL-3.
    05  MINOR-TOTAL    PICTURE S9(5)    VALUE ZERO.
    05  MAJOR-TOTAL    PICTURE S9(7)    VALUE ZERO.
    05  FINAL-TOTAL    PICTURE S9(9)    VALUE ZERO.
```

If data is simply to go from input to output, the DISPLAY usage clause should be used. (DISPLAY usage may be omitted, since this is a default option.)

INDEX. This option is discussed in the Table-Handling Section of the text.

PICTURE Clause

The PICTURE clause specifies the general characteristics and the detail description of an elementary item. The format is shown in figure 4.24.

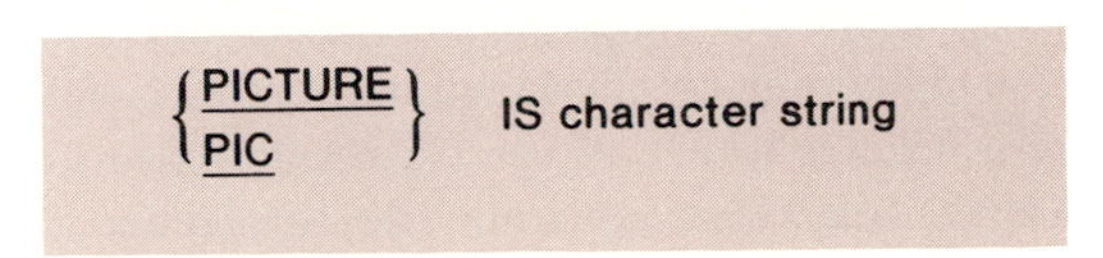

Figure 4.24 PICTURE Clause format.

Rules for the Use of PICTURE Clause are:

1. This clause is required in the description of every elementary item except those whose usages are COMPUTATIONAL-1 or COMPUTATIONAL-2. (Floating-point items have definite storage formats.)
2. The clause tells how many characters will be stored and describes the types of characters through the use of various symbols.
3. The clause is forbidden at the group level.
4. Numeric literals enclosed in parenthesis is a shorthand method of expressing the repetitively consecutive occurrence of a character. For example, X(10) is another way of writing XXXXXXXXXX.
5. All characters except P, V, and S are counted in the total size of any item.
6. CR and DB occupy two character positions in storage, and both may not appear in the same PICTURE clause.
7. A maximum of 30 characters is permitted in the clause. For example, PICTURE X(60) consists of 5 PICTURE characters, since only the actual characters appearing in the PICTURE clause are included in the count.
8. The characters S, V, CR, and DB may appear only once in a clause.

Figure 4.25 shows how to identify an item by its PICTURE clause. The meanings of some common PICTURE characters are given in figure 4.26.

Categories of Data

The categories of data that can be described with a PICTURE clause are

1. Alphabetic.
2. Numeric.
3. Alphanumeric.
4. Alphanumeric Edited.
5. Numeric Edited.

If the picture contains:	And also (possibly):	For example:	Then the item is called:	And will be used to store:
One or more Xs		XXX	Alphanumeric	Characters of any kind; letters, digits, special characters, or spaces
One or more As		A(35)	Alphabetic	Only letters or spaces
One or more 9s, but no editing symbols	S V P	S9(7)V99	Numeric	Only digits, and possibly an operational sign
One or more editing symbols; Z * $. , DB CR + − O B	9 V P	$ZZ,ZZZ.99	Report	Numeric data that is edited with spaces or certain special characters when the data is moved into the item

Figure 4.25 Identification of an item by its PICTURE.

The five categories of data items that can be described with a PICTURE clause are grouped into three classes: alphabetic, numeric, and alphanumeric. For alphabetic and numeric, the classes and categories are synonymous. The alphanumeric class includes the alphanumeric edited, numeric edited, and alphanumeric non-edited categories. Every elementary item belongs to one of the classes and also to one of the categories. The class of a group item at object time is treated as alphanumeric, regardless of the class of elementary items subordinate to that group item. In the following example, any processing involving the group item DATE-I will be treated as alphanumeric, despite the fact that the elementary items MONTH-I, DAY-I, and YEAR-I all have numeric PICTURE clauses.

```
05  DATE-I.
    10  MONTH-I     PICTURE 99.
    10  DAY-I       PICTURE 99.
    10  YEAR-I      PICTURE 99.
```

PICTURE Character	Meaning
X	Each X stands for one character of any kind—a letter, digit, special character, or space. The picture X(12) indicates that the item will contain twelve characters, but gives no indication of what characters they will be; all twelve could be spaces, or all could be digits, or there could be a mixture of various kinds of characters.
A	Each A stands for one letter or space.
9	Each 9 stands for one decimal digit. Numbers are always described in terms of the decimal digits they are the equivalent of—even when the data code is binary.
S	S indicates that the number has an operational sign. An "operational" sign tells the computer that the number is negative or positive; it is not a separate character that will print as "+" or "−".
V	V shows the location of an assumed decimal point in the number. An "assumed" decimal point is not a separate character in storage.
P	Each P stands for an assumed zero. Ps are used to position the assumed decimal point away from the actual number. For example, an item whose actual value is 25 will be treated as 25000 if its picture is 99PPPV; or as .00025 if its picture is VPPP99.

Figure 4.26 Meanings of some common PICTURE characters.

The following chart depicts the relationship of the class and categories of data items.

Level of Item	Class	Category
Elementary	Alphabetic	Alphabetic
	Numeric	Numeric
	Alphanumeric	Alphanumeric Alphanumeric Edited Numeric Edited
Group	Alphanumeric	Alphabetic Numeric Alphanumeric Alphanumeric Edited Numeric Edited

Alphabetic. An alphabetic item may contain any combination of the twenty-six letters of the alphabet and the space. No special characters or numerics are permitted in an alphabetic item. The permissible character in an alphabetic picture is A.

Numeric. A numeric item may contain any combination of the numerals 0–9; the item may have an operational sign. Permissible characters in a numeric picture are 9, V, P, and S.

Alphanumeric. An alphanumeric item may contain any combination of characters in the COBOL character set. A permissible character in an alphanumeric picture is X.

Alphanumeric Edited. An alphanumeric edited item is one whose picture clause is restricted to certain combinations of the following characters: A, X, 9, B, 0. To qualify as an alphanumeric edited item, one of the following conditions must exist. The PICTURE clause must contain at least:

1. One B and at least one X.
2. One 0 and at least one X.
3. One 0 and at least one A.

Numeric Edited. A numeric edited item is one whose PICTURE clause is restricted to certain combinations of the following characters: B, P, V, Z, 0, 9, ., *, +, −, CR, DB, $. The maximum number of digits in a numeric edited picture is 18.

PICTURE Characters

Nonedited PICTURE Clauses. A nonedited PICTURE clause may contain a combination of the characters as shown in figure 4.27. For examples, see figure 4.28.

Edited PICTURE Clauses. An edited PICTURE clause is used to describe items to be output on the printer (alphanumeric and numeric edited items).

Editing is used in the preparation of printed reports to give them a high degree of legibility and thereby greater usefulness. With proper planning, it is possible to suppress nonsignificant zeros, insert commas, insert decimal points, insert minus signs or credit symbols, and specify where suppressing of leading zeros should stop for numbers. Some examples of the editing applications of PICTURE clauses are shown in figure 4.29.

The characters and meanings of allowable editing characters in edited PICTURE clauses are shown in figures 4.30 and 4.31. (See also figures 4.32 and 4.33.) The reader should study these figures until the effects of the edit characters in the PICTURE clause are clear.

Insertion Characters

Insertion characters ,(comma), B(space), 0(zero) are editing characters that are inserted into the data for the purpose of improving the readability of the item.

Insertion characters are counted in determining the size of an item, and represent the position into which the character will be inserted.

The characters and meanings of the insertion characters are indicated in figure 4.34. (See also figure 4.35 for examples of various kinds of insertion editing.)

PICTURE Character	Meaning
9	The character 9 indicates that the position contains one decimal digit. Numbers are always described in terms of the decimal digits that they are equivalent to, even when the data is binary.
X	The character X indicates that the position can contain any type of character in the COBOL character set; a letter, digit, or special character.
V	The character V indicates the presence of an assumed decimal point. Since a numeric nonedited item may not contain an actual decimal point, an assumed decimal point provides the compiler with information concerning decimal alignment involved in computations. An "assumed decimal point" is not counted in the size of an elementary item and does not reserve any storage space.
P	The character P indicates the presence of an assumed zero. The Ps are used to position the assumed decimal point away from the actual number. For example, the actual value in storage is 15. It would be treated as 15000 if the PICTURE clause is 99PPP, or as .00015 if the PICTURE clause is VPPP99. The character V may be used or omitted when using the P character. When the V is used, it must be placed in the position of the assumed decimal point, to the left or the right of P or Ps that have been specified. The scaling position character P is not counted in the size of the data item.
S	The character S indicates the presence of an operational sign to the computer, either a positive or a negative number. It is not a separate character that will be printed as + or −. If used, S must be written as the leftmost character of the PICTURE clause. The presence of S is required where the USAGE clause is indicated as COMPUTATIONAL, since a sign appears in all binary numbers. The absence of S in a PICTURE clause will indicate a positive value. The operational sign is not counted in the size of an item.
A	The character A indicates the presence of a letter or space in an item. No special characters are permitted in a PICTURE clause with A picture.

Figure 4.27 Nonedited PICTURE characters and their meanings.

If the data item contains:	And the PICTURE clause is:	Then the data is interpreted as:	
3492	99V99	34.92	V indicates the location of the assumed decimal point.
169	9V99	1.69	
175	99V9	17.5	
254	V999	.254	
36985	9V9999	3.6985	
45694	9999V9	4569.4	
98745	99999	98745.	
246	999PPP	246000.	P indicates an assumed decimal scaling position and specifies the location of an assumed decimal point when the point is not within the data item. The use of P is generally avoided by programmers.
387	PP999	.00387	
487	999P	4870.	

Figure 4.28 Nonedited PICTURE examples.

	Source Area		Receiving Area	
	PICTURE	Data value	PICTURE	Edited data
1.	S99999	12345	−ZZ,ZZ9.99	12,345.00
2.	S99999V	00123	$ZZ,ZZ9.99	$123.00
3.	S9(5)	00100	$ZZ,ZZ9.99	$100.00
4.	S9(5)V	00000	−ZZ,ZZ9.99	0.00
5.	9(5)	00000	$ZZ,ZZZ.99	$.00
6.	9(5)	00000	$ZZ,ZZZ.ZZ	
7.	999V99	12345	$ZZ,ZZ9.99	$123.45
8.	V99999	12345	$ZZ,ZZ9.99	$0.12
9.	9(5)	12345	$**,**9.99	$12,345.00
10.	9(5)	00123	$**,**9.99	$***123.00
11.	9(5)	00000	$**,***.99	$******.00
12.	9(5)	00000	$**,***,**	$*********
13.	99V999	12345	$**,**9.99	$****12.34
14.	9(5)	12345	$$$,$$9.99	$12,345.00
15.	9(5)	00123	$$$,$$9.99	$123.00
16.	9(5)	00000	$$$,$$9.99	$0.00
17.	9(4)V9	12345	$$$,$$9.99	$1,234.50
18.	V9(5)	12345	$$$,$$9.99	$0.12
19.	S99999V	−12345	−ZZZZ9.99	−12345.00
20.	S9(5)V	12345	−ZZZZ9.99	12345.00
21.	S9(5)	−00123	−ZZZZ.99	−123.00
22.	S99999	12345	ZZZZ9.99	12345.00
23.	S9(5)	−12345	ZZZZ9.99−	12345.00−
24.	S9(5)	00123	−−−−−−−.99	123.00
25.	S9(5)	−00001	−−−−−−−.99	−1.00
26.	S9(5)	12345	+ZZZZZ.99	+12345.00
27.	S9(5)	−12345	+ZZZZZ.99	−12345.00
28.	S9(5)	12345	ZZZZZ.99+	12345.00+
29.	S9(5)	−12345	ZZZZZ.99	12345.00
30.	S9(5)	00123	++++++.99	+123.00
31.	S9(5)	00001	−−−−−−−.99	1.00
32.	9(5)	00123	++++++.99	+123.00
33.	9(5)	00123	−−−−−−−.99	123.00
34.	9(5)	12345	BB999.00	345.00
35.	9(5)	12345	00099.00	00045.00
36.	S9(5)	−12345	ZZZZZ.99CR	12345.00CR
37.	S9(5)	12345	$$$$$$.99CR	$12345.00

Figure 4.29 Editing applications of PICTURE Clauses—examples.

PICTURE Character	Meaning
Z	The character Z represents a digit-suppression character. 1. Each character Z represents a digit position. 2. All leading zeros appearing in positions represented by Zs are suppressed, leaving the positions blank. 3. Zero suppression is terminated when the actual or assumed decimal point is encountered. 4. A Z may appear to the right of the decimal point point only if all positions to the right are represented by Zs. 5. If all digit positions are represented by Zs, and the value of the data is zero, the entire area will be filled with blanks. 6. A Z character may not appear anywhere to the right of a 9 character. 7. Each Z is counted in the size of the item.
•	The character (.) represents an actual decimal point to be inserted in the printed output. 1. The decimal point is actually printed in the position indicated. 2. The source data is decimal aligned. 3. The character that appears to the right of the actual decimal point must consist of characters of one type (Z, *, 9, +, $, or−). 4. The character is counted in the size of a data item. 5. The actual decimal point may not be the last character in the PICTURE clause.
*	The asterisk (*) in the edited PICTURE clause is primarily used for protection of the amount in the printing of checks. 1. Each asterisk represents a digit position. 2. Leading nonsignificant zeros are replaced by asterisks. 3. Each field so defined will be replaced by asterisks until an actual or assumed decimal point is encountered. 4. An asterisk may appear to the right of the decimal point only if all digit positions are represented by asterisks. 5. If all digit positions are zero, the entire area will be filled with asterisks, except the actual decimal point. 6. The BLANK WHEN ZERO clause does not apply to any item having an asterisk (*) in its PICTURE. 7. An asterisk is counted in the size of an item.
CR DB	These character symbols are used as editing sign control symbols. These character symbols are printed *only if an item is negative.* They are called credit and debit symbols. 1. They may appear only at the right end of a PICTURE. 2. A positive value will blank out the characters, and only spaces will appear. 3. These symbols occupy 2 character positions and are counted in determining the size of an item.

Figure 4.30 Edited characters and their meanings.

PICTURE Character	Data Type	Specification	Additional Explanation
X	Alphanumeric	The associated position in the value will contain any character from the COBOL character set.	
A	Alphabetic	The associated position in the value will contain an alphabetic character or a space.	
9	Numeric or Numeric edited	The associated position in the value will contain any digit.	
V	Numeric	The decimal point in the value will be assumed to be at the location of the V. The V does not represent a character position.	
	Numeric edited	The associated position in the value will contain a point or a space.	A space will occur if the entire data item is suppressed.
$	Numeric edited	a. (simple insertion) The associated position in the value will contain a dollar sign. b. (floating insertion) The associated position in the value will contain a dollar sign, a digit, or a space.	The leftmost $ in a floating string does not represent a digit position. If the string of $ is specified only to the left of a decimal point, the rightmost $ in the picture corresponding to a position that precedes the leading nonzero digit in the value will be printed. A string of $ that extends to the right of a decimal point will have the same effect as a string to the left of the point unless the value is zero; in this case blanks will appear. All positions corresponding to $ positions to the right of the printed $ will contain digits; all to the left will contain blanks.
	Numeric edited	The associated position in the value will contain a comma, space, or dollar sign.	A comma included in a floating string is considered part of the floating string. A space or dollar sign could appear in the position in the value corresponding to the comma.
S	Numeric	A sign (+ or −) will be part of the value of the data item. The S does not represent a character position.	

Figure 4.31 PICTURE and edit characters.

If data moved to data-name is:	and the PICTURE clause is:	then the contents of data-name after move is:	
00000	ZZZZZ		Each Z in a character-string represents a leading numeric character position that is replaced by a space character when the content of that character position is zero.
39052	ZZZZZ	39052	
00006	ZZZZZ	6	
00295	ZZZZZ	295	
00005	ZZZ99	05	
00000	*****	*****	Each asterisk (*) in the character-string represents a leading numeric character position into which an * is placed when the content of that position is zero.
00820	*****	**820	
00858	***99	**858	
00075	**999	**075	
78963	*****	78963	

Figure 4.32 Edited PICTURE—examples.

Editing Symbol in PICTURE Character String	Result Data Item Positive or Zero	 Data Item Negative
+	+	−
−	Space	−
CR	2 spaces	CR
DB	2 spaces	DB

Figure 4.33 Editing sign control symbols and their results.

PICTURE Character	Meaning
,(comma) B(space) 0(zero)	1. The insertion character does not represent a digit position. 2. Zero Protection (Z) and Check Protection (*) indicates the replacement of insertion characters with spaces or asterisks if a significant digit or decimal point has not been encountered. 3. The comma, blank, or zero may appear with floating strings.

Figure 4.34 Insertion characters and their meanings.

PICTURE Character	Value of data	Edited result
99,999	12345	12,345
9,999,000	12345	2,345,000
99B999B000	1234	01 234 000
99B999B000	12345	12 345 000
99BBB999	123456	23 456

PICTURE Character	Value of data	Edited result
999.99	1.234	00.23
999.99	12.34	012.34
999.99	123.45	123.45
999.99	1234.5	234.50

PICTURE Character	Value of data	Edited result
999.99+	+6555.556	555.55+
+9999.99	−5555.555	−5555.55
9999.99−	+1234.56	1234.56
$999.99	−123.45	$123.45
−$999.99	−123.456	−$123.45
$9999.99CR	+123.45	$0123.45
$9999.99DB	−123.45	$0123.45DB

Figure 4.35 Simple insertion editing—examples.

PICTURE Character	Meaning
$(dollar sign) +(plus) −(minus)	These characters may appear in an edited PICTURE clause either in a floating string or singly as a fixed character. 1. As a fixed sign character, the + or − must appear as the first or last character (not both). 2. The plus sign (+) indicates that the sign of an item may be either plus or minus, depending on the algebraic values of the item. The plus or minus sign will be placed in the output area. 3. The minus sign (−) indicates that a minus sign for items will only be placed in the output area. If the item is positive, a blank will replace the minus sign. 4. As a fixed insertion character, the character $ may appear only once in a PICTURE clause. 5. Each character symbol is used in determining the size of the item.

Figure 4.36 Floating strings characters and their meanings.

Floating Strings

Floating strings are a series of continuous characters of either $, or +, or −, or a string composed of one or a repetition of one. Such characters may be interrupted by one or more insertion characters (comma, 0, B and/or V), or an actual decimal point. Floating strings begin with at least two consecutive occurrences of the characters to be floated. Some examples of floating strings follow:

```
$$,$$$,$$$
++++
--,---,--
$$$B$$$
++(8)V++
$$,$$$.$$
```

The floating string characters and their meanings are shown in figure 4.36.

Rules Governing the Use of Floating Strings

1. The floating string characters are inserted immediately to the left of the digit position indicated.
2. Blanks are placed in all positions to the left of the singly floating string character after insertion.
3. The presence of an actual or assumed decimal point in a floating string is treated as if all digit positions to the right of the decimal point where indicated by the PICTURE character 9 and BLANK WHEN ZERO clause were written for them.
4. A floating string need not constitute the entire picture.
5. When B (blank) or , (comma) or 0 (zero) appears to the right of the floating string, the character floats there to be as close to the leading digit as possible.
6. A comma may not be the last character in a PICTURE clause.

Some examples of floating string insertion editing appear in figure 4.37. Carefully study the additional examples and explanations that follow.

PICTURE Character	Value of data	Edited result
$$$$.99	12.34	$12.34
$$$$.99	1234	$234.00
$$$$.99	.1234	$.12
----.99	+12.34	12.34
----.99	−1.234	−1.23
$$99.99	1.234	$01.23

Figure 4.37 Floating strings insertion editing—examples.

The following examples illustrate a fixed insertion and a floating currency symbol using the symbol $.

If data moved to data-name is:	and the PICTURE clause is:	then, the contents of data-name after move is:
0000	$9999	$0000
6794	$9999	$6794
0008	$9999	$0008
0015	$ZZZZ	$ 15
003	$Z99	$ 03
0005	$$$$$	$5
1575	$$$$$	$1575
0004	$$$99	$04
00000	$****9	$****0
00225	$*****	$**225
4579	$999	$579

Each comma in the character-string represents a position into which a comma is inserted. This character position is counted in the size of the item. The comma cannot be the last character in the character-string. If the zero to the left of the symbol position has been suppressed, then the comma is not inserted at this position; the character that is replacing the zeros is inserted instead.

If data moved to data-name is:	and the PICTURE clause is:	then, the contents of data-name after move is:
00000	99,999	00,000
00000	ZZ,ZZZ	
00345	ZZ,ZZZ	345
02466	ZZ,ZZZ	2,466
02466	$ZZ,ZZZ	$ 2,466
00000	$ZZ,ZZZ	
00000	$**,***	*******
00000	$**,*99	$****00

Each period appearing in the character-string represents the decimal point for alignment purposes; it also represents a position into which the character period is inserted. The period is counted in the size of the item; it cannot be the last character in the PICTURE character-string. If the entire data item is suppressed to space characters, a space will replace the period. If the entire data item is suppressed to asterisks, an asterisk will not replace the period.

If data moved to data-name is:	and the PICTURE clause is:	then, the contents of data-name after move is:
000135	.999999	.000135
0013‸59	9999.99	0013.59
0004‸28	99.9999	04.2800
3‸9	999.99	003.90
1375	$$,$$$.99	$1,375.00
‸1375	$$,$$$.99	$.13
24675	$$,$$$.99	$4,675.00
00000	ZZZZZ.ZZ	
00000	***,***.**	******.**
‸05	$$.$$	$.05
‸00	$$.$$	
235‸07	$$$.$$	$35.07

‸represents assumed decimal point

Each B in the character-string represents a position into which the space character is inserted by the object program when data is placed in the field. The B is counted in the size of the field.

If data moved to data-name is:	and the PICTURE clause is:	then, the contents of data-name after move is:
A4892	XBXXXX	A 4892
CITYSTATE	XXXXBXXXXX	CITY STATE
FMLASTNAME	XBXBXXXXXXXX	F M LASTNAME
556086543	999B99B9999	556 08 6543
234‸45	999BV99	234 45

Each zero in the character-string represents a position into which the numeral zero is inserted by the object program when data is placed in the field. The 0 is counted in the size of the item.

If data moved to data-name is:	and the PICTURE clause is:	then, the contents of data-name after move is:
246	999000	246000
745	00999	00745
ABCXYZ	XXX0XXX	ABC0XYZ

When the minus sign (−) appears at either end of the character-string, the object program inserts a character describing the data as either positive or negative.

The minus sign itself will appear in the edited field only if the data being moved into the field is negative. If the data moving into the field is positive, a space character is inserted into the position indicated by the minus sign.

The minus sign can also be floated from the left end of the PICTURE by placing it in each leading numeric position to be suppressed. In the edited field each leading zero is replaced with a space, or a minus sign (depending on whether the sending data is negative or positive) will appear adjacent to the leftmost significant digit.

If data moved to data-name is:	and the PICTURE clause is:	then, the contents of data-name after move is:
9687	−9999	9687
−9687	−9999	−9687
4756	9999−	4756
−0756	9999−	0756−
12345	−−−−99	12345
00045	−−−−99	45
−00045	−−−−99	−45
−00045	−−−−−−	−45
000	−−−−−	
−000	−−−−−	

When the plus sign (+) appears at either end of the character-string, the object program inserts into the position either a plus or minus sign, describing the data as either positive or negative.

If the data moving into the edited field is positive, then a plus sign is inserted. If the data is negative, then a minus sign is inserted.

The plus sign can also be floated from the left end of the PICTURE by placing it in each leading numeric position to be suppressed. Each leading zero is replaced with a space, and either a plus sign or minus sign (depending on whether the sending data is positive or negative) will appear adjacent to the leftmost significant digit.

If data moved to data-name is:	and the PICTURE clause is:	then, the contents of data-name after move is:
5500	+9999	+5500
−5500	+9999	−5500
7689	9999+	7689+
−0987	9999+	0987−
12345	++++99	+12345
00045	++++99	+45
−00045	++++99	−45
000	++++	
−0000	++++	

When the symbol CR is written at the righthand end of the character-string, the object program inserts either CR or space characters, depending on whether the data is negative or positive.

If the data moving into the field is negative, a CR is inserted. If the data is positive, two space characters are inserted.

If data moved to data-name is:	and the PICTURE clause is:	then, the contents of data-name after move is:
123‸45	$999.99CR	$123.45
−123‸45	$999.99CR	$123.45CR
−123‸45	$***.99BCR	$123.45CR
−45‸6	$$,$$$.99CR	$45.60CR
4820‸33	$$,$$$.99CR	$4,820.33
−‸00	$$,$$$.99CR	$.00CR
0‸00	$$,$$$.99CR	$.00
0‸00	$*,***.99CR	$****.00

When the symbol DB is written into the righthand end of the character-string, the object program inserts either DB or space characters, depending on whether the data is negative or positive.

If the data moving into the field is negative, a DB is inserted; if the data is positive, two space characters are inserted.

If data moved to data-name is:	and the PICTURE clause is:	then, the contents of data-name after move is:
123‸45	$999.99DB	$123.45
−123‸45	$999.99DB	$123.45DB
−123‸45	$***.99BDB	$123.45DB
−45‸6	$$,$$$.99DB	$45.60DB
4820‸33	$$,$$$.99DB	$4,820.33
−‸00	$$,$$$.99DB	$.00DB
0‸00	$$,$$$.99DB	$.00
0‸00	$*,***.99DB	$****.00

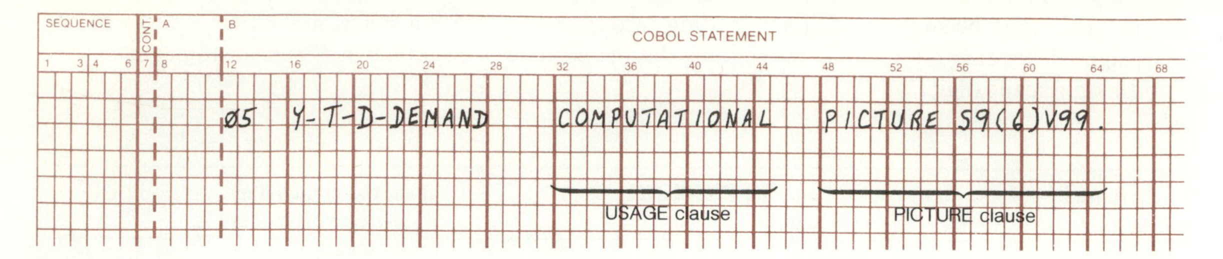

Figure 4.38 PICTURE and USAGE Clauses—example.

Relationship between PICTURE and USAGE Clauses
The usage of an item must be compatible with the PICTURE clause. The following kinds of items can have only DISPLAY usage: alphabetic, alphanumeric, alphanumeric edited, numeric edited, external decimal, and external floating point. Digits may have any usage: DISPLAY, COMPUTATIONAL, COMPUTATIONAL-1, -2, or -3, or INDEX. DISPLAY items may have any PICTURE clause; other than DISPLAY usage can have only numeric PICTURE clauses. (See figure 4.38.)

Rules Governing the Use of Edited PICTURE Clauses

1. There must be at least one digit position character in the clause.
2. If a fixed or floating string of plus or minus insertion characters are used, no other sign control character may be used.
3. The character to the left of an actual or assumed decimal point in the PICTURE clause (excluding the floating string of characters) are subject to the following restrictions.
   a. A Z may not follow 9, a floating string, or *.
   b. * may not follow 9, Z, or a floating string.
4. A floating string must begin with two consecutive characters (+, −, or $).
5. There may be only one type of floating string characters.
6. If the PICTURE clause does not contain 9s, BLANK WHEN ZERO is implied unless all the numeric positions contain asterisks. If the PICTURE clause does contain asterisks, and the area is zero, the area will be filled with asterisks.
7. The following restrictions apply to the characters to the right of the decimal point up to the end of the PICTURE (excluding insertion characters of +, −, CR, DB, if present).
   a. Only one type of digit character is permissible.
   b. If any of the characters appearing to the right of the decimal point is represented by +, −, Z, *, or $, then all the numeric characters in the PICTURE must be represented by the same characters.
8. The PICTURE character 9 can never appear to the left of the floating string or replacement character.
9. There cannot be a mixture of floating or replacement characters in an editing picture. They may appear as follows:
   a. An * or Z may appear with a fixed $.
   b. An * or Z may appear with a fixed leftmost + or fixed leftmost −.
   c. An * or Z may appear with a fixed rightmost + or fixed rightmost −.
   d. $ (fixed or floating string) may appear with fixed rightmost + or −.

VALUE Clause
For most items, one is not concerned about initial values. For example, to describe the items that make up an input record all that is necessary is to reserve an area in storage to receive the record. Specific values are going into these items when the record is read. The same conditions happen in the output record where all the values for the record are supplied by the program. These values will change for each record processed.

For this reason, COBOL bans the use of the VALUE clause for items in the File Section with the exception of its use in conjunction with the condition-name clause. (The condition-name, level-88 items will be discussed in the condition-name clause section.)

VALUE IS literal

Figure 4.39 VALUE Clause format.

The VALUE clause is permitted only in the description of elementary items, and only in the Working-Storage Section. Its general format is shown in figure 4.39. In some instances, one may want to specify an initial value of an item or a constant value that does not change within the program. For this type of programming, one would use the VALUE clause in the item description in the Working-Storage Section. One such instance is the setting up of constants such as

```
77  FICA-TAX    PICTURE V999    VALUE .067.
```

for the execution of the program whenever FICA-TAX is used. Another use of constants is the setting up of headings to appear in a report where the VALUE clause is used extensively.

Still another use of the VALUE clause is to have a certain value in a work area at the outset of the program execution. *Storage is not cleared before the object program is executed,* so one must not assume that those storage areas will contain blanks or zeros at the start of the program. It is the programmer's responsibility to clear these areas. A value must be specific such as the use of figurative constants ZERO and SPACE to set up an initial value such as

```
77  TRANSACTION-COUNT    PICTURE 999    VALUE ZERO.
```

The VALUE clause defines the value of constants, the initial value of working storage items, and the values associated with a condition-name.

Rules Governing the Use of the Value Clause

1. The size of the literal in the VALUE clause must be less than or equal to the size of an item as given in the PICTURE clause. All leading or trailing zeros reflected by Ps in a PICTURE clause must be included in the VALUE clause.
2. This clause is not permitted in the description of data items in the File Section other than condition-name entries at the 88 level.
3. When an initial value is not specified for an item in the Working-Storage Section, no assumption should be made regarding the original contents of the item.
4. A numeric literal must be used if the PICTURE clause designates a numeric item, as shown here:

```
77  DISCOUNT, USAGE COMPUTATIONAL,
    PICTURE SV99, VALUE .02.
```

5. A nonnumeric literal should be used if the item is alphabetic or alphanumeric, as shown here:

```
05  SUBSCRIPTION-BASIS, PICTURE X.
    88  REGULAR, VALUE '1'.
```

6. A figurative constant ZERO may be used in place of a numeric or nonnumeric literal. The number of zeros generated will be the same as the size specified in the PICTURE clause, as shown here:

```
77  TRANSACTION-COUNT, PICTURE 999,
        VALUE ZERO.
```

7. A figurative constant SPACE may be used as an initial value in the place of a nonnumeric literal. The number of blanks generated will depend upon the size of the item in the PICTURE clause.
8. The VALUE clause can only be specified for elementary items.
9. The VALUE clause must not be specified for any item whose size, explicit or implicit, is variable.
10. The VALUE clause must not be written in a record description entry that contains an OCCURS clause or REDEFINES clause, or an entry that is contained in an OCCURS or REDEFINES clause.

88 condition-name VALUE IS literal.

Figure 4.40 Condition-name clause format.

11. If the VALUE clause is written in an entry at the group level, the literal must be a figurative constant or nonnumeric literal, and the group area is initialized without consideration for the usage of the individual elementary or group items contained within this group. The VALUE clause cannot be specified at subordinate levels within this group.

Condition-Name Clause

A condition-name is a name assigned by the programmer to a particular value that may be assumed by a data item. The general format for the clause appears in figure 4.40. A condition-name is used only as an abbreviation for the relation condition. This relation condition stipulates that the associated condition variable is equal to one of the sets of variables to which that condition-name is associated. The result of the test is true if one of the values corresponding to the condition-name equals the value of its associated condition variable. For example, let MARRIED, SINGLE, WIDOW, WIDOWER, and DIVORCED correspond to the actual values 1, 2, 3, 4, 5, respectively, of a field called MARITAL-STATUS. The conditional statement

```
IF SINGLE . . .
```

would generate in the object program a test for the value of the condition variable MARITAL-STATUS against the value of "2."

Rules Governing the Use of Condition-Name Clauses

1. The condition-name is the name of the value of an item, not the name of the item itself. The item description entry complete with a PICTURE clause is required to describe the item.
2. Each condition-name must be unique or made unique through qualification.
3. A level 88 must be used in any condition-name entry.
4. A condition-name may pertain to an elementary item in a group item, with these exceptions: a level 66 item; a group containing items with descriptions that include JUSTIFIED, SYNCHRONIZED, or USAGE other than DISPLAY; or an index data item.
5. The condition-name is used in the Procedure Division's simple relational test statements.
6. In a condition-name clause, the VALUE clause is required. The VALUE clause and the condition-name itself are the only two entries permitted in the clause.

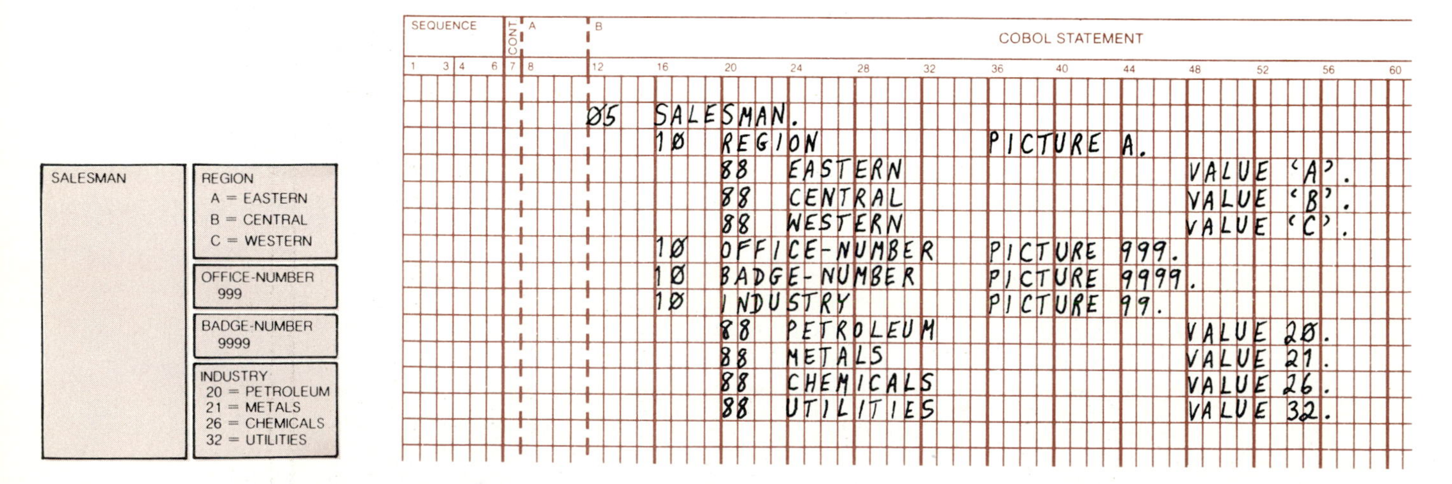

Figure 4.41 Condition-name clause entries—examples.

7. The condition-name must immediately follow the item with which it is associated.
8. The type of literal used must be consistent with the data type of the condition variable. For example, a numeric literal must be used if the item has a numeric picture, or a nonnumeric literal for alphabetic or alphanumeric pictures. The figurative constant zero may be used in place of a numeric or nonnumeric literal.

Some examples of condition-name clause entries appear in figure 4.41, along with the general breakdown of record levels. The use of level-88 (condition-names) entries will be discussed later in the Procedure Division to specify certain conditions for processing the data.

Working-Storage Section

The Working-Storage Section is composed of the section header, followed by data description entries for noncontiguous data items and/or record description entries. Each Working-Storage Section record-name must be unique since it cannot be qualified. Subordinate names need not be unique if they can be made unique by qualification.

The Working-Storage Section may contain descriptions of records which are not part of external data files, but which are developed and processed internally. This section is used to describe areas wherein intermediate results are stored temporarily at object time. The section is also used for descriptions of data to be used in the program. The section may be omitted if there aren't any constants or work areas needed in the program.

The Working-Storage Section is often used to provide headings for a report. Since this is the only section that is permitted to have VALUE clauses (outside of condition-name entries), report headings can be designed and moved to an output file description prior to a write operation. Output formats for detail lines to be printed can also be provided for in this section. In many programs, the Working-Storage Section is the largest unit in the Data Division.

Structure

The Working-Storage Section must begin with the header WORKING-STORAGE SECTION, followed by a period and a space on a line by itself. The section contains data entries for independent (noncontiguous) items and record items in that order. The general format for the Working-Storage Section appears in figure 4.42.

```
WORKING-STORAGE SECTION.

[ 77 noncontiguous item description entry ]  . . .

[ 01 record description entry ]  . . .
```

Figure 4.42 Working-Storage Section format.

Independent Items are items in the Working-Storage Section that bear no hierarchical relationship to one another and need not be grouped into records, provided that they do not need to be further subdivided.

1. These items must not be subdivided or themselves be a subdivision of another item.
2. These entries must precede record item entries (level numbers 01–49). (ANSI 1974 COBOL relaxed this rule. Level 77 items need no longer precede level 01 items in the Working-Storage Section.)
3. Each item must be defined in a separate record description as follows: Level number 77, Data Name, USAGE Clause (optional), VALUE clause (optional), and a PICTURE clause (required).
4. An OCCURS clause must not be used in describing an independent item.

Figure 4.43 provides a guide for coding these independent items.

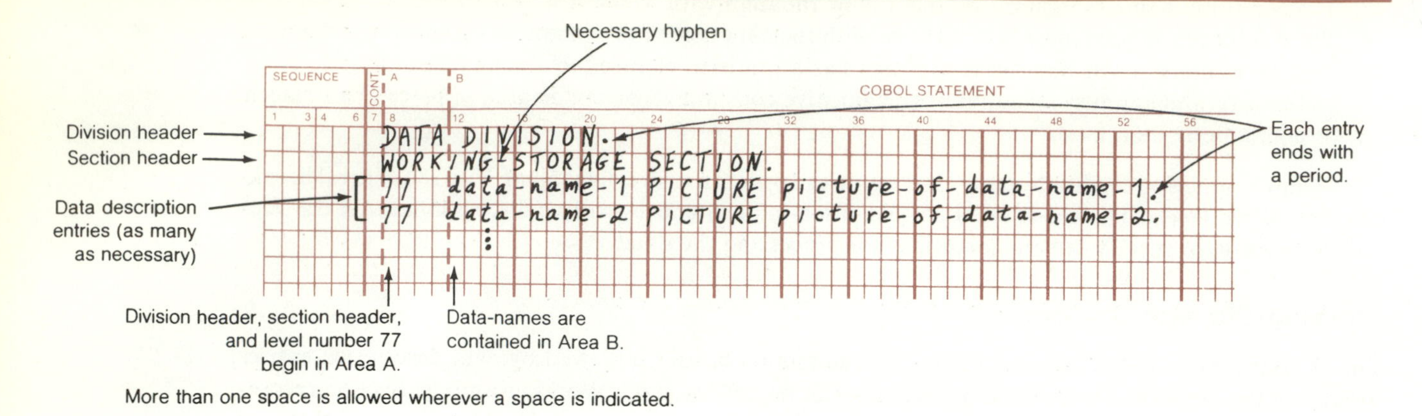

Figure 4.43 Guide for coding level 77 entries in the Working-Storage Section.

Other data description clauses are optional and can be used to complete the description of the item, if necessary.

These items are used primarily as temporary storage of an item pending the completion of a calculation, or to define a constant to be used in the program. Some examples appear in figure 4.44.

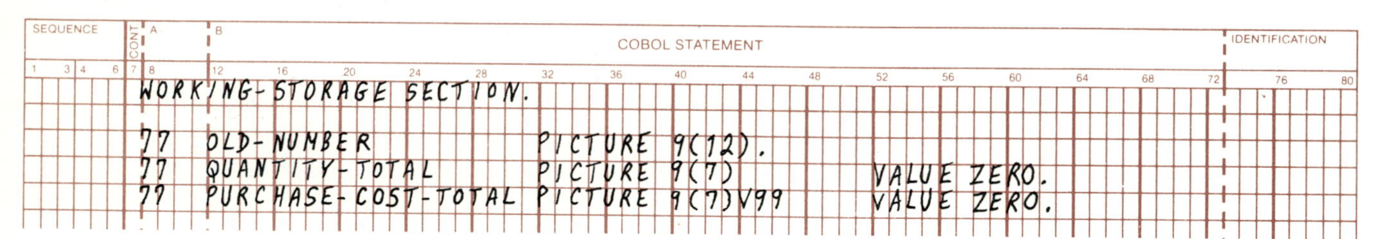

Figure 4.44 Level 77 entries: Working-Storage Section— example.

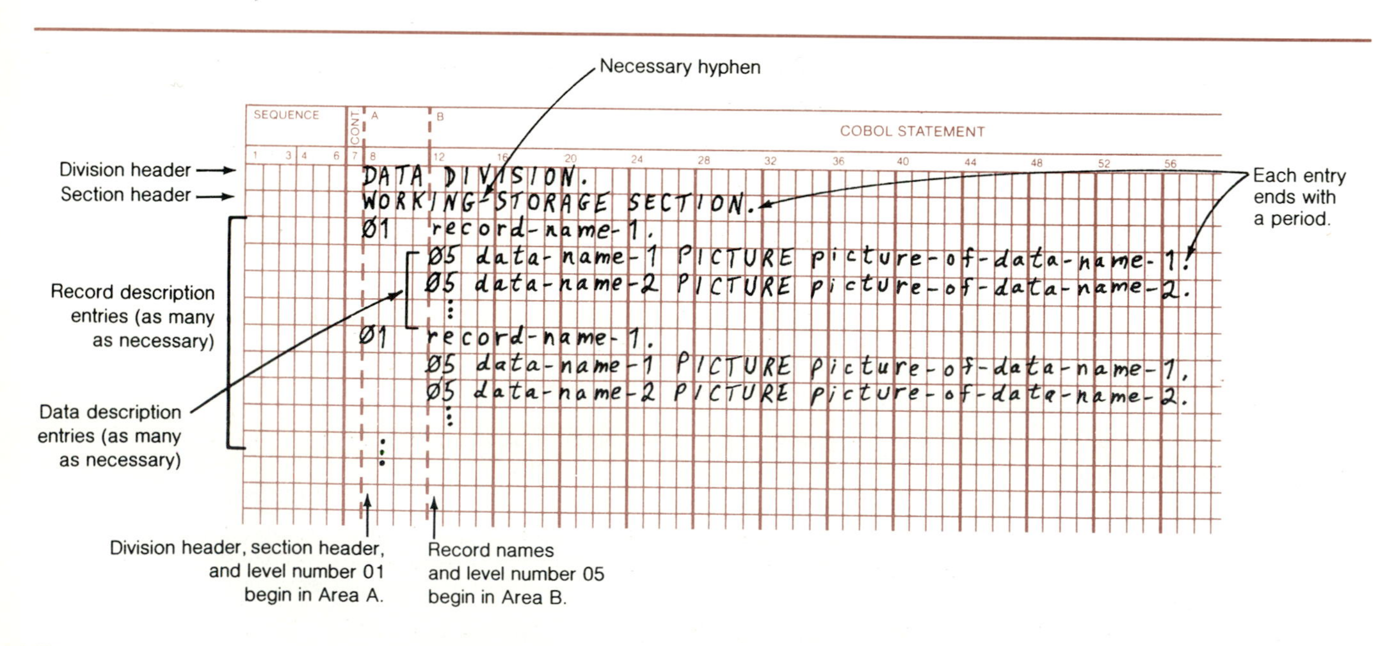

Figure 4.45 Guide for coding level 01 and 05 entries in the Working-Storage Section.

Record Items are data elements and constants in the Working-Storage Section among which there is a definite hierarchical relationship; they must be grouped together into records according to the rules for the formation of record descriptions. Figure 4.45 provides a guide for coding these entries.

All clauses that are used in record descriptions in the File Section can be used in record descriptions in the Working-Storage Section.

1. These items are subdivided into smaller units and bear a definite relationship to each other.
2. The entries used to describe these record items are identical to those used to describe a record in the File Section. However, one difference exists between records in the File Section and the entries in the Working-Storage Section. The entries in the File Section may be elementary items, but entries at the record level in the Working-Storage Section must be group items.

Record entries are often used for output heading and detail formats because of the ability to use the VALUE clause in the Working-Storage Section. Figure 4.46 shows their use in detail formats; the sample Data Division entries govern the processing in the diagram, which leads to the output file shown. Their use in heading formats is shown in figure 4.47, with a different arrangement of the detail format.

An internal value of an item in the Working-Storage Section may be specified by using the VALUE clause. The value is assumed by the item at the time of the execution of the program. Figures 4.48 and 4.49 show some examples. *No assumption can be made of the initial value of an item that has not been defined with a VALUE clause.*

Additional Data Division clauses, entries, examples, and uses will be found further along in the text.

Summary

The Data Division describes the data that the object program is to accept as input, to manipulate, to create, or to produce as output. The data to be processed falls into three categories:

1. The data in the files that are entering or leaving the internal storage area of the computer.
2. The data in the work areas of the computer that have been developed internally by the program.
3. Constant data that is to be used by the program.

The terms used in the Data Division are:

Item, which is a storage area used to contain a particular type of information. Items fall into three categories:

Elementary item—is the smallest unit available that is not divided into smaller units.

Group item—is a larger item composed of a named sequence of one or more elementary items. A referral to a group item refers to the entire area of elementary items.

Independent item—is an elementary item appearing in the Working-Storage Section that is not a record or part of a record that is usually used as work areas or to contain constant data.

Logical record, which is a group of related information, uniquely identifiable, and treated as a unit. It is commonly known as a data record.

Physical record, which is a physical unit of information whose size and record mode is convenient to a particular input or output device for the storage of data. Commonly referred to as a block.

File, which is composed of a series of related data records.

Label records, which usually contain information relative to the files that are stored in magnetic tape or direct access devices.

The function of the Data Division is to describe files and the records within the files, as well as items in working storage. The two most frequently used sections in the Data Division are the File Section and the Working-Storage Section.

The File Description entry represents the highest level of organization in the File Section. The File Section header is followed by a file description entry consisting of a level indicator (FD), a file-name,

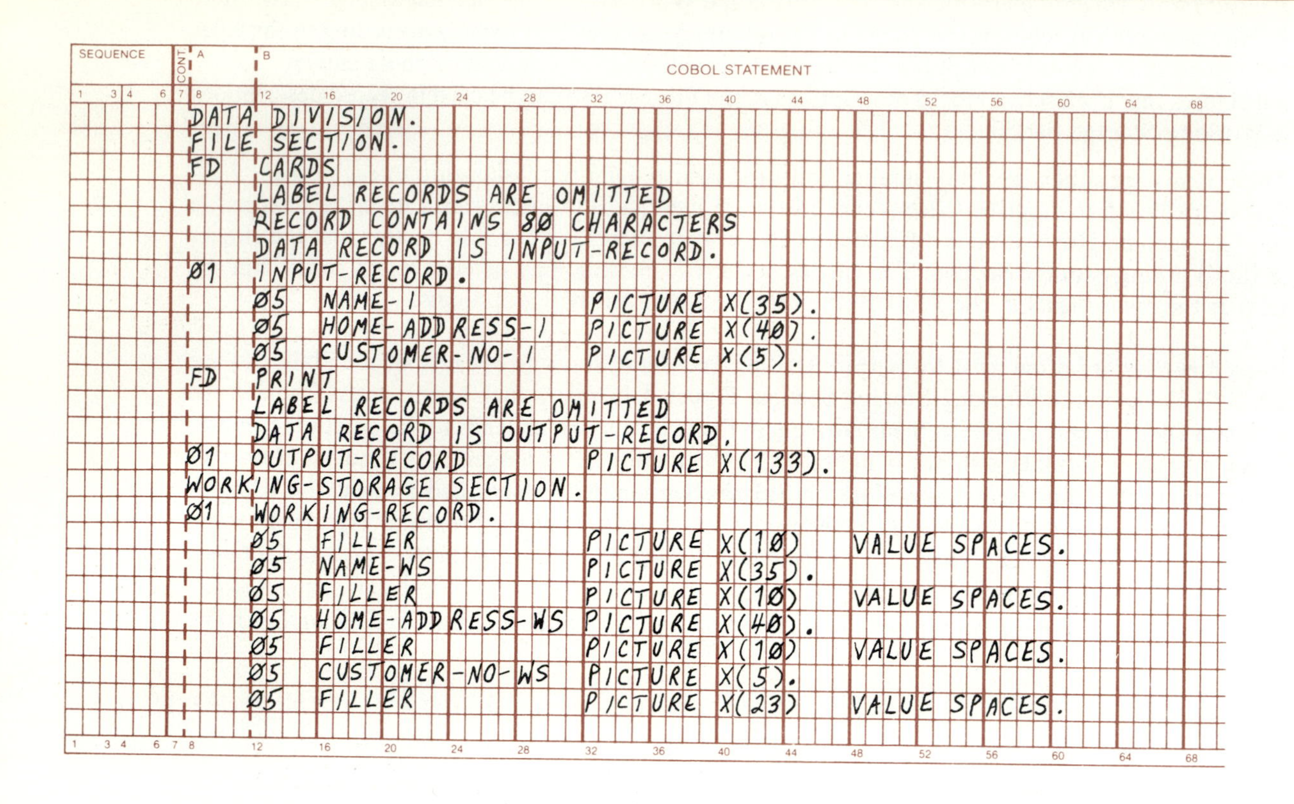

```
SEQUENCE  CONT. A   B                      COBOL STATEMENT

       DATA DIVISION.
       FILE SECTION.
       FD  CARDS
           LABEL RECORDS ARE OMITTED
           RECORD CONTAINS 8Ø CHARACTERS
           DATA RECORD IS INPUT-RECORD.
       Ø1  INPUT-RECORD.
           Ø5  NAME-1               PICTURE X(35).
           Ø5  HOME-ADDRESS-1       PICTURE X(4Ø).
           Ø5  CUSTOMER-NO-1        PICTURE X(5).
       FD  PRINT
           LABEL RECORDS ARE OMITTED
           DATA RECORD IS OUTPUT-RECORD.
       Ø1  OUTPUT-RECORD            PICTURE X(133).
       WORKING-STORAGE SECTION.
       Ø1  WORKING-RECORD.
           Ø5  FILLER               PICTURE X(1Ø)   VALUE SPACES.
           Ø5  NAME-WS              PICTURE X(35).
           Ø5  FILLER               PICTURE X(1Ø)   VALUE SPACES.
           Ø5  HOME-ADDRESS-WS      PICTURE X(4Ø).
           Ø5  FILLER               PICTURE X(1Ø)   VALUE SPACES.
           Ø5  CUSTOMER-NO-WS       PICTURE X(5).
           Ø5  FILLER               PICTURE X(23)   VALUE SPACES.
```

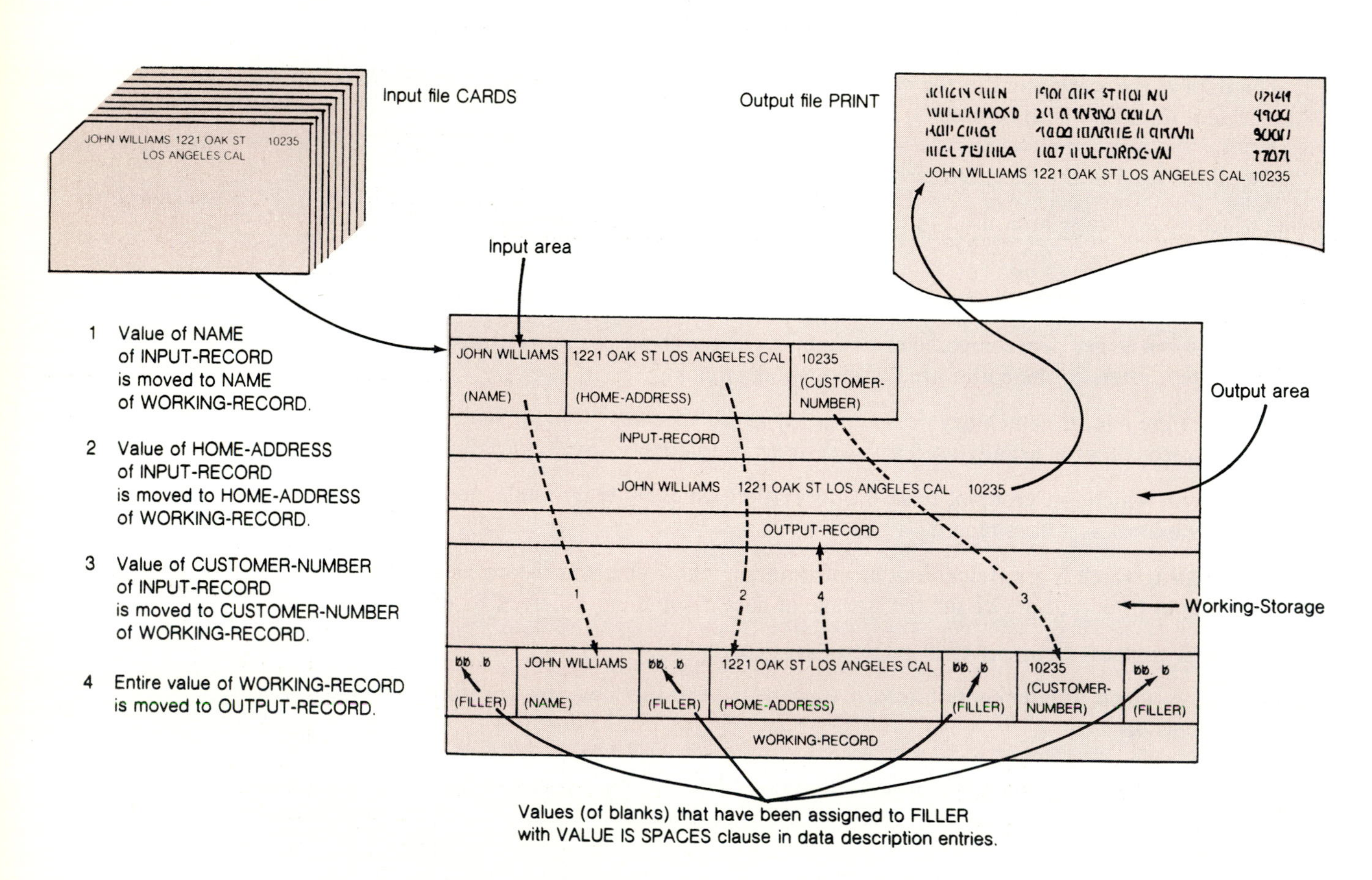

Figure 4.46 Data Division Entries for output of a file— sample.

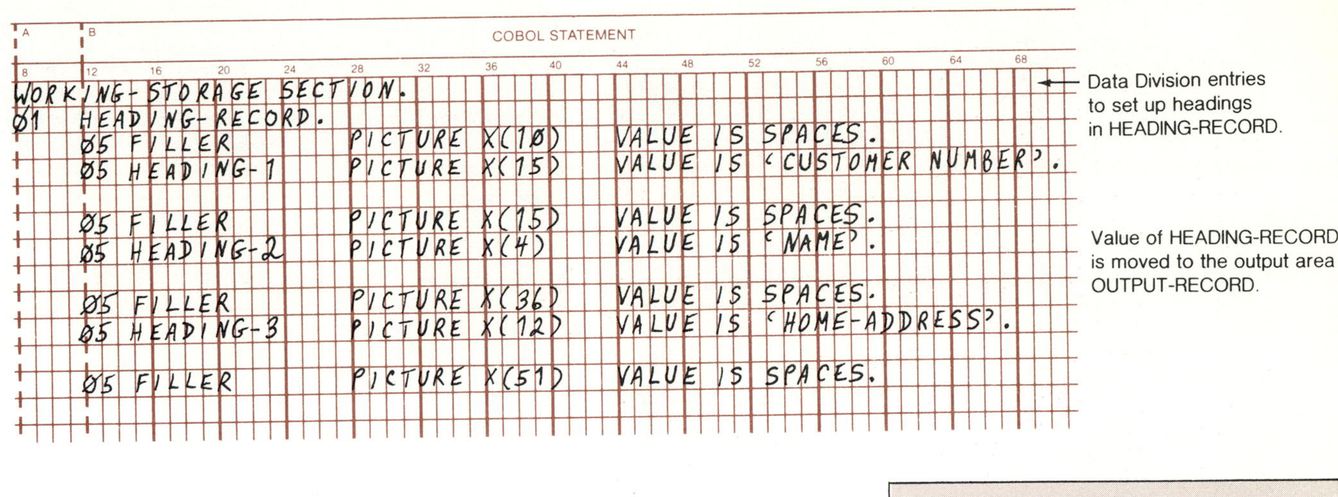
COBOL STATEMENT
WORKING-STORAGE SECTION.
Ø1 HEADING-RECORD.
Ø5 FILLER PICTURE X(1Ø) VALUE IS SPACES.
Ø5 HEADING-1 PICTURE X(15) VALUE IS 'CUSTOMER NUMBER'.
Ø5 FILLER PICTURE X(15) VALUE IS SPACES.
Ø5 HEADING-2 PICTURE X(4) VALUE IS 'NAME'.
Ø5 FILLER PICTURE X(36) VALUE IS SPACES.
Ø5 HEADING-3 PICTURE X(12) VALUE IS 'HOME-ADDRESS'.
Ø5 FILLER PICTURE X(51) VALUE IS SPACES.
Data Division entries to set up headings in HEADING-RECORD.
Value of HEADING-RECORD is moved to the output area OUTPUT-RECORD.

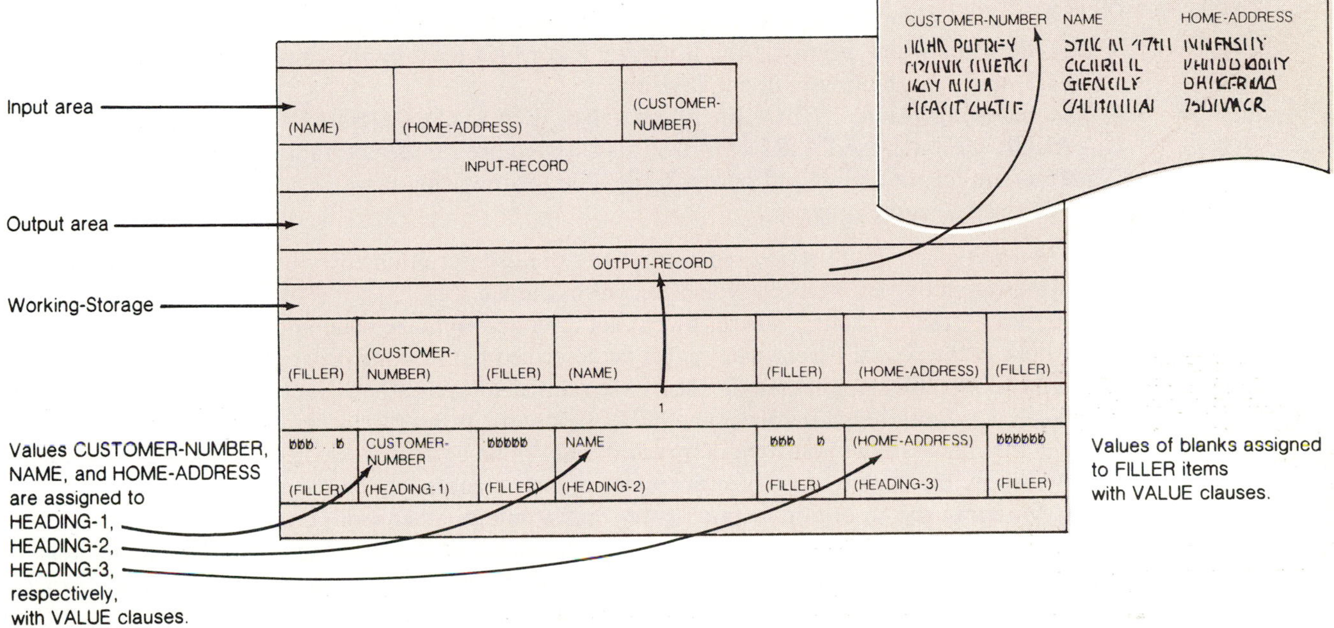
CUSTOMER-NUMBER
NAME
HOME-ADDRESS
Input area
(NAME)
(HOME-ADDRESS)
(CUSTOMER-NUMBER)
INPUT-RECORD
Output area
OUTPUT-RECORD
Working-Storage
(FILLER)
(CUSTOMER-NUMBER)
(FILLER)
(NAME)
(FILLER)
(HOME-ADDRESS)
(FILLER)
1
Values CUSTOMER-NUMBER, NAME, and HOME-ADDRESS are assigned to HEADING-1, HEADING-2, HEADING-3, respectively, with VALUE clauses.
CUSTOMER-NUMBER
(HEADING-1)
NAME
(HEADING-2)
(HOME-ADDRESS)
(HEADING-3)
Values of blanks assigned to FILLER items with VALUE clauses.

Figure 4.47 Working-Storage Section entries for heading—sample.

```
DATA DIVISION.
WORKING-STORAGE SECTION.
Ø1  INPUT-DATA.
    Ø5  FIELD-1,        PICTURE X(1Ø).
    Ø5  FIELD-2,        PICTURE X(5).
Ø1  OUTPUT-DATA.
    Ø5  FIELD-A,        PICTURE X(5).
    Ø5  FILLER,         PICTURE X(19),   VALUE SPACES.
    Ø5  FIELD-B,        PICTURE X(1Ø).
```

Figure 4.48 Levels 01 and 05 entries in Working-Storage Section—example.

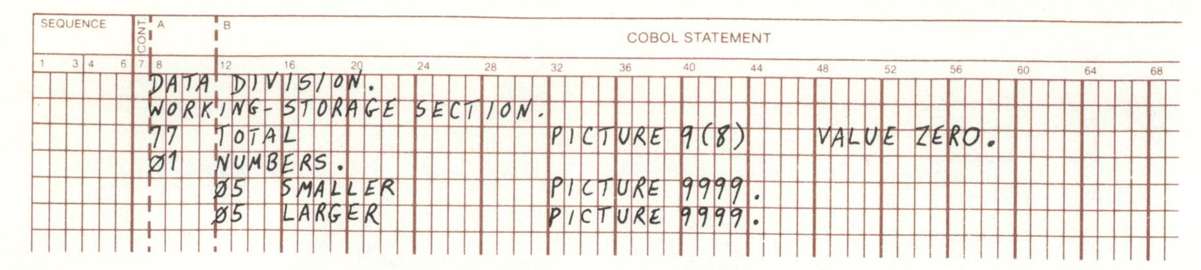

```
DATA DIVISION.
WORKING-STORAGE SECTION.
77  TOTAL                   PICTURE 9(8)    VALUE ZERO.
01  NUMBERS.
    05  SMALLER             PICTURE 9999.
    05  LARGER              PICTURE 9999.
```

Figure 4.49 Levels 77, 01, and 05 entries in Working-Storage Section—example.

and a series of independent clauses which specify the size of the logical and physical records, the presence or absence of label records, the value of label items, the name of the data records, and any other clauses that are necessary to the execution of the program. The file description entry contains:

1. The level indicator "FD," which is a reserved word and must begin at the A margin.
2. The file name which begins at the B margin and which must be the same as stipulated in the SELECT clause in the Environment Division.
3. The following clauses: BLOCK CONTAINS, RECORD CONTAINS, LABEL RECORDS, VALUE OF, and DATA RECORDS, may be required under certain conditions to explain the nature and form of input and output while other clauses only serve as documentation to assist the reader of the source program.

The format of the File Section requires that each file description entry be followed by one or more record description entries for each type of record in the file.

A record description entry consists of a set of data descriptions, which describe the characteristics of each item in the record. Every entry must be described in the same order in which it appears in the record and must indicate whether the items are related to each other. Each record description entry consists of a level number, a data-name or FILLER, and a series of clauses.

Level numbers are the first item of a record description entry and are used to structure the logical record to satisfy the need to specify subdivisions of data records for the purpose of data reference.

Every item in the record description entry must contain either a data-name or the reserved word FILLER immediately following the level number beginning at the B margin. The data-name must be unique or otherwise qualified and is used in the Procedure Division to refer to items individually. FILLER is a reserved word that may be used in place of data-name when the area contains information that is not necessary for the processing of the program.

The following are the most commonly used clauses that provide information about the data items.

The USAGE which describes the form in which the data is stored in the computer storage. The USAGE clause may be written at any level. If written at the group level, it applies to each elementary item. The forms of the USAGE clause are: DISPLAY, COMPUTATIONAL, COMPUTATIONAL-1 and COMPUTATIONAL-2, and COMPUTATIONAL-3.

The PICTURE clause specifies the general characteristics of the detail design of an elementary item. The categories of data that can be described with a PICTURE clause are as follows: *alphabetic items, numeric items, alphanumeric items, alphanumeric edited items, and numeric edited items.*

The PICTURE characters are divided into two groups:

1. *Nonedited PICTURE characters* may contain any combination of characters exclusive of the editing characters, and may be used in calculations. The *nonediting characters* are:
   - *X,* which represents any character in the COBOL character set.
   - *A,* which represents any alphabetic character and space.
   - *9,* which represents any numeral.
   - *V,* which represents an assumed decimal point.
   - *P,* which represents an assumed zero.
   - *S,* which represents the presence of an operational sign.

2. The *edited characters* are:

   Z, which represents a zero-suppression character.

   Each Z represents one digit.

   ., which represents an actual decimal point.

   , which represents a comma.

   *, which represents an asterisk used for check protection of amounts.

CR and DB characters are printed in the edited result only if the item is negative.

Insertion characters (,(comma), B(space), 0(zero)) are editing characters that are inserted in the edited result to improve the readability of the item.

Floating strings are a series of continuous characters of either $, or +, or −, or a string composed of one or a repetition of one such characters that are inserted in the edited result to improve the readability of the item.

The VALUE clause defines clauses or constants, the initial value of working storage items, and the values associated with condition-names. The VALUE clause is permitted in the description of elementary items in the Working-Storage Section with the exception of condition-names, which may have a VALUE clause in the File Section.

The condition-name is a name assigned by the program to a particular value that may be assumed by a data item. It must contain a level 88 entry and must immediately follow the item with which it is associated. The VALUE clause and the condition-name are the only entries permitted in the condition-name clause.

The Working-Storage Section is used to enter descriptions of records that are not part of external data files, but which are developed and processed internally. The Working-Storage Section is often used to provide headings for reports. This section contains entries for independent (noncontiguous) items and record description entries in that order.

Independent (noncontiguous) data items have no hierarchical relationship to each other and need not be grouped into records. Each item is defined with a level number 77 at the A margin, followed by a data-name at the B margin, PICTURE clause, and other optional clauses. These entries precede record description entries in the Working-Storage Section.

Record description items are data elements and constants in which there is a definite hierarchical relationship. They are grouped together in records according to the rules for the formation of record description entries.

Illustrative Program: Accounts Receivable Problem

Input

Field	Positions	
Entry Date	1–5	
Entry	6–7	
Customer Name	8–29	
Invoice Date	30–33	
Invoice Number	34–38	
Customer Number	39–43	
Location	44–48	
Blank	49–62	
Discount Allowed	63–67	XXX.XX
Amount Paid	68–73	XXXX.XX
Blank	74–80	

Calculations To Be Performed

1. Calculate Accounts Receivable = Amount Paid + Discount Allowed.
2. Final Totals for Accounts Receivable, Discount Allowed and Amount Paid.

Output

ACCOUNTS RECEIVABLE REGISTER

CUST. NO.	CUST. NAME	INV. NO.	ACCTS. REC.	DISCT. ALLOW.	AMT. PAID
67451	ACME MFG CO	345	$697.17	$13.67	$683.50
67452	AMERICAN STEEL CO	342	$1,398.93	$27.43	$1,371.50
67453	TAIYO CO LTD	447	$1,211.25	$23.75	$1,187.50
67454	ALLIS CHALMERS CO	451	$2,307.75	$45.25	$2,262.50
67455	XEROX CORP	435	$163.71	$3.21	$160.50
67456	GLOBE FORM CO	435	$229.50	$4.50	$225.00
67457	WATSON MFG CO	428	$113.73	$2.23	$111.50
67458	CALCOMP CORP	429	$165.75	$3.25	$162.50
67459	SHOP--RITE MARKETS	433	$168.30	$3.30	$165.00
67460	MICROSEAL CORP	440	$5.61	$.11	$5.50
67461	MITSUBISHI LTD	420	$2,305.20	$45.20	$2,260.00
67462	MARK KLEIN & SONS	431	$1,393.32	$27.32	$1,366.00
67463	HONEYWELL CORP	432	$11.73	$.23	$11.50
67464	SPERRY RAND CORP	449	$2,345.49	$45.99	$2,299.50
67465	WESTINGHOUSE CORP	460	$3,047.25	$59.75	$2,987.50
67466	GARRETT CORP	399	$184.62	$3.62	$181.00
67467	NANCY DOLL TOY CO	400	$22.95	$.45	$22.50
67468	RAMONAS FINE FOODS	430	$3,557.25	$69.75	$3,487.50
67469	EL CHOLOS	436	$1,795.71	$35.21	$1,760.50
67470	DATAMATION INC	437	$2,247.00	$374.50	$1,872.50
67471	MICROFICHE CORP	441	$2,555.10	$50.10	$2,505.00
67472	REALIST INC	389	$2,872.32	$56.32	$2,816.00
67473	EASTMAN KODAK CO	401	$2,311.32	$45.32	$2,266.00
67474	UNIVAC INC	410	$3,348.15	$65.65	$3,282.50
67475	AVCO CO	411	$5,015.85	$98.35	$4,917.50
67476	TRW SYSTEMS GROUP	412	$2,311.32	$45.32	$2,266.00
67477	BELL HELICOPTER CO	413	$2,878.95	$56.45	$2,822.50
67478	BOEING AEROSPACE CORP	414	$2,328.15	$45.65	$2,282.50
		TOTALS	$46,993.38	$1,251.88	$45,741.50**

Printer Spacing Chart

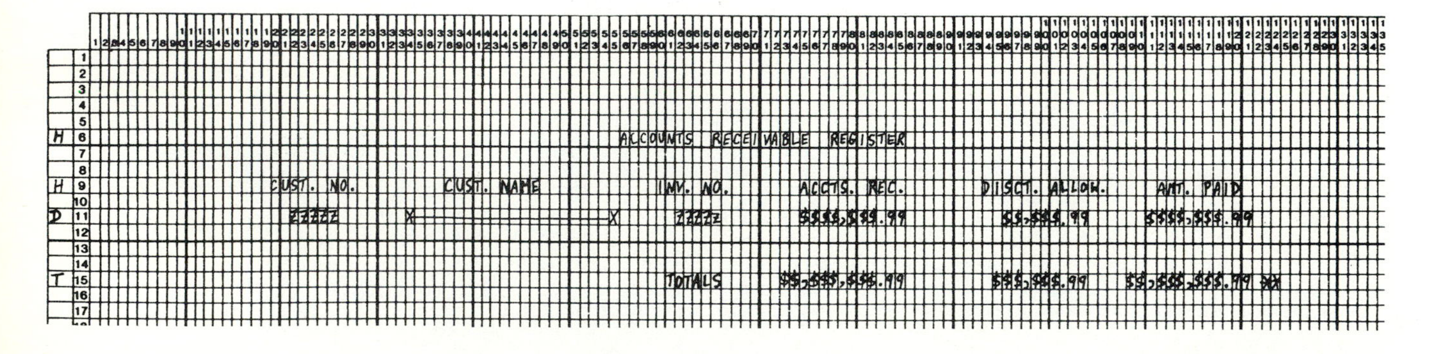

Hierarchy Chart

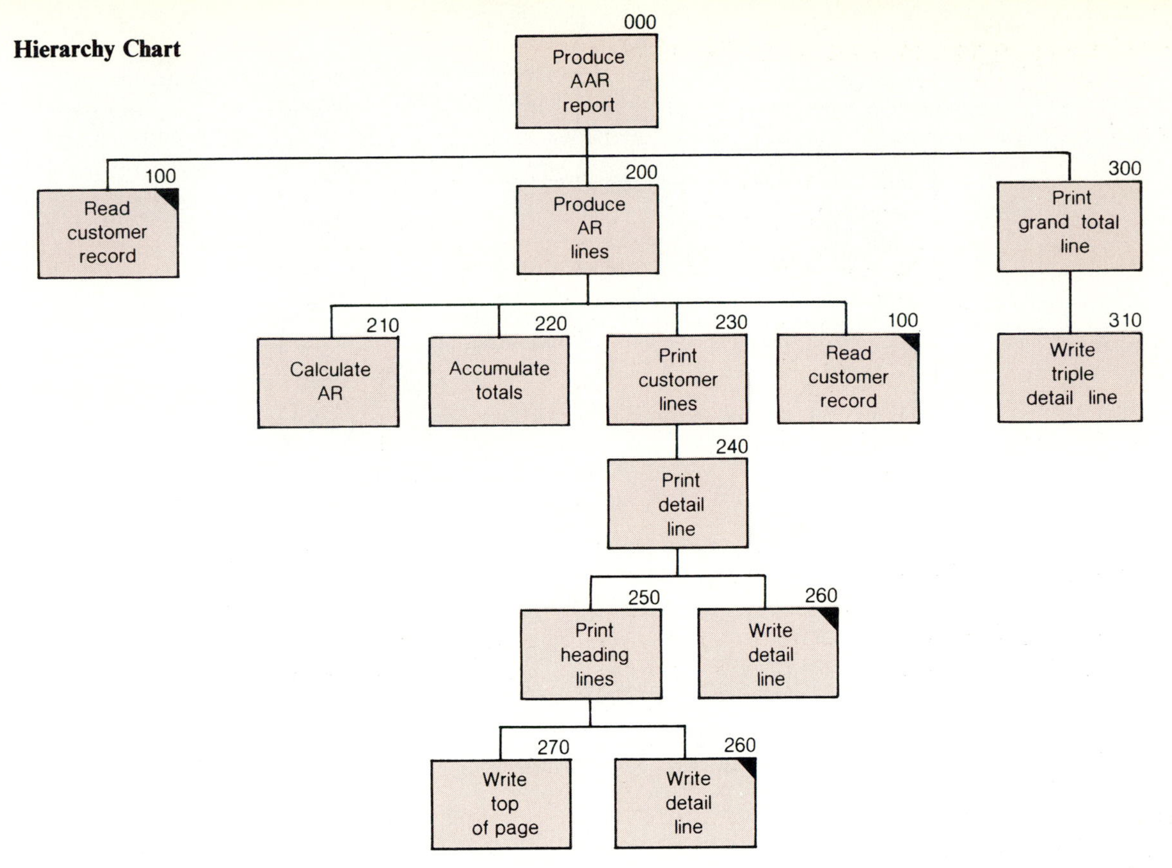

IPO Charts

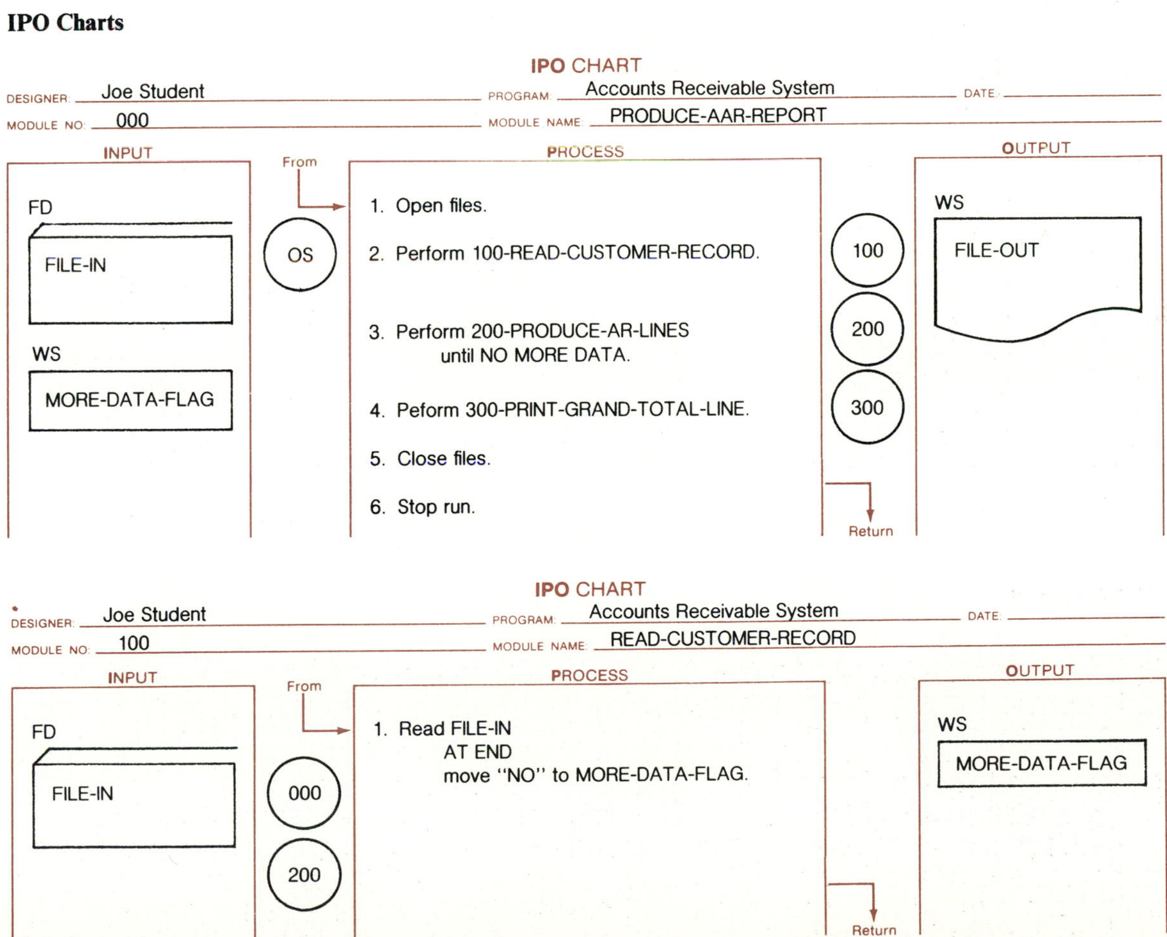

IPO CHART

DESIGNER: Joe Student PROGRAM: Accounts Receivable System DATE:

MODULE NO: 000 MODULE NAME: PRODUCE-AAR-REPORT

INPUT	From	PROCESS		OUTPUT
FD FILE-IN	OS	1. Open files.		WS FILE-OUT
		2. Perform 100-READ-CUSTOMER-RECORD.	100	
WS MORE-DATA-FLAG		3. Perform 200-PRODUCE-AR-LINES until NO MORE DATA.	200	
		4. Peform 300-PRINT-GRAND-TOTAL-LINE.	300	
		5. Close files.		
		6. Stop run.	Return	

IPO CHART

DESIGNER: Joe Student PROGRAM: Accounts Receivable System DATE:

MODULE NO: 100 MODULE NAME: READ-CUSTOMER-RECORD

INPUT	From	PROCESS		OUTPUT
FD FILE-IN	000 200	1. Read FILE-IN AT END move "NO" to MORE-DATA-FLAG.	Return	WS MORE-DATA-FLAG

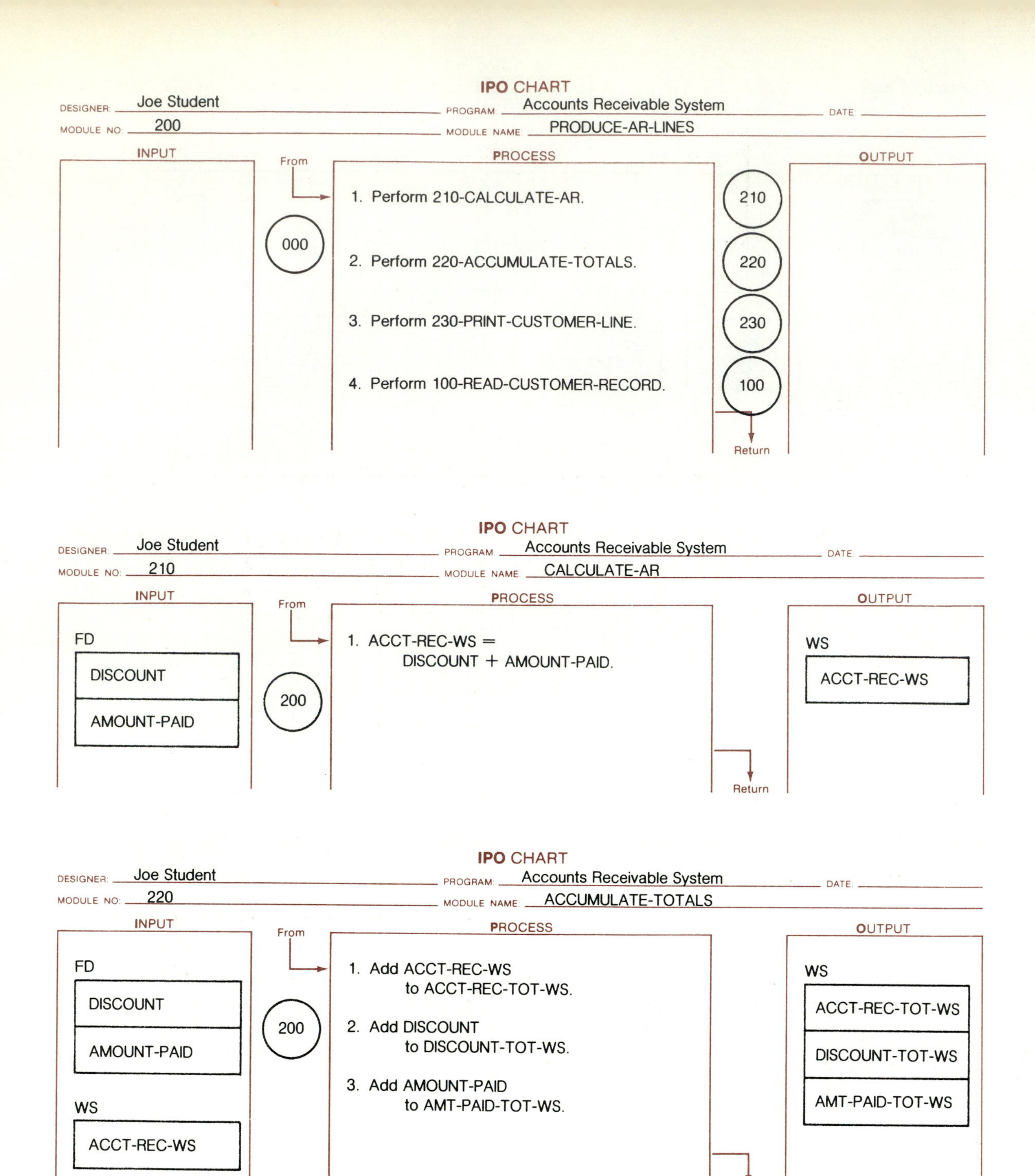
IPO CHART
DESIGNER: Joe Student
PROGRAM: Accounts Receivable System
DATE:
MODULE NO: 200
MODULE NAME: PRODUCE-AR-LINES
INPUT
From
000
PROCESS
1. Perform 210-CALCULATE-AR.
2. Perform 220-ACCUMULATE-TOTALS.
3. Perform 230-PRINT-CUSTOMER-LINE.
4. Perform 100-READ-CUSTOMER-RECORD.
210
220
230
100
Return
OUTPUT
IPO CHART
DESIGNER: Joe Student
PROGRAM: Accounts Receivable System
DATE:
MODULE NO: 210
MODULE NAME: CALCULATE-AR
INPUT
FD
DISCOUNT
AMOUNT-PAID
From
200
PROCESS
1. ACCT-REC-WS =
DISCOUNT + AMOUNT-PAID.
Return
OUTPUT
WS
ACCT-REC-WS
IPO CHART
DESIGNER: Joe Student
PROGRAM: Accounts Receivable System
DATE:
MODULE NO: 220
MODULE NAME: ACCUMULATE-TOTALS
INPUT
FD
DISCOUNT
AMOUNT-PAID
WS
ACCT-REC-WS
From
200
PROCESS
1. Add ACCT-REC-WS
to ACCT-REC-TOT-WS.
2. Add DISCOUNT
to DISCOUNT-TOT-WS.
3. Add AMOUNT-PAID
to AMT-PAID-TOT-WS.
Return
OUTPUT
WS
ACCT-REC-TOT-WS
DISCOUNT-TOT-WS
AMT-PAID-TOT-WS

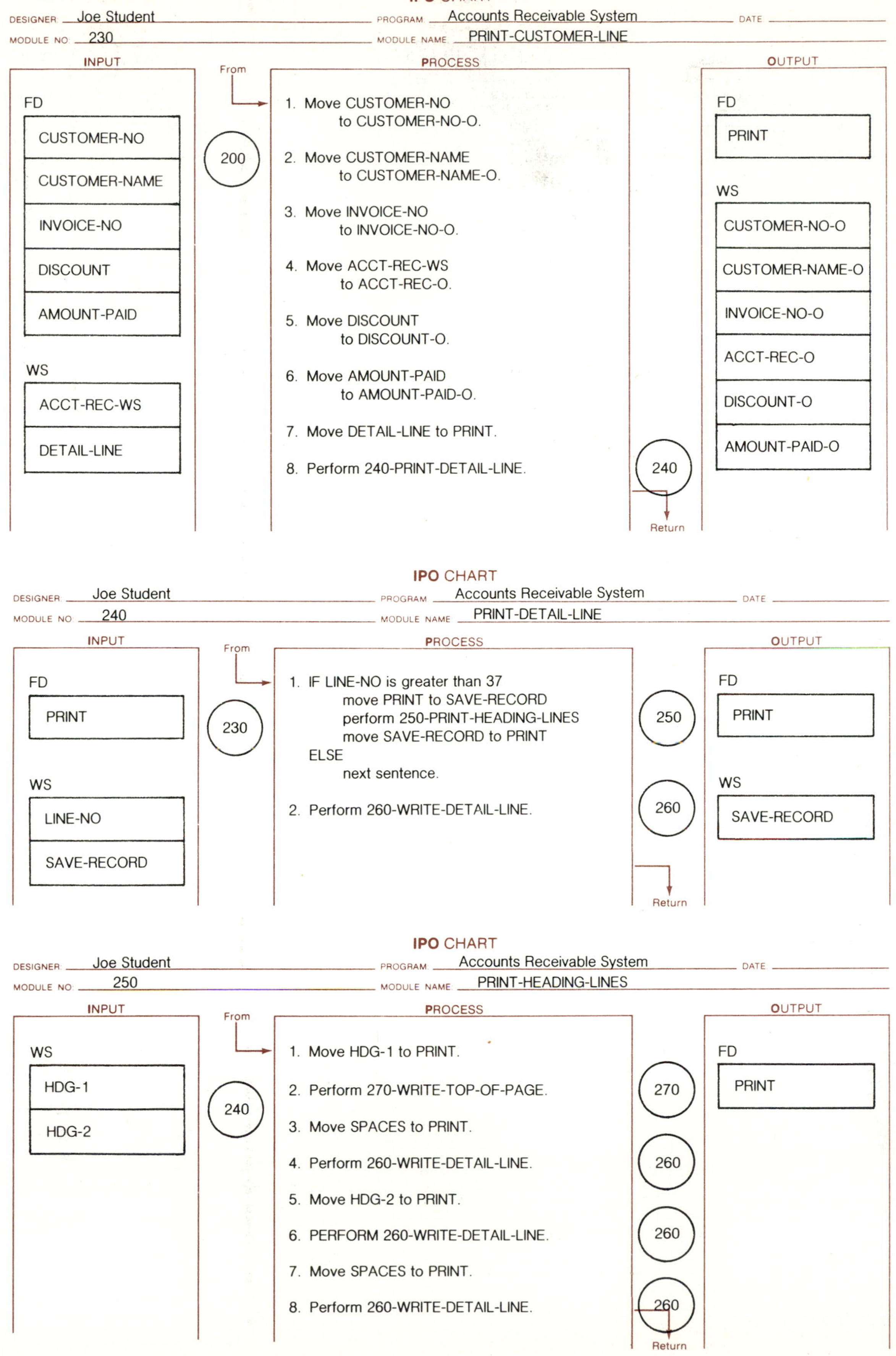

IPO CHART

DESIGNER: Joe Student — PROGRAM: Accounts Receivable System — DATE:

MODULE NO: 230 — MODULE NAME: PRINT-CUSTOMER-LINE

INPUT

FD
- CUSTOMER-NO
- CUSTOMER-NAME
- INVOICE-NO
- DISCOUNT
- AMOUNT-PAID

WS
- ACCT-REC-WS
- DETAIL-LINE

From: 200

PROCESS

1. Move CUSTOMER-NO to CUSTOMER-NO-O.
2. Move CUSTOMER-NAME to CUSTOMER-NAME-O.
3. Move INVOICE-NO to INVOICE-NO-O.
4. Move ACCT-REC-WS to ACCT-REC-O.
5. Move DISCOUNT to DISCOUNT-O.
6. Move AMOUNT-PAID to AMOUNT-PAID-O.
7. Move DETAIL-LINE to PRINT.
8. Perform 240-PRINT-DETAIL-LINE. (240)

Return

OUTPUT

FD
- PRINT

WS
- CUSTOMER-NO-O
- CUSTOMER-NAME-O
- INVOICE-NO-O
- ACCT-REC-O
- DISCOUNT-O
- AMOUNT-PAID-O

IPO CHART

DESIGNER: Joe Student — PROGRAM: Accounts Receivable System — DATE:

MODULE NO: 240 — MODULE NAME: PRINT-DETAIL-LINE

INPUT

FD
- PRINT

WS
- LINE-NO
- SAVE-RECORD

From: 230

PROCESS

1. IF LINE-NO is greater than 37
   move PRINT to SAVE-RECORD
   perform 250-PRINT-HEADING-LINES (250)
   move SAVE-RECORD to PRINT
   ELSE
   next sentence.
2. Perform 260-WRITE-DETAIL-LINE. (260)

Return

OUTPUT

FD
- PRINT

WS
- SAVE-RECORD

IPO CHART

DESIGNER: Joe Student — PROGRAM: Accounts Receivable System — DATE:

MODULE NO: 250 — MODULE NAME: PRINT-HEADING-LINES

INPUT

WS
- HDG-1
- HDG-2

From: 240

PROCESS

1. Move HDG-1 to PRINT.
2. Perform 270-WRITE-TOP-OF-PAGE. (270)
3. Move SPACES to PRINT.
4. Perform 260-WRITE-DETAIL-LINE. (260)
5. Move HDG-2 to PRINT.
6. PERFORM 260-WRITE-DETAIL-LINE. (260)
7. Move SPACES to PRINT.
8. Perform 260-WRITE-DETAIL-LINE. (260)

Return

OUTPUT

FD
- PRINT

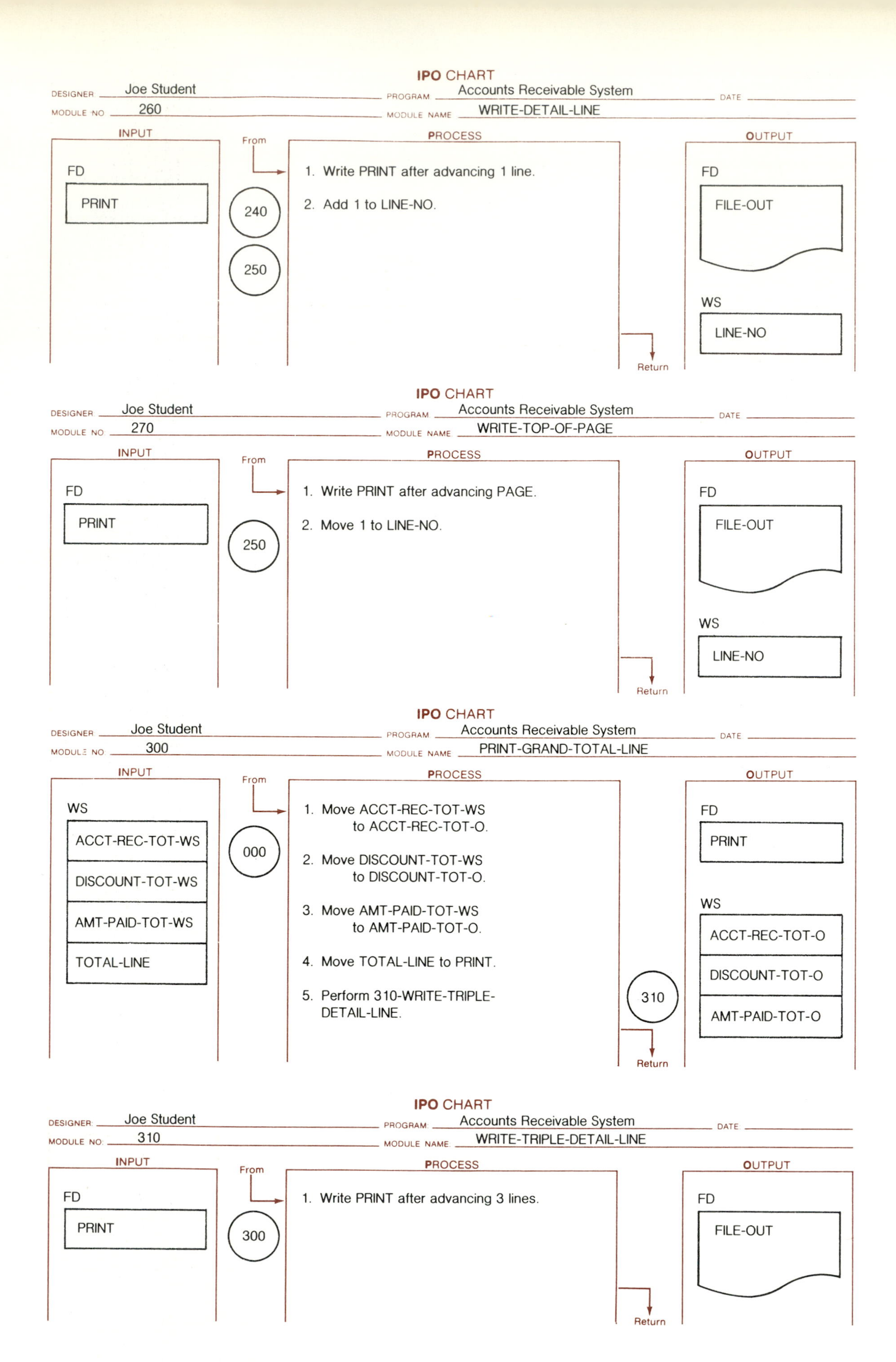

IPO CHART
DESIGNER Joe Student
PROGRAM Accounts Receivable System
DATE
MODULE NO 260
MODULE NAME WRITE-DETAIL-LINE
INPUT
FD
PRINT
From
240
250
PROCESS
1. Write PRINT after advancing 1 line.
2. Add 1 to LINE-NO.
Return
OUTPUT
FD
FILE-OUT
WS
LINE-NO
IPO CHART
DESIGNER Joe Student
PROGRAM Accounts Receivable System
DATE
MODULE NO 270
MODULE NAME WRITE-TOP-OF-PAGE
INPUT
FD
PRINT
From
250
PROCESS
1. Write PRINT after advancing PAGE.
2. Move 1 to LINE-NO.
Return
OUTPUT
FD
FILE-OUT
WS
LINE-NO
IPO CHART
DESIGNER Joe Student
PROGRAM Accounts Receivable System
DATE
MODULE NO 300
MODULE NAME PRINT-GRAND-TOTAL-LINE
INPUT
WS
ACCT-REC-TOT-WS
DISCOUNT-TOT-WS
AMT-PAID-TOT-WS
TOTAL-LINE
From
000
PROCESS
1. Move ACCT-REC-TOT-WS to ACCT-REC-TOT-O.
2. Move DISCOUNT-TOT-WS to DISCOUNT-TOT-O.
3. Move AMT-PAID-TOT-WS to AMT-PAID-TOT-O.
4. Move TOTAL-LINE to PRINT.
5. Perform 310-WRITE-TRIPLE-DETAIL-LINE.
310
Return
OUTPUT
FD
PRINT
WS
ACCT-REC-TOT-O
DISCOUNT-TOT-O
AMT-PAID-TOT-O
IPO CHART
DESIGNER Joe Student
PROGRAM Accounts Receivable System
DATE
MODULE NO 310
MODULE NAME WRITE-TRIPLE-DETAIL-LINE
INPUT
FD
PRINT
From
300
PROCESS
1. Write PRINT after advancing 3 lines.
Return
OUTPUT
FD
FILE-OUT

Flowcharts

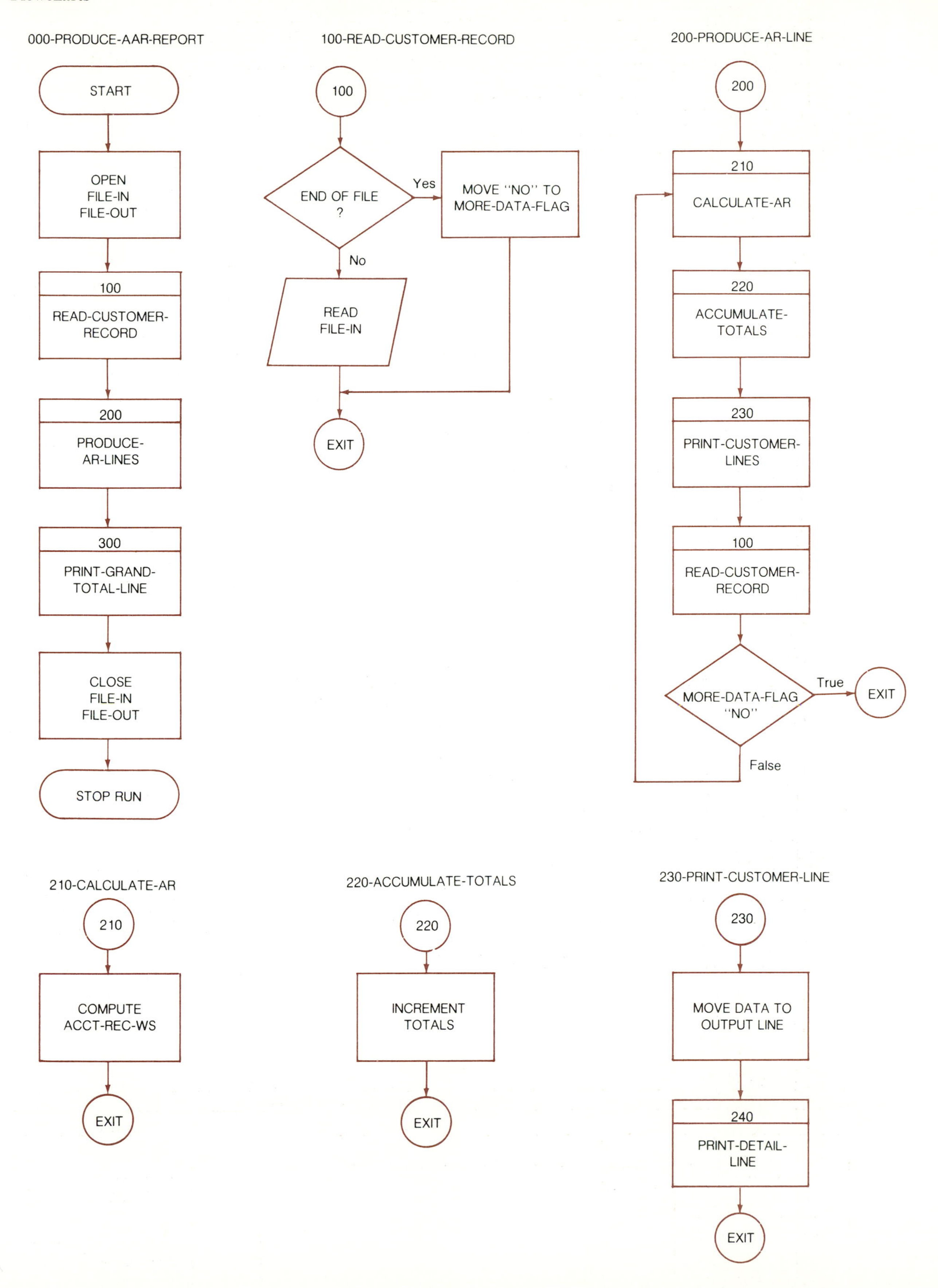
000-PRODUCE-AAR-REPORT
START
OPEN
FILE-IN
FILE-OUT
100
READ-CUSTOMER-
RECORD
200
PRODUCE-
AR-LINES
300
PRINT-GRAND-
TOTAL-LINE
CLOSE
FILE-IN
FILE-OUT
STOP RUN
100-READ-CUSTOMER-RECORD
100
END OF FILE
?
Yes
MOVE "NO" TO
MORE-DATA-FLAG
No
READ
FILE-IN
EXIT
200-PRODUCE-AR-LINE
200
210
CALCULATE-AR
220
ACCUMULATE-
TOTALS
230
PRINT-CUSTOMER-
LINES
100
READ-CUSTOMER-
RECORD
MORE-DATA-FLAG
"NO"
True
EXIT
False
210-CALCULATE-AR
210
COMPUTE
ACCT-REC-WS
EXIT
220-ACCUMULATE-TOTALS
220
INCREMENT
TOTALS
EXIT
230-PRINT-CUSTOMER-LINE
230
MOVE DATA TO
OUTPUT LINE
240
PRINT-DETAIL-
LINE
EXIT

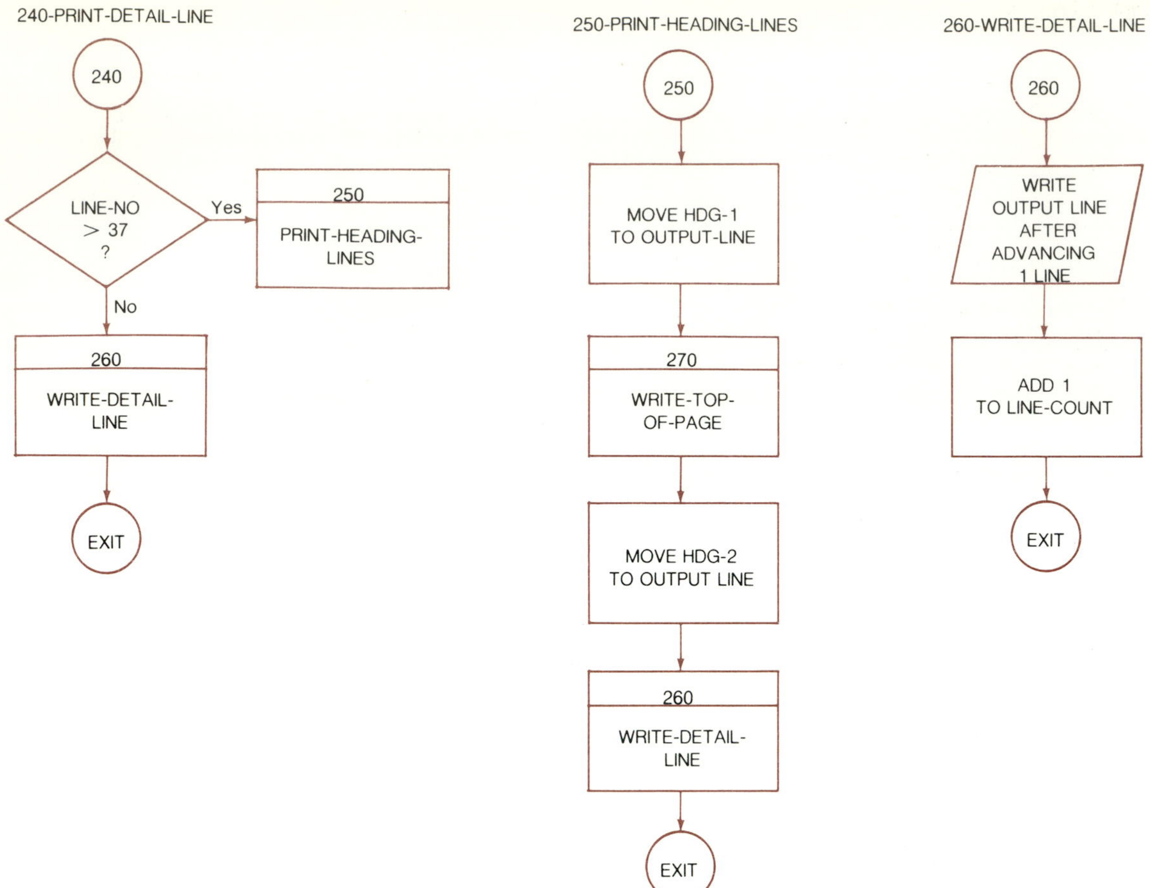
240-PRINT-DETAIL-LINE
240
LINE-NO
> 37
?
Yes
250
PRINT-HEADING-LINES
No
260
WRITE-DETAIL-LINE
EXIT
250-PRINT-HEADING-LINES
250
MOVE HDG-1
TO OUTPUT-LINE
270
WRITE-TOP-OF-PAGE
MOVE HDG-2
TO OUTPUT LINE
260
WRITE-DETAIL-LINE
EXIT
260-WRITE-DETAIL-LINE
260
WRITE
OUTPUT LINE
AFTER
ADVANCING
1 LINE
ADD 1
TO LINE-COUNT
EXIT

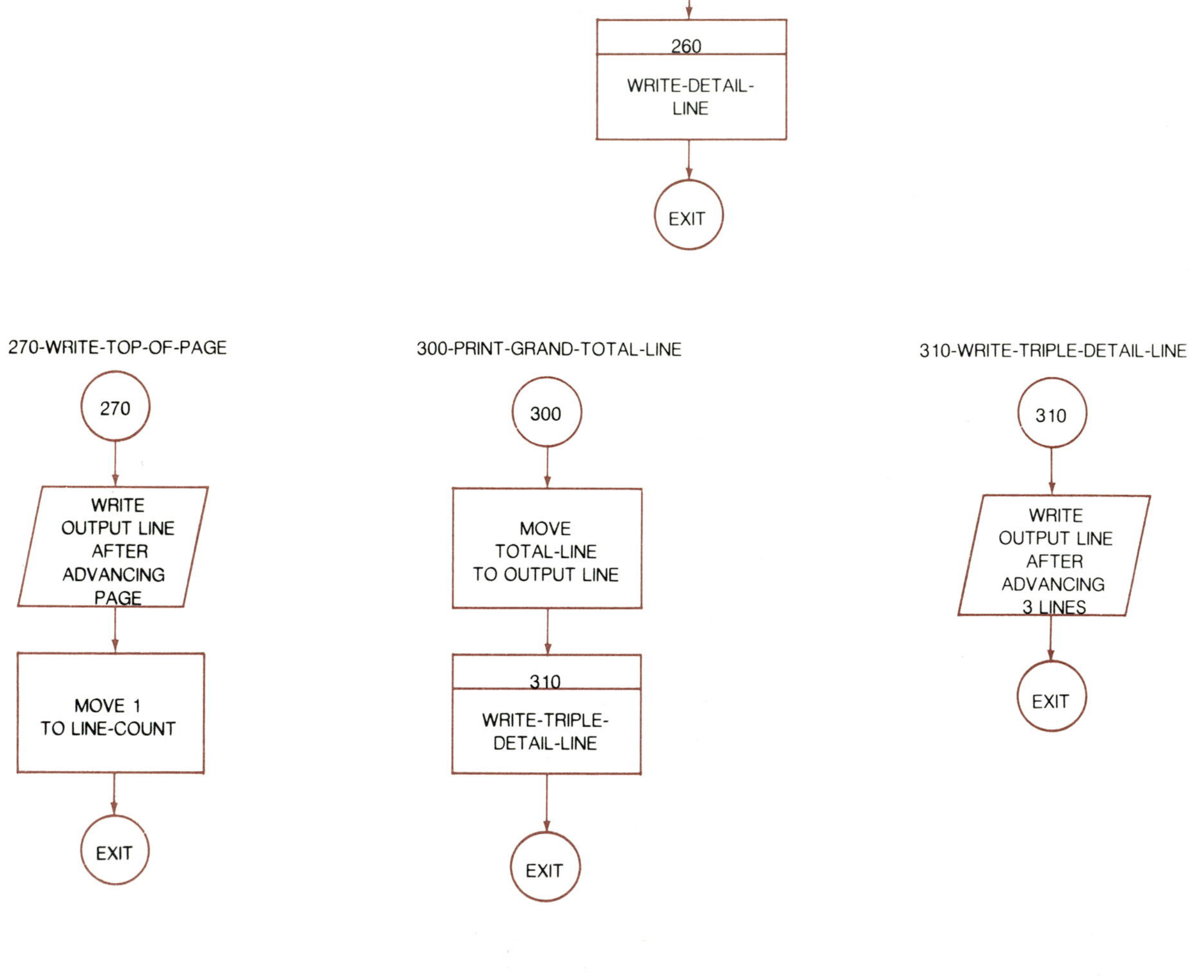
270-WRITE-TOP-OF-PAGE
270
WRITE
OUTPUT LINE
AFTER
ADVANCING
PAGE
MOVE 1
TO LINE-COUNT
EXIT
300-PRINT-GRAND-TOTAL-LINE
300
MOVE
TOTAL-LINE
TO OUTPUT LINE
310
WRITE-TRIPLE-DETAIL-LINE
EXIT
310-WRITE-TRIPLE-DETAIL-LINE
310
WRITE
OUTPUT LINE
AFTER
ADVANCING
3 LINES
EXIT

Source Program

```
 1
 2        *************************
 3
 4         IDENTIFICATION DIVISION.
 5
 6        *************************
 7
 8         PROGRAM-ID.
 9             ACCOUNTS-RECEIVABLE-REGISTER.
10         INSTALLATION.
11             WEST LOS ANGELES COLLEGE.
12         DATE-WRITTEN.
13             1 FEBRUARY 1982.
14
15        *PURPOSE.
16        *     THIS PROGRAM PREPARES AN ACCOUNTS RECEIVABLE REGISTER.
17
18        *************************
19
20         ENVIRONMENT DIVISION.
21
22        *************************
23
24         CONFIGURATION SECTION.
25
26         SOURCE-COMPUTER.
27             LEVEL-66-ASCII.
28         OBJECT-COMPUTER.
29             LEVEL-66-ASCII.
30
31         INPUT-OUTPUT SECTION.
32
33         FILE-CONTROL.
34             SELECT FILE-IN       ASSIGN TO C1-CARD-READER.
35             SELECT FILE-OUT      ASSIGN TO P1-PRINTER.
36
37        *************************
38
39         DATA DIVISION.
40
41        *************************
42
43         FILE SECTION.
44
45         FD  FILE-IN
46             CODE-SET IS GBCD
47             LABEL RECORDS ARE OMITTED
48             DATA RECORD IS CARDIN.
49
50         01  CARDIN.
51
52             03  ENTRY-DATE       PICTURE X(5).
53             03  ENTRI            PICTURE 9(2).
54             03  CUSTOMER-NAME    PICTURE X(21).
55             03  INVOICE-DATE     PICTURE X(5).
56             03  INVOICE-NO       PICTURE 9(5).
57             03  CUSTOMER-NO      PICTURE 9(5).
58             03  LOCATION         PICTURE 9(5).
59             03  FILLER           PICTURE X(14).
60             03  DISCOUNT         PICTURE 9(3)V99.
61             03  AMOUNT-PAID      PICTURE 9(4)V99.
62             03  FILLER           PICTURE X(7).
63
64         FD  FILE-OUT
65             CODE-SET IS GBCD
66             LABEL RECORDS ARE OMITTED
67             DATA RECORD IS PRINT.
68
69         01  PRINT                PICTURE X(132).
70
71         WORKING-STORAGE SECTION.
72
73         77  ACCT-REC-WS          PICTURE 9(5)V99 VALUE ZEROES.
74         77  ACCT-REC-TOT-WS      PICTURE 9(6)V99 VALUE ZEROES.
75         77  DISCOUNT-TOT-WS      PICTURE 9(5)V99 VALUE ZEROES.
76         77  AMT-PAID-TOT-WS      PICTURE 9(6)V99 VALUE ZEROES.
77         77  LINE-NO              PICTURE 9(2)    VALUE 38.
78
79         01  FLAGS.
80
81             03  MORE-DATA-FLAG   PICTURE X(3)    VALUE "YES".
82                 88  MORE-DATA                    VALUE "YES".
83                 88  NO-MORE-DATA                 VALUE "NO ".
84
85         01  HDG-1.
86
87             03  FILLER           PICTURE X(53)   VALUE SPACES.
88             03  FILLER           PICTURE X(30)   VALUE "ACCOUNTS RECEIVABL
89        -                                               "E REGISTER".
90
91         01  HDG-2.
92
93             03  FILLER           PICTURE X(17)   VALUE SPACES.
94             03  FILLER           PICTURE X(9)    VALUE "CUST. NO.".
95             03  FILLER           PICTURE X(10)   VALUE SPACES.
96             03  FILLER           PICTURE X(10)   VALUE "CUST. NAME".
97             03  FILLER           PICTURE X(11)   VALUE SPACES.
98             03  FILLER           PICTURE X(8)    VALUE "INV. NO.".
```

```
 99            03  FILLER            PICTURE X(7)     VALUE SPACES.
100            03  FILLER            PICTURE X(11)    VALUE "ACCTS. REC.".
101            03  FILLER            PICTURE X(8)     VALUE SPACES.
102            03  FILLER            PICTURE X(13)    VALUE "DISCT. ALLOW.".
103            03  FILLER            PICTURE X(5)     VALUE SPACES.
104            03  FILLER            PICTURE X(9)     VALUE "AMT. PAID".
105
106        01  DETAIL-LINE.
107
108            03  FILLER            PICTURE X(19)    VALUE SPACES.
109            03  CUSTOMER-NO-O     PICTURE Z(5).
110            03  FILLER            PICTURE X(7)     VALUE SPACES.
111            03  CUSTOMER-NAME-O   PICTURE X(21).
112            03  FILLER            PICTURE X(7)     VALUE SPACES.
113            03  INVOICE-NO-O      PICTURE Z(5).
114            03  FILLER            PICTURE X(8)     VALUE SPACES.
115            03  ACCT-REC-O        PICTURE $$$,$$$.99.
116            03  FILLER            PICTURE X(13)    VALUE SPACES.
117            03  DISCOUNT-O        PICTURE $$$$.99.
118            03  FILLER            PICTURE X(7)     VALUE SPACES.
119            03  AMOUNT-PAID-O     PICTURE $$,$$$.99.
120
121        01  TOTAL-LINE.
122
123            03  FILLER            PICTURE X(59)    VALUE SPACES.
124            03  FILLER            PICTURE X(6)     VALUE "TOTALS".
125            03  FILLER            PICTURE X(6)     VALUE SPACES.
126            03  ACCT-REC-TOT-O    PICTURE $,$$$,$$$.99.
127            03  FILLER            PICTURE X(10)    VALUE SPACES.
128            03  DISCOUNT-TOT-O    PICTURE $$$,$$$.99.
129            03  FILLER            PICTURE X(5)     VALUE SPACES.
130            03  AMT-PAID-TOT-O    PICTURE $$$$,$$$.99.
131            03  FILLER            PICTURE X(3)     VALUE "**".
132
133        01  SAVE-RECORD           PICTURE X(132).
134
135       **************************
136
137        PROCEDURE DIVISION.
138
139       **************************
140
141        000-PRODUCE-AAR-REPORT.
142
143            OPEN    INPUT   FILE-IN
144                    OUTPUT  FILE-OUT.
145            PERFORM 100-READ-CUSTOMER-RECORD.
146            PERFORM 200-PRODUCE-AR-LINES
147                    UNTIL NO-MORE-DATA.
148            PERFORM 300-PRINT-GRAND-TOTAL-LINE.
149            CLOSE   FILE-IN
150                    FILE-OUT.
151            STOP RUN.
152
153        100-READ-CUSTOMER-RECORD.
154
155            READ FILE-IN
156                AT END MOVE "NO" TO MORE-DATA-FLAG.
157
158        200-PRODUCE-AR-LINES.
159
160            PERFORM 210-CALCULATE-AR.
161            PERFORM 220-ACCUMULATE-TOTALS.
162            PERFORM 230-PRINT-CUSTOMER-LINE.
163            PERFORM 100-READ-CUSTOMER-RECORD.
164
165        210-CALCULATE-AR.
166
167            COMPUTE ACCT-REC-WS = DISCOUNT
168                + AMOUNT-PAID.
169
170        220-ACCUMULATE-TOTALS.
171
172            ADD ACCT-REC-WS TO ACCT-REC-TOT-WS.
173            ADD DISCOUNT TO DISCOUNT-TOT-WS.
174            ADD AMOUNT-PAID TO AMT-PAID-TOT-WS.
175
176        230-PRINT-CUSTOMER-LINE.
177
178            MOVE CUSTOMER-NO    TO CUSTOMER-NO-O.
179            MOVE CUSTOMER-NAME  TO CUSTOMER-NAME-O.
180            MOVE INVOICE-NO     TO INVOICE-NO-O.
181            MOVE ACCT-REC-WS    TO ACCT-REC-O.
182            MOVE DISCOUNT       TO DISCOUNT-O.
183            MOVE AMOUNT-PAID    TO AMOUNT-PAID-O.
184            MOVE DETAIL-LINE TO PRINT.
185            PERFORM 240-PRINT-DETAIL-LINE.
186
187        240-PRINT-DETAIL-LINE.
188
189            IF LINE-NO IS > 37
190                MOVE PRINT TO SAVE-RECORD
191                PERFORM 250-PRINT-HEADING-LINES
192                MOVE SAVE-RECORD TO PRINT
193            ELSE
194                NEXT SENTENCE.
195            PERFORM 260-WRITE-DETAIL-LINE.
196
197        250-PRINT-HEADING-LINES.
198
199            MOVE HDG-1 TO PRINT.
200            PERFORM 270-WRITE-TOP-OF-PAGE.
```

```
201           MOVE SPACES TO PRINT.
202           PERFORM 260-WRITE-DETAIL-LINE.
203           MOVE HDG-2 TO PRINT.
204           PERFORM 260-WRITE-DETAIL-LINE.
205           MOVE SPACES TO PRINT.
206           PERFORM 260-WRITE-DETAIL-LINE.
207
208       260-WRITE-DETAIL-LINE.
209
210           WRITE PRINT
211               AFTER ADVANCING 1 LINE.
212           ADD 1 TO LINE-NO.
213
214       270-WRITE-TOP-OF-PAGE.
215
216           WRITE PRINT
217               AFTER ADVANCING PAGE.
218           MOVE 1 TO LINE-NO.
219
220       300-PRINT-GRAND-TOTAL-LINE.
221
222           MOVE ACCT-REC-TOT-WS TO ACCT-REC-TOT-O.
223           MOVE DISCOUNT-TOT-WS TO DISCOUNT-TOT-O.
224           MOVE AMT-PAID-TOT-WS TO AMT-PAID-TOT-O.
225           MOVE TOTAL-LINE      TO PRINT.
226           PERFORM 310-WRITE-TRIPLE-DETAIL-LINE.
227
228       310-WRITE-TRIPLE-DETAIL-LINE.
229
230           WRITE PRINT
231               AFTER ADVANCING 3 LINES.

THERE WERE 231 SOURCE INPUT LINES.
THERE WERE NO DIAGNOSTICS.
```

ACCOUNTS RECEIVABLE REGISTER

CUST. NO.	CUST. NAME	INV. NO.	ACCTS. REC.	DISCT. ALLOW.	AMT. PAID
67451	ACME MFG CO	345	$697.17	$13.67	$683.50
67452	AMERICAN STEEL CO	342	$1,398.93	$27.43	$1,371.50
67453	TAIYO CO LTD	447	$1,211.25	$23.75	$1,187.50
67454	ALLIS CHALMERS CO	451	$2,307.75	$45.25	$2,262.50
67455	XEROX CORP	435	$163.71	$3.21	$160.50
67456	GLOBE FORM CO	435	$229.50	$4.50	$225.00
67457	WATSON MFG CO	428	$113.73	$2.23	$111.50
67458	CALCOMP CORP	429	$165.75	$3.25	$162.50
67459	SHOP--RITE MARKETS	433	$168.30	$3.30	$165.00
67460	MICROSEAL CORP	440	$5.61	$.11	$5.50
67461	MITSUBISHI LTD	420	$2,305.20	$45.20	$2,260.00
67462	MARK KLEIN & SONS	431	$1,393.32	$27.32	$1,366.00
67463	HONEYWELL CORP	432	$11.73	$.23	$11.50
67464	SPERRY RAND CORP	449	$2,345.49	$45.99	$2,299.50
67465	WESTINGHOUSE CORP	460	$3,047.25	$59.75	$2,987.50
67466	GARRETT CORP	399	$184.62	$3.62	$181.00
67467	NANCY DOLL TOY CO	400	$22.95	$.45	$22.50
67468	RAMONAS FINE FOODS	430	$3,557.25	$69.75	$3,487.50
67469	EL CHOLOS	436	$1,795.71	$35.21	$1,760.50
67470	DATAMATION INC	437	$2,247.00	$374.50	$1,872.50
67471	MICROFICHE CORP	441	$2,555.10	$50.10	$2,505.00
67472	REALIST INC	389	$2,872.32	$56.32	$2,816.00
67473	EASTMAN KODAK CO	401	$2,311.32	$45.32	$2,266.00
67474	UNIVAC INC	410	$3,348.15	$65.65	$3,282.50
67475	AVCO CO	411	$5,015.85	$98.35	$4,917.50
67476	TRW SYSTEMS GROUP	412	$2,311.32	$45.32	$2,266.00
67477	BELL HELICOPTER CO	413	$2,878.95	$56.45	$2,822.50
67478	BOEING AEROSPACE CORP	414	$2,328.15	$45.65	$2,282.50
		TOTALS	$46,993.38	$1,251.88	$45,741.50**

Questions for Review

1. Why is it important that data be stored properly?
2. What is the main function of the Data Division?
3. What are the three categories of data to be processed?
4. What must each entry contain?
5. What is an item? Briefly describe the three categories of items.
6. What is the distinction between the physical aspects of a file and the conceptual characteristics of the data contained in the file?
7. What is the relationship between a logical record, physical record, and a file?
8. What is the distinction between the record's external description and its internal content?
9. What is the function of the File Section?
10. What are the main functions of the Working-Storage Section?
11. What does the file description entry describe?
12. What does the file description entry contain?
13. When is the BLOCK CONTAINS clause used?
14. Briefly describe the main uses of the RECORD CONTAINS, LABEL RECORDS, VALUE OF, and DATA RECORDS clauses.
15. What does a record description entry consist of?
16. Why are items indented in record description entries?
17. What is the concept of levels and how are the various level numbers used?
18. What are condition-names and how are they used in the Procedure Division?
19. How is "FILLER" used, and does it always represent a blank area?
20. What are the rules governing the use of the USAGE clause? Explain the various types of USAGE clauses.
21. Describe the use of the PICTURE clause.
22. What are the five categories of data that can be described with a PICTURE clause?
23. What are non-edited PICTURE characters and their meanings? the edited characters and their meanings?
24. What is editing?
25. What are insertion characters? floating strings?
26. What is the relationship between the PICTURE and USAGE clauses?
27. What is the VALUE clause and how is it used?
28. What is a condition-name clause and how is it used?
29. What are the principal uses of the Working-Storage Section?

Matching Questions

Match each item with its proper description.

______ 1. Condition-name	A. Specifies the size and characteristics of a logical record and a physical record.
______ 2. File Section	B. Group of related information, uniquely identifiable, and treated as a unit.
______ 3. Record description entry	C. Used to give reports a high degree of legibility.
______ 4. Data Division	D. Describes records and non-contiguous data items which are not a part of external data files and are not developed and processed internally.
______ 5. File description entry	E. Used to indicate the organization of elementary items and group items.
______ 6. Physical record	F. Describes the structure of data files.
______ 7. Level numbers	G. Whose size and mode is convenient to a particular input or output device for the storage of data.
______ 8. Logical record	H. Describes characteristics of the information to be processed by the object program.
______ 9. Working-Storage Section	I. Assigned to items that may have various values.
______10. Editing	J. Describes the characteristics of each item in a record.

Exercises

Multiple Choice: Indicate the best *answer (question 1—42).*

1. The RECORD CONTAINS clause of the file description entry of a card file
   a. must state RECORD CONTAINS 80 CHARACTERS.
   b. should account for only the columns that are actually punched.
   c. may be omitted.
   d. None of the above.
2. In card files, label records are
   a. standard.
   b. prohibited.
   c. selected.
   d. optional.
3. Each clause of a file description ______________ written on a separate line.
   a. cannot be
   b. must be
   c. may be
   d. None of the above.
4. Descriptions of group items in the Working-Storage Section
   a. must not have VALUE clauses.
   b. must have VALUE clauses.
   c. need not have VALUE clauses.
   d. can selectively have VALUE clauses.
5. It is possible for a record description in the File Section to consist of only a level 01 entry; in that case, the level 01 entry should contain a level number, a name, and
   a. no descriptive clauses.
   b. a VALUE clause.
   c. no PICTURE clauses.
   d. a PICTURE clause.

6-7. Condition names are defined in the ______________ Division, and used in the ______________ Division.
   a. Identification
   b. Environment
   c. Data
   d. Procedure

8. Since an independent item is an elementary item, its description ______________ contain a PICTURE clause.
   a. can
   b. must, except when USAGE IS COMP-1 or COMP-2
   c. won't
   d. None of the above.
9. Description of group items ______________ PICTURE clauses.
   a. must have
   b. may have
   c. can selectively have
   d. must not have
10. The input data code is defined in a data item's
   a. VALUE clause.
   b. USAGE clause.
   c. PICTURE clause.
   d. Condition-name clauses.
11. Which PICTURE could produce the printed results ***17.25− and 2,675.25+ from source values −1725 and +267525?
   a. *,***.99−
   b. *,*99.99+
   c. ***99.99+
   d. *,*99.99−

12. Which PICTURE could produce the printed results ƀƀƀƀ8ƀ7ƀ6 and ƀƀ4ƀ3ƀ2ƀ1 from the source values 00876 and 04321? (ƀ denotes a blank character.)
   a. ZBB9B9B9
   b. 9B9B9B9B9
   c. ZB9B9B9B9
   d. ZZZB9B9B9
13. Which PICTURE could produce the printed results 3500ƀCR if the source value, 35, were negative?
   a. 99000CR
   b. 9999BCR
   c. 99990CR
   d. **99BCR
14. A PICTURE clause that specifies a numeric variable is
   a. PICTURE 9.9.
   b. PICTURE 99V99.
   c. PICTURE 9,999V9.
   d. PICTURE $$9.99.
15. Alignment of a decimal point is caused by
   a. only the PICTURE character . .
   b. only the PICTURE character V.
   c. both the PICTURE characters . and V.
   d. neither the PICTURE characters . and V.
16. The PICTURE that would allow printing of the value $65,495.75 is
   a. $$$,$$V.$$.
   b. $$,999.99.
   c. $99,999.00.
   d. $$$,$$$.99.
17. The file name in the file description entry
   a. must be a reserved word.
   b. comes before the level indicator.
   c. is programmer supplied.
   d. can be a PICTURE.
18. In the Data Division, file description entries are made up of
   a. item description entries.
   b. record description entries.
   c. description of independent items.
   d. None of the above.
19. Record description entries are made up of
   a. descriptions of independent items.
   b. file description items.
   c. item description entries.
   d. Any of the above.
20. If a file contains two types of records, how many record descriptions will be found below the file description entry?
   a. none at all
   b. only one
   c. at least two
   d. any number
21. In record descriptions, any item that is not further subdivided is
   a. an independent item.
   b. a non-group item.
   c. an elementary item.
   d. a group item.
22. Which of the five USAGE words may be omitted?
   a. DISPLAY
   b. COMPUTATIONAL
   c. COMPUTATIONAL-1 or -2
   d. COMPUTATIONAL-3

23. In a record description, an entry for a group item is followed by entries for the items that make it up. A group item comprises all the items described under it, until a level number equal to or less than the group item is encountered. This means that the items that make up a group item must have level numbers that are
    a. equal to the group item.
    b. less than the group item.
    c. identical to the group item.
    d. greater than the group item.
24. The word FILLER ______________ programmer-supplied name.
    a. is not a
    b. may be a
    c. is a reserved
    d. must be a
25. The usage specified for a group item
    a. has no bearing on the usage of the item in the group.
    b. applies to the first item in the group, but not the rest.
    c. applies to all the items in the group.
    d. applies to all the items in the group except the first.
26. If the USAGE clause is omitted for an elementary or group item,
    a. the item has no usage.
    b. the usage can be any of the five possibilities.
    c. it must be wrong.
    d. the usage is assumed to be DISPLAY.
27. Group items ______________ have PICTURE clauses.
    a. sometimes
    b. can
    c. never
    d. must
28. PICTURES are required in the item description entry of
    a. all elementary items.
    b. all elementary items except those that are only one character long.
    c. all elementary items except internal floating point items.
    d. None of the above.
29. Whenever a PICTURE contains the letter A, you can tell that the item is
    a. alphanumeric.
    b. a report item.
    c. alphabetic.
    d. numeric.
30. Pictures of alphanumeric items always contain the letter
    a. X
    b. Y
    c. Z
    d. A
31. Which kinds of items listed below could not be moved to alphanumeric receiving items?
    a. alphabetic
    b. EBCDIC
    c. floating point
    d. numeric
32. Numeric items are ______________ allowed to contain special characters.
    a. sometimes
    b. always
    c. not
    d. optionally
33. The value of the literal .85 is treated as (v is an assumed decimal point)
    a. v99
    b. .85
    c. 85%
    d. v85

34. If the souce value is 95v4 and the receiving picture is 999v99, the value of the receiving item following a numeric move will be
   a. 095.40
   b. 95.4
   c. 095v4
   d. 095v40
35. Given the PICTURE S9999, where is the assumed decimal point to be located?
   a. S9999.
   b. .S9999
   c. S99.99
   d. S.9999
36. If you have this example SPPP9(8), where is the assumed decimal point located?
   a. To the left of the last P?
   b. To the right of the 9?
   c. To the right of the last P?
   d. To the left of the last 9?
37. To move digits as in a numeric move, the PICTURE character that cannot be used is
   a. "V"
   b. "P"
   c. "9"
   d. "B"
38. A PICTURE ______________ contain both "." and a "V".
   a. may optionally
   b. can
   c. cannot
   d. can fractionally
39. Which statement below is correct?
   a. A "Z" can appear anywhere to the right of a 9.
   b. An "*" can appear anywhere to the right of a 9.
   c. A PICTURE can contain both an "*" and a "Z."
   d. A PICTURE cannot contain both an "*" and a "Z."
40. When CR, DB, or − is used in a PICTURE, what sign will be printed for a positive value?
   a. just the CR and DB
   b. each sign condition
   c. none
   d. +
41. When a + sign is used in a PICTURE, what sign will be printed for a negative value?
   a. +
   b. + or −
   c. none
   d. −
42. The largest amount that can be edited with the PICTURE $$$$ is
   a. 999
   b. 99
   c. no amount
   d. any amount
43. Match each item with its proper description

______	1. Group Item	A. Series of related data records.
______	2. Data Record	B. Named sequence of one or more elementary items.
______	3. Item	C. Physical record.
______	4. Block	D. Smallest item available.
______	5. Independent item	E. Item appearing in Working-Storage Section that is not a record or part of a record.
______	6. File	F. Field.
______	7. Elementary item	G. Logical record.

44. Match each clause with its proper description.

______	1. RECORD CONTAINS	A. Absence of labels.
______	2. BLOCK CONTAINS	B. Particularizes description of item in LABEL RECORDS clause.
______	3. LABEL RECORDS	C. Size of logical record.
______	4. VALUE OF	D. Number of records or characters in block.
______	5. DATA RECORDS	E. Number of records in file.

45. Match each level number group with its proper classification.

_____	1. 01	A. Subdivisions of group record items.
_____	2. 02-49	B. Condition entry.
_____	3. 66	C. Designates item as a record.
_____	4. 77	D. Used with RENAMES clause.
_____	5. 88	E. Independent elementary item in Working-Storage Section.

46. Match each USAGE clause with its proper description.

_____	1. Display	A. Short precision internal floating point format.
_____	2. Computational	B. Packed decimal format.
_____	3. Computational-1	C. One character per byte.
_____	4. Computational-2	D. Binary data items.
_____	5. Computational-3	E. Long precision internal floating point format.

47. Match each category of data with its proper description.

_____	1. Alphabetic	A. Numerals.
_____	2. Numeric	B. Combination of numerals and editing characters.
_____	3. Alphanumeric	C. Letters and space.
_____	4. Alphanumeric Edited	D. Any combination of characters in COBOL character set.
_____	5. Numeric Edited	E. Combination of any character plus the characters B or 0.

48. Match each non-editing character with its proper description.

_____	1. 9	A. Assumed decimal point.
_____	2. X	B. One decimal digit.
_____	3. V	C. Operational sign.
_____	4. P	D. Letter or space.
_____	5. S	E. Any type of character in the COBOL character set.
_____	6. A	F. Assumed zero.

49. Match each editing character with its proper description.

_____	1. Z	A. Credit symbol.
_____	2. .	B. Digit position.
_____	3. *	C. Credit symbol.
_____	4. CR	D. Actual decimal point.
_____	5. DB	E. Check protection.

50. Which of the following is a floating strings or insertion character? Use *F* for floating strings and *I* for insertion characters.

_____ $ _____ B _____ − _____ 0 _____ +

51. Match each clause with its proper description:

_____	1. PICTURE	A. Particular value that may be assumed by a data item.
_____	2. VALUE	B. General characteristics and size of an item.
_____	3. Condition-name	C. Form in which an item is stored.
_____	4. USAGE	D. Initial value of an item.

52. How will the following source data be interpreted?

	Source Data	**Picture**	**Interpretation**
a.	123	9V99	_____________
b.	−132	S999	_____________
c.	15671	999V999	_____________
d.	4071629	9(5)V99	_____________
e.	9263	V999	_____________
f.	+61	S99	_____________

53. Specify the actions that will take place when the source area data is moved to the corresponding receiving area.

	Source Area		**Receiving Area**	
	Picture	**Data Value**	**Picture**	**Data Value Interpretation**
a.	9999	1234	9(6)	_____________
b.	99V99	1234	999V99	_____________
c.	99V99	1234	99V999	_____________
d.	99V99	1234	9(4)V9(4)	_____________
e.	9(4)	1234	999	_____________
f.	999V9	1234	99V99	_____________
g.	999V9	1234	999	_____________
h.	999V999	123456	99V99	_____________
i.	99V999	12345	9(4)V99	_____________
j.	999V99	12345	99V999	_____________

54. Specify the actions that will take place when the source area data is moved to the corresponding receiving area.

	Source Area		Receiving Area	
	Picture	Data Value	Picture	Edited Data
a.	999V9	1234	999.9	
b.	9(5)V99	0001234	9(5).99	
c.	9(5)V99	0001234	Z(5).99	
d.	9(5)V99	0000123	ZZZ99.99	
e.	9(5)V99	0123456	99,999.99	
f.	9(5)V99	0123456	ZZ999.99	
g.	9(5)V99	0000000	ZZ,ZZZ.ZZ	
h.	9(5)V99	1234567	$99,999.99	
i.	S9(6)	+123456	+9(6)	
j.	S999	−123	−999	
k.	S9(4)	+0012	ZZZ9+	
l.	S99V99	−1234	99.99CR	
m.	S99V99	−1234	99.99DB	
n.	S99V99	+1234	99.99DB	
o.	9(5)V99	0234567	$$$,$$9.99	
p.	9(6)V99	00001234	***,***.99	
q.	S9999	+0123	++++9	
r.	S9999	−0123	++++9	
s.	S9999	+0001	−−−−9	
t.	9(9)	123456789	999B99B9999	

55. List the group and elementary items in the following.

a.

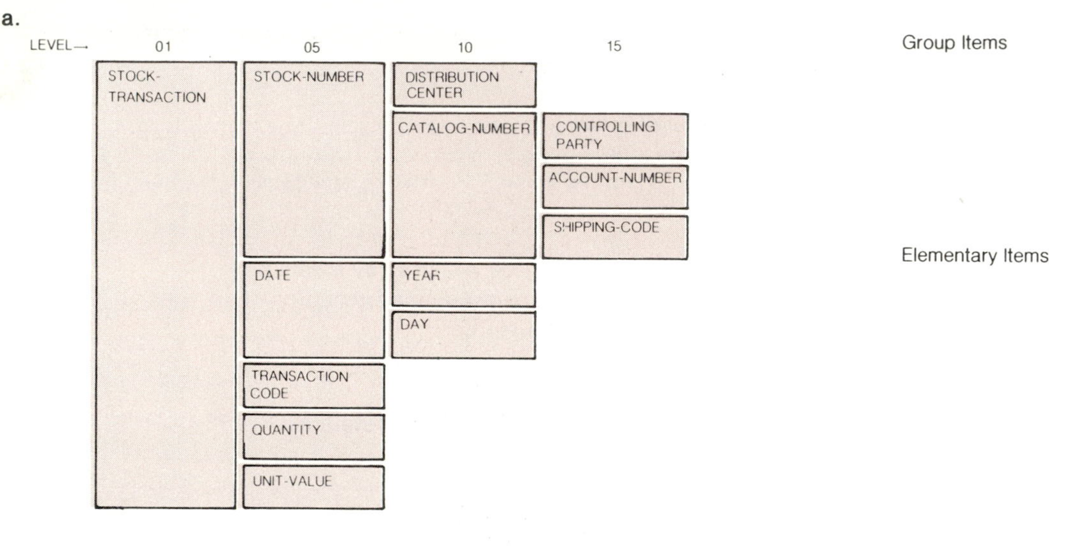

b.

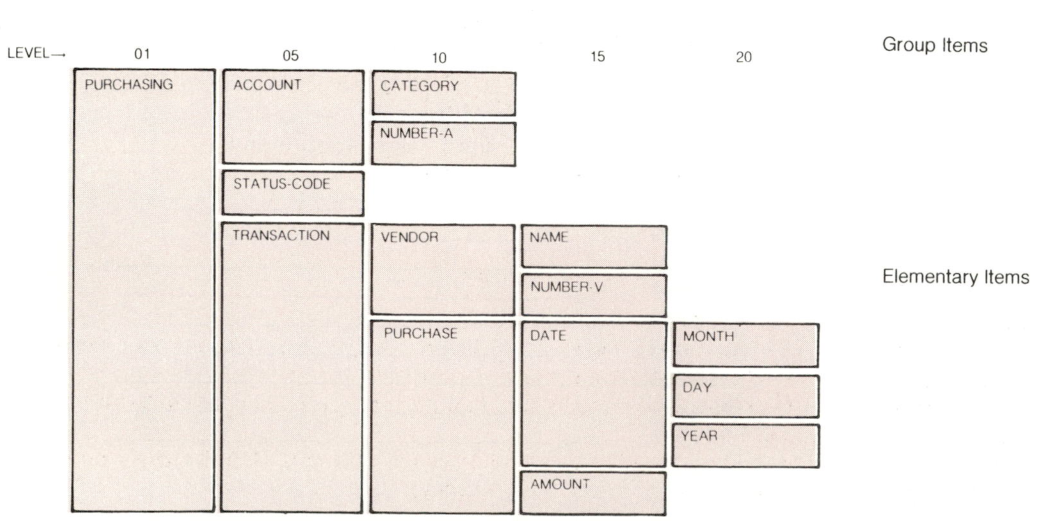

Problems

1. **Write the file description entry for a file whose name is EXPENSE-FILE.**
   The file is on magnetic tape, has standard label records and one type of data record called EXPENSE-RECORD. All records are of fixed length, and each record is preceded by a record length control field. There are twenty records per block.

2. **Write the record description entry for the following record. We are not concerned with the USAGE and PICTURE clauses of these items, just the level numbers, data-names or FILLER.**

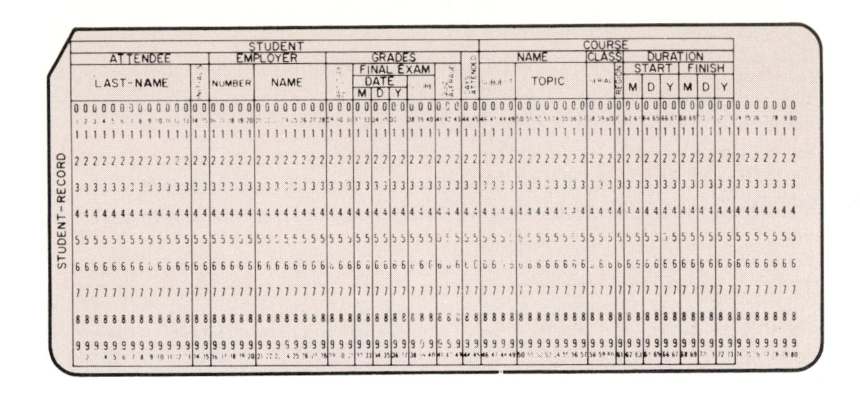

3. **Write the complete record description entry for the following input record including level numbers, data-names or FILLER, together with all necessary PICTURE and VALUE clauses.**

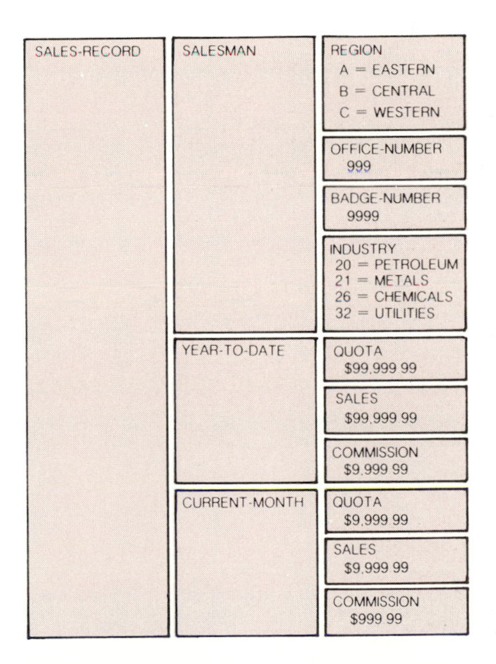

4. **In the Working-Storage Section, write the entries to set up the following:**
   a. A work area large enough to hold twenty-five alphanumeric characters.
   b. An independent item called DIFFERENCE to contain a sign, five digits, and stored in packed decimal format.
   c. A constant called LIMIT, whose value is 600 and is stored in a binary format.
   d. An alphanumeric constant which is to serve as a title of a report. The contents of the item are to be DEPRECIATION SCHEDULE and the item is to be named TITLE.
   e. A record to be called ADDRESS composed of STREET (20 alphanumeric characters), CITY (20 alphanumeric characters), STATE (5 alphanumeric characters) and ZIPCODE (5 digits).

5. **Write the necessary entries in the Working-Storage Section for the following headings:**

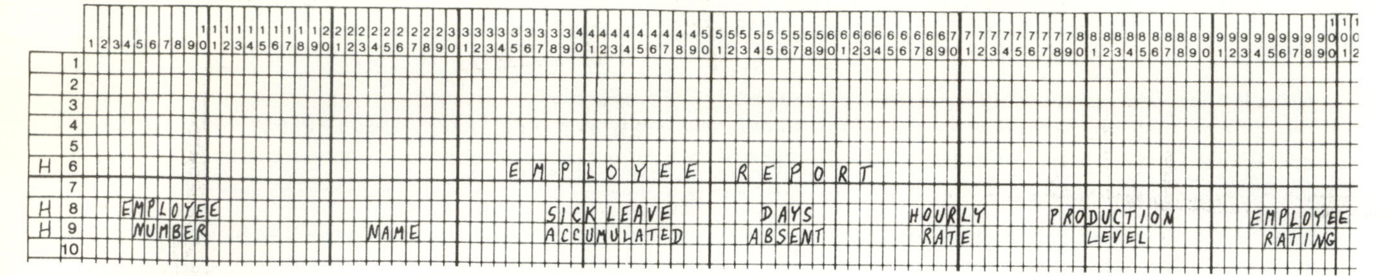

6. **Set up the Working-Storage Section for the following:**

   a. A percentage value of 25% to be used as the multiplier in a multiplication operation.
   b. A 4-digit counter containing an initial value of 1,000.
   c. A 5-position field to be used as a temporary storage area.
   d. A 3-digit page number which will be incremented for each new page printed. The initial value equals 001.
   e. A record containing six fields of fifteen characters each with the initial values equal to the names of the six New England States (Connecticut, Massachusetts, Maine, New Hampshire, Rhode Island and Vermont).
   f. Headings for a report as described in the following:

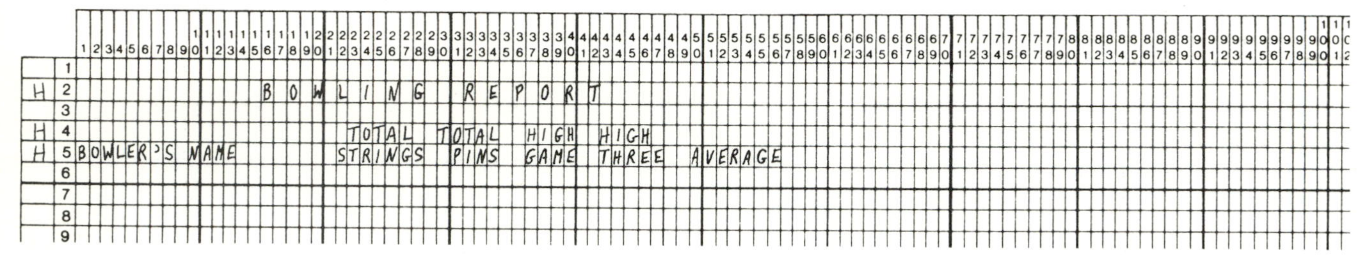

7. **Write the Identification, Environment and Data divisions for the following:**

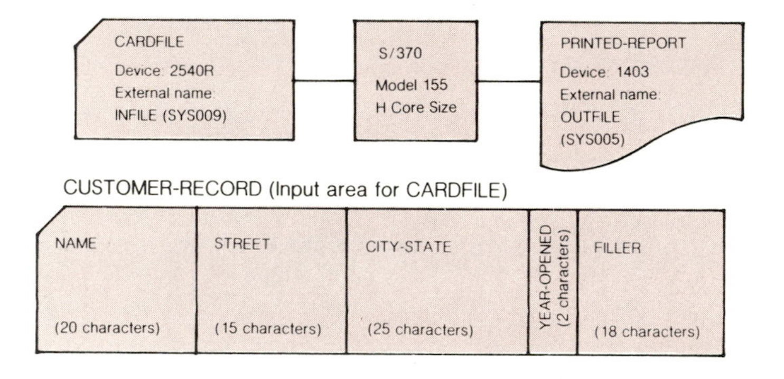

PRINT-RECORD (Output area for PRINTED-REPORT)

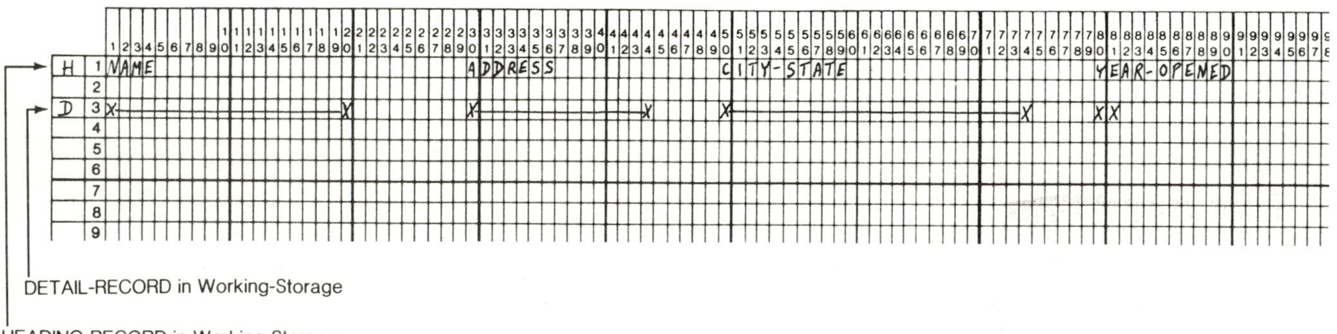

8. Write the Identification, Environment and Data divisions for the following:

Hardware

Computer	IBM 370 Model 155 with H Core size
Input	Magnetic Tape Unit Model 2400 External name RECBLES (SYS012).
Output	Printer Model 1403 External name ACCTLIST (SYS005).

Input File—RECEIVABLE
There are 50 records per block
as well as standard labels.

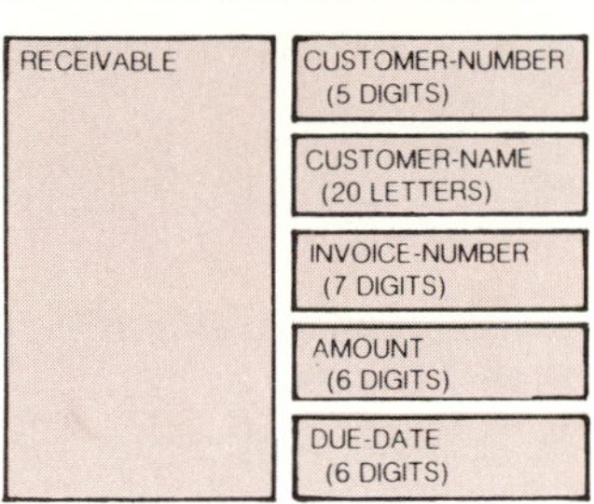

Output File—ACCOUNT

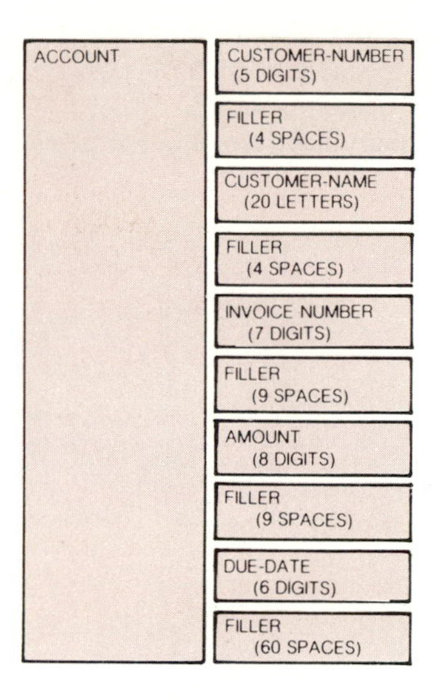

9. Write the Identification, Environment and Data divisions for the following:

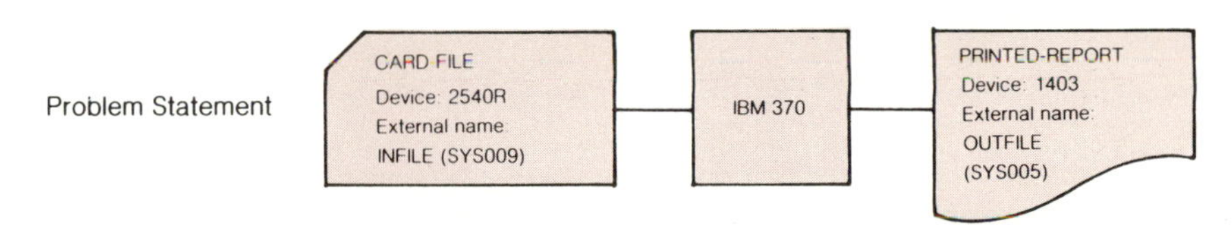

The system flowchart just presented shows the files and equipment to be used in this program. The forms of records in CARD-FILE and PRINTED-REPORT are illustrated as follows.

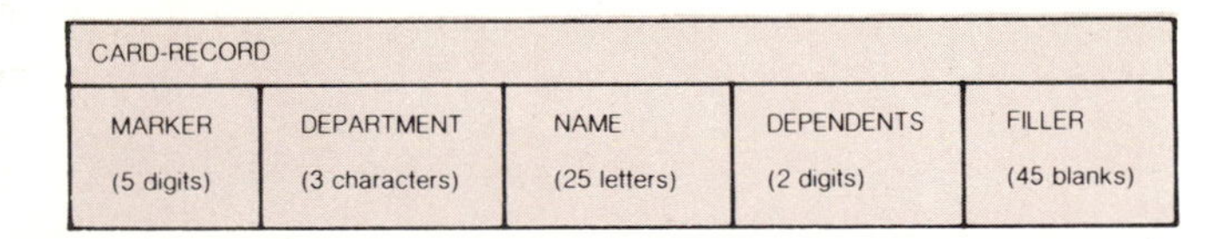

CARD-RECORD
MARKER (5 digits)
DEPARTMENT (3 characters)
NAME (25 letters)
DEPENDENTS (2 digits)
FILLER (45 blanks)

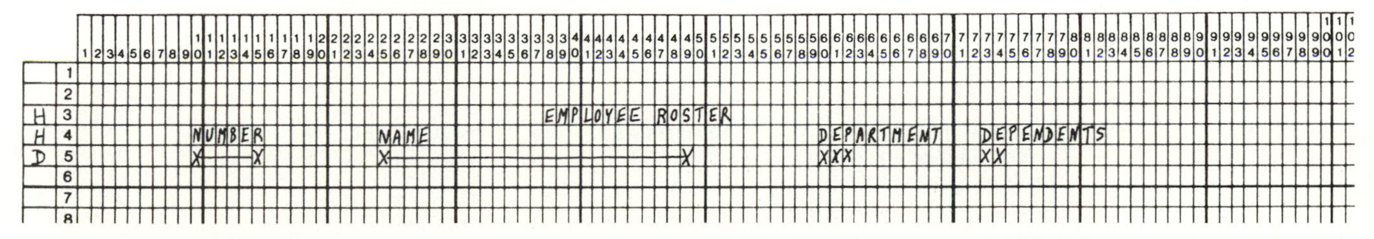

The file PRINTED-REPORT is to consist of a title and headings followed by a listing of the records in CARD-FILE with the data items rearranged as shown in the Printer Spacing Chart. Use the variables CARD-RECORD, PRINT-RECORD, WORK-RECORD-1, WORK-RECORD-2, and WORK-RECORD-3.

10. Write the Identification, Environment and Data divisions for the following:

Companies have master files of information that require changes and constant updating. One such change may be the deletion of discontinued items from the master file and the subsequent listing of the deleted items. In order to make these changes, a maintenance program should be written.

The following processing is involved in writing the maintenance program.

1. *Card-to-tape conversion.* Read a deck of delete cards and write the deleted items on a transaction file tape.
2. *File update.* Update the master file by passing the transaction file tape against the old master file deleting items according to the item number.
3. *Print.* Print those items which are deleted from the master file.

Hardware

Computer	IBM 370 Model 155 H Core Size
4 Magnetic Tape Units	Model 2400 (Tape-I-1, Tape I-2, Tape-O-1, Tape-O-2).
1 Card Reader	Model 2540(SYS009).
1 Printer	Model 1403(SYS005).

The following systems flowchart illustrates the two phases of the program.

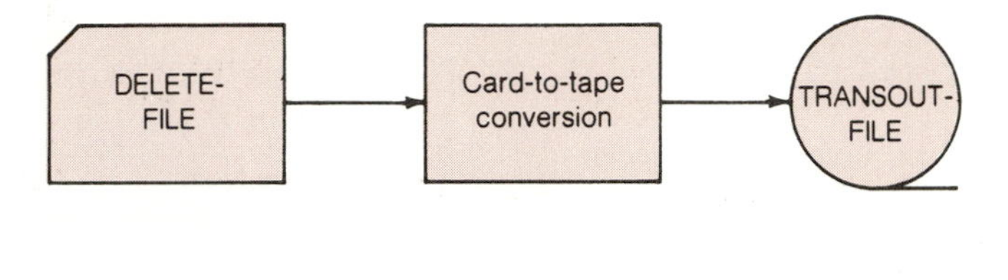

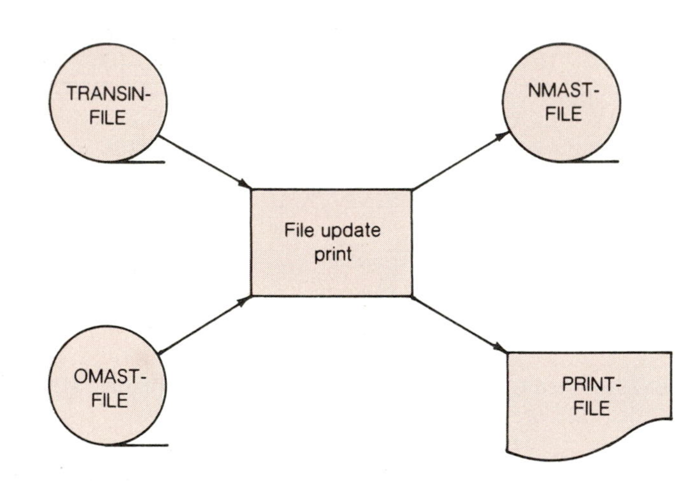

Data Division

File Section

DELETE-FILE	*Read from Card Reader*	*Record name is DELETE-RECORD*
Field Name	**Positions**	**Field Class**
Item number	1–5	Numeric
Delete date	6–11	Numeric
Description	12–24	Alphabetic

	Tape Unit	*Record name is*
TRANSOUT-FILE	*SYS012*	*TRANSOUT-RECORD*

FORMAT SAME AS DELETE-FILE

	Tape Unit	*Record name is*
TRANSIN-FILE	*SYS013*	*TRANSIN-RECORD*

FORMAT SAME AS TRANSOUT-FILE

	Tape Unit	*Record name is*
OMAST-FILE	*SYS014*	*OMAST-RECORD*

Field Name	**Positions**	**Field Class**
Item number	1-5	Numeric
Item description	6-18	Alphabetic
Quantity	19-22	Numeric
Balance	23-28	Numeric XXXX.XX

	Tape Unit	Record name is
NMAST-FILE	SYS015	NMAST-RECORD

FORMAT SAME AS OMAST-FILE
ALL TAPE RECORDS IN BLOCKS OF 4 RECORDS AND HAVE STANDARD LABELS.

PRINT-FILE	*LISTED REPORT 'DELETED ITEMS REPORT'*	*Record Name is PRINT-RECORD*

Field Name	**Positions**	**Field Class**
Delete Date	4-9	Numeric
Item number	14-18	Numeric
Item description	23-35	Alphabetic
Quantity	40-43	Numeric
Balance	49-56	Numeric Edited ($9999.99)

Working-Storage Section

Field Name	**Size**	**Field Class**
TOTAL	7	Numeric XXXXX.XX

Header information

Constant Value	**Positions**
DATE	5-8
ITEM-NO	13-19
DESCRIPTION	24-34
QUANTITY	39-46
BALANCE	50-56

Structured Programming

Chapter Outline

Chapter Objectives

The learning objectives of this chapter are:

1. To prepare you for the coding of programs using the latest techniques of structured programming including GO TO-less programming, top/down design, structure charts, pseudo coding, etc.

2. To explain how COBOL implements the basic control structures of structured programming so you can become familiar with the coding standards of structured programming.

Introduction

What is structured programming? A panacea for all programming ills? A solution to the rising costs of program preparation and maintenance? A set of rigid rules and no "GO TO" statements, thus removing all creativity from the art of programming? Or is it just a bunch of new names for old ideas and procedures that have always been around; procedures that you've more or less been using anyway in your programming?

The goal of structured programming is to change the process of programming from a frustrating, trial and error activity to a systematic, quality controlled activity. Structured programming also implies structured design and testing of a structured program made up of interdependent parts in a definite pattern of organization. However, to attain this goal of precision programming, the ideas of structured programming must be used constantly; not simply treated as good ideas to be used when convenient, but as basic principles that are always valid.

Structured programming applies a theoretical concept to the coding of computer programs. The main element of the concept is the use of basic structures to create complex programs. The application of structured programming in a commercial data processing environment requires a change in program design methods and a rigid application of standards to insure the best results. In addition, further difficulties arise from the use of COBOL, which does not provide all the basic structures directly.

The concepts behind structured programming and its related aspects are currently getting wide coverage and attention in the computer field. The concepts, to be useful, must be distilled into a practical approach that will fit the environment of the organization wishing to use structured programs.

A structured program tends to be much easier to understand than programs written in other styles. This capability of being more easily understood facilitates code checking, which in turn may reduce program testing and debugging time. This is true partly because structured programming concentrates on one of the most error-prone factors in programming, the logic.

A program that is easy to read and which is composed of well-defined segments tends to be simpler, faster, and less expensive to maintain. These benefits derive in part from the fact that since the program is to a significant extent its own documentation, the documentation is always up-to-date. This is seldom true with conventional methods. The benefits of structured programming may be summarized as follows:

Encourages programming discipline

Fewer errors

More easily modified and maintained

More nearly self-documenting

Programmer can control larger amount of code

Structured programming offers these benefits, but it should not be thought of as a panacea. Program development is still a demanding task requiring skill, effort, and creativity.

Objectives

The main objective of structured programming is to provide a single generalized method of program design, which yields the following benefits:

1. Program development by modules
2. Ease of coding
3. Speed of coding
4. Ease of debugging
5. Ease of maintenance

A

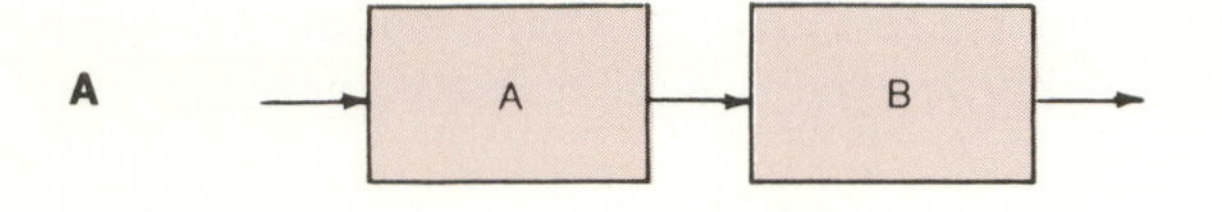

Sequence is represented by one function after the other as shown here. A and B are anything from single statements to complete modules. The combination of A followed by B is also a proper program since it has only one entry and one exit.

B

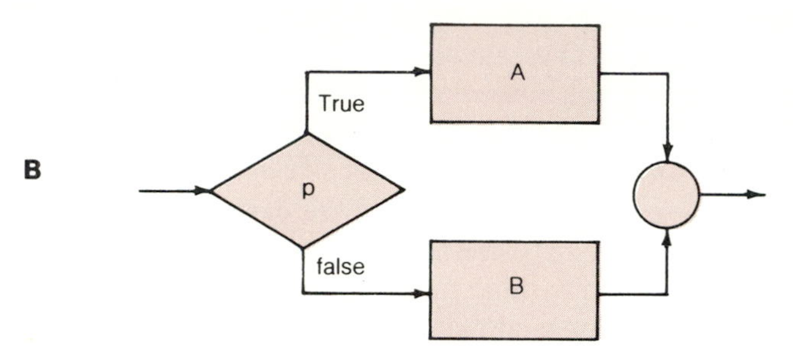

Selection is the choice between two actions based on a predicate; this is called the IFTHENELSE structure. In COBOL it is implemented with the IF statement, and the predicate called the condition. The usual flowchart for selection is shown at the left, where p is the predicate and A and B are the two functions.

C

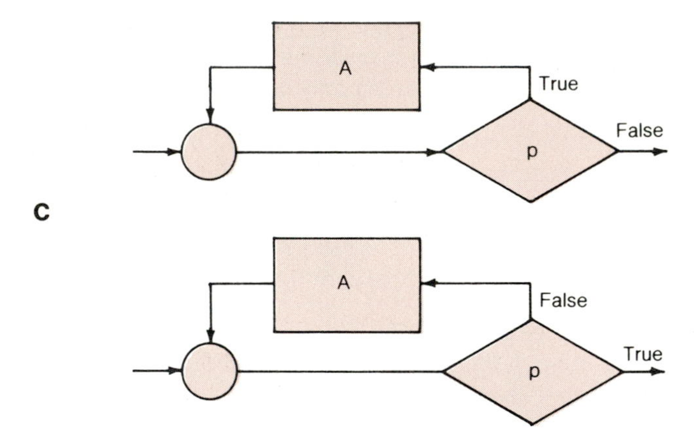

The iteration structure, used for the repeated execution of code is shown here in two forms. The first shows a flowchart for a loop known as a DOWHILE loop, which is implemented in some computer languages (PL/1, PASCAL). The second form shows the iteration structure for COBOL, PERFORM . . . UNTIL. Note the difference in the manner in which the test for exiting is made. In the DOWHILE, execution of the code is repeated WHILE the condition is true, and an exit is made when it becomes false. In the PERFORM . . . UNTIL, execution is repeated UNTIL a condition becomes true. In these flowcharts, p is the predicate and A is the controlled code.

Figure 5.1 Flowcharts for the control logic structures.

Structure

Structured programming is a style of programming in which the structure of a program (that is, the interrelationship of its parts) is made as clear as possible by restricting control logic to just three structures.

1. *Simple Sequence (Sequence).* In the absence of instructions to the contrary, statements are executed in the order in which they are written. See figure 5.1A.
2. *IFTHENELSE (Selection).* Combine this with statement brackets (begin and end) so that groups of statements can be included in the IF AND ELSE statements. In fact, the IF AND ELSE statements may themselves contain any of the three structures. See figure 5.1B.
3. *DOWHILE (Iteration)*—A loop control mechanism. See figure 5.1C.

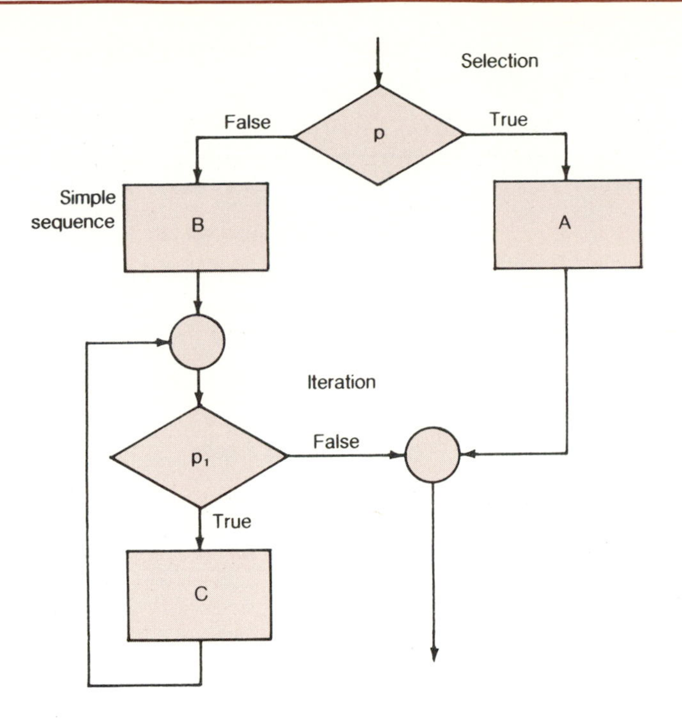

Figure 5.2 Nesting of control structures.

In a structured program, any program function can be performed using any of the aforementioned three basic control logic structures: simple sequence, selection, or iteration. Any kind of processing, any combination of decisions, any sort of logic, can be accommodated with any one of these control structures or a combination of them. Each structure is characterized by a simple and single point of transfer of control into the structure, and a single point of transfer of control out of the structure. These structures can be combined to form a program that is simple, in the sense that control flows from top to bottom or from beginning to end. There is no back tracking. The control structures can be nested as shown in figure 5.2, but they retain their characteristic of single/entry—single/exit.

Reducing program complexity can be thought of as a process of removing things from a program, e.g., obscure structure, complicated control paths, redundant and obsolete code, meaningless notes, etc. Improving program clarity can be thought of as a process of adding material features to a program, e.g., self-explanatory labels, important notes, code layout, and indentation that has information content for the reader, more levels of modularity, etc.

Much of a program's complexity arises from the fact that the program contains many jumps to other parts of a program—jumps both forward and backward in the code. These jumps make it difficult to follow the logic of the program and, moreover, make it difficult to ascertain at any given point of the program what existing conditions may be (such as, what the start of the variables is, and what other paths of the program have already been executed, etc.). Furthermore, as a program undergoes change during its development period, as it gets further debugged during its maintenance period, as it gets modified in subsequent new projects, its complexity grows alarmingly. New jumps are inserted, increasing that complexity. In some cases, new code is added because the programmer cannot find existing code that performs the desired functions, or isn't sure how the existing code works, or is afraid to disturb the existing code for fear of undoing another desirable function. The result, after many modifications, is a program that is nearly unintelligible. The program is by now shopworn; the time has come when it is better to throw the entire program out and start afresh.

Easy program readability requires that it not be necessary to turn a lot of pages in order to understand how something works. A structured program is composed of modules which may range from a few statements to a page of coding. Each module is allowed to have just one entry and one exit. A practical rule is that a module should not exceed a page of code, about fifty lines. In COBOL terms, a module can be a paragraph, section, subroutine, or code incorporated with a COPY. Such a module, assuming it has no infinite loops and no unreachable code, is called a *proper program.* When proper programs are combined using the three basic control logic structures, the result is a proper program. General rules for proper COBOL coding follow:

All paragraphs explicitly performed

- No 'perform A thru Z'
- No perform of section

No extra paragraph names

Paragraphs in order of execution

- Don't repeat the code
- Use meaningful names and clear code
- Remarks and overall comments good

New page for divisions, sections, modules

Separate paragraphs with blank lines

One statement per line

- Indent continuations and IF statements

One entry

One exit

No dead code

No endless loops

GO TO-Less Programming

Structured programming has occasionally been referred to as "GO TO-less Programming." A well-structured program gains an important part of its easy-to-read quality from the fact that it can be read in sequence, without "skipping around" from one part of the program to another. The "sequential ability," or "top-down readability," is beneficial because there is a definite limit to the amount of detail the human mind can encompass at one time. It is far easier to grasp completely what a statement does if its function can be understood in terms of just a few other statements, all of which are adjacent physically. The trouble with GO TO statements is that they generally defeat this purpose; in extreme cases, they can make a program essentially incomprehensible.

Here is an example of code involving GO TO statements:

```
PAR-1.
    IF A IS GREATER THAN .20, GO TO PAR-3.
    IF A IS GREATER THAN .10, GO TO PAR-2.
    MOVE 5 TO FLD-X.
    GO TO PAR-4.
PAR-2.
    MOVE 6 TO FLD-X.
    GO TO PAR-4.
PAR-3.
    MOVE 4 TO FLD-X.
PAR-4.
    ****************
```

To understand what is going on in the above program, the programmer would consider the conditional statements in the sequence presented. First, if the statement, "A IS GREATER THAN .20" is true, the program continues execution at PAR-3. Next, if the statement "A IS GREATER THAN .20" is false but the statement "A IS GREATER THAN .10" is true, the program continues execution at PAR-2. Only if both conditionals fail, does the program execute the next statement "MOVE 5 TO FLD-X" and then pass control to PAR-4.

To a COBOL-experienced eye, this does not seem to be a very difficult code structure, but if the eventual target, PAR-4, were several pages away (and if PAR-2 and PAR-3 were not so conveniently located) there would be a considerable amount of page flipping in order to discern the intended meaning of the program.

The problem would be eliminated or at least greatly simplified, if the program were organized so that there was greater "locality." In a sense, achieving this kind of correspondence between static placement of statements and dynamic flow depends on the vague concept of "program style." To illustrate how this program segment would look in GO TO-free form, it can be rewritten as follows:

```
IF A IS GREATER THAN .20
     MOVE 4 TO FLD-X
ELSE
     IF A IS GREATER THAN .10
          MOVE 6 TO FLD-X
     ELSE
          MOVE 5 TO FLD-X.
*************
```

The resulting value of FLD-X, after the set of checks on the value of A, is evident by the organization of the statements. Because there are no GO TO statements there are no labels and there is a direct correspondence between the static form of the program and the dynamic flow during execution.

There are uncommon situations where the use of GO TOs may improve readability compared with other ways of expressing a procedure. Such examples are exceptional, however, and do not usually occur in everyday programming. The impact of deviations from installation guidelines, such as using GO TOs in other than prescribed ways, should be given careful consideration before such deviations are permitted.

The blatant use of GO TO statements results in unnecessary and complex flow patterns leading to difficult debugging effort on the part of the programmer. No special effort is really required to "eliminate GO TOs," which sometimes has been misunderstood as the goal of structured programming. They just never occur when the standard control logic structures are used.

The fundamental difficulty with the GO TO statement is that it distracts the reader from the program by forcing him or her to examine the program in an unnatural way.

In structured programming in COBOL, GO TO* statements are used only in certain special instances. In place of GO TO statements, PERFORM** statements are used.

Program Design and Implementation

The mere removal of all of the GO TO statements will not of itself "structure" the program; in fact, even though GO TO-free programs are intrinsically easier to read and debug than their counterparts (programs with GO TO statements), the form and style of the expression of the algorithms is not explicitly changed by the avoidance of the GO TO statements. Structured programming is also concerned with ways of developing complicated program structures in an orderly fashion.

Structured COBOL solutions involve the use of three basic structures, SEQUENCE, SELECTION, and ITERATION. However, these structures by themselves are not enough to code a structured COBOL program. The techniques of program design and implementation must be applied to these structures before the structured program can be coded. Top-down programming and program modulation expand the use of these techniques to problem solutions.

Top-Down Design

The design of the program should proceed from top to bottom. Individual program modules should be as short as possible, preferably no longer than one page of machine output, to facilitate the partitioning of the logic into individual chunks that are easy to debug.

There are two major advantages of a strictly hierarchical form for a program. First, adhering to the hierarchical constraints forces the organization of the program along "natural" algorithm boundaries; individual program modules can be organized so that each performs some specific function. The result is that each module is easier to debug and so the entire program is easier to debug.

*A GO TO is merely a branch statement that is used as a device to have the program execution "jump" out of the normal sequence to a location specified by the label in the operand of the GO TO statement.

**The PERFORM statement transfers program control to the specified closed subroutine identified in the operand of the particular statement. Upon completion of the subroutine, control is returned to the program statement immediately following the PERFORM statement.

Structured programming is an aggregation of three basic ideas.

1. The beneficial properties of GO TO-less programming.
2. The application of management techniques to process from top to bottom design.
3. The idea of each program module being specifically related to each of the others.

Each level of programming is supported by the next lower level of the hierarchy. The program at the lower level manipulates the data and at the same time participates in generating a higher level of information for the next higher level of the hierarchy to manipulate.

For example, assume that a program is to be written to retrieve information from a file. A file may be considered a collection of records with each record residing on some section of the disk. Each record is, in turn, composed of words, each word is composed of bytes, and each byte is composed of bits. The hierarchy of resources needed to retrieve a single bit from the file is as follows:

Lowest Level	bit
↕	byte
	word
	record
Highest Level	file

A set of instructions would be written to manipulate files—in this case, to locate and extract particular records. The next level of instructions would operate solely on records—extracting words from records, retrieved by manipulating the program. In turn, the next level would be represented by a set of instructions that would extract bytes from words; at the top would be the program to extract bits from bytes.

Program Design

Data processing applications typically require a set of functions combined in a specified relationship. Applications design (also called system design) typically reflects the functions involved. For example, payroll applications involve functions such as:

1. *Create* payroll master file
2. *Update* payroll master file
3. *Edit* payroll (from gross to net)
4. *Write* payroll register
5. *Write* paychecks

Each function, in the aforementioned example, calls for a single, well-defined action (indicated by *italics*). To be more precise, a function is a transformation that usually receives input and returns output in a predictable manner.

Individual program designs (contrasted to system designs for entire applications) also require a set of functions combined in a specified relationship. However, program designs do not always reflect what functions are involved, or the relationship. In many cases, programs are designed by specifying the procedures to be performed in the form of a flowchart. For example, in a structure chart that updates a payroll master file (changes only, no additions or deletions—figure 5.3) the functions required are:

1. *Initialize* files, data, etc.
2. *Read* a transaction
3. *Read* a master record
4. *Update* old master record
5. *Write* new master record
6. *Print* line
7. *Print* error message
8. *Close* files, etc.

These program functions are all actions performed in a program where the overall function is to *update* a payroll master file.

The structure chart is part of the documentation for a program that updates a payroll master file. The structure chart shows the functions required for the program and their relationship. *Structured design* is a set of program design techniques and guidelines that are used to specify the functions in a

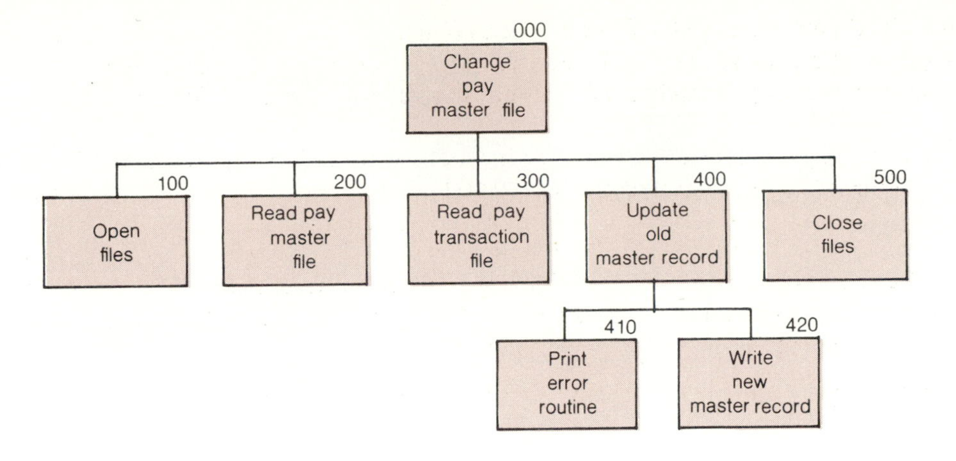

Figure 5.3 Structure chart—example.

program solution, the data processed by each function and the relationships among the functions. The overall result of the structured design process is documented in a structured chart that can be implemented using

Flowcharts and pseudo code to describe the procedure of each function,

Structured programming techniques to program each function, or

Top-down programming and modulation.

See the illustrations in figure 5.4.

The primary reason for using structured design is simplicity; that is, to define program solutions using single, well-defined functions that can be easily programmed, tested, and maintained.

Figure 5.4 Hierarchy, IPO, and flowcharts for updating a file.

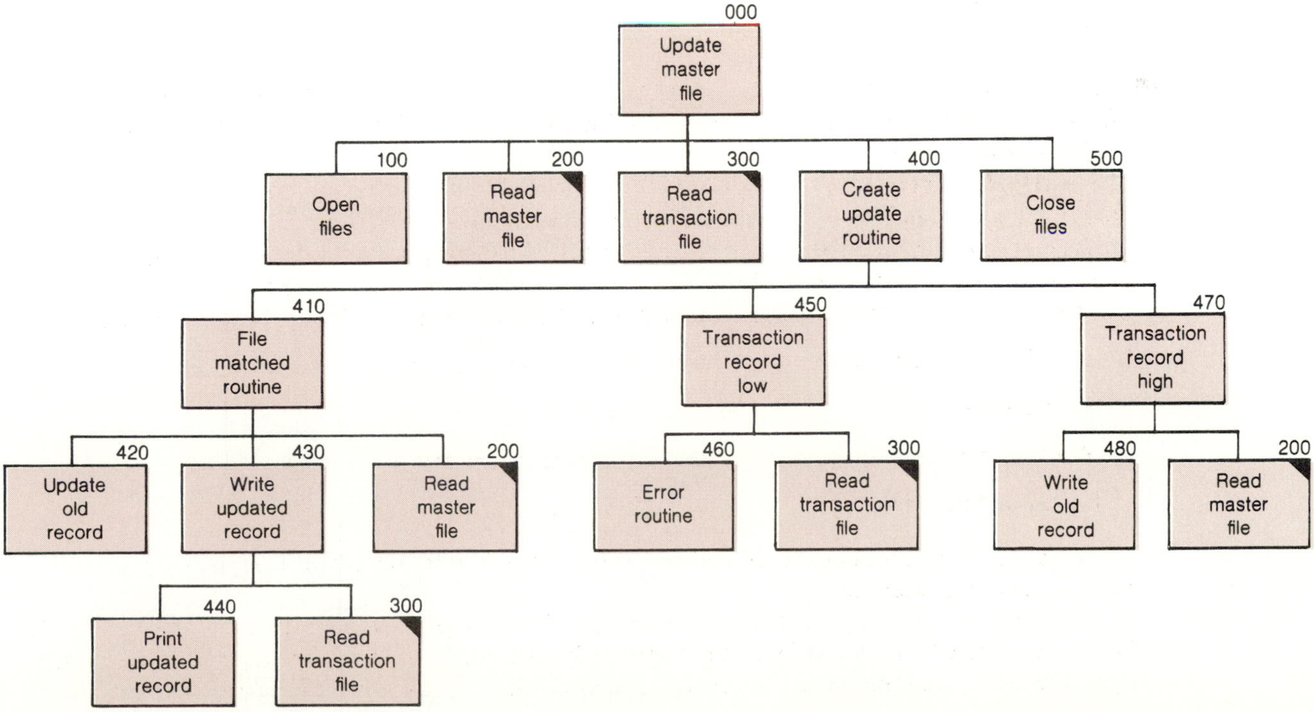

000 Update master file
100 Open files
200 Read master file
300 Read transaction file
400 Create update routine
500 Close files
410 File matched routine
450 Transaction record low
470 Transaction record high
420 Update old record
430 Write updated record
200 Read master file
460 Error routine
300 Read transaction file
480 Write old record
200 Read master file
440 Print updated record
300 Read transaction file

Figure 5.4 *(continued)*

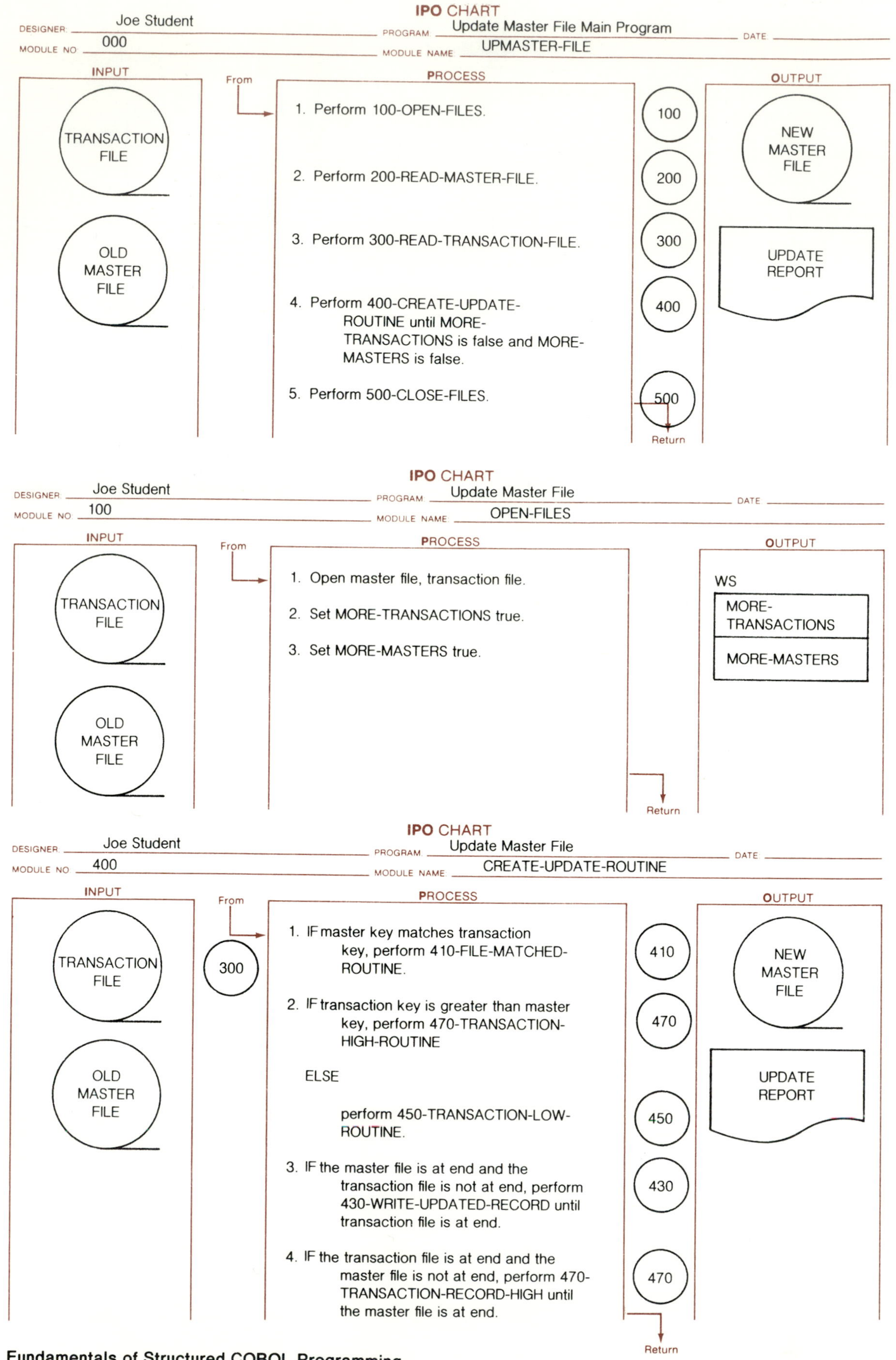

Figure 5.4 *(continued)*

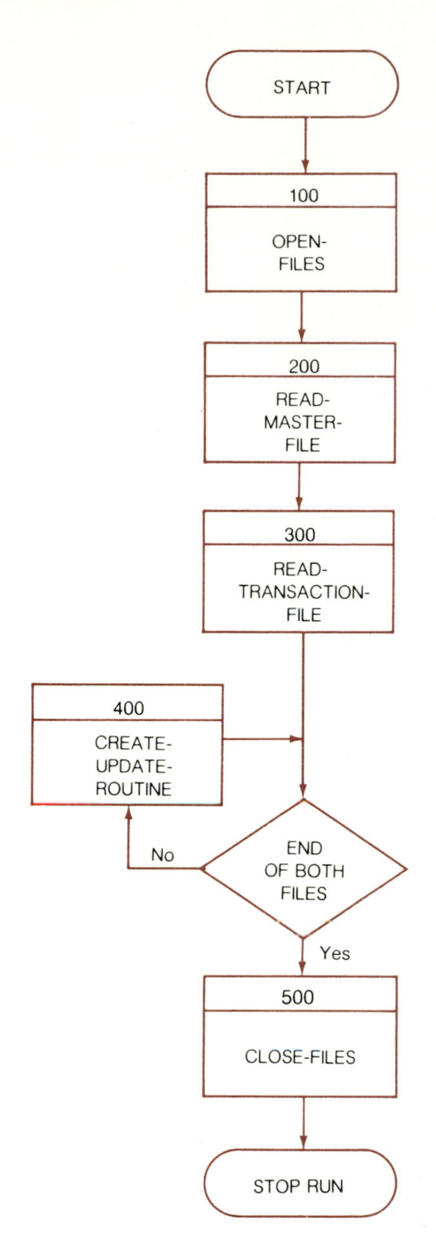

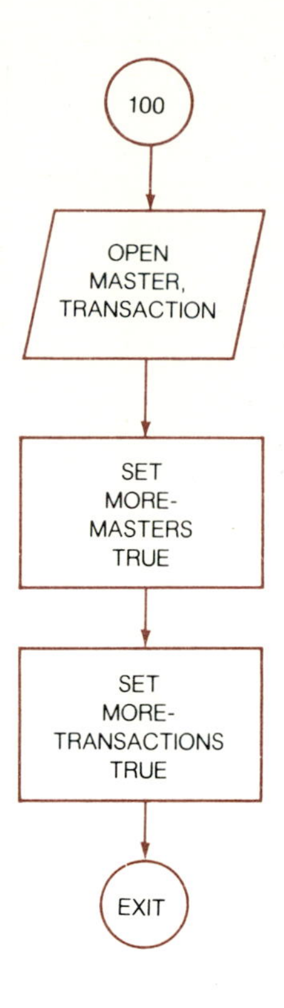

Figure 5.4 *(continued)*

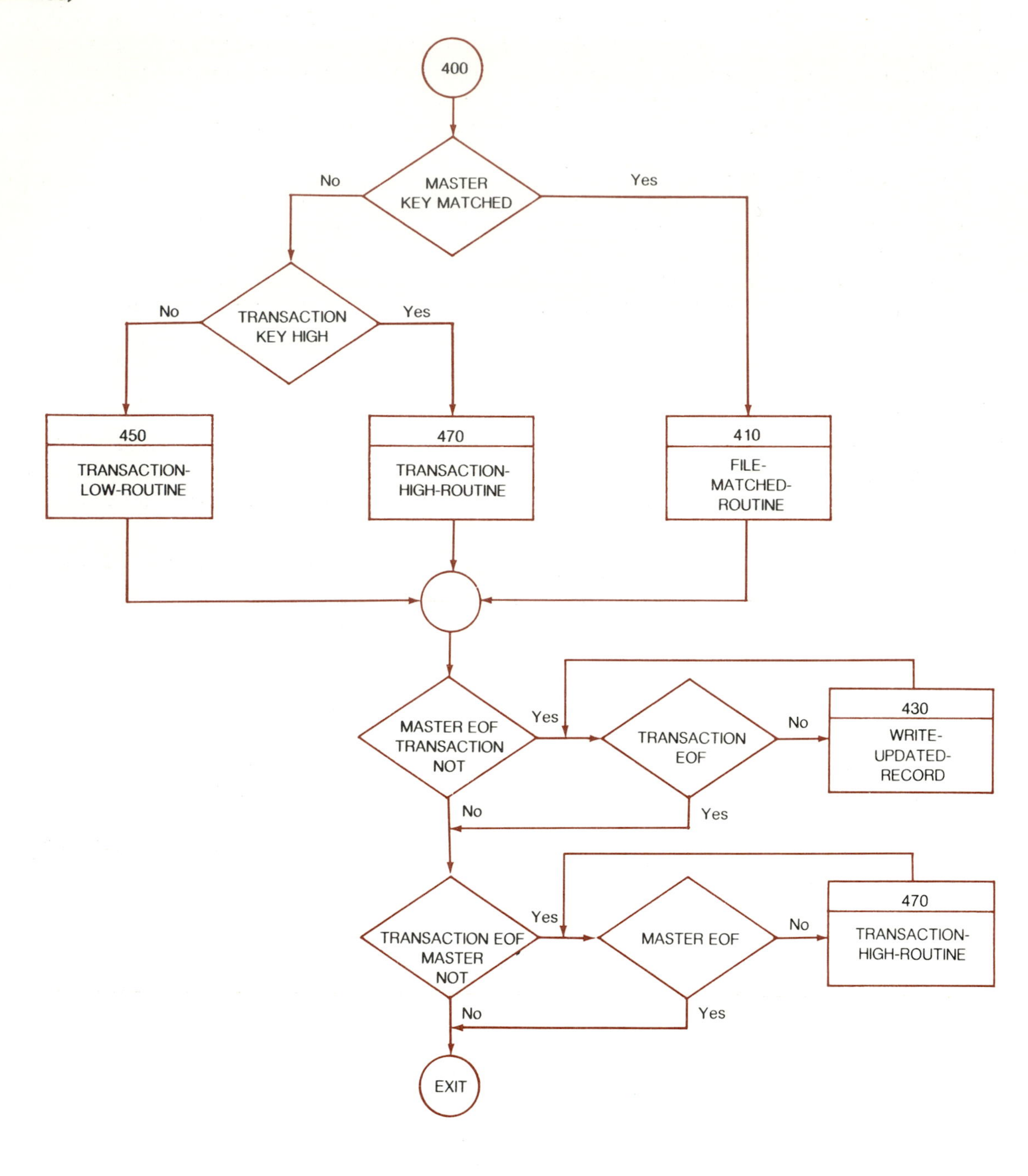

Note: Complete documentation of the system would require that an IPO and a program flowchart be made for each module on the hierarchical chart. Only three sets are shown. The student might do some of the remainder as an exercise. Also, the student might wish to write a program to update a real file. This would require that the documentation include the layout of the input files and print chart for the output report. Finally, after the program was tested and run, the sample output with the program would be added to the documentation.

The Design Procedure

Because structured design is a technique for designing programs, this assumes that a system design (application design) is already well defined, if not complete. Therefore, program requirements, record descriptions and so forth are already well defined. This is the point where the program design process begins. Starting with a well-defined problem statement, the structured design process involves a sequence of steps to define the functions required for the problem solution, functions that are described in terms of the problem itself. The steps in this process are performed in this order:

1. Sketch a *functional picture* of the problem.
2. Identify the *external, conceptual data streams,* both input(s) and output(s).
3. Identify the *major* external, conceptual data streams (one for input and one for output) and determine their *points of highest abstraction.*
4. Construct a *basic structure chart.*
5. Reiterate the aforementioned process until all functions are defined and the design process is complete.

Creating Structured Charts

Structured charts show the hierarchical structure of the various functions of the program according to the top/down design of structured programming. These charts serve as a basis for coding Procedure Division entries, as they show the overall logic of the program and the various levels of functions to be performed.

1. Module names should consist of one verb followed by one or two adjectives followed by one object. The module name together with the module number will form the paragraph name.
2. Because the module name together with the module number will form the paragraph name, the two should be no more than 30 characters.

Module Numbers

1. Module numbers should be written outside of the box.
2. The top level module number should be 000, which should be the overall function of the program.
3. All other modules should be broken down in 10s after the level 1 entries. (See figure 5.5.)

Using module numbers in the paragraphs in the Procedure Division will speed up the debugging process when tracing the execution of the object program.

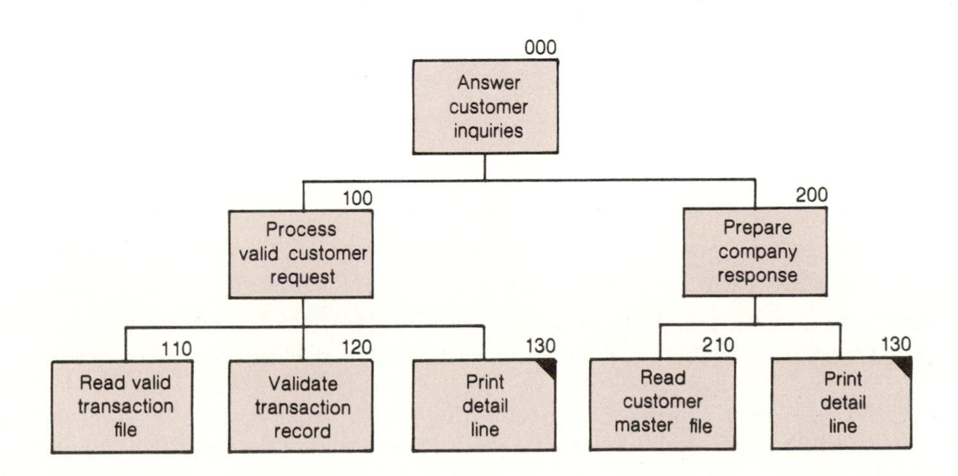

Figure 5.5 Structure chart—example.

Statements without Indentation

```
IF FINAL-MONTH
ADD LAST-INVOICE TO YEAR-TOTAL
IF STOP-ACCT
MOVE 12 TO STATUS-CODE
ELSE
MOVE 16 to STATUS-CODE
IF RENEW-ACCT
ADD REMAINING-AMT TO NEW-BALANCE
ELSE
ADD REMAINING-AMT TO FINAL-INVOICE
ELSE
ADD LAST-INVOICE TO MONTH-TOTAL.
```

Statements with Indentation

```
IF FINAL-MONTH
    ADD LAST-INVOICE TO YEAR-TOTAL
    IF STOP-ACCT
        MOVE 12 to STATUS-CODE
    ELSE
        MOVE 16 to STATUS-CODE
        IF RENEW-ACCT
            ADD REMAINING-AMT TO NEW-BALANCE
        ELSE
            ADD REMAINING-AMT TO FINAL-INVOICE
ELSE
    ADD LAST-INVOICE TO MONTH-TOTAL.
```

Identation can be a major benefit, as these skeleton programs show. Both do the same processing, but the second is far easier to understand and, therefore, to verify for correctness.

Figure 5.6 A nested IF statement, with and without indentation.

Indentation

The indentation of the program statements will make the structure more obvious to the reader of the program. This may make the program more difficult to code but it will greatly simplify the reading. When a program has to be maintained, it is the reading that becomes crucial to its success. Following such a practice also makes it much easier for another programmer to check a program for correctness.

The use of indentation is important because consistent indentation enhances readability so that the finished program exhibits in a pictorial way the relationships among statements. The basic idea is that all statements controlled by a control logic structure should be indented by a consistent amount, to show the scope of the control of the structure. In COBOL this means that the statements between the IF and the ELSE should be indented a consistent amount, and similarly for the statements between the ELSE and the next sentence. (See figure 5.6.)

Advantages

Use of the three classical structures of structured programming in their pure form results in inefficiency in two situations. This inefficiency is avoided through the use of a variant of the selection structure and a slight relaxation of the single/exit rule. The first situation is handled by computed GO TOs or switches (the CASE structure, where only one of a series of functions is to be performed, depending on the value of a variable). This is really a generalization of the selection process (IFTHENELSE) from a two-valued to a multivalued operation (fig. 5.7). The second situation arises when the programmer wishes to terminate a repetition block abnormally when the language does not explicitly allow this (fig. 5.8). Although such an abnormal termination violates the single/entry—single/exit rule of structured programming, it may produce a significant savings in space and time. If properly flagged, this practice maintains the spirit of structured programming.

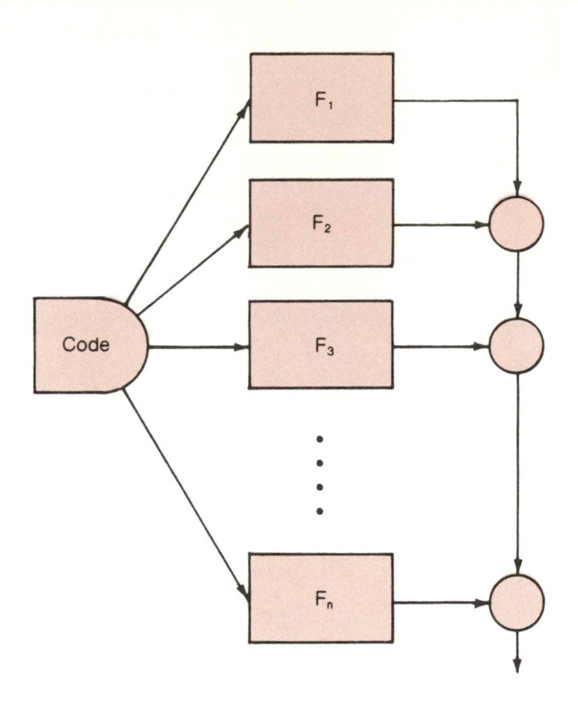

Figure 5.7 From a two-valued to a multi-valued operation.

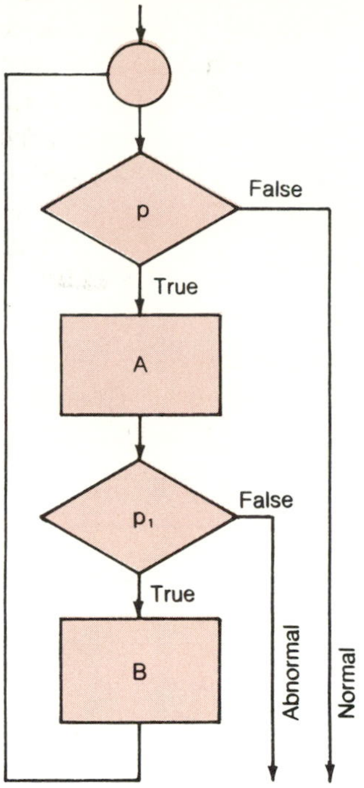

Figure 5.8 An abnormal termination of a repetition block.

Structured programming combined with some traditional coding practices, such as good annotation, descriptive labels, and judicious spacing in the source code, greatly clarifies source coding. This increased clarity, and the reduced complexity of structured programs are responsible for another advantage of structured programming: its correctness is more easily verified than that of unstructured programs. There are two senses in which this is true. First, because the flow of control is simpler in structured programs, the development and execution of test cases to adequately debug the program is simpler. Second, since the program is more understandable, its correctness is more easily verified by reading, that is, desk-checking. Compared to unstructured programs, structured programs are very easy to read and verify for correctness. The use of structured programming and of more desk-checking will, therefore, improve the quality of programs and reduce the cost of their development.

Using Pseudo Code

Pseudo code is an informal method of expressing structured programming logic. It is akin to COBOL but not bound by any formal language syntactical rules. The only conventions of pseudo code are those that pertain to the use of the structured figures and the indentation that aids in the visual perception of the logic. (In this text, flowcharts will be used in place of pseudo codes. However, pseudo codes are widely used and should be seriously examined for possible usage in COBOL programs.)

The primary purpose of pseudo code is to enable programmers to express their ideas in a natural form that uses English-language prose. The idea is that this will allow the programmer to concentrate on the logic solutions of the problem, instead of the form and constraints within which it must be stated. The intention is to help the programmer create an unambiguous solution to a problem.

As such, pseudo code may replace flowcharts and may be used instead of most program documentation. It can serve as a technical communication at all levels, from management review down to most detailed specifications from which code is directly written. (The purpose of stating the program design in pseudo code is "thought communication," not program specification.) (See figure 5.9.)

Regardless of whether or not a flowchart is created, pseudo code allows the programmer the freedom to express structured logic in an uninhibited manner, rather than in the rigid form of a programming language. Pseudo code expresses a program design that is both readable and easily converted to executable code.

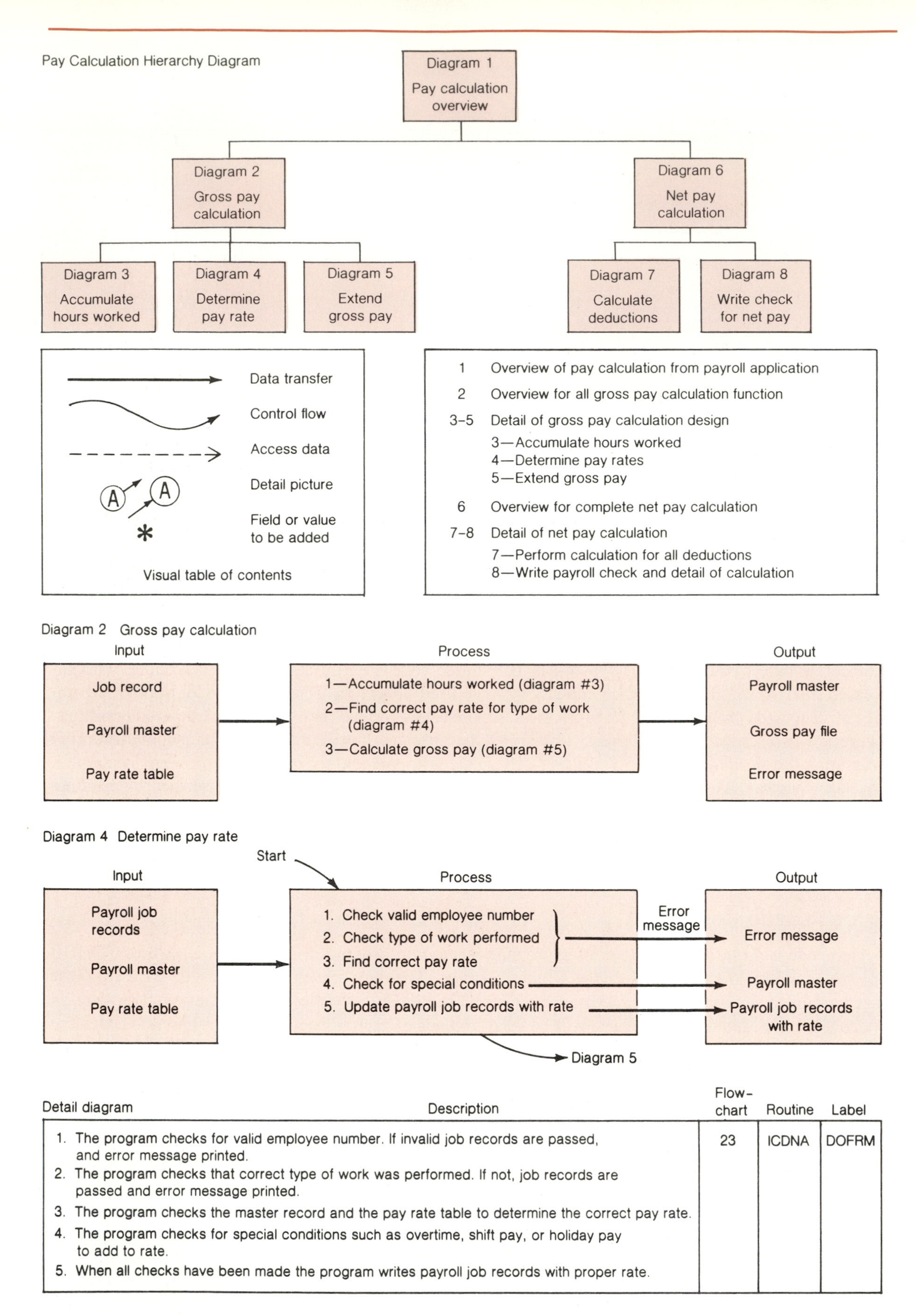

Detail diagram Description	Flow-chart	Routine	Label
1. The program checks for valid employee number. If invalid job records are passed, and error message printed. 2. The program checks that correct type of work was performed. If not, job records are passed and error message printed. 3. The program checks the master record and the pay rate table to determine the correct pay rate. 4. The program checks for special conditions such as overtime, shift pay, or holiday pay to add to rate. 5. When all checks have been made the program writes payroll job records with proper rate.	23	ICDNA	DOFRM

Figure 5.9 Conversion of hierarchy charts to IPO charts.

For example, consider the following pseudo code:

```
INITIALIZE THE PROGRAM
READ THE FIRST TEXT RECORD
DOWHILE THERE ARE MORE TEXT RECORDS
    DOWHILE THERE ARE MORE RECORDS IN THE TEXT RECORD
        EXTRACT THE NEXT TEXT WORD
        SEARCH THE WORD-TABLE FOR THE EXTRACTED WORD
        IF THE EXTRACTED WORD IS FOUND THEN
            INCREMENT THE WORD'S OCCURRENCE COUNT
        ELSE
            INSERT THE EXTRACTED WORD INTO THE TABLE AND SET COUNT TO 1.
        ENDIF
        INCREMENT THE WORDS-PROCESSED COUNT
    ENDDO (AT THE END OF THE TEXT RECORD)
    READ THE NEXT TEXT RECORD
ENDDO (WHEN ALL THE TEXT RECORDS HAVE BEEN READ)
PRINT THE TABLE AND SUMMARY INFORMATION
TERMINATE THE PROGRAM
```

The aforementioned relates to a word-frequency analysis program. It is designed to read and analyze a set of text records, preceded by a header card, extracting each individual word from the text records. The program creates and maintains a table, which maintains each unique word found in the text and a count of the number of times that the word occurs. When all text records have been processed, the contents of the table and appropriate statistics are printed.

The pseudo code describes each of the functions that are performed by the program. The programmer would have to translate the pseudo code into COBOL. That is, the programmer would have to rewrite this same program as follows:

```
PROCEDURE DIVISION.
WORD-FREQUENCY-MAIN-ROUTINE.
    PERFORM INITIALIZATION-PARAGRAPH.
    PERFORM TEXT-PROCESSING
        UNTIL THE LAST-RECORD-PROCESSED.
    PERFORM PRINT-TABLE-ROUTINE.
    PERFORM TERMINATION-PARAGRAPH.
    STOP RUN.
INITIALIZATION-PARAGRAPH.
    OPEN    INPUT    INPUT-FILE,
            OUTPUT   PRINT-FILE.
    READ INPUT-FILE INTO HEADING-LINE-TEXT,
        AT END PERFORM NO-INPUT-ERROR-ROUTINE.
    MOVE 0 TO WORDS-PROCESSED.
    PERFORM INITIALIZE-WORD-FREQUENCY-TABLE.
    READ INPUT-FILE,
        AT END MOVE '1' TO LAST-RECORD-PROCESSED-SWITCH.
TEXT-PROCESSING.
    PERFORM EXTRACT-NEXT-WORD-FROM-SOURCE.
    SEARCH TABLE-ENTRIES
        WHEN EXTRACTED-WORD IS NOT GREATER THAN
            TABLE-WORD (TABLE-INDEX)
            PERFORM INSERT-NEW-ENTRY-OR-INCR-COUNT.
PRINT-TABLE-ROUTINE.
    MOVE WORDS-PROCESSED TO HI-WORDS-PROCESSED.
    MOVE NUMBER-OF-UNIQUE-WORDS TO HI-NUMBER-OF-WORDS.
    MOVE 99 TO PRINT-LINE-COUNTER.
    PERFORM TABLE-PRINT-LOOP
        VARYING TABLE-INDEX FROM 1 BY 1
        UNTIL TABLE-INDEX = NUMBER-OF-ENTRIES.
TERMINATION-PARAGRAPH.
    CLOSE   INPUT-FILE,
            PRINT-FILE.
```

Note: The COBOL program on the preceding page is not complete. Many of the lower level paragraphs that are PERFORMed were omitted to conserve space. This example is meant to demonstrate the difference in pseudo coding and the actual COBOL programming codes.

The main goal of pseudo code is readability. It is an informal method of developing structured programs that solve the logic of the programmer's problem.

Pseudo code is a noncompliable quasi-code that requires program translation to a compliable language prior to execution. The primary reason for using pseudo code is to enable the programmer to express ideas in a highly readable material form using English-language prose.

Language Implementation

Computer programs can be written with a high degree of structure, which permits them to be more easily understood for testing, maintenance and modification. With structured programming, control branching is entirely standardized so that code can be read from top to bottom, without having to trace the branching logic as is typical for code generated in the past.

In structured programming, programmers must think deeper, but the end result is easier to read, understand, and maintain. Another advantage is the additional program design work that is required to produce such structured code. The programmer must think through the processing problem, not only writing down everything that is needed to be done, but writing it down in such a way that there are no after thoughts with subsequent jump-outs and jump-backs, and no indiscriminate use of a section of code from several sections because it "just happens" to do something at the time of the coding. Instead, the programmer must think through the control logic of the module completely at one time, in order to provide the proper structural framework for the control. This means that programs will be written in a much more uniform way because there is less freedom from arbitrary variety.

Such a program is much easier to understand than an unstructured logic jumble; readability has been improved. Because of its simplicity and clear logic, it minimizes the danger of the programmer's overlooking logical errors during implementation; reliability has been improved. And improved readability, in combination with the greater simplicity obtained by structured programming naturally leads to improved maintainability. Also, because structured code is simple, a programmer can control and understand a much larger amount of code.

Pseudo code expresses clearly and simply the logical solution to a data processing problem. The pseudo code is not compliable on a computer; that is, the programmer cannot submit the pseudo code directly to a computer for compilation and execution. The programmer must first translate this pseudo code into a machine-compliable, machine-executable programming language such as COBOL.

As the programmer gains experience in using structured programming techniques, he or she may successfully be able to code the solution in COBOL directly from the flowcharts omitting the pseudo code step. (However, don't forget the benefits inherent in the creation of pseudo code. Two such benefits are: (1) communication of the program logic to a non-technical person, and (2) the ability to state the logic unencumbered by the programming language rules and constraints.)

In COBOL, the structured programming techniques are primarily concerned with the Procedure Division and pertain basically to the application of the structured program figures. Four program figures are recommended for use in COBOL. They are:

1. SEQUENCE
2. IFTHENELSE (Selection)
3. DOWHILE (Iteration)
4. CASE (Multiway branches)

As we discuss each section of the Procedure Division, the relevant program figure will be explained and illustrated.

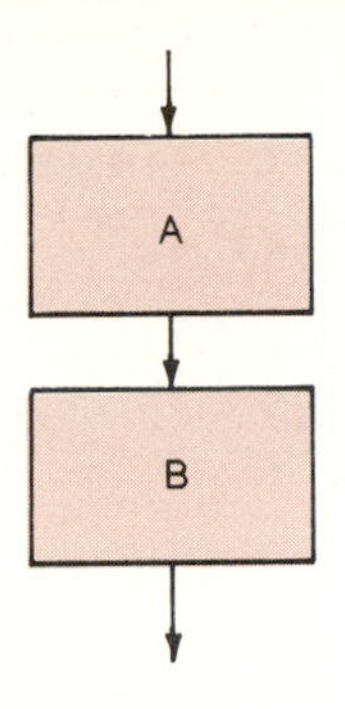

Figure 5.10 Flowchart for sequence program structure.

COBOL Implementation of Structured Programming

Once the principles of structured programming are understood, writing structured programs in COBOL is a matter of habitually following a few simple rules. The structure theorem states that any *proper program* can be written using only the control logic structure of *sequence, selection,* and *iteration (loop mechanism).* A proper program is defined as one that meets the following two requirements:

1. It has exactly one entry point and one exit point for program control.
2. There are paths from the entry to the exit that lead through every part of the program; this means that there are no infinite loops and no unreachable code. This requirement is, of course, no restriction but simply a statement that the structure theorem applies only to programs of consequence.

All control logic structures can be expressed in COBOL although in some cases not directly.

Basic Control Logic Structures

Sequence is simply a formularization of the idea that unless otherwise stated, program statements are executed in the order in which they appear in the program. This is true of all commonly used programming languages; it is not always realized that sequence is in fact a control logic structure.

Sequence is implemented in the COBOL language by simply writing statements in succession. (See figure 5.10.)

Selection is the choice between two actions based on a condition (predicate); this is called the IF-THENELSE structure.

In COBOL selection is implemented with the IF statement and the conditional statements. (See figure 5.11.)

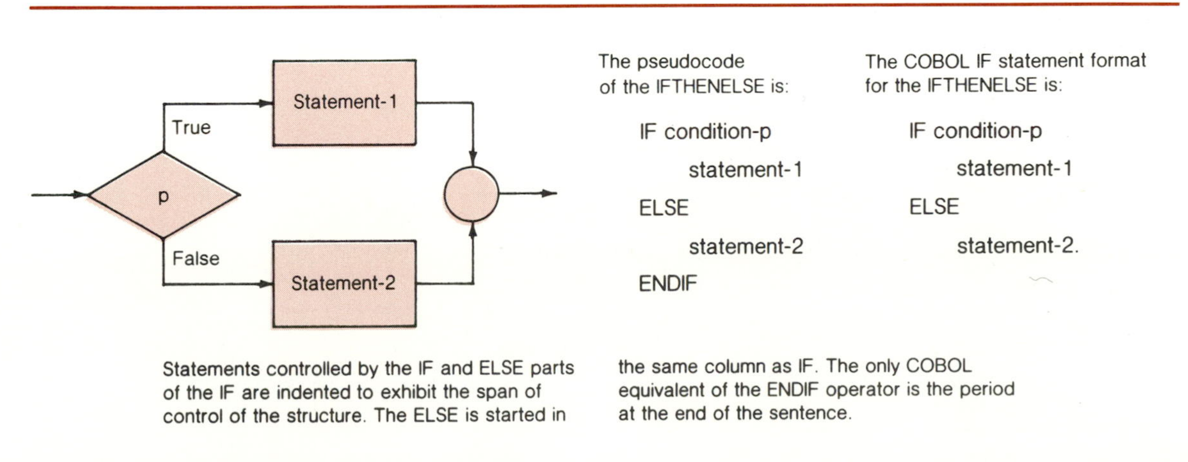

Statement-1
True
p
False
Statement-2
The pseudocode of the IFTHENELSE is:
IF condition-p
statement-1
ELSE
statement-2
ENDIF
The COBOL IF statement format for the IFTHENELSE is:
IF condition-p
statement-1
ELSE
statement-2.
Statements controlled by the IF and ELSE parts of the IF are indented to exhibit the span of control of the structure. The ELSE is started in the same column as IF. The only COBOL equivalent of the ENDIF operator is the period at the end of the sentence.

Figure 5.11 IFTHENELSE structure—example.

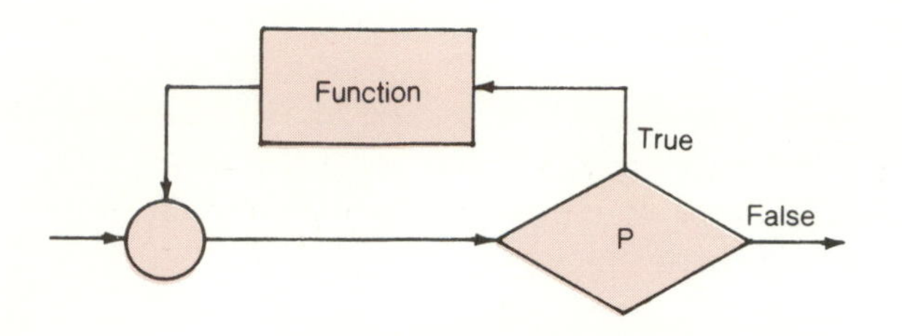

The pseudocode for the DOWHILE is:

DOWHILE p
function
ENDDO

The COBOL implementation of the DOWHILE involves the PERFORM verb with an UNTIL phrase.

PERFORM paragraph-name
UNTIL (NOT p)

The flowchart for the COBOL implementation of the DOWHILE is shown here. Since the PERFORM . . . UNTIL repeats execution of the named procedure until the condition is true, the opposite condition is tested, and the true and false labels on the predicate are reversed from the flowchart of the DOWHILE shown. In actual practice, it may be preferable to code the inverse operators. For example, if p is (A IS EQUAL TO B), then the inverse (A IS NOT EQUAL TO B), is used rather than (NOT [A IS EQUAL TO B]).

Figure 5.12 DOWHILE structure—example.

The *iteration* structure is used for repeated execution of code while a condition is true (or false). It is also referred to as the loop control. The iteration structure is called the DOWHILE structure (See figure 5.12.)

In COBOL, the DOWHILE structure is implemented with the PERFORM verb with the UNTIL option.

It is sometimes helpful—from both readability and efficiency standpoints—to have some way to express a multiway branch commonly referred to as the CASE structure. For example, it is necessary to execute appropriate routines based on a two-digit decimal code, and it is certainly possible to write 100 IF statements, or a compound statement with 99 ELSE IF statements, but common sense suggests that there is no reason to adhere so rigidly to the three basic structures. (See figure 5.13.)

The CASE structure uses the value of the variable to determine which of several routines is to be executed. Efficiency and convenience dictate reasonable use of language elements that may carry out logic functions in ways slightly different from those of the three basic structures.

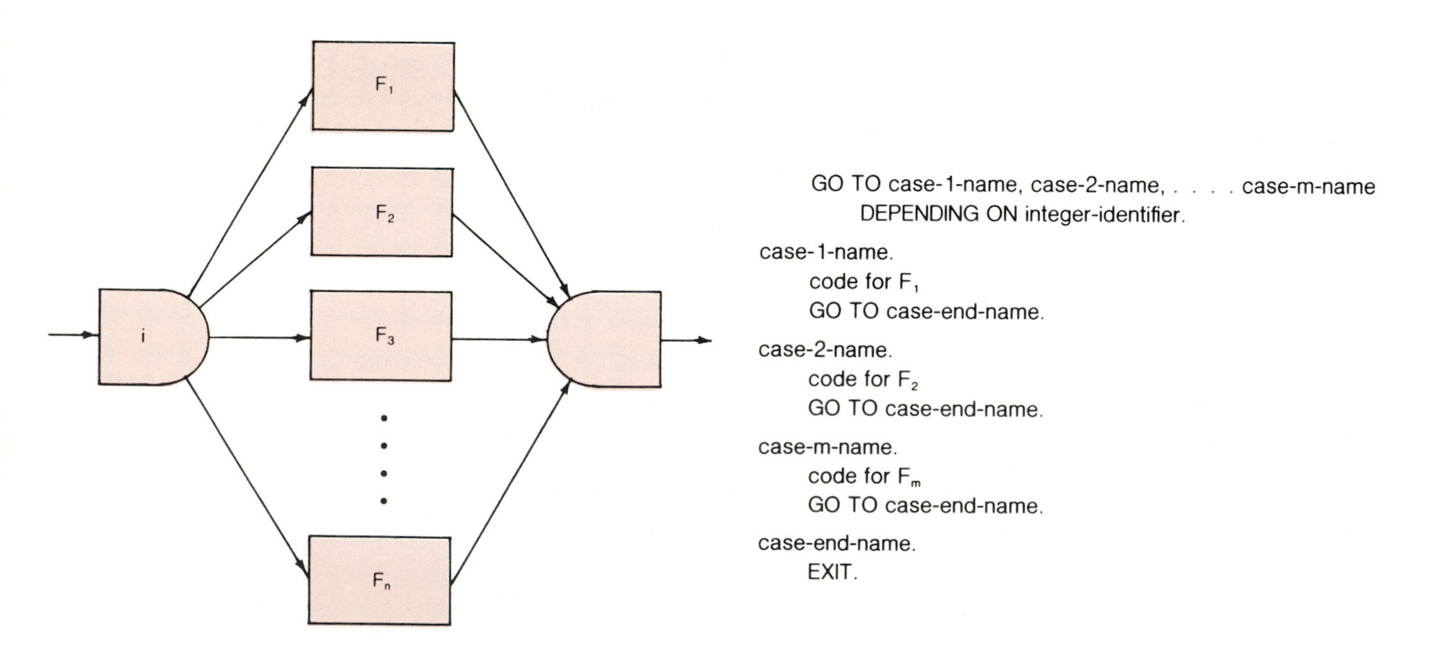

F1
F2
i
F3
Fn
GO TO case-1-name, case-2-name, ... case-m-name
DEPENDING ON integer-identifier.
case-1-name.
code for F1
GO TO case-end-name.
case-2-name.
code for F2
GO TO case-end-name.
case-m-name.
code for Fm
GO TO case-end-name.
case-end-name.
EXIT.

Figure 5.13 CASE structure—example with pseudo code.

Two of the standard programming structures—IFTHENELSE and DOWHILE—are implemented directly through COBOL statements. The CASE structure is simulated by conventional COBOL statements in particular structural forms.

Coding Standards

The standards described here represent recommendations based on experience to date in writing structured programs in COBOL. They are all devised to satisfy the same basic goals; to produce programs that are easy to read, easy to understand, and easy to maintain and modify.

Program Organization

These standards provide for the organization of a COBOL source program into a set of modules that are easily linked for compilation. (This use of the term *module* refers to a block of code that becomes part of a program for compilation directly as input to the computer, is COPYed into the program, or is a linked part of the COBOL compiler's primary data set. It is specifically not a reference to the COBOL reserved word "SEGMENT").

Any COBOL program requires a certain ordering of statements within the program. This ordering is further restricted to that of below for the sake of readability, clarity, and consistency. The first module of code in the program must contain the Identification and Environment Divisions. The second module should contain the Data Division, which may include COPY statements for detailed data specifications and is the only module in the program that may exceed one page in the source program listing—about fifty lines—including blank lines. Succeeding modules should contain the Procedure Division, which is made up of modules consisting of sections and paragraphs.

The construction of a module of the Procedure Division is as follows: At the head of the first module of the Procedure Division, the programmer will place the main line coding for the program. The code must be complete within this module. If the main line coding contains any PERFORMs, the placement of the PERFORMed routine is determined by the reason for coding the PERFORM.

1. If the PERFORM is used to invoke a large block of code, or code too complex to be physically inserted at the location of the PERFORM, then the PERFORMed paragraph will be placed in the next lower level module.
2. If the PERFORM statements are being used to implement a structured element (IFTHENELSE, or DOWHILE), then the PERFORMed paragraph is considered to be at the same logical level as the statements in the current module and, therefore, the PERFORMed paragraph will be placed in the current module after the main body of code.

If these paragraphs (the ones PERFORMed to implement structured elements) themselves PERFORM other paragraphs, then the "next-lower level" paragraphs will be placed in the current or lower module based upon the aforementioned paragraph placement rules. Thus, all code for a single logical level will be in the same module, and subordinate code will be in lower level modules. Hence, the number of modules in a program will indicate the level of complexity. Additionally, any PERFORMed paragraph will be found close to the PERFORMing statement and behind it, thereby continuing the philosophy that programs should be readable from the top down.

The modules of the Procedure Division should appear in some logical sequence, such as the order in which they are invoked, or a sequence that reflects their position in the logical hierarchy of the program according to the structure codes.

Coding Line Spacing

The following are *only suggestions.* No standardization of spacing and indentation has been developed as yet, and there seems to be little pressure for such standardization so long as consistent standards are followed within any one organization.

The key idea in devising helpful indentation is the production of programs in which the visual layout of the program elements aids the reader in understanding program relationship and function.

1. PERFORMed paragraphs will be separated from the main body of code above by blank line(s).
2. Logically noncontiguous paragraphs (other than those used in the CASE structure) will be separated by a blank line.

The free use of blank lines can exhibit more clearly that relationships exist among items so grouped.

Module Size

Each module of the Procedure Division should consist of no more than fifty lines of coding (approximately one page of the compiler listing).

Indenting and Formatting Standards

The purpose of these standards is to make all structured programs similar in format and to graphically highlight the structuring.

1. Paragraph names will begin in position 8 (except for the CASE variation) and will be the only coding on the line.
2. The first statement in every paragraph will begin in position 12. (Paragraph and section names are easier to locate if they are always written on a separate line, that is, if the first statement following a paragraph name begins on the next line.)
3. Any statement not subject to indentation rules will start in the same positions as the statements above it.
4. Only one statement per line should be written. Verbs are easier to locate, and programs are easier to correct and modify, if two statements are never written on the same line.
5. Any statements requiring more than one line should be indented on successive lines. Statements are easier to pick out if second and following lines of a continued statement are indented by some consistent amount of spaces, such as four spaces.
6. Options used with normal verbs will be coded separately on successive lines with indentation. These options are:

```
GIVING              AT END           UNTIL
ON SIZE ERROR       INVALID KEY      VARYING
DEPENDING ON        UPON             BEFORE
                                                    ADVANCING
AFTER               TALLYING         AFTER
TIMES               REPLACING
```

   All Sort Keys and Sort Options should be indented on separate lines. Important clauses of certain COBOL verbs are better set off if they are written on separate lines, indented by whatever amount of space is chosen as standard for continuation lines. For example, when the UNTIL and/or VARYING options are used with the PERFORM's, they can be written on separate lines and indented.

```
PERFORM PRINTOUT
    VARYING LINE-NUMBER FROM 1 BY 1
    UNTIL LINE-NUMBER IS GREATER THAN 50.
```

   The VARYING, AT END, and WHEN clauses of a SEARCH verb can be written on separate lines and indented:

```
SEARCH RATE-TABLE
    VARYING MAN-IDENTIFICATION
    AT END MOVE 'T' TO NO-FIND-SW
    WHEN RATE-MF = RATE-TABLE (RATE-INDEX)
```

7. Various statements are easier to read if appropriate portions of the statements are vertically aligned. For example,

```
OPEN     INPUT    TAPE-IN
                  FILE-IN
         OUTPUT   TAPE-OUT
                  FILE-OUT.
```

8. For multiple field operations (MOVE ZEROS TO FLD-A, FLD-G), place each successive field on a separate line directly below the preceding field. For example,

```
MOVE ZEROS TO FLD-A,
              FLD-G.
```

Characteristics of Structured COBOL Programming

1. Absence of GO TO statements and corresponding statement labels.
2. Inclusiveness of PERFORM statements, which result in the performance and return from another routine (closed subroutine).
3. One coding module per page.
4. One entry and one exit per module—enter at top and exit at bottom.
5. Special attention to nesting function (discussed later in text).
6. Special indentation and nesting of functions.
7. Use of names that convey intelligence about the data.
8. A limited number of statement types.
9. A reduction in the use of comments.

Conclusions

The first few programs using structured programming techniques may be somewhat difficult. However, once the learning phase is over, the coding should be faster using structured programming and associated techniques (e.g., top-down design, etc.) than with previous programmer methods. Proper use of the structured programming coding standards is the responsibility of the individual programmer with enforcement being provided through subsequent review of the programs. Study through the following problem, which is worked out using structured design.

Inventory Control Problem Using Structured Design

Problem Statement

A certain distributor company controls their product inventory with a data processing application. A record of each product distributed is contained on a direct access master file. Each day, orders received from customers are processed against the inventory master file to reflect that day's business on their product inventory. During this process, inventory master records are updated to show new quantities on hand, back-order situations (orders that can't be completely filled) are identified and reorder requirements (for any product whose quantity falls below a pre-specified level) are identified. Also during processing, sales statistics are updated by product and for classes of products and a report showing the day's activities is printed (the Order Status listing).

Functional Requirement Specifications

The Product Transaction File is in random sequence, with each transaction identifying a product, its quantity ordered and other related data.

Presented in the following illustrations are record layouts and a layout of the Order Status listing. Using these aids, design, using structured design techniques, a program solution to do the following:

1. For each transaction and its associated master record:
   a. Use the PRODUCT-NUMBER in the transaction record to find the product's master record. If the master record is not found, display an indication of the transaction PRODUCT-NUMBER on the console typewriter and continue with the next transaction.
   b. Reduce the master record quantity ON-HAND-M by the quantity on the transaction record.
   c. If ON-HAND-M plus ON-ORDER-M falls below the REORDER-LEVEL-M, issue a reorder indication on the listing and update ON-ORDER-M.
   d. Using the master record's UNIT-PRICE and the transaction quantity, update TOTAL-SALES on the master record.
   e. If ON-HAND-M (on the master record) goes to zero and a transaction quantity cannot be allocated, update BACK-ORDERS-M and indicate the quantity back ordered on the listing.
   f. Update the DATE-LAST-ACTIVE in the master record with the date in the transaction record.

2. For the Order Status listing:
   a. Print detail line as indicated in the following illustration. If no back-order or reorder conditions exist, blank out these positions.
   b. Accumulate a total of AMOUNT for each page and list as indicated.
   c. Accumulate a total for all transactions to be listed at the end of the report.
   d. At the end of the report, produce a TOTALS-BY-CATEGORY list as indicated in the illustration. Each master record contains a product class varying from 1 to 5. Build a table to accumulate sales by product code and update the table for each transaction. This table will then be listed under TOTALS-BY-CATEGORY.

Record Layouts

Transaction Record (Transaction for a Product)

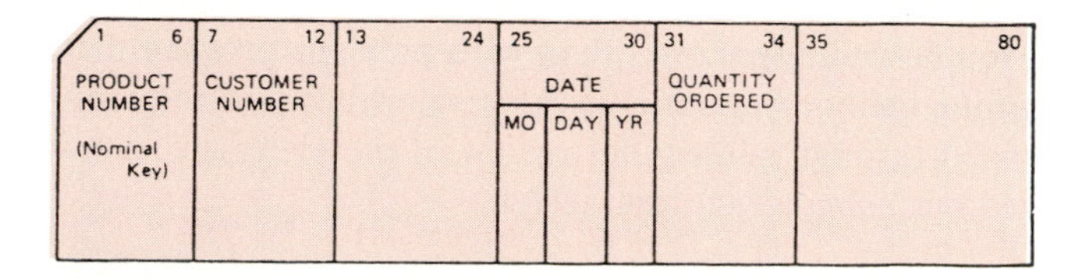

I-S-Record (Master Record for each Product)

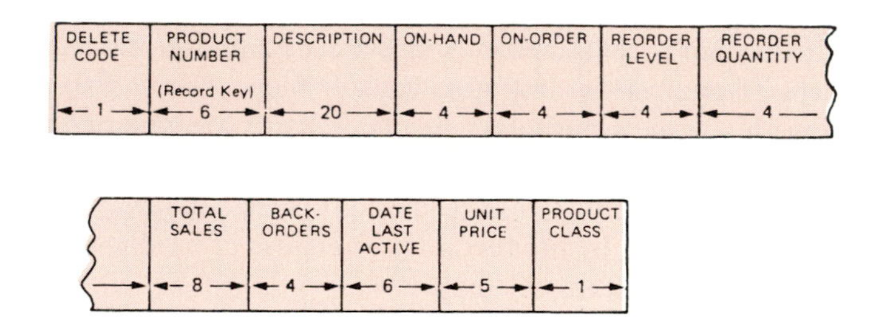

Order Status Listing

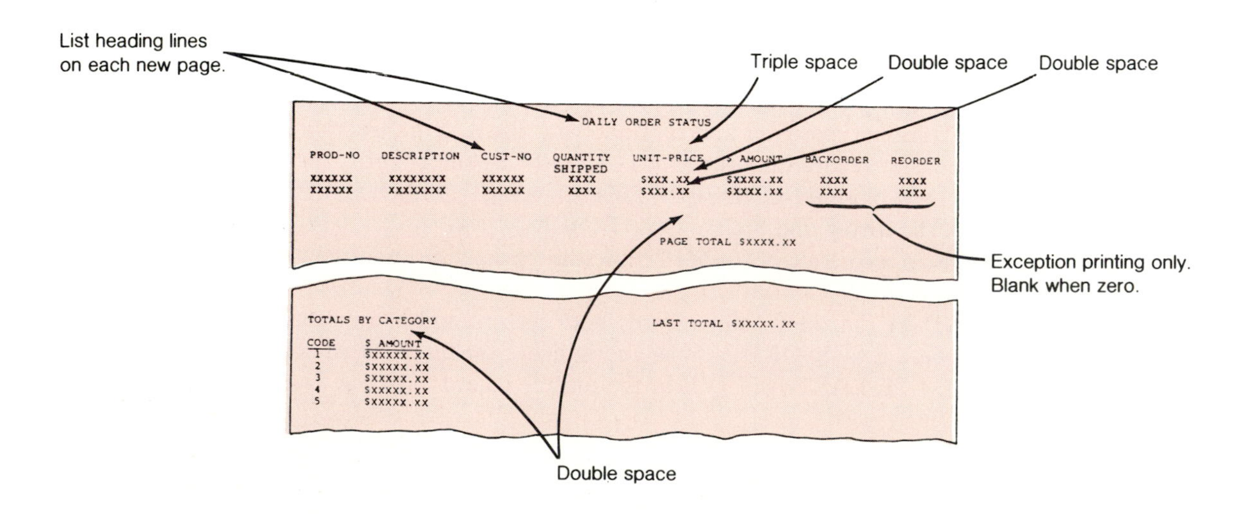

There are many variations with which one could view a functional sketch of INVENTORY CONTROL. One variation follows:

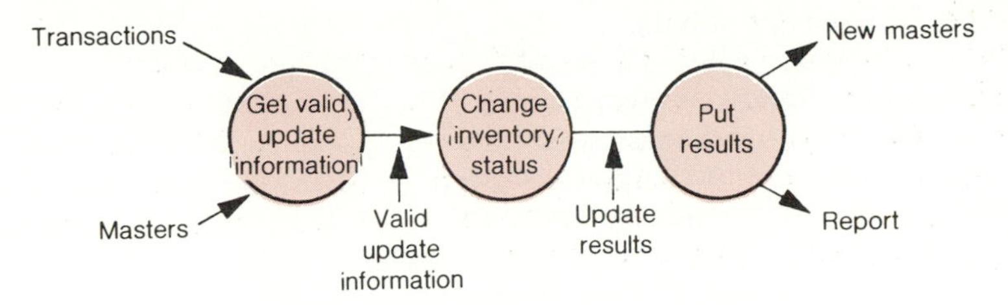

This functional sketch tends to simplify the fact that, in this problem there are two major input streams (transactions and masters) and two major output streams (updated masters and a printed report). A functional sketch that highlights these factors follows:

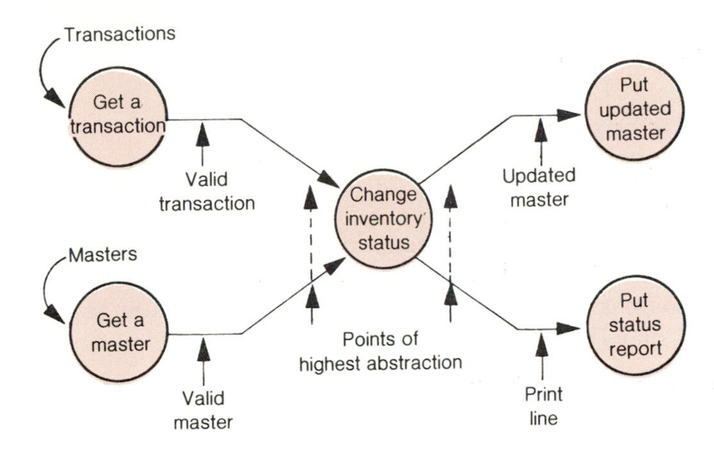

A basic structure chart that depicts the functions in the aforementioned sketch follows:

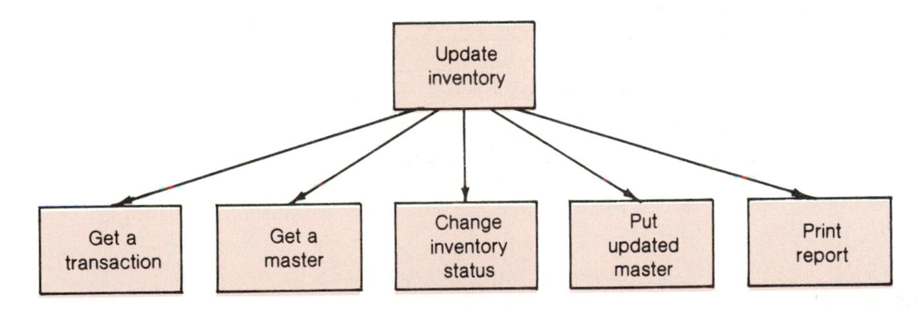

Expanding this structure chart to include subordinate functions, a final version of the design results:

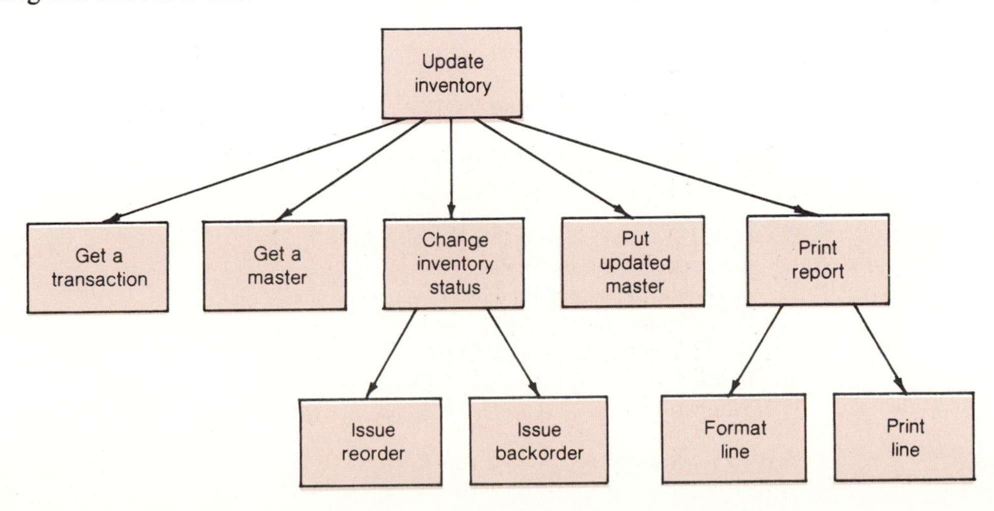

A solution for the top module pseudo code of INVENTORY-CONTROL follows:

```
BEGIN UPDATE-INVENTORY
    INCLUDE Initialization
    DOWHILE more transactions
        INVOKE Get-a-transaction
        IF more transactions THEN
            INVOKE Get-a-master
            IF master-found THEN
                INVOKE Change-inventory-status
                INVOKE Put-updated-master
                INVOKE Print-report
            ELSE
                Display no master found for product-number.
            ENDIF
        ELSE
            Null
        ENDIF
    ENDDO
    INCLUDE Termination
END UPDATE-INVENTORY
```

The solution should be detailed enough to show the control logic within the top module for INVENTORY CONTROL because that is the primary function of this module. Different function names may be used here. Notice that this solution does not use a "primary read function" before the DOWHILE; therefore, a conditional test for more transactions must follow the read function. Also notice the test for master found after INVOKE Get-a-master. This level of detail did not appear in the text structure, although it is a necessary check in the implementation.

The following is pseudo code for the lower-level modules in INVENTORY CONTROL.

```
BEGIN INITIALIZATION
    OPEN transaction and master files
    OPEN print file
    Set new-page switch ON
    Set more-transactions switch ON
    Set master-found switch ON
    Initialize page-total, final-total and line-count to zero
END INITIALIZATION
```

Note that other data initializations, etc. may become apparent requirements for the INITIALIZATION function as other modules are coded.

```
BEGIN GET-A-TRANSACTION
    Read transaction-file
    IF end-of-file THEN
        Set more-transactions OFF
    ELSE
        Null
    ENDIF
END GET-A-TRANSACTION.
```

Since GET-A-TRANSACTION is small it may be desirable to implement this function within UPDATE-INVENTORY when going to source code.

```
BEGIN GET-A-MASTER
    Set master-found switch ON
    Read indexed sequential master file using transaction key.
    IF no master is found THEN
        Set master-found switch OFF
    ELSE
        Null
    ENDIF
END GET-A-MASTER
```

GET-A-MASTER may also be included in its calling function when going to source language because it is small.

```
BEGIN CHANGE-INVENTORY-STATUS
    IF (quantity-on-hand > zero) THEN
        IF (quantity-on-hand − quantity ordered > zero) THEN
            update quantity-on-hand
            quantity-shipped = quantity-ordered
        ELSE
            quantity-shipped = quantity-on-hand
            quantity-on-hand = zero
            INVOKE Backorder
        ENDIF
    ELSE
        quantity-shipped = 0
        INVOKE Backorder
    ENDIF
    update date-last-active
    calculate value-of-sale
    update product-total-sales
    update sales by product-class
    IF ((quantity-on-hand + quantity-on-order) < reorder-level) THEN
        INVOKE Reorder
    ELSE
        Null
    ENDIF
END CHANGE-INVENTORY-STATUS
```

Notice above, how the pseudo code becomes more detailed. The Backorder and Reorder modules will be shown later after the pseudo code for all modules on this level.

```
BEGIN PUT-UPDATED-MASTER
    Write updated-master using product key.
END PUT-UPDATED-MASTER
```

The small amount of pseudo code just given indicates again the probability of placing this function's source code in the top module.

```
BEGIN PRINT-REPORT
    IF new-page is ON THEN
        move page total to output-line
        print output line
        update final-total
        set page-total to zero
        set new-page to OFF
        INVOKE Print-headings
    ELSE
        Null
    ENDIF
    move detail-line to output-line
    print output-line
    update page-total
    add 1 to line-count
    IF line-count = 40 THEN
        Set new-page ON
    ELSE
        Null
    ENDIF
END PRINT-REPORT
```

Notice that the pseudo code is somewhat different from that indicated in the structure chart.

```
BEGIN TERMINATION
    INVOKE Print-headings
    print final-total
    print totals-by-category table
    close files
END TERMINATION
```

The remaining three functions were referenced by second level modules.

```
BEGIN BACKORDER
    Calculate backorder quantity
    update total backorders on master
    move backorder-quantity to detail-output-line
END BACKORDER

BEGIN REORDER
    update on-order quantity on master
    move reorder quantity to output-line
END REORDER

BEGIN PRINT-HEADINGS
    move header-1 to output-line
    print output-line
    move header-2 to output-line
    print output-line
END PRINT-HEADINGS
```

The pseudo code will be used for coding the Procedure Division. The dashed line separates the top module from the rest of the program.

```
PROCEDURE DIVISION.
MAIN-PROCESS.
    PERFORM INITIALIZE.
    PERFORM READ-AND-PROCESS
        UNTIL END-OF-TRANSACTIONS.
    PERFORM TERMINATE.
    STOP RUN.

READ-AND-PROCESS.
    READ TRANS-FILE INTO TRANS-REC
        AT END MOVE '1' TO END-OF-TRANSACTIONS-SWITCH.
    IF NOT-END-OF-TRANSACTIONS
        PERFORM GET-A-MASTER
        IF MASTER-FOUND
            PERFORM CHANGE-INVENTORY-STATUS
            PERFORM PUT-UPDATED-MASTER
            PERFORM PRINT-REPORT
        ELSE
            MOVE '0' TO MASTER-FOUND-SWITCH
            DISPLAY 'NO MASTER FOUND FOR PRODUCT NO', PROD-NO
                UPON CONSOLE.

GET-A-MASTER.
    MOVE PROD-NO TO FILE-KEY.
    READ MASTER-INV INTO MASTER-REC
        INVALID KEY MOVE '1' TO MASTER-FOUND-SWITCH.
```

```
INITIALIZE.
    MOVE '0' TO END-OF-TRANSACTIONS-SWITCH.
    MOVE '0' TO MASTER-FOUND-SWITCH.
    OPEN     INPUT     TRANS-FILE
             I-O       MASTER-INV.
    DISPLAY 'INITIALIZE CALLED'
        UPON CONSOLE.

CHANGE-INVENTORY-STATUS.
    DISPLAY 'CHANGE-INVENTORY-STATUS CALLED',
        UPON CONSOLE.

PUT-UPDATED-MASTER.
    DISPLAY 'PUT-UPDATED-MASTER CALLED',
        UPON CONSOLE.

PRINT-REPORT.
    DISPLAY 'PRINT-REPORT CALLED',
        UPON CONSOLE.

TERMINATE.
    DISPLAY 'TERMINATE CALLED'.
        UPON CONSOLE.
    CLOSE      TRANS-FILE
               MASTER-INV.
```

Summary

Structured programming is the design, writing and testing of a program made up of interdependent parts in a definite pattern or organization. A structured program tends to be much easier to understand than programs written in other styles because of the use of only three basic structures: (1) *Simple sequence (sequence)*—In the absence of instructions to the contrary, statements are executed in the order that they are written; (2) *IFTHENELSE (selection)*—combine with statements (begin and end) so that groups of elements can be included in the THEN and ELSE statements; (3) *DOWHILE (repetition)*—a loop-control mechanism.

Each structure is characterized by a simple and single point of transfer of control into the structure and a single point of transfer of control out of the structure.

A *proper program* has no infinite loops and no unreachable code.

GO TO statements are used only in certain special instances. GO TO-free programs allow programs to be read in sequence from top to bottom without skipping around. GO TO statements are replaced by PERFORM statements.

The design of a program should proceed from top to bottom. Also, individual program modules should be as short as possible.

Structured design is a set of program design techniques and guidelines that are used to specify the functions in a program solution, the data processed by each function and the relationships among the functions. The overall result of the structured design is documented in a structured chart that can be implemented using flowcharts and/or pseudo code, structured programming techniques, top/down programming and modulation.

The structured design process involves a sequence of steps to define the functions required for the problem solution, functions that are described in terms of the problem itself.

Structured charts show the hierarchical structure of the various functions of the program according to the top/down design of structured programming. These charts serve as a basis for coding Procedure Division entries, as they show the overall logic of the program and the various levels of functions to be performed.

Module numbers are combined with module names on structured charts to form the paragraph names. Using module numbers in the Procedure Division will speed up the debugging process when tracing the execution of the object program.

The indentation of program statements will make the structures more obvious to the reader.

The main advantages of structured programming are: its correctness is more easily verified than that of unstructured programs because the flow of control is more easily verified than that of unstructured programs, and the program is more understandable, with its correctness easily verified by the reader.

Pseudo code is an informal method of expressing structured program logic. The primary purpose of pseudo code is to enable programmers to express their ideas in a natural form that uses English language prose. Pseudo code is a noncompilable quasi-code that requires program translation to a compilable language prior to execution. As the programmer gains experience using structured programming techniques, he/she may be able to code the solution directly in COBOL directly from the flowcharts omitting the pseudo code.

A proper program has the following requirements: (1) it has exactly one entry point and one exit point for program control; and (2) there are paths from the entry to the exit that lead through every part of the program.

All control logic structures can be expressed in COBOL; although in some cases not directly. They are: (1) *Sequence*—is implemented by simply writing statements in succession; (2) *Selection*—is implemented through the use of the IF statement and the conditional statements; and (3) *Iteration*—is implemented through the use of the PERFORM statement with the UNTIL option.

The CASE control logic structure is used to express a multiway branch and uses the value of a variable to determine which of several routines to execute. The CASE structure is simulated by conventional COBOL statements in particular structured form.

Every COBOL program requires a certain ordering of statements within the program. The first module (block of code) must contain the Identification and Environment Divisions. The second module should contain the Data Division. The construction of the Procedure Division is as follows: the first module of the Procedure Division shows the main line of coding containing the overall logic of the program. The PERFORMs in this main line coding are determined by the sequence of procedures to be performed according to levels. The modules of the Procedure Division should appear in some logical sequence such as the order in which they are invoked, in a sequence that reflects their position in the logical hierarchy of the processing according to the structured charts.

Helpful indentation in the production of programs in the form of visual layouts of the program elements aids the reader in understanding program relationships and functions. The purpose behind indentation is to make all structured programs similar in format, and also to highlight the structures graphically.

The main characteristics of structured programs are: (1) the absence of GO TO statements; (2) inclusiveness of PERFORM statements; (3) one coding module per page; (4) special indentations; (5) limited number of statement types; (6) intelligent names; and (7) reduction in the use of comments.

Coding should be faster using structured programming and associated techniques once the learning phase is over.

Illustrative Program: Two-Level Control Total Processing

Sample of Output Required

41	$203.37			
52	$110.00			
69	$134.65			
		1	$448.02	
18	$207.69			
32	$185.60			
		2	$393.29	
36	$194.15			
39	$121.40			
50	$51.80			
		3	$367.35	$1,208.66

Hierarchy Chart

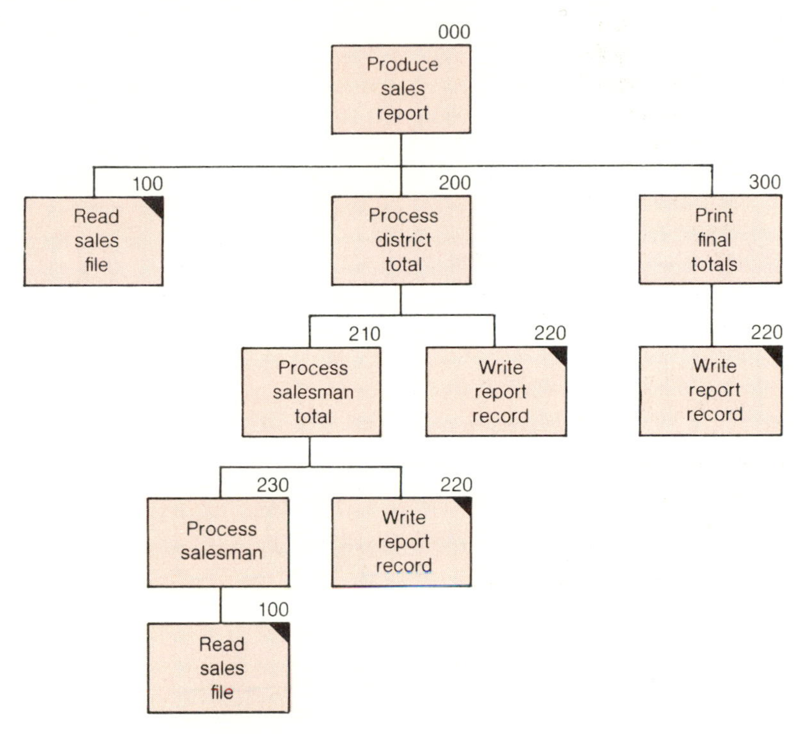

IPO Charts

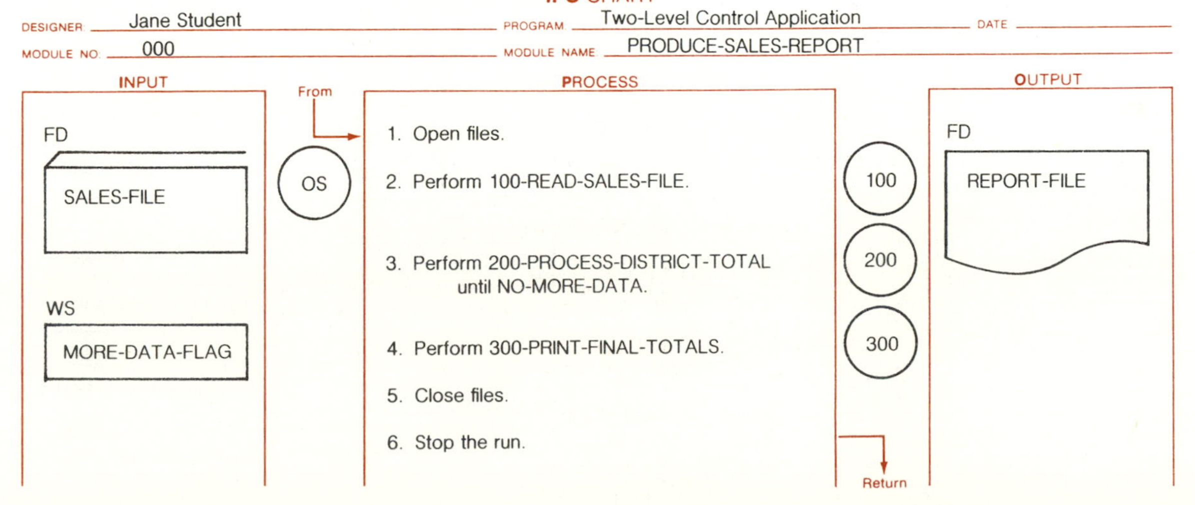

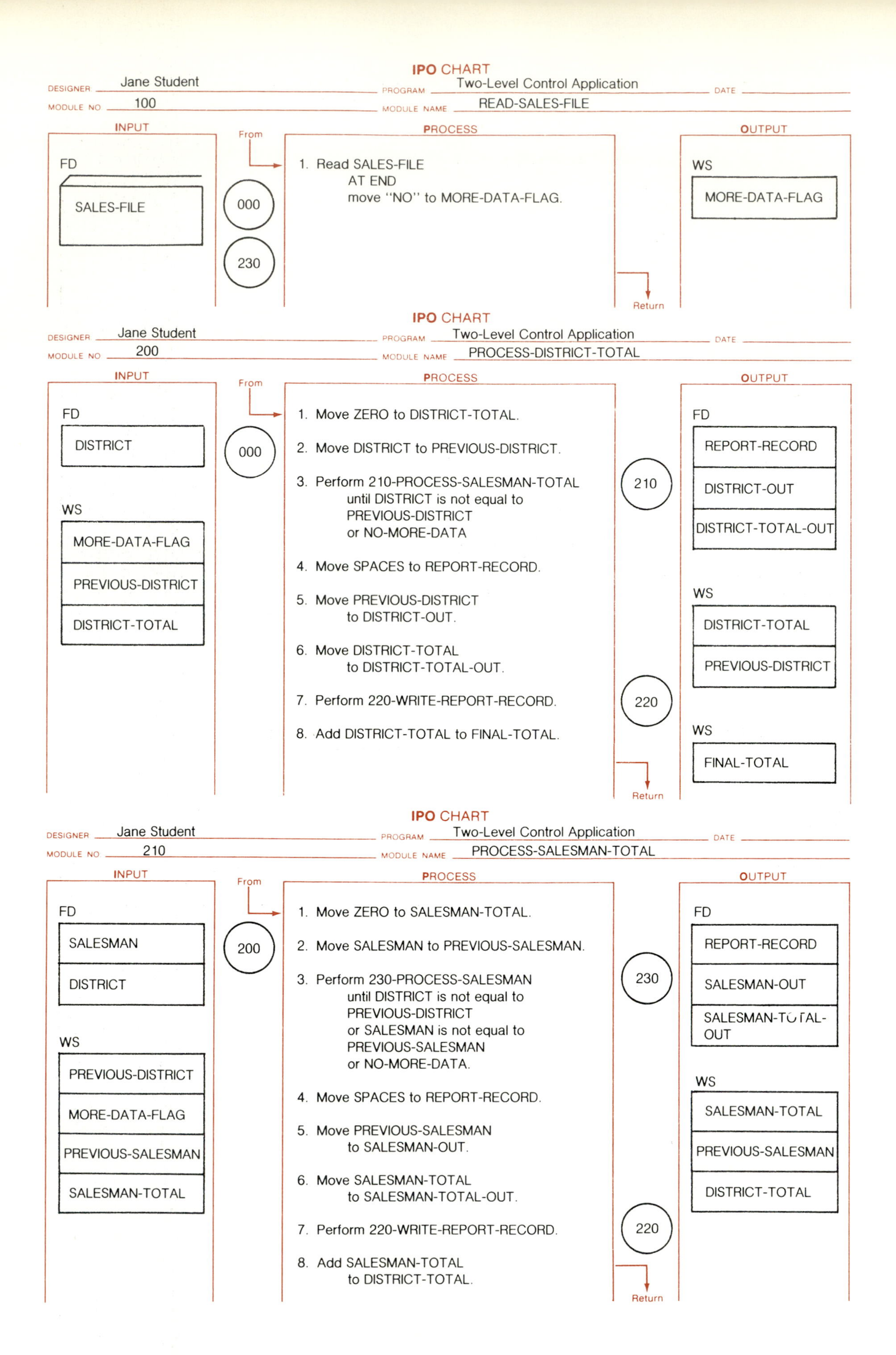

IPO CHART
DESIGNER Jane Student
PROGRAM Two-Level Control Application
DATE
MODULE NO 100
MODULE NAME READ-SALES-FILE
INPUT
FD
SALES-FILE
From
000
230
PROCESS
1. Read SALES-FILE
AT END
move "NO" to MORE-DATA-FLAG.
Return
OUTPUT
WS
MORE-DATA-FLAG
IPO CHART
DESIGNER Jane Student
PROGRAM Two-Level Control Application
DATE
MODULE NO 200
MODULE NAME PROCESS-DISTRICT-TOTAL
INPUT
FD
DISTRICT
WS
MORE-DATA-FLAG
PREVIOUS-DISTRICT
DISTRICT-TOTAL
From
000
PROCESS
1. Move ZERO to DISTRICT-TOTAL.
2. Move DISTRICT to PREVIOUS-DISTRICT.
3. Perform 210-PROCESS-SALESMAN-TOTAL
until DISTRICT is not equal to
PREVIOUS-DISTRICT
or NO-MORE-DATA
4. Move SPACES to REPORT-RECORD.
5. Move PREVIOUS-DISTRICT
to DISTRICT-OUT.
6. Move DISTRICT-TOTAL
to DISTRICT-TOTAL-OUT.
7. Perform 220-WRITE-REPORT-RECORD.
8. Add DISTRICT-TOTAL to FINAL-TOTAL.
210
220
Return
OUTPUT
FD
REPORT-RECORD
DISTRICT-OUT
DISTRICT-TOTAL-OUT
WS
DISTRICT-TOTAL
PREVIOUS-DISTRICT
WS
FINAL-TOTAL
IPO CHART
DESIGNER Jane Student
PROGRAM Two-Level Control Application
DATE
MODULE NO 210
MODULE NAME PROCESS-SALESMAN-TOTAL
INPUT
FD
SALESMAN
DISTRICT
WS
PREVIOUS-DISTRICT
MORE-DATA-FLAG
PREVIOUS-SALESMAN
SALESMAN-TOTAL
From
200
PROCESS
1. Move ZERO to SALESMAN-TOTAL.
2. Move SALESMAN to PREVIOUS-SALESMAN.
3. Perform 230-PROCESS-SALESMAN
until DISTRICT is not equal to
PREVIOUS-DISTRICT
or SALESMAN is not equal to
PREVIOUS-SALESMAN
or NO-MORE-DATA.
4. Move SPACES to REPORT-RECORD.
5. Move PREVIOUS-SALESMAN
to SALESMAN-OUT.
6. Move SALESMAN-TOTAL
to SALESMAN-TOTAL-OUT.
7. Perform 220-WRITE-REPORT-RECORD.
8. Add SALESMAN-TOTAL
to DISTRICT-TOTAL.
230
220
Return
OUTPUT
FD
REPORT-RECORD
SALESMAN-OUT
SALESMAN-TOTAL-OUT
WS
SALESMAN-TOTAL
PREVIOUS-SALESMAN
DISTRICT-TOTAL

IPO CHART

DESIGNER: Jane Student | PROGRAM: Two-Level Control Application | DATE:

MODULE NO: 220 | MODULE NAME: WRITE-REPORT-RECORD

INPUT

FD

REPORT-RECORD

From: 200, 210, 300

PROCESS

1. Write REPORT-RECORD.

Return

OUTPUT

FD

REPORT-FILE

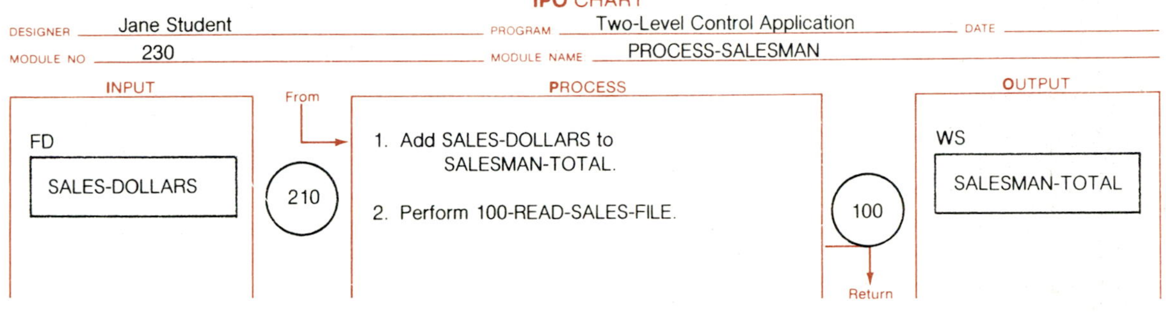

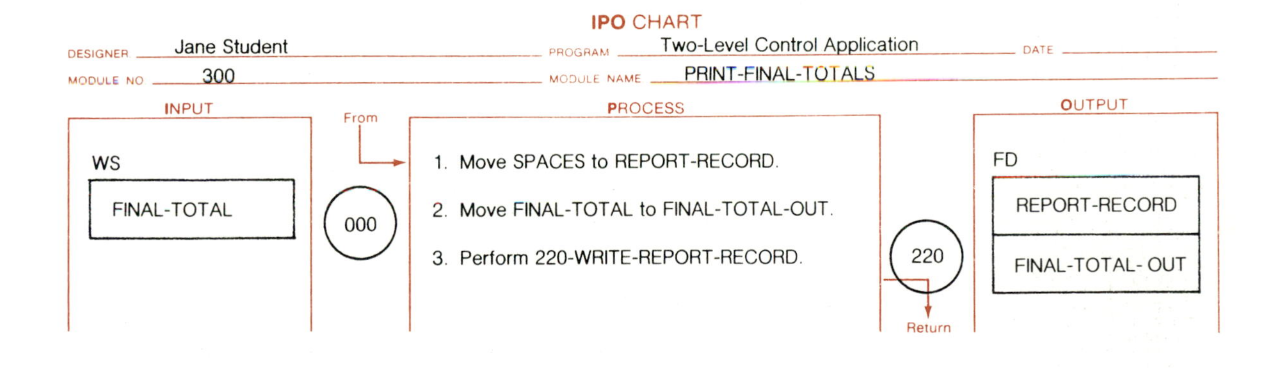

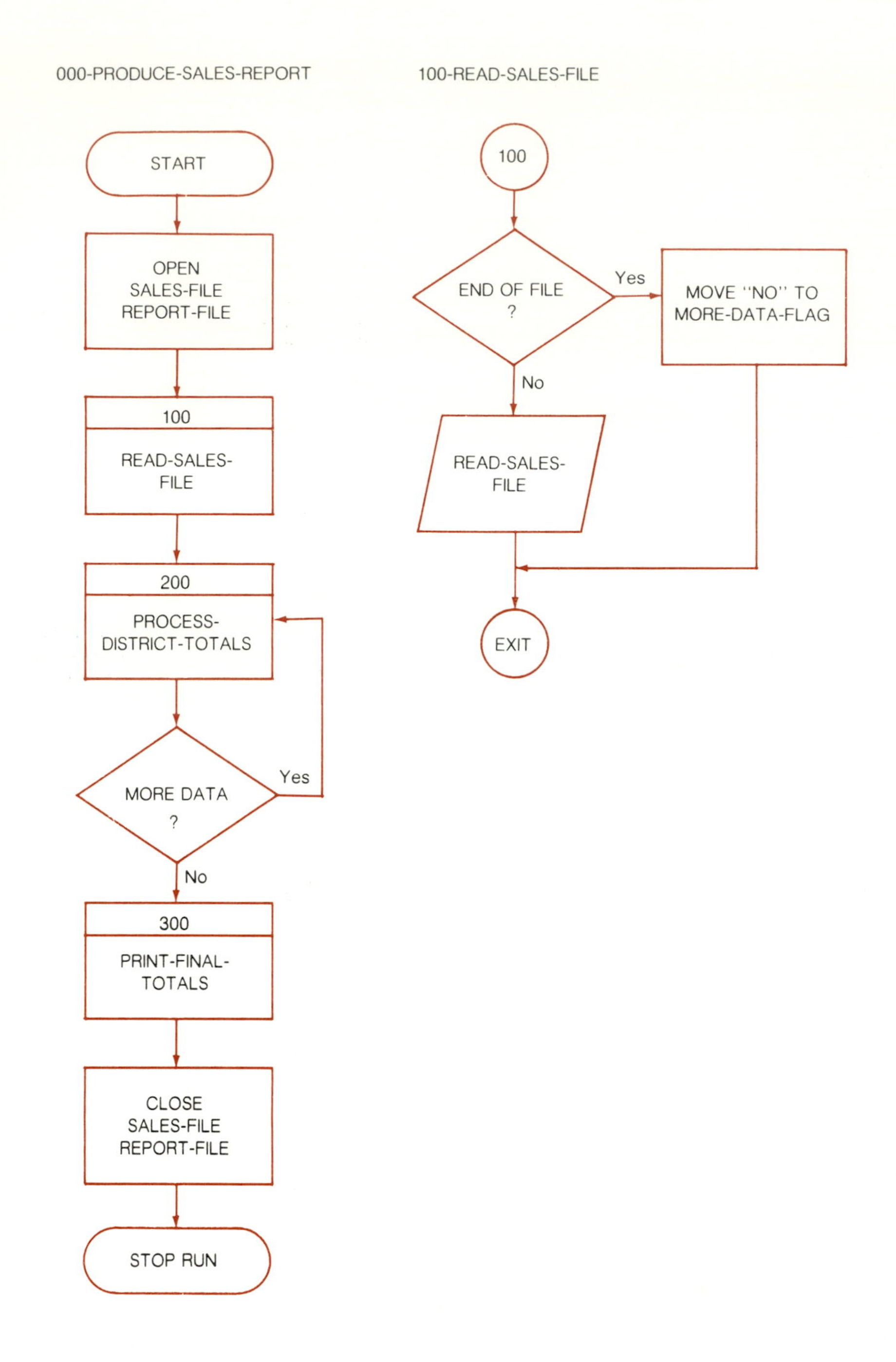
000-PRODUCE-SALES-REPORT
START
OPEN
SALES-FILE
REPORT-FILE
100
READ-SALES-
FILE
200
PROCESS-
DISTRICT-TOTALS
MORE DATA
?
Yes
No
300
PRINT-FINAL-
TOTALS
CLOSE
SALES-FILE
REPORT-FILE
STOP RUN
100-READ-SALES-FILE
100
END OF FILE
?
Yes
MOVE "NO" TO
MORE-DATA-FLAG
No
READ-SALES-
FILE
EXIT

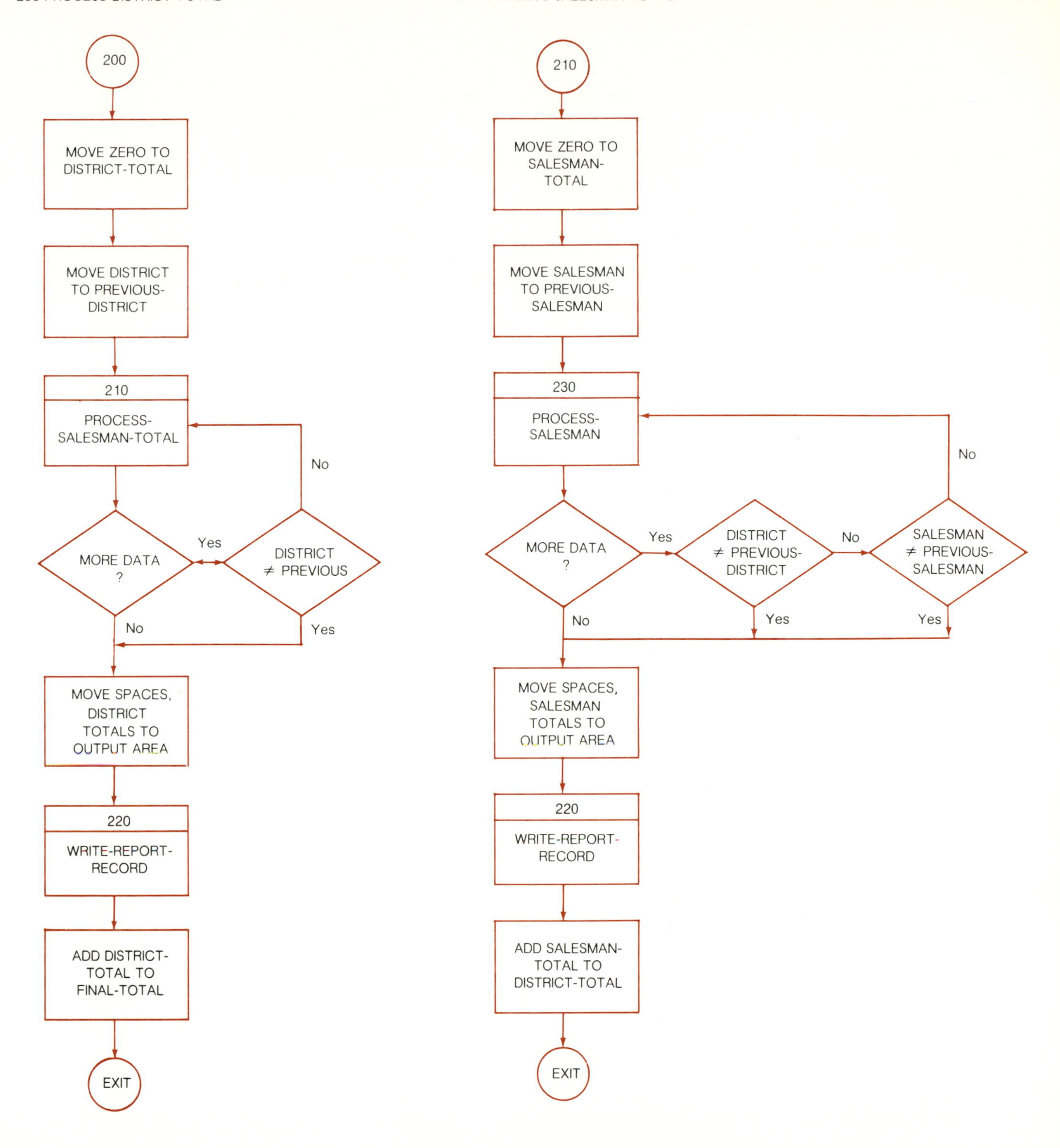
200-PROCESS-DISTRICT-TOTAL
200
MOVE ZERO TO DISTRICT-TOTAL
MOVE DISTRICT TO PREVIOUS-DISTRICT
210
PROCESS-SALESMAN-TOTAL
No
MORE DATA ?
Yes
DISTRICT ≠ PREVIOUS
No
Yes
MOVE SPACES, DISTRICT TOTALS TO OUTPUT AREA
220
WRITE-REPORT-RECORD
ADD DISTRICT-TOTAL TO FINAL-TOTAL
EXIT
210-PROCESS-SALESMAN-TOTAL
210
MOVE ZERO TO SALESMAN-TOTAL
MOVE SALESMAN TO PREVIOUS-SALESMAN
230
PROCESS-SALESMAN
No
MORE DATA ?
Yes
DISTRICT ≠ PREVIOUS-DISTRICT
No
SALESMAN ≠ PREVIOUS-SALESMAN
No
Yes
Yes
MOVE SPACES, SALESMAN TOTALS TO OUTPUT AREA
220
WRITE-REPORT-RECORD
ADD SALESMAN-TOTAL TO DISTRICT-TOTAL
EXIT

220-WRITE-REPORT-RECORD

230-PROCESS-SALESMAN

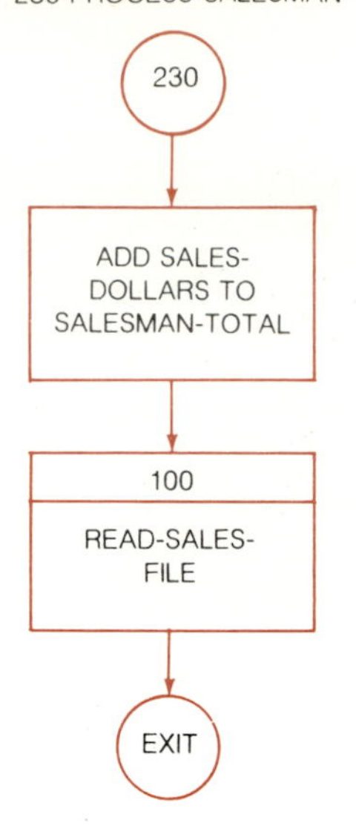

300-PRINT-FINAL-TOTALS

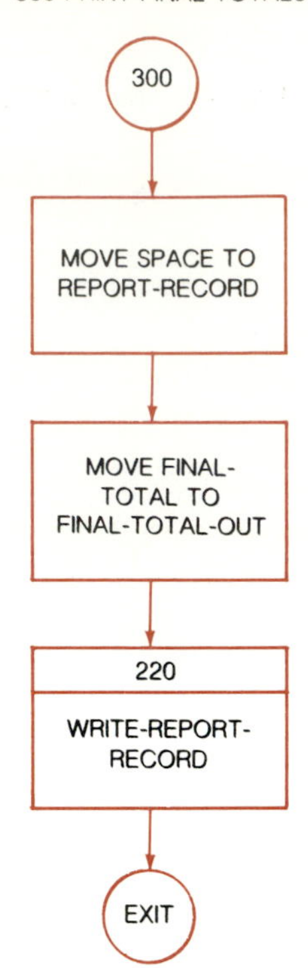

Source Program

```
IDENTIFICATION DIVISION.
PROGRAM-ID.
    TWOLEVEL.
ENVIRONMENT DIVISION.
INPUT-OUTPUT SECTION.
FILE-CONTROL.
    SELECT SALES-FILE ASSIGN TO SYS012-UR-2540R-S.
    SELECT REPORT-FILE ASSIGN TO SYS009-UR-1403-S.
DATA DIVISION.
FILE SECTION.
FD  SALES-FILE
    RECORD CONTAINS 80 CHARACTERS
    LABEL RECORDS ARE OMITTED
    DATA RECORD IS SALES-RECORD.
01  SALES-RECORD.
    05 SALESMAN                  PICTURE X(5).
    05 DISTRICT                  PICTURE XXX.
    05 SALES-DOLLARS             PICTURE 9(5)V99.
FD  REPORT-FILE
    RECORD CONTAINS 133 CHARACTERS
    LABEL RECORDS ARE OMITTED
    DATA RECORD IS REPORT-RECORD.
01  REPORT-RECORD.
    05 CARRIAGE-CONTROL          PICTURE X.
    05 SALESMAN-OUT              PICTURE ZZZZ9.
    05 FILLER                    PICTURE XXX.
    05 SALESMAN-TOTAL-OUT        PICTURE $$$,$$$,$$9.99.
    05 FILLER                    PICTURE X(8).
    05 DISTRICT-OUT              PICTURE ZZ9.
    05 FILLER                    PICTURE XXX.
    05 DISTRICT-TOTAL-OUT        PICTURE $$$,$$$,$$9.99.
    05 FILLER                    PICTURE X(8).
    05 FINAL-TOTAL-OUT           PICTURE $$$,$$$,$$9.99.
WORKING-STORAGE SECTION.
01  FLAGS.
    05 MORE-DATA-FLAG            PICTURE XXX       VALUE 'YES'.
       88 MORE-DATA                                VALUE 'YES'.
       88 NO-MORE-DATA                             VALUE 'NO'.
01  SAVE-ITEMS.
    05 PREVIOUS-SALESMAN         PICTURE X(5).
    05 PREVIOUS-DISTRICT         PICTURE XXX.
01  TOTALS COMPUTATIONAL-3.
    05 SALESMAN-TOTAL            PICTURE S9(8)V99 VALUE ZERO.
    05 DISTRICT-TOTAL            PICTURE S9(8)V99 VALUE ZERO.
    05 FINAL-TOTAL               PICTURE S9(8)V99 VALUE ZERO.
PROCEDURE DIVISION.
000-PRODUCE-SALES-REPORT.
     OPEN INPUT  SALES-FILE
          OUTPUT REPORT-FILE.
     PERFORM 100-READ-SALES-FILE.
     PERFORM 200-PROCESS-DISTRICT-TOTAL
          UNTIL NO-MORE-DATA.
     PERFORM 300-PRINT-FINAL-TOTALS.
     CLOSE   SALES-FILE,
             REPORT-FILE.
     STOP RUN.
100-READ-SALES-FILE.
     READ SALES-FILE
         AT END MOVE 'NO' TO MORE-DATA-FLAG.
200-PROCESS-DISTRICT-TOTAL.
     MOVE ZERO TO DISTRICT-TOTAL.
     MOVE DISTRICT TO PREVIOUS-DISTRICT.
     PERFORM 210-PROCESS-SALESMAN-TOTAL
         UNTIL DISTRICT IS NOT EQUAL TO PREVIOUS-DISTRICT
         OR NO-MORE-DATA.
     MOVE SPACES TO REPORT-RECORD.
     MOVE PREVIOUS-DISTRICT TO DISTRICT-OUT.
     MOVE DISTRICT-TOTAL TO DISTRICT-TOTAL-OUT.
     PERFORM 220-WRITE-REPORT-RECORD.
     ADD DISTRICT-TOTAL TO FINAL-TOTAL.
210-PROCESS-SALESMAN-TOTAL.
     MOVE ZERO TO SALESMAN-TOTAL.
     MOVE SALESMAN TO PREVIOUS-SALESMAN.
     PERFORM 230-PROCESS-SALESMAN
         UNTIL DISTRICT IS NOT EQUAL TO PREVIOUS-DISTRICT
         OR SALESMAN IS NOT EQUAL TO PREVIOUS-SALESMAN
         OR NO-MORE-DATA.
     MOVE SPACES TO REPORT-RECORD.
     MOVE PREVIOUS-SALESMAN TO SALESMAN-OUT.
     MOVE SALESMAN-TOTAL TO SALESMAN-TOTAL-OUT.
     PERFORM 220-WRITE-REPORT-RECORD.
     ADD SALESMAN-TOTAL TO DISTRICT-TOTAL.
220-WRITE-REPORT-RECORD.
     WRITE REPORT-RECORD
         AFTER ADVANCING 1 LINE.
230-PROCESS-SALESMAN.
     ADD SALES-DOLLARS TO SALESMAN-TOTAL.
     PERFORM 100-READ-SALES-FILE.
300-PRINT-FINAL-TOTALS.
     MOVE SPACES TO REPORT-RECORD.
     MOVE FINAL-TOTAL TO FINAL-TOTAL-OUT.
     PERFORM 220-WRITE-REPORT-RECORD.
```

```
   41  $203.37
   52  $110.00
   69  $134.65
                    1  $448.02
   18  $207.69
   32  $185.60
                    2  $393.29
   36  $194.15
   39  $121.40
   50   $51.80
                    3  $367.35   $1,208.66
```

Illustrative Program: Inquiry Response Application

Illustrative Master File

```
000108DESK                       0018500000160075010
000115CHAIR, FOLDING             0001810001270075100
000180LAMP, FLOOR                0003750000120075180
000181LAMP, DESK                 0002200001170075093
000200TYPEWRITER STAND           0002490000400074350
000309BOOKCASE, 5 SHELF          0004125000200075105
000310BOOKCASE, 4 SHELF          0003650000310075090
000311BOOKCASE, 3 SHELF          0002800000170075110
000480FILE CABINET, 4 DWR        0006180001000075130
000481FILE CABINET, 2 DWR        0003990000500075150
010684WASTEBASKET, GREEN         0000417000120075190
010686WASTEBASKET, GRAY          0000417001900075120
010687WASTEBASKET, BLUE          0000417000570075182
021732SOFA, LEATHER, BROWN       0035620000290075070
021739SOFA, LEATHER, RED         0035620000370075040
```

Illustrative Transaction File

```
00010875001
00018075001
00020075001
00025075001
000310 75001
00031075001
00048075140
00048175140
010.68575140
01069075150
02173975030
03194075150
```

Output

TRANSACTION		DESCRIPTION	UNIT PRICE	QOH	TOTAL COST	LAST ACTIVITY	
000108	75001	DESK	$185.00	16.00	$2,960.00	1975	010
000180	75001	LAMP, FLOOR	$37.50	12.00	$450.00	1975	180
000200	75001	NO ACTIVITY FOR THIS ITEM SINCE DATE IN INQUIRY					
000250	75001	NO MASTER FOR THIS STOCK NUMBER					
000310	7500	ALL ITEMS IN INQUIRY MUST BE NUMERIC					
000310	75001	BOOKCASE, 4 SHELF	$36.50	31.00	$1,131.50	1975	090
000480	75140	NO ACTIVITY FOR THIS ITEM SINCE DATE IN INQUIRY					
000481	75140	FILE CABINET, 2 DWR	$39.90	50.00	$1,995.00	1975	150
010.68	57514	ALL ITEMS IN INQUIRY MUST BE NUMERIC					
010690	75150	NO MASTER FOR THIS STOCK NUMBER					
021739	75030	SOFA, LEATHER, RED	$356.20	37.00	$13,179.40	1975	040
031940	75150	NO MASTER FOR THIS STOCK NUMBER					

Hierarchy Chart

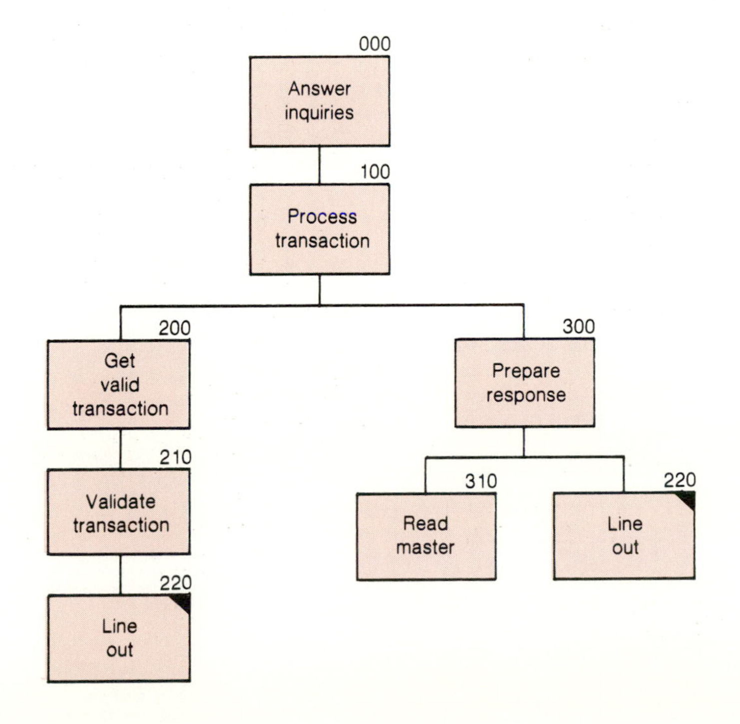

IPO Charts

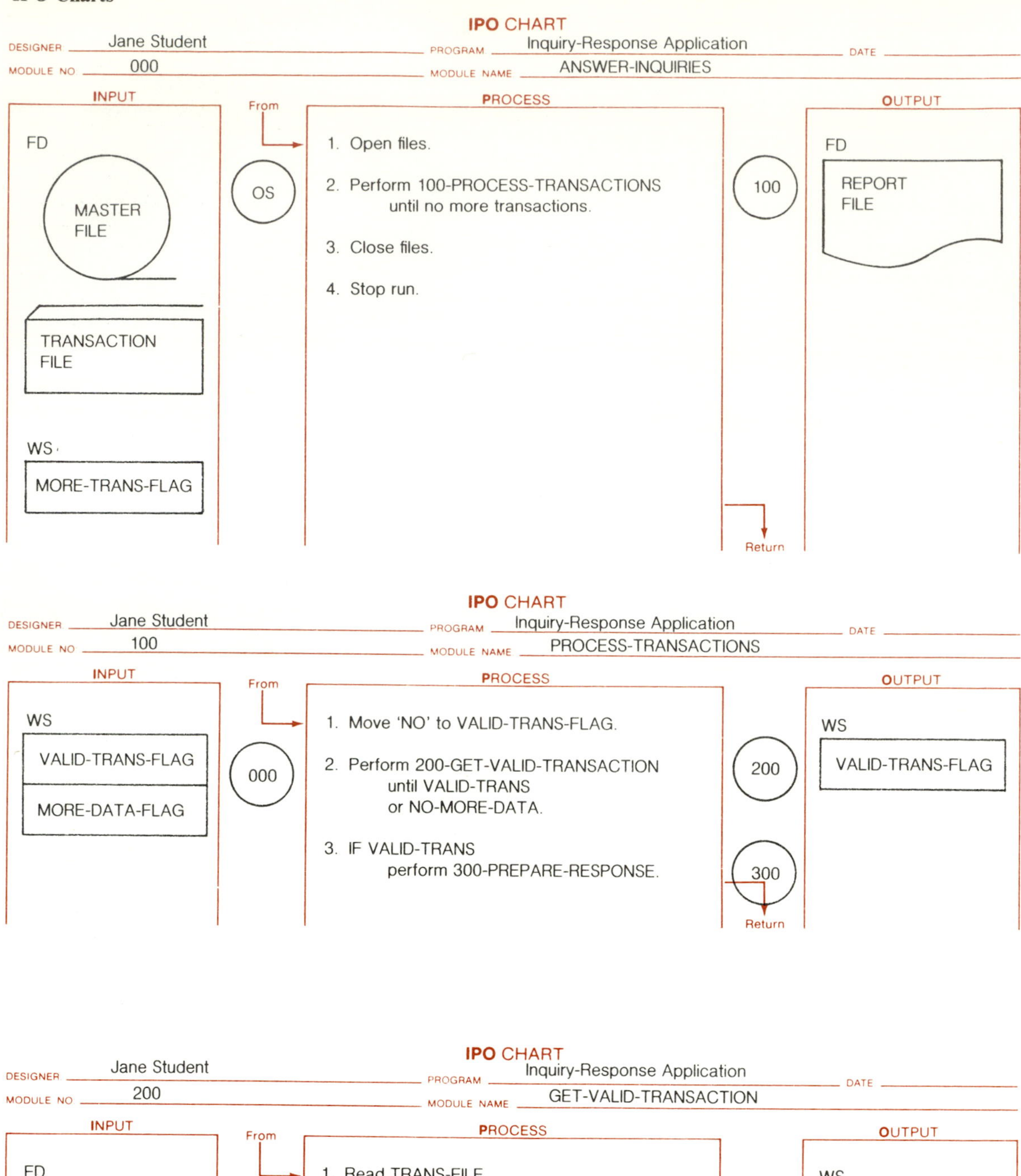
IPO CHART
DESIGNER Jane Student
PROGRAM Inquiry-Response Application
DATE
MODULE NO 000
MODULE NAME ANSWER-INQUIRIES
INPUT
FD
MASTER FILE
TRANSACTION FILE
WS
MORE-TRANS-FLAG
From
OS
PROCESS
1. Open files.
2. Perform 100-PROCESS-TRANSACTIONS until no more transactions.
3. Close files.
4. Stop run.
100
Return
OUTPUT
FD
REPORT FILE
IPO CHART
DESIGNER Jane Student
PROGRAM Inquiry-Response Application
DATE
MODULE NO 100
MODULE NAME PROCESS-TRANSACTIONS
INPUT
WS
VALID-TRANS-FLAG
MORE-DATA-FLAG
From
000
PROCESS
1. Move 'NO' to VALID-TRANS-FLAG.
2. Perform 200-GET-VALID-TRANSACTION until VALID-TRANS or NO-MORE-DATA.
3. IF VALID-TRANS perform 300-PREPARE-RESPONSE.
200
300
Return
OUTPUT
WS
VALID-TRANS-FLAG

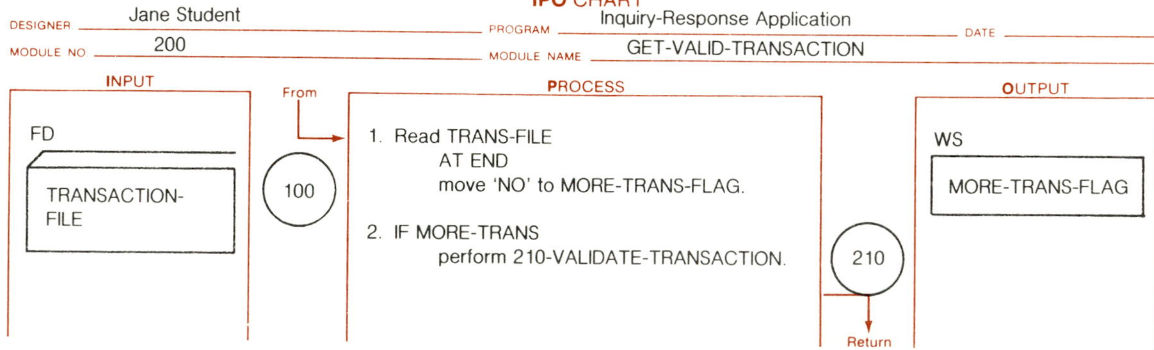
IPO CHART
DESIGNER Jane Student
PROGRAM Inquiry-Response Application
DATE
MODULE NO 200
MODULE NAME GET-VALID-TRANSACTION
INPUT
FD
TRANSACTION-FILE
From
100
PROCESS
1. Read TRANS-FILE AT END move 'NO' to MORE-TRANS-FLAG.
2. IF MORE-TRANS perform 210-VALIDATE-TRANSACTION.
210
Return
OUTPUT
WS
MORE-TRANS-FLAG

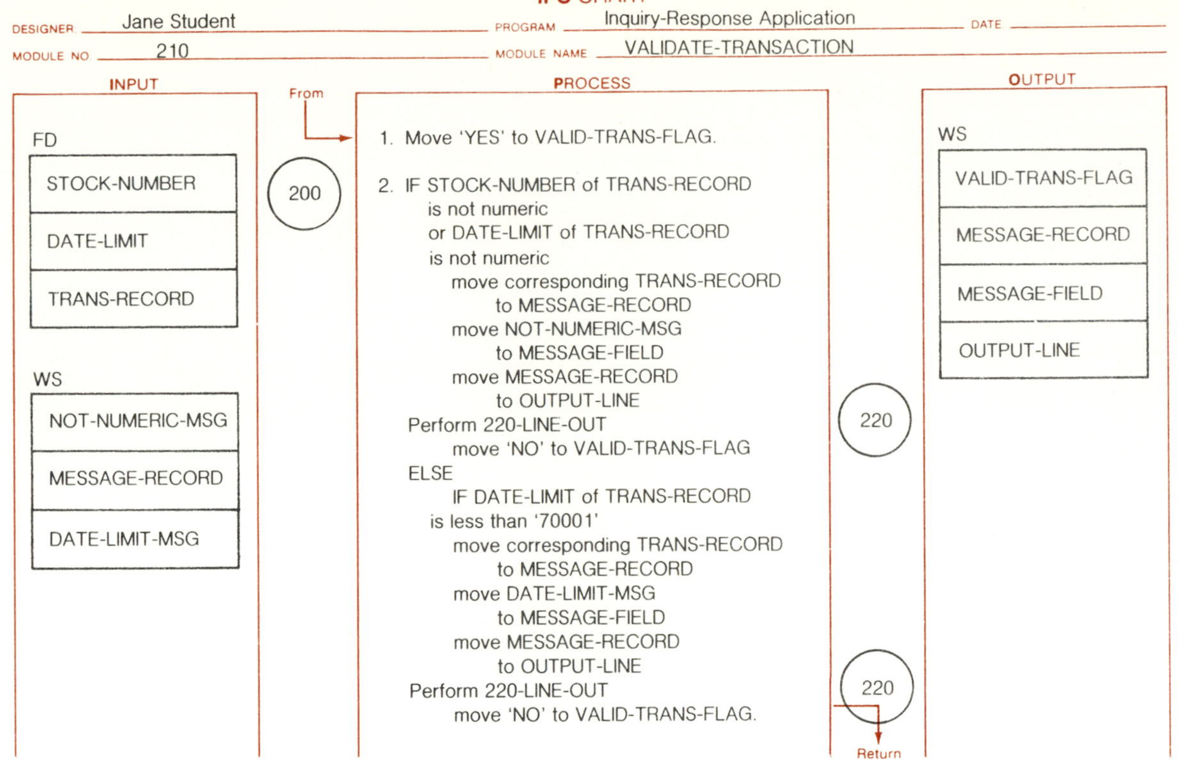
IPO CHART
DESIGNER Jane Student
PROGRAM Inquiry-Response Application
DATE
MODULE NO. 210
MODULE NAME VALIDATE-TRANSACTION
INPUT
FD
STOCK-NUMBER
DATE-LIMIT
TRANS-RECORD
WS
NOT-NUMERIC-MSG
MESSAGE-RECORD
DATE-LIMIT-MSG
From
200
PROCESS
1. Move 'YES' to VALID-TRANS-FLAG.
2. IF STOCK-NUMBER of TRANS-RECORD
is not numeric
or DATE-LIMIT of TRANS-RECORD
is not numeric
move corresponding TRANS-RECORD
to MESSAGE-RECORD
move NOT-NUMERIC-MSG
to MESSAGE-FIELD
move MESSAGE-RECORD
to OUTPUT-LINE
Perform 220-LINE-OUT
move 'NO' to VALID-TRANS-FLAG
ELSE
IF DATE-LIMIT of TRANS-RECORD
is less than '70001'
move corresponding TRANS-RECORD
to MESSAGE-RECORD
move DATE-LIMIT-MSG
to MESSAGE-FIELD
move MESSAGE-RECORD
to OUTPUT-LINE
Perform 220-LINE-OUT
move 'NO' to VALID-TRANS-FLAG.
220
220
Return
OUTPUT
WS
VALID-TRANS-FLAG
MESSAGE-RECORD
MESSAGE-FIELD
OUTPUT-LINE

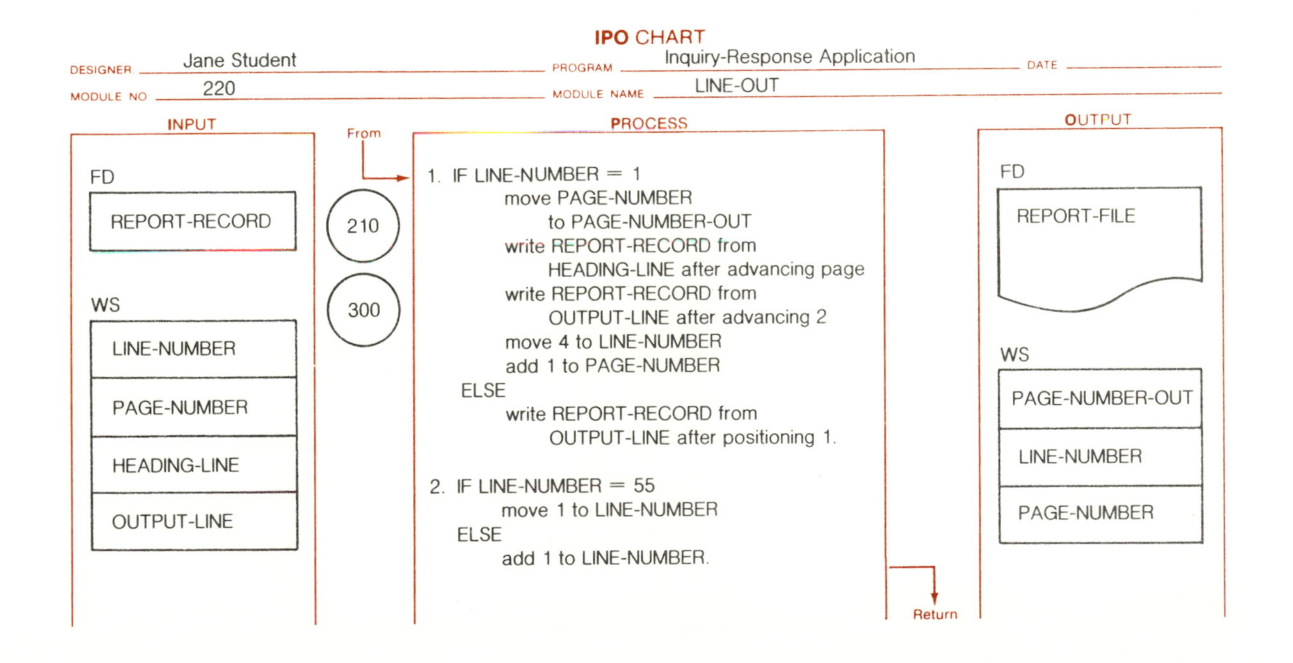
IPO CHART
DESIGNER Jane Student
PROGRAM Inquiry-Response Application
DATE
MODULE NO. 220
MODULE NAME LINE-OUT
INPUT
FD
REPORT-RECORD
WS
LINE-NUMBER
PAGE-NUMBER
HEADING-LINE
OUTPUT-LINE
From
210
300
PROCESS
1. IF LINE-NUMBER = 1
move PAGE-NUMBER
to PAGE-NUMBER-OUT
write REPORT-RECORD from
HEADING-LINE after advancing page
write REPORT-RECORD from
OUTPUT-LINE after advancing 2
move 4 to LINE-NUMBER
add 1 to PAGE-NUMBER
ELSE
write REPORT-RECORD from
OUTPUT-LINE after positioning 1.
2. IF LINE-NUMBER = 55
move 1 to LINE-NUMBER
ELSE
add 1 to LINE-NUMBER.
Return
OUTPUT
FD
REPORT-FILE
WS
PAGE-NUMBER-OUT
LINE-NUMBER
PAGE-NUMBER

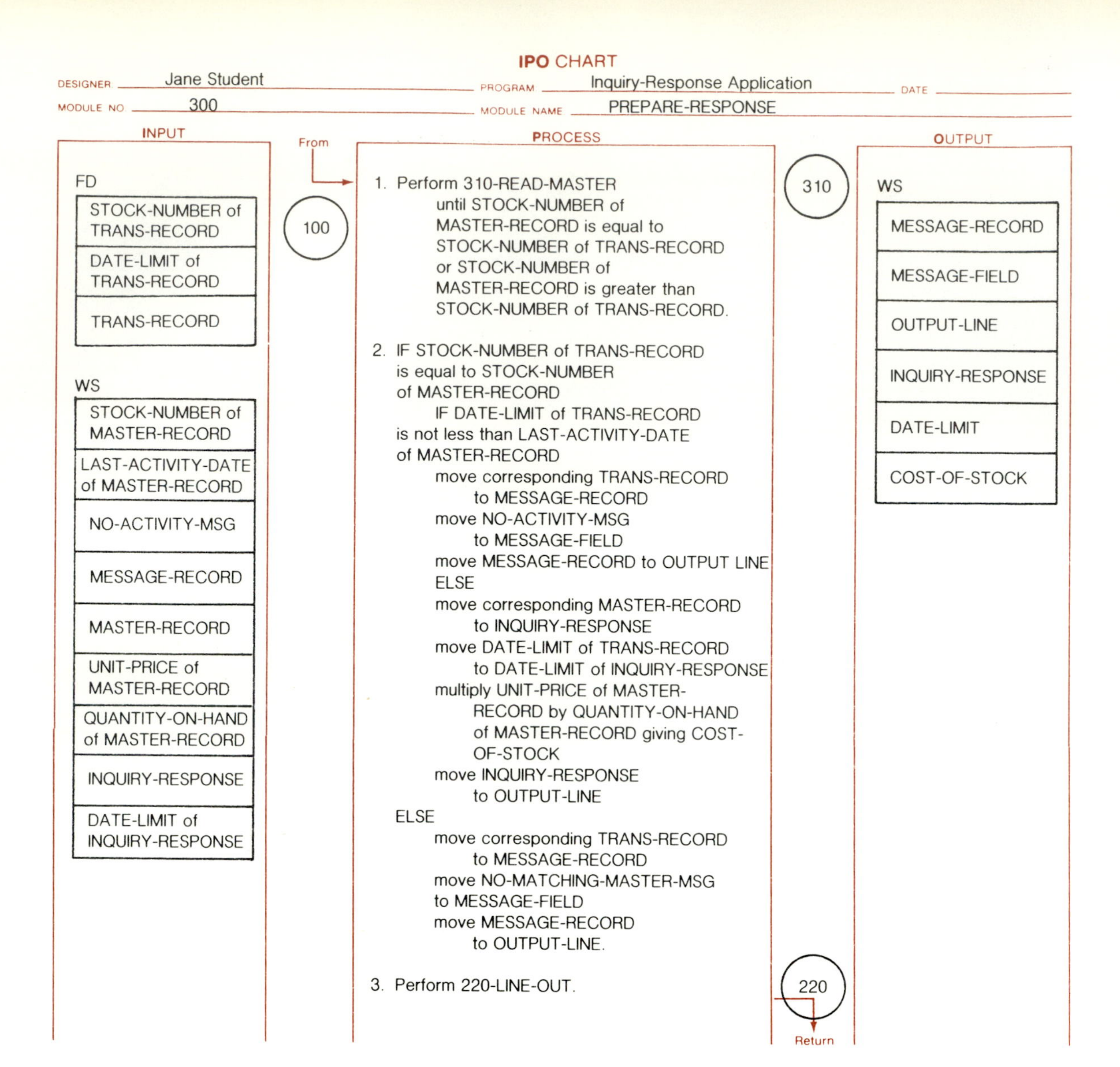
IPO CHART
DESIGNER Jane Student
PROGRAM Inquiry-Response Application
DATE
MODULE NO 300
MODULE NAME PREPARE-RESPONSE
INPUT
FD
STOCK-NUMBER of TRANS-RECORD
DATE-LIMIT of TRANS-RECORD
TRANS-RECORD
WS
STOCK-NUMBER of MASTER-RECORD
LAST-ACTIVITY-DATE of MASTER-RECORD
NO-ACTIVITY-MSG
MESSAGE-RECORD
MASTER-RECORD
UNIT-PRICE of MASTER-RECORD
QUANTITY-ON-HAND of MASTER-RECORD
INQUIRY-RESPONSE
DATE-LIMIT of INQUIRY-RESPONSE
From
100
PROCESS
1. Perform 310-READ-MASTER until STOCK-NUMBER of MASTER-RECORD is equal to STOCK-NUMBER of TRANS-RECORD or STOCK-NUMBER of MASTER-RECORD is greater than STOCK-NUMBER of TRANS-RECORD.
2. IF STOCK-NUMBER of TRANS-RECORD is equal to STOCK-NUMBER of MASTER-RECORD
IF DATE-LIMIT of TRANS-RECORD is not less than LAST-ACTIVITY-DATE of MASTER-RECORD
move corresponding TRANS-RECORD to MESSAGE-RECORD
move NO-ACTIVITY-MSG to MESSAGE-FIELD
move MESSAGE-RECORD to OUTPUT LINE
ELSE
move corresponding MASTER-RECORD to INQUIRY-RESPONSE
move DATE-LIMIT of TRANS-RECORD to DATE-LIMIT of INQUIRY-RESPONSE
multiply UNIT-PRICE of MASTER-RECORD by QUANTITY-ON-HAND of MASTER-RECORD giving COST-OF-STOCK
move INQUIRY-RESPONSE to OUTPUT-LINE
ELSE
move corresponding TRANS-RECORD to MESSAGE-RECORD
move NO-MATCHING-MASTER-MSG to MESSAGE-FIELD
move MESSAGE-RECORD to OUTPUT-LINE.
3. Perform 220-LINE-OUT.
310
220
Return
OUTPUT
WS
MESSAGE-RECORD
MESSAGE-FIELD
OUTPUT-LINE
INQUIRY-RESPONSE
DATE-LIMIT
COST-OF-STOCK

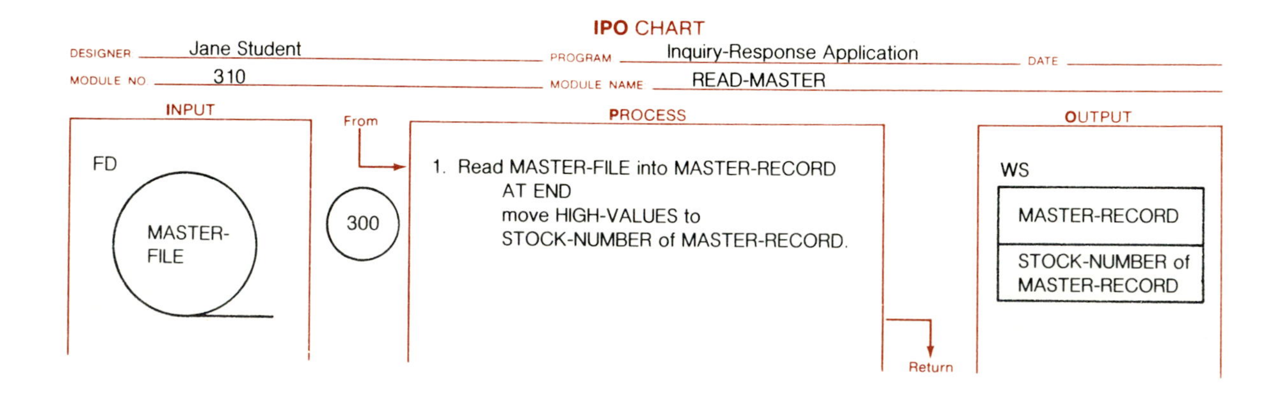
IPO CHART
DESIGNER Jane Student
PROGRAM Inquiry-Response Application
DATE
MODULE NO 310
MODULE NAME READ-MASTER
INPUT
FD
MASTER-FILE
From
300
PROCESS
1. Read MASTER-FILE into MASTER-RECORD
AT END
move HIGH-VALUES to STOCK-NUMBER of MASTER-RECORD.
Return
OUTPUT
WS
MASTER-RECORD
STOCK-NUMBER of MASTER-RECORD

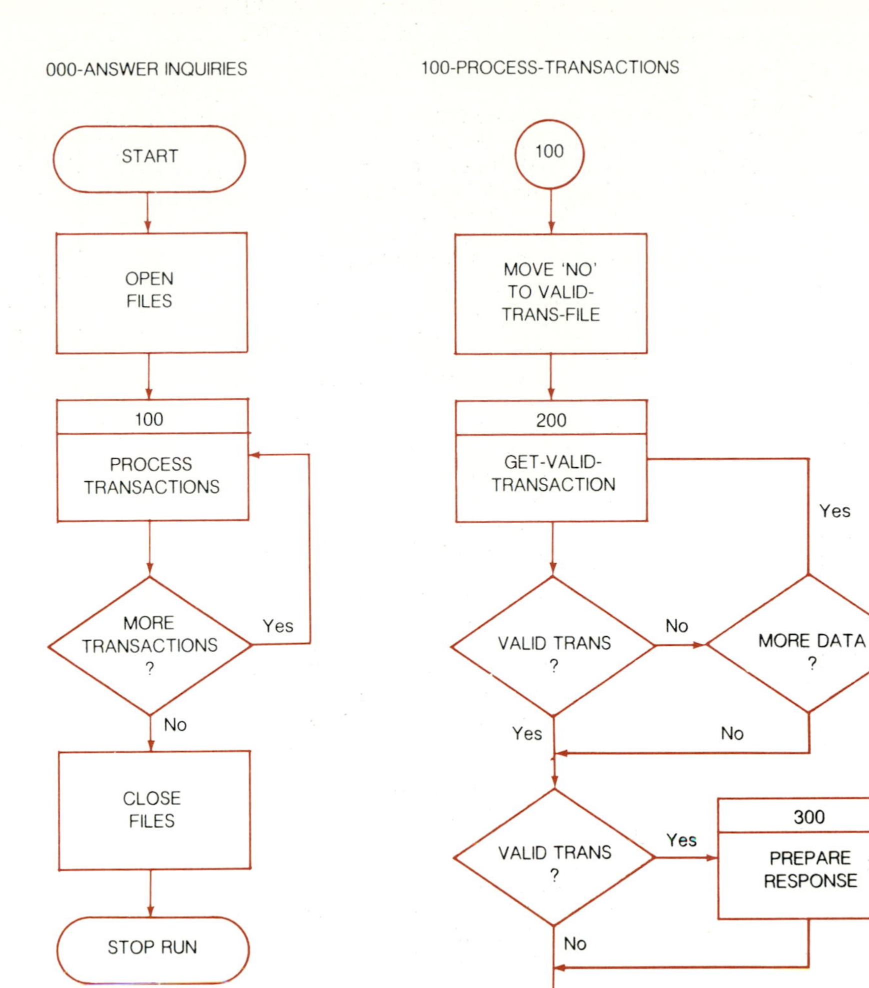
000-ANSWER INQUIRIES
START
OPEN FILES
100
PROCESS TRANSACTIONS
MORE TRANSACTIONS ?
Yes
No
CLOSE FILES
STOP RUN
100-PROCESS-TRANSACTIONS
100
MOVE 'NO' TO VALID-TRANS-FILE
200
GET-VALID-TRANSACTION
Yes
VALID TRANS ?
No
MORE DATA ?
Yes
No
VALID TRANS ?
Yes
300
PREPARE RESPONSE
No
EXIT

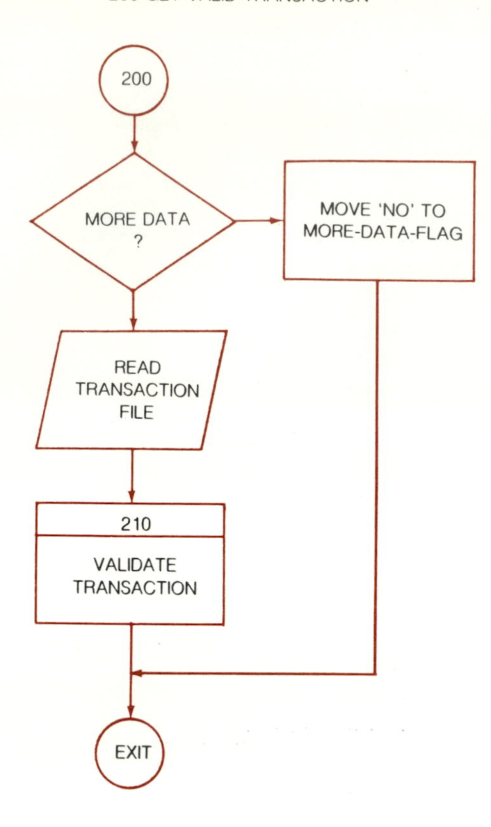
200-GET-VALID-TRANSACTION
200
MORE DATA
?
MOVE 'NO' TO
MORE-DATA-FLAG
READ
TRANSACTION
FILE
210
VALIDATE
TRANSACTION
EXIT

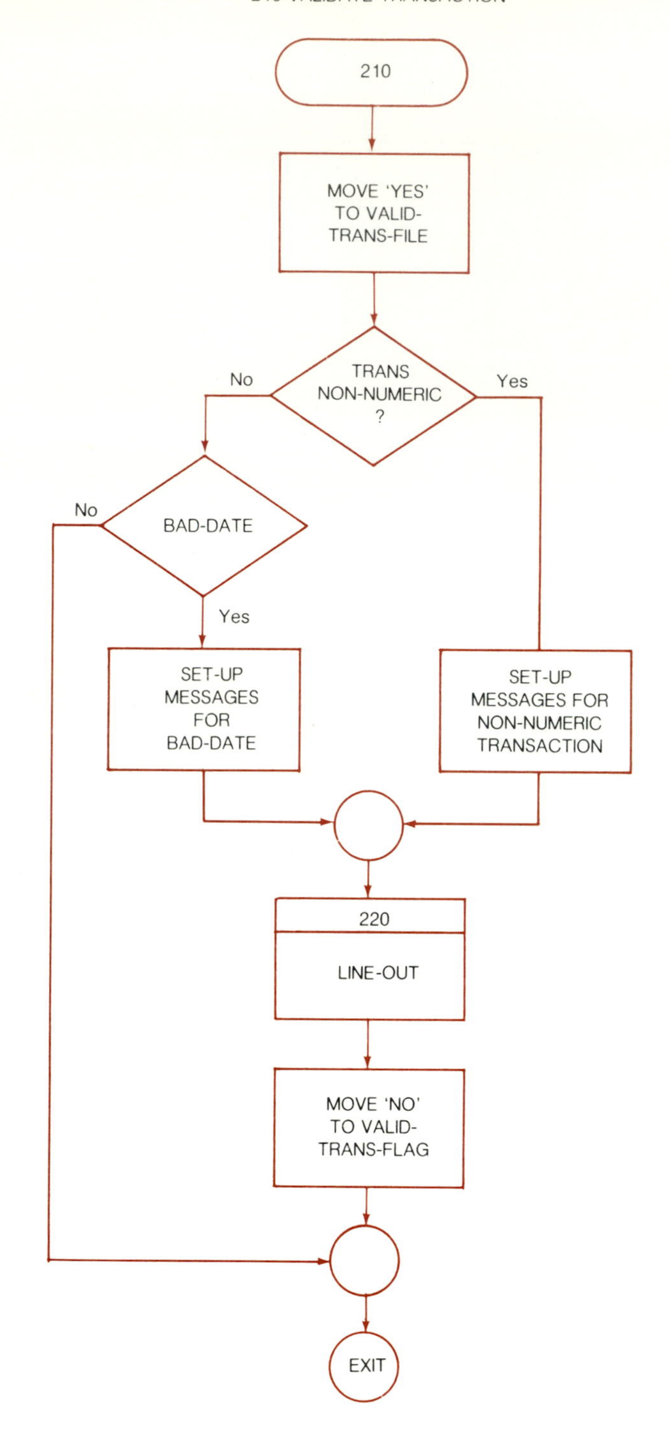
210-VALIDATE-TRANSACTION
210
MOVE 'YES'
TO VALID-
TRANS-FILE
TRANS
NON-NUMERIC
?
No
Yes
BAD-DATE
No
Yes
SET-UP
MESSAGES
FOR
BAD-DATE
SET-UP
MESSAGES FOR
NON-NUMERIC
TRANSACTION
220
LINE-OUT
MOVE 'NO'
TO VALID-
TRANS-FLAG
EXIT

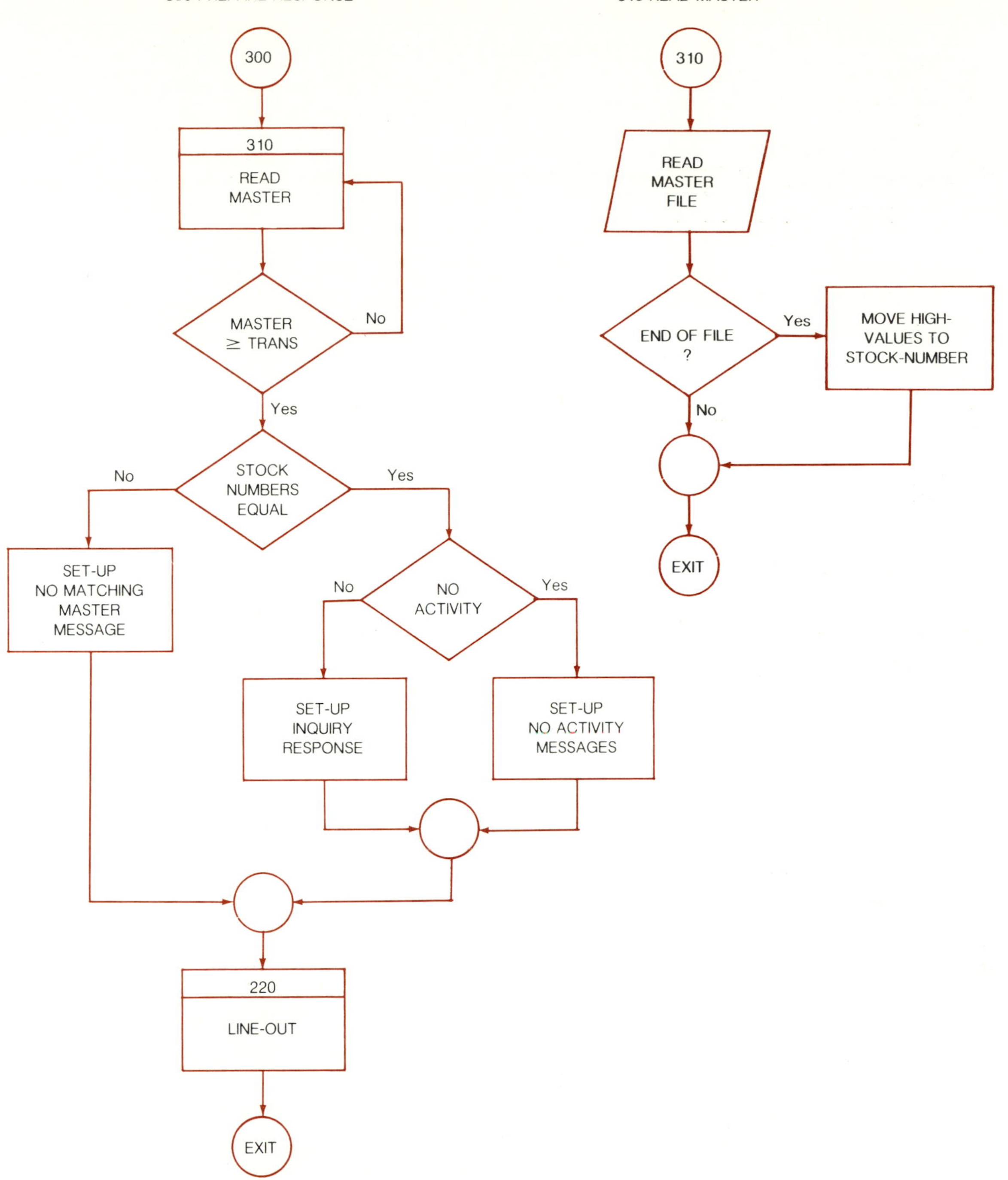
300-PREPARE-RESPONSE
300
310
READ
MASTER
MASTER
≥ TRANS
No
Yes
STOCK
NUMBERS
EQUAL
No
Yes
SET-UP
NO MATCHING
MASTER
MESSAGE
NO
ACTIVITY
No
Yes
SET-UP
INQUIRY
RESPONSE
SET-UP
NO ACTIVITY
MESSAGES
220
LINE-OUT
EXIT
310-READ-MASTER
310
READ
MASTER
FILE
END OF FILE
?
Yes
MOVE HIGH-
VALUES TO
STOCK-NUMBER
No
EXIT

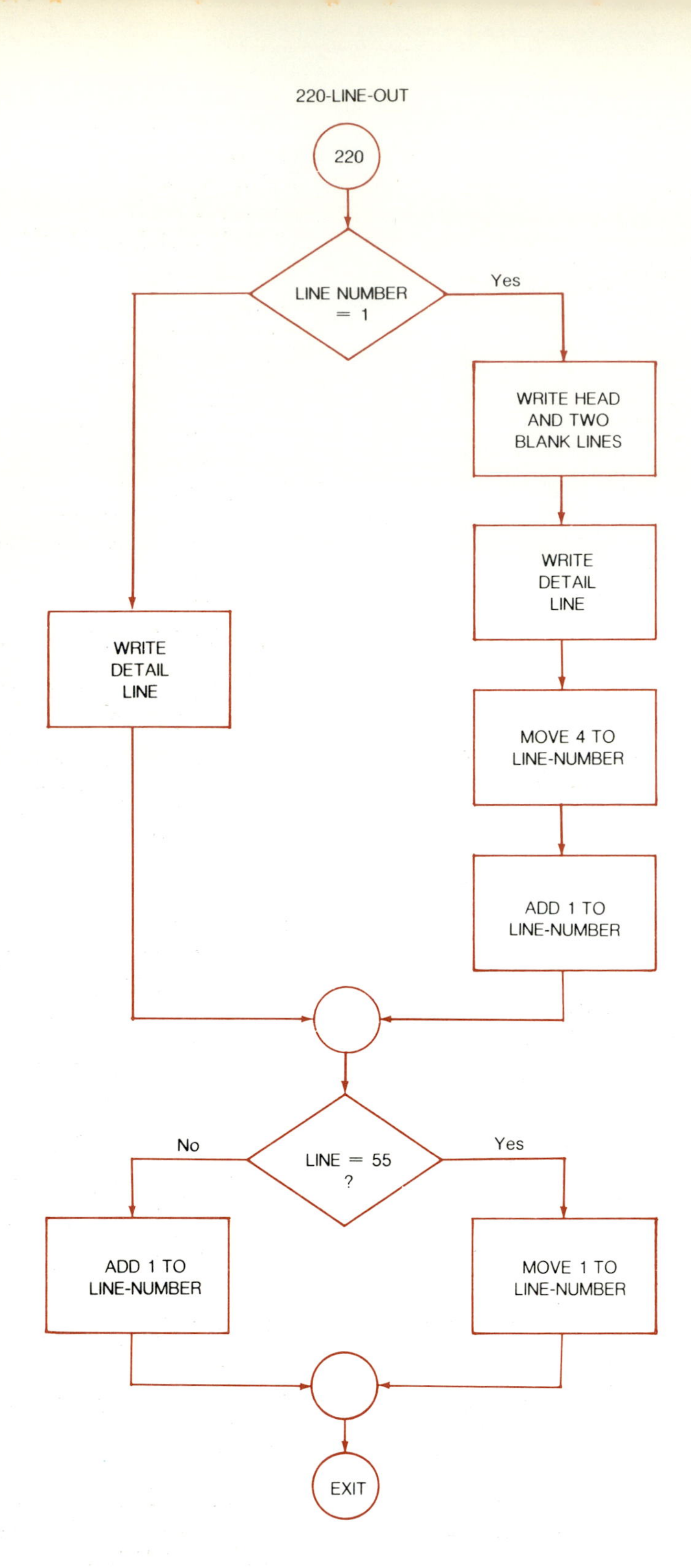
220-LINE-OUT
220
LINE NUMBER
= 1
Yes
WRITE HEAD
AND TWO
BLANK LINES
WRITE
DETAIL
LINE
WRITE
DETAIL
LINE
MOVE 4 TO
LINE-NUMBER
ADD 1 TO
LINE-NUMBER
No
LINE = 55
?
Yes
ADD 1 TO
LINE-NUMBER
MOVE 1 TO
LINE-NUMBER
EXIT

Source Program

```
IDENTIFICATION DIVISION.
PROGRAM-ID.
    STRUCTURED-PROGRAMMING-EXAMPLE.

ENVIRONMENT DIVISION.
INPUT-OUTPUT SECTION.
FILE-CONTROL.
    SELECT MASTER-FILE ASSIGN TO SYS011-UT-2400-S.
    SELECT TRANS-FILE ASSIGN TO SYS012-UR-2540R-S.
    SELECT REPORT-FILE ASSIGN TO SYS009-UR-1403-S.

DATA DIVISION.

FILE SECTION.

FD MASTER-FILE
    RECORD CONTAINS 80 CHARACTERS
    LABEL RECORDS ARE OMITTED
    DATA RECORD IS MASTER-RECORD-BUFFER.
01 MASTER-RECORD-BUFFER        PICTURE X(80).

FD TRANS-FILE
    RECORD CONTAINS 80 CHARACTERS
    RECORDING MODE IS F
    LABEL RECORDS ARE OMITTED
    DATA RECORD IS TRANS-RECORD.
01 TRANS-RECORD.
    05 STOCK-NUMBER             PICTURE X(6).
    05 DATE-LIMIT.
        10 YEAR                 PICTURE XX.
        10 DAY-OF-YEAR          PICTURE XXX.
    05 FILLER                   PICTURE X(69).
FD REPORT-FILE
    LABEL RECORDS ARE OMITTED.
    DATA RECORD IS REPORT-RECORD.
01 REPORT-RECORD.
    05 FORMS-CONTROL            PICTURE X.
    05 FILLER                   PICTURE X(132).

WORKING-STORAGE SECTION.

01 MASTER-RECORD.
    05 STOCK-NUMBER             PICTURE X(6)  VALUE LOW-VALUES.
    05 DESCRIPTION              PICTURE X(20).
    05 UNIT-PRICE               PICTURE 9(5)V99.
    05 QUANTITY-ON-HAND         PICTURE 9(5)V99.
    05 LAST-ACTIVITY-DATE.
        10 YEAR                 PICTURE XX.
        10 DAY-OF-YEAR          PICTURE XXX.
    05 FILLER                   PICTURE X(35).

01 FLAGS.
    05 MORE-TRANS-FLAG          PICTURE XXX   VALUE 'YES'.
        88 MORE-TRANS                         VALUE 'YES'.
        88 NO-MORE-TRANS                      VALUE 'NO'.
    05 VALID-TRANS-FLAG         PICTURE XXX.
        88 VALID-TRANS                        VALUE 'YES'.

01 MESSAGES.
    05 NO-ACTIVITY-MSG          PICTURE X(50)
        VALUE 'NO ACTIVITY FOR THIS ITEM SINCE DATE IN INQUIRY'.
    05 NO-MATCHING-MASTER MGS   PICTURE X(50)
        VALUE 'NO MASTER FOR THIS STOCK NUMBER'.
    05 NOT-NUMERIC-MSG          PICTURE X(50)
        VALUE 'ALL ITEMS IN INQUIRY MUST BE NUMERIC'.
    05 DATE-LIMIT-MSG           PICTURE X(50)
        VALUE 'DATE-LIMIT MUST NOT BE LESS THAN 70001'.

01 LINE-AND-PAGE-COUNTERS.
    05 LINE-NUMBER              PICTURE 99    VALUE 1.
    05 PAGE-NUMBER              PICTURE 999   VALUE 1.

01 OUTPUT-LINE.
    05 FORMS-CONTROL            PICTURE X.
    05 FILLER                   PICTURE X(132).

01 INQUIRY-RESPONSE.
    05 FORMS-CONTROL            PICTURE X.
    05 STOCK-NUMBER             PICTURE X(6).
    05 FILLER                   PICTURE XXX   VALUE SPACES.
    05 DATE-LIMIT               PICTURE X(5).
    05 FILLER                   PICTURE X(5)  VALUE SPACES.
    05 DESCRIPTION              PICTURE X(20).
    05 FILLER                   PICTURE XXX   VALUE SPACES.
    05 UNIT-PRICE               PICTURE $$$,$$9.99.
    05 FILLER                   PICTURE XXX   VALUE SPACES.
    05 QUANTITY-ON-HAND         PICTURE Z(4)9.99.
    05 FILLER                   PICTURE XXX   VALUE SPACES.
    05 COST-OF-STOCK            PICTURE $$,$$$,$$9.99.
    05 FILLER                   PICTURE XXX   VALUE SPACES.
    05 LAST-ACTIVITY-DATE.
        10 CENTURY              PICTURE XX    VALUE '19'.
        10 YEAR                 PICTURE XX.
        10 FILLER               PICTURE XX    VALUE SPACES.
        10 DAY-OF-YEAR          PICTURE XXX.
    05 FILLER                   PICTURE X(9)  VALUE SPACES.

01 MESSAGE-RECORD.
    05 FORMS-CONTROL            PICTURE X.
    05 STOCK-NUMBER             PICTURE X(6).
    05 FILLER                   PICTURE XXX   VALUE SPACES.
    05 DATE-LIMIT               PICTURE X(5).
    05 FILLER                   PICTURE X(5)  VALUE SPACES.
    05 MESSAGE-FIELD            PICTURE X(82).

01 HEADING-LINE.
    05 FORMS-CONTROL            PICTURE X.
    05 FILLER                   PICTURE X(46)
        VALUE'TRANSACTION           DESCRIPTION    UNIT'.
    05 FILLER                   PICTURE X(44)
        VALUE'PRICE    QOH        TOTAL COST LAST ACTIVITY'.
    05 PAGE-NUMBER-OUT          PICTURE Z(6)9.
```

```
PROCEDURE DIVISION.
000-ANSWER-INQUIRIES
    OPEN INPUT MASTER-FILE
               TRANS-FILE
         OUTPUT REPORT-FILE.
    PERFORM 100-PROCESS-TRANSACTIONS
         UNTIL NO-MORE-TRANS.
    CLOSE MASTER-FILE
          TRANS-FILE
          REPORT-FILE.
    STOP RUN.

100-PROCESS-TRANSACTION.
    MOVE 'NO' TO VALID-TRANS-FLAG.
    PERFORM 200-GET-VALID-TRANSACTION
        UNTIL VALID-TRANS
        OR NO-MORE-TRANS.
    IF VALID-TRANS

       PERFORM 300-PREPARE-RESPONSE
    ELSE
       NEXT SENTENCE.

200-GET-VALID-TRANSACTION.
    READ TRANS-FILE
        AT END MOVE 'NO' TO MORE-TRANS-FLAG.
    IF MORE-TRANS
       PERFORM 210-VALIDATE-TRANSACTION
    ELSE
        NEXT SENTENCE.

210-VALIDATE-TRANSACTION.
    MOVE 'YES' TO VALID-TRANS-FLAG.
    IF STOCK-NUMBER OF TRANS-RECORD IS NOT NUMERIC
        OR DATE-LIMIT OF TRANS-RECORD IS NOT NUMERIC

        MOVE CORRESPONDING TRANS-RECORD TO MESSAGE-RECORD
        MOVE NOT-NUMERIC-MSG TO MESSAGE-FIELD
        MOVE MESSAGE-RECORD TO OUTPUT-LINE
        PERFORM 220-LINE-OUT.
        MOVE 'NO' TO VALID-TRANS-FLAG
    ELSE
        IF DATE-LIMIT OF TRANS-RECORD IS LESS THAN '70001'

            MOVE CORRESPONDING TRANS-RECORD TO
                MESSAGE-RECORD
            MOVE DATE-LIMIT-MSG TO MESSAGE-FIELD
            MOVE MESSAGE-RECORD TO OUTPUT-LINE
            PERFORM 220-LINE-OUT
            MOVE 'NO' TO VALID-TRANS-FLAG
            ELSE
                NEXT SENTENCE.

300-PREPARE-RESPONSE.
    PERFORM 310-READ-MASTER
        UNTIL STOCK-NUMBER OF MASTER-RECORD
               IS EQUAL TO STOCK-NUMBER OF TRANS-RECORD
        OR STOCK-NUMBER OF MASTER-RECORD
               IS GREATER THAN STOCK-NUMBER OF TRANS-RECORD.

    IF STOCK-NUMBER OF TRANS-RECORD
       IS EQUAL TO STOCK-NUMBER OF MASTER-RECORD

       IF DATE-LIMIT OF TRANS-RECORD IS NOT LESS THAN
          LAST-ACTIVITY-DATE OF MASTER-RECORD

          MOVE CORRESPONDING TRANS-RECORD TO
              MESSAGE-RECORD
          MOVE NO-ACTIVITY-MSG TO MESSAGE-FIELD
          MOVE MESSAGE-RECORD TO OUTPUT-LINE
       ELSE
          MOVE CORRESPONDING MASTER-RECORD TO
              INQUIRY-RESPONSE
          MOVE DATE-LIMIT OF TRANS-RECORD TO
              DATE-LIMIT OF INQUIRY-RESPONSE
          MULTIPLY UNIT-PRICE OF MASTER-RECORD
              BY QUANTITY-ON-HAND OF MASTER-RECORD
              GIVING COST-OF-STOCK
          MOVE INQUIRY-RESPONSE TO OUTPUT-LINE
    ELSE
       MOVE CORRESPONDING TRANS-RECORD TO MESSAGE-RECORD
       MOVE NO-MATCHING-MASTER-MSG TO MESSAGE-FIELD
       MOVE MESSAGE-RECORD TO OUTPUT-LINE.
    PERFORM 220-LINE-OUT.
310-READ-MASTER.
    READ MASTER-FILE INTO MASTER-RECORD
        AT END MOVE HIGH-VALUES TO STOCK-NUMBER
            OF MASTER-RECORD.

220-LINE-OUT.
    IF LINE-NUMBER = 1

        MOVE PAGE-NUMBER TO PAGE-NUMBER-OUT
        WRITE REPORT-RECORD FROM HEADING-LINE
           AFTER ADVANCING PAGE
        WRITE REPORT-RECORD FROM OUTPUT-LINE
           AFTER ADVANCING 2
        MOVE 4 TO LINE-NUMBER
        ADD 1 TO PAGE-NUMBER
    ELSE
        WRITE REPORT-RECORD FROM OUTPUT-LINE
           AFTER ADVANCING 1.
    IF LINE-NUMBER = 55

        MOVE 1 TO LINE-NUMBER
    ELSE
        ADD 1 TO LINE-NUMBER.
```

TRANSACTION		DESCRIPTION	UNIT PRICE	QOH	TOTAL COST	LAST ACTIVITY	
000108	75001	DESK	$185.00	16.00	$2,960.00	1975	010
000180	75001	LAMP, FLOOR	$37.50	12.00	$450.00	1975	180
000200	75001	NO ACTIVITY FOR THIS ITEM SINCE DATE IN INQUIRY					
000250	75001	NO MASTER FOR THIS STOCK NUMBER					
000310	7500	ALL ITEMS IN INQUIRY MUST BE NUMERIC					
000310	75001	BOOKCASE, 4 SHELF	$36.50	31.00	$1,131.50	1975	090
000480	75140	NO ACTIVITY FOR THIS ITEM SINCE DATE IN INQUIRY					
000481	75140	FILE CABINET, 2 DWR	$39.90	50.00	$1,995.00	1975	150
010.68	57514	ALL ITEMS IN INQUIRY MUST BE NUMERIC					
010690	75150	NO MASTER FOR THIS STOCK NUMBER					
021739	75030	SOFA, LEATHER, RED	$356.20	37.00	$13,179.40	1975	040
031940	75150	NO MASTER FOR THIS STOCK NUMBER					

Questions for Review

1. Define structured programming.
2. What is a structured program and what difficulties does it create in COBOL programming?
3. What are the principal benefits of structured programming?
4. What are the main objectives of structured programming?
5. What is structured programming?
6. Briefly, what are the three basic structures of structured programming?
7. How are the basic structures combined into a program?
8. What is meant by "reduced program complexity" and "improved program clarity"?
9. What causes "jumps" within a program and what are the problems involved therein?
10. What is a segment?
11. What is a proper program?
12. What are the problems with using GO TO statements?
13. Why is it important that a program should proceed from top to bottom?
14. What are the three main ideas of structured programming?
15. What is a function?
16. What is a structured chart?
17. What is meant by structured design and what is its primary objective?
18. What are structured charts and what purpose do they serve?
19. What is the importance of indentation in a program?
20. What is "pseudo code" and what is its primary purpose?
21. What are the two requirements for a proper program?
22. What is the sequence control structure and how is it implemented in COBOL?
23. What is the selection control structure and how is it implemented in COBOL?
24. What is the iteration control structure and how is it implemented in COBOL?
25. What is the CASE structure?
26. What is the ordering of statements in a structured COBOL program?
27. What are the rules for construction of the Procedure Division?
28. How are blank lines used in coding line spacing?
29. List the rules for indenting and formatting structured programs.

Matching Questions

Match each item with its proper description.

______ 1. Top down design
______ 2. GO TO statements
______ 3. Selection
______ 4. Structured programming
______ 5. Segment
______ 6. CASE structure
______ 7. Sequence
______ 8. Indentation
______ 9. Proper program
______10. Iteration

A. Choice between two actions based on a condition.
B. Control logic restricted to three structures.
C. Used for repeated execution of code while a condition is true or false.
D. Exhibits in a pictorial way the relationships among statements.
E. Uses the value of a variable to determine which of several routines is to be executed.
F. A segment that has no infinite loops and no unreachable code.
G. A paragraph, section, subroutine, or code incorporated with a COPY.
H. Single point of transfer into or out of a structure.
I. Formularization of the idea that unless otherwise stated, statements are executed in the order in which they appear in a program.
J. Distracts reader from program by forcing him/her to examine a program in an unnatural way.

Exercises

Multiple Choice: Indicate the best *answer (questions 1-13).*

1. The objective(s) of structured programming are
   a. program development by segment.
   b. ease and speed of coding.
   c. ease of debugging.
   d. All of the above.
2. The following is a structure that is used in structured programming.
   a. GO TO
   b. DOWHILE
   c. PERFORM
   d. None of the above.
3. Each segment of a structured program
   a. must have at least one entry and may have multiple exits.
   b. may have many entries but only one exit.
   c. must have only one entry and one exit.
   d. may have many entries and many exits.
4. In structured programming,
   a. in place of PERFORM statements, GO TO statements are used.
   b. DOWHILE statements are used in place of PERFORM statements.
   c. in place of GO TO statements, PERFORM statements are used.
   d. GO TO statements are used with PERFORM statements.
5. The design of a proper program should proceed
   a. from top to bottom.
   b. in any logical sequence.
   c. from one paragraph to any paragraph in another segment.
   d. All of the above.
6. The structure chart
   a. is a set of program design techniques.
   b. shows the functions required for the program and their relationship.
   c. is a pseudo code that documents a program.
   d. All of the above.
7. The purpose of indentation of program statements is
   a. to make the structure more obvious to the reader.
   b. to exhibit in a pictorial way the relationships among statements.
   c. to enhance readability.
   d. All of the above.
8. Pseudo code
   a. enables programmers to express their ideas in a material form.
   b. allows the programmer to concentrate on the logic solutions of the problem.
   c. may replace flowcharts.
   d. All of the above.
9. A proper program
   a. has many entry and many exit points.
   b. has no paths from entry to exit.
   c. has exactly one entry point and one exit point for program control.
   d. may have infinite loops.
10. The choice between two actions based on a condition is called a(n)
    a. iteration structure.
    b. selection structure.
    c. sequence structure.
    d. CASE structure.
11. The construction of a segment of the Procedure Division is as follows:
    a. The PERFORM paragraph must appear in the first segment coding.
    b. The coding for a single logical level must be in several segments.
    c. The main line coding will appear in the first segment.
    d. It is not necessary for PERFORMed paragraphs to be close to the performing statements.

12. The formatting rule for indentation is as follows:
    a. Paragraph names will begin in position 12.
    b. More than one statement should be written on one line.
    c. The first statement in any program should begin in position 8.
    d. None of the above.
13. Some of the characteristics of structured COBOL programming are
    a. the presence of GO TO statements and corresponding statement labels.
    b. one or more entries and exits per segment.
    c. special indentation and nesting of functions.
    d. extensive use of comments.
14. Place the number in the second column (COBOL implementation) that correctly identifies the program structure.

Program Structure	*COBOL Implementation*
1. DOWHILE	_____ Writing statements in succession.
2. IFTHENELSE	_____ PERFORM with UNTIL option.
3. SEQUENCE	_____ IF statements.
4. CASE	_____ PERFORM with THRU option
	_____ GO TO with DEPENDING ON option.

15. **Rewrite the following statements using structured programming indenting and formatting standards.**
    a. IF TCODE IS EQUAL TO 20, ADD 7 TO AMOUNT, MOVE AMOUNT TO WORK-AMOUNT, ELSE SUBTRACT 10 FROM AMOUNT, MOVE AMOUNT TO EXTRA-WORK DIVIDE BALANCE BY 2 GIVING NEW-BAL.
    b. IF TRCODE IS EQUAL TO 50 NEXT SENTENCE ELSE ADD 100 TO NEW-AMT. MOVE NEW-AMT TO WORK-AMT.
    c. PERFORM DAILY-SALES-PROCESS VARYING DAY FROM 1 BY 1 UNTIL DAY IS GREATER THAN 7.
    d. OPEN INPUT QUANFILE, OUTPUT CALFILE, TFILE.
16. **Rewrite the following statements using structured programming indenting and formatting standards.**
    a. IF ACCT-NO = MACCOUNT-NO
       IF TCODE = 15
       ADD TRAN-AMT TO BALANCE
       ELSE SUBTRACT TRAN-AMT FROM BALANCE
       ELSE MOVE TRAN-FILE TO OUT-TRAN-FILE.
    b. MOVE SPACES TO FLD-A, FLD-B, FLD-C.
    c. IF TCODE = 9 AND AMT IS GREATER THAN 10 OR BALANCE IS LESS THAN ZERO ADD 700 TO EXCESS ELSE SUBTRACT 300 FROM OVER-DUE. MULTIPLY BALANCE BY 10 GIVING NEW-BALANCE.
17. **Rewrite the following statement in correct structured programming form.**
    PERFORM EDIT-ROUTINE UNTIL TRAN-NO IS EQUAL TO MAST-NO.
18. **What type of program structure is being used?**
    What is wrong with the indentation in the following statement? How can the statement be corrected to give the action represented by the indentation?

```
IF SIZE-CODE = 'M'
    IF TOTAL-PRICE IS GREATER THAN 500
        MOVE CHECK-PRICE-MESSAGE TO OUTPUT-LINE
        PERFORM PRINT-ROUTINE
ELSE
    MOVE BAD-SIZE-CODE-MESSAGE TO OUTPUT-LINE
    PERFORM PRINT-ROUTINE.
```

19. **Write a program segment using the CASE program structure based on the following information. Indicate each paragraph of the structure.**

```
IF TRAN-CODE = 1 GO TO DEPOSIT.
IF TRAN-CODE = 2 GO TO WITHDRAWAL.
IF TRAN-CODE = 3 GO TO INTEREST.
IF TRAN-CODE IS LESS THAN 1 OR GREATER THAN 3 GO TO NEXT SENTENCE.
```

Problem 1

Identify and label these control structures.

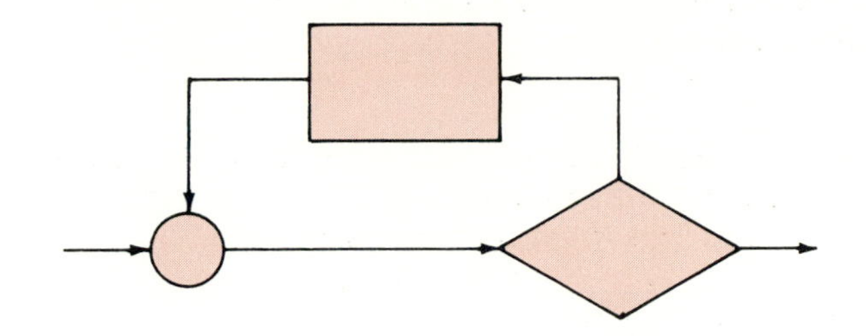

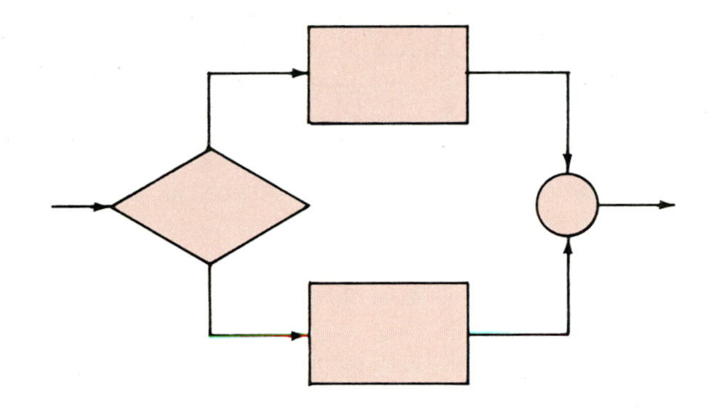

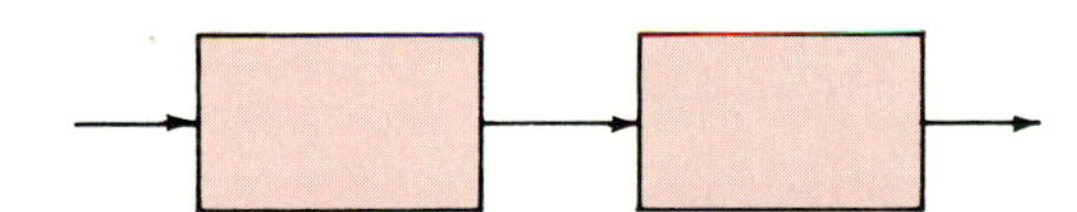

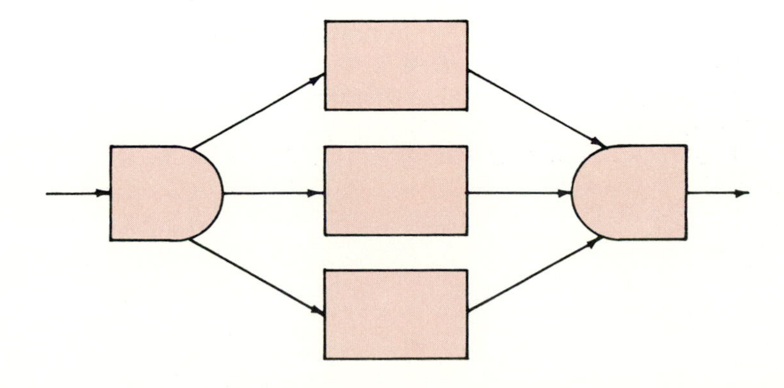

Problem 2

Using the flowchart and pseudo code just given, construct a flowchart using the CASE structure. Write the pseudo code for the CASE structure.

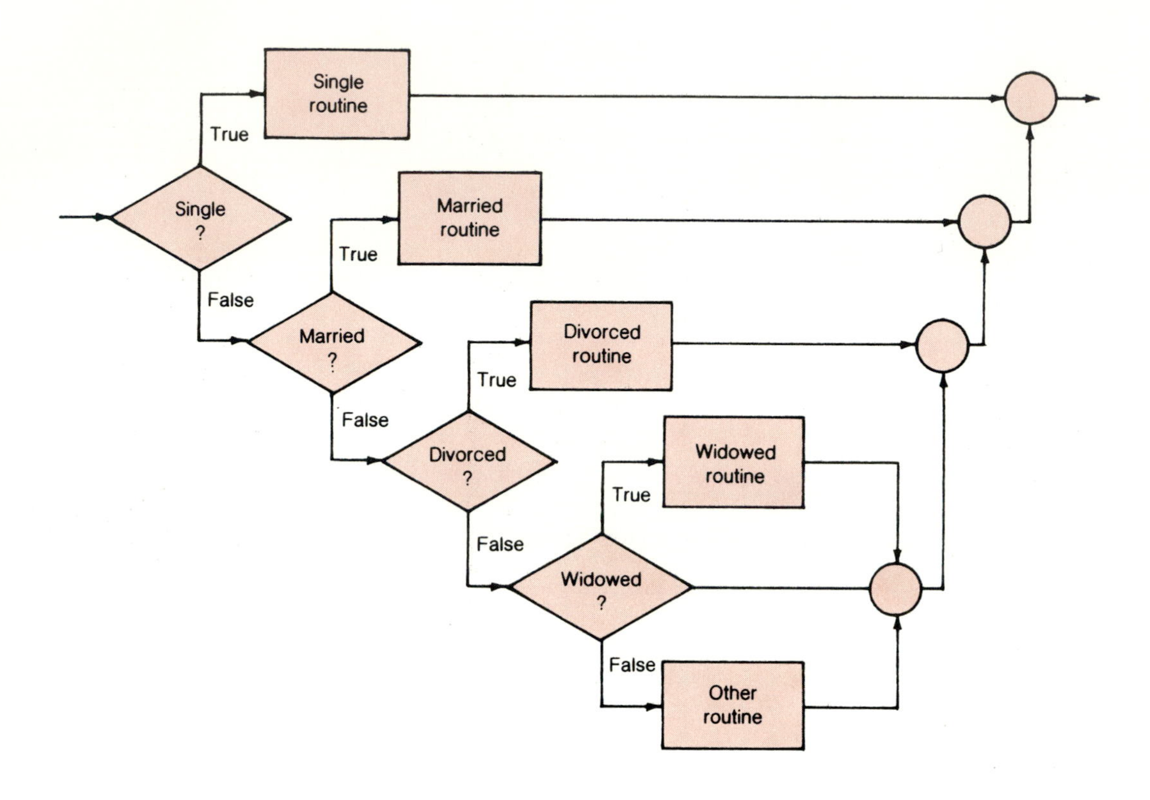


```
IF SINGLE THEN
     SINGLE ROUTINE
ELSE
     IF MARRIED THEN
          MARRIED ROUTINE
     ELSE
          IF DIVORCED THEN
               DIVORCED ROUTINE
          ELSE
               IF WIDOWED THEN
                    WIDOWED ROUTINE
               ELSE
                    OTHER ROUTINE
               ENDIF
          ENDIF
ENDIF
```

Problem 3

Write the pseudo code for the following flowchart.

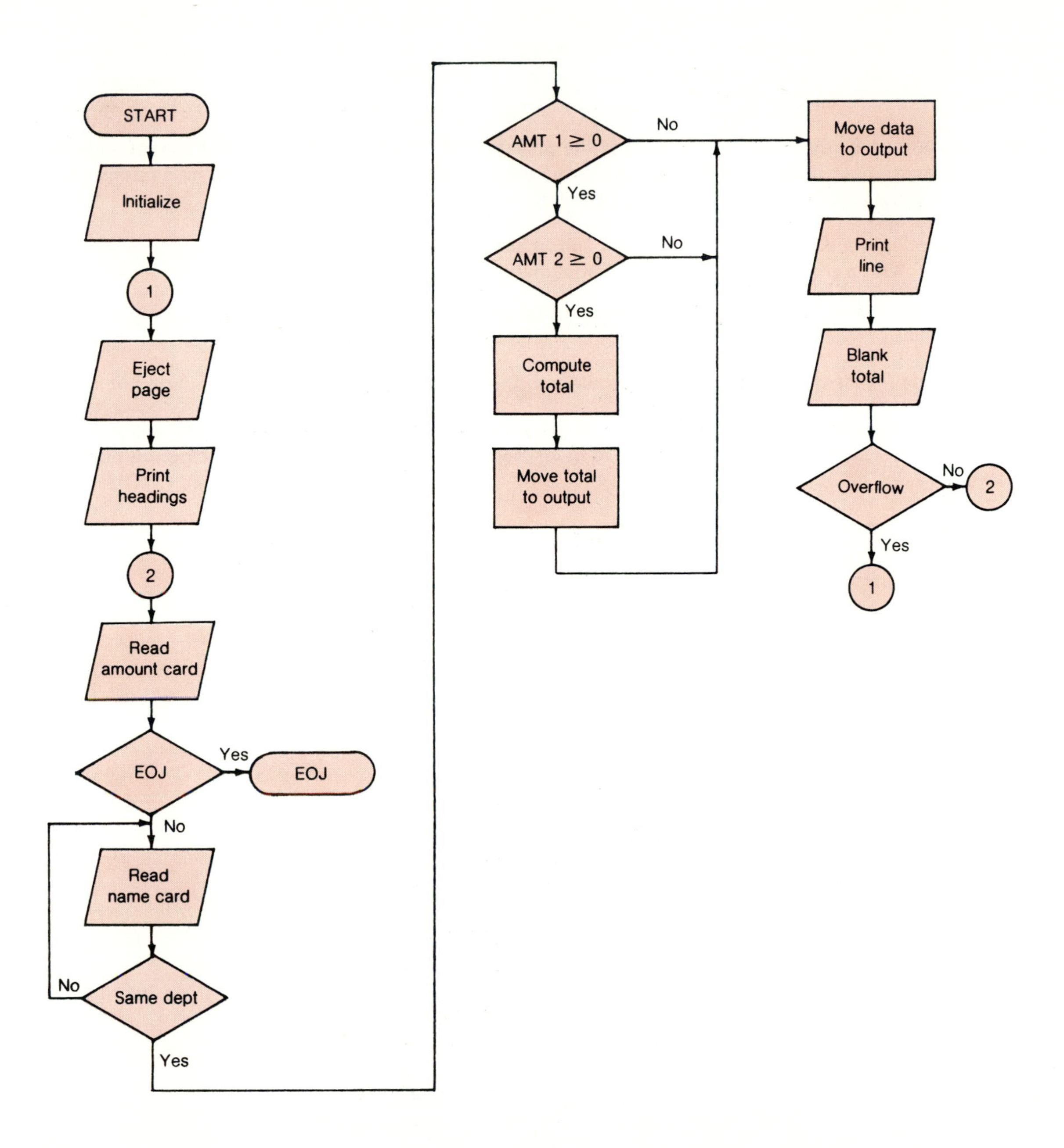

Problem 4

Code the following problem in pseudo code.

This Air Pollution problem concerns the recording of measurements of pollutants in the air near a large manufacturing plant.

General Description:

Air pollution measurements have been made near the smokestack of a large manufacturing plant every minute for a 24-hour period. Valid measurements range in value from one part per million (PPM) to 180,000 PPM of pollutant.

Functional Requirements Specifications:

1. **The measurements are stored as card images in the project TEST data set. There are ten measurements on each card. Each measurement takes seven positions on the card. A blank follows each of the measurements. The first measurement on each card image begins in card column 1.**

2. **Assume that the input measurements have been edited and all values are numeric and valid*, within the range of 1 to 180,000 PPM, with a zero value indicating an equipment malfunction.**

**Note:* The assumption of valid input is made to simplify the problem. Appropriate input checking should always be part of your problem solutions.

3. **Compute average (or mean) pollution values for each of the 24 hours for which no malfunction has occurred.**

4. **When an equipment malfunction occurs (represented by a zero measurement value), set the mean PPM value of the current hour to 1,000 PPM; bypass remaining measurements for the hour before starting the next hour's processing.**

5. **Print hourly mean values in the format shown below:**

HOUR	*MEAN PPM*
1	80864
2	86681
3	87540
4	75068
5	90222
6	87838
7	80653
8	105286
9	98699
10	82614
11	90411
12	1000
13	1000
14	86609
15	81348
16	87763
17	98166
18	86595
19	91963
20	88664
21	98657
22	91011
23	81726
24	92216

Procedure Division: Input / Output Operations

Chapter Outline

Chapter Objectives

1. To describe the formats and the various units of data used in the Procedure Division, thus enabling you to code this division properly.
2. To describe the functions, uses and formats of the various input / output COBOL verbs.
3. To explain the uses of the MOVE and PERFORM verbs in structured programming, thus enabling you to write simple listing programs.

Introduction

To some people, the Procedure Division represents "the program," for it consists of procedures which the computer is to follow in processing the data. The entries in this division are very similar to English.

The Procedure Division contains instructions that direct the data processing activities of the computer (see figure 6.1.); that specify the actions—such as input/output, data manipulation, arithmetic, and sequence control—that are required to process the data; and that control the sequence in which these actions are to be carried out.

The order in which the programmer codes the statements in this division will depend on the order of the data processing activities or the logic of the problem solution. (See figure 6.2.)

Units of Data

The following are units of expression that constitute the Procedure Division. These units may be combined to form larger units.

Procedure

A procedure is composed of a paragraph, or a group of successive paragraphs or a section, or a group of successive sections within the Procedure Division. See the following example:

```
DISTRICT-TOTAL-PROCESSING.
    MOVE ZERO TO DISTRICT-TOTAL.
    MOVE DISTRICT TO PREVIOUS-DISTRICT.
    PERFORM SALESMAN-TOTAL-PROCESSING
        UNTIL DISTRICT IS NOT EQUAL TO PREVIOUS-DISTRICT
        OR NO-MORE-DATA.
    MOVE SPACES TO REPORT-RECORD.
    MOVE PREVIOUS-DISTRICT TO DISTRICT-OUT.
    MOVE DISTRICT-TOTAL TO DISTRICT-TOTAL-OUT.
    WRITE REPORT-RECORD.
    ADD DISTRICT-TOTAL TO FINAL-TOTAL.

SALESMAN-TOTAL-PROCESSING.
    MOVE ZERO TO SALESMAN-TOTAL.
    MOVE SALESMAN TO PREVIOUS-SALESMAN.
    PERFORM PROCESS-AND-READ
        UNTIL DISTRICT IS NOT EQUAL TO PREVIOUS-DISTRICT
        OR SALESMAN IS NOT EQUAL TO PREVIOUS-SALESMAN
        OR NO-MORE-DATA.
    MOVE SPACES TO REPORT-RECORD.
    MOVE PREVIOUS-SALESMAN TO SALESMAN-OUT.
    MOVE SALESMAN-TOTAL TO SALESMAN-TOTAL-OUT.
    WRITE REPORT-RECORD.
    ADD SALESMAN-TOTAL TO DISTRICT-TOTAL.
```

If one paragraph is in a section, then all paragraphs must be in sections. A *procedure-name* is a word used to refer to a paragraph or section in the source program in which it occurs. It consists of a paragraph-name (which may be qualified), or a section-name.

The end of the Procedure Division and the physical end of the program is that physical position in a COBOL source program after which no further procedures appear.

Statement

A statement is a syntactically valid combination of words and symbols beginning with a COBOL verb. It begins with a verb and is completed by a combination of words that designate the data to be acted upon and that, at times, amplify the instructions. The portion of the statement following the verb consists of key words, optional words, and operands.

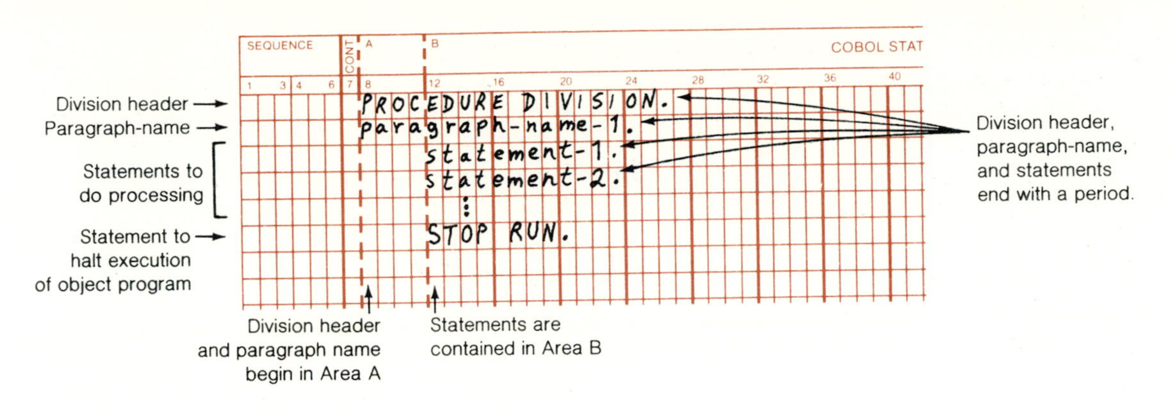

Figure 6.1 Guide for coding Procedure Division entries.

Figure 6.2 Procedure Division flowcharts—examples.

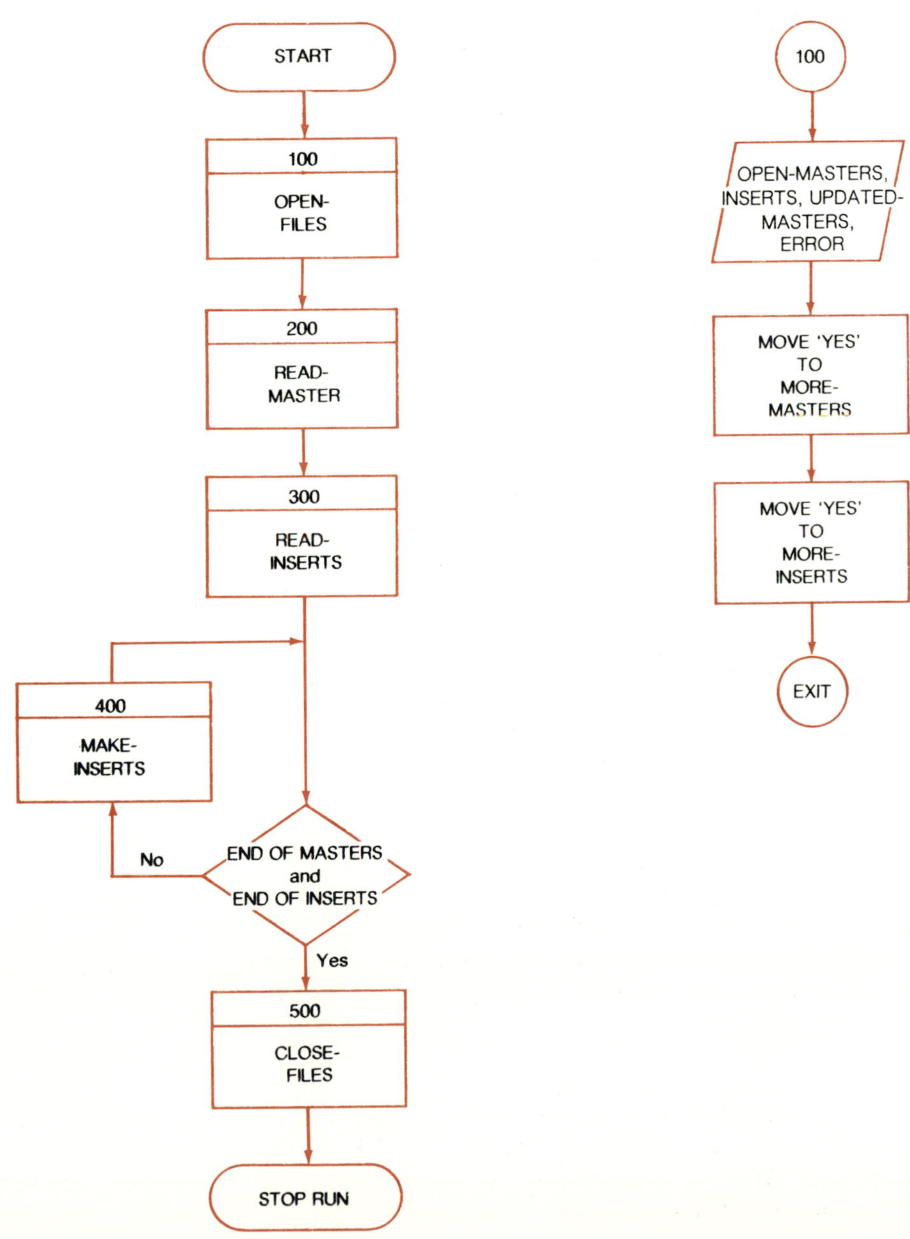

Figure 6.2 *(continued)*

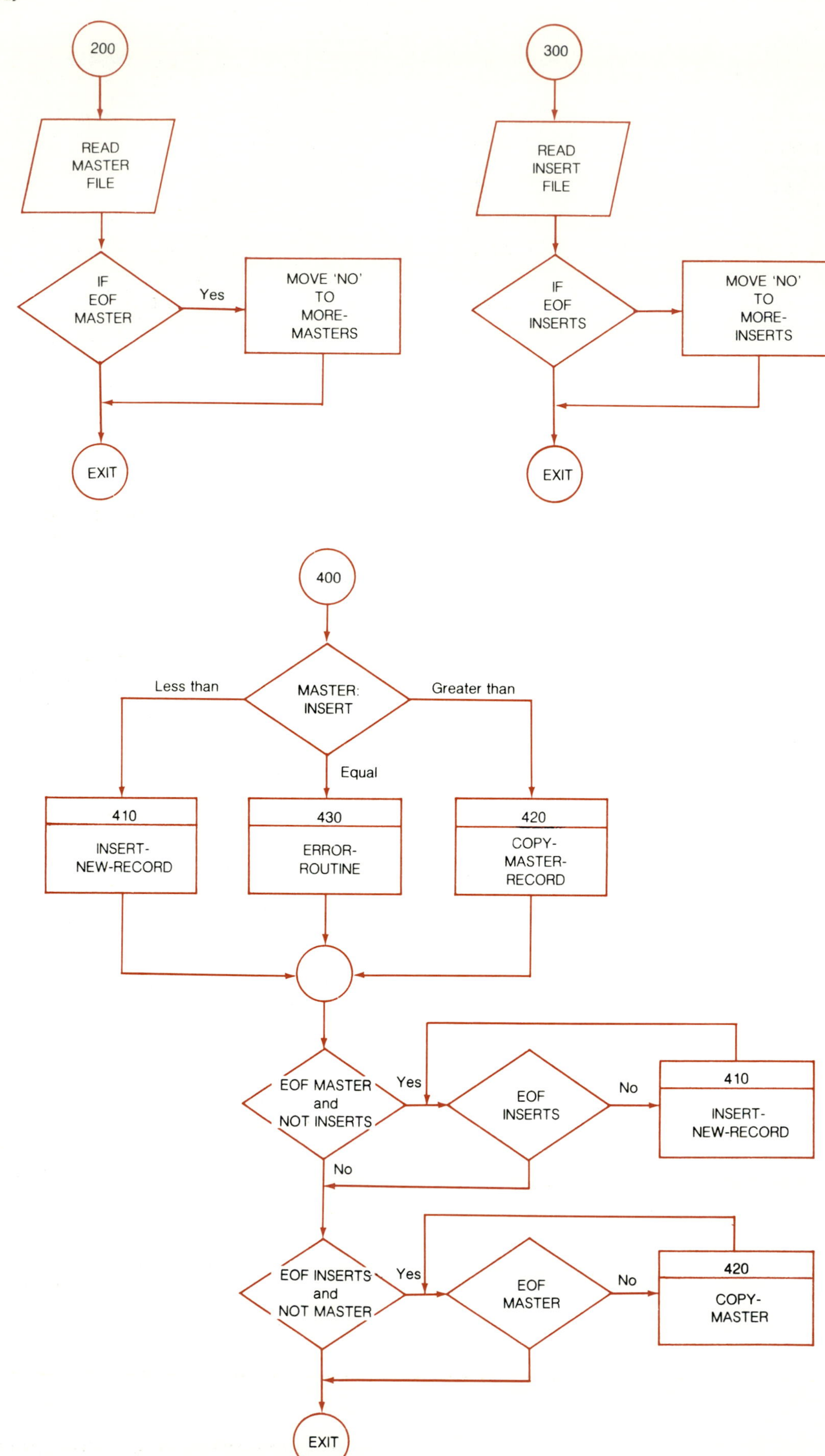

Figure 6.2 *(continued)*

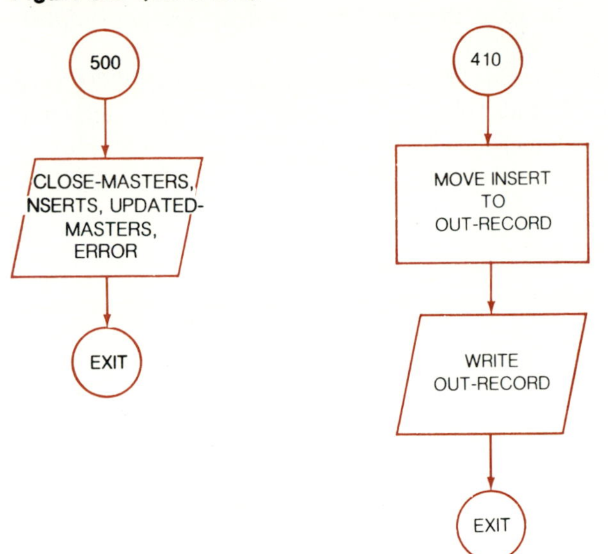

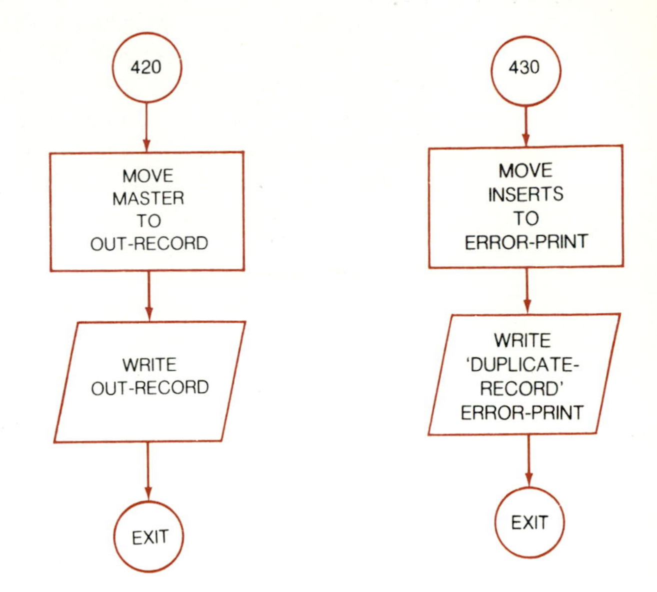


```
PROCEDURE DIVISION.
000-MAIN-ROUTINE.
    PERFORM 100-OPEN-FILES.
    PERFORM 200-READ-MASTER.
    PERFORM 300-READ-INSERTS.
    PERFORM 400-MAKE-INSERTS.
        UNTIL MORE-MASTERS-FLAG = 'NO'
        AND MORE-INSERTS-FLAG = 'NO'.
    PERFORM 500-CLOSE-FILES.

100-OPEN-FILES.
    OPEN INPUT    MASTERS,
                  INSERTS,
         OUTPUT   UPDATED-MASTERS,
                  ERROR.
    MOVE 'YES' TO MORE-MASTERS-FLAG
    MOVE 'YES' TO MORE-INSERTS-FLAG.

200-READ-MASTER.
    READ MASTERS
        AT END MOVE 'NO' TO MORE-MASTERS-FLAG.

300-READ-INSERTS.
    READ INSERTS
        AT END MOVE 'NO' TO MORE-INSERTS-FLAG.

400-MAKE-INSERTS.
    IF INSERT-NUMBER < MASTER-NUMBER
        PERFORM 410-INSERT-NEW-RECORD
    ELSE
        NEXT SENTENCE.
    IF INSERT-NUMBER > MASTER-NUMBER
        PERFORM 420-COPY-MASTER-RECORD
    ELSE
        NEXT SENTENCE.
    IF INSERT-NUMBER = MASTER-NUMBER
        PERFORM 430-ERROR-ROUTINE
    ELSE
        NEXT SENTENCE.
    IF MORE-MASTERS-FLAG = 'NO'
        AND MORE-INSERTS-FLAG = 'YES'
            PERFORM 410-INSERT-NEW-RECORD
            UNTIL MORE-INSERTS-FLAG = 'NO'
    ELSE
        NEXT SENTENCE.
    IF MORE-MASTERS-FLAG = 'YES'
        AND MORE-INSERTS-FLAG = 'NO'
            PERFORM 420-COPY-MASTER
            UNTIL MORE-MASTERS-FLAG = 'NO'.

410-INSERT-NEW-RECORD.
    MOVE INSERTS TO OUT-RECORD.
    WRITE OUT-RECORD.
    PERFORM 300-READ-INSERTS.

420-COPY-MASTER-RECORD.
    MOVE MASTER TO OUT-RECORD.
    WRITE OUT-RECORD.
    PERFORM 200-READ-MASTER.

430-ERROR-ROUTINE.
    MOVE INSERTS TO ERROR-PRINT
    WRITE 'DUPLICATE RECORD' ERROR-PRINT

500-CLOSE-FILES.
    CLOSE MASTERS
          INSERTS
          UPDATED-MASTERS
          ERROR.
```

Each statement has its own particular format that specifies the types of words required and the organization of the statement. It is necessary to include all words required in a statement and to present them to the compiler according to the prescribed structure. If these two requirements are not fulfilled, the compiler will not interpret the statements.

The statement is the basic unit of the Procedure Division. A statement consists of a COBOL verb or the words IF or ON followed by the appropriate operands (file-names, literals, data-names, etc.) and other COBOL words that are essential to the completion of the statement. COBOL statements may be compared to clauses in the English language. The statement may be one of three types: imperative, conditional, or compiler-directing.

Imperative Statement

An imperative statement indicates a specific unconditional action to be taken by the object program. See the following example:

```
MOVE CATALOG-NUMBER TO CONTROL-ITEM.
```

An imperative statement is any statement that is neither a conditional statement nor a compiler-directing statement. An imperative statement may consist of a sequence of imperative statements, each possibly separated from the next by a separator.

When the term 'imperative-statement' appears in the general format of statements, it refers to that sequence of consecutive imperative statements that must be ended by a period; or an ELSE phrase associated with a previous IF statement; or a WHEN phrase associated with a previous SEARCH statement.

An imperative statement directs the computer to perform certain specified actions. These actions are specified and unequivocal, and the computer does not have the option of not performing them. An example of an imperative statement follows:

```
SUBTRACT DEDUCTIONS FROM GROSS GIVING NET-PAY.
```

Conditional Statement

A conditional statement specifies that the truth value of a condition is to be determined and that the subsequent action of the object program is dependent upon this truth value.

```
IF MONTH EQUALS FIRST
    MOVE ZERO TO BALANCE
ELSE
    MOVE LAST-AMT TO BALANCE.
ADD INVOICE-AMT TO TOTAL
    ON SIZE ERROR
        PERFORM 400-SIZE-ERROR-ROUTINE.
```

The modification of an imperative statement permits the computer to perform an operation under certain conditions. If the programmer attaches one or more conditional statements to an imperative statement, then the entire statement becomes a conditional statement.

In a conditional statement, the stated action is performed only if the specified conditions are present. Some examples of conditional statements are found in a READ statement, an arithmetic statement with the ON SIZE ERROR option, and IF statements. These and other statements involving conditions will be discussed later in the chapter. Here is an example of a conditional statement:

```
IF AGE IS LESS THAN 21 PERFORM MINOR.
```

Compiler-Directing Statement

A compiler-directing statement consists of a compiler-directing verb and its operands. The compiler-directing verbs are COPY, ENTER, and USE. *A compiler-directing statement causes the compiler to take a specific action during compilation of the source program.* See the following example:

```
ENTER PARA-A.
```

These statements will be discussed in greater detail further along in the text.

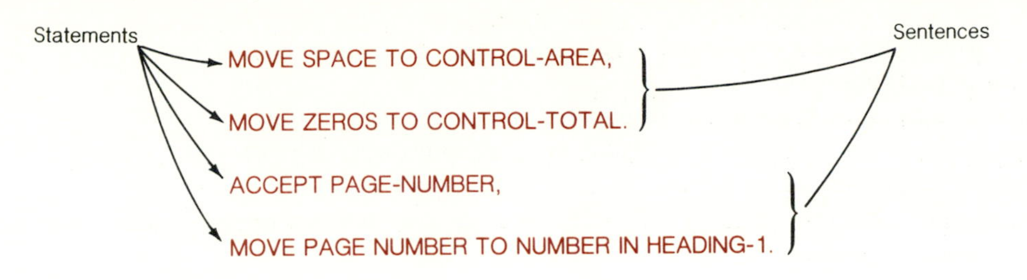

Figure 6.3 Difference-statement and sentence.

Sentence

A sentence is composed of one or more statements specifying action to be taken, and is terminated by a period followed by a space (fig. 6.3). Commas or semicolons may be used as separators between statements. When separators are used, they must be followed by a space. Separators improve readability, but their absence or presence has no effect upon the compilation of the object program. An example of a sentence follows:

```
ADD EARNINGS TO GROSS, PERFORM FICA-PROC.
```

Sentences may be either imperative, conditional, or compiler-directing. The following are imperative sentences:

```
MOVE COMPUTED-PAY TO NET-PAY.
ADD EARNING, OVERTIME
    GIVING WAGES.
MOVE TAX TO REPORT.
```

The following are conditional sentences:

```
IF CODE-1 IS EQUAL TO 4
    PERFORM PROCESS-PATH
ELSE
    NEXT SENTENCE.

IF OVERTIME
    MULTIPLY OVERTIME-HRS BY OVERTIME-RATE
        GIVING OVERTIME-PAY
    MULTIPLY REGULAR-HRS BY REGULAR-RATE
        GIVING REGULAR-PAY
    ADD OVERTIME-PAY, REGULAR-PAY
        GIVING TOTAL-PAY
ELSE
    MULTIPLY REGULAR-HRS BY REGULAR-RATE
        GIVING TOTAL-PAY.
```

Paragraph

In the Procedure Division, statements that are logically related are grouped into paragraphs. Every program must have at least one paragraph in the Procedure Division, and each paragraph must be identified by a paragraph-name preceding the first statement.

A paragraph consists of a paragraph-name followed by a period and a space, and by zero, one or more successive sentences. See the following example:

```
INSERT-NEW-RECORD.
    MOVE CARD-RECORD TO OUTPUT-RECORD.
    WRITE OUTPUT-RECORD.
    READ CARD-FILE
        AT END PERFORM FINISH-RUN.
```

A paragraph ends immediately before the next paragraph-name or section-name, or at the end of the Procedure Division.

Some of the rules for the use of paragraphs are:

1. Each paragraph must begin with a paragraph-name. Statements may be written on the same line as the paragraph-name.
2. A paragraph-name must not be duplicated within the same section.
3. Paragraph-names follow the same rules as data-names with one exception: a paragraph-name may be made up entirely of numerals.
4. A paragraph ends immediately before the next paragraph-name or section-name, or at the end of the Procedure Division.

Sections

The section in a COBOL program is the largest unit to which a procedure-name may be assigned.

1. A section is composed of one or more successive paragraphs.
2. A section must begin with a section header (a procedure-name) followed by a space and the word SECTION followed by a period. The section header must appear on a line by itself.
3. The Procedure Division need not be broken down into sections. Section usage is at the discretion of the programmer.
4. A section ends immediately before the next section name or at the end of the Procedure Division.
5. If sections are used in the Procedure Division then the entire division must be composed of sections.

A section consists of a section header, followed by a period and a space, and by zero, one or more successive paragraphs. A section ends immediately before the next section-name or at the end of the Procedure Division. See the following example:

```
CHECK-STATUS SECTION.
INPUT-ROUTINE.
    OPEN      INPUT      RAWMFILE.
    PERFORM PROCESS-INPUT-ROUTINE
        UNTIL NO-MORE-INPUT.
    CLOSE      RAWMFILE.
    GO TO CHECK-STATUS-EXIT.

PROCESS-INPUT-ROUTINE.
    IF ACCOUNT-STATUS IS EQUAL TO ZERO.
        NEXT SENTENCE
    ELSE
        RELEASE SRECORD FROM RAWMRECORD.
CHECK-STATUS-EXIT.      EXIT.

CHECK-BALANCE SECTION.
OUTPUT-ROUTINE.
    OPEN      OUTPUT      MFILE.
    PERFORM PROCESS-OUTPUT-ROUTINE
        UNTIL NO-MORE-OUTPUT.
    CLOSE      MFILE.
    GO TO CHECK-BALANCE-EXIT.

PROCESS-OUTPUT-ROUTINE.
    RETURN SFILE INTO MRECORD
        AT END MOVE 'NO' TO MORE-OUTPUT-FLAG.
    IF BALANCE IS EQUAL TO ZERO.
        DISPLAY 'ACCOUNT NUMBER' MNOO 'HAS A ZERO BALANCE'
            UPON PRINTER
  . ELSE
        WRITE MRECORD.
CHECK-BALANCE-EXIT.      EXIT.
```

It is important for the programmer to note that the only way a section can be ended is by either another section, or its appearance at the end of the Procedure Division. Since sections in the Procedure Division are created by the programmer, one section may run into another paragraph erroneously without the programmer being aware of it. Great care should be exercised by the programmer in the creation of sections, and they should only be used where absolutely necessary.

Summary: Organization of the Procedure Division

The Procedure Division consists of instructions that are written in statement form, and which may be combined to form sentences. Groups of sentences form paragraphs.

The Procedure Division generally consists of a series of paragraphs which may be optionally grouped into programmer-created sections. Each paragraph has a data-name and may consist of a varying number of entries (fig. 6.4).

The Procedure Division contains *procedures.*

Procedures are composed of paragraphs, groups of successive paragraphs, a section, or a group of successive sections within the Procedure Division. Paragraphs need not be grouped into sections. Execution begins with the first statement of the Procedure Division. Statements are then executed in the sequence in which they are written, unless altered by the program.

The end of the Procedure Division is the physical end of a COBOL program after which no further procedures may appear.

The term 'identifier' is defined as the word or words necessary to make a unique reference to a data item.

The Procedure Division will be coded using the structure charts as a guide for the overall logic of the program's procedures, and the flowchart for the individual processing steps. (Pseudo codes may also be used in conjunction with the flowcharts.)

Execution of the object program begins with the first statement of the Procedure Division.

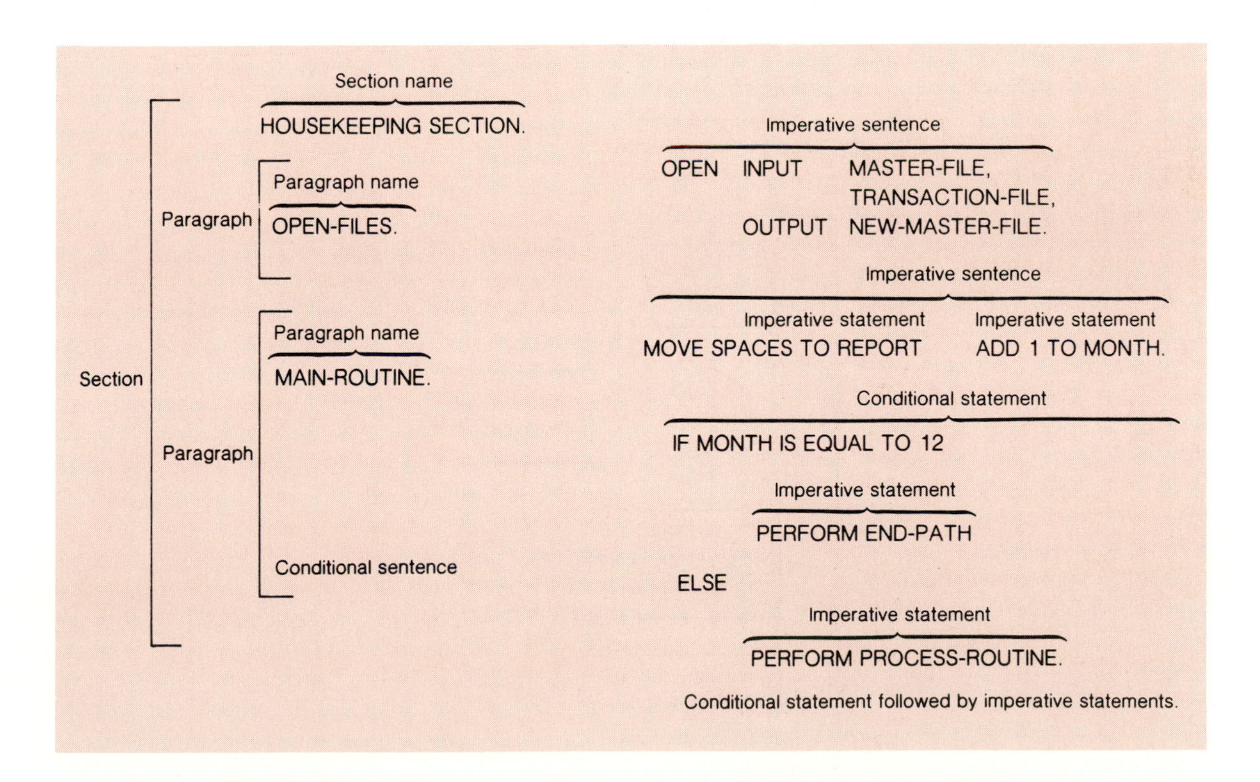

Figure 6.4 Procedure Division structure.

COBOL Verbs

COBOL Verbs are the basis of the Procedure Division of a source program. The organization of the division is based on the classification of COBOL verbs to be found below. (Other COBOL verbs will be discussed further along in the text.)

Input/Output	*Arithmetic*	*Data Manipulation*	*Sequence Control*	*Compiler-Directing*
ACCEPT	ADD	EXAMINE	GO TO	COPY
CLOSE	COMPUTE	INSPECT	PERFORM	ENTER
DISPLAY	DIVIDE	MOVE	STOP	USE
OPEN	MULTIPLY	STRING		
READ	SUBTRACT	UNSTRING		
WRITE				

Each of the verbs causes some event to take place either at compilation time or at program-execution time.

Input/Output Verbs

In data processing operations, the flow of data through a system is governed by an input/output system. The COBOL statements discussed in this section are used to initiate the flow of data that is stored on an external media device, such as punched cards, disk, or magnetic tape, and to govern the flow of low-volume information that is to be obtained from or sent to an input/output device, such as a console.

The programmer is concerned only with the use of individual records. The input/output system provides for operations such as the movement of data into internal storage, validity checking, and unblocking and blocking of physical records.

One of the important advantages of COBOL programming is the use of pretested input and output statements to get data into and out of data processing systems. The COBOL input and output verbs provide the means of storing data on an external device (magnetic tape, disk units, as well as on card readers, punches, and printers) and extracting such data from these external devices.

Four verbs, OPEN, READ, WRITE, and CLOSE, are used to specify the flow of data to and from files stored in an external media device. ACCEPT and DISPLAY are used in conjunction with low-

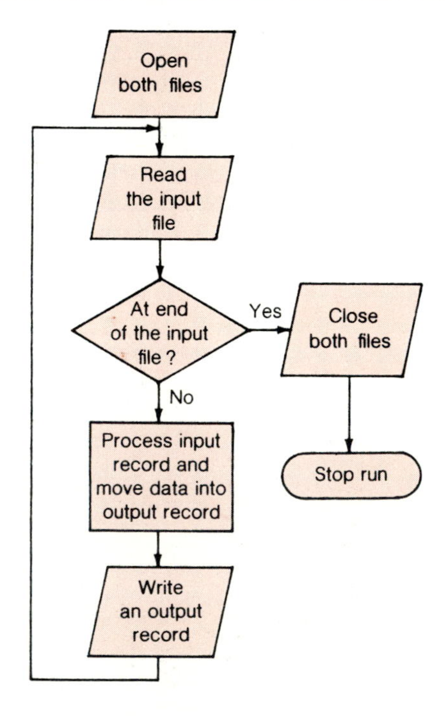

Figure 6.5 Overall logic input and output (sequential files).

```
        { INPUT file-name-1     [ , file-name-2 ] . . . }
OPEN    { OUTPUT file-name-3    [ , file-name-4 ] . . . }  . . .
        { I-O file-name-5       [ , file-name-6 ] . . . }
```

Figure 6.6 Format—OPEN statements.

volume data that has to be obtained or sent to a card reader, disk storage, console typewriter, or printer. (See figure 6.5.)

OPEN Statement

The statements in the Procedure Division are used to process data from record variables described in the Data Division. Before a file may be used, however, it must be prepared for processing. This is done at the beginning of the Procedure Division with an OPEN statement.

The OPEN statement makes one or more input or output files ready for reading or writing, checks or writes labels if needed, and prepares the storage areas to receive or send data (figs. 6.6 and 6.7).

Rules Governing the Use of the OPEN Statement

1. An OPEN statement must be specified for all files used in a COBOL program. The file must be designated as either INPUT, OUTPUT, or I-O (mass storage files).
2. An OPEN statement must be executed prior to any other input or output statement for a particular file. Prior to the successful execution of an OPEN statement for a given file, no statement that references that file may be executed. The successful execution of the OPEN statement determines the availability of the file and results in the file being in an open mode.
3. If the file has been closed during the processing, a second OPEN statement must be executed before the file can be used again.
4. The OPEN statement does not make input records available for processing, nor does it release output records to their respective devices. The successful execution of an OPEN statement makes the associated record area available to the program. A READ or WRITE statement respectively is required to perform the functions of making records available for processing or for output.
5. When a file is opened, such actions as checking and creating beginning file labels are done automatically for those files requiring such action.
6. An OPEN statement can name one or all the files to be processed by the program.
7. Each file that has been opened must be defined in the file description entry in the Data Division as well as the SELECT entry in the Environment Division.
8. At least one of the three optional clauses (INPUT, OUTPUT, or I-O) must be written.
9. The I-O option permits the opening of a mass storage file for both input and output operations. Since this option implies the existence of a file, it cannot be used if the mass storage file is being created.

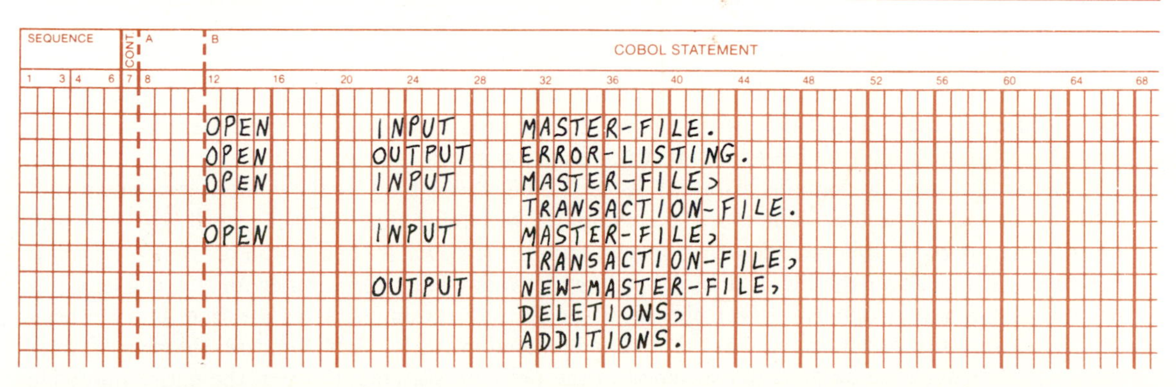

```
OPEN    INPUT     MASTER-FILE.
OPEN    OUTPUT    ERROR-LISTING.
OPEN    INPUT     MASTER-FILE,
                  TRANSACTION-FILE.
OPEN    INPUT     MASTER-FILE,
                  TRANSACTION-FILE,
        OUTPUT    NEW-MASTER-FILE,
                  DELETIONS,
                  ADDITIONS.
```

Figure 6.7 OPEN statement—example.

```
READ file-name RECORD  [ INTO identifier ]  [ ; AT END imperative-statement ]
```

Figure 6.8 Format—READ statement.

READ Statement

The READ statement makes available the next logical record for processing and allows the performance of one or more specified statements when the end of the file is detected. The general format for the READ statement appears in figure 6.8. Information contained in that record can subsequently be referred to in the program references to the appropriate variables. A READ statement must be executed before the data from a record can be processed.

Rules Governing the Use of the READ Statement

1. The data records are made available in the input block one at a time. The record is available in the input until the next READ or a CLOSE statement is executed.
2. If the file contains more than one type of record for a file being executed, the next record is made available regardless of type. If more than one record description is specified in the FD entry, it is the programmer's responsibility to recognize which record is in the input block at any one time, since these records automatically share the same storage area, one that is equivalent to an implicit redefinition of the area. The programmer cannot specify the type of record to be read because the format of the READ statement requires the name of a file, not a record.
3. The file must be opened before it can be read.

```
OPEN    INPUT    FORECAST-FILE
        OUTPUT   REPORT-FILE.
READ FORECAST-FILE;
    AT END MOVE 'NO' TO MORE-DATA-FLAG.
```

4. When the end of a volume is reached for multivolume files, such as tape files, volumes are automatically switched, the tape is rewound, and the next reel is read. All normal header and trailer labels are checked.
5. The INTO phrase is used to transmit a record from an input file to a Working-Storage variable or an output area associated with a previously opened file.

   The INTO option converts the READ statement into a READ and MOVE statement. The identifier must be the name of a Working-Storage Section entry or a previously opened record. The current record is now available in the input area as well as the area specified by the identifier.

```
READ MASTER-FILE INTO MASTER-WORK
    AT END PERFORM END-DATA-MASTER.
```

   If the format of the INTO area is different from the input area, the data is moved into that area in accordance with the rules for the MOVE statement *without* the CORRESPONDING option. *Note:* The largest record may be described in any 01 level entry; it need not be the first 01 entry. Using the INTO option, data is moved using the size of the largest record specified in the file description (FD) entry as the sending field size.

   The INTO phrase must not be used when the input file contains logical records of various sizes as indicated by their record descriptions. The storage area associated with the identifier and the record area associated with file-name must not be the same storage area.
6. The last record in a file contains a special code which is understood by the computer to mean that "this is the end of the file." Therefore, the computer will know that the last record has been processed when it reads this record with a special end-of-file code. Although the computer knows it has reached the end of the file, it doesn't know what to do. The programmer must specify an imperative statement in the AT END phrase of the READ statement after all records have been processed. In order to determine what should be included in the AT END phrase, the programmer must decide what is to be done (as specified in the problem statement) when the entire input file has been processed.

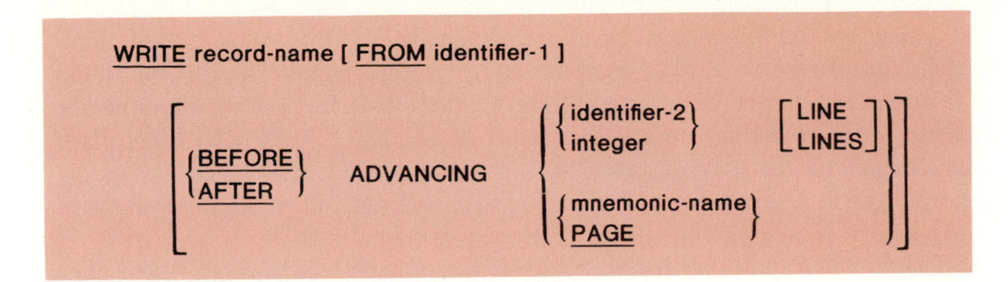

Figure 6.9 Format—WRITE statement.

7. An AT END phrase must be included in all READ statements for sequential input files. The statements following the records AT END up to the period are taken to be the end of file conditions. When the AT END clause is encountered, the last data record of the file has already been read.
8. Once the imperative statements in the AT END phrase have been executed for a file, any later referral to the file will constitute an error unless subsequent CLOSE and OPEN statements for that file are executed.

The READ statement does double duty. It not only makes a record available for processing from an input file, but also determines what to do when the end of the file is reached.

WRITE Statement

The WRITE statement releases a logical record for insertion in an output file. It can also be used for vertical positioning of lines within a logical page. (See figures 6.9 and 6.10.)

Rules Governing the Use of the WRITE Statement

1. If the records are blocked, the actual transfer of the data to the output block may not occur until later in the processing cycle when the output block is filled with the number of records specified in the file description entry in the Data Division.
2. When an end of volume is reached for multivolume files, such as magnetic tape files, volumes are switched, the tape is rewound, and the next reel is written. All normal standard header and trailer labels are written.
3. An OPEN statement must be executed prior to the execution of the first WRITE statement.
4. After the record has been released, the logical record named by the record-name may be no longer available for processing. All necessary processing of a record must be done prior to the WRITE statement.
5. The file associated with the record-name must be defined in the FD entry in the Data Division of the program. When a WRITE statement is executed, the record-name record is released to the output device.
6. The format requires a *record-name* rather than a file-name.

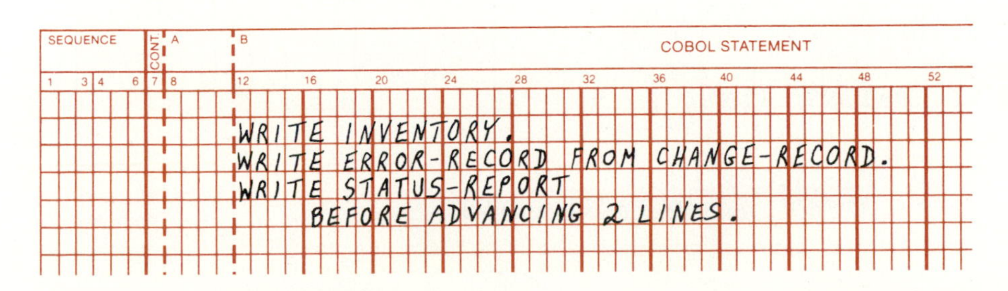

This will cause the report to be double spaced.

Figure 6.10 WRITE statement—examples.

7. The FROM option is used to transmit a record from Working-Storage variable or an input area associated with a previously opened file to an output area. When the FROM option is used, identifier-1 must not be the name of an item in the file containing the record-name. The FROM option converts the WRITE statement into a MOVE and WRITE statement. Identifier-1 must be the name of an item defined in the Working-Storage Section or in another FD. Moving takes place according to the rules specified for the MOVE statement without the CORRESPONDING option. After the execution of the WRITE statement with the FROM option, the information is still available in identifier-1 although it is no longer available in the record-name area.
8. The ADVANCING options allow control of the vertical position of each record on a printed page of a report.
   a. If the ADVANCING option is used with a WRITE statement, every WRITE statement for records associated with the same file must also contain one of these options. Automatic spacing is overriden by the ADVANCING option.
   b. If the ADVANCING option is not used, automatic advancing will be provided by the implementor so as to cause single spacing.
   c. When the ADVANCING option is used, the first character in each logical record of a file must be reserved by the programmer for control characters. Many computers use the first character print position for carriage control. In a printed report, if 132 characters are to be printed, PICTURE X(133) should be specified to allow for the control character. The compiler will generate instructions to insert the appropriate control character as the first character of a record. If the records are to be punched, the first character is used for pocket selection. PICTURE X(81) should be specified for punched output. It is the programmer's responsibility to see that the proper carriage-control tape is mounted on the printer prior to the execution of the program. Many newer printers do not use carriage control tapes.
9. *ADVANCING option*

   Identifier-2. If the identifier-2 option is specified, the printer page is advanced the number of lines equal to the current value of identifier-2. If the identifier is used, it must be the name of a nonnegative numeric elementary item (less than 100) described as an integer.

   Integer. If the integer option is specified, the printer page is advanced the number of lines equal to the value of the integer. The integer must be a nonnegative amount less than 100. The integer or the value of the data item referenced by identifier-2 may be zero.

   Mnemonic-Name. If the mnemonic-name option is specified, the printer page is advanced according to the rules specified by the implementor for that hardware device. The mnemonic-name must be defined as function-name in the SPECIAL-NAMES paragraph of the Environment Division. It is used to skip to channels 1–12 and to suppress spacing. It is also used for pocket selection for punched-card output files. (See figure 6.11.)

   If PAGE is specified, the record is presented in the logical page before or after (depending on the phrase used) the device is repositioned to the next logical page. (See figure 6.12.)

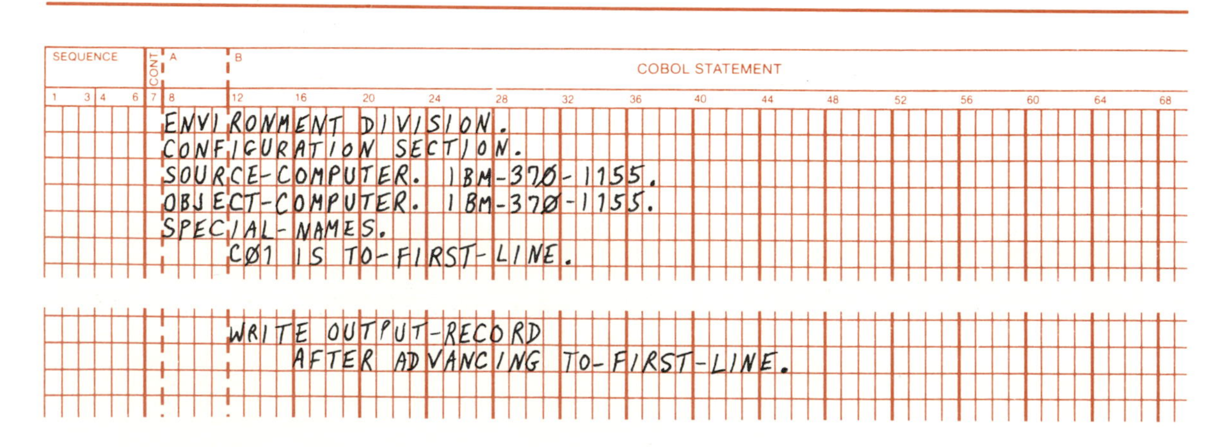

Figure 6.11 WRITE statement: mnemonic name—example.

Before Advancing. If the BEFORE ADVANCING option is used, the record is written *before* the printer page is advanced according to the preceding rules (figs. 6.13 and 6.14).

After Advancing. If the AFTER ADVANCING option is used, the record is written *after* the printer page is advanced according to the preceding rules (fig. 6.15).

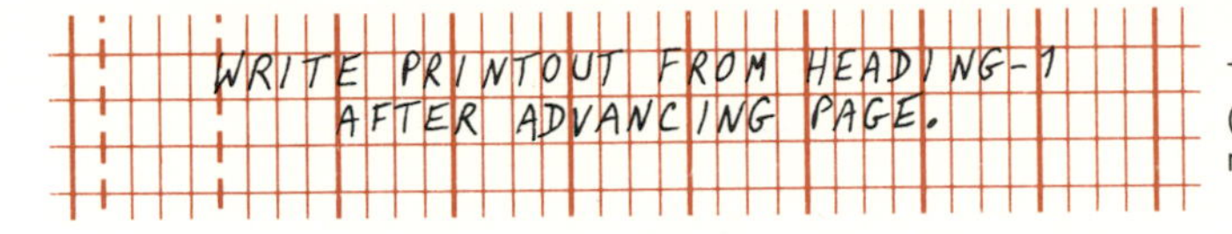

This coding will cause the first heading line (HEADING-1) to be printed at the top of the next page.

Figure 6.12 WRITE statement: PAGE option—example.

WRITE record-name
BEFORE ADVANCING integer LINES
Option

Figure 6.13 WRITE statement: BEFORE ADVANCING option—example.

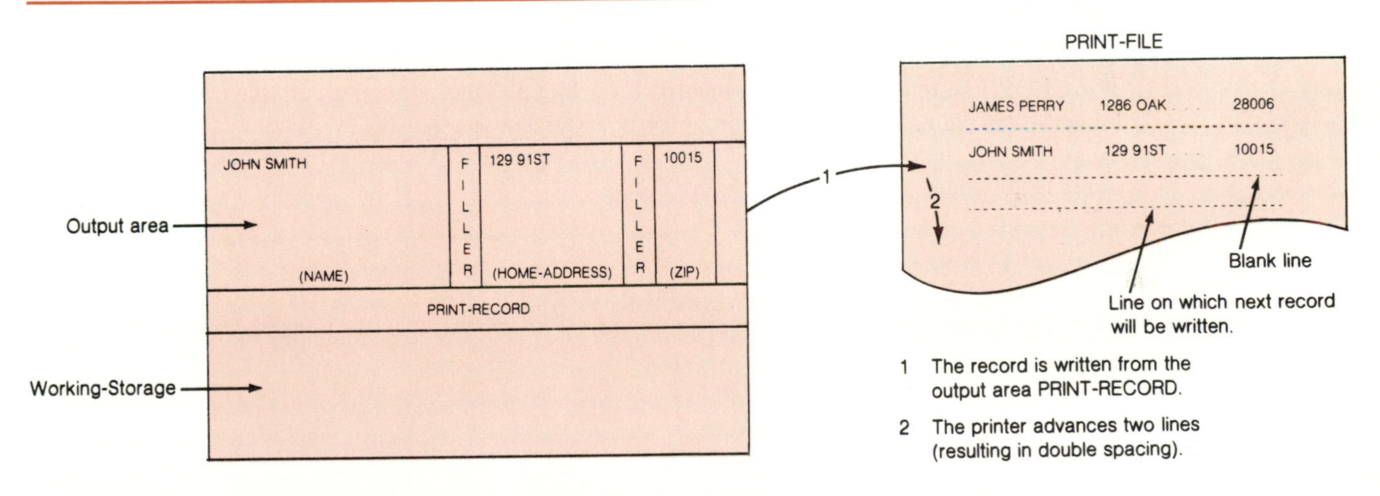

Figure 6.14 Execution of WRITE statement—before advancing option.

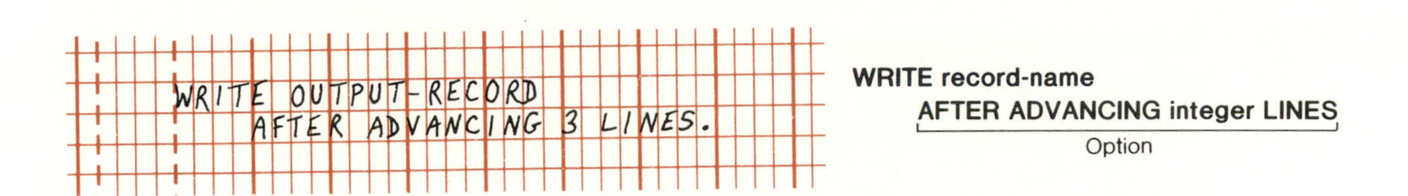

Figure 6.15 WRITE statement: AFTER ADVANCING option—example.

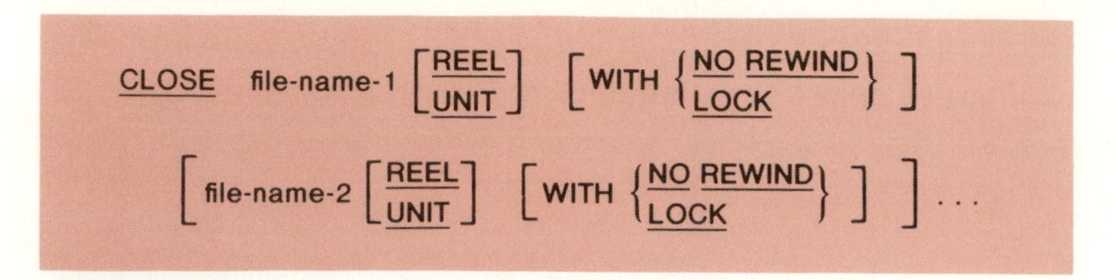

Figure 6.16 Format—CLOSE statement.

CLOSE Statement

All files used in a program should be closed by a CLOSE statement. Each file in a program may have separate CLOSE statements or a single CLOSE statement to close several files, both input or output. (There is no need to specify files as input or output as required in the OPEN statement.)

The CLOSE statement terminates the processing of reels/units and files with optional rewind and/or lock where applicable. (See figure 6.16.)

Rules Governing the Use of the CLOSE Statement

1. A CLOSE statement may be executed only for files that have been previously opened.
2. The file-name is the name of a file upon which the CLOSE statement is to operate. The file-name must be defined by a FD entry in the Data Division.
3. The REEL and WITH NO REWIND options apply only to files stored on magnetic tape devices and other devices to which these terms are applicable.
4. The UNIT option is applicable to mass storage devices in the sequential-access mode.
5. The LOCK option insures that the file cannot be opened during the execution of the object program.
6. The optional clauses (INPUT and OUTPUT) are not written for files closed. (See figure 6.17.)
7. If a CLOSE statement has been executed for a file, no other statement can be executed that references that file unless an intervening OPEN statement for that file is executed.
8. Following the successful execution of a CLOSE statement, the record area associated with file-name is no longer available. (See figure 6.18.)

```
CLOSE PAYROLL-MASTER.

CLOSE     PAYROLL-MASTER,
          INCOMING-CHANGES,
          ERROR-PRINT.
```

This will close all three files as specified under the CLOSE file-name-1 statement.

```
CLOSE     INVENTORY WITH NO REWIND.

CLOSE     INVENTORY WITH NO REWIND,
          INVENTORY-CHANGES.

CLOSE     NEW-INVENTORY-MASTER WITH LOCK,
          INVENTORY-MASTER WITH NO REWIND,
          INVENTORY-CHANGES.
```

Figure 6.17 CLOSE statement—examples.

INPUT-RECORD			
NAME		HOME-ADDRESS	
SUR (12 characters)	GIVEN (8 characters)	(40 characters)	(blank) (20 characters)

OUTPUT-RECORD	
LAST-NAME (12 characters)	ADDRESS-0 (40 characters)

The program flowchart shows the order of operations for the procedure division.

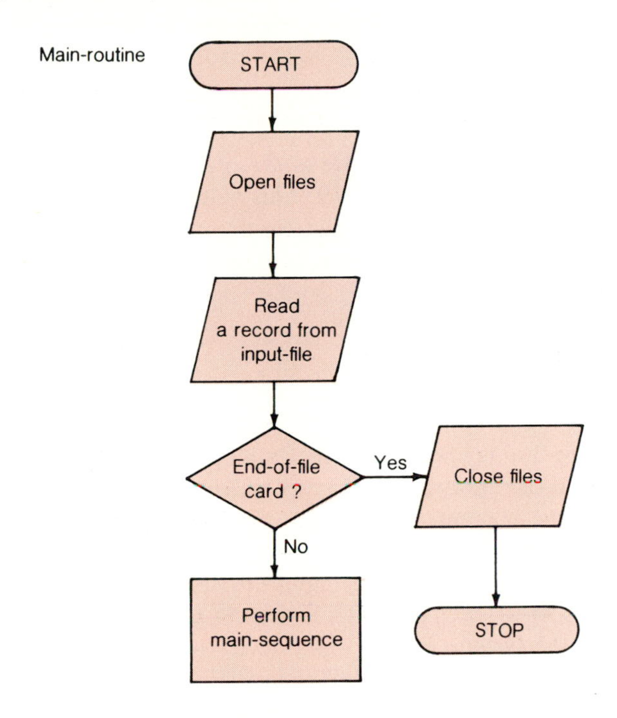

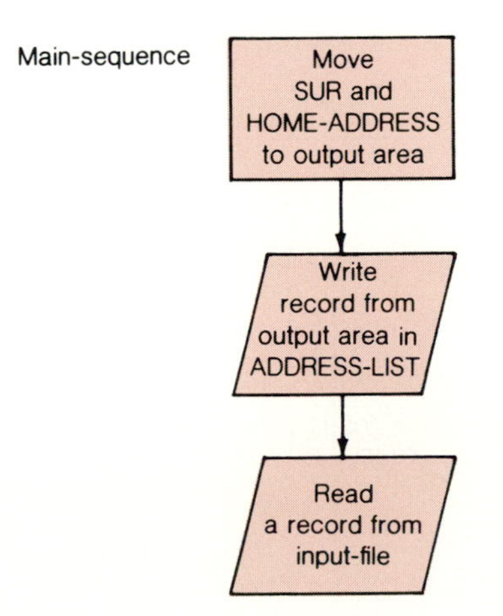


```
IDENTIFICATION DIVISION.
PROGRAM-ID.
    ADDRESS-LISTING.

ENVIRONMENT DIVISION.
CONFIGURATION SECTION.
SOURCE-COMPUTER.          IBM-370-1155.
OBJECT-COMPUTER.          IBM-370-1155.
INPUT-OUTPUT SECTION.
FILE-CONTROL.
    SELECT INPUT FILE           ASSIGN TO UR-2540R-S-INFILE.
    SELECT ADDRESS-LIST         ASSIGN TO UR-1403-S-OUTFILE.

DATA DIVISION.
FILE SECTION.
FD  INPUT-FILE
    LABEL RECORDS ARE OMITTED.
01  INPUT-RECORD.
    05  NAME.
        10  SUR                     PICTURE X(12).
        10  GIVEN                   PICTURE X(8).
    05  HOME-ADDRESS                PICTURE X(40).
    05  FILLER                      PICTURE X(20).

FD  ADDRESS-LIST
    LABEL RECORDS ARE OMITTED.
01  OUTPUT-RECORD.
    05  LAST-NAME                   PICTURE X(12).
    05  ADDRESS-O                   PICTURE X(40).

WORKING-STORAGE SECTION.
01  FLAGS.
    05  MORE-DATA-FLAG              PICTURE XXX       VALUE 'YES'.
        88  MORE-DATA                                 VALUE 'YES'.
        88  NO-MORE-DATA                              VALUE 'NO'.

PROCEDURE DIVISION.
MAIN-ROUTINE.
    OPEN     INPUT     INPUT-FILE
             OUTPUT    ADDRESS-LIST
    READ INPUT-FILE
        AT END MOVE 'NO' TO MORE-DATA-FLAG.
    PERFORM MAIN-SEQUENCE
        UNTIL NO-MORE-DATA.
    CLOSE       INPUT-FILE
                ADDRESS-LIST.

MAIN-SEQUENCE.
    MOVE SUR TO LAST-NAME.
    MOVE HOME-ADDRESS TO ADDRESS-O.
    WRITE OUTPUT-RECORD.
    READ INPUT-FILE
        AT END MOVE 'NO' to MORE-DATA-FLAG.
```

Figure 6.18 READ and WRITE statements—example.

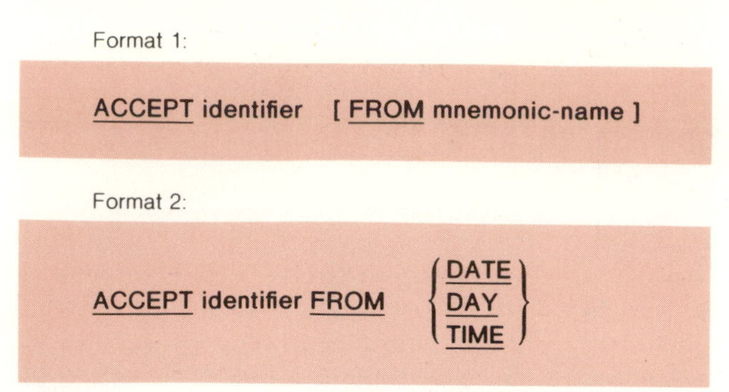

Format 1:
ACCEPT identifier [FROM mnemonic-name]
Format 2:
ACCEPT identifier FROM
DATE
DAY
TIME

Figure 6.19 Formats—ACCEPT statement.

ACCEPT Statement

The ACCEPT statement causes low-volume data to be made available to the specified data item. The two general formats for the ACCEPT statement appear in figure 6.19. This statement is used to get low-volume input, such as the current date, time, or an initial serial number or control totals; it is not used to read files of data. The input device might be a disk file, tape drive, card reader, console, terminal, etc.

Rules Governing the Use of the ACCEPT Statement

1. If the same input/output device is specified for the READ and ACCEPT statements, the results may be unpredictable.
2. The identifier will be described in the Working-Storage Section of the Data Division. The ACCEPT statement will cause the transfer of data from the hardware device specified, to the area specified by the identifier. This data replaces the previous contents of the area. (See figure 6.20.)
3. The *mnemonic-name* must be specified in the SPECIAL-NAMES paragraph of the Environment Division. The mnemonic-name may be either the systems logical input device, a card reader with an assumed input record size of 80 characters, or the CONSOLE which must not exceed 255 characters. If the FROM option is not specified, the systems logical input device is assumed.
4. If the hardware device specified is capable of transferring data of the same size as the receiving area, the transferred data is stored in the receiving data item. If the hardware device is not capable of transferring data of the same size as the receiving area item, then the following takes place:
   a. If the size of receiving area is greater than the transferred data, the transferred data is stored in the left portion of the receiving area, and additional data is requested.

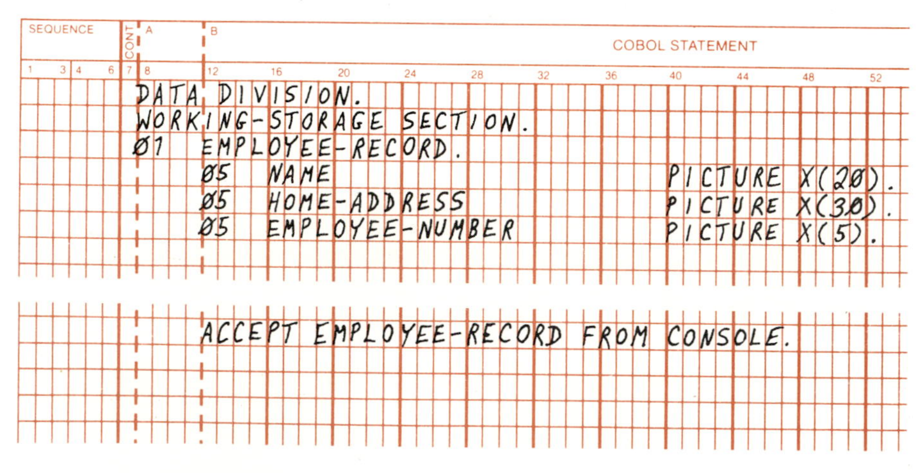

SEQUENCE
CONT
A
B
COBOL STATEMENT
DATA DIVISION.
WORKING-STORAGE SECTION.
Ø1 EMPLOYEE-RECORD.
Ø5 NAME PICTURE X(2Ø).
Ø5 HOME-ADDRESS PICTURE X(3Ø).
Ø5 EMPLOYEE-NUMBER PICTURE X(5).
ACCEPT EMPLOYEE-RECORD FROM CONSOLE.
This statement will allow values of elementary variables in EMPLOYEE-RECORD to be entered into Working-Storage through the console typewriter.

Figure 6.20 ACCEPT statement—CONSOLE option—example.

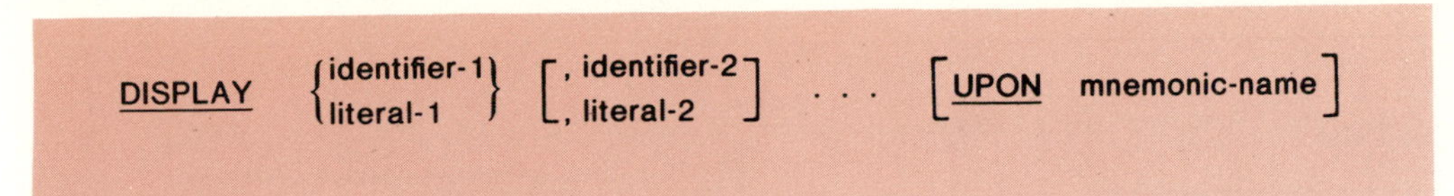

Figure 6.21 Format—DISPLAY statement.

   b. If the size of the receiving area is less than the transferred data, only the leftmost characters will be moved until the area is filled, with the excess character positions at the right being truncated.
5. The implementor will define for each hardware device the size of the data transfer. For example, if the data is in the form of punched cards, the input logical device will transfer eighty characters.

DISPLAY Statement

The DISPLAY statement is used for low-volume output, such as exception records, control totals or messages. The DISPLAY statement causes low-volume data to be transferred to an appropriate hardware device. (See figure 6.21.) This output device might be a disk file, tape drive, printer, console, or terminal.

When two or more items are displayed, no spaces are left between them. If a space is desired, either the figurative constant SPACE must be used, or a space must be included in a nonnumeric literal. For example: DISPLAY CUSTOMER-NAME, SPACE, TRANSACTION-AMOUNT, SPACE, EXCEPTION-CODE.

The quotation marks that enclose nonnumeric literals are not displayed. If quotation marks are to be part of the output, the figurative constant QUOTE must be used. For example: DISPLAY 'BEGIN' QUOTE 'PHASE 2' QUOTE. The data would be displayed as BEGIN 'PHASE 2'. (ANSI 1974 COBOL permits two contiguous quotation marks to be used to represent a single quotation mark character in a nonnumeric literal.)

Rules Governing the Use of the DISPLAY Statement

1. The contents of each operand will be transferred to the hardware device in the order listed.
2. If the same input/output device is used with both the WRITE and DISPLAY statements, the output resulting from the statements may not be in the same sequence as that in which the statements were encountered.
3. The mnemonic-name is associated with a hardware device in the SPECIAL-NAMES paragraph in the Environment Division.
4. The identifier may be either an elementary or a group item.
5. When a DISPLAY statement contains more than one operand, the size of the sending item is the sum of the sizes associated with the operands, and the value of the operands are transferred in the same sequence as that in which the operands are encountered.
6. Numeric or nonnumeric literals may be used. If the literal is numeric, it must be an unsigned integer. (See figure 6.22.)

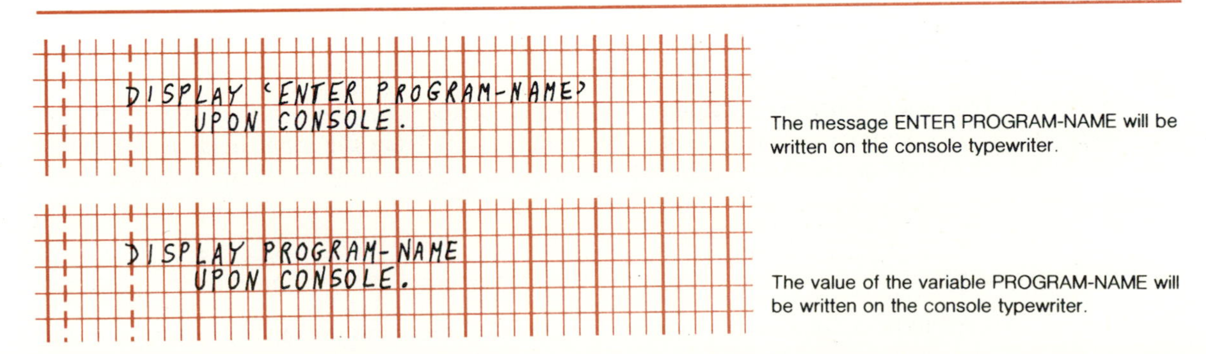

DISPLAY 'ENTER PROGRAM-NAME'
UPON CONSOLE.
The message ENTER PROGRAM-NAME will be written on the console typewriter.
DISPLAY PROGRAM-NAME
UPON CONSOLE.
The value of the variable PROGRAM-NAME will be written on the console typewriter.

Figure 6.22 DISPLAY statement—examples.

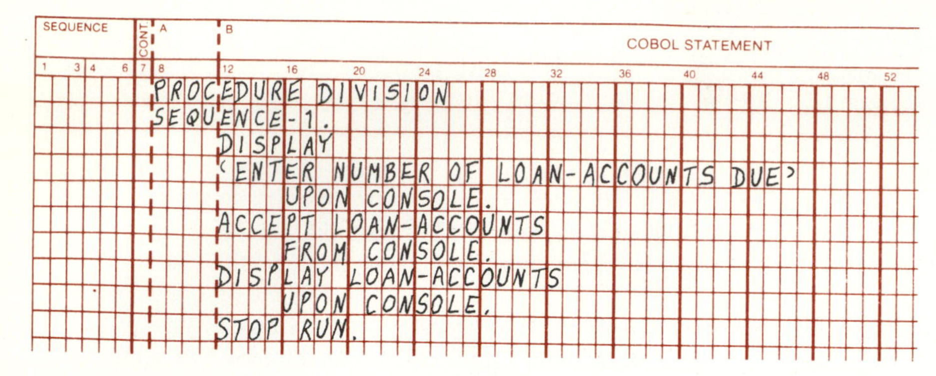

These entries will cause the following:

1. Write the message ENTER NUMBER OF LOAN-ACCOUNTS DUE on the console typewriter.
2. Allow the number of loan accounts that are due to be keyed into LOAN-ACCOUNTS.
3. Write the value of LOAN-ACCOUNTS on the console typewriter.
4. The STOP RUN statement will halt execution of the program.

Figure 6.23 DISPLAY and ACCEPT statements—examples.

7. Figurative constants, except ALL, may be used in DISPLAY statements. If a figurative constant is used as one of the operands, only a single occurrence of the figurative constant is displayed.
8. Any number of identifiers, literals, and figurative constants may be combined into one statement, but they must not exceed the specified maximum limit size. When more than one item is displayed, any spaces desired between multiple operands must be explicitly specified, either with designated spaces included in the literal or with the figurative constant SPACE between operands.
9. The implementor will define for each hardware device the size of the data transfer.
10. If the hardware device is capable of receiving the data of the same size being transferred, then the data is moved; otherwise, the following applies:
    a. If the size of the data item being transferred exceeds the size of the data that the hardware device is capable of receiving in a single transfer, the data, beginning with the leftmost character, is stored aligned to the left in the receiving hardware device, and additional data is requested.
    b. If the size of the data item that the hardware device is capable of receiving exceeds the size of the data being transferred, the transferred data is stored aligned to the left in the receiving hardware device.
11. If the UPON option is not used, the implementor's standard display device is assumed.

The ACCEPT and DISPLAY statements may be used as a communication device between the programmer and the computer operator. If the programmer wishes to display a message to the operator during the execution of the program, perhaps to enter a small amount of data so that the program can resume processing after it has halted, a DISPLAY statement can be coded the proper instruction displaying a message to the operator on what to do. If the programmer wishes the operator to reply by entering the data needed on the console, an ACCEPT statement must be coded.

Both the DISPLAY message and the response entered by the operator will be through the console. (See figure 6.23.)

Although the ACCEPT and DISPLAY statements provide one way to transfer data to and from the computer, they are used for low volume input and output such as codes or messages. A larger volume of data would be efficient on an input/output device such as a card reader, printer, tape drive, or disk drive.

In input/output operations, two additional statements, besides the input/output statements, are necessary: the MOVE statement that is used to move data items from the input fields to the output fields and the PERFORM statement that is used in structured programming to maintain the top/down design.

MOVE Statement

In order for a record to be outputted, the data must be transmitted from the input area to the output area. A MOVE statement is used for this purpose. When the MOVE statement is executed, the value of the variable is copied into the output area. The value of the variable remains the same but any pre-

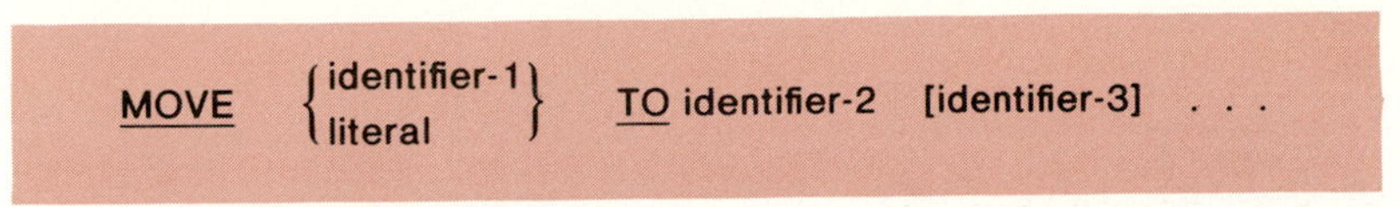

Figure 6.24 Format—MOVE statement.

vious value in the output area is destroyed. The variables specified in a MOVE statement may be either record variables or elementary variables. Values of one record variable may be moved to another record variable by a single MOVE statement, or they may be moved as elementary variables by more than one statement. Literals as well as variables can be specified in a MOVE statement.

MOVE is a data manipulation statement that moves the data from one area of storage to one or more other areas in storage. (See figure 6.24.) The MOVE statement converts the data (for instance from decimal to binary), if required, to fit the description of the receiving item. It also edits the data (inserts, deletes, or replaces characters) if the picture of the receiving item calls for it. (See figures 6.25 and 6.26.)

The MOVE statement transfers data, in accordance with the rules of editing, to one or more data areas.

Category of Sending Data Item		Category of Receiving Data Item		
		Alphabetic	Alphanumeric edited Alphanumeric	Numeric integer Numeric noninteger Numeric edited
Alphabetic		Legal	Legal	Illegal
Alphanumeric		Legal	Legal	Legal
Alphanumeric edited		Legal	Legal	Illegal
Numeric	Integer	Illegal	Legal	Legal
	Noninteger	Illegal	Illegal	Legal
Numeric edited		Illegal	Legal	Illegal

Data in this chart summarize the legality of the various types of MOVE statements.

Figure 6.25 Types of moves. Data in this chart summarize the legality of the various types of MOVE statements.

Type of Move	Receiving Item	Compiler Action During Move	Alignment	Padding if Necessary	Truncation if Necessary
Alphanumeric	Group	None	At left of value	On right with spaces	On right
	Alphabetic or alphanumeric	Any necessary conversion	At left of value	On right with spaces	On right
Numeric	External decimal or packed decimal	Any necessary conversion	At decimal point	On left and right with zeros	On left and right
Edit	Edited	Editing and any necessary conversion	At decimal point	On left and right with zeros (unless suppressed)	On left and right

Figure 6.26 Effects of types of moves.

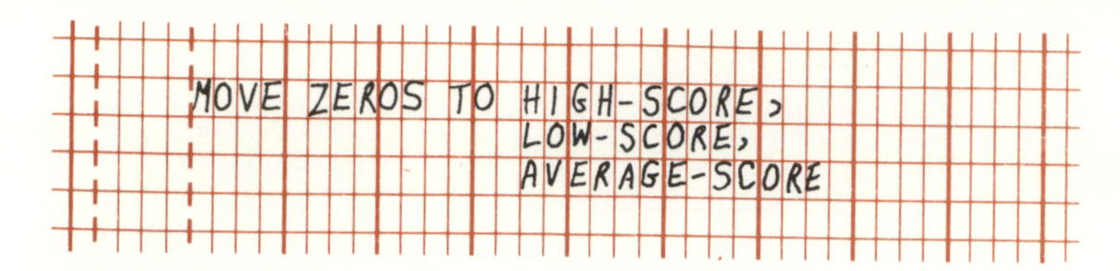

This statement will fill the entire area of HIGH-SCORE, LOW-SCORE, and AVERAGE-SCORE with zeros.

Figure 6.27 MOVE statement—example.

Rules Governing the Use of the MOVE Statement

1. Source data can be transferred to any number of receiving items. (See figure 6.27.)
2. When a group item is involved in a move, the data is moved without any regard for the level structure of the group items involved, and without editing. Thus, when a group item is present, the data being moved is treated simply as a sequence of alphanumeric characters and is placed in the receiving area in accordance with the rules for moving elementary nonnumeric items. If the size of the group item differs, the compiler will produce a warning message when the statement is encountered. Normally, when a group item is involved in a move, it is a group transfer, and the description of the two items are the same. (See figure 6.28.)

Figure 6.28 Individual move and group move—example.

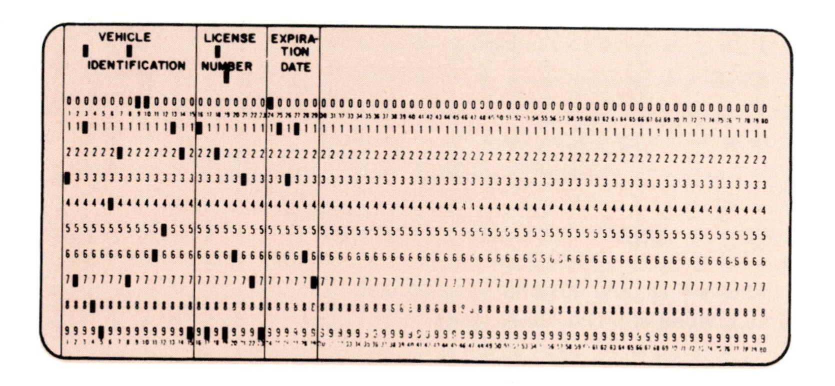

SEQUENCE | CONT. | A | B | COBOL STATEMENT

```
       DATA DIVISION.
       WORKING-STORAGE SECTION.
       01  CARD-DATA.
           05  VEHICLE                 PICTURE X(15).
           05  LICENSE                 PICTURE X(8).
           05  EXPIRATION              PICTURE 9(6).
       01  OUTPUT-LINE.
           05  AREA1                   PICTURE X(15).
           05  FILLER                  PICTURE X(5)    VALUE SPACES.
           05  AREA2                   PICTURE X(8).
           05  FILLER                  PICTURE X(5)    VALUE SPACES.
           05  AREA3                   PICTURE 9(6).
```

Figure 6.28 *(continued)*

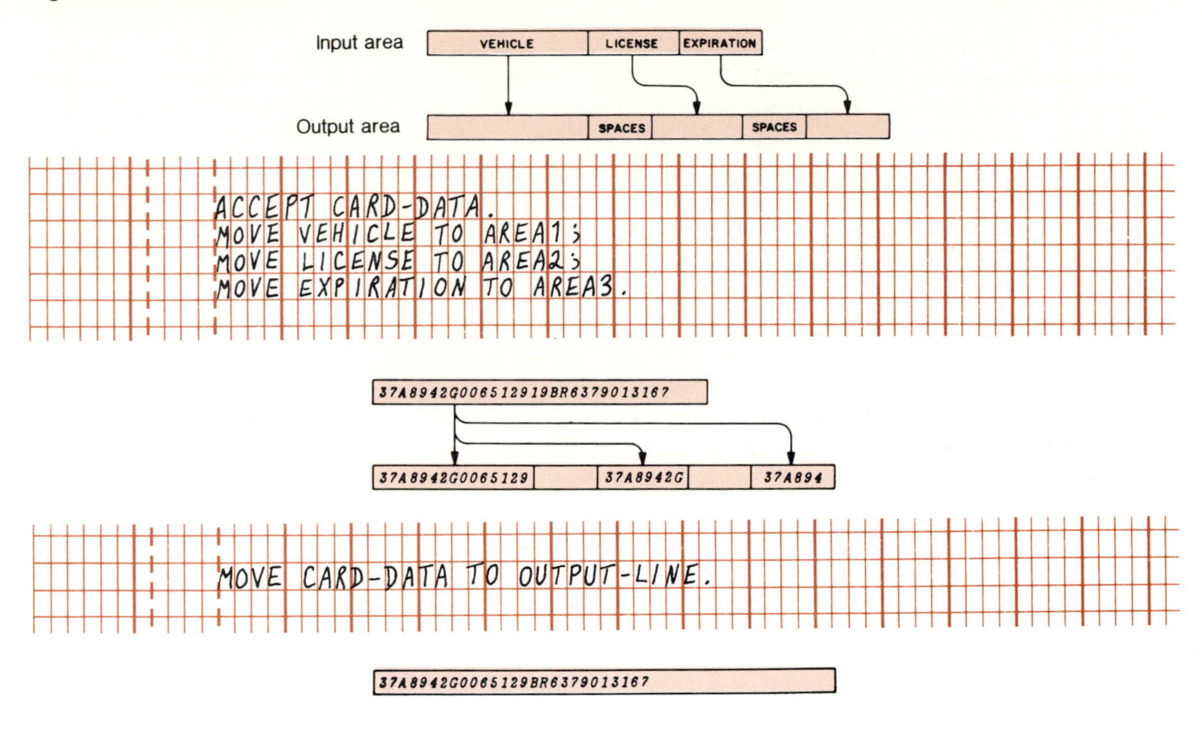

3. When both the source and receiving areas are elementary items, editing appropriate to the format of the receiving area takes place automatically after the MOVE instruction is executed. The type of editing depends upon whether the item is numeric or nonnumeric. (See figure 6.29.)
4. Numeric literals and the figurative constant ZERO belong to the numeric category (fig. 6.30). Nonnumeric literals and the figurative constant SPACE belong to the nonnumeric category.

Source Field		Receiving Field		
PICTURE	Value	PICTURE	Value before MOVE	Value after MOVE
99V99	1234	99V99	9876	1234
99V99	1234	99V9	987	123
9V9	12	99V999	98765	01200
XXX	A2B	XXXXX	Y9X8W	A2Bbb
9V99	123	99.99	87.65	01.23
AAAAAA	REPORT	AAA	JKL	REP

Figure 6.29 **Data movement—example.**

MOVE ZEROES TO FIELD-B.

	PICTURE	Before execution	After execution
FIELD-B	9999	0 1 2 3	0 0 0 0

Figure 6.30 **MOVE statement: numeric literal—example.**

Numeric Data

1. The data from the source area is aligned with respect to the decimal point (assumed or actual) in the receiving area. This alignment may result in the loss of leading or low-order digits (or both) if the source area is larger than the receiving area. Excess positions in the receiving area at either end will be filled with zeros. (See figure 6.31A and B.)
2. If the USAGE clause of the source and receiving fields differ, conversion to the representation specified in the receiving area takes place (fig. 6.31C).
3. If the receiving area specifies editing, zero suppression, insertion of dollar signs, commas, decimal points, etc., and decimal point alignment, all will take place in the receiving area (fig. 6.31D and E).
4. If no decimal point is specified, and the receiving area is larger than the sending area, right justification will take place and the blank left positions will become filled with zeros. (See figure 31F.)

A

MOVE 125.7 TO DOLLARS.

	PICTURE	Before execution	After execution
DOLLARS	9999V99	1234‸66	0125‸70

B

MOVE FIELD TO FIELD-1, FIELD-2, FIELD-3.

PICTURE	
FIELD	9(5)V999
FIELD-1	9V9(5)
FIELD-2	9(5)
FIELD-3	9(6)V99

Data in FIELD	Data in FIELD-1, FIELD-2, and FIELD-3 after the MOVE is executed	
23456‸789	FIELD-1	6‸78900
	FIELD-2	23456
	FIELD-3	023456‸78

C

MOVE FIELD-6 TO FIELD-7.

	PICTURE	Before execution	After execution
FIELD-6	999	157	157
FIELD-7	XXXX	AB24	157

D

MOVE AMOUNT TO AMOUNT-PR.

	PICTURE	Before execution	After execution
AMOUNT	9999V99	1258‸39	1258‸39
AMOUNT-PR	$9,999.99	$ 3,333.33	$1,258.39

E

MOVE AMOUNT-1 TO AMOUNT-1-OUT.

	PICTURE	Before execution	After execution
AMOUNT-1	9999V99	0000‸03	0000‸03
AMOUNT-1-OUT	$$,$$$.99	$2,219.00	$.03

F

MOVE AMOUNT-IN TO AMOUNT-OUT.

	PICTURE	Before execution	After execution
AMOUNT-IN	999PPP	238	238
AMOUNT-OUT	9(6)	129074	238000

MOVE TOTAL-1 TO TOTAL-2.

	PICTURE	Before execution	After execution
TOTAL-1	9(5)V99	14794‸23	14794‸23
TOTAL-2	999PP	438	147

‸ Indicates assumed decimal point.

Figure 6.31 MOVE statement: numeric data—examples.

A

MOVE FIELD-1 TO FIELD-2

	PICTURE	Before execution	After execution
FIELD-1	XXX	A B C	A B C
FIELD-2	XXXX	X Y W K	A B C

B

MOVE NAME TO FIELD-A, FIELD-B.

Data in NAME	Data in FIELD-A and FIELD-B after the MOVE is executed	
J O H N B R O W N 1502	FIELD-A	J O H N
	FIELD-B	J O H N B R O W N

C

MOVE '123' TO FIELD-5.

	PICTURE	Before execution	After execution
FIELD-5	XXXX	A B C D	1 2 3

D

MOVE NAME TO NAME-1.

Data in NAME	Data in NAME-1 after the MOVE is executed
J O H N	J O H N

E

MOVE FIELD-3 TO FIELD-4.

	PICTURE	Before execution	After execution
FIELD-3	XXX	A B C	A B C
FIELD-4	XX	X Y	A B

F

MOVE ACCOUNT-NO ACCT-NO-PR.

	PICTURE	Before execution	After execution
ACCOUNT-NO	XXXX	A 1 2 3	A 1 2 3
ACCT-NO-PR	XBXXX	A C D E	A 1 2 3

Figure 6.32 MOVE statement: nonnumeric data—examples.

Nonnumeric Data

1. The data from the source area is placed in the receiving area beginning at the left and continuing to the right, unless the field is specified as JUSTIFIED RIGHT, in which case the source data is placed in the right positions of the receiving area.
2. If the receiving area is not completely filled with data, the remaining positions are filled with spaces at the right or left for justified right items.
3. If the source field is longer than the receiving area, the move is terminated as soon as the receiving area is filled. Excess characters are truncated when the receiving area is filled. (See figure 6.32.)

Rules Governing Elementary Items MOVE Statements

1. A numeric edited, alphanumeric edited, the figurative constant SPACE, or any alphabetic data item must not be moved to a numeric edited data item.
2. A numeric literal, the figurative constant ZERO, a numeric data item, or a numeric edited data item must not be moved to an alphabetic item.
3. A numeric literal or a numeric data item whose implicit decimal point is not immediately to the right of the least significant digit must not be moved to an alphanumeric or alphanumeric edited item.
4. All other elementary moves are permissible and are performed in accordance with the rules previously mentioned.

The CORRESPONDING option will be discussed later in the text. (See figure 6.33.)

Since MOVE statements constitute such a large portion of COBOL programs, it is imperative that the programmer fully understand the effects of the various MOVE statement moves. The examples in this figure should help to further your understanding of the effects of MOVE statements.

The data referenced in the examples at the left are assumed to have the data images shown. (V appears in the table entries for some numeric items to illustrate assumed decimal point alignment; it does not actually appear in data items in the object program.)

Data-name	PICTURE
A	999999V
B	9999V99
E	X(10)
D	X(6)
C	999V9
F	X(10)

Statement	Sending item contents	Result
MOVE 1000 TO B.	1000	1000V00
MOVE 300 TO B.	300	0300V00
MOVE 1.235 TO B.	1.235	0001V23
MOVE C TO B.	024V5	0024V50
MOVE B TO C.	1234V56	234V5
MOVE F TO E.	AA1673BBCC	AA1673BBCC
MOVE D TO F.	AA1673	AA1673bbbb
MOVE E TO D.	AA1673BBCC	AA1673
MOVE "123456" to D.	123456	123456
MOVE A TO D.	007340V	007340
MOVE D TO A.	653000	653000V
MOVE ALL "Z" TO E.		ZZZZZZZZZZ
MOVE ALL "XY" TO E.		XYXYXYXYXY
MOVE ALL "XYZ" TO E.		XYZXYZXYZX
MOVE ZEROS TO C.		000V0
MOVE SPACES TO E.		bbbbbbbbbb

b denotes blank or space.

Here are some examples of the editing features of the MOVE statement.

Sending Item		Receiving Item	
PICTURE	Value of sending item	PICTURE	Value of receiving item
9999V99	567891	9999V99	567891
9999V99	567891	9999V9	56789
9V9	78	999V99	00780
XXX	M8N	XXXXX	M8Nbb
99V99	6789	999.99	067.89
AAAAAA	WARREN	AAA	WAR
99V99	6789	$ZZZ9.99	$bb67.89

b denotes space or blank

These examples illustrate the results of various MOVE statements. (V is shown to illustrate the assumed decimal point position.)

```
77  B  PICTURE 9999V99.
77  C  PICTURE 999V9.
77  D  PICTURE X (6).
```

MOVE	Sending value	Receiving value
1000 TO B	1000	1000V00
300 TO B	300	0300V00
1.235 TO B	1.235	0001V23
C TO B	024V5	0024V50
D TO B	123456	3456V00

Sending Item		Receiving Item	
PICTURE	Value	PICTURE	Resulting value
9 (5)	45678	$ZZ,ZZ9.99	$45,678.00
9 (3) V99	456.78	$ZZ,ZZ9.99	456.78
9 (3) V99	000.67	$ZZ,ZZ9.99	0.67
9 (3) V99	000.04	$ZZ,ZZZ.99	.04
9 (5)	00000	$ZZ,ZZZ.ZZ	
V9 (5)	.12345	$ZZ,ZZ9.99	0.12
9 (5)	12345	$**,**9.99	$12,345.00
9 (5)	67890	$$$,$$9.99	$67,890.00
9 (3) V99	678.90	$$$,$$9.99	$678.90
9 (5)	00000	$$$,$$9.99	$0.00
V9 (5)	.67890	$$$,$$9.99	$0.67
S9 (5) V	−56789.	−ZZZZ9.99	−56789.00
S9 (5)	+56789	−ZZZZ9.99	56789.00
S9 (5)	−56789	+ZZZZ9.99	−56789.00
S9 (5)	+56789	+ZZZZ9.99	+56789.00
S99V9 (3)	−56.789	−−−.99	−56.78
S9 (5)	−00567	ZZZZZ.99−	567.00−
S9 (5)	−56789	$$$$$$.99CR	$56789.00CR
S9 (5)	+56789	$$$$$$.99CR	$56789.00

A MOVE statement is considered to be nonelementary if the sending data item and/or the receiving data item is a group item. A nonelementary MOVE statement produces the same effect as if the sending and receiving data items were simple alphanumeric items.

Figure 6.33 MOVE statement—examples.

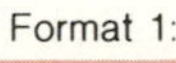

Format 1:

<u>PERFORM</u> procedure-name-1 [{ <u>THRU</u> / <u>THROUGH</u> } procedure name-2]

Figure 6.34 Format—PERFORM statement: format 1.

PERFORM Statement

PERFORM is a sequence control statement that causes a branch to a procedure or a series of procedures, and, following their execution, a return branch to the statement after the PERFORM statement. (See figure 6.34.) In other words, the procedure is treated as a subroutine, and the PERFORM statement in the main routine provides the linkage to and from the subroutine. The PERFORM statement is also used to control the execution of loops.

The PERFORM statement is useful when control must be transferred to a group of statements from several points in a program.

PERFORM has several different formats, which vary in complexity. In its simplest form, the procedure referred to is executed once each time the PERFORM is encountered. Other formats provide repetitive execution using one or more optional controls to control the "looping."

The DOWHILE Program Structure

The DOWHILE structure provides the basic loop capability. A COBOL paragraph or section is repeated while some condition is satisfied. The condition is tested prior to each execution of the paragraph or section, including the first.

The verb PERFORM with the UNTIL option is the COBOL implementation of the DOWHILE structure. Because the execution of the COBOL UNTIL option terminates on a "true" condition and the DOWHILE terminates on a "false" condition, it is necessary to code the inverse of the condition to loop while the condition is true. In the example below, this is shown (NOT condition).

The flowchart for the DOWHILE structure is shown in figure 6.35.

The COBOL format is:

```
PERFORM paragraph-name
    UNTIL (NOT condition)
```

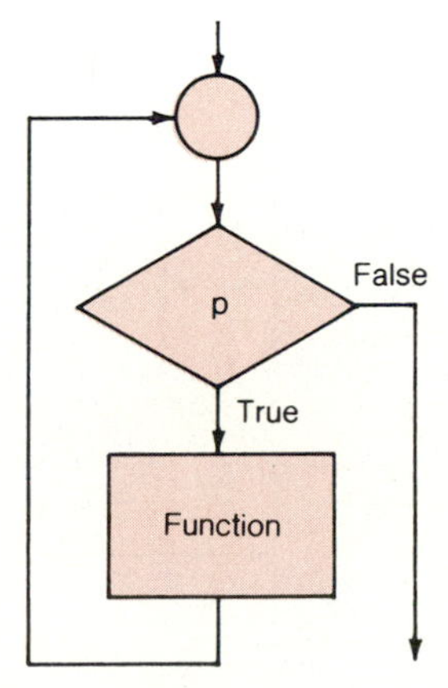

Figure 6.35 Flowchart for DOWHILE program structure.

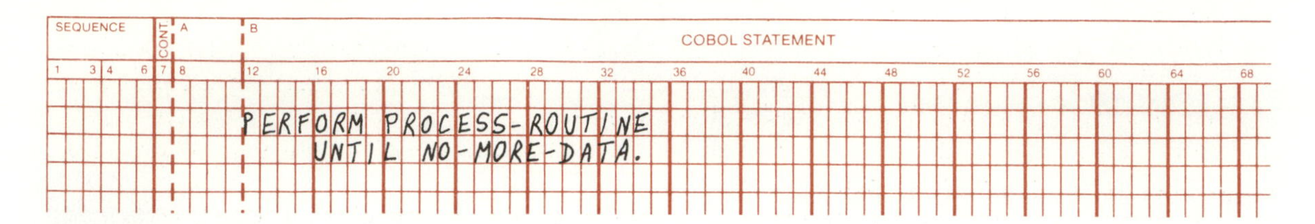

SEQUENCE
CONT.
A
B
COBOL STATEMENT
PERFORM PROCESS-ROUTINE
UNTIL NO-MORE-DATA.

Figure 6.36 PERFORM statement (DOWHILE structure—example).

It should be emphasized that the PERFORM with the UNTIL option carries out the test of the condition *before* executing the controlled function. If the condition in the UNTIL phrase is true when first encountered, the paragraph or section in the PERFORM statement is not executed at all. (See figure 6.36.)

The three formats of the PERFORM statement are used when it is desired not only to set up linkage to a routine, but also to control the number of times the routine is executed. Usually, these formats apply when "looping" is desired, that is, when a process is to be repeated until a terminal condition prevails.

There are a few instances in which the programmer needs to specify a literal number of repetitions of a procedure, such as two or ten or twenty-five times. More often the programmer will want to repeat a procedure a certain number of times for one record and a different number of times for another record depending on the value of some data item in the record.

There may be occasions where the programmer needs to repeat a loop until a certain result is reached or until a certain condition prevails regardless of whether it takes one or 100 performs.

Rules Governing the Use of the PERFORM Statement

1. When a procedure is performed, the PERFORM statement transfers the sequence control to the first statement in procedure-name-1 and also provides for the return of the control. The point at which the control is returned to the main program depends on the structure of the procedure being executed (fig. 6.37).
2. If procedure-name-1 is a paragraph-name, and procedure-name-2 is not specified, control is returned after the execution of the last statement of the procedure-name-1 paragraph.
3. If procedure-name-1 is the name of a section, and procedure-name-2 is not specified, control is returned after the last statement of the last paragraph in the procedure-name-1 section.
4. If procedure-name-2 is specified, control is transferred after the last statement of the procedure-name-2 paragraph.

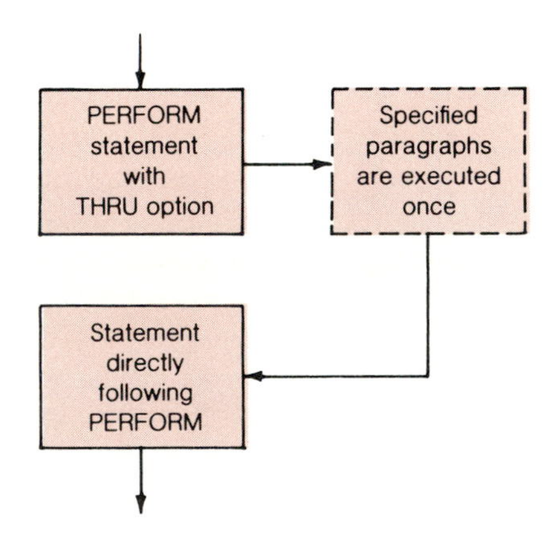

Figure 6.37 Flow of logic—PERFORM statement: format 1.

5. If procedure-name-2 is specified and is a section, control is transferred after the last statement of the procedure-name-2 section.
6. When procedure-name-2 is specified, the relationship between procedure-name-1 and procedure-name-2 must exist. Execution must proceed from procedure-name-1 throughout the last statement of procedure-name-2. GO TO and PERFORM statements are permitted between procedure-name-1 and the last statement in procedure-name-2 provided that the sequence ultimately returns prior to the final statement in procedure-name-2.
7. The last sentence in the procedure referred to must not contain an unconditional GO TO statement. If the logic of the procedure requires a conditional exit prior to the last sentence, the EXIT verb is used to satisfy the requirement. An EXIT statement consists solely of a paragraph-name and the word EXIT (see EXIT statement).
8. The procedure-name may be either a paragraph or section-name. The word SECTION is not required.
9. The procedure-name must not be the name of a procedure of which the PERFORM statement is a part.
10. A procedure-name can be referenced by more than one PERFORM statement.
11. Procedures to be performed can be outside the main program or can be part of the main routine so that they can be executed in line.
12. A referenced procedure may itself contain other PERFORM statements.
13. All procedures must be arranged in the order in which they are to be performed.
14. The words THRU and THROUGH are equivalent.

Basic PERFORM—Format-1

In a basic PERFORM statement, the procedure referenced is executed once, with control then passing to the first statement following the PERFORM statement (figs. 6.38 and 6.39). All statements in paragraphs or sections named in procedure-name-1 (through procedure-name-2) constitute the range and are executed before control is returned. (See figure 6.40.)

```
PAR-1.  MULTIPLY AMOUNT BY 300 GIVING TOTAL-AMOUNT.
        PERFORM CALCULATE.
        ADD 100 TO TOTAL.
        .
        .
CALCULATE.
        ADD 10 TO TOTAL.
        MOVE TOTAL TO NEW-TOTAL.
        SUBTRACT TOTAL FROM OLD-TOTAL.
MOVE-DATE.
        IF DATA IS EQUAL TO TODAY-DATE, MOVE TODAY-DATE TO OUTPUT-DATE.
        .
        .
```

In the above example, the statement PERFORM CALCULATE executes the three statements contained in the paragraph CALCULATE. The instructions are executed in the following sequence:

```
MULTIPLY AMOUNT BY 300 GIVING TOTAL-AMOUNT.
ADD 10 TO TOTAL.                  )
MOVE TOTAL TO NEW-TOTAL.          }   PERFORM CALCULATE
SUBTRACT TOTAL FROM OLD-TOTAL.    )
ADD 100 TO TOTAL.
.
.
```

Figure 6.38 PERFORM statement: format 1—example.

```
PERFORM COMPUTE-FICA
```

The paragraph COMPUTE-FICA will be executed; that is, PERFORMed and control will then pass to the statement following the PERFORM verb.

```
PERFORM COMPUTE-FICA THRU COMMON-CHECK-PRINT
      .
      .
      .
COMPUTE-FICA. . . .
COMPUTE-NET-PAY. . . .
COMMON-CHECK-PRINT. . . .
```

The range of procedures specified above is executed and control is passed to the statement following the PERFORM.

Figure 6.39 **PERFORM statement: format 1—example.**

```
PAR-1.   MOVE NEW-ACCT-NO TO ACCT-NO.
         PERFORM MOVEMENT THRU COMPUTATION.
         WRITE RECORD-OUT.
         .
         .
TEST-EQUALITY.
         IF TCODE = 1, PERFORM ROUTINE-1.
         IF TCODE = 2, PERFORM ROUTINE-2.
         IF TCODE = 3, PERFORM ROUTINE-3.
MOVEMENT. MOVE BALANCE TO NEW-BALANCE.
         MOVE TCODE TO NEW-CODE.
NEW-RECORD.
         MOVE 2 TO NEW-LL-CODE.
         MOVE DATE-1 TO NEW-DATE.
COMPUTATION.
         ADD AMOUNT TO BALANCE.
         SUBTRACT 35 FROM LOWER-LIMIT.
OUTPUT-ROUTINE.
         WRITE NEW-RECORD FROM OLD-AREA.
         ADD 1 TO COUNTER-1.
```

In the above example, the statement PERFORM MOVEMENT THRU COMPUTATION executes the six statements in the paragraphs entitled MOVEMENT, NEW-RECORD, and COMPUTATION. The instructions are executed in the following sequence:

```
MOVE NEW-ACCT-NO TO ACCT-NO.
MOVE BALANCE TO NEW-BALANCE.
MOVE TCODE TO NEW-CODE.
MOVE 2 TO NEW-LL-CODE.            PERFORM MOVEMENT
MOVE DATE-1 TO NEW-DATE.          THRU COMPUTATION.
ADD AMOUNT TO BALANCE.
SUBTRACT 35 FROM LOWER-LIMIT.
WRITE RECORD-OUT.
```

Figure 6.40 **PERFORM statement: format 1—example.**

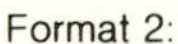

Format 2:

PERFORM procedure-name-1 [{THRU / THROUGH} procedure-name-2] {identifier-1 / integer-1} TIMES

Figure 6.41 Format—PERFORM statement: format 2 (TIMES option).

TIMES—Format-2

The TIMES option provides a means for performing a procedure a repetitive number of times, and for then returning control to the next statement after PERFORM. (See figures 6.41 and 6.42.)

The procedures are performed the number of times specified by integer-1 or by the initial value of the data item referenced by identifier-1 for that time. If at the time of execution of the PERFORM statement, the value of reference by identifier-1 is equal to zero or is negative, control passes to the next executable statement following the PERFORM statement. Following the execution of the procedure-specified number of times, control is transferred to the next executable statement following the PERFORM statement.

During the execution of the PERFORM statement, references to identifier-1 cannot alter the number of times the procedures are to be executed from that which was indicated by the initial value of identifier-1.

Rules Governing the Use of the TIMES Option

1. The number of times the procedure is to be performed is specified as a number or identifier.
2. If an identifier is used, it must have an integral value.

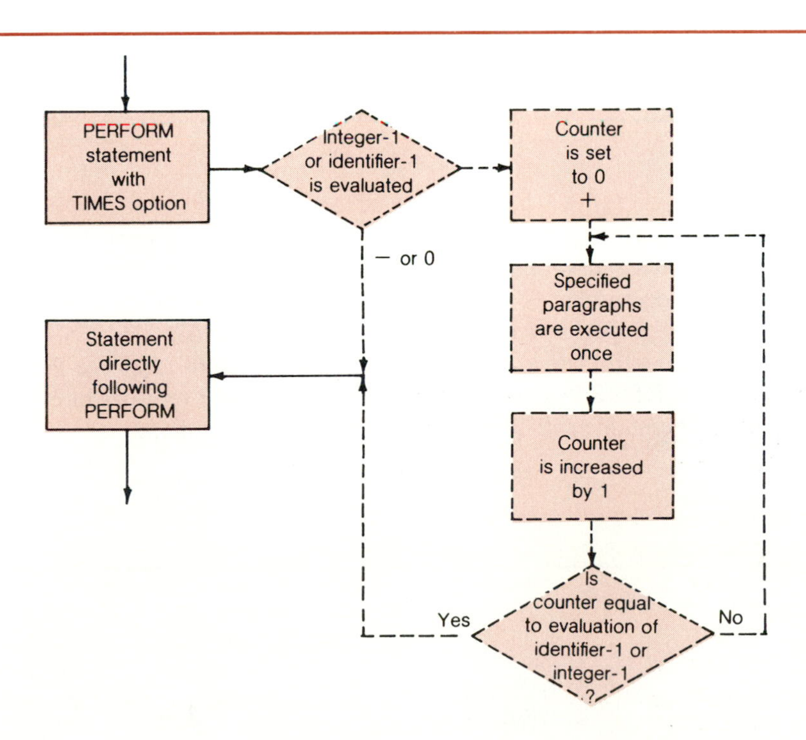

Figure 6.42 Flow of logic—PERFORM statement: format 2 (TIMES option).

PERFORM ITERATION Q-NUMBER TIMES.

The procedure ITERATION will be performed exactly the number of times as specified by the numeric quantity in the field Q-NUMBER, each time the PERFORM verb is executed. If the numeric quantity in the field Q-NUMBER is zero, then the procedure is not executed. Keep in mind that the numeric quantity in the field Q-NUMBER can be changed from time to time.

PERFORM A THROUGH E DETERMINED-NUMBER-OF TIMES.

The entire series of procedures beginning with A and ending with E will be executed exactly the number of times as specified by the numeric quantity in the field DETERMINED-NUMBER-OF, each time the PERFORM verb is executed.

PERFORM ITERATION 3 TIMES.

The procedure ITERATION will be executed exactly 3 times each time the PERFORM verb is executed.

Figure 6.43 PERFORM statement: format 2 (TIMES option)—example.

3. The identifier or the number must have a positive value. If the value of the identifier is zero or negative, control is transferred immediately to the next statement following the PERFORM statement.
4. When the TIMES option is used, a counter is set up. This counter is tested against the specified number of execution (times) before control is transferred to procedure-name-1. The counter is incremented by one after each execution, and the process is repeated until the value of the counter is equal to the number of times specified. At that point, control is passed to the next statement following the PERFORM statement. An initial value of zero will cause no execution of procedure-name-1. (See figure 6.43.)

UNTIL—Format-3

The UNTIL option operates in the same manner as the TIMES option, except that no counting takes place, and the PERFORM causes an evaluation of a specified test condition instead of testing the value of a counter against a specified number of executions. (See figures 6.44 and 6.45.)

The specified procedures are performed until the condition specified by the UNTIL phrase is true. When the condition is true, control is transferred to the next statement after the PERFORM. If the condition is true when the PERFORM statement is entered, no transfer to procedure-1 takes place and control is passed to the next statement following the PERFORM statement.

Format 3:

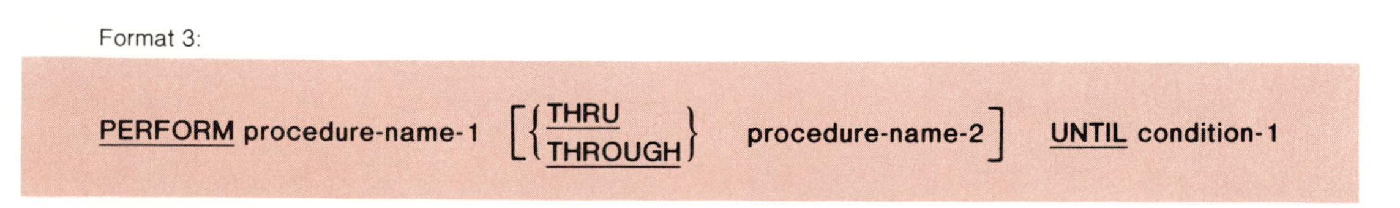

Figure 6.44 Format—PERFORM statement: format 3 (UNTIL option).

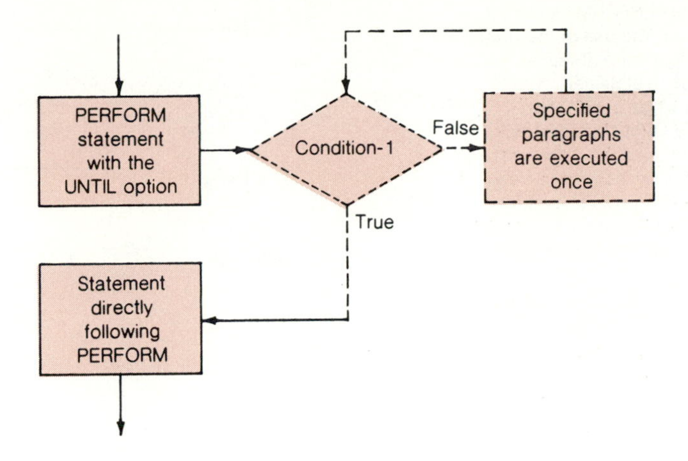

Figure 6.45 Flow of logic—PERFORM statement: format 3 (UNTIL option).

Rules Governing the Use of the UNTIL Option

1. Condition-1 may be a simple or compound expression.
2. Condition-1 is evaluated before the specified procedures are executed.
3. If condition-1 is true, control passes to the next statement after the PERFORM statement. The specified procedure is not executed.
4. If condition-1 is not true, control transfers to procedure-name-1.
5. The process is repeated until condition-1 is detected to be true.

(See figures 6.46 and 6.47.)

```
PERFORM EDIT-ROUTINE
    UNTIL KODE IS EQUAL TO '5'.
```

The procedure EDIT-ROUTINE will be executed until the data in the field KODE compares equal to 5.

```
PERFORM EDIT-ROUTINE
    UNTIL TRANSACTION-NUMBER IS GREATER THAN '12345'
    OR IS EQUAL TO KEY-NUMBER.
```

The procedure EDIT-ROUTINE will be executed until the data in the field TRANSACTION-NUMBER compares greater than 12345 or until the data compares equal to the data in the field KEY-NUMBER.

```
PERFORM A THROUGH G
    UNTIL V IS EQUAL TO M
    OR EQUAL TO N
    OR EQUAL TO P.
```

Figure 6.46 Format—PERFORM statement: format 3 (UNTIL option)—examples.

In its simplest form, the PERFORM statement is an imperative statement. The use of the UNTIL option, however, changes the PERFORM statement into a condition statement. The UNTIL option allows a single conditional termination of the PERFORM statement to be established. Since the condition for termination is tested before the referenced procedure is executed, the test may prevent the execution of the procedure. The flow diagram for the PERFORM . . . UNTIL statement is shown here.

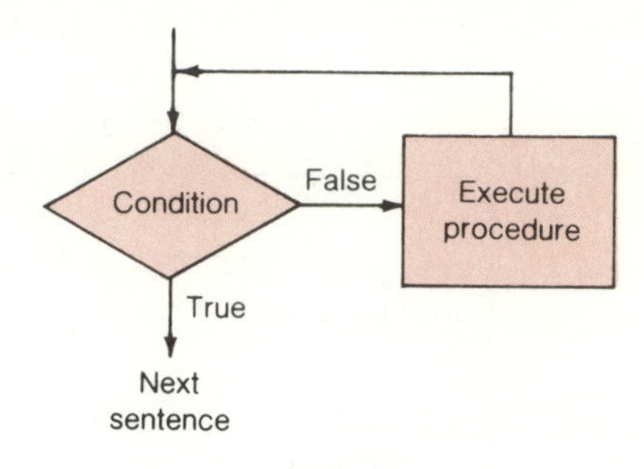


```
WORKING-STORAGE SECTION.
01  TRANSACTION-STATUS       PIC XXX.
    88  END-OF-TRANSACTIONS                 VALUE 'END'.
    .
    .
    MOVE SPACES TO TRANSACTION-STATUS.
    OPEN INPUT       TRANSACTION-FILE.
    READ TRANSACTION-FILE
        AT END MOVE 'END' TO TRANSACTION-STATUS.
    PERFORM TRANSACTION-PROCESS THRU NEXT-TRANSACTION
        UNTIL END-OF-TRANSACTIONS.
    CLOSE       TRANSACTION-FILE.

    .
    .
TRANSACTION-PROCESS.

    .
    .
NEXT-TRANSACTION.
    READ TRANSACTION-FILE
        AT END MOVE 'END' TO TRANSACTION-STATUS.
```

Figure 6.47 PERFORM statement: format 3 (UNTIL option)—example.

Summary

The Procedure Division contains instructions such as input/output, data manipulation, and arithmetic, which direct the data processing activities of the computer, and sequence control activities, which are required to process the data. The order in which the programmer codes these statements will depend on the logic of the problem solution.

The following are units of expression that constitute the Procedure Division:

A *procedure* is composed of a paragraph, or a group of successive paragraphs, or a section, or a group of successive sections within the Procedure Division.

A *procedure-name* is a word used to refer to a paragraph or a section.

The end of the Procedure Division and the physical end of the program is that physical position in a COBOL source program after which no further procedures appear.

A *statement* is a syntactically valid combination of words and symbols beginning with a COBOL verb. The statement is the basic unit of the Procedure Division and may be one of three types: (1) an *imperative* statement that indicates a specific unconditional action is to be taken by the object program, (2) a *conditional* statement that specifies that the truth value of a condition is to be determined and that the subsequent action of the object program is dependent upon this truth value, and (3) the *compiler-directing* statement that causes the compiler to take a specific action during the compilation of the source program.

A *sentence* is composed of one or more statements which may be either imperative, conditional, or compiler-directing, and which specifies actions to be taken, and is terminated by a period and followed by a space.

A *paragraph* consists of a paragraph-name followed by a period and a space, and by zero, one or more successive sentences. A paragraph ends immediately before the next paragraph or at the end of the Procedure Division.

A *section* consists of a section header followed by a period and a space, and by zero, one or more successive paragraphs. A section ends immediately before the next section or at the end of the Procedure Division.

The Procedure Division consists of procedures that are composed of paragraphs, groups of successive paragraphs, a section, or a group of successive sections. The end of the Procedure Division is the physical end of the COBOL program after which no further procedures can appear.

COBOL verbs are the basis of the Procedure Division of a source program. Each of these verbs causes some event to take place either at compilation time or at program execution time.

The input/output system provides for operations such as the movement of data into internal storage, validity checking, and unblocking and blocking of physical records. The COBOL input/output verbs provide a means of storing data on external devices and extracting such data from external devices. Four verbs, OPEN, READ, WRITE and CLOSE are used to specify the flow of data to and from files stored on an external medium. ACCEPT and DISPLAY verbs are used in conjunction with low-volume data that has to be obtained or sent to a card reader, disk storage device, console typewriter, or printer.

The OPEN statement makes one or more input or output files ready for reading or writing, checks or writes labels if needed, and prepares the storage areas to receive or send data. An OPEN statement must be executed prior to any other input/output statement for a particular file. The statement does not make input records available for processing nor does it release output records to their respective devices; it makes the associated record available to the program. A READ or WRITE statement respectively is required to perform the function of making records available for processing or output.

The READ statement makes the next logical record available for processing and allows for the performance of one or more specified statements when the end of a file is detected. A READ statement must be executed before the data from a record can be processed. Also, the file must be opened before it can be read. The INTO phrase is used to transmit a record from an input file to a Working-Storage variable or an output area associated with a previously opened file. The AT END phrase indicates that the last input record has been processed, and what action is to be taken following the processing of the input records.

The WRITE statement is used to release a logical record for insertion in an output file and for vertical positioning to line up within a logical page. An OPEN statement must be executed prior to the execution of the first WRITE statement. The FROM option is used to transmit a record from a Working-Storage variable or an input area associated with a previously opened file to an output area. The ADVANCING option allows control of the vertical positioning of each record on a printed page of a report. The BEFORE ADVANCING option is used if the record is to be written before the printed page is advanced the specified number of lines. The AFTER ADVANCING option is used if the record is to be written after the printed page is advanced the specified number of lines.

The CLOSE statement terminates the processing of reels/units and files with optional rewind and/or lock where applicable. All files used in a program should be closed. A CLOSE statement may be executed only for files that have been previously opened. If a file has been closed, no other statement can be executed that references that file unless an intervening OPEN statement for that file is executed.

The ACCEPT statement causes low-volume data to be made available to the specified data item. The implementor will define for each hardware device the size of the data transfer. The input device may be a disk file, tape drive, card reader, console, etc. This statement is used to get low-volume data such as current date, time, control totals, etc.

The DISPLAY statement causes low-volume data to be transferred to an appropriate hardware device. The implementor will define for each hardware device the amount of transfer. The contents of each operand will be transferred to the hardware in the order listed without any intervening space unless specified. Numeric and nonnumeric literals may be used. The DISPLAY statement may be used for such low-volume output as exception records, control totals and messages.

The ACCEPT and DISPLAY statements may be used as a communication device between the programmer and the computer operator.

A MOVE statement is a data manipulation statement that moves data from one data area to one or more storage areas. It is primarily used to move data from input to output. The value of the data remains the same in the input area but the previous values in the output areas are destroyed as the data is copied. Literals as well as variables may be moved. The MOVE statement may convert the data (decimal to binary), if required, to fit the description of the receiving field. This statement transfers data in accordance with the rules for editing (insertions, deletions or replacement of characters) if the picture of the receiving item calls for it.

A PERFORM statement is a sequence control statement that causes a branch to a procedure or a series of procedures and following the execution, a return branch to the statement after the PERFORM statement. The procedure is treated as a subroutine, with the PERFORM statement in the main routine providing the linkage to and from the subroutine, and is also used to control the execution of loops. The three formats of the PERFORM statement are used when it is desired not only to set up linkage of a routine but also to control the number of times the routine is executed. In the basic PERFORM-format-1, the procedure referenced is executed once, with control then passing to the first statement following the PERFORM statement. The TIMES option, format-2, provides a means for performing a procedure a repetitive number of times and then returning control to the next statement after the PERFORM statement. The UNTIL option, format-3, causes an evaluation of a specified test condition. The specified procedures are performed until the condition specified is true, then control is transferred to the next statement after the PERFORM statement.

Summary: COBOL Verbs

ACCEPT is an input verb that is used to read low-volume data from the system input device (which might be a disk file, tape drive, card reader, console, etc.). The information might be needed to initialize program switches, balance totals, serial numbers or other low-volume data. It is not used to read files of data (see READ).

CLOSE is an input-output verb that is used after the processing is finished to end the processing of a file. It terminates the processing of one or more previously opened data files. It also performs actions such as checking and creating end-of-file labels automatically when a file is closed (see OPEN).

DISPLAY is an output verb that puts data out on the system output device (which may be a disk drive, tape drive, printer, console, etc.). It is used for low-volume data such as exception records or messages to the computer operator. It is not used to write files of data (see WRITE).

MOVE is a data manipulation verb that moves data from one area of storage to another, converts the data (for instance, from decimal to binary) if required to fit the description of the receiving item, and edits the data (inserts, deletes, or replaces characters) if the picture of the receiving item calls for it.

OPEN is an input-output verb that makes one or more data files ready for reading or writing. Input files are named after the word INPUT; output files after OUTPUT. Actions such as checking and creating beginning-of-file labels are done automatically when a file is opened. Opening does not make input records available for processing; a READ statement is required for that. A file must be opened before a READ or WRITE statement can act on it. (See READ, WRITE, CLOSE.)

PERFORM is a sequence control verb that causes a branch to a procedure or a series of procedures and following their execution, a return branch to the statement after the PERFORM statement. This statement is also used to control the execution of loops. (PERFORM contrasts with GO TO, which causes a branch but not a return.)

READ is an input verb that makes a data record from an input file available for processing. For sequential files, such as tape and card files, the READ statement contains an AT END phrase, which specifies actions that are to be taken after the last record has been processed. A READ statement is valid only if the file is open (see OPEN).

WRITE is an output verb that releases a record for an output file. The actual transfer of this record to an output device may not occur until sometime later; in particular, if there are to be two or more records per block, the record may be held until there are enough records to fill a block. When the record is to be printed, the WRITE statement contains an AFTER or BEFORE ADVANCING phrase to specify the vertical spacing on the printer page. A WRITE statement is valid only if the file is open (see OPEN).

Illustrative Program: Simple Listing—Transaction Register

This program and all subsequent programs show only selected IPO charts and flowcharts. The charts selected are those that are likely to be most instructive. By this point you should understand these charts enough to know what would be contained in the charts not shown.

Job Definition

Print a report listing all items sold during a week. The selling of an item is known as a transaction, so the report is titled *Transaction Register.*

During the week, a transaction file is created. At end of each day, transaction records are punched into cards from information obtained from order forms during the day. To get the printed transaction report, list the information from all input records on the printed report.

Input

Sales transaction file consisting of 80-position records. The format of the input records is shown on this Record Layout Form:

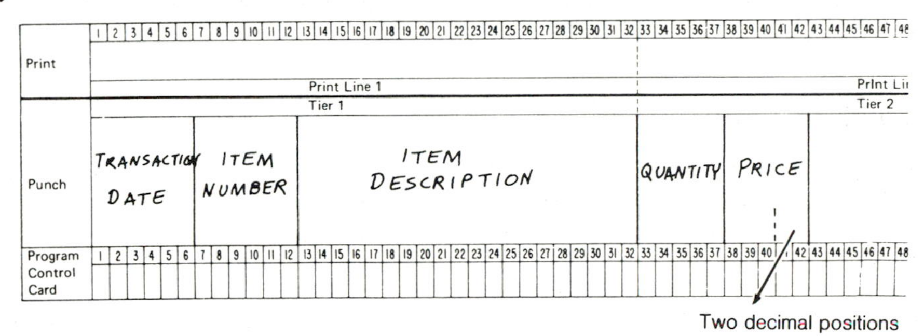

Output

A Transaction Register is to be printed as follows:

TRANSACTION REGISTER

TRANSACTION DATE	ITEM NO	DESCRIPTION	QUANTITY	UNIT PRICE
07/23/80	413010	CH001 BOX 100A FLUSH	10	4.90
07/23/80	412146	CH148 BREAKER 15A	100	.89
07/23/80	411116	1500 TWIN SOCKET B	500	1.12
07/24/80	503029	MOTOR 1/2 HP 60 CYC	2	146.78
07/24/80	317802	TERMINAL CLIP	100	5.12
07/24/80	326917	TERMINAL BAR	100	4.12
07/24/80	411121	1506 SOCT ADAPT BRN	400	.19
07/24/80	412997	CH173 BREAKER 30A	60	1.15
07/24/80	413088	CH176 BREAKER 60A	40	1.15
07/24/80	411174	C151 SIL SWITCH BRN	200	1.16
07/24/80	413090	CH005 BR BOX 150A	10	4.98
07/24/80	718326	FC803 FUSE 15A	200	.32

Printer Spacing Chart

This Printer Spacing Chart shows how the report is formatted:

Line	Type	Content
3	H	TRANSACTION REGISTER
5	H	TRANSACTION ITEM DESCRIPTION QUANTITY UNIT
6	H	DATE NO PRICE
8	D	XX/XX/XX XXXXXX X————X XXXXX XXX.XX
9–10		(DATE) (ITEMNO) (DESC) (QTY) (PRICE)
12		Single space detail lines

Hierarchy Chart

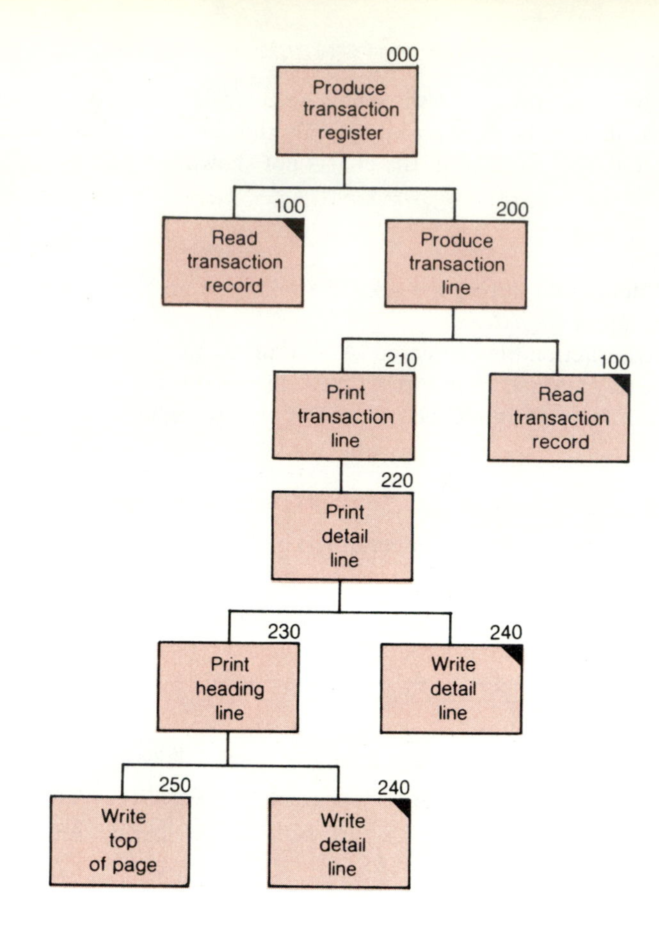
000
Produce transaction register
100
Read transaction record
200
Produce transaction line
210
Print transaction line
100
Read transaction record
220
Print detail line
230
Print heading line
240
Write detail line
250
Write top of page
240
Write detail line

IPO Charts

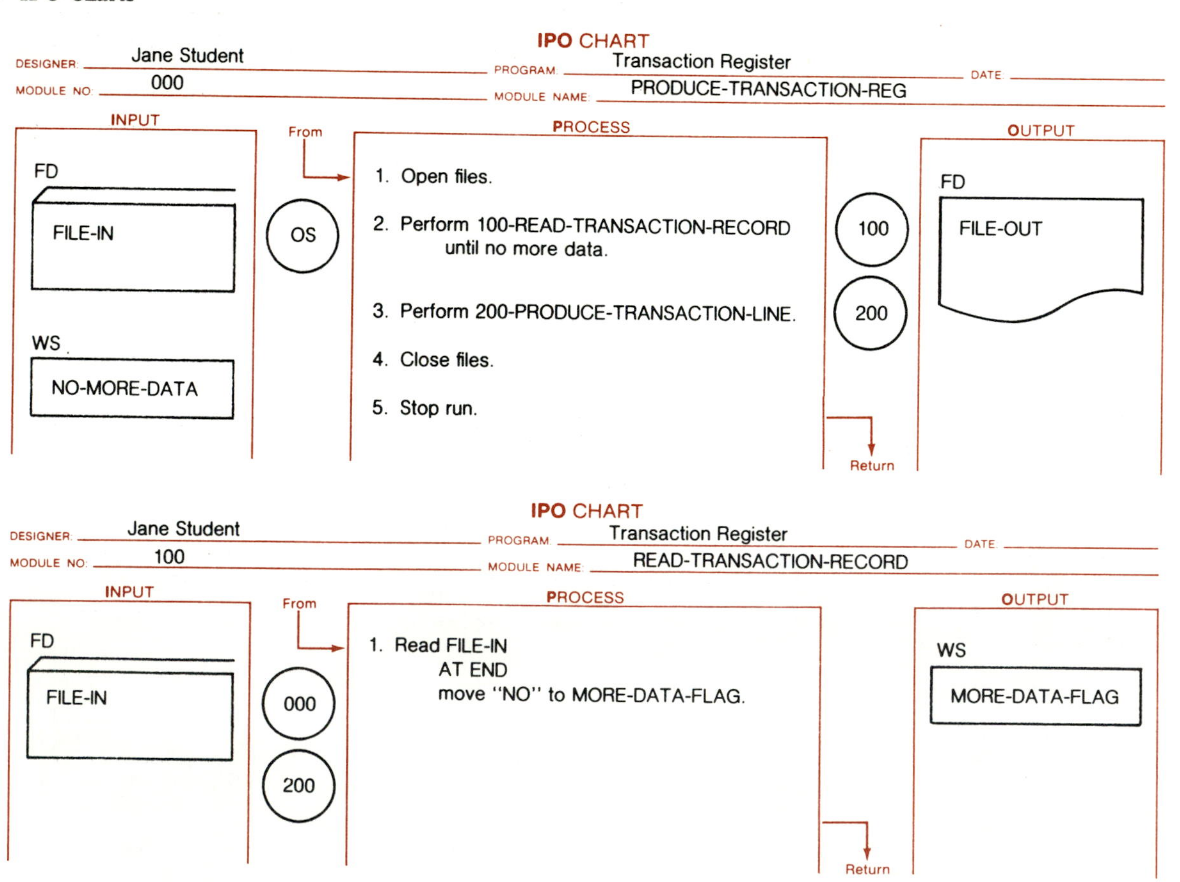
IPO CHART
DESIGNER: Jane Student
PROGRAM: Transaction Register
DATE:
MODULE NO: 000
MODULE NAME: PRODUCE-TRANSACTION-REG
INPUT
FD
FILE-IN
WS
NO-MORE-DATA
From
OS
PROCESS
1. Open files.
2. Perform 100-READ-TRANSACTION-RECORD until no more data.
3. Perform 200-PRODUCE-TRANSACTION-LINE.
4. Close files.
5. Stop run.
100
200
Return
OUTPUT
FD
FILE-OUT
IPO CHART
DESIGNER: Jane Student
PROGRAM: Transaction Register
DATE:
MODULE NO: 100
MODULE NAME: READ-TRANSACTION-RECORD
INPUT
FD
FILE-IN
From
000
200
PROCESS
1. Read FILE-IN
AT END
move "NO" to MORE-DATA-FLAG.
Return
OUTPUT
WS
MORE-DATA-FLAG

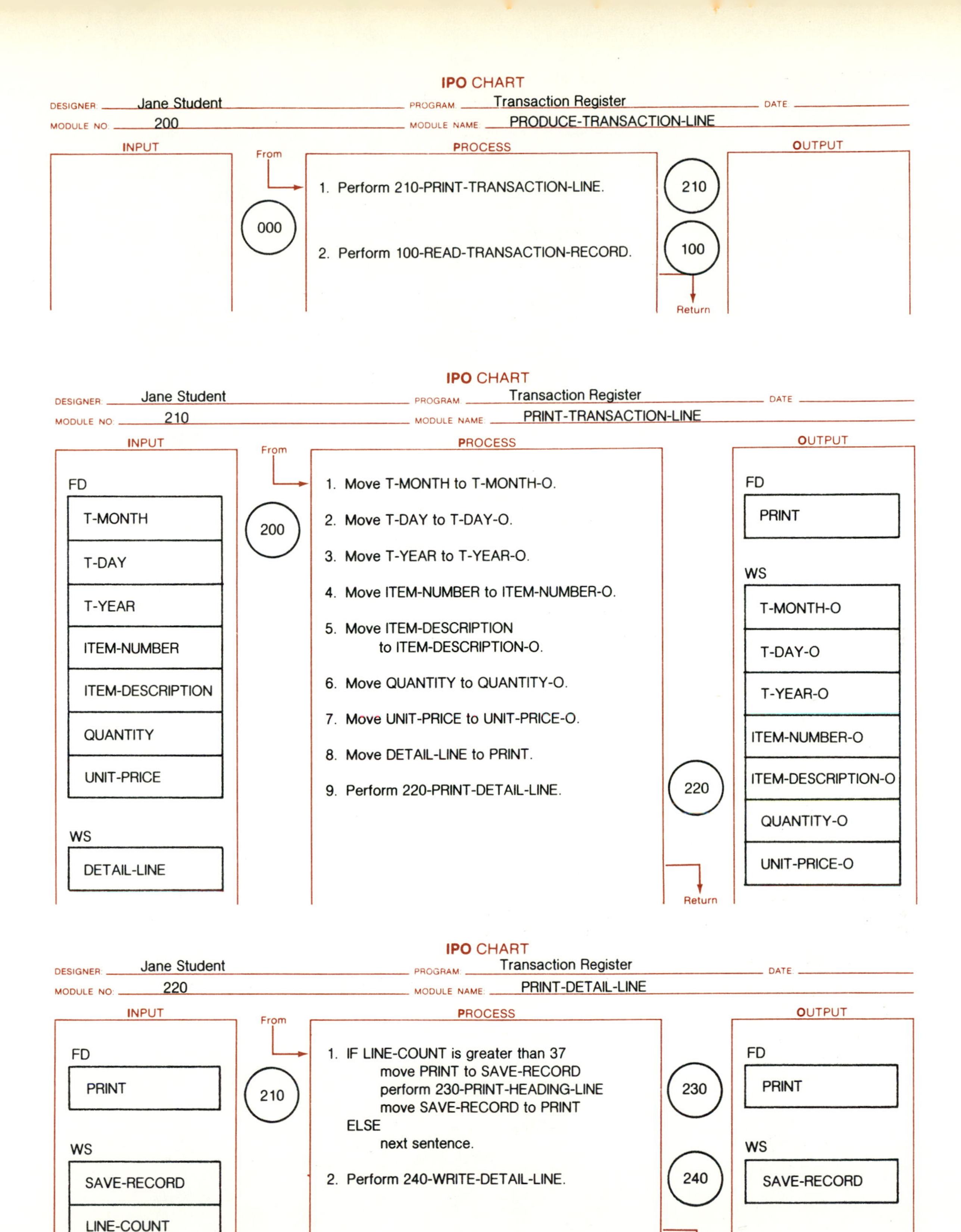
IPO CHART
DESIGNER Jane Student
PROGRAM Transaction Register
DATE
MODULE NO. 200
MODULE NAME PRODUCE-TRANSACTION-LINE
INPUT
From
PROCESS
OUTPUT
000
1. Perform 210-PRINT-TRANSACTION-LINE.
210
2. Perform 100-READ-TRANSACTION-RECORD.
100
Return
IPO CHART
DESIGNER Jane Student
PROGRAM Transaction Register
DATE
MODULE NO. 210
MODULE NAME PRINT-TRANSACTION-LINE
INPUT
From
PROCESS
OUTPUT
FD
T-MONTH
T-DAY
T-YEAR
ITEM-NUMBER
ITEM-DESCRIPTION
QUANTITY
UNIT-PRICE
WS
DETAIL-LINE
200
1. Move T-MONTH to T-MONTH-O.
2. Move T-DAY to T-DAY-O.
3. Move T-YEAR to T-YEAR-O.
4. Move ITEM-NUMBER to ITEM-NUMBER-O.
5. Move ITEM-DESCRIPTION to ITEM-DESCRIPTION-O.
6. Move QUANTITY to QUANTITY-O.
7. Move UNIT-PRICE to UNIT-PRICE-O.
8. Move DETAIL-LINE to PRINT.
9. Perform 220-PRINT-DETAIL-LINE.
220
Return
FD
PRINT
WS
T-MONTH-O
T-DAY-O
T-YEAR-O
ITEM-NUMBER-O
ITEM-DESCRIPTION-O
QUANTITY-O
UNIT-PRICE-O
IPO CHART
DESIGNER Jane Student
PROGRAM Transaction Register
DATE
MODULE NO. 220
MODULE NAME PRINT-DETAIL-LINE
INPUT
From
PROCESS
OUTPUT
FD
PRINT
WS
SAVE-RECORD
LINE-COUNT
210
1. IF LINE-COUNT is greater than 37
move PRINT to SAVE-RECORD
perform 230-PRINT-HEADING-LINE
move SAVE-RECORD to PRINT
ELSE
next sentence.
2. Perform 240-WRITE-DETAIL-LINE.
230
240
Return
FD
PRINT
WS
SAVE-RECORD

Flowcharts

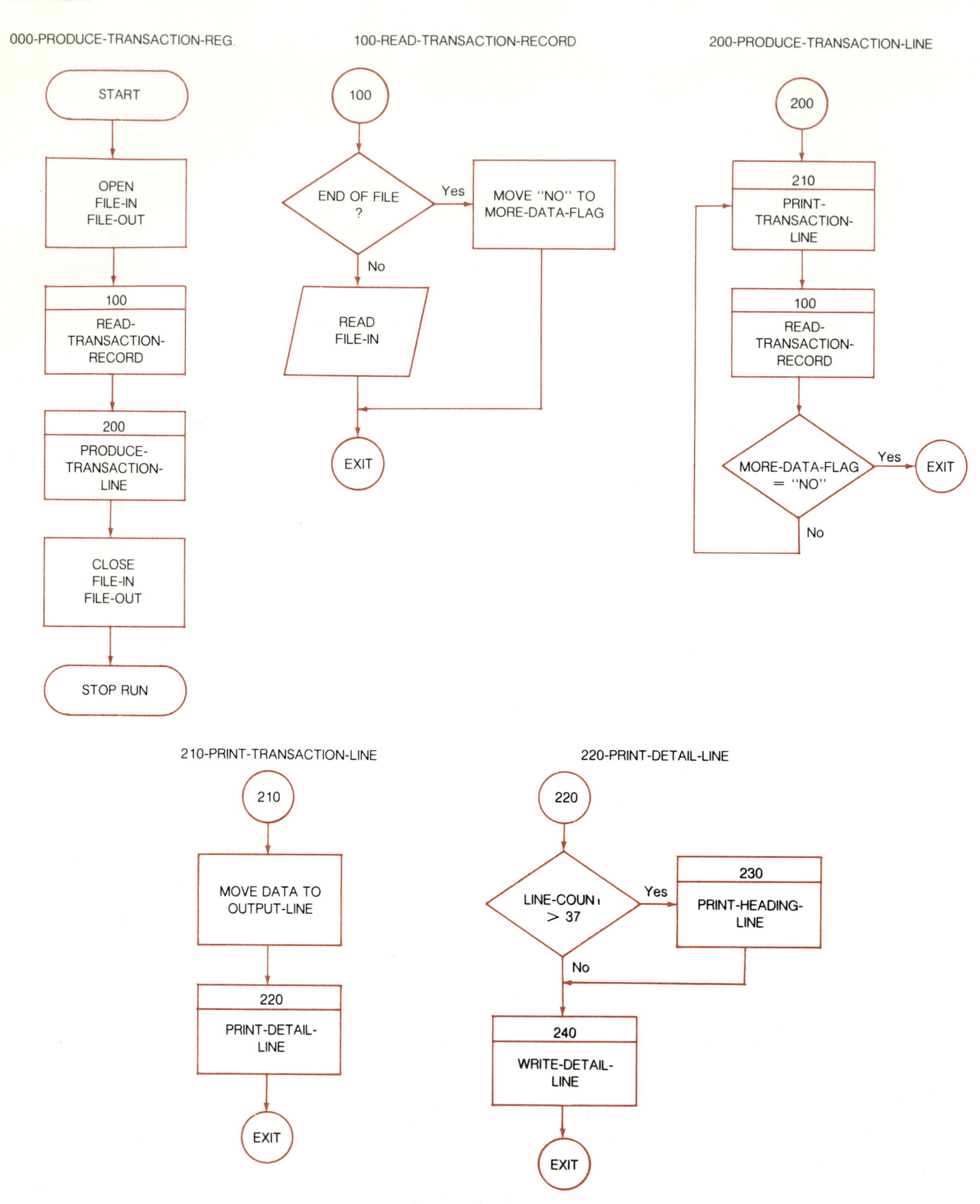

000-PRODUCE-TRANSACTION-REG.
START
OPEN
FILE-IN
FILE-OUT
100
READ-
TRANSACTION-
RECORD
200
PRODUCE-
TRANSACTION-
LINE
CLOSE
FILE-IN
FILE-OUT
STOP RUN
100-READ-TRANSACTION-RECORD
100
END OF FILE
?
Yes
MOVE "NO" TO
MORE-DATA-FLAG
No
READ
FILE-IN
EXIT
200-PRODUCE-TRANSACTION-LINE
200
210
PRINT-
TRANSACTION-
LINE
100
READ-
TRANSACTION-
RECORD
MORE-DATA-FLAG
= "NO"
Yes
EXIT
No
210-PRINT-TRANSACTION-LINE
210
MOVE DATA TO
OUTPUT-LINE
220
PRINT-DETAIL-
LINE
EXIT
220-PRINT-DETAIL-LINE
220
LINE-COUNT
> 37
Yes
230
PRINT-HEADING-
LINE
No
240
WRITE-DETAIL-
LINE
EXIT

Source Program

```
 1        *************************
 2
 3         IDENTIFICATION DIVISION.
 4
 5        *************************
 6
 7         PROGRAM-ID.
 8             TRANSACTION-REGISTER.
 9         INSTALLATION.
10             WEST LOS ANGELES COLLEGE.
11         DATE-WRITTEN.
12             3 FEBRUARY 1982.
13
14        *PURPOSE.
15        *    THIS PROGRAM PRINTS A REPORT LISTING ALL ITEMS SOLD DURING
16        *    A WEEK.
17
18        *************************
19
20         ENVIRONMENT DIVISION.
21
22        *************************
23
24         CONFIGURATION SECTION.
25
26         SOURCE-COMPUTER.
27             LEVEL-66-ASCII.
28         OBJECT-COMPUTER.
29             LEVEL-66-ASCII.
30
31         INPUT-OUTPUT SECTION.
32
33         FILE-CONTROL.
34             SELECT FILE-IN      ASSIGN TO C1-CARD-READER.
35             SELECT FILE-OUT     ASSIGN TO P1-PRINTER.
36
37        *************************
38
39         DATA DIVISION.
40
41        *************************
42
43         FILE SECTION.
44
45         FD  FILE-IN
46             CODE-SET IS GBCD
47             LABEL RECORDS ARE OMITTED
48             DATA RECORD IS CARDIN.
49
50         01  CARDIN.
51
52             03  TRANSACTION-DATE.
53                 05  T-MONTH          PICTURE 9(2).
54                 05  T-DAY            PICTURE 9(2).
55                 05  T-YEAR           PICTURE 9(2).
56             03  ITEM-NUMBER          PICTURE 9(6).
57             03  ITEM-DESCRIPTION     PICTURE X(20).
58             03  QUANTITY             PICTURE 9(5).
59             03  UNIT-PRICE           PICTURE 9(3)V99.
60             03  FILLER               PICTURE X(38).
61
62         FD  FILE-OUT
63             CODE-SET IS GBCD
64             LABEL RECORDS ARE OMITTED
65             DATA RECORD IS PRINT.
66
67         01  PRINT                    PICTURE X(132).
68
69         WORKING-STORAGE SECTION.
70
71         77  LINE-COUNT               PICTURE 9(2)    VALUE 38.
72
73         01  FLAGS.
74
75             03  MORE-DATA-FLAG       PICTURE X(3)    VALUE "YES".
76                 88  MORE-DATA                        VALUE "YES".
77                 88  NO-MORE-DATA                     VALUE "NO".
78
79         01  HDG-1.
80
81             03  FILLER               PICTURE X(37)   VALUE SPACES.
82             03  FILLER               PICTURE X(11)   VALUE "TRANSACTION".
83             03  FILLER               PICTURE X(10)   VALUE "  REGISTER".
84             03  FILLER               PICTURE X(74)   VALUE SPACES.
85
86         01  HDG-2.
87
88             03  FILLER               PICTURE X(9)    VALUE SPACES.
89             03  FILLER               PICTURE X(11)   VALUE "TRANSACTION".
90             03  FILLER               PICTURE X(6)    VALUE SPACES.
91             03  FILLER               PICTURE X(4)    VALUE "ITEM".
92             03  FILLER               PICTURE X(11)   VALUE SPACES.
93             03  FILLER               PICTURE X(11)   VALUE "DESCRIPTION".
94             03  FILLER               PICTURE X(11)   VALUE SPACES.
95             03  FILLER               PICTURE X(8)    VALUE "QUANTITY".
96             03  FILLER               PICTURE X(3)    VALUE SPACES.
97             03  FILLER               PICTURE X(4)    VALUE "UNIT".
98             03  FILLER               PICTURE X(54)   VALUE SPACES.
```

```
 99
100        01  HDG-3.
101
102            03  FILLER                  PICTURE X(12)    VALUE SPACES.
103            03  FILLER                  PICTURE X(4)     VALUE "DATE".
104            03  FILLER                  PICTURE X(11)    VALUE SPACES.
105            03  FILLER                  PICTURE X(2)     VALUE "NO".
106            03  FILLER                  PICTURE X(45)    VALUE SPACES.
107            03  FILLER                  PICTURE X(5)     VALUE "PRICE".
108            03  FILLER                  PICTURE X(53)    VALUE SPACES.
109
110        01  DETAIL-LINE.
111
112            03  FILLER                  PICTURE X(10)    VALUE SPACES.
113            03  T-MONTH-O               PICTURE 9(2).
114            03  FILLER                  PICTURE X(1)     VALUE "/".
115            03  T-DAY-O                 PICTURE 9(2).
116            03  FILLER                  PICTURE X(1)     VALUE "/".
117            03  T-YEAR-O                PICTURE 9(2).
118            03  FILLER                  PICTURE X(7)     VALUE SPACES.
119            03  ITEM-NUMBER-O           PICTURE 9(6).
120            03  FILLER                  PICTURE X(9)     VALUE SPACES.
121            03  ITEM-DESCRIPTION-O      PICTURE X(20).
122            03  FILLER                  PICTURE X(4)     VALUE SPACES.
123            03  QUANTITY-O              PICTURE ZZ,ZZ9.
124            03  FILLER                  PICTURE X(3)     VALUE SPACES.
125            03  UNIT-PRICE-O            PICTURE Z(3).99.
126            03  FILLER                  PICTURE X(53)    VALUE SPACES.
127
128        01  SAVE-RECORD                 PICTURE X(132).
129
130
131       *************************
132
133        PROCEDURE DIVISION.
134
135       *************************
136
137        000-PRODUCE-TRANSACTION-REG.
138
139            OPEN     INPUT    FILE-IN
140                     OUTPUT   FILE-OUT.
141            PERFORM 100-READ-TRANSACTION-RECORD.
142            PERFORM 200-PRODUCE-TRANSACTION-LINE
143                     UNTIL NO-MORE-DATA.
144            CLOSE    FILE-IN
145                     FILE-OUT.
146            STOP RUN.
147
148        100-READ-TRANSACTION-RECORD.
149
150            READ FILE-IN
151                AT END MOVE "NO " TO MORE-DATA-FLAG.
152
153        200-PRODUCE-TRANSACTION-LINE.
154
155            PERFORM 210-PRINT-TRANSACTION-LINE.
156            PERFORM 100-READ-TRANSACTION-RECORD.
157
158        210-PRINT-TRANSACTION-LINE.
159
160            MOVE T-MONTH              TO T-MONTH-O.
161            MOVE T-DAY                TO T-DAY-O.
162            MOVE T-YEAR               TO T-YEAR-O.
163            MOVE ITEM-NUMBER          TO ITEM-NUMBER-O.
164            MOVE ITEM-DESCRIPTION     TO ITEM-DESCRIPTION-O.
165            MOVE QUANTITY             TO QUANTITY-O.
166            MOVE UNIT-PRICE           TO UNIT-PRICE-O.
167            MOVE DETAIL-LINE          TO PRINT.
168            PERFORM 220-PRINT-DETAIL-LINE.
169
170        220-PRINT-DETAIL-LINE.
171
172            IF LINE-COUNT IS > 37
173                MOVE PRINT TO SAVE-RECORD
174                PERFORM 230-PRINT-HEADING-LINE
175                MOVE SAVE-RECORD TO PRINT
176            ELSE
177                NEXT SENTENCE.
178            PERFORM 240-WRITE-DETAIL-LINE.
179
180        230-PRINT-HEADING-LINE.
181
182            MOVE HDG-1 TO PRINT.
183            PERFORM 250-WRITE-TOP-OF-PAGE.
184            MOVE SPACES TO PRINT.
185            PERFORM 240-WRITE-DETAIL-LINE.
186            MOVE HDG-2 TO PRINT.
187            PERFORM 240-WRITE-DETAIL-LINE.
188            MOVE HDG-3 TO PRINT.
189            PERFORM 240-WRITE-DETAIL-LINE.
190            MOVE SPACES TO PRINT.
191            PERFORM 240-WRITE-DETAIL-LINE.
192
193        240-WRITE-DETAIL-LINE.
194
195            WRITE PRINT
196                AFTER ADVANCING 1 LINE.
197            ADD 1 TO LINE-COUNT.
198
199        250-WRITE-TOP-OF-PAGE.
200
```

```
201           WRITE PRINT
202               AFTER ADVANCING PAGE.
203           MOVE 1 TO LINE-COUNT.
204

THERE WERE 204 SOURCE INPUT LINES.
THERE WERE NO DIAGNOSTICS.
```

TRANSACTION REGISTER

TRANSACTION DATE	ITEM NO	DESCRIPTION	QUANTITY	UNIT PRICE
07/23/80	413010	CH001 BOX 100A FLUSH	10	4.90
07/23/80	412146	CH148 BREAKER 15A	100	.89
07/23/80	411116	1500 TWIN SOCKET B	500	1.12
07/24/80	503029	MOTOR 1/2 HP 60 CYC	2	146.78
07/24/80	317802	TERMINAL CLIP	100	5.12
07/24/80	326917	TERMINAL BAR	100	4.12
07/24/80	411121	1506 SOCT ADAPT BRN	400	.19
07/24/80	412997	CH173 BREAKER 30A	60	1.15
07/24/80	413088	CH176 BREAKER 60A	40	1.15
07/24/80	411174	C151 SIL SWITCH BRN	200	1.16
07/24/80	413090	CH005 BR BOX 150A	10	4.98
07/24/80	718326	FC803 FUSE 15A	200	.32

Questions for Review

1. What purpose does the Procedure Division serve?
2. What is a procedure and how is it referred to?
3. What is the physical end of a COBOL source program?
4. Define a statement and explain the three types of statements used in the Procedure Division.
5. Define a sentence, paragraph, and section.
6. Why is it important for the programmer to be careful in creating sections?
7. How is the Procedure Division organized?
8. How is the Procedure Division coded and where does execution of the object program begin?
9. What are the main functions of the input and output verbs? Give examples.
10. What are the main functions of the OPEN and READ statements and how are they related to each other?
11. What are the main functions of the WRITE and CLOSE statements?
12. Explain the use of the ADVANCING option of the WRITE statement.
13. What are the functions of the ACCEPT and DISPLAY statements?
14. How may the ACCEPT and DISPLAY statements be used as a communication device between the programmer and the computer operator?
15. What is the main function of the MOVE statement?
16. Explain the movement of numeric and nonnumeric data.
17. What is the main function of the PERFORM statement and how is it used?
18. What does the DOWHILE structure of structured programming provide and how is it implemented in COBOL programming?
19. What are the three formats of the PERFORM statement and when are they used?

Matching Questions

A.

Match each verb with its proper description.

_____ 1. CLOSE	A. Used for low-volume output.
_____ 2. DISPLAY	B. Makes the next logical record available for processing.
_____ 3. OPEN	C. Transfers data from one area of storage to one or more areas in storage.
_____ 4. PERFORM	D. A sequence control statement that causes a branch to a procedure or series of procedures, and, following their execution, a branch to the statement after the statement that caused the branch.
_____ 5. READ	E. Releases a logical record for insertion into an output file.
_____ 6. ACCEPT	F. Makes associated record areas available for processing.
_____ 7. WRITE	G. Terminates processing of a data file.
_____ 8. MOVE	H. Causes low-volume data to be made available to the specified data item.

B.

Match each item with its proper description.

_____ 1. Sentence	A. Causes compiler to take specific action during compilation of the source program.
_____ 2. Conditional statement	B. Composed of a paragraph, or a group of successive paragraphs, or a section, or a group of successive sections within the Procedure Division.
_____ 3. Statement	C. Specifies that the truth value of a condition is to be determined and that the subsequent action of the object program is dependent on this truth value.
_____ 4. Compiler-directing statement	D. Statements that are logically related and organized into groups.
_____ 5. Procedure	E. Indicates a specific unconditional action to be taken by the object program.
_____ 6. Section	F. Syntactically valid combination of words and symbols beginning with a COBOL verb.
_____ 7. Imperative statement	G. Composed of one or more statements specifying action to be taken and terminated by a period and a space.
_____ 8. Paragraph	H. Composed of one or more paragraphs.

Exercises

Multiple Choice: Indicate the best *answer (questions 1–25).*

1. The Procedure Division contains instructions that
   a. specify the actions to be performed.
   b. sequence which procedures are to be performed.
   c. specify data manipulation, input and output, and arithmetic actions.
   d. All of the above.
2. A statement is
   a. composed of paragraphs or sections.
   b. a syntactically valid combination of words and symbols beginning with a COBOL verb.
   c. always terminated by a period and a space.
   d. always on a line by itself.
3. A statement that specifies the truth value of a condition is called a(n)
   a. imperative statement.
   b. compiler-directing statement.
   c. conditional statement.
   d. procedure statement.
4. A paragraph ______________ have a paragraph header.
   a. cannot
   b. may
   c. must not
   d. must
5. The PERFORM verb has many advantages. The best advantages is that
   a. it allows you to set up subroutines.
   b. a frequently used routine need be coded only once and can be accessed from many locations within the program.
   c. it allows you to control with a minimum of writing.
   d. you cannot achieve these same functions with program switches.
6. In a PERFORM statement with the UNTIL option you ______________ change the value of the data whose condition is tested.
   a. may not
   b. must
   c. should not
   d. None of the above.
7. In a PERFORM statement with the UNTIL option, the word UNTIL is followed by a
   a. test condition.
   b. PERFORM statement.
   c. sequence control verb.
   d. STOP PERFORMANCE.
8. Each paragraph in the Procedure Division is called a procedure. A procedure must contain
   a. one sentence or less.
   b. one or more sentences.
   c. more than one sentence.
   d. All of the above.
9. An ACCEPT statement obtains up to eighty characters of data and moves this information into a
   a. display area.
   b. working-storage area.
   c. read files area.
   d. input-output area.
10. Which statement is correct?
    a. An OPEN statement cannot be executed prior to any other input-output statement for a file.
    b. The OPEN statement by itself makes an input record available for processing.
    c. You can open a file that is already open.
    d. You cannot open a file that is already open.

11. Can the following messages be displayed?
TYPE TWO FOUR-DIGIT NUMBERS AS A STRING OF EIGHT DIGITS NO SPACES. SMALLER NUMBER MUST BE FIRST
TO STOP RUN, TYPE 999999999.
Which is correct?
    a. The DISPLAY verb cannot be used because the message is too long.
    b. Alphanumeric messages cannot be displayed.
    c. Two DISPLAY statements can be used for the message, one for each line.
    d. The message is prohibited because it contains reserved words.
12. Controlling form spacing and skipping is the function of one of the formats of the following verb:
    a. WRITE
    b. PERFORM
    c. DISPLAY
    d. MOVE
13. Before we can read records from a file, we must ______________ the file.
    a. MOVE
    b. ACCEPT
    c. READ
    d. OPEN
14. Which of the following statements is *not* correct?
    a. In a CLOSE statement, files are not identified as "input" or "output" files.
    b. A CLOSE statement may only be executed for a file that is open.
    c. Any file that was opened must be closed before the run is stopped.
    d. Any file that was opened must be closed after the run is stopped.
15. A record name is specified in
    a. an OPEN statement.
    b. a READ statement.
    c. a PERFORM statement.
    d. a WRITE statement.
16. A file name is specified in
    a. a MOVE statement.
    b. a WRITE statement.
    c. a READ statement.
    d. a PERFORM statement.
17. Which of the following statements allows the console operator to key a name into storage through the console display station?
    a. READ NAME.
    b. ACCEPT NAME.
    c. DISPLAY NAME ON CONSOLE.
    d. ACCEPT NAME FROM CONSOLE.
18. You must define a location in working storage for
    a. a literal.
    b. a variable.
    c. a word enclosed in quotation marks.
    d. numbers not enclosed in quotation marks.
19. In order to write a literal on the console, the programmer must
    a. reserve a location in storage for the literal.
    b. always enclose the literal in quotation marks.
    c. reserve a location in the Environment Division.
    d. None of the above.
20. If you wish to display a message to the operator and accept his/her reply, you should code
    a. an ACCEPT statement first.
    b. a DISPLAY statement first.
    c. an OPEN statement first.
    d. a CLOSE statement first.
21. An eighty-character record named CARD-DATA has already been defined. Which of the entries below is the correct way to get the data from our card into CARD-DATA?
    a. ACCEPT DATA-IN, MOVE INTO CARD-DATA.
    b. ACCEPT; MOVE TO CARD-DATA.
    c. ACCEPT CARD-DATA.
    d. None of the above.

22. Now that we have the card's data in working storage (exercise 21), we want to display the data in order to check the validity of the data.
Which of the following statements is correct?
   a. We must first move it from CARD-DATA to an output area.
   b. We need only to write DISPLAY CARD-DATA UPON CONSOLE.
   c. We need to convert the data to DISPLAY usage.
   d. None of the above.
23. Paragraphs P1, P2, P3, and P4 are in linear sequence. Which of the following statements specifies execution of paragraphs P1, P3, and P4?
   a. PERFORM P1 P3 P4.
   b. PERFORM P1 THRU P4.
   c. PERFORM P1
      PERFORM P3 THRU P4.
   d. None of the above.
24. The problem of an "endless loop" in a PERFORM statement with an UNTIL option can be avoided by following this rule: In the procedure to be performed, you ______________ change the value of the data whose condition is being tested in the statement.
   a. must
   b. may
   c. may occasionally
   d. must not
25. In a PERFORM statement with the UNTIL option, if the test-condition is true to begin with, the procedure will
   a. be performed only once.
   b. not be performed at all.
   c. be repeated endlessly.
   d. be performed twice only.
26. PERFORM DISCOUNT 15 TIMES.
PERFORM DISCOUNT ITEMS TIMES.
Match the following words from the statements shown above with the terms from the PERFORM statement.

_____ 1. Procedure-name-1	A. 5
_____ 2. Identifier-1	B. DISCOUNT
_____ 3. Integer-1	C. ITEMS
	D. TIMES

27. PERFORM ALLOCATION
        UNTIL RESOURCE IS ZERO
        OR RESOURCE IS NEGATIVE
   a. If the value of RESOURCE is 5,000, will ALLOCATION be performed 5,000 times?
   b. What will happen if the original value of RESOURCE is 5,000 and its value is not changed by the ALLOCATION procedure?
28.
```
10  DATE-IN.
    15  MONTH-IN      PICTURE 99.
    15  DAY-IN        PICTURE 99.
    15  YEAR-IN       PICTURE 99.

05  INVOICE-DATE      PICTURE 9(6).

MOVE DATE-IN TO INVOICE-DATE.
```
Is this a valid MOVE statement? Give a reason for your answer.
29. A console operator will key fifteen characters, which may be both letters and digits, on the console keyboard. The first ten of these characters are to be printed in positions 25–34 of the output line on the printer. The remaining characters are to be printed in positions 10–14 of the output line.
Write the necessary Data and Procedure Division instructions for this problem.

30. *Procedure: READ-A-CARD.*

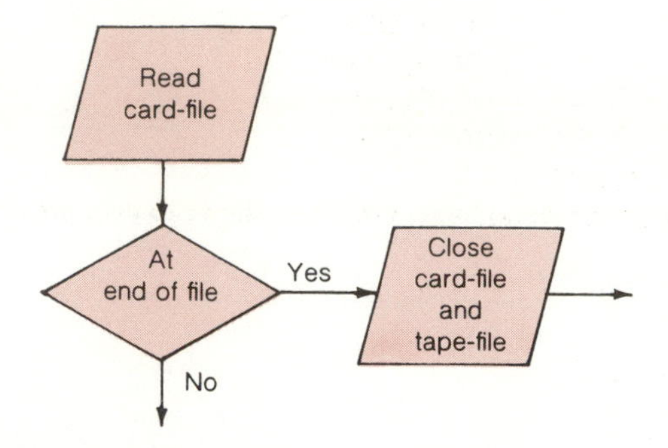

a. Write the READ entry that corresponds to the blocks above.
b. Complete the procedure READ-A-CARD by writing entries to
   1. move data into the output record.
   2. write the output record.

31. Match the effects below with the value of identifier-1 or integer-1 in the TIMES option of the PERFORM statement.

Value of identifier-1 or integer-1	*Effects*
_____ 1. −2	A. Invalid value (not an integer).
_____ 2. 2.8	B. Specified paragraph(s) will not be executed.
_____ 3. 26	C. Specified paragraph(s) will be executed −2 times.
_____ 4. 0	D. Specified paragraph(s) will be executed indicated number of times.

32. The hierarchy of entries in the Procedure Division beginning with the largest unit are:
    1. ______________________.   3. ______________________.
    2. ______________________.   4. ______________________.

33. Write Procedure Division statements for the following:
    a. Prepare a READ statement for a card file named CARD-INPUT with a record named CARD-REC-DATA. The termination paragraph is named LAST-RECORD.
    b. Prepare a WRITE statement for a printed report with a file name of PRINT-LIST and a record name of PRINT-IT, where each line is to be separated by two blank lines.
    c. Type on the console typewriter, the literal VALUE EXCEEDS BALANCE and the field defined in the Working-Storage Section as REMAINING-BALANCE.
    d. Prepare a CLOSE statement for the file named, INPUT-FILE, OUTPUT-FILE and PRINT-FILE.

34. Write the necessary WRITE statement with the ADVANCING option to accomplish the following:
    a. Advance the form three lines before a record is written.
    b. Advance the form three lines after a record is written.
    c. Double spacing.
    d. Triple spacing.
    e. Skipping to the first printing line of a new page before the line is printed.

35. Write the necessary Procedure Division entries for the following:
    a. Allow a value of CODE-DATA to be entered into Working-Storage through the console typewriter.
    b. Write the value of CODE-DATA on the console typewriter.

36. Match the result with the correct statement.

Statement	**Result**
_____ 1. DISPLAY DATE UPON CONSOLE.	A. The word DATE will be written on the console typewriter.
_____ 2. DISPLAY 'DATE' UPON CONSOLE.	B. The word 'DATE' will be written on the console typewriter.
	C. The value of the variable DATE will be written on the console typewriter.

37. Write the necessary Data Division and Procedure Division entries for the following:
    a. In the paragraph INIT, the following message is to be written out on the console typewriter ENTER OPERATION-CODE.
    b. A value of OPERATION-CODE is to be entered into storage through the console keyboard. Values of OPERATION-CODE have a form such as 107B, 509X; or 879G.
    c. The value of OPERATION-CODE is to be written on the console typewriter.

38. Using the partial Data Division entry that follows, write the necessary coding for the MOVE problem.

01 RECORD-1	01 RECORD-2
10 STOCK-NUMBER	10 STOCK-NUMBER
10 UNITS	10 DATA-MESSAGE
10 VALUE	10 QUANTITY
10 WEIGHT	10 UNITS

a. Transfer UNITS from RECORD-1 to RECORD-2.
b. Clear a field in Working-Storage Section called DATA-MESSAGE of previous field, and move information to area above in RECORD-2.

39. A name has been defined in the input record as follows:

```
05  NAME.
    10  LAST-NAME         PICTURE A(20).
    10  FIRST-INITIAL     PICTURE A.
    10  SECOND-INITIAL    PICTURE A.
```

a. Write the necessary entries to define a Working-Storage record called EDITED-NAME, with INITIAL-1 in the first position, followed by a period and space; then INITIAL-2 in the fourth position, with another period and space; finally twenty positions called SURNAME.
b. Write the MOVE statement to put each part of the name into their proper places in the Working-Storage record.
c. Write a MOVE statement to transfer EDITED-NAME to CUSTOMER-NAME. (Assume that CUSTOMER-NAME is a twenty-six-position alphanumeric item in the output record which has been previously defined.)

40. Indicate the RECEIVED DATA in the following:

	Source Data	Source Picture	Receiving Picture	Received Data
1.	8736	9999	9999	________
2.	8736	9999P	P9999	________
3.	8736	99V99	99V99	________
4.	8736	99V99	99.99	________
5.	8736	P9999	99.99	________
6.	8736	9999P	99.99	________
7.	8736	P9999	99V99	________
8.	8736	9999P	99V99	________
9.	8736	9999	$99.99	________
10.	8736	99V99	$99.999	________
11.	8736	9999P	99999.99	________
12.	ERROR	XXXXX	XXXXX	________
13.	ERROR	AAAAA	AAAAA	________
14.	ERROR	AAAAA	AAAAAAAA	________

Problem 1

Code the necessary entries in the Identification, Environment, Data, and Procedure divisions based on the following information. Create your own data names and use the system device names of your installation.

Input is as follows:

Printer Spacing Chart for LISTING

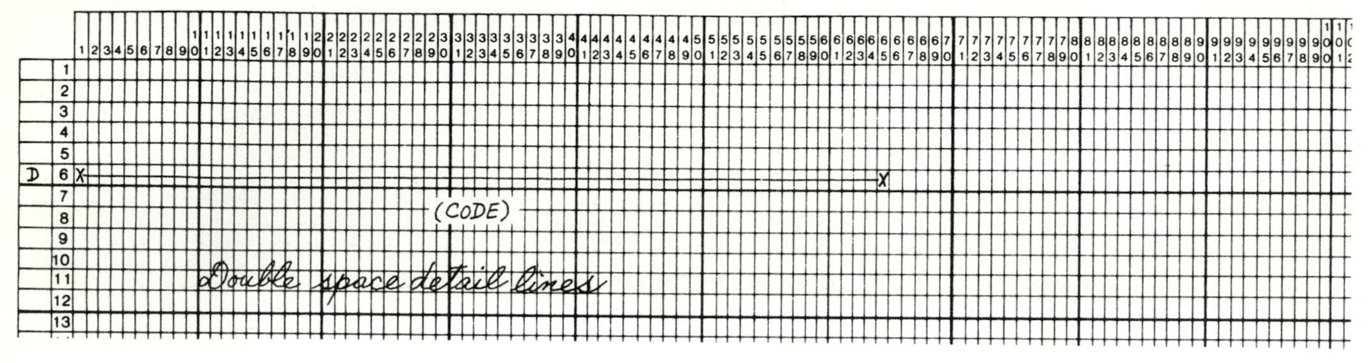

Output is as follows:

```
A B C D E F G H I J K L M N O P Q R S T U V W X Y Z
 A B C D E F G H I J K L M N O P Q R S T U V W X Y Z
0 1 2 3 4 5 6 7 8 9 0 1 2 3 4 5 6 7 8 9
 0 1 2 3 4 5 6 7 8 9 0 1 2 3 4 5 6 7 8 9
```

Problem 2

Use the flowchart, record layout, and printer spacing chart below to help you write a COBOL program that is responsible for printing a report that lists all the employees who were hired during the current calendar year.

The input is as follows: (5,0) indicates that there are five digit positions with no decimal positions; (3,A) indicates three alphabetic positions; and (25,A) indicates twenty-five alphabetic positions.

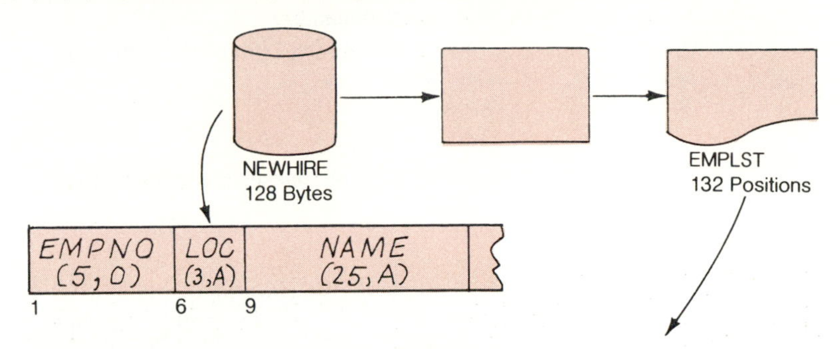

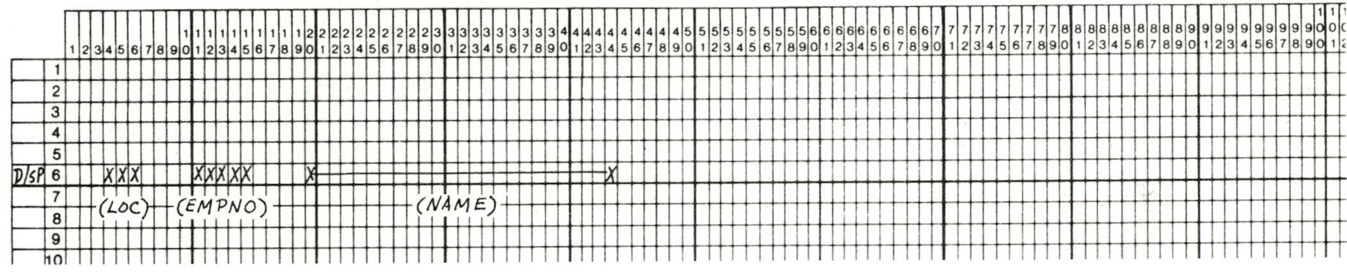

Output is as follows:

```
SWA    68832    ROBERT JONES
SWA    68322    JACK SMITH
STH    68832    HENRY KAHN
STH    68324    MARGARET KAISER
STH    68325    JUSTIN KRAMER
```

Problem 3

A sales analysis report is to be prepared, as follows, in the printer spacing chart.

The fields on the input file records are arranged as follows:

Position	Field
1-2	Salesperson number
3-8	Amount of sale (two decimal positions)
9-23	Customer name
30	Area

Using the above input specifications, code the necessary entries in all four divisions for this job, choosing your own file and field names.

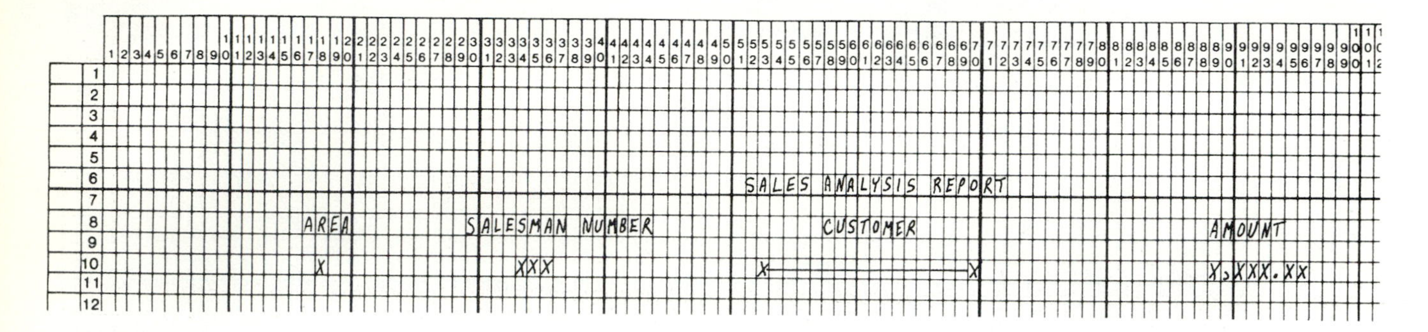

Output is as follows:

SALES ANALYSIS REPORT

AREA	SALESMAN NUMBER	CUSTOMER	AMOUNT
A	01	STEVEN LEWIS	398.64
A	02	DAVID MAIN	24.91
C	03	MICHAEL MELTON	9,641.11
D	04	JEAN MYERS	499.23
F	05	HAROLD OWENS	1,239.41

Problem 4

A report is to be prepared listing the quarterly earnings and social security tax on those earnings. The report is to be prepared as per the following printer spacing chart shown below.

The fields on the input file QTRFILE, are as follows:

Positions	Field Designation	Format
1-9	Social security number	
10-30	Name	
31-35	Social security tax	xxx.xx
36-41	Quarterly amount	xxxx.xx

Using the above input specifications, code the necessary entries in all divisions for this job. The output file is called SSLIST. Use the field names as per printer spacing chart.

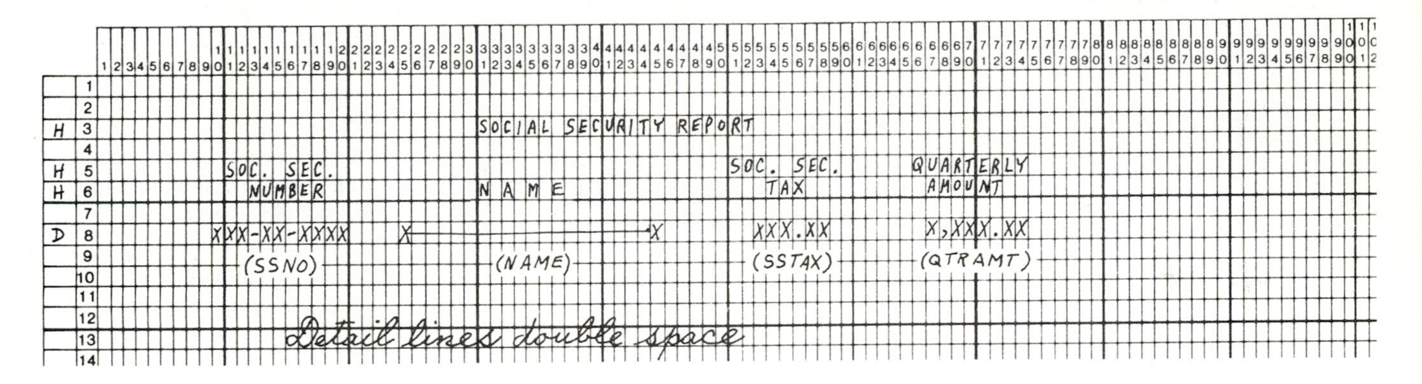

Output is as follows:

```
                         SOCIAL SECURITY REPORT

SOC. SEC.                                      SOC. SEC.      QUARTERLY
 NUMBER              N A M E                     TAX            AMOUNT

502-12-6934     RON PATTERSON                   313.67          627.34

419-63-8319     THOMAS PATRICK                  123.46          246.92

214-90-6184     MARIA PEREZ                      62.43          124.86

436-70-4125     LEE RICHARDSON                    9.79           19.58

383-80-7581     JOHN SANDERS                     47.74           95.48
```

Procedure Division: Arithmetic Operations

Chapter Outline

Chapter Objectives

The learning objectives of this chapter are:

1. To explain the arithmetic operations of the compiler and how it may be used to write more efficient arithmetic instructions.

2. To describe the various formats of the ADD, SUBTRACT, MULTIPLY and DIVIDE statements including the:
   a. GIVING option, to specify the variable in which the computed result will be stored.
   b. ROUNDED option, to specify that the computed result will be rounded to the nearest digit in the position corresponding to the rightmost position in the picture for the variable in which the result will be stored.
   c. SIZE ERROR option, to specify the action to be taken when the computed result exceeds the size of the variable in which the result is stored.

3. To describe the format of the COMPUTE statement and how to code arithmetic expressions using this statement.

4. To explain the operation of arithmetic expressions and the rules for the formation and evaluation of these expressions.

Introduction

In the previous chapter, the various techniques of structured programming were explained. The reasons for the use of these techniques were to more efficiently prepare programs, thereby reducing the costs of data processing operations. Programmers need to keep up with new techniques and approaches in order to solve future data processing problems.

Although the costs involved in program writing are high, it is the cost of testing and correcting (debugging) programs and maintaining and updating these programs that is taking a bigger and bigger portion of the data processing budget.

If the programmer is familiar with the actions of the compiler during the execution of arithmetic statements, more efficient source programs may be written.

Arithmetic Actions

Four major actions occur during the execution of an arithmetic statement: (1) the data values are prepared for calculations; (2) a raw result is calculated in a work area; (3) if desired, the size of the raw result is tested; and (4) the raw result is moved to the finished result item.

A detailed explanation of each point follows:

1. *The data values are prepared for calculation.* If necessary, the usage of the values is converted to a data code in which calculation is possible. Also, the sign bits are changed in certain cases.

   Arithmetic operations can be performed only with certain kinds of items. Specifically, computations may be done using only elementary numeric items and numeric literals. The computer can execute arithmetic operations either on binary items (COMPUTATION usage) or on packed decimal items (COMPUTATION-3 usage). Therefore, in order to execute an arithmetic statement, it is necessary to convert the data code of elementary numeric items whose usage is DISPLAY. Imagine that this statement appears in a COBOL program:

   ```
   ADD AMOUNT-W TO TOTAL-W.
   ```

   Here are the descriptions of the items to be added:

   ```
      05  AMOUNT-W      PIC S9(6)V99    USAGE COMP-3.
   77 TOTAL-W           PIC 9(8)V99     VALUE ZEROS.
   ```

   The items can be added because both are elementary items. TOTAL-W must be converted since it is a DISPLAY item; it will be converted to packed decimal (COMP-3) because that is the data code of AMOUNT-W.

   Although we typically say that an item is converted, it would be more accurate to say that the data from an item is converted to a different code. It works like this: the data is moved from the item to a work area, and converted there; then, the arithmetic operation is performed using the data in the work area. The contents of the original data item remain unchanged. For example, look at the following code:

   ```
   SUBTRACT STOCK-COUNT FROM BALANCE
       GIVING LOSS.
   ```

   If STOCK-COUNT is an external-decimal item (DISPLAY), and BALANCE is a packed-decimal item (COMP-3), the data from STOCK-COUNT is moved to a work area in which it is converted to packed-decimal. Then subtraction is done, using the data from the work area. If STOCK-COUNT is involved in later calculations in other procedures in the same program, the data from STOCK-COUNT will be converted again each time it is used in a calculation.

   The work area, by the way, is set up by the compiler, and not by the programmer. Also, the compiler generates all of the necessary instructions to cause conversion.

   Carefully note that the converted data is used only for one calculation. The next data conversion that is required will probably be done in the same work area, erasing the previously converted data.

   If repeated conversions are necessary in the calculations, the input item that is necessary in many calculations can be moved to a working-storage area with the usage of COMPUTATION-3 (packed decimal format). Thereafter, the working-storage item will be used in all calculations without any conversion. In this way, the data is converted only once, no matter how many times it is used in calculations.

The original input data in DISPLAY mode for output is still available in the storage input area.

If there is no 'S' in the item's picture, the absolute value of the item will be assumed each time the data of that particular item is involved in a calculation.

2. *A raw result is calculated in a work area.* At this step, the actual adding, subtracting, multiplying, dividing, or exponentiating is done. The "raw" result is the numerical outcome of the calculation. The compiler sets up the work areas needed for calculations. (The programmer need not define these areas in the Working-Storage Section.)

   The compiler makes the work area large enough to develop the result that the programmer wants—based on the operation (addition, subtraction, multiplication, division, or exponentiation) and the pictures of these items. For example, when two numbers both having the picture S999V99, are added, the work area need only contain six positions, but if the same numbers are multiplied, a ten-position work area is needed.

   Furthermore, the size of the work area is adjusted to take account of any shifting of data values that is needed to align decimal points. If the pictures of the three numbers to be added together were 99V99, 9V9999, and 9999V9, the work area would be nine positions long as calculated below:

```
     25v75
      8v0036
+  9981v2
  10014v9536
```

   The programmer can make the program more efficient and conserve storage space by trying to have the same number of decimal places in values that are to be added or subtracted.

3. *If desired, the size of the raw result is tested.* A test can be made to determine whether all of the significant integral digits will fit into the finished result item. The test is made only if an ON SIZE ERROR phrase follows the arithmetic statement.

   If all significant integral digits of the raw result will fit into the finished result item, the fourth arithmetic operation is performed—the raw result is moved to the finished result item. If they will not fit, the raw result is not moved; instead the execution of the arithmetic statement is suspended, and control goes to the statements that are written in the SIZE ERROR phrase. (The integral digits are those that appear to the left of the actual or assumed decimal point.) For example, an item whose picture is 999V99 has 3 integral places. And an item whose value is 3924v00895 has the integral digits 3924. Only the integral places are checked, not the decimal places; this is because it is normal to drop excess decimal digits.

4. *The raw result is moved from the work area to the finished result item.* The move will be a numeric move if the receiving item is numeric, or an edit move if the receiving item is a report item. The programmer *does not write a MOVE statement* to get the result into the finished result item. It is the compiler who generates the instructions that are needed to get the result into that item.

Methods of Computation

The computation and/or manipulation of numeric operands may be specified in COBOL through any of the following methods:

1. In arithmetic expressions within conditions. (This will be discussed under the Conditional Expressions section of the Procedure Division.)
2. With the arithmetic statements ADD, SUBTRACT, MULTIPLY, DIVIDE, and COMPUTE.
3. With the TALLYING phrase of the EXAMINE statement. (This will be discussed under the Data Manipulation statements section of the Procedure Division.)
4. With the TALLYING phrase of the INSPECT statement. (This will be discussed under the Data Manipulation statements section of the Procedure Division.)
5. With the VARYING phrase of the PERFORM statement. (This will be discussed under the Table Handling section of the Procedure Division.)
6. With the SUM counter. (This will be discussed in the Report Writer section.)

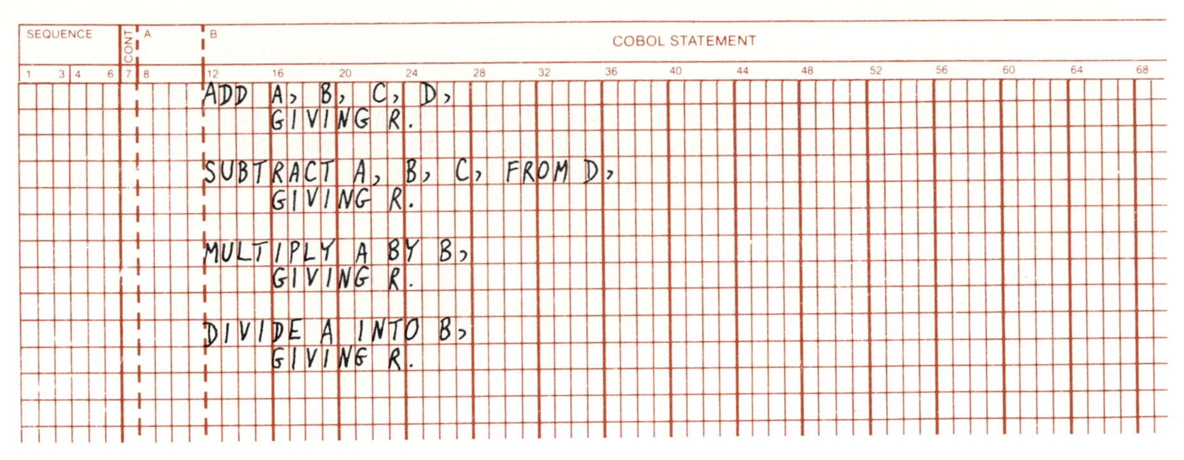

Figure 7.1 GIVING option—examples.

Arithmetic Statements

Five arithmetic verbs are provided for in COBOL to perform the necessary arithmetic functions: ADD, SUBTRACT, MULTIPLY, DIVIDE, and COMPUTE. ADD, SUBTRACT, MULTIPLY, and DIVIDE are arithmetic verbs used to perform individual arithmetic operations. The fifth verb, COMPUTE, permits the programmer to combine arithmetic operations into arithmetic expressions in a formula style using the various arithmetic operators.

Arithmetic expressions can be composed of an identifier of a numeric elementary item, a numeric literal, those identifiers and literals separated by arithmetic operators, or an arithmetic expression enclosed in parentheses.

Rules Governing the Use of Arithmetic Statements

1. All identifiers used in arithmetic statements must represent elementary numeric items that are defined in the Data Division.
2. The identifier that follows GIVING may contain editing symbols if it is not itself involved in the computation.
3. All literals used in arithmetic statements must be numeric.
4. The maximum size of a numeric literal or identifier is eighteen decimal digits.
5. The maximum size of a computation result is eighteen decimal digits after decimal alignment.
6. Decimal alignment is supplied automatically throughout the computation in accordance with individual PICTURE clauses of the results and operands.
7. The GIVING option applies to all arithmetic statements except the COMPUTE statement.

If the GIVING option is used, the value of the identifier following the word GIVING will be made equal to the calculated value of the arithmetic expression. This identifier may be an edited numeric item but must not be involved in the computation. *If the GIVING option is not used, the replaced operand in the arithmetic calculation must not be a literal.* Some examples of statements using the GIVING option appear in figure 7.1.

Common Options

For arithmetic statements, the calculated result may have more significant integral or fractional digits than those provided by the description of the resultant data item. Therefore, the arithmetic verbs offer the ROUNDED option and the SIZE ERROR option.

The following options are available in arithmetic statements: the ROUNDED option, the SIZE ERROR option, and the CORRESPONDING option. (The CORRESPONDING option will be discussed later in the text.)

In the following paragraphs, the term *resultant* is defined as that identifier associated with the result of the arithmetic operations.

ROUNDED Option

After decimal point alignment, the significant digits in a calculated result may extend to the right of the last digit of the resultant item's description (in COBOL, as in arithmetic, this situation is not regarded as an error condition). If the ROUNDED option is specified, the value stored in the resultant data item is determined as follows:

1. The excess digits on the right are dropped.
2. If the most significant digit of the excess digits exceeds four, the absolute magnitude of the retained value is increased by one in the least significant retained digit.
3. If the most significant digit of the excess digits does not exceed four, the retained value is unchanged.

The following shows the effect of specifying the ROUNDED option:

Results of Arithmetic Operation	PICTURE of Resultant	Value Stored in Resultant
3.14	S9V9	3.1
3.15	S9V9	3.2
−3.14	S9V9	−3.1
−3.15	S9V9	−3.2

The ROUNDED option is used when the number of places in the calculated result exceeds the number of places allocated for the sum, difference, product, quotient, or computed result. Here is an example.

```
MULTIPLY QUANTITY BY PRICE,
    GIVING AMOUNT, ROUNDED.
```

Rules Governing the Use of the ROUNDED Option

1. Truncation (dropping of excess digits) is determined by the identifier associated with the result.
2. The least significant digit in the result is increased by 1 if the most significant digit of the excess is greater than or equal to 5.
3. If the option is not specified, truncation occurs without rounding after decimal alignment.
4. Rounding of a computed negative result occurs by rounding the absolute value of the computed result and making the final result negative. Some examples of rounding and truncation are shown in figure 7.2.

	Item to Receive Calculated Result		
Calculated result	PICTURE	Value after rounding	Value after truncating
− 12.36	S99V9	− 12.4	− 12.3
8.432	9V9	8.4	8.4
35.6	99V9	35.6	35.6
65.6	99V	66	65
.0055	V999	.006	.005

Figure 7.2 Rounding or truncation of calculations.

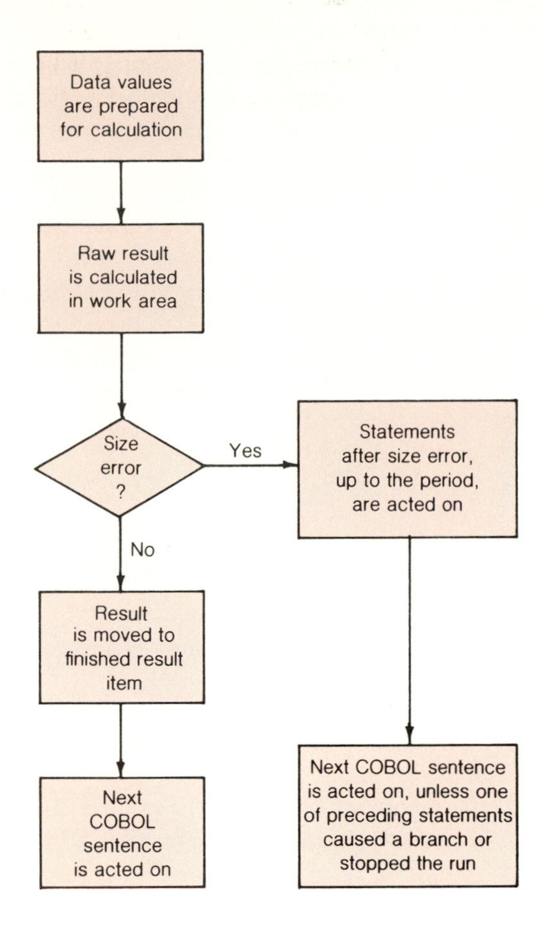

Figure 7.3 Flow of control through arithmetic statements that control on size error.

SIZE ERROR Option

If, after decimal point alignment, the value of the result exceeds the largest value that can be contained in the resultant, a size error condition exists. The following example shows the use of the ON SIZE ERROR option:

```
ADD OVERTIME TO REGULAR-EARNINGS
    ON SIZE ERROR, PERFORM OVERSIZE-ROUTINE.
```

Division by zero always causes a size error condition. The size error condition applies only to the final results of an arithmetic operation and does not apply to intermediate results, except the MULTIPLY and DIVIDE statements, in which case the size error condition applies to the intermediate results as well. If the ROUNDED option is specified, rounding takes place before checking for size error.

1. If the SIZE ERROR option is not specified and a size error condition occurs, the value of those resultant(s) affected is undefined. Values of resultant(s) for which no size error condition occurs are unaffected by size errors that occur for other resultant(s) during the execution of the same operation.
2. If the SIZE ERROR option is specified and a size error condition occurs, then the prior values of resultant(s) affected by the size errors are not altered. Values of the resultant(s) for which no size error condition occurs are not affected by size errors that occur during the execution of the same operation. After the completion of the operation, the imperative-statement in the SIZE ERROR option is executed. (See figure 7.3.)

Common Features

Arithmetic statements have several common features, such as (1) the data descriptions of the operands need not be the same; any necessary conversion and decimal point alignment is supplied throughout the calculation; and (2) the maximum size of each operand is eighteen decimal digits. The composite of operands, which is the hypothetical data item resulting from the superimposition of specified operands in a statement aligned on their decimal points, must not contain more than eighteen decimal digits.

Before arithmetic operations can be performed with an independent or elementary data item, that item must be defined as numeric. It may or may not have a sign specification S. It may or may not have a decimal point designation V. It must, however, be defined as numeric with the picture character 9.

Note: If a sign specification (S) does not appear in an arithmetic statement operand PICTURE, the operands involved will be treated as positive values. This may result in erroneous results in the calculation if there are any negative values involved. If the possibility of negative operands exists, a sign specification should be placed in the operands. Similarly, if the possibility of a negative resultant exists, a sign specification should be placed in the PICTURE of the resultant. This will permit the computer to recognize any negative values as well as positive values.

ADD Statement

The ADD statement is used to sum two or more numeric operands and store the result.

Each identifier of the ADD statement must refer to an elementary numeric item, except that the identifier following the word GIVING may refer either to an elementary numeric item or to an elementary numeric edited item.

When the ADD . . . TO statement is specified, the values of the operands preceding the word TO are added together. That sum is then added successively to the current value of each resultant following the word TO.

When the ADD . . . GIVING statement is specified, the value of the operands preceding the word GIVING are added together. That sum is then stored as a new value in each resultant following the word GIVING.

Examples:

```
ADD 0.5, RATE GIVING TOTAL.

ADD TOTAL-RECEIVED TO ON-HAND-QTY ROUNDED.

ADD VALUE-1, VALUE-2, GIVING VALUE-3,
    ON SIZE ERROR PERFORM ERROR-ROUTINE.
```

The ADD statement causes two or more numeric operands to be summed and the result to be stored.

Rules Governing the Use of the ADD Statement

1. An ADD statement must contain at least two addends (elementary numeric items).
2. When the TO option (format-1) is used, the values of the operands (literals and identifiers) preceding the word TO are added together. The sum is then added to the current value in each identifier-m, identifier-n, etc., and the resultant sum replaces the current values of identifier-m, identifier-n, etc. (See figures 7.4 and 7.5.)

   When the following statement is executed, the values of TOTAL-1 and TOTAL-2 and the value 100 will be added to the value of GRTOTAL; the sum of the four will be stored in GRTOTAL.

   ```
   ADD TOTAL-1, TOTAL-2, 100 TO GRTOTAL.
   ```

3. The resultant sums are not edited with the TO option.
4. The word GIVING may not be written in the same statement as TO.
5. When the GIVING option (format-2) is used, there must be at least two operands (literals and/or identifiers) preceding the word GIVING. The sum may be edited according to the item PICTURE and may be either an elementary numeric item or an elementary numeric edited item. (See figures 7.6 and 7.7.)

Format 1:

```
ADD {identifier-1} [identifier-2] . . . TO identifier—m [ ROUNDED ]
    {literal-1   } [literal-2   ]

    [ identifier—n [ ROUNDED ]] . . . [ ON SIZE ERROR imperative-statement ]
```

Figure 7.4 Format—ADD statement: format-1.

ADD 5 TO FIELD-A

Data in FIELD-A before the ADD is executed	Data in FIELD-A after the ADD is executed
0 0 1 5 0‸5	0 0 1 5 5‸5

ADD 125.25, FIELD-B, TO FIELD-C.

Data in the fields before the ADD is executed		Data in the fields after the ADD is executed	
FIELD-B	6 8‸5	FIELD-B	Unchanged
FIELD-C	0 1 0 0‸0 0	FIELD-C	0 2 9 3‸7 5

‸ represents assumed decimal point not inserted in data item.

Figure 7.5 ADD statement: format-1—example.

Format 2:

```
ADD {identifier-1} , {identifier-2} [, identifier-3] . . .
    {literal-1   }   {literal-2   } [, literal-3   ]

    GIVING identifier-m [ ROUNDED ] [, identifier-n [ ROUNDED ]] . . .

    [ ; ON SIZE ERROR imperative-statement ]
```

Figure 7.6 Format—ADD statement: format-2.

ADD FIELD-E, 5, FIELD-F GIVING FIELD-G.

Data in the fields before the ADD is executed		Data in the fields after the ADD is executed	
FIELD-E	‸0 0 0 2 5 0	FIELD-E	Unchanged
FIELD-F	1‸0 0	FIELD-F	Unchanged
FIELD-G	−0 0 5‸0 0	FIELD-G	0 0 0 6‸0 0

Figure 7.7 ADD statement: format-2—example.

When the GIVING option is used in an ADD statement, the reserved word TO is omitted. The following statement correctly specifies that the sum of FICA, FEDTAX, STATETAX, and INSURANCE is to be computed and stored in DEDUCTIONS.

```
ADD FICA, FEDTAX, STATETAX, INSURANCE,
    GIVING DEDUCTIONS.

ADD FICA, FEDTAX, STATETAX, INSURANCE,
    GIVING DEDUCTIONS ROUNDED,
    ON SIZE ERROR PERFORM SIZE-ERROR-ROUTINE.
```

If the ROUNDED and SIZE ERROR options are included in an ADD statement, the programmer might assume that the ROUNDED option would cause the sum to be rounded to the number of decimal places specified for the variable in which it is to be stored and, the SIZE ERROR option would specify the action to be taken if the sum exceeds the size specified for the variable in which it is to be stored.

6. Decimal points are aligned in all ADD operations.
7. The composite of all operands must not contain more than eighteen digits.
   a. In format-1 (TO option), the composite of operands is determined by using all operands in a statement.
   b. In format-2 (GIVING option), the composite of operands is determined by using all operands in a statement excluding the data items that follow the word GIVING.
8. The compiler insures that enough places are carried out so as not to lose any significant digits during execution.

SUBTRACT Statement

The SUBTRACT statement is used to subtract one, or the sum of two or more numeric data items from one or more items, and to set the value of one or more items equal to the results.

When the SUBTRACT statement without the GIVING option is specified (format-1), all literals or identifiers preceding the word FROM are added together; the total result is subtracted from each identifier following the word FROM, and the differences are stored as the new values of each successive resultant. The general format for the SUBTRACT statement in format-1 appears in figure 7.8. Here is an example of a SUBTRACT statement—format-1:

```
SUBTRACT CREDITS, PAYMENTS, FROM AMT-DUE.
```

When the SUBTRACT statement with the GIVING option is specified (format-2), all literals or identifiers preceding the word FROM are added together, the sum is subtracted from the literal or identifier named following the word FROM, and the result of the subtraction is stored as the new value of each successive resultant named in the GIVING phrase. The general format for the SUBTRACT statement in format-2 appears in figure 7.9.

Examples:

```
SUBTRACT UNION-DUES FROM ADJUSTED-PAY.

SUBTRACT RECEIPTS FROM ON-ORDER-QTY GIVING ADJ-ORDER-QTY,
    ON SIZE ERROR PERFORM ERROR-ROUTINE.

SUBTRACT A FROM B GIVING C.
```

The SUBTRACT statement is used to subtract one or the sum of two or more numeric data items from one or more specified items, and set the value of one or more items equal to the results.

Rules Governing the Use of the SUBTRACT Statement

1. All operands (literals and identifiers) must be elementary numeric items.
2. When the FROM option is used, all literals or identifiers preceding the word FROM are added together and the total is subtracted from the current value of identifier-m, and the difference replaces the value of identifier-m. (See figure 7.10.)
3. When the GIVING option is used, all literals or identifiers preceding the word FROM are added together; the sum is subtracted from literal-m or identifier-m and the difference replaces the value of identifier-n. An example of the SUBTRACT statement in format-2 appears in figure 7.11.

Format 1:

```
SUBTRACT  {identifier-1}  [identifier-2] . . .   FROM identifier-m  [ ROUNDED ]
          {literal-1   }  [literal-2   ]

     [identifier-n  [ ROUNDED ]] . . .

     [ ON SIZE ERROR imperative-statement ]
```

Figure 7.8 Format—SUBTRACT statement: format-1.

Format 2:

```
SUBTRACT  {literal-1   }  [, literal-2   ] . . .  FROM  {literal-m   }
          {identifier-1}  [, identifier-2]              {identifier-m}

     GIVING identifier-n  [ ROUNDED ] [, identifier-o [ ROUNDED ]] . . .

     [ ; ON SIZE ERROR imperative-statement ]
```

Figure 7.9 Format—SUBTRACT statement: format-2.

SUBTRACT −5.5 FROM INCREMENT.

	Before SUBTRACT	After SUBTRACT
INCREMENT	1 0 0	1 0 5

SUBTRACT −5.5 FROM INCREMENT ROUNDED.

	Before SUBTRACT	After SUBTRACT
INCREMENT	1 0 0	1 0 6

Figure 7.10 SUBTRACT statement: format-1—example.

SUBTRACT TOTAL-DEDUCTIONS FROM GROSS-PAY GIVING NET-PAY.

	Before SUBTRACT	After SUBTRACT
TOTAL-DEDUCTIONS	0 2 1 ‸ 2 2	Unchanged
GROSS-PAY	0 1 1 4 ‸ 7 6	Unchanged
NET-PAY	− 8 9 4 ‸ 0 0	0 0 9 3 ‸ 5 4

Figure 7.11 SUBTRACT statement: format-2—example.

4. The composite of all operands must not contain more than eighteen digits.
   a. In format-1 (FROM Option), the composite of operands is determined by using all operands in a statement.
   b. In format-2 (GIVING Option), the composite of operands is determined by using all operands in a statement excluding the data items that follow the word GIVING.
5. The compiler insures that enough places are carried out so as not to lose any significant digits during execution.

MULTIPLY Statement

The MULTIPLY statement is used to multiply numeric data items and to use the values of data items equal to the results.

When the MULTIPLY statement without the GIVING option is specified, the value of the identifier or literal referenced preceding the word BY is multiplied by the value of the identifier referenced following the word BY. The value of the latter identifier is then replaced by the product. Similar replacement occurs for each successive identifier named following the word BY.

When the MULTIPLY statement with the GIVING option is specified, the value of the identifier or literal multiplicand is multiplied by the value of the identifier or literal multiplier, and the product is stored in each successive identifier following the word GIVING.

Examples:

```
MULTIPLY A BY B GIVING C.

MULTIPLY 3.1416 BY R1.
```

The MULTIPLY statement causes numeric data items to be multiplied and sets the value of data items equal to the results.

Rules Governing the Use of the MULTIPLY Statement

1. All operands (literals and identifiers) must be elementary numeric items except that in format-2 (GIVING Option), each identifier following the word GIVING may be an elementary numeric edited item.

Format 1:

```
MULTIPLY  {identifier-1}  BY identifier-2  [ ROUNDED ]
          {literal-1   }

     [, identifier-3  [ ROUNDED ]] . . .

     [ ; ON SIZE ERROR imperative-statement ]
```

Figure 7.12 Format—MULTIPLY statement: format-1.

MULTIPLY BASE BY RATE.

	Before MULTIPLY	After MULTIPLY
BASE	− 500‸00	Unchanged
RATE	0000‸06	− 030‸00

Figure 7.13 MULTIPLY statement: format-1—example.

Format 2:

```
MULTIPLY {identifier-1} BY {identifier-2} GIVING identifier-3 [ ROUNDED ]
         {literal-1   }    {literal-2   }

    [, identifier-4  [ ROUNDED ]] . . .

    [ ; ON SIZE ERROR imperative-statement ]
```

Figure 7.14 Format—MULTIPLY statement: format-2.

MULTIPLY BASE BY RATE GIVING NEW-BASE.

	Before MULTIPLY	After MULTIPLY
BASE	− 5 0 0‸0 0	Unchanged
RATE	0 0 0 0‸0 6	Unchanged
NEW-BASE	− 7 0 0‸0 0	− 0 3 0‸0 0

Figure 7.15 MULTIPLY statement: format-2—example.

2. In format-1 (BY option), the value of identifier-1 or literal-1 is multiplied by the value of identifier-2 and the product replaces identifier-2. (See figures 7.12 and 7.13.)

   When the following example of a MULTIPLY statement—format-1 is executed, the product of the value stored in AMOUNT and the value stored in FACTOR is computed and then stored in AMOUNT.

```
MULTIPLY FACTOR BY AMOUNT.
```

   Assume the values 20v98 and 5v63 have been read into the variables PRICE and DISCOUNT, which are defined in the Working-Storage Section in the example of a MULTIPLY statement—format-1 that follows. (v denotes the assumed decimal point. It is not in the data item.) When the MULTIPLY statement is executed, the product of the values of DISCOUNT and PRICE is computed as 118v1174. The value for PRICE, in which the product will be stored, specifies that the value of PRICE is to have two decimal places to the right of an assumed decimal point. Consequently, the last two decimal places in the product are truncated and the value stored in PRICE as 118v12 and as the ROUNDED option was specified, and the digit to the right of the rightmost digit of the product is 5 or over.

```
WORKING-STORAGE SECTION.
77  PRICE          PIC 999V99.
77  DISCOUNT       PIC 9V99.
    .
    .
    .
    MULTIPLY DISCOUNT BY PRICE ROUNDED.
```

3. In format-2 (GIVING option), the value of identifier-1 or literal-1 is multiplied by identifier-2 or literal-2 and the product replaces identifier-3. (See figures 7.14 and 7.15.)

   In the following example of a MULTIPLY statement—format-2, the product of the values of WIDTH and HEIGHT will be computed and the value will be stored in SQUARE-FEET with the value rounded. The GIVING option and the ROUNDED option may be specified in the same

MULTIPLY statement; the reserved word ROUNDED follows the name of the variable in which the result is stored.

```
WORKING-STORAGE SECTION.
77  WIDTH            PIC 99V99.
77  HEIGHT           PIC 99V99.
77  SQUARE-FEET      PIC 9999V999.
    .
    .
    MULTIPLY WIDTH BY HEIGHT
        GIVING SQUARE-FEET ROUNDED.
```

4. The composite of operands, which is that hypothetical data item resulting from the superimposition of all data items of a given statement aligned on their decimal positions, must not contain more than eighteen digits.

DIVIDE Statement

The DIVIDE statement is used to divide one numeric data item into another and to set the value of a data item equal to the result.

When the DIVIDE . . . INTO statement is specified, the value of the identifier or literal divisor is divided into the value of the identifier referenced following the word INTO. The value of that dividend is replaced by this quotient; similar replacement occurs for the division and each successive dividend.

When the DIVIDE . . . INTO . . . GIVING statement is specified, the value of the identifier or literal divisor is divided into the identifier or literal dividend, and the resulting quotient is stored in the identifier referenced following the word GIVING. The dividend is not changed.

When the DIVIDE . . . BY . . . GIVING statement is specified, the value of the identifier or literal dividend named preceding the word BY is divided by the identifier or literal divisor named following the word BY, and the resulting quotient is stored in the identifiers following the word GIVING. The dividend is not changed.

The REMAINDER option is specified when a remainder from the division operation is desired. The remainder in COBOL is defined as the result of subtracting the product of the quotient and the divisor from the dividend. If ROUNDED is specified, the quotient used to calculate the remainder is an intermediate field that contains the quotient of the DIVIDE statement, truncated rather than rounded.

When the SIZE ERROR phrase is used in conjunction with the REMAINDER option, the following rules apply:

1. If SIZE ERROR occurs on the quotient, the contents of the quotient-resultant will not be changed, but the result of the divide will be used in computing the remainder.
2. If SIZE ERROR occurs on the remainder, the contents of the remainder-resultant will remain unchanged. However, as with other instances of multiple results of arithmetic statements, an analysis must be performed to determine which has occurred.

Examples:

```
DIVIDE TOTAL BY NUMBER GIVING AVERAGE.

DIVIDE 100.00 INTO K2H GIVING VALUE-1 ROUNDED.

DIVIDE A26 INTO R17K.
```

The DIVIDE statement divides one numeric data item into others and sets the value of data items equal to the quotient and the remainder.

Rules Governing the Use of the DIVIDE Statement

1. All operands (literals and identifiers) must represent elementary numeric items, although any identifier associated with the GIVING or REMAINDER phrase may be an elementary numeric-edited item.
2. The composite of operands, which is the hypothetical data item resulting from the superimposition of all data items (except the REMAINDER data item) of a given statement aligned on their decimal points, must not contain more than eighteen digits.

Format 1:

```
DIVIDE {identifier-1} INTO identifier-2 [ ROUNDED ]
       {literal-1   }

    [, identifier-3 [ ROUNDED ]] . . .

    [ ; ON SIZE ERROR imperative-statement ]
```

Figure 7.16 Format—DIVIDE statement: format-1.

DIVIDE 10 INTO QUANTITY.

	Before DIVIDE	After DIVIDE
QUANTITY	1 2 1‸5	0 1 2‸1

DIVIDE 10 INTO QUANTITY ROUNDED.

	Before DIVIDE	After DIVIDE
QUANTITY	1 2 1‸5	0 1 2‸2

Figure 7.17 DIVIDE statement: format-1—example.

3. When format-1 (INTO option) is used, the value of identifier-1 or literal-1 is divided into the value of identifier-2. The value of the dividend (identifier-2) is replaced by the quotient. (See figures 7.16 and 7.17.)

   If the values of FACTOR and RESULT are 02 and 125v00, respectively, after execution of the DIVIDE statement—format-1 that follows, the value stored in RESULT will be 062V50.

```
77  FACTOR      PIC 99.
77  RESULT      PIC 999V99.
    .
    .
    DIVIDE FACTOR INTO RESULT ROUNDED.
```

4. When format-2 (INTO . . . GIVING option) is used, the value of identifier-1 or literal-1 is divided into identifier-2 or literal-2 and the quotient is stored in identifier-3. (See figures 7.18 and 7.19.)

   The following is an example of a DIVIDE statement—format-2.

```
77  BALANCE     PIC 9999V99.
77  MONTHS      PIC 9V9.
77  PAYMENTS    PIC 9999V99.
    .
    .
    .
    DIVIDE MONTHS INTO BALANCE
        GIVING PAYMENTS ROUNDED.
```

Format 2:

```
DIVIDE {identifier-1} INTO {identifier-2} GIVING
       {literal-1   }      {literal-2   }

       identifier-3 [ ROUNDED ] [, identifier-4 [ ROUNDED ]] . . .

       [ ; ON SIZE ERROR imperative-statement ]
```

Figure 7.18 Format—DIVIDE statement: format-2.

DIVIDE 10 INTO QUANTITY GIVING AMOUNT.

	Before DIVIDE	After DIVIDE
QUANTITY	1 2 1 ^ 5	Unchanged
AMOUNT	0 0 0 ^ 0 0	0 1 2 ^ 1 5

Figure 7.19 DIVIDE statement: format-2—example.

5. When format-3 (BY . . . GIVING option) is used, the value of identifier-1 or literal-1 is divided by the value of identifier-2 or literal-2 and the quotient is stored in identifier-3. (See figure 7.20.)
   The following is an example of a DIVIDE statement—format-3.

```
DIVIDE DIVIDEND BY DIVISOR
    GIVING QUOTIENT ROUNDED.
```

6. When the BY or INTO option is used in conjunction with the GIVING and REMAINDER options, the value of identifier-1 or literal-1 is divided into or by identifier-2 or literal-2, and the quotient is stored in identifier-3 with the remainder being optionally stored in identifier-4. (A remainder is the result of subtracting the product of the quotient and the divisor from the dividend.) If the ROUNDED option is specified, the quotient is rounded after the remainder is determined. (See figures 7.21 and 7.22.)
   The REMAINDER option of the DIVIDE statement specifies that the remainder will be stored in the specified variable. (The remainder is the amount left over after one number cannot be divided into another number evenly.) When the following example of a DIVIDE statement—format-4 is executed, the remainder will be stored in LEFT-OVER.

```
DIVIDE PEOPLE INTO TOTAL
    GIVING SHARE
    REMAINDER LEFT-OVER.
```

   The following is an example of a DIVIDE statement—format-5

```
DIVIDE TOTAL-MINUTES BY 60
    GIVING HOURS
    REMAINDER MINUTES.
```

7. A division by zero will always result in a size error condition.

Note: In all four arithmetic statements (ADD, SUBTRACT, MULTIPLY, and DIVIDE), if the word GIVING is not used, the last operand (the one replaced by the result) must not be a literal.

Format 3:

```
DIVIDE  {identifier-1}  BY  {identifier-2}  GIVING
        {literal-1   }      {literal-2   }

        identifier-3  [ ROUNDED ] [, identifier-4  [ ROUNDED ]] . . .

      [ ; ON SIZE ERROR imperative-statement ]
```

Figure 7.20 Format—DIVIDE statement: format-3.

Format 4:

```
DIVIDE  {identifier-1}  INTO  {identifier-2}  GIVING
        {literal-1   }        {literal-2   }

        identifier-3  [ ROUNDED ]  REMAINDER  identifier-4

      [ ; ON SIZE ERROR imperative-statement ]
```

Figure 7.21 Format—DIVIDE statement: format-4.

Format 5:

```
DIVIDE  {identifier-1}  BY  {identifier-2}  GIVING
        {literal-1   }      {literal-2   }

        identifier-3  [ ROUNDED ]  REMAINDER  identifier-4

      [ ; ON SIZE ERROR imperative-statement ]
```

Figure 7.22 Format—DIVIDE statement: format-5.

COMPUTE Statement

The COMPUTE statement allows the programmer to combine arithmetic operations without the restrictions on composite of operands and/or receiving data items imposed by the arithmetic statements ADD, SUBTRACT, MULTIPLY, or DIVIDE.

The execution of a COMPUTE statement causes the resultants to receive the computed value of the arithmetic expression following the = symbol. In its simplest form, the arithmetic expression may consist of only a data-name or numeric literal.

Editing symbols must not be specified in any operand; except in an item that receives the calculated result, but is not used in the computation itself.

Examples:

```
COMPUTE GROSS-PAY ROUNDED = RATE * (BASE-HRS + (WORK-HRS -
    BASE-HRS) * PREMIUM-FACTOR).

COMPUTE QTY-ON-HAND = OLD-QTY + NO-RECVD - QTY-SHIPPED.

COMPUTE AVG-PERCENT-INCR = (END-PAY - START-PAY) /
    END-PAY * 100.

COMPUTE YTD-FICA = YTD-FICA + (GROSS-PAY * FICA-FACTOR).
```

COMPUTE identifier-1 [ROUNDED] [, identifier-2 [ROUNDED]] . . .

= arithmetic-expression [; ON SIZE ERROR imperative-statement]

Figure 7.23 Format—COMPUTE statement.

Exponentiation

The COMPUTE statement is the only statement available to the COBOL programmer that can specify exponentiation. The programmer should be aware of the pitfalls of exponentiation so that they can be avoided. The following uses of exponentiation are not allowed and may produce unpredictable results:

1. The value zero exponentiated by the value zero.
2. The value zero exponentiated by a negative value.
3. A negative value exponentiated by a non-integral value.

If a data item to be exponentiated can assume a negative value, a test (IF . . . NEGATIVE) should be arranged to bypass the exponentiation in case the value is negative.

The COMPUTE statement assigns to one or more data items the value of an arithmetic expression. (See figure 7.23.) Arithmetic expressions are discussed following this section.

Rules Governing the Use of the COMPUTE Statement

1. The arithmetic expression option permits the use of a meaningful combination of identifier, numeric literal, and figurative constant ZERO, joined by the operators in figure 7.24. This permits the pro-

Operator	Function
+	Addition
−	Subtraction
*	Multiplication
/	Division
**	Exponentiation

Figure 7.24 Arithmetic operators and their functions.

Hierarchy of Evaluation	Operator	Meaning	Example: Arithmetic Expression	Example: COBOL Expression
1	+	Unary plus sign	+2	+2
	−	Unary minus sign	−2	-2
2	**	Exponentiation	3^2	3 ** 2
3	*	Multiplication	3 × 2	3 * 2
	/	Division	3 ÷ 2	3 / 2
4	+	Addition	3 + 2	3 + 2
	−	Subtraction	3 − 2	3 - 2

1. Parentheses modify the order of evaluation; operations enclosed in parentheses are performed first, beginning with the innermost pair of parentheses.
2. When 2 operators of same level in hierarchy appear in the same expression, the operations are performed from left to right.
3. Every operator must be preceded by and followed by a space, except for unary signs.

Figure 7.25 Arithmetic operators and their meanings.

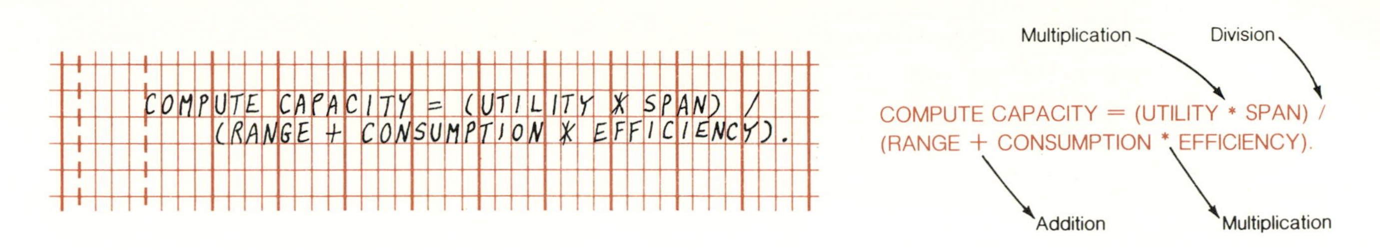

Figure 7.26 COMPUTE statement—example.

grammer to combine arithmetic operations without the restrictions imposed by the arithmetic statements ADD, SUBTRACT, MULTIPLY, and DIVIDE. (See figure 7.25.)

2. Operators must be preceded and followed by one or more spaces. (See figure 7.26.)
3. If exponentiation is desired, the COMPUTE statement must be used. The following is an example of a COMPUTE statement—exponentiation.

```
COMPUTE FACTOR = NUMBER ** 3.
```

4. An arithmetic expression consisting of a single identifier or literal provides a method of setting the values of identifier-1, identifier-2, etc., equal to the value of the single identifier or literal.

```
A   COMPUTE TOTAL = 5.
B   COMPUTE TOTAL = TOTAL B1.
C   COMPUTE RATES = 00000.
D   COMPUTE AMOUNT = QUANTITY.
```

The above expressions are used to set the values of identifier-1 equal to the value of a single identifier or literal as follows:

a. The variable TOTAL is set to the value of 5.
b. The variable TOTAL is set to the value of the variable TOTALB1.
c. The variable RATES is set to zeros.
d. The variable AMOUNT is set to the value of QUANTITY.

The result is the same as a move operation. Both of the following examples of a COMPUTE statement—single literal will accomplish the same thing: move a value of 3 to a variable named INDICATOR.

```
COMPUTE INDICATOR = 3.
        OR
MOVE 3 TO INDICATOR.
```

5. The value of the calculation's result must be written to the left of the equal sign as the item represented by identifier-1, identifier-2, etc.
6. Identifier-1, identifier-2, etc. must be elementary numeric items or elementary numeric-edited items. The calculated value is placed here and is edited according to the identifier-1, identifier-2, etc. picture. If more than one identifier is specified for the result of the operation, that is preceding =, the value of the arithmetic expression is computed, and then this value is stored as the new value of identifier-1, identifier-2, etc. in turn.

The following example of a COMPUTE statement will compute the value AMOUNT by adding TOTAL-1 and TOTAL-2 and multiplying the sum by the product of 2 and TOTAL-3. (Note: Parentheses could have been placed around [2 * TOTAL-3]. Although unnecessary because of COBOL's hierarchy of operations, this would, perhaps, make the statement more readable.

```
COMPUTE AMOUNT = (TOTAL-1 + TOTAL-2) * 2 * TOTAL-3.
```

7. The ROUNDED and SIZE ERROR options also apply to the COMPUTE statement.

   The following example of a COMPUTE statement—ROUNDED option will compute the value of the product of TAG and PERCENT and round it to the nearest cent and store it in the edited variable BILL.

```
77  TAG          PIC 99V99.
77  PERCENT      PIC V99.
77  BILL         PIC $99.99.
    .
    .
    .
    COMPUTE BILL ROUNDED = TAG * PERCENT.
```

Note: Since intermediate fields in a COMPUTE statement are given the same size as the final result field, and since the ON SIZE ERROR option applies only to the final result, it is possible for errors to occur when the intermediate results are larger than the final result field.

Arithmetic Expressions

An arithmetic expression can be an identifier of a numeric elementary item, a numeric literal, such identifiers and literals separated by arithmetic operators, two arithmetic expressions separated by an arithmetic operator, or an arithmetic expression enclosed in parentheses.

The COBOL arithmetic expression

```
D / B ** 2 + C
```

is equivalent to the algebraic expression $D/B^2 + C$.

Those identifiers and literals appearing in arithmetic expressions must represent either numeric elementary items or numeric literals on which the arithmetic may be performed.

It should be noted that identifier-1 may be used in the arithmetic expression; for instance, it is possible to write COMPUTE TOTAL = TOTAL + INCREASE − DECREASE. In this example, the original value of TOTAL is used to compute a new value to be placed into TOTAL. When identifier-1 is used this way, it must represent an elementary numeric item, and there is no editing of the result.

When a data-name or numeric literal is used as an arithmetic expression, the result is the same as if a MOVE had been written. For example, COMPUTE RESULT = 100 is equivalent to MOVE 100 TO RESULT.

Arithmetic Operators

There are five arithmetic operators that may be used in arithmetic expressions. They are represented by specific characters that must be preceded and followed by a space.

Arithmetic Operator	Meaning
+	addition
−	subtraction
*	multiplication
/	division
**	exponentiation

Formulation and Evaluation

The order of evaluation determines the precedence of arithmetic operators. The following listing shows the precedence of arithmetic operators in decreasing order of priority.

Class of Operator	Symbol(s)
parenthetical expression	( )
exponentiation	**
multiplication	*
division	/
addition	+
subtraction	−

The operators, multiplication and division, are of equal precedence, as are the operators addition and subtraction. A series of equal precedence will be evaluated from left to right.

Expression	Evaluation
A / B * C	Quotient of A/B multiplied by C.
A / B / C	Quotient of A/B divided by C.
A ** B ** C	Exponentiate by C that quantity obtained by raising A to the B power.

Parenthetical expressions have the highest precedence in COBOL, and any parenthetical expression within an arithmetic expression must be evaluated first. If parenthetical expressions are placed within parenthetical expressions, for example: A + ((C / D) * (E / F)) − 8, the innermost parenthetical expression C / D and E / F are evaluated first. Evaluation proceeds from the innermost to outermost parenthetical expression. The evaluation of C / D is multiplied by the evaluation of E / F.

Since the parenthetical expressions have the highest priority in evaluation, they may be used to modify the usual order of evaluation in an expression.

Expression	Evaluation
A / B + C	Add C to the quotient of A/B.
A / (B + C)	Divide A by the sum of B and C.
A / B * C	Multiply C by the quotient of A/B.
A / (B * C)	Divide A by the product of B times C.

Parenthetical expressions may also be inserted into arithmetic expressions to improve the readability of the expression even if the usual order of evaluation is not affected. This application of parentheses is especially useful in complex expressions or in expressions that require several lines of source program coding. For example, the following expression, which is unambiguous to the computer but not necessarily so to the reader, illustrates the normal precedence of operations:

```
A + B / C + D ** E * F − G
```

and is interpreted as if it were written

```
A + (B / C) + ((D ** E) * F) − G
```

and may be more comprehensible if it is actually written in this manner.

Evaluation Rules

1. Parentheses may be used in arithmetic expressions to specify the order in which elements are to be evaluated. Expressions within parentheses are evaluated first, and within nested parentheses, evaluation proceeds from the least inclusive set to the most inclusive set. When parentheses are not used, or when parenthetical expressions are at the same level of inclusiveness, the following hierarchy order of execution is implied:
   a. Exponentiation
   b. Multiplication and division
   c. Addition and subtraction
2. Parentheses are used either to eliminate ambiguities in logic where consecutive operations of the same hierarchical level appear or to modify the normal hierarchical sequence of execution in expressions where it is necessary to have some deviation from the normal procedure. When the sequence of execution is not specified by parentheses, the order of execution of consecutive operations of the same hierarchical level is from left to right.
3. An arithmetic expression may only begin with the symbol (, +, −, or a variable, and may only end with) or a variable. There must be a one-to-one correspondence between left and right parentheses of an arithmetic expression such that each left parenthesis is to the left of its corresponding right parenthesis.
4. Arithmetic expressions allow the programmer to combine arithmetic operations without the restrictions on the composition of operands and/or receiving data items.

```
COMPUTE VALUE ROUNDED = FACTOR * 5 - TOTAL ** 3.
```

In the above example of a COMPUTE statement—evaluation rules, the operations in the order in which they will be performed based on the hierarchy of evaluation, are,

** (exponentiation; TOTAL is cubed.)

* (multiplication; 5 and FACTOR are multiplied.)

— (subtraction; TOTAL^3 is subtracted from the product of 5 and FACTOR.)

Summary

The programmer needs to become familiar with the action of the compiler during the execution of arithmetic statements so that more efficient programs may be written. The major actions that occur during the execution of an arithmetic statement are: (1) The data values are prepared for calculation. If necessary, the usage of the values are converted to a data code in which calculations are possible. The compiler generates all the necessary instructions for conversion; (2) A raw result is calculated in a work area. The "raw" result is the numerical outcome of the calculation. The programmer need not define these areas as this is the responsibility of the compiler. The compiler sets up work areas and adjusts these areas to take into account any shifting of data values that is necessary to align decimal points; (3) If desired, the size of the raw results is tested. A test can be made to determine whether all of the significant integral digits will fit into the finished result item; and (4) The raw result is moved from the work area to the finished result item; The programmer does not write any MOVE statements, the compiler generates the instructions that are necessary to get the finished result.

The computation and/or manipulation of numeric operands may be specified in COBOL through arithmetic expressions within conditions, arithmetic statements (ADD, SUBTRACT, MULTIPLY, DIVIDE, COMPUTE), the TALLYING phrase of the EXAMINE statement, the TALLYING phrase of the INSPECT statement, the VARYING phrase of the PERFORM statement, and the SUM statement.

Five verbs are provided in COBOL to perform the necessary arithmetic functions: ADD, SUBTRACT, MULTIPLY, DIVIDE, and COMPUTE. COMPUTE permits the programmer to combine arithmetic operations into arithmetic expressions in a formula style using various arithmetic operations.

All arithmetic statements must use identifiers that represent elementary numeric items defined in the Data Division, except that the identifier that follows the GIVING option may contain editing symbols if not involved in the calculations. Other restrictions are that all literals used must be numeric, and that the maximum size of literals and results is not to exceed eighteen digits.

When the GIVING option is used, the identifier following GIVING is made equal to the calculated value of the expression.

The ROUNDED option is specified when the number of places in the calculated result exceeds the number of places allocated for the result. The excess digits on the right are dropped, and the least significant digit of the retained value is increased by one if the most significant digit of the excess digits exceeds four.

The SIZE ERROR condition exists if, after decimal alignment, the value of the result exceeds the largest value that can be contained in the resultant. If the ROUNDED option is specified, then rounding takes place before checking for the size error. If the SIZE ERROR is not specified and a size error condition occurs, the value of the resultant affected is undefined. If the option is specified and the size-error condition occurs, the prior value of resultants affected by the size error are not altered and the imperative statement following the SIZE ERROR phrase is executed.

The common features of arithmetic statements are as follows: the data description of the operands need not be the same, and the maximum size of each operand is not to exceed eighteen decimal digits. The arithmetic statement must be defined as numeric and may or may not have a sign specification, and it may or may not have a decimal point.

The ADD statement is used to sum two or more numeric operands and to store the result. When the ADD . . . TO statement is specified, the values of the operands preceding the word TO are added together and that sum is added successively to the current value of each resultant following the word TO. When the ADD . . . GIVING statement is specified, the value of the operands preceding the word GIVING are added together, and that sum is stored as the new value in each resultant following the word GIVING.

The SUBTRACT statement is used to subtract one, or the sum of two or more numeric data items from one or more items, and to set the value of one or more items equal to the results. When the SUBTRACT statement without the GIVING option is specified, all literals and identifiers preceding the word FROM are added together, and the total result is subtracted from each identifier following the word FROM and the differences are stored as the new values of each successive resultant. When the SUBTRACT statement with the GIVING option is specified, all literals and identifiers preceding the word FROM are added together; the sum is subtracted from the literal or identifier named following the word FROM, and the result of the subtraction is stored as the new values of each successive resultant named in the GIVING phrase.

The MULTIPLY statement is used to multiply numeric data items and to use the value of data items equal to the result. When the MULTIPLY statement without the GIVING option is specified, the value of the identifier or literal referenced preceding the word BY is multiplied by the value of the identifier following the word BY, and the value of the latter identifier is then replaced by the product. Similar replacements occur for each successive identifier following the word BY.

When the MULTIPLY statement with the GIVING option is specified, the value of the identifier or literal multiplicand is multiplied by the value of the identifier or literal multiplier and the product is stored in each successive identifier following the word GIVING.

The DIVIDE statement is used to divide one numeric data item into another and to set the value of a data item equal to the result. When the DIVIDE . . . INTO statement is specified, the value of the identifier or literal divisor is divided into the value of the identifier referenced following the word INTO, and the value of the dividend is replaced by the quotient; similar replacement occurs for the divisor and each successive dividend. When the DIVIDE . . . INTO . . . GIVING statement is specified, the value of the identifier or literal divisor is divided into the identifier or literal dividend and the resulting quotient is stored in the identifier referenced following the word GIVING. The dividend is not changed. When the DIVIDE . . . BY . . . GIVING statement is specified, the value of the identifier or literal dividend named preceding the word BY is divided by the identifier or literal divisor following the word BY, and the resulting quotient is stored in the identifiers following the word GIVING. The dividend is not changed.

The REMAINDER option is specified when a remainder from the division operations is desired. The remainder in COBOL is defined as the result of subtracting the product of the quotient and divisor from the dividend.

In all four arithmetic statements, if the word GIVING is not used, the last operand (the one replaced by the result) must not be a literal.

The COMPUTE statement allows the programmer to combine arithmetic operations without the restrictions on the composition of operands and/or receiving data items imposed by the arithmetic statements. The execution of a COMPUTE statement causes the resultant(s) to receive the computed value of the arithmetic expression following the equal (=) symbol. Editing symbols must not be specified in any operand except in an item that receives the calculated results, but which is not used in the computation itself. The COMPUTE statement is the only statement available to the COBOL programmer that can specify exponentiation. The COMPUTE statement assigns to one or more data items the value of an arithmetic expression.

An arithmetic expression can be an identifier of a numeric elementary item, a numeric literal, such identifiers and literals separated by arithmetic operators, two arithmetic expressions separated by an arithmetic operator or an arithmetic expression enclosed in parentheses. Those identifiers and literals appearing in arithmetic expressions must represent either numeric elementary items or numeric literals on which arithmetic may be performed.

There are five arithmetic operators that may be used in arithmetic expressions. They are represented by specific characters that must be preceded and followed by space.

The order of evaluation determines the precedence of arithmetic operations. The following is the precedence of arithmetic operations in decreasing order of priority: parenthetical expressions, exponentiation, multiplication and division, and addition and subtraction.

Summary: COBOL Verbs

ADD is an arithmetic verb that adds two or more numbers and puts the sum into data items after the word TO, unless there is a GIVING phrase. If GIVING is specified, the sum is put into the data item(s) after the word GIVING and it is edited according to the item's picture; the value of this item is not used in the addition.

COMPUTE is an arithmetic verb that computes the value of the data items, literals, or formula written to the right of the equal sign, and puts that value into the item named after the verb. It can add, subtract, multiply, divide, or exponentiate numbers, or combine these operations. COMPUTE can edit the result according to the receiving item's picture.

DIVIDE is an arithmetic verb that divides one number into another and puts the quotient into the data item named after the word INTO, or BY, unless there is a GIVING phrase. If a GIVING phrase is specified, the quotient is put into the data items named after the word GIVING and is edited according to the item's picture; the value of this item is not used in the division. The REMAINDER option may be specified when a remainder from the division operation is desired. The remainder is defined as the result of subtracting the product of the quotient and divisor from the dividend.

MULTIPLY is an arithmetic verb that multiplies one number by another and puts the product into the data item named after the word BY, unless there is a GIVING phrase. If GIVING is specified, the product is put into the data items after the word GIVING and is edited according to the item's picture; the value of this item is not used in the multiplication.

SUBTRACT is an arithmetic verb that subtracts one or more numbers from one or more numbers and puts the difference into the data items named after the word FROM, unless there is a GIVING phrase. If GIVING is specified, the difference is placed into the data items named after the word GIVING, and edited according to the item's picture; the value of this item is not used in the subtraction.

(See figure 7.27.)

Arithmetic Statements	Allowable Options			
	GIVING Variable Name	Variable Name ROUNDED	SIZE ERROR Statement	REMAINDER Variable Name
ADD {identifier-1 / numeric-literal-1} [identifier-2 / numeric-literal-2] . . . TO identifier-m	X	X	X	
SUBTRACT {identifier-1 / numeric-literal-1} [identifier-2 / numeric-literal-2] . . . FROM identifier-m	X	X	X	
MULTIPLY {identifier-1 / numeric-literal-1} BY identifier-2	X	X	X	
DIVIDE {identifier-1 / numeric-literal-1} [INTO / BY] identifier-2	X	X	X	X
COMPUTE identifier-1 = {identifier-2 / numeric-literal-1 / arithmetic-expression}		X	X	

*The reserved word TO is omitted when the GIVING option is specified.

Figure 7.27 Summary of arithmetic statements and their options.

Illustrative Program: Simple Calculations Accumulating Final Totals-Transaction Register

Job Definition

Print a report listing all sales transactions for a week. This report is to calculate the sales amount for each item and the final total of all sales items for the week. Sales amount (quantity sold times item price) is found in the input record.

Input

The Sales Transaction file consists of 80-position records. The format of the input records is as follows:

Positions	Field	Format
1–6	Date	mm/dd/yy
7–12	Item number	
13–32	Item description	
33–37	Quantity	
38–42	Price	xxx.xx

Processing

Multiply quantity times unit price to find sales amount.
Final total for all item sales for week.

Output

A Transaction Register is to be printed as follows:

```
                              TRANSACTION  REGISTER

TRANSACTION      ITEM           DESCRIPTION            QUANTITY   UNIT      SALES
   DATE           NO                                              PRICE     AMOUNT

 07/23/80       413010      CH001 BOX 100A FLUSH          10       4.90     $49.00
 07/23/80       412146      CH148 BREAKER 15A            100        .89     $89.00
 07/23/80       411116      1500 TWIN SOCKER B           500       1.12    $560.00
 07/24/80       503029      MOTOR 1/2 HP 60 CYC            2     146.78    $293.56
 07/24/80       317802      TERMINAL CLIP                100       5.12    $512.00
 07/24/80       326917      TERMINAL BAR                 100       4.12    $412.00
 07/24/80       411121      1506 COKT ADAPT BRN          400        .19     $76.00
 07/24/80       412997      CH173 BREAKER 30A             60       1.15     $69.00
 07/24/80       413088      CH176 BREAKER 60A             40       1.15     $46.00
 07/24/80       411174      C151 SIL SWITCH BRN          200       1.16    $232.00
 07/24/80       413090      CH005 BR BOX 150A             10       4.98     $49.80
 07/24/80       718326      FC803 FUSE 15A               200        .32     $64.00

                                          WEEKLY TOTAL                   $2,452.36
```

Hierarchy Chart

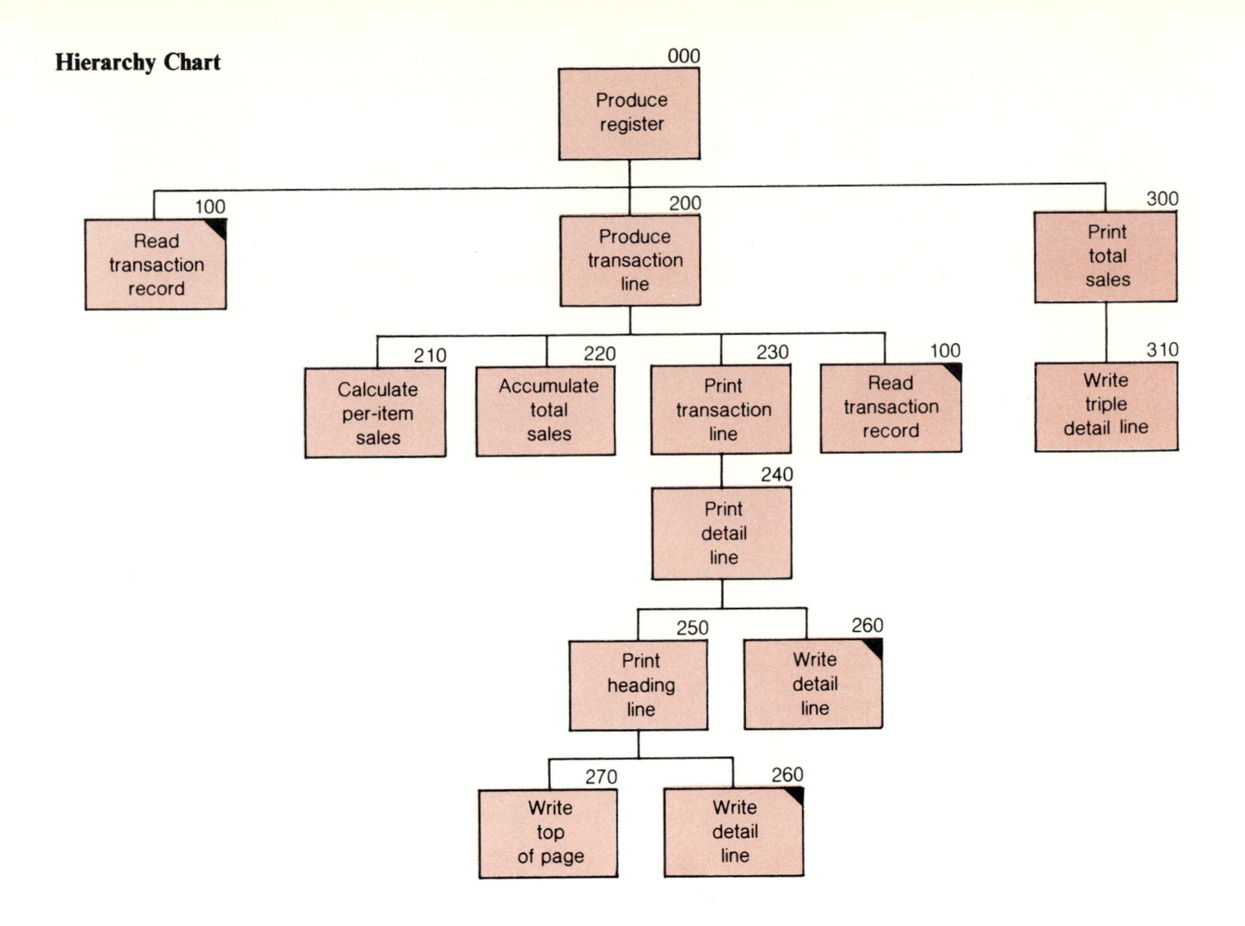

IPO Charts

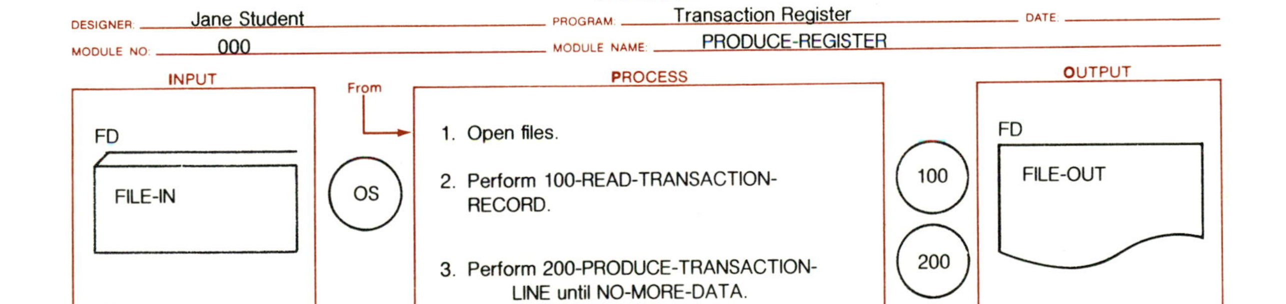

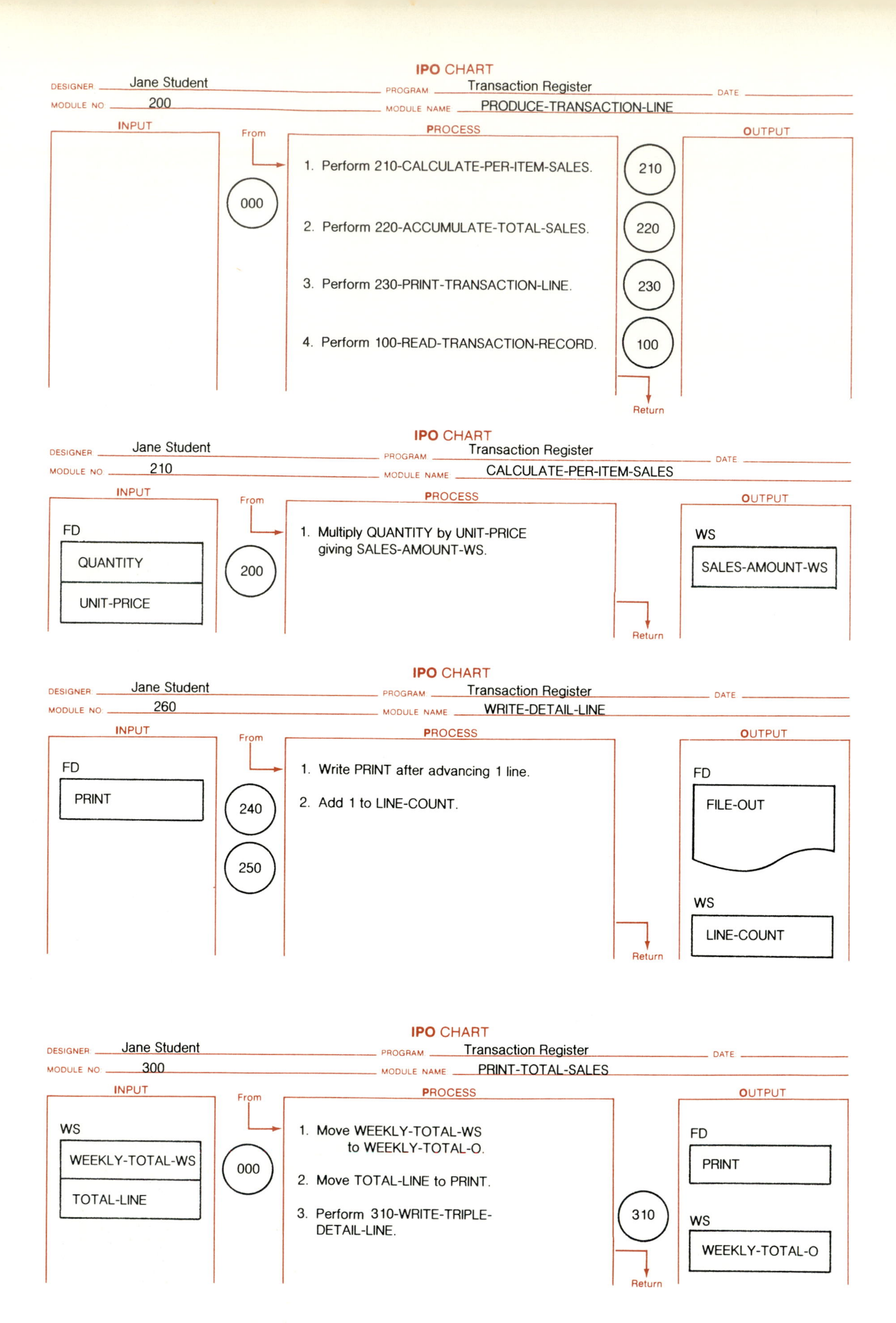

IPO CHART
DESIGNER Jane Student
PROGRAM Transaction Register
DATE
MODULE NO 200
MODULE NAME PRODUCE-TRANSACTION-LINE
INPUT
From
PROCESS
OUTPUT
000
1. Perform 210-CALCULATE-PER-ITEM-SALES.
210
2. Perform 220-ACCUMULATE-TOTAL-SALES.
220
3. Perform 230-PRINT-TRANSACTION-LINE.
230
4. Perform 100-READ-TRANSACTION-RECORD.
100
Return
IPO CHART
DESIGNER Jane Student
PROGRAM Transaction Register
DATE
MODULE NO 210
MODULE NAME CALCULATE-PER-ITEM-SALES
INPUT
From
PROCESS
OUTPUT
FD
QUANTITY
UNIT-PRICE
200
1. Multiply QUANTITY by UNIT-PRICE giving SALES-AMOUNT-WS.
WS
SALES-AMOUNT-WS
Return
IPO CHART
DESIGNER Jane Student
PROGRAM Transaction Register
DATE
MODULE NO 260
MODULE NAME WRITE-DETAIL-LINE
INPUT
From
PROCESS
OUTPUT
FD
PRINT
240
250
1. Write PRINT after advancing 1 line.
2. Add 1 to LINE-COUNT.
FD
FILE-OUT
WS
LINE-COUNT
Return
IPO CHART
DESIGNER Jane Student
PROGRAM Transaction Register
DATE
MODULE NO 300
MODULE NAME PRINT-TOTAL-SALES
INPUT
From
PROCESS
OUTPUT
WS
WEEKLY-TOTAL-WS
TOTAL-LINE
000
1. Move WEEKLY-TOTAL-WS to WEEKLY-TOTAL-O.
2. Move TOTAL-LINE to PRINT.
3. Perform 310-WRITE-TRIPLE-DETAIL-LINE.
310
FD
PRINT
WS
WEEKLY-TOTAL-O
Return

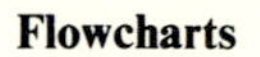

Flowcharts

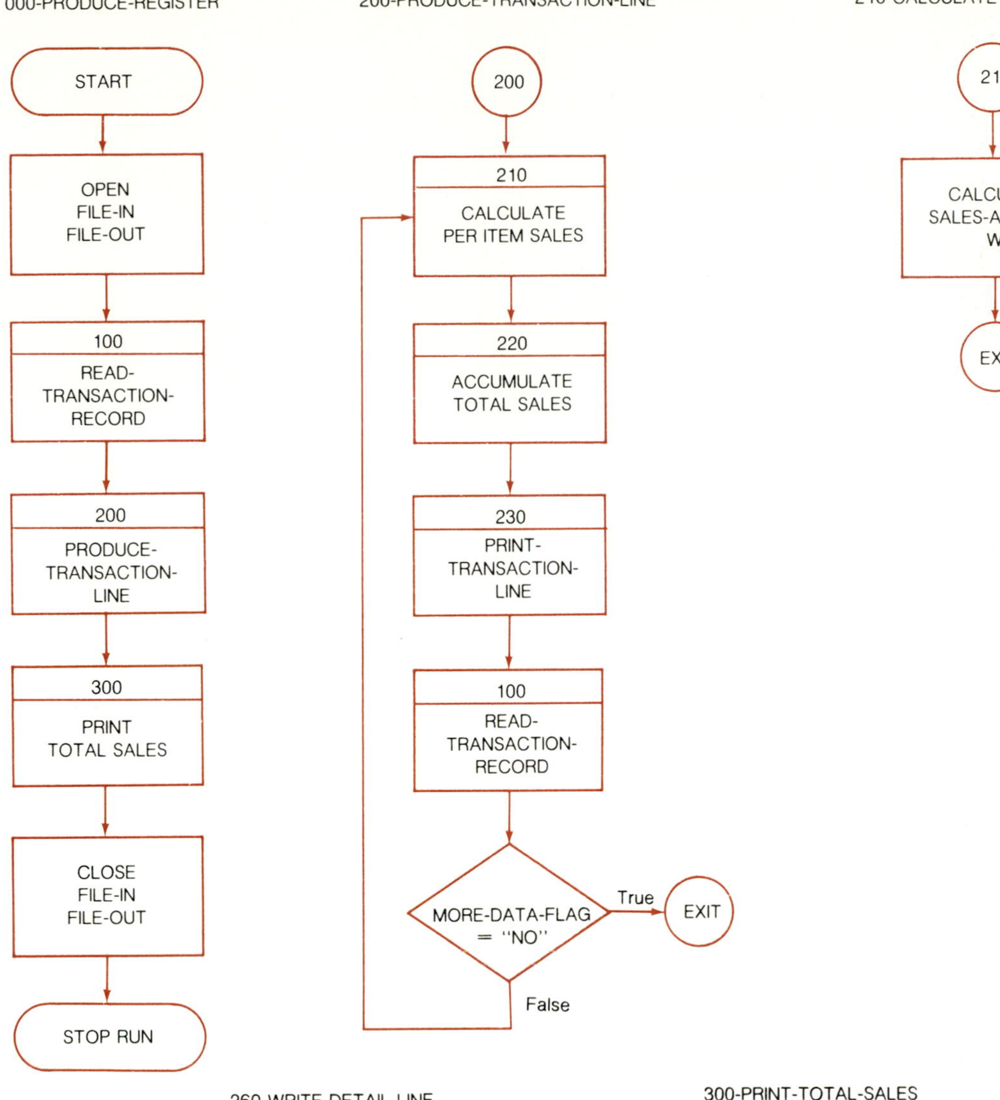
000-PRODUCE-REGISTER
START
OPEN FILE-IN FILE-OUT
100 READ-TRANSACTION-RECORD
200 PRODUCE-TRANSACTION-LINE
300 PRINT TOTAL SALES
CLOSE FILE-IN FILE-OUT
STOP RUN
200-PRODUCE-TRANSACTION-LINE
200
210 CALCULATE PER ITEM SALES
220 ACCUMULATE TOTAL SALES
230 PRINT-TRANSACTION-LINE
100 READ-TRANSACTION-RECORD
MORE-DATA-FLAG = "NO"
True
EXIT
False
210-CALCULATE-PER-ITEM-SALES
210
CALCULATE SALES-AMOUNT-WS
EXIT

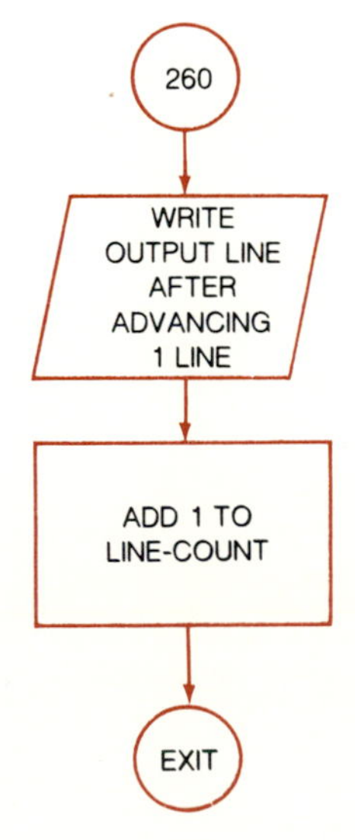
260-WRITE-DETAIL-LINE
260
WRITE OUTPUT LINE AFTER ADVANCING 1 LINE
ADD 1 TO LINE-COUNT
EXIT

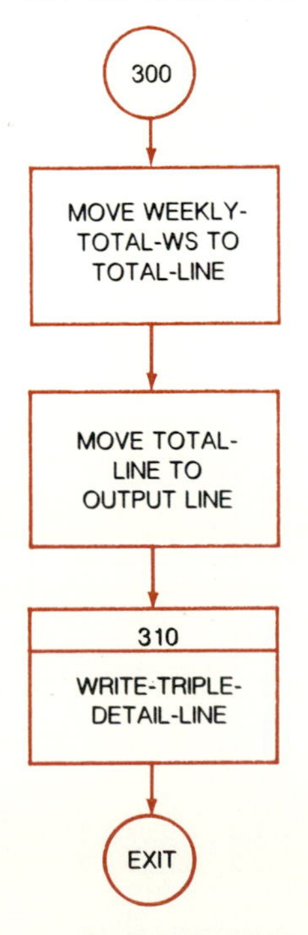
300-PRINT-TOTAL-SALES
300
MOVE WEEKLY-TOTAL-WS TO TOTAL-LINE
MOVE TOTAL-LINE TO OUTPUT LINE
310 WRITE-TRIPLE-DETAIL-LINE
EXIT

Source Program

```
 1        *************************
 2
 3         IDENTIFICATION DIVISION.
 4
 5        *************************
 6
 7         PROGRAM-ID.
 8             TRANSACTION-REGISTER-W-TOTALS.
 9         AUTHOR.
10             STEVE GONSOSKI.
11         INSTALLATION.
12             WEST LOS ANGELES COLLEGE.
13         DATE-WRITTEN.
14             3 FEBRUARY 1982.
15
16        *PURPOSE.
17        *    PRINT A REPORT LISTING ALL SALES TRANSACTIONS FOR A WEEK.
18        *    CALCULATING SALES AMOUNT PER ITEM AND TOTALING ALL SALES
19        *    ITEMS FOR THE WEEK.
20
21        *************************
22
23         ENVIRONMENT DIVISION.
24
25        *************************
26
27         CONFIGURATION SECTION.
28
29         SOURCE-COMPUTER.
30             LEVEL-66-ASCII.
31         OBJECT-COMPUTER.
32             LEVEL-66-ASCII.
33
34         INPUT-OUTPUT SECTION.
35
36         FILE-CONTROL.
37             SELECT FILE-IN       ASSIGN TO C1-CARD-READER.
38             SELECT FILE-OUT      ASSIGN TO P1-PRINTER.
39
40        *************************
41
42         DATA DIVISION.
43
44        *************************
45
46         FILE SECTION.
47
48         FD  FILE-IN
49             CODE-SET IS GBCD
50             LABEL RECORDS ARE OMITTED
51             DATA RECORD IS CARDIN.
52
53         01  CARDIN.
54
55             03  TRANSACTION-DATE.
56                 05  T-MONTH            PICTURE 9(2).
57                 05  T-DAY              PICTURE 9(2).
58                 05  T-YEAR             PICTURE 9(2).
59             03  ITEM-NUMBER            PICTURE 9(6).
60             03  ITEM-DESCRIPTION       PICTURE X(20).
61             03  QUANTITY               PICTURE 9(5).
62             03  UNIT-PRICE             PICTURE 9(3)V99.
63
64         FD  FILE-OUT
65             CODE-SET IS GBCD
66             LABEL RECORDS ARE OMITTED
67             DATA RECORD IS PRINT.
68
69         01  PRINT                      PICTURE X(132).
70
71         WORKING-STORAGE SECTION.
72
73         77  SALES-AMOUNT-WS            PICTURE 9(4)V99.
74         77  WEEKLY-TOTAL-WS            PICTURE 9(6)V99 VALUE ZERO.
75         77  LINE-COUNT                 PICTURE 9(2)    VALUE 38.
76
77         01  FLAGS.
78
79             03  MORE-DATA-FLAG         PICTURE X(3)    VALUE "YES".
80                 88  MORE-DATA                          VALUE "YES".
81                 88  NO-MORE-DATA                       VALUE "NO ".
82
83         01  HDG-1.
84
85             03  FILLER                 PICTURE X(37)   VALUE SPACES.
86             03  FILLER                 PICTURE X(11)   VALUE "TRANSACTION".
87             03  FILLER                 PICTURE X(10)   VALUE "  REGISTER".
88             03  FILLER                 PICTURE X(74)   VALUE SPACES.
89
90         01  HDG-2.
91
92             03  FILLER                 PICTURE X(9)    VALUE SPACES.
93             03  FILLER                 PICTURE X(11)   VALUE "TRANSACTION".
94             03  FILLER                 PICTURE X(6)    VALUE SPACES.
95             03  FILLER                 PICTURE X(4)    VALUE "ITEM".
96             03  FILLER                 PICTURE X(11)   VALUE SPACES.
```

```
 97          03  FILLER               PICTURE X(11)   VALUE "DESCRIPTION".
 98          03  FILLER               PICTURE X(11)   VALUE SPACES.
 99          03  FILLER               PICTURE X(8)    VALUE "QUANTITY".
100          03  FILLER               PICTURE X(3)    VALUE SPACES.
101          03  FILLER               PICTURE X(4)    VALUE "UNIT",
102          03  FILLER               PICTURE X(6)    VALUE SPACES.
103          03  FILLER               PICTURE X(5)    VALUE "SALES".
104          03  FILLER               PICTURE X(43)   VALUE SPACES.
105
106      01  HDG-3.
107
108          03  FILLER               PICTURE X(12)   VALUE SPACES.
109          03  FILLER               PICTURE X(4)    VALUE "DATE".
110          03  FILLER               PICTURE X(11)   VALUE SPACES.
111          03  FILLER               PICTURE X(2)    VALUE "NO".
112          03  FILLER               PICTURE X(45)   VALUE SPACES.
113          03  FILLER               PICTURE X(5)    VALUE "PRICE".
114          03  FILLER               PICTURE X(5)    VALUE SPACES.
115          03  FILLER               PICTURE X(6)    VALUE "AMOUNT".
116          03  FILLER               PICTURE X(42)   VALUE SPACES.
117
118      01  DETAIL-LINE.
119
120          03  FILLER               PICTURE X(10)   VALUE SPACES.
121          03  T-MONTH-O            PICTURE 9(2).
122          03  FILLER               PICTURE X(1)    VALUE "/".
123          03  T-DAY-O              PICTURE 9(2).
124          03  FILLER               PICTURE X(1)    VALUE "/".
125          03  T-YEAR-O             PICTURE 9(2).
126          03  FILLER               PICTURE X(7)    VALUE SPACES.
127          03  ITEM-NUMBER-O        PICTURE 9(6).
128          03  FILLER               PICTURE X(9)    VALUE SPACES.
129          03  ITEM-DESCRIPTION-O   PICTURE X(20).
130          03  FILLER               PICTURE X(5)    VALUE SPACES.
131          03  QUANTITY-O           PICTURE ZZZZ9.
132          03  FILLER               PICTURE X(3)    VALUE SPACES.
133          03  UNIT-PRICE-O         PICTURE ZZZ.99.
134          03  FILLER               PICTURE X(1)    VALUE SPACES.
135          03  SALES-AMOUNT-O       PICTURE $$$,$$9.99.
136          03  FILLER               PICTURE X(42)   VALUE SPACES.
137
138
139      01  TOTAL-LINE.
140
141          03  FILLER               PICTURE X(57)   VALUE SPACES.
142          03  FILLER               PICTURE X(7)    VALUE "WEEKLY".
143          03  FILLER               PICTURE X(5)    VALUE "TOTAL".
144          03  FILLER               PICTURE X(9)    VALUE SPACES.
145          03  WEEKLY-TOTAL-O       PICTURE $,$$$,$$$.99.
146          03  FILLER               PICTURE X(42)   VALUE SPACES.
147
148      01  SAVE-RECORD              PICTURE X(132).
149
150     *************************
151
152      PROCEDURE DIVISION.
153
154     *************************
155
156      000-PRODUCE-REGISTER.
157
158          OPEN    INPUT   FILE-IN
159                  OUTPUT  FILE-OUT.
160          PERFORM 100-READ-TRANSACTION-RECORD.
161          PERFORM 200-PRODUCE-TRANSACTION-LINE
162                  UNTIL NO-MORE-DATA.
163          PERFORM 300-PRINT-TOTAL-SALES.
164          CLOSE   FILE-IN
165                  FILE-OUT.
166          STOP RUN.
167
168      100-READ-TRANSACTION-RECORD.
169
170          READ FILE-IN
171              AT END MOVE "NO " TO MORE-DATA-FLAG.
172
173      200-PRODUCE-TRANSACTION-LINE.
174
175          PERFORM 210-CALCULATE-PER-ITEM-SALES.
176          PERFORM 220-ACCUMULATE-TOTAL-SALES.
177          PERFORM 230-PRINT-TRANSACTION-LINE.
178          PERFORM 100-READ-TRANSACTION-RECORD.
179
180      210-CALCULATE-PER-ITEM-SALES.
181
182          MULTIPLY QUANTITY BY UNIT-PRICE
183              GIVING SALES-AMOUNT-WS.
184
185      220-ACCUMULATE-TOTAL-SALES.
186
187          ADD SALES-AMOUNT-WS TO WEEKLY-TOTAL-WS.
188
189      230-PRINT-TRANSACTION-LINE.
190
191          MOVE T-MONTH            TO T-MONTH-O.
192          MOVE T-DAY              TO T-DAY-O.
193          MOVE T-YEAR             TO T-YEAR-O.
194          MOVE ITEM-NUMBER        TO ITEM-NUMBER-O.
195          MOVE ITEM-DESCRIPTION   TO ITEM-DESCRIPTION-O.
196          MOVE QUANTITY           TO QUANTITY-O.
197          MOVE UNIT-PRICE         TO UNIT-PRICE-O.
```

```
198           MOVE SALES-AMOUNT-WS  TO SALES-AMOUNT-O.
199           MOVE DETAIL-LINE TO PRINT.
200           PERFORM 240-PRINT-DETAIL-LINE.
201
202       240-PRINT-DETAIL-LINE.
203
204           IF LINE-COUNT IS > 37
205               MOVE PRINT TO SAVE-RECORD
206               PERFORM 250-PRINT-HEADING-LINE
207               MOVE SAVE-RECORD TO PRINT
208           ELSE
209               NEXT SENTENCE.
210           PERFORM 260-WRITE-DETAIL-LINE.
211
212       250-PRINT-HEADING-LINE.
213
214           MOVE HDG-1 TO PRINT.
215           PERFORM 270-WRITE-TOP-OF-PAGE.
216           MOVE SPACES TO PRINT.
217           PERFORM 260-WRITE-DETAIL-LINE.
218           MOVE HDG-2 TO PRINT.
219           PERFORM 260-WRITE-DETAIL-LINE.
220           MOVE HDG-3 TO PRINT.
221           PERFORM 260-WRITE-DETAIL-LINE.
222           MOVE SPACES TO PRINT.
223           PERFORM 260-WRITE-DETAIL-LINE.
224
225       260-WRITE-DETAIL-LINE.
226
227           WRITE PRINT
228               AFTER ADVANCING 1 LINE.
229           ADD 1 TO LINE-COUNT.
230
231       270-WRITE-TOP-OF-PAGE.
232
233           WRITE PRINT
234               AFTER ADVANCING PAGE.
235           MOVE 1 TO LINE-COUNT.
236
237       300-PRINT-TOTAL-SALES.
238
239           MOVE WEEKLY-TOTAL-WS TO WEEKLY-TOTAL-O.
240           MOVE TOTAL-LINE TO PRINT.
241           PERFORM 310-WRITE-TRIPLE-DETAIL-LINE.
242
243       310-WRITE-TRIPLE-DETAIL-LINE.
244
245           WRITE PRINT
246               AFTER ADVANCING 3 LINES.
247           ADD 3 TO LINE-COUNT.

THERE WERE 247 SOURCE INPUT LINES.
THERE WERE NO DIAGNOSTICS.
```

```
                                  TRANSACTION  REGISTER

       TRANSACTION      ITEM          DESCRIPTION           QUANTITY   UNIT       SALES
          DATE           NO                                            PRICE      AMOUNT

        07/23/80       413010       CH001 BOX 100A FLUSH         10     4.90      $49.00
        07/23/80       412146       CH148 BREAKER 15A           100      .89      $89.00
        07/23/80       411116       1500 TWIN SOCKER B          500     1.12     $560.00
        07/24/80       503029       MOTOR 1/2 HP 60 CYC           2   146.78     $293.56
        07/24/80       317802       TERMINAL CLIP               100     5.12     $512.00
        07/24/80       326917       TERMINAL BAR                100     4.12     $412.00
        07/24/80       411121       1506 COKT ADAPT BRN         400      .19      $76.00
        07/24/80       412997       CH173 BREAKER 30A            60     1.15      $69.00
        07/24/80       413088       CH176 BREAKER 60A            40     1.15      $46.00
        07/24/80       411174       C151 SIL SWITCH BRN         200     1.16     $232.00
        07/24/80       413090       CH005 BR BOX 150A            10     4.98      $49.80
        07/24/80       718326       FC803 FUSE 15A              200      .32      $64.00

                                                   WEEKLY TOTAL              $2,452.36
```

Questions for Review

1. Why is it important for the programmer to keep up with the state of the art and become familiar with new techniques to help solve data processing problems?
2. Briefly describe the four major actions during the execution of arithmetic statements.
3. What is an arithmetic expression?
4. How is the GIVING option used in arithmetic expressions?
5. How is the ROUNDED option used in arithmetic expressions?
6. How is the SIZE ERROR option used in arithmetic expressions?
7. What are the common features of arithmetic statements?
8. What is the function of the ADD statement and what are its options?
9. What is the function of the SUBTRACT statement and what are its options?
10. What is the function of the MULTIPLY statement and what are its options?
11. What is the function of the DIVIDE statement and what are its options?
12. When is the REMAINDER option specified?
13. What is the main purpose of the COMPUTE statement and how is it used?
14. What are the five arithmetic operators used in arithmetic expressions?
15. What is the order of evaluation of arithmetic expressions?

Matching Questions

Match each item with its proper description.

_____ 1. COMPUTE	A. Specifies accumulation of numeric values.
_____ 2. MULTIPLY	B. Made equal to the calculated value of the arithmetic expression.
_____ 3. Raw result	C. Requires use of an arithmetic expression.
_____ 4. ADD	D. Dropping of excess digits.
_____ 5. SIZE ERROR	E. Product of two numeric values.
_____ 6. SUBTRACT	F. Numerical outcome of the calculation.
_____ 7. GIVING	G. Determines the precedence of arithmetic operations.
_____ 8. DIVIDE	H. Exceeds the largest value that can be contained in the resultants.
_____ 9. ROUNDED	I. Difference between two numeric values.
_____10. Evaluation	J. Division of one numeric value by divisor.

Exercises

Multiple Choice: Indicate the best *answer (questions 1–28).*

1. In order to find the average sales per month for the first three months of the year, you could write
   a. COMPUTE AVERAGE = JAN + FEB + MAR / 3.
   b. COMPUTE AVERAGE = (JAN + FEB + MAR) / 3.
   c. ADD JAN FEB TO MAR GIVING TOTAL.
      DIVIDE 3 INTO TOTAL GIVING AVERAGE.
   d. All of the above.
2. An ADD or SUBTRACT statement can operate on
   a. more than two numbers.
   b. only one number.
   c. only two numbers.
   d. None of the above.
3. The test for SIZE ERROR is made ______________ the result is moved.
   a. after
   b. before
   c. while
   d. following
4. Select the valid MULTIPLY statement.
   a. MULTIPLY AMOUNT BY 5.
   b. MULTIPLY 5 BY AMOUNT.
   c. MULTIPLY 5 BY 5.
   d. MULTIPLY '5' BY AMOUNT.
5. Which of the following statements correctly specifies that the sum of FICA, STATETAX, and INSURANCE is to be computed and stored in DEDUCTIONS?
   a. ADD FICA STATETAX FEDTAX INSURANCE TO DEDUCTIONS.
   b. ADD FICA STATETAX FEDTAX TO INSURANCE GIVING DEDUCTIONS.
   c. ADD FICA STATETAX FEDTAX INSURANCE GIVING DEDUCTIONS.
   d. None of the above.
6. Which of the following SUBTRACT statements is in correct format?
   a. SUBTRACT 10, CHECK FROM ACCOUNT.
   b. SUBTRACT CHECK FROM 10, ACCOUNT.
   c. SUBTRACT CHECK FROM 10.
   d. All of the above.
7.
```
77  FACTOR    PICTURE 99.
77  RESULT    PICTURE 999V99.

        DIVIDE FACTOR INTO RESULT.
```
   If the values of FACTOR and RESULT were 02 and 12500 respectively, after the execution of the DIVIDE statement above, the value stored in
   a. FACTOR will be 062 ‸ 50.
   b. RESULT will be 625 ‸ 00.
   c. FACTOR will be 625 ‸ 00.
   d. RESULT will be 062 ‸ 50.
   ‸ denotes assumed decimal point.
8.
```
DIVIDE PEOPLE INTO TOTAL
    REMAINDER LEFT-OVER.
```
   When the statements above are executed, the REMAINDER will be
   a. lost.
   b. stored in TOTAL.
   c. stored in LEFT-OVER.
   d. None of the above.
9. According to the format of the COMPUTE statement, which of the following is correct?
   a. COMPUTE 5 = TOTAL.
   b. COMPUTE TOTAL = 5.
   c. COMPUTE TOTAL = 'TOTAL1'.
   d. COMPUTE TOTAL + 5 = TOTAL.

10. Arithmetic operations can be performed only with
    a. elementary alphanumeric items.
    b. group items.
    c. elementary numeric items.
    d. All of the above.
11. The only case in which both arithmetic items must be converted is when
    a. both are in binary format.
    b. one is in binary format and the other is in packed decimal format.
    c. both are in DISPLAY format.
    d. both are in packed decimal format.
12. A programming technique to avoid unnecessary conversion is to use the working storage item whose usage is COMPUTATIONAL-3 and to move the data to that item. This programming technique is an efficient way of treating an external decimal item,
    a. provided that the item is used in one calculation only.
    b. if the item is used in two or more calculations.
    c. no matter how many calculations are used.
    d. None of the above.
13. Since a raw result is calculated in a work area, it is
    a. necessary for the programmer to specify a work area in working storage.
    b. the compiler that sets up the work areas needed for calculation.
    c. the programmer who specifies the work area, as the compiler does not know the size of the operands.
    d. the programmer who supplies the size of the work area, and the compiler who sets it up.
14. A test can be made of the raw result by
    a. using a SIZE ERROR clause prior to the execution of the arithmetic statement.
    b. testing all digits including non-integral digits through the SIZE ERROR option.
    c. testing for SIZE ERROR before the raw result is moved to the result area.
    d. testing for SIZE ERROR after the raw result is moved to the result area.
15. The raw result is moved from the work area to the finished result record item by
    a. the programmer writing an appropriate MOVE statement.
    b. definition in the File Section.
    c. the compiler generating instructions that are needed to get the raw result into that area.
    d. None of the above.
16. Of all the arithmetic verbs, only the following permits the programmer to combine arithmetic operations into arithmetic expression in formular style.
    a. ADD
    b. COMPUTE
    c. MULTIPLY
    d. CALCULATE
17. The following options are available in arithmetic statements:
    a. ROUNDED
    b. SIZE ERROR
    c. GIVING
    d. All of the above.
18. The ROUNDED option is used when the number of places in the calculated result is
    a. greater than the number of places allowed for the result.
    b. less than the number of places allowed for the result.
    c. in packed decimal format.
    d. zero.
19. If the SIZE ERROR option is specified and a size error condition occurs,
    a. the prior values of resultant(s) are altered.
    b. values of the resultant(s) for which no size error occurred are altered.
    c. the values of the resultant(s) affected is undefined.
    d. after completion of the operation, the imperative-statement in the SIZE ERROR option is executed.
20. The ROUNDED option can be used
    a. only in arithmetic statements that contain a GIVING clause.
    b. in all arithmetic statements.
    c. in all arithmetic statements except the COMPUTE statement.
    d. in all arithmetic statements except the DIVIDE statement.

21. The ROUNDED option is always written
    a. at the end of the arithmetic statement.
    b. just ahead of the SIZE ERROR clause.
    c. right after the name of the finished result.
    d. at the end of the COMPUTE statement.
22. In an ADD statement,
    a. TO and GIVING options may be used together in the same statement.
    b. elementary numeric and elementary edited items may be used with the TO option.
    c. when the GIVING option is used, all elementary items are added, including the data item after the word GIVING.
    d. None of the above.
23. In a SUBTRACT statement,
    a. one or the sum of two or more numeric items may be subtracted from one or more items.
    b. the FROM and GIVING options may not be used in the same statement.
    c. when the FROM option is used, there may be no numeric literal preceding FROM.
    d. the data-name following GIVING must not be edited, as it is involved in the calculation.
24. In a MULTIPLY statement,
    a. if the GIVING option is not used, the word following BY may not be a numeric literal.
    b. if the GIVING option is not used, the product replaces the identifier following BY.
    c. when the GIVING option is specified, the product is placed in the identifier following GIVING.
    d. All of the above.
25. In the DIVIDE statement,
    a. when the DIVIDE . . . INTO statement is specified, the value of the identifier or literal is divided into the value of the identifier following the word DIVIDE.
    b. when the DIVIDE . . . INTO statement with the GIVING option is specified, the resulting quotient is stored in the item following the word INTO.
    c. when the DIVIDE . . . BY statement with the GIVING option is used, the quotient is stored in the identifier following GIVING.
    d. the REMAINDER option may not be used if the ROUNDED option is used.
26. In the COMPUTE statement,
    a. editing symbols may be used as operands involved in the calculation.
    b. exponentiation may be specified.
    c. numeric literals may not be used as operands.
    d. arithmetic verbs may be used as operands.
27. In an arithmetic expression, operands may be
    a. an identifier of a numeric elementary item.
    b. a numeric literal.
    c. two arithmetic expressions.
    d. All of the above.
28. Arithmetic operators
    a. may be used in arithmetic expressions.
    b. may be used with arithmetic verbs.
    c. must not be preceded by a space.
    d. must not be followed by a space.
29. WORKING-STORAGE SECTION.

```
77  WIDTH          PICTURE 99V99.
77  HEIGHT         PICTURE 99V99.
77  SQUARE-FEET    PICTURE 999V999.
```

    Write a MULTIPLY statement to compute the product of the values of WIDTH and HEIGHT and store the result in the data item SQUARE-FEET with the value rounded.
30. Write a statement to sum the values of DEPT1, DEPT2, DEPT3, and YEARLY and store this sum in YEARLY.
31. Write a statement specifying that the values of TOTAL-1 and TOTAL-2 are to be added and their sum rounded and stored in TOTAL. The STORE-NUMBER and TOTAL are to be output on the console display device.
32. Write a statement to divide PROFIT by 500. The rounded quotient is to be stored in PORTION and the remainder is to be stored in EXCESS.
33. List the operations in the order in which they will be performed in the following expression.
    COMPUTE ANSWER = FACTOR * 5 – TOTAL ** 3.
34. Write an arithmetic expression to add TOTAL-1 and TOTAL-2 and multiply the sum by the product of 2 and TOTAL-3.

35. Imagine that this statement appears in a COBOL expression.

ADD CHECK-AMOUNT TO FLOAT-TOTAL.

Here are the descriptions of the items to be added.

```
77  FLOAT-TOTAL            PICTURE S9(6)V99.
    05  CHECK-AMOUNT       PICTURE 9(8)V99 USAGE COMP-3.
```

1. Is it permissible to add these items? Why or why not?
2. Will it be necessary to convert the data code of either item? If so, which item, and what code must it be converted to?

36. Rewrite these two statements calling for the same calculations, adding ROUNDED in the correct place in each statement.

COMPUTE PRICE = AVERAGE * .75.

MULTIPLY AVERAGE BY .75 GIVING PRICE.

37. Write a COMPUTE statement based on the following information.

Four numbers (A, B, C, and D) are to be multiplied to get their product X. A, B, C, and D each contain six digits, that is, the picture of each item is S9(6). The picture of X is S9(18), which is the largest allowable size in a numeric item.

38. Shown below are two sets of entries. Which set of entries indicates the correct way to specify a rounded result or are all sets correct?

```
(1)      MULTIPLY QUANTITY BY PRICE GIVING AMOUNT ROUNDED.
     77  AMOUNT      PICTURE S9(4)V99      USAGE COMP-3.
(2)      MULTIPLY QUANTITY BY PRICE GIVING AMOUNT.
     77  AMOUNT      PICTURE S9(4)V99      USAGE COMP-3 ROUNDED.
```

39. List the precedence of arithmetic operators in decreasing order of priority.

40. Is statement 2 equivalent to statement 1? If not, give the correct equivalent DIVIDE statement.

(1) COMPUTE RATE = DISTANCE/TIME.

(2) DIVIDE DISTANCE INTO TIME GIVING RATE.

41. Arithmetic problems:

Data-Name	Picture	Data Values Before Execution	Data Values After Execution
FLD-A	S999V99	+10000	______
FLD-B	S999V999	+045550	______
FLD-C	S999V99	−12345	______
FLD-D	S9999	1234	______
FLD-E	S99V9999	+123456	______
FLD-F	S999V99	+90000	______
FLD-G	S999V9	+12345	______
FLD-H	S9V9	−45	______
FLD-I	S999V99	−32045	______
FLD-J	S99V99	+0475	______
FLD-K	S9999V9999	+46250000	______
FLD-L	S999V9	+4259	______
FLD-M	S999V99	−32007	______
FLD-N	S999V99	00000	______
FLD-O	S9999	4567	______
FLD-P	S9999V99	+123456	______

Arithmetic statements:

1. ADD FLD-A, FLD-B, GIVING FLD-C.
2. ADD FLD-A, FLD-B, FLD-H, FLD-I, FLD-J TO FLD-K.
3. SUBTRACT FLD-M FROM FLD-I ROUNDED.
4. MULTIPLY FLD-A BY FLD-B GIVING FLD-D ROUNDED.
5. DIVIDE FLD-A INTO FLD-B GIVING FLD-E ON SIZE ERROR PERFORM ERROR-RT.
6. MULTIPLY FLD-A BY FLD-F GIVING FLD-G ROUNDED ON SIZE ERROR PERFORM FIX-IT.
7. DIVIDE FLD-N INTO FLD-A GIVING FLD-P.
8. ADD FLD-O to FLD-O.

Required: In the DATA VALUES AFTER EXECUTION column, write the results of the arithmetic statements just presented.

42. Show the contents of each field after the calculation in the After Execution area.

a. ADD FLD-A to FLD-B.

Data-Name	Picture	Before Execution	Data Values After Execution
FLD-A	S99V99	+1234	____________
FLD-B	S99V99	−1200	____________

b. SUBTRACT FLD-A, FLD-B FROM FLD-C.

Data-Name	Picture	Before Execution	After Execution
FLD-A	S99V99	+1234	____________
FLD-B	S9999V99	−987654	____________
FLD-C	S9999V99	+123456	____________

c. MULTIPLY FLD-A BY FLD-B GIVING FLD-C.

Data-Name	Picture	Before Execution	After Execution
FLD-A	S99V99	+1234	____________
FLD-B	S9999V99	+98765	____________
FLD-C	S9(7)V9999	+1234567890	____________

d. DIVIDE FLD-A INTO FLD-B ROUNDED.

Data-Name	Picture	Before Execution	After Execution
FLD-A	S99V99	+1234	____________
FLD-B	S999V9	+9879	____________

e. ADD FLD-A, FLD-B TO FLD-C ON SIZE ERROR GO TO ERROR-ROUTINE.

Data-Name	Picture	Before Execution	After Execution
FLD-A	S99V99	+1234	____________
FLD-B	S99V99	+9876	
FLD-C	S999V99	+98765	____________
			____________

43. Prepare the following arithmetic statements.

a. Add the fields of GIANT and CONTAINER and place the result in CONTAINER.
b. Add the fields TOOL, TOTAL-NUMB, and NUMB, placing the result in TOTAL-NUMB.
c. Add the fields DATA-IN, PROD, and ROYALTY and place the result in GRAND-SUM.
d. Subtract the field QUANTITY from TOTAL-BALANCE and place the result in TOTAL-BALANCE.
e. Subtract the fields DATA-GIVEN, HOLD-DATA, CON-HOLD from TOTAL-HOLD and place the result in TOTAL-HOLD.
f. Subtract the field BALANCE-B from CON-NUMB and place the result in NEW-NUMB.

Problem 1

Use the flowchart, record layout, and printer spacing chart to help you write a COBOL program that is responsible for printing a report that lists information relative to the hours an employee worked during a particular pay period. You should accumulate the regular and overtime hours and print them at the end of the report.

Input is as follows:

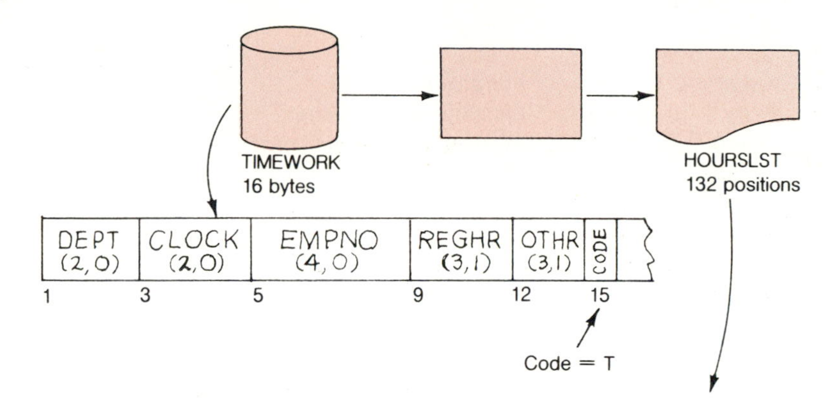

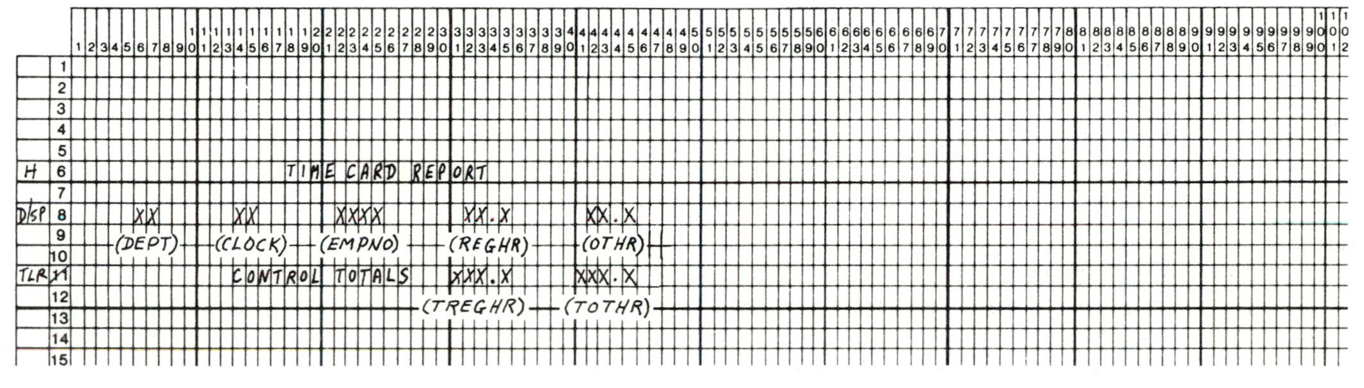

Output is as follows:

```
                        TIME CARD REPORT

 55         4          7214         40.0         4.0
 55         4          7392         35.5          .0
 55         9          7419         40.0        10.5
 74         9          4193         15.0          .0
 74        15          3284         20.5          .0
 99        15          1272         40.0        15.0
 99        15          1438         35.0          .0
 99        30          5277         40.0          .0

                 CONTROL TOTALS    266.0        29.5
```

Problem 2

Write a program that will print a salary table of monthly, yearly, daily, and hourly wages. The values to be entered are the initial monthly salary, the monthly limit of the table, and the monthly increments. For example, assume the three values 800, 1,200, and 50 are entered as input. The monthly initial value would be 800, 1,200 would be the monthly limit of the table, and 50 would be the monthly increments.

Input

Positions	Field Designation
1-5	Monthly salary—initial
6-10	Monthly salary—limit of the table
11-15	Monthly salary—increments

Processing

1. Compute the yearly, weekly, daily, and hourly salaries based on the monthly rate.
2. Assume a five-day week and an eight-hour day.
3. The output is to be printed as per the format shown in the following Printer Spacing Chart.

Output

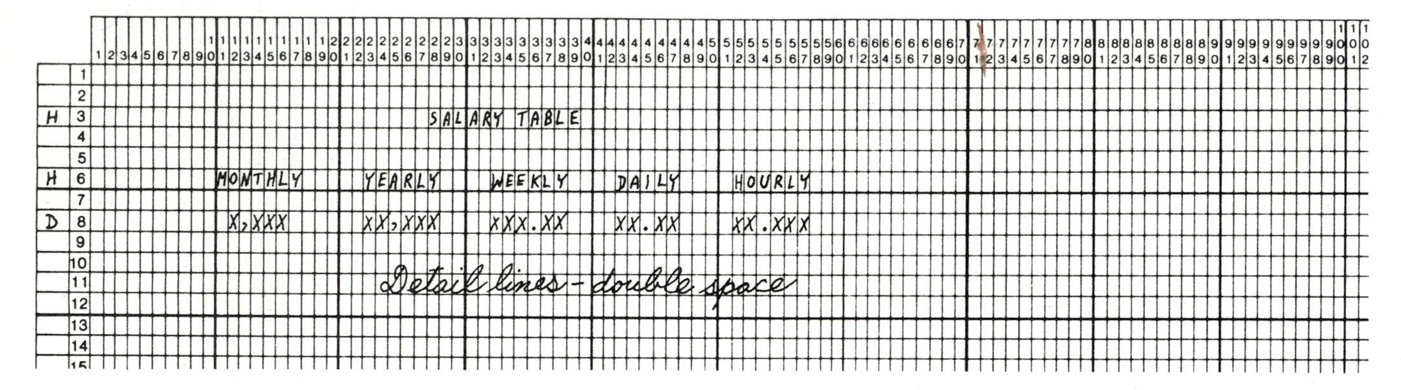

Output is as follows:

SALARY TABLE

MONTHLY	YEARLY	WEEKLY	DAILY	HOURLY
800	9,600	184.62	36.92	4.615
850	10,200	196.15	39.23	4.904
900	10,800	207.69	41.54	5.193
950	11,400	219.23	43.85	5.481
1,000	12,000	230.77	46.15	5.769
1,050	12,600	242.31	48.46	6.058
1,100	13,200	253.85	50.77	6.346
1,150	13,800	265.38	53.08	6.635
1,200	14,400	276.92	55.38	6.923

Problem 3

Write a COBOL program based on the following specifications.

General Description

A Monthly Commission Report is to be produced from an input card file.

Input

A file called SLSCDS; each record contains data on one transaction. The Record Layout follows.

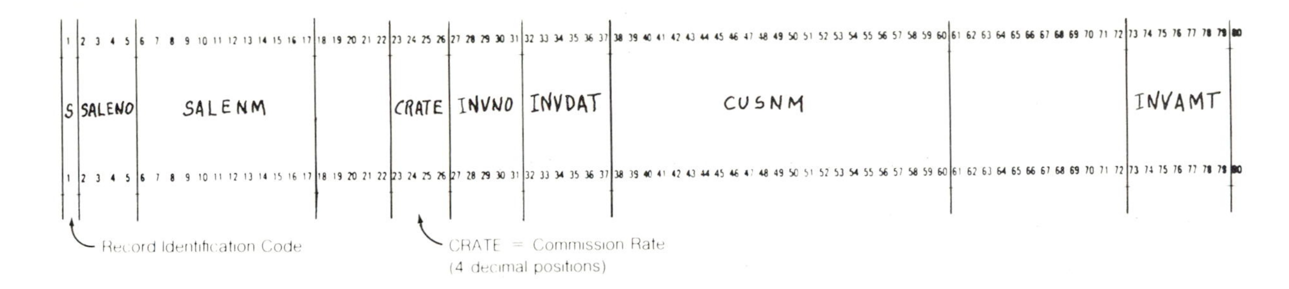

Calculations

For each record, Invoice Amount is multiplied by Commission Rate to give Commission Amount. A Final Total is printed at the end of the report.

Output

A printed report called COMREP is to be printed. The Printer Spacing Chart follows.

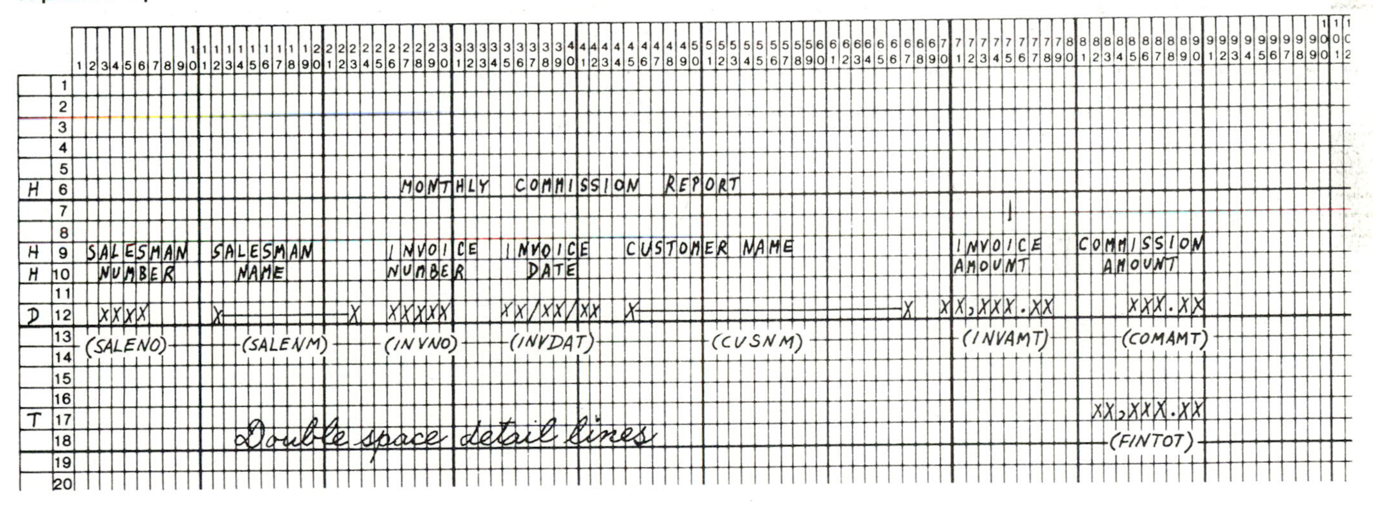

Output is as follows:

```
                              MONTHLY  COMMISSION  REPORT

SALESMAN   SALESMAN          INVOICE   INVOICE   CUSTOMER NAME              INVOICE    COMMISSION
 NUMBER      NAME            NUMBER     DATE                                 AMOUNT      AMOUNT

 1552      JOHN HOFFMAN      38164     4/19/80   PAUL FRIEDMAN              3,050.04      15.86

 1631      RICHARD KING      74719     9/23/80   BARBARA SMITH              3,050.04      43.62

 1679      LARRY HAM         54257     1/10/80   CARL JEFFERSON            19,468.32      99.96

 1741      PAULA LONDON      61906     8/20/80   HERBERT HOWARD                34.19       3.53

 1832      ED GRIFFEN        72393     9/30/80   RON MARTINEZ                 727.34     174.13

                                                                                         337.10
```

Problem 4

Write a COBOL program based on the following information:

Input

Field	Positions	Format
Account Number	1-6	XXXXXX
Principal	7-13	XXXXX.XX
Interest Rate	14-17	.XXXX
Monthly Payment	18-22	XXX.XX
Not Used	23-80	

Computations To Be Performed Each month the principal (unpaid balance) is multiplied by the annual interest rate. The resulting yearly interest must be divided by twelve to arrive at the monthly interest. The monthly mortgage payment consists of both interest and principal. When a monthly payment is received, the difference between the payment and the monthly interest reduces the principal. All calculations are rounded to two decimal positions.

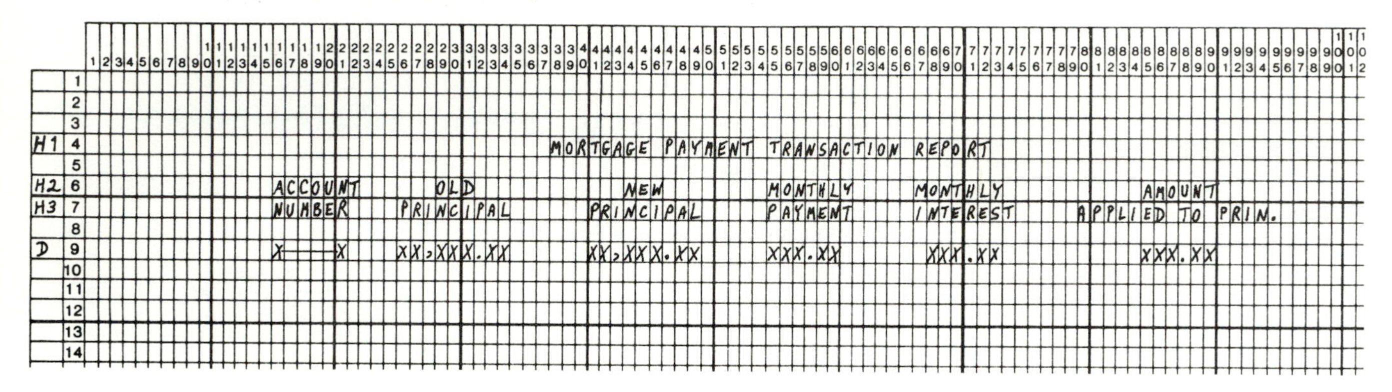

Output is as follows:

```
                    MORTGAGE PAYMENT TRANSACTION REPORT

ACCOUNT      OLD          NEW         MONTHLY    MONTHLY        AMOUNT
NUMBER    PRINCIPAL    PRINCIPAL      PAYMENT    INTEREST   APPLIED TO PRIN

123456    44,250.00    44,129.22      425.00     304.22         120.78
333255    33,375.00    33,306.20      298.25     229.45          68.80
013540    28,259.40    28,228.68      225.00     194.28          30.72
143689    30,000.00    29,931.25      275.00     206.25          68.75
208064    32,500.00    32,423.44      300.00     223.44          76.56
101325    58,650.00    58,603.22      450.00     403.22          46.78
```

8

Procedure Division: Sequence Control Statements

Chapter Outline

Chapter Objectives

The learning objectives of this chapter are:

1. To describe the flow of control through a COBOL program and how it may be altered to suit changing conditions.
2. To describe the use of the IF statement with its various conditional expressions that may be tested to transfer control based upon conditions affected by the data.
3. To describe the use of the GO TO statement with its DEPENDING ON option and how it may be used to simulate the CASE program structure.
4. To explain the purpose of the EXIT statement and the role it plays in the use of the PERFORM and GO TO statements.
5. To explain the use of the STOP statement and how it may be used to temporarily or permanently halt program execution.

Introduction

Another important category of procedural words is called "sequence control." These words enable the programmer to control the sequence in which other statements or procedures will be acted upon by the computer. The sequence control statements are designed to specify the sequence in which the various source program instructions are to be executed. Statements, sentences, and paragraphs of the Procedure Division are executed normally in the sequence in which they are written, except when a sequence control word is encountered.

Conditional Procedures

The use of conditions and conditional statements is essential in constructing COBOL programs. The conditional statement is the primary mechanism through which alternate paths of control can be developed within a program. (A condition is an assertion concerning the control of a data item.) The flow of control through a conditional statement depends upon the determination of the truth or falsity of the assertion at program execution.

The discussion of test-conditions leads naturally to a closer study of the flow of control through COBOL procedures. After all, the reason for having test-conditions is to permit control to flow along alternate procedural paths.

The COBOL compiler will cause COBOL procedural statements to be translated into actual machine language instructions. When we talk about flow of control, we are "playing computers," so to speak, and acting as though the COBOL statements had already been translated and are now being executed by the computer. In order to trace the flow of control, we must know such things as where the starting point is, what sequence is normally followed, and what statements cause deviations from the sequence.

Control normally flows from one statement to the next in the order in which they are written in the program. When control comes to the end of a procedure, it normally goes right to the next procedure in sequence.

Flow of Control

The way in which control flows through procedures in a COBOL program represents the sequence in which instructions in the object program will be executed. How control will flow will depend on the kinds of statements in the Procedure Division, along with their arrangement.

Starting Point

Control starts at the first statement of the first procedure in the Procedure Division, provided there are no declaratives. If there are declaratives, control starts at the first procedure after the END DECLARATIVES entry. (Declaratives are discussed later in the text.)

Sequence

Control automatically flows from one statement to the next in sequence, and from one paragraph to the next, except when

a GO TO statement causes a branch,

an IF statement causes control to jump over certain statements,

a PERFORM statement gives control temporarily to another procedure,

a STOP statement causes a delay in execution or terminates the run.

Conditional Expressions

Conditional expressions identify conditions that are tested to enable the object program to select between alternate paths of control depending upon the truth value of the conditions. Conditional conditions are specified in IF, PERFORM, and SEARCH statements. (SEARCH statements will be discussed in

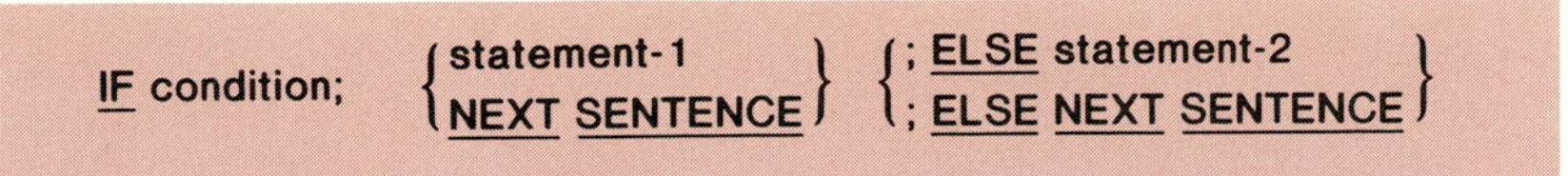

Figure 8.1 Format—CONDITIONAL statement.

the Table Handling section.) There are two categories of conditions associated with conditional expressions: simple conditions and compound conditions. Each may be enclosed within any number of paired parentheses, in which case its category is not changed.

A conditional expression contains one or more variables whose value may change during the course of the program. The conditional expression can be reduced to a single value that can be tested to determine to which of the alternate paths the program flow is to be taken. *A test condition is an expression that, taken as a whole, may be either true or false, depending on the circumstances existing when the expression is evaluated* (fig. 8.1).

Simple Conditions

The simple conditions are the relation, class, sign, and condition-name conditions. A simple condition has a truth value of true or false. The inclusion in parentheses of simple conditions does not change the simple truth value.

An IF statement causes a condition to be evaluated or tested, and an action to be taken based on whether the result of the test is true or false. (See figure 8.2.) Although IF is not a verb in the grammatical sense, it is regarded as such in COBOL. IF statements are used to evaluate test conditions. There are four types of test conditions: relation conditions, class conditions, sign conditions, and condition-name conditions.

Figure 8.2 Simple IF statement—example.

Basically, an IF statement is a simple IF statement whenever the IF verb occurs only once in the sentence. The simple IF statement represents a single level of procedure control. For example, the sentence form

IF **condition-1** **statement-1** ELSE **statement-2**

determines a choice in the flow of control:

True
Condition-1
False
ELSE
Statement-1
Statement-2
Next
sentence

The choice is present even if the NEXT SENTENCE option is used. The use of the NEXT SENTENCE option on the 'true' side of the condition yields:

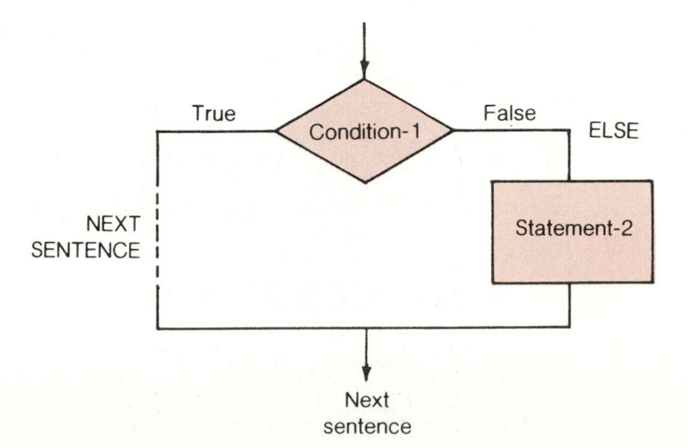

Figure 8.2 *(continued)*

The use of the NEXT SENTENCE option on the 'false' side of the condition is analogous. Rather than use the construction 'ELSE NEXT SENTENCE,' the entire ELSE phrase may be omitted, yielding the sentence form

IF **condition-1** **statement-1.**

and the flow of control:

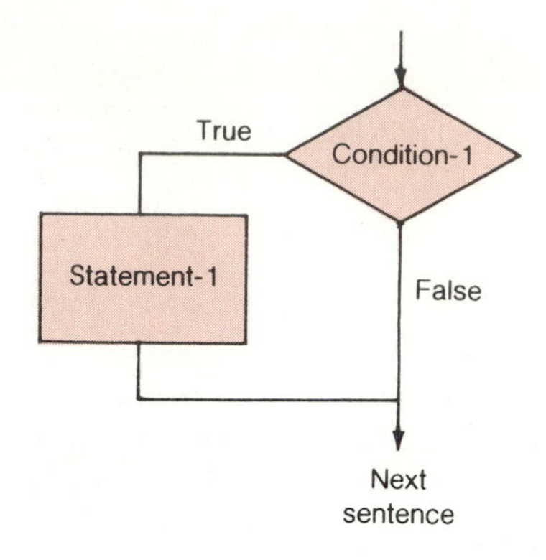

Although permissible, this omission of the ELSE phrase may be detrimental to the self-documenting capability (and therefore the maintainability) of the source program.

When the test condition is evaluated, the following action will take place.

1. If the condition is true, the statements immediately following the conditional expression are executed; control then passes to the next sentence.
2. If the condition is false, the statements immediately following ELSE are executed, or the next sentence, if the ELSE clause is omitted (figs. 8.3 and 8.4).
3. An IF statement must be terminated by a period and a space.
4. Any number of statements may follow the test condition. These statements are acted upon if the condition exists, and are skipped over if (a) the condition does not exist or (b) if they follow ELSE. (See figure 8.5.)
5. In a series of imperative statements executed when the condition is true, only the last statement may be an unconditional GO TO or STOP RUN statement; otherwise, the series of statements would contain statements into which control cannot flow. It is the programmer's responsibility to write the program steps in a logical sequence for execution.

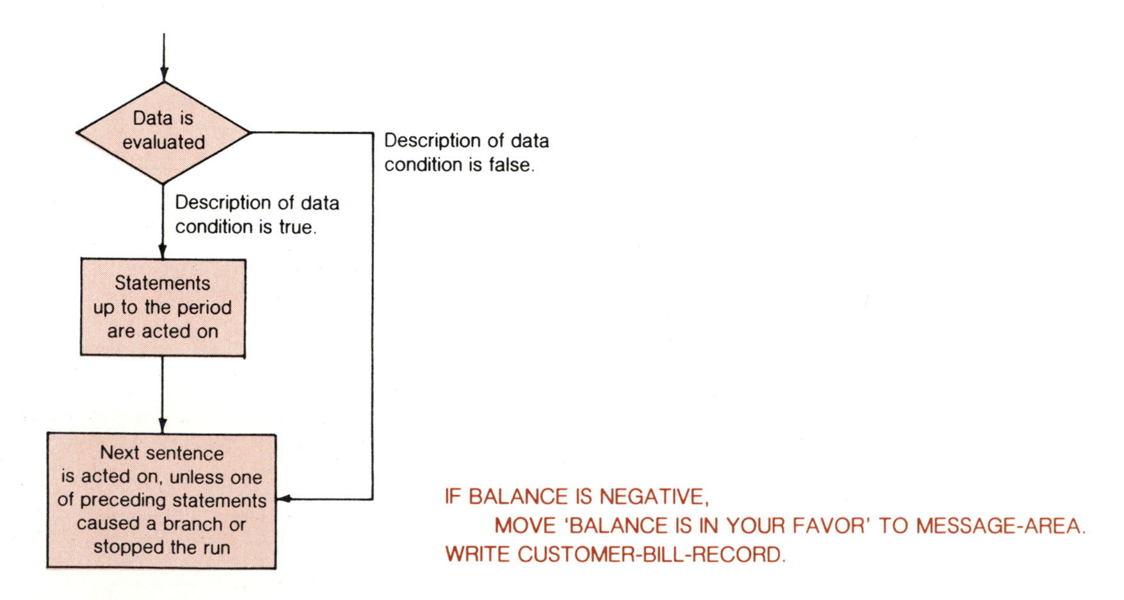

Figure 8.3 Flow of control through an IF statement that does not contain an ELSE statement.

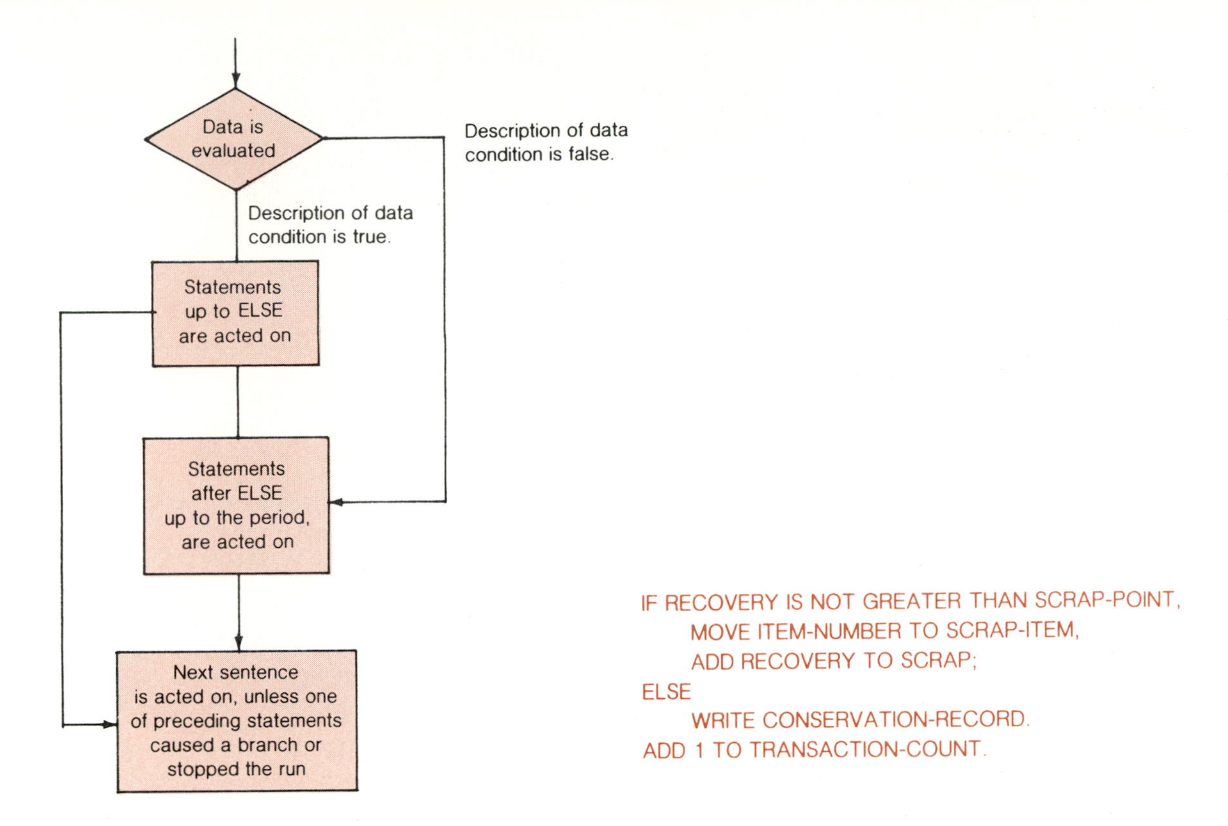


```
IF RECOVERY IS NOT GREATER THAN SCRAP-POINT,
      MOVE ITEM-NUMBER TO SCRAP-ITEM,
      ADD RECOVERY TO SCRAP;
ELSE
      WRITE CONSERVATION-RECORD.
ADD 1 TO TRANSACTION-COUNT.
```

Figure 8.4 Flow of control through an IF statement that contains an ELSE statement.

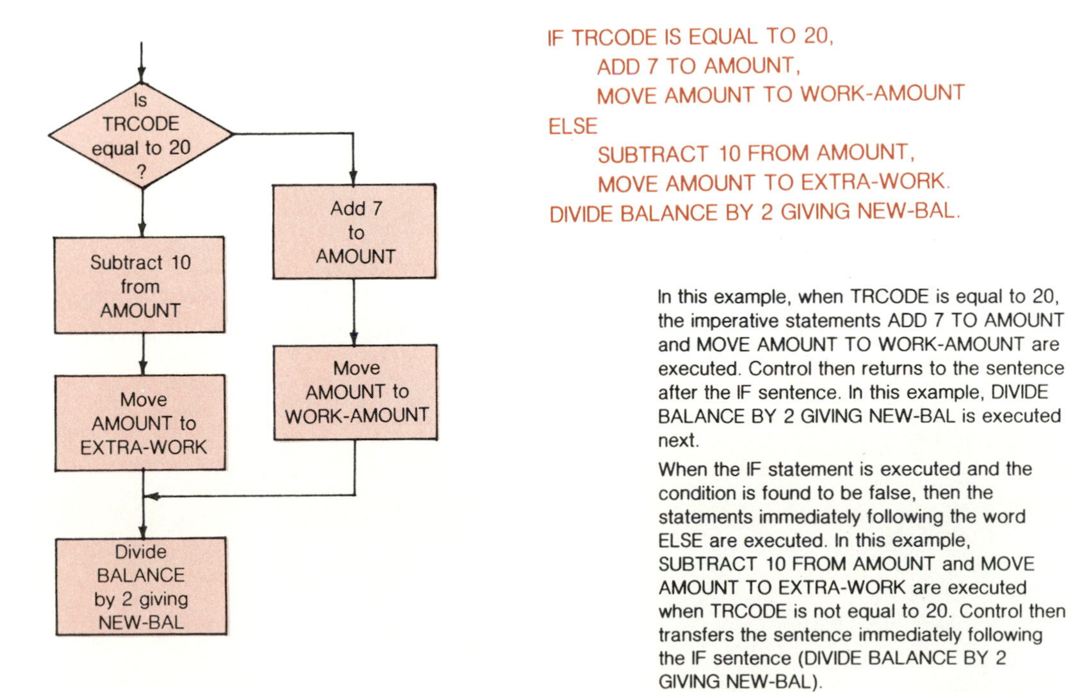


```
IF TRCODE IS EQUAL TO 20,
      ADD 7 TO AMOUNT,
      MOVE AMOUNT TO WORK-AMOUNT
ELSE
      SUBTRACT 10 FROM AMOUNT,
      MOVE AMOUNT TO EXTRA-WORK.
DIVIDE BALANCE BY 2 GIVING NEW-BAL.
```

In this example, when TRCODE is equal to 20, the imperative statements ADD 7 TO AMOUNT and MOVE AMOUNT TO WORK-AMOUNT are executed. Control then returns to the sentence after the IF sentence. In this example, DIVIDE BALANCE BY 2 GIVING NEW-BAL is executed next.

When the IF statement is executed and the condition is found to be false, then the statements immediately following the word ELSE are executed. In this example, SUBTRACT 10 FROM AMOUNT and MOVE AMOUNT TO EXTRA-WORK are executed when TRCODE is not equal to 20. Control then transfers the sentence immediately following the IF sentence (DIVIDE BALANCE BY 2 GIVING NEW-BAL).

Figure 8.5 IF statement—example.

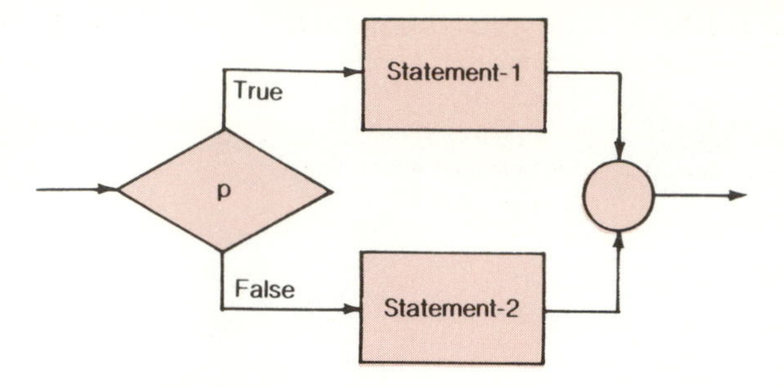

Figure 8.6 Flowchart for the IFTHENELSE structure.

The IFTHENELSE Program Structure—Coding Rules

The IFTHENELSE structure tests a single condition (predicate) to determine which of two function blocks will be executed. Alternately, the ELSE section may be replaced with ELSE NEXT SENTENCE when there isn't any function block to be executed.

The flowchart for this structure is detailed in figure 8.6.

Statements controlled by IF and ELSE parts of the IF statement are indented to exhibit the span of control of the structure. The word ELSE is started in the same columns as IF.

The structured programming coding rules for the IFTHENELSE structure follow:

1. Statements within the IF should be indented.
2. The ELSE should be coded on a line by itself in the same column as the IF.

```
IF    (condition)
      statement-1
      .
      statement-n
ELSE
      statement-2
      .
      statement-n.
```

3. When there are four or more statements (or any other number decided on) as a function of the IF, the following form should be used:

```
IF    (condition)
      PERFORM paragraph-1
ELSE
      PERFORM paragraph-2.
```

4. The condition portion of the IF will be enclosed in parentheses.
5. When there are no statements to follow the ELSE, ELSE NEXT SENTENCE will be coded fully.
6. A limit of levels of nested IF statements should be established. The general format for nested IF statements appears in figure 8.7.

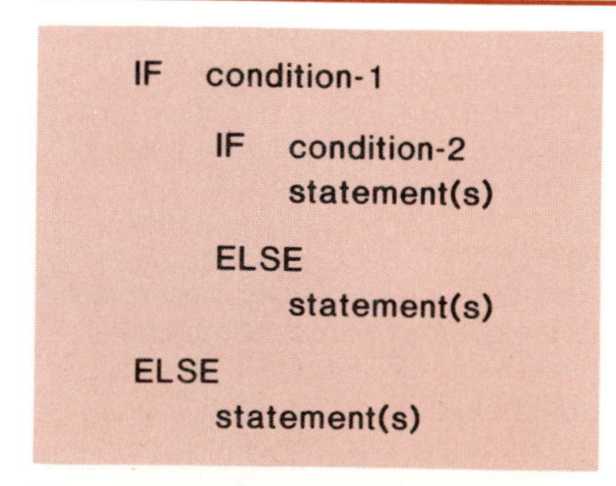

Figure 8.7 Nested IFs format.

7. If the IF or ELSE clause should include a nested IF, followed by one or more statements that should be executed without regard to evaluation of the nested IF (e.g., "statement-2" in the following example), then the nested IF must be PERFORMed.

```
    IF    (outer-condition)
          PERFORM inner-if-test-name
          statement-2
    ELSE
          statement-3
          .
          .
inner-if-test-name
    IF    (inner-condition)
          statement-1
    ELSE
          NEXT SENTENCE.
```

The statements controlled by the IF and ELSE parts of an IF statement are shown to be subordinate to the logic by indenting these statements by a consistent amount. If one tends to use nested IF statements sparingly, an indentation unit of four spaces would be reasonable. If one writes deeply nested IF statements, however, the indentation would need to be three or two positions, to avoid running out of space in the line. The following is an example of the nested IF:

Problem Statement

```
IF NOT A OR B IF C PERFORM Q ELSE IF D AND NOT E PERFORM R

ELSE PERFORM S ELSE PERFORM T.
```

Structured Instructions for Above Problem Statement

```
IF NOT A OR B
    IF C
        PERFORM G
    ELSE
        IF D AND NOT E
            PERFORM R
        ELSE
            PERFORM S
ELSE
    PERFORM T.
```

How to Simplify Nested IF's:

Keep compounds to a minimum

Avoid NOT's

Avoid implied subjects and operators

Restrict depth

No GO TO's

All on one page (use PERFORM's if necessary)

Indent

Use readable condition names

8. When the same field or condition is consecutively tested for more than three numeric values (or any other number decided on) using nested IF statements, the CASE structure should be used. In designing condition field values, the use of the CASE structure should be considered. (The CASE structure is discussed later.)

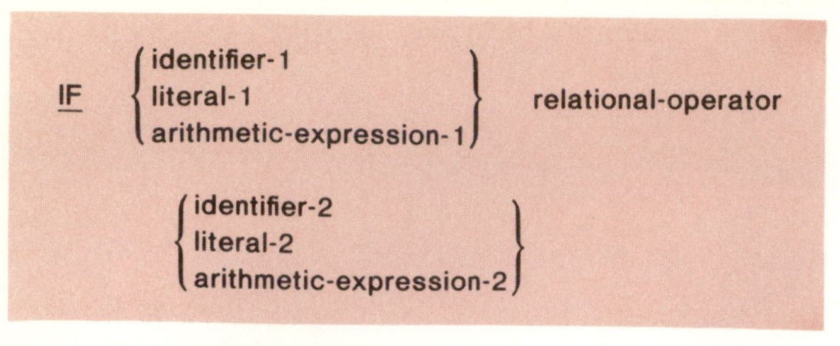

Figure 8.8 Format—relation condition.

Relational-Operator	Meaning
IS [NOT] GREATER THAN IS [NOT] >	Greater than or not greater than
IS [NOT] LESS THAN IS [NOT] <	Less than or not less than
IS [NOT] EQUAL TO IS [NOT] =	Equal to or not equal to

Figure 8.9 Relational operators and their meanings.

Relation Conditions

A relation condition causes a comparison of two operands, each of which may be the data item referenced by an identifier, a literal, or the value resulting from an arithmetic expression. A relation condition has a truth value of "true" if the relation exists between the operand. Comparison of two numeric operands is permitted regardless of the formats specified in their respective USAGE clauses. However, for all other comparisons, the operands must have the same usage. If either of the operands is a group item, the nonnumeric comparisons apply.

The first operands (identifier-1, literal-1, or arithmetic-expression-1) are called the subject of the condition; the second operands (identifier-2, literal-2, or arithmetic-expression-2) are called the object of the condition. (See figure 8.8.) The relation condition must contain at least one reference to a variable.

When used, NOT and the next key word or relation character are one relational operator that defines the comparison to be executed for truth value; e.g., NOT EQUAL is a truth test for an unequal comparison; NOT GREATER is a truth test for an equal or less comparison. The meaning of the relation operators appears in figure 8.9.

A test condition containing a single relational operator is called a relational condition. The result of a relational condition is either true or false. The test condition is a relational condition used to compare the value of a literal, or used to specify a relationship between two items. The appearance of a test condition in an IF statement causes the test condition to be evaluated.

There are two types of relational conditions: comparison of numeric operands and comparison of nonnumeric operands. Numeric operands are compared algebraically. For example, the value of 02 would be greater than −10, and −10 would be greater than −11. If a relational condition contains a numeric operand and a nonnumeric operand, it is treated as a nonnumeric comparison.

When a relational condition is evaluated, any unsigned value, other than zero, is considered positive. Zero is a unique value, and any preceding sign is ignored.

The second type of relational condition is the comparison of nonnumeric operands. Any comparison that includes a nonnumeric operand is a nonnumeric comparison. Nonnumeric comparisons are made in respect to the collating sequence of the particular computer.*

If the nonnumeric operands of a relational condition are the same, the characters in corresponding positions are compared from left to right. If an unequal pair of characters is encountered, the operand containing the character higher in the collating sequence is considered to be the greater operand.

When a relational condition with nonnumeric operands of unequal size is evaluated, the shorter operand is padded with blanks on the right to make it equal in length to the longer operand. The operands are then compared in the normal way beginning with the leftmost characters.

The logical operator NOT can be used with any relational operator. It must always be preceded by and followed by a space.

The phrases LESS THAN, EQUAL TO, and GREATER THAN are used in conditions in a COBOL statement to express a relationship between two variables or between a variable and a numeric literal.

*The *collating sequence* is the arrangement of all valid characters in the order of their relative precedence. The collating sequence of a computer is part of the computer design—each acceptable character has a predetermined place in the sequence. A collating sequence is used primarily in comparison operations.

First Operand		Second Operand				
		Group	Elementary			
			Alphanumeric	Alphabetic	Numeric	Literal
Group		C	C	C	C	C
Elementary	Alphanumeric	C	C	C	C	C
	Alphabetic	C	C	C	I	C
	Numeric	C	C	I	N	C
	Literal	C	C	C	C	I

C Compared logically (one character at a time, according to collating sequence of a particular computer)
N Compared algebraically (numeric values are compared)
I Invalid comparison

Example:

IF TOTAL GREATER THAN MAXIMUM PERFORM MESSAGE.

first operand (TOTAL) — operator (GREATER THAN) — second operand (MAXIMUM); first operand, operator, and second operand together form the condition

Explanation:

To use this chart, find the data type (determined by the picture) of the first operand in the column headed First Operand. Then find the data type of the second operand across the top of the figure opposite Second Operand. Extend imaginary lines into the figure from the data types of the first and second operands. In the block where these two lines intersect is a letter that tells you how the values are compared.

Figure 8.10 Types of valid comparisons.

A Relation Condition test involves the comparison of two data values, either of which may be an identifier, a literal, or an arithmetic expression. Either the relational operator symbol or relational operator may be used in the test. If the symbols are used, they must be preceded and followed by a space. NOT is used to specify the opposite of the expression.

Rules Governing Relation Condition Tests

1. The first operand is called the subject of the condition; the second operand is called the object of the condition. The subject and object may not both be literals.
2. Both operands must have the same USAGE, except when two numeric operands are involved. (See figure 8.10.)
3. Relational expression operators are used in relation conditions. The operators may be used in place of their names.

Comparison of Numeric Operands

For numeric operands, a comparison results in the determination that the algebraic value of one of the operands is less than, equal to or greater than the other. The operand length, in terms of the number of digits, is not significant. Zero is considered to represent a unique value regardless of the length, sign, or implied decimal point location.

Comparison of these operands is permitted regardless of the manner in which their usage is described. Unsigned numeric operands are considered to be positive for comparison purposes.

Numeric comparison is done *only* when a numeric item is compared to a numeric item. Differing usage of numeric items does not prevent a comparison of their values. This means that a binary item can be compared to a packed decimal item.

When usages are different, however, the computer will be instructed to convert one item or the other to make the usages the same. So, when a binary and packed decimal items are compared, the binary item is converted to packed decimal before the comparing begins.

This action is the same as the action of preparing data values for arithmetic operations. Numeric comparison is the same as arithmetic in another respect, namely, the computer can compare two packed decimals or it can compare two binary numbers. Conversion is required wherever the uses of the items are different, and whenever the external decimal items are compared.

Operand 1	Operand 2	Result of Comparison
1 2 5 ‸ 5 0	1 2 ‸ 4 0 1 0	125.50 is greater than 12.4010
1 5 1	5 1 ‸ 2 5	151 is greater than 51.25
1 0 0 0 ‸ 0	1 0 0 0 ‸ 1	1000.0 is less than 1000.1
1 ‸ 0 −	0 0 ‸ 0 +	−1.0 is less than +0.0
0 0 −	0 0 +	0 equals 0
3 4	0 1 2 8	34 is less than 0128
0 2 5	2 5	025 equals 25

Figure 8.11 Numeric operands comparisons—examples.

When the computer must convert data codes, additional instructions and more time is required to get the desired end result; thus the object program is less efficient. It should be quite clear that the relative efficiency of the program depends mainly on the characteristics of the data arithmetic being processed, and much less on the way a relation test or an arithmetic statement is written in the Procedure Division.

When two numbers are to be compared, the programmer must make the most efficient comparison; either make both items packed decimal, or if both items are external decimal, they both must be converted for a numeric comparison.

In numeric comparisons, the assumed decimal points of the values are aligned, and extra positions at either end of either value are filled with zeros. The end result is that the sizes of the values are made the same. This action is done in work areas set up by the compiler.

The actual comparison is based on two things: the sign of the number, and the magnitude of the number. Numbers that contain plus signs or no signs are considered positive; numbers that contain minus signs are negative. The value zero is a special case; zero has no magnitude, and its sign, if any, is disregarded.

Rules Governing the Numeric Items Comparison Tests

1. The test determines that the value of one of the items is GREATER THAN, LESS THAN, or EQUAL TO the other item, regardless of the length of the operands.
2. The items are algebraically compared after decimal point alignment.
3. Zero is considered a unique value regardless of its length, sign, or implied decimal-point location.
4. Numeric operands that do not have signs are considered positive values for purposes of comparison.
5. Comparison of numeric operands is permitted regardless of the manner in which their USAGE is described.

(See figure 8.11.)

Comparison of Nonnumeric Operands

For two nonnumeric operands, or one numeric (excluding the operational sign) and one nonnumeric operand, a comparison results in the determination that one of the operands is less than, equal to, or greater than the other with respect to a specified collating sequence of characters. The collating sequence of characters will depend upon the computer used. If one of the operands is specified as numeric, it must be an integer data item or an integer literal. Numeric and nonnumeric operands may be compared only when their usage is the same, explicitly or implicitly.

If the operands are of equal size, then characters in corresponding character positions are compared starting from the high-order end and continuing until either a pair of unequal characters is encountered or the low-order end of the item is reached, whichever comes first. The items are determined to be equal when the low-order end is reached.

The first encountered pair of unequal characters is compared for relative location in the collating sequence. The operand containing that character which is positioned higher in the collating sequence is determined to be the greater operand.

If the operands are of unequal size, comparison proceeds as though the shorter operand was extended on the right by sufficient spaces to make the operands equal. If this process exhausts the characters of the operand of the lesser size, then the operand of the lesser size is less than the operand of the larger size, unless the remainder of the operand of the larger size consists solely of spaces, in which case the two operands are equal.

Alphameric comparison is done in the same way that you might put words into alphabetical order. Values are compared character by character, proceeding from left to right.

Comparing continues until two characters are found that are not the same, or until the end of the items are reached. Characters are compared on the basis of the collating sequence of the particular computer.

It may be necessary to compare items that have different lengths. The rule is that the shorter item is thought of as being filled with blanks to the length of the longer item.

This rule is another way of saying that the comparison does not necessarily stop when the computer gets to the end of the shorter item. If all of the characters have been equal up to that point, the computer will look at the remaining characters, if any, in the longer item and compare them with blanks. So, if the remaining positions of the longer item contain blanks, the items are equal; but if the remaining positions contain any characters, the longer item has the greater value.

Rules Governing the Nonnumeric Items Comparison Tests

1. The test determines that one item is GREATER THAN, LESS THAN, or EQUAL TO the other item with respect to the specified collating sequence of characters for the particular computer. (In the IBM collating sequence, the numerals are in the highest category, followed by alphabetic characters, with the special characters the lowest of the group.)
2. Numeric and nonnumeric operands may be compared only if both items have the same USAGE, implicitly or explicitly.
3. The size of the operand is the total number of characters in the operand. All group items are considered in the nonnumeric operand group.
4. If both operands are of equal length in a nonnumeric comparison, the test proceeds from left (high-order position) to right (low-order position), and each character is compared to the corresponding character of the other item. The comparison of characters continues until an unequal condition is noted.

   If each individual character compared results in an equality, and the two items consist of the same number of characters, the items are considered equal.
5. If the operands are of differing lengths, the comparison proceeds as if the shorter item was filled with spaces until it is of the same length as the other operands.

(See figures 8.12 and 8.13.)

Operand 1	Operand 2	Result of Comparison
A B 9 5 4	A B 9 5 4	AB954 is equal to AB954
A 4 5 0	9 4 5 0	A450 is less than 9450
9 5 0	9 5 J	950 is greater than 95J
A B C D	S T U V	ABCD is less than STUV
M N O P Q	A B C	MNOPQ is greater than ABC
A B C	M N O P Q	ABC is less than MNOPQ
J B H N A L C A R N	J B H N A L C A N	JBHNALCARN is greater than JBHNALCAN
J K H N ƀ ƀ A L	J K H N ƀ ƀ A M	JKHNƀƀAL is less than KJHNƀƀAM
D E F	D E F G	DEF is less than DEFG
D E F	D E F ƀ	DEF is equal to DEFƀ

ƀ denotes blank character.

Figure 8.12 Nonnumeric operands comparisons— examples.

```
IF AMOUNT IS LESS THAN BALANCE
    MOVE SHIPMENT TO WORKSTORE
ELSE
    NEXT SENTENCE.
```

The MOVE is not executed on an "equal" or "greater" condition.

```
IF DATE-IN OF MASTER IS EQUAL TO TODAYS-DATE,
    PERFORM REVIEW
ELSE
    PERFORM NEXT-DETAIL.
```

On a "less" or "greater" condition the program performs NEXT- DETAIL.

Figure 8.13 Relational condition—example.

IF identifier is [NOT] { NUMERIC / ALPHABETIC }

Figure 8.14 Format—class condition.

Class Condition

The class condition determines whether the operand is numeric, that is, consists entirely of the characters 0, 1, 2, 3, . . . 9, with or without the operational sign, or alphabetic, that is, consists entirely of the characters A, B, C, . . . Z, space. The general format for the class conditions is shown in figure 8.14.

When used, NOT and the next key word specify one class condition that defines the class test to be executed for truth value; e.g., NOT NUMERIC is a truth test for determining that an operand is nonnumeric.

A possible use for class tests is to check the validity of certain data items. For example, it might be desired to determine whether an item that is supposed to contain numeric information actually contains digits. (Such an item would have a picture identifying it as a numeric item, but there is no automatic checking procedure to verify that data put into an item during the running of a program corresponds to the item's picture.)

Any item may be tested to determine whether its current content is either numeric or alphabetic. The usage of the item must be implicitly or explicitly DISPLAY. The ALPHABETIC test cannot be used with an item whose data description describes the item as numeric. The item being tested is determined to be alphabetic only if the contents consist of the alphabetic characters A through Z and the space.

The NUMERIC test cannot be used with an item whose data description describes the item as alphabetic or as a group item composed of elementary items whose data description indicates the presence of operational sign(s). If the data description of the item being tested does not indicate the presence of an operational sign, the item being tested is determined to be numeric only if the contents are numeric and an operational sign is not present.

Rules Governing the Use of the Class Condition

1. The Class Condition test is used to determine whether the data is numeric or alphabetic (fig 8.15).
2. *Numeric data* consists entirely of the numerals 0–9 with or without an operational sign. If the PICTURE clause of the record description of the identifier being tested does not contain an operational sign, the identifier is determined to be numeric only if the contents are numeric and an operational sign is not present.
3. A numeric test cannot be used with an item whose data descriptions describe items as alphabetic.
4. *Alphabetic data* consists of the characters A through Z plus the space character.

Type of Identifier	Valid Forms of the Class Test	
Alphabetic	ALPHABETIC	NOT ALPHABETIC
Alphanumeric	ALPHABETIC NUMERIC	NOT ALPHABETIC NOT NUMERIC
External-decimal	NUMERIC	NOT NUMERIC

Figure 8.15 Valid forms of class tests.

5. An alphabetic test cannot be used with an item whose record description is numeric. The following is an example of a class test:

```
IF ACTIVITY-RATING IS ALPHABETIC
    PERFORM HIGH-ACTIVITY-ANALYSIS
ELSE
    NEXT SENTENCE.
```

The following are examples of the class condition:

```
05  QUANTITY PICTURE S9999.
IF QUANTITY IS NUMERIC
    PERFORM PROCESS-1
ELSE
    PERFORM PROCESS-2.
```

In the example just presented, this class condition tests each character in QUANTITY for a numeric value and the rightmost character for a zone sign.

```
05  TCODE      PICTURE A.
IF TCODE IS ALPHABETIC
    MOVE MASTER TO NEW-MASTER
ELSE
    NEXT SENTENCE.
```

In the aforementioned, this class condition tests each character in TCODE for a letter of the alphabet or a space character.

Sign Condition

The test-condition sign determines whether or not the algebraic value of an arithmetic expression is less than, greater than, or equal to zero. (See figure 8.16.)

An operand is POSITIVE only if its value is greater than zero, NEGATIVE if its value is less than zero, and ZERO if its value is equal to zero. An operand whose value is zero is NOT POSITIVE and an operand whose value is zero is NOT NEGATIVE. The value zero is considered neither positive nor negative.

When used, NOT and the next key word specify one sign condition that defines the algebraic test to be executed for truth value; e.g., NOT ZERO is a truth test for a nonzero (positive or negative) value. (See figure 8.17.)

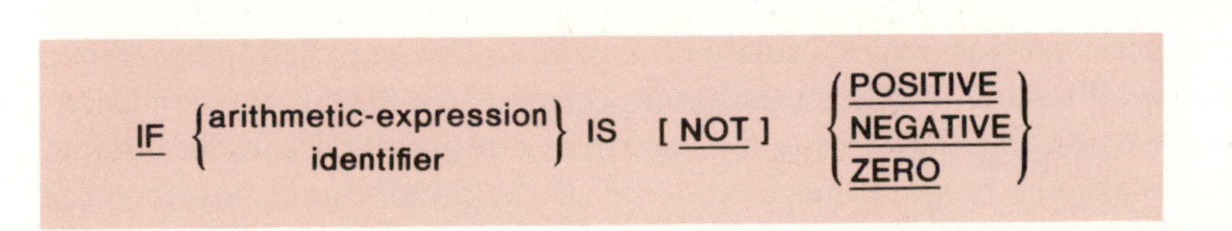

Figure 8.16 Format—sign condition.

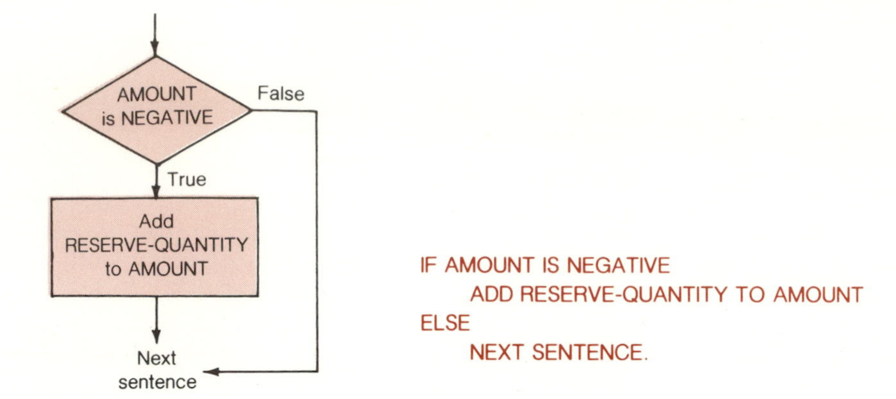
AMOUNT
is NEGATIVE
False
True
Add
RESERVE-QUANTITY
to AMOUNT
Next
sentence
IF AMOUNT IS NEGATIVE
ADD RESERVE-QUANTITY TO AMOUNT
ELSE
NEXT SENTENCE.

Figure 8.17 Sign test—example.

The following are examples of the sign condition:

```
IF AMOUNT IS NOT ZERO
    MOVE NEW-AMOUNT TO AMOUNT IN MASTER-RECORD
ELSE
    NEXT SENTENCE.
```

In the aforementioned, the sign condition tests the field AMOUNT for a nonzero value.

```
IF A * B IS POSITIVE
    PERFORM COMPUTATION
ELSE
    PERFORM ERROR-PROCEDURE.
```

In the aforementioned, the sign condition tests the product of A times B for a positive value.

```
IF BALANCE IS NEGATIVE
    PERFORM OVERDRAFT-PROCEDURE
ELSE
    PERFORM REGULAR-PROCEDURE.
```

In the aforementioned, the sign condition tests the field BALANCE for a negative value.

Rules Governing the Use of the Sign Condition

1. The value of zero is considered neither positive nor negative.
2. If an operand appears in a Sign Condition test, it must represent a numeric value. If the value is unsigned and not equal to zero, it is considered to be positive.

Condition-Name Condition

The relational condition is a type of condition. Another type of test condition that the programmer will find useful is the condition-name condition. In general, the condition-name test-condition is used in an IF statement that tests whether a variable has a specific value or one of a specific set of values.

When a condition-name test-condition is to be used in place of a relational condition, the condition names must be described in the Data Division, assigned specific values, and associated with their related variable.

In a condition-name test, the test-condition consists solely of a programmer-supplied name. Since a condition-name test is another way of testing whether a data item is equal to a literal, any condition-name test could be replaced by a relation test. The reason a programmer might use a condition-name test instead of a relation test is to make the program more readable. If the programmer has done the job correctly, the name of the condition will explain the meaning of the condition. A condition-name is defined in a level number 88 entry in the Data Division. Level-88 entries follow immediately after the

IF [NOT] condition-name

Figure 8.18 Format—condition-name test.

description of the entry to which they apply. There may, however, be more than one level-88 entry for an item.

Whenever a programmer wants to test whether the value of a data item is equal to a literal, he or she has the choice of using either a relation test or a condition-name test. The programmer's decision as to which to use would be based in part on the readability of the relation test. A relation test such as IF TEST-SCORE = 100 . . ., tells just as much as IF PERFECT-TEST-SCORE . . ., so the programmer would use the relation test. On the other hand, if the programmer had the choice of writing IF MARITAL-STATUS = 7 . . ., or IF DIVORCED . . ., he or she would certainly write IF DIVORCED. . . . In this instance then the preferred test is the condition-name test.

In a condition-name test-condition, a conditional variable is tested to determine whether or not its value is equal to one of the values associated with a condition-name. The general format for the condition-name test appears in figure 8.18.

Rules Governing the Use of a Condition-name Condition

1. The condition-name must be defined in a level-88 entry in the Data Division associated with the condition-name.
2. The condition-name condition test is an alternate way of expressing certain conditions that could be expressed by a simple relational condition. The rules for comparing a condition variable with a condition-name value are the same as those specified for relation conditions.
3. The test is true if the value corresponding to the condition-name equals the value of its associated condition variables.

(See figure 8.19.)

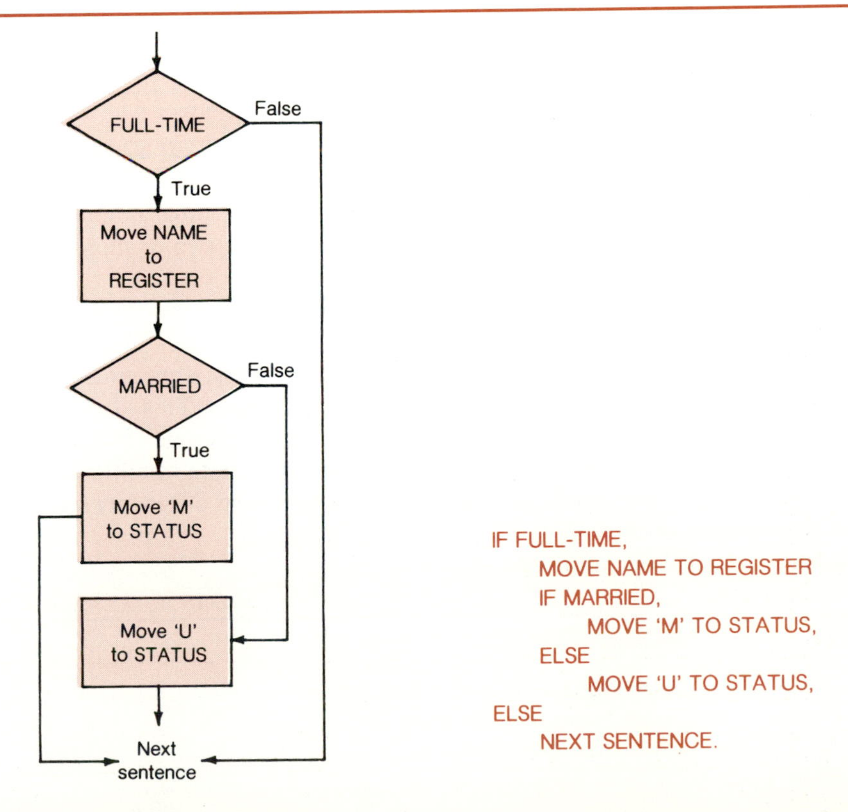


```
IF FULL-TIME,
      MOVE NAME TO REGISTER
      IF MARRIED,
            MOVE 'M' TO STATUS,
      ELSE
            MOVE 'U' TO STATUS,
ELSE
      NEXT SENTENCE.
```

Figure 8.19 Condition-name test—example.

Compound Conditional Expressions

Part of a programmer's task is to reduce a procedure to its most simple and efficient sequence. It is less efficient, of course, to ask the computer to make three tests when one test would supply the solution.

A compound condition is formed by combining simple conditions, combined conditions, and/or compound conditions with logical connectors (logical operators AND and OR) or negating these conditions with logical negation (the logical operator NOT). The truth value of a compound condition, whether parenthesized or not, is that truth value which results from the interaction of all stated logical operators, or the individual truth values of simple conditions, or the intermediate truth values of conditions logically connected or logically negated.

The logical operators and their meanings are:

Logical Operators	Meaning
AND	Logical conjunction; the truth value is "true" if both of the conjoined are true; "false" if one or both of the conjoined conditions is false.
OR	Logical inclusive OR; the truth value is "true" if one or both of the included conditions is true; "false" if both included conditions are false.
NOT	Logical negation or reversal of truth value; the true value is "true" if the condition is false; "false" if the condition is true.

The logical operators must be preceded by a space and followed by a space.

Negated Simple Conditions

A simple condition is negated through the use of the logical operator NOT. The negated simple condition affects the opposite truth value for a simple condition. Thus the truth value of negated simple condition is "true" if and only if the truth value of the simple condition is false; the truth value is "false" if and only if the truth value of the simple condition is true. The inclusion in parentheses of a negated simple condition does not change the truth value.

A compound condition is produced simply by tying together two or more test conditions with the words AND or OR. (The word NOT can be used as usual to give the opposite meaning to a condition.)

The word OR means that the compound condition is true if *either* of the test conditions joined by the logical operator OR is true.

The word AND means that the compound condition is only true if *all* of the test conditions joined by the logical operator AND are true.

Although compound conditions will often be very useful to the programmer in programming, it should be kept in mind that they will cause the computer to make two or more tests—and the programmer may want to avoid making two tests where one will do.

An IF statement causes a relational condition to be evaluated and the appropriate path of control to be selected based on whether the relational condition is true or false. The programmer might be required to write a program in which the basis for the selection of the appropriate path of control will depend on a set of circumstances instead of a single relational condition. This situation can be accounted for by using a series of IF statements, or a compound condition in a single IF statement.

A compound Conditional Expression consists of two or more simple conditions combined with logical operators AND and OR. These conditions are linked by AND and OR in any sequence which would produce the overall desired result.

Following are the logical operators and their meanings:

Logical Operator	Meaning
OR	Logical inclusive (either or both are true)
AND	Logical conjunction (both are true)
NOT	Logical negation (not true)

The logical operators must be preceded by a space and followed by a space.

Type of Operation	Operator (Operation Symbol)	Operation
Relational	IS GREATER THAN (>)	Is greater than
	IS LESS THAN (<)	Is less than
	IS EQUAL TO (=)	Is equal to
Logical	OR	Logical inclusive OR (either or both are true)
	AND	Logical conjunction (both are true)
	NOT	Logical negation

Figure 8.20 Relational and logical operators.

Rules for the Formation of Compound Conditional Expressions

1. Two or more simple conditions combined by AND/OR make up a compound condition. (See figure 8.20.)
2. The word OR is used to mean either or both. Thus the expression A OR B is true if A is true or B is true or both A and B are true.
3. The word AND is used to mean that both expressions must be true. Thus the expression A AND B is true only if both A and B are true.
4. The word NOT may be used to specify the opposite of the compound expression. Thus NOT A AND B is true if A is false and B is true.
5. Parentheses may be used to specify the sequence in which the conditions are to be evaluated. Parentheses must always appear as a pair. Logical evaluation begins with the innermost pair of parentheses and proceeds to the outermost pairs.
6. Three logical operators (AND, OR, or NOT) are used to combine simple statements in the same expression for the purpose of testing the condition of the expression.
7. If the sequence of evaluation is not specified by parentheses, the expression is evaluated in the following manner.
   a. Arithmetic expressions.
   b. Relational operators.
   c. [NOT] conditions.
   d. Conditions surrounding all ANDs are evaluated first, starting at the left and proceeding to the right.
   e. OR and its surrounding conditions are then evaluated, also proceeding from left to right.

(See figure 8.21 A, B, C, and D.)

Thus the expression A IS GREATER THAN B OR A IS EQUAL TO C AND D IS POSITIVE would be evaluated as if it were parenthesized as follows:

(A IS GREATER THAN B) OR (A IS EQUAL TO C) AND (D IS POSITIVE).

(See figure 8.22.)

Nested Conditional Expressions

Another kind of conditional sentence that is permitted in COBOL is one containing "nested" IF statements. The general idea of "nested" IFs is that one or more IFs appear within a sentence that begins with IF. It is thereby possible to make a series of decisions based on the outcome of previous decisions, and to take different courses of action, all within one sentence.

The rule about IFs that contain ELSE applies to nested IFs. The rule is: When control comes to ELSE, it goes to the next sentence. Thus, if the programmer wants control to leave a nested sequence and go to the next sentence, simply write ELSE followed by whatever statements are to be carried out on a "false" condition.

Figure 8.21 Compound conditional statements—examples.

A

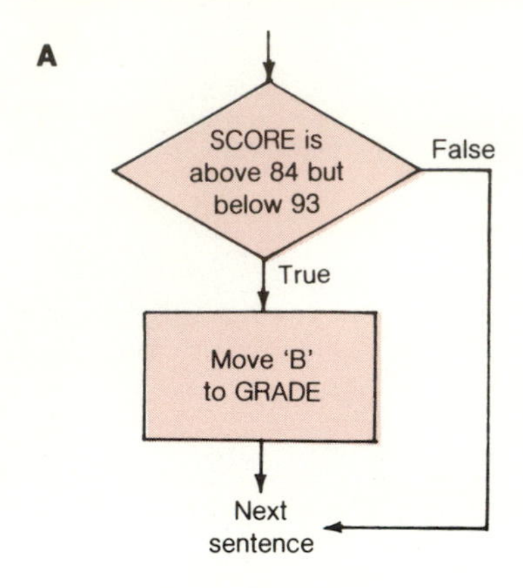


```
IF SCORE IS GREATER THAN 84
     AND SCORE IS LESS THAN 93,
     MOVE 'B' TO GRADE
ELSE
     NEXT SENTENCE.
```

B

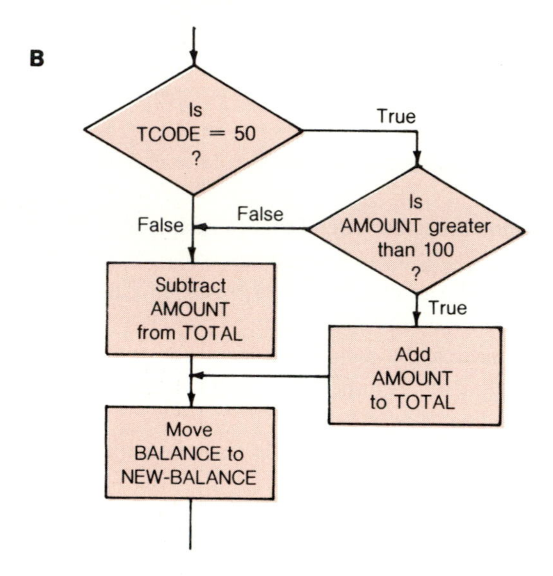

If AND is the only logical connective, then the compound condition is true only if each simple condition is true.

```
IF TCODE IS EQUAL TO 50
     AND AMOUNT IS GREATER THAN 100,
     ADD AMOUNT TO TOTAL,
ELSE
     SUBTRACT AMOUNT FROM TOTAL.
MOVE BALANCE TO NEW-BALANCE.
```

C

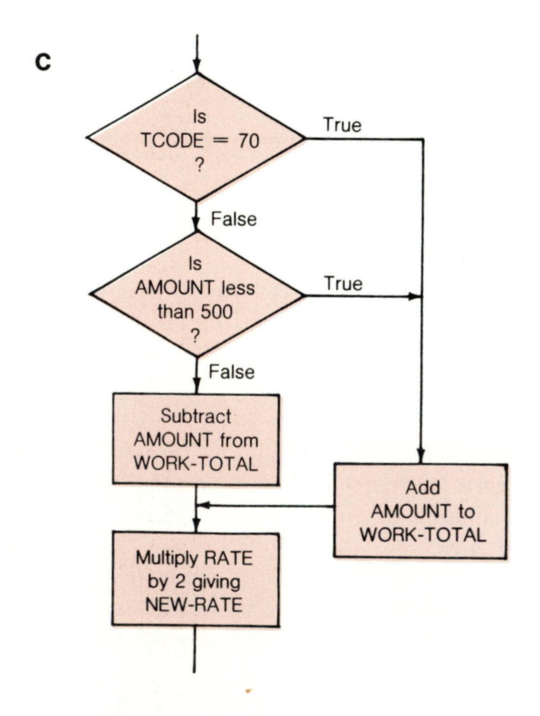

If OR is the only logical connective, then the compound condition is true if at least one of the simple conditions is true.

```
IF TCODE IS EQUAL TO 70,
     OR AMOUNT IS LESS THAN 500,
     ADD AMOUNT TO WORK-TOTAL,
ELSE
     SUBTRACT AMOUNT FROM WORK-TOTAL.
MULTIPLY RATE BY 2 GIVING NEW-RATE.
```

Figure 8.21 *(continued)*

D

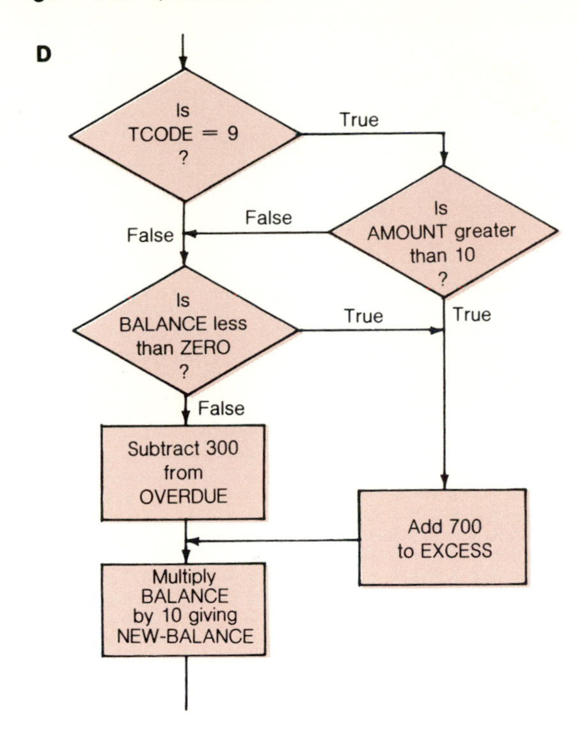

If both AND and OR are used, the conditions are evaluated proceeding left to right with each AND connective referring to the simple conditions before and after the word AND.

```
IF TCODE = 9,
    AND AMOUNT IS GREATER THAN 10,
    OR BALANCE IS LESS THAN ZERO,
    ADD 700 TO EXCESS,
ELSE
    SUBTRACT 300 FROM OVERDUE.
MULTIPLY BALANCE BY 10 GIVING NEW-BALANCE.
```

If condition A is:	If condition B is:	Then A AND B is:	Then A OR B is:	Then NOT A is:
True	True	True	True	False
True	False	False	True	False
False	True	False	True	True
False	False	False	False	True

The table summarizes the true or false value resulting from conditions having the logical connective AND, the logical connective OR, and the logical operator NOT.

Figure 8.22 AND/OR value table.

Naturally with nested IFs there will be more than one false condition. The rule here is: Control flows to the statement after ELSE from the last previous false condition that has not already been paired with an ELSE. This means that the ELSE statements for the first IF will come after the ELSE statements for the second IF.

Nested IFs can become devilishly complicated. Such statements are almost impossible for a reader to understand, and even harder for another programmer to change when program maintenance becomes necessary. What's more, even the original programmer is bound to have trouble debugging them if the results are not correct.

Any decisions made by nested IFs can also be made using common, ordinary, unnested IFs. And sometimes it is better to do exactly that.

A good rule is to keep the COBOL statements simple if possible. But if the program can't keep them simple, document them completely—with explanations of the processing they do (possible comments in the Procedure Division), and with flowcharts of the logic of the process.

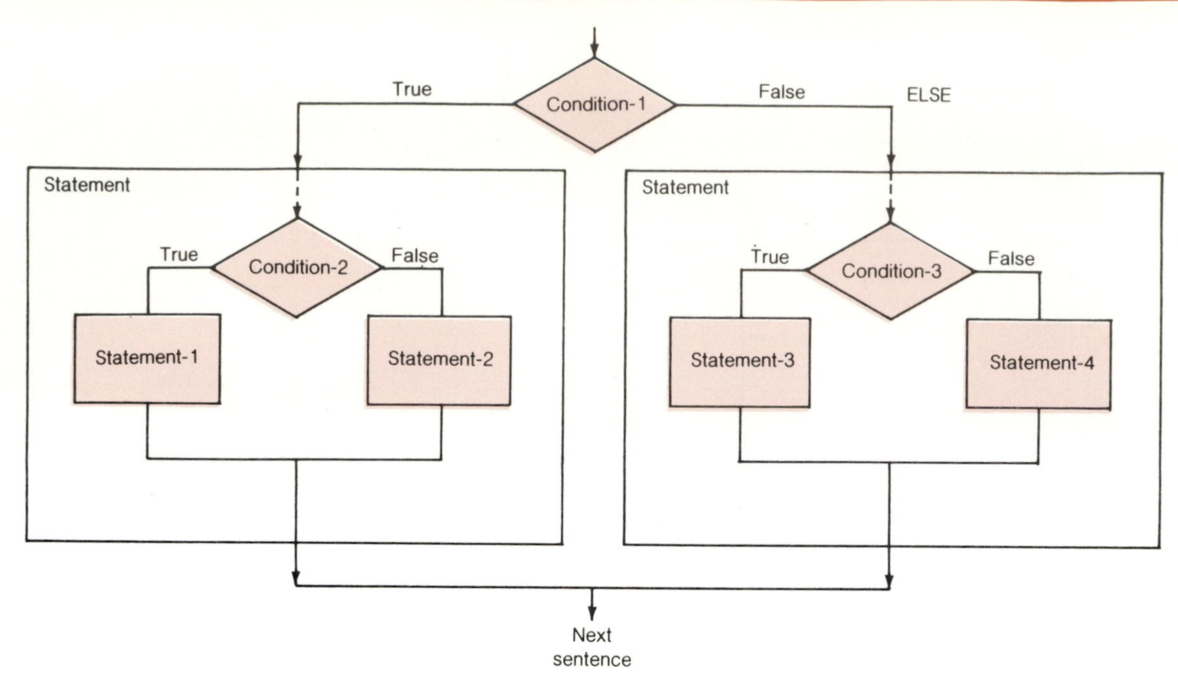

Figure 8.23 Flow structure.

Nested IFs establish more than one level of procedure control in a single sentence. Since each IF statement in the nesting implies a choice in the flow of control, a nested IF statement leads to two mutually exclusive flows of control. For example, the balanced nested IF statement follows:

```
IF    condition-1
      IF    condition-2
            Statement-1
      ELSE
            Statement-2
ELSE
      IF    condition-3
            Statement-3
      ELSE
            Statement-4.
```

The flow structure for this statement is in figure 8.23.

The inherent problem in programming such a statement and its resultant structure is not its logical complexity per se, but rather the rate at which the number of alternatives increases. A point is eventually reached at which the ability to mentally manipulate mutually exclusive alternatives is saturated. Confusion can be avoided if, and only if, each and every true path and false path is strictly stated. This is especially true whenever the ELSE NEXT SENTENCE might have been omitted. The temptation to remove the ELSE NEXT SENTENCE phrase should be resisted, even though the language rules permit such removal. The temptation is especially strong in a statement of the type that follows:

```
IF    condition-1
      IF    condition-2
            statement-1
      ELSE
            statement-2
ELSE
      NEXT SENTENCE.
```

In this case, the ELSE NEXT SENTENCE phrase serves as a definitive closure for the outermost IF statement. If these phrases are removed as redundant, it will be easy to lose control of the logic when attempting subsequent modifications.

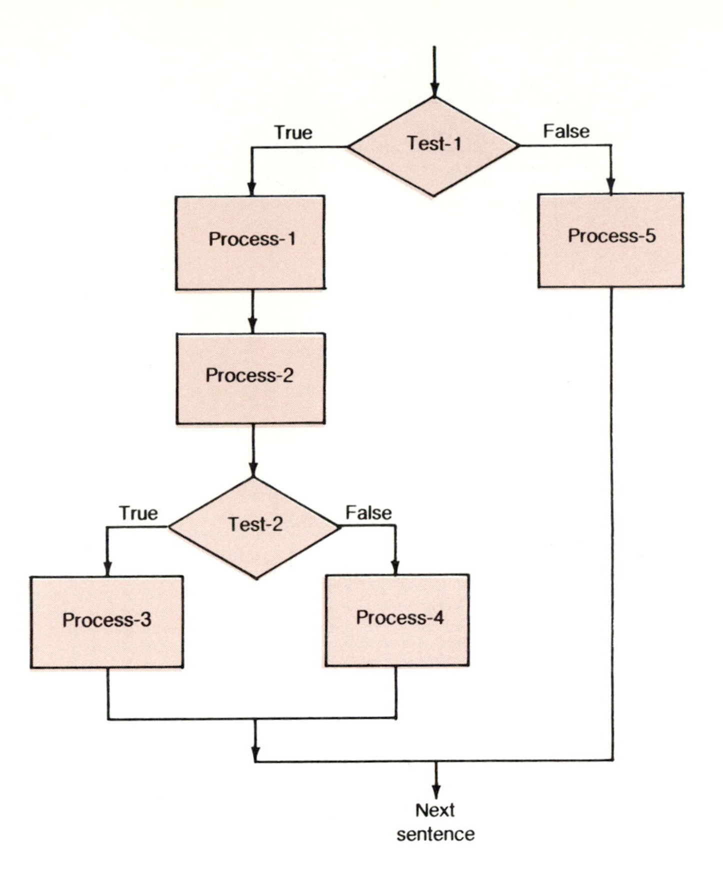

Figure 8.24 Flow structure.

When constructing IF statements, it must be remembered that each path in the logic must be composed of mutually exclusive procedures. For example, it is possible to write a valid nested IF statement for the following:

```
IF    condition-1
      process-1
      process-2
      IF    condition-2
            process-3
      ELSE
            process-4
ELSE
      process-5.
```

The flow structure for this statement is in figure 8.24.

It is not always possible, however, to make a straight-forward transformation of a flow diagram into an IF statement. The diagram cannot be represented as a single sentence without artificially introducing a redundant statement for process-4. One can get as far as

```
IF    condition-1
      process-1
      IF    condition-2
            process-2.
```

At this point, one is unable to affect a transfer of control *within* the sentence to the beginning of process-4. Since both process-3 and process-4 may consist of several statements, there is no way for the compiler

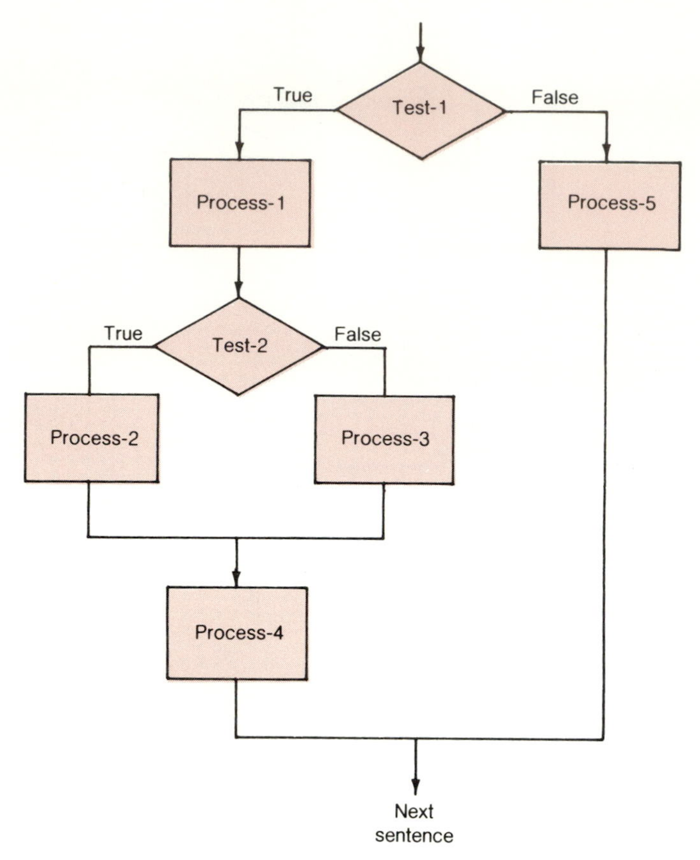

Figure 8.25 Flow structure.

to differentiate between them in the ELSE phrase. Either the flow must be recast into several or redundancy must be introduced, such as

```
IF    condition-1
      process-1
      IF    condition-2
            process-2
            process-4
      ELSE
            process-3
            process-4
ELSE
      process-5.
```

This flow structure is shown in figure 8.25.

If a conditional statement appears as statement-1 or as part of statement-1, it is said to be nested. Nesting a statement is like specifying a subordinate arithmetic expression enclosed in parentheses combined in a larger arithmetic expression. IF statements contained within are considered paired, with IF and ELSE combinations proceeding from left to right. Thus, any ELSE statement encountered must be considered to apply to the immediately preceding IF, if it has not already been paired with an ELSE.

Certain compilers may place some restrictions on the number and types of conditionals that can be nested. (See figure 8.26 A and B.)

(Note: *Selection* is a structured program theorem that is implemented by the IFTHENELSE structure. In COBOL, it is implemented with the IF statement and a stated condition is tested. Nested IFs and compound conditions for nested IFs are also used in structured programming.)

A

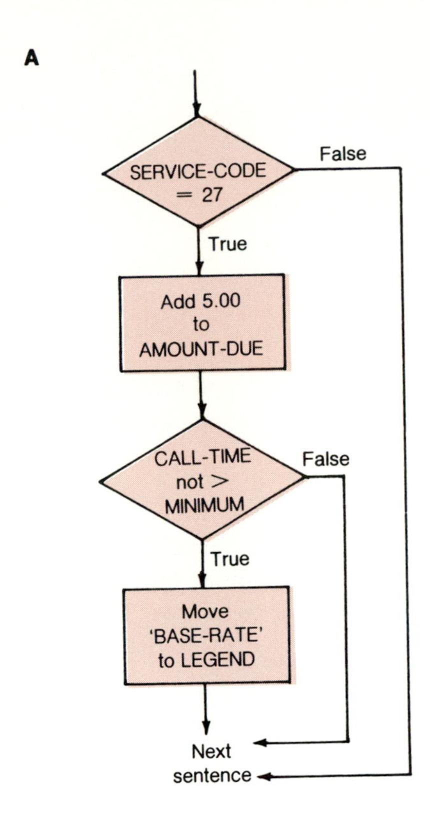


```
IF SERVICE-CODE IS EQUAL TO 27,
    ADD 5.00 TO AMOUNT-DUE
    IF CALL-TIME IS NOT GREATER THAN MINIMUM
        MOVE 'BASE RATE' TO LEGEND
    ELSE
        NEXT SENTENCE
ELSE
    NEXT SENTENCE.
```

B

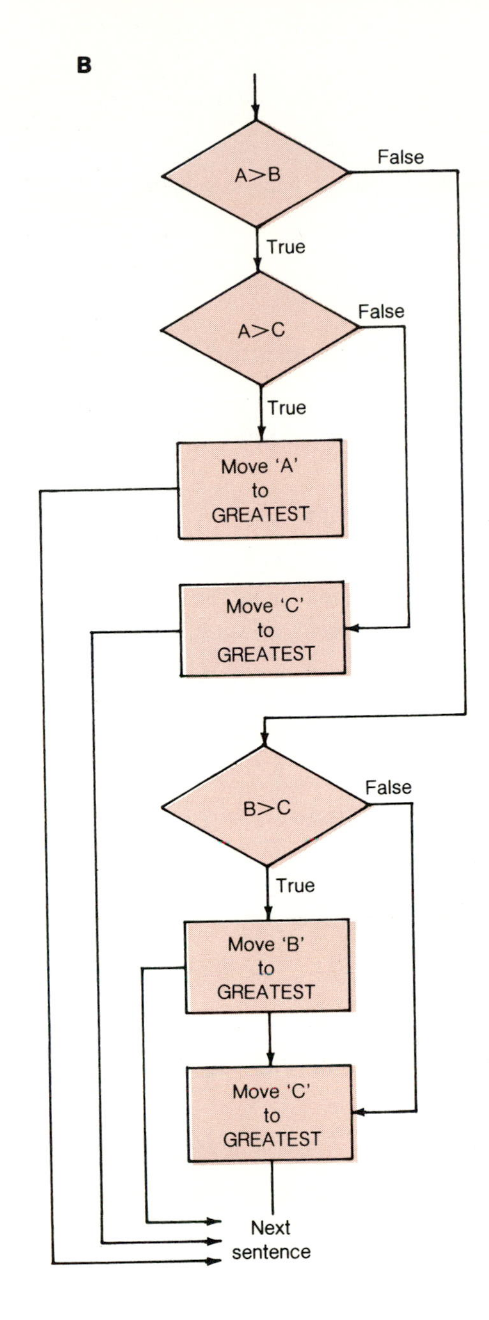


```
IF A IS GREATER THAN B,
    IF A IS GREATER THAN C,
        MOVE A TO GREATEST
    ELSE
        MOVE C TO GREATEST
ELSE
    IF B IS GREATER THAN C,
        MOVE B TO GREATEST
    ELSE
        MOVE C TO GREATEST.
```

Figure 8.26 Nested IF statement—examples.

. . . subject relational-operator object

{AND / OR} [NOT] relational-operator object . . .

Figure 8.27 Format: implied subject.

Abbreviated Combined Relation Conditions

Conditions involving full relation tests have three terms: a subject, a relation, and an object. COBOL allows the omission of some of these terms in certain forms of compound conditional expressions.

Implied Subjects

Many times a conditional expression will contain several simple relational conditions. (See figure 8.27.) These conditions may have the same subject. When a compound relational condition uses the same term as the subject of each relation, only the first occurrence of the subject need be written. The condition

```
AGE IS LESS THAN MAX-AGE AND AGE IS GREATER THAN 20.
```

could also be written as:

```
AGE IS LESS THAN MAX-AGE AND GREATER THAN 20.
```

As another illustration of this abbreviation, the condition

```
A EQUALS B OR EQUALS C AND IS GREATER THAN D.
```

is an abbreviation for

```
A EQUALS B OR A EQUALS C AND A IS GREATER THAN D.
```

which is equivalent to:

```
A EQUALS B OR (A EQUALS C AND A IS GREATER THAN D).
```

Rules Governing the Use of Implied Subjects

1. Only conditional expressions written as simple relational conditions may have implied subjects. SIGN and CLASS condition tests can never have implied subjects.
2. The first series of simple relational conditions must always consist of a subject, operator, and operation, all of which must be explicitly stated.
3. The subject may be implied only in a series of simple relational conditions connected by the logical operators AND and/or OR.
4. When the subject of a simple relational condition is implied, the subject used is the first subject to the left which is explicitly stated. For example, IF A = B OR =C OR D= E AND = F, A is the implied subject for C and D, while D is the implied subject of F since D is the first subject to the left of F.
5. When NOT is used in conjunction with a relational operator and an implied subject, the NOT is treated as a logical operator. For example, A IS GREATER THAN B AND NOT EQUAL TO C AND D is equivalent to A IS GREATER THAN B AND NOT A IS EQUAL TO C AND A IS EQUAL TO D.

The following is an example of an implied subject:

```
A = B OR NOT > C      (The subject, A, is implied.)
A = B OR NOT A > C    (The subject, A, is explicit.)
```

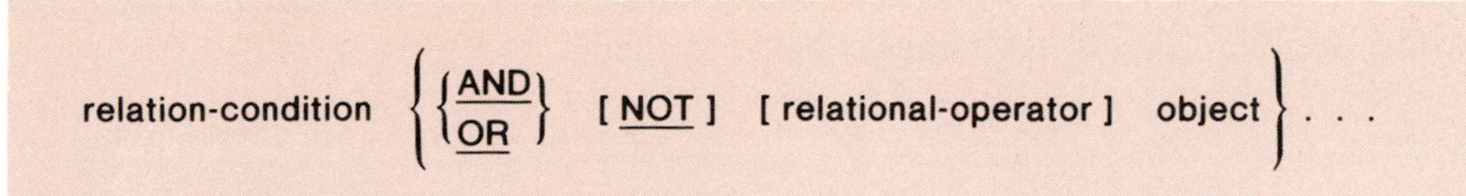

Figure 8.28 Format: implied subject and operator.

Implied Operators

Relational operators may be implied in a series of consecutive simple relational conditions in much the same way as that in which the subject can be implied. (If the operator is implied, then the subject is also assumed to be implied.) (See figure 8.28.)

When a compound relational condition uses the same subject and relation for a series of relations, only the first occurrence of the subject and the relation need be written.

This form of abbreviation is applicable regardless of the presence or absence of parentheses. For example, the condition

A = B OR C AND D

is equivalent to

A = B OR A = C AND A = D

which in turn is equivalent to

A = B OR (A = C AND A = D).

Rules Governing the Use of Implied Operators

1. A relational operator may be implied only in a simple relational condition when the subject is implied. SIGN and CLASS conditions can never be implied (do not have operators).
2. When an operator is implied, it is assumed to be the operator of the nearest completed stated simple condition to the left.

The following is an example of implied subject and operator:

A = B AND C	(Subject and relational-operator, A = , are implied.)
A = B AND A = C	(Subject and relational-operator, A = , are explicit.)

The following is an example of implied subject, and subject and operator:

A > B AND NOT < C AND D	(Subject, A, are implied in the second condition. Subject, A, and relational-operator, <, are implied in the third condition.)
A > B AND NOT A < C AND A < D	(Subject, A, and relational-operator, <, are explicit.)

Multiple Abbreviations

When relation conditions are written in consecutive sequence, any relation condition except the first may be abbreviated by (1) omitting the subject of the relation condition, or (2) omitting the subject and relation operator of the relation condition.

Within a sequence of relation conditions, both forms of abbreviations may be used. The effect of using them is as if the omitted subject were replaced by the last preceding stated subject or the omitted relational operator were replaced by the last preceding stated relational operator.

Example:

1. A = B AND > C OR D is equivalent to A = B AND A > C OR A > D.
2. A > B OR C AND < D is equivalent to A > B OR A > C AND A < D.

In using both types of abbreviations, any sequence of relation tests can occur in a sentence regardless of what verbs, keywords, or other types of tests appear between them.

Example:

```
IF    A = B
      MOVE X TO Y
ELSE
      IF    GREATER THAN C
            ADD M TO N
      ELSE
            NEXT SENTENCE.
```

is equivalent to

```
IF    A = B
      MOVE X TO Y
ELSE
      IF    A GREATER THAN C
            ADD M TO N
      ELSE
            NEXT SENTENCE.
```

Use of NOT in Abbreviations

The interpretation applied to the use of the word NOT in an abbreviated combined relation condition is as follows:

1. If the word or symbol immediately following NOT is GREATER, >, LESS, <, EQUAL, or =, then the NOT participates as part of the relation operator; otherwise
2. The NOT is interpreted as a logical operator and, therefore, the implied insertion of subject or relational operator results in a negated relation condition.

Some examples of abbreviated combined and negated combined relation conditions and expanded equivalents follow:

Abbreviated Combined Relation Condition:	Expanded Equivalent
A > B AND NOT > C OR D	((A > B) AND (A NOT > C) OR (A NOT > D)
A NOT EQUAL B OR C	(A NOT EQUAL B) OR (A NOT EQUAL C)
NOT A = B OR C	(NOT (A = B)) OR (A = C)
NOT (A GREATER B OR C)	NOT ((A GREATER B) OR (A > C))

Note: The use of implied subjects and relations should not be used if at all possible. Some programming standards speak strongly against their use as this can cause difficulty during program maintenance or modification.

GO TO, PERFORM, and STOP Statements

In addition to the IF statement, which permits different actions on the basis of the test-condition, there are other procedural words that can change the normal flow of control. They are GO TO and PERFORM, which cause branching, and STOP, which delays or halts the program.

A GO TO statement causes control to be branched unconditionally (or conditionally) to the first statement of a procedure. After a GO TO has caused control to branch to the beginning of a procedure, the normal flow of control is resumed.

GO TO and PERFORM statements cause control to flow to the beginning of a procedure, so the programmer must think in terms of *procedures* when using these statements. IF sentences, on the other hand, cause control either to flow through or to jump over certain statements. The difference is important in the logic of control flow.

A GO TO statement causes a branch to a procedure. The statement contains the name of the procedure to which a branch is desired. This name is given in the header entry of the procedure.

A GO TO statement differs from a PERFORM statement in that a PERFORM causes a branch to a procedure or a series of procedures, just as GO TO does. But after the procedure or procedures are acted upon, PERFORM causes a return branch to the statement after the PERFORM statement. (PERFORM statements are discussed in detail in chapter 6.)

The GO TO statement can be used for looping. When a GO TO statement is executed, control transfers to the given name, with execution continuing from that point.

PERFORM transfers control to the specified paragraph, executes it, then returns control to the statement following the cause of the transfer.

GO TO transfers control to the specified paragraph, executes it, then continues with the next paragraph.

When a GO TO statement with a DEPENDING ON option is executed, control is transferred to the paragraph whose position is represented by the value of the identifier. The identifier in the GO TO statement with the DEPENDING ON option must represent a positive or unsigned integer. If the integer is zero or larger than the number of paragraph names listed, the GO TO statement is ignored.

GO TO Statement

The subject of GO TO-less programming still stirs considerable controversy. Most structured program supporters claim that GO TO statements lead to difficulties in debugging, modifying, understanding, and proving programs. GO TO advocates argue that this statement, used correctly, need not lead to problems, and that it provides a natural, straightforward solution to common programming procedures.

No special effort is required to "eliminate GO TOs," which sometimes has mistakenly been considered the goal of structured programming. There are indeed good reasons for not wanting GO TOs, but no extra effort is required to "avoid" them; they just never occur when standard control logic structures are used. Naturally, if the chosen program language lacks essential control logic structures, they have to be simulated, and that does involve GO TOs. But this can be accomplished by carefully controlled means, such as the use of GO TO with the DEPENDING ON option in CASE structure problems.

There are situations in which the use of GO TOs may improve reading, compared with other ways of expressing a procedure. The GO TOs may be used to disable error routines that prevent further processing when certain types of interrupts occur. Good judgment should be used to determine whether the maintainability of the program is improved by using GO TOs in the particular situation.

The GO TO statement provides a means of transferring control conditionally or unconditionally from one part of the program to another. (The first of two general formats for the GO TO statement appears in figure 8.29.) The PERFORM statement also causes a branch out of normal sequence, but in addition provides a return to the program.

Rules Governing the Use of GO TO Statements

Unconditional

1. A procedure-name (name of paragraph or section) in the Procedure Division must follow the GO TO statement.
2. If the procedure-name is omitted, a paragraph-name must be assigned. The paragraph-name must be the only name in a paragraph, and must be modified by an ALTER statement prior to the execution of the GO TO statement.

Format 1:

```
GO TO procedure-name-1
```

Figure 8.29 GO TO statement: format 1.

3. If the procedure-name is omitted, and the GO TO sentence is not preset by an ALTER statement prior to its execution, erroneous processing will occur.
4. The GO TO statement can be used only as the final sentence in the sequence in which it appears.
5. A procedure-name may be the name of the procedure of which the GO TO statement is part. It is permissible to branch from a point in the paragraph back to the beginning of the procedure.

The following are GO TO statements—format-1—examples:

```
GO TO DETERMINE-TYPE-RECORD.
GO TO DEV-MR.
```

The CASE Program Structure

The CASE structure is used to select one of a set of functions for execution depending on the value of the integer-identifier whose range is from 1 to the number of procedure-names listed in the statement. For any value outside of the defined range, the statement is ignored. The use of the CASE structure is recommended when there are more than three (or any other number decided on for use with the IFTHENELSE statement) conditions. Three or less conditions are usually handled with nested IF statements.

COBOL offers no direct implementation of the CASE structure, but it may be simulated by using a combination of PERFORM and the GO TO statements with the DEPENDING ON option. The PERFORM statement must be written with the THRU option because the paragraphs that carry out the individual functions all end with a GO TO-paragraph-EXIT. Without the THRU option, the EXIT paragraph would not be branched correctly.

Where the "processing statements" are shown, it would also be quite possible to have PERFORM statements invoking out-of-line segments to do the processing.

The flowchart for the CASE structure is shown in figure 8.30.

The CASE statement is not part of conventional COBOL. It must be simulated as follows:

```
    GO TO case-1-name, case-2-name, . . . case-m-name
        DEPENDING ON integer-identifier.
case-1-name.
    code for F1
    GO TO case-end-name.
case-2-name.
    code for F2
    GO TO case-end-name.
case-m-name.
    code for Fm
    GO TO case-end-name.
case-end-name.
    EXIT.
```

Furthermore, the paragraph implementing the CASE statement must be placed out-of-line and executed via an inline PERFORM such as the following:

```
PERFORM case-paragraph THRU case-end-name.
```

The CASE statement cannot be implemented in COBOL without using the GO TO statement, but the problems associated with the GO TO statement are minimized by using it only in a highly controlled way. The implementation of the CASE structure is assured that transfer will be limited to paragraphs within this structure, and the processing paragraphs all transfer to the CASE EXIT. The problems associated with "skipping around" while reading a program are thereby avoided. (See figure 8.31.)

The CASE Structure

The COBOL simulation of the CASE structure is shown in the following example:

1. If the names of the paragraphs to be executed do not fit on one line, indent each continued line.
2. The DEPENDING ON clause will be coded on a separate line and will be indented.

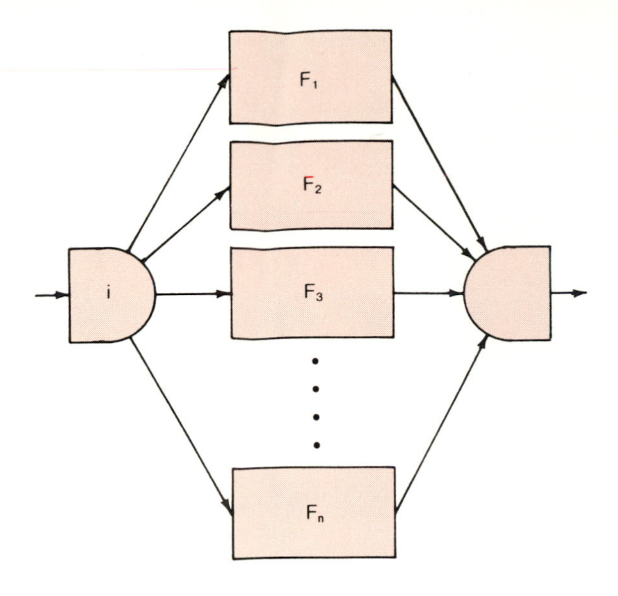

Figure 8.30 Flowchart for CASE program structure.

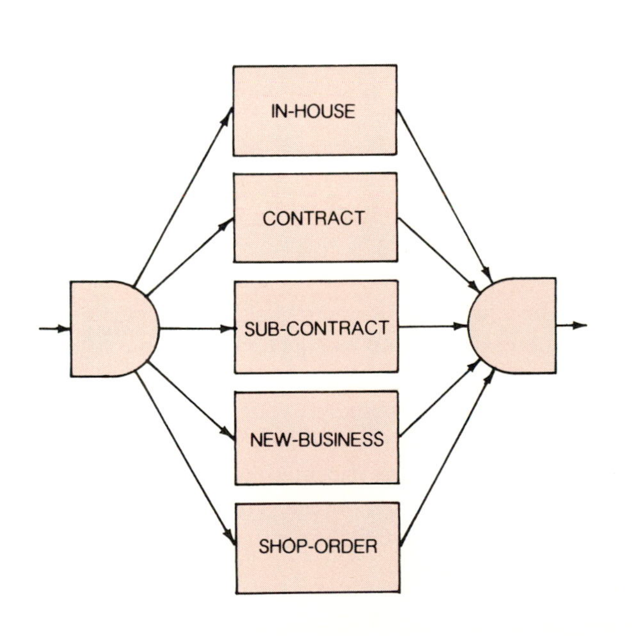


```
     PERFORM ACCOUNT-PROCESSING THRU
ACCOUNT-PROCESSING-EXIT.
     .
     .
ACCOUNT-PROCESSING.
     GO TO
          IN-HOUSE
          CONTRACT
          SUB-CONTRACT
          NEW-BUSINESS
          SHOP-ORDER
               DEPENDING ON ACCOUNT-CODE.
     error case processing statements.
     GO TO ACCOUNT-PROCESSING-EXIT.

IN-HOUSE.
     processing statements.
     GO TO ACCOUNT-PROCESSING-EXIT.

CONTRACT.
     processing statements.
     GO TO ACCOUNT-PROCESSING-EXIT.

.
.

SHOP-ORDER.
     processing statements.
     GO TO ACCOUNT-PROCESSING-EXIT.

ACCOUNT-PROCESSING-EXIT.
     EXIT.
```

Figure 8.31 CASE program structure—example.

3. The GO TO case-end-name statement, which is required to handle the situation where the value of the variable is greater than the greatest expected value or is less than one, will be coded on the next line and should be indented.
4. Each paragraph to be executed should be indented and must end with a GO TO statement that transfers control to the common end point of the CASE structure.
5. The case-end paragraph will follow the last of the function paragraphs and contain just the EXIT statement.

```
     PERFORM ACCOUNT-PROCESSING
          THRU ACCOUNT-PROCESSING-END.
          .
          .
ACCOUNT-PROCESSING
     GO TO IN-HOUSE, CONTRACT, SUB-CONTRACT,
          ESTIMATES, SHOP-ORDER
          DEPENDING ON ACCOUNT-CODE.
          PERFORM ERROR-ROUTINE.
          GO TO ACCOUNT-PROCESSING-END.
IN-HOUSE.
          .
     GO TO ACCOUNT-PROCESSING-END.
CONTRACT.
          .
          .
          .
     GO TO ACCOUNT-PROCESSING-END.
SHOP-ORDER.
          .
          .
     GO TO ACCOUNT-PROCESSING-END.
ACCOUNT-PROCESSING-END.
     EXIT.
```

Note: The CASE structure is the only allowable variance to the rule for using the THRU option of the PERFORM in structured programming techniques.

GO TO . . . DEPENDING ON Option

Any number of procedures can be named in the GO TO . . . DEPENDING ON statement. If the value of the data item is zero, or if it is greater than the number of named procedures, control automatically goes to the next sequential statement.

A GO TO . . . DEPENDING ON statement is the equivalent of a series of IF statements. It is, however, easier to write than a series of IF statements, and the instructions generated by the compiler will be somewhat more efficient than a series of relation tests would be.

The GO TO with the DEPENDING ON option is a way of analyzing any code number, and causes control to branch to procedures that correspond to various values of the number.

When a GO TO statement with the DEPENDING ON option is executed, control is transferred to one of a series of paragraphs depending on the value of the identifier. If the value of the identifier is zero or greater than the number of paragraph names specified, the GO TO statement is ignored and control goes to the next sentence in sequence.

DEPENDING ON (Conditional GO TO)

The DEPENDING ON option permits the multiple-branch type of operations according to the value of the current value of the identifier. (The general format for the GO TO statement with the DEPENDING ON option appears in figure 8.32.)

1. The identifier must have a positive integral value.
2. Control is passed to the 1st, 2nd. . . . . nth procedure-name as the value name of the identifier is 1, 2 . . . . (See figure 8.33.)
3. The identifier must have a range of values starting at 1 and continuing successively upward.

Format 2:

GO TO procedure-name-1 [procedure-name-2] . . . , procedure-name-n

DEPENDING ON identifier

Figure 8.32 Format—GO TO statement: format 2 (DEPENDING ON option).

4. If the value of the identifier is outside of the range of 1 through n, no branch occurs, and control passes to the first statement after the GO TO statement.

The following is an example of GO TO statement—format 2—DEPENDING ON option:

```
GO TO    RECEIPTS
         SHIPMENTS
         CUSTOMER-ORDERS
             DEPENDING ON TRANSACTION-CODE.
```

In the object program, if TRANSACTION-CODE contains the value 3, control will be transferred to CUSTOMER-ORDERS (the third procedure name in the series).

The following is another example of GO TO statement—format 2—DEPENDING ON option:

```
GO TO    DEPOSIT
         WITHDRAWAL
         INTEREST
             DEPENDING ON TRAN-CODE.
```

The above GO TO statement transfers control to DEPOSIT, WITHDRAWAL, or INTEREST depending on the number in TRAN-CODE. The following statements are equivalent to the above GO TO. . .DEPENDING ON statement.

```
IF TRAN-CODE = 1, GO TO DEPOSIT.
IF TRAN-CODE = 2, GO TO WITHDRAWAL.
IF TRAN-CODE = 3, GO TO INTEREST.
IF TRAN-CODE IS LESS THAN 1
    OR GREATER THAN 3,
    GO TO the next sentence
ELSE
    continue processing.
```

In a branching operation, after the transfer to the particular point in the program is executed, normal flow of control is resumed at the beginning of the particular procedure.

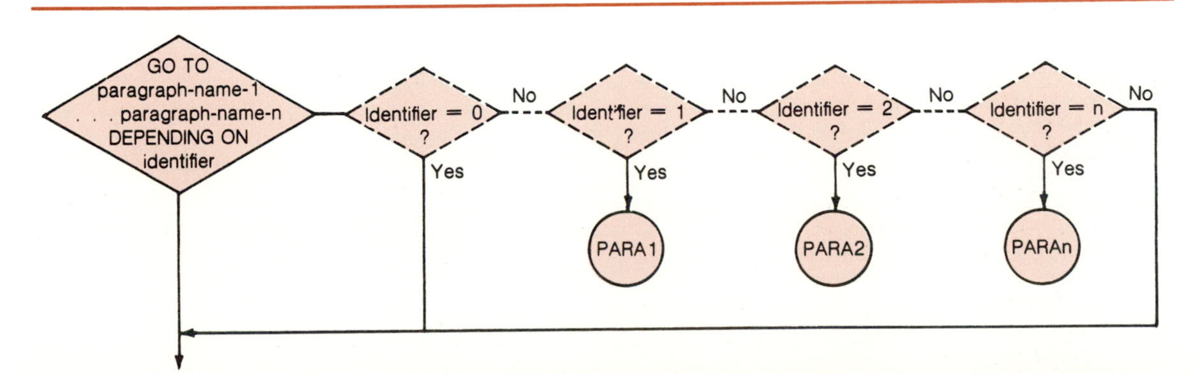

Figure 8.33 Flow of logic—GO TO statement: format 2 (DEPENDING ON option).

EXIT Statement

The EXIT statement is used when it is necessary to provide a common end point for a series of procedures. (The general format of the EXIT statement appears in figure 8.34.) The EXIT statement is also used when it is necessary to provide an ending point for a procedure that is executed by a PERFORM statement possibly having one or more conditional exits prior to the last sentence. The EXIT verb serves as an ending point common to all paths. The following is an example EXIT statement:

```
    PERFORM ANALYSIS-ROUTINE THRU FINISH-ANALYSIS.
         .
         .
         .
ANALYSIS-ROUTINE.
    COMPUTE RETURNS-RATIO = RETURNS/(ORDERS - FILLED +
        BACK-ORDERS-RETURNS).
    IF RETURNS-RATIO IS LESS THAN 20
        GO TO FINISH-ANALYSIS
    ELSE
        IF RETURNS-RATIO IS LESS THAN .33,
            ADD 1 TO HIGH-RATIO-COUNTER
            GO TO FINISH-ANALYSIS.
    PERFORM HIGH-RATIO-REPORT.
FINISH-ANALYSIS.
    EXIT.
```

Rules Governing the Use of the Exit Statement

1. The statement must appear in a source program as a one-word paragraph preceded by a paragraph-name.
2. In a PERFORM statement, the EXIT paragraph-name may be given as the object of the THRU option.
3. If the THRU option is used for the EXIT paragraph, a statement in the range of the PERFORM being executed may transfer to an EXIT paragraph, bypassing the remainder of the statements in the PERFORM range.
4. If the control reaches an EXIT paragraph and no associated PERFORM statement is used, control passes through the exit point to the first sentence of the next paragraph.

The following is another example EXIT statement:

```
PERFORM-1.      PERFORM TEST-CODE THRU EXIT-POINT.
SENTENCE-1.     .
                .
                .
TEST-CODE.
                IF TCODE = 12, GO TO PATH-B.
                IF TCODE = 15, GO TO PATH-C.
                IF TCODE = 16 GO TO EXIT-POINT.
PATH-A.
                .
                .
                GO TO EXIT-POINT.
PATH-B.         .
                .
                GO TO EXIT-POINT.
PATH-C.
                .
                .
EXIT-POINT.     EXIT.
```

paragraph-name. EXIT.

Figure 8.34 Format—EXIT statement.

In the example just given, the original PERFORM statement named PERFORM-1 executes the set of procedures from TEST-CODE through EXIT-POINT. In these procedures, the testing of TCODE results in the execution of one of four paths: PATH-A, PATH-B, PATH-C, and EXIT-POINT.

An EXIT statement serves only to enable the programmer to assign a procedure-name to a given point in a program. Such an EXIT statement has no other effect on the compilation or execution of the program.

Use of PERFORM, GO TO, and EXIT Statements

Although structured programming principles strongly reject the use of the GO TO statement, there are special circumstances where the GO TO statement can aid the programmer and still preserve the top-down theory of structured programming, such as, in the CASE structure program, where the PERFORM statement with the THRU option and the GO TO statement with the DEPENDING ON option are used. Immediately following the GO TO . . . DEPENDING ON statement there should be some statement to handle the error condition which may result from an argument out-of-range causing a non-branch to occur. Invalid results can be obtained if such a statement is not provided.

Since PERFORM error-routine will usually follow the GO TO . . . DEPENDING ON statement to handle invalid codes, the return from the error routine will be to the next statement following the PERFORM statement, which will cause the program to go through all paragraphs of the CASE structure. A GO TO statement directing the program to the EXIT paragraph should be written to avoid this.

This combination of PERFORM, GO TO, and EXIT statements can be used in conditions where it is necessary for the program to proceed to the end of the paragraph after a PERFORM statement is executed, such as in the following example:

```
    PERFORM PROCEDURE-1 THRU PROCEDURE-EXIT.
PROCEDURE-1.
    .
    IF A = B
        PERFORM PAR-1
        GO TO PROCEDURE-EXIT
    ELSE
        NEXT SENTENCE.
    IF A = C
        PERFORM PAR-2
        GO TO PROCEDURE-EXIT
    ELSE
        NEXT SENTENCE.
    .
    .
    .
    PROCEDURE-EXIT.
        EXIT.
```

The coding just mentioned will prevent the program from going through all IF statements. As soon as a match is found, the proper paragraph will be performed and the program will be directed to the end of the procedures.

This does not imply that GO TO statements should be used promiscuously, but strategic insertion of GO TO statements in crucial points in a program can save valuable time, and perhaps can prevent a record from being processed twice, especially after a PERFORM statement.

Figure 8.35 Format—STOP statement.

STOP Statement

There are two kinds of STOP statements. One stops the execution of the program permanently, the other temporarily. A permanent stop is indicated when the verb STOP is followed by the word RUN.

If the stop is temporary, that is, if execution is to be resumed after the computer operator takes some corrective steps, the verb STOP is followed by a literal. The literal will be output on the console. The literal is normally a message to the computer operator relative to the continuation of the program or its halt. Continuation of the program begins with the next executable statement in sequence.

The STOP statement permits the programmer to specify a temporary or permanent halt to the program. (The general format for the STOP statement appears in figure 8.35.)

STOP RUN

Rules Governing the Use of the STOP RUN Statement

1. The STOP RUN statement terminates the execution of a program.
2. Because of its terminal effect, the STOP RUN statement can only be used as a final statement in the sequence in which it appears; otherwise, the succeeding statements will never be executed.
3. All files should be closed before a STOP RUN statement is issued.
4. The actions following the execution of a STOP RUN statement depend upon the particular installation and/or a particular computer.

The following is an example of the STOP RUN statement:

```
STOP RUN.
```

STOP Literal

Rules Governing the Use of the STOP Literal Statement

1. The STOP Literal statement is used by the programmer to specify a temporary halt to the program.
2. When this statement is used, the specified literal will be displayed on the console at the time the stop occurs.
3. The program may be resumed only by operator intervention. A reply must be keyed in on the console to resume execution of the program.
4. Following the execution of the STOP Literal statement, continuation of the object program begins with the next sentence in sequence.
5. The literal may be numeric or nonnumeric, or it may be a figurative constant except ALL.

The following is an example of the STOP Literal statement:

```
STOP 'HALT 350—CONSULT RUN BOOK'.
```

Summary

The sequence control statements are designed to specify the sequence in which the various source program instructions are to be executed.

The conditional statement is the primary mechanism through which alternate paths of control can be developed within a program. A condition is an assertion concerning the control of a data item. The flow of control through a conditional statement depends upon the determination of the truth or falsity of the assertion at program execution.

The way in which control flows through procedures in a COBOL program represents the sequence in which the instructions in the object program will be executed. Control starts at the first statement in the Procedure Division provided there are no declaratives. If there are declaratives, execution starts after the END DECLARATIVES statement. Control flows from one statement to the next in sequence and from one program to the next except when a GO TO statement causes a branch, or an IF statement causes control to jump over certain statements, or a PERFORM statement gives temporary control to another procedure, or a STOP causes a delay in the execution or terminates the run.

A conditional expression may change the course of the program depending upon the circumstances existing when the expression is evaluated.

Simple conditions are the relation, class, sign, and condition-name conditions.

An IF statement causes a condition to be evaluated or tested, and an action to be taken, based on whether the result is true or false.

The IFTHENELSE structure tests a single condition to determine which of two function blocks will be executed. Statements controlled by IF and ELSE parts of the IF statement are indented to exhibit the span of control of the structure.

A test condition containing a single relational operation is called a relation condition. The test condition is a relation used to compare the value of a literal or used to specify a relationship between two items. The result of the relation condition is either true or false.

When used, NOT and the next key word or relation character are one relational operation that defines the comparison to be executed for truth values.

For numeric operands, a comparison results in the determination that the algebraic value of one of the operands is less than, equal to, or greater than the other. Numeric items must be compared to numeric items, and signs, if any, are considered in the comparison. Assumed decimal points are aligned and extra positions at either end of either data item are filled with zeros. The value zero has no magnitude, and its sign, if any, is disregarded.

For two nonnumeric operands, or one numeric (excluding the operational sign) and one nonnumeric operand, a comparison results in the determination that one operand is less than, equal to, or greater than the other with respect to a specified collating sequence of characters. The collating sequence will depend upon the computer used. If the operands are of equal size, the characters in corresponding positions are compared starting from the high-order and continuing until either a pair of unequal characters is encountered or the low-order end of the item is reached, whichever comes first. Items are determined to be equal when the low-order is reached. If operands are of unequal size, comparison proceeds as though the shorter operand was extended at the right with sufficient spaces to make the operands equal.

A class test condition is used to determine whether or not an item is alphabetic or numeric. The ALPHABETIC test cannot be used with items whose pictures are numeric. The item being tested is determined to be alphabetic only if the contents consist of the alphabetic characters A–Z and the space character. A numeric test cannot be used with items having alphabetic pictures, or which are group items. If the data description of the numeric item being tested does not indicate the presence of an operational sign, then the item being tested is determined to be numeric only if the contents are numeric and the operational sign is not present.

The sign test-condition determines whether or not the algebraic value of an arithmetic expression is less than, greater than, or equal to zero. An operand is POSITIVE only if its value is greater than zero, NEGATIVE if its value is less than zero, and ZERO if its value is equal to zero. The value zero is considered neither positive nor negative.

The condition-name test-condition is used in an IF statement that tests whether a variable has a specific value or one of a specific set of values. The condition-name must be described as a level-88 entry in the Data Division, assigned specific values, and associated with a related variable. The test-condition consists solely of a programmer-supplied name. A relation test may be used in place of the condition-name test. A programmer uses a condition-name test instead of a relation test to make the program more readable. In a condition-name test, a condition variable is tested to determine whether or not its value is equal to one of the values associated with the condition-name.

A compound condition is produced simply by tying together two or more test conditions with the words AND or OR. (The word NOT can be used as usual to give the opposite meaning to a condition.) The word OR means that the compound condition is true if *either* of the test conditions joined by the logical operator OR is true. The word AND means that the compound condition is true only if *all* of the test conditions joined by the logical operator AND are true. The conditions surrounding all ANDs are evaluated first, starting at the left and proceeding to the right. OR and its surrounding conditions are then evaluated, also proceeding from left to right. A simple condition is negated through the use of the logical operator NOT. The sequence of evaluation can be altered by parenthetical expressions.

Nested conditional expressions contain one or more IFs that appear in a sentence that begins with IF, thereby making it possible to make a series of decisions based on the outcome of the previous decisions, and to take different courses of action, all within one sentence. IF statements are considered paired with the closest ELSE phrase.

COBOL allows the omission of a subject, a relation, or both in certain forms of compound conditional expressions. A compound condition can use the same term as the subject of each relation, and only the first occurrence of the subject need be written. A compound statement can use the same subject and relation for a series of relations, and only the first occurrence of the subject and relation need be written. Both types (implied subjects and implied subjects and relations) may be used in the same conditional expressions with the subject and/or relation being replaced by the last preceding stated subject and/or relation. If the word immediately following NOT is a relation operator, then the NOT participates as part of the relation operator. Otherwise, the NOT is interpreted as a logical operator and, therefore, the implied insertion of a subject or relational operation results in a negated relation condition when used in conjunction with abbreviations.

The use of implied subjects and relations should be used sparingly or not at all, as this will cause problems in program maintenance and modification.

A GO TO statement causes control to branch unconditionally (or conditionally) to the first statement of a procedure. After a GO TO has caused control to branch to the beginning of a procedure, the normal flow of control is resumed.

GO TO and PERFORM statements cause control to flow to the beginning of a procedure, whereas the IF sentence causes control either to flow through or to jump over certain statements. A GO TO statement differs from a PERFORM statement in that a PERFORM causes a branch to a procedure or a series of procedures, but after the procedure or procedures are acted upon, PERFORM causes a return branch to the statement after the PERFORM statement.

The CASE structure is used to select one of a set of functions for executions depending on the value of the integer-identifier whose range is from 1 to the number of procedures listed in the statement. For any value outside of the defined range, the statement is ignored. CASE structure is simulated in COBOL by using the PERFORM statement with the THRU option and the GO TO statement with the DEPENDING ON option. The EXIT statement is used as a common exit point for the structure. The CASE structure is the only structure under structured programming rules to use the THRU option of the PERFORM statement and the GO TO verb with the DEPENDING ON option.

The GO TO . . . DEPENDING ON statement is the equivalent of a series of IF statements. It is a way of analyzing any code, and causes control to branch to procedures that correspond to the various values of the number. When the statement is executed, control is transferred to one of a series of procedures depending on the value of the identifier.

The EXIT statement is used when it is necessary to provide a common end point for a series of procedures. An EXIT statement serves only to enable the programmer to assign a procedure-name to a given point in a program. Such an EXIT statement has no effect on the compilation or the execution of the program.

In special circumstances, the use of PERFORM, GO TO, and EXIT statements combined can save valuable time in programming, and perhaps prevent a record from being processed twice.

There are two kinds of STOP statements. One stops the execution of the program permanently as indicated by following the word STOP with the word RUN; the other temporarily. If the STOP is temporary, that is, if execution is to be resumed after the computer operator takes some corrective action, the verb STOP is followed by a literal. The literal is normally a message output on the console relative to the continuation of the program after its halt.

Summary—COBOL Verbs

IF is a sequence control verb that causes one or more statements to be acted upon only if a certain condition exists, and also causes one or more other statements to be acted upon only if the condition does not exist.

The test-condition may be a relation test, sign test, condition-name test, or class test. Any number of *statements* may follow the *test-condition.* These are acted upon only if the condition exists, and are jumped over if the condition does not exist. In other words, statement-1 is acted upon first, statement-2, and so on. This means if statement-1 is a GO TO or STOP statement (for instance), then statement-2 is never acted upon. It is up to the programmer to make sure that the statements appear in a logical sequence.

Statements up to the word ELSE are acted upon only if the condition exists. Statements following ELSE up to the period are acted upon only if the condition does *not* exist. *Test-Conditions* compare two data values; examine the operational sign of a data item; check for a specific value in an item; and check the class of the data in an item. These tests are used in IF statements.

In a relation test, all combinations of operands are permitted, except two literals, two figurative constants, or a literal and a figurative constant. Condition-name must be defined in a level-88 entry in the Data Division.

GO TO is a sequence control verb that causes a branch to a procedure(s) in the program. The normal flow of control is resumed at the beginning of the specified procedure. The procedure-name must be either the name of a paragraph or the name of a section in the Procedure Division. The word SECTION is not used after a section name in the statement.

It is permissible to branch from some point in a procedure back to the beginning of the same procedure. In other words, procedure-name may be the name of the procedure that the GO TO is part of.

GO TO . . . DEPENDING ON Option. Any number of procedures can be named in the GO TO . . . DEPENDING ON statement. If the value of the data item is zero or if it is greater than the number of named procedures, control automatically goes to the next sequential statement.

A GO TO . . . DEPENDING ON statement is the equivalent of a series of IF statements. But it is easier to write than a series of IF statements, and the instructions generated by the compiler will be somewhat more efficient than those generated by a series of relation tests.

The GO TO with the DEPENDING ON option is a way of analyzing any code number and causing control to branch to procedures that correspond to the various values of the number.

When a GO TO statement with the DEPENDING ON option is executed, control is transferred to one of a series of paragraphs depending on the value of the identifier. If the value of the identifier is zero, or greater than the number of paragraph names specified, the GO TO statement is ignored and control goes to the next sentence in sequence.

The identifier must represent a positive or unsigned integer, with the value of the identifier representing the number of the paragraph names to which control is transferred.

The *EXIT* statement provides a common end point for a series of procedures. The EXIT statement must appear in a sentence by itself and must be the only sentence in the paragraph. The EXIT statement serves only to enable the user to assign a procedure-name to a given point in a program. Such an EXIT statement has no other effect on the compilation or execution of the program.

STOP is a sequence control verb that is used to temporarily halt or permanently terminate the program. To temporarily halt the program, the STOP literal format is used. The literal is communicated to the operator. Execution of the program is suspended until the operator uses the keyboard to type in some necessary data. Execution is then resumed at the statement following the STOP statement. The literal usually consists of a brief instruction to the operator.

If the STOP RUN statement is used, the control of the computer is turned over to the operating system control program or any other procedure established by the installation. In other words, when STOP RUN is acted upon, the computer does not halt, but rather goes on to some other job.

STOP RUN is used at the end of a job, or when an error in the data is so serious that it is impossible to continue the run.

If a STOP RUN statement appears in a consecutive sequence of imperative statements within a sentence, it must appear as the last statement in that sequence.

Illustrative Program: Payroll Register Problem

Input

Field	Positions	
Department	14–16	
Serial	17–21	
Gross Earnings	57–61	XXX.XX
Insurance	62–65	XX.XX
Federal Withholding Tax	69–72	XX.XX
State Withholding Tax	73–75	XX.XX
Miscellaneous Deductions	76–79	XX.XX
Code (Letter E)	80	

Calculations To Be Performed

1. FICA TAX = Gross Earnings × .067 (Rounded to 2 decimals)
2. NET EARNINGS = Gross Earnings − Insurance − FICA Tax − Federal Withholding Tax − State Withholding Tax − Miscellaneous Deductions.
3. Department Earnings value is the sum of the Net Earnings for each employee.
4. The Total Net Earnings is the sum of the Net Earnings for each department.

Output

Print a report as follows:

WEEKLY PAYROLL REGISTER

EMPLOYEE NO. DEPT.	SERIAL	GROSS EARNINGS	INSURANCE	TAX	FEDERAL WITHHOLDING	STATE WITHHOLDING	MISC. DEDUCT.	NET AMOUNT
9	16572	$115.78	$.00	$6.77	$8.90	$.84	$.00	$99.27
9	2A182	$84.17	$1.00	$4.92	$16.02	$1.19	$1.75	$59.29
9	970K5	$100.65	$3.00	$5.89	$12.57	$1.34	$.00	$77.85
9	63600	$116.40	$2.00	$6.81	$18.90	$1.85	$2.00	$84.84
							DEPT. EARNINGS	$321.25 *
10	3L715	$156.80	$2.00	$9.17	$22.46	$2.02	$.00	$121.15
10	38Y8J	$186.57	$.00	$10.91	$4.80	$.60	$.75	$169.51
10	W2923	$75.38	$1.25	$4.41	$9.75	$.78	$.00	$59.19
10	13E10	$84.17	$.00	$4.92	$8.90	$.84	$.00	$69.51
							DEPT. EARNINGS	$419.36 *
19	1P411	$100.65	$1.00	$5.89	$16.02	$1.19	$1.75	$74.80
19	64G12	$116.40	$3.00	$6.81	$12.57	$1.34	$.00	$92.68
19	13I43	$156.80	$7.00	$9.17	$18.90	$1.85	$7.00	$112.88
							DEPT. EARNINGS	$280.36 *
20	31B54	$186.57	$2.00	$10.91	$22.46	$2.02	$.00	$149.18
20	16X33	$75.38	$1.25	$4.41	$9.75	$.78	$.00	$59.19
20	137M5	$84.17	$.00	$4.92	$8.90	$.84	$.00	$69.51
							DEPT. EARNINGS	$277.88 *
29	V1388	$100.65	$1.00	$5.89	$16.02	$1.19	$1.75	$74.80
29	4C193	$116.40	$3.00	$6.81	$12.57	$1.34	$.00	$92.68
29	2N270	$156.80	$2.00	$9.17	$18.90	$1.85	$2.00	$122.88
							DEPT. EARNINGS	$290.36 *
30	52Q16	$186.57	$7.00	$10.91	$22.46	$2.02	$.00	$144.18
30	322D6	$67.70	$.00	$3.96	$4.80	$.60	$.75	$57.59
30	7U223	$75.18	$1.25	$4.40	$9.75	$.78	$.00	$59.00
30	2543S	$84.17	$.00	$4.92	$8.90	$.84	$.00	$69.51
							DEPT. EARNINGS	$330.28 *
39	2R598	$100.64	$1.00	$5.89	$16.02	$1.19	$1.75	$74.79
39	226T3	$116.40	$3.00	$6.81	$12.57	$1.34	$.00	$92.68
							DEPT. EARNINGS	$167.47 *
							TOTAL NET EARNINGS	$2,086.96 **

Printer Spacing Chart

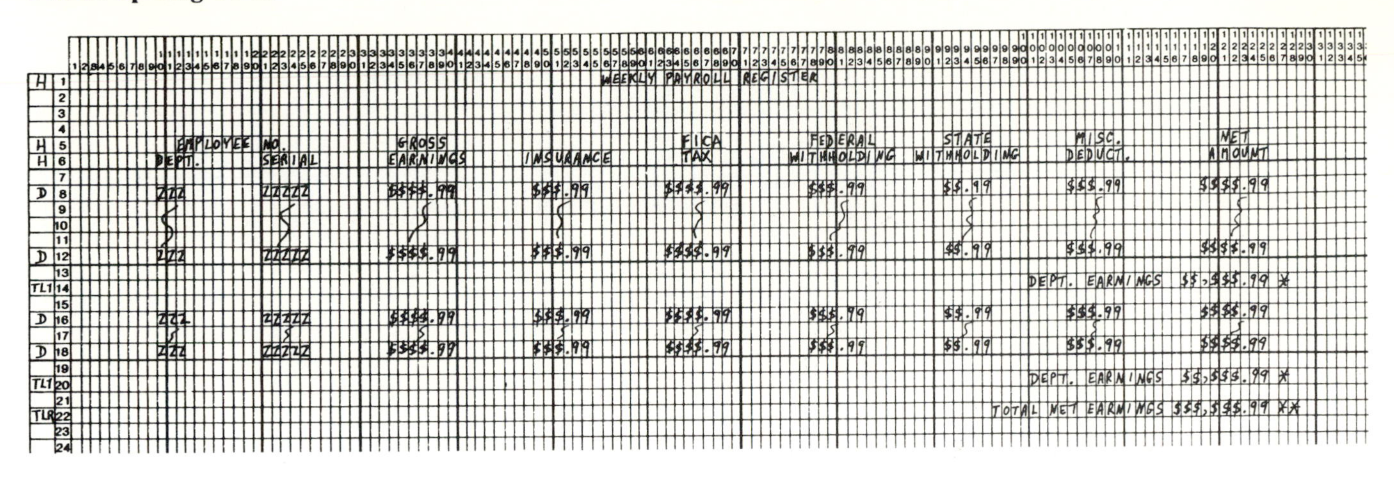

Hierarchy Chart

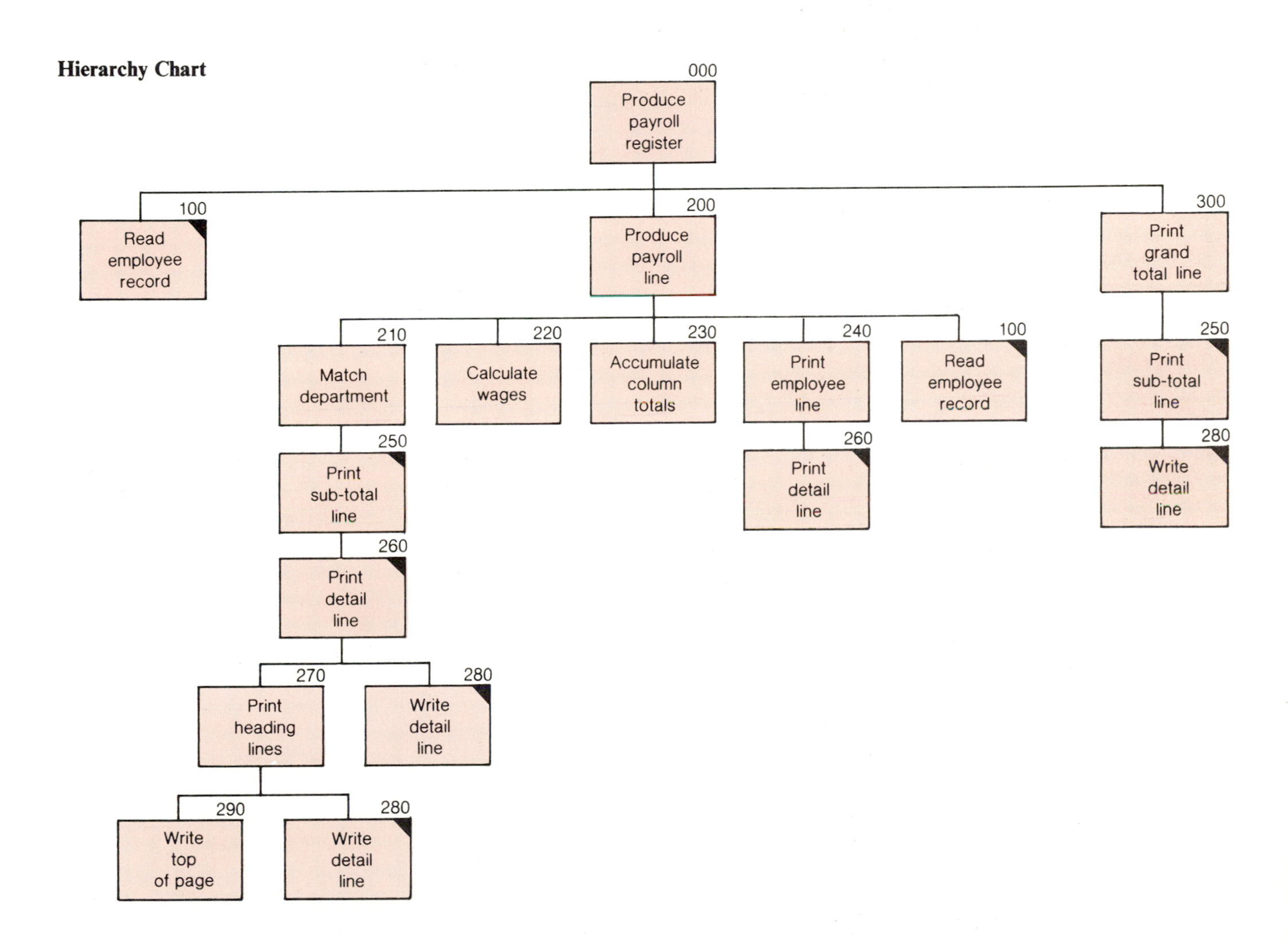

IPO Charts

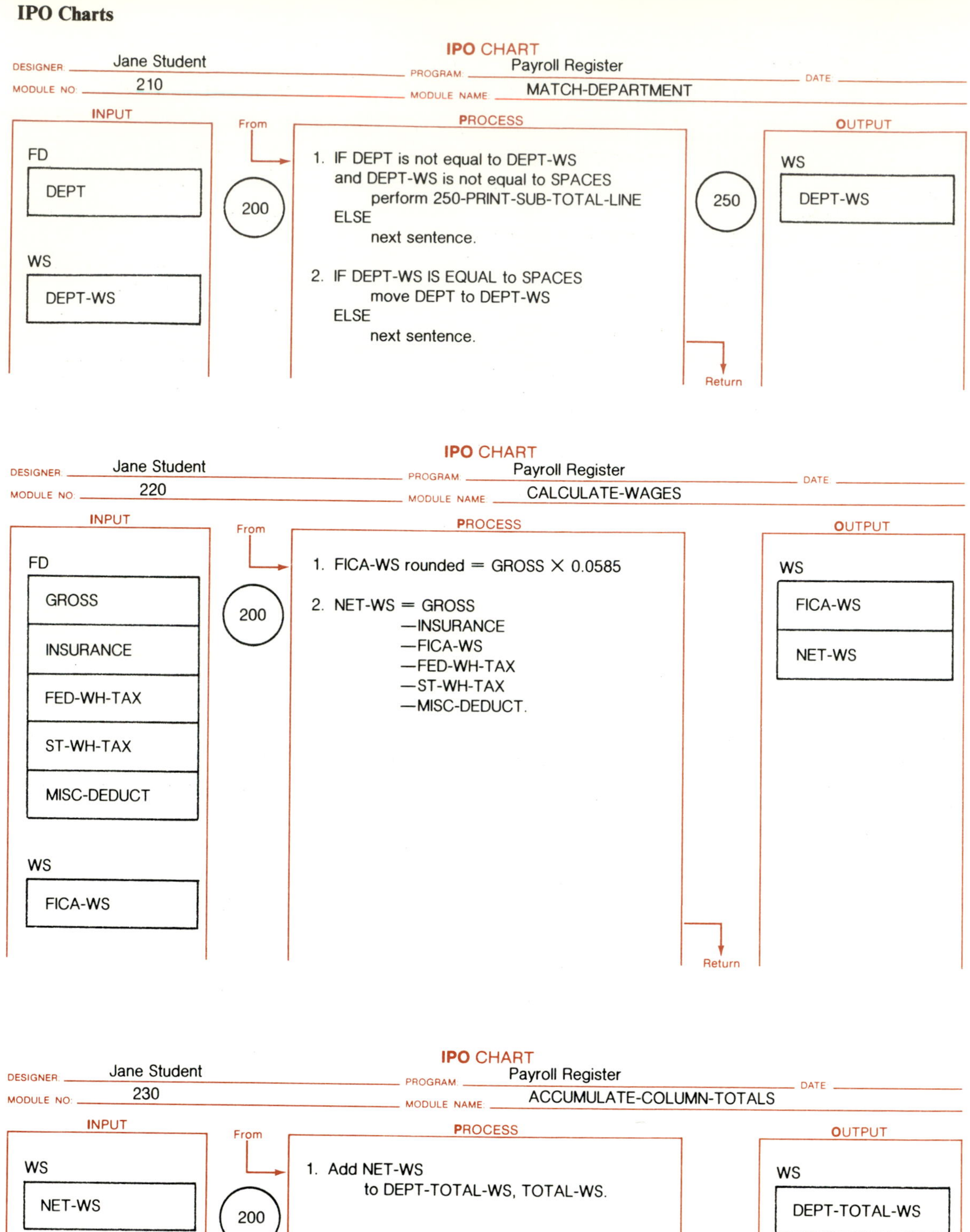

IPO CHART

DESIGNER: Jane Student — PROGRAM: Payroll Register — DATE:

MODULE NO: 210 — MODULE NAME: MATCH-DEPARTMENT

INPUT:
FD
- DEPT

WS
- DEPT-WS

From: 200

PROCESS:

1. IF DEPT is not equal to DEPT-WS
   and DEPT-WS is not equal to SPACES
   perform 250-PRINT-SUB-TOTAL-LINE
   ELSE
   next sentence.

2. IF DEPT-WS IS EQUAL to SPACES
   move DEPT to DEPT-WS
   ELSE
   next sentence.

250

Return

OUTPUT:
WS
- DEPT-WS

IPO CHART

DESIGNER: Jane Student — PROGRAM: Payroll Register — DATE:

MODULE NO: 220 — MODULE NAME: CALCULATE-WAGES

INPUT:
FD
- GROSS
- INSURANCE
- FED-WH-TAX
- ST-WH-TAX
- MISC-DEDUCT

WS
- FICA-WS

From: 200

PROCESS:

1. FICA-WS rounded = GROSS × 0.0585

2. NET-WS = GROSS
   —INSURANCE
   —FICA-WS
   —FED-WH-TAX
   —ST-WH-TAX
   —MISC-DEDUCT.

Return

OUTPUT:
WS
- FICA-WS
- NET-WS

IPO CHART

DESIGNER: Jane Student — PROGRAM: Payroll Register — DATE:

MODULE NO: 230 — MODULE NAME: ACCUMULATE-COLUMN-TOTALS

INPUT:
WS
- NET-WS

From: 200

PROCESS:

1. Add NET-WS
   to DEPT-TOTAL-WS, TOTAL-WS.

Return

OUTPUT:
WS
- DEPT-TOTAL-WS
- TOTAL-WS

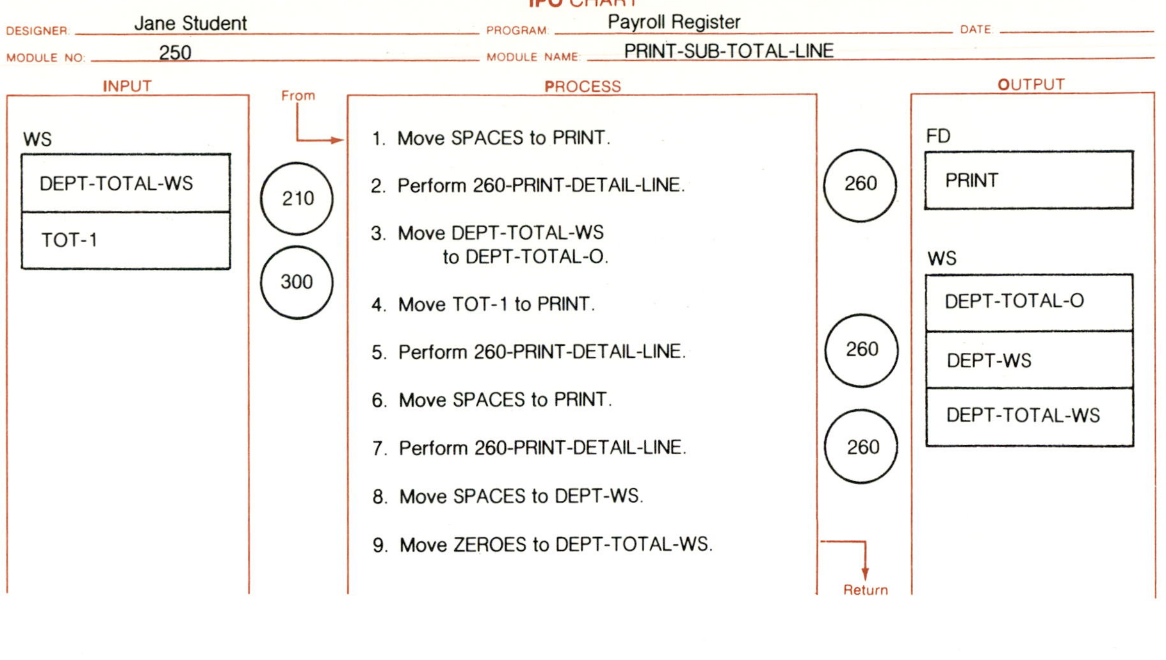

IPO CHART

DESIGNER: Jane Student
PROGRAM: Payroll Register
DATE:
MODULE NO: 250
MODULE NAME: PRINT-SUB-TOTAL-LINE

INPUT

WS

DEPT-TOTAL-WS
TOT-1

From: 210, 300

PROCESS

1. Move SPACES to PRINT.
2. Perform 260-PRINT-DETAIL-LINE. (260)
3. Move DEPT-TOTAL-WS to DEPT-TOTAL-O.
4. Move TOT-1 to PRINT.
5. Perform 260-PRINT-DETAIL-LINE. (260)
6. Move SPACES to PRINT.
7. Perform 260-PRINT-DETAIL-LINE. (260)
8. Move SPACES to DEPT-WS.
9. Move ZEROES to DEPT-TOTAL-WS.

Return

OUTPUT

FD

PRINT

WS

DEPT-TOTAL-O
DEPT-WS
DEPT-TOTAL-WS

IPO CHART

DESIGNER: Jane Student
PROGRAM: Payroll Register
DATE:
MODULE NO: 270
MODULE NAME: PRINT-HEADING-LINES

INPUT

WS

HDG-1
HDG-2
HDG-3

From: 260

PROCESS

1. Move HDG-1 to PRINT.
2. Perform 290-WRITE-TOP-OF-PAGE. (290)
3. Move HDG-2 to PRINT.
4. Perform 280-WRITE-DETAIL-LINE. (280)
5. Move HDG-3 to PRINT.
6. Perform 280-WRITE-DETAIL-LINE. (280)
7. Move SPACES to PRINT.
8. Perform 280-WRITE-DETAIL-LINE. (280)
9. Move SPACES to PRINT.
10. Perform 280-WRITE-DETAIL-LINE. (280)

Return

OUTPUT

FD

PRINT

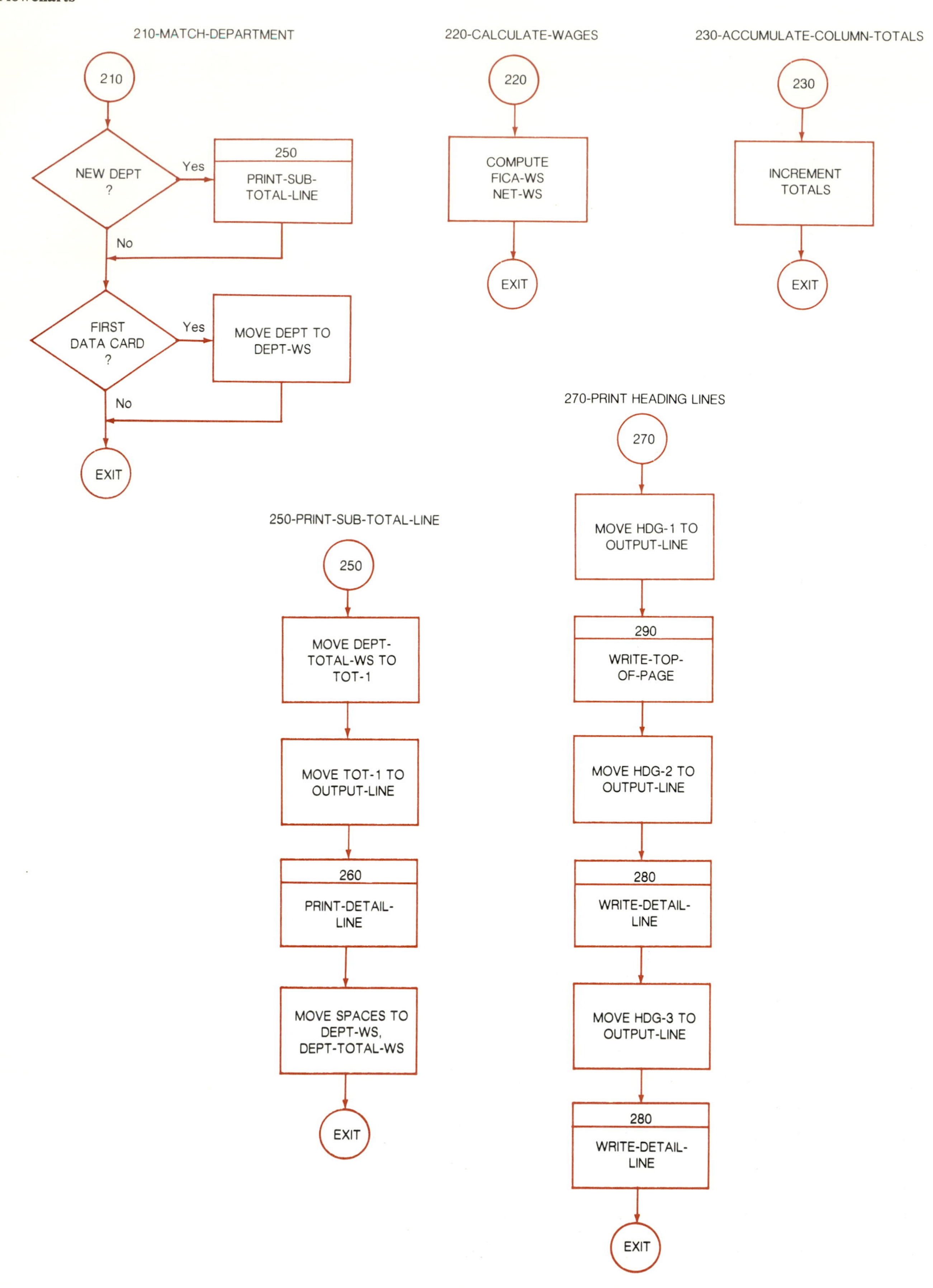
210-MATCH-DEPARTMENT
210
NEW DEPT ?
Yes
250
PRINT-SUB-TOTAL-LINE
No
FIRST DATA CARD ?
Yes
MOVE DEPT TO DEPT-WS
No
EXIT
220-CALCULATE-WAGES
220
COMPUTE FICA-WS NET-WS
EXIT
230-ACCUMULATE-COLUMN-TOTALS
230
INCREMENT TOTALS
EXIT
270-PRINT HEADING LINES
270
MOVE HDG-1 TO OUTPUT-LINE
290
WRITE-TOP-OF-PAGE
MOVE HDG-2 TO OUTPUT-LINE
280
WRITE-DETAIL-LINE
MOVE HDG-3 TO OUTPUT-LINE
280
WRITE-DETAIL-LINE
EXIT
250-PRINT-SUB-TOTAL-LINE
250
MOVE DEPT-TOTAL-WS TO TOT-1
MOVE TOT-1 TO OUTPUT-LINE
260
PRINT-DETAIL-LINE
MOVE SPACES TO DEPT-WS, DEPT-TOTAL-WS
EXIT

Source Program

```
 1        *************************
 2
 3         IDENTIFICATION DIVISION.
 4
 5        *************************
 6
 7         PROGRAM-ID.
 8             WEEKLY-PAYROLL-REGISTER.
 9         INSTALLATION.
10             WEST LOS ANGELES COLLEGE.
11         DATE-WRITTEN.
12             3 FEBRUARY 1982.
13
14        *PURPOSE.
15        *    THIS PROGRAM PREPARES A WEEKLY PAYROLL REGISTER.
16
17        *************************
18
19         ENVIRONMENT DIVISION.
20
21        *************************
22
23         CONFIGURATION SECTION.
24
25         SOURCE-COMPUTER.
26             LEVEL-66-ASCII.
27         OBJECT-COMPUTER.
28             LEVEL-66-ASCII.
29
30         INPUT-OUTPUT SECTION.
31
32         FILE-CONTROL.
33             SELECT FILE-IN      ASSIGN TO C1-CARD-READER.
34             SELECT FILE-OUT     ASSIGN TO P1-PRINTER.
35
36        *************************
37
38         DATA DIVISION.
39
40        *************************
41
42         FILE SECTION.
43
44         FD  FILE-IN
45             CODE-SET IS GBCD
46             LABEL RECORDS ARE OMITTED
47             DATA RECORD IS CARDIN.
48
49         01  CARDIN.
50
51             03  FILLER              PICTURE X(13).
52             03  DEPT                PICTURE X(3).
53             03  SERIAL              PICTURE X(5).
54             03  FILLER              PICTURE X(35).
55             03  GROSS               PICTURE 9(3)V99.
56             03  INSURANCE           PICTURE 9(2)V99.
57             03  FILLER              PICTURE X(3).
58             03  FED-WH-TAX          PICTURE 9(2)V99.
59             03  ST-WH-TAX           PICTURE 9(1)V99.
60             03  MISC-DEDUCT         PICTURE 9(2)V99.
61             03  KODE                PICTURE X(1).
62
63         FD  FILE-OUT
64             CODE-SET IS GBCD
65             LABEL RECORDS ARE OMITTED
66             DATA RECORD IS PRINT.
67
68         01  PRINT                   PICTURE X(132).
69
70         WORKING-STORAGE SECTION.
71
72         77  DEPT-WS                 PICTURE X(3)     VALUE SPACES.
73         77  FICA-WS                 PICTURE 9(3)V99  VALUE ZEROES.
74         77  NET-WS                  PICTURE 9(4)V99  VALUE ZEROES.
75         77  DEPT-TOTAL-WS           PICTURE 9(3)V99  VALUE ZEROES.
76         77  TOTAL-WS                PICTURE 9(5)V99  VALUE ZEROES.
77         77  LINE-COUNT-WS           PICTURE 9(2)     VALUE 38.
78
79         01  FLAGS.
80
81             03  MORE-DATA-FLAG  PICTURE X(3)     VALUE "YES".
82                 88  MORE-DATA                    VALUE "YES".
83                 88  NO-MORE-DATA                 VALUE "NO ".
84
85         01  HDG-1.
86
87             03  FILLER              PICTURE X(55)    VALUE SPACES.
88             03  FILLER              PICTURE X(14)    VALUE "WEEKLY PAYROLL".
89             03  FILLER              PICTURE X(9)     VALUE " REGISTER".
90             03  FILLER              PICTURE X(54)    VALUE SPACES.
91
92         01  HDG-2.
93
94             03  FILLER              PICTURE X(11)    VALUE SPACES.
95             03  FILLER              PICTURE X(12)    VALUE "EMPLOYEE NO.".
96             03  FILLER              PICTURE X(11)    VALUE SPACES.
97             03  FILLER              PICTURE X(5)     VALUE "GROSS".
```

```
 98           03  FILLER            PICTURE X(25)    VALUE SPACES.
 99           03  FILLER            PICTURE X(4)     VALUE SPACES.
100           03  FILLER            PICTURE X(9)     VALUE SPACES.
101           03  FILLER            PICTURE X(9)     VALUE "FEDERAL  ".
102           03  FILLER            PICTURE X(5)     VALUE SPACES.
103           03  FILLER            PICTURE X(5)     VALUE "STATE".
104           03  FILLER            PICTURE X(9)     VALUE SPACES.
105           03  FILLER            PICTURE X(5)     VALUE "MISC.".
106           03  FILLER            PICTURE X(10)    VALUE SPACES.
107           03  FILLER            PICTURE X(3)     VALUE "NET".
108           03  FILLER            PICTURE X(9)     VALUE SPACES.
109
110       01  HDG-3.
111
112           03  FILLER            PICTURE X(9)     VALUE SPACES.
113           03  FILLER            PICTURE X(5)     VALUE "DEPT.".
114           03  FILLER            PICTURE X(6)     VALUE SPACES.
115           03  FILLER            PICTURE X(6)     VALUE "SERIAL".
116           03  FILLER            PICTURE X(7)     VALUE SPACES.
117           03  FILLER            PICTURE X(8)     VALUE "EARNINGS".
118           03  FILLER            PICTURE X(6)     VALUE SPACES.
119           03  FILLER            PICTURE X(9)     VALUE "INSURANCE".
120           03  FILLER            PICTURE X(8)     VALUE SPACES.
121           03  FILLER            PICTURE X(3)     VALUE "TAX".
122           03  FILLER            PICTURE X(8)     VALUE SPACES.
123           03  FILLER            PICTURE X(13)    VALUE "WITHHOLDING  ".
124           03  FILLER            PICTURE X(11)    VALUE "WITHHOLDING".
125           03  FILLER            PICTURE X(5)     VALUE SPACES.
126           03  FILLER            PICTURE X(7)     VALUE "DEDUCT.".
127           03  FILLER            PICTURE X(8)     VALUE SPACES.
128           03  FILLER            PICTURE X(6)     VALUE "AMOUNT".
129           03  FILLER            PICTURE X(7)     VALUE SPACES.
130
131       01  DETAIL-LINE.
132
133           03  FILLER            PICTURE X(9)     VALUE SPACES.
134           03  DEPT-O            PICTURE Z(3).
135           03  FILLER            PICTURE X(8)     VALUE SPACES.
136           03  SERIAL-O          PICTURE X(5).
137           03  FILLER            PICTURE X(8)     VALUE SPACES.
138           03  GROSS-O           PICTURE $$$$.99.
139           03  FILLER            PICTURE X(8)     VALUE SPACES.
140           03  INSURANCE-O       PICTURE $$$.99.
141           03  FILLER            PICTURE X(8)     VALUE SPACES.
142           03  FICA-O            PICTURE $$$$.99.
143           03  FILLER            PICTURE X(8)     VALUE SPACES.
144           03  FED-WH-TAX-O      PICTURE $$$.99.
145           03  FILLER            PICTURE X(8)     VALUE SPACES.
146           03  ST-WH-TAX-O       PICTURE $$.99.
147           03  FILLER            PICTURE X(8)     VALUE SPACES.
148           03  MISC-DEDUCT-O     PICTURE $$$.99.
149           03  FILLER            PICTURE X(6)     VALUE SPACES.
150           03  NET-O             PICTURE $$,$$$.99.
151           03  FILLER            PICTURE X(7)     VALUE SPACES.
152
153       01  TOT-1.
154
155           03  FILLER            PICTURE X(100)   VALUE SPACES.
156           03  FILLER            PICTURE X(16)    VALUE "DEPT. EARNINGS  ".
157           03  DEPT-TOTAL-O      PICTURE $$,$$$.99.
158           03  FILLER            PICTURE X(7)     VALUE " *     ".
159
160       01  TOT-2.
161
162           03  FILLER            PICTURE X(96)    VALUE SPACES.
163           03  FILLER            PICTURE X(10)    VALUE "TOTAL NET ".
164           03  FILLER            PICTURE X(9)     VALUE "EARNINGS ".
165           03  TOTAL-O           PICTURE $$$,$$$.99.
166           03  FILLER            PICTURE X(7)     VALUE " **    ".
167
168       01  SAVE-RECORD           PICTURE X(132).
169
170      **************************
171
172       PROCEDURE DIVISION.
173
174      **************************
175
176       000-PRODUCE-PAYROLL-REGISTER.
177
178           OPEN    INPUT   FILE-IN
179                   OUTPUT  FILE-OUT.
180           PERFORM 100-READ-EMPLOYEE-RECORD.
181           PERFORM 200-PRODUCE-PAYROLL-LINE
182                   UNTIL NO-MORE-DATA.
183           PERFORM 300-PRINT-GRAND-TOTAL-LINE.
184           CLOSE   FILE-IN
185                   FILE-OUT.
186           STOP RUN.
187
188       100-READ-EMPLOYEE-RECORD.
189
190           READ FILE-IN
191               AT END MOVE "NO" TO MORE-DATA-FLAG.
192
193       200-PRODUCE-PAYROLL-LINE.
194
195           PERFORM 210-MATCH-DEPARTMENT.
196           PERFORM 220-CALCULATE-WAGES.
197           PERFORM 230-ACCUMULATE-COLUMN-TOTALS.
198           PERFORM 240-PRINT-EMPLOYEE-LINE.
```

```
199            PERFORM 100-READ-EMPLOYEE-RECORD.
200
201        210-MATCH-DEPARTMENT.
202
203            IF DEPT IS NOT = DEPT-WS
204                AND DEPT-WS IS NOT = "    "
205                PERFORM 250-PRINT-SUB-TOTAL-LINE
206            ELSE
207                NEXT SENTENCE.
208            IF DEPT-WS IS = "    "
209                MOVE DEPT TO DEPT-WS
210            ELSE
211                NEXT SENTENCE.
212
213        220-CALCULATE-WAGES.
214
215            COMPUTE FICA-WS ROUNDED = GROSS * 0.0585.
216            COMPUTE NET-WS = GROSS
217                  - INSURANCE
218                  - FICA-WS
219                  - FED-WH-TAX
220                  - ST-WH-TAX
221                  - MISC-DEDUCT.
222
223        230-ACCUMULATE-COLUMN-TOTALS.
224
225            ADD NET-WS TO DEPT-TOTAL-WS,
226                          TOTAL-WS.
227
228        240-PRINT-EMPLOYEE-LINE.
229
230            MOVE DEPT TO DEPT-O.
231            MOVE SERIAL TO SERIAL-O.
232            MOVE GROSS TO GROSS-O.
233            MOVE INSURANCE TO INSURANCE-O.
234            MOVE FICA-WS TO FICA-O.
235            MOVE FED-WH-TAX TO FED-WH-TAX-O.
236            MOVE ST-WH-TAX TO ST-WH-TAX-O.
237            MOVE MISC-DEDUCT TO MISC-DEDUCT-O.
238            MOVE NET-WS TO NET-O.
239            MOVE DETAIL-LINE TO PRINT.
240            PERFORM 260-PRINT-DETAIL-LINE.
241
242        250-PRINT-SUB-TOTAL-LINE.
243
244            MOVE SPACES TO PRINT.
245            PERFORM 260-PRINT-DETAIL-LINE.
246            MOVE DEPT-TOTAL-WS TO DEPT-TOTAL-O.
247            MOVE TOT-1 TO PRINT.
248            PERFORM 260-PRINT-DETAIL-LINE.
249            MOVE SPACES TO PRINT.
250            PERFORM 260-PRINT-DETAIL-LINE.
251            MOVE SPACES TO DEPT-WS.
252            MOVE ZEROES TO DEPT-TOTAL-WS.
253
254        260-PRINT-DETAIL-LINE.
255
256            IF LINE-COUNT-WS IS > 37
257                MOVE PRINT TO SAVE-RECORD
258                PERFORM 270-PRINT-HEADING-LINES
259                MOVE SAVE-RECORD TO PRINT
260            ELSE
261                NEXT SENTENCE.
262            PERFORM 280-WRITE-DETAIL-LINE.
263
264        270-PRINT-HEADING-LINES.
265
266            MOVE HDG-1 TO PRINT.
267            PERFORM 290-WRITE-TOP-OF-PAGE.
268            MOVE HDG-2 TO PRINT.
269            PERFORM 280-WRITE-DETAIL-LINE.
270            MOVE HDG-3 TO PRINT.
271            PERFORM 280-WRITE-DETAIL-LINE.
272            MOVE SPACES TO PRINT.
273            PERFORM 280-WRITE-DETAIL-LINE.
274            MOVE SPACES TO PRINT.
275            PERFORM 280-WRITE-DETAIL-LINE.
276
277        280-WRITE-DETAIL-LINE.
278
279            WRITE PRINT
280                AFTER ADVANCING 1 LINE.
281            ADD 1 TO LINE-COUNT-WS.
282
283        290-WRITE-TOP-OF-PAGE.
284
285            WRITE PRINT
286                AFTER ADVANCING PAGE.
287            MOVE 1 TO LINE-COUNT-WS.
288
289        300-PRINT-GRAND-TOTAL-LINE.
290
291            MOVE SPACES TO PRINT.
292            PERFORM 250-PRINT-SUB-TOTAL-LINE.
293            MOVE TOTAL-WS TO TOTAL-O.
294            MOVE TOT-2 TO PRINT.
295            PERFORM 280-WRITE-DETAIL-LINE.
296

THERE WERE 296 SOURCE INPUT LINES.
THERE WERE NO DIAGNOSTICS.
```

WEEKLY PAYROLL REGISTER

EMPLOYEE NO. DEPT.	SERIAL	GROSS EARNINGS	INSURANCE	TAX	FEDERAL WITHHOLDING	STATE WITHHOLDING	MISC. DEDUCT.	NET AMOUNT	
9	16572	$115.78	$.00	$6.77	$8.90	$.84	$.00	$99.27	
9	2A182	$84.17	$1.00	$4.92	$16.02	$1.19	$1.75	$59.29	
9	970K5	$100.65	$3.00	$5.89	$12.57	$1.34	$.00	$77.85	
9	63600	$116.40	$2.00	$6.81	$18.90	$1.85	$2.00	$84.84	
							DEPT. EARNINGS	$321.25	*
10	3L715	$156.80	$2.00	$9.17	$22.46	$2.02	$.00	$121.15	
10	38Y8J	$186.57	$.00	$10.91	$4.80	$.60	$.75	$169.51	
10	W2923	$75.38	$1.25	$4.41	$9.75	$.78	$.00	$59.19	
10	13E10	$84.17	$.00	$4.92	$8.90	$.84	$.00	$69.51	
							DEPT. EARNINGS	$419.36	*
19	1P411	$100.65	$1.00	$5.89	$16.02	$1.19	$1.75	$74.80	
19	64G12	$116.40	$3.00	$6.81	$12.57	$1.34	$.00	$92.68	
19	13I43	$156.80	$7.00	$9.17	$18.90	$1.85	$7.00	$112.88	
							DEPT. EARNINGS	$280.36	*
20	31B54	$186.57	$2.00	$10.91	$22.46	$2.02	$.00	$149.18	
20	16X33	$75.38	$1.25	$4.41	$9.75	$.78	$.00	$59.19	
20	137M5	$84.17	$.00	$4.92	$8.90	$.84	$.00	$69.51	
							DEPT. EARNINGS	$277.88	*
29	V1388	$100.65	$1.00	$5.89	$16.02	$1.19	$1.75	$74.80	
29	4C193	$116.40	$3.00	$6.81	$12.57	$1.34	$.00	$92.68	
29	2N270	$156.80	$2.00	$9.17	$18.90	$1.85	$2.00	$122.88	
							DEPT. EARNINGS	$290.36	*
30	52Q16	$186.57	$7.00	$10.91	$22.46	$2.02	$.00	$144.18	
30	322D6	$67.70	$.00	$3.96	$4.80	$.60	$.75	$57.59	
30	7U223	$75.18	$1.25	$4.40	$9.75	$.78	$.00	$59.00	
30	2543S	$84.17	$.00	$4.92	$8.90	$.84	$.00	$69.51	
							DEPT. EARNINGS	$330.28	*
39	2R598	$100.64	$1.00	$5.89	$16.02	$1.19	$1.75	$74.79	
39	226T3	$116.40	$3.00	$6.81	$12.57	$1.34	$.00	$92.68	
							DEPT. EARNINGS	$167.47	*
							TOTAL NET EARNINGS	$2,086.96	**

Questions for Review

1. What is the purpose of sequence control statements?
2. What is a conditional statement and how is it used to control the flow of data through a COBOL program?
3. What is the starting point in the Procedure Division?
4. How do the sequence control statements alter the sequence of program execution?
5. What is a test condition?
6. What is an IF statement?
7. What is a simple condition?
8. How is the IFTHENELSE structure used in COBOL programming?
9. What is a relation condition and how is it used?
10. What is the difference between comparison of numeric operands and nonnumeric operands?
11. What is a class test condition?
12. What is a sign test condition?
13. What is a condition-name condition?
14. What is a compound condition?
15. What are the logical operators and what are their meanings?
16. What is a nested condition?
17. What is an implied subject? an implied operator?
18. What is the function of the GO TO statement?
19. What is the difference between a GO TO statement and a PERFORM statement?
20. Explain the operation of the GO TO statement with the DEPENDING ON option, and how COBOL implements it in the CASE structure.
21. Why and when is the EXIT statement used?
22. What are the two types of STOP statements?

Matching Questions

Match each item with its proper description.

_____ 1. EXIT	A. Tests to see whether the item is alphabetic or numeric.
_____ 2. IF	B. Tests whether a variable has a specific value or one of a set of values.
_____ 3. GO TO	C. Tests a single condition to determine which of two function blocks will be executed.
_____ 4. Class condition	D. Evaluation of action whether true or false.
_____ 5. STOP	E. Value of item is less than, greater than, or equal to zero.
_____ 6. Condition-name condition	F. Transfers conditionally or unconditionally from one part of a program to another.
_____ 7. Relation condition	G. Selects one set of functions for execution depending on value of integer or identifier.
_____ 8. IFTHENELSE structure	H. Provides common end point for a series of procedures.
_____ 9. Sign condition	I. Compares values between two items.
_____10. CASE structure	J. Temporary or permanent halt to a program.

Exercises

Multiple Choice: Indicate the best *answer (questions 1-37).*

1. A GO TO statement causes a branch to a
   a. statement.
   b. sentence.
   c. procedure.
   d. division.
2. A permanent stop is indicated when the verb STOP is followed by
   a. a numeric literal.
   b. the word RUN.
   c. the word PROCESS.
   d. a nonnumeric literal.
3. Since a condition-name test is another way of testing whether a data item is equal to a literal, any condition-name test could be replaced by a
   a. relation test.
   b. sign test.
   c. class test.
   d. overflow test.
4. After a GO TO statement has caused control to branch to the beginning of a procedure,
   a. control immediately branches back to the statement after GO TO.
   b. the normal flow of control is resumed.
   c. control flows through that procedure and then returns to the GO TO.
   d. None of the above.
5. When an IF statement does not contain an ELSE statement, control jumps to the next
   a. statement.
   b. sentence.
   c. paragraph.
   d. procedure.
6. An EXIT statement must appear in a separate
   a. paragraph.
   b. statement.
   c. sentence.
   d. division.
7. When a GO TO with the DEPENDING ON option is executed,
   a. control is always transferred to the first paragraph listed.
   b. only one specific value of identifier will cause a transfer of control.
   c. control is transferred to the paragraph whose position is represented by the value of the identifier.
   d. control is always transferred to the next statement.
8.
```
IF STUDENT-NUMBER IS EQUAL TO ID-NUMBER
        PERFORM READ-CARD
ELSE
        NEXT SENTENCE.
```
   The COBOL statement just given specifies a test condition. The test condition is
   a. STUDENT-NUMBER IS EQUAL TO ID-NUMBER.
   b. IF STUDENT-NUMBER IS EQUAL TO ID-NUMBER.
   c. PERFORM READ-CARD.
   d. None of the above.
9. A test condition is
   a. a relation condition.
   b. used to compare the value of a variable with the value of a literal.
   c. used to specify a relationship between two items.
   d. All of the above.

10. IF COUNTER IS LESS THAN 1000
        PERFORM READ1
    ELSE
        PERFORM READ2.

    When the IF statement just given is executed,

    a. control will transfer to READ1 if counter is equal to 2000.
    b. the specified relational condition will be evaluated.
    c. if counter is equal to 500, the result of the relational condition will be false.
    d. None of the above.
11. **An example of a compound condition using OR is**
    a. NUMBER-1 OR GREATER THAN NUMBER-2.
    b. OR NUMBER-1 GREATER THAN NUMBER-2.
    c. NUMBER-1 GREATER THAN NUMBER-2 OR SWITCH EQUAL TO 1.
    d. None of the above.
12. (1) IF PAY-CODE EQUAL TO 2
            PERFORM WEEK-RATE
        ELSE
            NEXT SENTENCE.
    (2) IF WEEKLY
            PERFORM WEEK-RATE
        ELSE
            NEXT SENTENCE.

    In the two statements just given,

    a. WEEKLY is a relational condition.
    b. PAY-CODE is a condition-name.
    c. PAY-CODE is equivalent to WEEKLY.
    d. **None of the above.**
13. IF MARITAL-STATUS EQUAL TO 'S'
    IF DEPENDENTS NOT EQUAL TO 1
        PERFORM SPECIAL-CODE

    The coding just given represents

    a. a nested IF statement.
    b. a series of two simple IF statements.
    c. a compound condition.
    d. two nested IF statements.
14. To transfer control to a paragraph in a program and then return control to the statement directly following the point of transfer, you would use
    a. a GO TO statement.
    b. two PERFORM statements.
    c. a PERFORM statement.
    d. a relation condition.
15. In the sign test, zero is considered to be neither positive nor negative. This means that an item whose value is zero,
    a. must not have an operational sign.
    b. can have a sign but it is ignored.
    c. has a special + (plus) or − (minus) sign.
    d. cannot have a sign at all.
16. GO TO A B C DEPENDING ON FIELD.

    In the statement just given, FIELD must be

    a. an elementary numeric item.
    b. a group numeric item.
    c. an elementary alphanumeric item.
    d. a group alphanumeric item.
17. The GO TO statement with the DEPENDING ON option is a switching technique. Which following statement is correct?
    a. If the value of the data item is zero, or if its value is greater than the number of named procedures, control is passed to the next statement after the GO TO.
    b. If the value of the data item is 1, control will go to the second procedure named in the list.
    c. The GO TO statement lists the name of the procedure to which control does not branch.
    d. An unlimited number of procedures can be named, which need not correspond to the list named.

18. A valid relational condition can contain
   a. two literals as operands.
   b. an alphanumeric operand and an alphabetic operand.
   c. a numeric literal and an alphanumeric operand.
   d. a group item and a numeric operand.
19. GO TO TOTAL.
   In the statement just given, a restriction on TOTAL is that it must not contain a
   a. nested IF statement.
   b. PERFORM statement.
   c. GO TO statement.
   d. None of the above.
20. PERFORM TOTAL.
   In the statement just given, a restriction on TOTAL is that it must not contain a
   a. nested IF statement.
   b. PERFORM statement.
   c. GO TO statement that does not return control to TOTAL.
   d. None of the above.
21. A statement that gives control temporarily to another procedure is
   a. an IF statement.
   b. a PERFORM statement.
   c. a STOP statement.
   d. a GO TO statement.
22. When a test condition is evaluated, the following action takes place.
   a. If the condition is true, statements immediately following ELSE are executed.
   b. If the condition is true, statements immediately following the conditional expression are executed.
   c. A limited number of statements must follow the test condition.
   d. An unconditional GO TO or STOP RUN statement may appear anywhere in the series of statements executed if the conditional statement is true.
23. A simple condition is
   a. a relation condition.
   b. a class condition.
   c. a sign condition.
   d. All of the above.
24. An IFTHENELSE structure
   a. provides loop capability.
   b. selects one of a set of functions for execution depending on the value of the integer-identifier.
   c. tests a single condition to determine which of two function blocks will be executed.
   d. is writing statements in succession.
25. A relation condition can compare
   a. two numeric operands regardless of usage clauses.
   b. a group item and a nonnumeric operand.
   c. an alphabetic operand and an alphanumeric operand.
   d. All of the above.
26. The following phrases are used in relational operations.
   a. More than.
   b. Less than.
   c. Smaller than.
   d. All of the above.
27. The following is an *invalid* class condition test.
   a. Alphabetic.
   b. Alphanumeric.
   c. Numeric.
   d. None of the above.
28. An *invalid* sign condition test is
   a. POSITIVE.
   b. NEGATIVE.
   c. NUMERIC.
   d. ZERO.

29. A condition-name is defined in the Data Division with
    a. a level-66 entry.
    b. a level-88 entry.
    c. a level-77 entry.
    d. any level number from 1 through 49.
30. A compound condition is produced simply by tying two or more conditions with the words
    a. ALSO.
    b. ELSE.
    c. NOR.
    d. None of the above.
31. In a truth value of a conditional expression,
    a. the OR relation will be evaluated first.
    b. the AND relation will be evaluated first.
    c. OR and AND will be evaluated in the sequence written.
    d. OR and AND will be alternately evaluated.
32. The following is *not* a logical operator.
    a. AND.
    b. NO.
    c. NOT.
    d. OR.
33. COBOL allows the omission of certain terms in certain forms of compound expressions, such as
    a. the omission of objects.
    b. the omission of relations with different subjects.
    c. the omission of a subject.
    d. All of the above.
34. A GO TO statement
    a. causes control to be transferred back to the statement GO TO after execution of the procedure.
    b. causes control to be transferred to the first statement of a procedure.
    c. cannot be used for looping.
    d. None of the above.
35. The CASE structure is implemented in COBOL through the use of
    a. only PERFROM statements.
    b. a series of IF statements.
    c. a GO TO statement with the DEPENDING ON option.
    d. an IF statement with a GO TO statement.
36. A STOP statement with the literal option will cause the computer to
    a. stop temporarily and then resume after the computer operator takes some corrective action.
    b. stop permanently.
    c. print message and stop permanently.
    d. abort the program.
37. The EXIT statement is used to
    a. terminate a program.
    b. provide an end point for a series of procedures.
    c. terminate the compilation of a program.
    d. provide an exit point for a series of IF statements.
38. In this IF sentence, pick out the test condition and the statements to be acted on if the test condition is true. Also identify the test condition.

```
IF NET IS LESS THAN MINIMUM
      COMPUTE DEFICIT = MINIMUM − NET
ELSE
      NEXT SENTENCE.
```

39. The two following sentences do not mean the same thing. Explain the difference between them.

```
(1) IF SALES IS LESS THAN QUOTA
          MOVE 'BELOW' TO MEMO
    ELSE
          NEXT SENTENCE.
    ADD SALES TO YEAR-TO-DATE-SALES.
(2) IF SALES IS LESS THAN QUOTA
          MOVE 'BELOW' TO MEMO
          ADD SALES TO YEAR-TO-DATE-SALES
    ELSE
          NEXT SENTENCE.
```

40. Write the GO TO statement with the DEPENDING ON option, which will cause the same branches as the following series of IF statements.

IF SUB IS EQUAL TO 1 GO TO MANAGER
IF SUB IS EQUAL TO 2 GO TO ANALYST
IF SUB IS EQUAL TO 3 GO TO PROGRAMMER
IF SUB IS EQUAL TO 4 GO TO OPERATOR.

41. GO TO HOUR-RATE, WEEK-RATE, MONTH-RATE,
DEPENDING ON PAYCODE.

Match the effects of the statements just given with the assumed values of PAYCODE.

Assumed values of PAYCODE	*Action to be performed*
______ 1. 9	A. Control is transferred to HOUR-RATE.
______ 2. 3	B. Control is transferred to WEEK-RATE.
______ 3. 1	C. Control is transferred to MONTH-RATE.
______ 4. 0	D. Control is transferred to PAYCODE.
	E. Control is transferred to statement following GO TO.

42. Match the four types of simple test conditions on the left with the four names on the right.

______ 1. Relation test	A. ALPHABETIC
______ 2. Sign test	B. MARRIED
______ 3. Class test	C. IS NOT EQUAL TO
______ 4. Condition-name test	D. NEGATIVE

43.

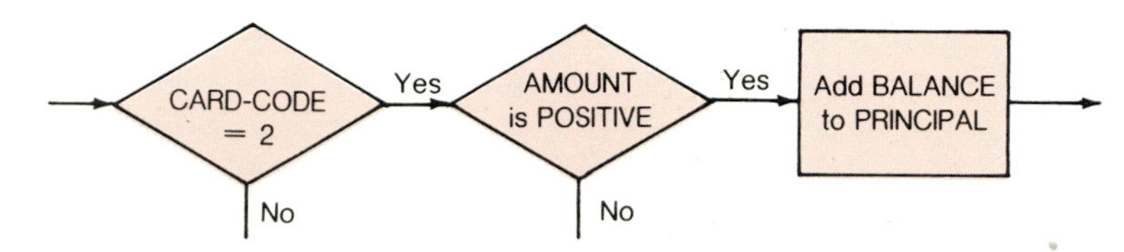

Select the proper coding for the procedures just given.

a. IF CARD-CODE IS EQUAL TO 2 OR AMOUNT IS POSITIVE, ADD BALANCE TO PRINCIPAL.
b. IF CARD-CODE IS EQUAL TO 2 AND AMOUNT IS POSITIVE, ADD BALANCE TO PRINCIPAL.

44. Write the conditional statements for the following:
   a. When the account number MAST-ACCT of the master file is not equal to the transaction account number, TRANS-ACCT, display 'SEQUENCE ERROR' TRANS-ACCT, on the console; otherwise continue regular processing.
   b. When the stock of parts, TOTAL-UNITS, is below the minimum REORDER-POINT, transfer the stock number PART-NO to the reorder report, REORDER-NUMBER; otherwise continue processing.
   c. When the line counter, LINE-COUNT, exceeds the specified number of lines per page, LINE-CONSTANT, control is passed to a page change routine, PAGE-CHANGE; otherwise a "1" (ONE-CONSTANT) is added to the line counter, LINE-COUNT.

45. For a given application, it is necessary to select all records that contain an item number equal to 41571, 58001 through 59720 or 64225.
Write the necessary procedural statements to accomplish the aforementioned.

46. A percentage of sales is offered to each salesman as a commission. The percentage differs if the item sold is class A, B, C or D. If the item sold is not class A, B, C or D, no commission is calculated and the program proceeds to the NO-COMM-ROUTINE.
Compute the sales commission based on the following rates.
CLASS A—equal to or less than 1,000, the commission is 6%.
—greater than 1,000 but less than 2,000, the commission is 7%.
—2,000 or greater, the commission is 10%.
CLASS B—less than 1,000, the commission is 4%.
—1,000 or greater, the commission is 6%.
CLASS C—all amounts, the commission is 4 1/2%.
CLASS D—all amounts, the commission is 5%.
The result of the computation is stored in COMMISSION.
Upon completion of the computation, proceed to NEXT-ROUTINE.
Write the necessary procedural statements for the aforementioned.

47. Using the following data values, solve the conditional statement by determining whether the tested condition is true or false.

A1 = 5	B1 = 3	C1 = 8	D1 = 15
A2 = 4	B2 = 3	C2 = 4	D2 = 15

Statement:
IF A1 = B1 OR (C1 IS LESS THAN D1 AND (B2 = C2 OR A1 IS GREATER THAN A2) and A2 = C2) OR C1 is GREATER THAN D2 and D1 is EQUAL TO D2.

48. **Write the single IF sentence that corresponds to the following flowchart. (Assume that all names have been defined. The decisions are condition-name tests.)**

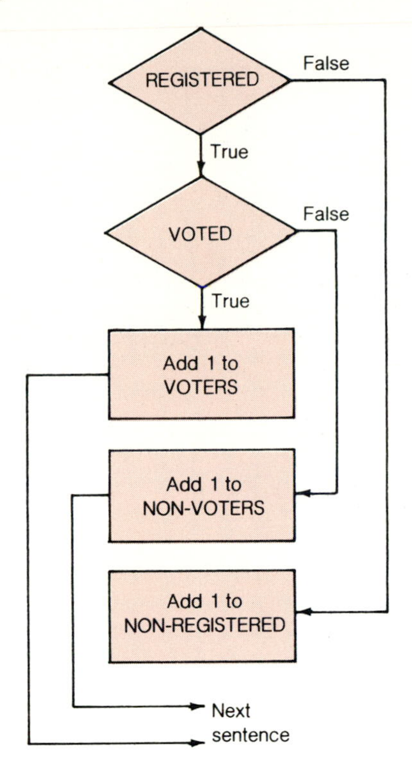

49. **Write the necessary procedural statements to accomplish the following:**

a.

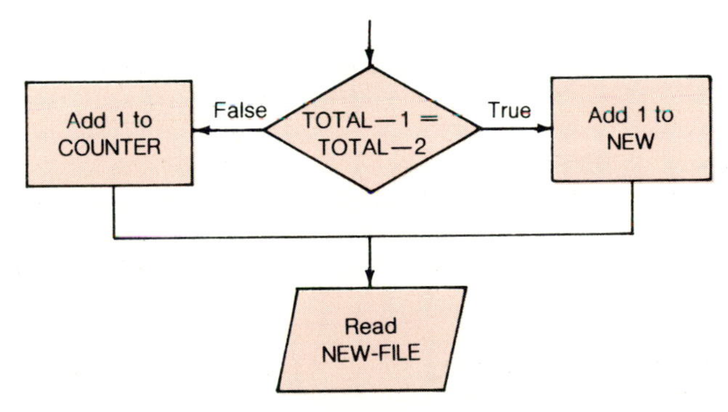

b.

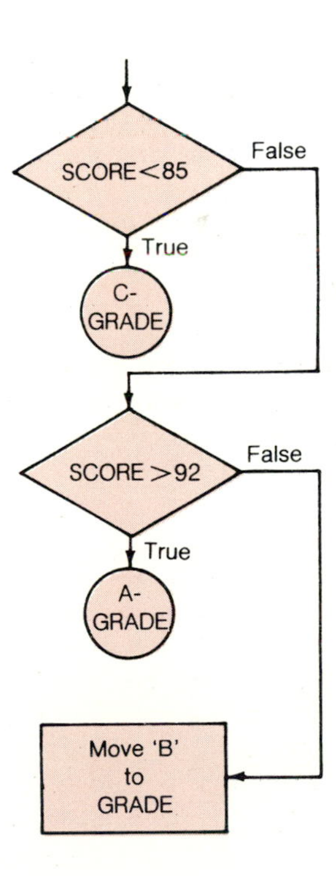

50. LOOP-1 SECTION.
SETUP-LOOP.

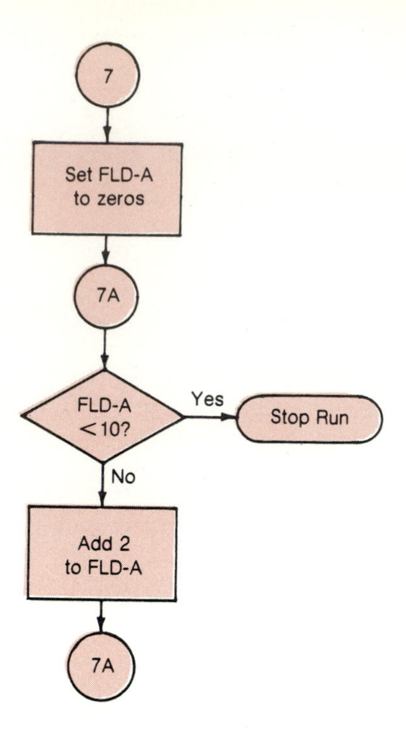

Write the Procedure Division entries for the above.

51. Largest number problem.

There are three unequal numbers labled NUM-A, NUM-B and NUM-C. Write the IF and MOVE statements necessary to move the largest number of the three to FLD-1, the next largest number to FLD-2 and the smallest number to FLD-3 and proceed to ROUT-X.

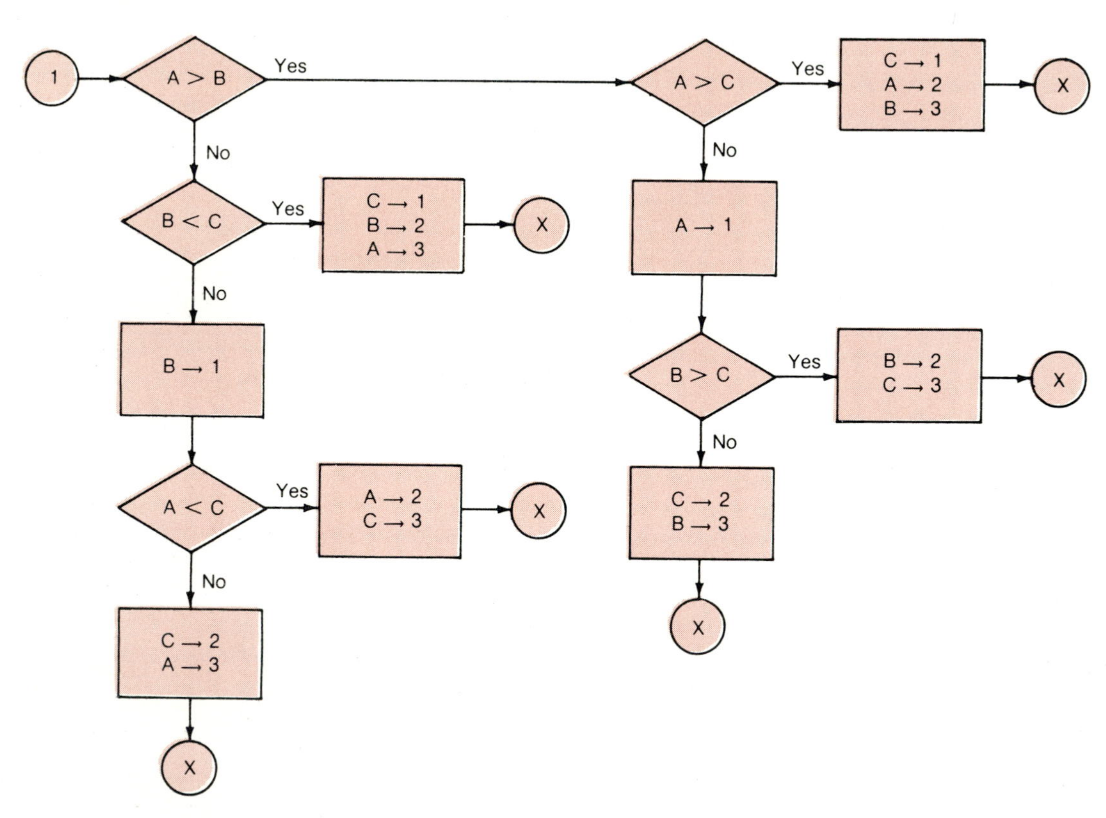

52. One of the many problems in payroll is the computation of the tax deduction for social security. The law (for 1982) states that 6.7% of the gross pay is to be taken out of each paycheck until a maximum of $2,170.80 (6.7% of $32,400) has been deducted. At this time, the deductions would cease. The flowchart at the right depicts a solution to this problem.

Write the Procedure Division statements to solve the flowchart at the right.

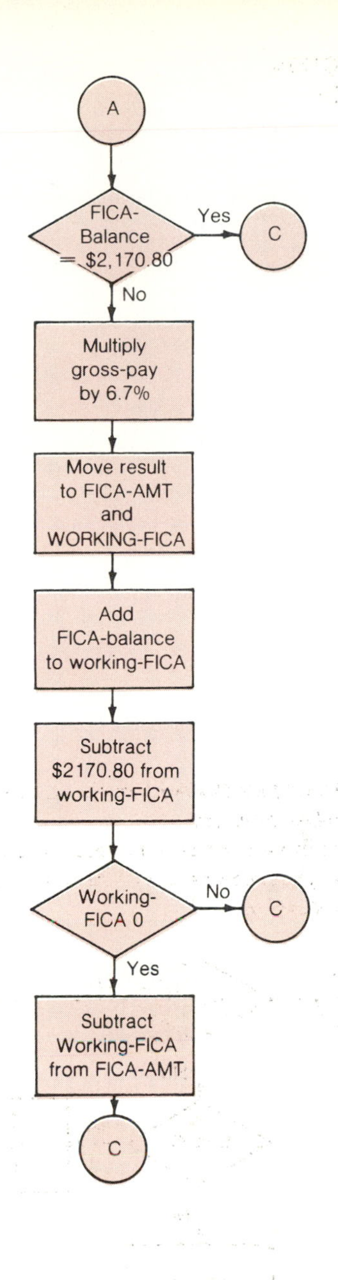

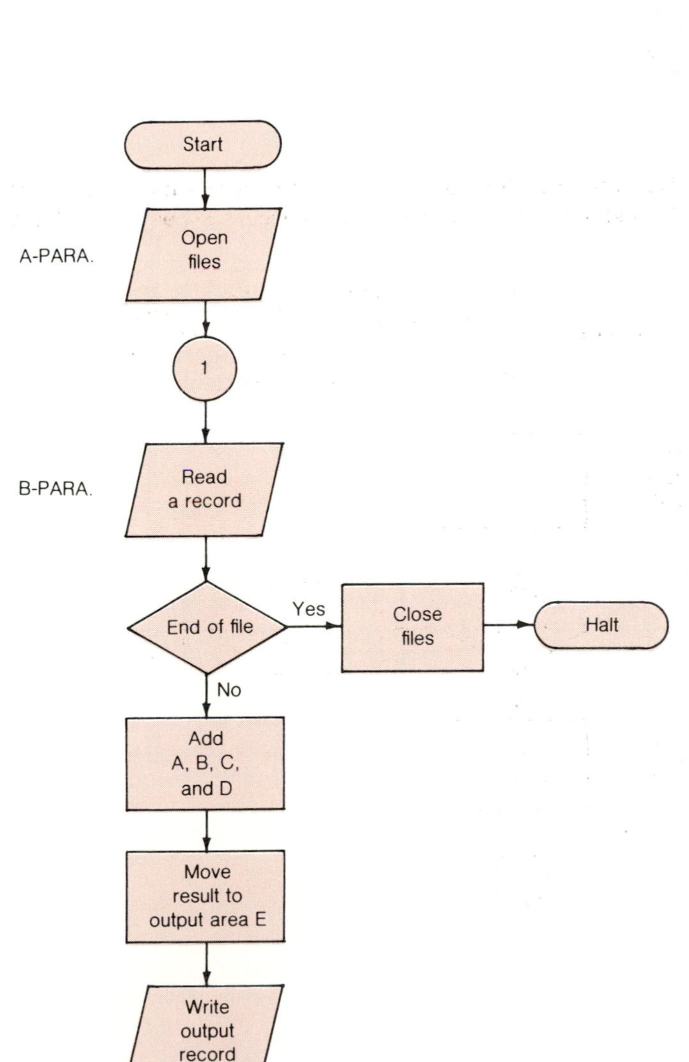

53. The following are the COBOL Procedure Division statements necessary to code the flowchart at the left. Rearrange them in their correct sequence and enter them on a COBOL coding form.

```
STOP RUN.
READ FILE-IN;
OPEN INPUT FILE-IN:
OUTPUT FILE-OUT.
AFTER ADVANCING 2 LINES.
A-PARA.
B-PARA.
GO TO B-PARA.
AT END CLOSE FILE-IN, FILE-OUT;
WRITE E-RECORD.
PROCEDURE DIVISION.
ADD A, B, C, D, GIVING E.
```

Problem 1

Code the necessary entries for the following:

General Description

Prepare a Billings Register report.

Input

A card file called ACCOUNTS; the record layout follows.

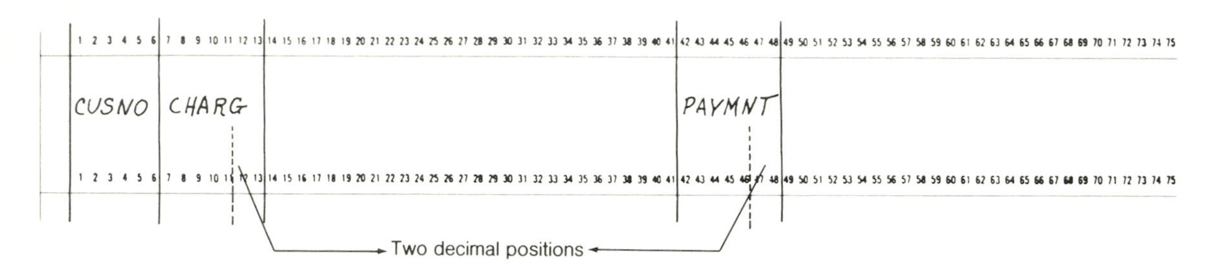

Calculations

Payments are to be compared with charges. If payments are less than charges, the message BILL is to be printed.

Output

Printed report called BILLINGS REGISTER. If payments are less than charges, print the message BILL on that detail line. The Printer Spacing Chart for BILLINGS follows.

Use your own names for files and other fields not defined on the Record Layout or Printer Spacing Chart.

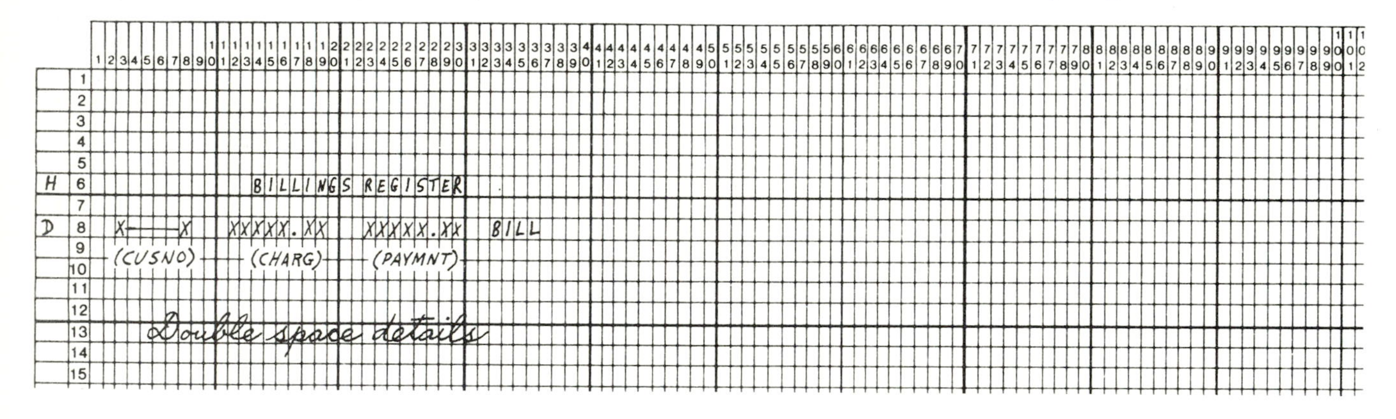

Output is as follows:

```
                BILLINGS REGISTER

245231      530.42      400.00    BILL

246134    5,000.01    4,000.00    BILL

251416      466.67      500.00

319874    2,500.25      500.00    BILL

431942   50,500.50   75,000.00
```

Problem 2

Using the flowchart, record layout, and the printer spacing chart below, write a COBOL program that is responsible for printing a directory of employee telephone numbers. The records are grouped by location so that a count of the number of employees in a given location can be printed. There is also a requirement to print the total number of phone numbers listed in the entire directory. For a given location, the location description should be printed only once.

The input is as follows:

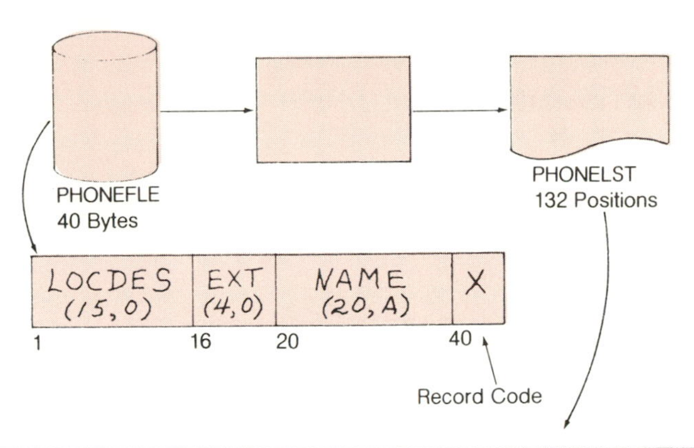

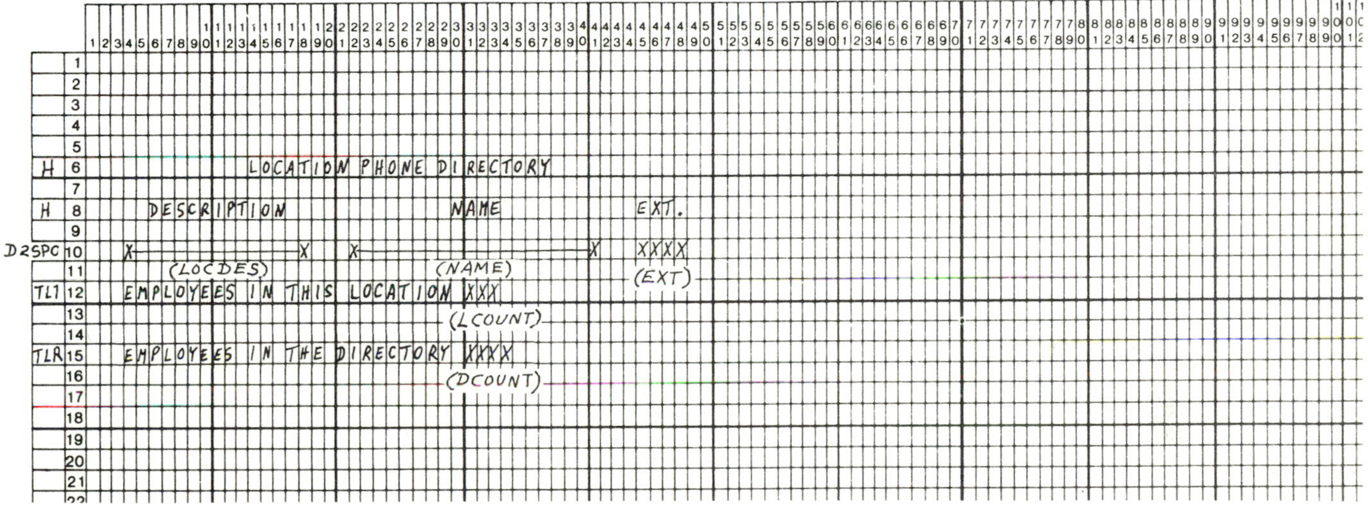

Output is as follows:

```
                 LOCATION PHONE DIRECTORY

   DESCRIPTION                NAME              EXT.

LOAD DEPARTMENT     JOHN CALDWELL               2680

                    BILL BROOKS                 2701

                    DAVID HAMILTON              2712

EMPLOYEES IN THIS LOCATION    3


PAYROLL SECTION     CHUNG LEE                   4014

                    LARRY HOOPER                4115

                    BENNY BROWN                 4216

                    JOHN JOHNSON                4229

                    ROBERT CARLSON              4238

EMPLOYEES IN THIS LOCATION    5


EMPLOYEES IN THE DIRECTORY     8
```

Problem 3

Code the necessary COBOL specifications for the following:

General Description

Prepare a Sales Analysis report.

Input

This is a file called SLSCDS. The Record Layout follows.

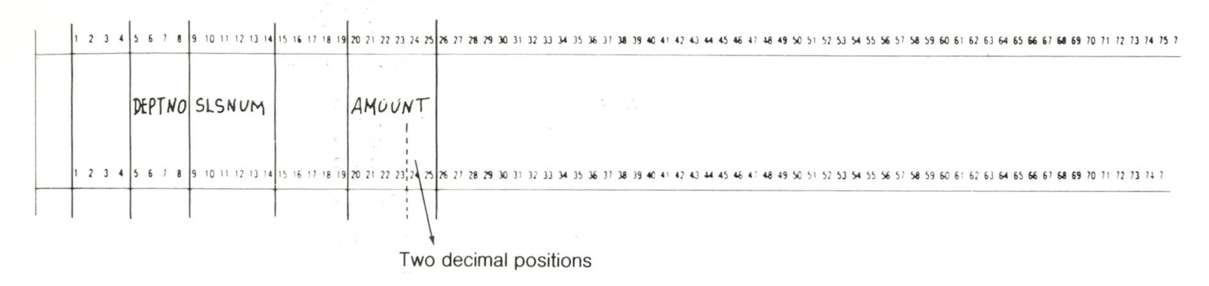

Calculations

For each salesman, accumulate total sales. For each department, accumulate total sales. Accumulate a final total for all departments.

AMOUNT + SLSAMT = SLSAMT
SLSAMT + DEPAMT = DEPAMT
DEPAMT + FINTOT = FINTOT

Output

A printed report called SLSRPT is to be output. The report is to be group-indicated by department and salesman. Print * beside each salesman total, ** beside each department total, and *** beside the final total. Print the current date and page number on each page. Print column headings DEPARTMENT, SALESMAN NO., and AMOUNT on each page.

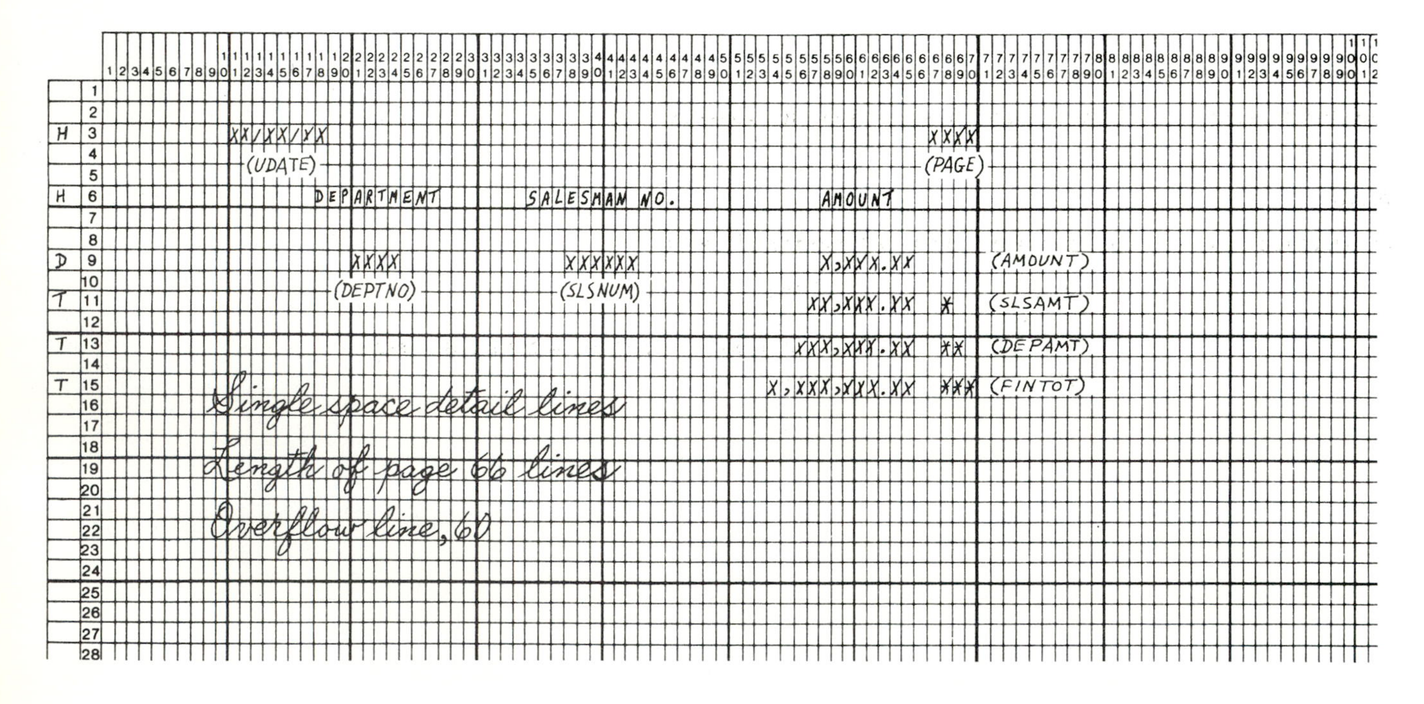

Output is as follows:

```
5/26/81                                                    PAGE     1

                  DEPARTMENT       SALESMAN NO.        AMOUNT

                     1002             791245          1,962.80
                                                      2,531.05
                                                        822.71

                                                      5,316.56  *

                                      789449          3,488.49
                                                        901.52

                                                      4,390.01  *

                                                      9,706.57  **

                     1050             691248            399.23
                                                      1,053.98
                                                         20.75

                                                      1,473.96  *

                                      729231          6,531.14
                                                      8,601.52
                                                        853.20
                                                         85.01

                                                     16,070.87  *

                                      798882          5,149.80

                                                      5,149.80  *

                                                     22,694.63  **

                                                     32,401.20  ***
```

Problem 4

Code the necessary COBOL instructions based on the information below.

Input

The input record is an 80-position record with the following information:

Positions	Field
1-5	Quantity
10-14	List price XXX.XX (two decimal positions)
20	Code

Calculations

Calculations are as per flowchart that follows.

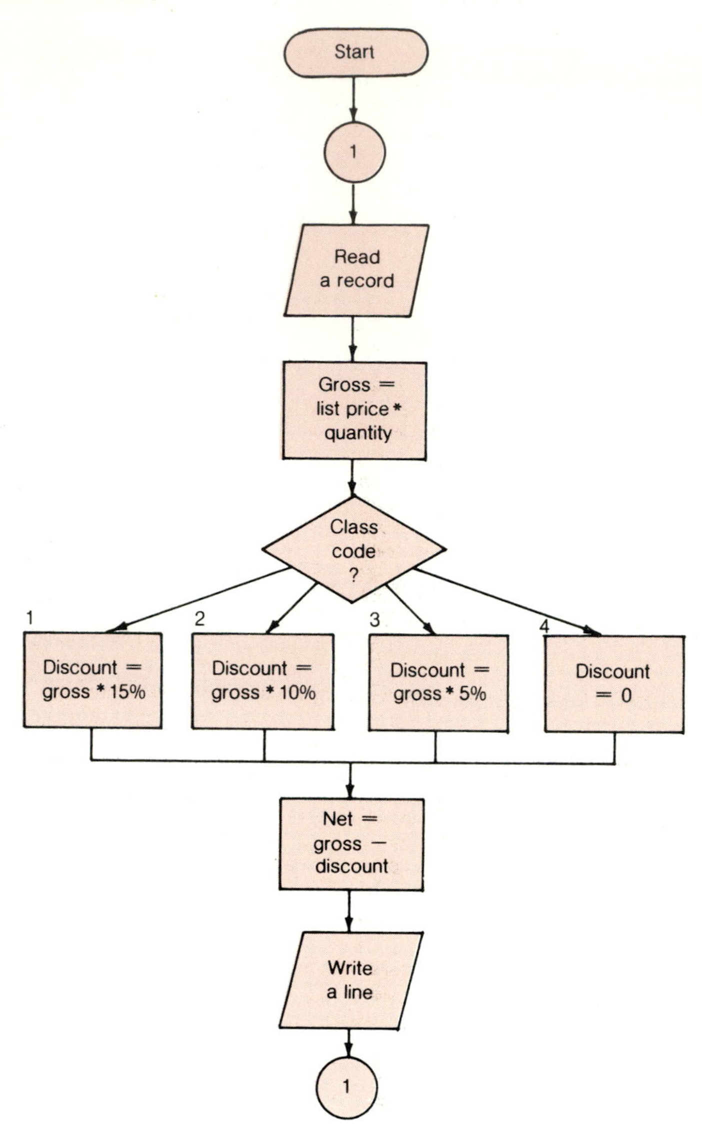

Output

Prepare a Printer Spacing Chart, leaving five spaces between each field and including all fields from input plus the calculated fields. Be sure to include all necessary headings.

Use your own data-names where necessary.

DISCOUNT REPORT

QUANTITY	LIST PR.	CODE	GROSS	DISCOUNT	NET
50	50.00	1	2,500.00	375.00	2,125.00
750	25.00	4	18,750.00	.00	18,750.00
1,500	15.30	3	22,950.00	1,147.50	21,802.50
12,000	9.63	2	115,560.00	11,556.00	104,004.00
500	1.20	2	600.00	60.00	540.00

Procedure Division: Data Manipulation Statements

Chapter Outline

Chapter Objectives

The learning objectives of this chapter are:

1. To identify the results to be obtained from the execution of data manipulation statements based upon the kind of statement and the characteristics of the data specified.
2. To describe the format and uses of the EXAMINE statement and its various options that can be used in data processing applications.
3. To describe the format of the INSPECT statement, the more powerful successor to the EXAMINE statement that is in ANSI 1974 COBOL.
4. To describe the formats and uses of the STRING and UNSTRING statements, which can be used to simplify the task of condensing data so that a larger volume of information can be stored in a data file. Conversely, when data is stored in a condensed format, it may be necessary to convert it into a noncondensed format for processing and/or display.

Introduction

Programs often require that data be moved from one area to another, and/or that operations be performed on the individual characters of a data set. Data manipulation verbs move data from one area to another within the computer, and the inspection of the data is explicit in the function of several of the COBOL verbs. The data manipulation verbs are:

1. The MOVE verb is used to move data from one area in storage to one or more areas within the computer.
2. The EXAMINE verb is used to inspect data with or without the movement of data.
3. The INSPECT verb is an improvement of the EXAMINE verb, and also much more powerful.
4. The STRING and UNSTRING verbs are used with character-strings to join several fields together into one field or to separate one field into several fields respectively.

EXAMINE Statement

Up to 1968, the data manipulation in COBOL was limited to two statements: MOVE and EXAMINE. (MOVE was discussed in chapter 6.) The EXAMINE statement is used to replace a given character and/or to count the number of times a given character appears in a data item. The EXAMINE statement may be used in the manners listed.

1. *To validate the input.* An input data item may have to be checked for validity, such as leading zeros in a numeric field. In some COBOL compilers, the absence of leading zeros (blanks) in a numeric field will cause a data exception error and terminate the program. The EXAMINE verb can be used to inspect the field and replace the leading blanks with zeros, thus allowing the program to continue.
2. *To translate information from one code to another.* Translation refers to the conversion of specified characters within a field to other specified characters. For instance, it may be necessary to translate fields in the Extended Binary-Coded Decimal Interchange Code (EBCDIC) to American National Standard Code for Information Interchange (ASCII) code. In this example, the fields can be translated using a series of EXAMINE statements with the REPLACING option.
3. *Free-form input processing.* The input data is frequently in the form of a series of character strings that have no strictly defined fields, but are separated by a distinct character such as a slash (/). For instance, it is necessary to print address labels where the name and address fields are of varying lengths. The problem is to determine the length of each field (where one field ends and another begins). A slash or any other unique character can be placed at the end of each field. The EXAMINE statement with its TALLYING option can be used to count the number of characters in each field (each character before the slash). TALLY would contain the length of each field.

NOTE: The INSPECT, STRING, and UNSTRING statements can perform the aforementioned in an improved manner.

The EXAMINE statement is used to replace a given character and/or to count the number of times it appears in a data item. The EXAMINE statement for format-1 is shown in figure 9.1.

Rules Governing the Use of the EXAMINE Statement

1. The EXAMINE statement may be applied only to an item whose USAGE IS DISPLAY.
2. Any literal used in the EXAMINE statement must be a number of characters associated with the class specified for the identifier. For example, if the class of the identifier is numeric, each literal in the statement must be numeric and may possess an operational sign. All must be single characters.
3. Nonnumeric literals must be single characters enclosed in quotation marks.
4. The examination of the data item begins with the first (leftmost) character of the data set and proceeds to the right. Each character in the data item specified by the identifier is examined in turn. If the data item is numeric, any operational sign associated with it will be ignored.
5. Figurative constants may be used in place of literal-2, with the exception of the figurative constant ALL.

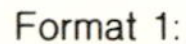

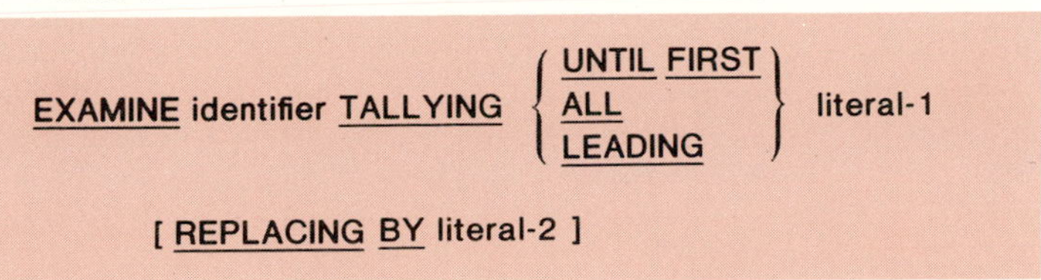

Figure 9.1 Format—EXAMINE statement: format 1.

6. *TALLYING option (format-1).* The TALLYING phrase of the EXAMINE statement is used to scan a data item, counting the number of occurrences of a given literal. The TALLYING phrase causes an integral count to be placed in TALLY (a special register set up by the compiler whose length is five decimal digits). (TALLY may be used in other procedural statements.) The significance of the count in TALLY depends upon which of the following options is specified.

Option	TALLY Value Represents
UNTIL FIRST	The number of occurrences of other characters not equal to the literal encountered prior to the first occurrence of the literal. (See figure 9.2.)
ALL	The number of occurrences of the literal throughout the item. (See figure 9.3.)
LEADING	The number of occurrences of the literal prior to encountering a character other than that literal. (See figure 9.4.)

```
EXAMINE ACCOUNT
    TALLYING UNTIL FIRST 'A'.
```

Data in ACCOUNT before EXAMINE	Data in ACCOUNT after EXAMINE	Contents of TALLY after EXAMINE
1296AA10	No change in data	000004
AA123456		000000
ƀƀƀA 1 2 3		000003
FRTS987123		000010
+A1234		000001

ƀ denotes blank character.

Figure 9.2 EXAMINE statement: format 1—example.

```
EXAMINE GROUP
    TALLYING ALL 9.
```

Data in GROUP before EXAMINE	Data in GROUP after EXAMINE	Contents of TALLY after EXAMINE
ABC9098719	No change in data	000003
****9999.56712		000004
1234567888		000000

Figure 9.3 EXAMINE statement: format 1—example.

```
EXAMINE PART-NUMBER
     TALLYING LEADING 'Z'.
```

Data in PART-NUMBER before EXAMINE	Data in PART-NUMBER after EXAMINE	Contents of TALLY after EXAMINE
ZZZZ90982		000004
ZZ123@ZZ	No change in data	000002
ZJOHNDOE		000001
YOUZZZZZQ		000000
12345		000000

Figure 9.4 EXAMINE statement: format 1—example.

Example:

```
WORKING-STORAGE SECTION.
01  EXAMINE-DATA.
    05  E-1     PICTURE A(10)     VALUE 'ABCADCADCD'.
    05  E-2     PICTURE 9(10)     VALUE 0150250350.
    .
    .
    .
PROCEDURE DIVISION.
EXAMINE-TEST-1.
    EXAMINE E-1
        TALLYING UNTIL FIRST 'D'.
    IF TALLY EQUAL TO 9
        PERFORM PASS
    ELSE
        PERFORM FAIL.
    .
    .
    .
EXAMINE-TEST-2.
    EXAMINE E-2
        TALLYING ALL 5
        REPLACING BY 0.
    IF TALLY EQUAL TO 3
        NEXT SENTENCE
    ELSE
        PERFORM FAIL.
```

7. *REPLACING option (format-2).* The REPLACING phrase of the EXAMINE statement is used to modify the value of an item by replacing certain characters in the original value with a new character. The REPLACING phrase for format-2 is shown in figure 9.5.

   When the REPLACING phrase is used, character replacement proceeds according to the options specified:

Option	Substitute Literal-2 for
ALL	Each occurrence of literal-1.
LEADING	All occurrences of literal-1 prior to the leftmost occurrence of any other character.
FIRST	The first occurrence of literal-1. (See figure 9.6.)
UNTIL FIRST	All occurrences of other characters prior to (but not including) the first occurrence of literal-1.

Format 2

```
                                       ⎧ ALL        ⎫
                                       ⎪ LEADING    ⎪
EXAMINE identifier REPLACING           ⎨ FIRST      ⎬  literal-1
                                       ⎩ UNTIL FIRST⎭

         BY literal-2
```

Figure 9.5 Format—EXAMINE statement—format 2.

Examples:

If a data item named TARGET had the value 'TO COME TO THE AID OF THE PARTY', then the statements:

1. EXAMINE TARGET
   REPLACING ALL 'O' by 'Z'.
   would result in a TARGET value of
   'TZ CZMD TZ THE AID ZF THE PARTY'.

2. EXAMINE TARGET
   REPLACING LEADING 'T' by 'Q'.
   would result in a TARGET value of
   'QO COME TO THE AID OF THE PARTY'.

3. EXAMINE TARGET
   REPLACING FIRST 'H' by 'Z'.
   would result in a TARGET value of
   'TO COM TO TZE AID OF THE PARTY'.

4. EXAMINE TARGET
   REPLACING UNTIL FIRST 'H' by 'Z'.
   would result in a TARGET value of
   'ZZZZZZZZZZZHE AID OF THE PARTY'.

5. When both the TALLYING and REPLACING phrases are specified, each character is tallied according to the aforementioned rules, and is also replaced with the literal specified in the REPLACING phrase. (See figures 9.7, 9.8, and 9.9.)
6. For nonnumeric data items, examination starts at the leftmost character and proceeds to the right. Each character in the data item specified by the identifier is examined in turn.
7. If a data item referred to by the EXAMINE statement is numeric, it must consist of numeric items and may possess an operational sign. Examination starts at the leftmost characters and proceeds to the right. Each character is examined in turn. When the character 'S' is used in the PICTURE character-string of the data item description to indicate the presence of an operational sign, the sign is ignored by the EXAMINE statement.

   The EXAMINE statement is used to manipulate the character within a field. To increase this capability, ANSI 1974 COBOL offers the INSPECT, STRING and UNSTRING statements.

```
EXAMINE LOCATION
    REPLACING FIRST 'E' BY ','.
```

Data in LOCATION before EXAMINE	Data in LOCATION after EXAMINE	Contents of TALLY after EXAMINE
EEENODATA	,EENODATA	Not affected
BILLEMARY	BILL,MARY	
SUEEBILLEMARY	SU,EBILLEMARY	

Figure 9.6 EXAMINE statement: format 2—example.

```
EXAMINE GROUP
    TALLYING ALL 9
    REPLACING BY '*'.
```

Data in GROUP before EXAMINE	Data in GROUP after EXAMINE	Contents of TALLY after EXAMINE
987612345	*87612345	000001
9999999	*******	000007
90909090	*0*0*0*0	000004

Figure 9.7 EXAMINE statement: format 2—example.

```
EXAMINE PART-NUMBER
    TALLYING LEADING 'M'
    REPLACING BY 'T'.
```

Data in PART-NUMBER before EXAMINE	Data in PART-NUMBER after EXAMINE	Contents of TALLY after EXAMINE
MMMM9876	TTTT9876	000004
MM@12@45MM	TT212@45MM	000002
MJOHNDOE	TJOHNDOE	000001
YOUMMM	YOUMMM	000000
909876	909876	000000

Figure 9.8 EXAMINE statement: format 2—example.

```
EXAMINE ACCOUNT
    TALLYING UNTIL FIRST 'A'
    REPLACING BY 'Z'.
```

Data in ACCOUNT before EXAMINE	Data in ACCOUNT after EXAMINE	Contents of TALLY after EXAMINE
09876AB@	ZZZZZAB@	000005
A8761AWE	A8761AWE	000000
JOHN01ZFFF	ZZZZZZZZZZ	000010

Figure 9.9 EXAMINE statement: format 2—example.

INSPECT Statement

The INSPECT statement increased the power of the EXAMINE statement. Although the three formats of the INSPECT statement allow for very complicated data manipulations, its execution is basically the same as that of the EXAMINE statement. Each field named is inspected from left to right, one character at a time, and specified characters are counted or replaced by other characters.

Differences between EXAMINE and INSPECT Statements

The major differences between the two statements are:

1. The INSPECT statement does not use a field TALLY to count the various characters in a field. Instead, it is the programmer's responsibility to create a tally field (in the Working-Storage Section of the Data Division or some other area in the Data Division) and supply this data name of the field after the word TALLYING.
2. The INSPECT statement permits the tallies and replacements to be named in series, so several can be done with one statement.
3. The BEFORE and AFTER options of the INSPECT statement permit some specifications that were not possible with the EXAMINE statement.
4. Whereas the EXAMINE statement permitted only single characters to be counted or replaced, INSPECT allows groups of characters to be counted and replaced.
5. All literals used with INSPECT must be alphanumeric (enclosed in quotation marks).

INSPECT is a COBOL statement that can be used in conjunction with data strings. It will examine the contents of a data item from left to right and will do any of the following: (1) count the number of occurrences of a particular character(s); or (2) replace this particular character(s) with a different character(s); or (3) both.

The INSPECT statement treats the value of each identifier as a character-string. Signed numeric identifiers will be treated as though they had been described as unsigned numeric identifiers during the execution of the INSPECT statement. All identifiers referenced in the INSPECT statement, with the exception of those identifiers that are to receive a tally, will be treated as though they had been described as alphanumeric during the execution of the INSPECT statement, even though they may actually be described as alphanumeric, alphanumeric edited, or unsigned numeric.

Rules Governing the Use of the INSPECT Statement

1. Inspection begins at the leftmost character position of the data item referenced by identifier-1 and proceeds from left to right to the rightmost character position.
2. The operands are considered in the order they are specified in the INSPECT statement, from left to right. The first character-string is compared to an equal number of contiguous characters, starting with the leftmost position in the data item referenced by identifier-1.
3. If no match occurs in the comparison of the first character-string, the comparison is repeated with each successive character-string until a match is found, or there is no next successive character string to be compared. When there is no successive character-string to be compared, the character position in the data item referenced by identifier-1 immediately to the right of the leftmost character position considered in the previous comparison cycle is considered as the leftmost character position, and the comparison cycle begins again with the first character-string to be replaced.
4. When a match occurs, the character position in the data item referenced by identifier-1 immediately to the right of the rightmost character position that participated in the match is considered the leftmost position of the data item referenced by identifier-1, and the comparison cycle continues beginning with the first character-string to be replaced or counted.
5. The comparison operation continues until the rightmost character position of the data item referenced by identifier-1 has participated in a match or has been considered as the leftmost character position. When this occurs, inspection is terminated.
6. *The TALLYING phrase* provides the ability to tally occurrences of strings of one or more characters in length. The TALLYING phrase for format-1 is shown in figure 9.10.

Format 1:

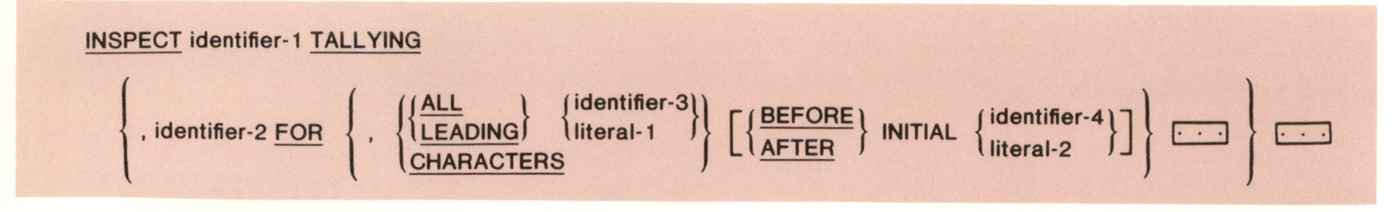

Figure 9.10 Format—INSPECT statement: format 1.

Example:

```
01  ID-1     PICTURE X(8)     VALUE 'ABCDEFGH'.
INSPECT ID-1
    TALLYING COUNTER-FIELD FOR ALL 'CDE'.
```

ABC would be compared to CDE. Since no match would occur, B would become the leftmost character position. BCD would be compared to CDE. Once again, there would be no match and C would then become the leftmost character position. A match would occur, COUNTER-FIELD would be increased by one, and F would become the leftmost character position, etc. Upon completion of the inspection, COUNTER-FIELD would have a value one greater than its value immediately before the execution of the INSPECT statement.

If the *CHARACTERS phrase* is specified, an implied one-character operand participates in the comparison cycle except that no comparison to the contents of the data referenced by identifier-1 takes place. This implied character is always considered to match the leftmost character of the contents of the data item referenced by identifier-1 participating in the current comparison cycle.

Example:

```
01  ID-1     PICTURE X(8)     VALUE 'ABCDEFGH'.
INSPECT ID-1
    TALLYING COUNTER-FIELD FOR CHARACTERS BEFORE INITIAL 'F'.
```

Upon completion of the inspection, COUNTER-FIELD would have a value of five.

7. *TALLYING . . . BEFORE option.* If the BEFORE phrase is specified, the character-string to be tallied participates only in those comparison cycles that involve that portion of the data item referenced by identifier-1 from its leftmost character position up to, but not including, the first occurrence within the data item of the character-string specified in the BEFORE phrase. The position of the first occurrence of the character specified in the BEFORE phrase is determined before the first cycle of the comparison cycle is begun. If there is no occurrence of the character-string specified in the BEFORE phrase within the contents of the data item referenced by identifier-1, then the comparison operation proceeds as though the BEFORE phrase had not been specified.

   Example:

```
77  ID-1    PICTURE X(15)    VALUE 'AFFFBCGHIFFFJKL'.
77  ID-2    PICTURE 99       VALUE 0.
77  ID-3    PICTURE 99       VALUE 0.
77  ID-4    PICTURE 99       VALUE 0.
INSPECT ID-1
    TALLYING ID-2 FOR CHARACTERS BEFORE 'I'.
INSPECT ID-1
    TALLYING ID-3 FOR ALL 'F'.
INSPECT ID-1
    TALLYING ID-4 FOR CHARACTERS BEFORE 'L'.
```

   Upon completion of the three inspections, ID-2, ID-3, and ID-4 would contain values of eight, six, and fourteen, respectively.

8. *TALLYING . . . AFTER option.* If the AFTER phrase is specified, the character-string to be tallied participates only in those comparison cycles that involve that portion of the data item referenced by identifier-1 from that character position immediately to the right of the rightmost character position of the first occurrence within the data item referenced by identifier-1 of the character-string specified in the AFTER phrase, to the rightmost character position of the data item referenced by identifier-1.

   The position of the first occurrence of the character-string specified in the AFTER phrase is determined before the first cycle of the comparison operation. If there is no occurrence of the character-string specified in the AFTER phrase within the data item referenced by identifier-1, then the character-string to be tallied is never eligible to participate in the comparison cycle.

   Example:

```
01  ID-1      PICTURE X(8)      VALUE 'ABCDEFGH'.
01  ID-2      PICTURE 99        VALUE 0.
INSPECT ID-1
    TALLYING ID-2 FOR CHARACTERS AFTER INITIAL 'F'.
```

   The existence of the character-string F in the sixth character position of ID-1 would be determined before the comparison cycle would be executed. Upon completion of the inspection, ID-2 would have a value of 02. The value of ID-2 would have been zero upon completion of the inspection if the INSPECT statement had been written as

```
INSPECT ID-1
    TALLYING ID-2 FOR CHARACTERS AFTER INITIAL 'Z'.
```

9. *ALL option.* If ALL is specified for tallying, the contents of the data item referenced by identifier-2 is incremented for each occurrence of the character-string matched within the contents of the data item referenced by identifier-1.

   Example:

```
01  ID-1      PICTURE X(14)     VALUE 'CCBADAQCCECCFA'.
01  ID-2      PICTURE 99        VALUE 5.
INSPECT ID-1
    TALLYING ID-2 FOR ALL 'A'.
```

   Upon completion of the inspection, the value of ID-2 would be eight.

10. *LEADING option.* If LEADING option is specified for tallying, each contiguous occurrence of the character-string matched in the contents of the data item referenced by identifier-1 would cause the contents of the data item referenced by identifier-2 to be incremented by one, provided that the leftmost occurrence of the character-string is at the point where comparison began in the first comparison cycle in which the character-string being tallied was eligible to participate.

   Example:

```
01  ID-1      PICTURE X(9)      VALUE 'AAABCAAEA'.
01  ID-2      PICTURE 99        VALUE 0.
INSPECT ID-1
    TALLYING ID-2 FOR LEADING 'A' AFTER INITIAL 'C'.
```

   Upon completion of the inspection, the value of ID-2 would be two. Following is an example of an INSPECT statement—format-1—TALLYING option:

```
INSPECT word
    TALLYING count FOR LEADING 'L' BEFORE INITIAL 'A',
        count-1 FOR LEADING 'A' BEFORE INITIAL 'L'.
where word = LARGE, count = 1, count-1 = 0.
where word = ANALYST, count = 0, count-1 = 1.
```

11. The *REPLACING phrase* operates in the same manner as the TALLYING phrase with the same options, except that instead of tallying occurrences of strings of one or more characters, they are

Format 2:

```
INSPECT identifier-1 REPLACING

{  CHARACTERS BY {identifier-6} [{BEFORE} INITIAL {identifier-7}]                                     }
                 {literal-4   }  {AFTER }          {literal-5   }
{{ {ALL    } {, {identifier-5} BY {identifier-6} [{BEFORE} INITIAL {identifier-7}]} ...} ...          }
{{ {LEADING} {  {literal-3   }    {literal-4   }  {AFTER }          {literal-5   }}     }             }
{{ {FIRST  } {                                                                  }     }             }
```

Figure 9.11 Format—INSPECT statement: format 2.

replaced by character-strings of equal length. The REPLACING phrase for format-1 is shown in figure 9.11.

Example:

```
01   ID-1        PICTURE X(8)        VALUE 'ABCDEFGH'.
INSPECT ID-1
     REPLACING ALL 'CDE' BY 'XYZ'.
```

Upon completion of the inspection, ID-1 would have a value of 'ABXYZFGH'.

```
01   ID-1        PICTURE X(8)        VALUE 'ABCDEFGH'.
INSPECT ID-1
     REPLACING CHARACTERS BY 'X'.
```

Upon completion of the inspection, ID-1 would have a value of 'XXXXXXXX'.

a. *REPLACING . . . BEFORE option.*

Example:

```
77   ID-1        PICTURE X(15)       VALUE 'AFFFBFGHIFFFJKL'.
INSPECT ID-1
     REPLACING ALL 'F' BY 'X' BEFORE 'H'.
```

would result in the value 'AXXXBXGHIFFJKL' for ID-1.

b. *REPLACING . . . AFTER option.*

Example:

```
01   ID-1        PICTURE X(8)        VALUE 'ABCDEFGH'.
INSPECT ID-1
     REPLACING CHARACTERS BY 'X' AFTER INITIAL 'F'.
```

Upon completion of the inspection, ID-1 would have a value of 'ABCDEFXX'.

c. *ALL option.*

Example:

```
01   ID-1        PICTURE X(14)       VALUE 'CCBADAQCCECCFA'.
INSPECT ID-1
     REPLACING ALL 'A' BY 'B'.
```

Upon completion of the inspection, the value of ID-1 would be 'CCBBDBQCCECCFB'.

d. *LEADING option.*

Example:

```
01   ID-1        PICTURE X(9)        VALUE 'AAABCAAEF'.
INSPECT ID-1
     REPLACING LEADING 'A' BY 'X' AFTER INITIAL 'C'.
```

Upon completion of the inspection, the value of ID-1 would be 'AAABCXXEF'.

e. *FIRST option.* This is an additional option available with the REPLACING phrase. The left-most occurrence of the matched data item is replaced.

Example:

```
01  ID-1     PICTURE X(11)     VALUE 'CCBADQCAEAF'.
INSPECT ID-1
    REPLACING FIRST 'A' BY 'B' AFTER INITIAL 'Q'.
```

Upon completion of the inspection, the value of ID-1 would be 'CCBADQCBEAF'.

Following are examples of an INSPECT statement—format-2—REPLACING option:

f. Assume that it is necessary to change all of the delimiters in CONDENSED-RECORD (following) from strokes (/) to asterisks (*).

```
DOE/JOHN/12345/MAIN/AVE./ANYTOWN/MO./12345*
```

The coding would be as follows:

```
INSPECT CONDENSED-RECORD
    REPLACING ALL '/' BY '*'.
```

after the execution of the aforementioned statement, the contents of CONDENSED-RECORD would be as follows:

```
DOE*JOHN*12345*MAIN*AVE.*ANYTOWN*MO.*12345*
```

g.
```
INSPECT word
    REPLACING ALL 'A' BY 'G' BEFORE INITIAL 'X'.
where word = ARXAX, word = GRXAX.
where word = HANDAX, word = HGNDGX.
```

h.
```
INSPECT word
    REPLACING ALL    'X' BY 'Y'
                     'B' BY 'Z'
                     'W' BY 'Q' AFTER INITIAL 'R'.
where word = RXXBQWY, word = RYYZQQY.
where word = YZACDWBR, word = YZACDWZR.
where word = RAWRXEB, word = RAQRYEZ.
```

i.
```
INSPECT word
    REPLACING CHARACTERS BY 'B' BEFORE INITIAL 'A'.
word before:  1 2   X Z A B C D
word after:   B B B B B A B C D
```

12. *Multiple replacement phrases.* The order in which a REPLACING phrase is specified can be significant particularly when there is a similarity in the character-strings that are to be replaced.

The following example demonstrates how the INSPECT statement would be executed when more than one phrase is specified for replacement.

```
05  ID-1     PICTURE X(14)     VALUE 'EEACEGKQGGKCAG'.
INSPECT ID-1
    REPLACING ALL    'A' BY 'E'
                     'C' BY 'D'
        LEADING      'E' BY 'F'
                     'G' BY 'H' AFTER INITIAL 'Q'
        FIRST        'K' BY 'L'.
```

Upon completion of the inspection, the value of ID-1 would be 'FFEDEGLQHHKDEG'.

13. *TALLYING and REPLACING options together.* A statement with both TALLYING and REPLACING options is interpreted and executed as though two successive INSPECT statements specifying the same identifier-1 had been written with one statement with TALLYING phrases identical to those specified in the TALLYING and REPLACING statement, and the other statement being a statement with REPLACING phrases identical to those specified in the TALLYING

Format 3:

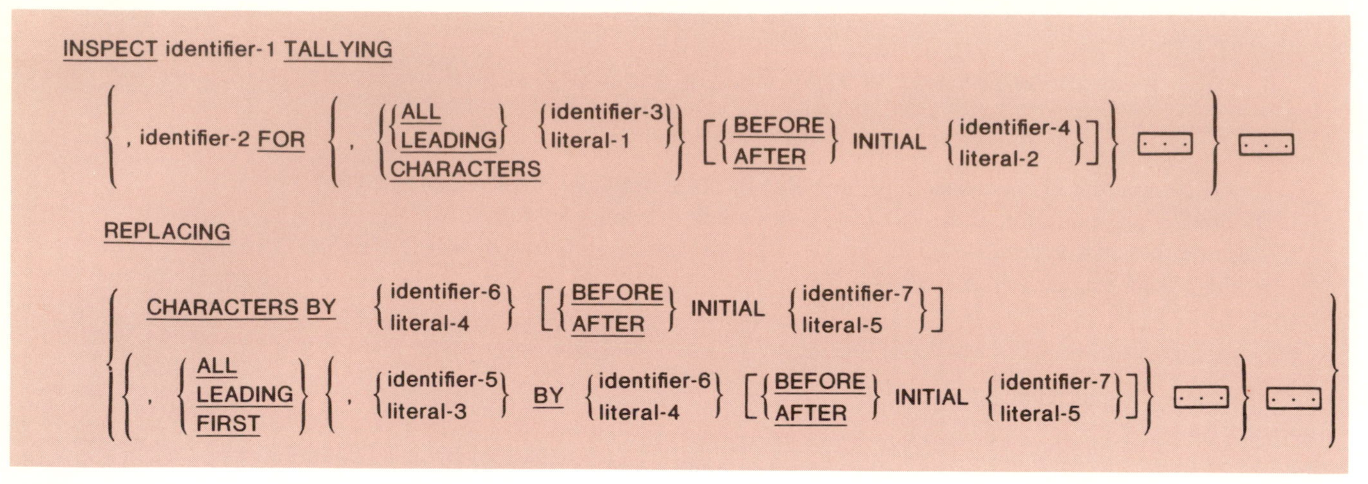

Figure 9.12 Format—INSPECT statement: format 3.

and REPLACING statement. The INSPECT statement for format-3 is shown in figure 9.12. The same general rules for matching and counting apply. Following are examples of an INSPECT statement—format-3—TALLYING and REPLACING options:

a. INSPECT word
   TALLYING count FOR ALL 'L'
   BEFORE REPLACING LEADING 'A' BY 'E' AFTER INITIAL 'L'.
   where word = CALLAR, count = 2, word = CALLAR.
   where word = SALAMI, count = 1, word = SALEMI.
   where word = LATTER, count = 1, word = LETTER.

b. INSPECT word
   TALLYING count FOR CHARACTERS AFTER INITIAL 'J'
   BEFORE REPLACING ALL 'A' BY 'B'.
   where word = ADJECTIVE, count = 6, word = BDJECTIVE.
   where word = JACK, count = 3, word = JBCK.
   where word = JUJMAB, count = 5, word = JUJMBB.

STRING and UNSTRING Statements

It is often desirable to condense data so that a larger volume of information can be stored in a data file. Conversely, when data is stored in a condensed format it may be necessary to convert it into a noncondensed format for processing or display. The STRING and UNSTRING statements can simplify these tasks.

The STRING statement causes characters from one or more data items to be transferred to a single data item. Each transfer is terminated because of the size of the sending and receiving data items, or because of the presence of a specific character(s) in the sending data item. Other features of the STRING statement are: no padding with blanks occurs; a pointer keeps track of the next available position in the receiving data item; and an ON OVERFLOW condition test can be included in the statement to signal the end of the receiving data item.

It is frequently necessary to transfer data from a condensed format to a noncondensed format. The UNSTRING statement can be used to perform this function.

STRING Statement

When a STRING statement is executed, several fields are joined together to form one field. This process is known as *concatenation.* The DELIMITED BY phrase allows the programmer to specify one or

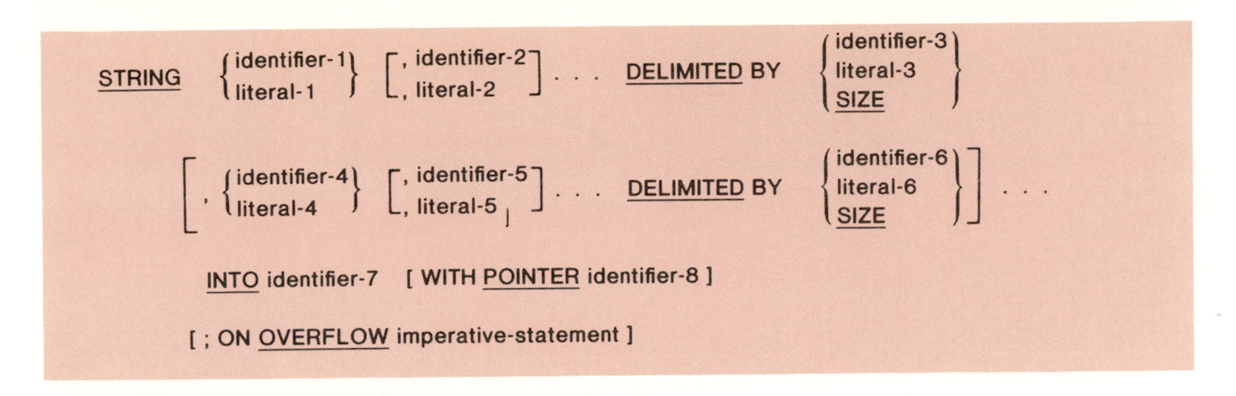

Figure 9.13 Format—STRING statement.

more characters to mark the end of the field to be strung. Only those characters in the sending field to the left of the delimiter are used in the STRING operation.

When the option DELIMITER BY SIZE is used, this means that the entire field is to be used in the STRING operation. If the delimiter is specified as a character (a nonnumeric literal enclosed in quotation marks), only the characters before the specified character are moved. (The specified character itself is not moved.)

When the POINTER option is used, the data is moved to the receiving field starting at the position indicated by the pointer field. (It is the programmer's responsibility to create a field for the pointer with an initial value.) Then every time a character is moved to the receiving field, the pointer field is increased by one. The initial value of the pointer field can never be less than one or more than the length of the receiving field.

The ON OVERFLOW option is used to signal (1) when the end of the receiving field is reached and when the sending field still contains characters waiting to be moved, or (2) when the pointer field does not contain an acceptable value (less than one or more than the length of the receiving field). Usually, the imperative statement that follows the ON OVERFLOW phrase is a statement that causes an overflow routine to be performed.

Explanation of the STRING Statement

In the STRING format, identifiers and literals 1, 2, 4 and 5 represent *sending items.* Identifiers and literals 3 and 6 indicate the character(s) *delimiting the move.* Identifier 7 represents the *receiving field.* Identifier 8 specifies the *pointer.* The general format for the STRING statement is shown in figure 9.13.

Characters are transferred from the sending fields to the receiving field in left-to-right order. The first named field (identifier-1) is moved to the leftmost positions of the receiving field. The next named complete sending field (identifier-2) is then moved immediately to the right of the filled positions of the receiving field. This process continues until either all of the sending fields have been transferred or the receiving field is full. The following illustration graphically describes the transfer of data. (See figure 9.14.)

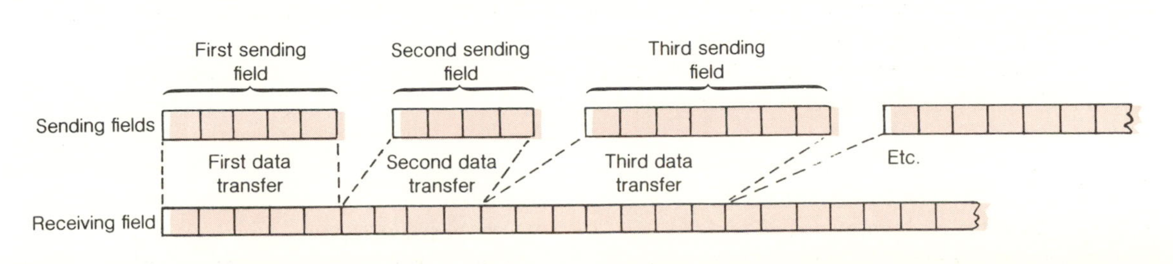

Figure 9.14 STRING statement: transfer of data.

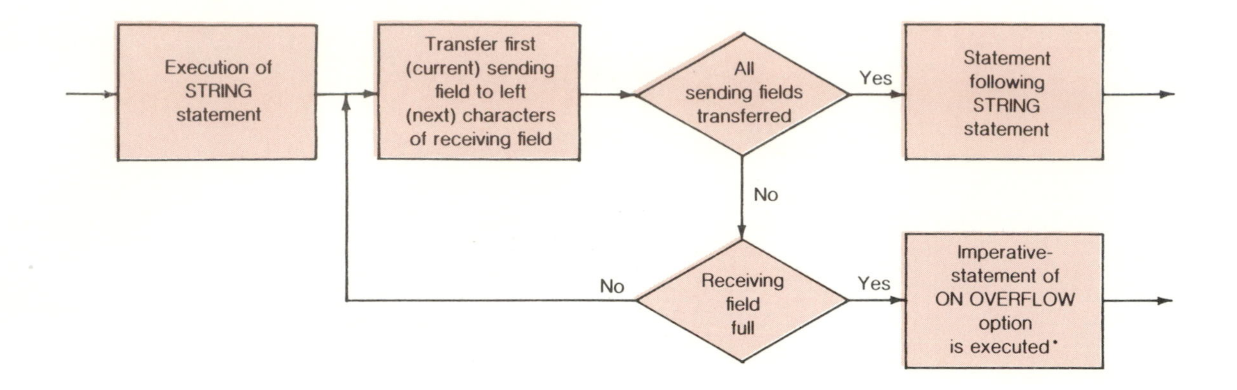

Figure 9.15 STRING statement: flow of logic.

Control passes to the imperative statement of the ON OVERFLOW option when the receiving field is full, and sending characters still remain in the sending fields, or the pointer contains an invalid value.

NOTE: Sending fields may be placed anywhere and in any order in the Data Division. The order in which they are named governs the order of the data transfer. If ON OVERFLOW is not specified, the next statement following STRING is executed when the receiving field is full, and the sending field still contains characters, or the pointer contains an invalid value. The flow of logic for the STRING statement is shown in figure 9.15.

The STRING statement provides juxtaposition (side by side) of the partial or complete contents of two or more data items into a single data item.

Rules Governing the Use of the STRING Statement

1. All identifiers must be described as usage of DISPLAY.
2. The DELIMITED BY phrase is required, which means that the transfer of data is to be stopped by a specified delimiter.
3. The receiving field must be an elementary data item with no editing symbols or the JUSTIFIED clause.
4. Each literal may be any figurative constant without the optional word ALL.
5. All literals must be described as nonnumeric literals (except the pointer), with a usage of DISPLAY.
6. The pointer must represent an elementary numeric data item of sufficient size to contain a value equal to the size plus one of the areas referenced by the receiving field. The symbol 'P' may not be used in the PICTURE character-string of the pointer.
7. When the sending field is an elementary numeric data item, it must be described as an integer without the symbol 'P' in its PICTURE character-string.
8. When a figurative constant is specified for a literal, it.refers to an implicit one-character data item whose usage is DISPLAY.
9. When the STRING statement is executed, the transfer of data is governed by the following rules:
   a. Those characters from the sending fields are transferred to the contents of the receiving field in accordance with the rules for alphanumeric to alphanumeric moves, except that no space filling will be provided.
   b. If the DELIMITED BY phrase is specified without the SIZE option, the contents of the data item referenced by the sending fields are transferred to the receiving data item in the sequence specified in the STRING statement beginning with the leftmost character and continuing from left to right until the end of the data item is reached, or until the character(s) specified as the delimiter is encountered. The delimiters are not transferred.
   c. If the DELIMITED BY phrase is specified with the SIZE option, the entire contents of the sending fields are transferred, in the sequence specified in the STRING statement, to the data

item in the receiving field until all the data has been transferred or the end of the data item referenced as the receiving field has been reached. Several extensive examples follow:

Example:

```
01  CUSTOMER-RECORD.
    05  NAME.
        10  LAST      PIC X(10).
        10  FIRST     PIC X(10).
    05  ADDRESS-1.
        10  NUMBER    PIC X(6).
        10  NAME-1    PIC X(10).
        10  NAME-2    PIC X(5).
    05  ADDRESS-2.
        10  CITY      PIC X(10).
        10  STATE     PIC X(10).
        10  ZIP       PIC 9(5).
```

The actual contents of the record are:

CUSTOMER RECORD

NAME		*ADDRESS-1*			*ADDRESS-2*		
LAST	*FIRST*	*NUMBER*	*NAME-1*	*NAME-2*	*CITY*	*STATE*	*ZIP*
Doe	John	123-45	Main	Ave.	Anytown	Mo.	12345

Consider the record description entries just listed for CUSTOMER-RECORD and the actual contents of the record shown. The total number of positions allocated for the record is 66, while for the particular record shown the total number of positions is 36.

Using the STRING statement and a data item called CONDENSED-RECORD the contents of CUSTOMER-RECORD can be condensed as follows:

```
77  CONDENSED-RECORD      PIC X(100).
              .
              .
              .
    STRING LAST '/' FIRST '/'
        NUMBER '/' NAME-1 '/'
        NAME-2 '/' CITY '/'
        STATE '/' ZIP
        DELIMITED BY ' '
        '*' DELIMITED BY SIZE
        INTO CONDENSED-RECORD.
```

This statement would produce the following result in the data item CONDENSED-RECORD.

```
DOE/JOHN/123-45/MAIN/AVE./ANYTOWN/MO./12345*
```

The total number of positions used to store the same information in CONDENSED-RECORD is 44 instead of 66, which represents a savings of one third.

The aforementioned statement specifies that each elementary item in CUSTOMER-RECORD is to be examined and transferred to CONDENSED-RECORD. Data transfer for the first and succeeding data items proceeds from left to right, character by character, until a blank character (delimiter) is encountered, or until the entire sending data item has been transferred.

The STRING statement specifies that a literal stroke (/) is to be transferred after each elementary item from CUSTOMER-RECORD is transferred.

The last character to be transferred is an asterisk used here to indicate the end of CUSTOMER-RECORD. The clause—'*' DELIMITED BY SIZE—specifies that the single character asterisk (*) is to be transferred to CONDENSED-RECORD until the size of the sending field (one character) is exhausted.

Example:

```
77  EMPLOYEE-MASTER       PIC X(20).
        .
        .
        .
01  EMPLOYEE-RECORD.
    05  NUMBER            PIC X(10).
    05  NAME.
        10  LAST          PIC X(10).
        10  FIRST         PIC X(10).
        10  MIDDLE        PIC X(10).
```

The actual contents of the record are:

EMPLOYEE RECORD

	NAME		
NUMBER	*LAST*	*FIRST*	*MIDDLE*
12345	Jones	John	J.

A STRING statement that will condense the above EMPLOYEE-RECORD in a data item called EMPLOYEE-MASTER follows. The sending data items are delimited by a single blank character and are separated by a stroke (/) in EMPLOYEE-MASTER.

```
STRING NUMBER '/' LAST '/'
    FIRST '/' MIDDLE '/'
    DELIMITED BY ' '
    INTO EMPLOYEE-MASTER.
```

The contents of EMPLOYEE-MASTER will look as follows after execution of the STRING statement.

```
12345/JONES/JOHN/J./
```

Example:

```
01  RECORD-IN.
    05  FIELD-A                   PIC X(5).
    05  FIELD-B                   PIC X(5).
    05  FIELD-C                   PIC X(5).

WORKING-STORAGE SECTION.
77  POINTER-W                     PIC 99.
77  CONDENSED-RECORD              PIC X(20).

1.      STRING     FIELD-A, FIELD-B, FIELD-C
            DELIMITED BY SIZE
            INTO CONDENSED-RECORD.

2.      STRING     FIELD-A, FIELD-B, FIELD-C
            DELIMITED BY '*'
            INTO CONDENSED-RECORD.

3.      STRING     FIELD-A, FIELD-B
            DELIMITED BY SIZE
            INTO CONDENSED-RECORD
            WITH POINTER POINTER-W.
```

The contents of all the fields after the execution of the STRING statements are:

					POINTER-W	
	FIELD-A	*FIELD-B*	*FIELD-C*	*CONDENSED-RECORD*	*BEFORE*	*AFTER*
1.	ABC*D	EF*GHI	JKLMN	ABC*DEF*GHIJKLMN	——	——
2.	ABC*D	EF*GHI	JKLMN	ABCEFJKLMN	——	——
3.	ABCDE	FGHIJ	—	ƀƀƀƀABCDEFGHIJƀƀƀƀƀƀ	5	15

ƀ denotes blank character.

UNSTRING Statement

The UNSTRING statement is basically the opposite of the STRING statement; instead of joining several fields into one field, the UNSTRING statement separates one field into several. The delimiter in the sending field tells where to separate the fields.

The COUNT IN option allows the programmer to keep a count of the number of characters placed in each field. The delimiter itself is not figured in this count. A count field can be specified for each of the receiving fields in the statement. (The count fields must be set up by the programmer.)

Like the STRING statement, the UNSTRING statement also permits a WITH POINTER option. If this option is specified, the UNSTRING operation begins with the position in the sending field indicated by the pointer field. Thus, if a pointer field contains an 8, the first seven characters of the sending field are ignored. As the UNSTRING statement is executed, the pointer field is increased by one so that it always contains the position of the next character to be processed. Remember that the pointer field must never be less than one or more than the length of the sending field.

The TALLYING IN option may be used to count the number of receiving fields operated on. The count includes all receiving fields used by the UNSTRING statement, even if no data was placed in a field, (that is, if two delimiters were found next to each other). Thus, if eight receiving fields are specified and only six are used before the end of the sending field is reached, the field specified in the TALLYING IN phrase will contain a value of 6.

It is often necessary to use several delimiters to unstring a field. In such cases, the OR phrase is used with the DELIMITED BY phrase in the UNSTRING statement. Any of the delimiters specified will then cause the string to be split. If it is necessary to keep track of which delimiter separated the two fields, the DELIMITER IN option can be used. Then, instead of being discarded, the delimiter is moved to the DELIMITER IN field.

The ALL option is another variation of the DELIMITED BY phrase. When ALL is used, any consecutive occurrences of the delimiter are treated as one delimiter. Whenever a figurative constant is used, the ALL option is automatically assumed. Thus, five consecutive spaces would be treated as one delimiter.

The two possible causes of overflow during an UNSTRING operation are:

1. The pointer field was improperly initialized; that is a value less than one or greater than the number of characters in the sending field, or
2. All possible receiving fields have been used, but some characters are still left in the sending field.

Without the ON OVERFLOW phrase, the program leaves the UNSTRING statement and continues with the next statement in sequence. With an ON OVERFLOW statement, the program does whatever the imperative-statement tells it to.

Explanation of the UNSTRING Statement

In the UNSTRING format, identifier-1 represents the *sending field.* Identifiers 2 and 3 or literals 1 and 2 indicate the character(s) *delimiting the move.* Identifiers 4 and 7 represent the *receiving fields.* Identifiers 6 and 9 represent the *COUNT IN field.* Identifier-10 represents the *pointer field.* Identifier-11 represents the *TALLYING IN field.* The general format for the UNSTRING statement is shown in figure 9.16.

Characters are transferred to the receiving fields from the sending field in left-to-right order. The first named receiving field (identifier-4) is completely filled with the leftmost characters of the sending field. The next receiving field (identifier-7) is then completely filled with the next group of characters from the sending field. This process continues until either all receiving fields are full, or the entire sending field has been transferred. Figure 9.17 graphically defines the transfer of data.

Control passes to the imperative statement of the ON OVERFLOW option when sending characters remain after all receiving fields are full. See the flow of logic for the UNSTRING statement in figure 9.18.

NOTE: Receiving fields may be placed anywhere and in any order in the Data Division. The order in which they are named governs the order of data transfer.

```
UNSTRING identifier-1

    [ DELIMITED BY [ ALL ] { identifier-2 }  [ , OR [ ALL ] { identifier-3 } ] . . . ]
                           { literal-1    }                 { literal-2    }

    INTO identifier-4 [ , DELIMITER IN identifier-5 ] [ , COUNT IN identifier-6 ]

      [ , identifier-7 [ , DELIMITER IN identifier-8 ] [ , COUNT IN identifier-9 ] ] . . .

    [ WITH POINTER identifier-10 ] [ TALLYING IN identifier-11 ]

    [ ; ON OVERFLOW imperative-statement ]
```

Figure 9.16 Format—UNSTRING statement.

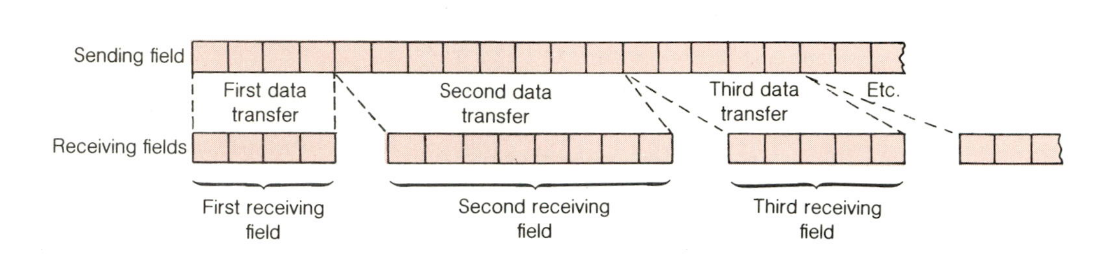

Figure 9.17 UNSTRING statement: transfer of data.

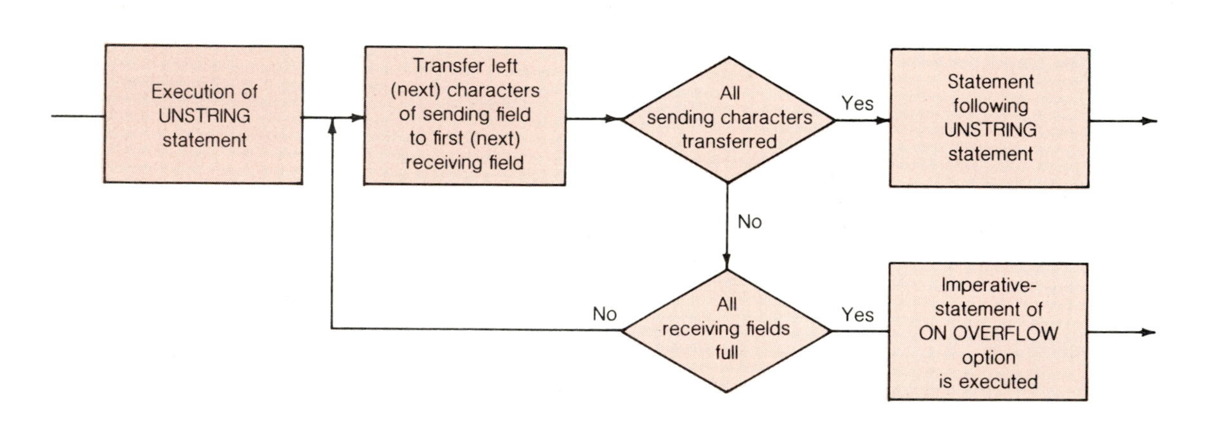

Figure 9.18 UNSTRING statement: flow of logic.

Example:

The UNSTRING statement can be used to transfer data from a condensed format to a noncondensed format. Consider the data item CONDENSED-RECORD:

```
DOE/JOHN/123-45/MAIN/AVE./ANYTOWN/MO./12345*
```

This information can be restored into its original format in the record CUSTOMER-RECORD.

```
01  CUSTOMER-RECORD.
    05  NAME.
        10  LAST        PIC X(10).
        10  FIRST       PIC X(10).
    05  ADDRESS-1.
        10  NUMBER      PIC X(6).
        10  NAME-1      PIC X(10).
        10  NAME-2      PIC X(5).
    05  ADDRESS-2.
        10  CITY        PIC X(10).
        10  STATE       PIC X(10).
        10  ZIP         PIC 9(5).
```

The UNSTRING statement could be used to achieve this result as follows:

```
UNSTRING CONDENSED-RECORD
    DELIMITED BY '/'
    INTO LAST FIRST NUMBER
        NAME-1 NAME-2 CITY
        STATE ZIP.
```

The UNSTRING statement causes contiguous data in a sending field to be separated and placed into multiple receiving fields.

Rules Governing the Use of the UNSTRING Statement

1. The sending field must be alphanumeric.
2. The receiving fields must be described as usage of DISPLAY. Receiving fields may be alphabetic, numeric or alphanumeric.
3. Each literal must be a nonnumeric literal. In addition, each literal may be any figurative constant without the optional word ALL.
4. The sending field and delimiters must be described as alphanumeric data items.
5. Receiving fields may be described as either alphabetic (except that the symbol 'B' may not be used in the PICTURE character-string), alphanumeric, or numeric (except that the symbol 'P' may not be used in the PICTURE character-string, and must be described as usage of DISPLAY.
6. COUNT IN, WITH POINTER, and TALLY IN fields must be described as elementary numeric integer data items (except that the symbol 'P' may not be used in the PICTURE character-string).
7. No identifier may name a level-88 entry.
8. The DELIMITER IN phrase and the COUNT IN phrase may be specified only if the DELIMITED BY phrase is specified.
9. COUNT IN represents the count of the number of characters within the data item referenced by the sending field isolated by the delimiters for the move to the receiving fields. This value does not include a count of the delimiter characters.
10. The data item referenced by POINTER contains a value that indicates the relative character position within the area defined by the sending field.
11. The data referenced by TALLYING IN is a counter that records the number of data items acted upon during the execution of an UNSTRING statement.
12. When a figurative constant is used as a delimiter, it stands for a single character nonnumeric literal.

    When the ALL phrase is specified, one occurrence of two or more contiguous occurrences of the delimiter (figurative constant or not), or the contents of the data item referenced by the delimiters, are treated as if it were only one occurrence, and this occurrence is not moved to the receiving data item according to the rules.

13. When any examination encounters two contiguous delimiters, the current receiving area is either space or zero, filled according to the description of the receiving area.
14. The delimiters can contain any character in the computer's character set.
15. The initialization of the contents of the data items associated with the POINTER phrase or the TALLYING phrase is the responsibility of the programmer.
16. Each delimiter represents one character. When a delimiter contains two or more characters, all of the characters must be present in contiguous positions of the sending item, and in the order given to be recognized as a delimiter.
17. When two or more delimiters are specified in the DELIMITED BY phrase, an 'OR' condition exists between them. Each delimiter is compared to the sending field. If a match occurs, the character(s) in the sending field is considered to be a single delimiter. No character(s) in the sending field can be considered a part of more than one delimiter.

    Each delimiter is applied to the sending field in the sequence specified in the UNSTRING statement.
18. When the UNSTRING statement is initiated, data is transferred from the sending item to the receiving area(s) according to the following rules:
    a. If the POINTER phrase is specified, the string of characters in the sending field is examined beginning with the relative character position indicated by the contents of the POINTER. (It is the programmer's responsibility to set the pointer before the UNSTRING statement is executed.) If the pointer is not specified, the string of characters is examined beginning with the leftmost character position.
    b. If the DELIMITED BY phrase is specified, the examination proceeds from left to right until a delimiter specified is encountered. If the DELIMITED BY phrase is not specified, the number of characters examined is equal to the size of the current receiving area.
    c. The characters thus examined (excluding the delimiter characters) are treated as elementary alphanumeric data items and are moved into the current receiving area according to the rules for the MOVE statement.
    d. If the DELIMITER IN phrase is specified, the delimiting character(s) are treated as an elementary alphanumeric data item and are moved into the receiving fields referenced by the delimiter according to the rules of the MOVE statement.
    e. If the COUNT IN phrase is specified, a value equal to the number of characters thus examined (excluding the delimiter character(s), if any) is moved in the COUNT IN area according to the rules for an elementary move.
    f. If the DELIMITED BY phrase is specified, the string of characters is further examined beginning with the first character to the right of the delimiter. If the DELIMITED BY phrase is not specified, the string of characters is further examined beginning with the character to the right of the last character transferred.
    g. After data is transferred to the receiving fields, the process is repeated until either all the characters are exhausted in the sending field or until there are no more receiving areas.
19. The contents of the POINTER will be incremented by one for each character examined in the sending field. When the execution of the UNSTRING statement with a POINTER phrase is completed, the contents of the data item referenced by the POINTER will contain a value equal to the initial value plus the number of characters examined in the sending item.
20. When the execution of the UNSTRING statement with a TALLYING IN phrase is completed, the contents of the data item referenced by TALLYING IN will contain a value equal to its initial value plus the number of data receiving items acted upon.
21. Either of the following conditions causes an overflow condition:
    a. An UNSTRING operation is initiated, and the value of the POINTER is less than one or greater than the size of the sending field.
    b. If, during execution of an UNSTRING statement, all data receiving areas have been acted upon, and the sending field still contains characters that have not been examined.
22. When an overflow condition exists, the UNSTRING operation is terminated. If an ON OVERFLOW phrase has been specified, the imperative statement included in the ON OVERFLOW phrase is executed. If the ON OVERFLOW phrase is not specified, control is transferred to the next executable statement.

Example:
Given the following information,

```
77  EMPLOYEE-MASTER        PIC X(20)
        VALUE '12345/JONES/JOHN/J./'

01  EMPLOYEE-RECORD.
    05  NUMBER             PIC 9(5).
    05  NAME.
        10  LAST           PIC X(10).
        10  FIRST          PIC X(10).
        10  MIDDLE         PIC X(10).
```

the following UNSTRING statement will cause data in a condensed format in a data item named EMPLOYEE-MASTER to be transferred to its original noncondensed format record called EMPLOYEE-RECORD.

```
UNSTRING EMPLOYEE-MASTER
    DELIMITED BY '/'
    INTO NUMBER LAST
        FIRST MIDDLE.
```

Example:

```
01  RECORD-IN.
    05  FIELD-A                    PIC X(5).
    05  FIELD-B                    PIC X(5).

WORKING-STORAGE SECTION.
77  CONDENSED-RECORD               PIC X(20).
77  COUNT-1                        PIC 99.
77  COUNT-2                        PIC 99.
77  POINTER-W                      PIC 99.
77  TALLY-W                        PIC 99.

1.      UNSTRING CONDENSED-RECORD
            DELIMITED BY '*'
            INTO FIELD-A FIELD-B.

2.      UNSTRING CONDENSED-RECORD
            DELIMITED BY '*'
            INTO FIELD-A COUNT IN COUNT-1
                FIELD-B COUNT IN COUNT-2
            WITH POINTER POINTER-W
            TALLYING IN TALLY-W.

3.      UNSTRING CONDENSED-RECORD
            DELIMITED BY '*' OR '/'
            INTO FIELD-A
                DELIMITER IN FIELD-B.

4.      UNSTRING CONDENSED-RECORD
            DELIMITED BY ALL '*'
            INTO FIELD-A FIELD-B.
```

The contents of each of the fields after the execution of the UNSTRING statements are:

					POINTER-W		
CONDENSED-RECORD	*FIELD-A*	*FIELD-B*	*COUNT-1*	*COUNT-2*	*BEFORE*	*AFTER*	*TALLY-W*
1. ABC*DEF	ABC	DEF	————	————	————	————	————
2. ABC*DEF	ABC	DEF	3	3	1	8	2
3. ABC/DEF*GH	ABC	/	————	————	————	————	————
4. ABC***DEF	ABC	DEF	————	————	————	————	————

Summary

Data manipulation verbs move data from one area to another within the computer, with the inspection of data being explicit in the function of several of the COBOL verbs.

The EXAMINE statement is used to replace a given character and/or count the number of times a given character appears in a data statement. The examination of the item starts at the left and proceeds to the right. The EXAMINE statement may be used in the following manner: (1) to validate the input, (2) to transfer information from one code to another, and (3) for free-form input processing where the data appears in a form of character strings.

The TALLYING option of the EXAMINE statement is used to scan a data item, counting the number of occurrences of a given character. The TALLYING phrase causes a five decimal digit special register to be set up by the compiler called TALLY to hold the count. TALLY may be used by other procedural statements in the following manner:

Option	The TALLY value represents
UNTIL FIRST	The number of occurrences of other characters not equal to the literal encountered prior to the first occurrence of the literal.
ALL	The number of occurrences of the literal throughout the item.
LEADING	The number of occurrences of the literal prior to encountering a character other than that literal.

The REPLACING phrase of the EXAMINE statement is used to modify the value of an item by replacing certain characters in the original value with new characters. The options are the same as those for the TALLYING option with the addition of the FIRST option where the replacement character is substituted for the first occurrence of the character to be replaced. In all options, the actions are the same except, instead of counting, the replacement character is substituted.

When both TALLYING and REPLACING phrases are used, each character is tallied and replaced according to the rules for tallying and replacing characters.

The INSPECT statement increased the powers of the EXAMINE statement. The execution of the INSPECT statement is basically the same as the EXAMINE statement. The major differences between the two statements are: (1) the INSPECT statement does not use a field called TALLY to count the various characters. It is the programmer's responsibility to create a tally field and to supply this name after the word TALLYING in the format; (2) the INSPECT statement permits the tallies and replacements to be named in series so several can be done with one statement; (3) the BEFORE and AFTER options of the INSPECT statement permit some specifications that are not possible with the EXAMINE statement; (4) whereas EXAMINE only permits a single character to be counted or replaced, INSPECT allows groups of characters to be counted and replaced; and (5) all literals used with the INSPECT statement must be alphanumeric.

The INSPECT statement is used in conjunction with character strings, and examines the contents of a data item from left to right and does any of the following: (1) count the number of occurrences of a particular character(s), or (2) replace the particular character(s) with a different character(s), or (3) both. The operands are considered in the order they are specified in the INSPECT statement starting with the leftmost position. The comparison of character-strings continues until a match is found, and the comparison further continues until the rightmost character position of the data item being inspected is reached.

The TALLYING phrase of the INSPECT statement provides the ability to tally occurrences of strings of one or more characters in a field specified by the programmer appearing after the word TALLYING.

The CHARACTERS phrase of the INSPECT statement implies a one character operand that participates in the operation without comparison. The implied character is always considered to match the leftmost character of the data item referenced.

If the BEFORE phrase is specified in the TALLYING option of the INSPECT statement, the character string to be tallied participates only on those comparison cycles that involve that portion of the data item referenced from its leftmost character position up to, but not including, the first occurrence with the data item of the character string referenced in the BEFORE phrase.

If the AFTER phrase is specified in the TALLYING option of the INSPECT statement, the character string to be tallied participates only in those comparison cycles that involve that portion of the data item referenced from that character position immediately to the right of the rightmost character of the first occurrence in the character string specified in the AFTER phrase prior to the rightmost position of the data item referenced.

The following options of the INSPECT statement are similar to those found in the TALLYING and REPLACING options of the EXAMINE statement. The ALL option is specified for all occurrences of a particular character. The LEADING option is used for all occurrences of a particular character up to the character specified. FIRST option is an additional option available with the REPLACING phrase whereby the leftmost occurrence of the matched data item is replaced.

The order in which a REPLACING phrase in an INSPECT statement is specified can be significant, particularly where there is a similarity in the character strings that are replaced.

The use of TALLYING and REPLACING options in one INSPECT statement is interpreted and executed as though there were two successive INSPECT statements specifying the same character string. The same general rules for counting and replacing apply.

The STRING statement causes characters from one or more data items to be transferred to a single data item. Each transfer is terminated because of the size of the sending and receiving data items, or because of the presence of a specific character(s) in the sending data item. Other features of the STRING statement are: no padding with blanks occur; a pointer keeps track of the next available position in the receiving area; and the ON OVERFLOW condition can be included in the statement to signal the end of the receiving data item.

When a STRING statement is executed, several fields are joined together (concatenation). The DELIMITER BY phrase allows the programmer to specify one or more characters to mark the end of the field to be strung. Only those characters in the sending field to the left of the delimiter are used in the STRING operation. The DELIMITER BY phrase is required.

The DELIMITER BY SIZE option of the STRING statement means that the entire field is to be used in the STRING operation. If the delimiter is specified as a character (a nonnumeric literal enclosed in quotation marks), only the characters before the specified character are moved. (The specified character [delimiter] is not moved.)

When the WITH POINTER option is specified in a STRING statement, the data is moved from the sending field to the receiving field starting at the point indicated by the pointer. (It is the programmer's responsibility to create a field for the pointer with an initial value.) The initial value can never be less than one or more than the length of the receiving field.

The ON OVERFLOW option of the STRING statement is used to signal (1) when the end of the receiving field is reached and when the sending field still contains characters to be moved, or (2) when the pointer field does not contain an acceptable value (less than one or more than the length of the receiving field). Usually, the imperative statement that follows ON OVERFLOW phrase is a statement that causes an overflow routine to be performed.

In the STRING statement, characters are transferred from the sending fields to the receiving field in a left-to-right order. The first sending field is moved first into the leftmost positions of the receiving field followed immediately by the next sending field. Delimiters define the length of the move. The process continues until all of the sending fields have been transferred or the receiving field is full. The sending fields may be placed anywhere and in any order in the Data Division, and the order in which they are named determines the order of the data transfer.

If the ON OVERFLOW option is not specified in the STRING statement, the next statement following STRING is executed when the receiving field is full and when the sending field still contains characters to be moved, or the pointer contains an invalid value.

The UNSTRING statement is basically the opposite of the STRING statement—instead of joining several fields into one, the UNSTRING separates one field into several. The delimiter in the sending field tells where to separate the fields.

The COUNT IN option of the UNSTRING statement allows the programmer to keep a count of the number of characters to be placed in each field. The delimiter itself is not figured in this count. Count fields can be set up by the programmer for each of the receiving fields in the UNSTRING statement.

The WITH POINTER option of the UNSTRING statement allows the programmer to set the position in the sending field where the transfer of data to the receiving field is to begin. The pointer must never be less than one or more than the length of the sending field.

The TALLYING IN option of the UNSTRING statement may be used to count the number of receiving fields acted upon.

If several delimiters are used to unstring a field, then the OR phrase may be used with the DELIMITER BY phrase in the UNSTRING statement.

The DELIMITER IN option of the UNSTRING statement is used when it is necessary to keep track of which delimiter separated two fields.

The two possible causes of overflow during an UNSTRING operation are: (1) the pointer field was improperly initialized, that is, a value less than one or greater than the number of characters in the sending field, or (2) all possible receiving fields have been used, but some characters still remain in the sending field.

When an overflow occurs and the ON OVERFLOW phrase is not specified, the program leaves the UNSTRING statement and continues with the next statement in sequence. With an ON OVERFLOW statement specified, the program does whatever the imperative statement tells it to.

In an UNSTRING statement, characters are transferred to the receiving fields from the sending field in left-to-right order. The first named receiving field is completely filled with the leftmost characters of the sending field, and then the next named receiving field is filled, etc., until all receiving fields are filled or the entire sending field has been transferred. Receiving fields may be placed anywhere and in any order in the Data Division. The order in which they are named determines the order of the data transfer.

The uses of the ANSI 1974 COBOL data movement elements of INSPECT, STRING, and UNSTRING are the same as those for the EXAMINE statement, namely input validation, translation, and free-form input processing. However, the new standards make these applications much easier.

Summary—COBOL Verbs

EXAMINE is a data manipulation verb that is used to replace a single given character and/or to count the number of times a single given character appears in a data item. A TALLY field (set up by the compiler) counts the number of occurrences. TALLY may be used in procedural statements. This verb was replaced in the ANSI 1974 COBOL by the INSPECT verb.

INSPECT is a data manipulation verb that can be used in conjunction with character strings. It can be used to count the number of occurrences of a particular character(s), or replace this particular character(s) with different character(s) or both. In addition to its TALLYING and REPLACING options, the BEFORE and AFTER TALLYING option allows the character string to be tallied or replaced before or after a given character.

STRING is a data manipulation verb that causes characters from one or more data items to be transferred to a single data item. Each transfer is terminated because of the size of the sending field and receiving data item, or because of the presence of a specific character(s) (delimiters) in the sending field. The transfer is from left to right in the receiving field. Other features of the STRING verb include: no padding with blanks occurs; a pointer keeps track of the next available position in the receiving data item; and an overflow test condition can be included in the statement to signal the end of the receiving data item.

UNSTRING is a data manipulation verb that causes contiguous data in a sending field to be separated and placed into multiple receiving fields. The characters are transferred to the receiving fields from the sending field in left-to-right order. The first named receiving field is completely filled first and then the next receiving field is filled, etc. This process continues until all receiving fields are filled or the entire sending field has been transferred. The delimiter in the sending field tells where to separate the fields. A count can be made of the number of characters transferred. A pointer will indicate in the sending field the leftmost point of data transfer. A field can be set up to keep track of the delimiters used when there is more than one used to separate fields. The number of fields transferred can be counted.

Illustrative Program: Control Levels—Group Indication—Transaction Register

Job Definition

Print a weekly sales transaction report that lists daily transactions, total sales for the day, and total sales for the week.

Input

This consists of a sales transaction file consisting of 80-position records. Records are arranged in ascending order by date. The format of the input records is shown on this record layout form:

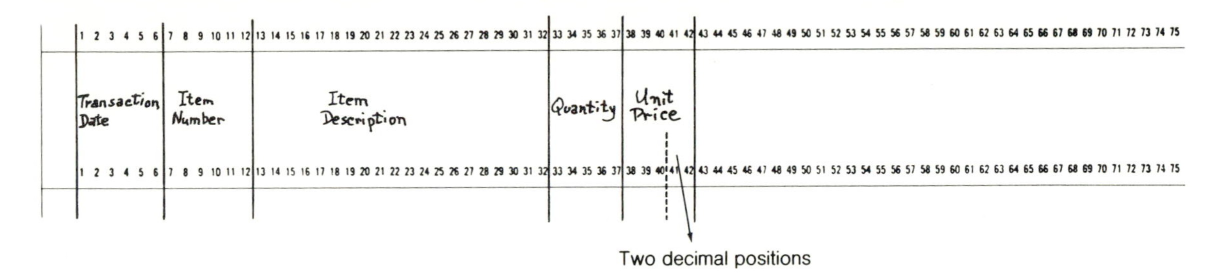

Processing

Multiply quantity times unit price to find sales amount.
Accumulate sales amount to find total item sales per day.
Accumulate total daily sales to find total weekly sales.

Output

A Transaction Register is to be printed as follows:

TRANSACTION REGISTER

TRANSACTION	ITEM	DESCRIPTION	QUANTITY	UNIT	SALES
07/23/80	413010	CH001 BOX 100A FLUSH	10	4.90	$49.00
	412146	CH148 BREAKER 15A	100	.89	$89.00
	411116	1500 TWIN SOCKET B	500	1.12	$560.00
					$698.00
07/24/80	503029	MOTOR 1/2 HP 60 CYC	2	146.78	$293.56
	317802	TERMINAL CLIP	100	5.12	$512.00
	326917	TERMINAL BAR	100	4.12	$412.00
	411121	1506 SOCKT ADAPT BRN	400	.19	$76.00
	412997	CH173 BREAKER 30A	60	1.15	$69.00
	413088	CH176 BREAKER 60A	40	1.15	$46.00
	411174	C151 SIL SWITCH BRN	200	1.16	$232.00
	413090	CH005 BR BOX 150A	10	4.98	$49.80
	718326	FC803 FUSE 15A	200	.32	$64.00
					$1,754.36
		WEEKLY TOTAL			$2,452.36

Printer Spacing Chart

This Printer Spacing Chart shows how the report is formatted:

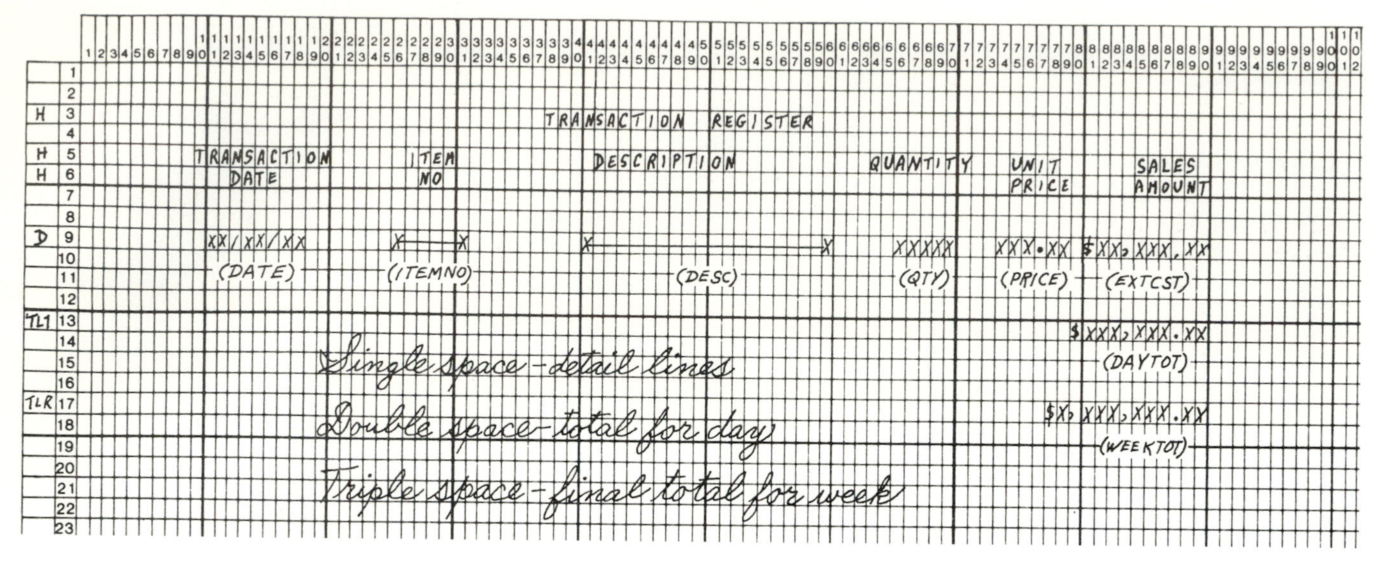

Hierarchy Chart

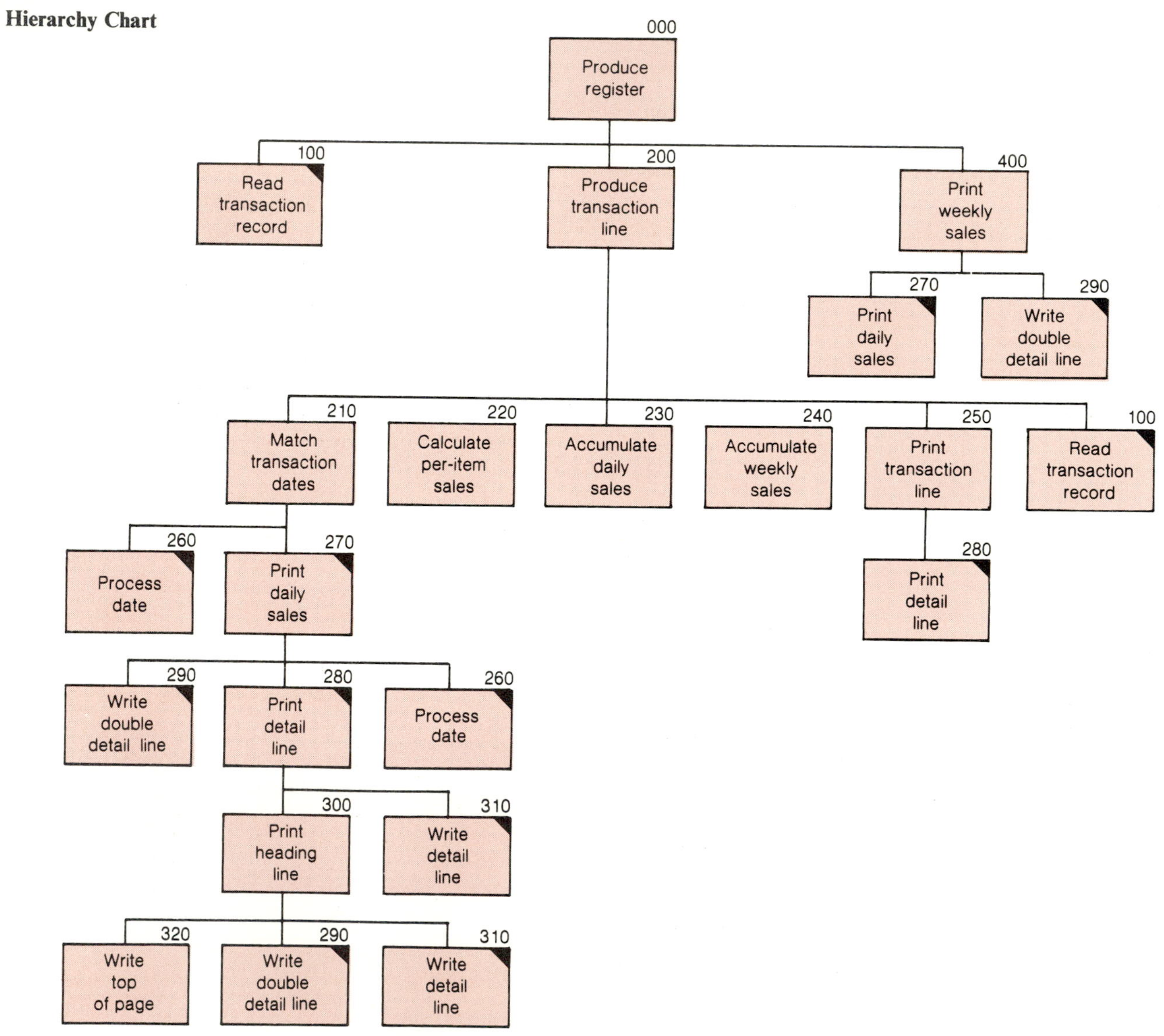

IPO Charts

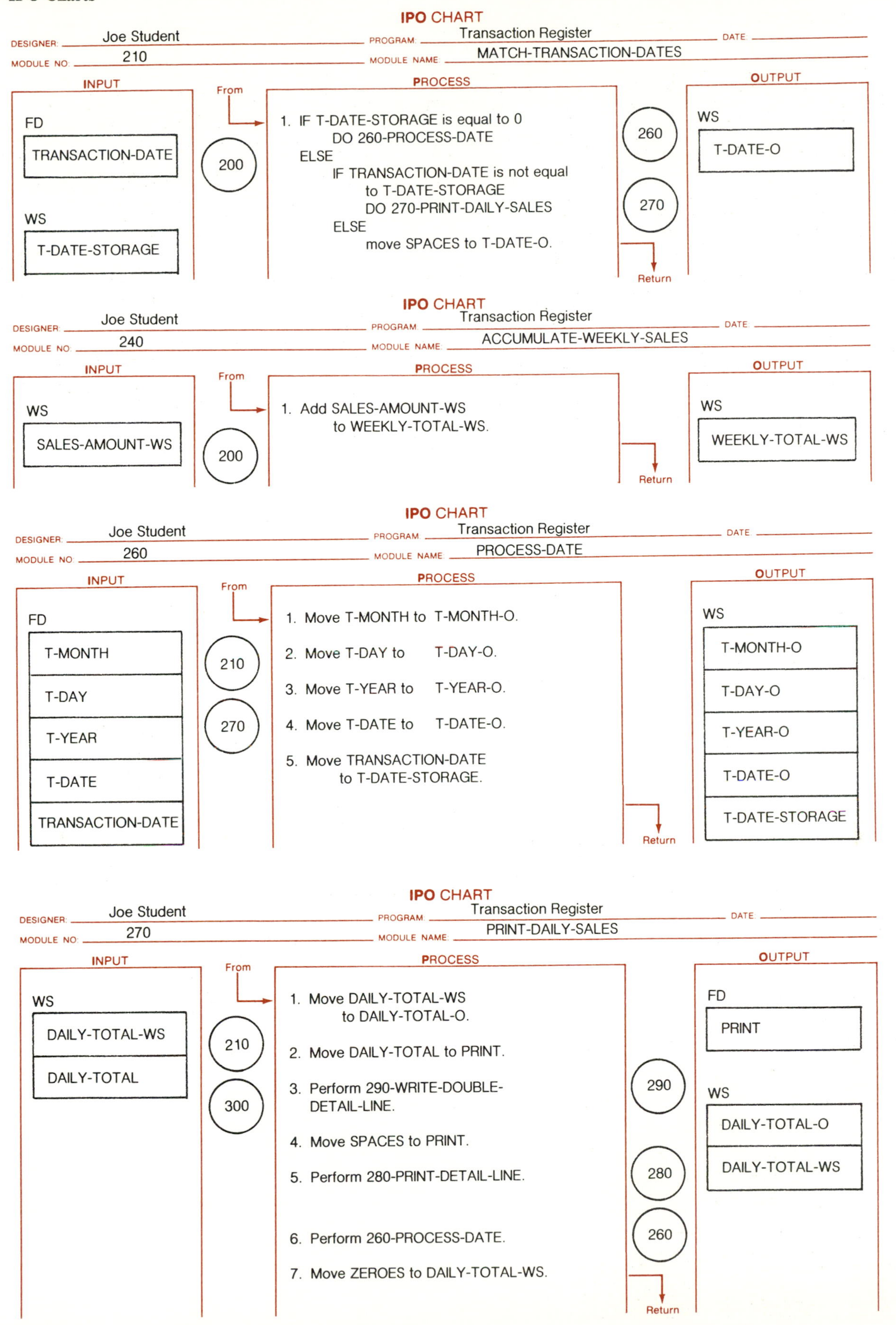

IPO CHART
DESIGNER: Joe Student
PROGRAM: Transaction Register
DATE:
MODULE NO: 210
MODULE NAME: MATCH-TRANSACTION-DATES
INPUT
FD
TRANSACTION-DATE
WS
T-DATE-STORAGE
From
200
PROCESS
1. IF T-DATE-STORAGE is equal to 0
DO 260-PROCESS-DATE
ELSE
IF TRANSACTION-DATE is not equal
to T-DATE-STORAGE
DO 270-PRINT-DAILY-SALES
ELSE
move SPACES to T-DATE-O.
260
270
Return
OUTPUT
WS
T-DATE-O
IPO CHART
DESIGNER: Joe Student
PROGRAM: Transaction Register
DATE:
MODULE NO: 240
MODULE NAME: ACCUMULATE-WEEKLY-SALES
INPUT
WS
SALES-AMOUNT-WS
From
200
PROCESS
1. Add SALES-AMOUNT-WS
to WEEKLY-TOTAL-WS.
Return
OUTPUT
WS
WEEKLY-TOTAL-WS
IPO CHART
DESIGNER: Joe Student
PROGRAM: Transaction Register
DATE:
MODULE NO: 260
MODULE NAME: PROCESS-DATE
INPUT
FD
T-MONTH
T-DAY
T-YEAR
T-DATE
TRANSACTION-DATE
From
210
270
PROCESS
1. Move T-MONTH to T-MONTH-O.
2. Move T-DAY to T-DAY-O.
3. Move T-YEAR to T-YEAR-O.
4. Move T-DATE to T-DATE-O.
5. Move TRANSACTION-DATE
to T-DATE-STORAGE.
Return
OUTPUT
WS
T-MONTH-O
T-DAY-O
T-YEAR-O
T-DATE-O
T-DATE-STORAGE
IPO CHART
DESIGNER: Joe Student
PROGRAM: Transaction Register
DATE:
MODULE NO: 270
MODULE NAME: PRINT-DAILY-SALES
INPUT
WS
DAILY-TOTAL-WS
DAILY-TOTAL
From
210
300
PROCESS
1. Move DAILY-TOTAL-WS
to DAILY-TOTAL-O.
2. Move DAILY-TOTAL to PRINT.
3. Perform 290-WRITE-DOUBLE-
DETAIL-LINE.
4. Move SPACES to PRINT.
5. Perform 280-PRINT-DETAIL-LINE.
6. Perform 260-PROCESS-DATE.
7. Move ZEROES to DAILY-TOTAL-WS.
290
280
260
Return
OUTPUT
FD
PRINT
WS
DAILY-TOTAL-O
DAILY-TOTAL-WS

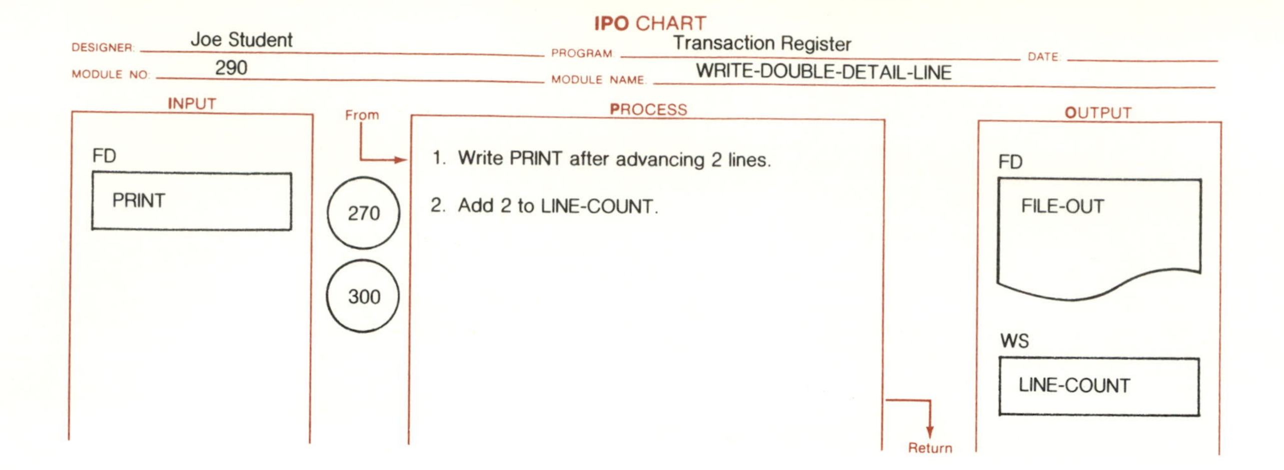
IPO CHART
DESIGNER Joe Student
PROGRAM Transaction Register
DATE
MODULE NO. 290
MODULE NAME WRITE-DOUBLE-DETAIL-LINE
INPUT
FD
PRINT
From
270
300
PROCESS
1. Write PRINT after advancing 2 lines.
2. Add 2 to LINE-COUNT.
Return
OUTPUT
FD
FILE-OUT
WS
LINE-COUNT

Flowcharts

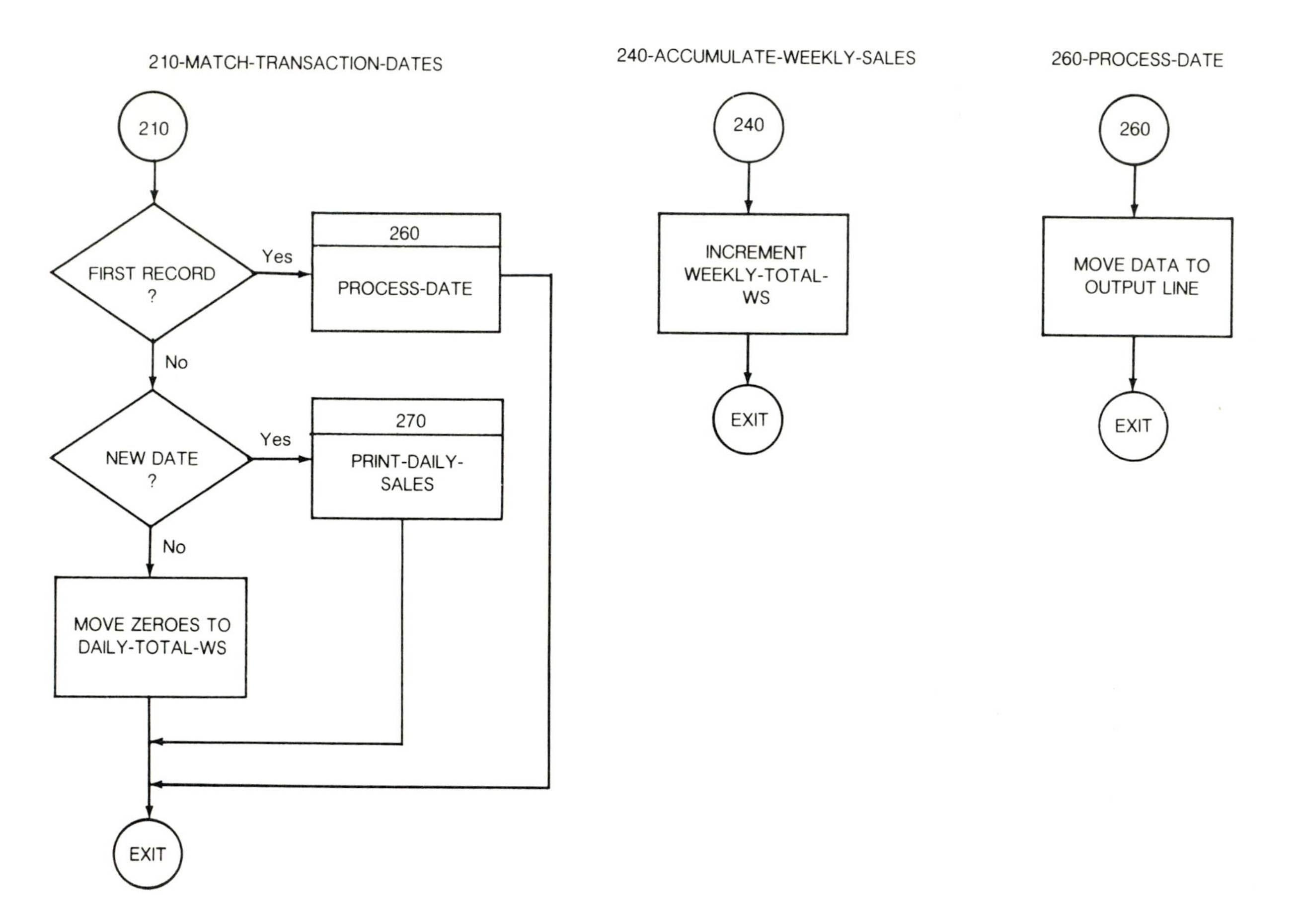
210-MATCH-TRANSACTION-DATES
210
FIRST RECORD ?
Yes
260
PROCESS-DATE
No
NEW DATE ?
Yes
270
PRINT-DAILY-SALES
No
MOVE ZEROES TO DAILY-TOTAL-WS
EXIT
240-ACCUMULATE-WEEKLY-SALES
240
INCREMENT WEEKLY-TOTAL-WS
EXIT
260-PROCESS-DATE
260
MOVE DATA TO OUTPUT LINE
EXIT

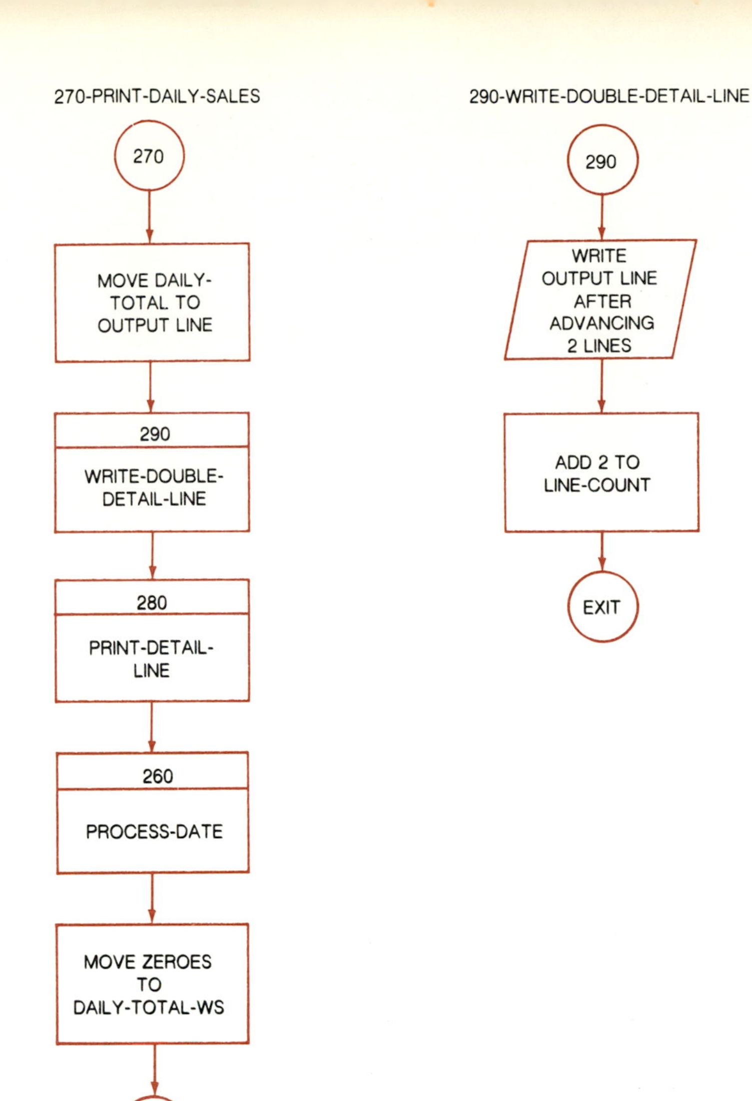

Source Program

```
 1        *************************
 2
 3         IDENTIFICATION DIVISION.
 4
 5        *************************
 6
 7         PROGRAM-ID.
 8             TRANSACTION-REGISTER.
 9         AUTHOR.
10             STEVE GONSOSKI.
11         INSTALLATION.
12             WEST LOS ANGELES COLLEGE.
13         DATE-WRITTEN.
14             3 FEBRUARY 1982.
15
16        *PURPOSE.
17        *    PRINTS A WEEKLY SALES TRANSACTION REPORT THAT LISTS DAILY
18        *    TRANSACTIONS, TOTAL SALES FOR THE DAY, AND TOTAL SALES FOR
19        *    THE WEEK.
20
21        *************************
22
23         ENVIRONMENT DIVISION.
24
25        *************************
26
27         CONFIGURATION SECTION.
28
29         SOURCE-COMPUTER.
30             LEVEL-66-ASCII.
31         OBJECT-COMPUTER.
32             LEVEL-66-ASCII.
33
```

```
34          INPUT-OUTPUT SECTION.
35
36          FILE-CONTROL.
37              SELECT FILE-IN       ASSIGN TO C1-CARD-READER.
38              SELECT FILE-OUT      ASSIGN TO P1-PRINTER.
39
40         ************************
41
42          DATA DIVISION.
43
44         ************************
45
46          FILE SECTION.
47
48          FD  FILE-IN
49              CODE-SET IS GBCD
50              LABEL RECORDS ARE OMITTED
51              DATA RECORD IS CARDIN.
52
53          01  CARDIN.
54
55              03  TRANSACTION-DATE.
56                  05  T-MONTH          PICTURE X(2).
57                  05  T-DAY            PICTURE X(2).
58                  05  T-YEAR           PICTURE X(2).
59              03  ITEM-NUMBER          PICTURE 9(6).
60              03  ITEM-DESCRIPTION     PICTURE X(20).
61              03  QUANTITY             PICTURE 9(5).
62              03  UNIT-PRICE           PICTURE 999V99.
63
64          FD  FILE-OUT
65              CODE-SET IS GBCD
66              LABEL RECORDS ARE OMITTED
67              DATA RECORD IS PRINT.
68
69          01  PRINT                    PICTURE X(132).
70
71          WORKING-STORAGE SECTION.
72
73          77  LINE-COUNT               PICTURE 9(2)     VALUE 38.
74          77  SALES-AMOUNT-WS          PICTURE 9(4)V99  VALUE ZERO.
75          77  DAILY-TOTAL-WS           PICTURE 9(6)V99  VALUE ZERO.
76          77  WEEKLY-TOTAL-WS          PICTURE 9(6)V99  VALUE ZERO.
77          77  T-DATE-STORAGE           PICTURE 9(6).
78
79          01  FLAGS.
80
81              03  MORE-DATA-FLAG       PICTURE X(3)     VALUE "YES".
82                  88  MORE-DATA                         VALUE "YES".
83                  88  NO-MORE-DATA                      VALUE "NO ".
84
85          01  T-DATE.
86
87              03  T-MONTH-O            PICTURE X(2).
88              03  FILLER               PICTURE X        VALUE "/".
89              03  T-DAY-O              PICTURE X(2).
90              03  FILLER               PICTURE X        VALUE "/".
91              03  T-YEAR-O             PICTURE X(2).
92
93          01  HDG-1.
94
95              03  FILLER               PICTURE X(37)    VALUE SPACES.
96              03  FILLER               PICTURE X(11)    VALUE "TRANSACTION".
97              03  FILLER               PICTURE X(10)    VALUE "  REGISTER".
98              03  FILLER               PICTURE X(74)    VALUE SPACES.
99
100         01  HDG-2.
101
102             03  FILLER               PICTURE X(9)     VALUE SPACES.
103             03  FILLER               PICTURE X(11)    VALUE "TRANSACTION".
104             03  FILLER               PICTURE X(6)     VALUE SPACES.
105             03  FILLER               PICTURE X(4)     VALUE "ITEM".
106             03  FILLER               PICTURE X(11)    VALUE SPACES.
107             03  FILLER               PICTURE X(11)    VALUE "DESCRIPTION".
108             03  FILLER               PICTURE X(11)    VALUE SPACES.
109             03  FILLER               PICTURE X(8)     VALUE "QUANTITY".
110             03  FILLER               PICTURE X(4)     VALUE SPACES.
111             03  FILLER               PICTURE X(4)     VALUE "UNIT".
112             03  FILLER               PICTURE X(5)     VALUE SPACES.
113             03  FILLER               PICTURE X(5)     VALUE "SALES".
114             03  FILLER               PICTURE X(43)    VALUE SPACES.
115
116         01  HDG-3.
117
118             03  FILLER               PICTURE X(12)    VALUE SPACES.
119             03  FILLER               PICTURE X(4)     VALUE "DATE".
120             03  FILLER               PICTURE X(11)    VALUE SPACES.
121             03  FILLER               PICTURE X(2)     VALUE "NO".
122             03  FILLER               PICTURE X(45)    VALUE SPACES.
123             03  FILLER               PICTURE X(5)     VALUE "PRICE".
124             03  FILLER               PICTURE X(5)     VALUE SPACES.
125             03  FILLER               PICTURE X(6)     VALUE "AMOUNT".
126             03  FILLER               PICTURE X(42)    VALUE SPACES.
127
128         01  DETAIL-LINE.
129
130             03  FILLER               PICTURE X(10)    VALUE SPACES.
131             03  T-DATE-O             PICTURE X(8).
132             03  FILLER               PICTURE X(7)     VALUE SPACES.
133             03  ITEM-NUMBER-O        PICTURE 9(6).
134             03  FILLER               PICTURE X(9)     VALUE SPACES.
```

```
135            03  ITEM-DESCRIPTION-O  PICTURE X(20).
136            03  FILLER              PICTURE X(5)     VALUE SPACES.
137            03  QUANTITY-O          PICTURE ZZZZ9.
138            03  FILLER              PICTURE X(3)     VALUE SPACES.
139            03  UNIT-PRICE-O        PICTURE ZZZ.99.
140            03  FILLER              PICTURE X(1)     VALUE SPACES.
141            03  SALES-AMOUNT-O      PICTURE $$$,$$9.99.
142            03  FILLER              PICTURE X(42)    VALUE SPACES.
143
144        01  DAILY-TOTAL.
145
146            03  FILLER              PICTURE X(78)    VALUE SPACES.
147            03  DAILY-TOTAL-O       PICTURE $,$$$,$$$.99.
148            03  FILLER              PICTURE X(42)    VALUE SPACES.
149
150        01  WEEKLY-TOTAL.
151
152            03  FILLER              PICTURE X(57)    VALUE SPACES.
153            03  FILLER              PICTURE X(7)     VALUE "WEEKLY".
154            03  FILLER              PICTURE X(5)     VALUE "TOTAL".
155            03  FILLER              PICTURE X(9)     VALUE SPACES.
156            03  WEEKLY-TOTAL-O      PICTURE $,$$$,$$$.99.
157            03  FILLER              PICTURE X(42)    VALUE SPACES.
158
159        01  SAVE-RECORD             PICTURE X(132).
160
161       ************************
162
163        PROCEDURE DIVISION.
164
165       ************************
166
167        000-PRODUCE-REGISTER.
168
169            OPEN    INPUT    FILE-IN
170                    OUTPUT   FILE-OUT.
171            PERFORM 100-READ-TRANSACTION-RECORD.
172            PERFORM 200-PRODUCE-TRANSACTION-LINE
173                    UNTIL NO-MORE-DATA.
174            PERFORM 300-PRINT-WEEKLY-SALES.
175            CLOSE   FILE-IN
176                    FILE-OUT.
177            STOP RUN.
178
179        100-READ-TRANSACTION-RECORD.
180
181            READ FILE-IN
182                AT END MOVE "NO " TO MORE-DATA-FLAG.
183
184        200-PRODUCE-TRANSACTION-LINE.
185
186            PERFORM 210-MATCH-TRANSACTION-DATES.
187            PERFORM 220-CALCULATE-PER-ITEM-SALES.
188            PERFORM 230-ACCUMULATE-DAILY-SALES.
189            PERFORM 240-ACCUMULATE-WEEKLY-SALES.
190            PERFORM 250-PRINT-TRANSACTION-LINE.
191            PERFORM 100-READ-TRANSACTION-RECORD.
192
193        210-MATCH-TRANSACTION-DATES.
194
195            IF T-DATE-STORAGE IS = 0
196                PERFORM 260-PROCESS-DATE
197                ELSE
198                IF TRANSACTION-DATE IS NOT = T-DATE-STORAGE
199                    PERFORM 270-PRINT-DAILY-SALES
200            ELSE
201                MOVE SPACES TO T-DATE-O.
202
203        220-CALCULATE-PER-ITEM-SALES.
204
205            MULTIPLY QUANTITY BY UNIT-PRICE
206                GIVING SALES-AMOUNT-WS.
207
208        230-ACCUMULATE-DAILY-SALES.
209
210            ADD SALES-AMOUNT-WS TO DAILY-TOTAL-WS.
211
212        240-ACCUMULATE-WEEKLY-SALES.
213
214            ADD SALES-AMOUNT-WS TO WEEKLY-TOTAL-WS.
215
216        250-PRINT-TRANSACTION-LINE.
217
218            MOVE ITEM-NUMBER        TO ITEM-NUMBER-O.
219            MOVE ITEM-DESCRIPTION   TO ITEM-DESCRIPTION-O.
220            MOVE QUANTITY           TO QUANTITY-O.
221            MOVE UNIT-PRICE         TO UNIT-PRICE-O.
222            MOVE SALES-AMOUNT-WS    TO SALES-AMOUNT-O.
223            MOVE DETAIL-LINE TO PRINT.
224            PERFORM 280-PRINT-DETAIL-LINE.
225
226        260-PROCESS-DATE.
227
228            MOVE T-MONTH            TO T-MONTH-O.
229            MOVE T-DAY              TO T-DAY-O.
230            MOVE T-YEAR             TO T-YEAR-O.
231            MOVE T-DATE             TO T-DATE-O.
232            MOVE TRANSACTION-DATE   TO T-DATE-STORAGE.
233
234        270-PRINT-DAILY-SALES.
235
```

```
236            MOVE DAILY-TOTAL-WS TO DAILY-TOTAL-O.
237            MOVE DAILY-TOTAL TO PRINT.
238            PERFORM 290-WRITE-DOUBLE-DETAIL-LINE.
239            MOVE SPACES TO PRINT.
240            PERFORM 280-PRINT-DETAIL-LINE.
241            PERFORM 260-PROCESS-DATE.
242            MOVE ZEROES TO DAILY-TOTAL-WS.
243
244        280-PRINT-DETAIL-LINE.
245
246            IF LINE-COUNT IS > 37
247                MOVE PRINT TO SAVE-RECORD
248                PERFORM 300-PRINT-HEADING-LINE
249                MOVE SAVE-RECORD TO PRINT
250            ELSE
251                NEXT SENTENCE.
252            PERFORM 310-WRITE-DETAIL-LINE.
253
254        290-WRITE-DOUBLE-DETAIL-LINE.
255
256            WRITE PRINT
257                AFTER ADVANCING 2 LINES.
258            ADD 2 TO LINE-COUNT.
259
260        300-PRINT-HEADING-LINE.
261
262            MOVE HDG-1 TO PRINT.
263            PERFORM 320-WRITE-TOP-OF-PAGE.
264            MOVE HDG-2 TO PRINT.
265            PERFORM 290-WRITE-DOUBLE-DETAIL-LINE.
266            MOVE SPACES TO PRINT.
267            PERFORM 310-WRITE-DETAIL-LINE.
268
269        310-WRITE-DETAIL-LINE.
270
271            WRITE PRINT
272                AFTER ADVANCING 1 LINE.
273            ADD 1 TO LINE-COUNT.
274
275        320-WRITE-TOP-OF-PAGE.
276
277            WRITE PRINT
278                AFTER ADVANCING PAGE.
279            MOVE 1 TO LINE-COUNT.
280
281        300-PRINT-WEEKLY-SALES.
282
283            PERFORM 270-PRINT-DAILY-SALES.
284            MOVE WEEKLY-TOTAL-WS TO WEEKLY-TOTAL-O.
285            MOVE WEEKLY-TOTAL TO PRINT.
286            PERFORM 290-WRITE-DOUBLE-DETAIL-LINE.

THERE WERE 286 SOURCE INPUT LINES.
THERE WERE NO DIAGNOSTICS.
```

```
                              TRANSACTION   REGISTER

  TRANSACTION       ITEM          DESCRIPTION              QUANTITY    UNIT       SALES

   07/23/80        413010       CH001 BOX 100A FLUSH          10       4.90      $49.00
                   412146       CH148 BREAKER 15A            100        .89      $89.00
                   411116       1500 TWIN SOCKET B           500       1.12     $560.00

                                                                                $698.00

   07/24/80        503029       MOTOR 1/2 HP 60 CYC            2     146.78     $293.56
                   317802       TERMINAL CLIP                100       5.12     $512.00
                   326917       TERMINAL BAR                 100       4.12     $412.00
                   411121       1506 SOCKT ADAPT BRN         400        .19      $76.00
                   412997       CH173 BREAKER 30A             60       1.15      $69.00
                   413088       CH176 BREAKER 60A             40       1.15      $46.00
                   411174       C151 SIL SWITCH BRN          200       1.16     $232.00
                   413090       CH005 BR BOX 150A             10       4.98      $49.80
                   718326       FC803 FUSE 15A               200        .32      $64.00

                                                                              $1,754.36

                                          WEEKLY TOTAL                        $2,452.36
```

Questions for Review

1. What is the main function of data manipulation statements?
2. List the data manipulation verbs and their main functions.
3. What is the main purpose of the EXAMINE statement and how is it used?
4. Briefly explain the TALLYING phrase of the EXAMINE statement and its various options.
5. Briefly explain the REPLACING phrase of the EXAMINE statement and its various options.
6. What is the main function of the INSPECT statement and what are the major differences between the EXAMINE and INSPECT statements?
7. Briefly explain the operation of the INSPECT statement.
8. Briefly explain the TALLYING phrase of the INSPECT statement and its various options.
9. What options of the INSPECT statement are similar to those found in the TALLYING and REPLACING options of the EXAMINE statement?
10. Why is it significant for the order in which multiple REPLACING phrases of the INSPECT statement to be specified, appear?
11. How are the uses of the TALLYING and REPLACING options of the INSPECT statement interpreted and executed?
12. What is the main function of the STRING statement?
13. What is the purpose of the STRING statement?
14. Briefly explain the various options available with the STRING statement.
15. Briefly explain the operation of the STRING statement.
16. What is the function of the UNSTRING statement?
17. What are the options of the UNSTRING statement?
18. What are the two possible causes of overflow during an UNSTRING operation?
19. What action takes place if the ON OVERFLOW phrase is not specified and overflow occurs?
20. Briefly explain the operation of the UNSTRING statement.

Matching Questions

Match each item with its proper description.

_____ 1. UNSTRING	A. Used to count the number of receiving fields operated on.
_____ 2. INSPECT	B. Used to replace a given character and/or count the number of times a given character appears in a data item.
_____ 3. DELIMITED BY	C. Indicates the starting position in a receiving field where data is to be moved.
_____ 4. EXAMINE	D. Modifies the value of an item by replacing certain original characters with new characters.
_____ 5. TALLYING IN	E. Several fields are joined together to form one field.
_____ 6. TALLYING	F. Used in conjunction with character strings.
_____ 7. POINTER	G. Allows the programmer to specify one or more characters to mark the end of a field to be strung.
_____ 8. STRING	H. Used to scan a data item counting the number of occurrences of a given literal.
_____ 9. REPLACING	I. Signals when the end of a receiving field is reached, and when the sending field still contains characters.
_____10. ON OVERFLOW	J. Separates one field into several fields.

Exercises

Multiple Choice: Indicate the best *answer.* (Questions 1-19)

1. The data manipulation verb that is used with character strings is
   a. EXAMINE.
   b. STRING.
   c. INSPECT.
   d. MOVE.
2. The EXAMINE statement may be used to
   a. validate input data.
   b. translate information from one code to another.
   c. do free-form input processing.
   d. All of the above.
3. The TALLYING option of the EXAMINE statement is used to
   a. count the number of occurrences of a given literal.
   b. modify the value of an item.
   c. replace certain characters in a data item.
   d. place the accumulation in a counter defined by the programmer.
4. The option of the REPLACING phrase that replaces all occurrences of a literal prior to the leftmost occurrence of any other character is
   a. ALL.
   b. FIRST.
   c. LEADING.
   d. UNTIL FIRST.
5. The INSPECT and EXAMINE statements are similar in that they both
   a. use a field called TALLY to count the various characters in a field.
   b. permit the tallies and replacements to be named in series.
   c. use the BEFORE and AFTER options that permit some modification of specifications.
   d. None of the above.
6. The INSPECT statement can be used to
   a. count the number of occurrences of a particular character.
   b. replace a particular character(s) with a different character(s).
   c. validate data.
   d. All of the above.
7. In the operation of an INSPECT statement,
   a. inspection begins at the rightmost character position of the data item referenced.
   b. the order of the operands need not be specified.
   c. if no match occurs in the comparison of the first character string, the comparison is repeated.
   d. the comparison continues until the lefthand character position referenced by the data item is reached.
8. If the CHARACTERS phrase is specified in the TALLY phrase,
   a. an implied literal participates in the comparison cycle.
   b. an implied one-character operand participates in the comparison cycle.
   c. an implied character is always considered to match the rightmost character of the data item referenced.
   d. All of the above.
9. The following is an allowable option of the REPLACING phrase.
   a. WHILE.
   b. DO.
   c. BEFORE.
   d. NOT.
10. A statement with both the TALLYING and REPLACING option together is interpreted as
    a. though one INSPECT statement with these options was executed.
    b. though two successive INSPECT statements specifying the same identifier have been coded.
    c. though the INSPECT statement with only one of the options can be executed.
    d. being nonexecutable.
11. The STRING statement causes characters from
    a. one sending field to be transferred to multiple data items.
    b. one or more data items to be transferred to a single data item.
    c. only one data item to be transferred to a single data item.
    d. None of the above.

12. Concatenation is the process of
    a. several fields joining together to form one field.
    b. one field being separated into several fields.
    c. several fields being joined to several other fields.
    d. None of the above.
13. The option that specifies that the entire field is to be used in a STRING operation is called
    a. DELIMITED BY POINTER.
    b. DELIMITED BY.
    c. DELIMITED BY CHARACTERS.
    d. DELIMITED BY SIZE.
14. When the POINTER option of the STRING statement is used,
    a. the initial value of the pointer can be zero.
    b. the initial value of the pointer is set by the compiler.
    c. it indicates the starting position of a receiving field.
    d. the initial value of the pointer is decreased by one every time a character is moved.
15. The ON OVERFLOW option of the STRING statement is used to signal
    a. when the end of the sending field is reached.
    b. when the pointer does not contain an acceptable value.
    c. when the receiving field still contains characters.
    d. None of the above.
16. The UNSTRING statement
    a. separates one field into several fields.
    b. joins several fields into one.
    c. joins several fields into corresponding fields.
    d. None of the above.
17. The COUNT IN phrase of the UNSTRING statement
    a. gives the number of fields involved in an operation.
    b. keeps count of the number of characters placed in each field.
    c. is set up by the compiler.
    d. All of the above.
18. If a pointer field in an UNSTRING statement contains a 9,
    a. only the first nine characters are moved.
    b. the receiving field will receive data starting in the tenth position.
    c. the first eight characters for the sending field are ignored.
    d. the pointer is decreased by one so it always contains the number of remaining characters to be moved.
19. If the DELIMITER BY phrase is specified in the UNSTRING statement, the string of characters is examined beginning with the
    a. first character to the left of the delimiter.
    b. first character to the right of the delimiter.
    c. character to the right of the last character transferred.
    d. None of the above.
20. For each of the following, fill in TALLY.
    a. EXAMINE FLD-A TALLYING UNTIL FIRST ZERO.
       FLD-A | 1 | 0 | 3 | 5 | 0 | 0 | 1 | TALLY | | | | | |

    b. EXAMINE FLD-B TALLYING ALL ZEROS.
       FLD-B | 1 | 0 | 3 | 5 | 0 | 0 | 1 | TALLY | | | | | |

    c. EXAMINE FLD-C TALLYING LEADING SPACES.
       FLD-C | | | | * | * | * | * | TALLY | | | | | |

    For each of the following, fill in the resulting data.

    d. EXAMINE FLD-D REPLACING ALL ZEROS BY QUOTES.
       FLD-D | 0 | 0 | 0 | 3 | 4 | 5 | 0 | FLD-D | | | | | | | |

    e. EXAMINE FLD-E REPLACING FIRST "O" BY "—".
       FLD-E | 5 | 7 | 3 | 0 | 5 | 1 | 4 | FLD-E | | | | | | | |

    f. EXAMINE FLD-F TALLYING LEADING ZEROS REPLACING BY SPACES.
       FLD-F | 0 | 0 | 0 | 0 | 6 | 7 | 0 | FLD-F | | | | | | | |
       TALLY | | | | | |

21. 77 STUDENT-MASTER-RECORD PICTURE X(40).

```
21. 77  STUDENT-MASTER-RECORD       PICTURE X(40).
            .
            .
            .
    01  STUDENT-RECORD.
        05  ID-NUMBER               PICTURE X(6).
        05  FILLER                  PICTURE X(7).
        05  FIRST                   PICTURE X(10).
        05  MIDDLE                  PICTURE X(10).
        05  LAST                    PICTURE X(12).
        05  FILLER                  PICTURE X(55).
```

a. Write a STRING statement that will condense the aforementioned STUDENT-RECORD in a data item called STUDENT-MASTER-RECORD. The sending data item should be delimited by a stroke (/) and should be separated by asterisks (*).

b. Write an UNSTRING statement assuming what is necessary to restore the information to its original form.

```
22. 01  STUDENT-RECORD.
        05  SOCIAL-SECURITY-NUMBER      PICTURE 9(9).
        05  FILLER                      PICTURE X(5).
        05  NAME.
            10  LAST                    PICTURE X(12).
            10  FIRST                   PICTURE X(12).
            10  INITIAL                 PICTURE X.
        05  FILLER                      PICTURE X(5).
        05  ADDRESS.
            10  STREET-NO               PICTURE X(6).
            10  STREET-NAME             PICTURE X(20).
            10  CITY                    PICTURE X(10).
            10  STATE                   PICTURE X(6).
            10  FILLER                  PICTURE XXX.
            10  ZIPCODE                 PICTURE X(5).
```

In the STUDENT-RECORD just given, using the STRING statement and a data item called CONDENSED-STUDENT-RECORD, the literal stroke (/) is to be transferred after each elementary item from STUDENT-RECORD is transferred. The last character to be transferred for each record should be indicated by the character dollar sign ($).

23. Assume that we wish to change the contents of CONDENSED-STUDENT-RECORD (exercise 22) to change all delimiters from stroke (/) to asterisk (*) and all dollar signs ($) to pound signs (#).
Write an INSPECT statement to accomplish the aforementioned.

24. Write a statement that will change each occurrence of the stroke (/) in EMPLOYEE-MASTER to a comma (,).

```
    77  EMPLOYEE-MASTER     PICTURE X(20).
        VALUE '12345/JONES/JOHN/J./'.
```

Problem 1

Use the flowchart, record layout, and printer spacing chart to help you write a COBOL program that prepares a report that totals the hours for a given job. Different employees, who may be working on several different jobs during the week, submit their time and number along with the job identification. All these records have been grouped by job number for this report.

Your calculations should include the accumulation of total hours for each job and a count of the number of active jobs.

The output should be according to the printer spacing chart. The job number should print only on the first record of a group that is group-indicated.

The input is as follows:

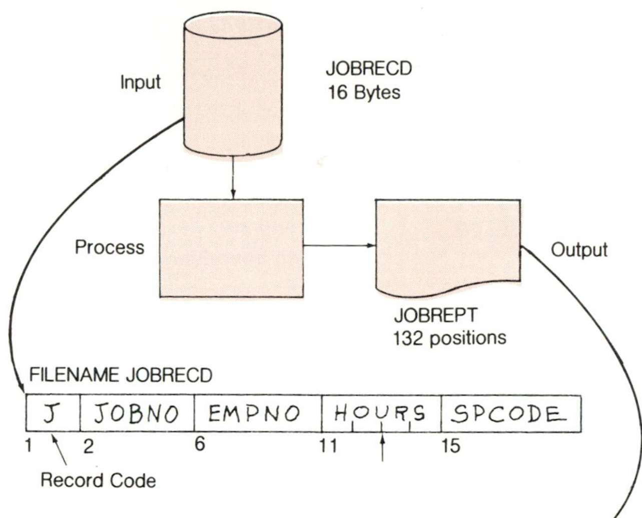

Filename = JOBREPT

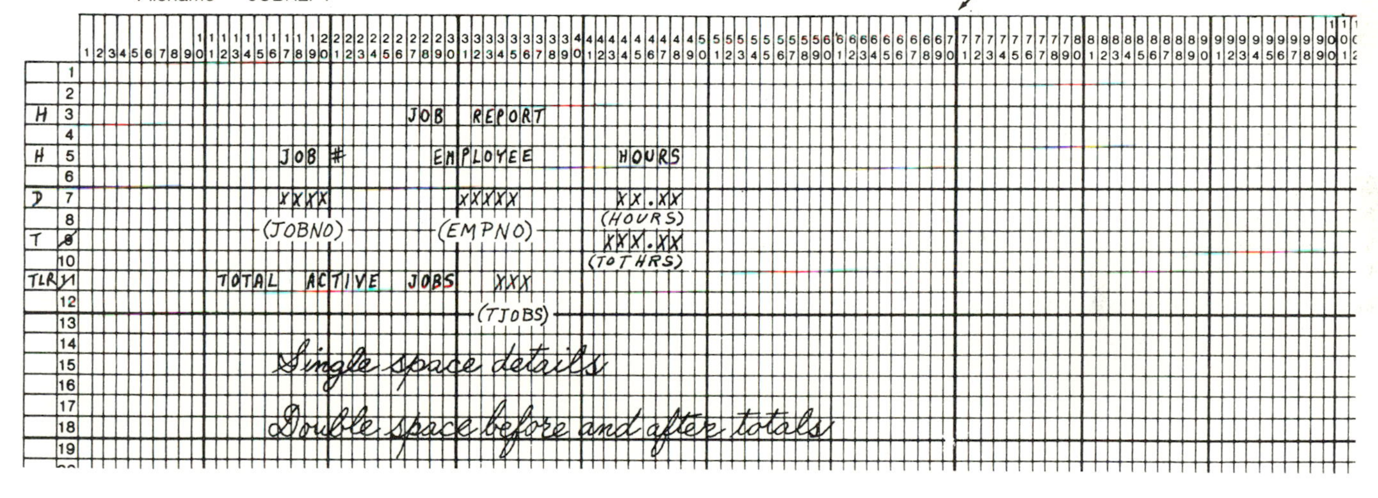

Output is as follows:

```
                    JOB REPORT

        JOB #        EMPLOYEE        HOURS

        0281           68312         12.50
        0281           59478         30.00


                                     42.50

        0283           67129         15.50
        0283           70049         42.30


                                     57.80

        0285           71312         40.30


                                     40.30

     TOTAL ACTIVE JOBS      5
```

Problem 2

Write a COBOL program for the following:

A payroll file consists of records containing the following information:

Positions	Fields
1-3	Department
4-8	Number
9-17	Social security number
18-22	Gross pay
23-27	Withholding tax
28-32	Social security tax
53-56	Deduction 1
57-60	Deduction 2
61-64	Deduction 3
65-68	Deduction 4
	All fields have two decimal positions

The net pay is to be computed using the following formula:

$$NET = GROSS - WITHTAX - SSTAX - D1 - D2 - D3 - D4$$

All deductions are to be taken from gross pay unless doing so causes the net pay amount to become zero or negative in value. If that happens, add back the last deduction, print an exception record on a special report, and bypass all further calculations for that record. Include messages on the exception report to indicate which, if any, deductions were actually taken.

Prepare a Printer Spacing Chart to represent heading and detail lines of the exception record report.
Create your own data names where necessary, and indicate necessary heading lines on the Printer Spacing Chart. In the output, include all information from the input, plus the net pay and any necessary messages.
Output is as follows:

```
         5/26/81                             PAYROLL   DEDUCTION   EXCESSES                             PAGE   1

DEPT. NOSOC. SEC. NO.     GROSS  WITH.TX.           DED 1  DED 2  DED 3   DED 4

 50   1600   564-49-1212 941.62  594.16  219.41  92.11  97.48  97.48    .00     ACTUAL PAY IS    35.94  INCLUDING D1
 62   2300   581-01-0334 877.77  587.70   58.97  81.99  94.19  94.19   9.28     ACTUAL PAY IS    54.92  INCLUDING D1, D2
 62   4400   495-14-9142  94.82   81.48    8.24   5.19   2.43   2.43   5.19     ACTUAL PAY IS     5.10  NO DEDUCTIONS TAKEN
```

Problem 3

The area in which all salesmen work is divided into three districts—A, B, and C. Some salesmen work only in one district, while others may work in parts of two or more districts.

For each salesman, the input file contains a record as follows. The amounts in the district fields show total weekly sales made by that salesman in each district. If the salesman did not work in a district or made no sales in that district, the field contains zeros.

Input

Positions	Field Designation	Format
1-25	Salesman name	
26-32	District A	XXXXX.XX
33-39	District B	XXXXX.XX
40-46	District C	XXXXX.XX
47-80	Not used	

Processing

1. A report is to be prepared showing the commissions earned in each district by each salesman.
2. The total commission is to be accumulated for each salesman and for each district.
3. The percentage of commission is as follows:
   a. 3 percent of gross sales $.01 to $1,000
   b. Plus 2 percent of gross sales $1,000.01 to $5,000
   c. Plus 1 percent of gross sales over $5,000

Output

The desired output shows two things as per the Printer Spacing Chart:

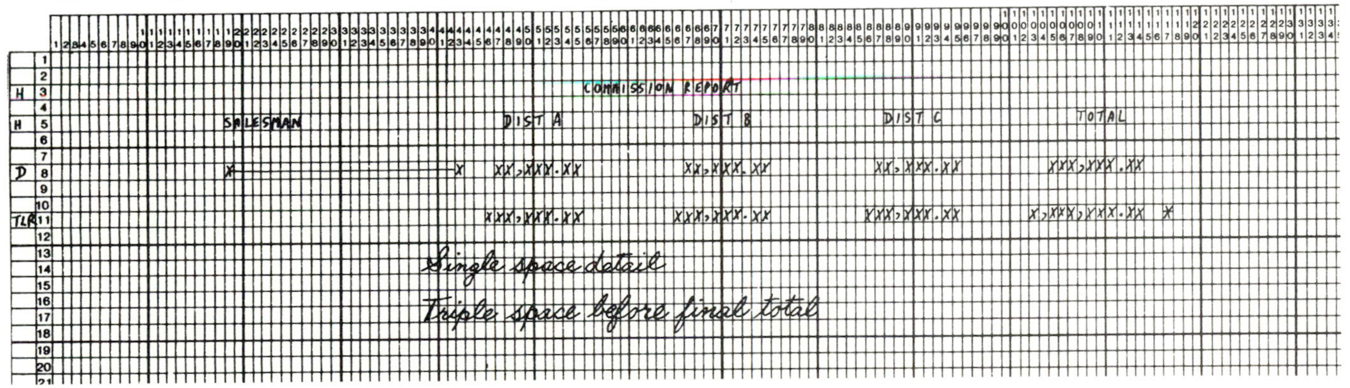

1. Total commission earned by each salesman by district and total for all districts
2. Total commissions paid for each district to all salesmen

   Output is as follows:

COMMISSION REPORT

SALESMAN	DIST A	DIST B	DIST C	TOTAL
HENRY HINES	28.52	120.00	.00	148.52
JACK SMITH	34.00	16.67	135.00	185.67
WALTER REID	13.80	.00	12.60	26.40
JANE DOE	41.85	44.41	151.79	238.05
CHARLES BROWN	139.29	886.29	711.23	1,736.81
	257.46 *	1,067.37 *	1,010.62 *	2,335.45 *

Problem 4

An invoice billing is to be prepared as per output sample.

Input

The input file consists of 80-position records with the following fields:

Positions	Field Designation	Format
1-5	Product number	XXXXX
6-11	Quantity	XXXXXX
12-17	Unit price	XXX.XXX
18-37	Description	
38-42	Invoice number	XXXXX
43-47	Customer number	XXXXX
48-67	Customer name	
68-80	Unused	

Processing

All answer fields should be rounded to two decimal places.

1. Compute sales amount = quantity $\times$ unit price.
2. All sales amounts are to be totaled for the same invoice number.
3. Compute discount amount and net amount due as follows:
   a. When total sales exceed $1,000, allow a 3 percent discount.
   b. When total sales are $1,000 or less, a 2 percent discount is allowed.
   c. Subtract discount amount from total sales amount to arrive at net amount due.
4. Customer Number, Customer Name, and Invoice Number are to be group-indicated.
5. The billing is to be printed in the following output format.

Output

The output format is as follows:

```
CUST NO.   CUSTOMER NAME        INVOICE NO.  PROD. NO.  QUANTITY UNIT PRICE    DESCRIPTION        SALES AMOUNT

  246      ACME HDWE CO., INC      24681       12345       651      4.751  HAMMER-BALL PEEN   EA    3,092.90
                                               24762        13    246.953  BOILER-STEAM       EA    3,210.39
                                               47672        11    189.752  WASHING MACHINE    EA    2,087.27
                                               67302       821      4.875  NAILS-STEEL WIRE   LB    4,002.38

                                                                             TOTAL SALES           12,392.94 *


                                                                             DISCOUNT ALLOWED          371.79


                                                                            NET AMOUNT DUE         12,021.15 **

 12481     E.C. MORGAN CO.         24682       15762       671       .752  LAG SCREWS         DZ      504.59
                                               38576        76      1.065  CLIPS-FILE         GR       80.94
                                               69251        52      6.521  PAINT              GL      339.09

                                                                             TOTAL SALES              924.62 *


                                                                             DISCOUNT ALLOWED           18.49


                                                                            NET AMOUNT DUE            906.13 **

 28762     WILLIAMS TOOL CO.       24683        7603     1,105       .151  NUTS HEX 1/8       DZ      166.86
                                                7603     1,105       .151  NUTS HEX 1/8       DZ      166.86
                                               39827        37    264.721  GRADERS            EA    9,794.68

                                                                             TOTAL SALES           10,128.40 *


                                                                             DISCOUNT ALLOWED          303.85


                                                                            NET AMOUNT DUE          9,824.55 **
```

Table Handling

Chapter Outline

Chapter Objectives

The learning objectives of this chapter are:

1. To describe the table handling features of COBOL and how it may be applied to data processing problems, including the uses of tables.
2. To differentiate between tables and arrays, and when each should be used in programming.
3. To describe the format and uses of the REDEFINES clause and the PERFORM statement with the VARYING option, and how they are used to set up tables at compilation and execution time.
4. To explain the operation of table searches and how the various formats of the SEARCH statement are used in serial and binary searches.
5. To describe the formats and functions of the SET statement and how it is used in a search operation.

Introduction

If you want to make a telephone call, you must first know the number to call. Imagine trying to obtain the number if no telephone directories or directory services were available. Therefore, to accomplish objectives such as this, similar items and types of information are grouped and organized so that they can be referenced easily and quickly.

A table is a collection of related items organized in such a way that each item of information can be referenced by its position within a table. (See figure 10.1.) A telephone directory consists of two tables of information, a name list arranged alphabetically, and a number list arranged in no apparent order. Each telephone number, however, occupies a position in the number list corresponding to the position of a particular name in the name list. (It is not the contents, but the relative position in the table, i.e., the seventh entry in both tables.)

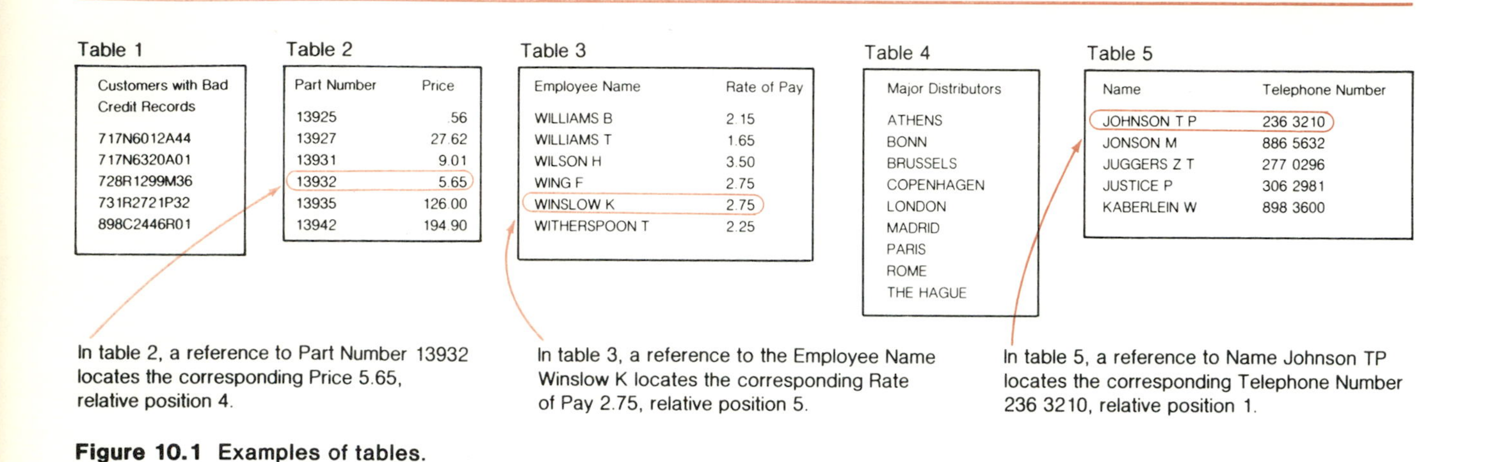

Figure 10.1 Examples of tables.

Each item within a table is called a *table element.* Thus, each name would be an element of the name table, while each number would be an element of the telephone number table.

If you wished to determine Ken Adams' telephone number, you would look through a list of names to locate KEN ADAMS. This procedure of checking the elements of a table one at a time to find a particular entry is called *searching* a table. The name KEN ADAMS is the search word and is known as the *argument.* The matching entry is the corresponding telephone number which is known as the *function.* (See figure 10.2.)

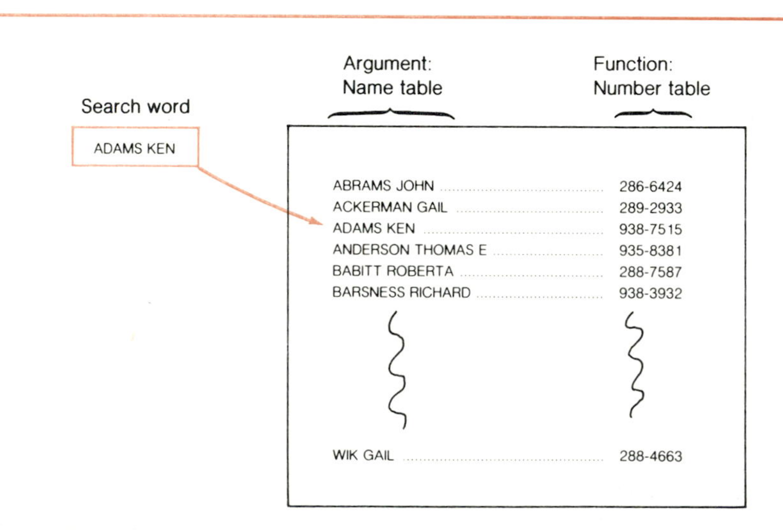

Figure 10.2 Searching a table.

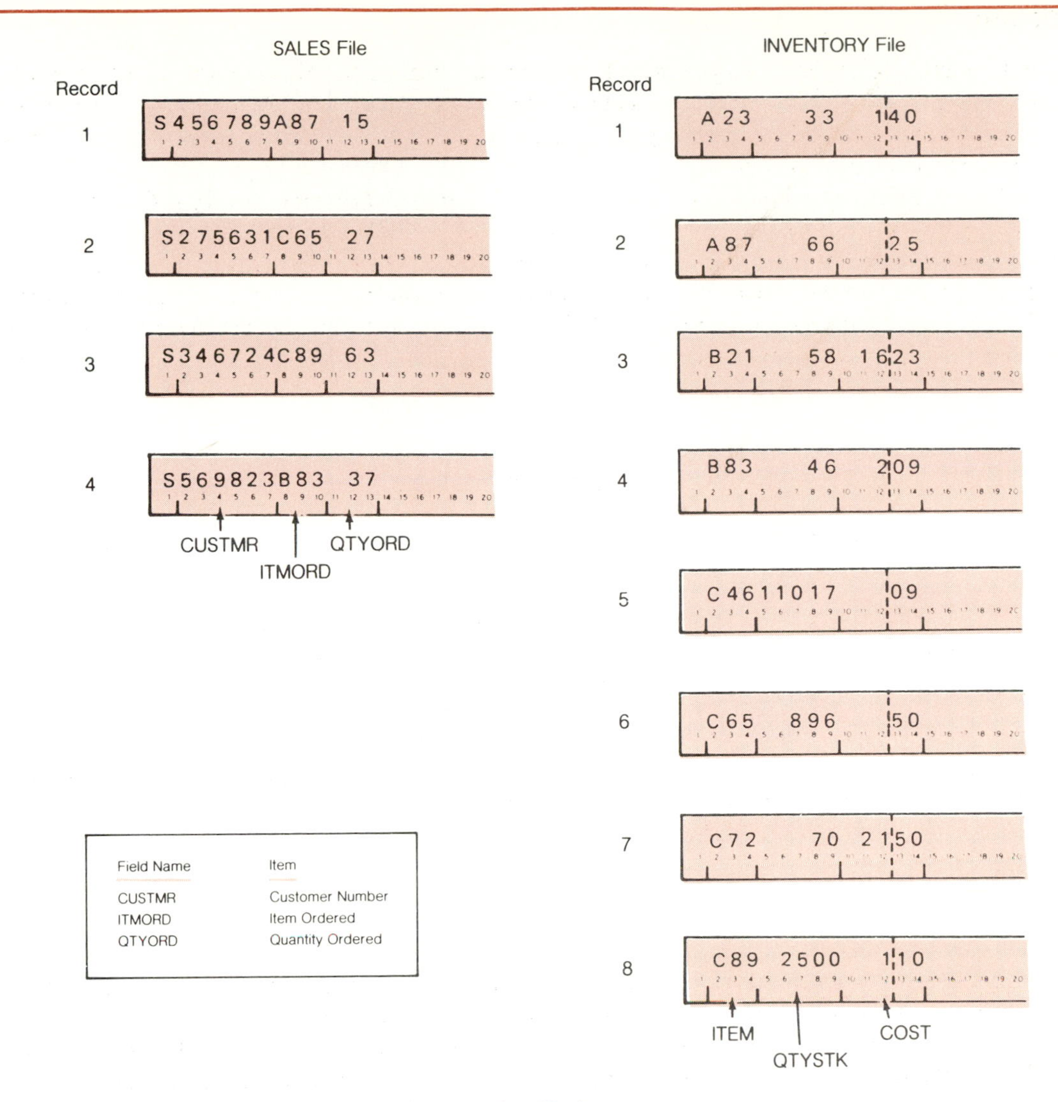

Figure 10.3 Data for determining if orders can be filled.

When two related tables are used, as in a telephone directory, only one table is searched (name table). When the search condition (in this case, an equal match) has been satisfied, the data in the corresponding element of the second table (number table) becomes available. Thus, the first table is used as a means of locating data in the second table.

A telephone directory is an example of tables containing organized information that must be referenced over and over again in our daily lives. Likewise, tables may be used to organize data that must be referenced repeatedly in data processing jobs as in the following example.

Assume that a customer has previously purchased various items from a company sales catalog. The sales file would contain records showing the customer's account number, the items ordered (each identified by a code), and how many of each item was ordered by the customer.

Furthermore, the company keeps an inventory file to contain data about each item that is carried in stock. A separate record is maintained for each item code, the quantity on hand, and the unit cost of each item. (See figure 10.3.)

Before any item is shipped, it is necessary to determine whether that item is still carried in stock. To do this, a clerk would spend time looking up each item ordered to see if that item is recorded in the inventory file. However, the same item will be ordered by many customers. Thus, the inventory file records would have to be referenced over and over again.

COBOL can search for the data in much less time by performing a table search function. A table would be set up in storage to contain all the items available.

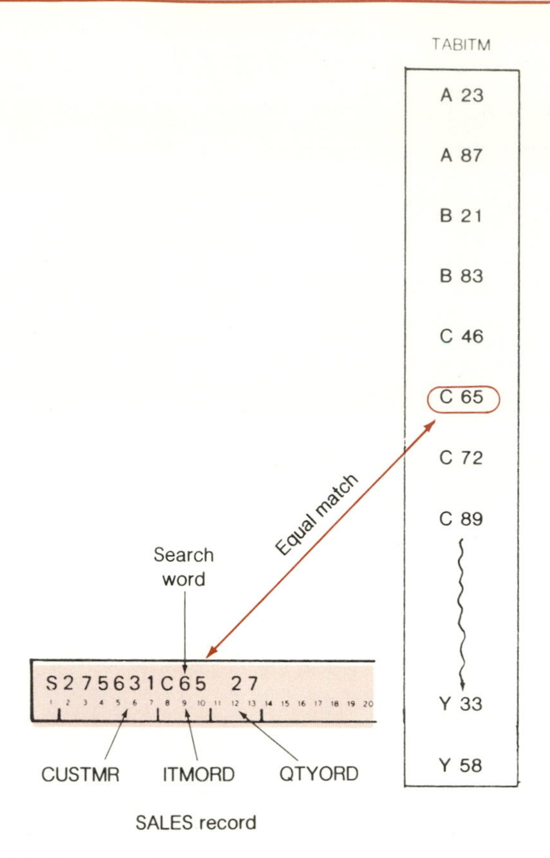

Figure 10.4 Searching a table for a particular data item.

A field in the sales record tells the program which item to look up. For every sales record read into the computer, the table is checked to see if the record read matches an entry in the table. (See figure 10.4.)

In addition to the searching for data quickly, use of table lookup can often reduce the number of instructions needed to do a job. All that need be done is to set up and define the table and to specify that the search operation is to be performed.

A program occasionally requires that one or more tables of values be stored in memory. These values can be referenced and used in computations or other procedural routines. For example, a tax percentage

Tax Table		
1 dependent	.175	(17.5%)
2 dependents	.164	
3 dependents	.159	
4 dependents	.153	
5 dependents	.147	
6 dependents	.141	
7 dependents	.127	
etc.		

Figure 10.5 Typical tax table—diagram.

table might be useful for calculating tax deductions during a payroll run. (See figure 10.5.) Such a table might be set up in the Working-Storage section of the Data Division as shown here:

```
WORKING-STORAGE SECTION.

01  TAX-TABLE.
    05  DEP-1      PICTURE V999      VALUE .175.
    05  DEP-2      PICTURE V999      VALUE .164.
    05  DEP-3      PICTURE V999      VALUE .159.
    05  DEP-4      PICTURE V999      VALUE .153.
    05  DEP-5      PICTURE V999      VALUE .147.
    05  DEP-6      PICTURE V999      VALUE .141.
    05  DEP-7      PICTURE V999      VALUE .127.
```

In the Procedure Division, the tax calculation routine might look like that shown here:

```
TAX-CALC.
    IF NO-DEP IS EQUAL TO 0, MULTIPLY DEP BY .179 GIVING
        TAX-AMT.
    IF NO-DEP IS EQUAL TO 1, MULTIPLY DEP BY DEP-1 GIVING
        TAX-AMT.
    IF NO-DEP IS EQUAL TO 2, MULTIPLY DEP BY DEP-2 GIVING
        TAX-AMT.
    IF NO-DEP IS EQUAL TO 3, MULTIPLY DEP BY DEP-3 GIVING
        TAX-AMT.
            Etc.
```

Notice that many IF statements must be written and executed, as there are values in the table. For larger tables, this would require a tremendous amount of writing, memory space, and execution time.

COBOL provides a Table Handling module that has the capability of defining tables of contiguous data, and to access an item relative to its position in the table. In order to discuss table handling features, it is necessary to define a few standard terms.

A *table* consists of a number of contiguous data items that are of like format. The individual data items in a table are called *table elements.* Within the COBOL programming language specifications, table elements are accessed by either *indexing* or *subscripting* techniques.

Subscripting is a method of specifying an occurrence number and to affix a *subscript* to the data-name of the table element. A *subscript* is an integer whose value identifies a particular element in a table.

Indexing is another method of specifying an occurrence number and to affix an *index* to the data-name of the table element. An *index* is a computer storage position or register whose content corresponds to an occurrence number.

Table Value refers to any one of the values (i.e., the value of any cell) contained in a table. The table itself simply consists of all values (cells) contained in it. In the tax table example, each tax percentage is a table value.

Search Argument is any number used to locate one of the values within the table. In the tax calculation example, the number of dependents constitutes the search argument.

To locate a value within a table, the search argument is used to scan the table until a table element associated with the search argument is found. If, in the tax example, the search is for the tax percentage to be used for an employee with five dependents (search argument), the search would produce the table element .147 which is associated with five dependents. The operation is called *table search* or *table lookup.*

The Table Handling Feature enables the COBOL programmer to process tables or lists of repeated data conveniently. A table may be set up with three dimensions; for example, three levels of subscripting or indexing can be handled. This Table Handling module provides a capability for defining tables of contiguous data items and for accessing an item relative to its position in the table. Language facility is provided for specifying how many times an item can be repeated. Each item may be identified through the use of a subscript or an index-name. Such a case exists when a group item described with an OCCURS clause contains another group item with an OCCURS clause, which in turn contains another group with an OCCURS clause. To make reference to any element within such a table, each level must be subscripted or indexed.

The Table Handling Feature provides a capability for accessing items in three-dimensional variable-length tables. The feature also provides the additional facilities for specifying ascending or descending keys, and permits searching of a table for an item satisfying a specified condition.

Tables

Tables are systematically arranged sets of information that are more limited in scope than other files. Examples are tables of freight rates, withholding percentages, prices, and conversion coefficients. Tables are used in computer programs much as they are used by clerks in manual systems. A document (or record) being processed provides a piece of information, such as an item stock number. The known information is then used to obtain another piece of information from a table by a search. Thus, the item stock number is used in looking up the item price in a table that contains the prices of all stock numbers.

Another use of tables is the verification of information with no requirement for retrieval of other data. For example, an input account code may be used to search a table containing all active account codes to determine whether or not the input code is valid.

A third use of tables entails updating, whereby entries may be changed during processing and then be returned to their places in the table. An example is the substitution of new prices in a price table. Another example is the posting of sales transactions to a summary table of department sales activity.

The advantage of using table format over a standard sequential file arrangement is that the tables are compact and may be completely contained in computer memory for random processing. Any member of the entire set of entries comprising the table may be retrieved for use either once or repeatedly during the course of processing by a program.

The entries in a table may be arguments, functions, pairs of arguments and functions, or pairs of functions and arguments. The arguments in a table are the values against which a search argument is compared when a search of table data is performed. A table of arguments may be used alone as in the validation example mentioned earlier in this section. More commonly, however, it has one or more associated function tables. A function table may be read into the computer together with the arguments table as a set of paired entries, or the tables may be read in separately.

To retrieve or access a function, the associated argument table is searched using a search argument and designated search criteria. When the argument entry that meets the criteria is found, its relative entry position in the argument table is noted. The desired function is then available in the same relative entry position in the function table. (See figure 10.6.)

The aforementioned figure (10.6) shows examples of table usage.

A shipping transaction file ① is sequenced by stock number within record type within customer number.

There is a type 1 record for each customer and a variable number of type 2 records.

The type 1 record shows the date of shipment and the address to which a shipment is made.

Each type 2 record shows a stock number for each item shipped, the quantity shipped, and a code assigned by a checker to each line item.

Line items are checked randomly by different checkers to control errors in both order makeup and shipping dock pilferage. The checker's codes are regenerated in a random pattern and are reassigned each processing cycle for coding integrity. The current authorized codes are in an argument table ② .

During processing of each line item transaction, the checker code is validated by searching the table of current authorized codes ③ , using the value from the transaction record as the search argument. Transactions without a valid checker code are returned to the shipping dock supervisor for rechecking of the order.

Item stock numbers and associated item prices are maintained in associated argument and function tables ④ . The stock number from the type 2 transaction record is used to obtain the item price from the price table ⑤. The price is then used in extending the item quantity to produce line item price.

A record for the line item is output for invoicing.

While tables are commonly used in combination with the processing of other files, they may be processed alone to produce a report or to update a table. Consider the processing of an employee hourly pay rate table to be updated by an across-the-board cost-of-living increase formula. The pay rate tables are in an unordered function table that is accessed by reference to an associated argument table of employee category codes in ascending sequence. The first search argument is a program constant or literal which is used to retrieve the first argument and function from the tables. The retrieval pay rate

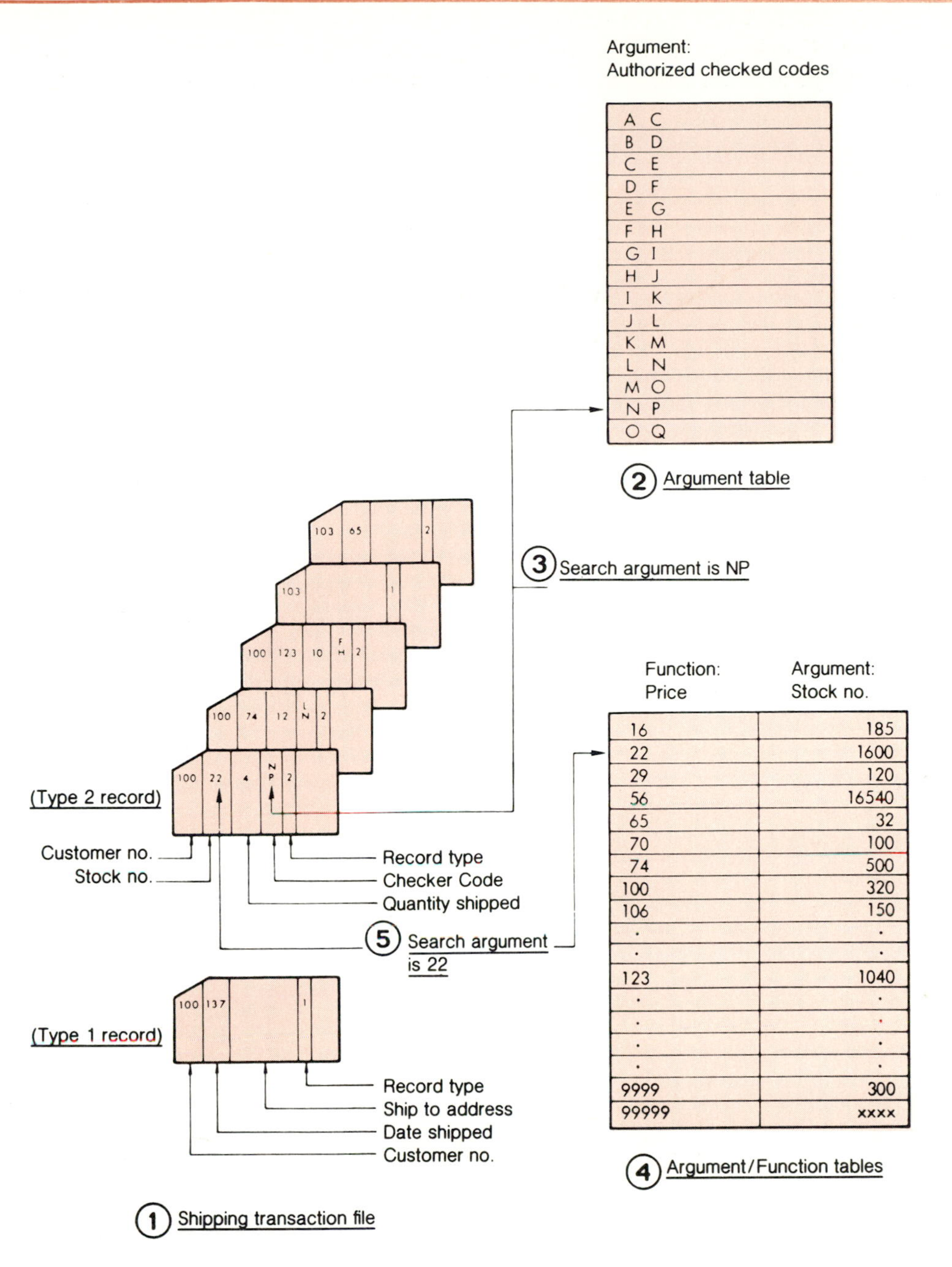

Figure 10.6 Table usage—example.

entry is updated by the increase formula and is restored to its original place in the function table. The retrieved argument is used in accessing the next higher entry in the argument table and its associated function from the function table. The process is continued successively for each entry in the table until there are no more entries. The updated function table is written in the file for future reference.

Arrays

An array is a continuous series of data fields, stored side by side, so they can be referenced as a group. In an array, each individual data field is called an *element*. Each element of the array has the same characteristics, that is, each contains data in the same format (alphanumeric or numeric), of the same length, and with the same number of decimal positions. (See figure 10.7.)

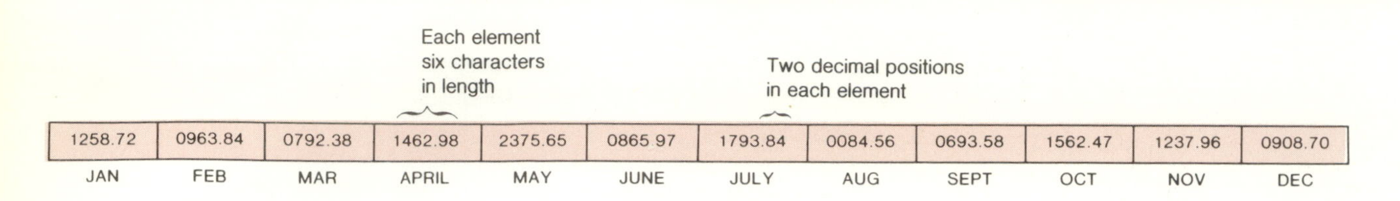

Figure 10.7 12-element numeric array.

An array is very similar to a table. Both arrays and tables are set up in a similar manner. The type of data that can be put into an array is the same as can be put into a table. The way data is arranged in storage is the same for tables and arrays; one element of the data follows another. However, the uses of tables and arrays differ considerably.

When to Use an Array Instead of a Table

In most cases, tables contain constant data, such as tax rates, shipping instructions, or discount rates. The constant data is then used for calculations or printing with variable transaction data. Arrays are generally used for variable data and totals that are used independently of the variable transaction data.

Arrays should be used instead of tables when it is necessary to reference all items at one time. Arrays should be used to directly reference a data item within a group of items, thus eliminating the necessity of doing a lookup based on a search word.

Tables and Arrays

A table or an array is a named set of data items arranged in some meaningful order. The distinction between a table and an array is that an array is composed of elementary data items having identical data descriptions, while a table may be composed of both elementary items and group items having differing data descriptions. Since the definition of array is a subset of the definition of table, the term "table" will be used in the discussions of table handling.

Successive item positions in a table have the ordinal position numbers 1, 2, 3, 4, . . . ; therefore, any particular item can be identified by its ordinal position number. To refer to any particular item in the table, the data-name and the desired item's position number are given. The data-name would be ambiguous if a specific position number were not given.

A table described in a single entry is said to be "one-dimensional." COBOL permits the use of one-, two- or three-dimensional tables, which are described, respectively, in one, two, or three data description entries. The number of occurrences in each dimension is specified via the OCCURS clause.

In a one-dimensional table, the number of occurrences of table items is specified in a single entry. For example,

```
05  FIELD-A          OCCURS . . .
```

In a two-dimensional table, the number of occurrences are in two entries: a group entry and a subordinate entry. The total number of occurrences of the table items is the product of the numbers specified in the two entries. For example,

```
05  MAJOR            OCCURS . . .
    10  MINOR        OCCURS . . .
```

In a three-dimensional table, the number of occurrences is specified in three entries; a group entry containing a subordinate group entry, which in turn contains another subordinate entry. The total number of occurrences of the elementary table items is the product of the numbers specified in the three entries that define the table. For example,

```
05  MAJOR                   OCCURS . . .
    10  INTERMEDIATE        OCCURS . . .
        15  MINOR           OCCURS . . .
```

The term "dimensions" refers to the spatial organization of the table. For example, a two-dimensional table has three occurrences specified in the group entry and four occurrences in the elementary entry. The total table then has twelve items, of which the first four make up the first group, the second four make up the second group, etc. The tenth item in the table is actually the second item of the third group. The whole table resembles a page that is ruled into three horizontal rows and four vertical columns, and the tenth item appears in the third row, second column:

	Col 1	Col 2	Col 3	Col 4
Row 1	o	o	o	o
Row 2	o	o	o	o
Row 3	o	o (tenth item)	o	o

The reference to any item is by its name, row number, and column number. In COBOL, such numbers are separated by a comma or space and enclosed in parentheses; that is, the tenth item position would be referred to by the data-name followed by (3,2). Such an expression is called a *subscript*.

Similarly, a three-dimensional table resembles a stack of ruled planes. The more inclusive group entry specifies the number of planes, the subordinate group specifies the number of rows in each plane, and the least inclusive entry specifies the number of columns in each row. Consider, for example, a table described as follows:

```
05  A                OCCURS 3 TIMES.
    10  B            OCCURS 3 TIMES.
        15  C        OCCURS 4 TIMES.
```

The table described may be visualized as a stack of three ruled planes:

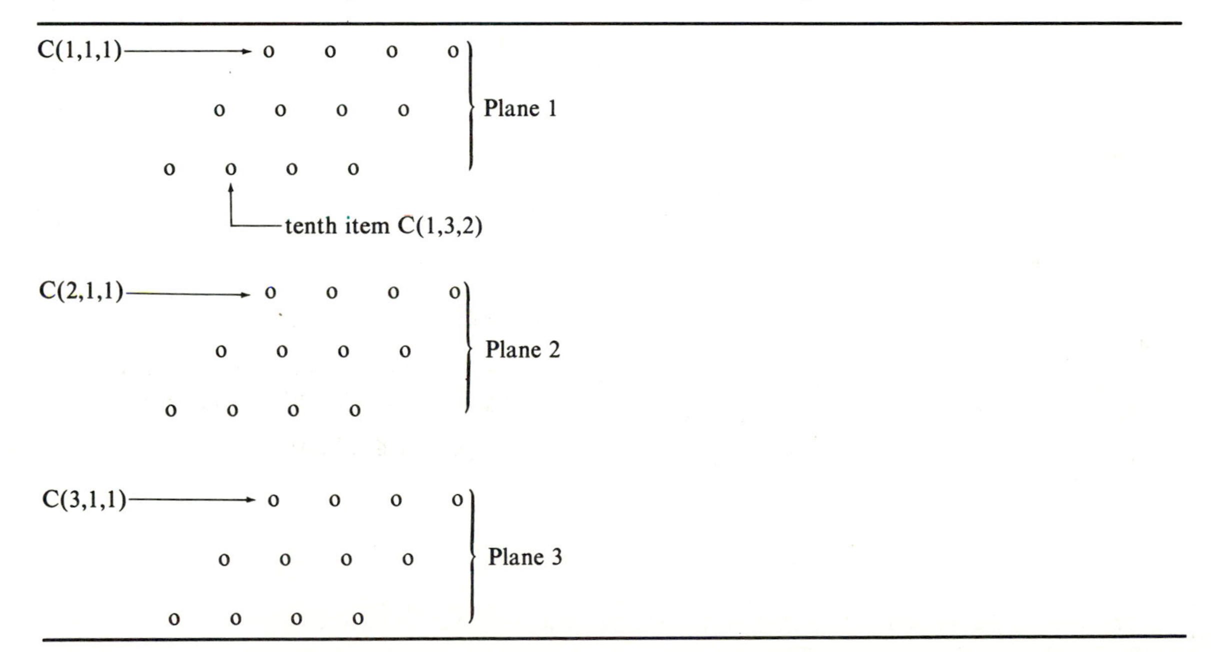

Note that items' positions in plane 1 correspond exactly to those in the prior example, except that the plane number (1) must also be specified within the subscript. The tenth item of this table is C(1,3,2).

COBOL permits table structures much more complicated than the examples just described, as long as no more than three dimensions are used. The minor (or only) OCCURS entry may be a group item. For example:

```
05  A                OCCURS 50 TIMES.
    10  B . . .
    10  C . . .
         .
         .
         .
```

In the definition of a two- or three-dimensional table, other entries may intervene within the hierarchy of OCCURS entries:

```
05  D                   OCCURS 50 TIMES.
    10  E . . .
    10  F               OCCURS 6 TIMES.
        15  G . . .
        15  H . . .
        15  I           OCCURS 3 TIMES.
            20  J . . .
            20  K . . .
        15  L           OCCURS 7 TIMES.
    10  M . . .
    10  N . . .
        15  P . . .
        15  Q . . .
            20  R       OCCURS 7 TIMES.
    10  S               OCCURS 2 TIMES.
```

Sometimes the number of significant items in a table varies throughout the execution of the object program. This variable number of occurrences may then be specified via the DEPENDING ON phrase of the OCCURS clause. Such a table can only be a one-dimensional table, can only appear once per record, and must be the last entry of the record.

Description of Table Handling

The table handling feature provides a capability for defining tables of contiguous data items and for accessing an item relative to its position in the table.

Data items may be accessed in up to three-dimensional variable-length tables. Additional facilities are needed to provide for specifying ascending or descending keys and for searching a dimension of a table for an item that satisfies a specific condition.

Tables of data are common components of business data processing problems. Although the items that make up a table could be described as contiguous data items, there are two reasons why this approach is not satisfactory. First, from a documentation standpoint, the underlying homogeneity of the items would not be readily apparent; and second, it would be difficult to make an individual element of such a table available if a decision is required to make one of these elements available at object program execution time.

Table Definition

Tables composed of contiguous data items are defined in COBOL by including the OCCURS clause in their data description entries. This clause specifies that the data item is to be repeated as many times as stated. The item is considered to be a table element and its name and description apply to each repetition or occurrence. Since each occurrence of a table element does not have to be assigned by a unique data-name, reference to a desired occurrence may be made simply by specifying the data-name of the table element together with the occurrence number of a desired table element.

To define a one-dimensional table, an OCCURS clause is used as part of the data description of the table element, but the OCCURS clause must not appear in the description of group items that contain the table element. An example of defining a one-dimensional table is:

```
01  TABLE-1.
    05  TABLE-ELEMENT OCCURS 20 TIMES.
        10  NAME . . .
        10  SSNO . . .
```

Defining a one-dimensional table within each occurrence of an element of another one-dimensional table calls for use of a two-dimensional table. To define a two-dimensional table, an OCCURS clause must appear in the group item that defines the table, and as part of an item which is subordinate to that group item. To define a three-dimensional table, an OCCURS clause should appear in the data description of the group item that defines the table and on two additional items (one of which is subordinate to the other). Note the following example of defining a three-dimensional table.

```
01  SALES-QUOTA-TABLE.
    05  REGION-TABLE                  OCCURS 5 TIMES.
        10  REGION-NAME                   PICTURE X(10).
        10  DISTRICT-TABLE            OCCURS 10 TIMES.
            15  DISTRICT-NAME             PICTURE X(10).
            15  OFFICE-TABLE          OCCURS 15 TIMES.
                20  OFFICE-NAME           PICTURE X(10).
                20  SALES-QUOTA           PICTURE 9(7).
```

Data Division for Table Handling

The OCCURS and USAGE clauses are included as part of the record description entry utilizing the table handling feature.

OCCURS Clause

The OCCURS clause is used to define tables and other homogeneous sets of data whose elements can be referred to by subscripting or indexing. The clause specifies the number of times that an item is repeated with no change in its USAGE or PICTURE clauses. The clause is used primarily in defining related sets of data such as tables, lists, matrixes, etc.

The OCCURS clause eliminates the need for separate entries for repeated data since it indicates the number of times a series of items with an identical form is repeated. In addition, it also supplies the required information for the application of subscripts and indexes.

The OCCURS clause is used in defining tables and other homogeneous sets of repeated sets of repeated data. (See figures 10.8 and 10.9.) Whenever the OCCURS clause is used, the data-name that is the subject of this entry must be either subscripted or indexed whenever it is referred to in a statement other than SEARCH. When subscripted, the subject refers to one occurrence within the table. When not subscripted (permitted only in a SEARCH statement), the subject represents the entire table element. (A table element consists of all occurrences of one table level.) Further, if the subject of this entry is the name of a group item, then all data-names belonging to the group must be subscripted or indexed whenever they are used as operands.

Format 1:

```
OCCURS integer-2 TIMES

  [ { ASCENDING  }  KEY IS data-name-2  [ data-name-3 ] . . . ] . . .
    { DESCENDING }

  [ INDEXED BY index-name-1  [ index-name-2 ] . . . ]
```

Format 2

```
OCCURS integer-1 TO integer-2 TIMES  [ DEPENDING ON data-name-1 ]

  [ { ASCENDING  }  KEY IS data-name-2  [ data-name-3 ] . . . ] . . .
    { DESCENDING }

  [ INDEXED BY index-name-1  [ index-name-2 ] . . . ]
```

Figure 10.8 Formats: OCCURS clause.

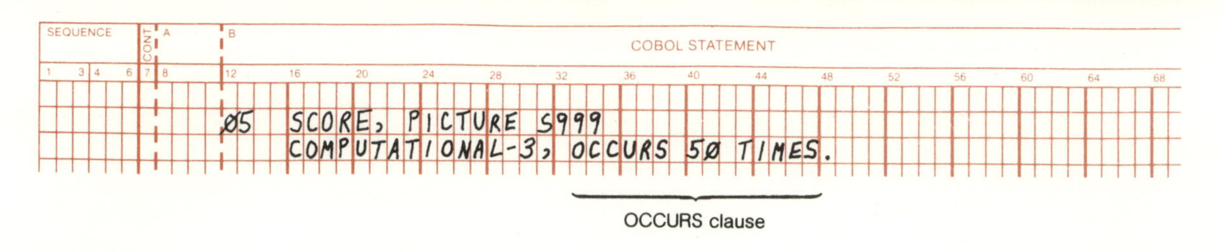

Figure 10.9 OCCURS clause—example.

Rules Governing the Use of the Occurs Clause

1. Record description clauses associated with an item that contains an OCCURS clause, apply to each repetition of the item being described.
2. Whenever the OCCURS clause is used, the data-name that is the defining name of the entry must be subscripted whenever used in the Procedure Division.
3. If the data-name is the name of a group item, then all data-names belonging to the group must be subscripted whenever used.
4. The OCCURS clause may not be used in the 01, 77, or 88 level of a record description entry.
5. The clause cannot describe an item whose size is variable.
   (See figures 10.10 and 10.11.)

PRODUCT NUMBER	QUANTITY				
9 9 9	9 9 9 ^ 9	9 9 9 ^ 9	9 9 9 ^ 9	9 9 9 ^ 9	9 9 9 ^ 9

```
05  PRODUCT-NUMBER       PICTURE 999.
05  QUANTITY             PICTURE 999V9 OCCURS 5 TIMES.
```

The definition of the first QUANTITY also applies to each of the four repetitions of QUANTITY. Thus, all five fields consist of four numeric characters, including one decimal position.

Figure 10.10 OCCURS clause—example.

PRODUCT NUMBER	WAREHOUSE-DATA LOC	BIN	QUANTITY						
9 9 9 9	1 7	6 2 9	9 9 9 9 9	1 8	1 5 8	9 9 9 9 9	1 8	1 6 0	9 9 9 9 9

```
05  PRODUCT-NUMBER         PICTURE 9999.
05  WAREHOUSE-DATA         OCCURS 3 TIMES ASCENDING KEYS ARE
                           LOCATION, BIN-NUMBER
    10  LOCATION           PICTURE 99.
    10  BIN-NUMBER         PICTURE 999.
    10  QUANTITY-ON-HAND   PICTURE 9(5).
```

In this example of data descriptions and data structure, the three occurrences of WAREHOUSE-DATA appear in ascending order according to the major key LOCATION and its minor key BIN-NUMBER.

Figure 10.11 OCCURS clause—example.

6. A record description entry that contains an OCCURS clause may not also contain a VALUE clause except for condition-name entries.
7. Integer-1 and integer-2 must be positive integers. Where both are used, the value of integer-1 must be less than the value of integer-2. The value of integer-1 may be zero, but integer-2 may not be zero.
8. Data-name-1, data-name-2 and data-name-3 may not be qualified.
9. In format-1, the value of integer-2 represents the exact number of occurrences. In format-2, the value of integer-2 represents the maximum number of occurrences.

DEPENDING ON Option

The DEPENDING ON option is used in format-2. This option is only required when the end of the occurrences cannot otherwise be determined. This indicates that the subject has a variable number of occurrences. This does not mean that the subject is variable, but rather that the number of times that the subject may be repeated is variable. The number of times is being controlled by the value of data-name-1 at object time.

Integer-1 represents the minimum number of occurrences while *integer-2* represents the maximum number of occurrences. Integer-1 cannot be zero in format-2.

Data-name-1 is the object of the DEPENDING ON option.

1. Must be described as a positive integer,
2. Must not exceed integer-2 in value,
3. May be qualified when necessary,
4. Must not be subscripted (that is, itself the subject of or an entry within a table), and
5. Must, if it appears in the same record as the table it controls, appear before the variable portion of the record.
   (See figure 10.12.)

SALE-TABLE		
SALE-ITEM	(1)	A21033
SALE-ITEM	(2)	A21455
SALE-ITEM	(3)	A22223
SALE-ITEM	(4)	A23762
SALE-ITEM	(5)	A24689
SALE-ITEM	(6)	A24643
SALE-ITEM	(7)	A29567
SALE-ITEM	(8)	J33468
SALE-ITEM	(9)	J24788
SALE-ITEM	(10)	J67011

```
1. WORKING-STORAGE SECTION.
   77      TABLE-SIZE      PICTURE 99.
   77      TABLE-VALUE     PICTURE X(6).
   77      SUBSCRIPT       PICTURE 99.
   01      SALE-TABLE.
        05      SALE-ITEM       OCCURS 10 TIMES      DEPENDING ON TABLE-SIZE
                               PICTURE X(6).

2. PROCEDURE DIVISION.
   INITIAL-ROUTINE.
       ACCEPT TABLE-SIZE   FROM SYSIN.
       PERFORM ACCEPT-TABLE-VALUE
           VARYING SUBSCRIPT FROM 1 BY 1
           UNTIL SUBSCRIPT IS GREATER THAN TABLE-SIZE
                 .
                 .
   ACCEPT-TABLE-VALUE.
       ACCEPT TABLE-VALUE FROM SYSIN.
       MOVE TABLE-VALUE TO SALE-ITEM (SUBSCRIPT).
```

1. Working-Storage Section entries to define the variable:
   TABLE-SIZE which is to contain the number of sale items,
   TABLE-VALUE to which a table value will be transmitted prior to being moved to a specific table element,
   SUBSCRIPT to be used as a subscript in referring to table elements.
   SALE-TABLE whose size is to be determined by TABLE-SIZE, and
2. Procedure Division entries to accept the table size and the table values.

Figure 10.12 Subscripting—example.

6. The DEPENDING ON phrase is required in the format-2 of the OCCURS clause.
7. A data description entry with an OCCURS DEPENDING ON clause may be followed within that record, only by entries subordinate to it (i.e., only the last part of the record may have a variable number of occurrences).
8. When a group item, having subordinate to it an entry that specifies format-2 of the OCCURS clause, is referenced, only part of the table area that is defined by the value of the operand of the DEPENDING ON phrase will be used in the operation. (The actual size of a variable length item is used, not the maximum size.)

KEY Option

The KEY option is used in conjunction with the INDEXED BY option in the execution of a SEARCH ALL statement. The option is used to indicate that the repeated data is arranged in ASCENDING or DESCENDING order according to the values contained in data-name-2, data-name-3, etc. The data-names are listed in descending order of significance.

If data-name-2 is the subject of the table entry, it is the only key that may be specified for the table, otherwise (1) all of the items identified by the data-name in the KEY IS phrase must be within the group item which is the subject of this entry, and (2) none of the items identified by the data-name in the KEY IS phrase can be described by an entry that either contains an OCCURS clause or is subordinate to an entry that contains an OCCURS clause.

INDEXED BY Option

The INDEXED BY option is required if the subject of this entry (the data name described by the OCCURS clause, or an item within this data-name, if it is a group item) is to be referred to by indexing. The index-name(s) identified by this clause is not defined elsewhere in the program since its allocation and format are dependent upon the system, and, not being data, cannot be associated with any data hierarchy.

Rules Governing the Use of Index-Names

1. Index-names must be unique words within a program.
2. An index-name must be initialized through a SET statement before it can be used. (See figure 10.13.)

USAGE IS INDEX Clause

The USAGE IS INDEX clause is used to specify the format of a data item in the computer storage. The clause permits the programmer to specify index-data items. (See figure 10.14.)

Rules Governing the Use of the USAGE IS INDEX Clause

1. The USAGE clause may be written at any level. If the USAGE clause is written at a group level, it applies to each elementary item in the group. The USAGE clause at an elementary level cannot contradict the USAGE clause of a group to which the item belongs, unless the group USAGE clause is not stated.
2. An elementary item described with the USAGE IS INDEX clause is called an index-data item. An index-data item is an elementary item (not necessarily connected with any table) that can be used to save index-name values for future reference. An index-data item must be assigned an index-name (e.g., [occurrence number-1]* entry length) through the SET statement. Such a value corresponds to an occurrence number in a table.
3. An index-data item can be referred to directly only in a SEARCH or SET statement or in a relation condition. An index-data item can be part of a group which is referred to in a MOVE or I/O statement, in which case no conversion will take place. (See figures 10.15 and 10.16.)

PRODUCT-NUMBER	WAREHOUSE-DATA								
	LOC	BIN	QUANTITY						
9999	17	629	99999	18	158	99999	18	160	99999

```
05  PRODUCT-NUMBER              PICTURE 9999.
05  WAREHOUSE-DATA              OCCURS 3 TIMES
                                ASCENDING KEYS ARE LOCATION,
                                BIN-NUMBER INDEXED BY
                                WAREHOUSE-INDEX.
    10  LOCATION                PICTURE 99.
    10  BIN-NUMBER              PICTURE 999.
    10  QUANTITY-ON-HAND        PICTURE 9(5).
```

The data description entries enable each of the three WAREHOUSE-DATA items to be accessed by indexing. In the Procedure Division, the index-name identifying the table element is enclosed in parentheses immediately following the last space of the table element data-name. This is an example of how the Procedure Division accesses the first WAREHOUSE-DATA item.

```
SET WAREHOUSE-INDEX to 1.
ADD QUANTITY-ON-HAND (WAREHOUSE-INDEX) TO
     TOTAL-QUANTITY.
MOVE WAREHOUSE-DATA (WAREHOUSE-INDEX) TO PRINT-AREA.
```

In order to access the third WAREHOUSE-DATA item, these Procedure Division statements are needed:

```
SET WAREHOUSE-INDEX to 3.
ADD QUANTITY-ON-HAND (WAREHOUSE-INDEX) TO
     TOTAL-QUAN-TITY.
MOVE WAREHOUSE-DATA (WAREHOUSE-INDEX) TO PRINT-AREA.
```

Notice that both the ADD and MOVE statements are exactly the same. Thus, the programmer would probably place the two statements in a subroutine and change the above procedure to the example below.

```
SET WAREHOUSE-INDEX TO 1.
PERFORM ADD-MOVE-ROUTINE.
SET WAREHOUSE-INDEX TO 2.
PERFORM ADD-MOVE-ROUTINE.

ADD-MOVE-ROUTINE.
    ADD QUANTITY-ON-HAND (WAREHOUSE INDEX) TO
         TOTAL-QUAN-TITY.
    MOVE WAREHOUSE-DATA (WAREHOUSE-INDEX) TO PRINT-AREA.
```

This example demonstrates the advantage of indexing when only one routine is written to process each element in a table.

Figure 10.13 **INDEXED BY, KEY options—example.**

[USAGE IS] INDEX

Figure 10.14 **Format: USAGE IS INDEX clause.**

The following data description entries describe a table containing five elements which are accessed by an index called TABLE 1-INDEX.

```
05  TABLE-ELEMENT OCCURS 5 TIMES INDEXED BY
        TABLE 1-INDEX.
    10  ACCOUNT-NUMBER        PICTURE X(5).
    10  BALANCE               PICTURE 9999V99.
```

In order to store the contents of the index (TABLE1-INDEX) temporarily, the programmer must move the contents of the index to an index data item. The index data item is defined in the Working-Storage Section as follows:

```
77  INDEX-STORAGE       USAGE IS INDEX.
```

Since an index is accessed only by a SET, SEARCH, and PERFORM statement, a SET statement must be used to store the contents of the index (TABLE1-INDEX) into the index data item (INDEX-STORAGE).

```
SET INDEX-STORAGE TO TABLE1-INDEX.
```

Thus, if TABLE1-INDEX contains the address or offset for the third table element in TABLE1, then the SET statement places the offset or occurrence number of the third table element into the index data item (INDEX-STORAGE). When the programmer wants to reset the index (TABLE1-INDEX) to the content of the index data item (INDEX-STORAGE), the following SET statement would be executed.

```
SET TABLE1-INDEX TO INDEX-STORAGE.
```

Figure 10.15 USAGE IS INDEX clause—example.

```
01  GROUP-1 USAGE IS INDEX.
    05  FIELD-A
    05  FIELD-B
    05  FIELD-C
```

If the USAGE IS INDEX clause describes a group item, all the elementary items in the group are index data items. The group itself is not an index data item. In the following data descriptions, FIELD-A, FIELD-B, and FIELD-C are index data items.

Figure 10.16 Group USAGE IS INDEX clause—example.

REDEFINES Clause

The REDEFINES clause specifies that the same area is to contain different data items (fig. 10.17). The entry gives another name and description to an item previously described. That is, the REDEFINES clause specifies the redefinition of a storage area, not of the items occupying the area.

The same area may be called by different names during the processing of the data. The area may contain different types of information and may be processed in a different manner under changing conditions. (See figure 10.18.)

level number data-name-1 <u>REDEFINES</u> data-name-2.

Figure 10.17 Format: REDEFINES clause.

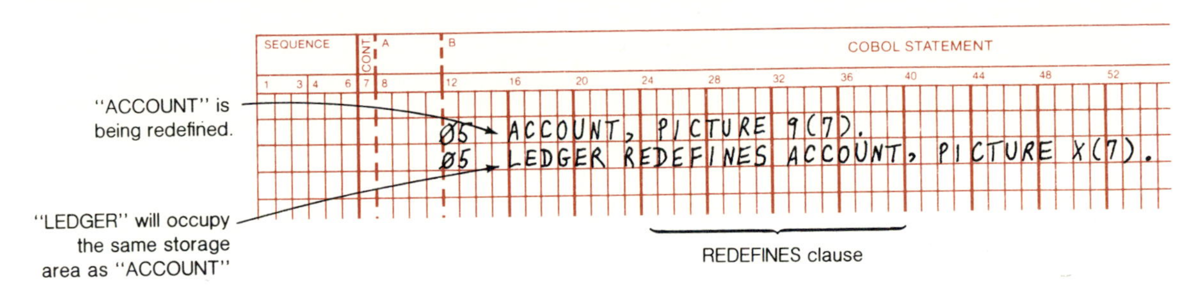

Figure 10.18 REDEFINES clause—example.

Rules Governing the Use of the REDEFINES Clause

1. The word REDEFINES must be written right after the data-name followed by the name of the item being redefined.
2. The level numbers of the two entries sharing the same area must be the same.
3. Redefines may be used at the 01 level in the working-storage section, but not in the file section.
4. The usage of data within the area cannot be redefined.
5. The redefinition starts at data-name-2 and ends when a level number that is less than or equal to that of data-name-2 is encountered. Between the data description of data-name-2 and data-name-1, there may be no entries having lower-level numbers than data-name-1 or -2.
6. A new storage area is not set aside by the redefinition. All descriptions of the area remain in effect.
7. The entries giving new descriptions of the area must immediately follow the entry that is being redefined.
8. A REDEFINES clause may be used for items subordinate to items that are also being redefined.
9. This entry should not contain any VALUE clauses.
   (See figures 10.19, 10.20, and 10.21.)

```
01  RECORD I.
    05  ACCOUNT-NUMBER                      PICTURE 9(6).
    05  AMOUNT                              PICTURE 9(5)V99.
    05  PERCENT                             PICTURE V999.
    05  PERCENT-OUT REDEFINES PERCENT       PICTURE 99V9.
    05  PRICE                               PICTURE 9V999.
```

In this example, a three character data item originally defined as PERCENT has been redefined as PERCENT-OUT. The redefinition with one decimal position enables the percentage to be printed as 36.5% instead of the decimal .365 which is used for internal calculation.

Figure 10.19 REDEFINES clause—example.

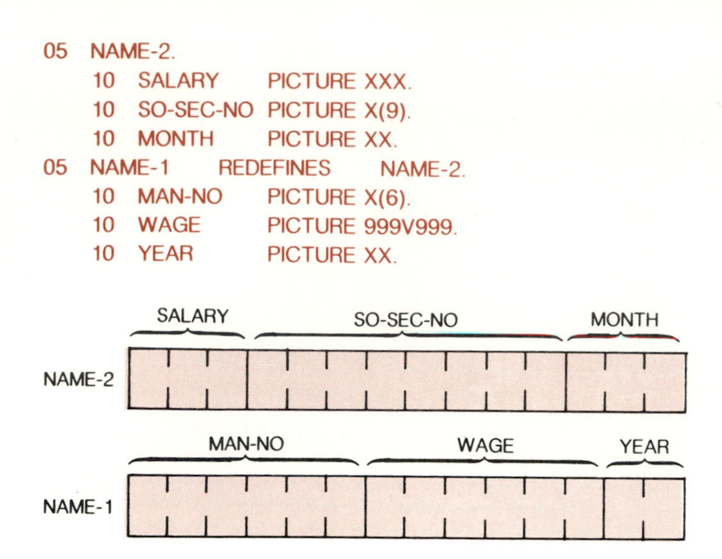
05 NAME-2.
10 SALARY PICTURE XXX.
10 SO-SEC-NO PICTURE X(9).
10 MONTH PICTURE XX.
05 NAME-1 REDEFINES NAME-2.
10 MAN-NO PICTURE X(6).
10 WAGE PICTURE 999V999.
10 YEAR PICTURE XX.
SALARY
SO-SEC-NO
MONTH
NAME-2
MAN-NO
WAGE
YEAR
NAME-1

Figure 10.20 REDEFINES clause—example.

```
05  REGULAR-EMPLOYEE.
    10  LOCATION                          PICTURE A(8).
    10  STATUS                            PICTURE X(4).
    10  SEMI-MONTHLY-PAY                  PICTURE 999V999.

05  TEMPORARY-EMPLOYEE REDEFINES REGULAR-EMPLOYEE.
    10  LOCATION                          PICTURE A(8).
    10  FILLER                            PICTURE X(6).
    10  HOURLY-PAY                        PICTURE 99V99.
    10  CODE-H REDEFINES HOURLY-PAY       PICTURE 9999.
```

Figure 10.21 REDEFINES clause—example.

Procedure Division for Table Handling

PERFORM Statement

In its simplest form, the PERFORM statement is an imperative statement. The use of the UNTIL phrase or the VARYING phrase, however, changes the PERFORM statement into a conditional statement. The VARYING and AFTER phrases allow the programmer to manipulate up to three control items and to establish separate conditional terminations for the manipulation of each item. (See figure 10.22.)

The PERFORM statement becomes a powerful mechanism for table handling when the VARYING phrase manipulates index-names or data-names used as subscripts in the performed procedure. The use of the FROM phrase and BY phrase allows a table to be created and table processing to begin at a specific table entry, and allows the processing of noncontiguous entries. Since three VARYING phrases may be specified, the PERFORM statement can manipulate all three dimensions of a table.

When an index-name is used in a VARYING phrase, the execution of the PERFORM statement causes the associated value to be manipulated in the same manner as a series of SET statements. Therefore, SET statements need not be included in the performed procedure to control the index-names specified in the PERFORM statement. (SET statements are discussed later in this section.)

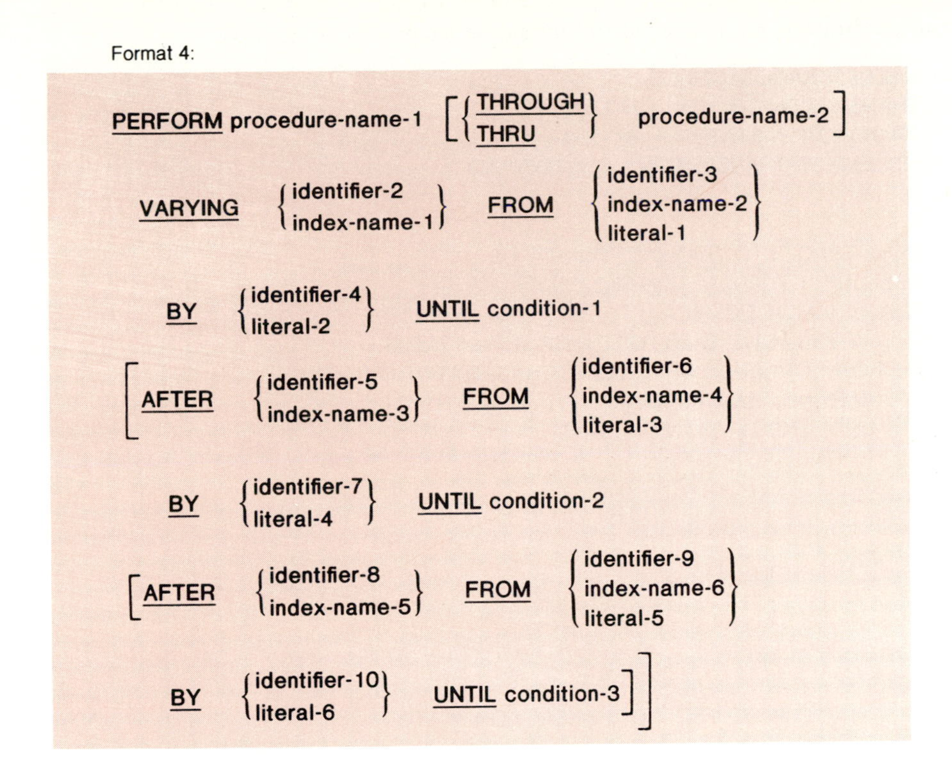

Figure 10.22 Format: PERFORM statement—format-4—VARYING option.

VARYING and AFTER Options

The VARYING and AFTER options of the PERFORM statement allow the programmer to manipulate up to three control items and to establish separate conditional terminations for the manipulation of each item. The termination conditions are all tested before the referenced procedure is executed, and the results of these tests may prevent the execution of the procedure. The flow diagram for a PERFORM statement using the VARYING phrase follows:

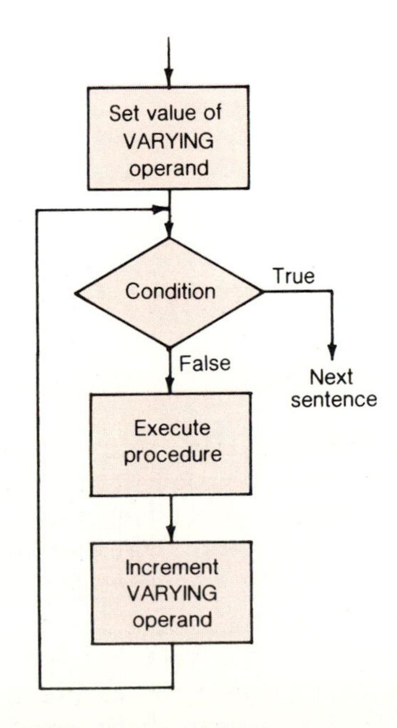

The condition whose truth terminates the PERFORM statement is often used to test the value of the varying operand, but the condition is not restricted to that single purpose. For example:

```
WORKING-STORAGE SECTION.
77  TRANSACTION-COUNT      PIC 9(5).
01  TRANSACTION-STATUS     PIC XXX.
    88  END-OF-TRANSACTION       VALUE 'END'.
.
.
    MOVE SPACES TO TRANSACTION-STATUS.
    OPEN INPUT TRANSACTION-FILE.
    READ TRANSACTION-FILE
        AT END MOVE 'END' TO TRANSACTION-STATUS.
    PERFORM TRANSACTION-PROCESS THRU NEXT-TRANSACTION
        VARYING TRANSACTION-COUNT FROM 0 BY 1
        UNTIL END-OF-TRANSACTIONS.
    CLOSE TRANSACTION-FILE.
    IF TRANSACTION-COUNT = 0 . . .
    .
    .
TRANSACTION-PROCESS
    .
    .
NEXT-TRANSACTION.
    READ TRANSACTION-FILE
        AT END MOVE 'END' TO TRANSACTION-STATUS.
    .
    .
```

When both the VARYING and AFTER phrases are used in the PERFORM statement, a hierarchy of tests is established. The execution of the statement becomes cyclic; that is, a varying operand at a lower level is varied through its full range for each step in the value of the next higher level varying operand. The flow diagram for a PERFORM statement with two varying operands follows:

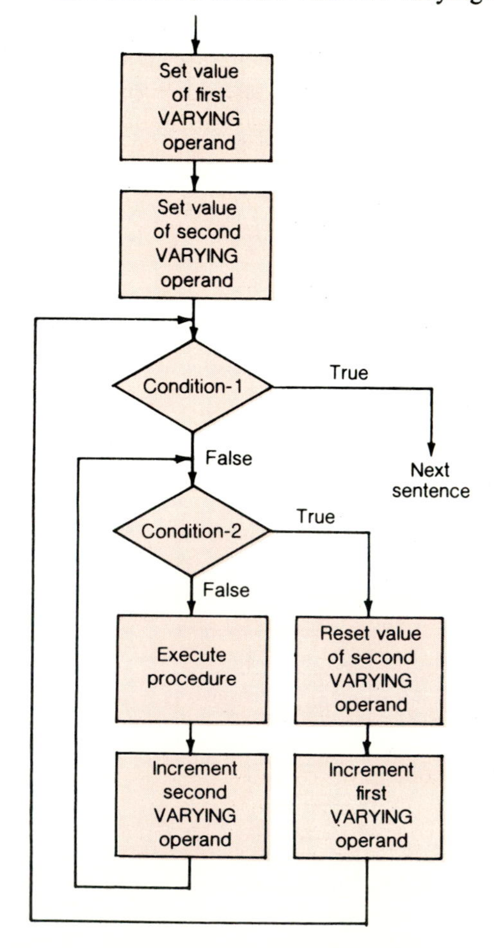

The flow diagram for a PERFORM statement with three varying operands follows:

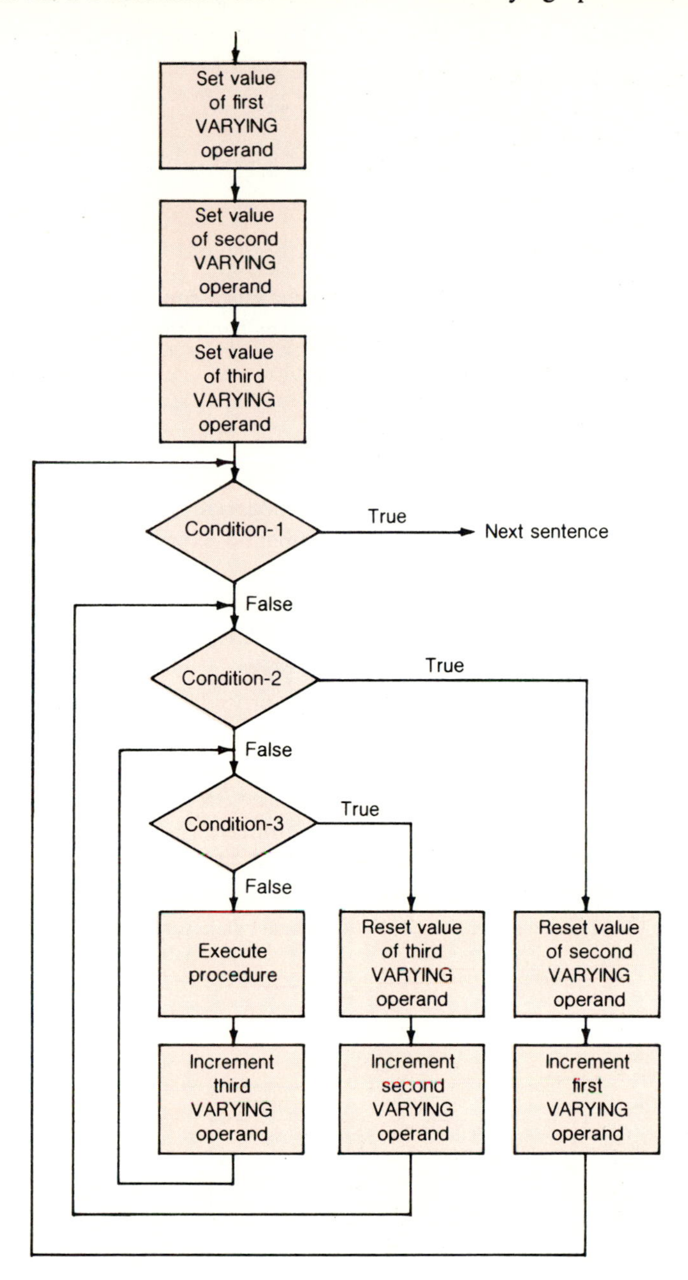

The cyclic nature of these forms of the PERFORM statement is useful in table handling. The controlling conditions can be used to test the subscript or index values. For example:

```
01  INFORMATION-ARRAY.
    05  PLANES                  OCCURS 50 TIMES INDEXED BY I.
        10  ROWS                OCCURS 42 TIMES INDEXED BY J.
            15  COLUMNS         OCCURS 12 TIMES INDEXED BY K.
                20  DATUM PIC . . .
.
.
.
*PROCESSING PLANES ACROSS ROWS.
    PERFORM DATUM-PROCESS
        VARYING I FROM 1 BY 1 UNTIL I > 50
        AFTER J FROM 1 BY 1 UNTIL J > 42
        AFTER K FROM 1 BY 1 UNTIL K > 12.
```

```
*PROCESSING PLANES DOWN COLUMNS.
     PERFORM DATUM-PROCESS
          VARYING I FROM 1 BY 1 UNTIL I > 50
          AFTER K FROM 1 BY 1 UNTIL K > 12
          AFTER J FROM 1 BY 1 UNTIL J > 42.
 .
 .
 .
DATUM-PROCESS.
     MOVE DATUM (I,J,K) . . .
```

The VARYING option is used to PERFORM a procedure repetitively, increasing or decreasing the value of one or more identifiers or index names once for each repetition until a specified condition is satisfied.

Rules Governing the Use of the VARYING Option

1. The option may be used to increase or decrease the value of one or more identifiers or index names depending upon whether the BY value is positive or negative. (See figure 10.23.)
2. The specified test condition may be a simple or compound expression.
3. The identifier, index name, or literal is set to the specified initial value (FROM) when commencing the PERFORM statement. Then condition-1 is evaluated (UNTIL).
4. If condition-1 is true, control passes to the next statement immediately following the PERFORM statement, and no execution of the procedures take place.

   If the statement is false, the procedure specified in procedure-name-1 through procedure-name-2 is executed once. The BY value is added to the index name or identifier and again causes condition-1 to be evaluated. This process continues until the conditional expression is found to be true, whereupon control passes to the next statement after the PERFORM statement.
5. The items used in BY and FROM must represent numeric values but need not be integers; such values may be positive, negative, or zero. (See figure 10.24.)
6. When more than one identifier is varied, the value of each identifier goes through the complete cycle (FROM, BY, UNTIL) each time that identifier-1 is altered by its BY value.

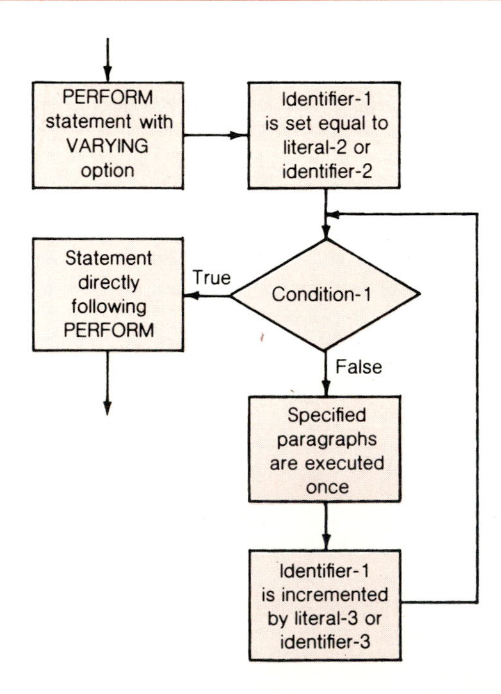

Figure 10.23 Flow of logic: PERFORM statement—format-4—VARYING option.

PERFORM 1 VARYING 2 FROM 3 BY 4 UNTIL 5

1. The name of the procedure that is to be performed.
2. The name of the base item.
3. The initial value of the base item—either the literal value, or the name of the data item that contains the value.
4. The amount by which the base item is to be increased each time the procedure is performed—either the literal amount, or the name of data item that contains the amount.
5. A condition which is tested to determine when to stop performing the procedure.

Figure 10.24 Using PERFORM statement—format-4—VARYING option.

7. Regardless of the number of identifiers being varied, as soon as test condition-1 is satisfied, control is transferred to the next statement after the PERFORM statement. (See figures 10.25 and 10.26.)

   Note: The PERFORM verb is used extensively in structured programming. The COBOL implementation of the DOWHILE structure involves the PERFORM verb with an UNTIL phrase. If indexing is used, the DOWHILE structure may be carried out using the PERFORM verb with the VARYING option.

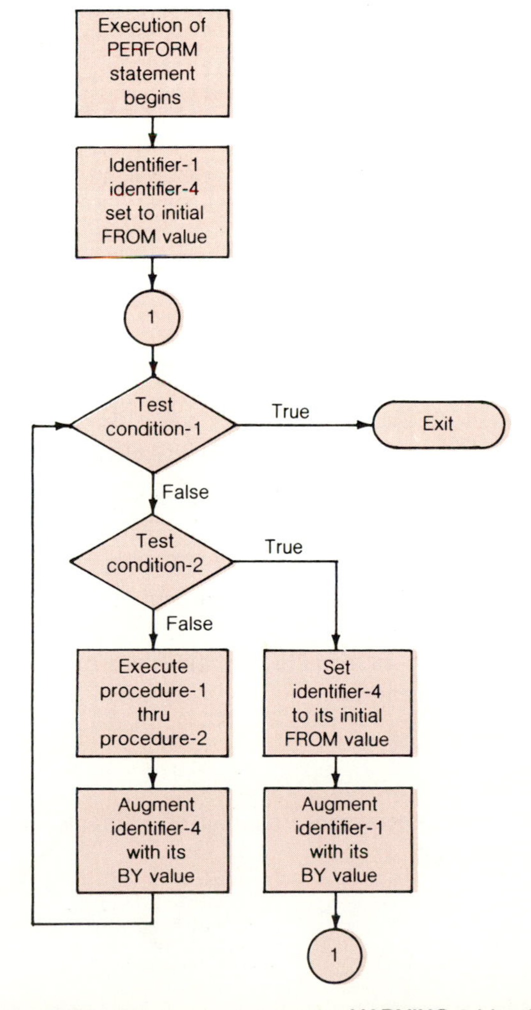

Figure 10.25 Logical flow: PERFORM statement—format-4—VARYING 2 identifiers.

```
PERFORM COMPUTE-FICA THRU CHECK-PRINT
     VARYING EMPLOYEE-COUNT FROM 0 BY 1
     UNTIL EMPLOYEE-COUNT IS EQUAL TO TOTAL-EMPLOYEE.
```

The previous example illustrates an item (EMPLOYEE-COUNT) that is set to 0 and incremented by 1 each time procedures COMPUTE-FICA through CHECK-PRINT are PERFORMed. If the item TOTAL-EMPLOYEE value is 5, the procedures are performed five times; that is, until EMPLOYEE-COUNT is incremented to 5.

Figure 10.26 **PERFORM statement—format-4—VARYING option—example.**

Reference to Table Items

Whenever the programmer refers to a table element, the reference must indicate which occurrence of the element is intended. For access to a one-dimensional table, the occurrence number of the desired element provides complete information. For access to tables of more than one dimension, an occurrence number must be supplied for each dimension of the table accessed. Thus, in the following example of a three-dimensional table, a reference to the fourth REGION-NAME would be complete, whereas a reference to the fourth DISTRICT-NAME would not. To refer to DISTRICT-NAME, which is an element of a two-dimensional table, the programmer must refer to, for example, the fourth DISTRICT-NAME within the fifth REGION-TABLE.

```
01  SALES-QUOTA-TABLE.
    05  REGION-TABLE          OCCURS 5 TIMES.
        10  REGION-NAME                PICTURE X(10).
        10  DISTRICT-TABLE    OCCURS 10 TIMES.
            15  DISTRICT-NAME          PICTURE X(10).
            15  OFFICE-TABLE OCCURS 15 TIMES.
                20  OFFICE-NAME        PICTURE X(10).
                20  SALES-QUOTA        PICTURE 9(7).
```

Referencing Tables

When a table is referenced, it is necessary to indicate, either by subscripting or by indexing, which element of the table is intended. The reference can be to any of the following items:

1. An individual table element (which may be either a group item or an elementary item).
2. An item within a group item that is a table element.
3. A condition-name associated with a table element.
4. A condition-name associated with an item within a group table element.

In COBOL, a table containing as many as three dimensions may be referenced using either subscripting or indexing. The general format is:

data-name   (a [, b [, c]])

where: a, b, and c each represents an occurrence number of a dimension of the table, expressed in the form of an index-name or a subscript.

An example of a three-dimensional table follows, with representative subscript references to table elements:

```
01  TABLE.
    05  STORE                          OCCURS 12 TIMES.
        10  STORE-NO . . .
        10  DEPARTMENT                 OCCURS 7 TIMES.
            15  DEPT-NO . . .
            15  SALE                   OCCURS 4 TIMES.
.
.
.
STORE (11)
STORE-NO (12)

DEPARTMENT (10,5)
DEPT-NO (12,7)

SALE (1,1,1)
SALE (12,7,4)
```

Rules for Subscripting and Indexing

1. Index-names may not be combined with data-name subscripts in a single data-name reference.
2. Where subscripting is not permitted, indexing is also not permitted.
3. Tables may have one, two, or three dimensions. Thus, a reference to a table element may require up to three subscripts or index-names.
4. The use of a data-name subscript in any reference to a table element or to an item within a table element will not cause any index-name to be altered by the object program.
5. A data-name used as a subscript may be neither subscripted nor indexed.
6. An occurrence number specified by a subscript or implied by an index-name must not be less than one or greater than the highest permissible occurrence number for the table element.
7. When indexing is used to reference a table, the INDEXED BY phrase must be employed in each OCCURS clause used to define the table. The SEARCH and SET statements appear in the Procedure Division only in connection with indexed table references.

Subscripting

The need often arises to have tables of information accessible to a source program for referencing. One method by which occurrence numbers may be specified is to append one or more subscripts to the data-name. Subscripting provides the facility for referring to data items in a list or a table that has been assigned individual values. *A subscript is an integer or integer data-name whose value specifies the occurrence number of an element in a table.*

Like all data tables, the tables must be described in the Data Division with an OCCURS clause to indicate the number of appearances of a particular item. The subscripts are used in the Procedure Division to reference a particular item in the table. If subscripting were not used, each item would have to be described in a separate entry.

The subscript can be represented either by a literal that is an integer or by a data-name that is defined elsewhere as a numeric elementary item with no character positions to the right of the assumed decimal point. In either case, the subscript is enclosed in parentheses and written immediately following the name of the table element (see fig. 10.27); examples of each are REGION-NAME (5) and REGION-NAME (FIVE) where FIVE is a data item defined elsewhere as integer-5. A table reference must include as many subscripts as there are dimensions in the table whose element is being referenced. That is, there must be a subscript for each OCCURS clause in the hierarchy containing the data-name itself. In the three-dimensional table example presented earlier, reference to REGION-NAME requires only one subscript, reference to DISTRICT-NAME requires two, and reference to OFFICE-NAME and SALES-QUOTA requires three.

When more than one subscript is required, each is written in order of successively less inclusive dimensions of the data organization. When a data-name is used as a subscript, it may be used to refer

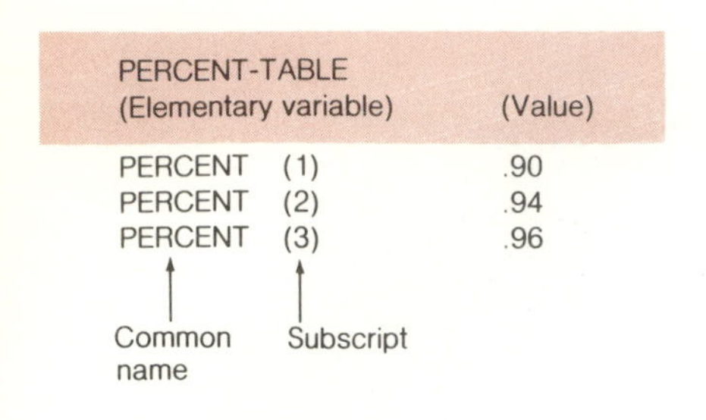

PERCENT-TABLE
(Elementary variable)
(Value)
PERCENT (1)
.90
PERCENT (2)
.94
PERCENT (3)
.96
Common name
Subscript


```
01  PERCENT-TABLE       USAGE COMP-3.
    05  PERCENT OCCURS 3 TIMES      PICTURE V99.
DISCOUNT-ROUTINE.
    IF GOOD,
        COMPUTE BILL ROUNDED = AMOUNT-WS * PERCENT (1)
    ELSE
        IF MEDIUM,
            COMPUTE BILL ROUNDED = AMOUNT-WS * PERCENT (2)
        ELSE
            COMPUTE BILL ROUNDED = AMOUNT-WS * PERCENT (3).
```

Figure 10.27 Subscripting—example.

to items in many different tables. These tables need not have elements of the same size. The data-name may also appear as the only subscript. It is also permissible to mix literal and data-names in subscripts; for example, DISTRICT-NAME (NEWKEY, 42).

Commas may be used to separate subscripts (or index-names). They are not required, however; spaces may be used to separate subscripts. Literal subscripts may be mixed with index-names when referencing a table item.

A data item is said to be "repeated" if the OCCURS clause is specified in the item's own description entry or in that of a group to which the item belongs. Any reference to a repeated item requires a subscript. The subscripts must be one-, two-, or three-dimensional, reflecting the number of OCCURS entries affecting the desired item. Do not use more than, or less than the correct number of subscripts. A data-name can be subscripted only if the item is repeated. If a conditional variable is repeated, the references to its condition-names also require subscripts.

The subscript formats for references to table items depend upon the dimensions of the table, as follows:

One dimensional: (position-number)

Two dimensional: (major, minor)

Three dimensional: (major, intermediate, minor)

Subscripts must be enclosed in parentheses, like those just shown, and commas or blanks must appear between the indicated items. In any reference to a table item, the parenthesized subscript must follow immediately after the terminal space of the table item's data-name. The name of a repeated item may require qualification in references. If so, the entire subscript follows the last qualifier.

Multilevel subscripts are always written from left to right in the following order: major, intermediate, and minor (that is plane, row, and column). The plane number, row number, and column number must be data-names or positive integers. If data-names are used, they must specify items that will have positive integral values when the object program is executed. Integers are used when the desired table positions are known at compilation time; data-names are used when the position depends upon data accessed or data developed in the object program. A data-name within a subscript may not itself be subscripted, but it may be qualified if necessary for uniqueness.

The lowest valid subscript is one. The highest valid subscript value is the maximum occurrence number of the item, as specified in the OCCURS clause. An invalid subscript will cause unpredictable results at program execution time.

A single data-name may be used as a subscript to refer to items in more than one table, and the tables need not have elements of the same size. The same data-name may be used to reference both a one-dimensional table and another two- or three-dimensional table.

An index data item may not be used as a subscript.

Rules Governing the Use of Subscripts

1. A subscript must always have a positive nonzero integral value whose value determines which item is being referenced within a table or list.

2. The subscript must be represented either by a numeric literal or a data-name that has an integral value.
3. Subscripts are enclosed in parentheses to the right of the subscripted data-name with an intervening space. If the subscripted data-name is qualified, the subscripts must appear to the right of all qualifiers.
4. If more than one level of subscript is present, the subscripts are separated by commas or spaces and arranged from right to left in increasing order of inclusiveness of the grouping within the table. Multiple subscripts are written with a single pair of parentheses separated by commas or spaces. A space should also separate the data-name from the subscripted expressions. *A maximum of three levels of subscripts is permitted.*
5. A subscripted data-name must be qualified when it is not unique in accordance with rules for qualification.
6. A subscript must always be used to reference an item that has an OCCURS clause or belongs to a group having an OCCURS clause. *Subscription may not be used with any undescribed data-name using an OCCURS clause.*
7. A programmer may refer to blocks (sets) of data within a table. The data-name is written, followed by the subscript of the particular block, plus any other subscript necessary to locate it. A complete table may be referenced just by using the name of the table.
8. Subscripts may not themselves be subscripted.
9. A data-name may not be subscripted when it is being used as
   a. A subscript or qualifier.
   b. A defining name of a record description entry.
   c. Data-name-2 in a REDEFINES clause.
   d. A data-name in a LABEL RECORDS clause.
   e. A data-name in the DEPENDING ON option of the OCCURS clause.
   (See figure 10.28.)

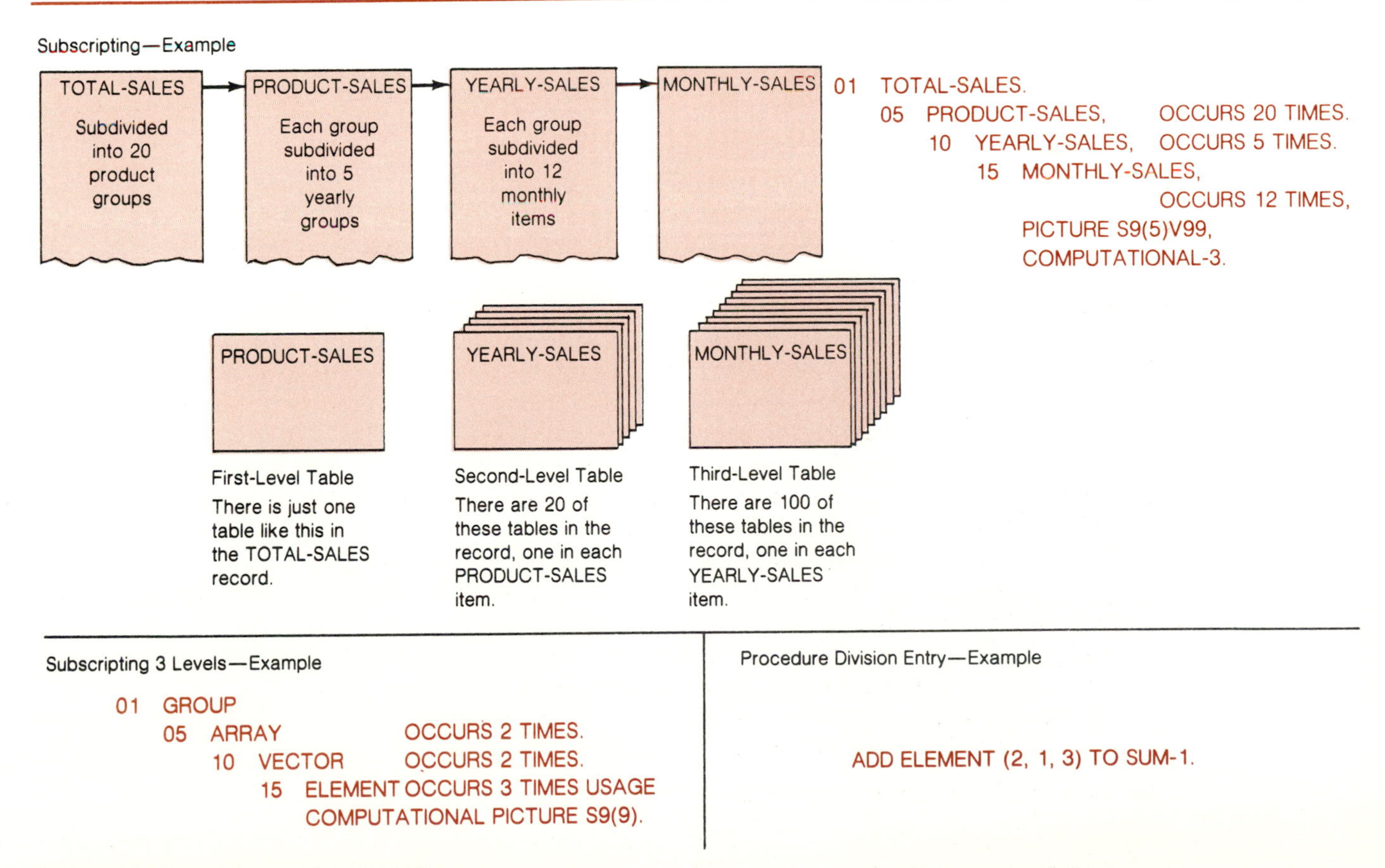

Figure 10.28 Subscripting—example.

Indexing

Another method of specifying occurrence numbers is to affix one or more index-names to an item whose data description includes an OCCURS clause by using the optional INDEXED BY phrase.

References are made to individual items in a table by specifying the name of the item followed by its related index-name in parentheses. The occurrence numbers required to complete the reference are obtained from the respective index-name; the index-name acting as a subscript. For references requiring more than one occurrence number, index-names and literals may be mixed, but index-names and data-name subscripts may not be mixed. Thus, if indexing is to be used, each OCCURS clause within the hierarchy must contain an INDEXED BY phrase. (See figure 10.29.)

Storage Layout for PARTY-TABLE

PARTY-TABLE	PARTY-CODE(1)	AGE-CODE (1, 1)	M-F-INFO (1, 1, 1)
			M-F-INFO (1, 1, 2)
		AGE-CODE (1, 2)	M-F-INFO (1, 2, 1)
			M-F-INFO (1, 2, 2)
		AGE-CODE (1, 3)	M-F-INFO (1, 3, 1)
			M-F-INFO (1, 3, 2)
	PARTY-CODE(2)	AGE-CODE (2, 1)	M-F-INFO (2, 1, 1)
			M-F-INFO (2, 1, 2)
		AGE-CODE (2, 2)	M-F-INFO (2, 2, 1)
			M-F-INFO (2, 2, 2)
		AGE-CODE (2, 3)	M-F-INFO (2, 3, 1)
			M-F-INFO (2, 3, 2)
	PARTY-CODE(3)	AGE-CODE (3, 1)	M-F-INFO (3, 1, 1)
			M-F-INFO (3, 1, 2)
		AGE-CODE (3, 2)	M-F-INFO (3, 2, 1)
			M-F-INFO (3, 2, 2)
		AGE-CODE (3, 3)	M-F-INFO (3, 3, 1)
			M-F-INFO (3, 3, 2)
	OCCURS 3 TIMES	OCCURS 3 TIMES	OCCURS 2 TIMES

```
01  PARTY-TABLE REDEFINES TABLE.
    05  PARTY-CODE          OCCURS 3 TIMES INDEXED BY PARTY.
        10  AGE-CODE        OCCURS 3 TIMES INDEXED BY AGE.
            15  M-F-INFO    OCCURS 2 TIMES INDEXED BY M-F
                PICTURE 9(7)V9 USAGE DISPLAY.
```

PARTY-TABLE contains three levels of indexing. Reference to elementary items within PARTY-TABLE is made by use of a name that is subscripted or indexed. A typical Procedure Division statement might be.

```
MOVE M-F-INFO (PARTY, AGE, M-F) TO M-F-RECORD.
```

Figure 10.29 Subscripting and indexing—example.

An index-name has no separate data description entry since its definition is hardware oriented, and it acts similarly to an index register* whose contents correspond to an occurrence number. Because of its contents, an index-name can be assigned to only one table. At object time execution, the contents of the index-name will correspond to an occurrence number for that specific dimension of the table with which the index-name was associated. An index-name must be initialized by a SET statement before being used as a table reference. Index-name values must not be less than one nor greater than the maximum occurrence number for the table element.

The value of an index-name can be modified only by using the PERFORM, SEARCH, and SET statements. If a data item is described with USAGE INDEX, the SET statement may be used to move data between the data item and an index-name. Data items described with USAGE INDEX are called *index data items.*

An index-name cannot be defined as part of a file; therefore, the index-name cannot be manipulated by input-output statements. However, a data item in a file can be described with USAGE INDEX; transfers may then be made without conversion, between these data items and index-names using the SET statement.

To clarify the difference between subscripting and indexing techniques, a simple table handling problem will be solved by each method. Consider the following two tables:

```
01  LIAB-RATES.
    05  TERR-L                                 OCCURS 9 TIMES.
        10  PREM            PIC 9(6).
        10  CLAS                               OCCURS 7 TIMES.
            15  L-LIMIT     PIC 9(6)           OCCURS 5 TIMES.
01  DAMG-RATES.
    05  TERR-D                                 OCCURS 9 TIMES.
        10  COMP-50D        PIC 9(6).
        10  COLL-50D        PIC 9(6).
        10  COMPOSIT        PIC 9(6)           OCCURS 196 TIMES.
```

These two tables are simplified versions of what might be used in developing automobile insurance premiums. If it is known that the territorial definition is the same for both liability and damage premiums, then both sets of territorial values can be referenced by the same data-name subscript.

Assume that an input file record of an individual policy is to be updated and the input record contains the following data:

```
TERR-IN
CLASS-IN
LIMIT-IN

CODE-IN
AGE-IN
```

where: TERR-IN is the territory number; CLASS-IN is a class code in the range of 1 to 7; and LIMIT-IN is a coverage limit code in the range of 1 to 5.

CODE-IN and AGE-IN are code numbers which, when multiplied, produce an occurrence number that points to 1 to 196 composite items (effectively, a table within a table).

Although the two tables have different dimensions, TERR-IN can be used to refer to the proper territorial occurrence in either table.

Example:

```
L-LIMIT (TERR-IN, CLASS-IN, LIMIT-IN) or
TERR-D (TERR-IN)
```

are both valid uses of the contents of TERR-IN as a subscript.

*An index register is a register whose contents may be used to modify an operand address during the execution of computer instructions so as to operate as a counter. An index register may be used for table lookup as a pointer.

The following procedure is also valid:

```
MULTIPLY CODE-IN BY AGE-IN
    GIVING SCRATCH.
MULTIPLY COMPOSIT (TERR-IN, SCRATCH)
    BY COLL-50D (TERR-IN)
    GIVING COLL-PREMIUM.
```

The first multiplication places the appropriate occurrence number for COMPOSIT into SCRATCH. Then the proper COMPOSIT element is multiplied by the $50-deductible collision rate for the territory to get a final collision premium into COLL-PREMIUM.

The above techniques describe some of the organizational and technical aspects of handling tables by the subscripting method.

If indexing rather than subscripting were used with the same table formats, the tables might be defined as follows:

```
01  LIAB-RATES.
    05  TERR-L                          OCCURS 9 TIMES INDEXED BY XTL.
        10  PREM            PIC 9(6).
        10  CLAS                        OCCURS 7 TIMES INDEXED BY XCD.
            15  L-LIMIT     PIC 9(6)    OCCURS 5 TIMES INDEXED BY XLF.

01  DAMG-RATES.
    05  TERR-D                          OCCURS 9 TIMES INDEXED BY XTD.
        10  COMP-50D        PIC 9(6).
        10  COLL-50D        PIC 9(6).
        10  COMPOSIT        PIC 9(6)    OCCURS 196 TIMES INDEXED BY XCF.
```

Indexing cannot be used to reference a table element unless the INDEXED BY phrase appears in the data description of the item and in any table group items to which the table element is subordinate. For example, COMPOSIT is INDEXED BY XCF. Thus, TERR-D (which has nine occurrences, each containing 196 occurrences of COMPOSIT) must also be indexed.

The handling of tables by indexing is similar to subscripting, except that index-names must be modified and controlled by means of SET and SEARCH statements. Data items that were used as subscripts in the subscripting example (such as TERR-IN, LIMIT-IN, SCRATCH, etc.) would have to be converted means of a SET statement.

Example:

```
SET XTL, XTD TO TERR-IN.
```

If the data-name TERR-IN is described by a USAGE IS INDEX clause, it must be moved without conversion to index-names XTL and XTD. If TERR-IN is a data-name not described by a USAGE IS INDEX clause, the compiler interprets the value in TERR-IN as a subscript value, which must be converted to the appropriate index value.

When one program invokes another program, the index-names in the called and calling programs always refer to separate indexes; therefore, each index-name must be initialized in the called program before referencing its associated array described in the Linkage Section. A calling program cannot SET a value in a called program's index-name, but a called program can use subscript values from a calling program if the subscripts are defined in the called program's Linkage Section.

Table Loading

Table data can be loaded into the computer at two different times; at the time the source program is compiled (compile time) or at the beginning of the object program execution (execution time).

Initial Values of Tables

In the Working-Storage Section, initial values of elements within tables are specified in one of the following ways:

1. *Compile Time Tables.* The table may be described as a record set of contiguous data description entries, each of which specifies the VALUE of an element, or part of an element of the table. In defining the record and its elements, any data description clause (USAGE, PICTURE, etc.) may be used to complete the definition where required. The hierarchical structure of the table is shown by using the REDEFINES entry and its subordinate entries. The subordinate entries (following the REDEFINES entry), which are repeated due to OCCURS clauses, must not contain VALUE clauses.

   Example:

```
01  MONTH-NAMES.
    05  FILLER,  PICTURE A(9), VALUE 'JANUARY'.
    05  FILLER,  PICTURE A(9), VALUE 'FEBRUARY'.
    05  FILLER,  PICTURE A(9), VALUE 'MARCH'.
    05  FILLER,  PICTURE A(9), VALUE 'APRIL'.
    05  FILLER,  PICTURE A(9), VALUE 'MAY'.
    05  FILLER,  PICTURE A(9), VALUE 'JUNE'.
    05  FILLER,  PICTURE A(9), VALUE 'JULY'.
    05  FILLER,  PICTURE A(9), VALUE 'AUGUST'.
    05  FILLER,  PICTURE A(9), VALUE 'SEPTEMBER'.
    05  FILLER,  PICTURE A(9), VALUE 'OCTOBER'.
    05  FILLER,  PICTURE A(9), VALUE 'NOVEMBER'.
    05  FILLER,  PICTURE A(9), VALUE 'DECEMBER'.
01  TABLE-OF-MONTHS REDEFINES MONTH-NAMES.
    05  MONTH,  PICTURE A(9), OCCURS 12 TIMES.
```

   The above coding shows how a month name table may be set up in the Working-Storage section at compile time. The data-name MONTH could be used as a subscript with the value being supplied by the numeric month (1–12) recorded in the input record. This subscript could be used to print the alphabetic month whenever the numeric month appears in the record.

   A Procedure Division statement such as:

```
MOVE MONTH (MO-IN) TO MONTH-OUT.
```

   would cause the alphabetic month subscripted by the numeric month from the input record to be moved to the output area.

   Example:

```
01  PRICE-TABLE.
    05  FILLER     PICTURE 99V99      VALUE 26.47.
    05  FILLER     PICTURE 99V99      VALUE 22.94.
    05  FILLER     PICTURE 99V99      VALUE 33.75.
    05  FILLER     PICTURE 99V99      VALUE 92.26.
    05  FILLER     PICTURE 99V99      VALUE 75.36.
01  PROD-PRICE TABLE REDEFINES PRICE-TABLE.
    05  PRICE      PICTURE 99V99          OCCURS 5 TIMES.
```

   A company has five products and sets up a compile-time table as above. The product number will serve as a subscript in accessing the correct price, as the following example will demonstrate.

   A Procedure Division entry may be coded as follows:

```
MULTIPLY UNITS-IN BY PRICE (PROD-NO)
    GIVING SALES-W.
```

   The PROD-NO field will appear in the input record and serve as the subscript.

2. *Execution Time Tables.* Usually a large table is loaded at execution time from a file of records onto a disk or main storage. The table should be loaded and read before the actual processing of data begins.

Example:

PERCENT-TABLE (Elementary variable)	(Value)
PERCENT (1)	.90
PERCENT (2)	.94
PERCENT (3)	.96

Common name ↑ (PERCENT) Subscript ↑ ((1), (2), (3))

```
WORKING-STORAGE SECTION.
77  E-VARIABLE                                  PICTURE V99.
01  PERCENT-TABLE        USAGE COMP-3.
    05  PERCENT      OCCURS 3 TIMES             PICTURE V99.
```

The Procedure Division for setting up the aforementioned table at execution time could contain the following coding:

```
ACCEPT PERCENT-TABLE FROM SYSIN.
```

When a table name is specified in an ACCEPT statement with the FROM SYSIN option, the number of characters specified in the picture for the table elements is transmitted to each element beginning with the element whose subscript is 1 and the first position.

In the aforementioned Procedure Division statement, the values of 90, 94, and 96 are transmitted to the elements of PERCENT-TABLE with subscripts 1, 2, and 3 respectively. Data in positions 7 through 80 will be disregarded.

The input record would contain information as follows starting in position 1 of the record: 909496.

If separate records are to be used for each element of the table, the following Procedure Division statements may be used:

```
ACCEPT E-VARIABLE FROM SYSIN.
MOVE E-VARIABLE TO PERCENT (1).
ACCEPT E-VARIABLE FROM SYSIN.
MOVE E-VARIABLE TO PERCENT (2).
ACCEPT E-VARIABLE FROM SYSIN.
MOVE E-VARIABLE TO PERCENT (3).
```

Note: When using the ACCEPT statement (or any other method) for loading tables at execution time, the data for the tables must precede any other data in the input file.

Tables may be stored permanently on disks or tapes, accessed by the program and brought into main storage as needed. Deletions and additions to these tables can be made by a separate computer run prior to the processing routine. A PERFORM statement with the VARYING option may be used to load tables at execution time.

The following is an example for loading and printing two-dimensional execution time tables:

Two-D-Table

Rows	Columns		
	1	2	3
1	11	21	31
2	12	22	32
3	13	23	33
4	14	24	34
5	15	25	35
6	16	26	36
7	17	27	37
8	18	28	38

```
WORKING-STORAGE SECTION.
77  COL                        PICTURE 9.
77  ROW                        PICTURE 9.
01  TWO-D-TABLE.
    05  COLUMN-TAB                                OCCURS 3 TIMES.
        10  ITEM-TAB           PICTURE 99         OCCURS 8 TIMES.

*READING AN ENTIRE TABLE
*DATA RECORD
 111213141516171821222324252627283132333435363738

    READ FILE-IN INTO TWO-D-TABLE
        AT END MOVE 'NO' TO MORE-DATA-FLAG.

*READING ONE COLUMN AT A TIME.
*DATA RECORDS
 1112131415161718
 2122232425262728
 3132333435363738

    PERFORM READ-COLUMN
        VARYING COL FROM 1 BY 1
        UNTIL COL IS GREATER THAN 3.

 READ-COLUMN.
    READ FILE-IN INTO COLUMN-TAB (COL)
        AT END MOVE 'NO' TO MORE-DATA-FLAG.

*PRINTING OUT BY COLUMNS.

 01  COLUMN-OUT.
     05  ITEM-OUT              PICTURE ZZZ99      OCCURS 8 TIMES.

     PERFORM PRINT-COLUMN
         VARYING COL FROM 1 BY 1
         UNTIL COL IS GREATER THAN 3.

 PRINT-COLUMN.
     PERFORM MOVE-COLUMN
         VARYING ROW FROM 1 BY 1
         UNTIL ROW IS GREATER THAN 8.
     WRITE PRINTOUT FROM COLUMN-OUT
         AFTER ADVANCING 2 LINES.

 MOVE-COLUMN.
     MOVE ITEM-TAB (COL, ROW) TO ITEM-OUT (ROW).

*PRINTING OUT BY ROWS.

 01  ROW-OUT.
     05  ITEM-OUT              PICTURE ZZZ99      OCCURS 3 TIMES.

     PERFORM PRINT-ROW
         VARYING ROW FROM 1 BY 1
         UNTIL ROW IS GREATER THAN 8.

 PRINT-ROW.
     PERFORM MOVE-ROW
         VARYING COL FROM 1 BY 1
         UNTIL COL IS GREATER THAN 3.
     WRITE PRINTOUT FROM ROW-OUT
         AFTER ADVANCING 2 LINES.
 MOVE-ROW.
     MOVE ITEM-TAB (COL-ROW) TO ITEM-OUT (COL).
```

The following is an example for loading and printing three-dimensional execution time tables.

Plane (1)			Plane (2)		
Column (1,1)	Column (1,2)	Column (1,3)	Column (2,1)	Column (2,2)	Column (2,3)
111	121	131	211	221	231
112	122	132	212	222	232
113	123	133	213	223	233
114	124	134	214	224	234
115	125	135	215	225	235
116	126	136	216	226	236
117	127	137	217	227	237
118	128	138	218	228	238

```
WORKING-STORAGE SECTION.
77  PLANE                       PICTURE 9.
77  COL                         PICTURE 9.
77  ROW                         PICTURE 9.
01  THREE-D-TABLE.
    05  PLANE-TAB                                OCCURS 2 TIMES.
        10  COLUMN-TAB                           OCCURS 3 TIMES.
            15  ITEM-TAB        PICTURE 999      OCCURS 8 TIMES.

*READING AN ENTIRE TABLE-CANNOT BE READ AS ONE RECORD IN
*PUNCHED CARDS.

*READING ONE PLANE AT A TIME.
*DATA RECORDS
 111112113114115116117118121122123124125126127128131132133134135136137138
 211212213214215216217218221222223224225226227228231232233234235236237238

    PERFORM READ-PLANE
        VARYING PLANE FROM 1 BY 1
        UNTIL PLANE IS GREATER THAN 2.
          .
          .
          .
 READ-PLANE.
    READ FILE-IN INTO PLANE-TAB (PLANE)
        AT END MOVE 'NO' TO MORE-DATA-FLAG.
*READING ONE COLUMN AT A TIME
*DATA RECORDS
 111112113114115116117118
 121122123124125126127128
 131132133134135136137138
 211212213214215216217218
 221222223224225226227228
 231232233234235236237238

    PERFORM READ-COLUMN
        VARYING PLANE FROM 1 BY 1
        UNTIL PLANE IS GREATER THAN 2
        AFTER COL FROM 1 BY 1
        UNTIL COL IS GREATER THAN 3.
          .
          .
          .
 READ-COLUMN.
    READ FILE-IN INTO COLUMN-TAB (PLANE, COL)
        AT END MOVE 'NO' TO MORE-DATA-FLAG.

*PRINTING OUT BY COLUMNS
```

```
01  COLUMN-OUT.
    05  ITEM-OUT                  PICTURE ZZZ99      OCCURS 8 TIMES.

    PERFORM PRINT-COLUMN
        VARYING PLANE FROM 1 BY 1
        UNTIL PLANE IS GREATER THAN 2
        AFTER COL FROM 1 BY 1
        UNTIL COL IS GREATER THAN 3.

PRINT-COLUMN.
    PERFORM MOVE-COLUMN
        VARYING ROW FROM 1 BY 1
        UNTIL ROW IS GREATER THAN 8.
    WRITE PRINTOUT FROM COLUMN-OUT
        AFTER ADVANCING 2 LINES.
        .
        .
        .
MOVE-COLUMN.
    MOVE ITEM-TAB (PLANE, COL, ROW) TO ITEM-OUT (ROW).

*PRINTING OUT BY ROWS.

01  ROW-OUT.
    05  ITEM-OUT                  PICTURE ZZZ99      OCCURS 3 TIMES.

    PERFORM PRINT-ROW
        VARYING PLANE FROM 1 BY 1
        UNTIL PLANE IS GREATER THAN 2
        AFTER ROW FROM 1 BY 1
        UNTIL ROW IS GREATER THAN 8.
        .
        .
        .
PRINT-ROW.
    PERFORM MOVE-ROW
        VARYING COL FROM 1 BY 1
        UNTIL COL IS GREATER THAN 3.
    WRITE PRINTOUT FROM ROW-OUT
        AFTER ADVANCING 2 LINES.
        .
        .
        .
MOVE-ROW.
    MOVE ITEM-TAB (PLANE, COL, ROW) TO ITEM-OUT (COL).
```

Regardless of the method for loading tables, all elements to be loaded into a table must have the same length and format. Each table must have a unique name and may or may not be sequentially organized in the type of lookup operation to be performed.

A one-dimensional table is shown in the following example.

The following is the record layout of each record in the file SALESFILE which is on a magnetic disk.

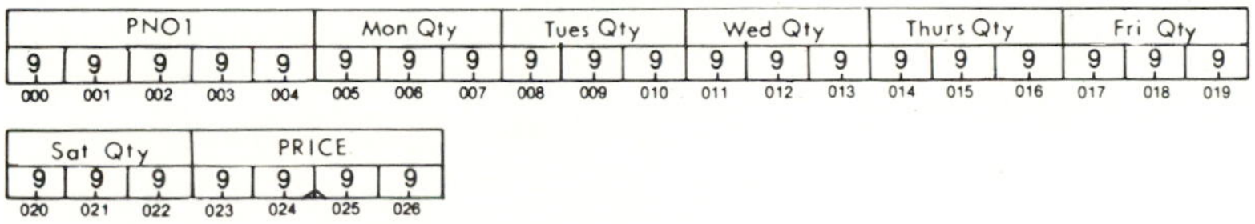

PNO1					Mon Qty			Tues Qty			Wed Qty			Thurs Qty			Fri Qty		
9	9	9	9	9	9	9	9	9	9	9	9	9	9	9	9	9	9	9	9
000	001	002	003	004	005	006	007	008	009	010	011	012	013	014	015	016	017	018	019

Sat Qty			PRICE			
9	9	9	9	9	9	9
020	021	022	023	024	025	026

The contents of the fields within the SALESFILE record are as follows:

PNO1: Product Number
QUANTITY: Quantity sold for each day
PRICE: Price per unit

Data from the records of SALESFILE is used to create records written onto SUMMFILE. The following is the record layout of each record in the file SUMMFILE.

PNO2					TOTAL QUANTITY				TOTAL-D-VALUE								Monday Dollar Value		
9	9	9	9	9	9	9	9	9	9	9	9	9	9	9	9	9	9	9	9
000	001	002	003	004	005	006	007	008	009	010	011	012	013	014	015	016	017	018	019

Monday Dollar Value				Tuesday Dollar Value							Wednesday Dollar Value							Thursday	
9	9	9	9	9	9	9	9	9	9	9	9	9	9	9	9	9	9	9	9
020	021	022	023	024	025	026	027	028	029	030	031	032	033	034	035	036	037	038	039

Thursday Dollar Value					Friday Dollar Value							Saturday Dollar Value						
9	9	9	9	9	9	9	9	9	9	9	9	9	9	9	9	9	9	9
040	041	042	043	044	045	046	047	048	049	050	051	052	053	054	055	056	057	058

The contents of the fields within the record of SUMMFILE are as follows.

PNO2: Product number

TOTAL-QUANTITY: Total quantity sold for one week.

TOTAL-D-VALUE: Dollar value of total quantity sold for one week.

D-VALUE: Dollar value of the quantity sold for each day.

The following flowchart illustrates the logic of the PERFORM VARYING statement in the following program.

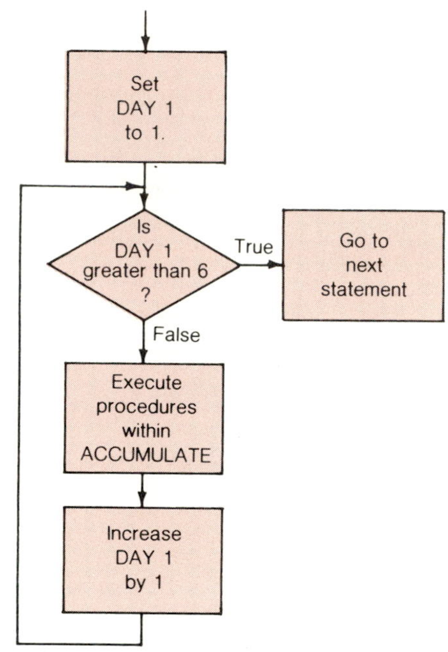


```
DATA DIVISION.
FILE SECTION.
FD  SALESFILE                    BLOCK CONTAINS 18 RECORDS
                                 RECORD CONTAINS 27 CHARACTERS
                                 LABEL RECORD IS STANDARD
                                 DATA RECORD IS SALESRECORD.
01  SALESRECORD.
    05  PNO1                     PICTURE 9(5).
    05  QUANTITY                 PICTURE 999        OCCURS 6 TIMES INDEXED BY DAY1.
    05  PRICE                    PICTURE 99V99.
FD  SUMMFILE                     BLOCK CONTAINS 8 RECORDS
                                 RECORD CONTAINS 59 CHARACTERS
                                 LABEL RECORD IS STANDARD
                                 DATA RECORD IS SUMMRECORD.

01  SUMMRECORD.
    05  PNO2                     PICTURE 9(5).
    05  TOTAL-QUANTITY           PICTURE 9999.
    05  TOTAL-D-VALUE            PICTURE 9(6)V99.
    05  D-VALUE                  PICTURE 9(5)V99    OCCURS 6 TIMES
                                                    INDEXED BY DAY2.
```

```
WORKING-STORAGE SECTION.
77  COUNTER                    PICTURE 9.
01  FLAGS.
    05  MORE-DATA-FLAG         PICTURE XXX     VALUE 'YES'.
        88  MORE-DATA                          VALUE 'YES'.
        88  NO-MORE-DATA                       VALUE 'NO'.

PROCEDURE DIVISION.
MAIN-ROUTINE.
    OPEN      INPUT     SALESFILE,
              OUTPUT    SUMMFILE.
    READ SALESFILE
        AT END MOVE 'NO' TO MORE-DATA-FLAG.
    PERFORM READ-IN
        UNTIL NO-MORE-DATA.
    CLOSE     SALESIFLE
              SUMMFILE
    STOP RUN.

READ-IN.
    MOVE PNO1 TO PNO2.
    MOVE ZERO TO TOTAL-QUANTITY.
    PERFORM ACCUMULATE
        VARYING DAY1 FROM 1 BY 1
        UNTIL DAY1 IS GREATER THAN 6.
    MULTIPLY TOTAL-QUANTITY BY PRICE
        GIVING TOTAL-D-VALUE.
    WRITE SUMMRECORD.
    READ SALESFILE
        AT END MOVE 'NO' TO MORE-DATA-FLAG.

ACCUMULATE.
    ADD QUANTITY (DAY1) TO TOTAL-QUANTITY.
    SET DAY2 TO DAY1.
    MULTIPLY QUANTITY (DAY1) BY PRICE
        GIVING D-VALUE (DAY2).
```

An example of a two-dimensional table is shown in the following diagram:

	DEPARTMENT No.	1st SALE	2nd SALE	3rd SALE	4th SALE	5th SALE
1st DEPARTMENT	1359	001293	029160	200015	025037	010035
2nd DEPARTMENT	1530	013000	000250	304000	002519	023410
3rd DEPARTMENT	2400	002459	002947	012044	002830	016625
4th DEPARTMENT	3594	000248	001590	002300	000122	003520

If the aforementioned two-dimensional table contains variable data, then the definition of the two-dimensional table would be as follows:

```
FILE SECTION.
01  TABLE-1.
    05  DEPARTMENT-ITEM                            OCCURS 4 TIMES    INDEXED BY DEPT.
        10  DEPARTMENT-NUMBER    PICTURE XXXX.
        10  SALE                 PICTURE 9999V99   OCCURS 5 TIMES    INDEXED BY DAY1.
```

In the aforementioned definition of the two-dimensional table, SALE is an element of the two-dimensional table. SALE occurs five times within each DEPARTMENT-ITEM. DEPARTMENT-

ITEM occurs four times within the table called TABLE-1. The INDEXED BY DEPT clause and INDEXED BY DAY1 are only needed when the indexing technique accesses the elements of the table.

The procedural statements required to add all SALE items and place the result of the addition into the field TOTAL-SALES follow:

```
PROCEDURE DIVISION.

    PERFORM DAILY-SALES-ACCUMULATION,
        VARYING DEPT FROM 1 BY 1
        UNTIL DEPT IS GREATER THAN 4
        AFTER DAY1 FROM 1 BY 1
        UNTIL DAY1 IS GREATER THAN 5.

DAILY-SALES-ACCUMULATION.
    ADD SALE (DEPT, DAY1) TO TOTAL-SALES.
```

An example of a three-dimensional table is shown in the following diagram:

	STORE-NO.	DEPARTMENT-NO.	DEPARTMENT 1st SALE	2nd SALE	3rd SALE	4th SALE	5th SALE
1st STORE	123	1359	001293	029160	200015	025037	010035
		1530	013000	000250	304000	002519	023410
		2400	002459	002947	012044	002830	016625
		3594	000248	001590	002300	000122	003520
2nd STORE	530	1280	002510	020050	001257	011120	025910
		3320	100245	001945	027933	123450	300020
		7730	002460	001024	003913	029433	002930
		9250	002200	001050	023910	000200	000391

If the aforementioned three-dimensional table contains variable data, then the definition of the three-dimensional table would be as follows:

```
FILE SECTION.
01  TABLE-2.
    05  STORE-ITEM                                   OCCURS 2 TIMES    INDEXED BY STORE.
        10  STORE-NO            PICTURE XXX.
        10  DEPARTMENT-ITEM                          OCCURS 4 TIMES    INDEXED BY DEPT.
            15  DEPARTMENT-NO   PICTURE XXXX.
            15  SALE            PICTURE 9999V99      OCCURS 5 TIMES    INDEXED BY DAY1.
```

In the aforementioned definition of the three-dimensional table, SALE is an element of the three-dimensional table. SALE occurs five times within each DEPARTMENT-ITEM. DEPARTMENT-ITEM occurs four times within each STORE-ITEM. STORE-ITEM occurs two times within the table called TABLE-2. The INDEXED BY STORE clause, the INDEXED BY DEPT clause, and the INDEXED BY DAY1 clause are only needed when the indexing technique accesses the elements of the table.

The procedural statements required to add all SALE items and place the result of the addition into the field TOTAL-SALE follow:

```
PERFORM DAILY-SALES-ACCUMULATION
     VARYING STORE FROM 1 BY 1
     UNTIL STORE IS GREATER THAN 2
     AFTER DEPT FROM 1 BY 1
     UNTIL DEPT IS GREATER THAN 4
     AFTER DAY1 FROM 1 BY 1
     UNTIL DAY1 IS GREATER THAN 5.

DAILY-SALES-ACCUMULATION.
     ADD SALE (STORE, DEPT, DAY1) TO TOTAL-SALES.
```

Table Searching

Data that has been arranged in the form of a table is often searched. In COBOL, the SEARCH statement, through its two options, provides facilities for producing serial and non-serial (binary, for example) searches. When using the SEARCH statement, an associated index-name or data-name may be varied. This statement also provides for the execution of imperative statements when certain conditions are true.

The SEARCH and SET statements may be used to facilitate table handling. In addition, there are special rules involving table-handling elements when they are used in relation conditions.

Relation Condition

Comparisons involving index-names and/or index-data items conform to the following rules:

1. The comparison of two index-names is actually the comparison of the corresponding occurrence numbers.
2. In the comparison of an index-name with a data item (other than index-data item) or in the comparison of an index-name with a literal, the occurrence number that corresponds to the value of the index-name is compared with the data item or literal.
3. In the comparison of an index-data item with an index-name or another index-data item, the actual values are compared without conversion.

   The result of the comparison of an index data item with any data item or literal not already specified is undefined.

SEARCH Statement

The SEARCH statement is used to search a table for a table element that satisfies the specified condition and to adjust the associated index-name to indicate the position of that table element.

There are two forms of the SEARCH statement, one for serial searches and the other, specifying the ALL keyword, for binary searches. Both forms of the SEARCH statement are conditional statements. Conditional control is exercised through the required WHEN phrase and the optional AT END phrase.

Only one level of a table (a table element) can be referenced with one SEARCH. There are two formats for the SEARCH statement. Format-1, SEARCH, is used for a serial search. Format-2, SEARCH ALL, is used for a binary search.

Format-1 SEARCH statements perform a serial search of a table element. If the programmer knows that the "found" condition will come after some intermediate point or some table element, to speed up execution, the SET statement can be used to set the index-names at that point and search only part of the table elements. If the number of table elements are large and must be searched from the first occurrence to the last, the use of format-2 (SEARCH ALL) is more efficient than format-1, since it uses a binary search technique that is much faster; however, the table must be ordered.

In format-2, the SEARCH ALL statement, the table must be ordered on the KEY(S) specified in the OCCURS clause.

Rules Governing the Use of the SEARCH Statement

1. Identifier-1 must not be subscripted or indexed, but its description must contain an OCCURS clause and an INDEXED BY clause.

```
SEARCH identifier-1 [VARYING {identifier-2   }]
                             {index-name-1   }

    [ ; AT END imperative-statement-1 ]

    ; WHEN condition-1   {imperative-statement-2}
                         {NEXT SENTENCE         }

    [; WHEN condition-2  {imperative-statement-3}] . . .
                         {NEXT SENTENCE         }
```

Figure 10.30 Format: SEARCH statement—format-1.

2. Identifier-2, when specified, must be described as USAGE IS INDEX, or as a numeric elementary item without any positions to the right of the assumed decimal point. Identifier-2 is incremented by the same amount, and at the same time, as the occurrence number.

Serial SEARCH Statement

The serial SEARCH statement is designed to search a one-dimensional array or a single dimension of a multidimensional array. The VARYING option may be used to step an index that is associated with the same table or another table. More than one WHEN phrase may be used to determine the termination condition for the search statement. The first format for the SEARCH statement is shown in figure 10.30.

The serial search does not initialize the controlling index value, but starts the search at its current value. The test for range limit on the controlling index is performed before the WHEN conditions are tested and may prevent the execution of the search. The flow of control in a serial SEARCH statement with an AT END phrase and two WHEN phrases is as follows:

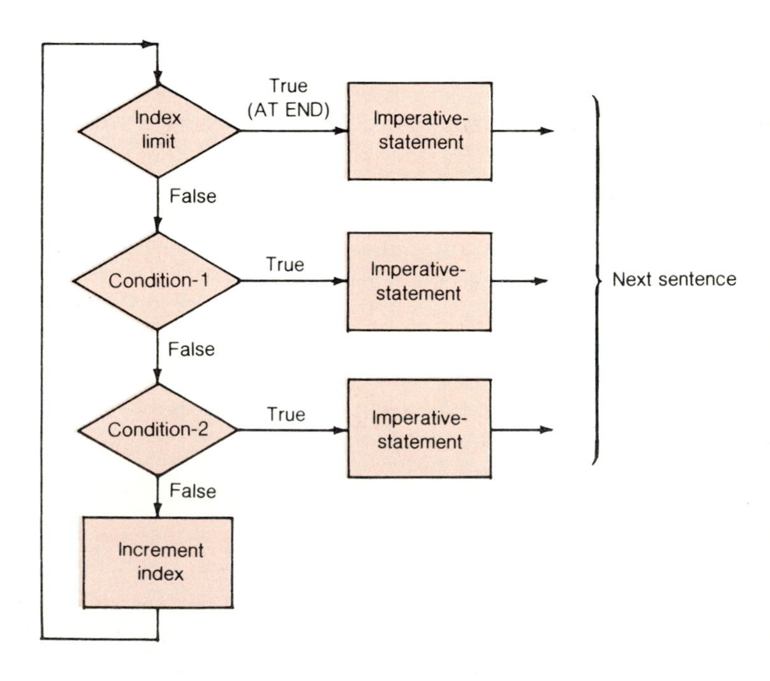

The imperative statements may contain GO TO statements that override the implicit transfer to the next sentence.

In a multidimensional array, the SEARCH statement may be used on any of the indexes. For example:

```
    .
    .
    .
    01  INFORMATION-ARRAY.
        05  PLANES              OCCURS 50 TIMES INDEXED BY I.
            10  ROWS            OCCURS 42 TIMES INDEXED BY J.
                15  COLUMNS     OCCURS 12 TIMES INDEXED BY K.
                    20    DATUM PIC . . .
    .
    .
    .
   *SEARCHING A ROW, COLUMN BY COLUMN.
        SET K TO 1.
        SEARCH COLUMNS
            AT END MOVE 'FAIL' TO SEARCH-FLAG
            WHEN DATUM (I,J,K) = TARGET-VALUE
                MOVE 'HIT' TO SEARCH-FLAG.
    .
    .
    .
   *SEARCHING A COLUMN, ROW BY ROW.
        SET J TO 1.
        SEARCH ROWS
            AT END MOVE 'FAIL' TO SEARCH-FLAG
            WHEN DATUM (I,J,K) = TARGET-VALUE
                MOVE 'HIT' TO SEARCH-FLAG.
```

Procedure for SEARCH Statement—Format-1

1. Upon execution of a SEARCH statement, a serial search takes place, starting with the current index setting.
2. If at the start of the search the value of the index-name associated with identifier-1 is not greater than the highest possible occurrence number for identifier-1, the following action takes place:
   a. The conditions in the WHEN option are evaluated in the order in which they are written.
   b. If none of the conditions is satisfied, the index-number for identifier-1 is incremented to reference the next table-element, and the search is repeated.
   c. If, upon evaluation, one of the WHEN conditions is satisfied, the search terminates immediately, and the imperative statement associated with that condition is executed. The index-name points to the table-element that satisfied the condition. (See figures 10.31 and 10.32.)
   d. If the end of the table is reached without the WHEN condition being satisfied, the search is terminated immediately, and if the AT END option is specified, the imperative statement is executed. If the AT END option is omitted, control passes to the next sentence. (See figures 10.33, 10.34, and 10.35.)

Figure 10.31 SEARCH statement—format-1—example.

Simple SEARCH Sentence

Format: SEARCH identifier WHEN condition imperative-statement.

Explanation: Beginning with the table element whose occurrence number corresponds to the current value of the index defined for identifier, table elements with the common name-identifier are tested serially by index until an element that satisfies condition is found. The imperative statement is then executed and control moves either to the sentence directly following the SEARCH sentence or the paragraph specified in a GO TO statement within the imperative statement.

Figure 10.31 *(continued)*

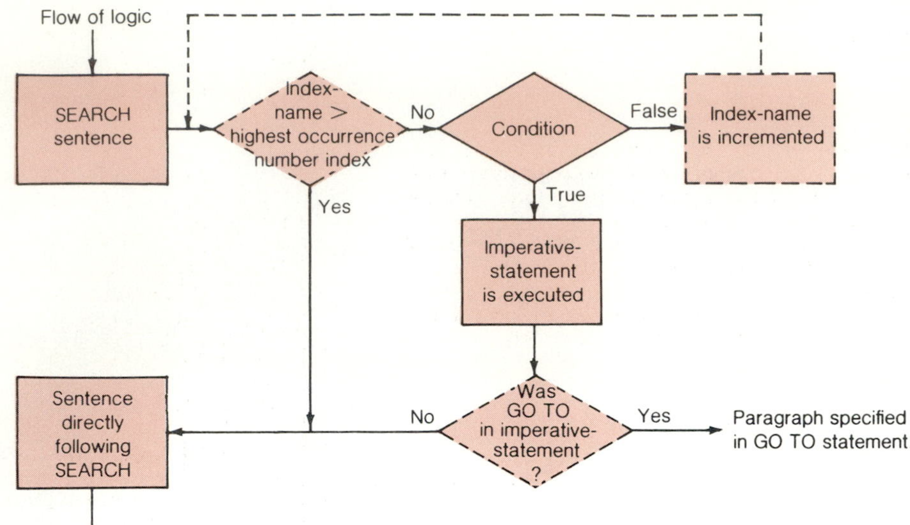

Rules

1. Identifier must be the common name of the table elements for which the OCCURS clause with the INDEXED BY option is specified.
2. Identifier must not be a level 01 variable.
3. Index-name (not specified in the SEARCH sentence but necessary for its execution) must be defined by the INDEXED BY option of the OCCURS clause in the data description entry for identifier.

(The dashed lines indicate logic that is done automatically by the compiler.)

Job:

Orders received by the mail-order warehouse are punched into cards in ORDER-FILE as shown below. As an order is processed, the program is to determine whether the item ordered is a sale item. If so, the price is to be computed as 90 percent of the regular price before the amount of the order is computed. The catalog numbers of sale items which have been punched one per card as shown below are treated as values of SALE-TABLE. The number of sale items for any one month will also be punched into a card to determine the desired size of SALE-TABLE for that month. The maximum number of sale items for any one month is 10.

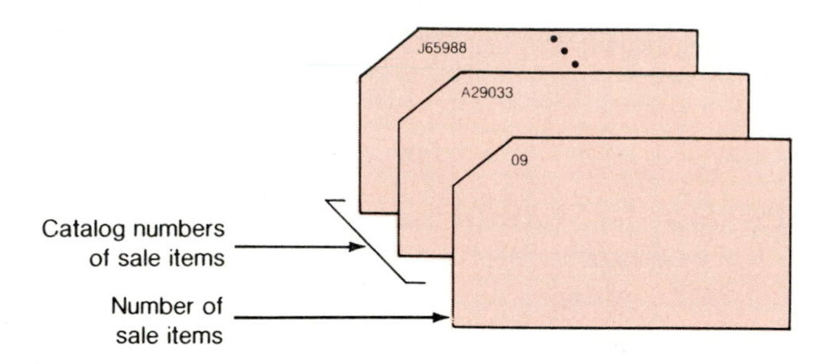

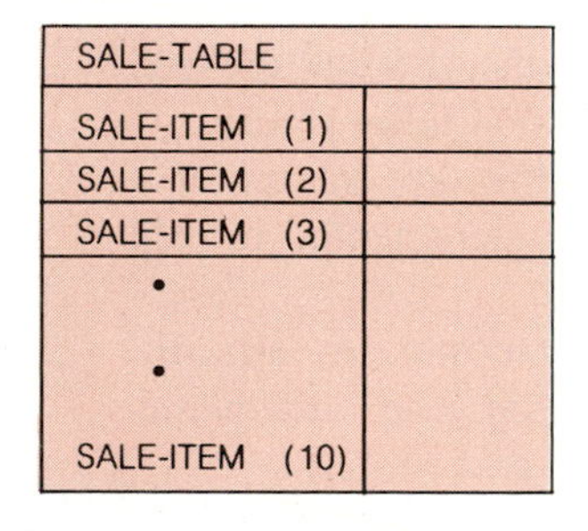

Card in ORDER-FILE

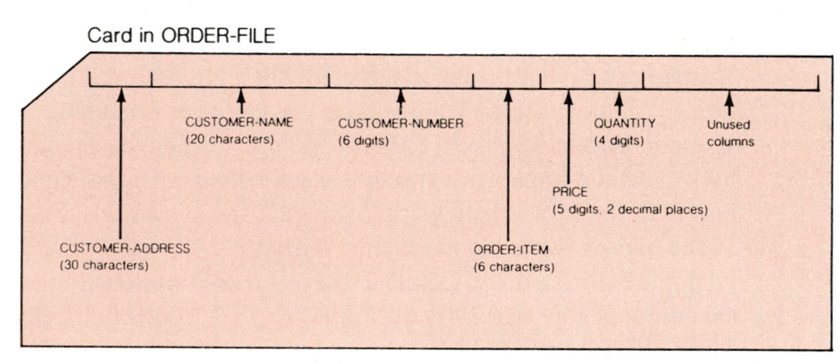

The coding is as follows:

```
01  SALE-TABLE.
    05  SALE-ITEM     OCCURS 10 TIMES
                DEPENDING ON TABLE-SIZE
                INDEXED BY S
                PICTURE X(6).
    SET S to 1.
    SEARCH SALE-ITEM
        WHEN ORDER-ITEM IS EQUAL TO SALE-ITEM (S)
        COMPUTE PRICE-WS ROUNDED = PRICE-WS * NINETY.
    COMPUTE AMOUNT = PRICE-WS * QUANTITY-WS.
```

Figure 10.32 SEARCH statement—format-1—example.

SEARCH sentence with the AT END Option

Format SEARCH identifier [AT END {imperative-statement-1 / NEXT SENTENCE}] WHEN condition imperative-statement-2.

Explanation Beginning with the table element whose occurrence number corresponds to the current value of the index defined for identifier, table elements are tested serially by index until an element that satisfies condition is found. Imperative-statement-2 is then executed and control moves either to the sentence directly following the SEARCH sentence or the paragraph specified in a GO TO statement within imperative-statement-2. If no table element satisfies condition, imperative-statement-1 is executed and control moves either to the sentence directly following the SEARCH sentence or the paragraph specified in a GO TO statement within imperative-statement-1.

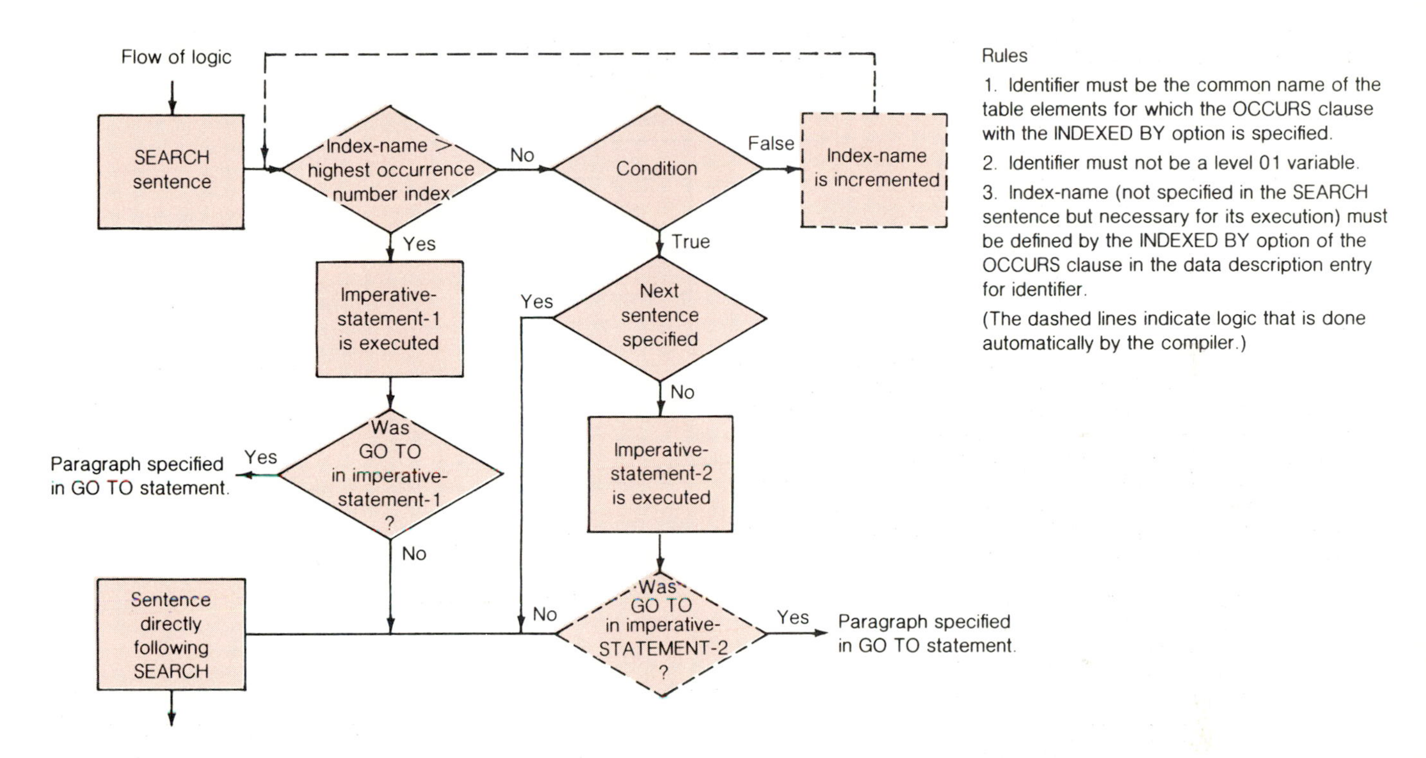

Rules

1. Identifier must be the common name of the table elements for which the OCCURS clause with the INDEXED BY option is specified.
2. Identifier must not be a level 01 variable.
3. Index-name (not specified in the SEARCH sentence but necessary for its execution) must be defined by the INDEXED BY option of the OCCURS clause in the data description entry for identifier.

(The dashed lines indicate logic that is done automatically by the compiler.)

Figure 10.33 SEARCH statement—AT END option—example.

Job:

If the item number in a transaction record matches a number in the discontinued table, a DISCONTINUED-DETAIL procedure is to be performed.

The table is shown below:

DISCONTINUED-TABLE		(Index D)
DISCONTINUED-ITEM-NUMBER	(1)	A17080
DISCONTINUED-ITEM-NUMBER	(2)	B92641
•		
•		
•		
•		
•		

The coding is as follows:

```
    SET D to 1.
TEST.
    SEARCH DISCONTINUED-ITEM-NUMBER
        AT END PERFORM DETAIL-PROC
        WHEN ITEM-NUMBER EQUAL TO
            DISCONTINUED-ITEM-NUMBER (D)
        PERFORM DISCONTINUED-DETAIL.
EXIT-POINT.
    EXIT.
```

Figure 10.34 SEARCH statement—AT END option—example.

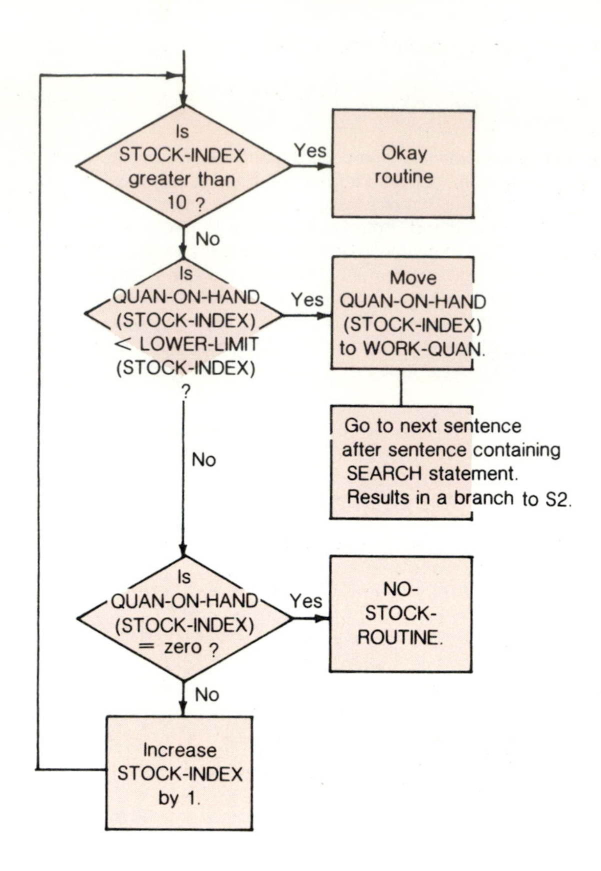

The table called STOCK-TABLE consists of 10 table elements (STOCK-ELEMENT). These are indexed by the index called STOCK-INDEX. SEARCH begins testing the fifth table element because the index associated with the table element was set to the occurrence number 5 before the SEARCH statement was executed. If the fifth table element does not meet either of the two conditions, the index (STOCK-INDEX) is incremented to the next occurrence number, 6. The search operation is then repeated, testing the fields in the sixth table element. The search continues with each successive table element, terminating when the end of the table is reached or a condition is met by the QUAN-ON-HAND field in one of the last, six table elements.

```
01  STOCK-TABLE.
    05  STOCK-ELEMENT OCCURS 10 TIMES
        INDEXED BY STOCK-INDEX.
        10  STOCK-NUMBER      PICTURE 9(5).
        10  PRICE             PICTURE 999V99.
        10  QUAN-ON-HAND      PICTURE 9999.
        10  BACK-ORDER-QUAN   PICTURE 9999.
        10  LOWER-LIMIT       PICTURE 999.

PROCEDURE DIVISION.
PAR-1.
    SET STOCK-INDEX TO 5.
    SEARCH STOCK-ELEMENT
        AT END PERFORM OKAY-ROUTINE,
        WHEN QUAN-ON-HAND (STOCK-INDEX) IS LESS THAN
            LOWER-LIMIT (STOCK-INDEX),
        MOVE QUAN-ON-HAND (STOCK-INDEX) TO WORK-QUAN
        WHEN QUAN-ON-HAND (STOCK-INDEX) EQUALS ZERO,
        PERFORM NO-STOCK-ROUTINE.

PAR-2.
    ADD 100 TO WORK-QUAN.
```

Figure 10.35 SEARCH statement—AT END option—example.

Condition-1 or condition-2 may be any condition as follows: a relation condition, class condition, condition-name condition, or sign condition.

When the VARYING option is specified, one of the following applies:

1. If index-name-1 is one of the indexes for identifier-1, index-name-1 is used for the search; otherwise, the first (or only) index-name is used for identifier-1.
2. If index-name-1 is an index for another table entry, then when the index-name for identifier-1 is incremented to represent the next occurrence of the table, index-name-1 is simultaneously incremented to represent the next occurrence of the table it indexes. (See figures 10.36 and 10.37.)
3. If identifier-2 is an index data item, it is incremented as the associated index is incremented.

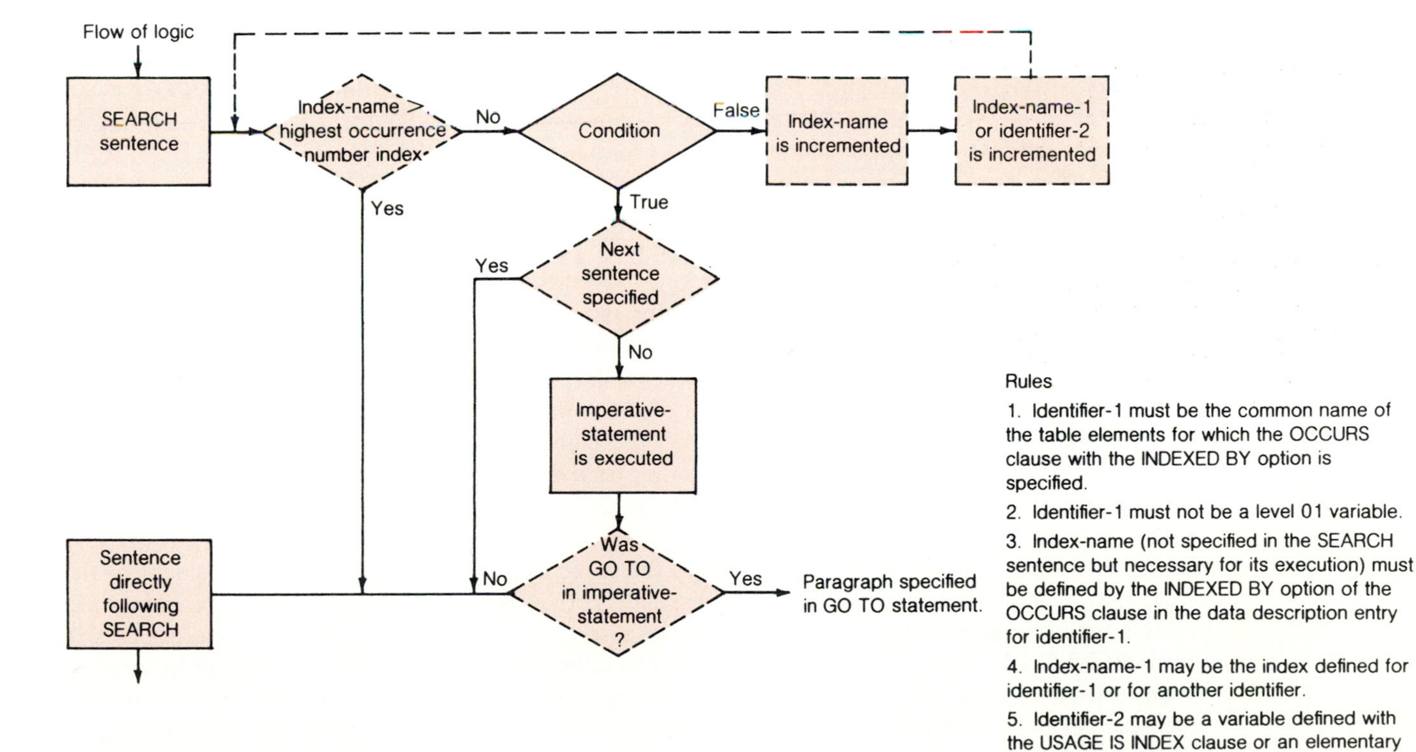

SEARCH Sentence with the VARYING Option
Format SEARCH identifier-1 [VARYING {index-name-1 / identifier-2}] WHEN condition {imperative-statement / NEXT SENTENCE}
Explanation Beginning with the table element whose occurrence number corresponds to the current value of the index defined for identifier-1, table elements with the common name identifier-1 are tested serially by index until an element that satisfies condition is found. Beginning with its value at the time the SEARCH sentence is executed the variable specified in the VARYING option is incremented each time the index defined for identifier-1 is incremented. When a table element satisfying condition is found, if NEXT SENTENCE is specified in the WHEN option, control moves to the sentence directly following the SEARCH sentence. Otherwise, the imperative statement is executed and control moves either to the sentence directly following the SEARCH sentence or the paragraph specified in a GO TO statement within the imperative statement.
Flow of logic
SEARCH sentence
Index-name > highest occurrence number index
No
Yes
Condition
False
True
Index-name is incremented
Index-name-1 or identifier-2 is incremented
Next sentence specified
Yes
No
Imperative-statement is executed
Was GO TO in imperative-statement ?
No
Yes
Paragraph specified in GO TO statement.
Sentence directly following SEARCH
Rules
1. Identifier-1 must be the common name of the table elements for which the OCCURS clause with the INDEXED BY option is specified.
2. Identifier-1 must not be a level 01 variable.
3. Index-name (not specified in the SEARCH sentence but necessary for its execution) must be defined by the INDEXED BY option of the OCCURS clause in the data description entry for identifier-1.
4. Index-name-1 may be the index defined for identifier-1 or for another identifier.
5. Identifier-2 may be a variable defined with the USAGE IS INDEX clause or an elementary variable that will have integer values.
(The dashed lines indicate logic that is done automatically by the compiler.)

Figure 10.36 SEARCH statement—VARYING option—example.

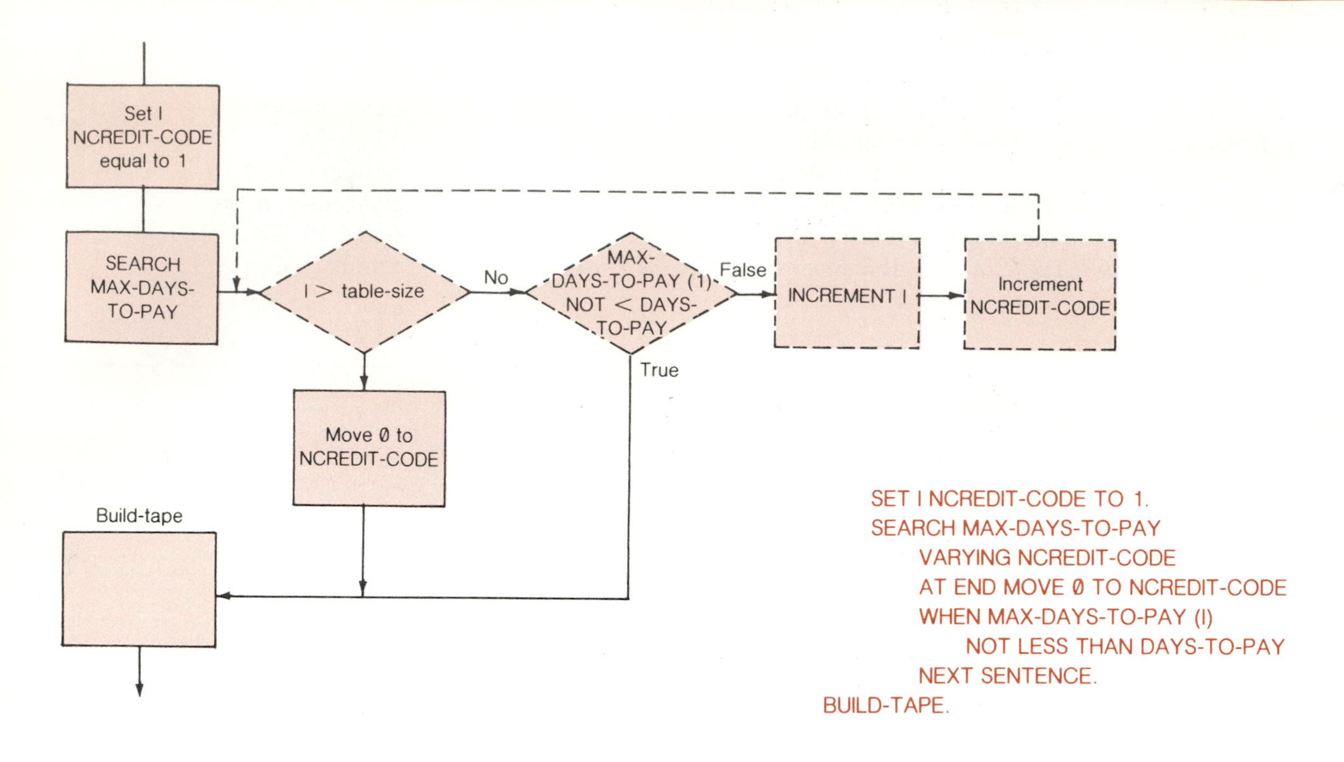
Set I
NCREDIT-CODE
equal to 1
SEARCH
MAX-DAYS-
TO-PAY
I > table-size
No
MAX-
DAYS-TO-PAY (1)
NOT < DAYS-
TO-PAY
False
True
INCREMENT I
Increment
NCREDIT-CODE
Move Ø to
NCREDIT-CODE
Build-tape
SET I NCREDIT-CODE TO 1.
SEARCH MAX-DAYS-TO-PAY
VARYING NCREDIT-CODE
AT END MOVE Ø TO NCREDIT-CODE
WHEN MAX-DAYS-TO-PAY (I)
NOT LESS THAN DAYS-TO-PAY
NEXT SENTENCE.
BUILD-TAPE.

Figure 10.37 SEARCH statement—VARYING option—example.

Binary SEARCH Statement

The binary SEARCH statement is designed to search a well-ordered, one-dimensional array.

The table referenced in the SEARCH ALL statement must not be defined with subscripting or indexing, but the data description of the table must contain both an OCCURS clause and an INDEXED BY phrase. The data description must also include a KEY phrase in the OCCURS clause.

The table must be well ordered when the SEARCH statement is executed. That is, every entry in the table must be in sorted order as defined by the KEY phrase specified for the table. The table should not contain duplicate key values. The key (or keys) specified in the KEY phrase must be tested by the condition contained in the WHEN phrase, and the test must be for a condition-name or for equality with a literal or a data item. Alternately, the condition may be a compound condition formed from simple conditions, with AND as the only connective. Any data-name that appears in the KEY phrase may appear as the subject or object of a test or as the name of the conditional variable with which the tested condition-name is associated. However, all preceding data-names in the KEY phrase must also be included within the WHEN condition. No other tests may appear within the WHEN phrase.

The initial setting of the index-name for the table is ignored and its setting is varied during the search operation in a manner that allows a "binary" search operation to be executed, with the restriction that at no time is it set to a value that exceeds the value that corresponds to the last element of the table, or that is less than the value that corresponds to the first element of the table. The index that controls the execution of the SEARCH statement is specified by the first index-name that appears in the INDEXED BY phrase that is associated with the table. If the WHEN condition cannot be satisfied for any setting of the index within this permitted range, then control is passed to an imperative statement when the AT END phrase appears, or to the next sentence when the AT END phrase is absent. In either case, the final setting of the index is not predictable. If the WHEN conditions can be satisfied, the index indicates the occurrence that allowed the WHEN condition to be satisfied, and control passes to the imperative statement associated with the WHEN phrase.

The imperative statements associated with the AT END and WHEN phrases may contain GO TO statements that override the implicit transfer to the next sentence.

The flow of control in a binary SEARCH statement is as follows:

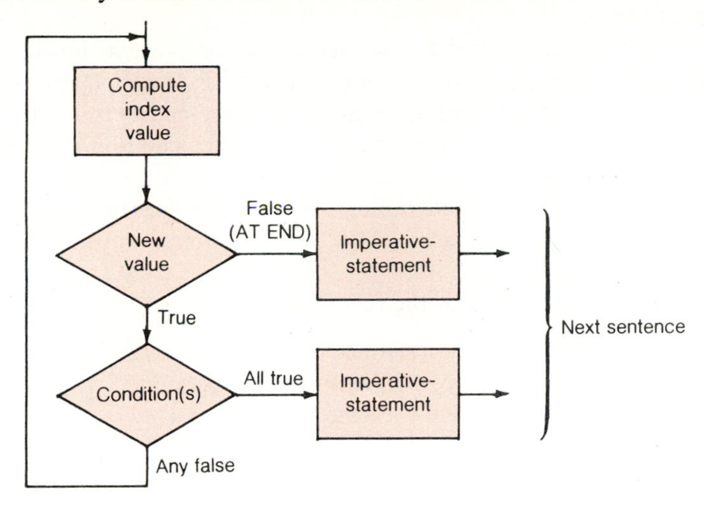

An example of a binary SEARCH statement follows:

```
        .
        .
        01  LEGAL-PART-NUMBERS      OCCURS 4000 TIMES
                ASCENDING KEY IS MASTER-PART-NUMBER
                INDEXED BY PART-INDEX.
            05  MASTER-PART-NUMBER      PIC . . .
        .
        .
            SEARCH ALL LEGAL-PART-NUMBERS
                AT END PERFORM PART-NUMBER-ERROR
                WHEN TRANSACTION-PART-NUMBER IS EQUAL TO
                    MASTER-PART-NUMBER (PART-INDEX)
                    PERFORM GOOD-PART-SERVICE.
```

Procedure for SEARCH Statement—Format-2

1. A nonserial type of search operation takes place. (The second format for the SEARCH statement is shown in figure 10.38.) When this occurs, the initial setting of the index-name for identifier-1 is ignored, and its setting is varied during the search. At no time is it less than the value that corre-

Format 2

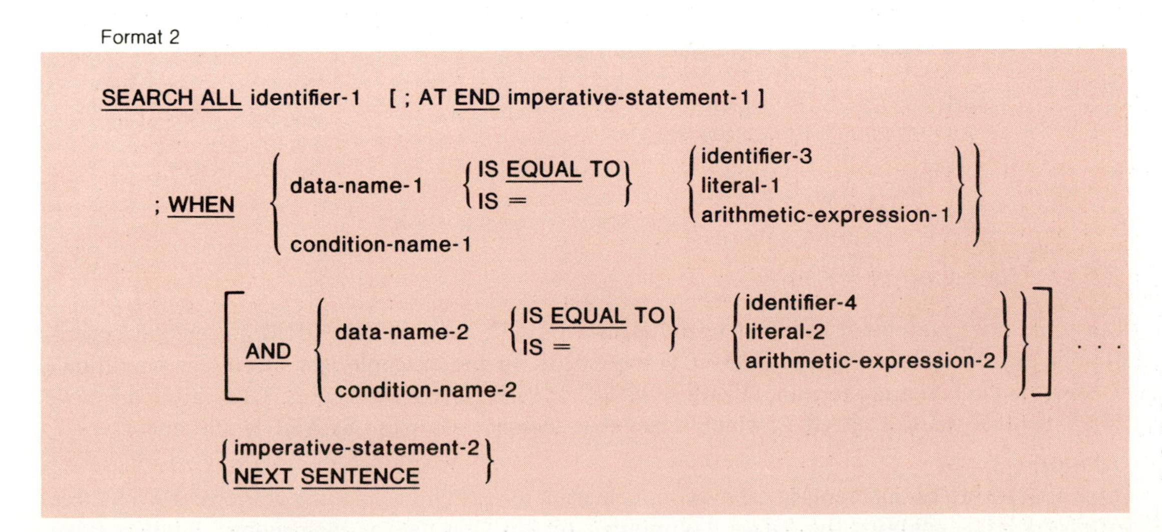

The required relational character '=' is not underlined to avoid confusion with other symbols.

Figure 10.38 Format: SEARCH statement—format-2.

sponds to the first element of the table, nor is it ever greater than the value that corresponds to the last element of the table.

   a. If WHEN cannot be satisfied for any setting of the index within the permitted range, control is passed to imperative statement-1 when the AT END option appears, or to the next sentence when this clause does not appear. In either case, the final setting of the index is unpredictable.
   b. If the index indicates an occurrence that allows the condition to be satisfied, control passes to imperative statement-2.
2. The first index-name assigned to identifier-1 will be used for the search.
3. The description of identifier-1 must contain the KEY option in its OCCURS clause.

Condition-1 must consist of one of the following:

1. *Relation Condition.* One of the data-names must appear in the KEY clause.
2. *Condition-name Condition.* The conditional variable associated with condition-name must be one of the names that appear in the KEY clause of identifier-1.
3. A compound condition from simple conditions of the types just described, with AND as the only connector.

The results of SEARCH ALL operation are predictable only when the data in the table is ordered as described by the ASCENDING/DESCENDING KEY clause associated with identifier-1.

Any data-name that appears in the KEY clause of identifier-1 may be tested in condition-1. However, all data-names in the KEY clause preceding the one to be tested must also be tested on condition-1. No other tests can be made on condition-1.

Example:

The table called STOCK-TABLE consists of 10 elements (STOCK-ELEMENT). These are indexed by the index called STOCK-INDEX. The SEARCH ALL statement begins testing the first table element with the index STOCK-INDEX set to the occurrence number 1. If the first table element does not have STOCK-NUMBER equal to TRAN-STOCK-NO, then STOCK-INDEX is incremented to the next occurrence number, 2. The search operation is then repeated with the testing of the STOCK-NUMBER in the second table element, continuing with each successive table element until the end of the table is reached or the condition is met.

```
01  STOCK-TABLE.
    05  STOCK-ELEMENT                    OCCURS 10 TIMES ASCENDING KEY IS
        STOCK-NUMBER                 INDEXED BY STOCK-INDEX.
        10  STOCK-NUMBER             PICTURE 9(5).
        10  PRICE                    PICTURE 999V99.
        10  QUAN-ON-HAND             PICTURE 9999.
        10  BACK-ORDER-QUAN          PICTURE 9999.
        10  LOWER-LIMIT              PICTURE 999.

PROCEDURE DIVISION.
PAR-1.
    SEARCH ALL STOCK-ELEMENT
        AT END PERFORM MISSING-ROUTINE
        WHEN STOCK-NUMBER (STOCK-INDEX) IS EQUAL TO
            TRAN-STOCK-NO
        MOVE STOCK-ELEMENT (STOCK-INDEX) TO WORK-AREA.
PAR-2.
    ADD QUANTITY TO TOTAL.
```

If an imperative-statement does not terminate with a GO TO or PERFORM statement, control passes to the sentence following the SEARCH statement. In the example just listed, the condition's imperative-statement does not terminate with a GO TO or PERFORM statement. Thus, after the move is executed, control passes to PAR-2, which is the next sentence after the SEARCH statement.

Example:

There are two tables that contain calendar information for the twelve months of the year. One table, JULIAN-VALUES, contains the Julian day number for the first day of each month; another table, INDEX-TABLE, contains month names ordered by ascending key of month numbers, as follows:

```
01  JULIAN-VALUES.
    05  JULIANS                         PIC X(36)     VALUE
        '001032060091121152182213244274305335'.
    05  JULIAN-TABLE REDEFINES JULIANS       OCCURS 12 TIMES
        INDEXED BY X2.
        10  FIRST-JULIAN                PIC 999.

01  INDEX-TABLE.
    05  CALENDAR.
        10  QUARTER-1                   PIC X(21)     VALUE
            'JAN0131FEB0228MAR0331'.
        10  QUARTER-2                   PIC X(21)     VALUE
            'APR0430MAY0531JUNE0630'.
        10  QUARTER-3                   PIC X(21)     VALUE
            'JUL0731AUG0831SEPT0930'.
        10  QUARTER-4                   PIC X(21)     VALUE
            'OCT1031NOV1130DEC1231'.
    05  CAL-TABLE REDEFINES CALENDAR         OCCURS 12 TIMES
        INDEXED BY X1
        ASCENDING KEY IS MONTH-NUM.
        10  CAL-ITEM.
            15  MONTH                   PIC XXX.
            15  MONTH-NUM               PIC 99.
            15  MAX-DAYS                PIC 99.
```

If each input record contains some month number, INPUT-MONTH-NO, and the matching month and beginning Julian day number are to be obtained from the tables for reporting purposes, then either a serial or binary search can be used.

For a binary search (format-2), the example procedures are:

```
SEARCH ALL CAL-TABLE
    AT END DISPLAY 'BAD INPUT MONTH'
        PERFORM ERROR-RTN
    WHEN MONTH-NUM (X1) = INPUT-MONTH-NO
        MOVE MONTH (X1) TO REPORT-MONTH
        SET X2 to X1
        MOVE FIRST-JULIAN (X2) TO BEGIN-JULIAN-DATE.
```

The index-name X1 requires no SET statement prior to a binary search since it is implicitly set upon entering the search operation. However, it is necessary that the table, CAL-TABLE, be ordered on a key, MONTH-SUM in the example, in order to use format-2 of the SEARCH statement. When the condition is satisfied, index-name X1 is left pointing to the table element that met the condition. Therefore, its value may be used in a SET statement to adjust index-name X2 so that a table element corresponding to the match in CAL-TABLE may be obtained from JULIAN-TABLE.

The advantage of the format-2 SEARCH statement is the relative speed of operation, which increases as the sizes of the tables to be searched increase. However, with format-2, data must be arranged in order of KEY values. Also, the results of the binary search are unpredictable if a table contains any duplicate items or if items are out of sequence.

The serial search technique (format-1) may be used if a binary search is not practical. It would be slower but, in the case of duplicate items, it would be possible to locate all table elements that satisfy a condition by continuing the search operation after first meeting the condition, and then setting the index-name up by 1.

The previous example with a serial search (format-1) is:

```
SET X1, X2 TO 1.

SEARCH CAL-TABLE
    VARYING X2
    AT END DISPLAY 'BAD INPUT MONTH'
    WHEN MONTH-NUM (X1) = INPUT-MONTH-NO
        MOVE MONTH (X1) TO REPORT-MONTH
        MOVE FIRST-JULIAN (X2) TO BEGIN-JULIAN-DATE.
```

Format 1:

```
SET {identifier-1    [, identifier-2]   ...}  TO {identifier-3  }
    {index-name-1    [, index-name-2]   ...}     {index-name-3  }
                                                 {integer-1     }
```

Format 2:

```
SET index-name-4  [, index-name-5]  ...  {UP BY  }   {identifier-4}
                                         {DOWN BY}   {integer-2   }
```

Figure 10.39 Formats: SET statement.

SET Statement

The SET statement is used to establish, modify, or preserve the value associated with an index-name. (Formats for the SET statement are shown in figure 10.39.) The SET statement may also be used to equate the contents of two index data items.

Uses of the SET Statement

Setting an Index-Name
The SET statement may be used to set the value associated with an index-name to the value of an integer literal, an integer data item, an index data item, or another index-name. In the first two cases, the value of the integer data item is converted into that special format of an index-name that represents the corresponding ordinal position number in the table associated with the index-name. In the third case, no conversion takes place since the index data item is assumed to be properly formatted. In the final case, the value of the sending index-name will be converted into the frame of reference of the receiving index-name. The index-names will therefore reference corresponding ordinal positions even though the two tables may have had different entry formats.

Modifying an Index-Name
The SET statement may be used to modify the present value of an index-name by the value of an integer literal or an integer data item. In both cases, the ordinal position number associated with the index-name is incremented (UP BY) or decremented (DOWN BY) the number of table positions that correspond to the value of the literal or data item.

Preserving an Index-Name
The SET statement may be used to save the value of an index-name in an integer data item or in an index data item. In the former case, the data item is set to that integer value that corresponds to the ordinal position number given by the present value associated with the index-name. In the latter case, the value associated with the index-name is stored into the index data item without conversion.

Restrictions on Table Values

Since transfers of information between index-names and index data items are made without conversion, use discretion when passing values from one index-name to another via an index data item. If both index-names appear in the same INDEXED BY phrase, no problems would arise. However, if the index-names appear in different INDEXED BY phrases, the two tables must have exactly the same descriptions in terms of entry sizes and maximum occurrence values.

The SET statement establishes reference points for table-handling operations by setting index-names associated with table-elements. The SET statement must be used when initializing index-name values before execution of a SEARCH statement. It may also be used to transfer values between index-names and other elementary data items.

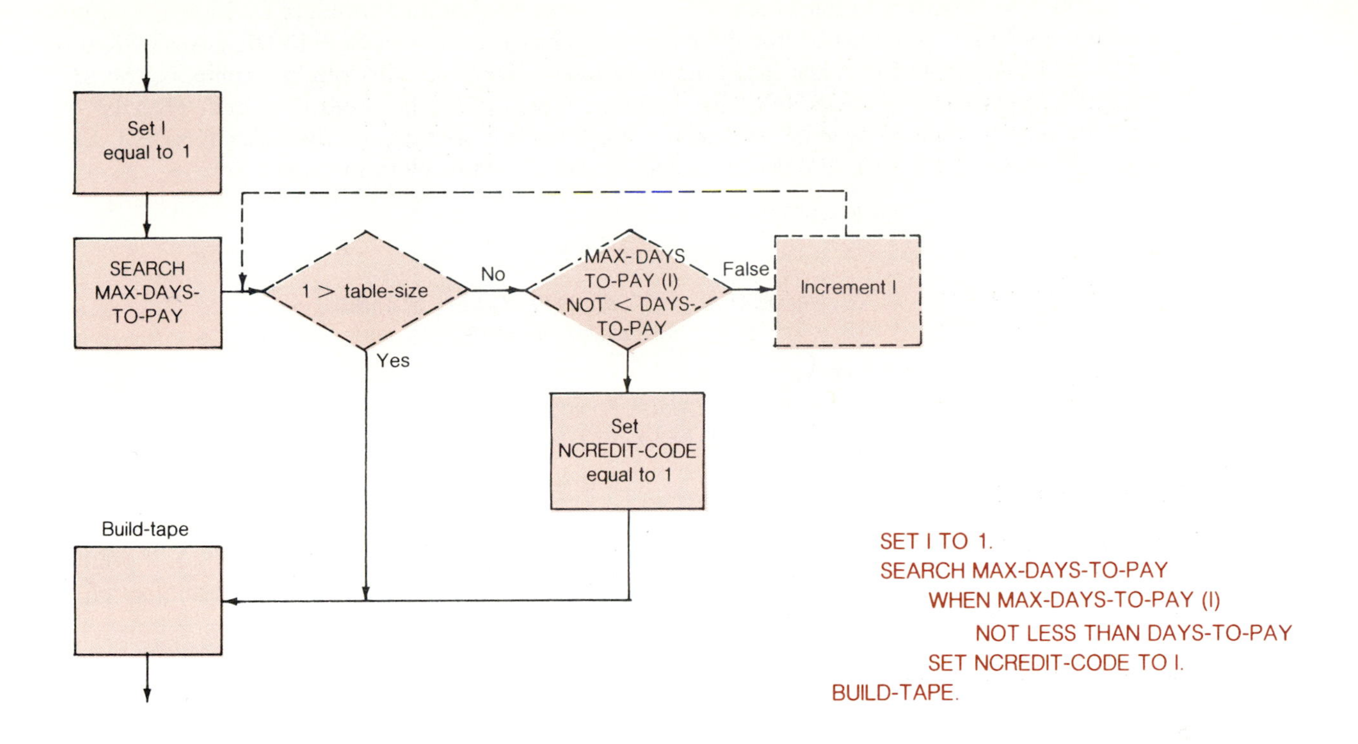

Figure 10.40 SET statement—example.

The SET statement is used to assign values to index-names and to index-data names. An *index-name* is the name given to the table associated with a table of elements. An index is assigned by including the INDEXED BY option of the OCCURS clause in the definition of the table element. An *index-data name* is an elementary data item described with the USAGE IS INDEX clause in the Data Division.

When the SET statement assigns to an index-name the value of a literal, identifier, or an index-name from another table element, it is set to an actual displacement from the beginning of the table element that corresponds to an occurrence number of an element in the associated table.

Rules Governing the Use of the SET Statement

All identifiers must name either index-data items or elementary items described as integers, except that identifier-4 must not name an index-data item. When a literal is used, it must be a positive integer. Index-names are considered related to a given table through the INDEXED BY option.

Format-1

When the SET statement is executed, one of the following actions takes place:

1. Index-name-1 is set to a value that corresponds to the same table-element to which either index-name-3, identifier-3, or integer-1 corresponds. If identifier-3 is an index-data item, or if index-name-3 is related to the same table as index-name-1, no conversion takes place. (See figure 10.40.)
2. If identifier-1 is an index-data item, it may be set to equal either the contents of index-name-3 or identifier-3 where Identifier-3 is also an index-data item. Integer-1 cannot be used in this case.
3. If identifier-1 is not an index-data item, it may be set only to an occurrence number that corresponds to the value of index-name-3. Neither identifier-3 nor integer-1 can be used in this case.

Example:

INDEX-A indexes a table of ten AMOUNT items. During the Procedure Division, the current contents of INDEX-A must be stored temporarily. Thus, the sentence SET INDEX-A-STORAGE TO INDEX-A is given. When it is time to restore INDEX-A to its original contents, the SET INDEX-A TO INDEX-A-STORAGE is given. Notice that the index-data-item (INDEX-A-STORAGE) is defined as a 77 level in the Working-Storage Section; the index (INDEX-A) is defined by the INDEXED BY clause associated with the table element (AMOUNT).

```
WORKING-STORAGE SECTION.
77  INDEX-A-STORAGE      USAGE IS INDEX.

    05  AMOUNT      PICTURE 99V9 OCCURS 10 TIMES
                    INDEXED BY INDEX-A.

PROCEDURE DIVISION.
    SET INDEX-A-STORAGE TO INDEX-A.
    SET INDEX-A TO INDEX-A-STORAGE.
```

Format-2

When the SET statement is executed, the content of index-number-4 (index-name-5, etc.), if present, is incremented (UP BY) or decremented (DOWN BY) a value that corresponds to the number of occurrence represented by the value of integer-2 or identifier-4.

The literal may be negative.

Following are three examples of SET statements and two Table Handling sample problems.

In the following example of a SET statement, the index Q2, is set to the value corresponding to the occurrence number represented by index Q1. If the index Q1 contains the offset for the second element in the table of QUANTITY elements, then the occurrence number of the index Q1 is 2. Thus, the SET statement changes the contents of the index Q2 to the offset for the second element in the table of LOWER-LIMIT elements. The offset contained in index Q2 now corresponds to the occurrence number of 2.

```
FILE SECTION.
    05  QUANTITY      PICTURE 999    OCCURS 5 TIMES INDEXED BY Q1.
    05  LOWER-LIMIT   PICTURE 99     OCCURS 5 TIMES INDEXED BY Q2.

PROCEDURE DIVISION.
    SET Q2 TO Q1.
    IF QUANTITY (Q1) IS LESS THAN LOWER-LIMIT (Q2)

        PERFORM UNDER-ROUTINE
    ELSE
        NEXT SENTENCE.
```

In the following SET statement example, the contents of index Q1 is incremented by 1. This permits the index to access each element in the table as the logic loops through the ADD statement five times.

```
FILE SECTION.

    05  QUANTITY    PICTURE 999    OCCURS 5 TIMES    INDEXED BY Q1.

PROCEDURE DIVISION.

    SET Q1 TO 1.
    PERFORM PAR-1
        UNTIL Q1 IS GREATER THAN 5.
PAR-1.
    ADD QUANTITY (Q1) TO TOTAL-QUANTITY.
    SET Q1 UP BY 1.
```

Example—SET statement.

```
SET IX-TABLE TO 3.
```

The index-name IX-TABLE is set to an index value that corresponds to occurrence number 3 for that table. If index-name IX-TABLE is not defined using an INDEXED BY phrase, the statement is illegal.

```
SET LOOKUP TO KEY-POINT.
```

If LOOKUP is an index-name, it is set to the occurrence number that corresponds to KEY-POINT. If KEY-POINT is an index-name that is not related to the same table as LOOKUP, an appropriate conversion is performed; otherwise, no conversion takes place. If LOOKUP is an index data item (that is, defined with a USAGE INDEX clause), it is set to the actual contents of KEY-POINT. KEY-POINT must be either an index-data item or an index-name or the statement is invalid.

If LOOKUP is neither an index-name nor an index data item, KEY-POINT must be an index-name. LOOKUP is then set to the occurrence value to which index-name KEY-POINT corresponds.

```
SET DEPT-INDEX UP BY KEY-JUMP.
```

The index-name DEPT-INDEX is incremented by a value that corresponds to the number of occurrences indicated by KEY-JUMP. That is, if the value of KEY-JUMP is three, DEPT-INDEX is incremented by a value that is equivalent to three occurrences.

Consider a hypothetical table of wholesale discount factors involved in determining the total price of merchandise orders, such as:

```
01  DISCOUNT-TABLE.
    05  RANGE-ENTRY           OCCURS 16 TIMES    INDEXED BY XRANGE.
        10  MAXRANGE                   PIC 9(6).
        10  CLASS-ITEM        OCCURS 12 TIMES
            INDEXED BY XCLASS          PIC 9(6).
```

Each wholesale order input record is computed against a basic price schedule catalog. An item in the wholesale order record, TOTAL-PRICE, is then compared against the discount schedule to determine which set of discount rates is to be applied to the particular order. Each occurrence of RANGE-ENTRY contains a field called MAXRANGE and twelve occurrences of CLASS-ITEM. MAXRANGE specifies the maximum order amount for which the accompanying set of CLASS-ITEM discount factors can be applied. Each CLASS-ITEM contains a discount factor for a particular class of merchandise. The coding would be:

```
SET XRANGE TO 1.

SEARCH RANGE-ENTRY
    AT END PERFORM ERROR-PROC
    WHEN MAXRANGE (XRANGE) > TOTAL-PRICE
        PERFORM RANGE-LOCATED-PROC.
```

When the search is completed, XRANGE will be set at the occurrence of RANGE-ENTRY that contains the appropriate set of discount factors.

Sample Table Handling program:

```
IDENTIFICATION DIVISION.
PROGRAM-ID.      TABLE-HANDLING-PROBLEM.
ENVIRONMENT DIVISION.
CONFIGURATION SECTION.
SOURCE-COMPUTER.     IBM-370-I155.
OBJECT-COMPUTER.     IBM-370-I155.
SPECIAL-NAMES.     CONSOLE IS TYPEWRITER.
INPUT-OUTPUT SECTION.
FILE-CONTROL.
    SELECT INFILE      ASSIGN TO UT-2400-S-INTAPE.
    SELECT OUTFILE     ASSIGN TO UR-1403-S-PRTOUT.
    SELECT INCARDS     ASSIGN TO UR-2540R-S-ICARDS.
```

```
DATA DIVISION.
FILE SECTION.
FD  INFILE     LABEL RECORDS ARE OMITTED.
01  TABLE                         PICTURE X(28200).
01  TABLE-2                       PICTURE X(1800).
FD  OUTFILE    LABEL RECORDS ARE OMITTED.
01  PRTLINE                       PICTURE X(133).
FD  INCARDS    LABEL RECORDS ARE OMITTED.
01  CARDS.
    05  STATE-NAME                PICTURE X(4).
    05  SEXCODE                   PICTURE 9.
    05  YEARCODE                  PICTURE 9(4).
    05  FILLER                    PICTURE X(71).
WORKING-STORAGE SECTION.
01  PRTAREA-20.
    05  FILLER                    PICTURE X          VALUE SPACES.
    05  YEARS-20                  PICTURE 9(4).
    05  FILLER                    PICTURE X(3)       VALUE SPACES.
    05  BIRTHS-20                 PICTURE 9(7).
    05  FILLER                    PICTURE X(3)       VALUE SPACES.
    05  DEATHS-20                 PICTURE 9(7).
    05  FILLER                    PICTURE X(108)     VALUE SPACES.
01  PRTAREA.
    05  FILLER                    PICTURE X          VALUE SPACES.
    05  YEAR                      PICTURE 9(4).
    05  FILLER                    PICTURE X(3)       VALUE SPACES.
    05  BIRTHS                    PICTURE 9(5).
    05  FILLER                    PICTURE X(3)       VALUE SPACES.
    05  DEATHS                    PICTURE 9(5).
    05  FILLER                    PICTURE X(112)     VALUE SPACES.
01  CENSUS-STATISTICS-TABLE.
    05  STATE-TABLE                   OCCURS 50 TIMES    INDEXED BY ST.
        10  STATE-ABBREV          PICTURE X(4).
        10  SEX                       OCCURS 2 TIMES     INDEXED BY SE.
            15  STATISTICS            OCCURS 20 TIMES ASCENDING KEY IS YEAR INDEXED BY YR.
                    20  YEAR      PICTURE 9(4).
                    20  BIRTHS    PICTURE 9(5).
                    20  DEATHS    PICTURE 9(5).
01  STATISTICS-LAST-20-YRS.
    05  SEX-20                        OCCURS 2 TIMES     INDEXED BY SE-20.
        10  STATE-20                  OCCURS 50 TIMES    INDEXED BY ST-20.
            15  YEARS-20          PICTURE 9(4).
            15  BIRTHS-20         PICTURE 9(7).
            15  DEATHS-20         PICTURE 9(7).
01  FLAGS.
    05  MORE-TABLE-FLAG           PICTURE XXX        VALUE 'YES'.
        88  MORE-TABLE                               VALUE 'YES'.
        88  NO-MORE-TABLE                            VALUE 'NO'.
    05  MORE-DATA-FLAG            PICTURE XXX        VALUE 'YES'.
        88  MORE-DATA                                VALUE 'YES'.
        88  NO-MORE-DATA                             VALUE 'NO'.
PROCEDURE DIVISION.
MAIN-ROUTINE
    OPEN      INPUT     INFILE
                        INCARDS
              OUTPUT    OUTFILE.
    READ INFILE INTO CENSUS-STATISTICS-TABLE
        AT END MOVE 'NO' TO MORE-TABLE-FLAG.
    READ INFILE INTO STATISTICS-LAST-20-YRS
        AT END MOVE 'NO' TO MORE-TABLE-FLAG.
```

```
    PERFORM READ-TABLE
        UNTIL NO-MORE-TABLE.
    READ INCARDS
        AT END MOVE 'NO' TO MORE-DATA-FLAG.
    PERFORM PROCESS-DATA
        UNTIL NO-MORE-DATA.
    CLOSE    INFILE
             INCARDS
             OUTFILE.
    STOP RUN.
READ-TABLE.

    READ INFILE INTO CENSUS-STATISTICS-TABLE
        AT END MOVE 'NO' TO MORE-TABLE-FLAG.
    READ INFILE INTO STATISTICS-LAST-20-YRS
        AT END MOVE 'NO' TO MORE-TABLE-FLAG.
PROCESS-DATA.
    PERFORM DETERMINE-ST.
    PERFORM DETERMINE-YR.
    READ INCARDS
        AT END MOVE 'NO' TO MORE-DATA-FLAG.

DETERMINE-ST.
    SET ST ST-20 TO 1.
    SEARCH STATE-TABLE
        VARYING ST-20
        AT END DISPLAY 'INCORRECT STATE' STATE-NAME
            UPON TYPEWRITER
            PERFORM PROCESS-DATA
        WHEN STATE-NAME = STATE-ABBREV (ST)
            NEXT SENTENCE.
    SET SE SE-20 TO SEXCODE.

DETERMINE-YR.
    SEARCH ALL STATISTICS
        AT END DISPLAY 'INCORRECT YEAR' YEARCODE
            UPON TYPEWRITER
        WHEN YEAR OF STATISTICS (ST, SE, YR) = YEARCODE
        PERFORM WRITE-RECORD.
EXIT-POINT.
    EXIT.

WRITE-RECORD.
    MOVE CORRESPONDING STATISTICS (ST, SE, YR) TO PRTAREA.
    WRITE PRTLINE FROM PRTAREA
        AFTER ADVANCING 3.
    MOVE CORRESPONDING STATE-20 (SE-20, ST-20) TO PRTAREA-20.
    WRITE PRTLINE FROM PRTAREA-20.
        AFTER ADVANCING 1.
```

The census bureau uses the program to compare:

1. The number of births and deaths that occurred in any of the 50 states in any of the past 20 years with
2. The total number of births and deaths that occurred in the same state over the entire 20-year period

The input, INCARDS, contains the specific information upon which the search of the table is to be conducted. INCARDS is formatted as follows:

STATE-NAME	a 4-character alphabetic abbreviation of the state name
SEXCODE	1 = male; 2 = female
YEARCODE	a 4-digit field in the range 1962 through 1981

A typical run might determine the number of females born in New York in 1973 as compared with the total number of females born in New York in the past 20 years.

Sample Table Handling program:

Problem Definition

The accessing of a table containing constant information is shown by the COBOL solution of a program processing data from a magnetic tape called READINGFILE. The following is the record layout of each record in the file READINGFILE.

The contents of the fields within a READINGFILE record are as follows:

RNO: account number

BEGIN-METER: beginning meter reading

END-METER: ending meter reading

RSC: service code

Data from the records of READINGFILE is used to create records written onto BILLFILE. The following is the record layout of each record in the file BILLFILE.

The contents of the fields within a BILLFILE record are as follows:

BNO: account number

BSC: service code

QUANTITY: quantity consumed

GROSS: gross dollar amount

NET: net dollar amount

The contents of the fields in a BILLFILE record are determined by the following formulas.

Quantity consumed = ending meter reading − beginning meter reading

Gross dollar amount = quantity consumed × charge rate (rounded)

Discount dollar amount = gross dollar amount × percent of discount

Net dollar amount = gross dollar amount − discount dollar amount.

The values of charge rate and percent of discount are determined from the following table.

RNO	BEGIN-METER	END-METER	RSC

Service code	Quantity consumed	Charge rate	Percent of discount
B	0100	$.055	01.5%
B	0500	$.045	05.3%
B	1000	$.032	05.3%
B	9999	$.027	05.3%
R	0050	$.060	01.5%
R	0100	$.058	03.2%
R	0200	$.051	03.2%
R	0500	$.047	03.2%
R	9999	$.040	03.2%

BNO	BSC	QUANTITY	GROSS	NET

If a zero results from subtracting beginning meter reading from ending meter reading or if the service code is not found within the table, then the READINGFILE record is to be output to the error file called BADFILE with the appropriate error code.

```
IDENTIFICATION DIVISION.
PROGRAM-ID. METER.

ENVIRONMENT DIVISION.
CONFIGURATION SECTION.
SOURCE-COMPUTER.     NCR-CENTURY-200.
OBJECT-COMPUTER.     NCR-CENTURY-200      MEMORY SIZE 3200 WORDS.

INPUT-OUTPUT SECTION.
FILE-CONTROL.
    SELECT READNGFILE       ASSIGN TO NCR-TYPE-45.
    SELECT BILLFILE         ASSIGN TO NCR-TYPE-45.
    SELECT BADFILE          ASSIGN TO NCR-TYPE-45.

*EACH FILE IS ASSIGNED TO A MAGNETIC TAPE UNIT TYPE 45.

DATA DIVISION.
FILE SECTION.
FD READNGFILE         BLOCK CONTAINS 100 RECORDS
                      RECORD CONTAINS 13 CHARACTERS
                      LABEL RECORD IS STANDARD.
01  READRECORD.
    05  RNO                     PICTURE 9999.
    05  BEGIN-METER             PICTURE 9999.
    05  END-METER               PICTURE 9999.
    05  RSC                     PICTURE X.

FD  BILLFILE                    BLOCK CONTAINS 90 RECORDS
                                RECORD CONTAINS 19 CHARACTERS
                                LABEL RECORD IS STANDARD.
01  BILLRECORD.
    05  BNO                     PICTURE 9999.
    05  BSC                     PICTURE X.
    05  QUANTITY                PICTURE 9999.
    05  GROSS                   PICTURE 999V99.
    05  NET                     PICTURE 999V99.

FD  BADFILE                     BLOCK CONTAINS 50 RECORDS
                                RECORD CONTAINS 14 CHARACTERS
                                LABEL RECORD IS STANDARD.
01  BADRECORD.
    05  BADATA                  PICTURE X(13).
    05  ECODE                   PICTURE X.
WORKING-STORAGE SECTION.
77  OPEN-CODE                   PICTURE 9         VALUE IS ZERO.
    88  BADFILE-NOT-OPEN                          VALUE IS ZERO.
    88  BADFILE-OPEN                              VALUE IS 1.
01  RATE-TABLE.
    05  FILLER                  PICTURE X(11)     VALUE 'B0100055015'.
    05  FILLER                  PICTURE X(11)     VALUE 'B0500045053'.
    05  FILLER                  PICTURE X(11)     VALUE 'B1000032053'.
    05  FILLER                  PICTURE X(11)     VALUE 'B9999027053'.
    05  FILLER                  PICTURE X(11)     VALUE 'R0050060015'.
    05  FILLER                  PICTURE X(11)     VALUE 'R0100058032'.
    05  FILLER                  PICTURE X(11)     VALUE 'R0200051032'.
    05  FILLER                  PICTURE X(11)     VALUE 'R0500047032'.
    05  FILLER                  PICTURE X(11)     VALUE 'R9999040032'.
```

```
01  RATE-TABLE1                REDEFINES RATE-TABLE.
    05  RATE-ITEM                  OCCURS 9 TIMES INDEXED BY T-INDEX.
        10  TSC                PICTURE X.
        10  TQUANTITY          PICTURE 9999.
        10  RATE               PICTURE V999.
        10  PERCENT            PICTURE V999.

01  FLAGS.
    05  MORE-DATA-FLAG         PICTURE XXX        VALUE 'YES'.
        88  MORE-DATA                             VALUE 'YES'.
        88  NO-MORE-DATA                          VALUE 'NO'.

PROCEDURE DIVISION.
MAIN-ROUTINE.
    OPEN      INPUT     READNGFILE
              OUTPUT    BILLFILE.
    READ READNGFILE
        AT END MOVE 'NO' TO MORE-DATA-FLAG.
    PERFORM READ-FILE
        UNTIL NO-MORE-DATA.
    PERFORM SEARCH-TABLE.
    CLOSE      READNGFILE
               BILLFILE.
    IF BADFILE-OPEN

        CLOSE BADFILE
    ELSE
        NEXT SENTENCE.
    STOP RUN.

SEARCH-TABLE.
    IF TSC (T-INDEX) = RSC
        AND TQUANTITY (T-INDEX) = QUANTITY

        PERFORM FOUND
    ELSE
        IF TSC (T-INDEX) = RSC
            AND TQUANTITY (T-INDEX) > QUANTITY

            PERFORM FOUND
        ELSE
            IF BADFILE-NOT-OPEN

                PERFORM OPEN-BAD
            ELSE

                NEXT SENTENCE.
        IF RSC = 'B' OR 'R'

            MOVE 'T' TO ECODE
            PERFORM CREATE-BADRECORD
        ELSE
            MOVE 'S' TO ECODE
            PERFORM CREATE-BADRECORD.

FOUND.
    MULTIPLY QUANTITY BY RATE (T-INDEX)
        GIVING GROSS ROUNDED.
    COMPUTE NET = GROSS * PERCENT (T-INDEX).
    WRITE BILLRECORD.
    PERFORM READ-FILE.
```

```
READ-FILE.
    IF END-METER IS LESS THAN BEGIN-METER

        COMPUTE QUANTITY = (1000 - BEGIN-METER) + END-METER
        PERFORM PROCESS-RECORD
    ELSE
        IF END-METER IS EQUAL TO BEGIN-METER

            PERFORM ZCODE
        ELSE
            SUBTRACT BEGIN-METER FROM END-METER
                GIVING QUANTITY
            PERFORM PROCESS-RECORD.
    READ READNGFILE
        AT END MOVE 'NO' TO MORE-DATA-FLAG.

ZCODE.
    IF BADFILE-NOT-OPEN

        PERFORM OPEN-BAD
    ELSE
        NEXT SENTENCE.
    MOVE 'Z' TO ECODE.
    PERFORM CREATE-BADRECORD.

PROCESS-RECORD.
    MOVE RNO TO BNO.
    MOVE RSC TO BSC.
    SET T-INDEX TO 1.

CREATE-BADRECORD.
    MOVE READRECORD TO BADATA.
    WRITE BADRECORD.

    PERFORM READ-FILE.

OPEN-BAD.
    OPEN OUTPUT     BADFILE.
    MOVE 1 TO OPEN-CODE.
```

Summary

A table is a collection of related items organized in such a way that each item of information can be referenced by its position within a table. Each item within a table is a *table element.*

The procedure in the checking of two elements of a table one at a time to find a particular entry is called *searching* a table. A search word is known as the *argument.* The matching entry is the corresponding entry which is known as the *function.*

When two tables are used, actually only one table is searched. The first table is used as a means of locating data in the second table. When the search condition has been satisfied, the data in the corresponding element of the second table becomes available.

In addition to speeding up the searching for data, the use of table lookup can often reduce the number of instructions needed to do a job. All that is needed to be done is to set up and define the table and to specify that the search operation is to be performed.

Subscripting is a method of specifying an occurrence number and affixing a subscript to the data-name of the table element. A *subscript* is an integer whose value identifies a particular element in a table.

Indexing is another method of specifying an occurrence number and affixing an index to the data-name of the table element. An *index* is a computer storage position or register whose contents correspond to an occurrence number.

Table values refer to any of the values contained in a table.

Search argument is any number used to locate one of the values within the table. The search judgment is used to scan the table until a table element associated with the search argument is found. The operation is called a *table search* or *table lookup*.

The table handling feature of COBOL enables the programmer to process tables or lists of repeated data conveniently. The table handling module provides a capability for defining tables of contiguous data items and for accessing an item relative to its position in the table. Each item may be identified through the use of a subscript or index. An OCCURS clause is used to specify how many times an item can be repeated.

The table handling feature provides a capability of accessing items from one- to three-dimensional variable-length tables. The feature also provides for specifying ascending or descending keys in the search of a table to satisfy a specified condition.

Tables are systematically arranged sets of information that are more limited in scope than other files. The uses of tables are: to use known information to obtain another piece of information; the verification of information with no requirement for retrieval of other data; and updating, where entries may be changed during processing and returned to their places in the table.

The advantage of using table format over a standard sequential file arrangement is that tables are compact and may be completely contained in computer memory for random processing. The entries in a table may be an argument, a function, pairs of arguments and functions, or pairs of functions and arguments.

An array is a continuous series of data fields, stored side by side, so that they can be referenced as a group. Arrays are similar to tables in the type of data that can be stored and the manner by which it is stored. The uses of tables and arrays differ considerably. Tables usually contain constant data that is used for calculation or printing with variable transaction data. Arrays are generally used for variable data and totals that are used independently of the variable transaction data. Arrays should be used instead of tables when it is necessary to reference all items at one time. The distinction between a table and an array is that an array is composed of elementary data items having identical data descriptions, while a table may be composed of both elementary and group items having differing data descriptions. Since the definition of array is a subset of the definition of table, the term "table" is used in the discussion of table handling.

Successive item positions in a table have the ordinal positions 1, 2, 3, 4 . . . , therefore any particular item can be identified by its ordinal position.

A table described in a single entry is said to be "one-dimensional"; COBOL permits the use of one-, two-, or three-dimensional tables. In one-, two-, or three-dimensional tables, the data description entries contain OCCURS clauses for the number of occurrences of each item in the table.

To define a one-dimensional table, an OCCURS clause is used as part of the data description of the table element, but the OCCURS clause must not appear in the description of the group items that contain the table element. To define a two-dimensional table, an OCCURS clause must appear in the group item that defines the table, and as part of an item which is subordinate to that group item. To define a three-dimensional table an OCCURS clause should appear in the data description of the group item that defines the table and on two additional items (one subordinate to the other).

The OCCURS and USAGE clauses are included as part of the record description entry utilizing the table handling feature. The OCCURS clause is used to define table and other homogeneous sets of repeated data. The data that is the subject of this entry must be either subscripted or indexed whenever it is referred to in a statement other than a SEARCH statement. When subscripted, the subject refers to one occurrence within a table. When not subscripted (permitted only in a SEARCH statement), the subject represents the entire table element. (A table element consists of all occurrences of one level of a table.)

The DEPENDING ON option of the OCCURS clause is only required when the number of occurrences cannot otherwise be determined. The number of times is controlled by the value of a data-name at object time.

The KEY option of the OCCURS clause is used in conjunction with the INDEXED BY option in the execution of a SEARCH ALL statement. The option is used to indicate that the repeated data is arranged in ASCENDING or DESCENDING order according to values contained in the data-name listed as describing the order of significance.

The INDEXED BY option of an OCCURS clause is required if the subject of this entry (the data name described by the OCCURS clause or an item within this data name, if it is a group item) is to be referred to by indexing. The index-name(s) identified by this clause is not defined elsewhere in the program, since its allocation and format are dependent upon the system, and not being data, cannot be associated with any data element.

The USAGE IS INDEX clause is used to specify the format of a data item in the computer storage. The clause permits the programmer to specify an index-data item. An *index-data item* is an elementary item that can be used to save index-name values for future reference. An index-data item must be assigned an index-name (occurrence number) through a SET statement.

A REDEFINES clause specifies that the same area is to contain different data items. This clause specifies the redefinition of a storage area, not the items occupying the area.

The PERFORM statement becomes a powerful mechanism for table handling when the VARYING phrase manipulates index-names or data names used as subscripts in the performed procedure. The use of the FROM and BY phrases allows a table to be created and table processing to begin at a specified table entry, and also allows for the processing of noncontiguous entries. Since three VARYING phrases may be specified, the PERFORM statement can manipulate all three dimensions of a table.

The VARYING and AFTER options of the PERFORM statement allow the programmer to manipulate up to three control items and establish separate conditional terminations of the manipulation of each item. When both the VARYING and AFTER phrases are used in the PERFORM statement, a hierarchy of tests is established. The execution of the statement becomes cyclic, that is, a VARYING operand at a lower level is varied through its full range for each step in the values of the next higher level VARYING operand. The cyclic nature of these forms of the PERFORM statement is useful in table handling because the controlling conditions can be used to test the subscript or index-names. The VARYING option is used to PERFORM a procedure repeatedly, increasing or decreasing the values of one or more identifiers or index-names once for each repetition until a specified condition is reached.

The reference to a table must indicate which occurrence of the element is intended. For access to a one-dimensional table, the occurrence number of the desired element provides complete information. For access to a table of more than one dimension, an occurrence number must be supplied for each dimension of the table. The reference should indicate either by subscripting or indexing which element of the table is intended. A table may contain as many as three dimensions that may be referenced.

A subscript is an integer whose value specifies the occurrence number of an element. Subscripts are used in the Procedure Division to reference a particular item in a table. The subscript can be represented either by a literal that is an integer or by a data-name that is defined elsewhere as a numeric elementary item with no character positions to the right of the assumed decimal point. The subscript is enclosed in parentheses and written immediately following the name of the table element. A table reference may include as many subscripts as there are dimensions in the table written in less important dimensions of the table organization. The lowest level of subscript is 1 and the highest number is the maximum occurrence number of the item specified in the OCCURS clause.

An index-name is affixed to a data item whose description includes an OCCURS clause with the optional phrase INDEXED BY. Reference is made to individual items in a table by specifying the name of the item followed by its related index-name in parentheses. Index-names may be mixed with literals but not with subscripts. An index-name has no separate data description entry since its definition is hardware oriented. An index-name can be assigned to only one table. An index-name must be initialized by a SET statement before it is used. Index-name values must not be less than one or greater than the maximum occurrence number for the table element. The index-name will correspond to an occurrence number for that specific dimension of the table with which the index-name is associated. The value of an index-name can be modified only by using the PERFORM, SEARCH and SET statements.

Index data items are data items described with the USAGE IS INDEX clause, and may be used by the SET statement to move data between the data items and index-names. Transfers are made without conversions between data items and index-names using the SET statement.

Table data can be loaded into the computer at two different times: at the time the source program is compiled (compile time) or at the beginning of the object program execution (execution time).

In the Working-Storage section, initial values of elements within a table are specified in one of the following ways:

Compile Time Tables may be described as a record set of contiguous data entries, each of which specifies the VALUE of an element, or part of an element. Any data description clause (USAGE, PICTURE, etc.) may be used to complete the entry. The hierarchical structure of the table is shown by the programmer in a REDEFINES entry. Subordinate entries following the REDEFINES clause must not contain any VALUE entries.

Execution Time Tables are usually large tables loaded at execution time from a file of records either onto a disk or main storage. The tables are loaded before the actual processing of data is to begin. These tables may be stored permanently with deletions and additions made to the tables through separate computer runs. A PERFORM statement with the VARYING option may be used to load tables at execution time.

The SEARCH statement, through its two options, provides the facilities for producing serial or non-serial (binary) searches. When using the SEARCH statement, an associated index-name or data name may be used. This statement also provides for the execution of imperative statements when certain conditions are true. The SEARCH and SET statements may be used to facilitate table handling.

Comparisons involving index-names and/or index data items actually compare the corresponding occurrence numbers.

The SEARCH statement is used to search a table for a table element that satisfies a specified condition and adjusts the associated index-name to indicate the specified position of that table element. There are two forms of the SEARCH statement, one for serial searches and the other specifying the ALL keyword for binary searches. Both forms are conditional statements. Conditional control is exercised through the required WHEN phrase and the optional AT END phrase. Only one level of a table (table element) can be referenced with one search. Format-1 is used for serial searches. Format-2, SEARCH ALL, is used for binary searches.

Format-1 of the SEARCH statement is used to perform a serial search of a table element of a one-dimensional table or a single dimension of a multidimensional table. The VARYING option may be used to step an index that is associated with the same table or another table. More than one WHEN phrase may be used to determine the termination condition for the search. If it is known that the "found" condition will come after some intermediate point or some table element, the SET statement can be used to set index-names at that point to speed up the operation to search only that part of the table.

Format-2 of the SEARCH statement is a more efficient form than format-1, since it uses a binary search technique that is much faster. However, the table must be ordered. The binary SEARCH statement is designed to search a well-ordered one-dimensional table. The table referenced in the SEARCH statement must not be defined with subscripting or indexing, but the data descriptions of the table must contain both an OCCURS clause and an INDEXED BY phrase. The data description must also include a KEY phrase in the OCCURS clause. The table must be well-ordered when the SEARCH ALL statement is used.

In both formats-1 and -2 of the SEARCH statement, when an unsuccessful statement occurs, the control is passed to the AT END phrase, which may contain an imperative sentence, or to the next sentence if the AT END phrase is absent.

The SET statement is used to establish, modify, or preserve the value associated with an index-name. This statement may be used to equate the contents of two index data items. The uses of the SET statement are:

Setting an index-name. The SET statement may be used to set the value associated with an index-name to the value of an integer literal, integer data item, or index data item, or another index-name.

Modifying an index-name. The SET statement may be used to modify the present value of an index-name by the value of an integer literal in an integer data item. In both cases, the ordinal pointer number associated with the index-name is incremented (UP BY) or decremented (DOWN BY) the number of the positions that correspond to the value of the literal or data item.

Preserving an index-name. The SET statement may be used to save the value of an index-name or an integer data item or an index data item.

Since transfers of information between index-names and index data items are made without conversion, discretion must be used when processing values from one index-name to another in index data items.

Illustrative Program: Subscript Problem

Input

Field	Positions	
Customer Number	2–5	
Item Number	6–10	
Item Cost	21–24	XX.XX
Department	30	
Customer Name	32–50	

Calculations to be Performed

Calculate the total bill for each customer by applying the appropriate discount for the department, that is, multiply the item cost by the discount factor for the department to arrive at the charge price.

Output

Print reports as follows:

D I S C O U N T   T A B L E

DEPARTMENT	DISCOUNT
1	05%
2	07%
3	10%
4	15%
5	06%
6	22%
7	12%
8	09%
9	20%

C U S T O M E R   R E P O R T

CUST. NO.	CUST. NAME	DEPT.	ITEM NO.	ITEM COST	DISCT. PER.	DISCT. AMT.	CHARGE
152	J. LANGDON	8	87653	$24.75	09%	$2.23	$22.52
152	J. LANGDON	6	64025	$9.45	22%	$2.08	$7.37
152	J. LANGDON	4	41915	$13.70	15%	$2.06	$11.64
152	J. LANGDON	1	17410	$2.51	05%	$.13	$2.38
							$43.91 *
2468	L. MORRISEY	1	18520	$3.75	05%	$.19	$3.56
2468	L. MORRISEY	2	20012	$4.20	07%	$.29	$3.91
2468	L. MORRISEY	3	31572	$10.15	10%	$1.02	$9.13
2468	L. MORRISEY	4	48792	$37.50	15%	$5.63	$31.87
2468	L. MORRISEY	5	50407	$15.15	06%	$.91	$14.24
2468	L. MORRISEY	6	61575	$20.10	22%	$4.42	$15.68
2468	L. MORRISEY	7	79204	$51.70	12%	$6.20	$45.50
2468	L. MORRISEY	8	85075	$37.84	09%	$3.41	$34.43
2468	L. MORRISEY	9	98476	$87.94	20%	$17.59	$70.35
							$228.67 *
3451	M. JACKSON	3	37847	$27.90	10%	$2.79	$25.11
3451	M. JACKSON	5	58492	$68.50	06%	$4.11	$64.39
3451	M. JACKSON	6	60010	$20.40	22%	$4.49	$15.91
3451	M. JACKSON	8	85260	$78.52	09%	$7.07	$71.45
3451	M. JACKSON	9	90520	$27.52	20%	$5.50	$22.02
							$198.88 *
4512	S. LEVITT	2	24680	$30.50	07%	$2.14	$28.36
4512	S. LEVITT	5	56784	$52.53	06%	$3.15	$49.38
4512	S. LEVITT	6	60410	$12.15	22%	$2.67	$9.48
4512	S. LEVITT	7	78952	$89.25	12%	$10.71	$78.54
4512	S. LEVITT	8	85278	$49.75	09%	$4.48	$45.27
4512	S. LEVITT	8	87492	$64.25	09%	$5.78	$58.47

C U S T O M E R   R E P O R T

CUST. NO.	CUST. NAME	DEPT.	ITEM NO.	ITEM COST	DISCT. PER.	DISCT. AMT.	CHARGE
4512	S. LEVITT	9	97204	$84.75	20%	$16.95	$67.80
							$337.30 *
5417	K. CONKLIN	1	13579	$35.72	05%	$1.79	$33.93
5417	K. CONKLIN	2	24615	$18.75	07%	$1.31	$17.44
5417	K. CONKLIN	3	34928	$37.45	10%	$3.75	$33.70
5417	K. CONKLIN	4	48527	$87.50	15%	$13.13	$74.37
5417	K. CONKLIN	5	50150	$18.95	06%	$1.14	$17.81
5417	K. CONKLIN	5	54652	$38.92	06%	$2.34	$36.58
5417	K. CONKLIN	5	59765	$98.95	06%	$5.94	$93.01
5417	K. CONKLIN	7	71572	$18.95	12%	$2.27	$16.68
5417	K. CONKLIN	8	85175	$80.10	09%	$7.21	$72.89
5417	K. CONKLIN	9	90275	$4.60	20%	$.92	$3.68
5417	K. CONKLIN	9	91572	$18.57	20%	$3.71	$14.86
5417	K. CONKLIN	9	97576	$84.95	20%	$16.99	$67.96
							$482.91 *
6213	Z. HAMPTON	1	15792	$64.25	05%	$3.21	$61.04
6213	Z. HAMPTON	1	19975	$98.75	05%	$4.94	$93.81
6213	Z. HAMPTON	3	34576	$51.15	10%	$5.12	$46.03
6213	Z. HAMPTON	4	49512	$85.20	15%	$12.78	$72.42
							$273.30 *
7545	M. LARSON	1	14676	$38.45	05%	$1.92	$36.53
7545	M. LARSON	1	18592	$82.51	05%	$4.13	$78.38
7545	M. LARSON	1	19994	$98.98	05%	$4.95	$94.03
7545	M. LARSON	2	21214	$15.15	07%	$1.06	$14.09
7545	M. LARSON	3	37515	$82.12	10%	$8.21	$73.91
7545	M. LARSON	3	38592	$96.15	10%	$9.62	$86.53
7545	M. LARSON	4	48485	$87.14	15%	$13.07	$74.07
7545	M. LARSON	5	52762	$37.92	06%	$2.28	$35.64
7545	M. LARSON	5	57684	$80.15	06%	$4.81	$75.34
7545	M. LARSON	7	79015	$96.25	12%	$11.55	$84.70
7545	M. LARSON	8	80123	$5.60	09%	$.50	$5.10
7545	M. LARSON	8	82462	$20.15	09%	$1.81	$18.34
7545	M. LARSON	9	91520	$18.15	20%	$3.63	$14.52
7545	M. LARSON	9	93715	$40.15	20%	$8.03	$32.12
							$723.30 *

Printer Spacing Chart

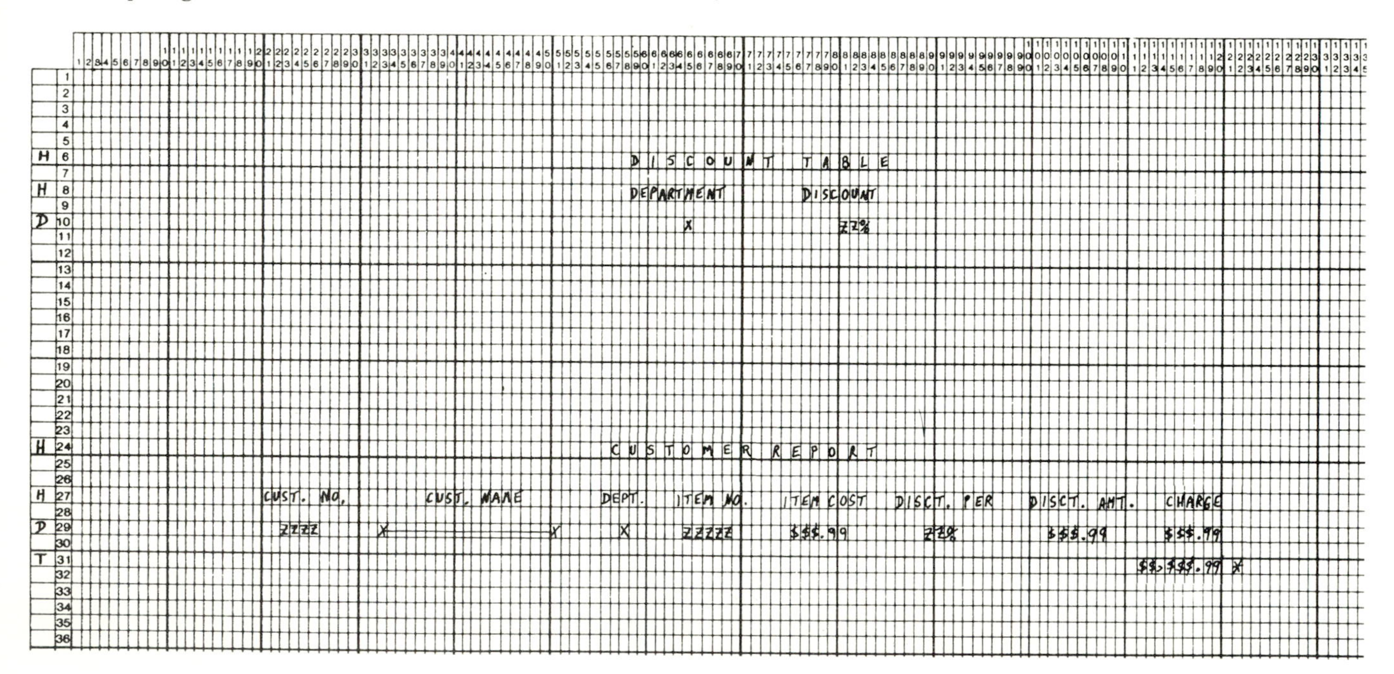

Hierarchy Chart

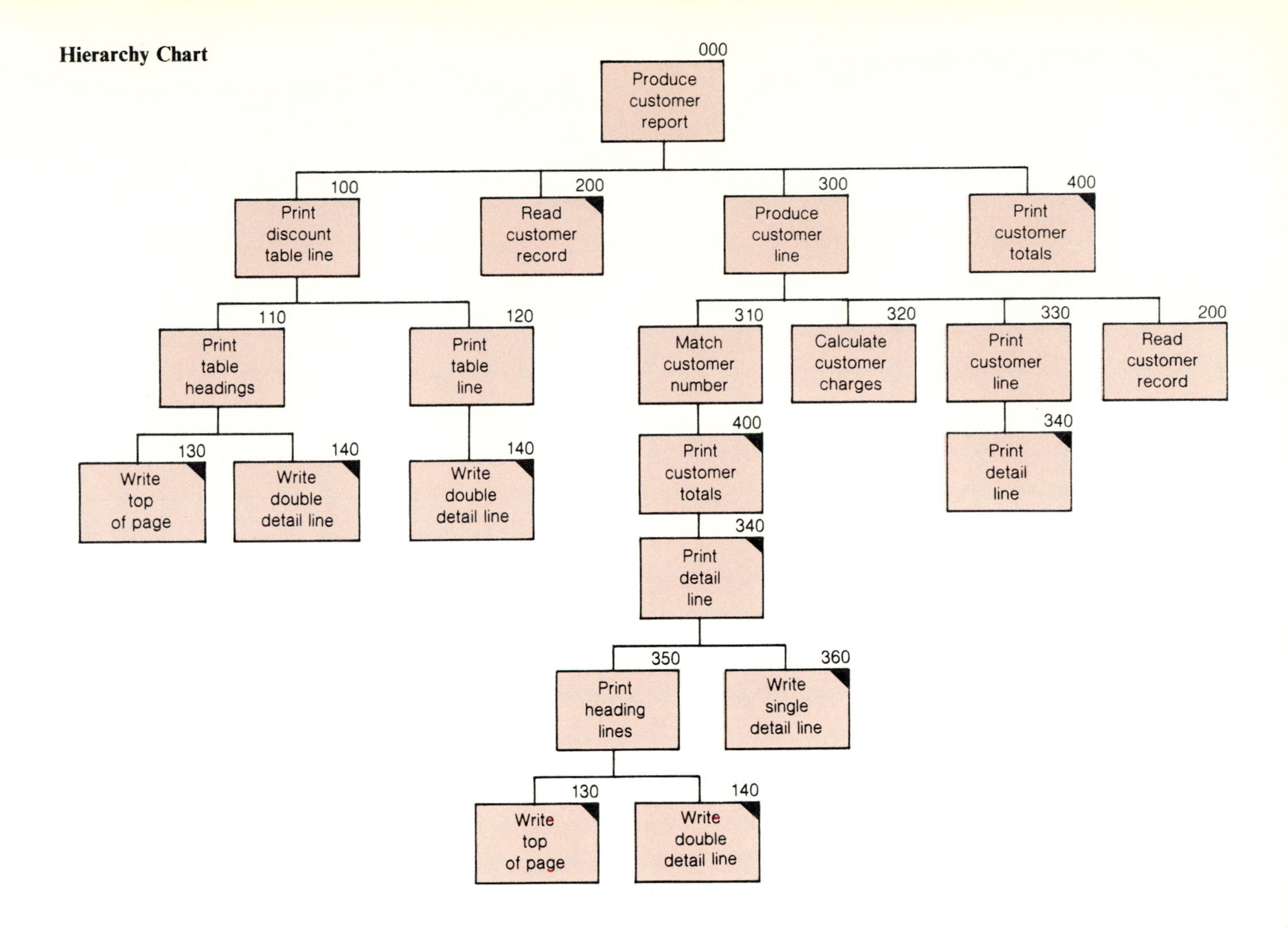

IPO Charts

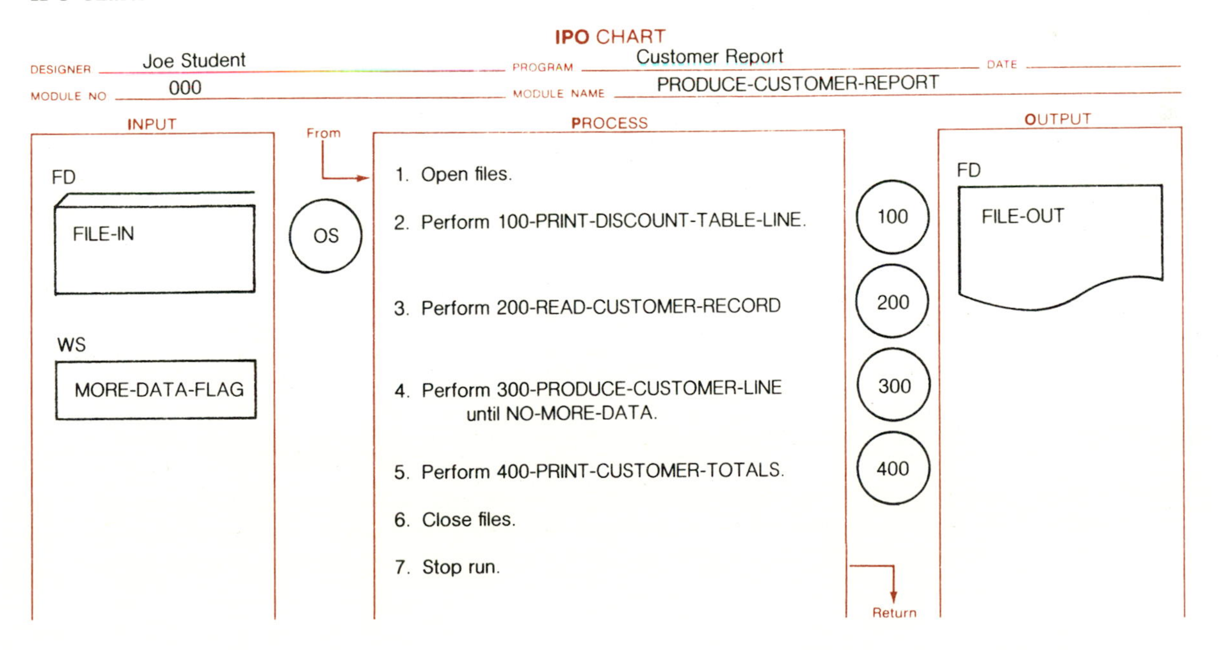

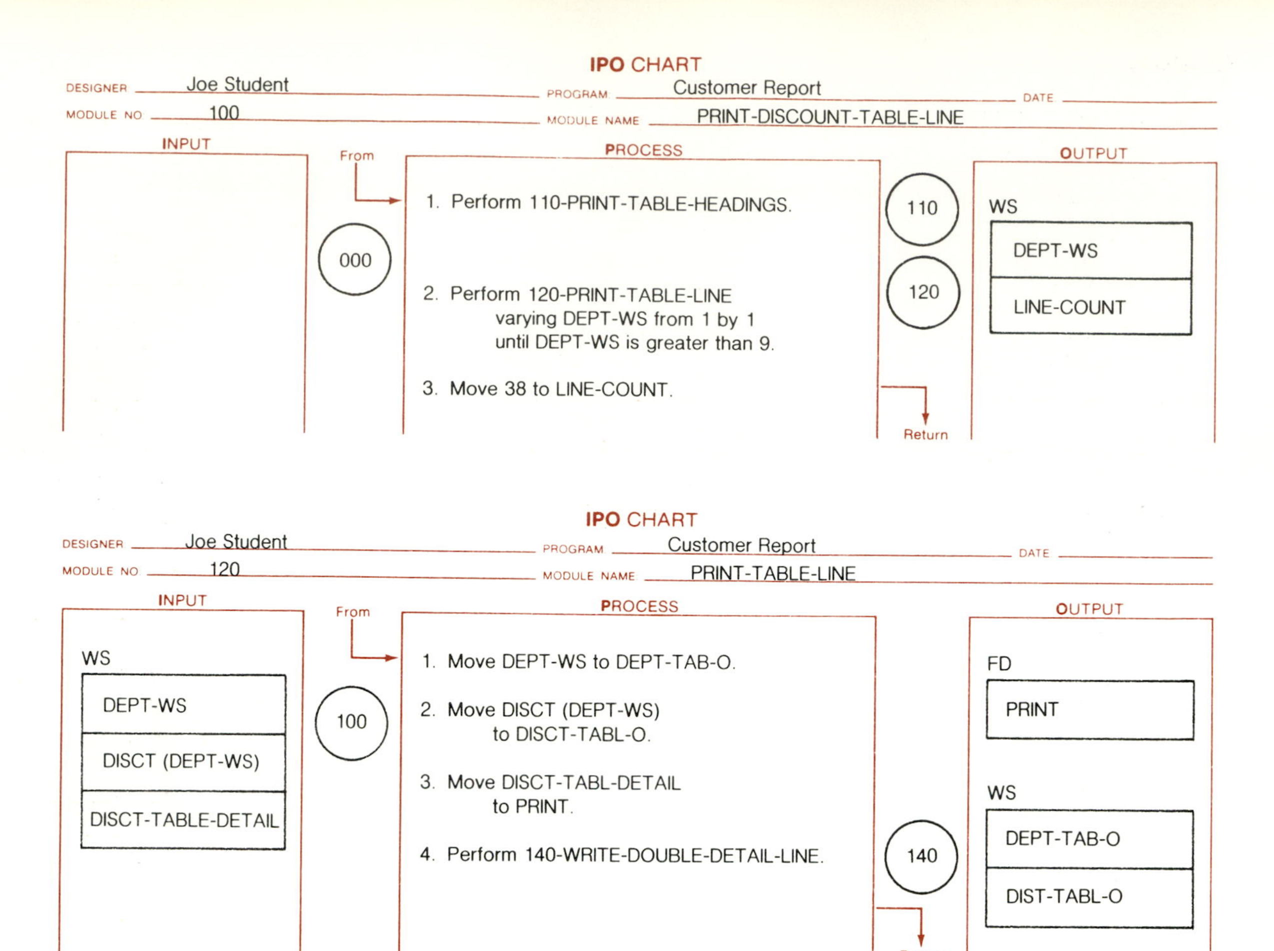
IPO CHART

DESIGNER: Joe Student
PROGRAM: Customer Report
DATE:
MODULE NO: 100
MODULE NAME: PRINT-DISCOUNT-TABLE-LINE

INPUT	From	PROCESS		OUTPUT
	000	1. Perform 110-PRINT-TABLE-HEADINGS.	110	WS
		2. Perform 120-PRINT-TABLE-LINE varying DEPT-WS from 1 by 1 until DEPT-WS is greater than 9.	120	DEPT-WS
		3. Move 38 to LINE-COUNT.	Return	LINE-COUNT

IPO CHART

DESIGNER: Joe Student
PROGRAM: Customer Report
DATE:
MODULE NO: 120
MODULE NAME: PRINT-TABLE-LINE

INPUT	From	PROCESS		OUTPUT
WS		1. Move DEPT-WS to DEPT-TAB-O.		FD
DEPT-WS	100	2. Move DISCT (DEPT-WS) to DISCT-TABL-O.		PRINT
DISCT (DEPT-WS)		3. Move DISCT-TABL-DETAIL to PRINT.		WS
DISCT-TABLE-DETAIL		4. Perform 140-WRITE-DOUBLE-DETAIL-LINE.	140	DEPT-TAB-O
			Return	DIST-TABL-O

Flowcharts

000-PRODUCE-CUSTOMER-REPORT

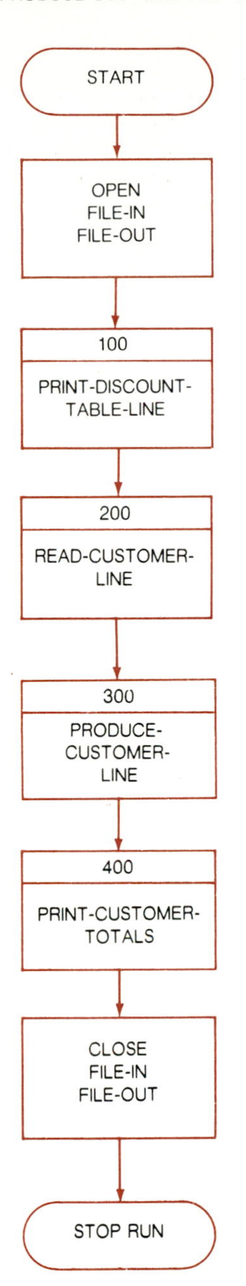

100-PRINT-DISCOUNT-TABLE-LINE

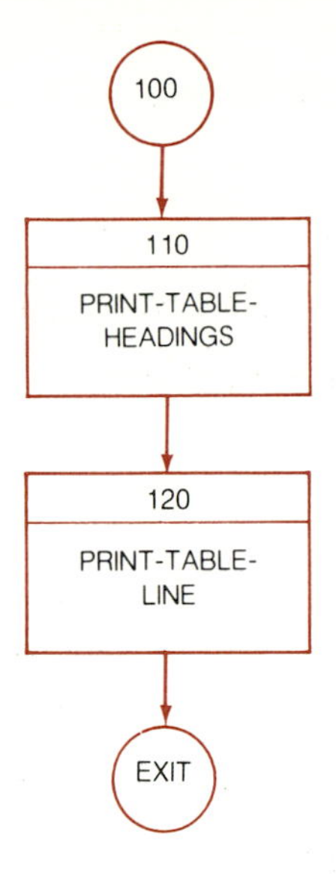

120-PRINT-TABLE-LINE

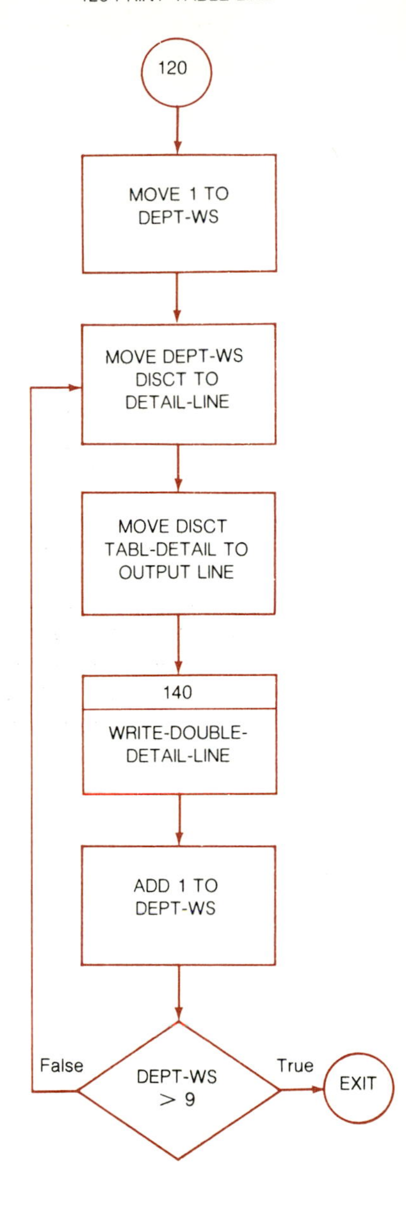

Source Program

```
 1        *************************
 2
 3         IDENTIFICATION DIVISION.
 4
 5        *************************
 6
 7         PROGRAM-ID.
 8             SUBSCRIPT-PROBLEM.
 9         INSTALLATION.
10             WEST LOS ANGELES COLLEGE.
11         DATE-WRITTEN.
12             08 FEBRUARY 1982.
13
14        *PURPOSE.
15        *    THIS PROGRAM PREPARES A DISCOUNT TABLE AND CUSTOMER REPORT,
16        *        CALCULATING THE TOTAL BILL FOR EACH CUSTOMER.
17
```

```
 18        *************************
 19
 20         ENVIRONMENT DIVISION.
 21
 22        *************************
 23
 24         CONFIGURATION SECTION.
 25
 26         SOURCE-COMPUTER.
 27             LEVEL-66-ASCII.
 28         OBJECT-COMPUTER.
 29             LEVEL-66-ASCII.
 30
 31         INPUT-OUTPUT SECTION.
 32
 33         FILE-CONTROL.
 34             SELECT FILE-IN      ASSIGN TO C1-CARD-READER.
 35             SELECT FILE-OUT     ASSIGN TO P1-PRINTER.
 36
 37        *************************
 38
 39         DATA DIVISION.
 40
 41
 42        *************************
 43
 44
 45         FILE SECTION.
 46
 47         FD  FILE-IN
 48             CODE-SET IS GBCD
 49             LABEL RECORDS ARE OMITTED
 50             DATA RECORD IS CARDIN.
 51
 52         01  CARDIN.
 53
 54             03  FILLER                PICTURE X.
 55             03  CUSTOMER-NUMBER       PICTURE 9(4).
 56             03  ITEM-NUMBER           PICTURE 9(5).
 57             03  FILLER                PICTURE X(10).
 58             03  ITEM-COST             PICTURE 99V99.
 59             03  FILLER                PICTURE X(5).
 60             03  DEPARTMENT            PICTURE 9.
 61             03  FILLER                PICTURE X.
 62             03  CUSTOMER-NAME         PICTURE X(19).
 63             03  FILLER                PICTURE X(30).
 64
 65         FD  FILE-OUT
 66             CODE-SET IS GBCD
 67             LABEL RECORDS ARE OMITTED
 68             DATA RECORD IS PRINT.
 69
 70         01  PRINT                     PICTURE X(132).
 71
 72         WORKING-STORAGE SECTION.
 73
 74         77  DEPT-WS                   PICTURE 9(2)     VALUE 1.
 75         77  CUST-NO-WS                PICTURE 9(4)     VALUE ZEROES.
 76         77  DISCT-WS                  PICTURE 99V99.
 77         77  CHARGE-WS                 PICTURE 99V99.
 78         77  TOTL-CHG-WS               PICTURE 9999V99  VALUE ZEROES.
 79         77  LINE-COUNT                PICTURE 99.
 80
 81         01  FLAGS.
 82
 83             03  MORE-DATA-FLAG        PICTURE XXX      VALUE "YES".
 84                 88  MORE-DATA                          VALUE "YES".
 85                 88  NO-MORE-DATA                       VALUE "NO ".
 86
 87         01  TABLE-DISCOUNT.
 88
 89             03  DISCT-TAB.
 90
 91                 05  FILLER            PICTURE V99      VALUE .05.
 92                 05  FILLER            PICTURE V99      VALUE .07.
 93                 05  FILLER            PICTURE V99      VALUE .10.
 94                 05  FILLER            PICTURE V99      VALUE .15.
 95                 05  FILLER            PICTURE V99      VALUE .06.
 96                 05  FILLER            PICTURE V99      VALUE .22.
 97                 05  FILLER            PICTURE V99      VALUE .12.
 98                 05  FILLER            PICTURE V99      VALUE .09.
 99                 05  FILLER            PICTURE V99      VALUE .20.
100
101             03  DISCT-TABLE REDEFINES DISCT-TAB.
102
103                 05  DISCT             PICTURE V99      OCCURS 9 TIMES.
104
105         01  HEADING-D-1.
106
107             03  FILLER                PICTURE X(58)    VALUE SPACES.
108             03  FILLER                PICTURE X(18)    VALUE "D I S C O U N
109        -                                                 "T   ".
110             03  FILLER                PICTURE X(9)     VALUE "T A B L E".
111             03  FILLER                PICTURE X(47)    VALUE SPACES.
112
113         01  HEADING-D-2.
114
115             03  FILLER                PICTURE X(58)    VALUE SPACES.
116             03  FILLER                PICTURE X(10)    VALUE "DEPARTMENT".
117             03  FILLER                PICTURE X(8)     VALUE SPACES.
118             03  FILLER                PICTURE X(8)     VALUE "DISCOUNT".
119             03  FILLER                PICTURE X(48)    VALUE SPACES.
120
```

```
121        01  HEADING-C-1.
122
123            03  FILLER                PICTURE X(56)   VALUE SPACES.
124            03  FILLER                PICTURE X(17)   VALUE "C U S T O M E
125    -                                                       "R  ".
126            03  FILLER                PICTURE X(11)   VALUE "R E P O R T".
127            03  FILLER                PICTURE X(48)   VALUE SPACES.
128
129        01  HEADING-C-2.
130
131            03  FILLER                PICTURE X(20)   VALUE SPACES.
132            03  FILLER                PICTURE X(9)    VALUE "CUST. NO.".
133            03  FILLER                PICTURE X(8)    VALUE SPACES.
134            03  FILLER                PICTURE X(10)   VALUE "CUST. NAME".
135            03  FILLER                PICTURE X(8)    VALUE SPACES.
136            03  FILLER                PICTURE X(5)    VALUE "DEPT.".
137            03  FILLER                PICTURE X(3)    VALUE SPACES.
138            03  FILLER                PICTURE X(8)    VALUE "ITEM NO.".
139            03  FILLER                PICTURE X(3)    VALUE SPACES.
140            03  FILLER                PICTURE X(9)    VALUE "ITEM COST".
141            03  FILLER                PICTURE X(3)    VALUE SPACES.
142            03  FILLER                PICTURE X(11)   VALUE "DISCT. PER.".
143            03  FILLER                PICTURE X(3)    VALUE SPACES.
144            03  FILLER                PICTURE X(11)   VALUE "DISCT. AMT.".
145            03  FILLER                PICTURE X(3)    VALUE SPACES.
146            03  FILLER                PICTURE X(6)    VALUE "CHARGE".
147            03  FILLER                PICTURE X(12)   VALUE SPACES.
148
149        01  DISCT-TABL-DETAIL.
150
151            03  FILLER                PICTURE X(60)   VALUE SPACES.
152            03  DEPT-TAB-O            PICTURE Z9.
153            03  FILLER                PICTURE X(18)   VALUE SPACES.
154            03  DISCT-TABL-O          PICTURE VZ9.
155            03  FILLER                PICTURE X       VALUE "%".
156            03  FILLER                PICTURE X(48)   VALUE SPACES.
157
158        01  CUST-RPRT-DETAIL.
159
160            03  FILLER                PICTURE X(22)   VALUE SPACES.
161            03  CUSTOMER-NUMBER-O     PICTURE Z(4).
162            03  FILLER                PICTURE X(6)    VALUE SPACES.
163            03  CUSTOMER-NAME-O       PICTURE X(19).
164            03  FILLER                PICTURE X(6)    VALUE SPACES.
165            03  DEPARTMENT-O          PICTURE 9.
166            03  FILLER                PICTURE X(6)    VALUE SPACES.
167            03  ITEM-NUMBER-O         PICTURE Z(5).
168            03  FILLER                PICTURE X(6)    VALUE SPACES.
169            03  ITEM-COST-O           PICTURE $$$.99.
170            03  FILLER                PICTURE X(8)    VALUE SPACES.
171            03  DISCT-O               PICTURE VZ9.
172            03  FILLER                PICTURE X       VALUE "%".
173            03  FILLER                PICTURE X(10)   VALUE SPACES.
174            03  DISCT-AMT-O           PICTURE $$$.99.
175            03  FILLER                PICTURE X(6)    VALUE SPACES.
176            03  CHARGE-O              PICTURE $$$.99.
177            03  FILLER                PICTURE X(12)   VALUE SPACES.
178
179        01  CUST-RPRT-TOTAL.
180
181            03  FILLER                PICTURE X(111)  VALUE SPACES.
182            03  TOTAL-CHARGE-O        PICTURE $$,$$$.99.
183            03  FILLER                PICTURE X(2)    VALUE " *".
184            03  FILLER                PICTURE X(10)   VALUE SPACES.
185
186        01  SAVE-RECORD               PICTURE X(132).
187
188       *************************
189
190        PROCEDURE DIVISION.
191
192       *************************
193
194        000-PRODUCE-CUSTOMER-REPORT.
195            OPEN    INPUT   FILE-IN
196                    OUTPUT  FILE-OUT.
197            PERFORM 100-PRINT-DISCOUNT-TABLE-LINE.
198            PERFORM 200-READ-CUSTOMER-RECORD.
199            PERFORM 300-PRODUCE-CUSTOMER-LINE
200                    UNTIL NO-MORE-DATA.
201            PERFORM 400-PRINT-CUSTOMER-TOTALS.
202            CLOSE   FILE-IN
203                    FILE-OUT.
204            STOP RUN.
205
206        100-PRINT-DISCOUNT-TABLE-LINE.
207
208            PERFORM 110-PRINT-TABLE-HEADINGS.
209            PERFORM 120-PRINT-TABLE-LINE
210                VARYING DEPT-WS FROM 1 BY 1
211                UNTIL DEPT-WS > 9.
212            MOVE 38 TO LINE-COUNT.
213
214        110-PRINT-TABLE-HEADINGS.
215
216            MOVE HEADING-D-1 TO PRINT.
217            PERFORM 130-WRITE-TOP-OF-PAGE.
218            MOVE HEADING-D-2 TO PRINT.
219            PERFORM 140-WRITE-DOUBLE-DETAIL-LINE.
220
```

```
221         120-PRINT-TABLE-LINE.
222
223             MOVE DEPT-WS TO DEPT-TAB-O.
224             MOVE DISCT(DEPT-WS) TO DISCT-TABL-O.
225             MOVE DISCT-TABL-DETAIL TO PRINT.
226             PERFORM 140-WRITE-DOUBLE-DETAIL-LINE.
227
228         130-WRITE-TOP-OF-PAGE.
229
230             WRITE PRINT
231                 AFTER ADVANCING PAGE.
232             MOVE 1 TO LINE-COUNT.
233
234         140-WRITE-DOUBLE-DETAIL-LINE.
235
236             WRITE PRINT
237                 AFTER ADVANCING 2 LINES.
238             ADD 2 TO LINE-COUNT.
239
240         200-READ-CUSTOMER-RECORD.
241
242             READ FILE-IN
243                 AT END MOVE "NO " TO MORE-DATA-FLAG.
244
245         300-PRODUCE-CUSTOMER-LINE.
246
247             PERFORM 310-MATCH-CUSTOMER-NUMBER.
248             PERFORM 320-CALCULATE-CUSTOMER-CHARGES.
249             PERFORM 330-PRINT-CUSTOMER-LINE.
250             PERFORM 200-READ-CUSTOMER-RECORD.
251
252         310-MATCH-CUSTOMER-NUMBER.
253
254             IF CUST-NO-WS IS = 0
255                 MOVE CUSTOMER-NUMBER TO CUST-NO-WS
256             ELSE
257                 NEXT SENTENCE.
258             IF CUSTOMER-NUMBER IS > CUST-NO-WS
259                 PERFORM 400-PRINT-CUSTOMER-TOTALS
260             ELSE
261                 NEXT SENTENCE.
262
263         320-CALCULATE-CUSTOMER-CHARGES.
264
265             COMPUTE DISCT-WS ROUNDED =
266                 ITEM-COST * DISCT(DEPARTMENT).
267             COMPUTE CHARGE-WS ROUNDED =
268                 ITEM-COST - DISCT-WS.
269             ADD CHARGE-WS TO TOTL-CHG-WS.
270
271         330-PRINT-CUSTOMER-LINE.
272
273             MOVE CUSTOMER-NUMBER     TO CUSTOMER-NUMBER-O.
274             MOVE ITEM-NUMBER         TO ITEM-NUMBER-O.
275             MOVE ITEM-COST           TO ITEM-COST-O.
276             MOVE DEPARTMENT          TO DEPARTMENT-O.
277             MOVE CUSTOMER-NAME       TO CUSTOMER-NAME-O.
278             MOVE DISCT (DEPARTMENT)  TO DISCT-O.
279             MOVE DISCT-WS            TO DISCT-AMT-O.
280             MOVE CHARGE-WS           TO CHARGE-O.
281             MOVE CUST-RPRT-DETAIL    TO PRINT.
282             PERFORM 340-PRINT-DETAIL-LINE.
283
284         340-PRINT-DETAIL-LINE.
285
286             IF LINE-COUNT IS > 37
287                 MOVE PRINT TO SAVE-RECORD
288                 PERFORM 350-PRINT-HEADING-LINES
289                 MOVE SAVE-RECORD TO PRINT
290             ELSE
291                 NEXT SENTENCE.
292             PERFORM 360-WRITE-SINGLE-DETAIL-LINE.
293
294         350-PRINT-HEADING-LINES.
295
296             MOVE HEADING-C-1 TO PRINT.
297             PERFORM 130-WRITE-TOP-OF-PAGE.
298             MOVE HEADING-C-2 TO PRINT.
299             PERFORM 140-WRITE-DOUBLE-DETAIL-LINE.
300             MOVE SPACES TO PRINT.
301             PERFORM 140-WRITE-DOUBLE-DETAIL-LINE.
302
303         360-WRITE-SINGLE-DETAIL-LINE.
304
305             WRITE PRINT
306                 AFTER ADVANCING 1 LINE.
307             ADD 1 TO LINE-COUNT.
308
309         400-PRINT-CUSTOMER-TOTALS.
310
311             MOVE SPACES TO PRINT.
312             PERFORM 340-PRINT-DETAIL-LINE.
313             MOVE TOTL-CHG-WS TO TOTAL-CHARGE-O.
314             MOVE CUST-RPRT-TOTAL TO PRINT.
315             PERFORM 340-PRINT-DETAIL-LINE.
316             MOVE SPACES TO PRINT.
317             PERFORM 340-PRINT-DETAIL-LINE.
318             MOVE CUSTOMER-NUMBER TO CUST-NO-WS.
319             MOVE ZEROES TO TOTL-CHG-WS.
320
```

D I S C O U N T   T A B L E

DEPARTMENT	DISCOUNT
1	05%
2	07%
3	10%
4	15%
5	06%
6	22%
7	12%
8	09%
9	20%

C U S T O M E R   R E P O R T

CUST. NO.	CUST. NAME	DEPT.	ITEM NO.	ITEM COST	DISCT. PER.	DISCT. AMT.	CHARGE
152	J. LANGDON	8	87653	$24.75	09%	$2.23	$22.52
152	J. LANGDON	6	64025	$9.45	22%	$2.08	$7.37
152	J. LANGDON	4	41915	$13.70	15%	$2.06	$11.64
152	J. LANGDON	1	17410	$2.51	05%	$.13	$2.38
							$43.91 *
2468	L. MORRISEY	1	18520	$3.75	05%	$.19	$3.56
2468	L. MORRISEY	2	20012	$4.20	07%	$.29	$3.91
2468	L. MORRISEY	3	31572	$10.15	10%	$1.02	$9.13
2468	L. MORRISEY	4	48792	$37.50	15%	$5.63	$31.87
2468	L. MORRISEY	5	50407	$15.15	06%	$.91	$14.24
2468	L. MORRISEY	6	61575	$20.10	22%	$4.42	$15.68
2468	L. MORRISEY	7	79204	$51.70	12%	$6.20	$45.50
2468	L. MORRISEY	8	85075	$37.84	09%	$3.41	$34.43
2468	L. MORRISEY	9	98476	$87.94	20%	$17.59	$70.35
							$228.67 *
3451	M. JACKSON	3	37847	$27.90	10%	$2.79	$25.11
3451	M. JACKSON	5	58492	$68.50	06%	$4.11	$64.39
3451	M. JACKSON	6	60010	$20.40	22%	$4.49	$15.91
3451	M. JACKSON	8	85260	$78.52	09%	$7.07	$71.45
3451	M. JACKSON	9	90520	$27.52	20%	$5.50	$22.02
							$198.88 *
4512	S. LEVITT	2	24680	$30.50	07%	$2.14	$28.36
4512	S. LEVITT	5	56784	$52.53	06%	$3.15	$49.38
4512	S. LEVITT	6	60410	$12.15	22%	$2.67	$9.48
4512	S. LEVITT	7	78952	$89.25	12%	$10.71	$78.54
4512	S. LEVITT	8	85278	$49.75	09%	$4.48	$45.27
4512	S. LEVITT	8	87492	$64.25	09%	$5.78	$58.47
4512	S. LEVITT	9	97204	$84.75	20%	$16.95	$67.80
							$337.30 *
5417	K. CONKLIN	1	13579	$35.72	05%	$1.79	$33.93
5417	K. CONKLIN	2	24615	$18.75	07%	$1.31	$17.44
5417	K. CONKLIN	3	34928	$37.45	10%	$3.75	$33.70
5417	K. CONKLIN	4	48527	$87.50	15%	$13.13	$74.37
5417	K. CONKLIN	5	50150	$18.95	06%	$1.14	$17.81
5417	K. CONKLIN	5	54652	$38.92	06%	$2.34	$36.58
5417	K. CONKLIN	5	59765	$98.95	06%	$5.94	$93.01
5417	K. CONKLIN	7	71572	$18.95	12%	$2.27	$16.68
5417	K. CONKLIN	8	85175	$80.10	09%	$7.21	$72.89
5417	K. CONKLIN	9	90275	$4.60	20%	$.92	$3.68
5417	K. CONKLIN	9	91572	$18.57	20%	$3.71	$14.86
5417	K. CONKLIN	9	97576	$84.95	20%	$16.99	$67.96
							$482.91 *
6213	Z. HAMPTON	1	15792	$64.25	05%	$3.21	$61.04
6213	Z. HAMPTON	1	19975	$98.75	05%	$4.94	$93.81
6213	Z. HAMPTON	3	34576	$51.15	10%	$5.12	$46.03
6213	Z. HAMPTON	4	49512	$85.20	15%	$12.78	$72.42
							$273.30 *
7545	M. LARSON	1	14676	$38.45	05%	$1.92	$36.53
7545	M. LARSON	1	18592	$82.51	05%	$4.13	$78.38
7545	M. LARSON	1	19994	$98.98	05%	$4.95	$94.03
7545	M. LARSON	2	21214	$15.15	07%	$1.06	$14.09
7545	M. LARSON	3	37515	$82.12	10%	$8.21	$73.91
7545	M. LARSON	3	38592	$96.15	10%	$9.62	$86.53
7545	M. LARSON	4	48485	$87.14	15%	$13.07	$74.07
7545	M. LARSON	5	52762	$37.92	06%	$2.28	$35.64
7545	M. LARSON	5	57684	$80.15	06%	$4.81	$75.34
7545	M. LARSON	7	79015	$96.25	12%	$11.55	$84.70
7545	M. LARSON	8	80123	$5.60	09%	$.50	$5.10
7545	M. LARSON	8	82462	$20.15	09%	$1.81	$18.34
7545	M. LARSON	9	91520	$18.15	20%	$3.63	$14.52
7545	M. LARSON	9	93715	$40.15	20%	$8.03	$32.12
							$723.30 *

Job Definition

Print a report listing the name of the taxpayer with the proper tax deduction for miles driven during the year based on the table.

Input

A record for each taxpayer contains the following information:

Positions	Field Designation
1–9	Social security number
10–29	Name
36–41	Mileage driven

Processing

Set up a compile time table for the tax deductions based on the following information:

State Gasoline Tax Table

Nonbusiness Miles Driven	Tax Rate
Under 3,000	$12
3,000 to 3,499	19
3,500 to 3,999	22
4,000 to 4,499	25
4,500 to 4,999	28
5,000 to 5,499	30
5,500 to 5,999	33
6,000 to 6,499	36
6,500 to 6,999	39
7,000 to 7,499	42
7,500 to 7,999	45
8,000 to 8,499	48
8,500 to 8,999	51
9,000 to 9,499	53
9,500 to 9,999	56
10,000 to 10,999	61
11,000 to 11,999	67
12,000 to 12,999	72
13,000 to 13,999	78
14,000 to 14,999	84
15,000 to 15,999	90
16,000 to 16,999	95
17,000 to 17,999	101
18,000 to 18,999	107
19,000 to 19,999	113
20,000 miles*	116

Using the aforementioned table, determine the tax deduction for the number of miles driven.

Perform a table lookup to determine the tax deduction based on the number of miles driven.

*For over 20,000 miles, use table amounts for total miles driven. For example, for 25,000 miles, add the deduction for 5,000 to the deduction for 20,000 miles.

Output

A Tax Deduction Report is to be printed as follows.

TAX DEDUCTION REPORT

SOC. SEC. NUMBER		MILES DRIVEN	GASOLINE TAX DED
556-98-3021	BELL,ERNEST	75,362	$438.00
556-94-2447	CHATMAN,CARLA E	13,471	$ 78.00
011-30-9013	DINIZ,FRANCISCO J	6,314	$ 36.00
330-52-8528	HODGES,MICHAEL A	36,293	$211.00
556-13-8842	JOHNSON,NIMROD J	17,954	$101.00
560-66-7910	NELSON,RAYMOND R	27,901	$161.00
556-96-4996	POWELL,BOBBIE J	104,727	$608.00
573-02-1310	SANFORD,MICHAEL M	400	$ 12.00
573-98-9084	TAKAYAMA,GARY M	13,017	$ 78.00
563-96-2106	TRAVIS,HIRAM S	8,520	$ 51.00
573-93-5201	WAGGONER,LARRY	18,791	$107.00
574-70-2075	WINGERT,JOSEPH R	12,001	$ 72.00

Printer Spacing Chart

This Printer Spacing Chart shows how the report is formatted:

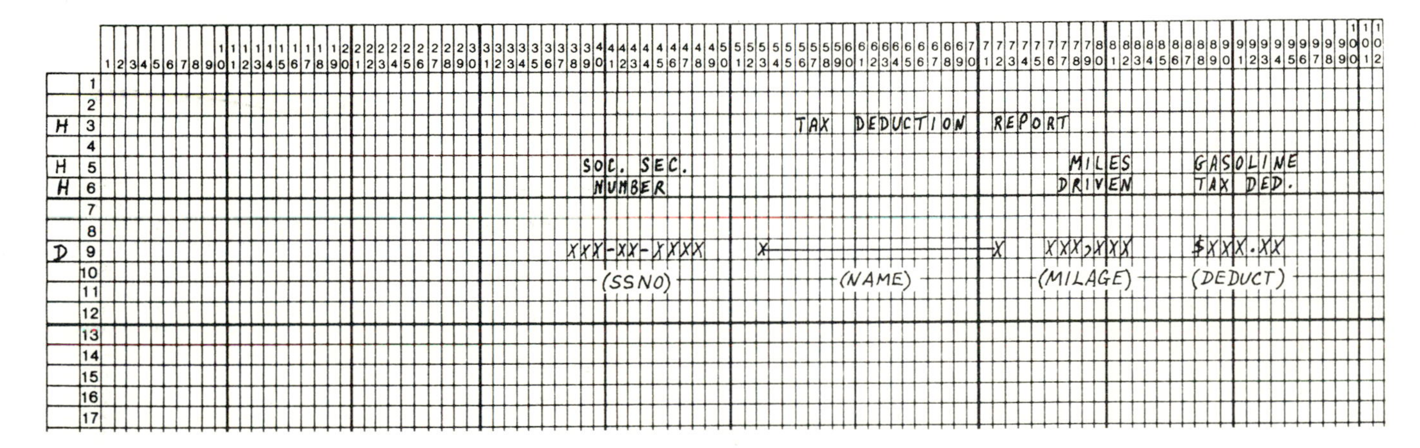

Hierarchy Chart

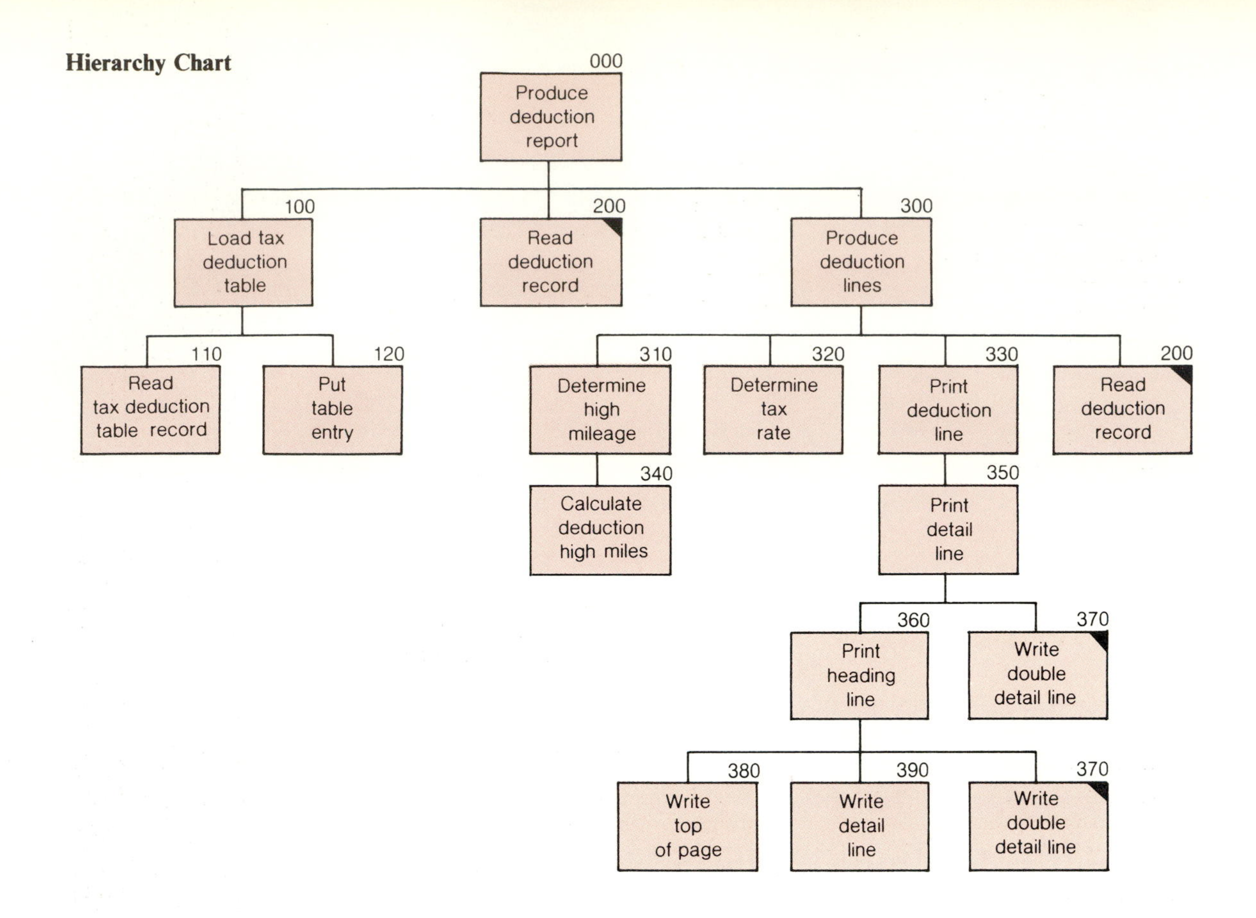

IPO Charts

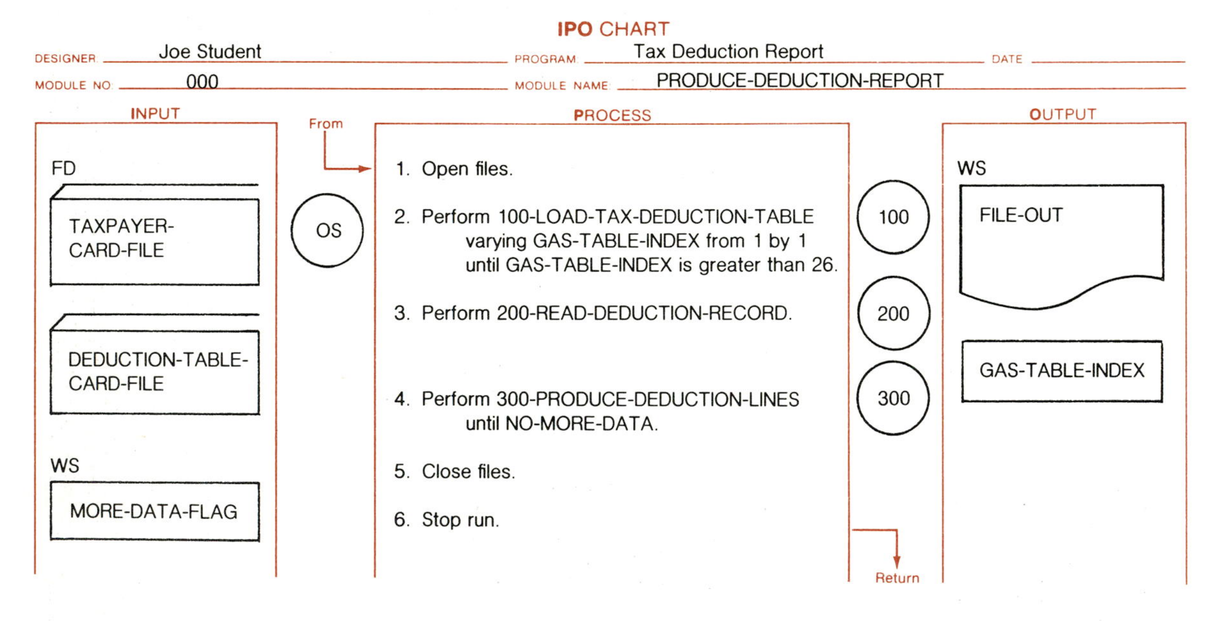

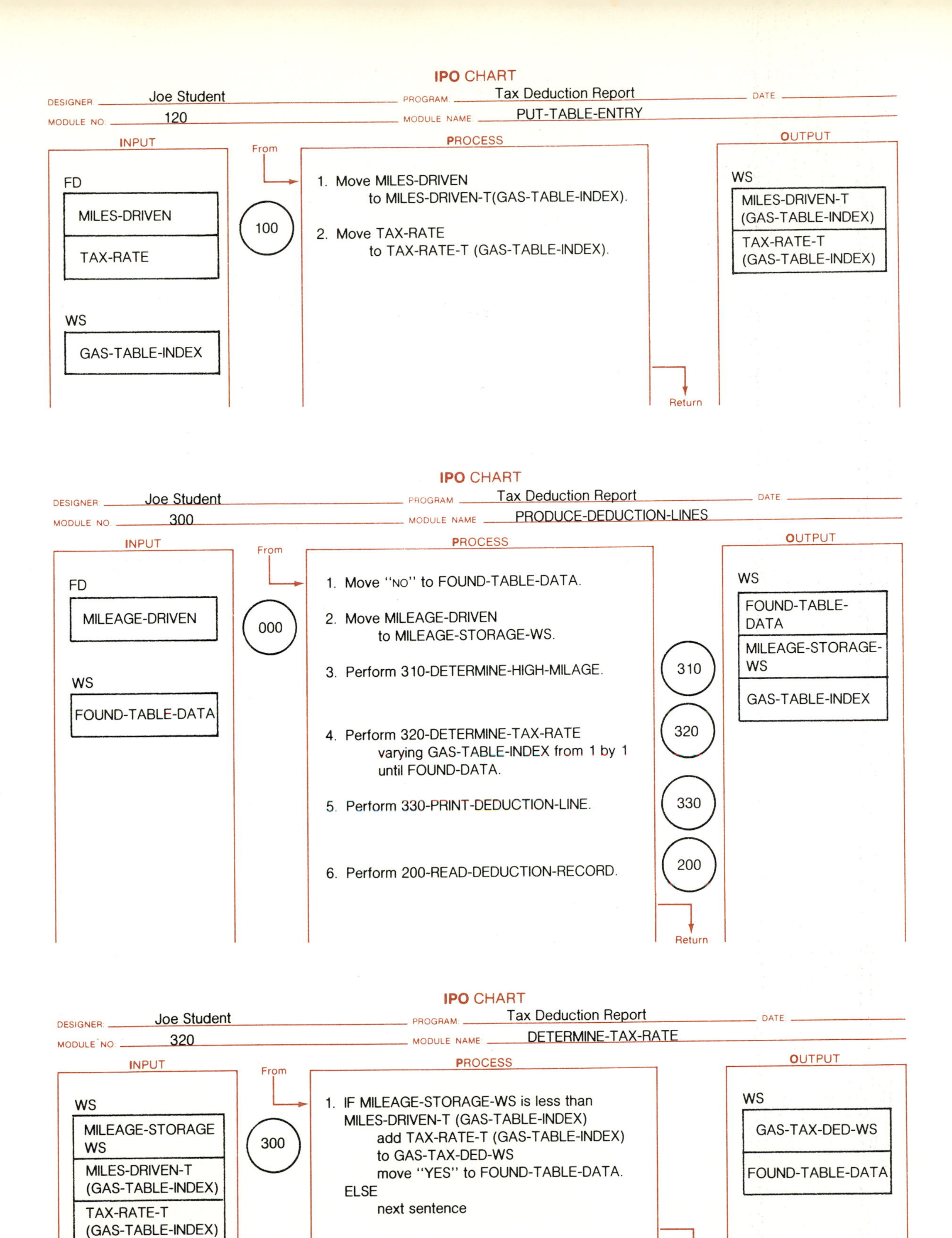
IPO CHART
DESIGNER Joe Student
PROGRAM Tax Deduction Report
DATE
MODULE NO. 120
MODULE NAME PUT-TABLE-ENTRY
INPUT
FD
MILES-DRIVEN
TAX-RATE
WS
GAS-TABLE-INDEX
From
100
PROCESS
1. Move MILES-DRIVEN
to MILES-DRIVEN-T(GAS-TABLE-INDEX).
2. Move TAX-RATE
to TAX-RATE-T (GAS-TABLE-INDEX).
Return
OUTPUT
WS
MILES-DRIVEN-T
(GAS-TABLE-INDEX)
TAX-RATE-T
(GAS-TABLE-INDEX)
IPO CHART
DESIGNER Joe Student
PROGRAM Tax Deduction Report
DATE
MODULE NO. 300
MODULE NAME PRODUCE-DEDUCTION-LINES
INPUT
FD
MILEAGE-DRIVEN
WS
FOUND-TABLE-DATA
From
000
PROCESS
1. Move "NO" to FOUND-TABLE-DATA.
2. Move MILEAGE-DRIVEN
to MILEAGE-STORAGE-WS.
3. Perform 310-DETERMINE-HIGH-MILAGE.
310
4. Perform 320-DETERMINE-TAX-RATE
varying GAS-TABLE-INDEX from 1 by 1
until FOUND-DATA.
320
5. Perform 330-PRINT-DEDUCTION-LINE.
330
6. Perform 200-READ-DEDUCTION-RECORD.
200
Return
OUTPUT
WS
FOUND-TABLE-
DATA
MILEAGE-STORAGE-
WS
GAS-TABLE-INDEX
IPO CHART
DESIGNER Joe Student
PROGRAM Tax Deduction Report
DATE
MODULE NO. 320
MODULE NAME DETERMINE-TAX-RATE
INPUT
WS
MILEAGE-STORAGE
WS
MILES-DRIVEN-T
(GAS-TABLE-INDEX)
TAX-RATE-T
(GAS-TABLE-INDEX)
From
300
PROCESS
1. IF MILEAGE-STORAGE-WS is less than
MILES-DRIVEN-T (GAS-TABLE-INDEX)
add TAX-RATE-T (GAS-TABLE-INDEX)
to GAS-TAX-DED-WS
move "YES" to FOUND-TABLE-DATA.
ELSE
next sentence
Return
OUTPUT
WS
GAS-TAX-DED-WS
FOUND-TABLE-DATA

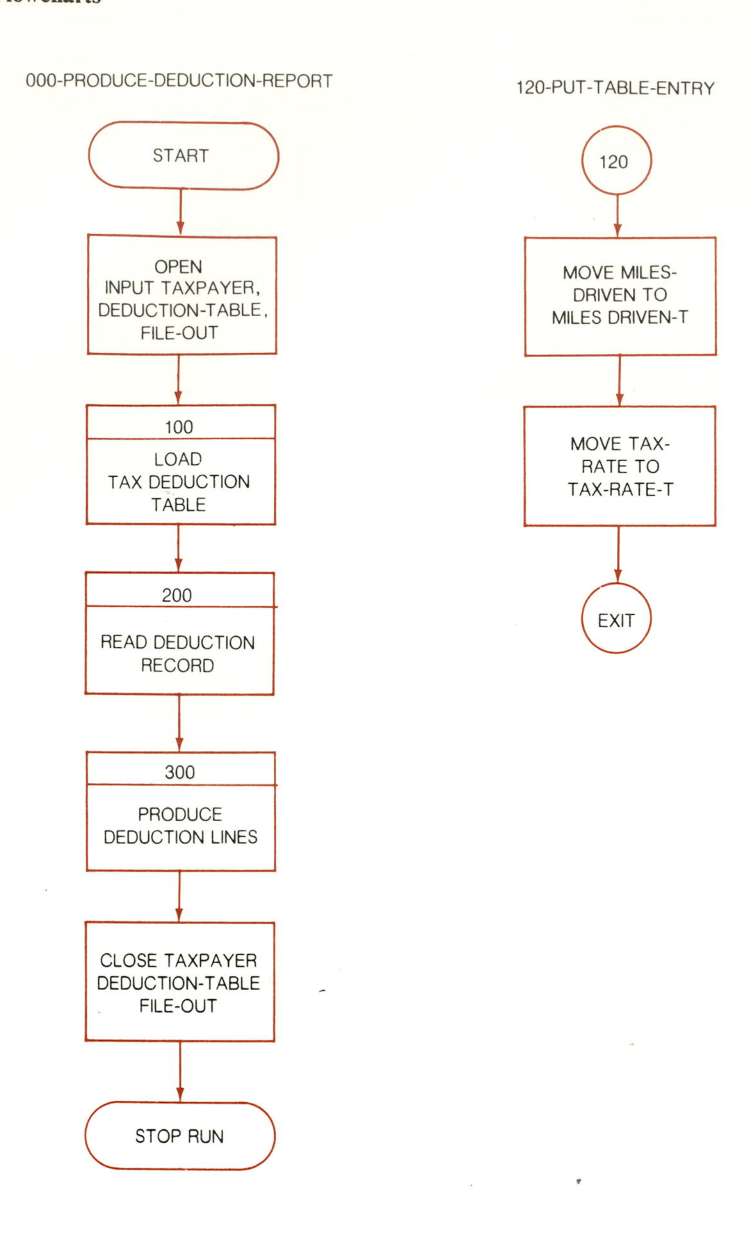
000-PRODUCE-DEDUCTION-REPORT
START
OPEN INPUT TAXPAYER, DEDUCTION-TABLE, FILE-OUT
100
LOAD TAX DEDUCTION TABLE
200
READ DEDUCTION RECORD
300
PRODUCE DEDUCTION LINES
CLOSE TAXPAYER DEDUCTION-TABLE FILE-OUT
STOP RUN
120-PUT-TABLE-ENTRY
120
MOVE MILES-DRIVEN TO MILES DRIVEN-T
MOVE TAX-RATE TO TAX-RATE-T
EXIT

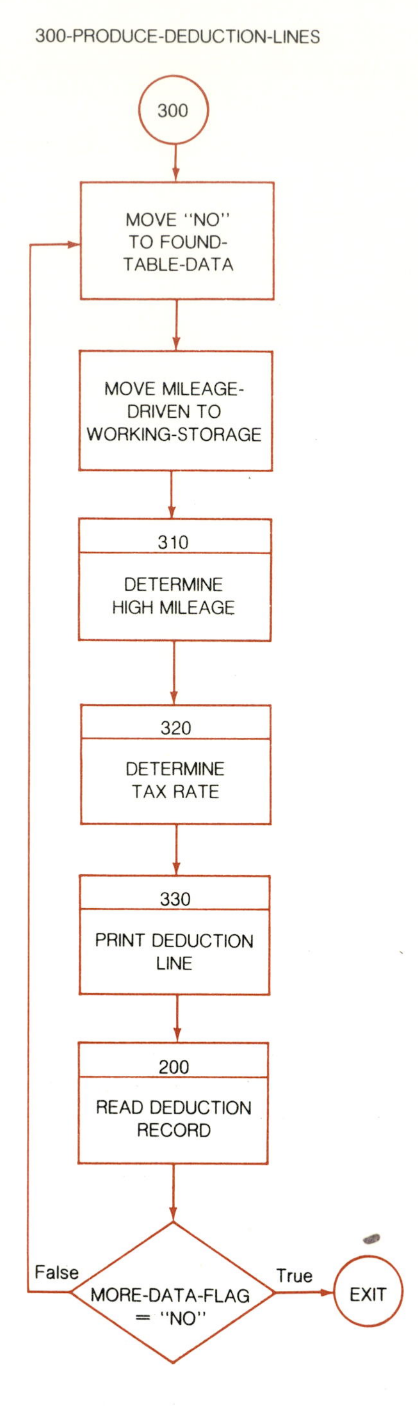
300-PRODUCE-DEDUCTION-LINES
300
MOVE "NO" TO FOUND-TABLE-DATA
MOVE MILEAGE-DRIVEN TO WORKING-STORAGE
310
DETERMINE HIGH MILEAGE
320
DETERMINE TAX RATE
330
PRINT DEDUCTION LINE
200
READ DEDUCTION RECORD
False
MORE-DATA-FLAG = "NO"
True
EXIT

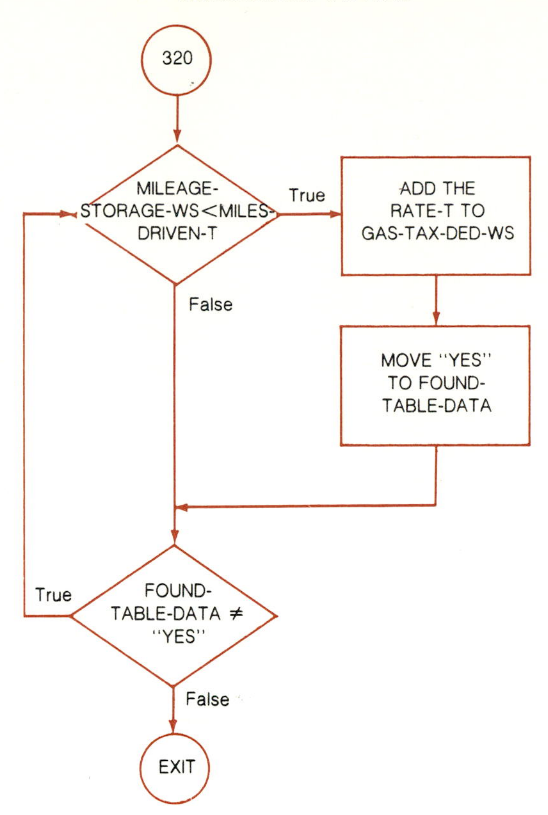

Source Program

```
 1        **************************
 2
 3         IDENTIFICATION DIVISION.
 4
 5        **************************
 6
 7         PROGRAM-ID.
 8             TAX-DEDUCTION-REPORT.
 9         AUTHOR.
10             STEVE GONSOSKI.
11         INSTALLATION.
12             WEST LOS ANGELES COLLEGE.
13         DATE-WRITTEN.
14             3 FEBRUARY 1982.
15
16        *PURPOSE.
17        *    THIS PROGRAM PRINTS A REPORT LISTING THE NAME OF THE TAXPAYER
18        *    WITH THE PROPER TAX DEDUCTION FOR MILES DRIVEN DURING THE
19        *    YEAR BASED ON THE TABLE.
20
21        **************************
22
23         ENVIRONMENT DIVISION.
24
25        **************************
26
27         CONFIGURATION SECTION.
28
29         SOURCE-COMPUTER.
30             LEVEL-66-ASCII.
31         OBJECT-COMPUTER.
32             LEVEL-66-ASCII.
33
34         INPUT-OUTPUT SECTION.
35
36         FILE-CONTROL.
37             SELECT TAXPAYER          ASSIGN TO C1-CARD-READER.
38             SELECT DEDUCTION-TABLE   ASSIGN TO C2-CARD-READER.
39             SELECT FILE-OUT          ASSIGN TO P1-PRINTER.
40
```

```
41        ***************************
42
43         DATA DIVISION.
44
45        ***************************
46
47         FILE SECTION.
48
49         FD  TAXPAYER
50
51             CODE-SET IS GBCD
52             LABEL RECORDS ARE OMITTED
53             DATA RECORD IS TAX-PAYER.
54
55         01  TAX-PAYER.
56
57             03  SOCIAL-SECURITY-NUMBER.
58                 05  SS-1                PICTURE 9(3).
59                 05  SS-2                PICTURE 9(2).
60                 05  SS-3                PICTURE 9(4).
61             03  TAXPAYER-NAME           PICTURE X(20).
62             03  FILLER                  PICTURE X(6).
63             03  MILEAGE-DRIVEN          PICTURE 9(6).
64             03  FILLER                  PICTURE X(39).
65
66         FD  DEDUCTION-TABLE
67
68             CODE-SET IS GBCD
69             LABEL RECORDS ARE OMITTED
70             DATA RECORD IS TABLE-ENTRY.
71
72         01  TABLE-ENTRY.
73
74             03  MILES-DRIVEN            PICTURE 9(5).
75             03  TAX-RATE                PICTURE 9(3).
76             03  FILLER                  PICTURE X(72).
77
78         FD  FILE-OUT
79
80             CODE-SET IS GBCD
81             LABEL RECORDS ARE OMITTED
82             DATA RECORD IS PRINT.
83
84         01  PRINT                       PICTURE X(132).
85
86         WORKING-STORAGE SECTION.
87
88         77  MILEAGE-STORAGE-WS          PICTURE 9(6).
89         77  LINE-COUNT                  PICTURE 9(2)    VALUE 38.
90         77  TAX-RATE-WS                 PICTURE 9(3).
91         77  OVER-TWENTY-WS              PICTURE 9(1).
92         77  EXTRA-MILES-WS              PICTURE 9(5).
93         77  GAS-TAX-DED-WS              PICTURE 9(3)V99.
94         77  GAS-TABLE-INDEX             PICTURE 9(2).
95
96         01  FLAGS.
97
98             03  MORE-DATA-FLAG          PICTURE X(3)    VALUE "YES".
99                 88  MORE-DATA                           VALUE "YES".
100                88  NO-MORE-DATA                        VALUE "NO ".
101            03  MORE-TABLE-DATA         PICTURE X(3)    VALUE "YES".
102                88  MORE-TABLE                          VALUE "YES".
103                88  NO-MORE-TABLE-DATA                  VALUE "NO ".
104            03  FOUND-TABLE-DATA        PICTURE X(3)    VALUE "NO ".
105                88  FOUND-DATA                          VALUE "YES".
106                88  NOT-FOUND-DATA                      VALUE "NO ".
107
108        01  TAX-DEDUCTION.
109
110            03  GAS-TABLE
111                    OCCURS 26 TIMES.
112                05  MILES-DRIVEN-T      PICTURE 9(5).
113                05  TAX-RATE-T          PICTURE 9(3).
114                05  FILLER              PICTURE X(72).
115
116        01  HDG-1.
117
118            03  FILLER                  PICTURE X(55)   VALUE SPACES.
119            03  FILLER                  PICTURE X(23)   VALUE "TAX  DEDUCTION
120       -                                                    "  REPORT".
121            03  FILLER                  PICTURE X(54)   VALUE SPACES.
122
123        01  HDG-2.
124
125            03  FILLER                  PICTURE X(38)   VALUE SPACES.
126            03  FILLER                  PICTURE X(9)    VALUE "SOC. SEC.".
127            03  FILLER                  PICTURE X(30)   VALUE SPACES.
128            03  FILLER                  PICTURE X(5)    VALUE "MILES".
129            03  FILLER                  PICTURE X(5)    VALUE SPACES.
130            03  FILLER                  PICTURE X(8)    VALUE "GASOLINE".
131            03  FILLER                  PICTURE X(37)   VALUE SPACES.
132
133        01  HDG-3.
134
135            03  FILLER                  PICTURE X(39)   VALUE SPACES.
136            03  FILLER                  PICTURE X(6)    VALUE "NUMBER".
137            03  FILLER                  PICTURE X(31)   VALUE SPACES.
138            03  FILLER                  PICTURE X(6)    VALUE "DRIVEN".
139            03  FILLER                  PICTURE X(5)    VALUE SPACES.
140            03  FILLER                  PICTURE X(7)    VALUE "TAX DED".
141            03  FILLER                  PICTURE X(38)   VALUE SPACES.
```

```
142
143        01  DETAIL-LINE.
144
145            03  FILLER                 PICTURE X(37)    VALUE SPACES.
146            03  SS-1-O                 PICTURE 9(3).
147            03  FILLER                 PICTURE X(1)     VALUE "-".
148            03  SS-2-O                 PICTURE 9(2).
149            03  FILLER                 PICTURE X(1)     VALUE "-".
150            03  SS-3-O                 PICTURE 9(4).
151            03  FILLER                 PICTURE X(4)     VALUE SPACES.
152            03  TAXPAYER-NAME-O        PICTURE X(20).
153            03  FILLER                 PICTURE X(3)     VALUE SPACES.
154            03  MILEAGE-DRIVEN-O       PICTURE ZZZ,ZZ9.
155            03  FILLER                 PICTURE X(5)     VALUE SPACES.
156            03  GAS-TAX-DEDUCTION-O PICTURE $ZZ9.99.
157            03  FILLER                 PICTURE X(38)    VALUE SPACES.
158
159        01  SAVE-RECORD                PICTURE X(132).
160
161       *************************
162
163        PROCEDURE DIVISION.
164
165       *************************
166
167        000-PRODUCE-DEDUCTION-REPORT.
168
169            OPEN    INPUT   TAXPAYER,
170                            DEDUCTION-TABLE,
171                    OUTPUT  FILE-OUT.
172            PERFORM 100-LOAD-TAX-DEDUCTION-TABLE
173                    VARYING GAS-TABLE-INDEX FROM 1 BY 1
174                    UNTIL GAS-TABLE-INDEX IS GREATER THAN 26.
175            PERFORM 200-READ-DEDUCTION-RECORD.
176            PERFORM 300-PRODUCE-DEDUCTION-LINES
177                    UNTIL NO-MORE-DATA.
178            CLOSE   TAXPAYER,
179                    DEDUCTION-TABLE,
180                    FILE-OUT.
181            STOP RUN.
182
183        100-LOAD-TAX-DEDUCTION-TABLE.
184
185            PERFORM 110-READ-TAX-DEDUCTION-TABLE.
186            PERFORM 120-PUT-TABLE-ENTRY.
187
188        110-READ-TAX-DEDUCTION-TABLE.
189
190            READ DEDUCTION-TABLE
191                AT END MOVE "NO " TO MORE-TABLE-DATA.
192
193        120-PUT-TABLE-ENTRY.
194
195            MOVE MILES-DRIVEN TO MILES-DRIVEN-T(GAS-TABLE-INDEX).
196            MOVE TAX-RATE      TO TAX-RATE-T    (GAS-TABLE-INDEX).
197
198        200-READ-DEDUCTION-RECORD.
199
200            READ TAXPAYER
201                AT END MOVE "NO " TO MORE-DATA-FLAG.
202
203        300-PRODUCE-DEDUCTION-LINES.
204
205            MOVE "NO " TO FOUND-TABLE-DATA.
206            MOVE MILEAGE-DRIVEN TO MILEAGE-STORAGE-WS.
207            PERFORM 310-DETERMINE-HIGH-MILEAGE.
208            PERFORM 320-DETERMINE-TAX-RATE
209                VARYING GAS-TABLE-INDEX FROM 1 BY 1
210                UNTIL FOUND-DATA.
211            PERFORM 330-PRINT-DEDUCTION-LINE.
212            PERFORM 200-READ-DEDUCTION-RECORD.
213
214        310-DETERMINE-HIGH-MILEAGE.
215
216            MOVE ZERO TO GAS-TAX-DED-WS.
217            IF MILEAGE-DRIVEN IS > 20000
218                PERFORM 340-CALCULATE-DED-HIGH-MILES
219            ELSE
220                NEXT SENTENCE.
221
222        320-DETERMINE-TAX-RATE.
223
224            IF MILEAGE-STORAGE-WS IS < MILES-DRIVEN-T(GAS-TABLE-INDEX)
225                ADD TAX-RATE-T (GAS-TABLE-INDEX) TO GAS-TAX-DED-WS
226                MOVE "YES" TO FOUND-TABLE-DATA
227            ELSE
228                NEXT SENTENCE.
229
230        330-PRINT-DEDUCTION-LINE.
231
232            MOVE SS-1                  TO SS-1-O.
233            MOVE SS-2                  TO SS-2-O.
234            MOVE SS-3                  TO SS-3-O.
235            MOVE TAXPAYER-NAME         TO TAXPAYER-NAME-O.
236            MOVE MILEAGE-DRIVEN        TO MILEAGE-DRIVEN-O.
237            MOVE GAS-TAX-DED-WS        TO GAS-TAX-DEDUCTION-O.
238            MOVE DETAIL-LINE           TO PRINT.
239            PERFORM 350-PRINT-DETAIL-LINE.
240
```

```
241        340-CALCULATE-DED-HIGH-MILES.
242
243            DIVIDE MILEAGE-DRIVEN BY 20000
244                GIVING OVER-TWENTY-WS.
245            COMPUTE EXTRA-MILES-WS =
246                MILEAGE-DRIVEN - (20000 * OVER-TWENTY-WS)
247            MOVE EXTRA-MILES-WS TO MILEAGE-STORAGE-WS.
248            MULTIPLY OVER-TWENTY-WS BY TAX-RATE-T (26)
249                GIVING GAS-TAX-DED-WS.
250
251        350-PRINT-DETAIL-LINE.
252
253            IF LINE-COUNT IS > 37
254                MOVE PRINT TO SAVE-RECORD
255                PERFORM 360-PRINT-HEADING-LINE
256                MOVE SAVE-RECORD TO PRINT
257            ELSE
258                NEXT SENTENCE.
259            PERFORM 370-WRITE-DOUBLE-DETAIL-LINE.
260
261        360-PRINT-HEADING-LINE.
262
263            MOVE HDG-1 TO PRINT.
264            PERFORM 380-WRITE-TOP-OF-PAGE.
265            MOVE SPACES TO PRINT.
266            PERFORM 390-WRITE-DETAIL-LINE.
267            MOVE HDG-2 TO PRINT.
268            PERFORM 390-WRITE-DETAIL-LINE.
269            MOVE HDG-3 TO PRINT.
270            PERFORM 390-WRITE-DETAIL-LINE.
271            MOVE SPACES TO PRINT.
272            PERFORM 370-WRITE-DOUBLE-DETAIL-LINE.
273
274        370-WRITE-DOUBLE-DETAIL-LINE.
275
276            WRITE PRINT
277                AFTER ADVANCING 2 LINES.
278            ADD 2 TO LINE-COUNT.
279
280        380-WRITE-TOP-OF-PAGE.
281
282            WRITE PRINT
283                AFTER ADVANCING PAGE.
284            MOVE 1 TO LINE-COUNT.
285
286        390-WRITE-DETAIL-LINE.
287
288            WRITE PRINT
289                AFTER ADVANCING 1 LINE.
290            ADD 1 TO LINE-COUNT.

THERE WERE 290 SOURCE INPUT LINES.
THERE WERE NO DIAGNOSTICS.
```

```
                         TAX  DEDUCTION  REPORT

     SOC. SEC.                                   MILES      GASOLINE
      NUMBER                                     DRIVEN     TAX DED


     556-98-3021    BELL,ERNEST                  75,362     $438.00

     556-94-2447    CHATMAN,CARLA E              13,471     $ 78.00

     011-30-9013    DINIZ,FRANCISCO J             6,314     $ 36.00

     330-52-8528    HODGES,MICHAEL A             36,293     $211.00

     556-13-8842    JOHNSON,NIMROD J             17,954     $101.00

     560-66-7910    NELSON,RAYMOND R             27,901     $161.00

     556-96-4996    POWELL,BOBBIE J             104,727     $608.00

     573-02-1310    SANFORD,MICHAEL M               400     $ 12.00

     573-98-9084    TAKAYAMA,GARY M              13,017     $ 78.00

     563-96-2106    TRAVIS,HIRAM S                8,520     $ 51.00

     573-93-5201    WAGGONER,LARRY               18,791     $107.00

     574-70-2075    WINGERT,JOSEPH R             12,001     $ 72.00
```

Illustrative Program: Sales Report

The BG Company sells their three products in different geographical areas located throughout the United States. A report is to be prepared showing the total sales of each product as well as the total sales for each product by salesman, territory, and area and a final total for all fields.

Arrays are to be used for the three products as well as for the different levels of totals. Totals are to be "rolled" into other totals at each different level.

Input

Positions	Field	
1–6	Date	
7–10	Salesman number	
11–14	Territory number	
15–16	Area number	
30–36	Product 1 sales	XX,XXX.XX
37–43	Product 2 sales	XX,XXX.XX
44–50	Product 3 sales	XX,XXX.XX

Calculations

The sales for each salesman are to be totaled. The sales for each product as well as the salesman total are to be totaled by salesman number, territory number, area number, and a final total for all fields.

Output

The input fields are to be outputted in the same sequence, as well as the total field.

Level totals for salesman, territory, area, and final are to be outputted.
Leave five spaces between each field and provide the necessary headings.

Required

Code the necessary COBOL forms to accomplish the above.

Output is as follows:

DATE	SALESMAN NUMBER	TERRITORY NUMBER	AREA NUMBER	PRODUCT 1 SALES	PRODUCT 2 SALES	PRODUCT 3 SALES	TOTAL SALES
01/11/81	5104	30	21	3,500.00	45,000.00	32,005.00	80,505.00
01/22/81	5104	30	21	2,950.50	3,598.55	1,110.10	88,164.15
01/24/81	5104	30	21	590.30	1,000.00	2,000.08	91,754.53
	SALESMAN TOTALS			7,040.80	49,598.55	35,115.18	91,754.53
01/26/81	2134	30	21	1,953.00	2,553.30	3,280.50	99,541.33
01/28/81	2134	30	21	591.10	5,410.03	6,028.00	111,570.46
	SALESMAN TOTALS			2,544.10	7,963.33	9,308.50	19,815.93
	TERRITORY TOTALS			9,584.90	57,561.88	44,423.68	111,570.46
	AREA TOTALS			9,584.90	57,561.88	44,423.68	111,570.46
01/13/81	3007	60	22	555.55	6,532.10	7,755.00	126,413.11
01/19/81	3007	60	22	98.00	1,014.44	444.47	127,970.02
	SALESMAN TOTALS			653.55	7,546.54	8,199.47	16,399.56
01/27/81	1510	60	22	888.33	100.50	1,875.00	130,833.85
01/27/81	1510	60	22	97.50	1,500.00	2,350.00	134,781.35
	SALESMAN TOTALS			985.83	1,600.50	4,225.00	6,811.33
	TERRITORY TOTALS			1,639.38	9,147.04	12,424.47	23,210.89
01/17/81	1980	80	22	7,000.00	9,000.00	4,500.00	155,281.35
01/20/81	1980	80	22	175.00	1,950.00	8,510.00	165,916.35
01/22/81	1980	80	22	5,777.00	594.40	819.40	173,107.15
	SALESMAN TOTALS			12,952.00	11,544.40	13,829.40	38,325.80
	TERRITORY TOTALS			12,952.00	11,544.40	13,829.40	38,325.80
	AREA TOTALS			14,591.38	20,691.44	26,253.87	61,536.69
	FINAL TOTALS			24,176.28	78,253.32	70,677.55	173,107.15

Hierarchy Chart

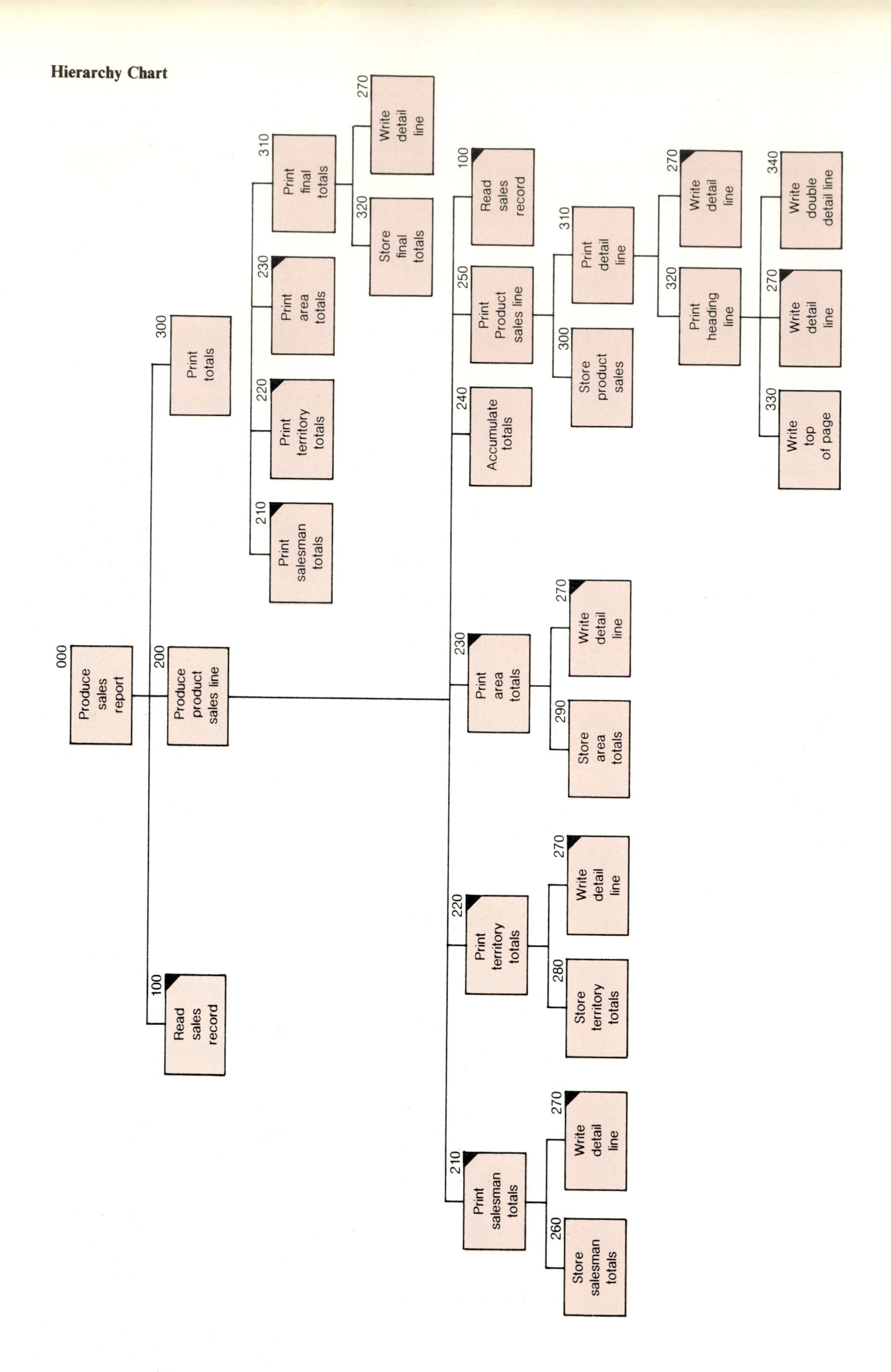

IPO Charts

IPO CHART

DESIGNER Joe Student PROGRAM Sales Report DATE

MODULE NO 000 MODULE NAME PRODUCE-SALES-REPORT

INPUT	From	PROCESS		OUTPUT
FD: CARD-FILE	OS	1. Open files.		FD: FILE-OUT
		2. Perform 100-READ-SALES-RECORD.	100	
		3. Perform 200-PRODUCE-PRODUCT-SALES-LINE until NO-MORE-DATA.	200	
WS: MORE-DATA-FLAG		4. Perform 300-PRINT-TOTALS.	300	
		5. Close files.		
		6. Stop run.	Return	

IPO CHART

DESIGNER Joe Student PROGRAM Sales Report DATE

MODULE NO 200 MODULE NAME PRODUCE-PRODUCT-SALES-LINE

INPUT:

FD
- SALESMAN-NUMBER
- TERRITORY-NUMBER
- AREA-NUMBER

WS
- FIRST-RECORD-FLAG
- SALESMAN-NUMBER-SAVED
- TERRITORY-NUMBER-SAVED
- AREA-NUMBER-SAVED

From: 000

PROCESS:

```
1. IF FIRST-RECORD
         move SALESMAN-NUMBER
              to SALESMAN-NUMBER-SAVED
         move TERRITORY-NUMBER
              to TERRITORY-NUMBER-SAVED
         move AREA-NUMBER
              to AREA-NUMBER-SAVED
         move "NO" to FIRST-RECORD-FLAG
   ELSE
         next sentence.

2. IF SALESMAN-NUMBER
   not = SALESMAN-NUMBER-SAVED
         perform 210-PRINT-SALESMAN-TOTALS          (210)
         move SALESMAN-NUMBER
              to SALESMAN-NUMBER-SAVED
   ELSE
         next sentence.

3. IF TERRITORY-NUMBER
   not = TERRITORY-NUMBER-SAVED
         perform 220-PRINT-TERRITORY-TOTALS         (220)
         move TERRITORY NUMBER
   to TERRITORY-NUMBER-SAVED
   ELSE
         next sentence.

4. IF AREA-NUMBER
   not = AREA-NUMBER-SAVED
         perform 230-PRINT-AREA-TOTALS              (230)
         move AREA-NUMBER
   to AREA-NUMBER-SAVED
   ELSE
         next sentence.

5. Perform 240-ACCUMULATE-TOTALS                    (240)
         varying PRODUCT-SUBSCRIPT
         from 1 by 1
         until PRODUCT-SUBSCRIPT
         is greater than 3.

6. Perform 250-PRINT-PRODUCT-SALES-LINE.            (250)

7. Perform 100-READ-SALES-RECORD.                   (100)
```

Return

OUTPUT:

WS
- SALESMAN-NUMBER-SAVED
- TERRITORY-NUMBER-SAVED
- AREA-NUMBER-SAVED
- FIRST-RECORD-FLAG
- PRODUCT-SUBSCRIPT

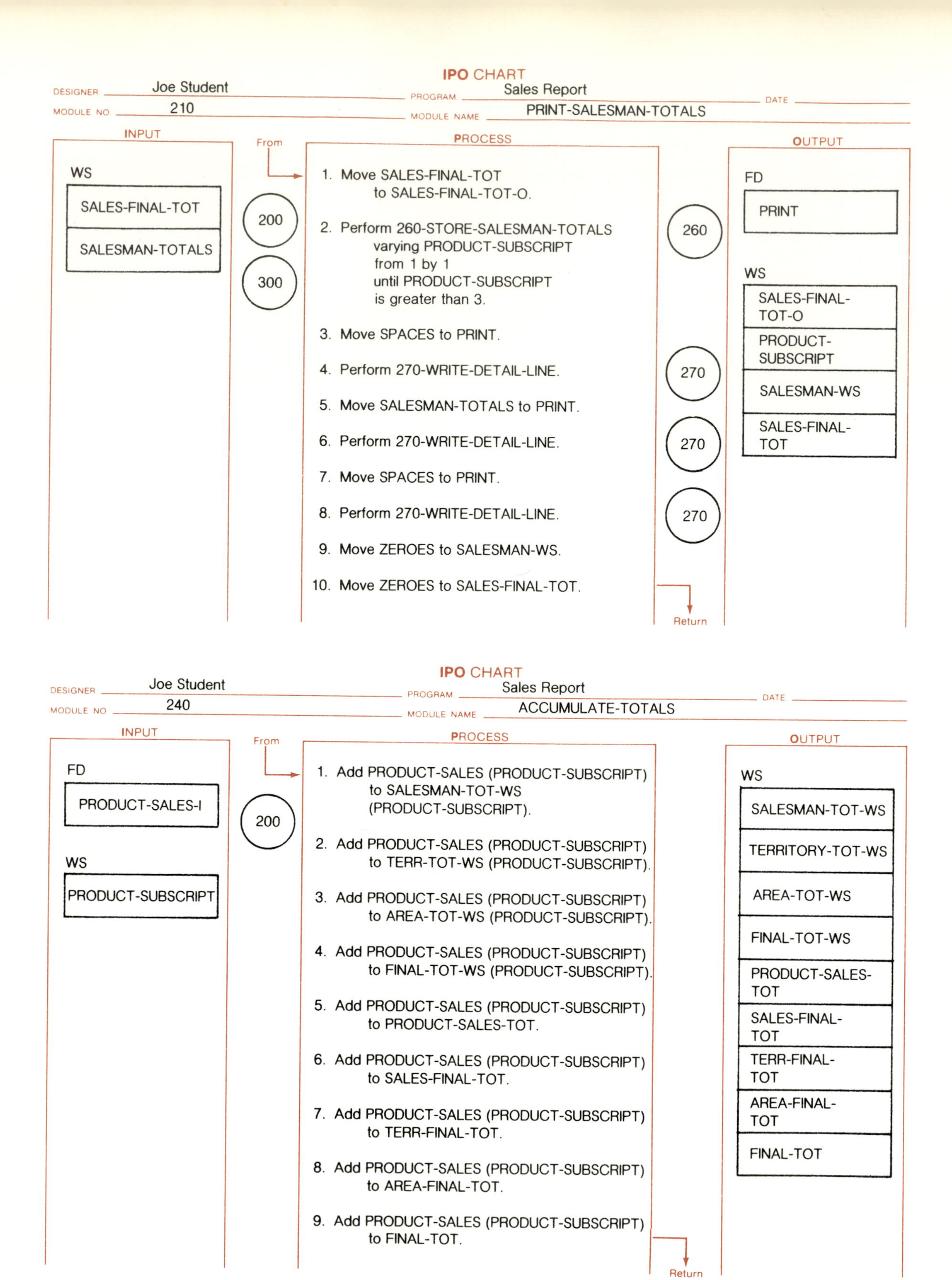

IPO CHART
DESIGNER Joe Student
PROGRAM Sales Report
DATE
MODULE NO. 210
MODULE NAME PRINT-SALESMAN-TOTALS
INPUT
WS
SALES-FINAL-TOT
SALESMAN-TOTALS
From
200
300
PROCESS
1. Move SALES-FINAL-TOT to SALES-FINAL-TOT-O.
2. Perform 260-STORE-SALESMAN-TOTALS varying PRODUCT-SUBSCRIPT from 1 by 1 until PRODUCT-SUBSCRIPT is greater than 3.
3. Move SPACES to PRINT.
4. Perform 270-WRITE-DETAIL-LINE.
5. Move SALESMAN-TOTALS to PRINT.
6. Perform 270-WRITE-DETAIL-LINE.
7. Move SPACES to PRINT.
8. Perform 270-WRITE-DETAIL-LINE.
9. Move ZEROES to SALESMAN-WS.
10. Move ZEROES to SALES-FINAL-TOT.
260
270
270
270
Return
OUTPUT
FD
PRINT
WS
SALES-FINAL-TOT-O
PRODUCT-SUBSCRIPT
SALESMAN-WS
SALES-FINAL-TOT
IPO CHART
DESIGNER Joe Student
PROGRAM Sales Report
DATE
MODULE NO. 240
MODULE NAME ACCUMULATE-TOTALS
INPUT
FD
PRODUCT-SALES-I
WS
PRODUCT-SUBSCRIPT
From
200
PROCESS
1. Add PRODUCT-SALES (PRODUCT-SUBSCRIPT) to SALESMAN-TOT-WS (PRODUCT-SUBSCRIPT).
2. Add PRODUCT-SALES (PRODUCT-SUBSCRIPT) to TERR-TOT-WS (PRODUCT-SUBSCRIPT).
3. Add PRODUCT-SALES (PRODUCT-SUBSCRIPT) to AREA-TOT-WS (PRODUCT-SUBSCRIPT).
4. Add PRODUCT-SALES (PRODUCT-SUBSCRIPT) to FINAL-TOT-WS (PRODUCT-SUBSCRIPT).
5. Add PRODUCT-SALES (PRODUCT-SUBSCRIPT) to PRODUCT-SALES-TOT.
6. Add PRODUCT-SALES (PRODUCT-SUBSCRIPT) to SALES-FINAL-TOT.
7. Add PRODUCT-SALES (PRODUCT-SUBSCRIPT) to TERR-FINAL-TOT.
8. Add PRODUCT-SALES (PRODUCT-SUBSCRIPT) to AREA-FINAL-TOT.
9. Add PRODUCT-SALES (PRODUCT-SUBSCRIPT) to FINAL-TOT.
Return
OUTPUT
WS
SALESMAN-TOT-WS
TERRITORY-TOT-WS
AREA-TOT-WS
FINAL-TOT-WS
PRODUCT-SALES-TOT
SALES-FINAL-TOT
TERR-FINAL-TOT
AREA-FINAL-TOT
FINAL-TOT

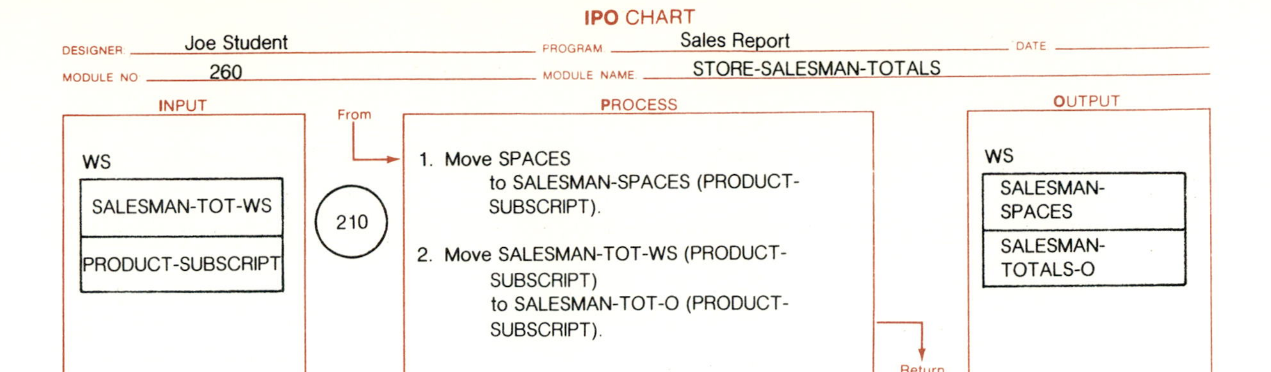
IPO CHART
DESIGNER Joe Student
PROGRAM Sales Report
DATE
MODULE NO 260
MODULE NAME STORE-SALESMAN-TOTALS
INPUT
WS
SALESMAN-TOT-WS
PRODUCT-SUBSCRIPT
From
210
PROCESS
1. Move SPACES
to SALESMAN-SPACES (PRODUCT-SUBSCRIPT).
2. Move SALESMAN-TOT-WS (PRODUCT-SUBSCRIPT)
to SALESMAN-TOT-O (PRODUCT-SUBSCRIPT).
Return
OUTPUT
WS
SALESMAN-SPACES
SALESMAN-TOTALS-O

Flowcharts

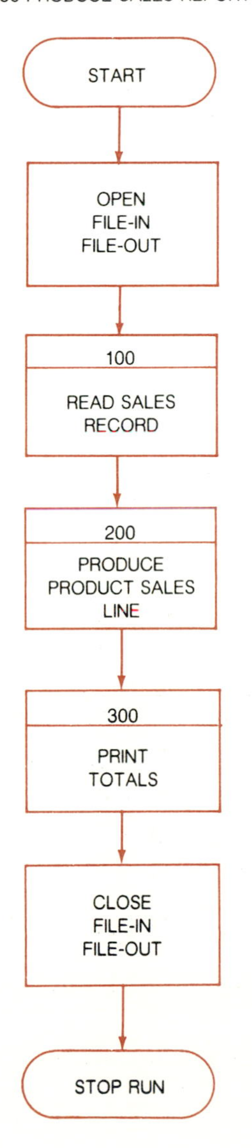
000-PRODUCE-SALES-REPORT
START
OPEN
FILE-IN
FILE-OUT
100
READ SALES RECORD
200
PRODUCE PRODUCT SALES LINE
300
PRINT TOTALS
CLOSE
FILE-IN
FILE-OUT
STOP RUN

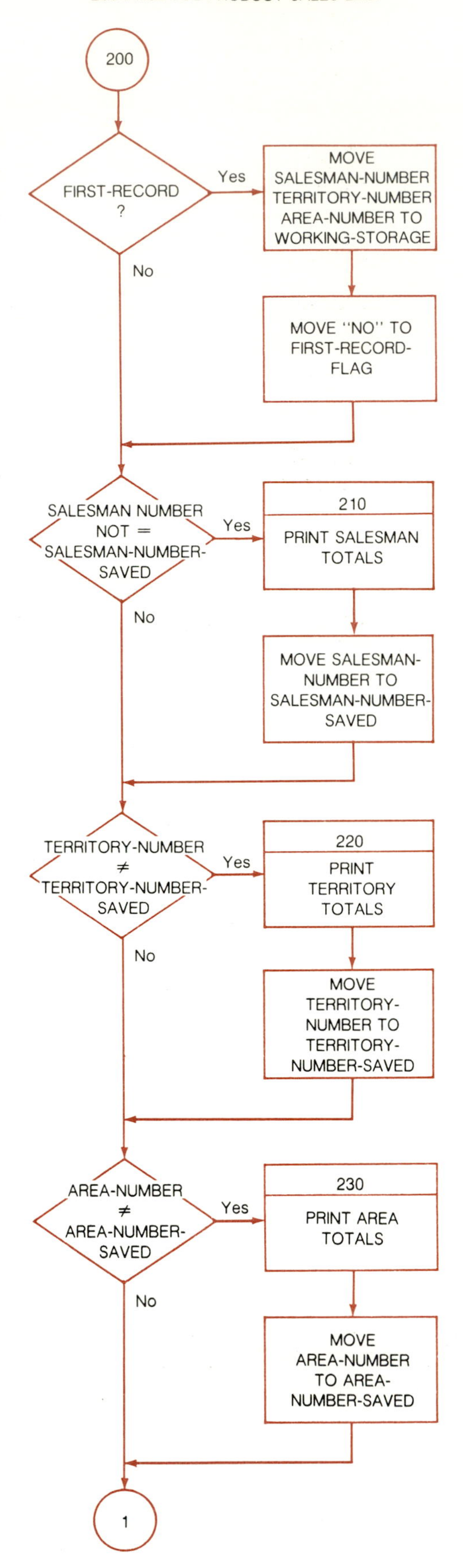
200-PRODUCE-PRODUCT-SALES-LINE
200
FIRST-RECORD ?
Yes
MOVE SALESMAN-NUMBER TERRITORY-NUMBER AREA-NUMBER TO WORKING-STORAGE
MOVE "NO" TO FIRST-RECORD-FLAG
No
SALESMAN NUMBER NOT = SALESMAN-NUMBER-SAVED
Yes
210
PRINT SALESMAN TOTALS
MOVE SALESMAN-NUMBER TO SALESMAN-NUMBER-SAVED
No
TERRITORY-NUMBER ≠ TERRITORY-NUMBER-SAVED
Yes
220
PRINT TERRITORY TOTALS
MOVE TERRITORY-NUMBER TO TERRITORY-NUMBER-SAVED
No
AREA-NUMBER ≠ AREA-NUMBER-SAVED
Yes
230
PRINT AREA TOTALS
MOVE AREA-NUMBER TO AREA-NUMBER-SAVED
No
1

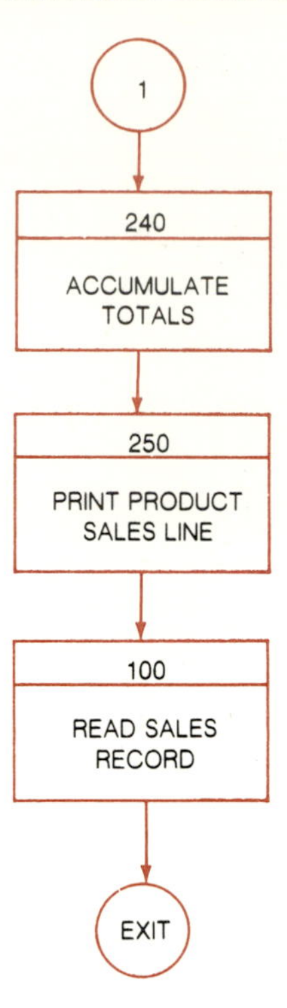
200-PRODUCE-PRODUCT-SALES-LINE (CONT)
1
240
ACCUMULATE TOTALS
250
PRINT PRODUCT SALES LINE
100
READ SALES RECORD
EXIT

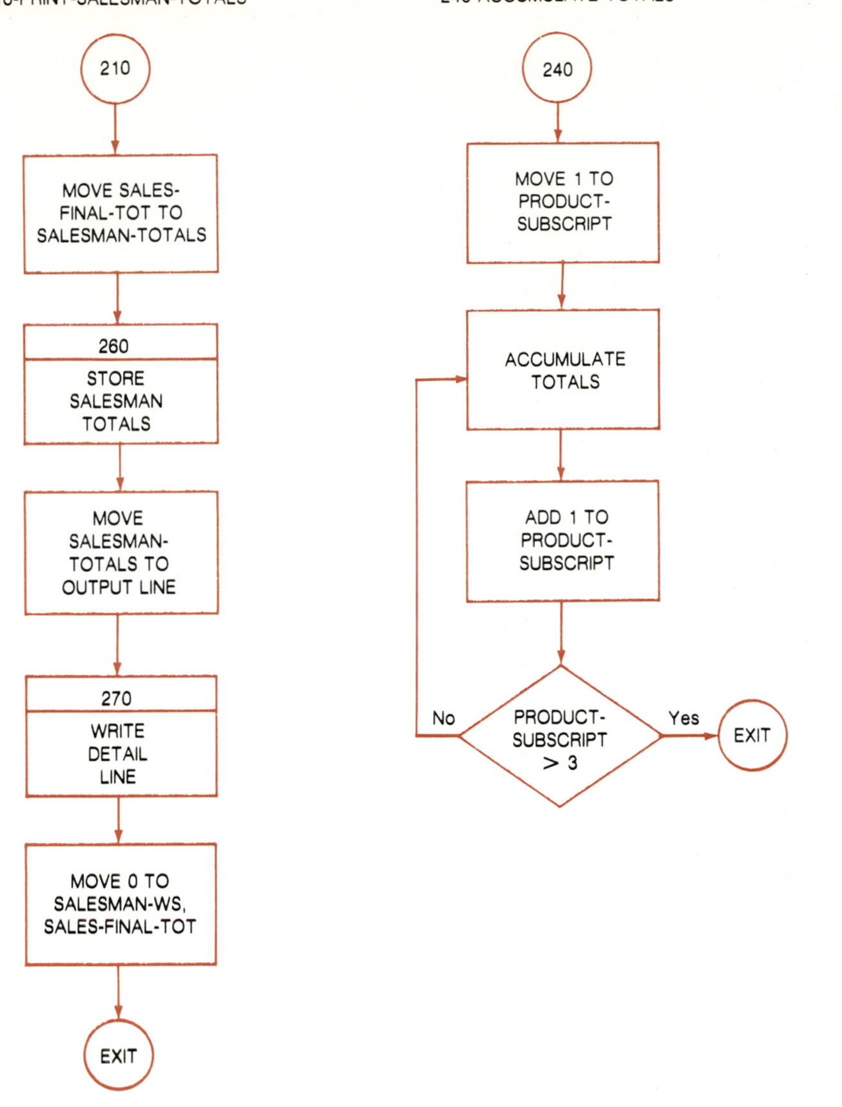

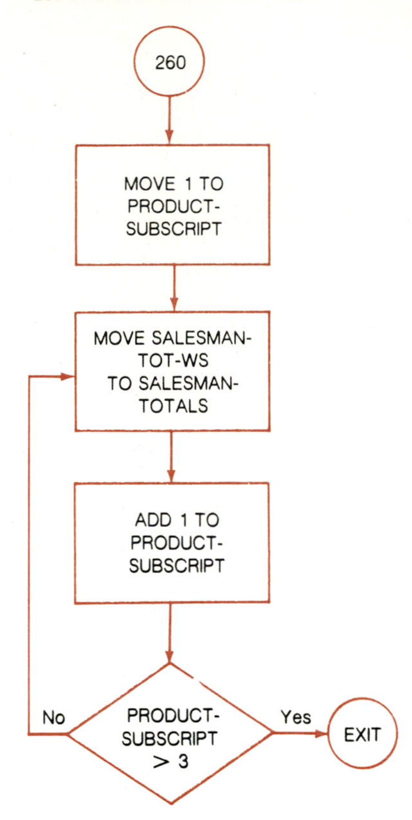

Source Program

```
 1       *************************
 2
 3        IDENTIFICATION DIVISION.
 4
 5       *************************
 6
 7        PROGRAM-ID.
 8            PRODUCT-SALES-REPORT.
 9        AUTHOR.
10            STEVE GONSOSKI.
11        INSTALLATION.
12            WEST LOS ANGELES COLLEGE.
13        DATE-WRITTEN.
14            3 FEBRUARY 1982.
15
16       *PURPOSE.
17       *    THIS PROGRAM PRINTS A REPORT SHOWING THE TOTAL SALES OF EACH
18       *    PRODUCT AS WELL AS THE TOTAL SALES FOR EACH PRODUCT BY
19       *    SALESMAN, TERRITORY, AREA AND A FINAL TOTAL FOR ALL FIELDS.
20
21       *************************
22
23        ENVIRONMENT DIVISION.
24
25       *************************
26
27        CONFIGURATION SECTION.
28
29        SOURCE-COMPUTER.
30            LEVEL-66-ASCII.
31        OBJECT-COMPUTER.
32            LEVEL-66-ASCII.
33
```

```
34        INPUT-OUTPUT SECTION.
35
36        FILE-CONTROL.
37            SELECT FILE-IN      ASSIGN TO C1-CARD-READER.
38            SELECT FILE-OUT     ASSIGN TO P1-PRINTER.
39
40       *************************
41
42        DATA DIVISION.
43
44       *************************
45
46        FILE SECTION.
47
48        FD  FILE-IN
49            CODE-SET IS GBCD
50            LABEL RECORDS ARE OMITTED
51            DATA RECORD IS CARDIN.
52
53        01  CARDIN.
54
55        03  TRANSACTION-DATE.
56                05  T-MONTH           PICTURE 9(2).
57                05  T-DAY             PICTURE 9(2).
58                05  T-YEAR            PICTURE 9(2).
59            03  SALESMAN-NUMBER       PICTURE 9(4).
60            03  TERRITORY-NUMBER      PICTURE 9(4).
61            03  AREA-NUMBER           PICTURE 9(2).
62            03  FILLER                PICTURE X(13).
63            03  PRODUCT-SALES-I.
64                05  PRODUCT-SALES     PICTURE 9(5)V99
65                                          OCCURS 3 TIMES.
66            03  FILLER                PICTURE X(30).
67
68        FD  FILE-OUT
69            CODE-SET IS GBCD
70            LABEL RECORDS ARE OMITTED
71            DATA RECORD IS PRINT.
72
73        01  PRINT                     PICTURE X(132).
74
75        WORKING-STORAGE SECTION.
76
77        77  LINE-COUNT-WS             PICTURE 99       VALUE 38.
78        77  TERRITORY-NUMBER-SAVED    PICTURE 9(4).
79        77  AREA-NUMBER-SAVED         PICTURE 9(2).
80        77  SALESMAN-NUMBER-SAVED     PICTURE 9(4).
81        77  SALES-FINAL-TOT           PICTURE 9(6)V99.
82        77  TERR-FINAL-TOT            PICTURE 9(6)V99.
83        77  AREA-FINAL-TOT            PICTURE 9(6)V99.
84        77  FINAL-TOT                 PICTURE 9(6)V99.
85        77  PRODUCT-SALES-TOT         PICTURE 9(6)V99.
86        77  PRODUCT-SUBSCRIPT         PICTURE 9(2).
87
88        01  SALESMAN-WS.
89
90            03  SALESMAN-TOT-WS       PICTURE 9(5)V99
91                                          OCCURS 3 TIMES.
92
93        01  TERRITORY-WS.
94
95            03  TERR-TOT-WS           PICTURE 9(5)V99
96                                          OCCURS 3 TIMES.
97
98        01  AREA-WS.
99
100           03  AREA-TOT-WS           PICTURE 9(5)V99
101                                         OCCURS 3 TIMES.
102
103       01  FINAL-WS.
104
105           03  FINAL-TOT-WS          PICTURE 9(5)V99
106                                         OCCURS 3 TIMES.
107
108       01  FLAGS.
109
110           03  MORE-DATA-FLAG        PICTURE X(3)     VALUE "YES".
111               88  MORE-DATA                          VALUE "YES".
112               88  NO-MORE-DATA                       VALUE "NO ".
113           03  FIRST-RECORD-FLAG     PICTURE X(3)     VALUE "YES".
114               88  FIRST-RECORD                       VALUE "YES".
115               88  NON-FIRST-REC                      VALUE "NO".
116
117       01  HDG-1.
118
119           03  FILLER                PICTURE X(28)    VALUE SPACES.
120           03  FILLER                PICTURE X(8)     VALUE "SALESMAN".
121           03  FILLER                PICTURE X(5)     VALUE SPACES.
122           03  FILLER                PICTURE X(9)     VALUE "TERRITORY".
123           03  FILLER                PICTURE X(6)     VALUE SPACES.
124           03  FILLER                PICTURE X(4)     VALUE "AREA".
125           03  FILLER                PICTURE X(9)     VALUE SPACES.
126           03  FILLER                PICTURE X(9)     VALUE "PRODUCT 1".
127           03  FILLER                PICTURE X(5)     VALUE SPACES.
128           03  FILLER                PICTURE X(9)     VALUE "PRODUCT 2".
129           03  FILLER                PICTURE X(5)     VALUE SPACES.
130           03  FILLER                PICTURE X(9)     VALUE "PRODUCT 3".
131           03  FILLER                PICTURE X(8)     VALUE SPACES.
132           03  FILLER                PICTURE X(5)     VALUE "TOTAL".
133           03  FILLER                PICTURE X(13)    VALUE SPACES.
134
```

```
135        01  HDG-2.
136
137            03  FILLER               PICTURE X(19)    VALUE SPACES.
138            03  FILLER               PICTURE X(4)     VALUE "DATE".
139            03  FILLER               PICTURE X(6)     VALUE SPACES.
140            03  FILLER               PICTURE X(6)     VALUE "NUMBER".
141            03  FILLER               PICTURE X(7)     VALUE SPACES.
142            03  FILLER               PICTURE X(6)     VALUE "NUMBER".
143            03  FILLER               PICTURE X(6)     VALUE SPACES.
144            03  FILLER               PICTURE X(6)     VALUE "NUMBER".
145            03  FILLER               PICTURE X(10)    VALUE SPACES.
146            03  FILLER               PICTURE X(5)     VALUE "SALES".
147            03  FILLER               PICTURE X(9)     VALUE SPACES.
148            03  FILLER               PICTURE X(5)     VALUE "SALES".
149            03  FILLER               PICTURE X(9)     VALUE SPACES.
150            03  FILLER               PICTURE X(5)     VALUE "SALES".
151            03  FILLER               PICTURE X(10)    VALUE SPACES.
152            03  FILLER               PICTURE X(5)     VALUE "SALES".
153            03  FILLER               PICTURE X(14)    VALUE SPACES.
154
155        01  DETAIL-LINE.
156
157            03  FILLER               PICTURE X(17)    VALUE SPACES.
158            03  T-MONTH-O            PICTURE 9(2).
159            03  FILLER               PICTURE X(1)     VALUE "/".
160            03  T-DAY-O              PICTURE 9(2).
161            03  FILLER               PICTURE X(1)     VALUE "/".
162            03  T-YEAR-O             PICTURE 9(2).
163            03  FILLER               PICTURE X(5)     VALUE SPACES.
164            03  SALESMAN-NUMBER-O    PICTURE Z(4).
165            03  FILLER               PICTURE X(9)     VALUE SPACES.
166            03  TERRITORY-NUMBER-O   PICTURE Z(4).
167            03  FILLER               PICTURE X(9)     VALUE SPACES.
168            03  AREA-NUMBER-O        PICTURE Z(2).
169            03  FILLER               PICTURE X(10)    VALUE SPACES.
170            03  SALES-O                    OCCURS 3 TIMES.
171                05  PRODUCT-SALES-O  PICTURE ZZ,ZZ9.99.
172                05  SALES-SPACES     PICTURE X(5).
173            03  PRODUCT-SALES-TOT-O  PICTURE ZZZ,ZZ9.99.
174            03  FILLER               PICTURE X(12)    VALUE SPACES.
175
176        01  SALESMAN-TOTALS.
177
178            03  FILLER               PICTURE X(33)    VALUE SPACES.
179            03  FILLER               PICTURE X(9)     VALUE "SALESMAN ".
180            03  FILLER               PICTURE X(6)     VALUE "TOTALS".
181            03  FILLER               PICTURE X(20)    VALUE SPACES.
182            03  SALESMAN-TOTALS-O          OCCURS 3 TIMES.
183                05  SALESMAN-TOT-O   PICTURE ZZ,ZZ9.99.
184                05  SALESMAN-SPACES  PICTURE X(5).
185            03  SALES-FINAL-TOT-O    PICTURE ZZZ,ZZ9.99.
186            03  FILLER               PICTURE X(12)    VALUE SPACES.
187
188        01  TERRITORY-TOTALS.
189
190            03  FILLER               PICTURE X(33)    VALUE SPACES.
191            03  FILLER               PICTURE X(10)    VALUE "TERRITORY ".
192            03  FILLER               PICTURE X(6)     VALUE "TOTALS".
193            03  FILLER               PICTURE X(19)    VALUE SPACES.
194            03  TERRITORY-TOTALS-O         OCCURS 3 TIMES.
195                05  TERR-TOT-O       PICTURE ZZ,ZZ9.99.
196                05  TERR-SPACES      PICTURE X(5).
197            03  TERR-FINAL-TOT-O     PICTURE ZZZ,ZZ9.99.
198            03  FILLER               PICTURE X(12)    VALUE SPACES.
199
200        01  AREA-TOTALS.
201
202            03  FILLER               PICTURE X(33)    VALUE SPACES.
203            03  FILLER               PICTURE X(11)    VALUE "AREA TOTALS".
204            03  FILLER               PICTURE X(24)    VALUE SPACES.
205            03  AREA-TOTALS-O              OCCURS 3 TIMES.
206                05  AREA-TOT-O       PICTURE ZZ,ZZ9.99.
207                05  AREA-SPACES      PICTURE X(5).
208            03  AREA-FINAL-TOT-O     PICTURE ZZZ,ZZ9.99.
209            03  FILLER               PICTURE X(12)    VALUE SPACES.
210
211        01  FINAL-TOTALS.
212
213            03  FILLER               PICTURE X(33)    VALUE SPACES.
214            03  FILLER               PICTURE X(12)    VALUE "FINAL TOTALS".
215            03  FILLER               PICTURE X(23)    VALUE SPACES.
216            03  FINAL-TOTALS-O             OCCURS 3 TIMES.
217                05  FINAL-TOT-O      PICTURE ZZ,ZZ9.99.
218                05  FINAL-SPACES     PICTURE X(5).
219            03  FINAL-TOTAL-O        PICTURE ZZZ,ZZ9.99.
220            03  FILLER               PICTURE X(12)    VALUE SPACES.
221
222        01  SAVE-RECORD              PICTURE X(132).
223
224       *************************
225
226        PROCEDURE DIVISION.
227
228       *************************
229
230        000-PRODUCE-SALES-REPORT.
231
232            OPEN    INPUT   FILE-IN
233                    OUTPUT  FILE-OUT.
234            PERFORM 100-READ-SALES-RECORD.
235            PERFORM 200-PRODUCE-PRODUCT-SALES-LINE
236                    UNTIL NO-MORE-DATA.
```

```
237              PERFORM 300-PRINT-TOTALS.
238              CLOSE   FILE-IN
239                      FILE-OUT.
240              STOP RUN.
241
242          100-READ-SALES-RECORD.
243
244              READ FILE-IN
245                  AT END MOVE "NO " TO MORE-DATA-FLAG.
246
247          200-PRODUCE-PRODUCT-SALES-LINE.
248
249              IF FIRST-RECORD
250                  MOVE SALESMAN-NUMBER     TO SALESMAN-NUMBER-SAVED
251                  MOVE TERRITORY-NUMBER    TO TERRITORY-NUMBER-SAVED
252                  MOVE AREA-NUMBER         TO AREA-NUMBER-SAVED
253                  MOVE "NO "               TO FIRST-RECORD-FLAG
254              ELSE
255                  NEXT SENTENCE.
256              IF SALESMAN-NUMBER IS NOT = SALESMAN-NUMBER-SAVED
257                  PERFORM 210-PRINT-SALESMAN-TOTALS
258                  MOVE SALESMAN-NUMBER TO SALESMAN-NUMBER-SAVED
259              ELSE
260                  NEXT SENTENCE.
261              IF TERRITORY-NUMBER IS NOT = TERRITORY-NUMBER-SAVED
262                  PERFORM 220-PRINT-TERRITORY-TOTALS
263                  MOVE TERRITORY-NUMBER TO TERRITORY-NUMBER-SAVED
264              ELSE
265                  NEXT SENTENCE.
266              IF AREA-NUMBER IS NOT = AREA-NUMBER-SAVED
267                  PERFORM 230-PRINT-AREA-TOTALS
268                  MOVE AREA-NUMBER TO AREA-NUMBER-SAVED
269              ELSE
270                  NEXT SENTENCE.
271              PERFORM 240-ACCUMULATE-TOTALS
272                  VARYING PRODUCT-SUBSCRIPT FROM 1 BY 1
273                  UNTIL PRODUCT-SUBSCRIPT IS GREATER THAN 3.
274              PERFORM 250-PRINT-PRODUCT-SALES-LINE.
275              PERFORM 100-READ-SALES-RECORD.
276
277          210-PRINT-SALESMAN-TOTALS.
278
279              MOVE SALES-FINAL-TOT TO SALES-FINAL-TOT-O.
280              PERFORM 260-STORE-SALESMAN-TOTALS
281                  VARYING PRODUCT-SUBSCRIPT FROM 1 BY 1
282                  UNTIL PRODUCT-SUBSCRIPT IS GREATER THAN 3.
283              MOVE SPACES TO PRINT.
284              PERFORM 270-WRITE-DETAIL-LINE.
285              MOVE SALESMAN-TOTALS TO PRINT.
286              PERFORM 270-WRITE-DETAIL-LINE.
287              MOVE SPACES TO PRINT.
288              PERFORM 270-WRITE-DETAIL-LINE.
289              MOVE ZEROES TO SALESMAN-WS.
290              MOVE ZEROES TO SALES-FINAL-TOT.
291
292          220-PRINT-TERRITORY-TOTALS.
293
294              MOVE TERR-FINAL-TOT TO TERR-FINAL-TOT-O.
295              PERFORM 280-STORE-TERRITORY-TOTALS
296                  VARYING PRODUCT-SUBSCRIPT FROM 1 BY 1
297                  UNTIL PRODUCT-SUBSCRIPT IS GREATER THAN 3.
298              MOVE TERRITORY-TOTALS TO PRINT.
299              PERFORM 270-WRITE-DETAIL-LINE.
300              MOVE SPACES TO PRINT.
301              PERFORM 270-WRITE-DETAIL-LINE.
302              MOVE ZEROES TO TERRITORY-WS.
303              MOVE ZEROES TO TERR-FINAL-TOT.
304
305          230-PRINT-AREA-TOTALS.
306
307              MOVE AREA-FINAL-TOT TO AREA-FINAL-TOT-O.
308              PERFORM 290-STORE-AREA-TOTALS
309                  VARYING PRODUCT-SUBSCRIPT FROM 1 BY 1
310                  UNTIL PRODUCT-SUBSCRIPT IS GREATER THAN 3.
311              MOVE AREA-TOTALS TO PRINT.
312              PERFORM 270-WRITE-DETAIL-LINE.
313              MOVE SPACES TO PRINT.
314              PERFORM 270-WRITE-DETAIL-LINE.
315              MOVE ZEROES TO AREA-WS.
316              MOVE ZEROES TO AREA-FINAL-TOT.
317
318          240-ACCUMULATE-TOTALS.
319
320              ADD PRODUCT-SALES (PRODUCT-SUBSCRIPT)
321                  TO SALESMAN-TOT-WS (PRODUCT-SUBSCRIPT).
322              ADD PRODUCT-SALES (PRODUCT-SUBSCRIPT)
323                  TO TERR-TOT-WS (PRODUCT-SUBSCRIPT).
324              ADD PRODUCT-SALES (PRODUCT-SUBSCRIPT)
325                  TO AREA-TOT-WS (PRODUCT-SUBSCRIPT).
326              ADD PRODUCT-SALES (PRODUCT-SUBSCRIPT)
327                  TO FINAL-TOT-WS (PRODUCT-SUBSCRIPT).
328              ADD PRODUCT-SALES (PRODUCT-SUBSCRIPT)
329                  TO PRODUCT-SALES-TOT.
330              ADD PRODUCT-SALES (PRODUCT-SUBSCRIPT)
331                  TO SALES-FINAL-TOT.
332              ADD PRODUCT-SALES (PRODUCT-SUBSCRIPT)
333                  TO TERR-FINAL-TOT.
334              ADD PRODUCT-SALES (PRODUCT-SUBSCRIPT)
335                  TO AREA-FINAL-TOT.
336              ADD PRODUCT-SALES (PRODUCT-SUBSCRIPT)
337                  TO FINAL-TOT.
338
```

```
339        250-PRINT-PRODUCT-SALES-LINE.
340
341            MOVE T-MONTH              TO T-MONTH-O.
342            MOVE T-DAY                TO T-DAY-O.
343            MOVE T-YEAR               TO T-YEAR-O.
344            MOVE SALESMAN-NUMBER      TO SALESMAN-NUMBER-O.
345            MOVE TERRITORY-NUMBER     TO TERRITORY-NUMBER-O.
346            MOVE AREA-NUMBER          TO AREA-NUMBER-O.
347            MOVE PRODUCT-SALES-TOT    TO PRODUCT-SALES-TOT-O.
348            PERFORM 300-STORE-PRODUCT-SALES
349                VARYING PRODUCT-SUBSCRIPT FROM 1 BY 1
350                UNTIL PRODUCT-SUBSCRIPT IS GREATER THAN 3.
351            MOVE DETAIL-LINE TO PRINT.
352            PERFORM 310-PRINT-DETAIL-LINE.
353
354        260-STORE-SALESMAN-TOTALS.
355
356            MOVE SPACES TO SALESMAN-SPACES (PRODUCT-SUBSCRIPT).
357            MOVE SALESMAN-TOT-WS (PRODUCT-SUBSCRIPT)
358                TO SALESMAN-TOT-O (PRODUCT-SUBSCRIPT).
359
360        270-WRITE-DETAIL-LINE.
361
362            WRITE PRINT
363                AFTER ADVANCING 1 LINE.
364            ADD 1 TO LINE-COUNT-WS.
365
366        280-STORE-TERRITORY-TOTALS.
367
368            MOVE SPACES TO TERR-SPACES (PRODUCT-SUBSCRIPT).
369            MOVE TERR-TOT-WS (PRODUCT-SUBSCRIPT)
370                TO TERR-TOT-O (PRODUCT-SUBSCRIPT).
371
372        290-STORE-AREA-TOTALS.
373
374            MOVE SPACES TO AREA-SPACES (PRODUCT-SUBSCRIPT).
375            MOVE AREA-TOT-WS (PRODUCT-SUBSCRIPT)
376                TO AREA-TOT-O (PRODUCT-SUBSCRIPT).
377
378        300-STORE-PRODUCT-SALES.
379
380            MOVE SPACES TO SALES-SPACES (PRODUCT-SUBSCRIPT).
381            MOVE PRODUCT-SALES (PRODUCT-SUBSCRIPT)
382                TO PRODUCT-SALES-O (PRODUCT-SUBSCRIPT).
383
384        310-PRINT-DETAIL-LINE.
385
386            IF LINE-COUNT-WS IS > 37
387                MOVE PRINT TO SAVE-RECORD
388                PERFORM 320-PRINT-HEADING-LINE
389                MOVE SAVE-RECORD TO PRINT
390            ELSE
391                NEXT SENTENCE.
392                PERFORM 270-WRITE-DETAIL-LINE.
393
394        320-PRINT-HEADING-LINE.
395
396            MOVE HDG-1 TO PRINT.
397            PERFORM 330-WRITE-TOP-OF-PAGE.
398            MOVE HDG-2 TO PRINT.
399            PERFORM 270-WRITE-DETAIL-LINE.
400            MOVE SPACES TO PRINT.
401            PERFORM 340-WRITE-DOUBLE-DETAIL-LINE.
402
403        330-WRITE-TOP-OF-PAGE.
404
405            WRITE PRINT
406                AFTER ADVANCING PAGE.
407            MOVE 1 TO LINE-COUNT-WS.
408
409        340-WRITE-DOUBLE-DETAIL-LINE.
410
411            WRITE PRINT
412                AFTER ADVANCING 2 LINES.
413            ADD 2 TO LINE-COUNT-WS.
414
415        300-PRINT-TOTALS.
416
417            PERFORM 210-PRINT-SALESMAN-TOTALS.
418            PERFORM 220-PRINT-TERRITORY-TOTALS.
419            PERFORM 230-PRINT-AREA-TOTALS.
420            PERFORM 310-PRINT-FINAL-TOTALS.
421
422        310-PRINT-FINAL-TOTALS.
423
424            MOVE FINAL-TOT TO FINAL-TOTAL-O.
425            PERFORM 320-STORE-FINAL-TOTALS
426                VARYING PRODUCT-SUBSCRIPT FROM 1 BY 1
427                UNTIL PRODUCT-SUBSCRIPT IS GREATER THAN 3.
428            MOVE FINAL-TOTALS TO PRINT.
429            PERFORM 270-WRITE-DETAIL-LINE.
430
431        320-STORE-FINAL-TOTALS.
432
433            MOVE SPACES TO FINAL-SPACES (PRODUCT-SUBSCRIPT).
434            MOVE FINAL-TOT-WS (PRODUCT-SUBSCRIPT)
435                TO FINAL-TOT-O (PRODUCT-SUBSCRIPT).

THERE WERE 435 SOURCE INPUT LINES.
THERE WERE NO DIAGNOSTICS.
```

DATE	SALESMAN NUMBER	TERRITORY NUMBER	AREA NUMBER	PRODUCT 1 SALES	PRODUCT 2 SALES	PRODUCT 3 SALES	TOTAL SALES
01/11/81	5104	30	21	3,500.00	45,000.00	32,005.00	80,505.00
01/22/81	5104	30	21	2,950.50	3,598.55	1,110.10	88,164.15
01/24/81	5104	30	21	590.30	1,000.00	2,000.08	91,754.53
	SALESMAN TOTALS			7,040.80	49,598.55	35,115.18	91,754.53
01/26/81	2134	30	21	1,953.00	2,553.30	3,280.50	99,541.33
01/28/81	2134	30	21	591.10	5,410.03	6,028.00	111,570.46
	SALESMAN TOTALS			2,544.10	7,963.33	9,308.50	19,815.93
	TERRITORY TOTALS			9,584.90	57,561.88	44,423.68	111,570.46
	AREA TOTALS			9,584.90	57,561.88	44,423.68	111,570.46
01/13/81	3007	60	22	555.55	6,532.10	7,755.00	126,413.11
01/19/81	3007	60	22	98.00	1,014.44	444.47	127,970.02
	SALESMAN TOTALS			653.55	7,546.54	8,199.47	16,399.56
01/27/81	1510	60	22	888.33	100.50	1,875.00	130,833.85
01/27/81	1510	60	22	97.50	1,500.00	2,350.00	134,781.35
	SALESMAN TOTALS			985.83	1,600.50	4,225.00	6,811.33
	TERRITORY TOTALS			1,639.38	9,147.04	12,424.47	23,210.89
01/17/81	1980	80	22	7,000.00	9,000.00	4,500.00	155,281.35
01/20/81	1980	80	22	175.00	1,950.00	8,510.00	165,916.35
01/22/81	1980	80	22	5,777.00	594.40	819.40	173,107.15
	SALESMAN TOTALS			12,952.00	11,544.40	13,829.40	38,325.80
	TERRITORY TOTALS			12,952.00	11,544.40	13,829.40	38,325.80
	AREA TOTALS			14,591.38	20,691.44	26,253.87	61,536.69
	FINAL TOTALS			24,176.28	78,253.32	70,677.55	173,107.15

Questions for Review

1. Define table, table element, searching, argument, and function.
2. Briefly describe the following standard terms of table handling: subscripting, indexing, table values, search arguments, and table lookup.
3. What are the principal functions of the table handling feature?
4. What are the main uses of tables?
5. What is the advantage of using table format over a standard sequential file?
6. How may the entries in a table be defined?
7. What are arrays, in what ways are they similar to tables, and in what ways do they differ?
8. How are items identified in a table?
9. How is a one-dimensional, two-dimensional, and three-dimensional table defined?
10. What is the function of the OCCURS clause? Explain the use of the DEPENDING ON, KEY, and INDEXED BY options.
11. Explain the use of the USAGE IS INDEX clause.
12. What is the function of the REDEFINES clause?
13. How is the PERFORM statement with the VARYING phrase used in table handling?
14. How does the VARYING and AFTER options of the PERFORM statement allow the programmer to manipulate tables?
15. How is a reference to a table indicated?
16. What is a subscript and how is it used in table referencing?
17. How is indexing used in table handling?
18. Briefly describe the two methods of loading tables.
19. How is a table searched?
20. Define the SEARCH statement and its two forms.
21. What is the function of the SET statement and how is it used in table handling?
22. What are the restrictions on table values?

Matching Questions

Match each item with its proper description.

_____ 1. INDEXED BY
_____ 2. Argument
_____ 3. Table
_____ 4. OCCURS
_____ 5. Index
_____ 6. Array
_____ 7. DEPENDING ON
_____ 8. Function
_____ 9. KEY
_____10. Subscript

A. Used to define tables and other homogeneous sets of data whose elements can be referenced by subscripting or indexing.
B. A continuous series of data fields stored side by side, so they can be referenced as a group.
C. Matches entry in a corresponding position.
D. Phrase of OCCURS clause used when the end of the occurrence cannot be otherwise determined.
E. An integer whose value identifies a particular element within a table.
F. A collection of related information organized in such a way so that each item can be referenced by its position.
G. Computer storage position or register whose contents corresponds to an occurrence number.
H. Search word.
I. Required if the subject of the entry is to be referred to by indexing.
J. Used in conjunction with the INDEXED BY option in the execution of the SEARCH ALL statement.

Match each item with its proper description.

_____ 1. AT END
_____ 2. REDEFINES
_____ 3. Execution-time
_____ 4. SET
_____ 5. USAGE IS INDEX
_____ 6. SEARCH
_____ 7. WHEN
_____ 8. VARYING
_____ 9. Compile-time
_____10. SEARCH ALL

A. Used to specify the format of the data-item in the computer storage so that index-data items are specified.
B. Manipulates index-names or data-names used as subscripts in the PERFORM procedure.
C. Table element that specifies the condition and adjusts the associated index-name to indicate the position of the table element.
D. Hierarchical structure of the table.
E. Binary search.
F. Search is terminated and imperative statement is executed.
G. Specifies that the same area is to contain different data items.
H. A large table loaded from a file of records onto a disk or main storage.
I. Used to establish, modify, or preserve the value associated with an index-name.
J. Condition to be satisfied in a search.

Exercises

Multiple Choice: Indicate the best *answer (questions 1–31).*

1. An argument is sometimes called a(n)
   a. function.
   b. search word.
   c. index.
   d. subscript.
2. Tables are sets of
   a. unrelated variables.
   b. sequential files.
   c. related variables.
   d. noncontiguous data items.
3. A use of tables involves
   a. using known information to obtain another piece of information.
   b. verification of information with no retrieval of information.
   c. updating information.
   d. All of the above.
4. The entries in a table may be
   a. arguments.
   b. functions.
   c. pairs of arguments and functions.
   d. All of the above.
5. An array is
   a. not a continuous series of data fields.
   b. used in the same manner as tables.
   c. referenced as a group.
   d. All of the above.
6. COBOL allows up to
   a. five dimensions of tables.
   b. three dimensions of tables.
   c. two dimensions of tables.
   d. an unlimited number of dimensions of tables.
7. The OCCURS clause
   a. may be used with 01 level items.
   b. can describe an item whose size is variable.
   c. defines tables and other homogeneous sets of repeated sets of data.
   d. does not require subscripts or indexes of the items that it is associated with.
8. The OCCURS clause is used with the following option, if the number of occurrences cannot be determined.
   a. KEY.
   b. INDEXED BY.
   c. REDEFINES.
   d. DEPENDING ON.
9. The KEY option is used
   a. with the INDEXED BY option in the execution of a SEARCH ALL statement.
   b. to indicate that repeated sets of data are organized in ASCENDING or DESCENDING order.
   c. to indicate that the data names are listed in descending order of significance.
   d. All of the above.
10. The INDEXED BY option is required
    a. if the subject of this entry is to be referenced to by indexing.
    b. if the subject of this entry is to be referenced to by subscripting.
    c. if the subject of this entry is to be referenced to by both subscripting and indexing.
    d. All of the above.
11. The USAGE IS INDEX clause may
    a. be used at the elementary level only.
    b. contradict the USAGE clause of a group of which it is part of.
    c. be written only at the group level.
    d. None of the above.

12. The REDEFINES clause
    a. specifies the redefinition of items occupying two areas.
    b. redefines a storage area.
    c. requires that the redefining area be the same as the area it redefines.
    d. sets aside a new storage area.
13. The VARYING and AFTER phrases of the PERFORM statement allow the programmer to
    a. manipulate up to five control items.
    b. establish one control condition for the manipulation of all items.
    c. establish separate conditions for the manipulation of each item.
    d. None of the above.
14. A reference can be made to the following item.
    a. An individual table element.
    b. An item within a group item that is a table element.
    c. A condition-name associated with a table element.
    d. All of the above.
15. The following relate to subscripting and indexing:
    a. Index-names may be combined with data-name subscripts in the same statement.
    b. Indexing may be used where subscripting is prohibited.
    c. A data-name used as subscript may be neither subscripted nor indexed.
    d. All of the above.
16. A subscript
    a. may have a negative value.
    b. may have a decimal point.
    c. must be an integer.
    d. may be zero.
17. Indexing
    a. specifies occurrence numbers.
    b. must be used with a data item that is included in an OCCURS clause.
    c. must be used with the optional INDEXED BY phrase.
    d. All of the above.
18. The REDEFINES clause is used in the setting up of the following tables:
    a. Compile-time tables.
    b. Execution-time tables.
    c. File tables.
    d. Preexecution-time tables.
19. The following options are used with the SEARCH statement.
    a. INDEXED BY.
    b. DEPENDING ON.
    c. WHEN.
    d. PERFORM.
20. A binary search statement
    a. searches a table in any sequence.
    b. is defined with subscripting and indexing.
    c. must include a KEY phrase.
    d. may search multidimensional tables.
21. A SET statement may be used to
    a. set an index-name.
    b. modify an index-name.
    c. preserve an index-name.
    d. All of the above.
22. A subscript may be
    a. any integer including zero.
    b. any variable.
    c. any variable with an integer value.
    d. any integer including negative values.
23. A variable may be defined as a subscript with a picture of
    a. Xs.
    b. 9s.
    c. 9s and a V specifying a decimal place.
    d. All of the above.

24. A value may be assigned to a variable in the Procedure Division by
    a. accepting the value through the console keyboard.
    b. reading the value from a file.
    c. moving a literal to a variable.
    d. All of the above.
25. When the DEPENDING ON option of the OCCURS clause is specified, the integer must be
    a. at least one.
    b. the maximum number of elements the table will have during any execution.
    c. the same as the highest occurrence number the table will have during execution.
    d. All of the above.
26. The variable specified in the DEPENDING ON option of the OCCURS clause may be
    a. qualified.
    b. subscripted.
    c. both A and B.
    d. None of the above.
27. An index-name
    a. must be defined with a picture of 9s.
    b. may be defined with a picture of Xs.
    c. is defined by its appearance in the INDEXED BY option.
    d. is defined in the USAGE IS INDEX clause.
28. When a SEARCH statement is executed,
    a. the index is automatically set to 1.
    b. the search begins with the element whose occurrence number corresponds to the current index value.
    c. the index is set the maximum number of occurrences in the table referenced.
    d. the index number is set to zero.
29. One of the actions of the PERFORM statement with the VARYING option is to change the value of an item. In the following, which item will be changed?

```
PERFORM RETIREMENT-COMPUTATION
    VARYING AGE FROM AGE-WHEN-HIRED
    BY 1 UNTIL AGE IS EQUAL TO 65.
```

    a. AGE.
    b. AGE-WHEN-HIRED.
    c. RETIREMENT-COMPUTATION.
    d. None of the above.
30. After a procedure has been performed, the value of a base item is increased by a specified amount. Either the literal amount or the name of an item containing the amount in the PERFORM . . . VARYING statement is written after the word
    a. UNTIL.
    b. FROM.
    c. BY.
    d. VARYING.
31. The number of items that a table may contain is
    a. standardized, and the same in all programs.
    b. determined by the programmer.
    c. determined by the system.
    d. determined by the job control cards.
32.

```
01  PERCENT-TABLE       USAGE COMP-3.
    05  PERCENT         PICTURE V99     OCCURS 3 TIMES.
```

Match the correct facts about PERCENT-TABLE with the appropriate portion of the Data Division entries.

_____ 1. USAGE-COMP-3	A. PERCENT-TABLE will have two elements.
_____ 2. OCCURS 3 TIMES	B. PERCENT-TABLE will have three elements.
_____ 3. PICTURE V99	C. Each element of PERCENT-TABLE will have two digits.
	D. Each element of PERCENT-TABLE will have three digits.
	E. Each element will be stored as packed decimal.

33. WORKING-STORAGE SECTION.

```
01  DAYS-OF-THE-WEEK.
    05  FILLER      PICTURE A(9)      VALUE 'SUNDAY'.
    05  FILLER      PICTURE A(9)      VALUE 'MONDAY'.
    05  FILLER      PICTURE A(9)      VALUE 'TUESDAY'.
    05  FILLER      PICTURE A(9)      VALUE 'WEDNESDAY'.
    05  FILLER      PICTURE A(9)      VALUE 'THURSDAY'.
    05  FILLER      PICTURE A(9)      VALUE 'FRIDAY'.
    05  FILLER      PICTURE A(9)      VALUE 'SATURDAY'.
```

a. Write the REDEFINES statement WEEK, which will allow us to use the table.

b. Is this a compile-time or execution-time table?

34. **Here is one way in which the table of wage rates can be structured. The man number and wage rate for each employee are stored together in the table. In this example, there are seventy-five employees on the payroll.**

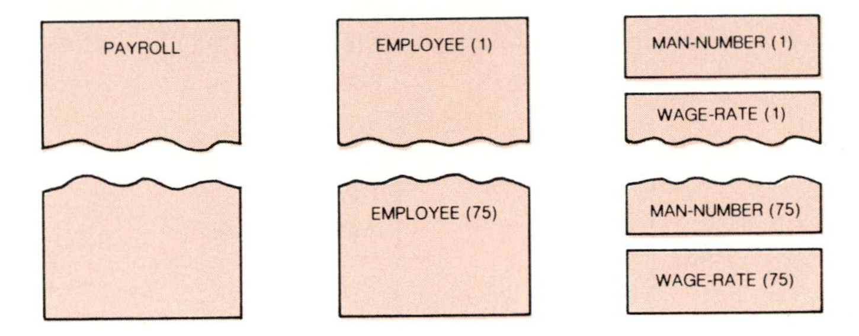

Which set of the following entries corresponds to the record structure just diagrammed?

a.
```
01  PAYROLL.
    05  EMPLOYEE.
        10  MAN-NUMBER OCCURS 75 TIMES.
                PICTURE 9(5).
        10  WAGE-RATE OCCURS 75 TIMES.
                PICTURE 9(5).
```

b.
```
01  PAYROLL.
    05  EMPLOYEE OCCURS 75 TIMES.
        10  MAN-NUMBER, PICTURE 9(5).
        10  WAGE-RATE, PICTURE 9V999.
```

35. **In COBOL, our thinking must proceed from the largest to the smallest item; therefore, in planning our description of this record, we must begin with the record as a whole. This drawing suggests how we may think of the structure of this record and its progressive subdivision.**

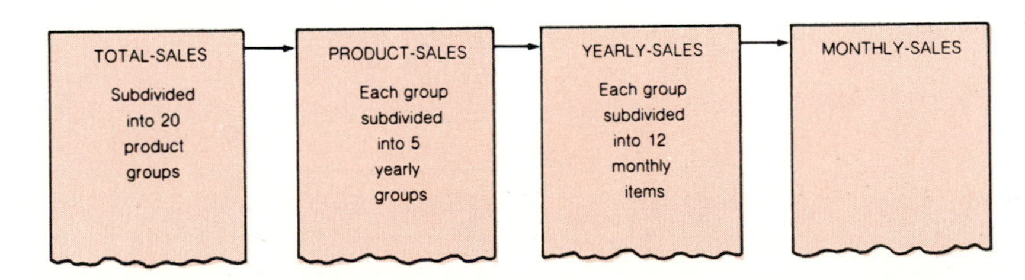

Here is the start of the record description. See if you can complete it. The elementary item, MONTHLY-SALES, consists of a sign plus seven digits, including two decimal places, and is stored in packed-decimal format.

```
01  TOTAL-SALES.
    05  PRODUCT-SALES, OCCURS 20 TIMES.
```

36. A typical tax table follows:

Dependents	Rate
1	.175
2	.164
3	.159
4	.153
5	.147
6	.141
7	.121

In the Working-Storage Section,

a. Set up table.
b. Redefine the table describing it as one value entry repeated a certain number of times.
c. Write the procedural statement using subscripts to multiply GROSS by the fourth element in the table to arrive at TAXAMT.

37. In the following insurance table, the entries for ages are 1 to 65 (Premium XXX.XX).

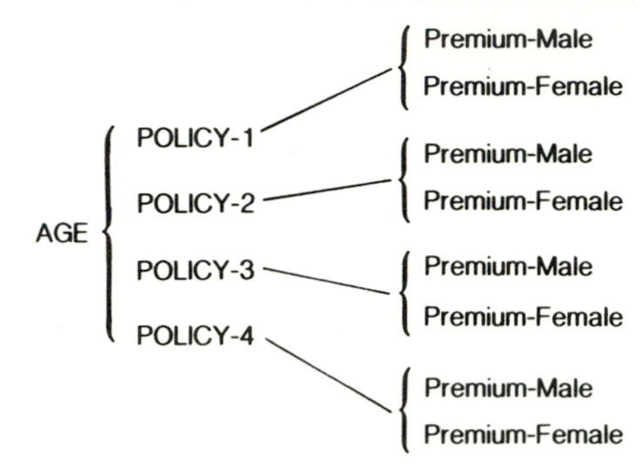

It is necessary to reference the premium for a female, age 64, for policy 2. Write the necessary Data and Procedure Division entries to accomplish the aforementioned.

38. A deck of sales records (in random sequence) is read and the total from each record is accumulated in the appropriate counter depending upon the value (01-50) recorded in the STATE-CODE field of the record. At the end of the file, the totals of each of the fifty records is to be printed along with the state code.
Write the necessary Data and Procedure Division entries to accomplish the following:

a. Set up a table for fifty items. Each item is to contain eight positions to represent a sales total counter for each of the fifty states.
b. Print a report indicating the state number and total.

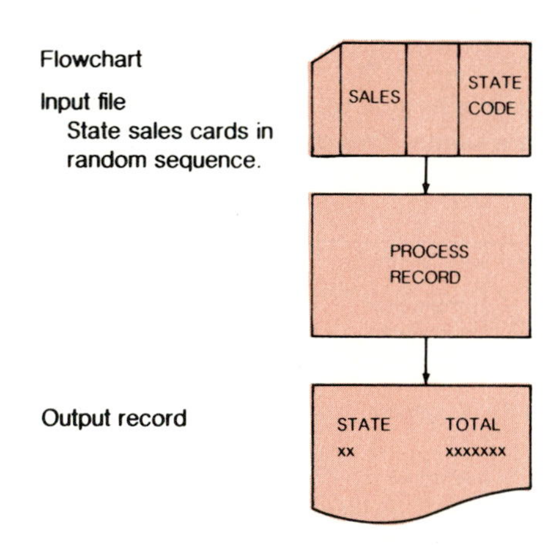

39.
```
05  A OCCURS 5 TIMES.
    10  B OCCURS 4 TIMES.
        15  C PICTURE 99.
```
It is necessary to reference C in the 4th B in the 5th A. Which of the following entries are correct? (There is more than one correct answer.)

a. C IN B IN A (5,4)
b. C (5,4) IN B IN A
c. C IN B (5,4)
d. C IN A (5,4)
e. C (4) IN A (5)

40. Assume you wish to print a header line as follows:

```
01  HEADING-1.
    05  H1              PICTURE X(8) VALUE IS 'ITEM-NO.'.
    05  FILLER          PICTURE X(5).
    05  DESC            PICTURE X(8).
    05  FILLER          PICTURE X(5).
    05  H2              PICTURE X(6) VALUE IS 'TOTAL.'.
    05  FILLER          PICTURE X(5).
    05  DATEI
        10  MONTHI      PICTURE X(9).
        10  FILLER      PICTURE X(3).
        10  DAYI        PICTURE 99.
        10  FILLER      PICTURE XX.
        10  YEARI       PICTURE 99.
```

In another work area, you have the following record:

```
01  WORK-AREA.
    05  ITEM-NO.        PICTURE 99.
    05  MONTH-CODE      PICTURE 99.
    05  DAY-CODE        PICTURE 99.
    05  YEAR-CODE       PICTURE 99.
```

Write the necessary procedural entries to fill in the header-line as follows:

a. ITEM-NO can vary from 01 through 45. Depending upon its contents, DESC is set to one of the following:

01—PENCIL #2	04—PAINTSET	07—STAPLES	10—MISC.
02—PENCIL #3	05—BNDPAPERS	08—CARBONS	
03—PENCIL #4	06—STAPLERS	09—ERASERS	

b. MONTH is to be set depending upon the contents of MONTH-CODE:

01—JANUARY	04—APRIL	07—JULY	10—OCTOBER
02—FEBRUARY	05—MAY	08—AUGUST	11—NOVEMBER
03—MARCH	06—JUNE	09—SEPTEMBER	12—DECEMBER

c. DAY-CODE and YEAR-CODE are to be inserted into DAYI and YEARI.

41. In an insurance premium run, the monthly rates are determined by the risk class. Here we have a situation where there is a wide and irregular gap between the argument assigned to one table value and the argument assigned to the table value that follows. This means that the risk class code could not be used directly as a subscript.

Write the necessary Data and Procedure Division entries to

a. set up the table in the Working-Storage Section and

b. conduct a step-by-step search to locate the appropriate premium rate.

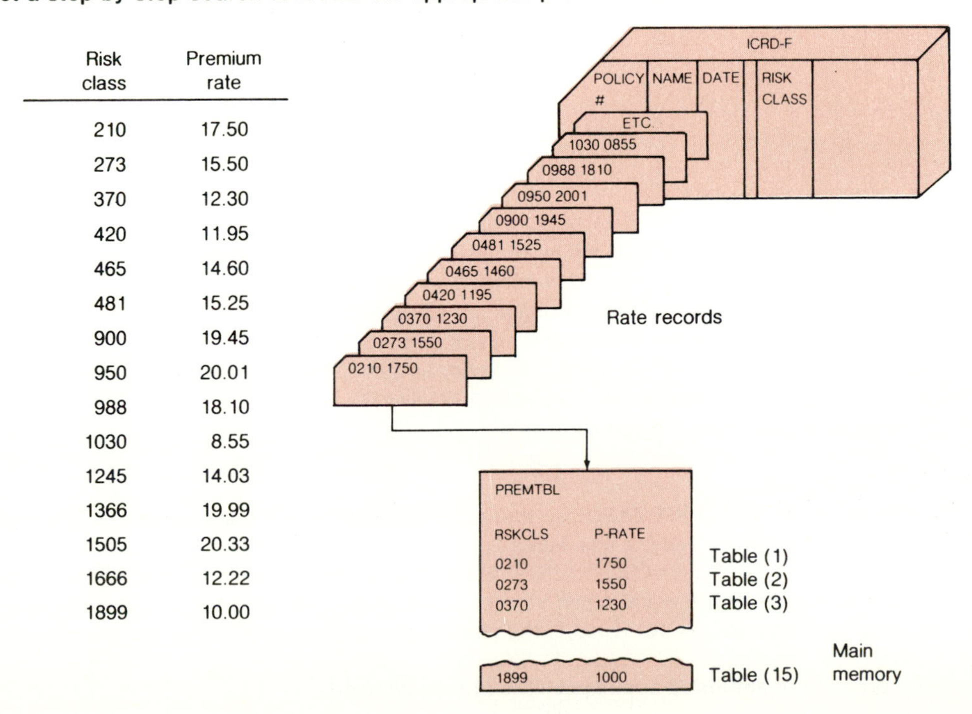

Risk class	Premium rate
210	17.50
273	15.50
370	12.30
420	11.95
465	14.60
481	15.25
900	19.45
950	20.01
988	18.10
1030	8.55
1245	14.03
1366	19.99
1505	20.33
1666	12.22
1899	10.00

42. Given the following information:

```
01  TRANS-RECORD.
    05  CITY-ITEM                                     OCCURS 2 TIMES.
        10  STORE-ITEM                                OCCURS 4 TIMES.
            15  DEPT-NO          PICTURE X(5).
            15  DAILY-SALES      PICTURE 9999V99      OCCURS 3 TIMES.
77  DAYW                         PICTURE 99.
77  DEPT                         PICTURE 99.
77  STORE                        PICTURE 99.
```

Required:

a. Write the procedural statements to add DAILY-SALES to TOTAL-SALES for each item using STORE, DEPT and DAY as subscripts.

b. Using the INDEXED BY option, write the necessary Data and Procedure Division entries to add DAILY-SALES to TOTAL-SALES for each item using STORE-INDEX, DEPT-INDEX and DAY-INDEX as indexes.

43. Given the following information:

```
05  PRODUCT-NO              PICTURE 9999.
05  WAREHOUSE-DATA              OCCURS 3 TIMES
                            ASCENDING KEY IS QTY-ON-HAND
                            INDEXED BY WAREHOUSE-ITEM.
    10  LOCATION            PICTURE 99.
    10  BIN-NO              PICTURE 999.
    10  QTY-ON-HAND         PICTURE 9(5).
```

Write the necessary procedural statements to search the aforementioned to locate QTY-ON-HAND equal to 100. If the item is found, the program is to branch to FOUND-ITEM where the quantity will be moved to an output area (OUTPUT-FIELD). If the item is not found, the program is to branch to NOT-IN-TABLE-ROUTINE.

44. Orders for merchandise are received and recorded as follows:

ORDER-ITEM

Field	Positions	
Code	1	
Stock Number	2-7	
Date	8-13	
Customer Number	14-19	
Customer Name	20-35	
Quantity	36-39	
Price	40-44	XXX.XX
Blank	45-80	

Set up a table for DISCOUNT-TABLE which can be used for items subject to discount. The stock number and discount rate are punched into cards and are treated as values of the DISCOUNT-TABLE as follows:

```
DISCOUNT-TABLE
    DISCOUNT-ITEM (1)           .87
    DISCOUNT-ITEM (2)           .79
    DISCOUNT-ITEM (3)           .85
          .                      .
          .                      .
    DISCOUNT-ITEM (15)          .69
```

The total number of discount items for each month will be recorded to determine the size of the DISCOUNT-TABLE for that month. The maximum number of discount items for one month is fifteen.

a. In the Working-Storage Section, write the entries to define the variables,
   TABLE-MAXIMUM—which is to contain the number of discount items.
   TABLE-VALUE—to which a table value will be transmitted prior to being moved to a specific table element.
   DISCOUNT-TABLE—whose size is determined by the TABLE-MAXIMUM.
   DISCOUNT—to be used as a subscript in referring to table elements.

b. Write the necessary Procedure Division entries to accept the TABLE-MAXIMUM and the DISCOUNT-TABLE.

45. Using the information in exercise 44,

a. Code the record description for DISCOUNT-TABLE so that a SEARCH statement can be used for the text of discount items. (Specify an index for the table. Use D as an index.)

b. Using the SET statement at one, search the DISCOUNT-TABLE until you find a stock number equal to a discount item number, then multiply the quantity of the equal stock number by the price and the discount percentage to arrive at a charge price. If an equal stock number is not found in the search, multiply the quantity by the price to arrive at the charge price.
   Write the necessary procedural entries to accomplish the aforementioned.

Problem 1

There are two related tables containing the names and telephone numbers of subscribers to a certain system of purchasing. You are to set up two related tables so that a search can be made using the name as the search argument and the telephone number as the function. There are ten subscribers.

The output should be a listing including the name of the subscriber and the telephone number. An indication should be made if the subscriber is not in the table.

The table is recorded in the following positions.

Positions	Field
1-11	Name
12-18	Telephone number
19-29	Name
30-36	Telephone number
37-47	Name
48-54	Telephone number
55-65	Name
66-72	Telephone Number

The input record contains only one field.

1-11	Name

Code the necessary COBOL specifications. There is only one area code.

Output is as follows:

```
     NAME           TELEPHONE NUMBER

JOHNSON T P            236-3210

JONSON M               886-5632

JUGGERS Z T            277-0296

JUSTICE P              306-2981

KAMBERLEIN             SUBSCRIBER NOT IN TABLE
```

Problem 2

To compute the weekly pay for each employee, the rate of pay as found in table TABPAY is multiplied by the number of hours worked (a three-position field with one decimal position). The weekly pay amount is a five-position field with two decimals after rounding.

Using the following specifications, code the necessary COBOL instructions.

Table

TABEMP	TABPAY
1062	5.25
1063	6.25
1064	5.40
1065	4.50
1066	5.50
1067	4.75
1068	6.50
1069	4.50
1070	5.00
1071	8.00

Input

Positions	Field
1-4	Employee number
35-37	Hours XX.X

Calculations

Perform a search of the table using employee number as the search word. After a successful search, perform the necessary calculations.

Output

Leave five positions between each field. Output the employee number, hours worked, rate of pay, and gross pay.

Output is as follows:

EMPLOYEE NUMBER	HOURS WORKED	RATE OF PAY	GROSS PAY
1062	15.5	5.25	81.38
1065	32.5	4.50	146.25
1066	40.0	5.50	220.00
1068	5.0	6.50	32.50
1071	25.0	8.00	200.00

Problem 3

To compute the correct amount of an invoice, the amount of tax must be added to the purchase price. To do this, a table is set up as follows, and the correct tax is added to the purchase price.

Table

Tax is calculated at 6¢ per $1.00 and each fraction over $1.00 is based on the following rates. (There is no tax on any amount under $1.00.)

Fraction over $1.00	Rate
.01-.19	$.01
.20-.39	.02
.40-.59	.03
.60-.79	.04
.80-.99	.05

Input

Positions	Field	
1-5	Customer number	
10-14	Purchase price	(XXX.XX)

Calculations

Search the table for the correct amount. Add this amount to the purchase price to determine the total price.

Output

Print a report showing the customer number, purchase price, tax, and total price. Leave five spaces between each field. Be sure to put in the proper headings.

Output is as follows:

CUSTOMER NUMBER	PURCHASE PRICE	TAX	TOTAL PRICE
10865	1.49	.09	1.58
12850	.52	.00	.52
22560	1.68	.10	1.78
25647	1.21	.08	1.29
26841	1.85	.11	1.96
36875	123.56	7.41	130.97
47250	258.63	15.52	274.15
78250	653.24	39.20	692.44

Problem 4

The EZ Money Corp. manufactures five products as follows:

Product Number	Description
1	Soap
2	Bleach
3	Detergent
4	Cleanser
5	Powder

The unit cost price and unit selling price for each product is as follows:

Product Number	Unit Cost	Unit Selling Price
1	3.75	5.25
2	2.38	4.25
3	5.67	7.25
4	3.19	5.00
5	2.76	4.55

A report is to be prepared showing the following information:

Product Report

Prod. No. Description Units Cost Sales Profit

Input

Positions	Field
1-6	Date
7	Product number
8-12	Units sold

Required

Two arrays are to be set up to perform the following (one alternating array for the description and cost and one array for the selling price):

1. The product number is to serve as the index.
2. The description of the product is to be looked up in the array.
3. The units sold are to be multiplied by the unit cost to arrive at the cost.
4. The units sold are to be multiplied by the unit selling price to arrive at sales.
5. The cost is subtracted from the sales, giving the profit.

   All arrays are to be loaded at compile time.

   Output is as follows:

PRODUCT NUMBER	DESCRIPTION	UNITS	COST	SALES	PROFIT
1	SOAP	250	937.50	1,312.50	375.00
5	POWDER	30	82.80	136.50	53.70
3	DETERGENT	100	567.00	725.00	158.00
4	CLEANSER	20	63.80	100.00	36.20
2	BLEACH	15	35.70	63.75	28.05
3	DETERGENT	157	890.19	1,138.25	248.06
1	SOAP	79	296.25	414.75	118.50
4	CLEANSER	524	1,671.56	2,620.00	948.44
5	POWDER	17	46.92	77.35	30.43

Sort-Merge Feature

Chapter Outline

Chapter Objectives

The learning objectives of this chapter are:

1. To describe the sort and merge features of COBOL and how they may be applied to data processing problems.
2. To differentiate between sorting and merging and when each should be used in programming.
3. To explain the operation of the SORT and MERGE statements and how they are used to order files or to combine two or more files that are already ordered into one ordered file.
4. To explain the operation and function of input procedures that are used to process records prior to the sorting process, and also output procedures that are used to process records after the sorting or merging process.
5. To describe the format and use of the RELEASE and RETURN statements that are used in the input and output procedures respectively.

Introductory Concepts

Sorting

Much data processing depends upon the order in which records appear on the files being processed. Such processing often depends upon the order of the records being sequenced according to the values of one or more fields that appear in each of the records. The fields upon which the ordering depends are called "keys" of the file.

Since data in its original form seldom occurs in well-ordered sequences, a technique, sorting, is provided by which the programmer can impose the desired ordering upon the records in a file. A sorting procedure manipulates an input file whose records are in an indeterminate sequence, and produces an output file containing the same set of data records rearranged into the desired sequence. The number of records in the input file is usually unknown at the inception of the sort procedure and is generally not relevant to the sorting procedure.

Merging

Some data processing operations are performed on a group of records that are distributed on several files. If all of those several files are themselves well ordered by the same rules, their contents may be merged into one composite file, which is itself well ordered. A merging procedure manipulates two or more well-ordered input files and produces an output file containing the total set of data records in one well-ordered sequence.

Ordering

The sequencing of the output file in a sorting or merging procedure is governed by the values of one or more fields in each of the records being ordered. These fields, the keys, must appear in the same position relative to the start of the record, in every record being sorted or merged.

The order in which the keys are specified to the sort or merge procedure determines their hierarchical relationship. The first key named, the major key, is the most significant field. Each successive key specified is of decreasing significance until the last key, the most minor key, is reached.

Each key may also be specified as determining an ascending ordering or descending ordering. Ascending ordering means that those records with lower values of that key will appear in the output prior to the records with higher values for that key. Descending ordering implies the inverse result.

Ordering is accomplished by comparing, from major to most minor, the corresponding keys of two records until an inequality of value is found. The output order is then determined by the ascending or descending rule that applies to that particular key field. If there is no inequality of value in any corresponding pair of keys, the ordering is determined by the sort or merge procedure.

Description of Sort-Merge

The sort-merge feature provides the capability to order one or more files of records, or to combine two or more identically ordered files of records, according to a set of programmer-specified keys contained within each record. Optionally, a programmer may apply some special processing to each of the individual records by input or output procedures. This special processing may be applied before and/or after the records are ordered by the sort (see figure 11.1) or after the records have been combined by the merge.

Record Ordering

The ability to arrange records into a particular order is a common requirement of a data processing programmer. The sort and merge features of COBOL provide facilities to assist in meeting this requirement.

While both are concerned with record ordering, the functions and capabilities of the SORT and MERGE statements are different in a number of respects. *The SORT statement will produce an ordered file from one or more files that may be completely unordered in the sort sequence, whereas the*

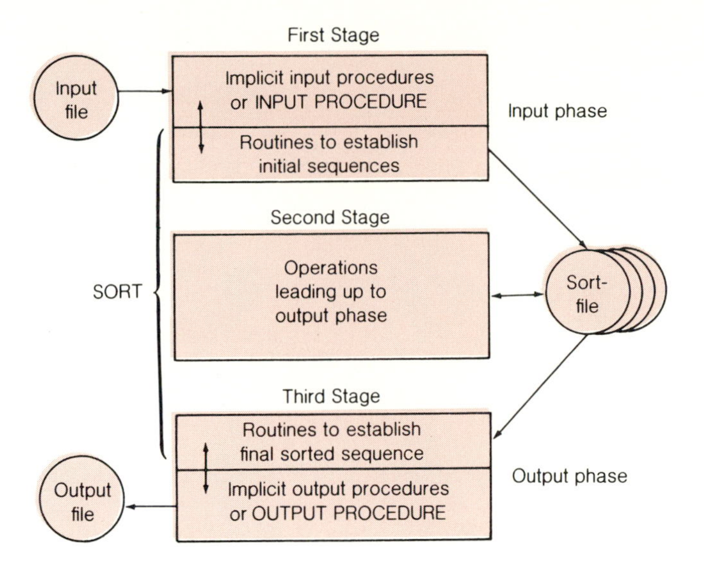

Figure 11.1 Schematic—sort operation.

MERGE statement can only produce an ordered file from two or more files, each of which is already ordered in the specified sequence.

In many applications, it is necessary to apply some special processing to the contents of the sort or merge file(s) before or after sorting or merging. This special processing may consist of addition, deletion, creation, altering, editing, or some other modification of the individual records in the file. The COBOL sort-merge feature allows the programmer to express these procedures in the COBOL language. A COBOL program may contain any number of sorts and merges, and each of them may have its own independent special procedures. The sort-merge feature automatically causes execution of these procedures in such a way that extra passes over the sort or merge files are not required.

Relationship with File Input-Output

The files specified in the USING and GIVING phrases of the SORT and MERGE statements must be described explicitly or implicitly in the FILE-CONTROL paragraph as having sequential organization. No input-output statement may be executed for the file named in the sort-merge file description.

Program Organization

The COBOL language contains two verbs which initiate the SORT and MERGE ordering procedures. Several additional language features are associated with both of these verbs. The RELEASE verb may be used with the execution of a SORT verb and the RETURN verb may be used with the execution of SORT and MERGE verbs. A special form of the SELECT clause in the FILE-CONTROL paragraph of the Environment Division is associated with the intermediate working files of the sort procedure and the implicitly working file of the merge procedure. In addition, those files are described by a special type of file description (SD rather than FD) in the Data Division.

The SORT verb invokes the execution of a set of sorting procedures contained within the standard software library. These procedures operate with the COBOL object program to perform the sort. During the execution of the sorting function, the object program and sorting procedures combine in the organization as pictured in figure 11.2. The programmer can include several SORT statements in the source program. If several SORT statements are present, they are completely independent of each other.

The MERGE verb invokes the execution of a set of merging procedures contained within the standard software library. These procedures operate with the COBOL object program to perform the merge. During the execution of the merging function, the object program and merging procedures combine in

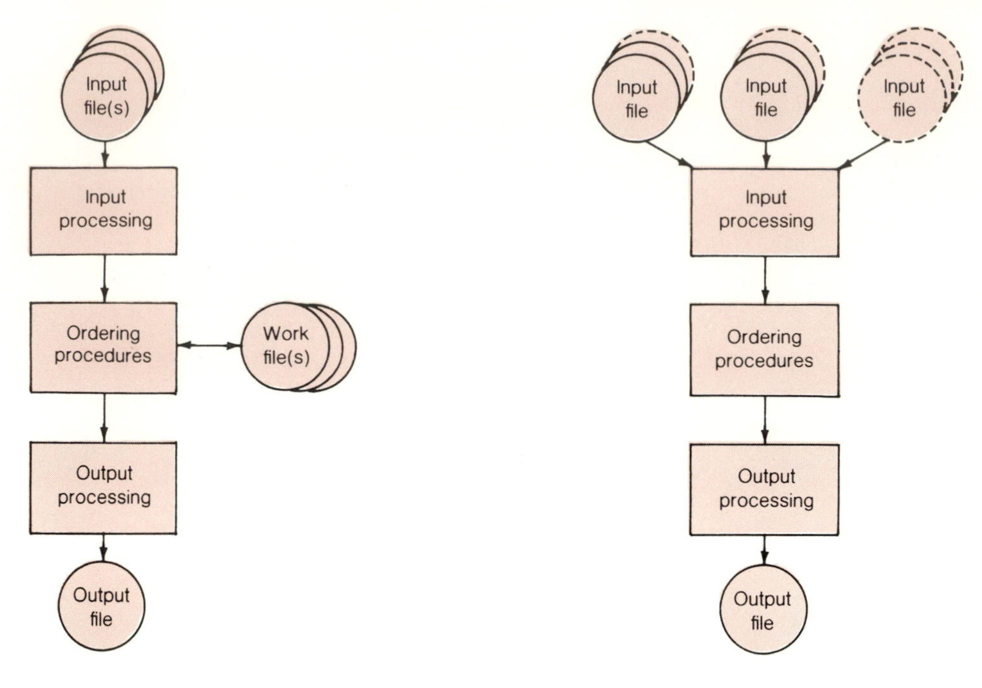

Figure 11.2 Sort program organization.

Figure 11.3 Merge program organization.

the organization as pictured in figure 11.3. The programmer can include several MERGE statements in the source program. If several MERGE statements are present, they are completely independent of each other.

SORT Statement

Because sorting constitutes a large percentage of the workload in a business data processing system, an efficient sort program is a highly necessary part of that system.

The purpose of the SORT statement is to invoke the execution of a sorting procedure.

The SORT statement arranges a collection of records into a sequence determined by the programmer. Each SORT statement has its own internal working file, called a *sort file,* created and used during the sorting operation. It may be either magnetic disk or magnetic tape. The sort file itself is referred to and accessed only by the SORT statement.

The Sort Feature provides the COBOL programmer with a convenient access to the sorting capacity of the Sort/Merge program by including a COBOL SORT statement and specifying other necessary elements of the Sort Feature in his or her program. The Sort Feature provides the capability for sorting a file of records according to a set of user-specified keys within each record (see figure 11.4). Sorting operations for fixed or variable records of varying modes of data representation can be specified by the programmer. Optionally, the programmer may apply some special processing which may consist of addition, deletion, creation, alteration, editing, or other modification of individual records by input or output procedures. This special processing allows the programmer to summarize, delete, shorten, or otherwise alter the records being sorted during the initial or final phases of the sort. (See figure 11.5.)

Basic Elements of the Sort Feature

To use the Sort Feature, the COBOL programmer must provide additional information in the Environment, Data, and Procedure Divisions of the source program. The basic elements of the Sort Feature are the SORT statement in the Procedure Division and the Sort-File-Description (SD) entry, with its associated record description entries in the Data Division.

1. The programmer must name in the Environment Division, with SELECT sentences for all files to be used as input to and output from Sort-file.

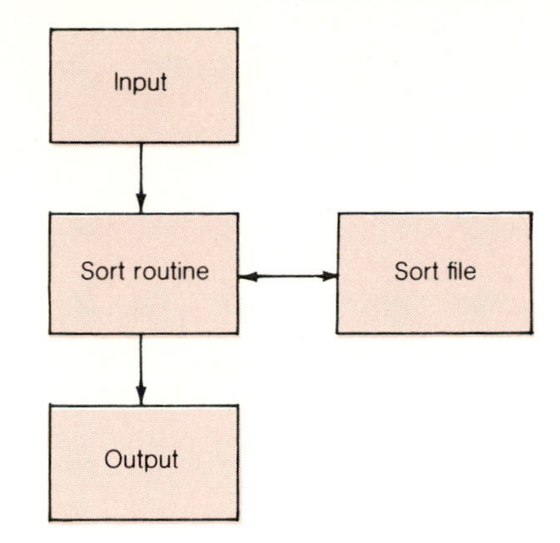

In this illustration the execution of the SORT statement performs the following functions:

1. Opens the input, sort, and output files.
2. Transcribes all of the input file to the sort file.
3. Sorts the file according to the prescribed specifications.
4. Merges the output of the sort to the output file.
5. Closes the input, sort, and output files.

Figure 11.4 Sort with no INPUT or OUTPUT procedure—example.

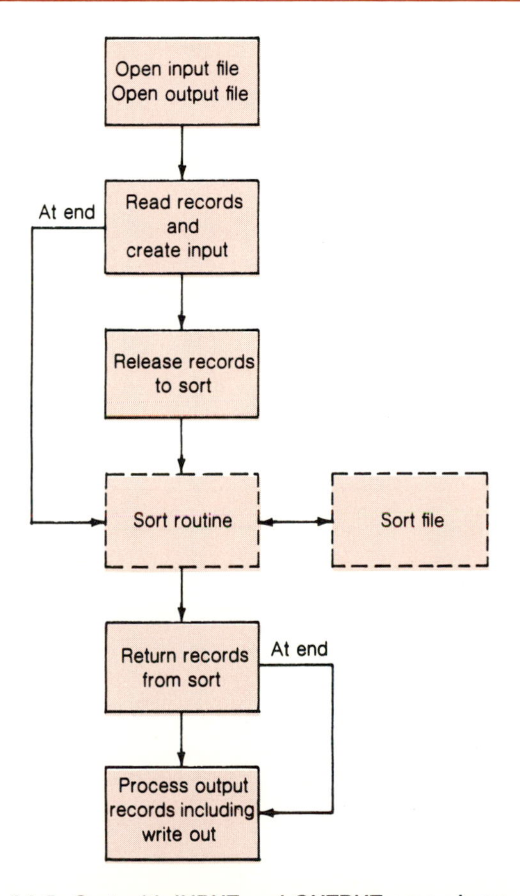

In this operation the following functions are performed:

1. When a SORT statement is executed, control is transferred to the INPUT procedures.
2. In the INPUT procedures, the programmer opens the input file(s) to the sort. Records are read, created, and released to the sort. When the input data is exhausted, the file is to be closed.
3. The sort is performed when all records to be sorted have been passed to the sort by the RELEASE verb and the last logical instruction of the INPUT procedure has been executed.
4. On the final merge pass of the sort, control is given to the programmer to perform OUTPUT procedures in which the output file is opened, sorted records are processed, and the output file is closed.
5. The sort terminates when all records have been sorted and passed by the RETURN verb to the OUTPUT procedures and the last logical instruction of the OUTPUT procedures have been executed.

Figure 11.5 Sort with INPUT and OUTPUT procedures—example.

2. In the Data Division, the programmer must write file description entries (FD) for all files that are to provide input and output to the sort program. In addition, a Sort-File-Description entry (SD) must be written describing the records to be sorted, including the sorting-key fields. The record description entry associated with the Sort-File-Description may be considered as redefining the records being sorted.
3. The programmer must specify the records being sorted, the sort-key names to be sorted on, and whether the sort is to be in ascending or descending sequence or a mixture of both, that is, sort-

keys may be specified as ascending or descending, independent of one another, and the sequence of the records will conform to the mixture specified, and whether the records are to be processed before and/or after the sort. The programmer writes a SORT statement in the Procedure Division specifying all of the aforementioned options.

4. The sort-work files are provided by the Sort/Merge program to serve as intermediate work files during the sorting process.

   The SORT statement in the Procedure Division is the primary element of a source program that performs one or more sorting operations. A sorting operation is based on the sort-keys named in the SORT statement. A sort-key specifies the fields within a record on which the file is sorted. Sort-keys are defined in the record description clauses associated with the Sort-File-Description (SD) entry. The term sorting operation means not only the manipulation of the Sort Program of sort-work files on the basis of sort-keys designated by the COBOL program, but it also includes the method of making available to, and returning records from, these sort-work files. A *sort-work file* is a collection of records that is involved in the sorting operation as it exists in intermediate device(s). Records are made available to the Sort/Merge program by the USING or INPUT PROCEDURE options of the SORT statement. Sorted records are returned from the Sort/Merge program by either the GIVING or OUTPUT PROCEDURE options of the SORT statement.

Environment Division for Sort-Merge/Input-Output Section

FILE-CONTROL Paragraph

The FILE-CONTROL paragraph for sort-merge names each file and allows for specification of other file-related information. (The format for the FILE-CONTROL paragraph is shown in figure 11.6.) This paragraph identifies the file medium and permits particular hardware assignments. SELECT entries must be written in the FILE-CONTROL paragraph of the Input-Output Section for all files used within input and output procedures, as well as files specified in the USING or GIVING options of the SORT statement in the Procedure Division. The SELECT statement may be specified for the sort-file. (The format for the SELECT statement is shown in figure 11.7.) The file-name identifies the sort-file to the compiler.

```
FILE-CONTROL.   [ file-control-entry ]   . . .
```

Figure 11.6 Format—FILE-CONTROL paragraph.

```
SELECT file-name ASSIGN TO implementor-name-1   [ , implementor-name-2 ]   . . .  .
```

Figure 11.7 Format—SELECT statement.

Rules Governing the Use of the SELECT Clause

1. Each sort or merge file described in the Data Division must be named once and only once as file-name in the FILE-CONTROL paragraph. Each sort or merge file specified in the file control entry must have a sort-merge file description entry in the Data Division.
2. Since file-name represents a sort or merge file, only the ASSIGN clause is permitted to follow file-name in the FILE-CONTROL paragraph. The ASSIGN clause specifies the association of the sort or merge file referenced by file-name to a storage medium.

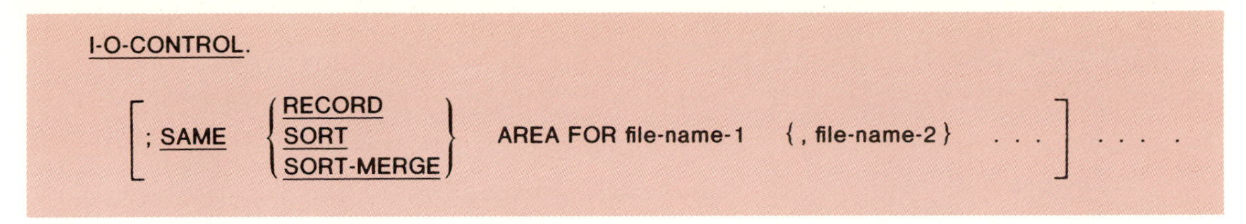

Figure 11.8 Format—I-O-CONTROL paragraph.

I-O-CONTROL Paragraph

The I-O-CONTROL paragraph for sort-merge specifies the memory area that is to be shared by different files. (The format for the I-O-CONTROL paragraph is shown in figure 11.8.)

The SAME RECORD AREA clause specifies that two or more files are to use the same memory area for processing the current logical record. All of the files may be opened at the same time. A logical record in the SAME RECORD AREA is considered as a logical record of each opened output file whose file-name appears in this SAME RECORD AREA clause, and of the most recently read input file whose file-name appears in this SAME RECORD AREA clause. This is equivalent to implicit redefinition of the area; that is, records are aligned on the leftmost character position.

If the SAME SORT AREA or SAME SORT-MERGE AREA is used, at least one of the file-names must represent a sort file or a merge file. Files that do not represent sort or merge files may also be named in the clause. This clause specifies that memory is shared as follows:

1. The SAME SORT AREA or SAME SORT-MERGE AREA clause specifies a memory area that will be made available for use in sorting or merging each sort or merge file named. Therefore, any memory area allocated for the sorting or merging of a sort or merge file is available for reuse in sorting or merging any of the other sort or merge files.
2. In addition, memory areas assigned to files that do not represent sort or merge files may be allocated as needed for sorting or merging the sort or merge files named in the SAME SORT AREA or SAME SORT-MERGE AREA clause.
3. Files other than sort or merge files do not share the same memory area with each other. If the programmer wishes these files to share the same memory area with each other, a SAME AREA or SAME RECORD AREA clause naming these files must be included in the program.
4. During the execution of a SORT or MERGE statement that refers to a sort or merge file named in this clause, any non-sort-merge files named in this clause must not be open.

Rules Governing the Use of the I-O-CONTROL Paragraph

1. The I-O-CONTROL paragraph for sort-merge is optional.
2. In the SAME AREA clause, SORT and SORT-MERGE are equivalent.
3. If the SAME SORT AREA or SAME SORT-MERGE AREA clause is used, at least one of the file-names must represent a sort or merge file. Files that do not represent sort or merge files may also be named in the clause.
4. The three formats of the SAME clause (SAME RECORD AREA, SAME SORT AREA, SAME SORT-MERGE AREA) are considered separately in the following:

   More than one SAME clause may be included in a program. However:

   a. A file-name must not appear in more than one SAME RECORD AREA clause.
   b. A file-name that represents a sort or merge file must not appear in more than one SAME SORT AREA or SAME SORT-MERGE AREA clause.
   c. If a file-name that does not represent a sort or merge file appears in a SAME AREA clause and in one or more SAME SORT AREA or SAME SORT-MERGE AREA clauses, then all of the files named in that SAME AREA clause must be named in that SAME SORT AREA or SAME SORT-MERGE AREA clause(s).
5. The files referenced in the SAME SORT AREA, SAME SORT-MERGE AREA, or SAME RECORD AREA clauses are not all required to have the same organization or access.

```
SD file-name

    [ ; RECORD CONTAINS  [ integer-1 TO ]  integer-2 CHARACTERS ]

    [ ; DATA  { RECORD IS    }  data-name-1  [ , data-name-2 ]  . . . ] .
              { RECORDS ARE }
```

Figure 11.9 Format—sort-file-description entry.

Data Division for Sort-Merge

In the Data Division, the programmer must include file description entries for files to be sorted or merged, sort-file description entries for sort-work files, and record description entries for each.

File Section

An SD file description is used to provide information about the size and the names of the data records associated with the file to be sorted or merged. (The format for the Sort-File Description entry is shown in figure 11.9.) There are no label procedures that the programmer can control, and the rules for blocking and internal storage are peculiar to the SORT and MERGE statements.

Sort-File-Description Entry

The Sort-File-Description entry furnishes information concerning the physical structure, identification, and record-names of the file to be sorted or merged.

SD is a COBOL reserved word that must appear at the A margin. The level indicator SD identifies the beginning of the sort-merge file description and must precede the file-name. The clauses that follow the name of the file are optional and their order of appearance is not significant. One or more record description entries must follow the sort-merge file description entry. However, no input-output statements may be executed for this file.

File-Name is a programmer-supplied name that is used in the Sort-Merge feature. The SORT or MERGE statement in the Procedure Division specifies this name. All rules for data-names apply, and it must not be qualified or subscripted.

Data Record and Record Contains clauses follow the same rules for File Section entries in the Data Division. (See Data Record and Record Contains entries in the Data Division.)

The format of the *Record Description entry* will vary according to the type of item being described. All rules for Record Description entries in the Data Division apply to this entry. (See Record Description entry in the Data Division.)

Sort-Keys

Sort-Keys are identified by data-names assigned to each field involved in the sorting or merging operation.

Rules Governing the Use of Sort-Keys

1. The keys must be physically located in the same position and must have the same data format in every logical record of the sort-file.
2. Key items must not contain an OCCURS clause or be subordinate to entries that contain an OCCURS clause.
3. A maximum of twelve keys may be specified. The total length of all the keys must not exceed 256 bytes.
4. All keys must be at a fixed displacement from the beginning of a record; that is, they cannot be located after a variable table in a record.
5. All key fields must be located within the first 4,092 bytes of a logical record.
6. The data-names describing the keys may be qualified.

Sort-Key Evaluation
When the values of a key in a pair of sort file records are compared, one value is found to be greater than, equal to, or less than the other according to the rules given under the Relation Condition paragraph in the Procedure Division. The key comparison determines the order of the records in the sort output.

1. If all corresponding key values of a pair of sort file records are identical, the record that appeared first in the input to the sorting procedure will precede the other in the output.
2. If any key values of a pair of sort file records are unequal, the output sequence is based upon the most significant key item for which inequality is found.
   a. If that key item is governed by the ASCENDING option, the record with the lower value will precede the other in the output.
   b. If the item is governed by the DESCENDING option, the record with the higher value will precede the other in the output.

All comparisons are made on the basis of the collating sequence of that particular computer unless the COLLATING SEQUENCE phrase of the SORT statement is used to specify a different sequence. (COLLATING SEQUENCE will be discussed under the SORT statement further in this section.)

Procedure Division for Sort-Merge

Sort Procedures

The COBOL procedural statements that are available for use in a sort program are as follows:

1. The SORT statement for performing a sorting operation on a collection of records.
2. The RELEASE statement for transferring records to the initial phase of a sort operation.
3. The RETURN statement for obtaining sorted records from the final phase of a sort operation.

The Procedure Division must contain a SORT statement to describe the sorting operation and, optionally, any necessary input and output procedures. The procedure-names constituting the input and output procedures are specified in the SORT statements.

The Procedure Division may contain more than one SORT statement appearing anywhere except in the Declaratives portion or in the input or output procedures associated with the SORT statement.

Sort Input Processing

A choice must be made between having the sorting process handle the input processing of the file being sorted or having the programmer's program specify the input processing procedures. In most cases the former technique is more efficient, but the latter may be necessary in order to accomplish selective editing of the input records. If such editing would result in the deletion of a significant portion of the input file, then the input procedure technique would be more appropriate.

SORT Statement

The SORT statement creates a sort file by executing input procedures or by transferring records from another file; sorts the records in the sort file on a set of specified keys; and, in the final phase of the sort operation, makes available each record from the sort file, in sorted order, to some output procedure or to an output file. (The format for the SORT statement is shown in figure 11.10.)

Functions of the SORT Statement

1. Directs the sorting operation to obtain the necessary records to be sorted either from an INPUT PROCEDURE or the USING file.
2. The records are then sorted on a set of specified keys.
3. After the last sort is completed, the sorted records are made available to either an OUTPUT PROCEDURE or to the GIVING file.

```
SORT file-name-1 ON {ASCENDING } KEY data-name-1 [, data-name-2] ...
                    {DESCENDING}

     [ON {ASCENDING } KEY data-name-3 [, data-name-4] ... ] ...
         {DESCENDING}

  [COLLATING SEQUENCE IS alphabet-name]

  {INPUT PROCEDURE IS section-name-1 [{THROUGH} section-name-2]}
  {                                  [{THRU   }               ]}
  {USING file-name-2 [, file-name-3] ...                        }

  {OUTPUT PROCEDURE IS section-name-3 [{THROUGH} section-name-4]}
  {                                   [{THRU   }               ]}
  {GIVING file-name-4                                            }
```

Figure 11.10 Format—SORT statement.

An example of the SORT statement follows:

```
SORT SALES-RECORDS
    ON ASCENDING KEY CUSTOMER-NUMBER
        DESCENDING KEY DATE-IN
    USING FN-1
    GIVING FN-2.
```

File-Name-1

This is the name given in a Sort-File-Description entry in the Data Division that describes the records to be sorted.

Ascending or Descending

One of these clauses must be included (both may be included) to specify the sequence of records to be sorted. The sequence is applicable to all sort-keys immediately following the clause. More than one data-name may be specified for sorting after the ASCENDING or DESCENDING word. Every data-name used must have been described in a clause associated with the Sort-File-Description entry. Sort-keys are always listed from left to right in order of decreasing significance regardless of whether they are ASCENDING or DESCENDING. The sort-keys must be specified in the logical sequence in which the records are to be sorted.

When the ASCENDING clause is used, the sorted sequence is from the lowest value of the key to the highest value, according to the collating sequence for the COBOL character set.

When the DESCENDING clause is used, the sorting sequence is from the highest value of the key to the lowest, according to the collating sequence for the COBOL character set.

Data-Name

This is the data-name assigned to each sort-key; it is required in every statement.

Rules Governing the Use of Data-Names

1. More than one data-name may be specified after the ASCENDING or DESCENDING options.
2. The same data-name must not be used twice in the same SORT statement.
3. Every data-name must have been defined in a Record Description entry associated with the Sort Description entry.
4. The sort-keys are specified in the desired order of sorting.

Rules Governing the Use of the SORT Statement

1. File-name-1 must be described in a sort-merge file description entry in the Data Division.
2. Section-name-1 represents the name of an input procedure. Section-name-3 represents the name of an output procedure.
3. File-name-2, file-name-3, and file-name-4 must be described in a file description entry (not in a sort-merge file description entry) in the Data Division.
4. Data-name-1, data-name-2, data-name-3, and data-name-4 are KEY data-names and are subject to the following rules:
   a. The data items identified by KEY data-names must be described in records associated with file-name-1.
   b. KEY data-names may be qualified.
   c. The data items identified by KEY data-names must not be variable-length items.
   d. If file-name-1 has more than one record description, then the data items identified by KEY data-names need be described in only one of the record descriptions.
   e. None of the data items identified by KEY data-names can be described by an entry which either contains an OCCURS clause or is subordinate to an entry which contains an OCCURS clause.
5. The words THRU and THROUGH are equivalent.
6. SORT statements may appear anywhere except in the declarative portion of the Procedure Division or in an input or output procedure associated with a SORT or MERGE statement.
7. No more than one file-name from a multiple file reel can appear in the SORT statement.
8. The data-names following the word KEY are listed from left to right in the SORT statement in order of decreasing significance without regard to how they are divided into KEY phrases. When the ASCENDING phrase is specified, the sorted sequence will be from the lowest value of the contents of the data items identified by the KEY data-names to the highest value, according to the rules for comparison of operands in a relation condition.

   When the DESCENDING phrase is specified, the sorted sequence will be from the highest value of the contents of the data items identified by the KEY data-names to the lowest value, according to the rules for comparison of operands in a relation condition.
9. The collating sequence that applies to the comparison of the nonnumeric key data items specified is determined in the following order of precedence: first, the collating sequence established by the COLLATING SEQUENCE phrase, if specified, in the SORT statement; and second, the collating sequence established in the program collating sequence.
10. The input procedure must consist of one or more sections that appear contiguously in a source program and do not form part of any output procedure. If INPUT PROCEDURE is specified, control is passed to the input procedure before file-name-1 is sequenced by the SORT statement. The compiler inserts a return mechanism at the end of the last section in the input procedure, and when control passes the last statement in the input procedure, the records that have been released to file-name-1 are sorted.
11. The output procedure must consist of one or more sections that appear contiguously in a source program and do not form part of any input procedure. If OUTPUT PROCEDURE is specified, control passes to it after file-name-1 has been sequenced by the SORT statement. The compiler inserts a return mechanism at the end of the last section in the output procedure. And when control passes to the last statement in the output procedure, the return mechanism provides for termination of the sort, and passes control to the next executable statement after the SORT statement. Before entering the output procedure, the sort procedure reaches a point at which it can select the next record in sorted order when requested. The RETURN statements in the output procedure are the requests for the next record.

    An example of the SORT program with USING and GIVING options follows:

```
ENVIRONMENT DIVISION.
      .
      .
      .
INPUT-OUTPUT SECTION
FILE-CONTROL.
      SELECT INPUT-FILE
            ASSIGN TO INPUT-MASTER.
```

```
    SELECT SORT-FILE
        ASSIGN TO COLLATE-FILE.
    SELECT OUTPUT-FILE
        ASSIGN TO OUTPUT-MASTER.
    .
    .
    .
DATA DIVISION.
FILE SECTION.
FD  INPUT-FILE . . .
01  . . .
    .
    .
SD  SORT-FILE . . .
01  . . .
    .
    .
FD  OUTPUT-FILE . . .
01  . . .
    .
    .
PROCEDURE DIVISION.
SORT-CALL.
    SORT SORT-FILE
        ON . . .
        USING INPUT-FILE
        GIVING OUTPUT-FILE.
    STOP RUN.
```

Note that these files may have multiple record types and sizes, provided the sort file and USING file have the same records, the sort file and GIVING file have the same records, and key descriptions and positions are equivalent for all record types.

Sort Program with Input and Output Procedures

For many sort applications, it is necessary to apply some processing to the content of a sort file. This special processing may consist of addition, deletion, creation, alteration, editing, or some other modification of the individual records. The special processing may be necessary before, after, or both before and after the records are reordered by the sort. The COBOL sort allows the programmer to express these procedures in the COBOL language and to specify at which point, before or after the sort, they are to be executed. When the procedures are executed before the sort, they are called *input procedures.* When the procedures are executed after the sort, they are called *output procedures.*

A COBOL sort may contain any number of sorts; each may have its own independent special procedures. The SORT feature automatically executes these procedures at the specified point in such a way that extra passes over the sort file are not required.

Before the SORT operation is executed, the programmer specifies input procedures in which records are read and operated on. In the input procedures, the RELEASE statement creates the sort file. At the conclusion of the input procedures those records that have been output by the RELEASE statement comprise the sort file.

At the end of the input procedures, the SORT statement arranges the entire set of records in the sort file according to the keys specified in the SORT statement.

After the records have been sorted, they are available for output procedures. In these output procedures, the RETURN statement makes the next record available in sorted order from the sort file. At the end of the output procedures are the records that have been made available by the RETURN statement for further processing. This processing must include the writing of records into the output file.

INPUT PROCEDURE

The presence of an INPUT PROCEDURE indicates that the programmer has written an input procedure to process the records before they are sorted. This procedure is included in the Procedure Division in one or more sections. This procedure passes one record at a time to the Sort Feature after it has completed its processing.

When the INPUT PROCEDURE option is specified, the programmer is responsible for all the input processing for the sorting processing. An input procedure must consist of one or more sections.

Since the input procedure is invoked by the sorting procedure via the same techniques used in the PERFORM statement execution, the structure of the input procedure must follow the same basic rules as a set of sections that are the object of a PERFORM statement. Control must not be passed to the input procedure except through the execution of a SORT statement. The input procedure may contain any procedures needed to select, create, or modify records for input to the sorting process.

Section-name-1 is the name of the first or only section in the main program that contains the input procedures. This section is required if the INPUT PROCEDURE is used.

Section-name-2 is the name of the last section and is required if the procedure terminates in a section other than that in which it was started.

The INPUT PROCEDURE consists of one or more sections that are written into a source program.

Rules Governing the Use of the INPUT PROCEDURE

1. The INPUT PROCEDURE can include any statements needed to select, create, or modify records.
2. Control must not be passed on to an INPUT PROCEDURE unless the related SORT statement is executed, because the RELEASE statement in the INPUT PROCEDURE has no meaning unless it is controlled by a SORT statement.
3. An INPUT PROCEDURE must not contain a SORT statement, a MERGE statement, or a RETURN statement.
4. Any files used as sort-work files may not be opened or referred to by an INPUT PROCEDURE.
5. The INPUT PROCEDURE must build the records to be sorted one at a time. The record must have been described and assigned a data-name in the record description entry associated with the Sort Description entry.
6. The INPUT PROCEDURE must make the record available to the sorting operation after it has been processed. A RELEASE statement is used for this purpose.
7. After all the records have been released to the sorting operation, the INPUT PROCEDURE must transfer control to the last statement in the INPUT PROCEDURE to terminate the procedure.
8. The INPUT PROCEDURE must not contain any transfers of control to points outside the INPUT PROCEDURE.
   The remainder of the Procedure Division must not contain any transfer of control to points inside the input procedures (with the exception of the return of control from a Declarative section).

If an INPUT PROCEDURE is specified, control is passed to the INPUT PROCEDURE when the SORT program input phase is ready to receive the first record. The compiler inserts a return mechanism at the end of the last section of the INPUT PROCEDURE, and when control passes the last statement in the INPUT PROCEDURE, the records that have been released to file-name-1 are sorted. The RELEASE statement transfers records from the INPUT PROCEDURE to the input phase of the sort operation. (See RELEASE statement.)

USING

The USING option indicates that the records to be sorted are all in one file and are to be passed to the sorting operation as one unit when the SORT statement is executed. If the option is used, all records to be sorted must be in the same files.

If the USING phrase is specified, all the records in file-name-2 and file-name-3 are transferred automatically to file-name-1. When the SORT statement is executed, file-name-2 and file-name-3 must not be open. The SORT statement automatically initiates the processing of, makes available the logical records for, and terminates the processing of file-name-2 and file-name-3. The terminating function for all files is performed as if a CLOSE statement, without optional phrases, had been executed for each file. The SORT statement also automatically performs the implicit functions of moving the records from the file area of file-name-2 and file-name-3 to the file area for file-name-1 and releasing records to the initial phrase of the sort operation.

If this option is used, all records in file-name-2 and file-name-3 are transferred to file-name-1. At the time of execution of the SORT statement, file-name-2 and file-name-3 must not be open. File-name-2 and file-name-3 must be standard sequential files. *For the USING option, the compiler will automatically OPEN, READ, RELEASE, and CLOSE file-name-2 and file-name-3 without the programmer specifying these functions.*

Sort Output Processing

A choice must be made between having the sorting process handle the output processing of the newly sorted file or having the programmer's program specify the output processing procedures. The choice is not as easily made as in the case of the input procedure since it involves the assessment between the efficiency of giving the sorting process more memory to work with and the total system efficiency of embedding the processing of the output file within an output procedure. Any requirement for a copy of the file for later processing by other procedures may also be a consideration.

OUTPUT PROCEDURE

This procedure indicates that the programmer has written an OUTPUT PROCEDURE to process the sorted records. This procedure is included in the Procedure Division in the form of one or more sections. The procedure returns records one at a time from the Sort Feature after they have been sorted.

Section-name-3 is the name of the first or only section in the main program that contains the OUTPUT PROCEDURE. This section is required if the output procedures are used.

Section-name-4 is the name of the last section that contains OUTPUT procedures and is required if the procedure terminates in a section other than the one in which it started.

The OUTPUT PROCEDURE consists of one or more sections that are written into a source program. When the OUTPUT PROCEDURE is specified, final output file processing for the sorting process is the responsibility of the object program. Since the OUTPUT PROCEDURE is invoked by the sorting procedure via the same techniques used in the execution of a PERFORM statement, the structure of the OUTPUT PROCEDUREs must follow the same basic rules as sets of sections that are the object of a PERFORM statement. Control must not be passed to the OUTPUT PROCEDURE except through the execution of a SORT statement. The OUTPUT PROCEDURE may contain any procedures needed to select, modify, or copy the records that are being returned from the sorting process.

Three general restrictions apply to the procedural statements within the OUTPUT PROCEDURE:

1. An OUTPUT PROCEDURE must not contain a SORT statement, a MERGE statement, or a RELEASE statement.
2. The OUTPUT PROCEDURE must not contain ALTER statements or transfers of control to points outside the OUTPUT PROCEDURE; that is, GO TO and PERFORM statements in the OUTPUT PROCEDURE are not permitted to refer to procedure-names outside the output procedure. COBOL statements that will cause an implied transfer of control to USE procedures are allowed.
3. The remainder of the Procedure Division must not contain ALTER statements or transfers of control to points inside the output procedure; i.e., GO TO and PERFORM statements in the remainder of the Procedure Division are not permitted to refer to procedure-names within the output procedure.

Rules Governing the Use of the OUTPUT PROCEDURE

1. The OUTPUT PROCEDURE can include any statement necessary to select, modify, or copy the sorted records being returned one at a time in sorted order from the sort-file.
2. Control must not be passed to the OUTPUT PROCEDURE, except when a related SORT statement is being executed, because RETURN statements in the OUTPUT PROCEDURE have no meaning unless they are controlled by the SORT statement.
3. The OUTPUT PROCEDURE must not contain any SORT statements.
4. Any files used as sort-files may not be opened or referred to by the OUTPUT PROCEDURE.
5. Records must be obtained one at a time from the Sort/Merge program over the RETURN statement. Once a record is returned, the previously returned record is no longer available.
6. The OUTPUT PROCEDURE must manipulate the returned sorted record by referring to the data record that has been described and assigned a data-name in the record description entry. If the records are to be written on an output file, the programmer must provide the appropriate OPEN statement prior to the execution of the SORT statement or the OUTPUT PROCEDURE itself.
7. The OUTPUT PROCEDURE must not contain any transfer of control outside the OUTPUT PROCEDURE.
8. The remainder of the Procedure Division must not contain any transfers of control to points inside the OUTPUT PROCEDURE (with the exception of the return of control from a Declaratives section).

9. After all the records have been returned by the Sort Feature, and the OUTPUT PROCEDURE attempts to execute another RETURN statement, the AT END clause of the RETURN statement will be executed. The AT END clause should direct control of the program to the last statement of the OUTPUT PROCEDURE to terminate the OUTPUT PROCEDURE.

If an OUTPUT PROCEDURE is specified, control passes to it after file-name-1 has been placed in sequence by the SORT statement. The compiler inserts a return mechanism at the end of the last section of the OUTPUT PROCEDURE. When control passes the last statement in the OUTPUT PROCEDURE, the return mechanism provides for the termination of the sort and then passes control to the next statement after the SORT statement.

When all records are sorted, control is passed to the OUTPUT PROCEDURE. The RETURN statement in the OUTPUT PROCEDURE is a request for the next record. (See RETURN statement.)

GIVING

If the GIVING phrase is specified, all the sorted records in file-name-1 are automatically written on file-name-4 as the implied output procedure for this SORT statement. When the SORT statement is executed, file-name-4 must be open. The SORT statement automatically initiates the processing of, releases the logical records to, and terminates the processing of file-name-4. The terminating function is performed as if a CLOSE statement, without optional phrases, had been executed for the file. The SORT statement automatically performs the implicit functions of the return of the sorted records from the final phase of the sort operation and moves the records from the file area for file-name-1 to the file area for file-name-4. An example of the SORT program—USING and GIVING options—follows:

```
ENVIRONMENT DIVISION.
    .
    .
INPUT-OUTPUT SECTION.
FILE-CONTROL.
    SELECT RAWTFILE
        ASSIGN TO . . .
    SELECT SWFILE
        ASSIGN TO . . .
    SELECT TRANFILE
        ASSIGN TO . . .

DATA DIVISION.
FILE SECTION.
FD  RAWTFILE
    LABEL RECORD IS STANDARD.
01  RAWTRECORD      PICTURE X(13).

SD  SWFILE
    RECORD CONTAINS 13 CHARACTERS.
01  SWRECORD.
    05  TNO         PICTURE X(6).
    05  FILLER      PICTURE X(6).
    05  TCODE       PICTURE X.

FD  TRANFILE
    LABEL RECORD IS STANDARD.
01  TRANRECORD      PICTURE X(13).

PROCEDURE DIVISION.
BEGIN.
    SORT SWFILE
        ON ASCENDING KEY TNO, TCODE
        USING RAWFILE
        GIVING TRANFILE.
    STOP RUN.
```

Note: The SORT and MERGE statements and INPUT and OUTPUT PROCEDURES are permitted anywhere in the Procedure Division except in the Declaratives Section.

RELEASE record-name [FROM identifier]

Figure 11.11 Format—RELEASE statement.

RELEASE Statement

The RELEASE statement is used to transfer logical input records from an input procedure to the initial phase of a sorting operation. (The format for the RELEASE statement is shown in figure 11.11.) The RELEASE statement may appear only in an input procedure; every input procedure must contain at least one RELEASE statement.

The record-name referenced by the RELEASE statement must be the name of a record defined within a sort file; that is, a file described with an SD level indicator. If the sort file description contains more than one record description, and if the record descriptions define records of different sizes, a separate RELEASE statement must be specified for each record size. If the FROM phrase is used, the contents of "identifier" must be the name of a data item in working-storage or of an input record area. If the format of identifier is different from that of the record name, then moving takes place according to the rules specified for the MOVE statement without the CORRESPONDING option. The information in the sort record area is no longer available, but the information in the identifier area is available. It is illegal to use the same name for both the record-name and the identifier or for the two names to reference the same memory area.

After the RELEASE statement is executed, the contents of record-name are no longer available to the COBOL procedure. The execution of a RELEASE statement causes the contents of record-name (after the contents of identifier have been moved to it in the FROM phrase) to be made available to the initial phase of the sort process. When control passes from the input procedure, the sort file consists of all those records that were placed in it by the execution of RELEASE statements. No OPEN, READ, WRITE, or CLOSE statements may be given for the sort file.

Rules Governing the Use of the RELEASE Statement

1. A RELEASE statement may only be used within the range of an INPUT PROCEDURE.
2. If the INPUT PROCEDURE option is specified, the RELEASE statement must be included within the given set of procedures.
3. Sort-record-name must be the name of a logical record in the associated sort-file-description entry and may be qualified.
4. If the FROM option is used, the contents of the identifier data area are moved to the record-name, then the contents of sort-record-name are released to the sort-file. Moving takes place according to the rules of the MOVE statement without the CORRESPONDING option. The information in the record area is no longer available, but the information in the data area associated with the identifier is available.
5. After the RELEASE statement is executed, the logical record is no longer available. When control passes from the INPUT PROCEDURE, the file consists of all those records that were placed in it by the execution of the RELEASE statement.
6. Record-name and identifier must not refer to the same storage area.

The following are examples of a RELEASE statement:

```
RELEASE INPUT-RECORD.

RELEASE RECORD-ONE.
```

RETURN Statement

The RETURN statement is used to obtain logical output records from a sort operation and transfer them to an output procedure. (The format for the RETURN statement is shown in figure 11.12.) The RETURN statement may appear only in an output procedure, and every output procedure must contain at least one RETURN statement.

The file-name referenced in the RETURN statement must be described with an SD level indicator and must be the same file-name that is referenced in the SORT statement currently being executed.

RETURN file-name RECORD [INTO identifier] ; AT END imperative-statement

Figure 11.12 Format—RETURN statement.

The INTO phrase may be used only when the file referred to by file-name contains just one type of record. The identifier must be the name of a data item in working-storage or of an output record area. If the format of the identifier differs from that of the input record, then moving is performed according to the rules specified for the MOVE statement without the CORRESPONDING option. When the INTO phrase is used, the logical record is still available in the sort file's record area.

The execution of the RETURN statement causes the next record in sorted order (according to the keys listed in the SORT statement) to be made available for processing in the record area associated with the sort file.

After the contents of the sort file are exhausted, the next execution of the RETURN statement will result in the execution of the imperative statement in the AT END phrase.

After execution of the AT END imperative-statement, no RETURN statement may be executed within the current output procedure.

No OPEN, READ, WRITE, or CLOSE statements may be given for the sort file.

Rules Governing the Use of the RETURN Statement

1. File-name must be described by a Sort File Description entry in the Data Division.
2. A RETURN statement may be used only within the range of an OUTPUT PROCEDURE associated with a SORT statement for a file-name.
3. The INTO option may be used only when the input file contains just one type of record. The storage area associated with the identifier and the storage area which is the record area associated with file-name must not be the same storage area.
4. The identifier must be the name of a Working-Storage area or an output record area. Use of the INTO option has the same affect as the MOVE statement for alphanumeric items.
5. The imperative statement in the AT END phrase specifies the action to be taken when all the sorted records have been obtained for the sorting operation.
6. When a file consists of more than one type of logical record, these records automatically share the storage area. This is equivalent to saying that there exists an implicit redefinition of the area, and only the information that is present in the current record is available.
7. After the execution of the imperative statement in the AT END phrase, no RETURN statement may be executed within the OUTPUT PROCEDURE.

The following are RETURN statement examples:

```
RETURN FILE-ONE
    AT END PERFORM END-PROGRAM.

RETURN FILE-SORT
    AT END PERFORM LAST-PROC.
```

Control of INPUT and OUTPUT PROCEDURES

The INPUT and OUTPUT PROCEDURES function in a manner similar to option 1 of the PERFORM statement; for example, naming a section in an INPUT PROCEDURE clause causes execution of that section during the sorting operation to proceed as though that section had been the subject of a PERFORM statement. As with the PERFORM statement, the execution of the section is terminated after execution of its last statement.

Return is back to the next statement after the INPUT or OUTPUT PROCEDURES have been executed. The EXIT verb may be used as a common exit point for conditional exits from INPUT or OUTPUT PROCEDURES. If the EXIT verb is used, it must appear as the last paragraph of the INPUT or OUTPUT PROCEDURES (see EXIT statement in Procedure Division).

Flow of Control

Any sequence of procedural statements may be executed before or after the sort statement is executed. When the SORT statement is executed, the sorting process receives control.

If an input procedure has been specified, the sort transfers control to the input procedure to start processing records. The input procedure is responsible for opening the input file and reading the logical input records. Each time a record is ready for the sort file, the input procedure will execute a RELEASE statement, causing the sort to place the record in the sort file. Control then passes to the statement following the RELEASE statement. The input procedure continues reading and releasing until all the input records have been given to the sort, at which time the input procedure is responsible for closing the input file. Control must pass to the exit point of the input procedure, thereby returning control to the sorting process. An example of a SORT program—INPUT PROCEDURE and GIVING options—follows:

```
ENVIRONMENT DIVISION.
    .
    .
    .
INPUT-OUTPUT SECTION.
FILE-CONTROL.
    SELECT NET-FILE-IN
        ASSIGN TO . . .
    SELECT NET-FILE
        ASSIGN TO . . .
    SELECT NET-FILE-OUT
        ASSIGN TO . . .

DATA DIVISION.
FILE SECTION.
FD  NET-FILE-IN
    LABEL RECORDS OMITTED
    DATA RECORD IS NET-CARD-IN.
01  NET-CARD-IN.
    05  EMPL-NO-IN              PICTURE 9(6).
    05  DEPT-IN                 PICTURE 99.
    05  NET-SALES-IN            PICTURE 9(7)V99.
    05  NAME-ADDR-IN            PICTURE X(55).

SD  NET-FILE
    DATA RECORD IS SALES-RECORD.
01  SALES-RECORD.
    05  EMPL-NO                 PICTURE 9(6).
    05  DEPT                    PICTURE 99.
    05  NET-SALES               PICTURE 9(7)V99.
    05  NAME-ADDR               PICTURE X(55).

FD  NET-FILE-OUT
    LABEL RECORDS OMITTED.
    05  EMPL-NO-OUT             PICTURE 9(6).
    05  DEPT-OUT                PICTURE 99.
    05  NET-SALES-OUT           PICTURE 9(7)V99.
    05  NAME-ADDR-OUT           PICTURE X(55).

WORKING-STORAGE SECTION.
01  FLAGS.
    05  MORE-INPUT-FLAG         PICTURE XXX      VALUE 'YES'.
        88  MORE-INPUT                           VALUE 'YES'.
        88  NO-MORE-INPUT                        VALUE 'NO'.
    05  MORE-TRANS-FLAG         PICTURE XXX      VALUE 'YES'.
        88  MORE-TRANS                           VALUE 'YES'.
        88  NO-MORE-TRANS                        VALUE 'NO'.
```

```
PROCEDURE DIVISION.
ELIM-DEPT-7-9-PRINTOUT SECTION.
    SORT NET-FILE
        ON ASCENDING KEY DEPT
            DESCENDING KEY NET-SALES
        INPUT PROCEDURE IS SCREEN-DEPT
        GIVING NET-FILE-OUT.
    PERFORM CHECK-RESULTS.
    STOP RUN.

SCREEN-DEPT SECTION.
INPUT-ROUTINE.
    OPEN    INPUT    NET-FILE-IN.
    READ NET-FILE-IN
        AT END MOVE 'NO' TO MORE-INPUT-FLAG.
    PERFORM PROCESS-INPUT-ROUTINE
        UNTIL NO-MORE-INPUT.
    CLOSE NET-FILE-IN
    GO TO INPUT-EXIT.

PROCESS-INPUT-ROUTINE.
    DISPLAY EMPL-NO-IN, DEPT-IN, NET-SALES-IN, NAME-ADDR-IN
        UPON PRINTER.
    IF DEPT-IN IS EQUAL TO 7 OR 9
        NEXT SENTENCE
    ELSE
        MOVE NET-CARD-IN TO SALES-RECORD
        RELEASE SALES-RECORD.
    READ NET-FILE-IN
        AT END MOVE 'NO' TO MORE-INPUT-FLAG.

INPUT-EXIT.
    EXIT.

CHECK-RESULTS SECTION.
CHECK-ROUTINE.
    OPEN    INPUT    NET-FILE-OUT
    READ NET-FILE-OUT
        AT END MOVE 'NO' TO MORE-TRANS-FLAG.
    PERFORM PROCESS-CHECK-ROUTINE
        UNTIL NO-MORE-TRANS.
    CLOSE    NET-FILE-OUT.
    GO TO OUTPUT-EXIT.

PROCESS-CHECK-ROUTINE.
    DISPLAY EMPL-NO-OUT, NET-SALES-OUT, NAME-ADDR-OUT
        UPON PRINTER.
    READ NET-FILE-OUT
        AT END MOVE 'NO' TO MORE-TRANS-FLAG.

OUTPUT-EXIT.
    EXIT.
```

The program just presented illustrates a sort based on a sales contest. The records to be sorted contain data on salesperson, name and address, employee number, department number, and precalculated net sales for the contest period.

The salesperson with the highest net sales in each department wins a prize, and smaller prizes are awarded for second highest sales, third highest, etc. The order of the SORT is (1) by department, the lowest numbered first (ASCENDING KEY DEPT); and (2) by net sales within each department, the highest net sales first (DESCENDING KEY NET-SALES).

The records for the employees of departments 7 and 9 are eliminated in an input procedure (SCREEN-DEPT) before sorting begins. The remaining records are then sorted, and the output is placed on another file for use in a later job step.

A simple input procedure might be organized as follows:

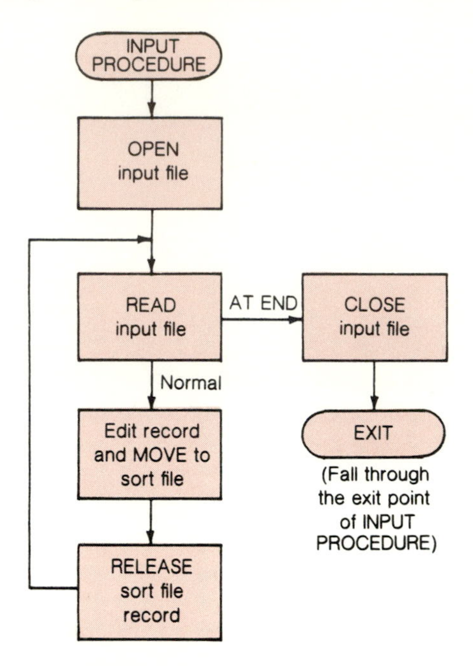

The sorting process orders the records, up to the point of determining which record goes first in the final output sequence. If an output procedure has been specified, the sorting process at this point transfers control to the output procedure. The output procedure is responsible for opening its output file, if any, and obtaining records in the final sequence by means of the RETURN statement. When the output procedure has disposed of each record, it returns the next record, and thus continues to return and process them. After the last record has been returned, the sorting process will cause control to pass to the AT END phrase the next time a RETURN statement is executed. The output procedure is then responsible for closing its output file, if any, and allowing control to pass to its exit point. At this point the sorting process terminates its own procedures. Control then passes to the statement following the SORT statement. A simple output procedure might be organized as follows:

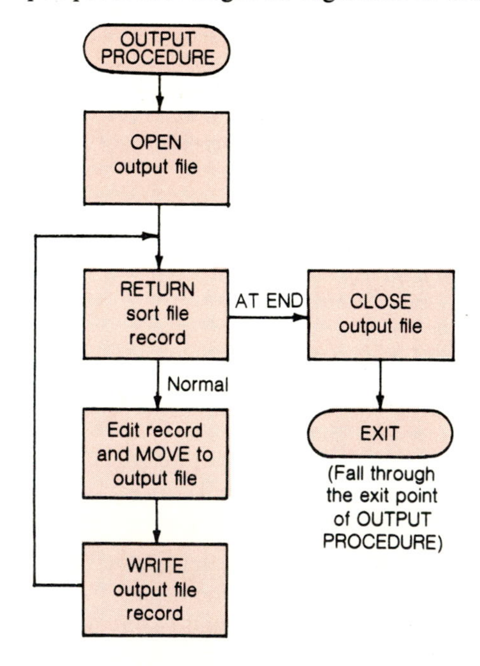

In effect, the sequence of events just described also applies when the USING or GIVING option is used, except that the input or output procedure becomes implicit, rather than specified in detail by the programmer.

An example of the SORT program—INPUT PROCEDURE and OUTPUT PROCEDURE options—follows:

```
ENVIRONMENT DIVISION.
        .
        .
        .
INPUT-OUTPUT SECTION.
FILE-CONTROL.
    SELECT RAWMFILE
        ASSIGN TO . . .
    SELECT SFILE
        ASSIGN TO . . .
    SELECT MFILE
        ASSIGN TO . . .

DATA DIVISION.
FILE SECTION.
FD  RAWMFILE
    LABEL RECORD IS STANDARD.
01  RAWMRECORD.
    05  FILLER                          PICTURE X(26).
    05  ACCOUNT-STATUS                  PICTURE X.
    05  FILLER                          PICTURE X(11).

SD  SFILE
    RECORD CONTAINS 38 CHARACTERS.
01  SRECORD.
    05  MNO                             PICTURE X(6).
    05  FILLER                          PICTURE X(32).

FD  MFILE
    LABEL RECORD IS STANDARD.
01  MRECORD.
    05  MNOO                            PICTURE X(6).
    05  BALANCE                         PICTURE 9(5)V99.
    05  FILLER                          PICTURE X(25).

WORKING-STORAGE SECTION.
01  FLAGS.
    05  MORE-INPUT-FLAG                 PICTURE XXX       VALUE 'YES'.
        88  MORE-INPUT                                    VALUE 'YES'.
        88  NO-MORE-INPUT                                 VALUE 'NO'.
    05  MORE-OUTPUT-FLAG                PICTURE XXX       VALUE 'YES'.
        88  MORE-OUTPUT                                   VALUE 'YES'.
        88  NO-MORE-OUTPUT                                VALUE 'NO'.

PROCEDURE DIVISION.
BEGIN SECTION.
    SORT SFILE
        ON ASCENDING KEY MNO
        INPUT PROCEDURE IS CHECK-STATUS
        OUTPUT PROCEDURE IS CHECK-BALANCE.
    STOP RUN.

CHECK-STATUS SECTION.
INPUT-ROUTINE.
    OPEN    INPUT    RAWMFILE.
    READ RAWMFILE
        AT END MOVE 'NO' TO MORE-INPUT-FLAG.
    PERFORM PROCESS-INPUT-ROUTINE
        UNTIL NO-MORE-INPUT.
    CLOSE RAWMFILE.
    GO TO CHECK-STATUS-EXIT.
```

```
PROCESS-INPUT-ROUTINE.
    IF ACCOUNT-STATUS IS EQUAL TO ZERO
        NEXT SENTENCE
    ELSE
        RELEASE SRECORD FROM RAWMRECORD.
    READ RAWMFILE
        AT END MOVE 'NO' TO MORE-INPUT-FLAG.

CHECK-STATUS-EXIT.
    EXIT.

CHECK-BALANCE SECTION.
OUTPUT-ROUTINE.
    OPEN    OUTPUT    MFILE.
    RETURN SFILE INTO MRECORD
        AT END MOVE 'NO' TO MORE-OUTPUT-FLAG.
    PERFORM PROCESS-OUTPUT-ROUTINE
        UNTIL NO-MORE-OUTPUT.
    CLOSE    MFILE.
    GO TO CHECK-BALANCE-EXIT.

PROCESS-OUTPUT-ROUTINE.
    IF BALANCE IS EQUAL TO ZERO
        DISPLAY 'ACCOUNT NUMBER' MNOO 'HAS A ZERO BALANCE'
            UPON PRINTER
    ELSE
        WRITE MRECORD.
    RETURN SFILE INTO MRECORD
        AT END MOVE 'NO' TO MORE-OUTPUT-FLAG.

CHECK-BALANCE-EXIT.
    EXIT.
```

An example of the SORT program—INPUT PROCEDURE and OUTPUT PROCEDURE options—follows:

```
ENVIRONMENT DIVISION.
    .
    .
    .
INPUT-OUTPUT SECTION.
FILE-CONTROL.
    SELECT INFILE
        ASSIGN TO . . .
    SELECT SORTFILE
        ASSIGN TO . . .
    SELECT OUTFILE
        ASSIGN TO . . .

DATA DIVISION.
FILE SECTION.
FD  INFILE
    LABEL RECORDS OMITTED
    DATA RECORD IS INREC.
01  INREC                              PICTURE X(80).

SD  SORTFILE
    DATA RECORD IS SRTREC.
01  SRTREC.
    05  KEY1                           PICTURE X(5).
    05  KEY2                           PICTURE X(12).
    05  KEY3                           PICTURE X(15).
    05  FILLER                         PICTURE X(20).
    05  KEY4                           PICTURE X(28).
```

```
FD  OUTFILE
    LABEL RECORDS STANDARD
    DATA RECORD IS OUTREC.
01  OUTREC                          PICTURE X(80).

WORKING-STORAGE SECTION.
01  FLAGS.
    05  MORE-INPUT-FLAG             PICTURE XXX     VALUE 'YES'.
        88  MORE-INPUT                              VALUE 'YES'.
        88  NO-MORE-INPUT                           VALUE 'NO'.
    05  MORE-OUTPUT-FLAG            PICTURE XXX     VALUE 'YES'.
        88  MORE-OUTPUT                             VALUE 'YES'.
        88  NO-MORE-OUTPUT                          VALUE 'NO'.

PROCEDURE DIVISION.
BEGIN SECTION.
    DISPLAY 'START EXAMPLE'
        UPON CONSOLE.
    SORT SORTFILE
        ON DESCENDING KEY KEY1, KEY2
            ASCENDING KEY KEY3,
            DESCENDING KEY KEY4
        INPUT PROCEDURE IS INP-PROC
        OUTPUT PROCEDURE IS OUT-PROC.
    STOP RUN.

INP-PROC SECTION.
INPUT-ROUTINE.
    OPEN     INPUT     INFILE.
    READ INFILE
        AT END MOVE 'NO' TO MORE-INPUT-FLAG.
    PERFORM PROCESS-INPUT-ROUTINE
        UNTIL NO-MORE-INPUT.
    CLOSE INFILE.
    DISPLAY 'END OF INPUT'
        UPON CONSOLE.
    GO TO INP-PROC-EXIT.

PROCESS-INPUT-ROUTINE.
    MOVE INREC TO SRTREC
    RELEASE SRTREC.

INP-PROC-EXIT.
    EXIT.

OUT-PROC SECTION.
OUTPUT-ROUTINE.
    OPEN     OUTPUT     OUTFILE.
    RETURN SORTFILE
        AT END MOVE 'NO' TO MORE-OUTPUT-FLAG.
    PERFORM PROCESS-OUTPUT-ROUTINE
        UNTIL NO-MORE-OUTPUT.
    CLOSE OUTFILE.
    DISPLAY 'END OF OUTPUT'
        UPON CONSOLE.
    GO TO OUT-PROC-EXIT.

PROCESS-OUTPUT-ROUTINE.
    MOVE SRTREC TO OUTREC.
    WRITE OUTREC.
    RETURN SORTFILE
        AT END MOVE 'NO' TO MORE-OUTPUT-FLAG.

OUT-PROC-EXIT.
    EXIT.
```

```
MERGE file-name-1 ON {ASCENDING | DESCENDING} KEY data-name-1 [ , data-name-2 ] . . .

        [ON {ASCENDING | DESCENDING} KEY data-name-3 [ , data-name-4 ] . . . ] . . .

   [ COLLATING SEQUENCE IS alphabet-name ]

    USING file-name-2, file-name-3 [ , file-name-4 ] . . .

   { OUTPUT PROCEDURE IS section-name-1 [ {THROUGH | THRU} section-name-2 ] }
   { GIVING file-name-5                                                     }
```

Figure 11.13 Format—MERGE statement.

MERGE Statement

The MERGE statement has the same options as the SORT statement with the exception of the INPUT PROCEDURE, which is not available with the MERGE statement. Environment, Data, and Procedure Division statements are the same as the SORT statement, except that the MERGE statement is used in the Procedure Division in place of the SORT statement.

The MERGE statement combines two or more identically sequenced files on a set of specified keys, and during the process makes records available, in merged order, to an output procedure or to an output file. (The format for the MERGE statement is shown in figure 11.13.)

Rules Governing the Use of the MERGE Statement

1. The MERGE statement will merge all records contained on file-name-2, file-name-3, and file-name-4. The files referenced in the MERGE statement must be open when the MERGE statement is executed. These files are automatically opened and closed by the merge operation with all implicit functions performed, such as the execution of any associated USE procedures. The terminating function for all files is performed as if a CLOSE statement, without optional phrases, had been executed for each file.
2. The data-names following the word KEY are listed from left to right in the MERGE statement in order of decreasing significance and without regard to how they are divided into KEY phrases. In the format, data-name-1 is the major key, data-name-2 is the next most significant key, etc. The same rules for ASCENDING and DESCENDING phrases apply as in the SORT statement except the files are being merged instead of sorted.
3. The COLLATING SEQUENCE phrase operates in the same manner as with the SORT statement.
4. The operation of the OUTPUT PROCEDURE is the same as the SORT statement. There are no INPUT PROCEDUREs in the MERGE statement.
5. If the GIVING phrase is specified, all the merged records in file-name-1 are automatically written on file-name-5 as the implied output procedure for the MERGE statement.
6. File-name-1 must be described in a sort-merge file description entry in the Data Division.
7. Section-1 represents the name of an output procedure.
8. File-name-2, file-name-3, file-name-4, and file-name-5 must be described in a file description entry, not in a sort-merge file description entry, in the Data Division.
9. The words THRU and THROUGH are equivalent.
10. Data-name-1, data-name-2, data-name-3, and data-name-4 are KEY data-names and are subject to the same rules that apply for the SORT statement KEY data-names.
11. No more than one file-name from a multiple file reel can appear in the MERGE statement.
12. File-names must not be repeated within the MERGE statement.

13. MERGE statements may appear anywhere except the declarative portion of the Procedure Division or in an input procedure or output procedure associated with a SORT or MERGE statement.

An example of the MERGE program—USING and GIVING options—follows:

```
ENVIRONMENT DIVISION.
      .
      .
      .
INPUT-OUTPUT SECTION.
FILE-CONTROL.
    SELECT INPUT-FILE-1
        ASSIGN TO . . .
    SELECT INPUT-FILE-2
        ASSIGN TO . . .
    SELECT MERGE-FILE
        ASSIGN TO . . .
    SELECT OUTPUT-FILE
        ASSIGN TO . . .
DATA DIVISION.
FILE SECTION.
FD  INPUT-FILE-1
      .
      .
01    . . .
      .
      .
FD  INPUT-FILE-2
      .
      .
01    . . .
      .
      .
SD  MERGE-FILE
      .
      .
01    . . .
      .
      .
FD  OUTPUT-FILE
      .
      .
01    . . .
      .
      .
PROCEDURE DIVISION.
MERGE-CALL.
    MERGE MERGE-FILE
        ON . . .
        USING  INPUT-FILE-1,
               INPUT-FILE-2
        GIVING OUTPUT-FILE.
    STOP RUN.
```

Note that these files may have multiple record types and sizes, provided the merge file and USING file have the same records, the merge file and GIVING file have the same records, and key descriptions and positions are equivalent for all record types.

Summary

Much data processing depends upon the order in which records appear in the files being processed. A sorting procedure manipulates an input file whose records are in an indeterminate sequence and produces an output file containing the same set of data records rearranged into the desired sequence.

A merging procedure manipulates two or more well-ordered input files and produces an output file containing the total set of data records in one well-ordered sequence.

The sequencing of the output file in a sorting or merging procedure is governed by the values of one or more fields in each record being ordered called *keys,* which appear in the same position relative to the start of the record. The order in which the keys are specified in sort or merge procedures determine their hierarchical relationship. Ascending ordering means that those records with lower values of that key will appear in the output prior to the record with higher values for that key. Descending order implies the inverse result.

The sort-merge procedure provides the capability to order one or more files of records, or to combine two or more identically ordered files of records, according to a set of programmer-specified key(s). Optionally, a programmer may apply some special processing to each of the individual records either before and/or after the record has been ordered by the sort or after the records have been combined by the merge.

The functions and capabilities of the SORT and MERGE statements are different in a number of respects. The SORT statement will produce an ordered file from one or more files that may be completely unordered in that sort sequence, whereas the MERGE statement can only produce an ordered file from two or more files, each of which is already ordered in the specified sequence.

The files specified in the USING and GIVING phrases of SORT and MERGE statements must be described in FILE-CONTROL paragraphs as being of sequential organization. No input/output statements may be executed for the file named in the Sort-Merge file description.

SORT and MERGE verbs initiate ordering procedures. The RELEASE verb may be used with the execution of the SORT verb and the RETURN verb may be used with the execution of the SORT and MERGE verbs.

The purpose of the SORT statement is to invoke the execution of a sorting procedure. Each SORT statement has its own internal working file called sort file, created and used during the sorting operation. This file is only accessed and referred to by the SORT statement.

To use the SORT feature, the programmer must provide additional information in the Environment, Data, and Procedure divisions of the source program. The basic elements are the SORT statement in the Procedure Division and the Sort-File-Description (SD) entry, with associated record description entries in the Data Division.

The FILE-CONTROL paragraph of the Environment Division names each file and allows for specification of other related information. SELECT entries must be written for all files used with input and output procedures as well as files specified in the USING and GIVING options of the SORT and MERGE statements.

The I-O-CONTROL paragraph is optional and specifies the memory area that is to be shared by different files. The SAME RECORD AREA clause specifies that two or more files are to use the same memory area for processing the current logical record. The SAME SORT AREA or SAME SORT-MERGE AREA clauses specify a memory area that will be made available for use in sorting or merging each sort or merge file named.

In the Data Division, the programmer must include file description entries for files to be sorted or merged, sort-file-description entries for sort work files, and record description entries for each. An SD file description is used to provide information about the size and the name of the data record associated with the file to be sorted or merged.

The Sort-File-Description entry furnishes information concerning the physical structure and record-names of all records to be sorted or merged. One or more record description entries must follow the sort-merge file description entry. However, no input/output statements may be executed for this file.

The values in key comparisons determine the order of the record in the sort output. All comparisons are made on the basis of the collating sequence of that particular computer unless the COLLATING SEQUENCE phrase of the SORT statement is used to specify a different sequence.

The COBOL procedural statements that are available for use in a sort program are: the SORT statement for performing a sorting operation on a collection of unordered records; the RELEASE statement for transferring records to the initial phase of a sort operation; and the RETURN statement for obtaining records from the final phase of a sort operation.

The Procedure Division must contain one or more SORT statements to indicate the sorting operation and, optionally, any necessary input and output procedures. The procedure-names constituting the input and output procedures are specified in the SORT statement.

A choice is made by the programmer of whether to have the sorting process handle the input processing of the file being sorted or of having the programmer specify the input processing procedure.

The SORT statement creates a sort file by executing input procedures or by transferring records from another file, sorts the records in the sort file on a set of specified keys, and in the final phase of the sort operation, makes accessible each record from the sort file, in sorted order, to some output procedure or output file.

Procedures that are executed before the sort are specified in the INPUT PROCEDURE. Procedures that are executed after the sort are specified in the OUTPUT PROCEDURE.

When the INPUT PROCEDURE is specified, the programmer is responsible for all input processing for the sorting process. An input procedure may consist of one or more sections. Since the input procedure is invoked via the same technique used in the PERFORM statement, it must conform to the same rules as applied to the PERFORM statement. Control passes to the input procedure only through the execution of a SORT statement. The input procedure may contain any procedures needed to select, create, or modify records for input to the sorting process. The input procedure makes records available to the sorting operation through the RELEASE statement.

The USING option indicates that the records to be sorted are all in one file and are passed to the sorting operation as one unit when the SORT statement is executed. If this option is used, all records must be in the same file. For the USING option, the compiler will automatically OPEN, READ, RELEASE, and CLOSE file-name-2 and file-name-3 without the programmer specifying these functions.

A choice must be made between having the sorting process handle the output processing of the newly sorted file or having the programmer's program specify the output processing procedures.

When the OUTPUT PROCEDURE is specified, the programmer must specify the output processing of the newly sorted file. Since the output procedure is invoked by the sorting procedures using the same techniques used in the execution of the PERFORM statement, the structure of the output procedure must follow the same basic rules applied to the PERFORM statement. Control can only be passed to the output procedure through the execution of the SORT statement. The output procedure may contain any procedures needed to select, modify, or copy the records that are being returned from the sorting process. Records are obtained one at a time from the SORT/MERGE process through the RETURN statement.

If the GIVING phrase is specified, all the sorted records in file-name-1 are automatically written in file-name-4 as the implied procedure for the SORT statement. The SORT statement automatically initiates the processing of, releases the logical records to, and terminates the processing of file-name-4.

The RELEASE statement is used to transfer logical input records from an input procedure prior to the initial phase of a sorting operation. The RELEASE statement may appear only in an input procedure, and each input procedure must contain at least one RELEASE statement. The record-name referenced by the RELEASE statement must be the name of a record defined within a sort file; that is, a file described with a SD level indicator. If the FROM phrase is used, the contents of "identifier" must be the name of a data item in working-storage or of an input record area. After the RELEASE statement is executed, the contents of record-name is no longer available to the COBOL procedure.

The RETURN statement is used to obtain logical output records from a sort operation and to transfer them to an output procedure. The RETURN statement may appear only in an output procedure, and every output procedure must contain at least one RETURN statement. The file-name in the RETURN statement must be described with an SD level indicator and must be the same file-name that was referred in the SORT statement currently being executed. The INTO phrase may be used only when the file referred to by file-name contains just one type of record. The identifier must be the name of a data item in working-storage or of an output record area. The execution of the RETURN statement causes the next record sorted to be made available for processing in the record area associated with the sort file. After the contents of the sort file are exhausted, the next execution of the RETURN statement will result in the execution of the imperative statement in the AT END phrase. After execution of the AT END imperative statement, no RETURN statement may be executed within the current output procedure.

No OPEN, READ, WRITE, or CLOSE statements may be given for a sort file.

If an input procedure is specified, the SORT statement transfers control to start processing records. The input procedure is responsible for opening the input file and reading the logical record. The input procedure continues reading and releasing all input records that have been given to the sort through the RELEASE statement. The input procedure is responsible for closing the input file after all records have been released. Control is returned to the sorting process. The sorting process orders the records according to the programmer-specified keys. If an output procedure is specified, the sorting process at this point transfers control to output procedures. The output procedure is responsible for opening its output file, if any, and obtaining records in the final sequence by the use of the RETURN statement. After the last record has been returned, the sorting process causes control to pass to the AT END phrase the next time a RETURN statement is executed. The output procedure is responsible for closing its output file, if any, and allowing control to pass to its exit point. The sorting process then terminates its own procedures. At this point, control passes to the statement following the SORT statement.

In effect, the sequence of events just described also applies when the USING or GIVING option is used, except that the input or output procedure becomes implicit, rather than specified in detail by the programmer.

The MERGE statement has the same options as the SORT statement, with the exception of the INPUT PROCEDURE, which is not available to the MERGE statement. Environment, Data, and Procedure Division statements are the same as the SORT statement except that the MERGE statement is used in the Procedure Division in place of the SORT statement.

The MERGE statement combines two or more identically sequenced files on a set of specified keys, and during the process makes records available, in merged order, to an output procedure or to an output file.

Questions for Review

1. What is a sorting procedure? a merging procedure?
2. What is a key and what is its role in the sort and merge procedures?
3. What does the Sort-Merge procedure provide?
4. How do the functions and capabilities of the SORT and MERGE statements differ?
5. Where must the files specified in the USING and GIVING phrases of the SORT and MERGE statements be described?
6. What verbs initiate ordered procedures and what verbs are used with each?
7. What is the purpose of the SORT verb?
8. What is the purpose of the MERGE verb?
9. What is the purpose of the SORT statement?
10. What are the basic elements of the SORT feature?
11. Briefly describe the necessary entries in the Environment Division for the SORT and MERGE features.
12. What entries must be written in the Data Division for the SORT-MERGE feature?
13. What information does the Sort-File-Description entry provide?
14. How are Sort-Keys evaluated?
15. What are the procedural statements that are available for the SORT, and what are their main purposes?
16. What choice must be made by the programmer regarding input processing?
17. Briefly explain the operation of the SORT statement.
18. What is an INPUT PROCEDURE and how is it specified?
19. What does the USING option indicate?
20. What choice must be made by the programmer regarding output processing?
21. What is an OUTPUT PROCEDURE and how is it specified?
22. What does the GIVING option indicate?
23. What is the function of the RELEASE statement and how is it used in SORT procedures?
24. What is the function of the RETURN statement and how is it used in SORT-MERGE procedures?
25. Briefly explain the operation of INPUT PROCEDURES and OUTPUT PROCEDURES.
26. Briefly explain the operation of the MERGE statement.

Matching Questions

Match each item with its proper description.

_____ 1. DESCENDING
_____ 2. Sorting procedure
_____ 3. Sort-File-Description
_____ 4. Sort-Keys
_____ 5. Ordering
_____ 6. SORT statement
_____ 7. Sort-Work-File
_____ 8. Merging procedure
_____ 9. ASCENDING
_____ 10. Merge statement
_____ 11. SAME RECORD AREA

A. Collection of records that is involved in the sorting operation as it exists in intermediate devices.
B. Produces an ordered file from one or more files that may be completely unordered.
C. Manipulates two or more well-ordered files and produces an output file containing the total set of records in one well-ordered sequence.
D. Specifies that two or more files are to use the same memory area for processing the current logical record.
E. Data-names assigned to each field involved in the sorting operation.
F. Produce an ordered file from two or more files, each of which is ordered in a particular sequence.
G. Manipulates an input file whose records are in an indeterminate sequence, and produces an output file containing the same set of records arranged in a desired sequence.
H. Lowest value of a key to its highest value according to the collating sequence of the particular computer.
I. Comparing keys and then determining the sequence as determined by the ASCENDING or DESCENDING rule, whichever is applicable.
J. Furnishes information concerning the physical structure identification and record-names of files to be sorted.
K. Highest value of the key to the lowest according to the collating sequence of the particular computer.

Match each item with its proper description.

_____ 1. RELEASE
_____ 2. ASSIGN
_____ 3. OUTPUT PROCEDURE
_____ 4. SD
_____ 5. USING
_____ 6. I-O-Control
_____ 7. INPUT PROCEDURE
_____ 8. RETURN
_____ 9. GIVING
_____ 10. COLLATING SEQUENCE

A. Specifies memory areas to be shared by different files.
B. Processes records before they are sorted.
C. Indicates that the records to be sorted are all in one file and are to be passed to the sorting operation as one unit.
D. Ordering sequence for a particular computer.
E. All sorted records are automatically written on an output file.
F. Transfers logical input records from an INPUT PROCEDURE to the initial phase of a sorting operation.
G. Identifies the beginning of the Sort-Merge file described and must precede file-name.
H. Processes records after they are sorted.
I. Obtains logical output records from a sort operation and transfers them to an OUTPUT PROCEDURE.
J. Specifies the association of the sort or merge file referenced by file-name to a storage medium.

Exercises

Multiple Choice: Indicate the best *answer (questions 1–19).*

1. A sorting procedure
   a. manipulates an output file whose records are not ordered.
   b. combines two files in a well-ordered sequence.
   c. manipulates an input file whose records are in an indeterminate order.
   d. None of the above.
2. A merging procedure
   a. manipulates two or more unordered input files.
   b. manipulates one input file in a well-ordered sequence.
   c. manipulates two or more well-ordered output files.
   d. None of the above.

3. In the ordering of a file, keys
    a. may be specified in any order unrelated to their hierarchical relationship.
    b. may appear in different positions in each record.
    c. that are the most significant are specified last.
    d. None of the above.
4. The Sort-Merge feature provides the capability to
    a. order one or more files of records.
    b. combine two or more identically ordered files according to their specified keys.
    c. apply some special processing to each of the individual records.
    d. All of the above.
5. The files specified in the USING and GIVING phrases of the SORT and MERGE statements must be described explicitly or implicitly in the FILE-CONTROL paragraph as having
    a. random organization.
    b. sequential organization.
    c. relative organization.
    d. indexed sequential organization.
6. The FILE-CONTROL paragraph may specify all files in the
    a. INPUT PROCEDURE.
    b. OUTPUT PROCEDURE.
    c. Sort-Files.
    d. All of the above.
7. The Sort-File-Description entry furnishes information regarding the
    a. files in OUTPUT PROCEDUREs.
    b. files in INPUT PROCEDUREs.
    c. physical structure of the file to be sorted or merged.
    d. All of the above.
8. A statement for transferring records to the initial phase of a sort operation is the
    a. SORT statement.
    b. RELEASE statement.
    c. RETURN statement.
    d. GIVING statement.
9. The SORT statement
    a. creates a sort file by executing input procedures.
    b. transfers records from another file.
    c. sorts the records in a sort file on a set of specified keys.
    d. All of the above.
10. An INPUT PROCEDURE indicates
    a. that the programmer has written a procedure to process records after they have been sorted.
    b. a procedure that is included in the Procedure Division in one or more sections.
    c. that the programmer is not responsible for all input processing.
    d. that one file at a time will be passed to the Sort feature.
11. The USING option indicates that
    a. the records to be sorted are in one file.
    b. the records to be sorted are in more than one file.
    c. the programmer must issue instructions to transfer the records to the Sort feature.
    d. the programmer must issue instructions to terminate the processing.
12. An OUTPUT PROCEDURE
    a. may contain a SORT statement.
    b. must contain ALTER statements.
    c. may transfer control outside the OUTPUT PROCEDURE.
    d. All of the above.
13. The GIVING phrase
    a. specifies that all records to be sorted are in one file.
    b. requires the programmer to write OPEN statements for the file.
    c. transfers all sorted records to an output file.
    d. All of the above.
14. The RELEASE statement
    a. may appear in an OUTPUT PROCEDURE.
    b. must appear in an INPUT PROCEDURE.
    c. may appear in a Sort-File-Description entry.
    d. after execution, still has access to the output logical record.

15. The RETURN statement
    a. is used to obtain logical output records from a sort operation.
    b. transfers output records to the OUTPUT PROCEDURE.
    c. must be included in the OUTPUT PROCEDURE.
    d. All of the above.
16. After the contents of the sort file are exhausted, the next execution of the RETURN statement will cause the
    a. GIVING option to be executed.
    b. execution of the AT END imperative statement.
    c. USING option to be executed.
    d. OUTPUT PROCEDURE to be executed.
17. The INPUT PROCEDURE and OUTPUT PROCEDURE function in a manner similar to a
    a. MOVE statement.
    b. GO TO statement.
    c. PERFORM statement.
    d. READ statement.
18. The statement in the INPUT PROCEDURE that causes a record to be placed in the sort file is the
    a. RELEASE statement.
    b. RETURN statement.
    c. GIVING statement.
    d. USING statement.
19. The MERGE statement
    a. has the same options as the SORT statement with the exception of INPUT PROCEDURE.
    b. combines two or more files identically sequenced.
    c. makes records available, in merged sequence, to an output file.
    d. All of the above.
20. Given the following information:

TAPE FORMAT

IRG	Dept. No.	Clock No.	Name	SS#	YTD gross	YTD W/T	YTD fica	IRG
	2 pos.	3 pos.	20 pos.	9 pos.	7 pos.	6 pos.	5 pos.	

52 character logical record, fixed length
10 logical records/block

Write a program to
    a. Input the tape record.
    b. Sort the record on Social Security Number sequence.
    c. Output the tape record in the same format as the input record.
21. Given the following information:

CARD FORMAT

Dept. No.	Clock No.		Name	Soc. Sec. No.		Current gross	Current W/T	Current fica	code
1-2	3-5	6	7 26	27-35		62-68	69-74	75-79	80

TAPE FORMAT

IRG	Dept. No.	Clock No.	Name	Soc. Sec. No.	Gross	W/T	Fica	IRG
	2 pos.	3 pos.	20 pos.	9 pos.	7 pos.	6 pos.	5 pos.	

52 characters per logical record, fixed length
5 records per block

Write the program to accomplish the following:
    a. Input the card record.
    b. Sort record to Department Number and Clock Number sequence.
    c. Write card record on tape output.

Problem 1

A STUDENT-MASTER file contains records in the following format:

Positions	Field
1-9	Social Security Number
10-30	Student Name
31-54	Street Address
55-66	City
67-68	State
69-73	Zipcode

Mailing lables are to be produced so that registration information can be mailed to each student. The records to be sorted in zipcode sequence are to be outputted as follows:

```
NAME
STREET ADDRESS
CITY          STATE      ZIPCODE
```

Write the necessary entries to produce the aforementioned.

Problem 2

A report is to be printed listing all checks in check number sequence. After the checks have been sorted, they are to be listed on the printer in the output form that follows.

Job Definition

A report is to be printed listing all returned checks and designating them as either regular or special.

Input

The input file consists of records as follows:

Positions	Field Designation
1-5	Check number
6-10	Check amount (xxx.xx)
11-21	Last name
22	Initial
23	Code

Output

A Returned Check Register is to be printed as follows:

```
                 RETURNED CHECK REGISTER

      NAME        INIT   CHKAMT     CHKNO    CODE

   HOLMES          A      97.50     1567      R
   FRANKLIN        B      55.10     1569      R
   STEVENS         S      73.50     1570      R
   WALKER          F     127.75     1571      S
   JONES           D      33.00     1573      R
   ROSEMAN         B      18.75     1574      S
   MORGAN          P      47.50     1577      S
   GREEN           H     108.15     1578      R
   MITCHELL        T      74.00     1579      R
   PENNEY          J      29.00     1582      R
   MORGAN          R     144.50     1583      S
   CRAIG           A      82.50     1588      R
   WARD            M     119.75     1590      R
```

This Printer Spacing Chart shows how the report is formatted:

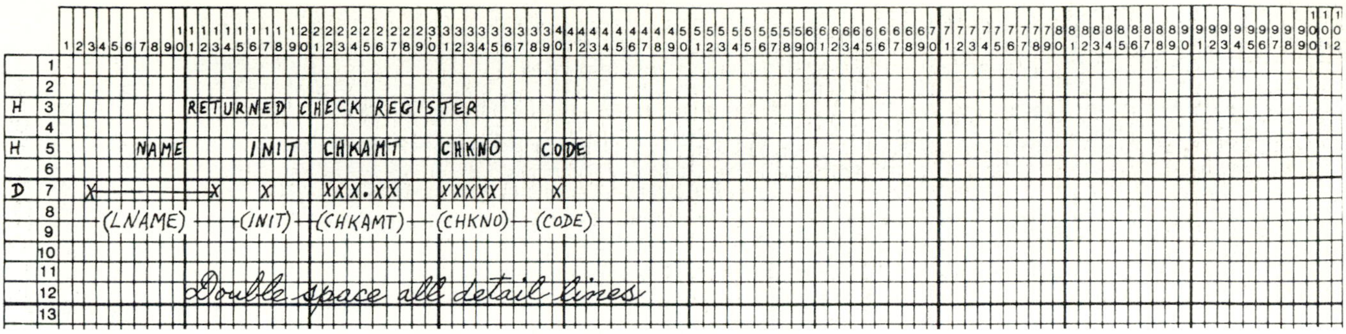

```
H  3            RETURNED CHECK REGISTER
H  5       NAME      INIT  CHKAMT    CHKNO     CODE
D  7  X----------X  X    XXX.XX    XXXXX     X
   8   (LNAME)     (INIT) (CHKAMT)  (CHKNO)  (CODE)
  12            Double space all detail lines
```

Problem 3

Write the necessary INPUT and OUTPUT PROCEDUREs to sort the payroll file to employee number sequence within each department so that the checks may be distributed by department as specified in the following:
A payroll file consists of records containing the following information:

Positions	Fields
1-3	Department
4-8	Number
9-17	Social security number
18-22	Gross pay
23-27	Withholding tax
28-32	Social security tax
53-56	Deduction 1
57-60	Deduction 2
61-64	Deduction 3
65-68	Deduction 4
	All fields have two decimal positions

The net pay is to be computed using the following formula:
NET = GROSS − WITHTAX − SSTAX − D1 − D2 − D3 − D4
All deductions are to be taken from gross pay unless doing so causes the net pay amount to become zero or negative in value. If that happens, add back the last deduction, print an exception record on a special report, and bypass all further calculations for that record. Include messages on the exception report to indicate which, if any, deductions were actually taken.
Prepare a Printer Spacing Chart to represent heading and detail lines of the exception record report. Include the UDATE and PAGE fields on the heading lines.
Create your own data names where necessary, and the necessary heading lines are to be indicated on the Printer Spacing Chart. In the output, include all information from the input plus the net pay and any necessary messages.
Output is as follows:

```
          5/26/81                       PAYROLL   DEDUCTION   EXCESSES                            PAGE   1

  DEPT. NOSOC. SEC. NO.      GROSS   WITH.TX.          DED 1  DED 2  DED 3   DED 4
   50   1600   564-49-1212  941.62  594.16  219.41   92.11  97.48  97.48    .00      ACTUAL PAY IS   35.94  INCLUDING D1
   62   2300   581-01-0334  877.77  587.70   58.97   81.99  94.19  94.19   9.28      ACTUAL PAY IS   54.92  INCLUDING D1, D2
   62   4400   495-14-9142   94.82   81.48    8.24    5.19   2.43   2.43   5.19      ACTUAL PAY IS    5.10  NO DEDUCTIONS TAKEN
```

Problem 4

Given the following information:

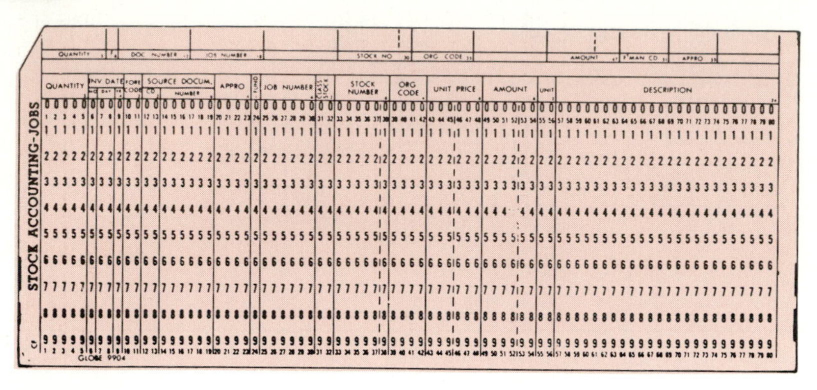

Write the necessary program to accomplish the following:

a. Read in the input records.
b. Multiply quantity by unit price giving amount.
c. Sorting the records into stock number sequence.
d. Output a report on the printer as follows:

Record Positions	Field	Print Positions	
31-32	Class Stock	3-4	
33-38	Stock Number	6-11	
55-56	Unit	13-14	
57-80	Description	16-39	
1-5	Quantity	41-46	XX,XXX
43-48	Unit Price	48-54	$XXX.XX
	Amount	56-65	$XX,XXX.XX

Processing to be performed:

1. Read in the input records.
2. Multiply quantity by unit price giving amount.
3. Sort the records into stock number sequence.
4. Output a record on the printer as format above.
5. Final total of all amounts.

Report Writer Feature

Chapter Outline

Chapter Objectives

The learning objectives of this chapter are:

1. To describe and explain the operation of the Report Writer feature and how it may be applied to data processing problems.
2. To demonstrate how it is easier to specify the physical appearance of a report rather than require specifications of the detailed procedures necessary to produce the report.
3. To describe the format of the necessary entries in the Data Division to produce the desired output.
4. To explain the various functions and operations of SUM counters in the Report Writer.
5. To describe the necessary Procedure Division statements of INITIATE, GENERATE, and TERMINATE and how they are used to produce the necessary output report.
6. To explain the function and operation of the USE BEFORE REPORTING declarative and how the SUPPRESS statement may be used within the declarative to inhibit the printing of lines.

Introduction

A report represents a pictorial organization of data. To present a report, the physical aspects of the report format must be differentiated from the conceptual characteristics of the data to be included in the report. In defining the physical aspects of the report format, consideration must be given to the width and length of the report medium, to the individual page structure, to the type of hardware device on which the report is finally to be written. Structure controls are established to insure that the report format is maintained.

To define the conceptual characteristics of the data, i.e., the logical organization of the report itself, the concept of level structure is used. Each report may be divided into respective report groups, which, in turn are subdivided into a sequence of items. Level structure permits the programmer to refer to an entire report-name, a major report group, a minor report group, an elementary item within a report group, etc.

To create the report, the approach taken is to define the types of report groups that must be considered in presenting data in a formal manner. Types may be defined as heading groups, footing groups, control groups, or detail print groups. A report group describes a set of data that is to be considered as an individual unit, irrespective of its physical format structure. The unit may be the representation of a data record, a set of constant report headings, or a series of variable control totals. The description of the report group is a separate entity. The report group may extend over several actual lines of a page and may have a descriptive heading above it, which sometimes is necessary in order to produce the desired output report format.

The Report Writer Feature provides the facility for producing reports by specifying the physical appearance of a report rather than requiring specification of the detailed procedures necessary to produce that report. The programmer can specify the format of the printed report in the Data Division, thereby minimizing the amount of Procedure Division coding he or she would have to write to create the report.

A hierarchy is used in defining the logical organization of the report. Each report is divided into report groups, which in turn are divided into a sequence of items. Such a hierarchical structure permits explicit reference to a report group with implicit reference to other levels in the hierarchy. A report group contains one or more items to be presented on one or more lines.

The specification for the format of the printed output together with any necessary control totals and control headings can be written into the program. The detailed report group items are the basic elements of this report. The necessary data for the detail group items can be supplied from sources outside of the report or by the summation of data items within the report. This summation is the process of adding either the individual data items or other control totals.

Additional information in the form of control headings and control totals can be printed with the detail group items. Control totals and control headings occur automatically when the machine senses a control break. When the value of a specified item used for control purposes changes, a control break occurs.

Report headings at the beginning of the report, as well as totals at the end of the report, can be printed as the report is being prepared. Individual page headings at the top of each page, and totals at the bottom of each sheet, may also be printed concurrently. As the program is being executed, line and page counters are incremented automatically and are used by the program to print the various headings, as well as to control the skipping and spacing of the printed items. Data is added, and the totals are printed automatically.

The Report Writer option can print the report as the information is being processed and can also put the data into intermediate storage where it may be used for subsequent off-line printing in a specified format.

A printed report consists of the information reported in the format in which it is printed. Several reports may be printed from the same program. A special two-character identification is necessary to specify each individual report.

At program execution time the report is put in the specified format, the data to be accumulated is added, the necessary totals are printed, counters are incremented and reset, and each line and page is printed. Thus the programmer need not be concerned with any of the details of the operations.

Description of the Report Writer

The Report Writer feature provides the facility for producing reports by specifying the physical appearance of a report instead of specifying the detailed procedures necessary to produce the report.

The Report Writer feature places emphasis on the organization, format, and contents of an output report. Although a report can be produced using the standard COBOL language, the Report Writer language characteristics provide a more concise method for report structuring and report production. Much of the Procedure Division coding which would normally be supplied by the programmer is instead provided automatically by the Report Writer Control System (henceforth referred to as RWCS). Thus, the programmer is relieved of writing procedures for moving data, constructing print lines, counting lines on a page, numbering pages, producing heading and footing lines, recognizing the end of logical data subdivisions, updating SUM counters, etc. All of these operations are accomplished by the RWCS from source language statements that appear primarily in the Report Section of the Data Division of the source program.

Data movement to a report description is directed by the Report Section clauses SOURCE, SUM, and VALUE. Fields of data are positioned on a print line by means of the COLUMN NUMBER clause. The PAGE clause specifies the length of the page, the size of the heading and footing areas, and the size of the area in which the detail lines will appear. Data items may be specified to form a control hierarchy. During the execution of a GENERATE statement, the RWCS uses the hierarchy to check automatically for control breaks. When a control break occurs, summary information (SUM counters) can be presented.

Report Format

A report may consist of any meaningful combination of the following syntax selections:

Report Heading	(one for each report)
Page Heading	(one format for each report)
Overflow Heading	(one format for each report)
Control Heading	(one format for each control level)
Detail	(no limit for each report)
Control Footing	(one format for each control level)
Overflow Footing	(one format for each report)
Page Footing	(one format for each report)
Report Footing	(one format for each report)

In the Data Division, the programmer provides the necessary data-names and describes the formats of the report to be produced. In the Procedure Division the programmer writes the necessary statements that provide the desired reports.

Data Division

The Report Writer Feature allows the programmer to describe the report pictorially in the Data Division, thereby minimizing the amount of Procedure Division coding necessary. The programmer must write in the File Section of the Data Division a description of all names and formats of the reports to be produced. In addition, a complete file and record description of the input data must also be written in this section. A Report Section must be added at the end of the Data Division to define the format of each finished report. A report may be written in two files at the same time.

In COBOL, each report is described in the Report Section of the Data Division. The programmer specifies the intended format for each of the headings, footings, and detail lines in the report, as well as all sources of data. A report may utilize data described in the File Section, Working-Storage Section, and Linkage Section. In addition, the programmer specifies the overall organization and intended page layout of the report.

```
FD file-name

  [; BLOCK CONTAINS  [ integer-1 TO ]  integer-2  {RECORDS   }]
                                                   {CHARACTERS}

  [; RECORD CONTAINS  [ integer-3 TO ]  integer-4 CHARACTERS]

   ; LABEL  {RECORD IS   }  {STANDARD}
            {RECORDS ARE }  {OMITTED }

  [; VALUE OF implementor-name-1 IS  {data-name-1}
                                     {literal-1  }

          [, implementor-name-2 IS  {data-name-2}]  . . .]
                                    {literal-2  }

  [; CODE-SET IS alphabet-name]

   ;{REPORT IS  }   report-name-1  [ , report-name-2 ]  . . . .
    {REPORTS ARE}
```

Figure 12.1 Format—file description entry.

The compiler provides the following functions in the object program:

1. Vertical format control, including line counting, page counting, and production of page headings and footings.
2. Detection of control breaks.
3. Production of control headings and footings.
4. Accumulation of SUM counters to any number of control levels.
5. Execution of programmer-defined procedures before presentation of nondetail report groups.
6. Production of overflow headings and footings.

File Section

File Description

The file description entry furnishes the necessary information concerning the identification, physical structure, and record-names pertaining to the file. (The format for the file description entry is shown in figure 12.1.) A detailed discussion of the clauses, with the exception of the Report clause, will be found in the Data Division (chap. 4).

REPORT Clause

A REPORT clause is required in the FD entry to list the name of the report to be produced. (The format for the REPORT clause is shown in figure 12.2.) The name or names of each report to be produced appears in this clause. The sequence of the names is not important, and these reports may be of

```
{REPORT IS  }   report-name-1  [ report-name-2 ]  . . .
{REPORTS ARE}
```

Figure 12.2 Format—REPORT clause.

different sizes, formats, etc. The Report Name(s) must be the same name as appears in the Report Section, since the REPORT clause references the description entries with their associated file description entry. An example of the REPORT clause follows:

```
ENVIRONMENT DIVISION.
    .
    .
INPUT-OUTPUT SECTION.
FILE-CONTROL.
    SELECT FILE-1
        ASSIGN TO . . .
    SELECT FILE-2
        ASSIGN TO . . .

DATA DIVISION.
FILE SECTION.
FD  FILE-1
    RECORD CONTAINS 121 CHARACTERS
    REPORT IS REPORT-A.

FD  FILE-2
    RECORD CONTAINS 101 CHARACTERS
    REPORT IS REPORT-A.
```

For each GENERATE statement in the Procedure Division, the records for REPORT-A will be written on FILE-1 and FILE-2, respectively. The records on FILE-2 will not contain columns 102 through 121 of the corresponding records on FILE-1.

The REPORT clause specifies the names of reports that make up a report file. Report-name(s) are the unique names of the report and must be specified in the REPORT clause of the file description entry for the file in which the report is to be written.

Rules Governing the Use of the REPORT Clause

1. Each report-name specified in a REPORT clause must be the subject of a report description entry in the Report Section. The order of appearance of the report-name is not significant.
2. A report-name must appear in only one REPORT clause.
3. The subject of a file description entry that specifies a REPORT clause may only be referred to by the OPEN OUTPUT, OPEN EXTEND, and CLOSE statements.
4. The presence of more than one report-name in a REPORT clause indicates that the file contains more than one report.

Report Section

In the Report Section, the description of each report must begin with a report description entry (RD entry) and be followed by the entries that describe the report groups within the report.

Skeletal Format of the Report Section

The Report Section consists of two types of entries for each report; one describes the physical aspects of the report format, while the other describes the conceptual characteristics of the items that make up the report and their relation to the report format.

The definition of each report includes these two types of entries:

1. The RD entry specifies the basic page layout and the overall organization of the report.
2. Report group description entries give the detailed formats of all elements of the report and the sources of all information for the report.

An RD entry in the Report Section is analogous to an FD entry in the File Section; it is the highest level of hierarchical organization for the report. The report-name specified in each RD entry must be unique.

A level 01 report group description entry is analogous to a level 01 data record description entry in the File Section. A level 01 report item is called a *report group*. Normally, the hierarchical definition of the report group is completed with a series of subordinate entries with level numbers 02–49; however, it is permissible to define an entire report group in a single 01 entry.

An item with no subordinate items (even if its level is 01) is an elementary item. Any report item whose entry is followed by subordinate entries is a group item.

Since several reports may be defined in the Report Section, the skeletal format of the Report Section is as follows:

```
REPORT SECTION.
RD  report-name-1 . . .                 )
01  report-group-name . . .             |   Complete
    05 . . .                            |   description of
    .                                   |   first report
    .                                   |
    .                                   )
01 . . .
.
.
.
RD  report-name-2 . . .
01 . . .
.
.
.
RD  report-name-n . . .
    .
    .
    .
```

Within each report group, items to be printed must be described from left to right. If the report group contains multiple lines, they must be described in order from top to bottom.

The length of each line is determined by the programmer. In the formatting of a print line, spaces are assumed except where a specific item is to be printed. (In a data record, on the other hand, every character position must be described.)

Report Description (RD) Entry

In addition to naming the report, the RD entry defines the format of each page of the report by specifying the vertical boundaries of the region within which each type of report group may be printed. (The format for the Report Description Entry is shown in figure 12.3.) The RD entry also specifies the control data items. When the report is produced, changes in the values of the control data items cause the detail information of the report to be processed in groups called *control groups.*

```
RD report-name

    [ ; CODE literal-1 ]

    [ ; { CONTROL IS    } { data-name-1  [ , data-name-2 ] . . .                    } ]
        { CONTROLS ARE  } { FINAL  [ , data-name-1  [ , data-name-2 ] . . . ]       }

    [ ; PAGE  [ LIMIT IS   ]  integer-1  [ LINE  ]  [ , HEADING integer-2 ]
              [ LIMITS ARE ]             [ LINES ]

            [ , FIRST DETAIL integer-3 ]  [ , LAST DETAIL integer-4 ]

            [ , FOOTING integer-5 ] ] .
```

Figure 12.3 Format—report description entry.

Each report named in the REPORT(S) clause of an FD entry in the File Section must be the subject of an RD entry in the Report Section. Furthermore, each report in the Report Section must be named in one, and only one, FD entry.

The Report Section must contain at least one report description entry. This entry contains information pertaining to the overall format of a report named in the File Section, and is uniquely identified by the level indicator RD. The characteristics of the report page are provided by describing the number of physical lines per page and the limits for presenting the specified headings, footings, and details within a page structure. RD is the reserved word for the level indicator, and each report named in a FD entry in the File Section must be defined by an RD entry.

RD Entries

The description of each report begins with an RD entry. Except for the level indicator (RD) and the report-name, all clauses in an RD entry are optional.

The optional clauses in an RD entry are:

Clause	Function
CODE	To assign unique letters or digits to label each line of this report on intermediate storage. (The code characters do not appear in the printed report.)
CONTROL(S)	To specify data-names of control items, in the order from most significant to least significant.
PAGE	To specify the maximum number of lines per page and the physical format of a page of the report.
HEADING	To specify the line number at which page or overflow headings may begin.
FIRST DETAIL	To specify the line number at which detail and control lines may begin.
LAST DETAIL	To specify the line number beyond which detail and control heading lines must not be printed.
FOOTING	To specify the line number beyond which control footing lines must not be printed. Also, to specify the line number on or before which a page or overflow footing must not be printed.

Rules Governing the Use of the Report Description Entry

1. The report-name must appear in one, and only one, REPORT(S) clause.
2. The order of entry of the clauses following the report-name is not significant.
3. Report-name is the highest permissible qualifier that may be specified for LINE-COUNTER, PAGE-COUNTER, and all data-names defined within the Report Section.
4. One or more report group description entries must follow the report description entry.

CODE Clause

The CODE clause is used to specify identifying characters that are added to each line produced. (The format for the CODE clause is shown in figure 12.4.) A two-character literal that identifies each print line is specified as belonging to a specific report.

```
CODE literal-1
```

Figure 12.4 Format—CODE clause.

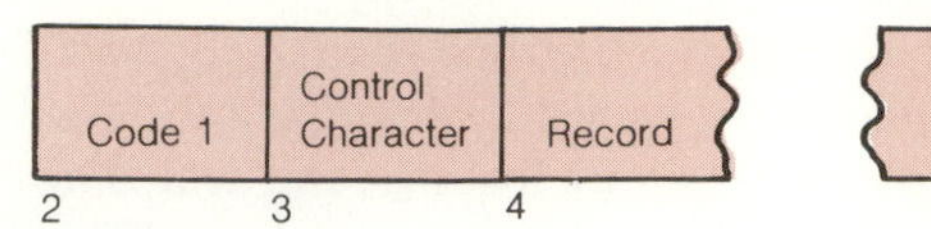

When the programmer wishes to write a report from a file, he/she needs merely to read a record, check the first two characters for the desired code, and have it printed if the desired code is found. The record should be printed starting from the fourth character.

Figure 12.5 Format of a report record with the CODE clause specified.

Rules Governing the Use of the CODE Clause

1. Literal-1 is a two-character nonnumeric literal.
2. If the CODE clause is specified for any report in a file, it must be specified for all reports in the same file.
3. When the CODE clause is specified, literal-1 is automatically placed in the first two character positions of each Report Writer logical record.
4. The positions occupied by literal-1 are not included in the description of the print line, but are included in the logical record size. (The format for a Report Record with the CODE clause specified is shown in figure 12.5.)

An example of REPORT and CODE clauses follows:

```
DATA DIVISION.
FILE SECTION.
FD  RPT-OUT-FILE
    RECORDS CONTAIN 122 CHARACTERS
    LABEL RECORDS ARE STANDARD
    REPORTS ARE REP-FILE-A REP-FILE-B.
    .
    .
    .
REPORT SECTION.
RD  REP-FILE-A
    CODE 'AA'
    .
    .
    .
RD  REP-FILE-B
    CODE 'BB'
    .
    .
    .
```

CONTROL Clause

Control Data Items

Each control heading or footing is associated with a specific control data item. A control item may be any item described in the File Section, Working-Storage Section, or Linkage Section.

Control items are related to the report by a list of control data-names specified in the CONTROL(S) clause of the RD entry. When control items are specified, the reporting procedures in the object program automatically monitor all control items for changes in value.

The most significant possible control level is associated with the reserved word FINAL, which may be optionally specified in the RD entry's CONTROL(S) clause and in a control heading and/or a control footing report group description entry.

Any control item may be associated with a specific control heading and/or a specific control footing report group. (Control report groups may be specified for each control item, for none, or for any subset of control items.) Control footings may call for automatic accumulation of SUM counters.

```
{ CONTROL IS    } { data-name-1  [ , data-name-2 ] . . .
{ CONTROLS ARE  } { FINAL  [ , data-name-1  [ , data-name-2 ]  . . . ] }
```

Figure 12.6 Format—CONTROL clause.

Control heading report groups are presented in the following hierarchical order:

FINAL CONTROL HEADING

MAJOR CONTROL HEADING

.

.

.

MINOR CONTROL HEADING

Control footing report groups are presented in the following hierarchical order:

MINOR CONTROL FOOTING

.

.

.

MAJOR CONTROL FOOTING

FINAL CONTROL FOOTING

A control break is recognized whenever a control item has changed in value between the execution of the previous GENERATE statement and the current GENERATE statement.

If the item producing a control break is not the least significant (rightmost) item in the list of all control items, then a control break has occurred at all less significant levels as well.

A control break causes the following automatic actions:

1. Rolling forward and crossfooting SUM counters.
2. Presentation of control footing up through the control break level.
3. Resetting SUM counters to zero, up through the control break level, unless inhibited by a RESET phrase.
4. Presentation of control headings from the control break level down through the least significant control level.

The CONTROL clause is used to specify the different levels of controls to be applied to the report. (The format of the CONTROL clause is shown in figure 12.6.) *A control is a data item that is tested each time a detail group is generated.* Controls govern the basic format of the report. When a control break occurs, special actions will be taken before the next line of the report is printed. Controls are listed in hierarchical order, proceeding from the most important down to the least important. Thus, by specifying HEADING and FOOTING controls, the programmer is able to instruct the Report Writer to produce the report in the desired sequence and format.

Rules Governing the Use of the CONTROL Clause

1. A control is a data item whose value is tested each time a detail group item is to be printed.
2. If the test indicates a change in the value of the data item, a control break occurs, and special action is taken before the detail line is printed.

3. The special action to be taken depends upon what the programmer has stipulated. When controls are tested, the highest control level specified is tested first, then the second-highest level, etc. When a control break is indicated for a higher level, an implied lower control break occurs as well. A control heading or control totals or neither may be defined for each control break by the programmer.

   An example of a CONTROL clause is:

```
RD . . . CONTROLS ARE YEAR MONTH WEEK DAY-I
```

4. The control footing and headings that are defined are printed prior to printing the original referenced data. They are printed in the following sequence: lowest-level control footing, next-higher-level footing, etc. up to and including the control footing for the level at which the control break occurred; then the control heading for that level, then the next-lower control heading, etc. down to and including the minor control heading; then the detail line is printed.

   If, in the course of printing control headings and footing, an end-of-page condition is detected, the current page is ejected, and a new page is begun. If the associated report groups are given, a page footing and/or a page heading is also printed.
5. The levels of control are indicated by the order in which they are written. The identifiers specify the hierarchy of controls. Data-Name-1 is the major control, data-name-2 is the intermediate control, data-name-3 is the minor control, etc.
6. FINAL is the exception to the rule that controls the data items. It is the highest control level possible. A control break by FINAL occurs at the beginning and end of the report only. All implied lower-level control breaks are taken at the same time as FINAL.
7. All level totals are printed in the sequence in which they are written with the exception of the FINAL total.
8. The CONTROL clause is required when CONTROL FOOTING or CONTROL HEADING is specified.

PAGE Clause

The PAGE clause is used to describe the physical format of a page of the report. (The format for the PAGE clause is shown in figure 12.7.) The PAGE clause specifies the specific line control to be maintained within the logical presentation of a page. The PAGE clause is required when the page format is to be controlled by the Report Writer.

The fixed data-names, PAGE-COUNTER and LINE-COUNTER are numeric counters automatically generated by the Report Writer based on the presence of specific entries, and do not require any data description clauses. The description of these two counters is included here for the purpose of explaining their effect on the overall report format.

PAGE-COUNTER

A PAGE-COUNTER is implicitly provided for each report. It is primarily used as a SOURCE data item within page/overflow heading or page/overflow footing report groups, in order to provide consecutive page numbers for the report.

The initial value of the page counter is one. Its value is automatically incremented by one each time a page break occurs. (The increment follows production of any page or overflow footing, but precedes production of any page or overflow heading.)

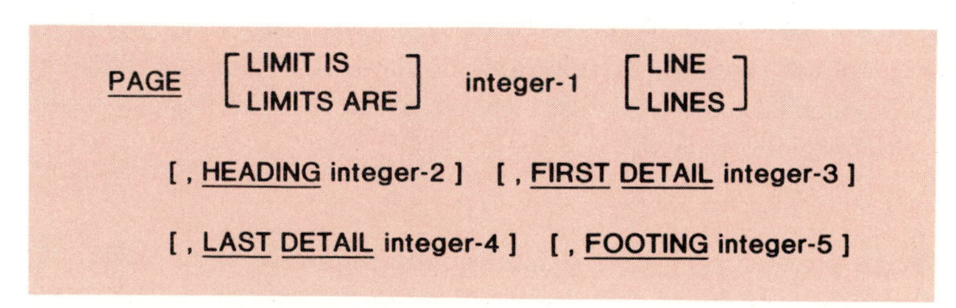

Figure 12.7 Format—PAGE clause.

The reserved word PAGE-COUNTER is a name for a special register that is generated for each report description entry in the Report Section of the Data Division. The implicit description is that of an unsigned integer that must be capable of representing a range of values from 1 through 999999. The usage is defined by the implementor. The value in PAGE-COUNTER is maintained by the RWCS and is used by the program to number the pages of a report. The value in PAGE-COUNTER may be altered by Procedure Division statements.

PAGE-COUNTER may be referenced in a SOURCE clause or in the Procedure Division to access the page counter value. The report-name may be used as a qualifier for PAGE-COUNTER; such qualification is necessary whenever the Report Section includes more than one report.

Normally, Procedure Division statements should not change the value of a page counter. However, a Procedure Division statement may change the starting value of a page counter if an initial page number other than one (1) is desired.

The maximum size of a PAGE-COUNTER is based on the size specified in the PICTURE clause associated with an elementary item whose SOURCE is PAGE-COUNTER.

Rules Governing the Use of the PAGE-COUNTER Clause

1. PAGE-COUNTER is the reserved word used to reference a special register that is automatically created for each report specified in the Report Section.
2. In the Report Section, a reference to PAGE-COUNTER can only appear in a SOURCE clause. Outside of the Report Section, PAGE-COUNTER may be used in any context in which a data-name of an integral value can appear.
3. If more than one PAGE-COUNTER exists in a program, then PAGE-COUNTER must be qualified by a report-name whenever it is referenced in the Procedure Division.
4. Execution of the INITIATE statement causes the RWCS to set the PAGE-COUNTER of the referenced report to one (1).
5. PAGE-COUNTER is automatically incremented by one (1) each time the RWCS executes a page advance.
6. PAGE-COUNTER may be altered by Procedure Division statements.

LINE-COUNTER

A LINE-COUNTER is implicitly provided for each report. It is used by the generated reporting procedures to recognize page (and overflow) breaks, and to control vertical page format.

The LINE-COUNTER is automatically set to zero initially, and is reset to zero whenever a page break occurs. It is automatically set, reset, and incremented on the basis of values specified in the LINE NUMBER and NEXT GROUP clauses in the respective report groups. It is automatically tested on the basis of values specified in the PAGE clause of the RD entry.

A page break occurs whenever a relative LINE NUMBER or relative NEXT GROUP value causes the line counter to exceed the value specified in the PAGE clause.

The reserved word LINE-COUNTER is a name for a special register that is generated for each report description entry in the Report Section of the Data Division. The implicit description is that of an unsigned integer that must be capable of representing a range of values from 0 through 999999. The usage is defined by the implementor. The value in LINE-COUNTER is maintained by the RWCS, and is used to determine the vertical positioning of a report. The value in LINE-COUNTER may be accessed by Procedure Statements; however, only the RWCS can change the value of LINE-COUNTER.

LINE-COUNTER may be referred to if it is necessary to access the line counter contents. The report-name may be used as a qualifier for LINE-COUNTER; such qualification is necessary whenever the Report Section includes more than one report.

If the last line produced had no relevant NEXT GROUP clause, the line counter value would be the number of the last line printed. Otherwise, the line counter value is the number of the last line skipped.

Procedure Division statements should never change the value of a line counter. Otherwise, an unpredictable loss of page format control may occur.

Rules Governing the Use of the LINE-COUNTER Clause

1. LINE-COUNTER is the reserved word used to reference a special register that is automatically created for each report specified in the Report Section.

2. A reference to LINE-COUNTER in the Report Section can only appear in a SOURCE clause. Outside the Report Section, LINE-COUNTER may be used in any context in which a data-name of integral value might appear. However, only the RWCS can change the contents of LINE-COUNTER.
3. If more than one LINE-COUNTER exists in a program, it must be qualified by a report-name whenever it is referenced in the Procedure Division.
4. Execution of an INITIATE statement causes the RWCS to set the LINE-COUNTER of the referenced report to zero (0). The RWCS also automatically resets LINE-COUNTER to zero each time it executes a page advance.
5. The value of LINE-COUNTER is not affected by the processing of non-printable report groups, nor by the processing of a printable report group whose printing is suppressed by means of the SUPPRESS statement.
6. At the time each print line is presented, the value of LINE-COUNTER represents the line number on which the print line is presented. The value of LINE-COUNTER after the presentation of a report group is governed by the presentation rules for the report group.

In the Report Section, neither a sum counter nor the special registers LINE-COUNTER can be used as a subscript.

The PAGE clause defines the length of a page and the vertical subdivisions within which report groups are presented.

The format of the PAGE clause, shown in figure 12.7, is explained as follows:

LIMIT(S). LIMIT IS and LIMITS ARE are optional words and need not be included in the clause.

Integer-1. The integer-1 LINE(S) clause is required to specify the depth of the report page; the depth of the report page may or may not be equal to the physical perforated continuous form often associated in a report with the page length. The size of the fixed data-name LINE-COUNTER is the maximum numeric size based on integer-1 LINE(S) required for counter to prevent overflow. LINE and LINES are optional words and need not be included in the clause.

HEADING integer-2. This is the first line number of the first heading print group. No print group will start preceding integer-2. Integer-2 is the first line upon which anything is printed.

FIRST DETAIL integer-3. This is the first line number of the first normal print group, that is body; no DETAIL print group will start before integer-3.

LAST DETAIL integer-4. This is the last line number of the last normal print group, that is body; no DETAIL print group will extend beyond integer-4.

FOOTING integer-5. The last line number of the last CONTROL FOOTING print group is specified by integer-5. No CONTROL FOOTING print group will extend beyond integer-5. PAGE FOOTING print groups will follow integer-5. (See figure 12.8.)

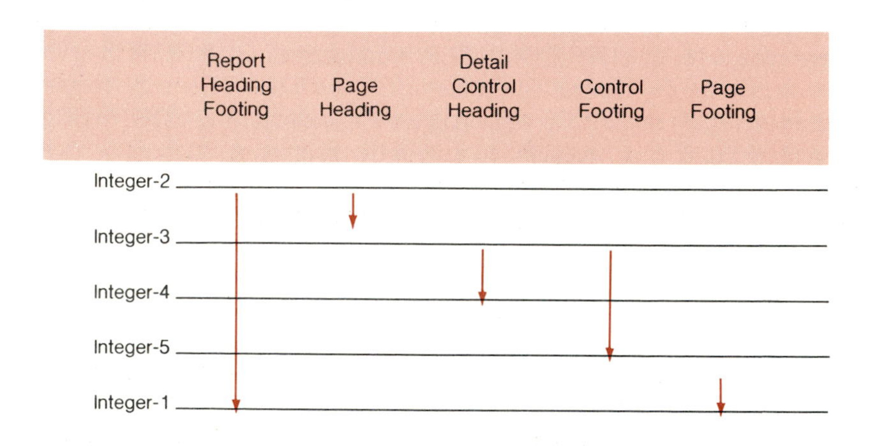

Figure 12.8 Page format when the PAGE clause is specified.

Report Groups that May Be Presented in the Region	First Line Number of the Region	Last Line Number of the Region
REPORT HEADING described with NEXT GROUP NEXT PAGE REPORT FOOTING described with LINE integer-1 NEXT PAGE	Integer-2	Integer-1
REPORT HEADING not described with NEXT GROUP NEXT PAGE PAGE HEADING	Integer-2	Integer-3 minus 1
CONTROL HEADING DETAIL	Integer-3	Integer-4
CONTROL FOOTING	Integer-3	Integer-5
PAGE FOOTING REPORT FOOTING not described with LINE integer-1 NEXT PAGE	Integer-5 plus 1	Integer-1

Figure 12.9 Page regions table.

Note: The following implicit control is assumed for omitted specifications:

1. If HEADING integer-2 is omitted, integer-2 is considered to be equivalent to the value 1, that is, LINE NUMBER one.
2. If FIRST DETAIL integer-3 is omitted, integer-3 is considered to be equivalent to the value of integer-2.
3. If LAST DETAIL integer-4 is omitted, integer-4 is considered to be equivalent to the value of integer-5.
4. If FOOTING integer-5 is omitted, integer-5 is considered to be equivalent to the value of integer-4. If both LAST DETAIL integer-4 and FOOTING integer-5 are omitted, integer-4 and integer-5 are both considered to be equivalent to the value of integer-1.

Page regions that are established by the PAGE clause are described in figure 12.9.

Report Groups

Each integral unit of data represented in a report, such as a page heading or footing, control heading or footing, or detail line is called a *report group.* A report group may consist of one or several actual lines in the printed report. In the Report Section, the first entry of each report must be level 01.

The TYPE clause is a required part of each level 01 report group description entry. The TYPE clause identifies the report group as detail or as report, page, overflow, or control heading or footing. Each report must contain at least one TYPE DETAIL report group. All other types are optional. A given heading type may be used with or without the corresponding footing, or vice versa. A report may have several distinct detail report groups or control heading or footing report groups, but no more than one of each of the other types.

Report Group Description Entry

A report group may be a set of data made up of several print lines with many data items, or it may consist of one print line with one data item. (The format for the Report Group Description Entry is shown in figure 12.10.) Report groups may exist within report groups—all or each capable of a reference by a GENERATE or a USE statement in the Procedure Division. A description of a set of data becomes a report group by the presence of a level number and a TYPE description. The level number gives the

Figure 12.10 Format—report group description entry.

Format 1:

```
01      [data-name-1]

[; LINE NUMBER IS  {integer-1   [ ON NEXT PAGE ]}]
                   {PLUS integer-2              }

[; NEXT GROUP IS   {integer-3      }]
                   {PLUS integer-4 }
                   {NEXT PAGE      }

                {{REPORT HEADING}                                  }
                {{RH            }                                  }
                {                                                  }
                {{PAGE HEADING}                                    }
                {{PH          }                                    }
                {                                                  }
                {{CONTROL HEADING}          {data-name-2}          }
                {{CH             }          {FINAL      }          }
                {                                                  }
    ; TYPE is   {{DETAIL}                                          }
                {{DE    }                                          }
                {                                                  }
                {{CONTROL FOOTING}          {data-name-3}          }
                {{CF             }          {FINAL      }          }
                {                                                  }
                {{PAGE FOOTING}                                    }
                {{PF          }                                    }
                {                                                  }
                {{REPORT FOOTING}                                  }
                {{RF            }                                  }

[;  [ USAGE IS ]  DISPLAY] .
```

Format 2:

```
level-number [ data-name-1 ]

[; LINE NUMBER IS  {integer-1 [ ON NEXT PAGE ]}.]
                   {PLUS integer-2            }

[;  [ USAGE IS ]  DISPLAY] .
```

depth of the group and the TYPE clause describes the purpose of the report group presentation. If report groups exist within report groups, all must have the same TYPE descriptions. Including a data-name with the entry permits the group to be referred to by a GENERATE or a USE statement in the Procedure Division. At object program time, report groups are created as a result of the GENERATE statement.

The Report Group Description Entry defines the characteristics for a report group, whether it be a series of lines, one line, or an elementary item. The placement of an item in relation to the entire report group, the hierarchy of a particular group within a report group, the format descriptions of all items, and any control factors associated with the group—all are defined in the entry. The system of level numbers is employed here to indicate elementary items and group items within the range of 01–49.

Pictorially to the programmer, a report group is a line or a series of lines, initially consisting of all SPACES; its length is determined by the compiler based on the environmental specifications. Within the framework of a report, the order of report groups specified is not significant. Within the framework

Figure 12.10 *(continued)*

Format 3:

```
level-number [ data-name-1 ]

    [ ; BLANK WHEN ZERO ]

    [ ; GROUP INDICATE ]

    [ ;  { JUSTIFIED }  RIGHT ]
         { JUST      }

    [ ; LINE NUMBER IS  { integer-1  [ ON NEXT PAGE ] } ]
                        { PLUS integer-2              }

    [ ; COLUMN NUMBER IS integer-3 ]

      ;  { PICTURE }  IS character-string
         { PIC     }

  { ; SOURCE IS identifier-1                                            }
  {                                                                     }
  { ; VALUE IS literal                                                  }
  {                                                                     }
  { { ; SUM identifier-2  [ , identifier-3 ] . . .                      }
  {                                                                     }
  {      [ UPON data-name-2  [ , data-name-3 ]  . . . ] } . . .         }
  {                                                                     }
  {      [ RESET ON  { data-name-4 } ]                                  }
  {                  { FINAL       }                                    }

    [ ; [ USAGE IS ]  DISPLAY ] .
```

of a report group, the programmer describes the presented elements consecutively from left to right, and then from top to bottom. The description of a report group is analogous to the data record, which consists of a set of entries defining the characteristics of the included elements. However, in the report group, SPACES are associated except where a specified entry is indicated for presentation, whereas in the data record, each character position must be defined.

The Report Group Description Entry specifies the characteristics of a particular report group and of the individual data-names within a report group. A report group may be comprised of one or more report groups. Each report group is described by a hierarchy of entries similar to the description of the data record. There are three types of report groups: heading, detail, and footing.

A report group is considered to be one unit of the report consisting of a line or a series of lines that are printed (or not printed) under certain conditions.

The report groups that will make up the report are described following the RD entry. The description of each report group begins with a report group description entry; that is, an entry that contains a level number 01 and a TYPE clause. Subordinate to the report group description entry, group and elementary entries that further describe the characteristics of the report group may be included.

Typically, the description of a report includes two or more level 01 report group entries, each followed by a hierarchy of subordinate entries. Depending upon a number of factors, most clauses (except the level-number clause) are optional. In most entries, the data-name is optional and is normally omitted. A data-name is specified in level 01 detail report group entries and in nondetail report groups that are referenced by the BEFORE REPORTING phrase of the USE statement.

At the 01 report group level, the following clause is required:

Clause	Function
TYPE	To specify the purpose of this report group (detail, page or control heading, etc.).

The optional clauses in a report group entry are:

Clause	Function
LINE NUMBER	To specify vertical spacing that is to precede production of this report group.
NEXT GROUP	To specify vertical spacing that is to follow production of this report group.
USAGE	To declare the usage of printable items.
COLUMN NUMBER	To indicate that this item is to be printed, and to specify its horizontal position on the line.
GROUP INDICATE	To cause a repetitive item to be printed only at the top of the page and just after each control break.
JUSTIFIED RIGHT	To override normal left justification when this item is edited for output.
PICTURE	To specify the desired output format for this item.
RESET	To specify the control break where a SUM counter is to be reset to zero.
BLANK WHEN ZERO	To cause this item's value to be "spaces" when the SOURCE or SUM with which it is associated has the value zero.
SOURCE, SUM, or VALUE	To specify the source of data for this item: 1. SOURCE—a data item. 2. SUM—a SUM counter. 3. VALUE—a literal.

Rules Governing the Use of the Report Group Description Entry.

1. The report group description entry can appear only in the Report Section.
2. Except for the data-name clause, which, when present, must immediately follow the level-number, the clauses may be written in any sequence.
3. In format-2 the level-number may be any integer from 02 to 49 inclusive. In format-3 the level-number may be any integer from 02 to 49 inclusive.
4. The description of a report group may consist of one, two or three hierarchic levels:
   a. The first entry that describes a report group must be a format-1 entry.
   b. Both format-2 and format-3 entries may be immediately subordinate to a format-1 entry.
   c. At least one format-3 entry must be immediately subordinate to a format-2 entry.
   d. Format-3 entries must be elementary.
5. In a format-1 entry, data-name-1 is required only when:
   a. A DETAIL group is referenced by a GENERATE statement.
   b. A DETAIL group is referenced by the UPON phrase of a SUM clause.
   c. A report group is referenced in a USE BEFORE REPORTING sentence.
   d. The name of a CONTROL FOOTING report group is used to qualify a reference to a sum counter.
6. A format-2 entry must contain at least one optional clause.
7. In a format-2 entry, data-name-1 is optional. If present, it may be used only to qualify a sum counter reference.
8. In the Report Section, the USAGE clause is used only to declare the usage of printable items.
   a. If the USAGE clause appears in a format-3 entry, that entry must define a printable item.
   b. If the USAGE clause appears in a format-1 or format-2 entry, at least one subordinate entry must define a printable item.

9. An entry that contains a LINE NUMBER clause must not have a subordinate entry that also contains a LINE NUMBER clause.
10. In format-3:
    a. A GROUP INDICATE clause may appear only in a TYPE DETAIL report group.
    b. A SUM clause may appear only in a TYPE CONTROL FOOTING report group.
    c. An entry that contains a COLUMN NUMBER clause but no LINE NUMBER clause must be subordinate to an entry that contains a LINE NUMBER clause.
    d. Data-name-1 is optional but may be specified in any entry. Data-name-1, however, may be referenced only if the entry defines a sum counter.
    e. A LINE NUMBER clause must not be the only clause specified.
    f. An entry that contains a VALUE clause must also have a COLUMN NUMBER clause.
11. The following table shows all permissible clause combinations for a format-3 entry. The table is read from left to right along the selected row.
    An "M" indicates that the presence of the clause is mandatory.
    A "P" indicates that the presence of the clause is permitted, but not required.
    A blank indicates that the clause is not permitted.

Clauses

PIC	COLUMN	SOURCE	SUM	VALUE	JUST	BLANK WHEN ZERO	GROUP INDICATE	USAGE	LINE
M			M						P
M	M		M			P		P	P
M	P	M			P		P	P	P
M	P	M				P	P	P	P
M	M			M	P		P	P	P

Data-Name Clause
The data-name clause specifies the name of the data being described.

Rules Governing the Use of the Data-Name Clause

1. In the Report Section, a data-name is not required in a data description entry; the word FILLER must not be used.
2. In the Report Section, data-name must be specified in the following cases:
    a. When the data-name represents a report group to be referred to by a GENERATE statement or a USE statement in the Procedure Division.
    b. When reference is to be made to the sum counter in the Procedure Division or the Report Section.
    c. When a DETAIL report group is referenced in the UPON phrase of the SUM clause.
    d. When the data-name is required to provide sum counter qualification.

LINE NUMBER Clause
The LINE NUMBER clause specifies vertical positioning information for its report group. (The format for the LINE NUMBER clause is shown in figure 12.11.)

```
LINE NUMBER IS  { integer-1  [ ON NEXT PAGE ] }
                { PLUS integer-2              }
```

Figure 12.11 Format—LINE NUMBER clause.

Rules Governing the Use of the LINE NUMBER Clause

1. A LINE NUMBER clause must be specified to establish each print line of a report group.
2. The RWCS affects the vertical positioning specified by a LINE NUMBER clause before presenting the print line established by that LINE NUMBER clause.
3. Integer-1 specifies an absolute line number. An absolute line number specifies the line number on which the print line is presented.
4. Integer-2 specifies a relative line number. If a relative LINE NUMBER clause is not the first LINE NUMBER clause in the report group description entry, then the line number on which its print line is presented is determined by calculating the sum of the line number on which the previous print line of the report group was presented and integer-2 of the relative LINE NUMBER clause.
5. The NEXT PAGE phrase specifies that the report group is to be presented beginning on the indicated line number on a new page.
6. Integer-1 and integer-2 must not exceed three significant digits in length.
7. Within a given report group description entry, an entry that contains a LINE NUMBER clause must not contain a subordinate entry that also contains a LINE NUMBER clause.
8. Within a given report group description entry, all absolute LINE NUMBER clauses must precede all relative LINE NUMBER clauses.
9. Within a given report group description entry, successive absolute LINE NUMBER clauses must specify integers that are in ascending order. The integers need not be consecutive.
10. If the PAGE clause is omitted from a given report description entry, only the relative LINE NUMBER clauses can be specified in any report group description entry within that report.
11. Within a given report group description entry, a NEXT PAGE phrase can appear only once and, if present, must be in the first LINE NUMBER clause in that report group description entry.
12. Every entry that defines a printable item must either contain a LINE NUMBER clause or be subordinate to an entry that contains a LINE NUMBER clause.
13. The first LINE NUMBER clause specified within a PAGE FOOTING report group must be an absolute LINE NUMBER clause.

NEXT GROUP Clause

The NEXT GROUP clause specifies information for vertical positioning of a page following the presentation of the last line of a report group. (The format for the NEXT GROUP clause is shown in figure 12.12.)

Rules Governing the Use of the NEXT GROUP Clause

1. Any positioning of the page specified by the NEXT GROUP clause takes place after the presentation of the report group in which the clause appears.
2. The vertical positioning information supplied by the NEXT GROUP clause is interpreted by the RWCS along with information from the TYPE and PAGE clauses, and the value in LINE-COUNTER, to determine a new value for LINE-COUNTER.
3. The NEXT GROUP clause is ignored by the RWCS when it is specified on a CONTROL FOOTING report group that is at a level other than the highest level at which a control break is detected.
4. The NEXT GROUP clause of a body group refers to the next body group to be presented, and therefore can affect the location at which the next body group is presented.
5. A report group entry must not contain a NEXT GROUP clause unless the description of that report group contains at least one LINE NUMBER clause.
6. Integer-1 and integer-2 must not exceed three significant digits in length.

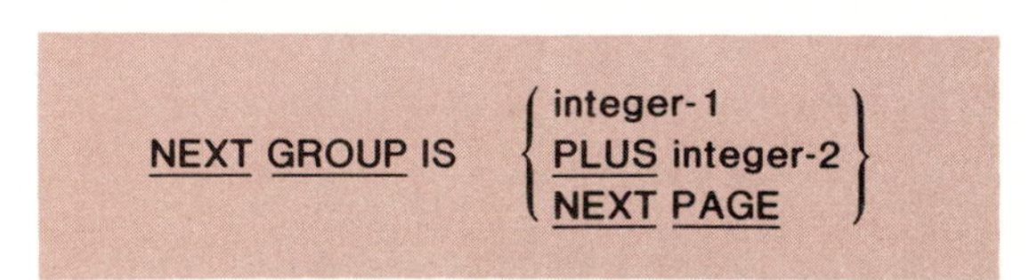

Figure 12.12 Format—NEXT GROUP clause.

7. If the PAGE clause is omitted from the report description entry, only a relative NEXT GROUP clause may be specified in any report group description entry within that report.
8. The NEXT PAGE phrase of the NEXT GROUP clause must not be specified in a PAGE FOOTING report group.
9. The NEXT GROUP clause must not be specified in a REPORT FOOTING report group or in a PAGE HEADING report group.

An example of the NEXT GROUP clause follows:

```
RD  EXPENSE-REPORT CONTROLS ARE FINAL, MONTH, DAYI
        .
        .
        .
01  TYPE CONTROL FOOTING DAYI
    LINE PLUS 1 NEXT GROUP NEXT PAGE
        .
        .
        .
01  TYPE CONTROL FOOTING MONTH
    LINE PLUS 1 NEXT GROUP NEXT PAGE.
        .
        .
        .
    (Execution Output)

EXPENSE REPORT
        .
        .
        .
January 31 . . . 29.30
    (Output for CF DAYI)

January total . . . 131.40
    (Output for CF MONTH)

Note: The NEXT GROUP NEXT PAGE clause for the
control footing DAYI is not activated.
```

TYPE Clause

The TYPE clause specifies the particular type of report group that is described in this entry and indicates the time at which the report group is to be generated. (The format for the TYPE clause is shown in figure 12.13.) Abbreviations may be used in the TYPE clause.

Rules Governing the Use of the TYPE Clause

1. The level number 01 identifies a particular report group to be generated as output and the TYPE clause indicates the time for the generation of this report group.
2. If the report group is described as other than TYPE DETAIL, its generation is an automatic Report Writer function.
3. If the report group is described with the TYPE DETAIL clause, the Procedure Division statement—GENERATE data-name—directs the Report Writer to produce the named report group.
4. Nothing precedes a REPORT HEADING entry, and nothing follows a REPORT FOOTING entry within a report.
5. A FINAL type control break may be designated only once for CONTROL HEADING or CONTROL FOOTING entries within a particular report group.
6. CONTROL HEADING report groups appear with the current values of any indicated SOURCE data items before the DETAIL report groups of the CONTROL group are produced.
7. CONTROL FOOTING report groups appear with the previous value of any indicated CONTROL SOURCE data items just after the DETAIL report groups of the CONTROL groups have been produced.

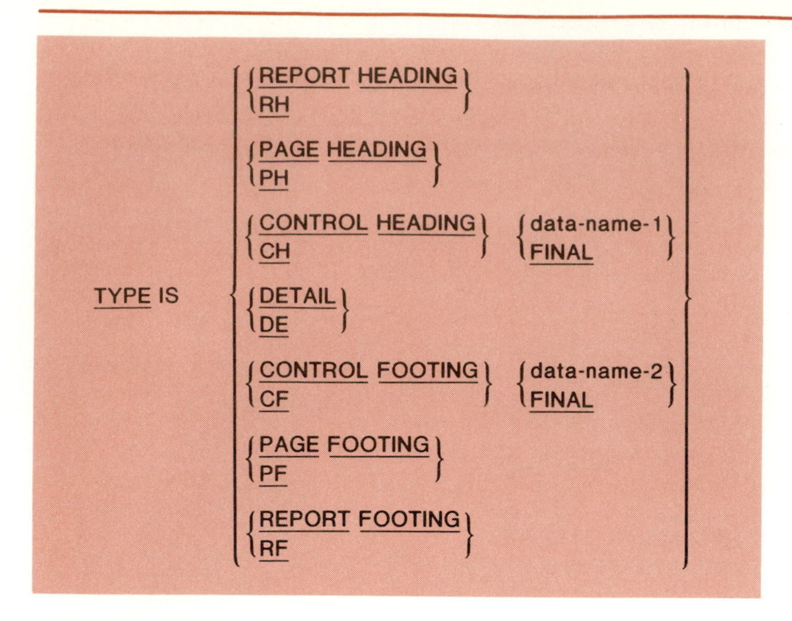

Figure 12.13 Format—TYPE clause.

8. The USE procedure specified for a CONTROL FOOTING report group refers to:
   a. Source data items specified in the CONTROL clause affect the previous value of the data item.
   b. Source data items not specified in the CONTROL clause affect the current value of the item. These report groups appear whenever a control break is noted.
9. LINE NUMBER determines the absolute or relative position of the CONTROL report group exclusive of the other HEADING and FOOTING report group. The following is a HEADING or FOOTING report group sequence:

```
REPORT HEADING      (one occurrence only)
PAGE HEADING
      .
      .
      .
CONTROL HEADING
DETAIL
CONTROL FOOTING
      .
      .
      .
PAGE FOOTING
REPORT FOOTING      (one occurrence only)
```

10. PAGE HEADING and PAGE FOOTING report groups may be specified only if a PAGE clause is specified in the corresponding report group description entry.
11. TYPE clause data-names may not be subscripted or indexed.
12. In CONTROL FOOTING, PAGE HEADING, PAGE FOOTING, and REPORT FOOTING report groups, SOURCE clauses and USE statements may not reference:
    a. Group items containing control data items.
    b. Data items subordinate to a control data item.
    c. A redefinition or renaming of any part of a control data item.

In PAGE HEADING and PAGE FOOTING report groups, SOURCE clauses and USE statements must not reference control data-names.

Report Heading (RH). The REPORT HEADING entry indicates a report group that is produced only once, at the beginning of the report, during the execution of the first GENERATE statement. Only one REPORT HEADING is permitted in a report. SOURCE clauses used in REPORT HEADING report group items refer to the value of data items at the time the first GENERATE statement is executed.

Page Heading (PH). The PAGE HEADING entry indicates a report group that is automatically produced at the beginning of each page of the report. Only one page heading is permitted for each report. The page heading is printed on the first page after the REPORT HEADING is specified.

Control Heading (CH). The CONTROL HEADING entry indicates a report group that is printed at the beginning of a control group for a designated identifier, or, in the case of FINAL, is produced once before the first control at the initiation of a report during the execution of the first GENERATE statement. There can be only one report group of this type for each identifier and for the FINAL specified in a report. In order to produce any CONTROL HEADING report groups, a control break must occur. SOURCE clauses used in TYPE CONTROL HEADING FINAL report groups refer to the value of the items at the time the first GENERATE statement is executed. The following is a CONTROL HEADING report groups sequence:

```
Final Control Heading     (one occurrence only)
Major Control Heading
     .
     .
     .
Minor Control Heading
```

Identifier-n as well as FINAL, must be one of the identifiers described in the CONTROL clause in the report description entry.

Detail (DE). The DETAIL entry indicates a report group that is produced for each GENERATE statement in the Procedure Division. (The data-name specified in the 01 level is referred to by the GENERATE statement. This name must be unique.) There is no limit to the number of DETAIL report groups that may be included in a report.

Control Footing (CF). The CONTROL FOOTING entry indicates a report group that is produced at the end of a control group for a designated identifier or one that is produced once at the termination of a report ending in a FINAL control group. There can be only one report group for each identifier and for the FINAL entry specified in a report. In order to produce any CONTROL FOOTING report group, a control break must occur. SOURCE clauses used in TYPE CONTROL FOOTING FINAL report groups refer to the values of the items at the time the TERMINATE statement is executed.

Page Footing (PF). The PAGE FOOTING entry indicates a report group that is automatically produced at the bottom of each page of the report. There can be only one report group of this type in a report.

Report Footing (RF). The REPORT FOOTING entry indicates a report group that is produced only once at the termination of a report. There can be only one report group of this type in a report. SOURCE clauses used in TYPE REPORT FOOTING report groups refer to the value of the items at the time the TERMINATE statement is executed. The following is a CONTROL FOOTING report groups sequence:

```
Minor Control Footing
     .
     .
     .
Major Control Footing
Final Control Footing     (one occurrence only)
```

USAGE Clause

DISPLAY is the only option that may be specified for elementary or group items in a report group description entry. (See "USAGE" clause in the Data Division, chap. 4.)

COLUMN Clause

The COLUMN clause indicates the absolute column number in the printed page of the high-order (leftmost) character of an elementary item. (The format for the COLUMN clause is shown in figure 12.14.) Integer-1 must be a positive integer. The clause can only be given at the elementary level within a report group.

COLUMN NUMBER IS integer-1

Figure 12.14 Format—COLUMN clause.

GROUP INDICATE

Figure 12.15 Format—GROUP INDICATE clause.

The COLUMN clause indicates that the *leftmost* character of the elementary item is placed in the position specified by the integer. If the column number is not indicated, the elementary item, though included in the description of the report group, is suppressed when the report group is produced at object time.

GROUP INDICATE Clause

The GROUP INDICATE clause specifies that the elementary item is to be produced only on the first occurrence of the item after any CONTROL or PAGE break. This clause must be specified only at the elementary level within a DETAIL report group. (The format for the GROUP INDICATE clause is shown in figure 12.15.)

The elementary item is not only group-indicated on the first DETAIL report group containing the item after a control break, but is also indicated on the first DETAIL report group containing the item on a new page, even though a control break did not occur. The following sample shows the GROUP INDICATE clause and resultant execution output:

```
REPORT SECTION.
    .
    .
01  DETAIL-LINE TYPE IS DETAIL LINE
    NUMBER IS PLUS 1.
    05  COLUMN IS 2 GROUP INDICATE
        PICTURE IS A(9) SOURCE IS
        MONTHNAME OF RECORD-AREA (MONTH)
    .
    .
              (Execution Output)
    .
    .
JANUARY     15     A00 . . .
                   A02 . . .

PURCHASES AND COST . . .

JANUARY     21     A03 . . .
                   A03 . . .
```

GROUP INDICATE items are printed after page and control breaks.

The GROUP INDICATE clause may only appear in a DETAIL report group entry that defines a printable item (contains a COLUMN and PICTURE clause.)

JUSTIFIED Clause

The same rules are applicable to the use of the JUSTIFIED clause in a report group description as discussed in the Data Division. (See "JUSTIFIED" clause in the Data Division, chap. 4.)

PICTURE Clause

The same rules are applicable to the use of the PICTURE clause in a report group description as discussed in the Data Division. (See "PICTURE" clause in the Data Division, chap. 4.)

BLANK WHEN ZERO Clause

The same rules are applicable to the use of the BLANK WHEN ZERO clause in a report group description as described in the Data Division. (See BLANK WHEN ZERO clause in the Data Division, chap. 4.)

SOURCE IS identifier-1

Figure 12.16 Format—SOURCE clause.

SOURCE Clause

The SOURCE clause indicates a data item to be used as the source for this report item. (The format for the SOURCE clause is shown in figure 12.16.) The item is presented according to the PICTURE and COLUMN clauses in this elementary item entry.

The SOURCE clause has two functions:

1. To specify a data item that is to be printed.
2. To specify a data item that is to be summed in a CONTROL FOOTING report group. The following are examples of COLUMN and SOURCE clauses:

```
05  COLUMN 10    PICTURE 9(6)            SOURCE IS UNITS-IN.
05  COLUMN 25    PICTURE ZZZ,ZZZ.99      SOURCE IS AMOUNT-IN.
05  COLUMN 85    PICTURE ZZZZ            SOURCE IS PAGE-COUNTER.
05  COLUMN 35    PICTURE 9(5).99         SOURCE IS TAB-ITEM (PROD).
```

SUM Counter Manipulation

A function of the Report Writer that must be clarified to avoid producing inefficient object code is the manipulation of SUM counters. There are three distinct types of SUM counter manipulation: subtotalling, rolling forward, and crossfooting. The following sections contain definitions and illustrations of the types of manipulation just presented.

Subtotalling

Subtotalling is the most basic type of SUM counter manipulation. In this method, a SUM counter is augmented by the value of the SUM operand for each execution of a GENERATE statement of the TYPE DETAIL report group, which contains the SOURCE counterpart of the SUM operand. For example:

```
01  DETAIL-1          TYPE DE              LINE PLUS 1.
    05  SOURCE   IS   COST.
    .
    .
    .
01  MINOR             TYPE CF MINR         LINE PLUS 1.
    05  SCTR-1        COLUMN 50     PIC Z(6).99     SUM COST.
    .
    .
    .
01  INTERMEDIATE      TYPE CF INTRM        LINE PLUS 1.
    05  SCTR-2        COLUMN 50     PIC Z(6).99     SUM COST.
    .
    .
    .
01  MAJOR             TYPE CF MAJR         LINE PLUS 1.
    05  SCTR-3        COLUMN 50     PIC Z(6).99         SUM COST.
    .
    .
    .
01  FIN-TOT           TYPE CF FINAL        LINE PLUS 1  NEXT GROUP NEXT PAGE.
    05  SCTR-4        COLUMN 50     PIC Z(6).99         SUM COST.
    .
    .
    .
```

At each execution of a GENERATE DETAIL-1, the value of COST will be added into SUM counters SCTR-1, SCTR-2, SCTR-3, and SCTR-4. When a control break occurs, no "rolling totals" of counters is necessary since all counters are effectively "subtotalled." The only remaining actions to be performed are:

1. Presenting the controlling footing report groups from the least inclusive (MINOR) up through the control footing representing the control break level.
2. Resetting the corresponding SUM counters to zero after each control footing is presented.

Rolling Forward

Rolling forward is a type of SUM counter manipulation in which SUM counters defined in control footing report groups of lower control levels are added to SUM counters defined in control footing report groups of higher control levels during control break processing.

In the previous example, for instance, the identical results may be obtained more efficiently by the "rolling forward" of the SUM counters. For example:

```
01  DETAIL-1            TYPE DE              LINE PLUS 1.
    05  SOURCE  IS      COST.
    .
    .
    .
01  MINOR               TYPE CF MINR         LINE PLUS 1.
    05  SCTR-1          COLUMN 50      PIC Z(6).99        SUM COST.
    .
    .
    .
01  INTERMEDIATE        TYPE CF INTRM        LINE PLUS 1.
    05  SCTR-2          COLUMN 50      PIC Z(6).99        SUM SCTR-1.
    .
    .
    .
01  MAJOR               TYPE CF MAJR         LINE PLUS 1.
    05  SCTR-3          COLUMN 50      PIC Z(6).99        SUM SCTR-2.
    .
    .
    .
01  FIN-TOT             TYPE CF FINAL        LINE PLUS 1  NEXT  GROUP
                                                          NEXT PAGE.
    05  SCTR-4          COLUMN 50      PIC Z(6).99        SUM SCTR-3.
    .
    .
    .
```

The following sequence of events occurs in the example just presented:

1. At each execution of a GENERATE DETAIL-1 statement, the value of COST is added into SUM counter SCTR-1 (subtotalling).
2. When a control break occurs on control data-name MINR, the control footing report group called MINOR is presented; then SUM counter SCTR-1 is added (rolled forward) to SUM counter SCTR-2.
3. When a control break occurs at a higher control break level, the control footing report groups are presented in sequence from the inclusive (MINOR) up to and including the control footing at which the control break occurred. After each control footing is presented, the SUM counters for that report group are rolled forward to corresponding SUM counters in higher level control footing report groups.

Thus, the subtotalling operation occurs only at the least inclusive (MINOR) control break level. The remaining SUM counters are augmented only when control break processing takes place.

The following is an example of SUM clause—rolling forward totals:

```
RD . . .
    CONTROLS ARE YEAR-I, MONTH-I, WEEK-I, DAY-I.
    .
    .
    .
```

Method 1:

```
01  TYPE CONTROL FOOTING DAY-I LINE PLUS 1.
    05  COLUMN 50    PICTURE Z(6).99    SUM COST.

01  TYPE CONTROL FOOTING WEEK-I LINE PLUS 1.
    05  COLUMN 50    PICTURE Z(6).99    SUM COST.

01  TYPE CONTROL FOOTING MONTH-I LINE PLUS 1.
    05  COLUMN 50    PICTURE Z(6).99    SUM COST.

01  TYPE CONTROL FOOTING YEAR-I LINE PLUS 1.
    05  COLUMN 50    PICTURE Z(6).99    SUM COST.
```

Method 2:

```
01  TYPE CONTROL FOOTING DAY-I LINE PLUS 1.
    05  C COLUMN 50    PICTURE Z(6).99    SUM COST.

01  TYPE CONTROL FOOTING WEEK-I LINE PLUS 1.
    05  D COLUMN 50    PICTURE Z(6).99    SUM C.

01  TYPE CONTROL FOOTING MONTH-I LINE PLUS 1.
    05  E COLUMN 50    PICTURE Z(6).99    SUM D.

01  TYPE CONTROL FOOTING YEAR-I LINE PLUS 1.
    05  COLUMN 50      PICTURE Z(6).99    SUM E.
```

Method 2 will execute faster and will be more accurate as we are checking the totals being printed to a control total for the year. One addition will be performed for each day, one more for each week, and one for each month. In method 1, four additions will be performed for each day.

Crossfooting

Crossfooting is a type of SUM counter manipulation in which SUM counters defined in a given control footing report group are added to other SUM counters in the same report group during control break processing. For example:

```
01  DETAIL-1    TYPE DE    LINE PLUS 1.
    05  SOURCE IS COST-1.
    05  SOURCE IS COST-2.
    .
    .
    .
01  MINOR TYPE CF MINR LINE PLUS 1.
    05  SCTR-1    COLUMN 50    PIC Z(6).99    SUM COST-1.
    05  SCTR-2    COLUMN 60    PIC Z(6).99    SUM COST-2.
    05  SCTR-3    COLUMN 70    PIC Z(9).99    SUM SCTR-1, SCTR-2.
    .
    .
    .
01  INTERMEDIATE    TYPE CF    INTRM LINE PLUS 1.
    05  SCTR-4    COLUMN 50    PIC Z(6).99    SUM SCTR-1.
    05  SCTR-5    COLUMN 60    PIC Z(6).99    SUM SCTR-2.
    05  SCTR-6    COLUMN 70    PIC Z(9).99    SUM SCTR-4, SCTR-5.
    .
    .
    .
```

The following sequence of events occurs in the sequence just presented:

1. At each execution of a GENERATE DETAIL-1 statement, SUM counters SCTR-1 and SCTR-2 are augmented by the corresponding values of COST-1 and COST-2 (subtotalling).
2. When a control break occurs for the control footing report group called MINOR, SUM counters SCTR-1 and SCTR-2 are added into SUM counter SCTR-3 before the report group is presented (crossfooting).
3. After the report group called MINOR is presented, SUM counters SCTR-1 and SCTR-2 are added into SUM counters SCTR-4 and SCTR-5, respectively (rolled totals).
4. SUM counters SCTR-1, SCTR-2, and SCTR-3 are reset to zero.
5. When a control break occurs for the control footing report group called INTERMEDIATE, SUM counters SCTR-4 and SCTR-5 are added into SUM counter SCTR-6 before the report group is presented (crossfooting).

SUM Clause

The SUM clause establishes a sum counter and names the data items to be summed. (The format for the SUM clause is shown in figure 12.17.)

Rules Governing the Use of the SUM Clause

1. The SUM clause establishes a sum counter. The sum counter is a numeric data item with an optional sign. At object program execution, the RWCS adds directly into the sum counter each of the values contained in identifier-1 and identifier-2. This addition is performed under the rules of the ADD statement. An example of the SUM clause follows:

```
RD  REPORT-1 CONTROLS ARE MINR . . .
    01  DETAIL-1 TYPE DE     LINE PLUS 1.
        05  SOURCE IS COST.
        .
        .
        .
    01  DETAIL-2 TYPE DE     LINE PLUS 1.
        05  SOURCE IS COST.
        .
        .
        .
    01  MINOR TYPE CF MINR LINE PLUS 1.
        05  SCTR-1 COLUMN 50 PIC Z(6).99 SUM COST.
        .
        .
        .
```

   For each execution of either a GENERATE DETAIL-1 or a GENERATE DETAIL-2 statement, the SUM counter SCTR-1 will be augmented by the value of COST since it is the object of a SOURCE IS clause in both TYPE DETAIL report groups.
2. The size of the sum counter is equal to the number of receiving character positions specified by the PICTURE clause that accompanies the SUM clause in the description of the elementary item.

```
{SUM identifier-1  [ , identifier-2 ] . . .
     [UPON data-name-1  [ , data-name-2 ] . . .]} . . .
     [RESET ON  {data-name-3}]
                {FINAL      }
```

Figure 12.17 Format—SUM clause.

3. Only one sum counter exists for an elementary report entry regardless of the number of SUM clauses specified in the elementary report entry.
4. If the elementary report entry for a printable item contains a SUM clause, the sum counter serves as a source data item. The RWCS moves the data contained in the sum counter, according to the rules of the MOVE statement, to the printable item for presentation.
5. If a data-name appears as the subject of an elementary report entry that contains a SUM clause, the data-name is the name of the sum counter; the data-name is not the name of the printable item that the entry may also define.

   It is permissible for Procedure Division statements to alter the contents of sum counters.
6. Addition of the identifiers into sum counters is performed by the RWCS during the execution of GENERATE and TERMINATE statements. The three categories of sum counter incrementing are called subtotalling, crossfooting, and rolling forward. Subtotalling is accomplished during execution of GENERATE statements only, after any control break processing, but before processing of the DETAIL report group. Crossfooting and rolling forward are accomplished during the processing of CONTROL FOOTING report groups.
7. The UPON phrase provides the capability to accomplish selective subtotalling for the DETAIL report groups named in the phrase. The UPON phrase of the SUM clause may be used to selectively augment a given SUM counter. An example of the SUM clause—UPON phrase—follows:

```
01  MINOR TYPE CF MINR LINE PLUS 1.
    05  SCTR-1 COLUMN 50 PIC Z(6).99 SUM COST UPON DETAIL-1.
    .
    .
    .
```

   In this example, the definition just given indicates that SUM counter SCTR-1 will be augmented only when a GENERATE statement is executed for DETAIL-1.
8. The RWCS adds each individual addend into the sum counter at a time that depends upon the characteristics of the addend.
   a. When the addend is a sum counter defined in the same CONTROL FOOTING report group, then the accumulation of that addend into the sum counter is termed crossfooting. Crossfooting occurs when a control break takes place and at the time the CONTROL FOOTING report group is processed.

      Crossfooting is performed according to the sequence in which sum counters are defined within the CONTROL FOOTING report group. That is, all crossfooting into the first sum counter defined in the CONTROL FOOTING report is completed, and then all crossfooting into the second sum counter defined in the CONTROL FOOTING report group is completed. This procedure is repeated until all crossfooting operations are completed.
   b. When the addend is a sum counter defined in a lower level CONTROL FOOTING report group, the accumulation of that addend into the sum counter is termed *rolling forward.* A sum counter in a lower level CONTROL FOOTING report group is rolled forward when a control break occurs and at the time that the lower level CONTROL FOOTING report group is processed.
   c. When the addend is not a sum counter, the accumulation into a sum counter of such an addend is called subtotalling. If the SUM clause contains the UPON phrase, the addends are subtotalled when a GENERATE statement for the designated DETAIL report group is executed. If the SUM clause does not contain the UPON phrase, the addends that are not sum counters are subtotalled when any GENERATE data-name statement is executed for the report in which the SUM clause appears.
9. If two or more of the identifiers specify the same addend, then the addend is added into the sum counter as many times as the addend is referenced in the SUM clause. It is permissible for two or more of the data-names to specify the same DETAIL report group. When a GENERATE data-name statement for such a DETAIL report group is given, the incrementing occurs repeatedly, as many times as data-name appears in the UPON phrase.
10. For the subtotalling that occurs when a GENERATE report-name statement is executed, refer to the GENERATE statement in this section.

11. In the absence of an explicit RESET phrase, the RWCS will set a sum counter to zero when the RWCS is processing the CONTROL FOOTING report group within which the sum counter is defined. If an explicit RESET phrase is specified, the RWCS will set the sum counter to zero when the RWCS is processing the designated level of the control hierarchy. Examples of the SUM clause—RESET phrase—follow:

```
REPORT SECTION.
     .
     .
     .
   01  MINOR-CONTROL          TYPE CF MINOR . . .
       05  LINE PLUS 1.
           10  A COLUMN 1     PIC Z(6).99     SUM COST RESET ON INTERMEDIATE.
     .
     .
     .
   01  INTERMEDIATE-CONTROL TYPE CF INTERMEDIATE . . .
       05  LINE PLUS 1.
           10  B COLUMN 1     PIC Z(6).99     SUM COST RESET ON FINAL.
     .
     .
     .
   01  FINAL-CONTROL          TYPE CF FINAL . . .
       05  LINE PLUS 1.
     .
     .
     .
```

In the examples just presented, the MINOR SUM counter will be reset on an INTERMEDIATE control break. The INTERMEDIATE SUM counter will be reset on FINAL.

Sum counters are initially set to zero by the RWCS during the execution of the INITIATE statement for the report containing the sum counter.

12. Identifier-1 and identifier-2 must be defined as numeric data items. When defined in the Report Section, identifier-1 and identifier-2 must be the name of sum counters.

    If the UPON phrase is omitted, any identifiers in the associated SUM clause which are themselves sum counters must be defined either in the same report group that contains this SUM clause or in a report group that is at a lower level in the control hierarchy of this report.

    If the UPON phrase is specified, any identifiers in the associated SUM clause must not be sum counters.
13. Data-name-1 and data-name-2 must be the names of DETAIL report groups described in the same report as the CONTROL FOOTING report group in which the SUM clause appears. Data-name-1 and data-name-2 may be qualified by a report-name.
14. A SUM clause can appear only in the description of a CONTROL FOOTING report group.
15. Data-name-3 must be one of the data-names specified in the CONTROL clause for this report. Data-name-3 must not be a lower-level control than the associated control for the report group in which the RESET phrase appears.

    FINAL, if specified in the RESET phrase, must also appear in the CONTROL clause for this report.
16. The highest permissible qualifier of a sum counter is the report-name.
17. Counters are reset automatically to zero unless the explicit RESET phrase is given, specifying resetting based on a higher-level control than the associated control for the report group. The RESET phrase may be used for programming the totalling of identifiers where subtotals of identifiers may be desired without automatic resetting upon printing the report group.

VALUE Clause

The VALUE clause causes the report data item to assign the specified value each time its report group is presented only if the elementary item entry does not contain a GROUP INDICATE clause. (The format for the VALUE clause is shown in figure 12.18.) If the GROUP INDICATE clause is present,

VALUE IS literal-1

Figure 12.18 Format—VALUE clause.

and a given object time condition exists, the item will assume the specified value. (See "GROUP INDICATE" rules.)

Procedure Division

The production of a report is controlled in the Procedure Division with four report writing statements:

1. INITIATE
2. GENERATE
3. SUPPRESS
4. TERMINATE

In addition, the BEFORE REPORTING phrase of the USE statement may also be used to control the production of a report.

The SUPPRESS statement inhibits the presentation of a report group and may be specified only in a USE BEFORE REPORTING procedure.

The relationship of the above statements to other Procedure Division statements is illustrated by the following chart of a simple reporting program:

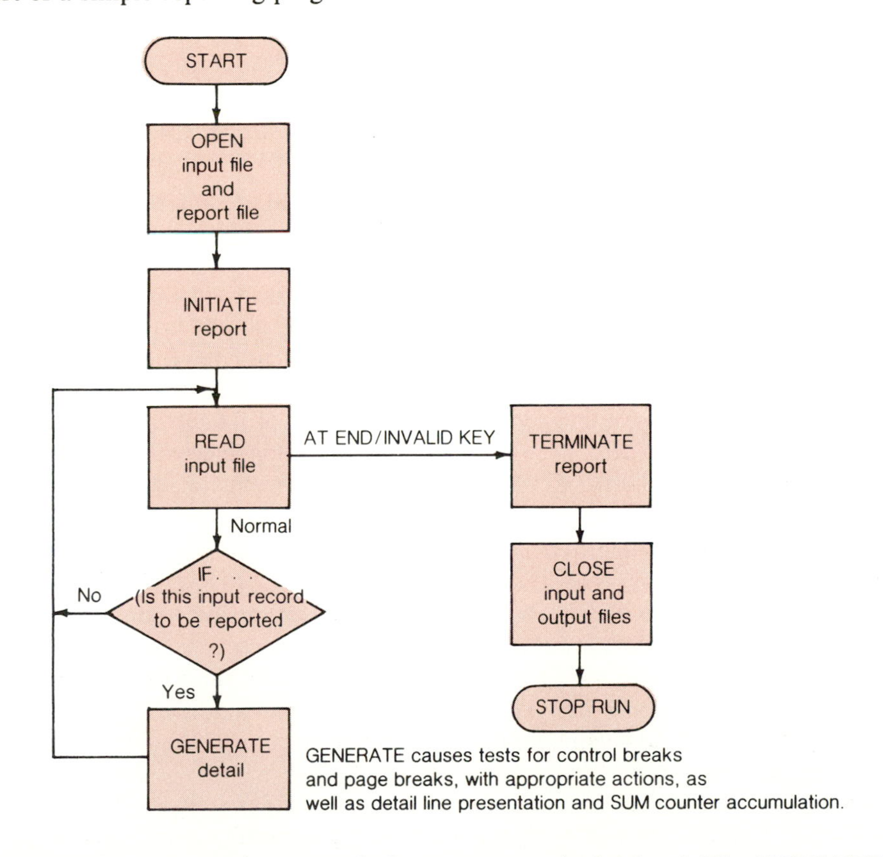

Before a GENERATE statement is executed, the report must be initiated. The INITIATE statement causes initial housekeeping values to be established.

The GENERATE statement provides for all aspects of report editing, writing, and housekeeping. But GENERATE in itself makes no provision for reading input data or deciding when detail lines should be produced. Instead, the programmer explicitly obtains each input record via COBOL statements such as the READ statement.

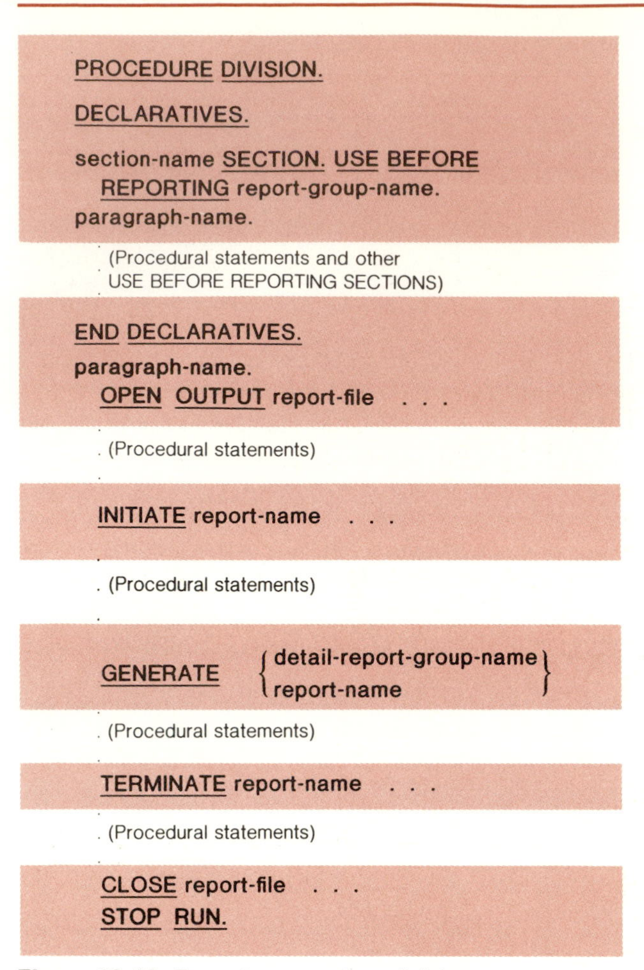

PROCEDURE DIVISION.
DECLARATIVES.
section-name SECTION. USE BEFORE
REPORTING report-group-name.
paragraph-name.
(Procedural statements and other
USE BEFORE REPORTING SECTIONS)
END DECLARATIVES.
paragraph-name.
OPEN OUTPUT report-file . . .
(Procedural statements)
INITIATE report-name . . .
(Procedural statements)
GENERATE {detail-report-group-name / report-name}
(Procedural statements)
TERMINATE report-name . . .
(Procedural statements)
CLOSE report-file . . .
STOP RUN.

Figure 12.19 Format—procedure division.

The INITIATE, GENERATE, and TERMINATE statements are required entries in the Procedure Division for the production of a report (or reports). The Declarative Section statement USE BEFORE REPORTING allows the programmer to gain control during the generation of any report group.

1. Procedural statements in USE BEFORE REPORTING must conform to the restrictions stipulated under DECLARATIVES and USE Declarative described in the Procedure Division section of this text. USE BEFORE REPORTING procedures permit manipulation or alteration of report-group-elements immediately prior to printing. Refer to Guides to Use of the Report Writer described later in this section.
2. All other features of ANS COBOL as contained in this text may be used in a program containing the Report Writer feature. Complete freedom in processing data, etc., is allowed before, during, and after producing a report.

When the last GENERATE statement has been executed, the report must be terminated. The TERMINATE statement causes final control footings and report footings to be presented.

The immediate destination of a report is always a file specified in the File Section of the Data Division. The file must be explicitly opened prior to execution of the report's INITIATE statement, and the file must be explicitly closed after the TERMINATE statement. The report writing statements implicitly perform whatever writing is required for the report.

In a program that utilizes the Report Writer feature, records are read and data is manipulated by programmer instructions in the Procedure Division prior to entering the report phase. (The format for the Procedure Division is shown in figure 12.19.) A report is produced by the execution of the INITIATE, GENERATE, and TERMINATE statements in the Procedure Division. The INITIATE statement initializes all counters associated with the Report Writer; the GENERATE statement is used each time a detailed portion of the report is to be produced; and the TERMINATE statement is used to end the report. The Report Writer feature allows additional manipulation of data by means of a USE BEFORE REPORTING declarative in the Declaratives Section of the Procedure Division.

INITIATE Statement

The INITIATE statement causes the Report Writer Control System to begin the processing of a report. (The format for the INITIATE statement is shown in figure 12.20.)

Rules Governing the Use of the INITIATE Statement

1. Each report name must be defined by a report description entry in the Report Section of the Data Division.

INITIATE report-name-1 [, report-name-2] . . .

Figure 12.20 Format—INITIATE statement.

2. The INITIATE statement resets all data-name entries that contain SUM clauses associated with the report. The Report Writer controls for all the TYPE report groups that are associated with the report are set up in their respective order.
3. The PAGE-COUNTER, if specified, is set to 1 prior to or during the execution of the INITIATE statement. If a different starting value for the PAGE-COUNTER other than 1 is desired, the programmer may reset this counter with a statement in the Procedure Division following the INITIATE statement.
4. The LINE-COUNTER, if specified, is set to zero prior to or during the execution of the INITIATE statement.
5. The INITIATE statement does not open the file with which the report is associated. An OPEN statement for the file must be specified. The INITIATE statement performs Report Writer functions for individually described programs analogous to the input-output functions that the OPEN statement performs for individually described files.
6. A second INITIATE statement for a particular report-name may not be executed unless a TERMINATE statement has been executed for that report-name subsequent to the first INITIATE statement.

GENERATE Statement

The GENERATE statement directs the RWCS to produce a report in accordance with the report description that was specified in the Report Section of the Data Division. (The format for the GENERATE statement is shown in figure 12.21.)

Rules Governing the Use of the GENERATE Statement

1. Data-name represents a TYPE DETAIL report group or an RD (report description) entry.
2. If data-name is the name of a TYPE DETAIL report group, the GENERATE statement performs all the automatic operations of the Report Writer and produces an actual output detail report group in the output. This is called *detail reporting.*
3. If report-name is the name of a RD entry, the GENERATE statement does all the automatic operations of the Report Writer and updates the footing report group(s) within a particular report group without producing an actual detail report group associated with the report. In this case, all SUM counters associated with the report descriptions are algebraically incremented each time a GENERATE statement is executed. This is called *summary reporting.* If more than one TYPE DETAIL group is specified, all SUM counters are algebraically incremented each time a GENERATE statement is executed.
4. The GENERATE statement, implicit in both detail and summary reporting, produces the following automatic operations (if defined):
   a. Steps and tests the LINE-COUNTER and/or PAGE-COUNTER to produce appropriate PAGE-FOOTING and/or PAGE HEADING report groups.
   b. Recognizes any specified control breaks to produce appropriate CONTROL FOOTING and/or CONTROL HEADING report groups.

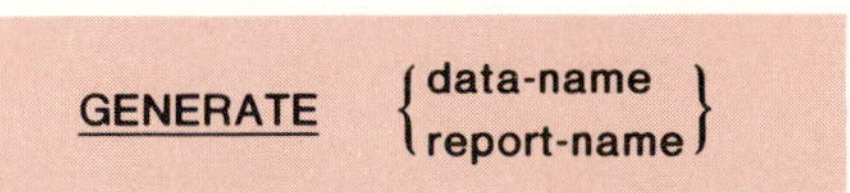

Figure 12.21 Format—GENERATE statement.

c. Accumulates into the SUM counters all specified identifiers. Resets the SUM counters on an associated control break. Performs an updating procedure between control break levels for each set of SUM counters.
d. Executes any specified routines defined by a USE statement before generation of the associated report groups.

5. During the execution of the first GENERATE statement, the following report groups associated with the report, if specified, are produced in the following order:
   a. REPORT HEADING report group.
   b. PAGE HEADING report group.
   c. All CONTROL HEADING report groups in this order: FINAL, major to minor.
   d. The DETAIL report group, if specified, in the GENERATE statement.
6. If a control break is recognized at the time of the execution of a GENERATE statement (other than the first that is executed for the report), all CONTROL FOOTING report groups specified for the report are produced from the minor group up to and including the report group specified for the identifier which caused the control break. Then the CONTROL HEADING report group(s) specified for the report, from the report group specified for the identifier that causes the control break, down to the minor report group, are produced in that order. The DETAIL report group specified in the GENERATE statement is then produced.
7. Data is moved to the data item in the report group description entry of the Report Section and is edited under control of the Report Writer according to the same rules for movement and editing as described for the MOVE statement. (See "MOVE" statement in the Procedure Division.)

If no GENERATE statements have been executed for a report during the interval between the execution of an INITIATE statement and a TERMINATE statement for that report, the TERMINATE statement does not cause the RWCS to perform any of the related processing.

TERMINATE Statement

The TERMINATE statement causes the RWCS to complete the processing of the specified reports. (The format for the TERMINATE statement is shown in figure 12.22.)

Rules Governing the Use of the TERMINATE Statement

1. Each report-name given in a TERMINATE statement must be defined by an RD entry in the Data Division.
2. The TERMINATE statement produces all the control footings associated with this report, as if a control break had just occurred at the highest level, and completes the Report Writer functions for the named report. The TERMINATE statement also produces the last page footings and report footing report groups associated with this report.
3. Appropriate PAGE HEADING or FOOTING report groups are prepared in their respective order for the report description.
4. A second TERMINATE statement for a particular file may not be executed unless a second INITIATE statement has been executed for the report-name. If a TERMINATE statement has been executed for a report, a GENERATE statement for that report must not be executed unless an intervening INITIATE statement for that report is executed.
5. The TERMINATE statement does not close the file with which the report is associated. A CLOSE statement for the file must be given by the programmer. The TERMINATE statement performs Report Writer functions for individually described report programs analogous to the input-output functions that the CLOSE statement performs for individually described files.

```
TERMINATE report-name-1  [ report-name-2 ]  . . .
```

Figure 12.22 Format—TERMINATE statement.

SUPPRESS PRINTING

Figure 12.23 Format—SUPPRESS statement.

6. SOURCE clauses used in TYPE CONTROL FOOTING FINAL or TYPE REPORT FOOTING report groups refer to the values of the items during the execution of the TERMINATE statement.

SUPPRESS Statement

The SUPPRESS statement causes the RWCS to inhibit the presentation of a report group. (The format for the SUPPRESS statement is shown in figure 12.23.)

Rules Governing the Use of the SUPPRESS Statement

1. The SUPPRESS statement may only appear in a USE BEFORE REPORTING procedure.
2. The SUPPRESS statement inhibits presentation only for the report group named in the USE procedure within which the SUPPRESS statement appears.
3. The SUPPRESS statement must be executed each time the presentation of the report group is to be inhibited.
4. When the SUPPRESS statement is executed, the RWCS is instructed to inhibit the processing of the following report group functions:
   a. The presentation of the print line of the report group.
   b. The processing of all LINE clauses in the report group.
   c. The processing of the NEXT GROUP clause in the report group.
   d. The adjustment of LINE-COUNTER.

USE BEFORE REPORTING Statement

A USE BEFORE REPORTING declarative statement may be written in the Declaratives Section of the Procedure Division if the programmer wishes to alter or manipulate the data before it is presented in the report. (The format for the USE BEFORE REPORTING declarative is shown in figure 12.24.)

The USE statement specifies Procedure Division statements that are executed just before a report group named in the Report Section of the Data Division is produced.

Rules Governing the Use of a USE BEFORE REPORTING Statement

1. A USE statement, when present, must immediately follow a section header in the Declaratives portion of the Procedure Division, and must be followed by a period and a space. The remainder of the section must consist of one or more procedural paragraphs that define the procedures to be processed.
2. Identifier represents a report group named in the Report Section of the Data Division. The identifier must not be used in more than one USE statement. The identifier must be qualified by the report-name if not unique.
3. No Report Writer statement (INITIATE, GENERATE or TERMINATE) may be written in a procedural paragraph or a paragraph following the USE sentence in the Declaratives Section.
4. The USE statement itself is never executed; rather, it defines the conditions calling for the execution of the USE procedures.

USE BEFORE REPORTING identifier.

Figure 12.24 Format—USE BEFORE REPORTING declarative.

5. The designated procedures are executed by the Report Writer just before the named report is produced, regardless of page or control breaks associated with report groups. The report group may be any type except DETAIL.
6. Within a USE procedure, there must not be any reference to any nondeclarative procedure. Conversely, in the nondeclarative portion, there must be no reference to procedure-names that appear in the Declaratives portion, except that PERFORM statements may refer to USE declaratives or to procedures associated with USE declaratives.

The USE BEFORE REPORTING procedures must not change the contents of control data items or alter subscripts or indexes used in referencing the controls.

The following is an example of the USE BEFORE REPORTING declarative:

```
WORKING-STORAGE SECTION.
77  A                                       PICTURE 9(6)V99     VALUE 0.
77  B                                       PICTURE 9(6)V99     VALUE 0.

REPORT SECTION.
    .
    .
    .
01  MINOR-CONTROL              TYPE CF      MINOR . . .
    05  LINE PLUS 1.
        10  COLUMN 1                        PICTURE Z(6).99     SOURCE IS A.
    .
    .
    .
01  INTERMEDIATE-CONTROL       TYPE CF      INTERMEDIATE . . .
    05  LINE PLUS 1.
        10  COLUMN 1                        PICTURE Z(6).99     SOURCE IS B.
    .
    .
    .
01  FINAL-CONTROL              TYPE CF      FINAL . . .
    .
    .
    .
PROCEDURE DIVISION.
DECLARATIVES.
INT-CTL SECTION.
    USE BEFORE REPORTING INTERMEDIATE-CONTROL.
PARA-1.
    MOVE ZEROS TO A.
FIN-CTL SECTION.
    USE BEFORE REPORTING FINAL-CONTROL.
PARA-2.
    MOVE ZEROS TO B.
END DECLARATIVES.
    .
    .
    .
BEGIN.
    READ INPUT-FILE.
        AT END . . .
    COMPUTE A = A + COST.
    COMPUTE B = B + COST.
    GENERATE . . .
    .
    .
```

The programming just presented manually resets the SUM counters. In previous SUM counter examples, the counters were automatically reset.

The following coding example illustrates the use and operation of the USE BEFORE REPORTING declarative section. Part 1 of the example shows the definition of the two CONTROL FOOTING report groups in the Report Section of the Data Division. Part 2 of the example shows a method of using the USE BEFORE REPORTING declarative. In this example, the USE BEFORE REPORTING declarative section in Part 2 will be executed before MINOR control footing (described in Part 1) is produced.

Part 1

```
01  MINOR TYPE CONTROL FOOTING C-1 LINE PLUS 2.
    05  A     COLUMN 3     PICTURE $$$.99          SUM P
        BLANK WHEN ZERO.

    05  B     COLUMN 10    PICTURE $$$99.99CR      SUM Q.

    05        COLUMN 25    PICTURE ** *** **       SOURCE E.

01  TYPE     CONTROL FOOTING C-2    LINE PLUS 1.
    05        COLUMN  5    PICTURE $$$$$.99        SUM A
        BLANK WHEN ZERO.
```

P, Q, and E can be defined in any of the File, Working-Storage, or Linkage Sections. C-1 and C-2 are control data items in the CONTROLS clause.

Part 2

```
PROCEDURE DIVISION.
DECLARATIVES.
RW SECTION.
    USE BEFORE REPORTING MINOR.

MINOR-PARA.

    IF A = 0
        AND B = 0
        SUPPRESS
    ELSE
        NEXT SENTENCE.
END DECLARATIVES.
```

When a control break occurs for C-1, the following operations are performed:

1. The RW SECTION is executed.
2. A test is made whether to SUPPRESS printing. If the condition tested is true, steps 3 and 4 are bypassed.
3. The print line is constructed under control of the Report Writer as follows:
   a. moving the correct character to the first part of record (to cause a double-space);
   b. moving the sum counters to the print line (edited); and
   c. moving E to the print line (edited).
4. The line is printed and 2 is added to the LINE-COUNTER.
5. The sum counter A is added to the sum counter defined in the second report group.
6. The sum counters in the first report group are set to zero.

Guides to the Use of the Report Writer

The Report Writer is a valuable feature of the COBOL language, making possible the automatic generation of a report, or reports, in either detail or summary form. The data processing potential of the Report Writer is greatly enhanced in that it may be combined with any other COBOL language element, e.g., SORT or MERGE verb, ENTER verb, input-output verbs, computation, data-manipulation and other procedural statements. However, the actual processing of the report is automatic. The construction of print-lines, "crossfooting" and "rolling" of sum counters, control break-testing and printing, heading and footing generation, etc., require no programming effort once the report has been defined in the Report Section.

The following is an example of the Report Writer program with the Sort program:

```
ENVIRONMENT DIVISION.
    .
    .
    .
INPUT-OUTPUT SECTION.
FILE-CONTROL.
    SELECT INPUT-FILE
        ASSIGN TO INPUT-MASTER.
    SELECT SORT-FILE
        ASSIGN TO COLLATE-FILE.
    SELECT REPORT-OUTPUT
        ASSIGN TO OUTPUT-MASTER.

DATA DIVISION.
FILE SECTION.
FD  INPUT-FILE
    .
    .
01  . . .
    .
    .
SD  SORT-FILE
    .
    .
01  . . .
    .
    .
FD  REPORT-OUTPUT
    REPORT IS XYZ
    .
    .
    .
WORKING-STORAGE SECTION.
    .
    .
    .
01  FLAGS.
    05  MORE-INPUT-FLAG      PICTURE XXX      VALUE 'YES'.
        88  MORE-INPUT                        VALUE 'YES'.
        88  NO-MORE-INPUT                     VALUE 'NO'.
    .
    .
    .
REPORT SECTION.
RD  XYZ
    .
    .
    .
01  DETAIL-LINE     TYPE DETAIL . . .
    .
    .
    .
PROCEDURE DIVISION.
SORT-CALL SECTION.
DRIVER.
    SORT SORT-FILE
        ON . . .
        USING INPUT-FILE
        OUTPUT PROCEDURE IS EDIT.
    STOP RUN.
```

```
EDIT SECTION.
STARTUP.
    OPEN      OUTPUT      REPORT-OUTPUT.
    INITIATE XYZ.
    RETURN SORT-FILE RECORD
        AT END MOVE 'NO' TO MORE-INPUT-FLAG.
    PERFORM LOOP
        UNTIL NO-MORE-INPUT.
    TERMINATE XYZ.
    CLOSE REPORT-OUTPUT.
    GO TO LOOP-EXIT.

LOOP.
    GENERATE DETAIL-LINE.
    RETURN SORT-FILE RECORD
        AT END MOVE 'NO' TO MORE-INPUT-FLAG.

LOOP-EXIT.
    EXIT.
```

The Sort program just presented entails the use of an output procedure to deliver a report (on any suitable device) rather than an output tape.

The following is an example of the Report Writer program with the Merge program:

```
ENVIRONMENT DIVISION.
    .
    .
INPUT-OUTPUT SECTION.
FILE-CONTROL.
    SELECT INPUT-FILE-1
        ASSIGN TO . . .
    SELECT INPUT-FILE-2
        ASSIGN TO . . .
    SELECT MERGE-FILE
        ASSIGN TO . . .
    SELECT REPORT-OUTPUT
        ASSIGN TO . . .

DATA DIVISION.
FILE SECTION.
FD  INPUT-FILE-1
    .
    .
01  . . .
    .
    .
FD  INPUT-FILE-2
    .
    .
01  . . .
    .
    .
SD  MERGE-FILE
    .
    .
01  . . .
    .
    .
FD  REPORT-OUT
    REPORT IS XYZ
    .
    .
01  . . .
    .
    .
```

```
WORKING-STORAGE SECTION.
    .
    .
    .
01  FLAGS.
    05  MORE-INPUT-FLAG     PICTURE XXX     VALUE 'YES'.
        88  MORE-INPUT                      VALUE 'YES'.
        88  NO-MORE-INPUT                   VALUE 'NO'.
    .
    .
    .
REPORT SECTION.
RD  XYZ
    .
    .
    .
01  DETAIL-LINE     TYPE DETAIL . . .
    .
    .
    .
PROCEDURE DIVISION.
MERGE-CALL SECTION.
DRIVER.
    MERGE MERGE-FILE
        ON . . .
        USING INPUT-FILE-1,
            INPUT-FILE-2
        OUTPUT PROCEDURE IS EDIT.
    STOP RUN.

EDIT SECTION.
STARTUP.
    OPEN    OUTPUT    REPORT-OUT.
    INITIATE XYZ.
    RETURN MERGE-FILE RECORD
        AT END MOVE 'NO' TO MORE-INPUT-FLAG.
    PERFORM LOOP
        UNTIL NO-MORE-INPUT.
    TERMINATE XYZ.
    CLOSE REPORT-OUT.
    GO TO LOOP-EXIT.

LOOP.
    GENERATE DETAIL-LINE.
    RETURN MERGE-FILE RECORD
        AT END MOVE 'NO' TO MORE-INPUT-FLAG.

LOOP-EXIT.
    EXIT.
```

The merge program just presented entails the use of an output procedure to deliver a report (on any suitable device) rather than an output tape.

A report definition specifies the page and line format, and describes the source data to be printed and the numerical data to be summed. The page and line formats are defined in the Report Section with the clauses CONTROL, PAGE, TYPE, LINE, NEXT GROUP, COLUMN, and GROUP INDICATE. How these clauses should be defined for a particular report is best determined by referring to a print spacing chart. The source data which is to appear on each print-line is referred to by a SOURCE clause that causes it to be moved to the print buffer as the report-group is generated. The numerical data to be summed is referred to by a SUM clause that causes it to be added to a counter and printed, with or without editing, as defined in the associated PICTURE clause.

Summary

The Report Writer feature provides the facility for producing reports by specifying the physical appearance of a report rather than requiring specifications of detailed procedures necessary to produce that report. The programmer can specify the format of the printed report in the Data Division, thereby minimizing the amount of Procedure Division coding he or she would have to write to create the report. All operations of the Report Writer, such as procedures for moving data, constructing print lines, counting lines on a page, numbering pages, producing heading and footing lines, recognizing the end of logical data subdivisions, updating SUM counters, etc. are accomplished by the Report Writer Control System (RWCS) from source language statements that appear primarily in the Report Section of the Data Division of the source program.

In the Data Division, the programmer provides the necessary data-names and describes the format of the report to be produced. In the Procedure Division, the programmer writes the necessary statements that provide the desired results.

The programmer must write in the File Section of the Data Division, a description of all names and formats of the report to be produced. In addition, a complete file and record description of the input data must also be written in this section. A Report Section must be added at the end of the Data Division to define the format of each finished report. More than one report may be written on two files at the same time.

A Report clause is required in the file description (FD) entry to list the name of the report(s) to be produced.

In the Report Section, the description of each report must begin with a report description (RD) entry, and be followed by entries that describe the report groups within the report. The Report Section consists of two types of entries; one describes the physical aspect of the report format, while the other describes the conceptual characteristics of the items that make up the report and their relation to the report format.

The RD entry specifies the basic page layout and the overall organization of the report. The RD entry is analogous to the FD entry in the File Section.

The Report Group Description entry gives the detailed formats of all elements of the report and the sources of all information for that report.

A level 01 report group description entry is analogous to a level 01 data description entry in the File Section and is called a report group.

The Report Section must contain at least one report description entry. This entry contains information pertaining to the overall format of the report named in the File Section and is uniquely identified by the level indicator RD. Each report named in the REPORT(S) clause of a FD entry in the File Section must be the subject of an RD entry in the Report Section.

The REPORT clause specifies the names of reports that make up a report file. The report names are unique names that must be specified in the REPORT clause of the file description entry for the file in which this report is to be written.

The CODE clause is used to specify identifying characters that are added to each line produced. The two-character literal identifies each print line as belonging to a specific report.

Each control heading or footing is associated with a specific control data item described in the File, Working-Storage, or Linkage Section. Control items are related to the report by a list of control data-names entered in declining order of importance in the CONTROL(S) clause of the RD entry. The most significant possible control level is associated with the reserved word FINAL. A control break is recognized whenever a control item has changed in value between the execution of the previous GENERATE statement and the current GENERATE statement.

The PAGE clause is used to describe the physical format of a page of the report. The PAGE clause is required when the page format is to be controlled by the Report Writer.

The fixed data-names, PAGE-COUNTER and LINE-COUNTER, are numeric counters automatically generated by the Report Writer based on the presence of specific entries and do not require any data description clauses.

A page counter is implicitly provided for each report. It is primarily used as a SOURCE data item within page/overflow heading or page/overflow footing report groups, to provide consecutive page numbers for the report. Its initial value is 1 and is automatically incremented by 1 each time a page break occurs. It may be altered by a Procedure Division statement and may be referenced in a SOURCE clause.

A line counter is implicitly provided for each report. It is used by the generated reporting procedures to recognize page (and overflow) breaks, and to control the vertical page format. The line number is automatically set to zero initially, and is reset to zero whenever a page break occurs. It is automatically set, reset, and incremented on the basis of values specified in the LINE NUMBER and NEXT GROUP clauses in the respective report groups. Procedure Division statements should never change the value of a line counter.

Each integral unit of data represented in a report, such as page heading or footing, or detail lines is called a report group. The report group may consist of one or several actual lines in the printed report. Each report group must contain at least one TYPE DETAIL report group.

The Report Group Description entry defines the characteristics of a report group and of the individual data-names within a report group. There are three types of report groups: heading, detail and footing. The report groups that make up the report are described fully in the RD entry. The description of each report group begins with one report group description entry that contains a level number 01 and a TYPE clause, followed by a group of subdivision entries.

The data-name clause specifies the name of the data being described. It must be specified when referred to by a GENERATE statement, a USE statement, or any other statement.

The LINE NUMBER clause specifies the vertical positioning information for the report. Integer-1 specifies an absolute number while integer-2 specifies a relative line number. The NEXT PAGE phrase specifies that the report group is to be presented beginning in the indicated line number on a new page.

The NEXT GROUP clause specifies information for vertical positioning of a page following the presentation of the last line of a report group.

The TYPE clause specifies the particular type of report group that is described in this entry and indicates the time at which this report group is to be generated.

The REPORT HEADING group entry indicates a report group that is produced only once at the beginning of the report during the execution of the first GENERATE statement.

The PAGE HEADING entry indicates a report group that is automatically produced at the beginning of each page of the report.

The CONTROL HEADING entry indicates a report group that is printed at the beginning of a control group of a designated identifier, or in the case of FINAL, is produced once before the first control at the initiation of a report during the execution of the first GENERATE statement.

The DETAIL entry indicates a report group that is produced for each GENERATE statement. The data-name is specified in the 01 level, and is referred to by the GENERATE statement.

The CONTROL FOOTING entry indicates a report group that is produced either at the end of a control group for a designated identifier, or once at the termination of a report ending in a FINAL control group.

The PAGE FOOTING entry indicates a report group that is automatically produced at the bottom of each page of the report.

The REPORT FOOTING entry indicates a report group that is produced only once at the termination of the report at the time the TERMINATE statement is executed.

DISPLAY is the only USAGE option that may be specified for elementary or group items.

The COLUMN clause indicates that the leftmost character of the elementary item is placed in a position specified by the integer.

The GROUP INDICATE clause specifies that an elementary item is to be produced only on the first occurrence of an item after any CONTROL or PAGE break.

The same rules apply to the JUSTIFIED, BLANK WHEN ZERO, and PICTURE clauses as in the Data Division.

The SOURCE clause indicates a data item to be used as a source for this report item. It specifies a data item that is to be printed or a data item that is to be summed in a CONTROL FOOTING report group.

The three distinct types of SUM counters manipulations are subtotalling, rolling forward, and crossfooting.

In subtotalling, a SUM counter is augmented by the value of the SUM operand for each execution of a GENERATE statement of the TYPE DETAIL report group which contains the SOURCE counterpart of the SUM operand.

In rolling forward, the SUM counters defined in control footing groups of lower control levels are added to SUM counters defined in control footing report groups of higher control levels during control break processing.

Crossfooting is a type of SUM counter manipulation in which SUM counters defined in a given control control footing report group are added to other SUM counters in the same report group during control break processing.

The SUM clause establishes a SUM counter and names the data items to be summed. It is permissible for Procedure Division statements to alter the contents of the SUM counter. The UPON phrase provides the capability to accomplish selective subtotalling for the DETAIL report groups named in the phrase. SUM counters are initially set to zero by the RWCS during the execution of the INITIATE statement for the report containing the SUM counters. Counters are reset automatically to zero unless the explicit RESET phrase is given, specifying resetting based on a higher level of control than the associated control for the report group.

The VALUE clause causes the report data item to assign the specified value each time its report group is presented only if the elementary item entry does not contain a GROUP INDICATE clause.

The INITIATE statement causes the RWCS to begin the processing of a report. It resets all data-name entries that contain SUM clauses associated with the report. The PAGE-COUNTER, if specified, is set to 1. The LINE-COUNTER, if specified, is set to 0. The INITIATE statement does not open the file. An OPEN statement for the file must be specified.

The GENERATE statement directs the RWCS to produce a report in accordance with the report description that was specified in the Report Section of the Data Division. It performs all the automatic operations of the Report Writer and produces an actual output detail report group. If the data-name is the subject of the GENERATE statement, all detail operations are performed. If report-name is the subject of the GENERATE statement, summary processing is performed.

The TERMINATE statement causes the RWCS to complete the processing of the specified report. The TERMINATE statement does not close the file with which the report is associated. A CLOSE statement for the file must be given by the programmer.

The SUPPRESS statement causes the RWCS to inhibit the presentation of a report group. It may only appear in the USE BEFORE REPORTING procedure. It inhibits the presentation only for the group in which the SUPPRESS statement appears.

The USE BEFORE REPORTING declarative statement may be written in the Declaratives Section of the Procedure Division if the programmer wishes to alter or manipulate the data before it is presented in the report. The USE statement, when present, must immediately follow a section header in the Declaratives portion of the Procedure Division. The USE statement itself is never executed; rather it defines the conditions calling for the execution of the USE procedure.

The Report Writer feature may be combined with any other COBOL element such as the Sort-Merge feature.

Illustrative Program: Quarterly Expenditures Report

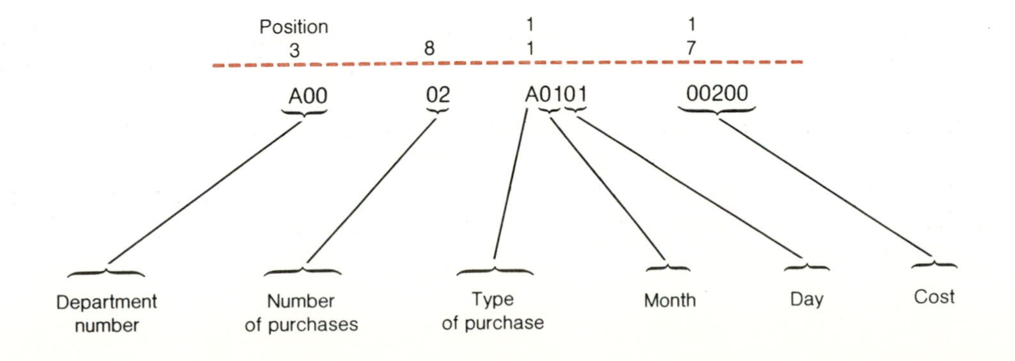

```
 1      *************************
 2
 3       IDENTIFICATION DIVISION.
 4
 5      *************************
 6
 7       PROGRAM-ID.
 8           ACME.
 9       AJTHOR.
10           WAN-JU CHIEN.
11       INSTALLATION.
12           WEST LOS ANGELES COLLEGE.
13       DATE-WRITTEN.
14           3 FEBRUARY 1982.
15      *
16      *PURPOSE.
17           REPORT WAS PRODUCED 3Y THE REPORT WRITER FEATURE.
18      *************************
19
20       ENVIRONMENT DIVISION.
21
22      *************************
23
24       CONFIGURATION SECTION.
25
26       SOURCE-COMPUTER.
27           LEVEL-66-ASCII.
28       OBJECT-COMPUTER.
29           LEVEL-66-ASCII.
30
31       INPUT-OUTPUT SECTION.
32
33       FILE-CONTROL.
34           SELECT FILE-IN      ASSIGN TO C1-CARD-READER.
35           SELECT FILE-OUT     ASSIGN TO P1-PRINTER.
36
37      *************************
38
39       DATA DIVISION.
40
41      *************************
42
43       FILE SECTION.
44
45       FD  FILE-IN
46           CODE-SET IS GBCD
47           LABEL RECORDS ARE OMITTED
48           DATA RECORD IS CARDIN.
49
50       01  CARDIN.
51
52           02  FILLER          PICTURE AA.
53           02  DEPT            PICTURE XXX.
54           02  FILLER          PICTURE AA.
55           02  NO-PURCHASE     PICTURE 99.
56           02  FILLER          PICTURE A.
57           02  TYPE-PURCHASE   PICTURE A.
58           02  MONTH           PICTURE 99.
59           02  DAY-I           PICTURE 99.
60           02  FILLER          PICTURE A.
61           02  COST            PICTURE 999V99.
62           02  FILLER          PICTURE X(59).
63       FD  FILE-OUT
64           CODE-SET IS GBCD
65           LABEL RECORDS ARE OMITTED
66           REPORT IS EXPENSE-REPORT.
67
68       WORKING-STORAGE SECTION.
69
70       77  SAVED-MONTH PICTURE 99     VALUE 1.
71       77  SAVED-DAY   PICTURE 99     VALUE 0.
72       77  CONTINUED   PICTURE X(11) VALUE SPACE.
73       01  MONTHS.
74           02  RECORD-MONTH.
75               03  FILLER PICTURE A(9) VALUE IS "JANUARY  ".
76               03  FILLER PICTURE A(9) VALUE IS "FEBRUARY ".
77               03  FILLER PICTURE A(9) VALUE IS "MARCH    ".
78               03  FILLER PICTURE A(9) VALUE IS "APRIL    ".
79               03  FILLER PICTURE A(9) VALUE IS "MAY      ".
80               03  FILLER PICTURE A(9) VALUE IS "JUNE     ".
81               03  FILLER PICTURE A(9) VALUE IS "JULY     ".
82               03  FILLER PICTURE A(9) VALUE IS "AUGUST   ".
83               03  FILLER PICTURE A(9) VALUE IS "SEPTEMBER".
84               03  FILLER PICTURE A(9) VALUE IS "OCTOBER  ".
85               03  FILLER PICTURE A(9) VALUE IS "NOVEMBER ".
86               03  FILLER PICTURE A(9) VALUE IS "DECEMBER ".
87           02  RECORD-AREA REDEFINES RECORD-MONTH OCCURS 12 TIMES.
88               03  MONTHNAME PICTURE A(9).
89
90       01  FLAGS.
91
92           03  MORE-DATA-FLAG  PICTURE XXX      VALUE "YES".
93               88  MORE-DATA                    VALUE "YES".
94               88  NO-MORE-DATA                 VALUE "NO ".
95
96       REPORT SECTION.
97
```

```
 98        RD  EXPENSE-REPORT
 99            CONTROLS ARE FINAL, MONTH, DAY-I
100            PAGE              44 LINES
101            HEADING           1
102            FIRST DETAIL      9
103            LAST DETAIL       33
104            FOOTING           37.
105        01  TYPE REPORT HEADING.
106            02  LINE 1 COLUMN 27 PICTURE A(26)
107                VALUE IS "ACME MANUFACTURING COMPANY".
108            02  LINE 3 COLUMN 26 PICTURE A(29)
109                VALUE IS "QUARTERLY EXPENDITURES REPORT".
110
111        01  PAGE-HEAD TYPE PAGE HEADING.
112            02  LINE NUMBER 5.
113                03  COLUMN 30 PICTURE A(9)
114                    SOURCE MONTHNAME OF RECORD-AREA(MONTH).
115                03  COLUMN 39 PICTURE A(12) VALUE IS "EXPENDITURES".
116                03  COLUMN 52 PICTURE X(11) SOURCE CONTINUED.
117            02  LINE NUMBER 7.
118                03      COLUMN 2 PICTURE X(35)
119                    VALUE IS "MONTH      DAY     DEPT  NO-PURCHASES".
120                03  COLUMN 40 PICTURE X(33)
121                    VALUE IS "TYPE          COST  CUMULATIVE-COST".
122
123        01  DETAIL-LINE TYPE DETAIL LINE PLUS 1.
124            02  COLUMN  2 GROUP INDICATE PICTURE A(9)
125                SOURCE MONTHNAME OF RECORD-AREA(MONTH).
126            02  COLUMN 13      PIC 99        SOURCE DAY-I GROUP INDICATE.
127            02  COLUMN 19      PIC XXX       SOURCE DEPT.
128            02  COLUMN 31      PIC Z9        SOURCE NO-PURCHASE.
129            02  COLUMN 42      PIC A         SOURCE TYPE-PURCHASE.
130            02  COLUMN 50      PIC ZZ9.99    SOURCE COST.
131        01  TYPE CONTROL FOOTING DAY-I.
132            02  LINE PLUS 2.
133                03  COLUMN  2 PIC X(22)    VALUE "PURCHASES AND COST FOR".
134                03  COLUMN 24 PIC Z(2)     SOURCE SAVED-MONTH.
135                03  COLUMN 26 PIC X        VALUE "-".
136                03  COLUMN 27 PIC 99       SOURCE SAVED-DAY.
137                03  COLUMN 30 PIC ZZ9      SUM NO-PURCHASE.
138                03  MIN COLUMN 49 PIC $$$9.99 SUM COST.
139                03  COLUMN 65 PIC $$$$9.99 SUM COST RESET ON FINAL.
140            02  LINE PLUS 1 COLUMN 2 PIC X(71) VALUE ALL "*".
141            02  LINE PLUS 1 COLUMN 2 PIC X(70) VALUE SPACES.
142        01  TYPE CONTROL FOOTING MONTH LINE PLUS 2 NEXT GROUP NEXT PAGE.
143            02  COLUMN 16      PIC A(14)    VALUE "TOTAL COST FOR".
144            02  COLUMN 31      PIC A(9)
145                SOURCE MONTHNAME OF RECORD-AREA(SAVED-MONTH).
146            02  COLUMN 40      PIC A(3)     VALUE "WAS".
147            02  INT COLUMN 49 PIC $$$9.99 SUM MIN.
148        01  TYPE CONTROL FOOTING FINAL LINE PLUS 2.
149            02  COLUMN 16      PIC       A(26)
150                VALUE IS "TOTAL COST FOR QUARTERS".
151            02  COLUMN 47      PIC $$,$$9.99  SUM INT.
152        01  TYPE PAGE FOOTING LINE 37.
153            02  COLUMN 59      PIC X(12)    VALUE "REPORT-PAGE-".
154            02  COLUMN 71      PIC 9(2)     SOURCE PAGE-COUNTER.
155        01  TYPE REPORT FOOTING.
156            02  LINE PLUS 1 COLUMN 32 PIC A(13) VALUE IS "END OF REPORT".
157
158        PROCEDURE DIVISION.
159
160        DECLARATIVES.
161        PAGE-HEAD-RTN SECTION.
162                USE BEFORE REPORTING PAGE-HEAD.
163        TEST.
164            IF NO-MORE-DATA
165                SUPPRESS PRINTING
166            ELSE
167                IF MONTH EQUALS SAVED-MONTH
168                    MOVE "(CONTINUED)" TO CONTINUED
169                ELSE
170                    MOVE SPACES TO CONTINUED
171                    MOVE MONTH TO SAVED-MONTH.
172        END DECLARATIVES.
173
174        MAIN-ROUTINE SECTION.
175        MAIN-PAR.
176            OPEN INPUT  FILE-IN
177                 OUTPUT FILE-OUT.
178            INITIATE EXPENSE-REPORT.
179            READ FILE-IN
180                AT END MOVE "NO" TO MORE-DATA-FLAG.
181            PERFORM WRITE-REPORT
182                UNTIL NO-MORE-DATA.
183            PERFORM DONE.
184        WRITE-REPORT.
185            GENERATE DETAIL-LINE.
186            MOVE DAY-I TO SAVED-DAY.
187            READ FILE-IN
188                AT END MOVE "NO" TO MORE-DATA-FLAG.
189        DONE.
190            TERMINATE EXPENSE-REPORT.
191            CLOSE   FILE-IN
192                    FILE-OUT.
193            STOP RUN.

THERE WERE 193 SOURCE INPUT LINES.
THERE WERE NO DIAGNOSTICS.
```

```
                         ACME MANUFACTURING COMPANY

                       QUARTERLY EXPENDITJRES REPORT

                           JANUARY  EXPENDITURES

MONTH      DAY     DEPT  NO-PURCHASES   TYPE        COST  CUMULATIVE-COST

JANUARY     01     A00          2         A          2.00
                   A02          1         A          1.00
                   A02          2         C         16.00

PURCHASES AND COST FOR 1-01    5                   $19.00           $19.00
**************************************************************************

JANUARY     02     A01          2         B          2.00
                   A04         10         A         10.00
                   A04         10         C         80.00

PURCHASES AND COST FOR 1-02   22                   $92.00          $111.00
**************************************************************************

JANUARY     05     A01          2         B          2.00

PURCHASES AND COST FOR 1-05    2                    $2.00          $113.00
**************************************************************************

JANUARY     08     A01         10         A         10.00
                   A01          8         B         12.48
                   A01         20         D         38.40

PURCHASES AND COST FOR 1-08   38                   $60.88          $173.88
**************************************************************************

JANUARY     13     A00          4         B          6.24
                   A00          1         C          8.00

PURCHASES AND COST FOR 1-13    5                   $14.24          $188.12
**************************************************************************

JANUARY     15     A00         10         D         19.20
                   A02          1         C          8.00

PURCHASES AND COST FOR 1-15   11                   $27.20          $215.32
**************************************************************************

JANUARY     21     A03         10         E         30.00
                   A03         10         F         25.00
                   A03         10         G         50.00

PURCHASES AND COST FOR 1-21   30                  $105.00          $320.32
**************************************************************************

JANUARY     23     A00          5         A          5.00

PURCHASES AND COST FOR 1-23    5                    $5.00          $325.32
**************************************************************************

JANUARY     26     A04          5         A          5.00
                   A04          5         B          7.80

PURCHASES AND COST FOR 1-26   10                   $12.80          $338.12
**************************************************************************

JANUARY     27     A00          6         B          9.36
                   A00         15         C        120.00

PURCHASES AND COST FOR 1-27   21                  $129.36          $467.48
**************************************************************************

JANUARY     30     A00          2         B          3.12
                   A02         10         A         10.00
                   A02          1         C          8.00
                   A04          1         B         23.40
                   A04         10         C         80.00

PURCHASES AND COST FOR 1-30   24                  $124.52          $592.00
**************************************************************************

JANUARY     31     A00          1         A          1.00
                   A04          6         A          6.00

PURCHASES AND COST FOR 1-31    7                    $7.00          $599.00
**************************************************************************

              TOTAL COST FOR JANUARY WAS          $599.00
              TOTAL COST FOR QUARTERS             $599.00


                                                          REPORT-PAGE-04
                              END OF REPORT
```

Output

```
(1)                           ACME MANUFACTURING COMPANY

                            QUARTERLY EXPENDITJRES REPORT

(2)                            JANUARY  EXPENDITURES

           MONTH      DAY    DEPT  NO-PURCHASES   TYPE      COST   CUMULATIVE-COST

(3)        JANUARY     01    A00         2         A        2.00
(4)                          A02         1         A        1.00
                             A02         2         C       16.00

(5)        PURCHASES AND COST FOR 1-01   5                $19.00            $19.00
           ***********************************************************************

           JANUARY     02    A01         2         B        2.00
                             A04        10         A       10.00
                             A04        10         C       80.00

           PURCHASES AND COST FOR 1-02  22                $92.00           $111.00
           ***********************************************************************

           JANUARY     05    A01         2         B        2.00

           PURCHASES AND COST FOR 1-05   2                 $2.00           $113.00
           ***********************************************************************

           JANUARY     08    A01        10         A       10.00
                             A01         8         B       12.48
                             A01        20         D       38.40

           PURCHASES AND COST FOR 1-08  38                $60.88           $173.88
           ***********************************************************************


(6)                                                               REPORT-PAGE-01




                               JANUARY  EXPENDITURES (CONTINUED)

           MONTH      DAY    DEPT  NO-PURCHASES   TYPE      COST   CUMULATIVE-COST

           JANUARY     13    A00         4         B        6.24
                             A00         1         C        8.00

           PURCHASES AND COST FOR 1-13   5                $14.24           $188.12
           ***********************************************************************

           JANUARY     15    A00        10         D       19.20
                             A02         1         C        8.00

           PURCHASES AND COST FOR 1-15  11                $27.20           $215.32
           ***********************************************************************

           JANUARY     21    A03        10         E       30.00
                             A03        10         F       25.00
                             A03        10         G       50.00

           PURCHASES AND COST FOR 1-21  30               $105.00           $320.32
           ***********************************************************************

           JANUARY     23    A00         5         A        5.00

           PURCHASES AND COST FOR 1-23   5                 $5.00           $325.32
           ***********************************************************************

(3)        JANUARY     26    A04         5         A        5.00


                                                                  REPORT-PAGE-02
```

```
                              JANUARY  EXPENDITURES (CONTINUED)
      MONTH       DAY      DEPT   NO-PURCHASES    TYPE          COST   CUMULATIVE-COST

      JANUARY      26      A04          5          B            7.80
(4)   PURCHASES AND COST FOR 1-26      10                     $12.80           $338.12
(5)   ****************************************************************************

      JANUARY      27      A00          6          B            9.36
                           A00         15          C          120.00

      PURCHASES AND COST FOR 1-27      21                    $129.36           $467.48
      ****************************************************************************

      JANUARY      30      A00          2          B            3.12
                           A02         10          A           10.00
                           A02          1          C            8.00
                           A04          1          B           23.40
                           A04         10          C           80.00

      PURCHASES AND COST FOR 1-30      24                    $124.52           $592.00
      ****************************************************************************

      JANUARY      31      A00          1          A            1.00
                           A04          6          A            6.00

      PURCHASES AND COST FOR 1-31       7                      $7.00           $599.00
      ****************************************************************************

(7)                 TOTAL COST FOR JANUARY WAS               $599.00
                                                                          REPORT-PAGE-03



(8)                 TOTAL COST FOR QUARTERS                  $599.00

      ~~~~~~~~~~~~~~~~~~~~~~~~~~~~~~~~~~~~~~~~~~~~~~~~~~~~~~~~~~~~~~~~~~~~~~~~~~~~

                                                                          REPORT-PAGE-04
(9)                              END OF REPORT
```

Key Relating Report to Report Writer Source Program

① is the report heading resulting from source lines 105–109.
② is the page heading resulting from source lines 111–121.
③ is the detail line resulting from source lines 123–130 (note that since it is the first detail line after a control break, the fields defined with "group indicate," lines 124–126, appear).
④ is a detail line resulting from the same source lines as 3. In this case, however, the fields described as "group indicate" do not appear (since the control break did not immediately precede the detail line).
⑤ is the control footing for (DAY-I) resulting from source lines 131–141.
⑥ is the page footing resulting from source lines 152–154.
⑦ is the control footing (for MONTH) resulting from source lines 142–147.
⑧ is the control footing (for FINAL) resulting from source lines 148–151.
⑨ is the report footing resulting from source lines 155–156.

Lines 161–171 of the example illustrate a use of "USE BEFORE REPORTING" declarative. The effect of the source is that each time a new page is started, a test is made to determine if the end of the report has been reached so excessive printing of the headings can be suppressed, or if the new page is being started because a change in MONTH has been recognized (the definition for the control footing for MONTH specifies "NEXT GROUP NEXT PAGE") or because the physical limits of the page were exhausted. The calculation involved sets up a fixed ("PAGE GROUP") which is referenced by a SOURCE clause in the PAGE FOOTING description. Consequently, two page counters can be maintained; one indicating physical pages and one indicating logical pages.

Questions for Review

1. What are the important items to be considered when presenting a report?
2. What are the report groups that must be considered in presenting data in a format manner?
3. What is the purpose of the Report Writer feature?
4. Describe the principal features of the Report Writer.
5. What is meant by a control break?
6. Briefly describe the operation of the Report Writer.
7. What does the programmer provide in the Data Division?
8. What is the purpose of the REPORT clause?
9. What does the Report Section contain?
10. What is the CODE clause used for?
11. How is the CONTROLS clause used in the Report Writer feature?
12. What is the importance of the PAGE clause?
13. Describe the operation of the PAGE-COUNTER and LINE-COUNTER.
14. What is a Report Group?
15. What is a Report Group Description entry?
16. Briefly describe the data-name, LINE-NUMBER and NEXT GROUP clauses.
17. What is a TYPE clause? Briefly describe the different types of TYPE clauses.
18. Explain the uses of the following clauses: COLUMN, GROUP INDICATE, RESET, SOURCE.
19. Briefly describe the three types of SUM counter manipulations.
20. What is the purpose of the SUM clause and how does it operate on data?
21. How is the VALUE clause used?
22. Explain the purposes and uses of the INITIATE, GENERATE, and TERMINATE statements.
23. Explain the purpose and use of the SUPPRESS statement.
24. What are the uses of the USE BEFORE REPORTING declarative?

Matching Questions

A. *Match each item with its proper description.*

_____ 1. LINE-COUNTER	A. Statement executed before Report Group.
_____ 2. Report Section	B. Provides much of the Procedure Division coding normally supplied by the programmer.
_____ 3. SUPPRESS	C. Stops processing of a report.
_____ 4. Report Writer Control System (RWCS)	D. Characteristics of a particular Report Group.
_____ 5. INITIATE	E. Produces a report.
_____ 6. TERMINATE	F. Information pertaining to the overall format of a report.
_____ 7. Report Group Description entry	G. Specifications for each report.
_____ 8. Report Description entry	H. Begins processing of a report.
_____ 9. USE	I. Causes the Report Writer to inhibit presentation of a Report Group.
_____10. GENERATE	J. Generates procedures to receive page and overflow breaks and control vertical page format.

B. *Match each clause with its proper description.*

_____ 1. REPORT	A. Physical format of a report.
_____ 2. CODE	B. Line control following current report group.
_____ 3. CONTROL	C. First occurrence after control or page break.
_____ 4. PAGE	D. Name of report to be produced.
_____ 5. LINE	E. Automatic summation of data.
_____ 6. NEXT GROUP	F. Value to be tested.
_____ 7. TYPE	G. More than one report from a file.
_____ 8. GROUP INDICATE	H. Time that report group is to be generated.
_____ 9. SOURCE	I. Absolute or relative number.
_____10. SUM	J. Item to be used for report item.

Exercises

Multiple Choice: Indicate the best *answer (questions 1-26).*

1. The Report Writer feature provides the facility for producing reports by
   a. requiring specific detailed procedures to produce the report.
   b. specifying the format in the Procedure Division.
   c. specifying the physical appearance of the report.
   d. All of the above.
2. The programmer is responsible in the Report Writer for writing instructions to
   a. move data.
   b. construct print lines.
   c. produce heading and footing lines.
   d. None of the above.
3. The compiler provides
   a. vertical format control.
   b. detection of control breaks.
   c. accumulation of SUM counters.
   d. All of the above.
4. A Report clause
   a. is optional in the Report Writer feature.
   b. must appear in the FD entry.
   c. describes the report to be produced.
   d. must contain the same name that appears in a FD entry.
5. The Report Section consists of
   a. the Report clause.
   b. the FD entries.
   c. record description entries.
   d. None of the above.
6. A report group is written at the
   a. RD level.
   b. 01 level.
   c. 77 level.
   d. FD level.
7. The optional clauses in the RD entry are
   a. CONTROL.
   b. PAGE.
   c. HEADING.
   d. All of the above.
8. The CODE clause is
   a. a one-character literal that identifies all print lines.
   b. a one-character literal that appears in the last character position of a logical record.
   c. a numeric literal.
   d. None of the above.
9. The most significant possible control level is
   a. major.
   b. final.
   c. minor.
   d. intermediate.
10. A control is a data item that is to be tested each time a
    a. detail group is generated.
    b. heading group is generated.
    c. total group is generated.
    d. footing group is generated.
11. A PAGE clause
    a. is never required.
    b. specifies the specific line control to be maintained within a report.
    c. describes the logical format of a page.
    d. All of the above.

12. The PAGE-COUNTER
    a. has an initial value of one.
    b. has a value that is incremented by one each time a page break occurs.
    c. is a name for a special register that is generated for the report.
    d. All of the above.
13. The LINE-COUNTER
    a. is automatically set to 1 initially.
    b. is automatically tested on the basis of values specified in the PAGE clause.
    c. should be changed by Procedure Division statements.
    d. All of the above.
14. The following clause is required in the Report Group Description entry.
    a. REPORT
    b. CONTROL
    c. TYPE
    d. PAGE
15. The following clause specifies the purpose of the report group.
    a. USAGE
    b. PICTURE
    c. TYPE
    d. NEXT GROUP
16. The NEXT GROUP clause specifies
    a. information for vertical positioning following the last line of a report group.
    b. positioning to take place before the report group in which the clause appears.
    c. a page heading report group.
    d. All of the above.
17. The following report group may appear in a TYPE clause.
    a. DETAIL HEADING
    b. CONTROL FOOTING
    c. DETAIL FOOTING
    d. GROUP HEADING
18. The COLUMN clause
    a. indicates the leftmost character of all elementary items.
    b. specifies the relative column number in the printed page.
    c. may be a negative integer.
    d. may be written at the group level.
19. The clause that specifies that an elementary item is to be produced only once at the first occurrence of an item after any control break or page break is the
    a. JUSTIFIED clause.
    b. SOURCE clause.
    c. GROUP INDICATE clause.
    d. CONTROL clause.
20. The technique that in SUM counter manipulation adds lower level control SUM counters to higher control levels during control break processing is
    a. subtotalling.
    b. rolling forward.
    c. cross footing.
    d. control footing.
21. A SUM clause
    a. does not establish a SUM counter.
    b. may be in more than one elementary report entry.
    c. cannot be altered by Procedure Division statements.
    d. names the data items to be summed.
22. The INITIATE statement
    a. causes the Report Writer to begin the processing of a report.
    b. resets all data-name entries that contain SUM counters in the report.
    c. sets the PAGE-COUNTER to 1.
    d. All of the above.
23. The GENERATE statement
    a. may reference a TYPE CONTROL FOOTING report group.
    b. does not recognize any control breaks.
    c. may reference a report-name.
    d. All of the above.

24. The TERMINATE statement
    a. closes the file with which the report is associated.
    b. causes the Report Writer to complete processing of the specified reports.
    c. will not provide last page footing and report footing groups associated with the report.
    d. None of the above.
25. The SUPPRESS statement
    a. may appear in a TYPE DETAIL statement.
    b. inhibits presentation of all report groups.
    c. must be executed each time the presentation of the report group is to be inhibited.
    d. All of the above.
26. The USE BEFORE REPORTING declarative
    a. specifies Procedure Division statements that are to be executed before a report group named.
    b. must immediately follow a section header in the Declaratives Section.
    c. must be followed by a period and a space.
    d. All of the above.
27. The input fields of a record are

Positions	**Field**	
5-8	DEPTNO	
9-14	SLSNUM	
20-25	AMOUNT	XXXX.XX

    For every salesman accumulate total sales.
    For every sales department accumulate total sales.
    Accumulate a final total.
    Assume that all fields were properly defined.
    REQUIRED: 1. Code the CONTROL clause for the aforementioned.
    2. Using the rolling forward technique, code the necessary entries to accumulate the proper totals based on the aforementioned information.
28. Write a declarative to suppress printing of FINAL control group if the code is "A."
29. Using the following information, write the necessary entries in the Data and Procedure Divisions to prepare the following report based on the following information. Use the crossfooting technique.
    The records of the Himargin Department Stores contain the following information:

	STORE-NO.	DEPARTMENT-NO.	DEPARTMENT 1st SALE	2nd SALE	3rd SALE	4th SALE	5th SALE
1st STORE	123	1359	001293	029160	200015	025037	010035
		1530	013000	000250	304000	002519	023410
		2400	002459	002947	012044	002830	016625
		3594	000248	001590	002300	000122	003520
2nd STORE	530	1280	002510	020050	001257	011120	025910
		3320	100245	001945	027933	123450	300020
		7730	002460	001024	003913	029433	002930
		9250	002200	001050	023910	000200	000391

A report is to be prepared showing the total sales for each day as well as the total for each department for each day and a total for all days for all stores.

There are two stores and there are four departments in each store.

Input

Columns	Field	
1-3	Store number	
4-7	Department number	
8-13	Sales 1st day	XXXX.XX
14-19	Sales 2nd day	XXXX.XX
20-25	Sales 3rd day	XXXX.XX
26-31	Sales 4th day	XXXX.XX
32-37	Sales 5th day	XXXX.XX

Calculations

Add all sales for each department. Determine totals for each store by day and totals.

Required

Arrays are to be set up for the five days' sales. Arrays are to be set up for the two stores. Department totals are to be added to store totals. The records are in sequence by department number within store number.

Code the necessary COBOL entries to accomplish the aforementioned. Use the Report Writer feature.

Output is as follows.

STORE NUMBER	DEPARTMENT NUMBER	SALES 1ST DAY	SALES 2ND DAY	SALES 3RD DAY	SALES 4TH DAY	SALES 5TH DAY	TOTAL SALES
123	1359	12.93	291.60	2,000.15	250.37	111.35	2,666.40
123	1530	130.00	2.50	3,040.00	25.19	234.10	3,431.79
123	2400	24.59	29.47	120.44	28.30	166.25	369.05
123	3594	2.48	15.90	23.00	1.22	35.20	77.80
	STORE TOTALS	170.00	339.47	5,183.59	305.08	546.90	6,545.04
530	1280	25.10	200.50	12.57	111.20	259.10	608.47
530	3320	1,002.45	19.45	279.33	1,234.50	3,000.20	5,535.93
530	7730	24.60	10.24	39.13	294.33	29.30	397.60
530	9250	22.00	10.50	239.10	22.00	3.91	297.51
	STORE TOTALS	1,074.15	240.69	570.13	1,662.03	3,292.51	6,839.51

Problem 1

Prepare the Report Writer entries for the following:

REPORT

```
                    CUSTOMER NAME AND ADDRESS LISTING

ARTSON           H V     123 WOOD LANE        DE MONES, CAL.       1
BELEBOR          G       784 GRAND DRIVE      SEMMDALE, VA.        2
BREIGHT          H M     NEW SPRING BLVD      HEER, MD.            3
CALIPHANDER      A C     STRETCH BLVD         MITTAK, ALA.         4
DIERR            D       1 MADISON ROAD       HEAROLD, N.M.        5
```

PRINTER SPACING CHART

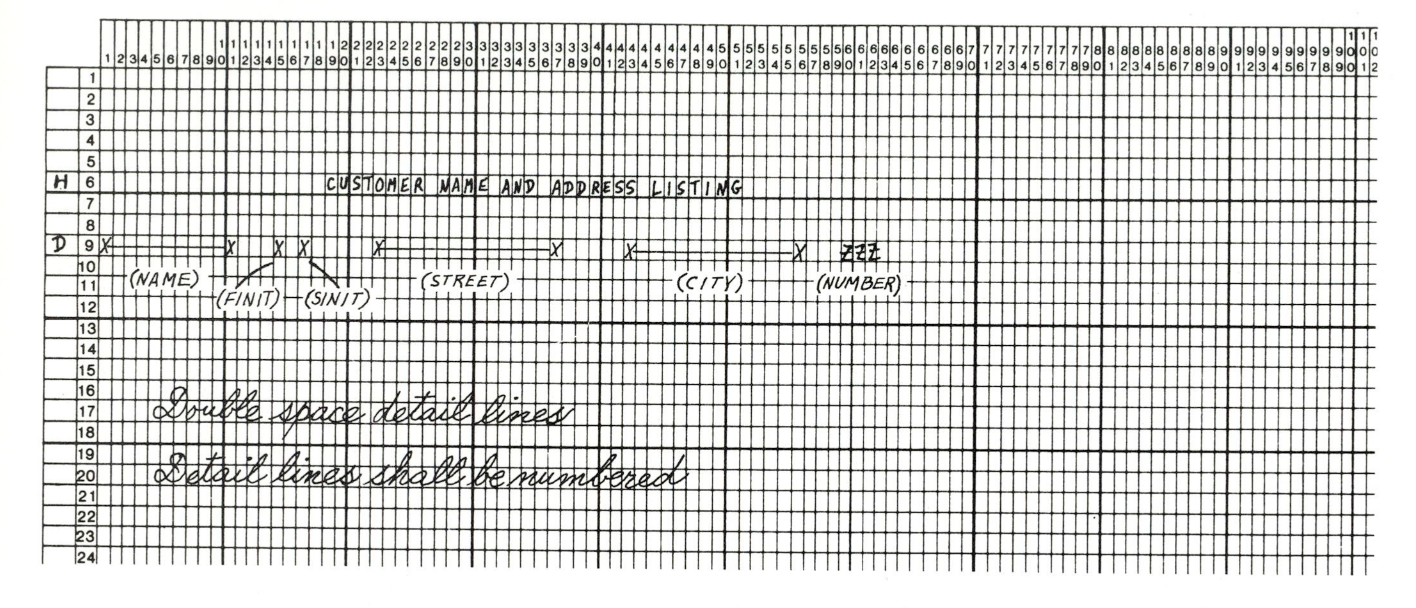

INPUT RECORD

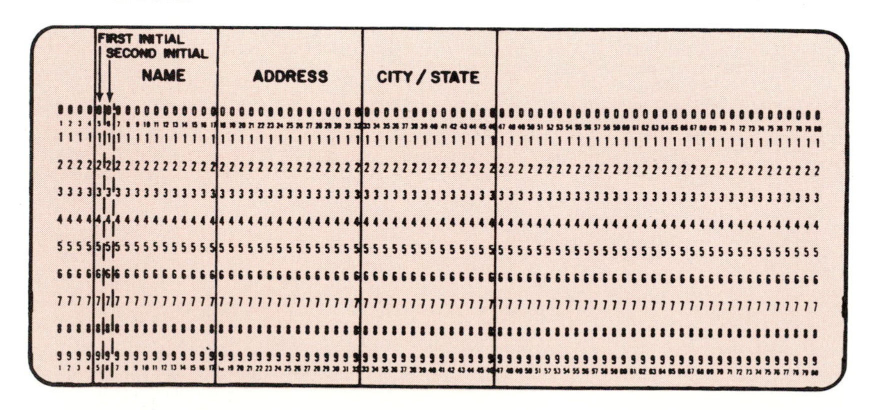

Problem 2

Prepare the Report Writer entries for the following:

Input

Field	Positions
Date	1-6
Location	7-9
Department	10-13
Employee Number	14-17
Employee Name	18-32 First initial, second initial, name.
Group Insurance	33-36 XX.XX
Union Dues	37-40 XX.XX
Bonds	41-44 XX.XX
United Fund	45-48 XX.XX
Pension	49-52 XX.XX
Major Medical	53-56 XX.XX
Other	57-60 XX.XX
Total Deductions	61-65 XXX.XX

Calculations

Records are in sequence by departments within locations.

a. Totals by departments for Group Insurance, Union Dues, Bonds, United Fund, Pension, Major Medical, Other and Total Deductions.

b. Totals by location for Group Insurance, Union Dues, Bonds, United Fund, Pension, Major Medical, Other and Total Deductions.

Output

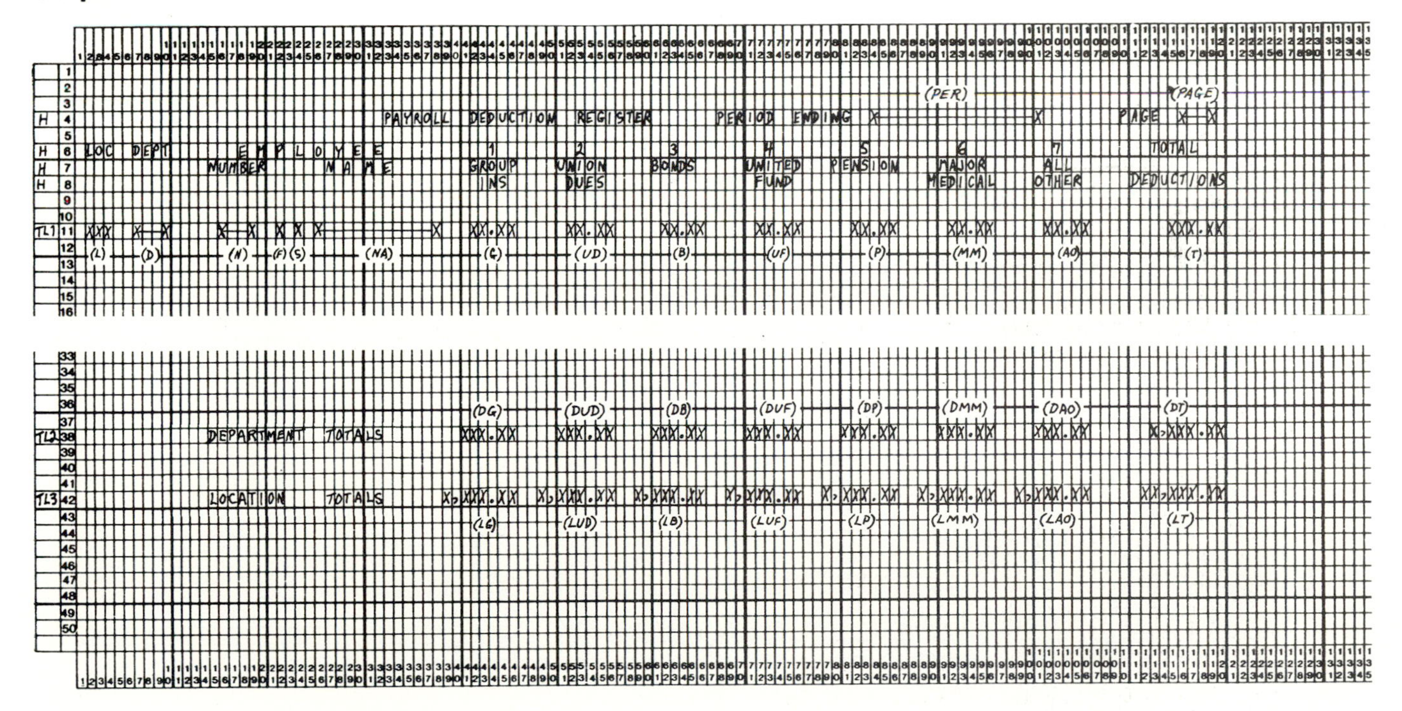

Problem 3

Prepare the Report Writer entries for the following:

Input

Field	Positions	
Date	1-6	xx/xx/xx
Customer Number	7-10	xxx.xx
Old Balance	11-15	xxx.xx
Purchases	16-20	xxx.xx
Payments	21-25	xxx.xx
New Balance	26-30	xxx.xx

Calculations:

a. Totals by customer number for Old Balance, Purchases, Payments and New Balance.
b. Final totals for Old Balance, Purchases, Payments and New Balance.

Chart

```
H    6            CHARGE ACCOUNT STATUS REPORT                     P. XXXX
                                                                   (PAGE)
H    8  CUSTOMER         OLD          PURCHASES      PAYMENTS         NEW
H    9   NUMBER        BALANCE                                      BALANCE
TL1 12   X  X         $XXX.XXCR      $XXX.XX        $XXX.XX         $XXX.XXCR
        (CUST)        (OLDBAL)       (PURC)         (PAY)           (NEWBAL)
TLR 16  TOTALS      $XX,XXX.XX     $XX,XXX.XX     $XX,XXX.XX     $XX,XXX.XX
                      (TOLD)         (TPUR)         (TPAY)          (TNEW)
```

Problem 4

Using the following information, write a program to accomplish the following:

a. Read the input record.
b. Calculate the net pay by multiplying hours by the rate.
c. Sort records into Department and Serial Number sequence.
d. Punch card using the same format as the input record, including net pay.
e. Write report per format using the Report Writer Feature.

Input

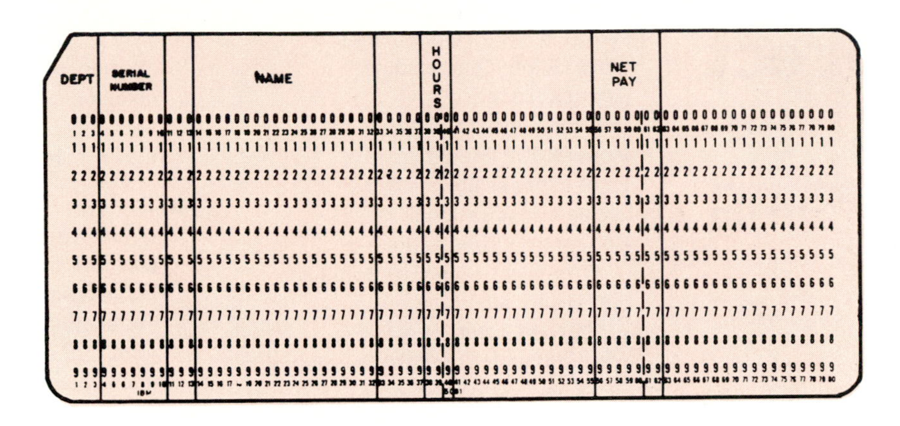

Report

DEPT.	SERIAL NUMBER	N A M E		HOURS WORKED	NET PAY
1	1234560	SMITH	JW	40.1	285.15
1	1892750	JONES	RA	39.6	152.16
1	8929016	MAUS	JB	62.5	182.55
					619.86
2	0238648	GOLDMAN	H	31.7	100.25
2	0333367	WOLFE	DJ	9.5	26.60
					126.85

Chart

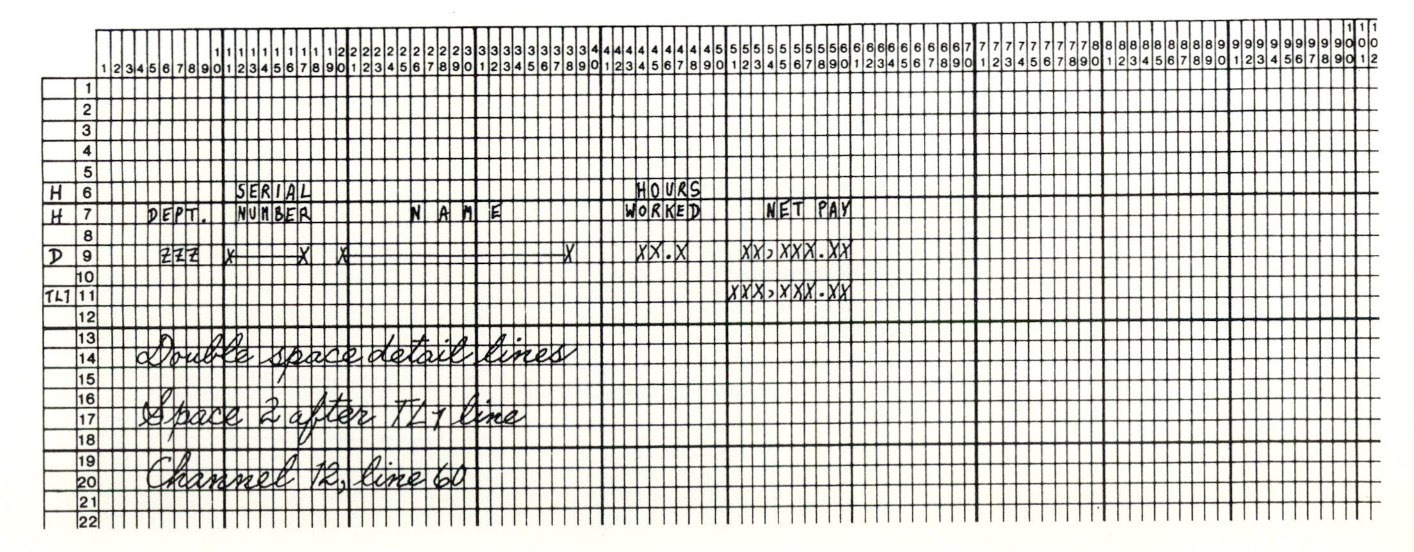

13

Direct Access Processing

Chapter Outline

Chapter Objectives

The learning objectives of this chapter are:

1. To describe the operation of the different file organizations stored on mass storage devices such as standard sequential organization, indexed sequential organization, and relative organization.
2. To explain the various terms used in direct access processing.
3. To discuss the various factors involved in determining the correct file organization for a particular file on a mass storage device.
4. To describe the different methods of file organization and how to program them for the creation of a file; accessing records sequentially and randomly or both from these files; updating the files, and replacing records in the files.

Introduction

"Inline" processing denotes the ability of the data processing system to process data as soon as it becomes available. This implies that the input data does not have to be sorted in any manner, manipulated, or edited before it can be entered into a system, whether the input consists of transactions of a single application or of many applications.

Direct-access mass storage devices have made inline processing feasible for many applications. (See figure 13.1.) While sorting transactions are still advantageous before certain processing runs, in most instances the necessity for presorting has been eliminated. The ability to process data inline provides solutions to problems that heretofore were thought impractical.

Direct-access mass storage enables the programmer to maintain current records of diversified applications and to process nonsequential and intermixed data for multiple application areas. The term *direct-access* implies access at random by multiple users of data (files, programs, subroutines, programming aids) involving mass storage devices. These storage devices differ in physical appearance, capacity, and speed, but are functionally similar in terms of data recording, checking, and programming. The direct-access devices used for mass memory storage are disk storage, drum, and data cells. (See figure 13.2.) The disk storage is the most popular of the mass storage devices in use today. (See figure 13.3.)

Terminology

Direct-access terminology and concepts are a prerequisite to an understanding of programming using direct-access devices. (See figure 13.4.)

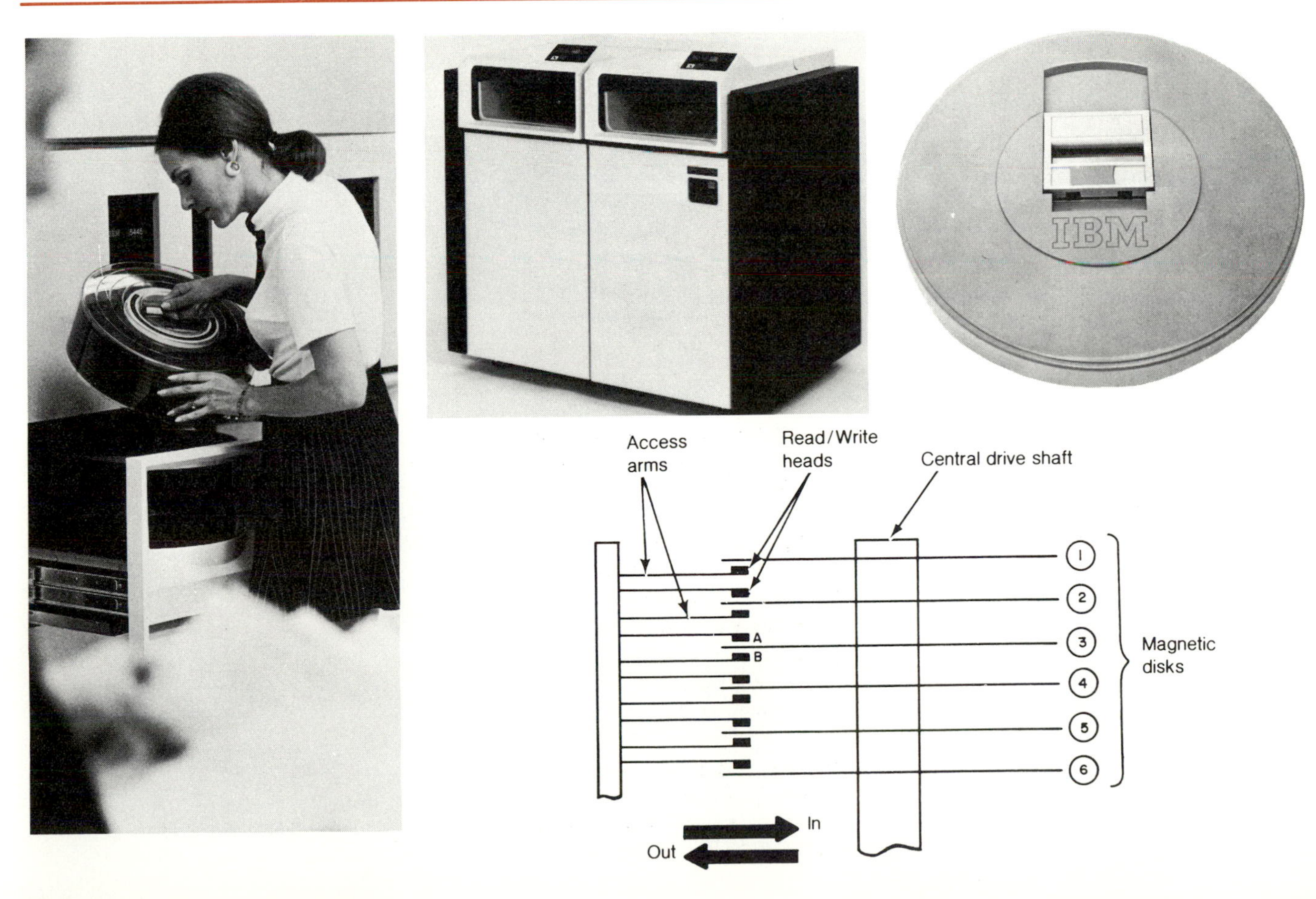

Figure 13.1 Mass storage devices.

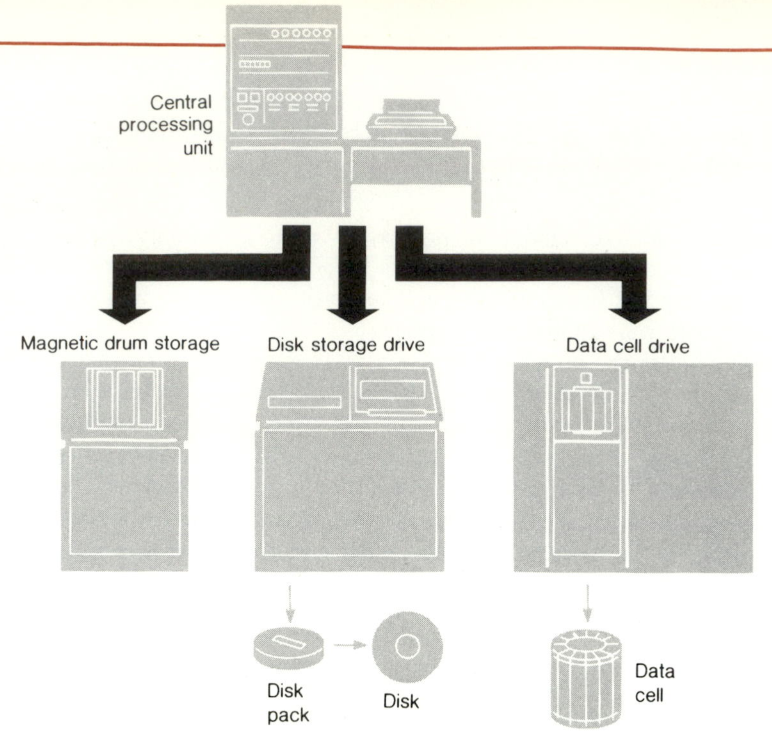

Figure 13.2 Mass storage devices—types.

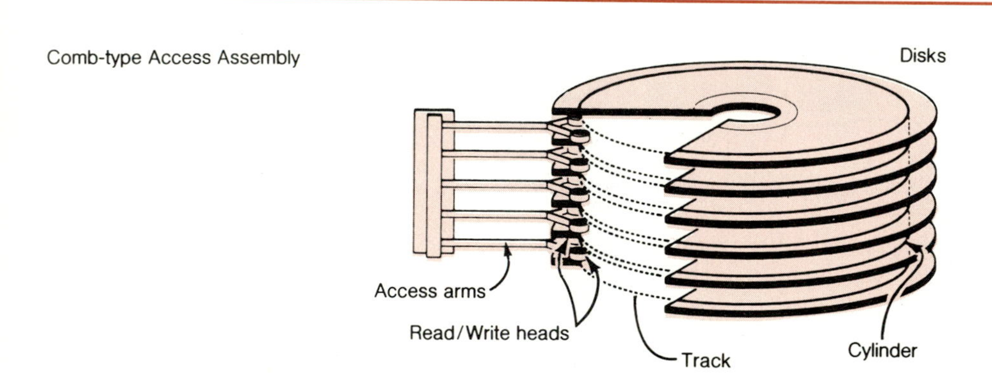

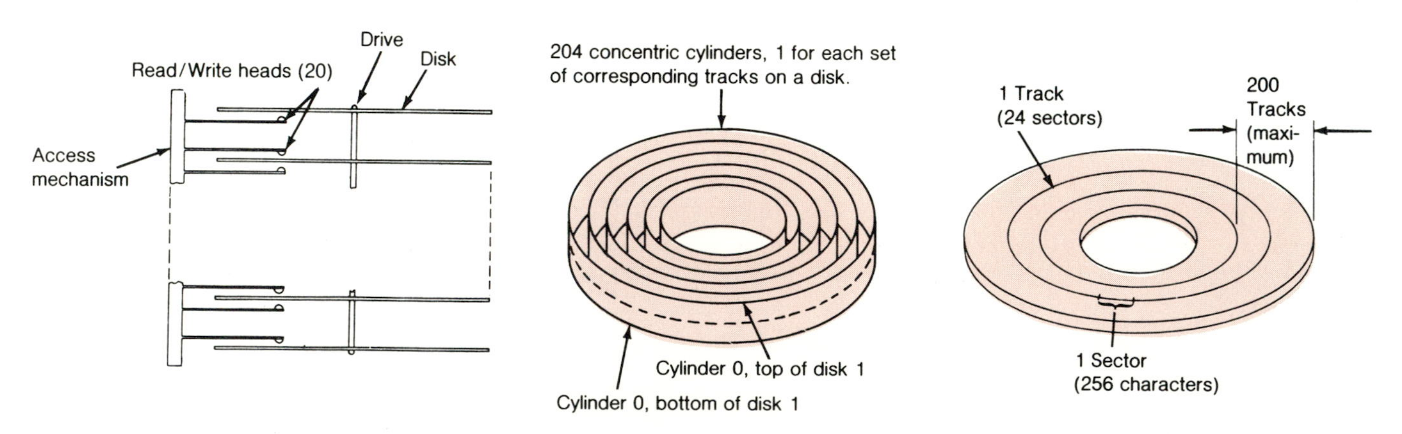

Figure 13.3 Disk storage device.

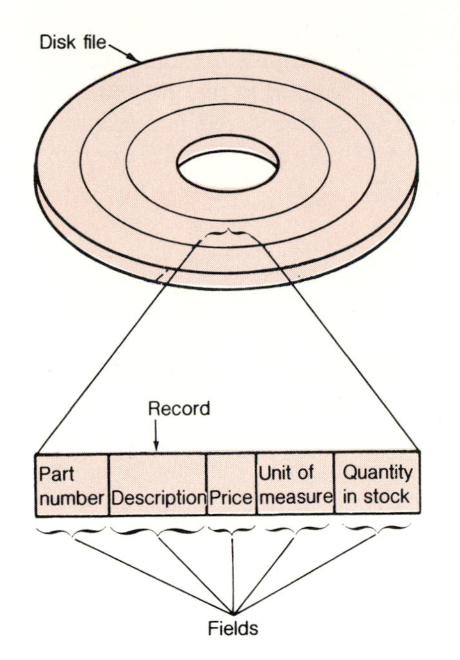

Field: An area on a record reserved and used for a particular item of information.

Record: A group of related fields.

File: A group of related records.

Figure 13.4 File, record, field example.

Direct-Access Storage Device (DASD)

A direct-access storage device (DASD) is one in which each physical record has a discrete location and a unique address. Thus records can be stored on a DASD in such a way that the location of any one record can be determined without extensive searching. Records can be accessed directly as well as serially.

File

The term file *can mean a physical unit (a DASD, for instance), or an organized collection of related information.* In this text, the latter definition applies. An inventory, for example, contains all of the data concerning a particular inventory. It may occupy several physical units, or part of one physical unit. The term *data set* is often used instead of *file* to describe an organized collection of related information.

Record

The term *record* can also mean a physical unit or a logical unit. *A logical record may be defined as a collection of data related to a common identifier.* An inventory file, for example, would contain a record (logical record) for each part number in the inventory. A physical record consists of one or more logical records. The term *block* is equivalent to the term *physical record.* On a DASD, certain "nondata" information required by the control unit of the device is recorded in the same record area as the physical record. This nondata information and the physical record may be referred to as a whole with the term *data record.*

Key

Each logical record contains a control field(s) or key(s) that uniquely identify it. The key(s) of the inventory record, for example, would probably be the part numbers.

Introduction to File Organization

File organization refers to the logical relationships among the various records that constitute the file, particularly with respect to the means of identification and access to any specific record.

Records in a file must be logically organized so that they can be efficiently retrieved for processing. This chapter discusses some factors to be considered in selecting a method of organization and the COBOL processing of files under the various methods of file organization.

Data File Characteristics

The inherent characteristics of the file must be considered in selecting an efficient method of organization.

Volatility. This term refers to the addition and deletion of records from a file. A static file is one that has a low percentage of additions and deletions, while a volatile file is one that has a high rate of additions and deletions. No matter how the file is organized, additions and deletions are of significant concern, but, with some methods of organization, they can be handled more efficiently than others.

Activity. The percentage of activity is one of the factors to be considered when discussing data file characteristics. If a low percentage of the records are to be processed on a run, the file should probably be organized in such a way that any record can be quickly located without having to look at all the records in the file.

How to distribute the activity should also be a consideration. With some methods of organization, some records can be located more quickly than others. The records processed most frequently should certainly be the ones that can be located most quickly.

The amount of activity also makes a difference. An active file (that is, one which is frequently referred to) must be organized very carefully, since the time involved in locating records may amount to an appreciable period of time. At the other extreme, an inactive file may be referred to so infrequently that the time required to locate records is immaterial.

Size. A file so large that it cannot all be online (available to the system) at one time must be organized and processed in certain ways. A file may be so small that the method of organization makes little difference, since the time required to process it is very short no matter how it is organized.

The growth potential of the file is also a consideration. Files are usually planned on the basis of their anticipated growth over a period of time. Initial planning must also consider how growth that exceeds this size will eventually be handled.

Processing Characteristics

The distinction between the organization of a master file and the order of input detail records processed against that file is important. In *sequential processing,* the input transactions are grouped together, sorted into the same sequence as the master file, and the resulting batch is then processed against the master file. Sequential processing is the most efficient means of processing when tape and cards are used to store master files. Direct-access storage devices are also very efficient sequential processors, especially when the percentage of activity against the master file is high.

Non-sequential processing (random processing) is the processing of detail transactions against a master file in whatever order they occur. With direct access devices, non-sequential processing can be very efficient, since a file can be organized in such a way that any record can be quickly located.

It is possible, on a run, to process the input transactions against more than one file. This saves setup and sorting time. It may also minimize control problems, since the transactions are handled less frequently.

The use of Direct Access Storage Devices (DASD) to store a master file makes it possible to choose the processing method that will suit the application best. Thus some applications can be processed sequentially, while those in which the time required to sort, or where the delay associated with batching is undesirable can be processed nonsequentially. Real savings in overall job time can only be made by combining runs in which each input affects several master files; the details can be processed sequentially against a primary file and non-sequentially against the secondary files, all in a single run. This is the basis of inline processing. (See figures 13.5 and 13.6.)

Sequential Processing

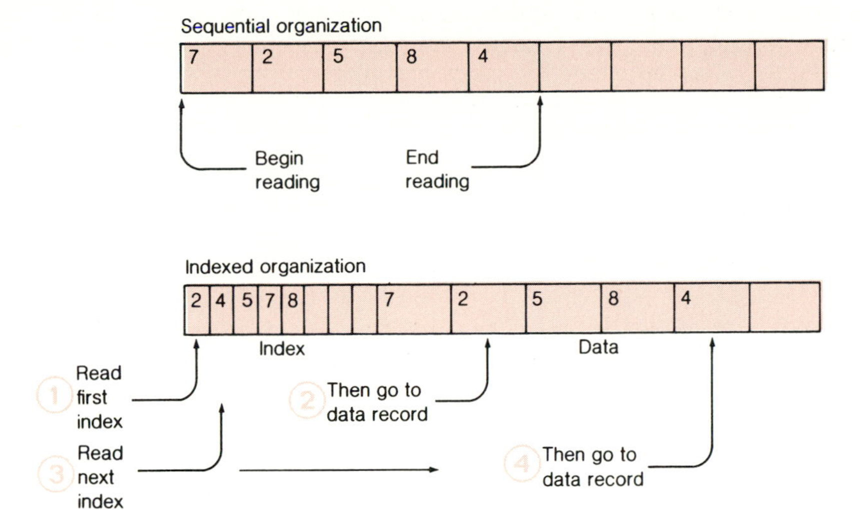

Random Processing

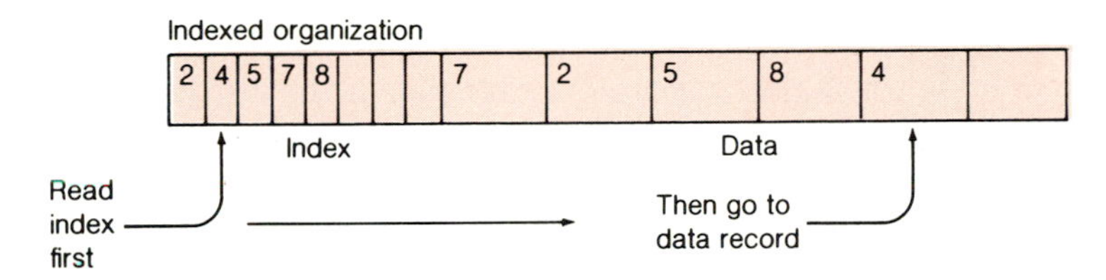

Figure 13.5 Sequential and random processing examples.

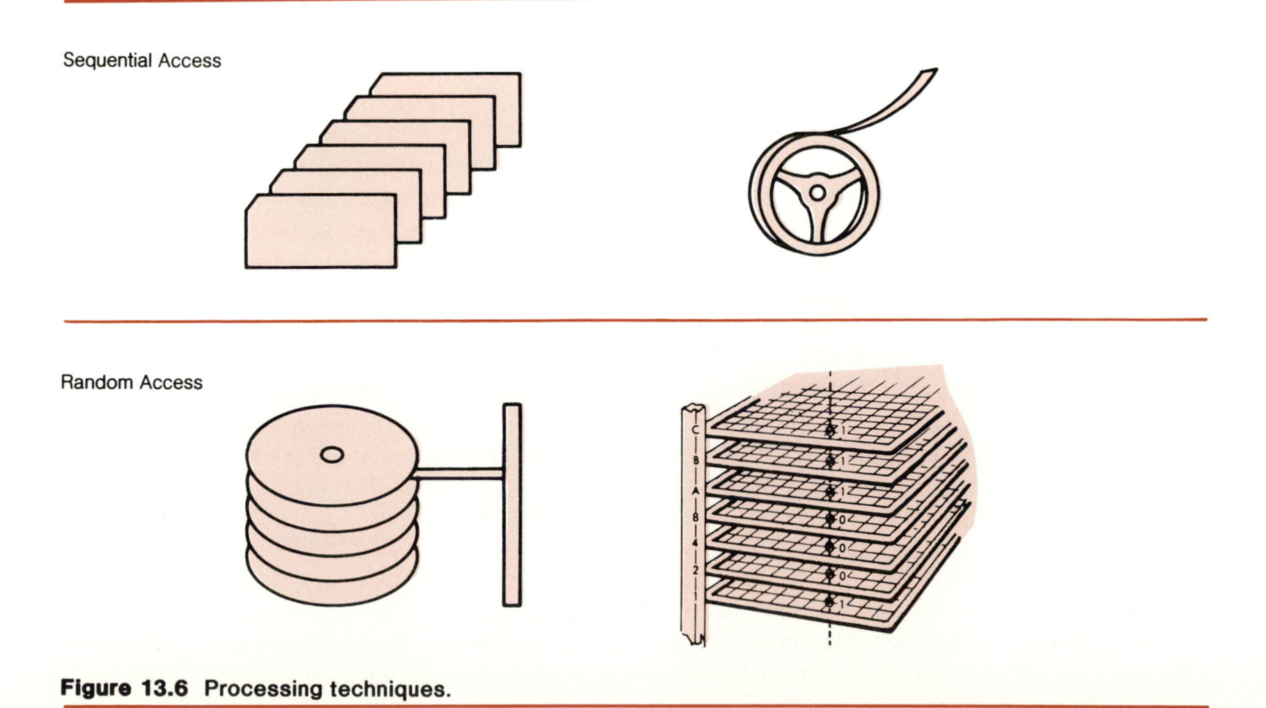

Figure 13.6 Processing techniques.

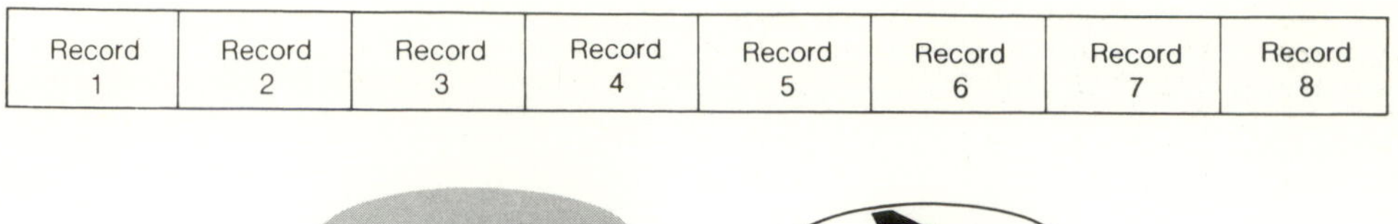

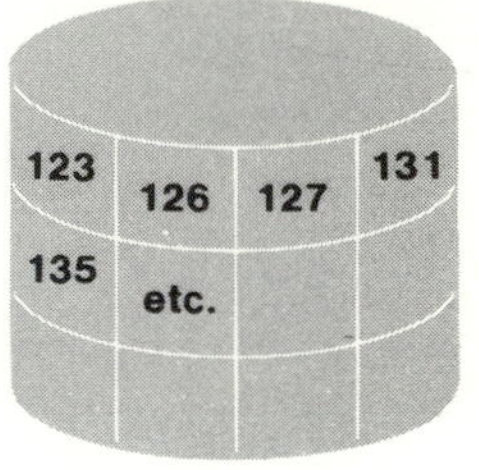

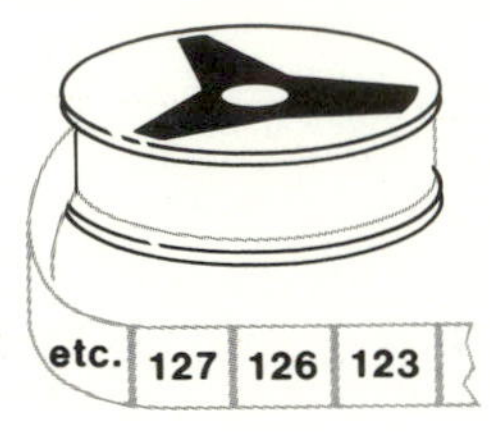

Figure 13.7 Sequentially organized data file—example.

Methods of Organization

The most popular methods of organization for direct access devices are briefly described as follows:

Sequential Organization. In a sequential file, records are organized solely on the basis of their successive physical locations in the file. The records are generally, but not necessarily, in sequence according to their keys (control numbers) as well as in physical sequence. The records are usually read or updated in the same order in which they appear. For example, the hundredth record is usually read only after the first ninety-nine have been read.

Individual records cannot be located quickly. Records usually cannot be deleted or added unless the entire file is rewritten. This organization is generally used when most records are processed each time the file is read. (See figure 13.7.)

Indexed Sequential Organization. An indexed sequential file is similar to a sequential file in that rapid sequential processing is possible. Indexed sequential organization, however, by references to indexes associated with the file, makes it also possible to quickly locate individual records for nonsequential processing. Moreover, a separate area of the file is set aside for additions; this obviates a rewrite of the entire file, a process that would usually be necessary when adding records to a sequential file. Although the added records are not physically in key sequence, the indexes are referred to in order to retrieve the added records in key sequence, thus making rapid sequential processing possible.

In this method of organization, the programming system has control over the location of the individual records. The programmer, therefore, need do very little I/O programming; the programming system does almost all of it, since the characteristics of the file are known. (See figure 13.8.)

Relative Organization. A relative file consists of records that are identified by relative record numbers. The file may be thought of as composed of a serial string of areas, each capable of holding a logical record. Each of these areas is denominated by a relative record number. Records are stored and retrieved based on this number. For example, the tenth record is the one addressed by relative record number ten and is in the tenth record area, whether or not records have been written in the first through the ninth record areas.

Relative file organization is permitted only on random mass storage devices. (See figure 13.9.)

In the succeeding sections, each data file organization will be discussed in detail as to the use of these file organizations, as well as the COBOL programming statements required to create, update, add and delete records to the various data file organizations.

Description of File Input-Output

File input-output provides the capability for transferring data from an external recording device into computer memory (input), or from memory to an external recording device (output) with a minimum of concern for the physical characteristics of the recording device or the processes required to complete the transfer.

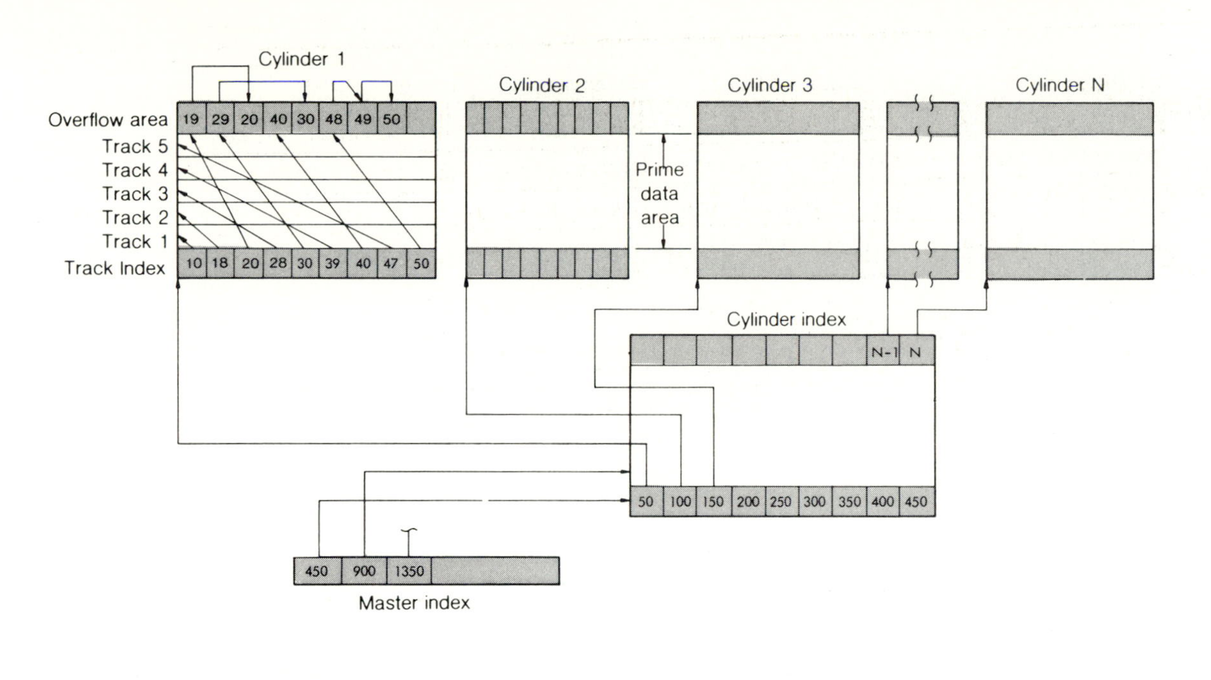

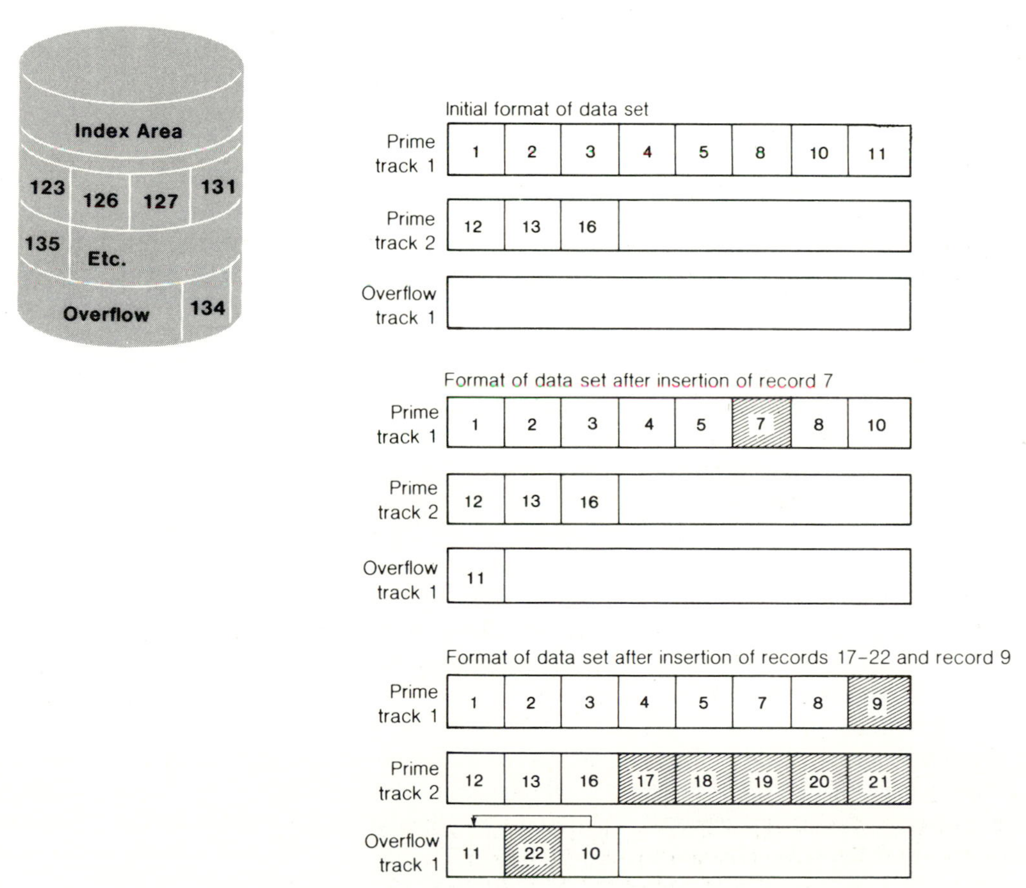

Figure 13.8 **Indexed organized data file—examples.**

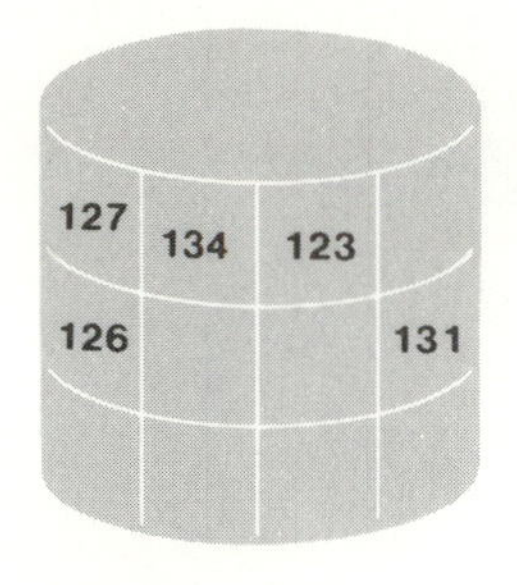

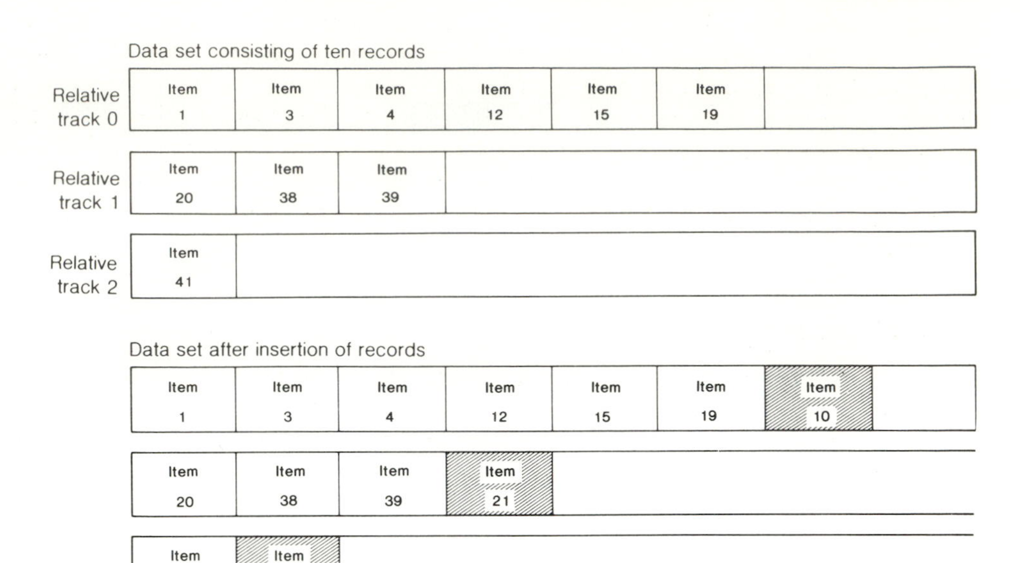

Figure 13.9 Relative organized data file—example.

Input-Output Organization

Sequential files are organized such that each record in the file except the first has a unique predecessor record, and each record except the last has a unique successor record. These predecessor-successor relationships are established by the order of WRITE statements when the file is created. Once established, the predecessor-successor relationships do not change except in the case where records are added to the end of the file.

A file whose organization is indexed is a random mass storage file in which data records can be accessed by the value of a key(s). A record description may include one or more key data items, each of which is associated with an index. Each index provides a logical path to the data records according to the contents of a data item within each record which is the record key for that index.

The data item named in the RECORD KEY clause of the file control entry for a file is the prime record key for that file. For the purposes of inserting, updating, and deleting records in a file, each record is identified solely by the value of its prime record key. This value must, therefore, be unique and must not be changed when the record is updated.

A data item named in the ALTERNATE RECORD KEY clause of the file control entry for a file is an alternate record key for that file. The value of an alternate key may be non-unique if the DUPLICATES phrase is specified for it. These keys provide alternate access paths for the retrieval of records from the file.

Relative file organization is permitted only on random mass storage devices. A relative file consists of records which are identified by relative record numbers. The file may be thought of as composed of a serial string of areas, each capable of holding a logical record. Each of these areas is denominated by a relative record number. Records are stored and retrieved based on this number. For example, the tenth record is the one addressed by relative record number 10 and is in the tenth record area, whether or not records have been written in the first through the ninth record areas.

Access Modes

In the sequential-access mode, the sequence in which records are accessed is the order in which the records were originally written (if organization is sequential), the ascending order of the record key values (if organization is indexed) or the ascending order of the relative record numbers of all records within the file (if organization is relative). The order of records retrieval within a set of records having duplicate record key values is the order in which the records were written into the set.

In the random-access mode, the sequence in which records are accessed is controlled by the programmer. The desired record is accessed by placing the value of its record key(s) in a record key data item(s) (if organization is indexed), or by placing its relative record number in a relative key data item (if organization is relative). Random access is not allowed with sequential organization.

In the dynamic-access mode, the programmer may alternate from sequential access to random access using appropriate forms of input-output statements. Dynamic access is not allowed with sequential organization.

Current Record Pointer

The current record pointer is a conceptual entity used in this text to facilitate specification of the next record to be accessed within a given file. The concept of current record pointer has no meaning for a file opened in the output mode. The setting of the current record pointer is affected only by the OPEN, READ, and START statements.

Input-Output Status

If the FILE STATUS clause is specified in a file control entry, a value is placed into the specified two-character data item during the execution of a CLOSE, DELETE, OPEN, READ, REWRITE, START, or WRITE statement and before any applicable USE procedure is executed, to indicate to the COBOL program the status of that input-output operation.

Status Key 1

The leftmost character position of the FILE STATUS data item is known as status key 1 and is set to indicate one of the following conditions upon completion of the input-output operations:

0—indicates Successful Completion

1—indicates At End

2—indicates Invalid Key

3—indicates Permanent Error

9—indicates Implementor Defined

Status Key 2

The rightmost character position of the FILE STATUS data item is known as status key 2 and is used to further describe the results of the input-output operations.

(For further information concerning the status keys, consult the appendix).

Error Processing

When an error occurs during an input-output operation, the programmer can exercise a degree of control by specifying USE procedures that check the FILE STATUS data item associated with the file, and modify it in order to communicate to the input-output control system the desired disposition of the error.

Data File Organizations

Data file organizations refers to the physical arrangement of data records within a file. To give the programmer maximum flexibility in reading and writing data sets from mass storage devices, the following methods of data organization are most commonly used for disk operations: Sequential (Standard), Indexed Sequential, and Relative.

Record 1	Record 2	Record 3	Record 4	Record 5

Figure 13.10 Sequentially organized data set.

Sequential Organization

In a sequential file, records are organized solely on the basis of their successive physical locations in the file. (See figure 13.10.) The records are written one after the other—track by track, cylinder by cylinder—at successively higher locations. The records are usually, but not necessarily, in sequence according to their keys (control numbers). The records are usually read or updated in the same sequence as that in which they appear. (See figure 13.11.)

A disk file can be organized and processed like a card file. Such a disk file is called a *sequential file*. The sequence of the file can be determined by control fields, such as employee number or a customer number, or the records may be in no particular sequence. Consecutive processing means that the records are processed one after another in the physical order in which they occur.

An example of a sequential file is an employee master file arranged in employee number order and containing information about each employee. When this file is used for processing, such as for payroll checks, the records are processed consecutively. The lowest employee number is processed first and so on until the last record, the highest employee number, is processed.

A sequential file may span multiple disk volumes. (A volume refers to one disk pack. A multivolume file is a file that is contained on more than one disk pack.) A multivolume file, however, affects the processing of the file.

When the sequential file is created, the records are written onto a disk for the first time. The records in a sequential file are placed on the disk consecutively, that is, they are written on the disk in the order in which they are read. All tracks in one cylinder are filled first, then all tracks in the next cylinder, and so on until the whole file is placed on the disk.

In the example in figure 13.11, each record is 128 positions (bytes) long. Since each track contains 6,144 bytes of data, 48 records can be written on each track and 96 records can be written on each cylinder. The numbers on the tracks in the example correspond to the number and position of each record.

Updating a sequential file that is located on a mass storage device is more efficient than updating a file located on a magnetic tape or punched cards. After the file is opened, the record can be read,

Writing Records on a Disk

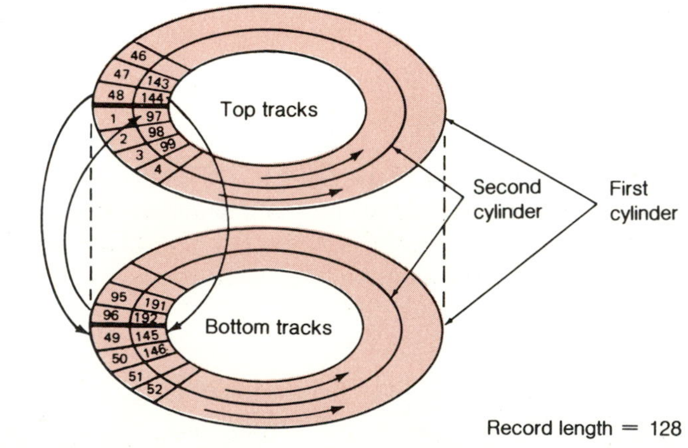

Top tracks

Bottom tracks

Second cylinder

First cylinder

Record length = 128

Figure 13.11 Sequential file organization—example.

```
SELECT  [ OPTIONAL ]  file-name

    ASSIGN TO implementor-name-1  [ , implementor-name-2 ] . . .

  [ ; RESERVE integer-1  [ AREA  ] ]
                         [ AREAS ]

  [ ; ORGANIZATION IS SEQUENTIAL ]

  [ ; ACCESS MODE IS SEQUENTIAL ]

  [ ; FILE STATUS IS data-name-1 ]
```

Figure 13.12 Format—File Control entry—sequential files.

updated, and written back in the same location in the file without creating a new file. Thus the file can be used for both input and output activities without the necessity for opening and closing files between operations.

A file on a mass storage device may have the same standard sequential data file organization as any of the unit record or magnetic tape files. The mass storage file, however, may be differently organized so that any record may be accessed merely by specifying the key(s) or unique field that tells the system where the desired record is located. This differs from standard sequential organization in that the desired records can be accessed at random without accessing all previous records.

Nonsequential processing of a sequential file is, at best, very inefficient. If it is done infrequently, the time required to locate the records may not matter. There are several ways to program nonsequential processing, with significant differences in the time required. The slowest way is to read the records sequentially until the desired one is located. On the average, half of the files would have to be read. A sequential search takes less time if the records are formatted with keys. The search is done in Search Key High or Equal at the speed of one revolution per track. When the search condition is satisfied, the corresponding record is read.

Additions and deletions require a complete rewrite of a sequential file. Therefore, sequential organization is used in direct-access storage devices primarily for tables and intermediate storage rather than for master files. Its use is recommended for master files, only, if there is a high percentage of activity and if virtually all processing is sequential.

Creating a Standard Sequential Disk File

ENVIRONMENT DIVISION. In order to specify that a file will be on a mass storage device, certain entries must be specified in the File Control Entry. The format for File Control Entry is shown in figure 13.12.

The file control entry names a file and may specify other file-related information.

Rules Governing the Use of the File Control Entry.

1. The SELECT clause must be specified first in the file control entry. The clauses that follow the SELECT clause may appear in any order. Each file described in the Data Division must be named once and only once as the file-name in the FILE-CONTROL paragraph. Each file specified in the file control entry must have a file description entry in the Data Division.
2. The ASSIGN clause specifies the association of the file referenced by file-name to a storage medium.
3. The RESERVE clause allows the programmer to specify the number of input-output areas allocated. If the RESERVE clause is specified, the number of input-output areas allocated is equal to the value of integer-1. If the RESERVE clause is not specified, the number of input-output areas allocated is specified by the implementor.

4. The ORGANIZATION clause specifies the logical structure of a file. The file organization is established at the time a file is created and cannot subsequently be changed. When the ORGANIZATION clause is not specified, the ORGANIZATION IS SEQUENTIAL clause is implied.
5. Records in the file are accessed in the sequence dictated by the organization. This sequence is specified by predecessor-successor record relationships established by the execution of WRITE statements when the file is created or extended. If the ACCESS MODE is not specified, the ACCESS MODE IS SEQUENTIAL clause is implied.
6. When the FILE STATUS clause is specified, a value will be moved by the operating system into the data item specified by data-name-1 after the execution of every statement that references the file either explicitly or implicitly. This value indicates the status of the statement's execution. Data-name-1 must be defined in the Data Division as a two-character data item of the category alphanumeric and must not be defined in the File Section. Data-name-1 may be qualified.

Data Division. The Data Division entries required for sequential disk files are as follows:

FD. File-name of the disk file.

Block Contains. This clause is required if records are blocked.

Label Records Are Standard. All files in direct-access devices must have standard labels.

Example Environment and Data division entries for a sequential disk file follow:

```
ENVIRONMENT DIVISION.
        .
        .
        .
INPUT-OUTPUT SECTION.
FILE-CONTROL.
    SELECT DISK-SEQ
        ASSIGN TO . . .
        ORGANIZATION IS SEQUENTIAL
        ACCESS MODE IS SEQUENTIAL.
        .
        .
        .
DATA DIVISION.
FILE SECTION.
FD  DISK-SEQ
    BLOCK CONTAINS 4 RECORDS
    LABEL RECORDS ARE STANDARD.
```

Procedure Division. To create a sequential file, the programmer must use an output file and a WRITE statement.

WRITE Statement. The WRITE statement without the ADVANCING option is used when a sequential file is being created on disk. (The format and an example for the WRITE statement are shown in figure 13.13.) The amount of space reserved for a file is usually specified in job-control cards according to the operating system in a particular installation.

The following operations may be performed on sequential files:

WRITE Accesses the space immediately following that area into which the previous logical record was written and places the contents of the specified record in that place. The file must be open for output or extension.

WRITE record-name [FROM identifier-1]

WRITE DISKFACT FROM INRECORD.

Figure 13.13 Format—WRITE statement and example.

READ Accesses the next logical record on the file and makes the contents of that record available in the file record area. If no "next" record exists, an AT END condition exists. The file must be open for input or input-output.

REWRITE Replaces the logical record accessed by the previous input-output operation (which must have been a successful READ) with the contents of the specified record. The logical record being replaced must be equal in size to the record specified in the REWRITE statement (the replacement record). REWRITE can be executed only on files that are allocated to random mass storage and are open for input-output.

REWRITE Statement

The REWRITE statement logically replaces a record existing in a mass storage file. (The format and an example of the REWRITE statement are shown in figure 13.14.)

Rules Governing the Use of the REWRITE Statement

1. The file associated with record-name must be a mass storage file and must be open in the I-O mode at the time of execution of this statement.
2. The last input-output statement executed for the associated file prior to the execution of the REWRITE statement must have been a successfully executed READ statement. The Mass Storage Control System (MSCS) logically replaces the record that was accessed by the READ statement.
3. The number of character positions in the record referenced by record-name must be equal to the number of character positions in the record being replaced.
4. The logical record released by a successful execution of the REWRITE statement is no longer available in the record area.
5. The execution of a REWRITE statement with the FROM phrase is equivalent to the execution of:

   MOVE identifier TO record-name

   followed by the execution of the same REWRITE statement without the FROM phrase. The contents of the record area prior to the execution of the implicit MOVE statement have no effect on the execution of the REWRITE statement.
6. The current record pointer is not affected by the execution of a REWRITE statement.
7. The execution of the REWRITE statement causes the value of a FILE STATUS data item, if any, associated with the file to be updated.

REWRITE record-name [FROM identifier]

REWRITE STUDENT-RECORD
FROM WORK-RECORD.

In the instruction, WORK-RECORD is moved to STUDENT-RECORD, and then placed back into the file.

Figure 13.14 Format—REWRITE statement and example.

The following is a program for creating a sequential disk file from a tape file:

```
ENVIRONMENT DIVISION.
      .
      .
      .
INPUT-OUTPUT SECTION.
FILE-CONTROL.
    SELECT DISKFILE.
        ASSIGN TO . . .
        ORGANIZATION IS SEQUENTIAL.
        ACCESS IS SEQUENTIAL.
    SELECT TAPEFILE
        ASSIGN TO . . .

DATA DIVISION.
FILE SECTION.
FD  DISKFILE
    BLOCK CONTAINS 5 RECORDS
    LABEL RECORDS ARE STANDARD.
01  DISK-RECORD                         PICTURE X(80).

FD  TAPEFILE
    LABEL RECORDS OMITTED.
01  TAPELIST                            PICTURE X(80).

WORKING-STORAGE SECTION.
01  FLAGS.
    05  MORE-DATA-FLAG                  PICTURE XXX        VALUE 'YES'.
        88  MORE-DATA                                      VALUE 'YES'.
        88  NO-MORE-DATA                                   VALUE 'NO'.

PROCEDURE DIVISION.
MAIN-ROUTINE.
    OPEN      INPUT     TAPEFILE
              OUTPUT    DISKFILE.
    READ TAPEFILE INTO DISK-RECORD
        AT END MOVE 'NO' TO MORE-DATA-FLAG.
    PERFORM PROCESS-ROUTINE
        UNTIL NO-MORE-DATA.
    PERFORM TERMINATION-ROUTINE.

PROCESS-ROUTINE.
    WRITE DISK-RECORD.
    READ TAPEFILE INTO DISK-RECORD
        AT END MOVE 'NO' TO MORE-DATA-FLAG.

TERMINATION-ROUTINE.
    CLOSE     TAPEFILE
              DISKFILE.
    STOP RUN.
```

The following example shows how to create sequential disk files.

Data from a disk called QUANFILE is to be used to create data records to be written onto one of two disk files called CALFILE and TFILE. The following is the record layout of each record in the input file QUANFILE.

ACCTNO					QMONTH		QDAY		QUAN1				QUAN2				QUAN3 →		
X	X	X	X	X	9	9	9	9	9	9	9	S	9	9	9	S	9	9	9
000	001	002	003	004	005	006	007	008	009	010	011	012	013	014	015	016	017	018	019

← QUAN3
S
020

The contents of the data fields within the record of QUANFILE are as follows:

ACCTNO: Account number	QUANT1: Quantity 1
QMONTH: Month	QUANT2: Quantity 2
QDAY: Day	QUANT3: Quantity 3

The record layout of each record in the output file, CALFILE, is as follows:

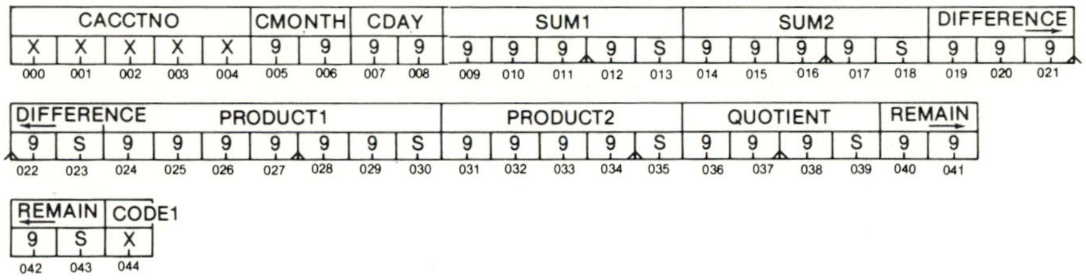

Field	Picture characters	Positions
CACCTNO	X X X X X	000–004
CMONTH	9 9	005–006
CDAY	9 9	007–008
SUM1	9 9 9 ^ 9 S	009–013
SUM2	9 9 9 ^ 9 S	014–018
DIFFERENCE	9 9 9 ^ 9 S	019–023
PRODUCT1	9 9 9 9 ^ 9 9 S	024–030
PRODUCT2	9 9 9 9 ^ S	031–035
QUOTIENT	9 9 ^ 9 S	036–039
REMAIN	9 9 9 S	040–043
CODE1	X	044

The contents of the data fields within the record of CALFILE are as follows:

CACCTNO: Account number
CMONTH: Month
CDAY: Day
SUM1: Sum 1
SUM2: Sum 2
DIFFENCE: Difference
PRODUCT1: Product 1
PRODUCT2: Product 2
QUOTIENT: Quotient
REMAIN: Remainder
CODE1: Code

If at least one of the three fields, quantity 1, quantity 2, and quantity 3, in the QUANFILE record is greater than 10 or equal to 10, then one record is created and written onto CALFILE. The following formulas determine the contents of the fields in the record of CALFILE.

Sum 1 = quantity 1 + quantity 2
Sum 2 = quantity 3 + 15
Difference = quantity 3 − quantity 2
Product 1 = quantity 3 × quantity 2
Product 2 = quantity 3 × quantity 2 (rounded to nearest whole number)

$$\text{Quotient} = \frac{\text{quantity 1} + \text{quantity 3}}{\text{quantity 2}}$$ with the remainder placed in REMAIN.

Code = A if quantity is less than 50
Code = X if quantity 2 is greater than 50
Code = E if quantity is equal to 50.

If the three fields, quantity 1, quantity 2, and quantity 3, in the QUANFILE record are all less than 10, then one record is created and written onto TFILE. Each record in the output file TFILE has the same structure as each record in the output file CALFILE.

The program for this example is:

```
IDENTIFICATION DIVISION.
PROGRAM-ID.
    CALPROGRAM.
AUTHOR.
    C FEINGOLD.
DATE-WRITTEN.     JUNE 15 1982.
DATE-COMPILED.
*PURPOSE.         CREATE TWO DISK FILES SEQUENTIALLY.
```

```
ENVIRONMENT DIVISION.
CONFIGURATION SECTION.
SOURCE-COMPUTER . . .
OBJECT-COMPUTER . . .
INPUT-OUTPUT SECTION.
FILE-CONTROL.
    SELECT QUANFILE
        ASSIGN TO . . .
    SELECT CALFILE
        ASSIGN TO . . .
        ORGANIZATION IS SEQUENTIAL.
        ACCESS IS SEQUENTIAL.
    SELECT TFILE
        ASSIGN TO . . .
        ORGANIZATION IS SEQUENTIAL.
        ACCESS IS SEQUENTIAL.

DATA DIVISION.
FILE SECTION.
FD  QUANFILE
    BLOCK CONTAINS 24 RECORDS
    RECORD CONTAINS 21 CHARACTERS
    LABEL RECORD IS STANDARD
    DATA RECORD IS QUANRECORD.
01  QUANRECORD.
    05  ACCTNO              PICTURE X(5).
    05  QMONTH              PICTURE 99.
    05  QDAY                PICTURE 99.
    05  QUAN1               PICTURE S99V99.
    05  QUAN2               PICTURE S99V99.
    05  QUAN3               PICTURE S99V99.

FD  CALFILE
    BLOCK CONTAINS 11 RECORDS
    RECORD CONTAINS 45 CHARACTERS
    LABEL RECORD IS STANDARD
    DATA RECORD IS CALRECORD.
01  CALRECORD.
    05  CACCTNO             PICTURE X(5).
    05  CMONTH              PICTURE 99.
    05  CDAY                PICTURE 99.
    05  SUM1                PICTURE S999V99.
    05  SUM2                PICTURE S999V99.
    05  DIFFENCE            PICTURE S999V99.
    05  PRODUCT1            PICTURE S9999V999.
    05  PRODUCT2            PICTURE S9999V9.
    05  QUOTIENT            PICTURE S99V99.
    05  REMAIN              PICTURE S9999.
    05  CODE1               PICTURE X.

FD  TFILE
    BLOCK CONTAINS 11 RECORDS
    LABEL RECORD IS STANDARD
    DATA RECORD IS TRECORD.
01  TRECORD                 PICTURE X(45).

WORKING-STORAGE SECTION.
77  WSUM                    PICTURE S999V99.
01  FLAGS.
    05  MORE-DATA-FLAG      PICTURE XXX       VALUE 'YES'.
        88  MORE-DATA                         VALUE 'YES'.
        88  NO-MORE-DATA                      VALUE 'NO'.
```

```
PROCEDURE DIVISION.
MAIN-ROUTINE.
    OPEN     INPUT    QUANFILE
             OUTPUT   CALFILE
                      TFILE.
    READ QUANFILE
        AT END MOVE 'NO' TO MORE-DATA-FLAG.
    PERFORM PROCESS-ROUTINE
        UNTIL NO-MORE-DATA.
    CLOSE    QUANFILE
             CALFILE
             TFILE.
    STOP RUN.

PROCESS-ROUTINE.
    MOVE ACCTNO TO CACCTNO.
    MOVE QMONTH TO CMONTH.
    MOVE QDAY TO CDAY.
    ADD QUAN1, QUAN2,
        GIVING SUM1.
    ADD QUAN3, 15
        GIVING SUM2.
    SUBTRACT QUAN2 FROM QUAN3
        GIVING DIFFERENCE.
    MULTIPLY QUAN3 BY QUAN2
        GIVING PRODUCT1.
    MULTIPLY QUAN3 BY QUAN2
        GIVING PRODUCT2 ROUNDED.
    ADD QUAN1, QUAN3
        GIVING WSUM.
    DIVIDE WSUM BY QUAN2
        GIVING QUOTIENT
        REMAINDER REMAIN.
    IF QUAN2 IS LESS THAN 50
        MOVE 'A' TO CODE1
        NEXT SENTENCE
    ELSE
        IF QUAN2 IS GREATER THAN 50
            MOVE 'X' TO CODE1
        ELSE
            MOVE 'E' TO CODE1.
    IF QUAN1 IS LESS THAN 10
        AND QUAN2 IS LESS THAN 10
        AND QUAN3 IS LESS THAN 10
        WRITE TRECORD FROM CALRECORD
    ELSE
        WRITE CALRECORD.
    READ QUANFILE
        AT END MOVE 'NO' TO MORE-DATA-FLAG.
```

Updating a Standard Sequential File

After a sequential file has been created, it may be maintained by updating the records and placing them back in the file. This can be accomplished only if the file has been opened with the I-O option. The I-O phrase permits the opening of a mass storage file for both input and output operations.

After a sequential file has been opened as I-O, a record is accessed just as in an input file. A programmer would access a record with a READ statement. After updating, the record is placed back into a sequential file as though it were being placed in an output file. The programmer would use a REWRITE statement.

The following is a program to update a sequential disk file:

```
IDENTIFICATION DIVISION.
PROGRAM-ID.
    SEQUENTIAL-DISK-UPDATE.

ENVIRONMENT DIVISION.
CONFIGURATION SECTION.
SOURCE-COMPUTER . . .
OBJECT-COMPUTER . . .
INPUT-OUTPUT SECTION.
FILE-CONTROL.
    SELECT DISKFILE
        ASSIGN TO . . .
        ORGANIZATION IS SEQUENTIAL
        ACCESS IS SEQUENTIAL.
    SELECT CARDFILE
        ASSIGN TO . . .

DATA DIVISION.
FILE SECTION.
FD  DISKFILE
    LABEL RECORDS ARE STANDARD
    BLOCK CONTAINS 5 RECORDS.
01  DISK-RECORD.
    05  PERSONAL.
        10  NAME                  PICTURE X(21).
        10  CUSTOMER-NO           PICTURE X(6).
        10  ADDRESS-HOME.
            15  STREET            PICTURE X(15).
            15  CITYSTATE         PICTURE X(15).
    05  PAYRECORD.
        10  YEAR-OPENED           PICTURE 99.
        10  MAXIMUM-CREDIT        PICTURE S9999V99     USAGE COMP-3.
        10  MAXIMUM-BILL          PICTURE S9999V99     USAGE COMP-3.
        10  BALANCE-DUE           PICTURE S9999V99     USAGE COMP-3.
        10  PAYCODE               PICTURE 9.
FD  CARDFILE
    LABEL RECORDS OMITTED.
01  INPUT-CARDS.
    05  C-NUMBER                  PICTURE X(6).
    05  STREET-NEW                PICTURE X(15).
    05  FILLER                    PICTURE X(59).

WORKING-STORAGE SECTION.
01  FLAGS.
    05  MORE-DATA-FLAG            PICTURE XXX     VALUE 'YES'.
        88  MORE-DATA                             VALUE 'YES'.
        88  NO-MORE-DATA                          VALUE 'NO'.

PROCEDURE DIVISION.
MAIN-ROUTINE.
    OPEN     INPUT    CARDFILE
             I-O      DISKFILE.
    READ CARDFILE
        AT END MOVE 'NO' TO MORE-DATA-FLAG.
    READ DISKFILE
        AT END MOVE 'NO' TO MORE-DATA-FLAG.
    PERFORM MATCH THRU MATCH-EXIT
        UNTIL NO-MORE-DATA.
    CLOSE    CARDFILE
             DISKFILE.
    STOP RUN.
```

```
MATCH.
    IF C-NUMBER > CUSTOMER-NO
        READ DISKFILE
            AT END MOVE 'NO' TO MORE-DATA-FLAG
        GO TO MATCH-EXIT
    ELSE
        NEXT SENTENCE.
    IF C-NUMBER < CUSTOMER-NO
        DISPLAY C-NUMBER 'NOT FOUND IN FILE'
            UPON CONSOLE
        READ CARDFILE
            AT END MOVE 'NO' TO MORE-DATA-FLAG.
        GO TO MATCH-EXIT
    ELSE
        NEXT SENTENCE.
    MOVE STREET-NEW TO STREET
    WRITE DISK-RECORD.
    READ CARDFILE
        AT END MOVE 'NO' TO MORE-DATA-FLAG.
    READ DISKFILE
        AT END MOVE 'NO' TO MORE-DATA-FLAG.

MATCH-EXIT.
    EXIT.
```

Note: This program updates a sequential disk file by inserting address changes to previous records.

Indexed Sequential Organization

An indexed sequential organization file is a sequential file with indexes that permit rapid access to individual records as well as rapid sequential processing. (See figures 13.15, 13.16, and 13.17.)

In some data-processing applications the programmer may not want to process the file consecutively. Consecutive processing is time consuming if the programmer only wants to process certain records in the file. It is faster to skip the records not needed in a job and to process only the required ones. An indexed file allows this type of processing.

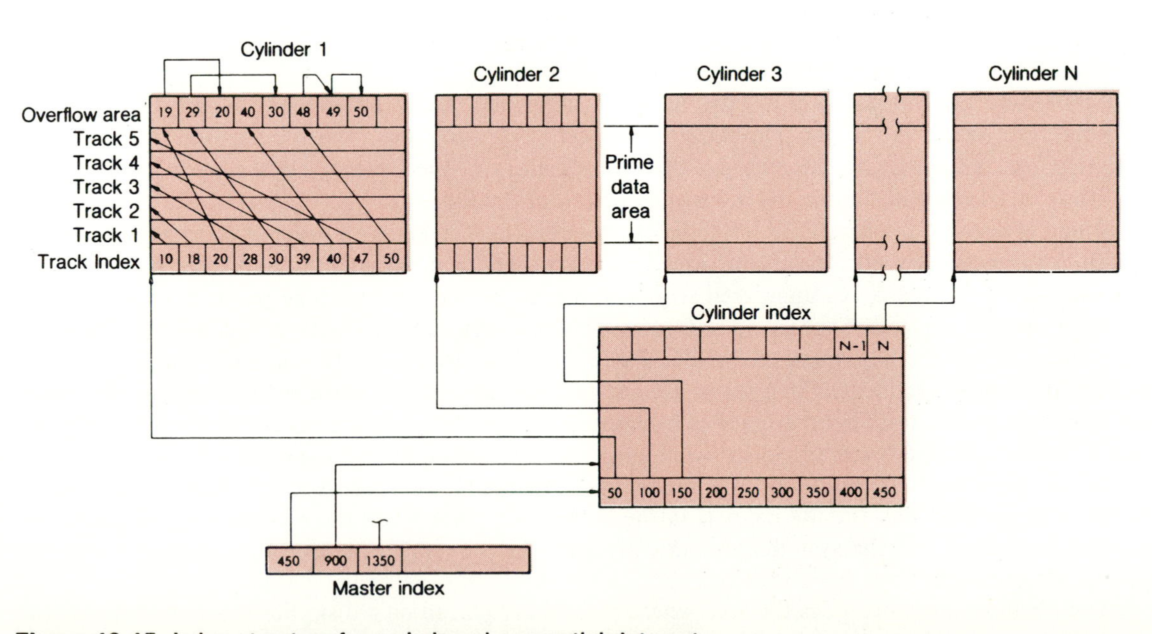

Figure 13.15 Index structure for an indexed sequential data set.

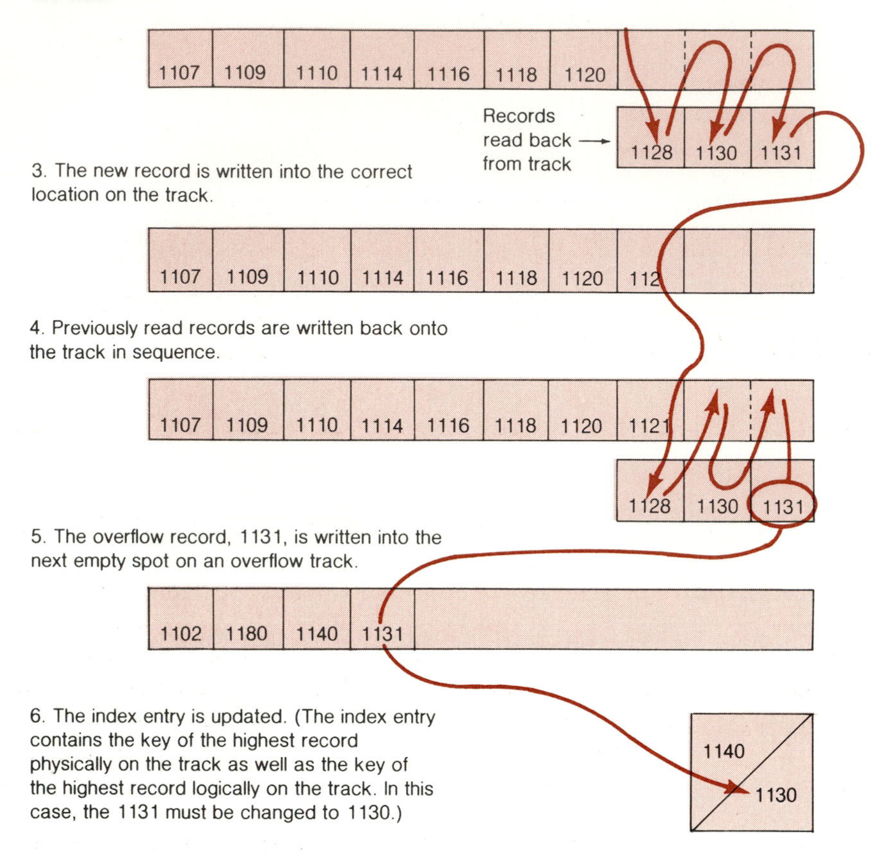
1. Records in an indexed file are stored on tracks. The compiler must first determine the track on which the record is to be added, 1121 in this example, could be inserted with the key in logical sequence.
2. All records on the track that have keys higher than the key of the record to be added are read.
1107
1109
1110
1114
1116
1118
1120
Records read back from track
1128
1130
1131
3. The new record is written into the correct location on the track.
1107
1109
1110
1114
1116
1118
1120
4. Previously read records are written back onto the track in sequence.
1107
1109
1110
1114
1116
1118
1120
1128
1130
1131
5. The overflow record, 1131, is written into the next empty spot on an overflow track.
1102
1180
1140
1131
6. The index entry is updated. (The index entry contains the key of the highest record physically on the track as well as the key of the highest record logically on the track. In this case, the 1131 must be changed to 1130.)
1140
1130

Figure 13.16 Indexed sequential organization—example.

An indexed file is organized into two parts: an index and the data records. The index contains an entry for each record in the file. The programmer can go to the index, find the location of the record, go to that location, and find the record wanted. Each entry in the file index describes a record in the file. There is an entry in the file index for each record in the file.

When creating an indexed file for COBOL, the records can be in an ordered or an unordered sequence. An ordered sequence means the records are arranged in order according to some major control field used as the key field. An unordered sequence means the records are in no particular order.

An inventory file loaded according to frequency of use is an example of an unordered file. The most active items are at the beginning of the file. When the file is used to write customers' orders, most of the records needed are located in a small area of the file, rather than scattered throughout the entire file. This reduces the total time it takes to process the records because the access mechanism does not have to move back and forth across the whole disk to access the required records.

When an indexed file is created, the file index is created as the records are written on the disk. If the file is an ordered file, the file index is in the correct sequence when the records are written. If the file is an unordered file, the system automatically sorts the file index into ascending sequence after all the records in the file have been loaded.

The file index area precedes the area where records are placed on a disk. For example, suppose the file index for a certain file requires five tracks. The file index entries would be written on the first five

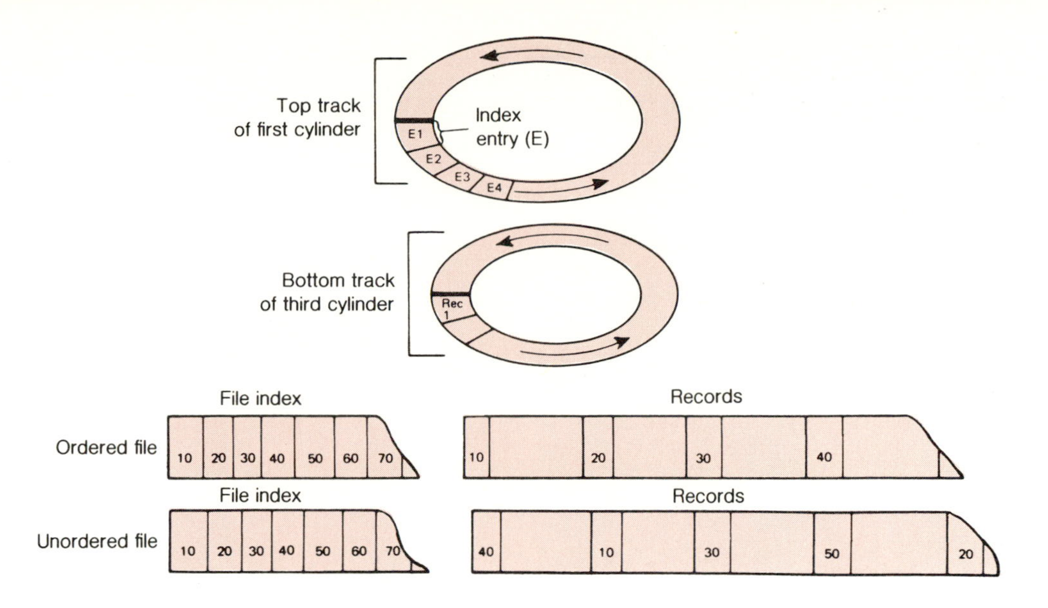

Figure 13.17 Indexed file organization—example.

tracks of the file. Records would be written beginning in the first sector of the sixth track. Both the file index area and the record area must start at the beginning of a track.

The indexes are created and written by the system as the file is created or organized. Keys provided by the programmer precede each block of data and are used to provide the index. An indexed sequential file is similar to a sequential file; however, by referring to the indexes maintained with the file, it is possible to quickly locate individual records for random processing. Moreover, a separate area can be set aside for additions, making it unnecessary to rewrite the entire file, a process that would be required for sequential processing. Although the records are not maintained in key sequence, the indexes are referred to in order to retrieve the added records in key sequence, thus making rapid sequential processing possible. (See figure 13.18.)

The programming system has control over the location of the individual records in this method of organization. The programmer need do very little input or output programming; the programming system does most of it since the characteristics of the file are known.

Indexed sequential organization gives the programmer greater flexibility in the operations that can be performed on the data file. The programmer is provided with the facilities for reading or writing records in a manner similar to that for sequential organization. The programmer can also read or write individual records whose keys may be in any order, and can add logical records with new keys. The system locates the proper position in the data file for the new record and makes all the necessary adjustments to the indexes.

The indexed sequential file must be stored on a direct-access device. Just as with standard sequential files, the indexed sequential files must be created sequentially. Identify-data in the control field of the record, called *keys,* must be in ascending sequence in succeeding records. As the records are written into the file, the system creates indexes based on the key or control field in each record to make possible quick location of any record in the file. Thus any record in an indexed sequential file may be accessed by specifying the appropriate key.

In addition to quick access of any record, an advantage of the indexed sequential file is that records may be added to any part of the file after it has been created, and the system will keep all records in logical sequence, although some records may technically be in a special "overflow" area. In accessing the file sequentially, the system will access records in logical sequence by key rather than in physical sequence by position on the device. (See figure 13.19.)

An indexed sequential file is a sequential file with indexes that permit rapid access to individual records as well as rapid sequential processing. An indexed sequential file has three distinct areas: a prime area, indexes, and an overflow area. (See figure 13.20.)

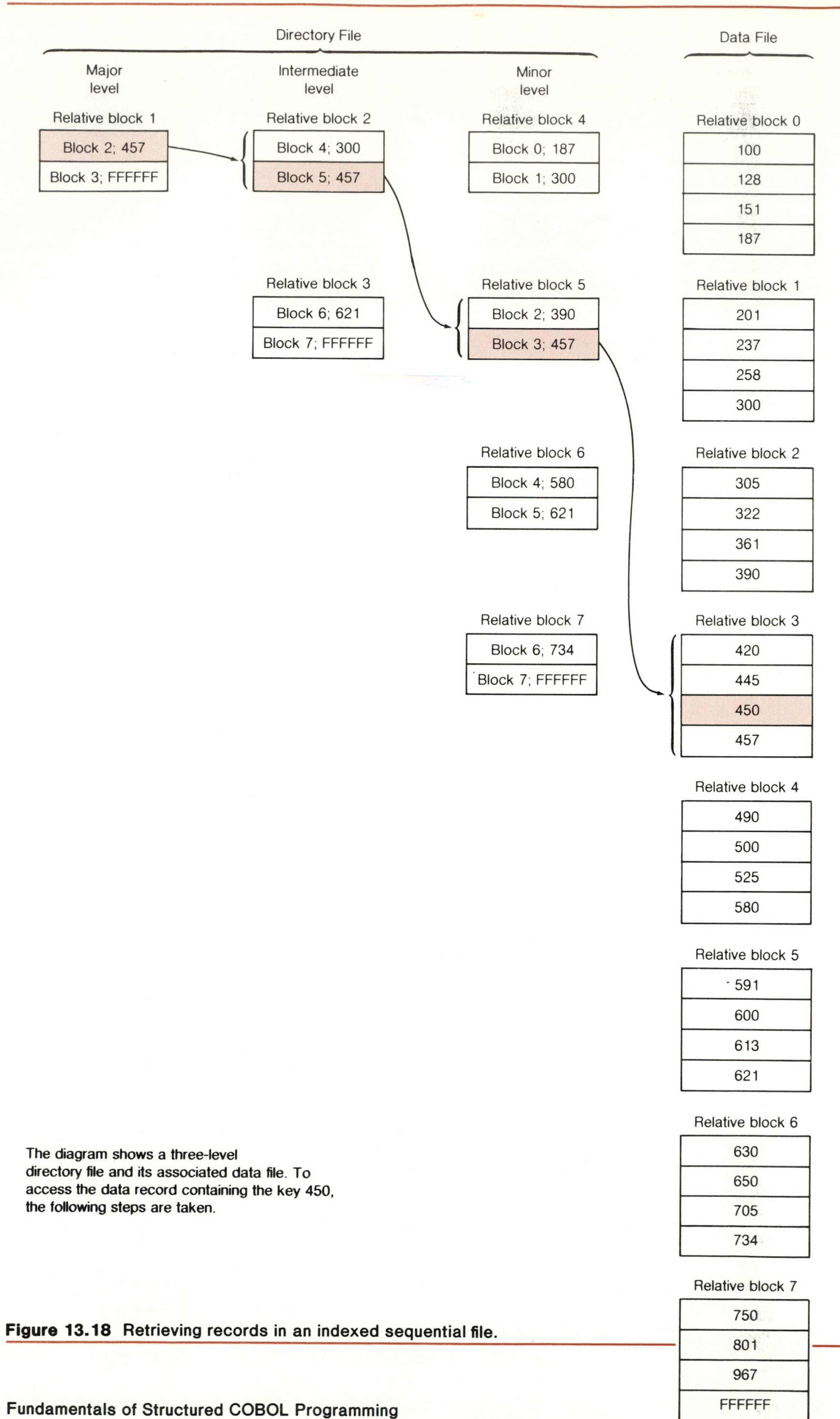

Figure 13.18 Retrieving records in an indexed sequential file.

Figure 13.19. Addition of records to an indexed sequential data set.

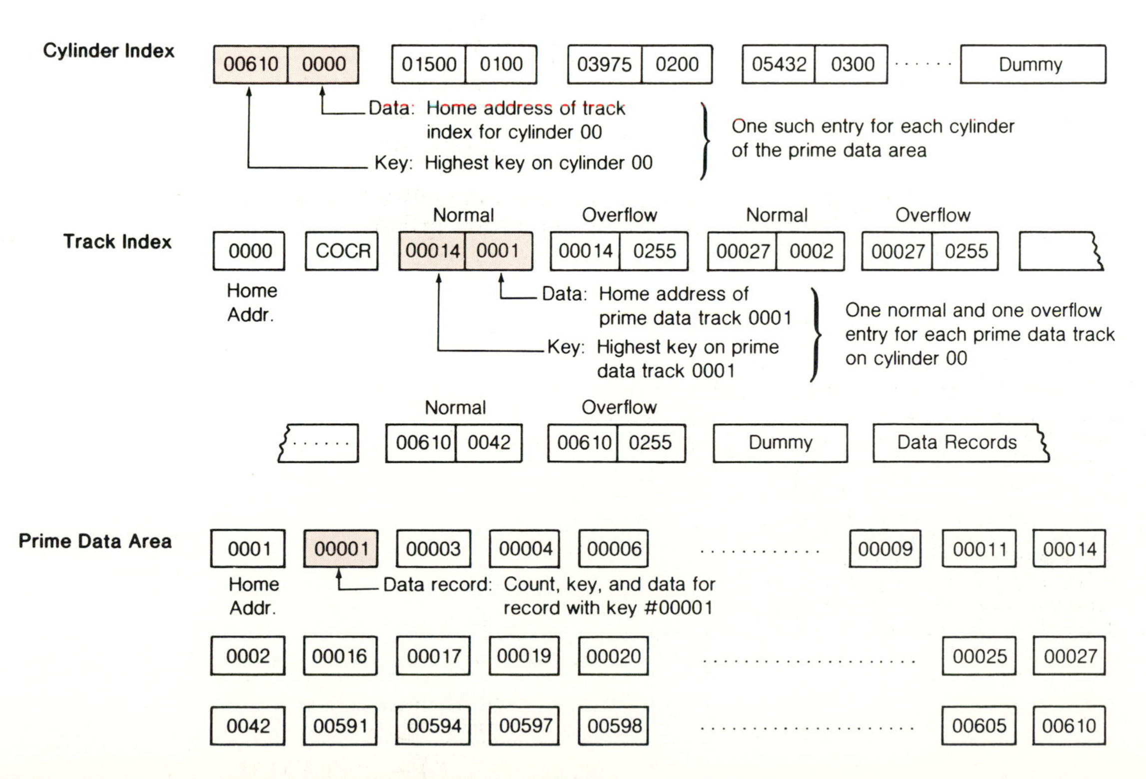

Figure 13.20 An indexed sequential file with no additions.

Prime Area

The prime area is the area in which records are written when the file is created or subsequently reorganized. Additions to the file may also be written in the prime area. The prime area may span multiple volumes and consist of several noncontiguous areas. The records in the prime area are in key sequence.

Prime records must be formatted with keys, and they may be blocked or unblocked. If blocked, each logical record contains its key, and the key area contains the key of the highest record in the block.

Indexes

There are two or more indexes of different levels. They are created and written by the operating system when the file is created or reorganized.

Track Index

This is the lowest index level and is always present. Its entries point to data records. There is one track address for each cylinder in the prime area. It is always written on the first track(s) of the cylinder that it indexes. There is a pair of entries for each prime data track in the cylinder containing the home address of the prime track and the key of the highest record in the track (normal entry), and the overflow area. The last entry of each track index is a dummy entry indicating the end of the index. The rest of the index track contains prime records if there is enough room for them.

Cylinder Index

This is a higher index level and is always present. Its entries point to track indexes. There is one cylinder index for the file. It may reside on a different type of DASD than the rest of the file. It consists of one entry for each cylinder in the prime area, followed by a dummy entry. The entries are formatted in the same fashion as the track index entries. The key area contains the key of the highest record in the cylinder to which the entry points. The data area contains the Home Address of the track index for that cylinder.

Overflow Area

A certain number of whole tracks, as specified by the programmer, are reserved in each cylinder for overflow data from prime tracks in that cylinder. (See figures 13.21 and 13.22.)

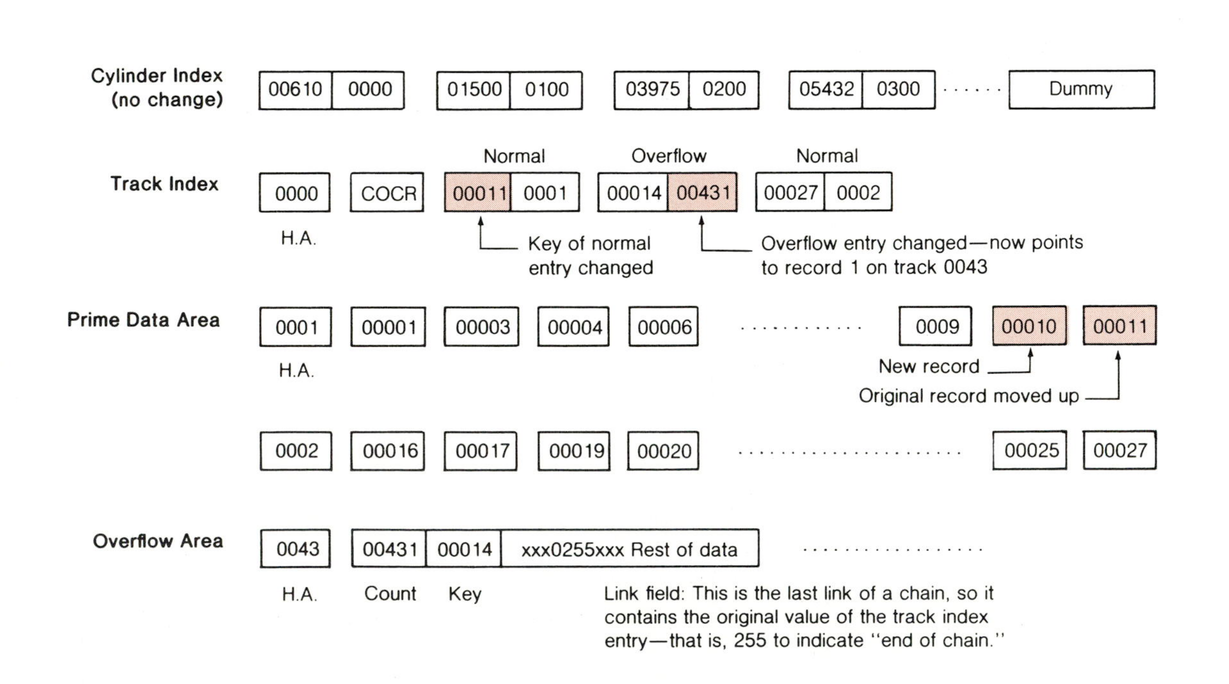

Figure 13.21 An indexed sequential file after the first addition to a prime track.

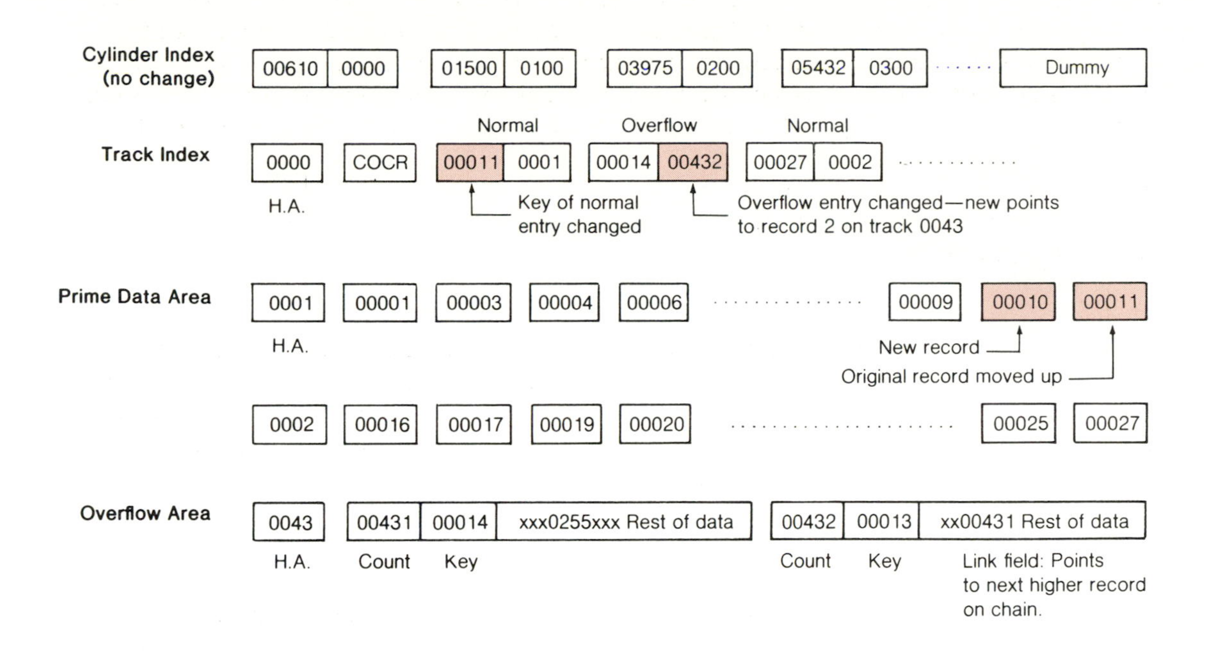

Figure 13.22 An indexed sequential file after subsequent additions to a track.

Creating an Indexed Sequential File

An indexed sequential file, which must be allocated in the execution activity to *two* random mass storage files (one for the data, the other for the index), is organized such that each record is uniquely identified by the value of a key within the record. In the RECORD KEY phrase of the SELECT clause, the source program specifies one of the data items within one of the records associated with the file as the record key data item. Each attempt to access a record based on the record key item causes a search of the index file for a key that matches the current contents of the record key data item in the file record area. The matching record in turn points to the location of the associated data record.

Environment Division. In order to specify that a file will be on a mass storage device, certain entries must be specified in the File-Control entry. (The format for File-Control entry—indexed files is shown in figure 13.23.)

An example of a File-Control entry for indexed files follows:

```
ENVIRONMENT DIVISION.
    .
    .
    .
INPUT-OUTPUT SECTION.
FILE-CONTROL.
    SELECT FILE-IN
        ASSIGN TO . . .
    SELECT PAYROLL-REGISTER
        ASSIGN TO . . .
        ORGANIZATION IS INDEXED
        ACCESS MODE IS SEQUENTIAL
        RECORD KEY IS SOCIAL-SEC-NO.
```

In addition to the normal file control entries, the following are additional entries required for indexed files.

1. The RESERVE clause allows the programmer to specify the number of input-output areas allocated.

```
SELECT file-name

    ASSIGN TO implementor-name-1  [ , implementor-name-2 ] . . .

  [ ; RESERVE integer-1  [ AREA  ] ]
                         [ AREAS ]

    ; ORGANIZATION IS INDEXED

  [                      { SEQUENTIAL } ]
  [ ; ACCESS MODE IS     { RANDOM     } ]
  [                      { DYNAMIC    } ]

    ; RECORD KEY IS data-name-1

  [ ; ALTERNATE RECORD KEY IS data-name-2  [ with DUPLICATES ] ] . . .

  [ ; FILE STATUS IS data-name-3 ]
```

Figure 13.23 Format—File Control entry—indexed files.

2. The ORGANIZATION clause specifies the logical structure of a file. The file organization is established at the time a file is created and cannot subsequently be changed. ORGANIZATION IS INDEXED is used for all indexed files.
3. The ACCESS MODE clause defines the manner in which the records or a file are to be accessed. An indexed file may be accessed in the sequential, random, or dynamic mode. When the access mode is sequential, records in the file are accessed in the sequence dictated by the file organization. For indexed files this sequence is the order of ascending record key values within a given key of reference. The next logical record is made available from the file when the READ statement is executed, or the next logical record is placed into the file when a WRITE statement is executed. ACCESS IS SEQUENTIAL may be applied to files assigned to magnetic tape, unit record, or direct-access devices.

   If the access mode is random, storage and retrieval are based on the value of the record key data item, which indicates the record to be accessed. To do a random search, it is necessary to specify where, in the record being searched for, the key can be found, and also where in main storage the particular key that corresponds to the record to be retrieved can be found. The RECORD KEY clause answers the first requirement and the NOMINAL KEY clause answers the second requirement. (The NOMINAL KEY clause is IBM terminology and may not be found in all COBOL systems.) The record key identifies a field in the disk record and the nominal key is defined in the Working-Storage Section.

   When the access mode is dynamic, records in the file may be accessed sequentially and/or randomly.

   If the ACCESS MODE clause is not specified, the ACCESS MODE IS SEQUENTIAL is implied. However, for documentation purposes, it is recommended that this clause be specified even if the ACCESS MODE IS SEQUENTIAL.
4. The RECORD KEY clause specifies the record key that is the prime record key for the file. The values of the prime record key must be unique among records of the file. The prime record key provides an access path to records in an indexed file.

   The RECORD KEY clause is used to access an indexed file. It specifies the elementary variable within the file record that identifies the record and is required for all records stored in indexed files. Any unique elementary variable in the record associated with the indexed file can be specified as the RECORD KEY.

An example of the program for the RECORD KEY clause follows:

```
01  WAGE-RECORD.
    05  EMPLOYEE-NUMBER     PICTURE 9(5).
    05  HOURS.
        10  REGULAR         PICTURE 999.
        10  OVERTIME        PICTURE 999.
    05  WAGES.
        10  REGULAR         PICTURE 999V99.
        10  OVERTIME        PICTURE 999V99.
    05  FILLER              PICTURE X(57).
```

```
SELECT WAGE-FILE
    ASSIGN TO . . .
    ORGANIZATION IS INDEXED
    ACCESS MODE IS SEQUENTIAL
    RECORD KEY IS EMPLOYEE-NUMBER.
```

Data-name-1 may be qualified. The data items referenced by data-name-1 must each be defined as a data item of the alphanumeric category within a record description entry associated with that file-name. Data-name-1 can describe an item whose size is variable.

5. An ALTERNATE RECORD KEY clause specifies a record key that is an alternate record key for the file. This alternate record key provides an alternate access path to records in an indexed file.

   The DUPLICATES phrase specifies that the value of the associated alternate record key may be duplicated within any of the records in the file. If the DUPLICATES phrase is not specified, the value of the associated alternate record key must not be duplicated among any of the records in the file.

   The same rules apply to data-name-2 as to data-name-1 as to qualification, definition, and size. The data descriptions of data-name-1 and data-name-2 as well as their relative locations within a record must be the same as that used when the file was created.

6. When the FILE STATUS clause is specified, a value will be moved by the operating system into the data item specified by data-name-3 after the execution of every statement that references that file either explicitly or implicitly. This value indicates the status of the statement's execution. Data-name-3 must be defined in the Data Division as a two-character data item of the alphanumeric category.

Data Division. The Data Division entries required for indexed sequential disk files are as follows.

FD. Name of disk file.

Block Contains. This clause required of all records that are blocked.

Label Records Are Standard. All files on direct-access devices must have standard labels.

Procedure Division. To create an indexed sequential file, you must use an output file and the INVALID KEY option in any WRITE statement.

WRITE Statement. When the WRITE statement is being used to refer to an indexed sequential file, the optional word KEY must be used. (The format for the WRITE statement is shown in figure 13.24.)

The INVALID KEY option is activated if the key field of the record being written does not contain a greater value than the key field of the record just written; that is, a key out of sequence or a duplicate

---

```
WRITE record-name [ FROM identifier-1 ]
    INVALID KEY imperative-statement
```

When the WRITE statement is used to refer to an indexed sequential file, the optional COBOL word KEY is also used as indicated in this format.

Figure 13.24 Format—WRITE statement—indexed files.

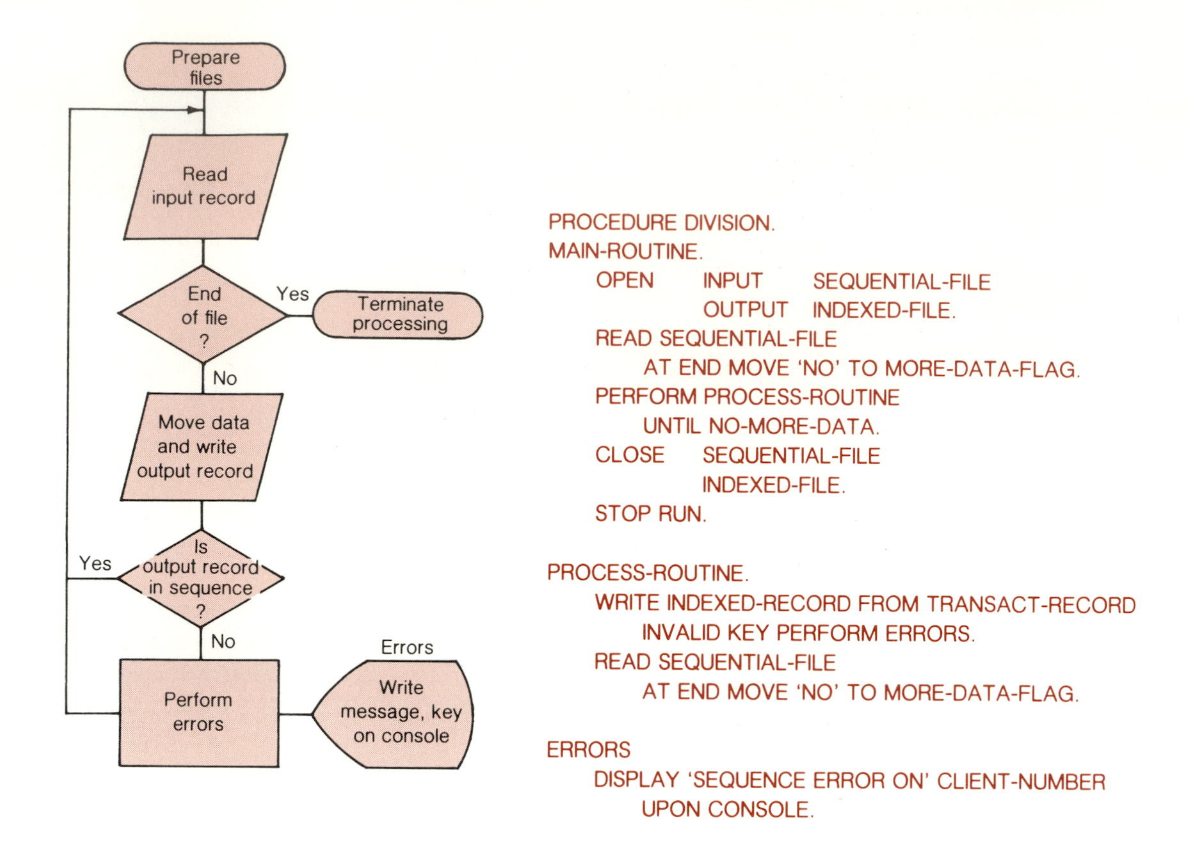
Prepare files
Read input record
End of file ?
Yes
Terminate processing
No
Move data and write output record
Is output record in sequence ?
Yes
No
Perform errors
Errors
Write message, key on console
PROCEDURE DIVISION.
MAIN-ROUTINE.
OPEN INPUT SEQUENTIAL-FILE
OUTPUT INDEXED-FILE.
READ SEQUENTIAL-FILE
AT END MOVE 'NO' TO MORE-DATA-FLAG.
PERFORM PROCESS-ROUTINE
UNTIL NO-MORE-DATA.
CLOSE SEQUENTIAL-FILE
INDEXED-FILE.
STOP RUN.
PROCESS-ROUTINE.
WRITE INDEXED-RECORD FROM TRANSACT-RECORD
INVALID KEY PERFORM ERRORS.
READ SEQUENTIAL-FILE
AT END MOVE 'NO' TO MORE-DATA-FLAG.
ERRORS
DISPLAY 'SEQUENCE ERROR ON' CLIENT-NUMBER
UPON CONSOLE.

Figure 13.25 Creation of indexed sequential file—example.

key may generate an INVALID KEY option. An imperative statement after the word INVALID KEY directs the program to an error procedure.

When the WRITE statement with the INVALID KEY option is executed, the key of the record is checked for correct sequence before the record is written. If the INVALID KEY option is activated in an attempt to write a record into an indexed sequential file, that record is not placed into the indexed sequential file. Subsequent records, however, may still be placed into the file. The record with the INVALID KEY may be placed into a sequential file, if desired, for subsequent individual checking of the key. The imperative statement following the words INVALID KEY could direct the program to the proper routine to accomplish this. (See figure 13.25.)

The following example shows the creation of an indexed sequential file:

```
IDENTIFICATION DIVISION.
PROGRAM-ID.     CREATEIS.

ENVIRONMENT DIVISION.
CONFIGURATION SECTION.
SOURCE-COMPUTER . . .
OBJECT-COMPUTER . . .
INPUT-OUTPUT SECTION.
    SELECT DISKFILE
        ASSIGN TO . . .
    SELECT IND-SEQ
        ASSIGN TO . . .
        ORGANIZATION IS INDEXED
        ACCESS MODE IS SEQUENTIAL
        RECORD KEY IS RECORDID-IS.
```

```
DATA DIVISION.
FILE SECTION.
FD  DISKFILE
    LABEL RECORDS ARE STANDARD
    BLOCK CONTAINS 5 RECORDS.
01  DISK-RECORD.
    05  PERSONAL-IN.
        10  NAME-IN              PICTURE X(21).
        10  CUSTNO-IN            PICTURE X(6).
        10  HOMEADDR-IN.
            15  STREET-IN        PICTURE X(15).
            15  CITYST-IN        PICTURE X(15).
    05  PAYRECORD-IN.
        10  YEAROPEN-IN          PICTURE 99.
        10  MAXCRED-IN           PICTURE 9999V99    USAGE COMP-3.
        10  MAXBILL-IN           PICTURE 9999V99    USAGE COMP-3.
        10  BALDUE-IN            PICTURE 9999V99    USAGE COMP-3.
        10  PAYCODE-IN           PICTURE 9.
FD  IND-SEQ
    LABEL RECORDS ARE STANDARD
    BLOCK CONTAINS 5 RECORDS.
01  RECORD-IS.
    05  PERSONAL-IS.
        10  NAME-IS              PICTURE X(21).
        10  RECORDID-IS          PICTURE X(6).
        10  HOMEADDR-IS.
            15  STREET-IS        PICTURE X(15).
            15  CITYST-IS        PICTURE X(15).
    05  PAYRECORD-IS.
        15  YEAROPEN-IS          PICTURE 99.
        15  MAXCRED-IS           PICTURE 9999V99    USAGE COMP-3.
        15  MAXBILL-IS           PICTURE 9999V99    USAGE COMP-3.
        15  BALDUE-IS            PICTURE 9999V99    USAGE COMP-3.
        15  PAYCODE-IS           PICTURE 9.
WORKING-STORAGE SECTION.
01  FLAGS.
    05  MORE-DATA-FLAG           PICTURE XXX      VALUE 'YES'.
        88  MORE-DATA                             VALUE 'YES'.
        88  NO-MORE-DATA                          VALUE 'NO'.
PROCEDURE DIVISION.
MAIN-ROUTINE.
    OPEN    INPUT    DISKFILE
            OUTPUT   IND-SEQ.
    READ DISKFILE
        AT END MOVE 'NO' TO MORE-DATA-FLAG.
    PERFORM PROCESS-ROUTINE
        UNTIL NO-MORE-DATA.
    DISPLAY 'INDEXED-FILE CREATED'
        UPON CONSOLE.
    CLOSE   DISKFILE
            IND-SEQ.
    STOP RUN.
PROCESS-ROUTINE.
    MOVE NAME-IN            TO NAME-IS.
    MOVE CUSTNO-IN          TO RECORDID-IS.
    MOVE STREET-IN          TO STREET-IS.
    MOVE CITYST-IN          TO CITYST-IS.
    MOVE PAYRECORD-IN       TO PAYRECORD-IS.
```

```
WRITE RECORD-IS
    INVALID KEY DISPLAY CUSTNO-IN
        'RECORD NOT WRITTEN'
        UPON CONSOLE.
READ DISKFILE
    AT END MOVE 'NO' TO MORE-DATA-FLAG.
```

Following is another example for the creation of an indexed sequential file:

```
 IDENTIFICATION DIVISION.
 PROGRAM-ID.      CREATIS.
*PURPOSE.         ILLUSTRATE CREATION OF INDEXED SEQUENTIAL FILE.

 ENVIRONMENT DIVISION.
 CONFIGURATION SECTION.
 SOURCE-COMPUTER . . .
 OBJECT-COMPUTER . . .
 INPUT-OUTPUT SECTION.
     SELECT FILE-IS
         ASSIGN TO . . .
         ORGANIZATION IS INDEXED
         ACCESS MODE IS SEQUENTIAL
         RECORD KEY IS REC-ID.
     SELECT CARD-FILE
         ASSIGN TO . . .

 DATA DIVISION.
 FILE SECTION.
 FD  FILE-IS
     BLOCK CONTAINS 5 RECORDS
     LABEL RECORDS ARE STANDARD
     DATA RECORD IS DISK.
 01  DISK.
     05  DELETE-CODE        PICTURE X.
     05  REC-ID             PICTURE 9(10).
     05  DISK-FLD1          PICTURE X(10).
     05  DISK-NAME          PICTURE X(20).
     05  DISK-BAL           PICTURE 99999V99.
     05  FILLER             PICTURE X(52).

 FD  CARD-FILE
     LABEL RECORDS ARE OMITTED
     DATA RECORD IS CARDS.
 01  CARDS.
     05  KEY-ID             PICTURE 9(10).
     05  CD-NAME            PICTURE X(20).
     05  CD-BAL             PICTURE 99999V99.
     05  FILLER             PICTURE X(43).

 WORKING-STORAGE SECTION.
 01  FLAGS.
     05  MORE-DATA-FLAG     PICTURE XXX        VALUE 'YES'.
         88  MORE-DATA                         VALUE 'YES'.
         88  NO-MORE-DATA                      VALUE 'NO'.

 PROCEDURE DIVISION.
 MAIN-ROUTINE.
     OPEN     INPUT    CARD-FILE
              OUTPUT   FILE-IS.
     READ CARD-FILE
         AT END MOVE 'NO' TO MORE-DATA-FLAG.
     PERFORM PROCESS-ROUTINE
         UNTIL NO-MORE-DATA.
```

```
     DISPLAY 'END OF JOB'
          UPON CONSOLE.
     CLOSE    CARD-FILE
              FILE-IS.
     STOP RUN.

PROCESS-ROUTINE.
     MOVE LOW-VALUE       TO DELETE-CODE.
     MOVE KEY-ID          TO REC-ID.
     MOVE CD-NAME         TO DISK-NAME.
     MOVE CD-BAL          TO DISK-BAL.
     WRITE DISK
          INVALID KEY PERFORM ERR.
     READ CARD-FILE
          AT END MOVE 'NO' TO MORE-DATA-FLAG.

ERR.
     DISPLAY 'DUPLICATE KEY OR SEQ-ERR FOR KEY NO' REC-ID
          UPON CONSOLE.
```

Sequentially Accessing an Indexed Sequential File

It is frequently necessary to access records sequentially from an indexed sequential file. These records may be accessed sequentially, record by record, until the file is closed. This program would be similar to a program for accessing a standard sequential file, with the exception of the File-Control paragraph entries describing the file as indexed sequential.

The following example shows how to access records sequentially in an indexed sequential file (Environment and Procedure Division entries only):

```
ENVIRONMENT DIVISION.
     .
     .
     .
INPUT-OUTPUT SECTION.
FILE-CONTROL.
     SELECT STUDENT-FILE
          ASSIGN TO . . .
          ORGANIZATION IS INDEXED
          ACCESS MODE IS SEQUENTIAL
          RECORD KEY IS STUDENT-NUMBER.
     SELECT PRINTOUT
          ASSIGN TO . . .

PROCEDURE DIVISION.
MAIN-ROUTINE.
     OPEN     INPUT     STUDENT-FILE
              OUTPUT    PRINTOUT.
     READ STUDENT-FILE
          AT END MOVE 'NO' TO MORE-DATA-FLAG.
     PERFORM PROCESS-ROUTINE
          UNTIL NO-MORE-DATA.
     CLOSE    STUDENT-FILE
              PRINTOUT.
     STOP RUN.

PROCESS-ROUTINE.
     MOVE STUDENT-NUMBER      TO S-NUMBER.
     MOVE CORRESPONDING STUDENT-DATA TO PRINTDATA.
     WRITE PRINTOUT FROM PRINTDATA
          AFTER ADVANCING 2 LINES.
     READ STUDENT-FILE
          AT END MOVE 'NO' TO MORE-DATA-FLAG.
```

Records in an indexed sequential file may be accessed at a record that is not the first record in a file, continuing until the file is closed. This is not possible in a standard sequential file, but can be accomplished in an indexed sequential file by specifying the desired beginning key, and positioning the file at that desired key prior to the accessing of the record.

Environment Division. The Environment Division entries for an indexed sequential file from which all records will be sequentially accessed are identical to the entries required when an indexed sequential file is to be created.

Data Division. The Data Division entries for an indexed sequential file from which all records will be sequentially accessed are similar to entries previously described for standard and indexed sequential programs.

Procedure Division. The Procedure Division coding is similar to the standard sequential accessing of records.

Whenever sequential access of records from an indexed sequential file will begin at some other record than the beginning of the file, the NOMINAL KEY clause must be specified in addition to the RECORD KEY clause in the File-Control paragraph of the Environment Division.

Nominal Key Clause. A NOMINAL KEY clause is used with indexed sequential files and specifies any elementary variable described in the Working-Storage Section of the program.

The following example shows RECORD KEY and NOMINAL KEY clauses:

```
DATA DIVISION.
FILE SECTION.
FD  INDEXED-FILE
    LABEL RECORDS ARE STANDARD
    BLOCK CONTAINS 5 RECORDS.
01  IF-RECORD.
    05  FILLER          PICTURE X.
    05  RECORD-ID       PICTURE 9(10).
    05  FIELD-1         PICTURE X(20).
    05  FIELD-2         PICTURE X(10).
    05  FILLER          PICTURE X(62).
WORKING-STORAGE SECTION.
77  RECORD-NUMBER   PICTURE 9(10).
_____________________________________

    RECORD KEY IS RECORD-ID
    NOMINAL KEY IS RECORD-NUMBER
```

In order to sequentially access the records in an indexed sequential file beginning at some record other than the first record of a file, the value of the NOMINAL KEY variable must be set equal to the RECORD KEY of the record at which the sequential access is to begin. Since the NOMINAL KEY variable must be given a value before accessing of records begins, a value must be set in Working-Storage through the execution of a MOVE, ACCEPT, or VALUE clause.

After the NOMINAL KEY variable has been set to the value of the RECORD KEY variable of the record at which accessing of records is to begin, the indexed sequential file must be positioned so that a READ statement will access the desired record. A START statement is used to position the file. Absence of the START statement causes the searching of a file to start at the first record.

START Statement

The START statement provides a basis for logical positioning within an index file, for subsequent retrieval of records. (The format for the START statement is shown in figure 13.26.)

Rules Governing the Use of the START Statement

1. File-name must be the name of an indexed file with sequential or dynamic access and must be open in the INPUT or I-O mode at the time that the START statement is executed.

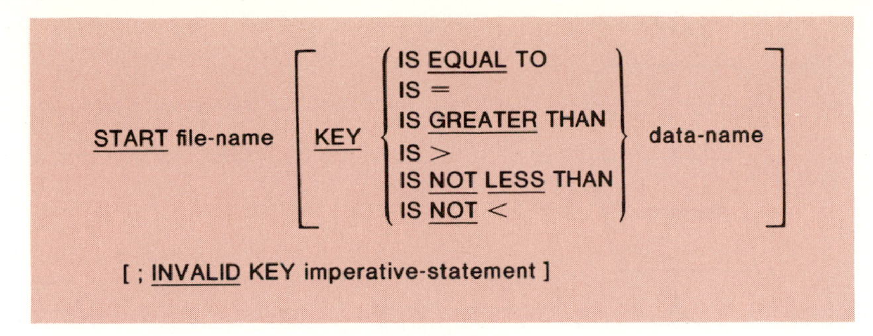

Figure 13.26 Format—START statement.

The required relational characters >, <, and = are not underlined to avoid confusion with other symbols such as ≥ (greater than or equal to).

2. The type of comparison specified by the relational operator in the KEY phrase occurs between a key associated with a record in the file referenced by file-name and a data item. The current record pointer is positioned to the first logical record currently existing in the file whose key satisfies the comparison.

   If the comparison is not satisfied by any record in the file, an INVALID KEY condition exists, the execution of the START statement is unsuccessful, and position of the current record pointer is undefined.

   If the KEY phrase is specified, the comparison described uses the data item referenced by data-name. (Data-name may be qualified.)

   If the KEY phrase is not specified, the comparison described uses the data item referenced in the RECORD KEY clause associated with the file-name.
3. Upon completion of the successful execution of the START statement, a key of reference is established and used in subsequent READ statements as follows:
   a. If the KEY phrase is not specified, the prime record key specified for file-name becomes the key of reference.
   b. If the KEY phrase is specified, and data-name is specified as record key for file-name, the record key becomes the key of reference.
   c. If the KEY phrase is specified, and data-name is not specified as a record key for file-name, the record key whose leftmost character position corresponds to the leftmost character position of the data item specified by data-name, becomes the key of reference.

      The following example shows: START statement to position file before accessing an indexed file sequentially—KEY phrase option.

```
MOVE 'A208787' TO PATIENT-NUMBER.

START PATIENT-FILE
      KEY IS NOT LESS THAN PATIENT-NUMBER
      INVALID KEY PERFORM TERM-ROUTINE.
```

   In the programming just given, processing would begin with key A208787. If there were no record with key A208787, processing would start with the first record greater than A208787. If the START statement wasn't used, processing would begin with the first record in the file.
4. If the KEY phrase is not specified, the relational operator 'IS EQUAL TO' is implied.

   The following example shows: START statement to position file before accessing an indexed file sequentially—no KEY phrase option.

```
NOMINAL KEY IS PATIENT-NUMBER

MOVE 'A208787' TO PATIENT-NUMBER.

START PATIENT-FILE
      INVALID KEY PERFORM TERM-ROUTINE.
```

5. If the execution of the START statement is not successful, the key of reference is undefined. (See figure 13.27.)

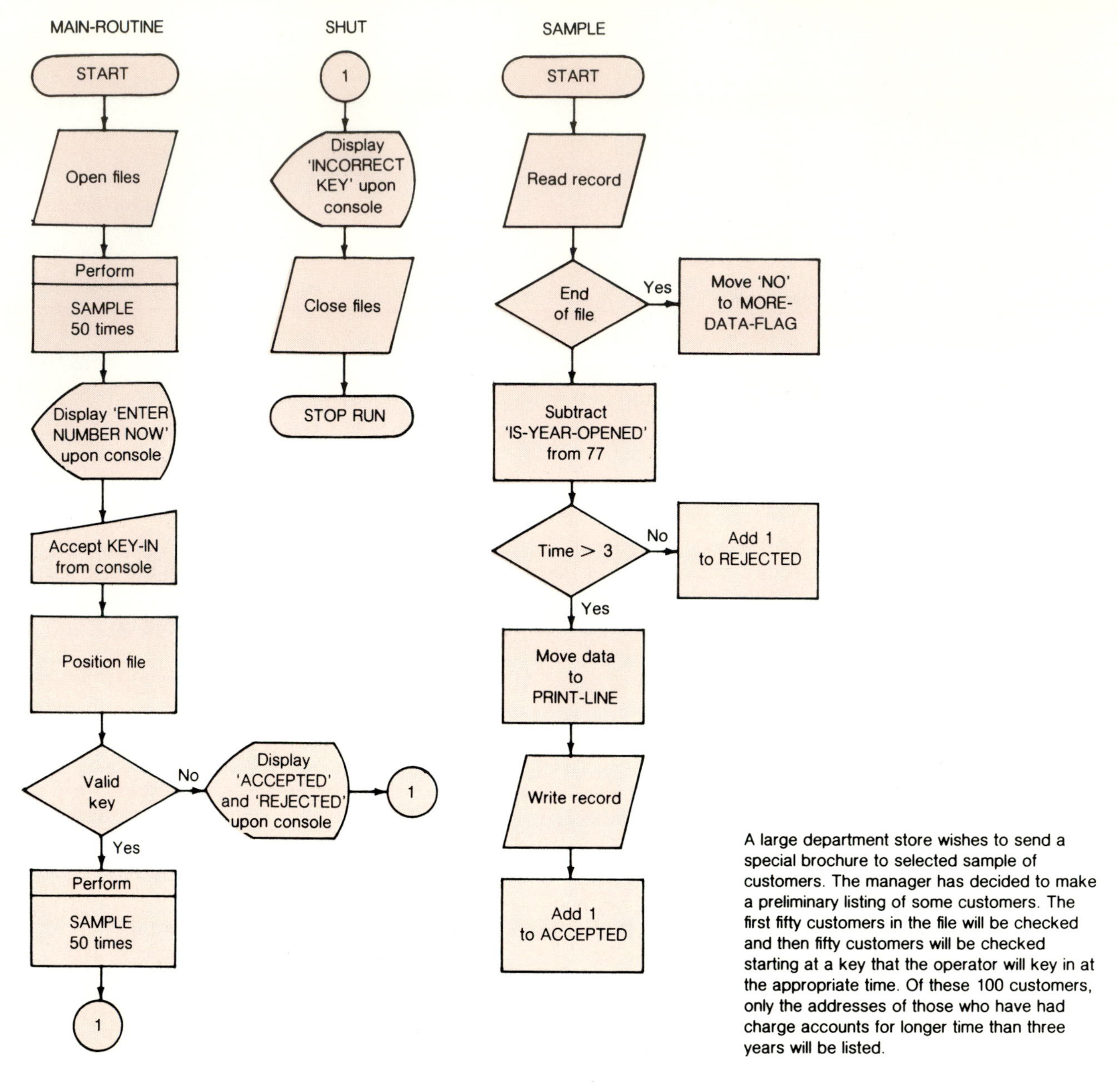

A large department store wishes to send a special brochure to selected sample of customers. The manager has decided to make a preliminary listing of some customers. The first fifty customers in the file will be checked and then fifty customers will be checked starting at a key that the operator will key in at the appropriate time. Of these 100 customers, only the addresses of those who have had charge accounts for longer time than three years will be listed.

Figure 13.27 Program accessing records sequentially from an indexed file starting at any point—example.

The program accompanying the example in figure 13.27 follows:

```
IDENTIFICATION DIVISION.
PROGRAM-ID.      SEQ-ACCESS.

ENVIRONMENT DIVISION.
CONFIGURATION SECTION.
SOURCE-COMPUTER . . .
OBJECT-COMPUTER . . .
INPUT-OUTPUT SECTION.
FILE-CONTROL.
    SELECT IND-SEQ
        ASSIGN TO . . .
        ORGANIZATION IS INDEXED
        ACCESS MODE IS SEQUENTIAL
```

```
            RECORD KEY IS RECORD-ID
            NOMINAL KEY IS KEY-IN.
        SELECT PRINT-FILE
            ASSIGN TO . . .

DATA DIVISION.
FILE SECTION.
FD  IND-SEQ
        LABEL RECORDS ARE STANDARD
        BLOCK CONTAINS 5 RECORDS.
01  DISK-RECORD.
    05  IS-PERSONAL.
        10  IS-NAME                 PICTURE X(21).
        10  RECORD-ID               PICTURE X(6).
        10  IS-ADDRESS.
            15  IS-STREET           PICTURE X(15).
            15  IS-CITYST           PICTURE X(15).
    05  IS-PAYRECORD.
        10  IS-YEAR-OPEN            PICTURE 99.
        10  IS-MAXCRED              PICTURE 9999V99                    USAGE COMP-3.
        10  IS-MAXBILL              PICTURE 9999V99                    USAGE COMP-3.
        10  IS-BALDUE               PICTURE 9999V99                    USAGE COMP-3.
        10  IS-PAYCODE              PICTURE 9.
            88  BAD                                    VALUE 1.
            88  POOR                                   VALUE 2.
            88  SLOW                                   VALUE 3.
            88  AVERAGE                                VALUE 4.
            88  GOOD                                   VALUE 5.
            88  EXCELLENT                              VALUE 6.
            88  NONE                                   VALUE 7.

FD  PRINT-FILE
        LABEL RECORDS ARE OMITTED.
01  OUTGO                           PICTURE X(133).

WORKING-STORAGE SECTION.
77  TIME-WS                         PICTURE 99.
77  KEY-IN                          PICTURE X(6).
77  ACCEPTED                        PICTURE 999        VALUE ZEROS     USAGE COMP-3.
77  REJECTED                        PICTURE 999        VALUE ZEROS     USAGE COMP-3.
01  PRINT-LINE.
    05  FILLER                      PICTURE X(9)       VALUE SPACES.
    05  FILLER                      PICTURE XX         VALUE '19'.
    05  YEARS                       PICTURE 99.
    05  FILLER                      PICTURE X(10)      VALUE SPACES.
    05  IDENT                       PICTURE X(6).
    05  FILLER                      PICTURE X(10)      VALUE SPACES.
    05  CUSTOMER                    PICTURE X(20).
    05  FILLER                      PICTURE X(15)      VALUE SPACES.
    05  O-STREET                    PICTURE X(15).
    05  FILLER                      PICTURE X(10)      VALUE SPACES.
    05  O-CITYST                    PICTURE X(15).
    05  FILLER                      PICTURE X(19)      VALUE SPACES.
01  FLAGS.
    05  MORE-DATA-FLAG              PICTURE XXX        VALUE 'YES'.
        88  MORE-DATA                                  VALUE 'YES'.
        88  NO-MORE-DATA                               VALUE 'NO'.

PROCEDURE DIVISION.
MAIN-ROUTINE.
    OPEN    INPUT     IND-SEQ
            OUTPUT    PRINT-FILE.
    PERFORM SAMPLE 50 TIMES.
```

```
     DISPLAY 'ENTER NUMBER NOW'
         UPON CONSOLE.
     ACCEPT KEY-IN FROM CONSOLE.
     START IND-SEQ
         INVALID KEY DISPLAY 'INCORRECT KEY' UPON CONSOLE
         PERFORM SHUT.
     PERFORM SAMPLE 50 TIMES.
     PERFORM SHUT.

 SAMPLE.
     READ IND-SEQ
         AT END MOVE 'NO' TO MORE-DATA-FLAG.
     SUBTRACT IS-YEAR-OPEN FROM 82
         GIVING TIME-WS.
     IF TIME-WS IS GREATER THAN 3
         MOVE IS-YEAR        TO    YEARS
         MOVE RECORD-ID      TO    IDENT
         MOVE IS-NAME        TO    CUSTOMER
         MOVE IS-STREET      TO    O-STREET
         MOVE IS-CITYST      TO    O-CITYST
         WRITE OUTGO FROM PRINT-LINE
             AFTER ADVANCING 1 LINES
         ADD 1 TO ACCEPTED
     ELSE
         ADD 1 TO REJECTED.

 SHUT.
     DISPLAY 'ACCEPTED 'ACCEPTED 'REJECTED 'REJECTED
         UPON CONSOLE.
     CLOSE    IND-SEQ
              PRINT-FILE.
     STOP RUN.
```

DELETE Statement

The DELETE statement logically removes a record from a mass storage file. (The format for the DELETE statement is shown in figure 13.28.)

Rules Governing the Use of the DELETE Statement

1. The associated file must be open in the I-O mode at the time of the execution of this statement. For files in the sequential access mode, the last input-output statement executed for file-name prior to the execution of the DELETE statement must have been a successfully executed READ statement. The MSCS logically removes from the file the record that was accessed by the READ statement.

   For a file in random or dynamic access mode, the MSCS logically removes from the file the record identified by the contents of the prime record key data item associated with file-name. If the file does not contain the record specified by the key, an INVALID KEY condition exists.
2. The INVALID KEY phrase must not be specified for a DELETE statement that references a file that is in a sequential access mode. The INVALID KEY phrase must be specified for a DELETE statement that references a file that is not in a sequential access mode and for which an applicable USE procedure is not specified.
3. After the successful execution of a DELETE statement, the identified record has been logically

```
DELETE file-name RECORD   [ ; INVALID KEY imperative-statement ]
```

Figure 13.28 Format—DELETE statement.

removed from the file and can no longer be accessed. The execution of a DELETE statement does not affect the contents of the record area associated with file-name.

4. The current record pointer is not affected by the execution of a DELETE statement.
5. The execution of the DELETE statement causes the value of the specified FILE STATUS data item, if any, associated with file-name, to be updated.

An example of the DELETE statement follows:

```
IF    TRANS-COD IS EQUAL TO 'D'
      PERFORM DELETE-PROC
ELSE
      PERFORM PROCESS-PROC.
.
.
.
DELETE-PROC.
      MOVE TRANS-NO TO DISK-NO.
      DISPLAY 'ITEM NUMBER ' DISK-NO ' DELETED FROM FILE'
            UPON CONSOLE.
      DELETE DISK-FILE
            INVALID KEY
                  DISPLAY DISK-NO ' NOT IN FILE '
                        UPON CONSOLE.
.
.
.
```

Sequential Mode

The following operations may be performed on an indexed file accessed in the sequential mode:

WRITE	Accesses the space immediately following that location into which the previous logical record was written, and places the contents of the specified record in that space. If the contents of the record key data item are less than or equal to the key of the previously written logical record, or if writing the logical record would exceed the boundaries of the allocated file space, an INVALID KEY condition exists. The file must be open for output.
DELETE	Removes the logical record accessed by the previous input-output operation (which must have been a successful READ statement) from the file. The file must be open for input-output.
START	Positions the file such that the key of the next logical record accessed will be based on the comparison specified in the START statement, in relation to the contents of the record key data item. If the comparison is not satisfied by any logical record on the file, an INVALID KEY condition will exist. The file must be open for input or input-output.
READ	Accesses the next logical record on the file and makes the contents of that record available in the file record area. If there is no 'next' logical record, an AT END condition exists. The file must be open for input or input-output.
REWRITE	Replaces the logical record accessed by the previous input-output operation (which must have been a successful READ statement) with the contents of the specified record. If the contents of the record key data item are not equal to the key of the last record read, an INVALID KEY condition exists. The logical records being replaced must be equal in size to the record specified in the REWRITE statement (the replacement record). The file must be open for input-output.

Accessing an Indexed Sequential File at Random

Environment Division. The SELECT, ASSIGN, RECORD KEY, and NOMINAL KEY clauses are used in the same manner as sequentially accessing an indexed sequential file.

The ACCESS IS RANDOM option is required for the random access of records from a file. Storage and retrieval are based on the RECORD KEY or NOMINAL KEY associated with each record. The RECORD KEY must be specified in the File-Control paragraph when sequential access is to begin at some record other than the first record of the file or when the ACCESS IS RANDOM option is specified.

When the records are accessed at random, the NOMINAL KEY field must be set to the value of the RECORD KEY field of the desired record before a READ statement is executed for the file. The NOMINAL KEY in Working-Storage will contain the key of the desired record.

Any READ statement that refers to a randomly accessed indexed sequential file must include the INVALID KEY option. Following are the entries necessary to access a record at random from an indexed file:

```
SELECT RANDOM-FILE
    ASSIGN TO . . .
    ORGANIZATION IS INDEXED
    ACCESS MODE IS RANDOM
    RECORD KEY IS ID-FIELD
    NOMINAL KEY IS W-S-ID.

READ RANDOM-FILE
    INVALID KEY PERFORM SPECIAL-CASES.
```

The end-of-file record is not checked when a file is randomly accessed. The INVALID KEY option of a READ statement is activated when no record with a control field equal to the value of the NOMINAL KEY variable can be found in the file.

The following program illustrates the random accessing of an indexed file.

The random accessing of a file having the indexed sequential type of organization is illustrated by the COBOL solution of a program having an input file called TRFILE on disk. The records of TRFILE have been sorted by customer number. The following is the record layout of each record in TRFILE.

CNO					DEPT				STYLE			AMOUNT						DATE →	
X	X	X	X	X	9	9	9	9	9	9	9	9	9	9	9	9	9	9	9
000	001	002	003	004	005	006	007	008	009	010	011	012	013	014	015	016	017	018	019

← DATE				CLERK			
9	9	9	9	9	9	9	9
020	021	022	023	024	025	026	027

The contents of the fields within the TRFILE record are as follows:

CNO: Customer number
DEPT: Department number
STYLE: Style number
AMOUNT: Amount of purchase
DATE: Date of transaction
CLERK: Clerk number

A disk file called DESFILE has its records sorted by style number. Each record of DESFILE contains descriptive information concerning a specified style number. The following is the record layout of each record in DESFILE.

ODATE				STYLE			DES-DATA →												
							DESC →												
9	9	9	9	9	9	9	X	X	X	X	X	X	X	X	X	X	X	X	X
000	001	002	003	004	005	006	007	008	009	010	011	012	013	014	015	016	017	018	019

← DES-DATA →																			
← DESC																	MCODE →		
X	X	X	X	X	X	X	X	X	X	X	X	X	X	X	X	X	9	9	9
020	021	022	023	024	025	026	027	028	029	030	031	032	033	034	035	036	037	038	039

← DES-DATA							
← MCODE			PRICE				
9	9	9	9	9	9	9	9
040	041	042	043	044	045	046	047

The contents of the fields within the DESFILE record are as follows:

ODATE: Original order date
STYLE: Style number
DESC: Description
MCODE: Manufacturer code
PRICE: Unit price

Using the style number from the TRFILE record, the DESFILE record having the same style number is to be located. Fields from the DESFILE record are to be combined with fields from the TRFILE record to create a record for output to a disk file called TR2FILE. The following is the record layout of each record written onto TR2FILE.

TRDATA-1 →																			
CNO					DEPT				STYLE			AMOUNT						DATE →	
X	X	X	X	X	9	9	9	9	9	9	9	9	9	9	9	9	9	9	9
000	001	002	003	004	005	006	007	008	009	010	011	012	013	014	015	016	017	018	019

← TRDATA-1								TRDATA-2 →											
← DATE				CLERK				DESC →											
9	9	9	9	9	9	9	9	X	X	X	X	X	X	X	X	X	X	X	X
020	021	022	023	024	025	026	027	028	029	030	031	032	033	034	035	036	037	038	039

← TRDATA-2 →																			
← DESC																		MCODE →	
X	X	X	X	X	X	X	X	X	X	X	X	X	X	X	X	X	X	9	9
040	041	042	043	044	045	046	047	048	049	050	051	052	053	054	055	056	057	058	059

← TRDATA-2								
← MCODE				PRICE				
9	9	9	9	9	9	9	9	9
060	061	062	063	064	065	066	067	068

The contents of the fields within TR2FILE record are as follows:

CNO: Customer number
DEPT: Department number
STYLE: Style number
AMOUNT: Amount of purchase
CLERK: Clerk number
DESC: Description
MCODE: Manufacturer code
PRICE: Unit price

If a style number equal to the TRFILE style number cannot be found in the description file DESFILE, then the TRFILE record is to be written onto an error file called EFILE. The following is the record layout of each record written onto EFILE.

CNO					DEPT				STYLE			AMOUNT						DATE →	
X	X	X	X	X	9	9	9	9	9	9	9	9	9	9	9	9	9	9	9
000	001	002	003	004	005	006	007	008	009	010	011	012	013	014	015	016	017	018	019

← DATE				CLERK			
9	9	9	9	9	9	9	9
020	021	022	023	024	025	026	027

The contents of the fields within the EFILE record are as follows:

CNO: Customer number
DEPT: Department number
STYLE: Style number
AMOUNT: Amount of purchase
DATE: Date of transaction
CLERK: Clerk number

Following is the program for this example:

```
IDENTIFICATION DIVISION.
PROGRAM-ID.    ISPROGRAM.

ENVIRONMENT DIVISION.
CONFIGURATION SECTION.
SOURCE-COMPUTER . . .
OBJECT-COMPUTER . . .
INPUT-OUTPUT SECTION.
```

```
FILE-CONTROL.
    SELECT DYSTEM
        ASSIGN TO . . .
        ORGANIZATION IS INDEXED
        ACCESS MODE IS RANDOM
        RECORD KEY IS DSTYLE
        NOMINAL KEY IS STYLE-W.
    SELECT TRFILE
        ASSIGN TO . . .
    SELECT TR2FILE
        ASSIGN TO . . .
    SELECT EFILE
        ASSIGN TO . . .

DATA DIVISION.
FILE SECTION.
FD  DYSTEM
    BLOCK CONTAINS 10 RECORDS
    RECORD CONTAINS 48 CHARACTERS
    LABEL RECORD IS STANDARD
    DATA RECORD IS DESRECORD.
01  DESRECORD.
    05  FILLER                  PICTURE XXXX.
    05  DSTYLE                  PICTURE 999.
    05  DES-DATA                PICTURE X(41).

FD  TRFILE
    BLOCK CONTAINS 18 RECORDS
    RECORD CONTAINS 28 CHARACTERS
    LABEL RECORD IS STANDARD
    DATA RECORD IS TRECORD.
01  TRECORD.
    05  FILLER                  PICTURE X(9).
    05  STYLE                   PICTURE 999.
    05  FILLER                  PICTURE X(16).

FD  TR2FILE
    BLOCK CONTAINS 7 RECORDS
    RECORD CONTAINS 69 CHARACTERS
    LABEL RECORD IS STANDARD
    DATA RECORD IS TR2RECORD.
01  TR2RECORD.
    05  TRDATA-1                PICTURE X(28).
    05  TRDATA-2                PICTURE X(41).

FD  EFILE
    BLOCK CONTAINS 18 RECORDS
    RECORD CONTAINS 28 CHARACTERS
    LABEL RECORD IS STANDARD
    DATA RECORD IS ERECORD.
01  ERECORD                     PICTURE X(28).

WORKING-STORAGE SECTION.
01  NOMINAL-KEY-FIELD.
    05  STYLE-W                 PICTURE 999.
01  FLAGS.
    05  MORE-DATA-FLAG          PICTURE XXX         VALUE 'YES'.
        88  MORE-DATA                               VALUE 'YES'.
        88  NO-MORE-DATA                            VALUE 'NO'.
```

```
PROCEDURE DIVISION.
MAIN-ROUTINE.
    OPEN      INPUT     DYSTEM
                        TRFILE
              OUTPUT    TR2FILE
                        EFILE.
    PERFORM READ-TRFILE.
    PERFORM PROCESS-ROUTINE THRU PROCESS-EXIT
        UNTIL NO-MORE-DATA.
    CLOSE     DYSTEM
              TRFILE
              TR2FILE
              EFILE.
    STOP RUN.

PROCESS-ROUTINE.
    READ DYSTEM
        INVALID KEY
            MOVE TRECORD TO ERECORD
            WRITE ERECORD
            PERFORM READ-TRFILE
            GO TO PROCESS-EXIT.
    MOVE DES-DATA TO TRDATA-2.
    MOVE TRECORD TO TRDATA-1.
    WRITE TR2RECORD.
    PERFORM READ-TRFILE.

PROCESS-EXIT.
    EXIT.

READ-TRFILE.
    READ TRFILE
        AT END MOVE 'NO' TO MORE-DATA-FLAG.
    MOVE STYLE TO STYLE-W.
```

Adding Records to an Indexed Sequential File at Random

When adding records to an indexed sequential file, it is not necessary to recreate an indexed sequential file as it was with standard sequential files. Any record with a RECORD KEY that is not currently in the indexed sequential file may be added to it.

Environment Division. SELECT, ASSIGN, RECORD KEY, NOMINAL KEY, and ACCESS IS RANDOM clauses similar to those clauses used in indexed sequential file random accessing are required.

Data Division. Similar entries for indexed sequential files are accessed at random.

Procedure Division. When records are to be added to an indexed sequential file, the file must have random access and be opened as an I-O file. The INVALID KEY option of the WRITE statement is activated if the NOMINAL KEY field associated with the record duplicates the RECORD KEY field of a record already in the file.

The following is an example for randomly adding records to an indexed file (Environment and Procedure Division entries only):

```
ENVIRONMENT DIVISION.
    .
    .
INPUT-OUTPUT SECTION.
FILE-CONTROL.
    SELECT STUDENT-MASTER
        ASSIGN TO . . .
        ORGANIZATION IS INDEXED
        ACCESS MODE IS RANDOM
        RECORD KEY IS STUDENT-NUM
        NOMINAL KEY IS KEY-NUMBER.
```

```
        SELECT UPDATE-DATA
            ASSIGN TO . . .

    PROCEDURE DIVISION.
    MAIN-ROUTINE.
        OPEN     INPUT    UPDATE-DATA
                 I-O      STUDENT-MASTER.
        READ UPDATE-DATA
            AT END MOVE 'NO' TO MORE-DATA-FLAG.
        PERFORM PROCESS-ROUTINE
            UNTIL NO-MORE-DATA.
        CLOSE    UPDATE-DATA
                 STUDENT-MASTER.
        STOP RUN.

    PROCESS-ROUTINE.
        MOVE CARD-NUMBER TO KEY-NUMBER.
        WRITE STUDENT-DATA FROM TRANSFER
            INVALID KEY
                PERFORM BAD-KEY.
        READ UPDATE-DATA
            AT END MOVE 'NO' TO MORE-DATA-FLAG.

    BAD-KEY.
        DISPLAY KEY-NUMBER ' BAD KEY '
            UPON CONSOLE.
```

Updating and Replacing Records in an Indexed Sequential File at Random

Environment Division. Same entries as randomly *adding* records to an indexed sequential file.

Data Division. Same entries as randomly *adding* records to an indexed sequential file.

Procedure Division. Every *READ* statement that refers to a randomly accessed indexed sequential file must have the INVALID KEY option. The INVALID KEY option of a READ statement refers to a randomly accessed indexed sequential file opened as I-O is activated under the same circumstances as the INVALID KEY option of a READ statement that refers to randomly accessed files opened as INPUT. The INVALID KEY option is activated when no RECORD KEY equal to the current value of the NOMINAL KEY variable can be located in the file.

Before a READ statement can be executed for a randomly accessed indexed sequential file, the NOMINAL KEY variable must be set to the desired value, and the file must be opened as I-O or INPUT.

After a READ statement is executed for an indexed sequential file opened as I-O, the accessed record may be updated and placed back in the same position in the file. An updated record is placed back into an indexed sequential file with a REWRITE statement. The next input or output statement for an indexed sequential file opened as I-O after a READ statement may be a statement to place the record back into the file with a REWRITE statement. The key fields should not be altered between a READ and a REWRITE statement.

The READ statement and its associated REWRITE statement may be separated by any number of statements provided that they are not separated by any other input or output statements that refer to the indexed sequential file.

REWRITE Statement. The function of the REWRITE statement is to place a logical record on a direct-access device with a specified record, if the contents of the associated key(s) are found to be valid. (The format for the REWRITE statement is shown in figure 13.29.)

```
REWRITE record-name   [ FROM identifier ]   [ ; INVALID KEY imperative-statement ]
```

Figure 13.29 Format—REWRITE statement.

The READ statement for a file whose access mode is sequential must be executed before a REWRITE statement for the file can be executed. A REWRITE statement can be executed only for indexed sequential files opened as I-O. If the ACCESS IS RANDOM option is specified for the file, the NOMINAL KEY must be set to the desired value prior to the execution of the REWRITE statement. The record-name is the name of the logical record in the Data Division.

The record-name must be associated with an indexed sequential file that was opened as I-O. The record that is placed back into the file is the last record accessed by the READ statement referring to that file.

If the INVALID KEY option of a READ statement is activated, no record has been accessed. The next input or output statement for an indexed sequential file after the INVALID KEY of a READ statement is activated could be another READ statement to access a different record.

The following program is designed to change records at random in an indexed sequential file (Environment and Procedure Division entries only):

```
ENVIRONMENT DIVISION.
      .
      .
INPUT-OUTPUT SECTION.
FILE-CONTROL.
    SELECT STUDENT-MASTER-FILE
        ASSIGN TO . . .
        ORGANIZATION IS INDEXED
        ACCESS MODE IS RANDOM
        RECORD KEY IS S-NUMBER
        NOMINAL KEY IS KEY-NUMBER.
    SELECT UPDATE-DATA-FILE
        ASSIGN TO . . .

PROCEDURE DIVISION.
MAIN-ROUTINE.
    OPEN     INPUT     UPDATE-DATA-FILE
             I-O       STUDENT-MASTER-FILE.
    READ UPDATE-DATA-FILE
        AT END MOVE 'NO' TO MORE-DATA-FLAG.
    PERFORM PROCESS-ROUTINE THRU PROCESS-ROUTINE-EXIT
        UNTIL NO-MORE-DATA.
    CLOSE    UPDATE-DATA-FILE
             STUDENT-MASTER-FILE.
    STOP RUN.

PROCESS-ROUTINE.
    MOVE C-NUMBER      TO      KEY-NUMBER.
    READ STUDENT-MASTER-FILE
        INVALID KEY
            PERFORM BAD-KEY
            GO TO PROCESS-ROUTINE-EXIT.
    MOVE C-STREET      TO      STREET.
    MOVE C-CITY        TO      CITY.
    MOVE C-STATE       TO      STATE.
    REWRITE STUDENT-DATA-RECORD.
    READ UPDATE-DATA-FILE
        AT END MOVE 'NO' TO MORE-DATA-FLAG.

PROCESS-ROUTINE-EXIT.
    EXIT.

BAD-KEY.
    DISPLAY KEY-NUMBER 'BAD KEY'
        UPON CONSOLE.
```

The Toluca Community College, which has a problem with students who move frequently, uses the program just presented to update addresses in its indexed sequential master file.

The following program is designed to add, delete, and update records at random in an indexed sequential file.

The random accessing of a master file having an indexed sequential type of organization is illustrated by the COBOL solution of a program having a transaction file called TFILE on disk. The records of TFILE have not been sorted. One or more transaction records may have the same customer number. The following is the layout of each record in TFILE.

TNO						TAMOUNT					TCODE	
X	X	X	X	X	X	9	9	9	9	9	9	9
000	001	002	003	004	005	006	007	008	009	010	011	012

The contents of the data fields within the TFILE records are as follows:

TNO: Customer number
TAMOUNT: Transaction amount
TCODE: Transaction code.

The type of transaction record is determined by the content of the transaction code field as follows:

1 indicates an insertion transaction record.
2 indicates a deposit transaction record.
3 indicates an interest transaction record.
4 indicates a withdrawal transaction record.
5 indicates a deletion transaction record.

A disk file record called MPILE is a master file whose records have been sorted by an ascending customer number. Each master record has a unique customer number. The structure of each record in this master file follows:

MNO						BALANCE						LTCODE	LTAMOUNT						
X	X	X	X	X	X	9	9	9	9	9	9	9	9	9	9	9	9	9	9
000	001	002	003	004	005	006	007	008	009	010	011	012	013	014	015	016	017	018	019

LTDATE						MDATA											
9	9	9	9	9	9	X	X	X	X	X	X	X	X	X	X	X	X
020	021	022	023	024	025	026	027	028	029	030	031	032	033	034	035	036	037

The contents of the data fields within the MFILE record are as follows:

MNO: Customer number
BALANCE: Balance
LTCODE: Last transaction code
LTAMOUNT: Last transaction amount
LTDATE: Last transaction date in the order day, month, year
MDATA: Miscellaneous data

The *master file* called MFILE must be *reorganized as a data file of an indexed sequential system called MSYSTEM.* Using the customer number from the transaction record, the master record having the same customer number is to be located. A master record is to be updated by the transaction record having the same customer number. The updating actions to be taken are determined by the transaction code.

An *insertion transaction* is processed by inserting a new record into the master file. The inserted record is created by placing the transaction customer number into the master customer number field. The transaction amount is placed into both the current balance and the last transaction amount fields of the master record. The transaction code is placed into the last transaction code field. The data is placed into the last transaction date field. Zero characters are placed into the remaining fields of the master record.

A *deposit or interest transaction* record updates the master record by adding the amount of the transaction to the current balance in the master record. The transaction code is placed into the last transaction code field. The transaction amount is placed into the last transaction amount field. The date is placed into the last transaction date field.

A *withdrawal transaction* record updates the master record by subtracting the amount of the transaction from the current balance in the master record. The transaction code is placed into the last transaction code field. The transaction amount is placed into the last transaction amount field. The date is placed into the last transaction date field.

A *deletion transaction* is processed by deleting the record from the master file.

A record is created and output to an error file called ERROR-FILE whenever one of the six error conditions occur.

The following is the record layout of each record in ERROR-FILE.

ENO						EAMOUNT							ETCODE	ECODE
X	X	X	X	X	X	9	9	9	9	9 ^	9	9	9	X
000	001	002	003	004	005	006	007	008	009	010	011	012	013	014

The contents of the data fields within the ERROR-FILE record are as follows:

ENO: Customer number
EAMOUNT: Transaction amount or balance if master record is being deleted
ETCODE: Transaction code
ECODE: Error code

The value of the error code is determined by the type of error condition. The error code for each of the six conditions are as follows:

1. The error code is equal to the letter D when a master record has been deleted.
2. The error code is equal to the letter E when a deletion record cannot update the master file.
3. The error code is equal to the letter I when a master record already exists for the customer number in an insertion record.
4. The error code is equal to the letter J when an insertion record cannot update the master file.
5. The error code is equal to the letter M when a master record does not exist for the customer number in a deposit, interest, withdrawal, or deletion transaction record.
6. The error code is equal to the letter C when a transaction record does not have a transaction code equal to 1, 2, 3, 4, or 5.

The program for this example follows:

```
IDENTIFICATION DIVISION.
PROGRAM-ID.     ISUPDATE.

ENVIRONMENT DIVISION.
CONFIGURATION SECTION.
SOURCE-COMPUTER . . .
OBJECT-COMPUTER . . .
INPUT-OUTPUT SECTION.
FILE-CONTROL.
    SELECT MSYSTEM
        ASSIGN TO . . .
        ORGANIZATION IS INDEXED
        ACCESS MODE IS RANDOM
        RECORD KEY IS MNO
        NOMINAL KEY IS WNO.
    SELECT TFILE
        ASSIGN TO . . .
    SELECT ERRORFILE
        ASSIGN TO . . .

DATA DIVISION.
FILE SECTION.
FD  MSYSTEM
    BLOCK CONTAINS 13 RECORDS
    RECORD CONTAINS 38 CHARACTERS
    LABEL RECORD IS STANDARD
    DATA RECORD IS MRECORD.
01  MRECORD.
    05  MNO                       PICTURE X(6).
    05  BALANCE                   PICTURE 9(5)V99.
    05  LTCODE                    PICTURE 9.
    05  LTAMOUNT                  PICTURE 9999V99.
    05  LTDATE                    PICTURE X(6).
    05  FILLER                    PICTURE X(12).
```

```
FD  TFILE
    BLOCK CONTAINS 39 RECORDS
    RECORD CONTAINS 13 CHARACTERS
    LABEL RECORD IS STANDARD
    DATA RECORD IS TRECORD.
01  TRECORD.
    05  TNO                         PICTURE X(6).
    05  TAMOUNT                     PICTURE 9999V99.
    05  TCODE                       PICTURE 9.

FD  ERRORFILE
    BLOCK CONTAINS 34 RECORDS
    RECORD CONTAINS 15 CHARACTERS
    LABEL RECORD IS STANDARD
    DATA RECORD IS ERECORD.
01  ERECORD.
    05  ENO                         PICTURE X(6).
    05  EAMOUNT                     PICTURE 9(5)V99.
    05  ETCODE                      PICTURE 9.
    05  ECODE                       PICTURE X.

WORKING-STORAGE SECTION.
01  WRECORD.
    05  WNO                         PICTURE X(6).
    05  WBALANCE                    PICTURE 9(5)V99.
    05  WLTCODE                     PICTURE 9.
    05  WLTAMOUNT                   PICTURE 9999V99.
    05  FILLER                      PICTURE X(12)      VALUE ZEROS.
01  FLAGS.
    05  MORE-DATA-FLAG              PICTURE XXX        VALUE 'YES'.
        88  MORE-DATA                                  VALUE 'YES'.
        88  NO-MORE-DATA                               VALUE 'NO'.

PROCEDURE DIVISION.
MAIN-ROUTINE.
    OPEN    INPUT    TFILE
            I-O      MSYSTEM
            OUTPUT   ERRORFILE.
    READ TFILE
        AT END MOVE 'NO' TO MORE-DATA-FLAG.
    PERFORM PROCESS-ROUTINE THRU PROCESS-ROUTINE-EXIT
        UNTIL NO-MORE-DATA.
    CLOSE   TFILE
            MSYSTEM
            ERRORFILE.
    STOP RUN.
    PROCESS-ROUTINE.
        GO TO   INSERTION,
                UPDATE, UPDATE, UPDATE,
                DELETION,
                DEPENDING ON TCODE.
        PERFORM TRANSERR.
        READ TFILE
            AT END MOVE 'NO' TO MORE-DATA-FLAG.
        GO TO PROCESS-ROUTINE-EXIT.

    INSERTION.
        MOVE TNO            TO WNO.
        MOVE TAMOUNT        TO WBALANCE, WLTAMOUNT.
        MOVE VDATE          TO LTDATE.
        MOVE TCODE          TO WLTCODE.
```

```
    WRITE MRECORD FROM WRECORD
        INVALID KEY PERFORM CHECK-FLAG.
    GO TO PROCESS-ROUTINE-EXIT.

UPDATE.
    READ MSYSTEM
        INVALID KEY  PERFORM BADTRAN
                     GO TO PROCESS-ROUTINE-EXIT.
    IF TCODE = 2 OR 3
        ADD TAMOUNT TO BALANCE
    ELSE
        SUBTRACT TAMOUNT FROM BALANCE.
    MOVE TCODE              TO LTCODE.
    MOVE TAMOUNT            TO LTAMOUNT.
    MOVE VDATE              TO LTDATE.
    REWRITE MRECORD.
    GO TO PROCESS-ROUTINE-EXIT.

DELETION.
    MOVE 'D'                TO ECODE.
    MOVE BALANCE            TO EAMOUNT.
    DISPLAY 'CUSTOMER NUMBER ' TNO ' DELETED FROM FILE '
        UPON CONSOLE.
    DELETE MSYSTEM RECORD
        INVALID KEY
            DISPLAY 'CUSTOMER NUMBER ' TNO ' NOT IN FILE '
                UPON CONSOLE.
    GO TO PROCESS-ROUTINE-EXIT.

PROCESS-ROUTINE-EXIT.
    EXIT.

TRANSERR.
    MOVE 'C'                TO ECODE.
    PERFORM E1 THRU E2.

CHECK-FLAG.
    IF MYSYSTEM $ISCONTFLY = 49
        MOVE 'I'            TO ECODE
    ELSE
        MOVE 'J'            TO ECODE.
    PERFORM E1 THRU E2.

E1.
    MOVE TAMOUNT            TO EAMOUNT.
E2.
    MOVE TNO                TO ENO.
    MOVE TCODE              TO ETCODE.
    WRITE ERECORD.
```

The following program is for random retrieval and updating of an indexed sequential file. This program randomly updates an existing indexed sequential file. The READ IS-FILE statement causes a search of indexes for an equal comparison between the NOMINAL KEY obtained from the input record and the RECORD KEY of the I-O file. If an equal comparison occurs, the record is updated, and the details of this update are printed. If a matching record is not found, the invalid key branch is taken.

```
IDENTIFICATION DIVISION.
PROGRAM-ID.     RANDOMIS.
*PURPOSE.       ILLUSTRATE RANDOM RETRIEVAL FROM IS-FILE.

ENVIRONMENT DIVISION.
CONFIGURATION SECTION.
SOURCE-COMPUTER . . .
OBJECT-COMPUTER . . .
```

```
INPUT-OUTPUT SECTION.
FILE-CONTROL.
    SELECT IS-FILE
        ASSIGN TO . . .
        ORGANIZATION IS INDEXED
        ACCESS MODE IS RANDOM
        RECORD KEY IS REC-ID
        NOMINAL KEY IS KEY-ID.
    SELECT CARD-FILE
        ASSIGN TO . . .
    SELECT PRINT-FILE
        ASSIGN TO . . .

DATA DIVISION.
FILE SECTION.
FD  IS-FILE
    BLOCK CONTAINS 5 RECORDS
    RECORD CONTAINS 100 CHARACTERS
    LABEL RECORDS ARE STANDARD
    DATA RECORD IS DISK.
01  DISK.
    05  DELETE-CODE            PICTURE X.
    05  REC-ID                 PICTURE 9(10).
    05  DISK-FLD1              PICTURE X(10).
    05  DISK-NAME              PICTURE X(20).
    05  DISK-BAL               PICTURE 99999V99.
    05  FILLER                 PICTURE X(52).

FD  CARD-FILE
    LABEL RECORDS ARE OMITTED
    DATA RECORD IS CARDS.
01  CARDS.
    05  KEY-IDA                PICTURE 9(10).
    05  CD-NAME                PICTURE X(20).
    05  CD-AMT                 PICTURE 99999V99.
    05  FILLER                 PICTURE X(43).

FD  PRINT-FILE
    LABEL RECORDS ARE OMITTED
    DATA RECORD IS PRINTER.
01  PRINTER.
    05  FILLER                 PICTURE X.
    05  FORMSC                 PICTURE X.
    05  PRINT-ID               PICTURE X(10).
    05  FILLER                 PICTURE X(10).
    05  PRINT-NAME             PICTURE X(20).
    05  FILLER                 PICTURE X(10).
    05  PRINT-BAL              PICTURE $ZZZ,999.99.
    05  FILLER                 PICTURE X(10).
    05  PRINT-AMT              PICTURE $ZZZ,ZZZ.99.
    05  FILLER                 PICTURE X(10).
    05  PRINT-NEW-BAL          PICTURE $ZZZ,ZZZ.99.
    05  FILLER                 PICTURE X(28).
```

```
WORKING-STORAGE SECTION.
77  KEY-ID                          PICTURE 9(10).
01  FLAGS.
    05  MORE-DATA-FLAG              PICTURE XXX        VALUE 'YES'.
        88  MORE-DATA                                  VALUE 'YES'.
        88  NO-MORE-DATA                               VALUE 'NO'.

PROCEDURE DIVISION.
MAIN-ROUTINE.
    OPEN    INPUT    CARD-FILE
            OUTPUT   PRINT-FILE
            I-O      IS-FILE.
    READ CARD-FILE
        AT END MOVE 'NO' TO MORE-DATA-FLAG.
    PERFORM PROCESS-ROUTINE THRU PROCESS-ROUTINE-EXIT
        UNTIL NO-MORE-DATA.
    CLOSE    CARD-FILE
             PRINT-FILE
             IS-FILE.
    DISPLAY ' END OF JOB '
        UPON CONSOLE.
    STOP RUN.

PROCESS-ROUTINE.
    MOVE SPACES         TO PRINTER.
    MOVE KEY-IDA        TO KEY-ID.
    READ IS-FILE
        INVALID KEY
            PERFORM NO-RECORD
            GO TO PROCESS-ROUTINE-EXIT.
    MOVE REC-ID         TO PRINT-ID.
    MOVE DISK-NAME      TO PRINT-NAME.
    MOVE DISK-BAL       TO PRINT-BAL.
    MOVE CD-AMT         TO DISK-BAL.
    MOVE DISK-BAL       TO PRINT-NEW-BAL.
    REWRITE DISK
        INVALID KEY
            PERFORM NO-RECORD
            GO TO PROCESS-ROUTINE-EXIT.
    WRITE PRINTER
        AFTER ADVANCING 2 LINES.
    READ CARD-FILE
        AT END MOVE 'NO' TO MORE-DATA-FLAG.

PROCESS-ROUTINE-EXIT.
    EXIT.

NO-RECORD.
    DISPLAY 'NO RECORD FOUND FOR ' KEY-ID
        UPON CONSOLE.
    READ CARD-FILE
        AT END MOVE 'NO' TO MORE-DATA-FLAG.
```

Random Mode
The following operations may be performed on an indexed file accessed in the random mode:

WRITE	Places the contents of the specified record into a record space situated such that the content of the record key data item is greater than the key of the preceding logical record and less than the key of the succeeding logical record. If the contents of the record key data item are equal to the key of an existing logical record, or if writing the logical record exceeds the boundaries of the allocated file space, an INVALID KEY condition exists. The file must be open for output or input-output.
DELETE	Removes the logical record identified by the contents of the record key data item from the file. If no such logical record exists on the file, an INVALID KEY condition exists. The file must be open for input-output.
READ	Accesses the logical record identified by the contents of the record key data item and makes the contents of that record available in the file record area. If no such logical record exists on the file, an INVALID KEY condition exists. The file must be open for input or input-output.
REWRITE	Replaces the logical record identified by the contents of the record key data item with the contents of the specified record. If no such record exists on the file, an INVALID KEY condition exists. The logical record being replaced must be equal in size to the record specified in the REWRITE statement (the replacement record). The file must be open for input-output.

Dynamic-Access Mode
In the dynamic-access mode, the programmer may alternate from sequential access to random access using appropriate forms of input-output statements. Dynamic access is not allowed with sequential organization.

The following operations may be performed on an indexed file in the dynamic mode:

WRITE	Places the contents of the specified record into record space situated such that the content of the record key data item is greater than the key of the preceding logical record and less than the key of the succeeding logical record. If the contents of the record key data item are equal to the key of the existing logical record, or if writing the logical record would exceed the boundaries of the allocated file space, an INVALID KEY condition exists. The file must be open for output or input-output.
DELETE	Removes the logical record identified by the contents of the record key data item from the file. If no such logical record exists on the file, an INVALID KEY condition exists. The file must be open for input-output.
START	Positions the file such that the next logical record accessed will be based on the comparison specified in the START statement, in relation to the contents of the record key data item. If the comparison is not satisfied by any logical record on the file, an INVALID KEY condition exists. The file must be open for input or input-output.
READ NEXT	If the file was positioned by an OPEN or START statement, and the logical record to which it was positioned is still accessible (not having been deleted), the input-output control system makes the contents of that logical record available in the file record area. Otherwise, the input-output control system accesses the next logical record on the file and makes the contents of that record available in the file record area. If no 'next' logical record exists on the file, an AT END condition exists. The file must be open for input or input-output.
READ	Accesses the logical record identified by the contents of the record key data item and makes the contents of that logical record available in the file record area. If no such record exists on the file, an INVALID KEY condition exists. The file must be open for input or input-output.
REWRITE	Replaces the logical record identified by the contents of the record key data item with the contents of the specified record. If no such logical record exists on the file, an INVALID KEY condition exists. The logical record being replaced must be equal in size to the record specified in the REWRITE statement (the replacement record). The file must be open for input-output.

```
READ file-name  [ NEXT ]  RECORD  [ INTO identifier ]

    [ ; AT END imperative-statement ]
```

Figure 13.30 Format—READ NEXT statement.

READ NEXT Statement
The NEXT phrase must be specified for files in the dynamic access mode. The READ statement with the NEXT phrase specified causes the next logical record to be retrieved from that file. (The format for the READ NEXT statement is shown in figure 13.30.)

Relative Organization

Relative file organization permits accessing of records of a mass storage device in either a random or a sequential manner. Each record in a relative file is uniquely identified by an integer value greater than zero that specifies the record's logical ordinal position in the file.

Relative organization does not use an index or record key to identify each record in a file. The relative file consists of records that are identified by relative record numbers. The file may be thought of as composed of a serial string of areas, each capable of holding a logical record. Each of these areas is denominated by a relative record number. Records are stored and retrieved based on this number. For example, the tenth record is the one addressed by relative record number 10, and is in the tenth record area, whether or not records have been written in the first through ninth record areas.

Access Modes

In the *sequential* access mode, the sequence in which records are accessed is the ascending order of the relative record numbers of all records that currently exist within the file.

In the *random* access mode, the sequence in which records are accessed is controlled by the programmer. The desired record is accessed by placing its relative record number in a relative key data item.

In the *dynamic* access mode, the programmer may change at will from sequential access to random access using appropriate forms of input-output statements.

The relative file organization can be best used when the record identification key can be used as a record number or can be easily converted to a record number.

Environment Division. The SELECT, ASSIGN, RESERVE, ACCESS MODE, and FILE STATUS clauses are written in the same manner as indexed files with the exception of the ORGANIZATION and RELATIVE KEY. (The format for the File-Control entry—relative organization is shown in figure 13.31.)

```
SELECT file-name

  ASSIGN TO implementor-name-1  [ , implementor-name-2 ] . . .

  [ ; RESERVE integer-1  [ AREA  ] ]
                         [ AREAS ]

  ; ORGANIZATION IS RELATIVE

  [                  { SEQUENTIAL  [ , RELATIVE KEY IS data-name-1 ] } ]
  [ ; ACCESS MODE IS { { RANDOM  }                                   } ]
  [                  { { DYNAMIC }   , RELATIVE KEY IS data-name-1   } ]

  [ ; FILE STATUS IS data-name-2 ]  .
```

Figure 13.31 Format—File Control entry—relative organization.

The ORGANIZATION IS RELATIVE statement is used for all files that are relatively organized.

The RELATIVE KEY clause applies only to files with relative organization. The relative key data item contains the logical ordinal position of the record in the file. The first logical record has a relative record number of 1, and subsequent logical records have relative record numbers of 2, 3, 4, etc. The format of the RELATIVE KEY clause is

RELATIVE KEY IS data-name-1

Data-name-1 must not be in a record description entry for that file. The value contained must be an unsigned integer in the range of 1 . . .

The ACCESS IS SEQUENTIAL clause is optional; however, if specified, data-name-1 will contain the current relative record number.

When the relative file is opened in the output mode, the file may be created by one of the following:

1. If the ACCESS MODE IS SEQUENTIAL, the WRITE statement causes a record to be released by the system. The first record will have a relative record number of 1 and subsequent records released will have relative numbers of 2, 3, 4, etc. If the RELATIVE KEY data item has been specified in the SELECT clause for the associated file, the relative record number of the record just released will be placed in the RELATIVE KEY data item during execution of the WRITE statement.
2. If the ACCESS MODE IS RANDOM, prior to the execution of the WRITE statement, the value of the RELATIVE KEY data item must be initialized in the program, while the relative record number to be associated with the record is then released by the WRITE statement.

When a relative file is opened in the I-O mode and the ACCESS MODE IS RANDOM, records are to be inserted in the associated file. The value of the RELATIVE KEY data item must be initialized by the program, with the record to be associated with relative record number in the record area. Execution of the WRITE statement then causes the contents of the record area to be released.

Data Division. The same entries as indexed sequential are to be used.

Procedure Division. The same entries may be used as for indexed sequential organization.

A program to create and sequentially access a relative file follows:

```
IDENTIFICATION DIVISION.
PROGRAM-ID.      REL-IO.
*      1. TO ILLUSTRATE CREATION OF A RELATIVE FILE IN RANDOM MODE.
*
*      2. TO ILLUSTRATE SEQUENTIAL READ OF A RELATIVE FILE.

ENVIRONMENT DIVISION.
CONFIGURATION SECTION.
SOURCE-COMPUTER . . .
OBJECT-COMPUTER . . .
INPUT-OUTPUT SECTION.
FILE-CONTROL.
    SELECT CARD-FILE
        ASSIGN TO . . .
    SELECT RELATIVE-CR
        ASSIGN TO . . .
        ORGANIZATION IS RELATIVE
        ACCESS MODE IS RANDOM
        RELATIVE KEY IS WORK-KEY.
    SELECT RELATIVE-RD
        ASSIGN TO . . .
        ORGANIZATION IS RELATIVE
        ACCESS MODE IS SEQUENTIAL.
    SELECT PRINT-FILE
        ASSIGN TO . . .

I-O-CONTROL.
    SAME AREA FOR RELATIVE-CR, RELATIVE-RD.
```

```
DATA DIVISION.
FILE SECTION.
FD  CARD-FILE
    LABEL RECORDS ARE OMITTED
    DATA RECORD IS CARD-RECORD.
01  CARD-RECORD.
    05  CARD-KEY.
        10  CARD-REGION       PICTURE 9.
        10  CARD-ACCTNO       PICTURE 999.
    05  CARD-SALESMAN         PICTURE X(30).
    05  CARD-LASTYEAR         PICTURE 9(7)V99.
    05  CARD-TYPE             PICTURE XX.
    05  FILLER                PICTURE X(35).
FD  RELATIVE-CR
    LABEL RECORDS OMITTED
    RECORD CONTAINS 60 CHARACTERS
    DATA RECORD IS REL-CR-RECORD.
01  REL-CR-RECORD.
    05  REL-CR-SALESMAN       PICTURE X(30).
    05  REL-CR-TYPE           PICTURE XX.
    05  REL-CR-LASTYEAR       PICTURE 9(7)V99.
    05  FILLER                PICTURE X(19).

FD  RELATIVE-RD
    LABEL RECORDS ARE OMITTED
    RECORD CONTAINS 60 CHARACTERS
    DATA RECORD IS REL-RD-RECORD.
01  REL-RD-RECORD.
    05  REL-RD-IMAGE          PICTURE X(60).

FD  PRINT-FILE
    LABEL RECORDS ARE OMITTED
    DATA RECORD IS PRINT-RECORD.
01  PRINT-RECORD              PICTURE X(133).

WORKING-STORAGE SECTION.
77  WORK-KEY                  PICTURE 9(4).
01  FLAGS.
    05  MORE-INPUT-FLAG       PICTURE XXX    VALUE 'YES'.
        88  MORE-INPUT                       VALUE 'YES'.
        88  NO-MORE-INPUT                    VALUE 'NO'.
    05  MORE-TRANS-FLAG       PICTURE XXX    VALUE 'YES'.
        88  MORE-TRANS                       VALUE 'YES'.
        88  NO-MORE-TRANS                    VALUE 'NO'.

PROCEDURE DIVISION.
MAIN-ROUTINE.
    OPEN    INPUT    CARD-FILE
            OUTPUT   RELATIVE-CR.
    READ CARD-FILE
        AT END MOVE 'NO' TO MORE-INPUT-FLAG.
    PERFORM CREATE-RELATIVE
        UNTIL NO-MORE-INPUT.
    CLOSE   CARD-FILE
            RELATIVE-CR.
    OPEN    INPUT    RELATIVE-RD
            OUTPUT   PRINT-FILE.
    READ RELATIVE-RD
        AT END MOVE 'NO' TO MORE-TRANS-FLAG.
    PERFORM READ-RELATIVE
        UNTIL NO-MORE-TRANS.
    CLOSE   RELATIVE-RD
            PRINT-FILE.
    STOP RUN.
```

```
CREATE-RELATIVE.
    MOVE CARD-KEY               TO WORK-KEY.
    MOVE CARD-SALESMAN          TO REL-CR-SALESMAN.
    MOVE CARD-LASTYEAR          TO REL-CR-LASTYEAR.
    MOVE CARD-TYPE              TO REL-CR-TYPE.
    WRITE REL-CR-RECORD
        INVALID KEY
            PERFORM WRITE-ERROR.
    READ CARD-FILE
        AT END MOVE 'NO' TO MORE-INPUT-FLAG.

WRITE-ERROR.
    DISPLAY ' INVALID KEY ' CARD-KEY
        UPON CONSOLE.
    CLOSE    CARD-FILE
             RELATIVE-CR.
    STOP RUN.

READ-RELATIVE.
    WRITE PRINT-RECORD FROM REL-RD-RECORD
        AFTER ADVANCING 1 LINES.
    READ RELATIVE-RD
        AT END MOVE 'NO' TO MORE-TRANS-FLAG.
```

Sequential Mode

The following operations may be performed on a relative file accessed in the sequential mode:

WRITE	Accesses the next record location on the file, places the contents of the specified record in that location, and places the relative record number of the record in the relative key data item. If there is no 'next' record location, an INVALID KEY condition exists. The file must be open as output.
DELETE	Removes the logical record accessed by the previous input-output operation (which must have been a successful READ statement) from that file. The file must be open for input-output.
START	Positions the file such that the relative record number of the next logical record accessed will be based on the comparison specified in the START statement, in relation to the contents of the relative key data item. If the comparison is not satisfied by any logical record on the file, an INVALID KEY condition exists. The file must be open for input or input-output.
READ	Accesses the next logical record on the file, makes the contents of that record available in the file record area, and places the relative record number of the record in the relative key data item. The number of record locations traversed to access the next logical record is immaterial. If there is no 'next' logical record, an AT END condition exists. The file must be open for input or input-output.
REWRITE	Replaces the logical record accessed by the previous input-output operation (which must have been a successful READ statement) with the contents of the specified record. The file must be open for input-output.

Random Mode
The following operations may be performed on a relative file in the random mode:

WRITE	Places the contents of the specified record in the record location identified by the contents of the relative key data item. If the specified record location is outside the boundaries of the allocated file space, or if a logical record already occupies that location, an INVALID KEY condition exists. The file must be open for output or input-output.
DELETE	Removes the logical record identified by the contents of the relative key data item from the file. If the specified record location is outside the boundaries of the allocated file space, or if a logical record does not occupy that location, an INVALID KEY condition exists. The file must be open for input-output.
READ	Accesses the logical record identified by the contents of the relative key data item and makes the contents of that record available in the file record area. If the specified record location is outside the boundaries of the allocated file space, or if a logical record does not occupy that location, an INVALID KEY condition exists. The file must be open for input or input-output.
REWRITE	Replaces the logical record identified by the contents of the relative key data item with the contents of the specified record. If the specified record location is outside the boundaries of the allocated file space, or if a logical record does not occupy that record location, an INVALID KEY condition exists. The logical record being replaced must be equal in size to the record specified in the REWRITE statement (the replacement record). The file must be open for input- output.

Dynamic Mode
The following operations may be performed on a relative file accessed in the dynamic mode:

WRITE	Places the contents of the specified record in the record location identified by the contents of the relative key data item. If the specified location is outside the boundaries of the allocated file space, or if a logical record already occupies that location, an INVALID KEY condition exists. The file must be open for output or input-output.
DELETE	Removes the logical record identified by the contents of the relative key data item from the file. If the specified record is outside the boundaries of the allocated file space, or if a logical record does not occupy the specified record location, an INVALID KEY condition exists. The file must be open for input-output.
START	Positions the file at a record location such that the next logical record accessed will be based on the comparison specified in the START statement, in relation to the contents of the relative key data item. If the comparison is not satisfied by any logical record on the file, an INVALID KEY condition exists. The file must be open for input or input-output.
READ NEXT	If the file was positioned by an OPEN or START statement, and the record location to which it was positioned is still occupied by a logical record (the logical record not having been deleted), the input-output control system makes the contents of that logical record available in the file record area and places the relative record number of the record in the relative key data item. Otherwise, the input-output control system accesses the next logical record on the file, makes the contents of that record available in the file record area, and places the relative record number of the record in the relative key data item. If there is no 'next' logical record, an AT END condition exists. The file must be open for input or input-output.
READ	Accesses the logical record identified by the contents of the relative key data item and makes the contents of that record available in the file record area. If the specified record location is outside the boundaries of the allocated file space, or if a logical record does not occupy that location, an INVALID KEY condition exists. The file must be open for input or input-output.
REWRITE	Replaces the logical record identified by the contents of the relative key data item with the contents of the specified record. If the specified record location is outside the boundaries of the allocated file space, or if that record does not contain a logical record, an INVALID KEY condition exists. The logical record being replaced must be equal in size to the record specified in the REWRITE statement (the replacement record). The file must be open for input-output.

Summary

Direct-access mass storage enables the programmer to maintain current records of diversified applications, as well as process nonsequential and intermixed data for multiple access. The term direct-access implies access at random by multiple users of data (files, programs, subroutines, programming aids) involving mass storage devices.

A direct-access storage device (DASD) is one in which each physical record has a discrete location and a unique address.

The term "file" can mean a physical record (a DASD, for instance) or an organized collection of related information. The term file is also referred to as a data set.

The term "record" can also mean a physical record or a logical unit. A logical record may be defined as a collection of data related to a common identifier. A physical record consists of one or more logical records. The term block is equivalent to the term physical record.

A key is a control field that uniquely identifies a record.

File organization refers to the logical relationships among the various records that constitute the file, particularly with respect to the means of identification and access to any specific record. Records in a file must be logically organized so that they can be retrieved efficiently for processing.

The inherent characteristics of the file must be considered in selecting an efficient method of organization. Volatility refers to the additions and deletions of records from a file. The percentage of activity is another factor to be considered. The size of the file is an important consideration in the organization of the file.

The manner in which the file is to be processed has a bearing on its organization. In sequential processing, the input transactions are grouped together, sorted into the same sequence as the master file, and the resulting batch is then processed against the master file. Random processing is the processing of detail transactions against a master file in whatever order they occur.

Files may be organized in a sequential, indexed sequential, or relative manner. In a sequential file, records are organized solely on the basis of their successive physical location in the file. An indexed sequential file is similar to a sequential file in that rapid sequential file processing is possible. An indexed sequential organization, however, by references associated with the file, also makes it possible to quickly locate individual records for non-sequential processing. A relative file consists of records that are identified by relative record numbers. Records are stored and retrieved based on this number.

File input-output devices provide the capability for transferring data from an external recording device into computer memory (input), or from memory to an external recording device (output) with minimum concern for the physical characteristics of the recording device in the processes required to complete the transfer. Sequential files are organized such that each record in the file except the first has a unique predecessor record and each record except the last has a unique successor record. Indexed organized files contain records that may be accessed by the value of a key. A relative file consists of records that are identified by relative record numbers. The file may be thought of as composed of a serial string of areas, each capable of holding a logical record.

In the sequential-access mode, the sequence in which records are accessed is: the order in which the records were originally written, if the sequence was sequential; the ascending order of the record key values, if the organization was indexed; or the ascending order of the relative record numbers of all records written within the file, if the organization was relative.

In the random-access mode, the sequence in which records are accessed is controlled by the programmer. The desired record is accessed by placing the value of its record key in a record key data item (if organization is indexed), or by placing its relative number in a relative key data item (if the organization is relative). Random access is not allowed with sequential organization.

In a dynamic-access mode, the programmer may alternate from sequential access to random access using appropriate forms of input-output statements. Dynamic access is not allowed with sequential organization.

The current record pointer is a conceptual entry used to facilitate specification of the next record to be accessed within a given file. The setting of the current record pointer is affected only by the OPEN, READ, and START statements.

The FILE STATUS clause places a specified two-character data item during the execution of a CLOSE, DELETE, OPEN, READ, REWRITE, START, or WRITE statement before any USE procedure is executed, to indicate to the COBOL program the status of that input-output operation. The leftmost character of the FILE STATUS data item is known as status key 1 and is set to indicate a condition upon completion of the input-output operation. The rightmost character position of the FILE STATUS data item is known as status key 2 and is used to further describe the results of an input-output operation.

When an error occurs during an input-output operation, the programmer can exercise a degree of control by specifying USE procedures that check the FILE STATUS data item associated with the file, and modify it in order to communicate to the input-output control system the desired disposition of the error.

Data file organization refers to the physical arrangement of data-records within a file. In sequential files, records are organized solely on the basis of their successive physical locations in the file.

The creation of a standard sequential file includes certain entries in the Environment, Data, and Procedure divisions. In the Environment Division, the programmer uses the SELECT, ASSIGN, RESERVE, ORGANIZATION, ACCESS MODE, and FILE STATUS clauses to specify the necessary entries.

In the Data Division, the name of the file on disk, as well as the BLOCK CONTAINS and LABEL RECORDS clauses are written.

In the Procedure Division, to create a sequential file, the programmer must use an output file and a WRITE statement. The WRITE statement without the ADVANCING option is used when a sequential file is created on a disk. WRITE, READ, and REWRITE statements may be used when processing sequential files. The WRITE statement accesses the space immediately following that area into which the previous logical record was written and places the contents of the specified record in that place.

The READ statement accesses the next logical record and makes the contents of that record available in the file record area.

The REWRITE statement replaces the logical record accessed by the previous input-output operation with the contents of the specified record.

Records are maintained in a sequential file by updating the record and placing them back into the file.

An indexed sequential organization file is a sequential file with indexes that permit rapid access to individual records as well as rapid sequential processing. The programming system has control over the location of the individual records in this method of organization.

The prime area is the area in which records are written when the file is created or subsequently reorganized. Prime records must be formatted with keys.

There are two or more indexes of different levels when the indexed sequential file is created. The track index is the lowest level of index and is always present. Its entries point to data records. The cylinder index is the highest level of index and its entries point to track indexes.

The overflow area is a certain number of whole tracks specified by the programmer that are reserved in each cylinder for overflow data from prime tracks in that cylinder.

An indexed sequential file, which must be allocated in the execution activity to two random mass storage files (one for the data and the other for the index), is organized such that each record is uniquely identified by the value of a key within the record. The key is specified in the RECORD KEY phrase of the SELECT clause in the Environment Division.

In addition to the normal entries in the Environment Division, these additional entries are necessary for Indexed Sequential organization: The ORGANIZATION IS INDEXED is used for all indexed files. The ACCESS MODE defines the manner in which the records are to be accessed. The RECORD KEY clause specifies the key for the record being searched. The NOMINAL KEY clause specifies the particular key in mass storage that corresponds to the record to be retrieved.

When the access mode is DYNAMIC, records in the file may be accessed sequentially and/or randomly.

An ALTERNATE RECORD KEY clause specifies a record key that is an alternate record key for the file.

The DUPLICATES phrase specifies that the values of the associated alternate record key may be duplicated within any of the records.

The Data Division entries for an indexed sequential file are the same as those for sequential files.

In the Procedure Division, the WRITE statement with the INVALID KEY option is used to create the indexed sequential files.

An indexed sequential file may be sequentially accessed in the same manner as a standard sequential file. The entries in the Environment, Data, and Procedure Divisions are the same as those for accessing a sequential file.

The NOMINAL KEY clause is used to sequentially access the records in an indexed sequential file beginning at some other record than the first. A key value is defined in the NOMINAL KEY clause before accessing begins.

The START statement provides a basis for logical positioning within an index file for the subsequent retrieval of records.

The DELETE statement logically removes a record from a mass storage file.

The following operations may be performed in an indexed file accessed in a sequential mode: WRITE, DELETE, START, READ, and REWRITE.

The Environment Division entries for accessing an indexed sequential file at random are the same as that of sequentially accessing an indexed file, with the exception that the ACCESS IS RANDOM option is required.

When adding records to an indexed sequential file, it is not as necessary to recreate an index file as it was with standard sequential files. Any record with a RECORD KEY that is not currently in the indexed sequential file may be added to it. The Environment and Data Division entries are the same as those required in accessing an indexed sequential file at random. The Procedure Division must have the file opened as an I-O file.

To update or replace records in a randomly indexed sequential file, the same entries are required as adding records to a randomly indexed file, with the exception that the REWRITE statement is used to place the record in the indexed sequential file.

The following operations may be performed on an indexed sequential file accessed in the random access mode: WRITE, DELETE, READ, and REWRITE.

The following operations may be performed on an indexed sequential file accessed in the dynamic access mode: WRITE, DELETE, START, READ NEXT, READ, and REWRITE.

The READ statement with the NEXT phrase specification causes the next logical record to be retrieved from that file. It must be specified for files in the dynamic access mode.

Relative file organization permits accessing of records of a mass storage device either in random or sequential manners. Each record in a relative file is uniquely identified by an integer value greater than zero that specifies the record's logical ordinal position in the file. The same access modes may be used for relative files as for indexed sequential files, with the exception that the ORGANIZATION IS RELATIVE and the key is RELATIVE KEY. The RELATIVE KEY contains the logical ordinal position of the record in the file. The entries necessary for adding records are the same as those for indexed sequential files.

The following operations may be performed on a relative file accessed in the sequential mode: WRITE, DELETE, START, READ, and REWRITE.

The following operations may be performed on a relative file accessed in the random mode: WRITE, DELETE, READ, and REWRITE.

The following operations may be performed on a relative file accessed in the dynamic access mode: WRITE, DELETE, START, READ NEXT, READ, and REWRITE.

Illustrative Program: Creating an Indexed Disk File

Note: Since the first two programs that follow are meant to show only the basic concepts used for disk reads and writes, IPO charts are not shown for them. They are shown for the third program, which incorporates these concepts in an application of greater complexity.

Job Definition

An index disk file is to be created from an 80-position record.

Input

An 80-position record is to be used as input.

Processing

An 80-position indexed disk record is to be created.

Output

The created indexed disk file record is to have the same format as the input record.

Hierarchy Chart

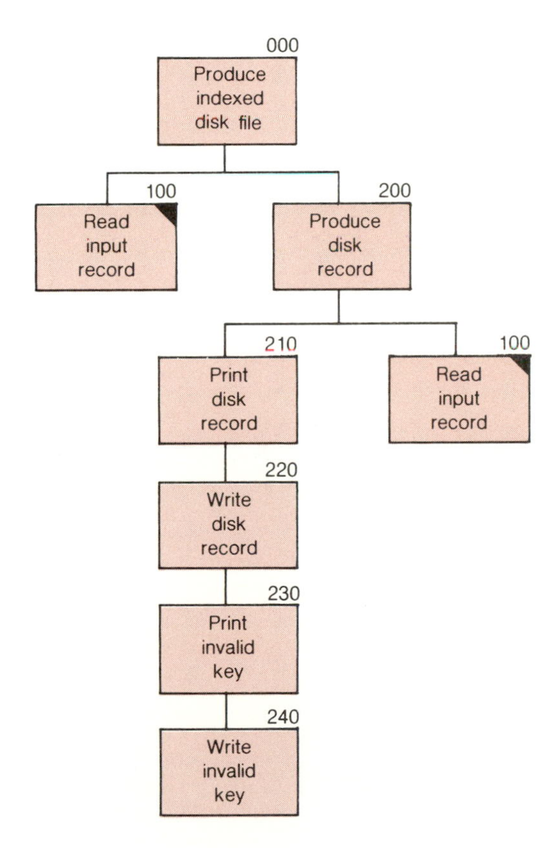

Flowcharts

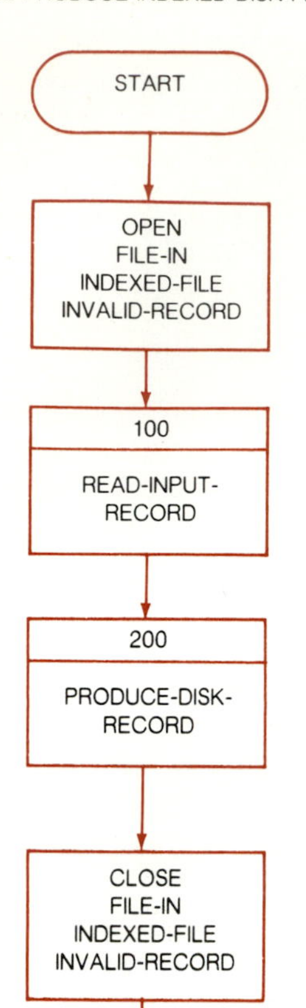
000-PRODUCE-INDEXED-DISK-FILE
START
OPEN
FILE-IN
INDEXED-FILE
INVALID-RECORD
100
READ-INPUT-
RECORD
200
PRODUCE-DISK-
RECORD
CLOSE
FILE-IN
INDEXED-FILE
INVALID-RECORD
STOP RUN

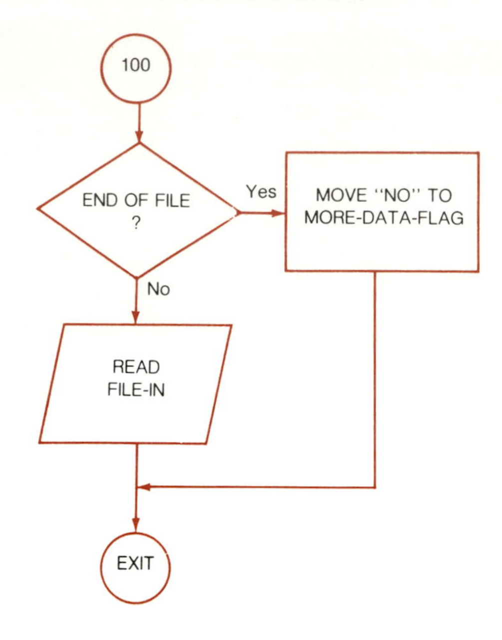
100-READ-INPUT-RECORD
100
END OF FILE
?
Yes
MOVE "NO" TO
MORE-DATA-FLAG
No
READ
FILE-IN
EXIT

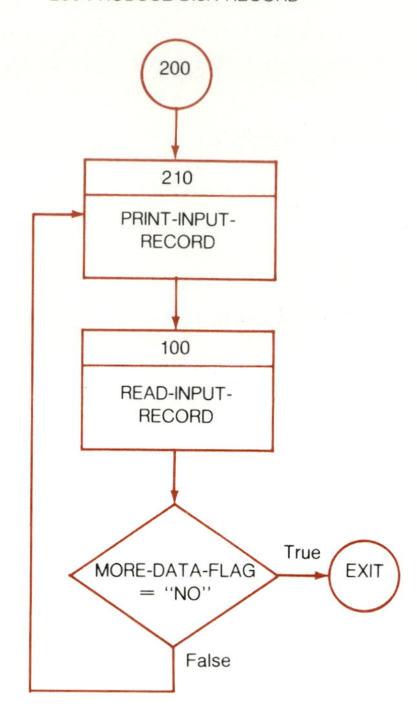
200-PRODUCE-DISK-RECORD
200
210
PRINT-INPUT-
RECORD
100
READ-INPUT-
RECORD
MORE-DATA-FLAG
= "NO"
True
EXIT
False

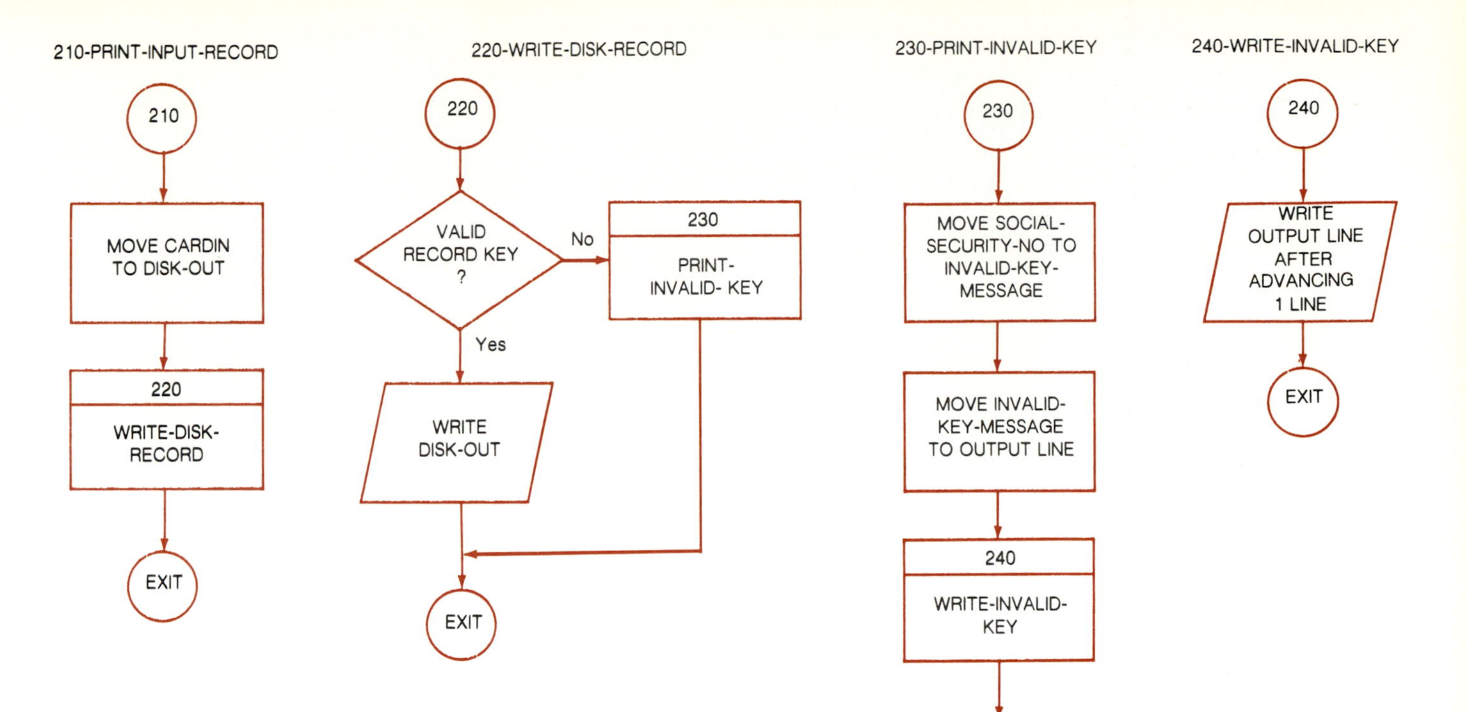

Source Program

```
 1        **************************
 2
 3         IDENTIFICATION DIVISION.
 4
 5        **************************
 6
 7         PROGRAM-ID.
 8              CREATING-AN-INDEXED-DISK-FILE.
 9         AUTHOR.
10              STEVE GONSOSKI.
11         INSTALLATION.
12              WEST LOS ANGELES COLLEGE.
13         DATE-WRITTEN.
14              3 FEBRUARY 1982.
15
16        *PURPOSE.
17        *     CREATING AN INDEXED FILE FROM AN 80-COLUMN CARD.
18
19        **************************
20
21         ENVIRONMENT DIVISION.
22
23        **************************
24
25         CONFIGURATION SECTION.
26
27         SOURCE-COMPUTER.
28              LEVEL-66-ASCII.
29         OBJECT-COMPUTER.
30              LEVEL-66-ASCII.
31
32         INPUT-OUTPUT SECTION.
33
34         FILE-CONTROL.
35              SELECT FILE-IN ASSIGN TO C1-CARD-READER.
36              SELECT INDEXED-FILE
37                   ASSIGN TO DATA-FILE INDEXED-FILE
38                   ORGANIZATION IS INDEXED
39                   ACCESS MODE IS SEQUENTIAL
40                   RECORD KEY IS SOCIAL-SECURITY-NO.
41              SELECT INVALID-RECORD ASSIGN TO P1-PRINTER.
42
43        **************************
44
45         DATA DIVISION.
46
47        **************************
48
49         FILE SECTION.
50
51         FD  FILE-IN
52              CODE-SET IS GBCD
53              RECORD CONTAINS 80 CHARACTERS
54              LABEL RECORDS ARE OMITTED
55              DATA RECORD IS CARDIN.
56
```

```
 57        01  CARDIN.
 58
 59            03  RECORD-IN            PICTURE X(80).
 60
 61        FD  INDEXED-FILE
 62            RECORD CONTAINS 80 CHARACTERS
 63            BLOCK CONTAINS 2 TO 4 RECORDS
 64            LABEL RECORDS ARE STANDARD
 65            DATA RECORD IS DISK-OUT.
 66
 67        01  DISK-OUT.
 68
 69            03  RECORD-PART-1        PICTURE X(25).
 70            03  SOCIAL-SECURITY-NO   PICTURE X(9).
 71            03  RECORD-PART-2        PICTURE X(46).
 72
 73        FD  INVALID-RECORD
 74            CODE-SET IS GBCD
 75            RECORD CONTAINS 80 CHARACTERS
 76            LABEL RECORDS ARE OMITTED
 77            DATA RECORD IS PRINT.
 78
 79        01  PRINT                    PICTURE X(132).
 80
 81        WORKING-STORAGE SECTION.
 82
 83        01  FLAGS.
 84
 85            03  MORE-DATA-FLAG       PICTURE X(3)     VALUE "YES".
 86                88  MORE-DATA                         VALUE "YES".
 87                88  NO-MORE-DATA                      VALUE "NO ".
 88
 89        01  INVALID-KEY-MESSAGE.
 90
 91            03  SOCIAL-SECURITY-O    PICTURE X(9).
 92            03  FILLER               PICTURE X(2)     VALUE SPACES.
 93            03  FILLER               PICTURE X(18)    VALUE "RECORD NOT WRI
 94       -                                                "TTEN".
 95            03  FILLER               PICTURE X(103)   VALUE SPACES.
 96
 97       *************************
 98
 99        PROCEDURE DIVISION.
100
101       *************************
102
103        000-PRODUCE-INDEXED-DISK-FILE.
104
105            OPEN     INPUT    FILE-IN
106                     OUTPUT   INDEXED-FILE
107                              INVALID-RECORD.
108            PERFORM 100-READ-INPUT-RECORD.
109            PERFORM 200-PRODUCE-DISK-RECORD
110                     UNTIL NO-MORE-DATA.
111            CLOSE    FILE-IN
112                     INDEXED-FILE
113                     INVALID-RECORD.
114            STOP RUN.
115
116        100-READ-INPUT-RECORD.
117
118            READ FILE-IN
119                AT END MOVE "NO " TO MORE-DATA-FLAG.
120
121        200-PRODUCE-DISK-RECORD.
122
123            PERFORM 210-PRINT-INPUT-RECORD.
124            PERFORM 100-READ-INPUT-RECORD.
125
126        210-PRINT-INPUT-RECORD.
127
128            MOVE CARDIN TO DISK-OUT.
129            PERFORM 220-WRITE-DISK-RECORD.
130
131        220-WRITE-DISK-RECORD.
132
133            WRITE DISK-OUT
134                INVALID KEY
135                PERFORM 230-PRINT-INVALID-KEY.
136
137        230-PRINT-INVALID-KEY.
138
139            MOVE SOCIAL-SECURITY-NO TO SOCIAL-SECURITY-O.
140            MOVE INVALID-KEY-MESSAGE TO PRINT.
141            PERFORM 240-WRITE-INVALID-KEY.
142
143        240-WRITE-INVALID-KEY.
144
145            WRITE PRINT
146                AFTER ADVANCING 1 LINE.

THERE WERE 146 SOURCE INPUT LINES.
THERE WERE NO DIAGNOSTICS.
```

Illustrative Program: List Indexed Disk Records

Job Definition

List the data of the record stored on disk in the previous program to assure that the record was stored properly.

Input

An 80-position record is to be used.

Processing

Print out the contents of an 80-position indexed disk record.

Output

The 80-position indexed disk record is to be printed as follows:

```
75925FOX WILLIAM              13009529415130303219678851 2
42252JOHNSON BEN              54301222204000000370002340 0
21663WASHINGTON GEORGE        54301222304067000376672379 2
21074MONTGOMERY ALEX          54301222404133670383332418 2
37185SMITH JOSEPH             54301222504200330390002457 2
15296BROWN WALLACE            54301222604267000396702496 2
42307DUNIGAN HENRY            54301222704333500403352535 1
37418JONES WILLIAM            54301222804666670436672730 0
63529DELANEY JERRY            54301222904866670456672847 0
42630HALLECK FRANCES          54301223004000330370002340 2
63741REID PATRICIA            54301223104007000370672344 1
15375JACKSON KENNETH          54301223211188331790076545 2
63852ALEXANDER CHARLES        54301223304933330463332886 0
37963HALL GEORGE              54301224005666670560003315 0
21074SIPLE CHARLES            54301224105666670563333315 0
15185GOODMAN HENRY            54301224206033330603333529 5
12345CAMM FRED J              54301224306333330633333705 0
87307DENTON TERRENCE          54301224807666670766674485 0
15141GOODSALL PHILLIP         54301224908666670866675070 0
42902SAWYER DAVID             55632020113215212193677730 9
21472YOUNG SAMUEL             55716778212231882414207155 6
87524HEPNER ELMER             55910929913338172648607802 8
63708HORNE ALBERT             57820114111281231249936599 5
```

The Printer Spacing Chart shows how the report is formatted:

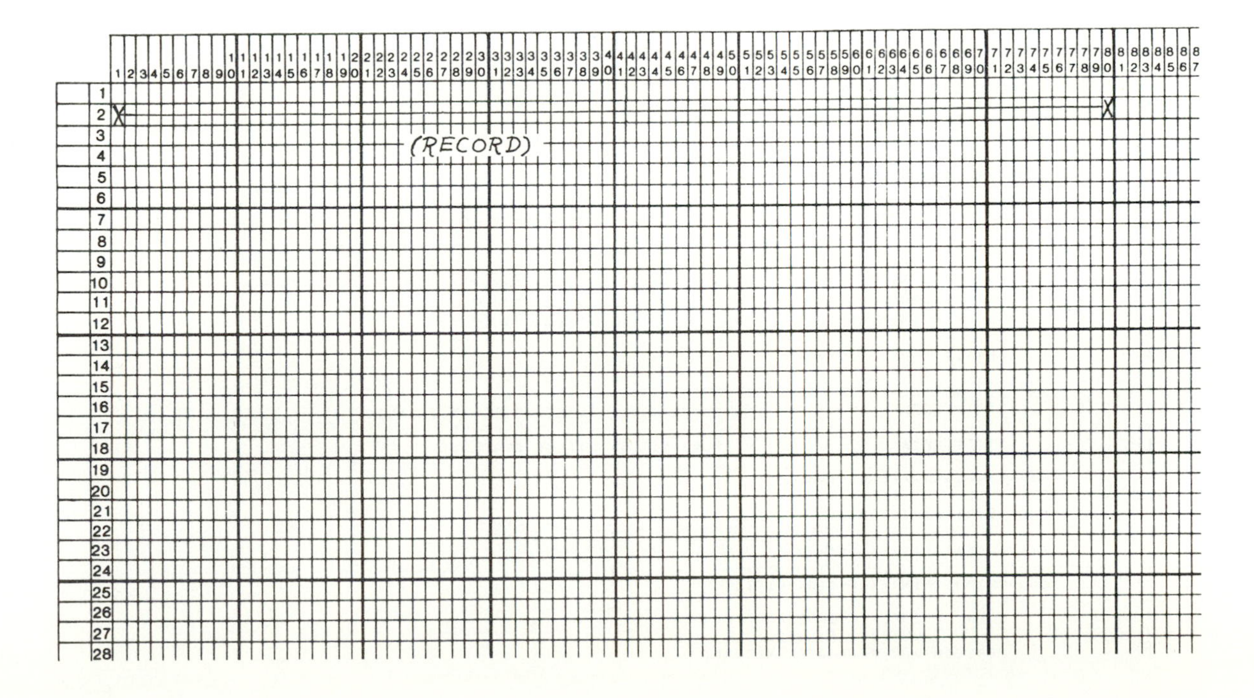

Source Program

```
 1        *************************
 2
 3         IDENTIFICATION DIVISION.
 4
 5        *************************
 6
 7         PROGRAM-ID.
 8             LISTING-INDEXED-DISK-FILE.
 9         AUTHOR.
10             STEVE GONSOSKI.
11         INSTALLATION.
12             WEST LOS ANGELES COLLEGE.
13         DATE-WRITTEN.
14             3 FEBRUARY 1982.
15
16        *PURPOSE.
17        *    LIST THE DATA STORED ON DISK FROM A PREVIOUS PROGRAM TO
18        *    ASSURE THE RECORDS WERE STORED PROPERLY.
19
20        *************************
21
22         ENVIRONMENT DIVISION.
23
24        *************************
25
26         CONFIGURATION SECTION.
27
28         SOURCE-COMPUTER.
29             LEVEL-66-ASCII.
30         OBJECT-COMPUTER.
31             LEVEL-66-ASCII.
32
33         INPUT-OUTPUT SECTION.
34
35         FILE-CONTROL.
36             SELECT INDEXED-FILE
37                 ASSIGN TO DATA-FILE INDEX-FILE
38                 ORGANIZATION IS INDEXED
39                 ACCESS MODE IS SEQUENTIAL
40                 RECORD KEY IS SOCIAL-SECURITY-NO.
41             SELECT FILE-OUT ASSIGN TO P1-PRINTER.
42
43        *************************
44
45         DATA DIVISION.
46
47        *************************
48
49         FILE SECTION.
50
51         FD  INDEXED-FILE
52             RECORD CONTAINS 80 CHARACTERS
53             BLOCK CONTAINS 2 TO 4 RECORDS
54             LABEL RECORDS ARE STANDARD
55             DATA RECORD IS DISK-RECORD.
56
57         01  DISK-RECORD.
58
59             03  RECORD-PART-1        PICTURE X(25).
60             03  SOCIAL-SECURITY-NO   PICTURE X(9).
61             03  RECORD-PART-2        PICTURE X(46).
62
63         FD  FILE-OUT
64             CODE-SET IS GBCD
65             LABEL RECORDS ARE OMITTED
66             DATA RECORD IS PRINT.
67
68         01  PRINT                    PICTURE X(132).
69
70         WORKING-STORAGE SECTION.
71
72         77  LINE-COUNT-WS            PICTURE 9(2)     VALUE 38.
73
74         01  FLAGS.
75
76             03  MORE-DATA-FLAG       PICTURE X(3)     VALUE "YES".
77                 88  MORE-DATA                         VALUE "YES".
78                 88  NO-MORE-DATA                      VALUE "NO ".
79
80         01  INVALID-KEY-MESSAGE.
81
82             03  SOCIAL-SECURITY-O    PICTURE X(9).
83             03  FILLER               PICTURE X(18)    VALUE " IS AN INVALID
84        -                                                   " KEY".
85             03  FILLER               PICTURE X(105)   VALUE SPACES.
86
87        *************************
88
89         PROCEDURE DIVISION.
90
91        *************************
92
93         000-PRODUCE-INDEXED-DISK-LIST.
94
95             OPEN    INPUT   INDEXED-FILE
96                     OUTPUT  FILE-OUT.
97             PERFORM 100-READ-DISK-RECORD.
```

```
 98           PERFORM 200-PRODUCE-INDEX-DISK-LINES
 99                   UNTIL NO-MORE-DATA.
100           CLOSE   INDEXED-FILE
101                   FILE-OUT.
102           STOP RUN.
103
104       100-READ-DISK-RECORD.
105
106           READ INDEXED-FILE
107               AT END MOVE "NO " TO MORE-DATA-FLAG.
108
109       200-PRODUCE-INDEX-DISK-LINES.
110
111           PERFORM 210-PRINT-INDEX-DISK-LINE.
112           PERFORM 100-READ-DISK-RECORD.
113
114       210-PRINT-INDEX-DISK-LINE.
115
116           MOVE DISK-RECORD TO PRINT.
117           PERFORM 220-PRINT-DETAIL-LINE.
118
119       220-PRINT-DETAIL-LINE.
120
121           IF  LINE-COUNT-WS IS > 37
122               PERFORM 230-WRITE-TOP-OF-PAGE
123           ELSE
124               PERFORM 240-WRITE-DETAIL-LINE.
125
126       230-WRITE-TOP-OF-PAGE.
127
128           WRITE PRINT
129               AFTER ADVANCING PAGE.
130           MOVE 1 TO LINE-COUNT-WS.
131
132       240-WRITE-DETAIL-LINE.
133
134           WRITE PRINT
135               AFTER ADVANCING 1 LINE.
136           ADD 1 TO LINE-COUNT-WS.

THERE WERE 136 SOURCE INPUT LINES.
THERE WERE NO DIAGNOSTICS.
```

```
75925FOX WILLIAM           1300952941513030321967885 12
42252JOHNSON BEN           54301222204000000370002340 0
21663WASHINGTON GEORGE     543012223040670003766723792
21074MONTGOMERY ALEX       543012224041336703833324182
37185SMITH JOSEPH          543012225042003303900024572
15296BROWN WALLACE         543012226042670003967024962
42307DUNIGAN HENRY         543012227043335004033525351
37418JONES WILLIAM         543012228046666704366727300
63529DELANEY JERRY         543012229048666704566728470
42630HALLECK FRANCES       543012230040003303700023402
63741REID PATRICIA         543012231040070003706723441
15375JACKSON KENNETH       543012232111883317900765452
63852ALEXANDER CHARLES     543012233049333304633328860
37963HALL GEORGE           543012240056666705600033150
21074SIPLE CHARLES         543012241056666705633333150
15185GOODMAN HENRY         543012242060333306033335295
12345CAMM FRED J           543012243063333306333337050
87307DENTON TERRENCE       543012248076666707666744850
15141GOODSALL PHILLIP      543012249086666708666750700
42902SAWYER DAVID          556320201132152121936777309
21472YOUNG SAMUEL          557167782122318824142071556
87524HEPNER ELMER          559109299133381726486078028
63708HORNE ALBERT          578201141112812312499365995
```

Illustrative Program: Disk Updating—Matching Records—Updated Records

Job Definition

An indexed disk record is to be updated with a current file. Both files are in social security number sequence. A report is to be printed showing the new updated record and at the same time the indexed disk file is to be updated.

If no matching record is found, it should be so indicated.

Input

The indexed disk file contains the following fields:

Positions	Field Designation
1–2	Department number
3–5	Clock number
6–25	Name
26–34	Social security number
35–41	Old year-to-date gross earnings (XXXXX.XX)
42–47	Old year-to-date withholding tax (XXXX.XX)
48–53	Old year-to-date FICA (XXXX.XX)

The input record contains the following fields:

Positions	Field Designation
1–2	Department number
3–5	Clock number
26–34	Social security number
62–68	Current gross earnings (XXXXX.XX)
69–74	Current withholding tax (XXXX.XX)
75–78	Current FICA (XX.XX)
80	Code 1

Processing

On matching fields (social security number) the following operations are to be performed:

1. The current gross earning is to be added to the old year-to-date gross earnings to find the updated year-to-date gross earnings.
2. The current withholding tax is to be added to the old year-to-date withholding tax to find the updated year-to-date withholding tax.
3. The current FICA tax is to be added to the old year-to-date FICA tax.

If the input record does not match the index disk record, a message is to be printed. The indexed disk file will be updated.

Output

```
75925FOX WILLIAM          1300952942269545482950889545
364208841 IS A BAD READ KEY
364208841 IS A BAD WRITE KEY
75925FOX WILLIAM          3642088412844035580473892905
42252JOHNSON BEN          5430122220600000055500235170
21663WASHINGTON GEORGE    5430122230610050056500239109
21074MONTGOMERY ALEX      5430122240620050057500243029
37185SMITH JOSEPH         5430122250630050058500246948
15296BROWN WALLACE        5430122260640050059505250868
42307DUNIGAN HENRY        5430122270650025060502254777
37418JONES WILLIAM        5430122280700000065500274365
63529DELANEY JERRY        5430122290730000068500286123
42630HALLECK FRANCES      5430122300600050055500235190
63741REID PATRICIA        5430122310601050055600235582
15375JACKSON KENNETH      5430122321678250268510657792
37963HALL GEORGE          5430122400850000084000333157
21074SIPLE CHARLES        5430122410850000084500333157
15185GOODMAN HENRY        5430122420905000090500354714
543012246 IS A BAD READ KEY
543012246 IS A BAD WRITE KEY
15185GOODMAN HENRY        5430122461221667122167356566
87307DENTON TERRENCE      5430122481150000115000450742
15141GOODSALL PHILLIP     5430122491300000130000509535
42902SAWYER DAVID         5563202011982282329050776955
21472YOUNG SAMUEL         5571677821837786362130719155
87524HEPNER ELMER         5591092992000725397290784181
63708HORNE ALBERT         5782011411692184187490663249
```

Hierarchy Chart

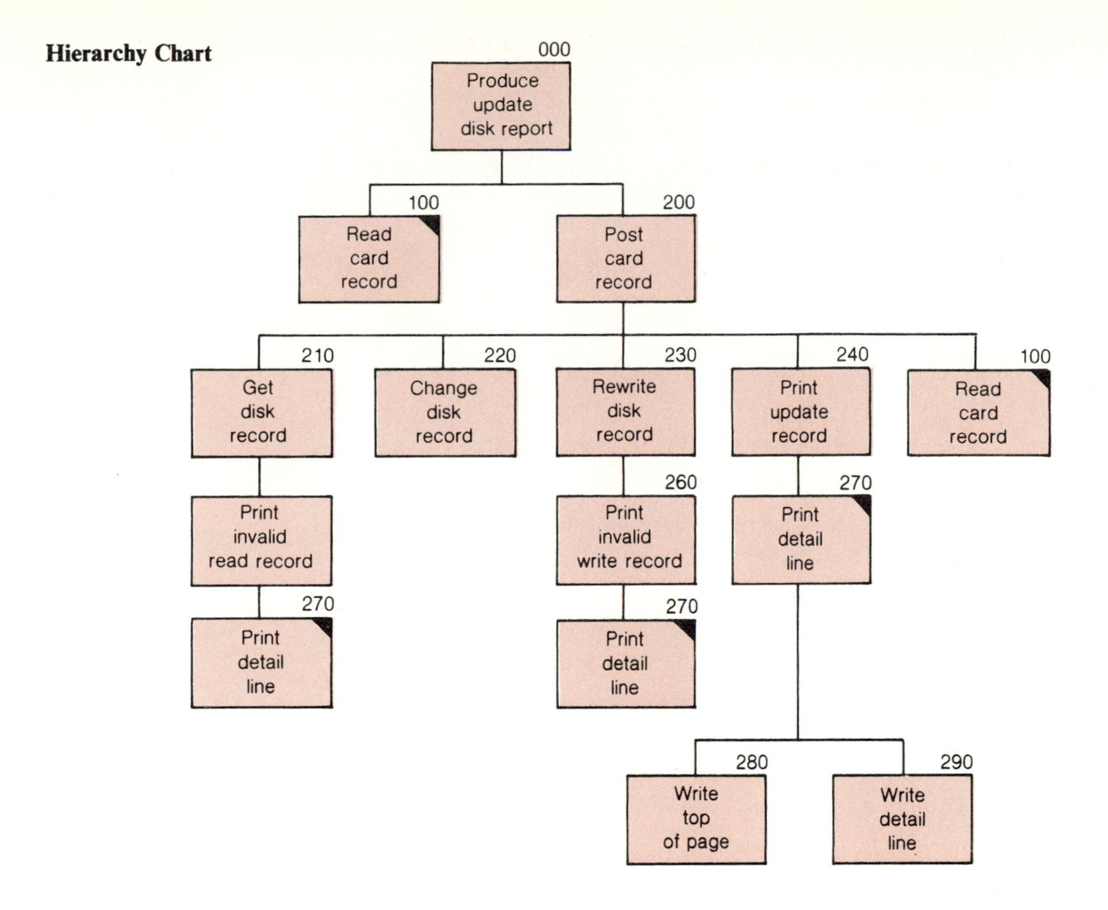

IPO Charts

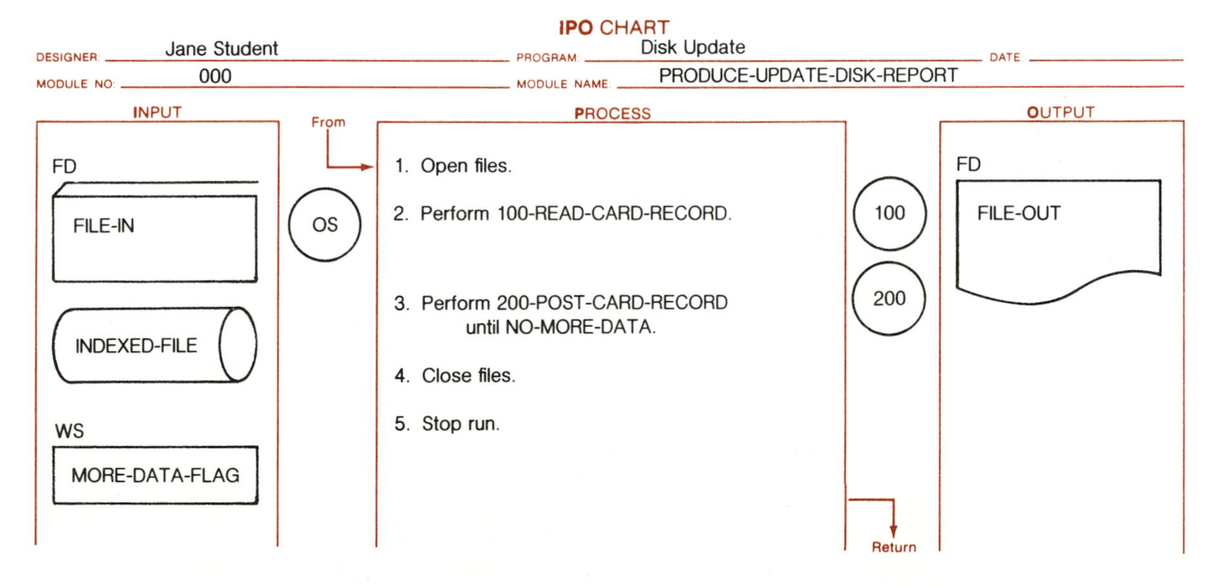

IPO CHART

DESIGNER: Jane Student | PROGRAM: Disk Update | DATE:

MODULE NO: 200 | MODULE NAME: POST-CARD-RECORD

INPUT	From	PROCESS		OUTPUT
	000	1. Perform 210-GET-DISK-RECORD.	210	
		2. Perform 220-CHANGE-DISK-RECORD.	220	
		3. Perform 230-REWRITE-DISK-RECORD.	230	
		4. Perform 240-PRINT-UPDATE-RECORD.	240	
		5. Perform 100-READ-CARD-RECORD.	100	
			Return	

IPO CHART

DESIGNER: Jane Student | PROGRAM: Disk Update | DATE:

MODULE NO: 210 | MODULE NAME: GET-DISK-RECORD

INPUT	From	PROCESS		OUTPUT
FD SOC-SEC-C	200	1. Move SOC-SEC-C to SOC-SEC-D.		FD SOC-SEC-D
INDEXED-FILE		2. Read INDEXED-FILE if key invalid perform 250-PRINT-INVALID-READ-RECORD.	250	
			Return	

IPO CHART

DESIGNER: Jane Student | PROGRAM: Disk Update | DATE:

MODULE NO: 230 | MODULE NAME: REWRITE-DISK-RECORD

INPUT	From	PROCESS		OUTPUT
FD DISK-RECORD	200	1. Rewrite DISK-RECORD if invalid key perform 260-PRINT-INVALID-WRITE-RECORD.	260	FD INDEXED-FILE
			Return	

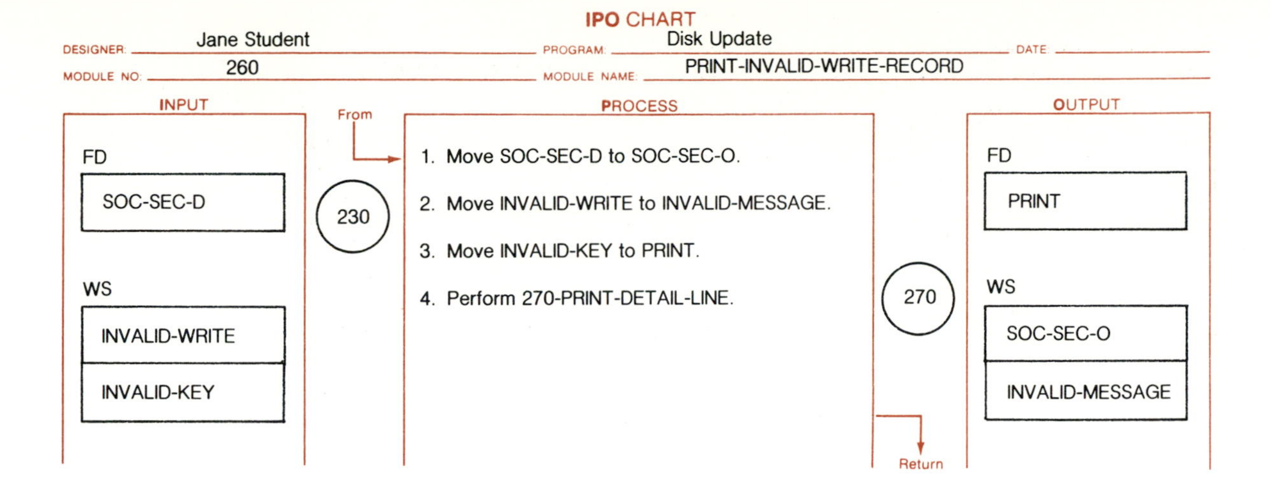
IPO CHART
DESIGNER: Jane Student
PROGRAM: Disk Update
DATE:
MODULE NO: 260
MODULE NAME: PRINT-INVALID-WRITE-RECORD
INPUT
FD
SOC-SEC-D
WS
INVALID-WRITE
INVALID-KEY
From
230
PROCESS
1. Move SOC-SEC-D to SOC-SEC-O.
2. Move INVALID-WRITE to INVALID-MESSAGE.
3. Move INVALID-KEY to PRINT.
4. Perform 270-PRINT-DETAIL-LINE.
270
Return
OUTPUT
FD
PRINT
WS
SOC-SEC-O
INVALID-MESSAGE

Flowcharts

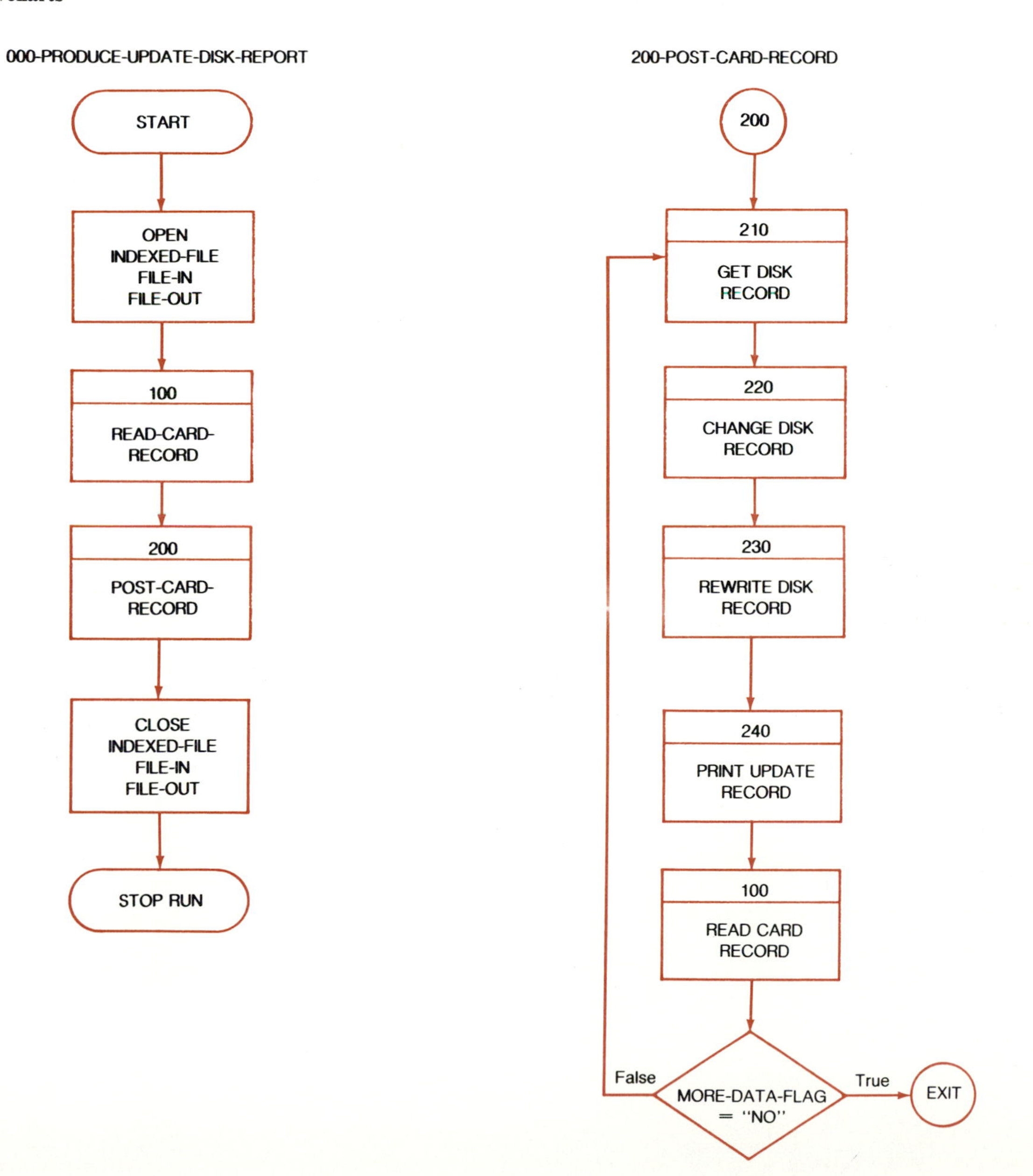
000-PRODUCE-UPDATE-DISK-REPORT
START
OPEN INDEXED-FILE FILE-IN FILE-OUT
100
READ-CARD-RECORD
200
POST-CARD-RECORD
CLOSE INDEXED-FILE FILE-IN FILE-OUT
STOP RUN
200-POST-CARD-RECORD
200
210
GET DISK RECORD
220
CHANGE DISK RECORD
230
REWRITE DISK RECORD
240
PRINT UPDATE RECORD
100
READ CARD RECORD
False
MORE-DATA-FLAG = "NO"
True
EXIT

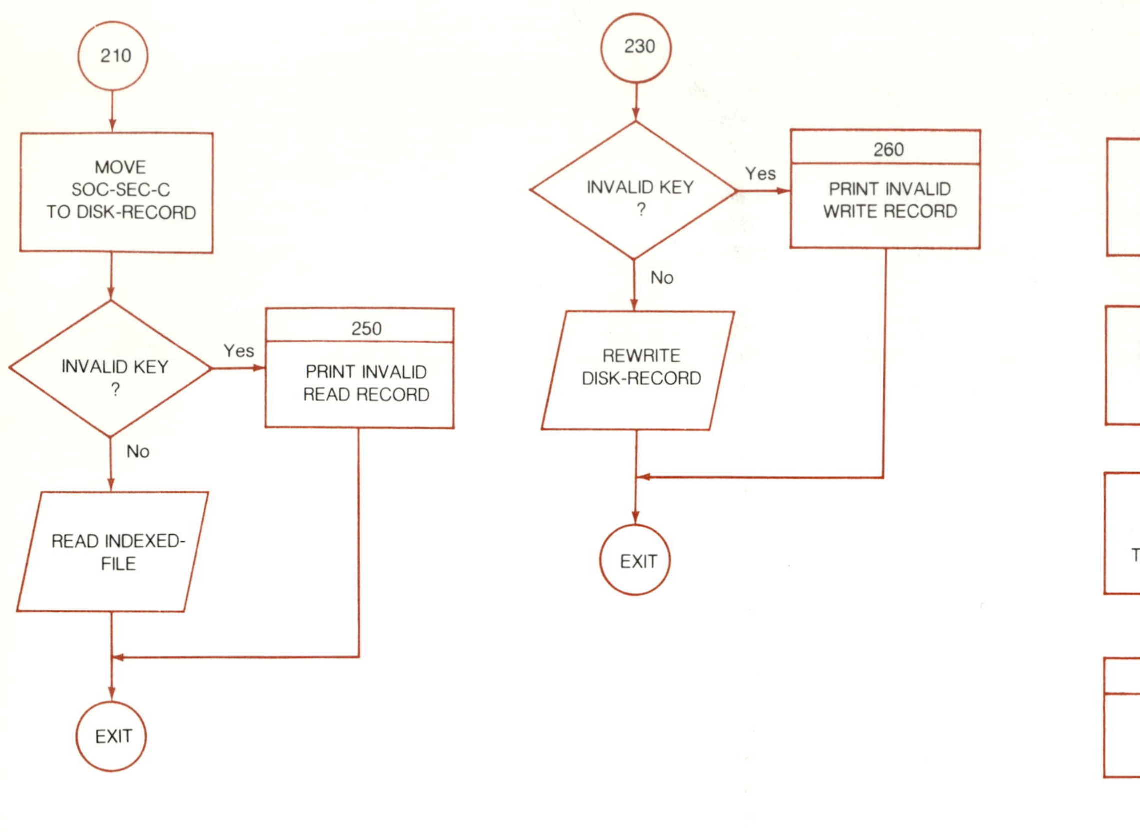

210-GET-DISK-RECORD
210
MOVE SOC-SEC-C TO DISK-RECORD
INVALID KEY ?
Yes
250
PRINT INVALID READ RECORD
No
READ INDEXED-FILE
EXIT
230-REWRITE-DISK-RECORD
230
INVALID KEY ?
Yes
260
PRINT INVALID WRITE RECORD
No
REWRITE DISK-RECORD
EXIT

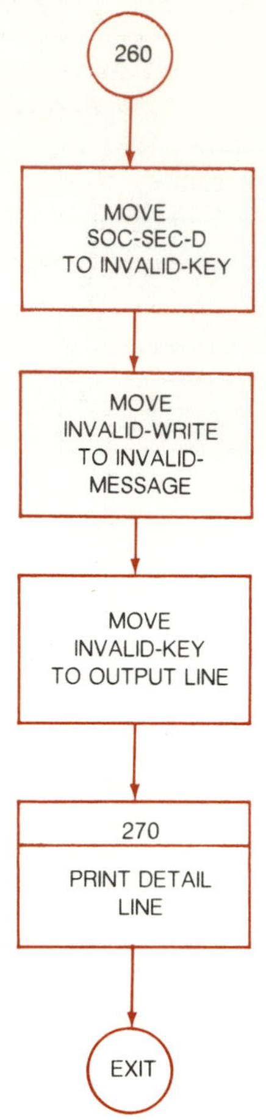

260-PRINT-INVALID-WRITE-RECORD
260
MOVE SOC-SEC-D TO INVALID-KEY
MOVE INVALID-WRITE TO INVALID-MESSAGE
MOVE INVALID-KEY TO OUTPUT LINE
270
PRINT DETAIL LINE
EXIT

Source Program

```
 1        *************************
 2
 3         IDENTIFICATION DIVISION.
 4
 5        *************************
 6
 7         PROGRAM-ID.
 8             DISK-UPDATING.
 9         AUTHOR.
10             STEVE GONSOSKI.
11         INSTALLATION.
12             WEST LOS ANGELES COLLEGE.
13         DATE-WRITTEN.
14             3 FEBRUARY 1982.
15
16        *PURPOSE.
17        *    AN INDEXED DISK RECORD IS TO BE UPDATED WITH A CURRENT CARD
18        *    FILE.
19
20        *************************
21
22         ENVIRONMENT DIVISION.
23
24        *************************
25
26         CONFIGURATION SECTION.
27
28         SOURCE-COMPUTER.
29             LEVEL-66-ASCII.
30         OBJECT-COMPUTER.
31             LEVEL-66-ASCII.
32
33         INPUT-OUTPUT SECTION.
34
35         FILE-CONTROL.
36             SELECT FILE-IN ASSIGN TO C1-CARD-READER.
37             SELECT INDEXED-FILE
38                 ASSIGN TO DATA-FILE INDEXED-FILE
39                 ORGANIZATION IS INDEXED
40                 ACCESS MODE IS RANDOM
41                 RECORD KEY IS SOC-SEC-D.
42             SELECT FILE-OUT ASSIGN TO P1-PRINTER.
43
44        *************************
45
46         DATA DIVISION.
47
48        *************************
49
50         FILE SECTION.
51
52         FD  FILE-IN
53             CODE-SET IS GBCD
54             LABEL RECORDS ARE OMITTED
55             DATA RECORD IS CARDIN.
56
57         01  CARDIN.
58
59             03  DEPARTMENT-NUMBER-C PICTURE 9(2).
60             03  CLOCK-NUMBER-C      PICTURE 9(3).
61             03  FILLER              PICTURE X(20).
62             03  SOC-SEC-C           PICTURE X(9).
63             03  FILLER              PICTURE X(27).
64             03  CUR-GROSS-EARN-C    PICTURE 9(5)V99.
65             03  CUR-WITH-TAX-C      PICTURE 9(4)V99.
66             03  CUR-FICA-C          PICTURE 9(2)V99.
67             03  FILLER              PICTURE X(1).
68             03  CODE-C              PICTURE X(1).
69
70         FD  INDEXED-FILE
71             RECORD CONTAINS 80 CHARACTERS
72             BLOCK CONTAINS 2 TO 4 RECORDS
73             LABEL RECORDS ARE STANDARD
74             DATA RECORD IS DISK-RECORD.
75
76         01  DISK-RECORD.
77
78             03  DEPARTMENT-NUMBER-D PICTURE 9(2).
79             03  CLOCK-NUMBER-D      PICTURE 9(3).
80             03  NAME-D              PICTURE X(20).
81             03  SOC-SEC-D           PICTURE X(9).
82             03  O-YTD-GROSS-EARN-D  PICTURE 9(5)V99.
83             03  O-YTD-WITH-TAX-D    PICTURE 9(4)V99.
84             03  O-YTD-FICA-D        PICTURE 9(4)V99.
85             03  FILLER              PICTURE X(27).
86
87         FD  FILE-OUT
88             CODE-SET IS GBCD
89             LABEL RECORDS ARE OMITTED
90             DATA RECORD IS PRINT.
91
92         01  PRINT.
93
94             03  OUTPUT-LINE         PICTURE X(80).
95             03  FILLER              PICTURE X(52).
96
```

```
 97        WORKING-STORAGE SECTION.
 98
 99        77  LINE-COUNT-WS            PICTURE 9(2)    VALUE 38.
100
101        01  FLAGS.
102
103            03  MORE-DATA-FLAG       PICTURE X(3)    VALUE "YES".
104                88  MORE-DATA                        VALUE "YES".
105                88  NO-MORE-DATA                     VALUE "NO ".
106
107        01  INVALID-KEY-MESSAGES.
108
109            03  INVALID-READ.
110                05  FILLER           PICTURE X(19)   VALUE " IS A BAD READ
111       -                                                 " KEY".
112                05  FILLER           PICTURE X(104)  VALUE SPACES.
113            03  INVALID-WRITE.
114                05  FILLER           PICTURE X(19)   VALUE " IS A BAD WRIT
115       -                                                 "E KEY".
116                05  FILLER           PICTURE X(104)  VALUE SPACES.
117
118        01  INVALID-KEY.
119
120            03  SOC-SEC-O            PICTURE X(9).
121            03  INVALID-MESSAGE      PICTURE X(123).
122
123        01  SAVE-RECORD              PICTURE X(132).
124
125       *************************
126
127        PROCEDURE DIVISION.
128
129       *************************
130
131        000-PRODUCE-UPDATE-DISK-REPORT.
132
133            OPEN    I-O      INDEXED-FILE
134                    INPUT    FILE-IN
135                    OUTPUT   FILE-OUT.
136            PERFORM 100-READ-CARD-RECORD.
137            PERFORM 200-POST-CARD-RECORD
138                    UNTIL NO-MORE-DATA.
139            CLOSE   INDEXED-FILE
140                    FILE-IN
141                    FILE-OUT.
142            STOP RUN.
143
144        100-READ-CARD-RECORD.
145
146            READ FILE-IN
147                AT END MOVE "NO " TO MORE-DATA-FLAG.
148
149        200-POST-CARD-RECORD.
150
151            PERFORM 210-GET-DISK-RECORD.
152            PERFORM 220-CHANGE-DISK-RECORD.
153            PERFORM 230-REWRITE-DISK-RECORD.
154            PERFORM 240-PRINT-UPDATE-RECORD.
155            PERFORM 100-READ-CARD-RECORD.
156
157        210-GET-DISK-RECORD.
158
159            MOVE SOC-SEC-C TO SOC-SEC-D.
160            READ INDEXED-FILE
161                INVALID KEY
162                PERFORM 250-PRINT-INVALID-READ-RECORD.
163
164        220-CHANGE-DISK-RECORD.
165
166            ADD CUR-GROSS-EARN-C     TO O-YTD-GROSS-EARN-D.
167            ADD CUR-WITH-TAX-C       TO O-YTD-WITH-TAX-D.
168            ADD CUR-FICA-C           TO O-YTD-FICA-D.
169
170        230-REWRITE-DISK-RECORD.
171
172            REWRITE DISK-RECORD
173                INVALID KEY
174                PERFORM 260-PRINT-INVALID-WRITE-RECORD.
175
176        240-PRINT-UPDATE-RECORD.
177
178            MOVE SPACES TO PRINT.
179            MOVE DISK-RECORD TO OUTPUT-LINE.
180            PERFORM 270-PRINT-DETAIL-LINE.
181
182        250-PRINT-INVALID-READ-RECORD.
183
184            MOVE SOC-SEC-D TO SOC-SEC-O.
185            MOVE INVALID-READ TO INVALID-MESSAGE.
186            MOVE INVALID-KEY TO PRINT.
187            PERFORM 270-PRINT-DETAIL-LINE.
188
189        260-PRINT-INVALID-WRITE-RECORD.
190
191            MOVE SOC-SEC-D TO SOC-SEC-O.
192            MOVE INVALID-WRITE TO INVALID-MESSAGE.
193            MOVE INVALID-KEY TO PRINT.
194            PERFORM 270-PRINT-DETAIL-LINE.
195
```

```
196        270-PRINT-DETAIL-LINE.
197
198            IF LINE-COUNT-WS IS > 37
199                MOVE PRINT TO SAVE-RECORD
200                MOVE SPACES TO PRINT
201                PERFORM 280-WRITE-TOP-OF-PAGE
202                MOVE SAVE-RECORD TO PRINT
203            ELSE
204                NEXT SENTENCE.
205            PERFORM 290-WRITE-DETAIL-LINE.
206
207        280-WRITE-TOP-OF-PAGE.
208
209            WRITE PRINT
210                AFTER ADVANCING PAGE.
211            MOVE 1 TO LINE-COUNT-WS.
212
213        290-WRITE-DETAIL-LINE.
214
215            WRITE PRINT
216                AFTER ADVANCING 1 LINE.
217            ADD 1 TO LINE-COUNT-WS.

THERE WERE 217 SOURCE INPUT LINES.
THERE WERE NO DIAGNOSTICS.
```

```
75925FOX WILLIAM          130095294226954548295088954 5
364208841 IS A BAD READ KEY
364208841 IS A BAD WRITE KEY
75925FOX WILLIAM          3642088412844035580473892905
42252JOHNSON BEN          5430122220600000055500235170
21663WASHINGTON GEORGE    5430122230610050056500239109
21074MONTGOMERY ALEX      5430122240620050057500243029
37185SMITH JOSEPH         5430122250630050058500246948
15296BROWN WALLACE        5430122260640050059505250868
42307DUNIGAN HENRY        5430122270650025060502254777
37418JONES WILLIAM        5430122280700000065500274365
63529DELANEY JERRY        5430122290730000068500286123
42630HALLECK FRANCES      5430122300600050055500235190
63741REID PATRICIA        5430122310601050055600235582
15375JACKSON KENNETH      5430122321678250268510657792
37963HALL GEORGE          5430122400850000084000333157
21074SIPLE CHARLES        5430122410850000084500333157
15185GOODMAN HENRY        5430122420905000090500354714
543012246 IS A BAD READ KEY
543012246 IS A BAD WRITE KEY
15185GOODMAN HENRY        5430122461221667122167356566
87307DENTON TERRENCE      5430122481150000115000450742
15141GOODSALL PHILLIP     5430122491300000130000509535
42902SAWYER DAVID         5563202011982282329050776955
21472YOUNG SAMUEL         5571677821837786362130719155
87524HEPNER ELMER         5591092992000725397290784181
63708HORNE ALBERT         5782011411692184187490663249
```

Questions for Review

1. Explain the "inline" processing technique and how it is used with mass storage devices.
2. Briefly define the following terms: Direct Access, Direct Access Storage Device, file, physical and logical record, key.
3. What is meant by file organization?
4. What are the inherent characteristics of the file that must be considered in selecting an efficient method of organization?
5. What are the methods of processing direct access files?
6. Briefly describe the different methods of file organization.
7. What capability does file input-output devices provide?
8. How are sequential, indexed, and relative files organized?
9. What are the RECORD KEY and ALTERNATE RECORD KEY clauses used for?
10. Briefly describe the sequential, random, and dynamic access modes.
11. What is the current record pointer?
12. What is the main purpose of the FILE STATUS clause and how is it used in error processing?
13. What is meant by data file organization?
14. What Environment Division entries are necessary to create a standard sequential file?
15. What Data and Procedure division entries are necessary to create a standard sequential file?
16. What input/output verbs are used to process standard sequential files?
17. How is an indexed sequential file organized?
18. What additional entries are necessary in the Environment Division for indexed sequential files?
19. What entries are required in the Data and Procedure Divisions for the creation of indexed sequential files?
20. How is an indexed sequential file sequentially accessed?
21. What are the uses of the NOMINAL KEY clause? START statement? DELETE statement?
22. What operations may be performed on an indexed sequential file accessed in the sequential mode?
23. What entries are required in the Environment Division for randomly accessing an indexed sequential file?
24. How are records added to a randomly indexed sequential file?
25. How are records updated or replaced in a randomly indexed sequential file?
26. What operations may be performed on a randomly accessed indexed sequential file?
27. What is the dynamic access mode and what operations may be performed on an indexed sequential file in this mode?
28. Explain the use of the READ NEXT statement.
29. How is a relative file organized and what operations may be performed on a relative file?

Matching Questions

A. *Match each item with its proper description.*

_____ 1. Key	A. Records are organized solely on the basis of their successive physical locations in the file.
_____ 2. Relative organization	B. Input transactions are grouped together, sorted into the same sequence as master file, and the resulting batch is then processed against the master file.
_____ 3. Direct access storage device	C. A collection of information related to a common identifier.
_____ 4. Random processing	D. A physical unit or a collection of records.
_____ 5. Indexed sequential organization	E. A control field that uniquely identifies a record.
_____ 6. File	F. Each physical record has a discrete location and a unique address.
_____ 7. Sequential processing	G. Sequential file with indexes that permit rapid access to records.
_____ 8. Record	H. Each record uniquely identified by an integer value greater than zero which specifies the record's ordinal position in the file.
_____ 9. Sequential organization	I. Processing of detail transactions against the master file regardless of sequence.
_____10. WRITE	J. Accesses space immediately following that area into which the previous record was written and puts the contents of the specified record in that space.

B. *Match each item with its proper description.*

______ 1. READ NEXT	A. Certain number of whole tracks reserved for each cylinder for data that will not fit into the prime area.
______ 2. Prime area	B. Used to position the file.
______ 3. INVALID KEY	C. Replaces the logical record accessed by previous input/output operation with the contents of the specified record.
______ 4. Dynamic access	D. Records are written when the file is created or subsequently reorganized.
______ 5. READ	E. Created and written by the operating system when the indexed sequential file is created or reorganized.
______ 6. Indexes	F. Removes identified record from file.
______ 7. START	G. Alternates from sequential access to random access using appropriate forms of input/output statements.
______ 8. Overflow area	H. Accesses the next logical record in the file and makes the contents of that record available in the file record area.
______ 9. REWRITE	I. Activated when no record with a control field equal to key is found.
______10. DELETE	J. Makes next logical record available.

C. *Match each clause with its proper description.*

______ 1. ORGANIZATION	A. Control entry for a file.
______ 2. FILE STATUS	B. Provides alternate paths for the retrieval of records from a file.
______ 3. ACCESS MODE	C. Specifies the logical structure of a file.
______ 4. RECORD KEY	D. Initiates the processing of sequentially indexed sequential files at a specified record.
______ 5. RELATIVE KEY	E. A two-character data item that indicates to the COBOL program the status of that input/output operation.
______ 6. RESERVE	F. Contains the logical ordinal position of the record in the file.
______ 7. ALTERNATE RECORD KEY	G. Allows the programmer to specify the number of input/output areas to be allotted.
______ 8. NOMINAL KEY	H. Specifies how records will be retrieved.

Exercises

Multiple Choice: Indicate the best *answer (questions 1-35).*

1. Direct access implies
   a. access at random by multiple users of data.
   b. sequential access by few users of data.
   c. random access by one user.
   d. None of the above.
2. An organized collection of record information is known as a(n)
   a. record.
   b. direct access device.
   c. file.
   d. organization.
3. The control field that uniquely identifies an item is known as a(n)
   a. record.
   b. prime area.
   c. index.
   d. key.
4. The inherent characteristics of a file are
   a. volatility.
   b. activity.
   c. size.
   d. All of the above.

5. The input transactions are sorted and grouped together and processed against the master file in
   a. sequential processing.
   b. relative processing.
   c. random processing.
   d. dynamic processing.
6. A method of file organization where records are retrieved by their logical ordinal position in the file is known as
   a. sequential organization.
   b. relative organization.
   c. indexed sequential organization.
   d. direct organization.
7. The prime key for a control entry in a record is
   a. RECORD KEY.
   b. RELATIVE KEY.
   c. NOMINAL KEY.
   d. ACTUAL KEY.
8. Relative file organization is permitted on
   a. magnetic tape units.
   b. random mass storage units.
   c. punched card files.
   d. All of the above.
9. If the FILE STATUS clause is specified in a file control entry, a value is placed in the specified two-character data item during the execution of a(n)
   a. CLOSE statement.
   b. OPEN statement.
   c. READ statement.
   d. All of the above.
10. If the programmer wishes additional input/output areas allotted, the following clause in the Environment Division would be specified:
    a. ASSIGN.
    b. SELECT.
    c. RESERVE.
    d. None of the above.
11. The logical structure of a file is specified by the following clause:
    a. ORGANIZATION.
    b. ACCESS MODE.
    c. FILE STATUS.
    d. RECORD KEY.
12. The statement that replaces the logical record accessed by the previous input/output operation is the
    a. READ statement.
    b. REWRITE statement.
    c. WRITE statement.
    d. READ NEXT statement.
13. To update a sequential file, the file must be opened as
    a. I-O.
    b. INPUT.
    c. OUTPUT.
    d. DYNAMIC.
14. In a sequential indexed file,
    a. the programmer has control over the location of individual records.
    b. the programmer must do most of the input/output programming.
    c. the programmer has no flexibility in the operation that can be performed on the data.
    d. None of the above.
15. The area in which records are initially written on an indexed sequential file is the
    a. main area.
    b. prime area.
    c. overflow area.
    d. core area.

16. The clause that is used to access an indexed sequential file is the
    a. ACCESS MODE clause.
    b. RECORD KEY clause.
    c. ALTERNATE RECORD KEY clause.
    d. NOMINAL KEY clause.
17. The statement that provides a basis for logical positioning within an indexed sequential file is the
    a. READ statement.
    b. WRITE statement.
    c. START statement.
    d. READ NEXT statement.
18. The following operations may be performed on an indexed file in the dynamic mode.
    a. DELETE.
    b. START.
    c. READ NEXT.
    d. All of the above.
19. Relative organization
    a. uses a record key to identify each record in the file.
    b. uses an index to identify each record in the file.
    c. permits records to be accessed randomly only.
    d. None of the above.
20. To create any file you must use
    a. an output file.
    b. a READ statement.
    c. a REWRITE statement.
    d. All of the above.
21. OPEN I-O UP-FILE.
    The above statement would be acceptable if UPFILE were
    a. on a magnetic tape.
    b. on a disk.
    c. not yet created.
    d. in punched cards.
22. The RECORD KEY clause specifies an elementary variable within the record associated with the indexed file that contains the unique key. The RECORD KEY clause could specify the
    a. zip-code for a file of magazine subscribers.
    b. machine-serial-number for a file of service records on electric typewriters.
    c. last-name for a file of telephone users in New York.
    d. None of the above.
23. The RECORD KEY clause must be specified for
    a. any indexed sequential file.
    b. every file in a mass storage device.
    c. every relative file.
    d. every sequential file.
24.
```
SELECT INDEXED-FILE
       ASSIGN TO DISK
       RECORD KEY IS RECORD-ID.
```
    In the File-Control entry above,
    a. RECORD-ID is an elementary variable that is defined in the Working-Storage Section.
    b. standard sequential organization is used.
    c. DISK is the file-name.
    d. None of the above.
25. Which of the following must be included in a File-Control entry for an indexed sequential file?
    a. INVALID KEY option.
    b. AT END option.
    c. LABEL RECORDS clause.
    d. ASSIGN clause.
26. For which key below will the INVALID KEY option of the WRITE statement be activated? (Assume records are being read in the following sequence.)
    a. 079326.
    b. 079327.
    c. 084719.
    d. 084719.

27. If the INVALID KEY option of the WRITE statement is activated in an attempt to write a specific record,
    a. the record for which the option was activated will not have been written into the file.
    b. the file will contain the record that was a duplicate or out of sequence.
    c. the file must be recreated to eliminate the invalid record.
    d. None of the above.
28. In order to sequentially access records from an indexed sequential file opened as INPUT, you would have to use
    a. the INVALID KEY option in the WRITE statement referring to the file.
    b. the AT END option in the READ statement referring to the file.
    c. the INVALID KEY option in the READ statement referring to the file.
    d. None of the above.
29. The NOMINAL KEY clause specifies when
    a. an indexed sequential file is to be created.
    b. all records are to be sequentially accessed from the last half of an indexed sequential file.
    c. all records to be sequentially accessed from an indexed sequential file.
    d. All of the above.
30. RECORD KEY IS MAN-NUMBER.

    According to the clause above, the variable MAN-NUMBER
    a. contains the key or control field of the record.
    b. is an elementary variable within the record associated with the indexed sequential file.
    c. has been described in the File Section.
    d. All of the above.
31.
```
NOMINAL KEY IS LOCATION.
      .
      .
      .
MOVE 1811 TO LOCATION.
      .
      .
      .
READ POSITION-FILE
     AT END PERFORM TERMINATION-PROC.
```
    In order to begin sequential processing of records from the indexed sequential file POSITION-FILE with the record whose key is 1811, a START statement should be placed in the following sequence:
    a. before the MOVE statement.
    b. between the MOVE statement and the READ statement.
    c. after the READ statement.
    d. in the AT END phrase.
32. If you wish to access four records in a row beginning with the key 87108 and then four records in a row beginning with the key 98411, you would execute
    a. a START statement, then a READ statement four times, then another START statement, then a READ statement four times.
    b. eight START statements and eight READ statements.
    c. two START statements and two READ statements.
    d. None of the above.
33. The ACCESS MODE clause is
    a. optional for sequentially accessed files.
    b. required for randomly accessed files.
    c. omitted, then ACCESS IS SEQUENTIAL is implied.
    d. All of the above.
34. ACCESS MODE IS RANDOM.

    This form of an ACCESS MODE clause can refer to a file on
    a. a mass storage device.
    b. a magnetic tape unit.
    c. a cassette.
    d. punched cards.
35. Which of the entries below would be required to update all records in an indexed sequential file with sequential access?
    a. START statement.
    b. REWRITE statement.
    c. ACCESS IS SEQUENTIAL clause.
    d. NOMINAL KEY clause.

36. Match the correct facts with each portion of a File-Control paragraph. (There may be more than one matching item.)

_____ 1. SELECT clause
_____ 2. ASSIGN clause
_____ 3. RECORD KEY clause

A. Identifies the file-name.
B. Specifies an elementary variable within the file record that identifies the record.
C. Links the file to a device.
D. Is required for all indexed sequential files.
E. Is required for all files.

37. Match the entries in a program to use an indexed sequential file for sequential input of all records with the divisions in which the clauses would be written. (There may be more than one matching item.)

_____ 1. Identification Division
_____ 2. Environment Division
_____ 3. Data Division
_____ 4. Procedure Division

A. LABEL RECORDS clause.
B. PROGRAM-ID paragraph.
C. INVALID KEY option.
D. BLOCK clause.
E. ASSIGN clause.
F. RECORD KEY clause.
G. OPEN statement.
H. SELECT clause.
I. AT END option.

38. Match the following (there may be more than one matching item):

_____ 1. RECORD KEY
_____ 2. NOMINAL KEY

A. Required whenever sequential access of records in an indexed sequential file is to begin at a record other than the first record in the file.
B. Required in every file control entry for an indexed sequential file.
C. Specifies a variable within the record associated with an indexed sequential file.
D. Specifies a variable described in the Working-Storage Section of a program that accesses an indexed sequential file.

39. Match the following:

_____ 1. START statement
_____ 2. READ statement

A. Accesses a record from the indexed sequential file.
B. Positions the indexed sequential file at the desired record.
C. Sets the value of the NOMINAL KEY variable.

40.
```
MOVE NUMERO TO NOM-KEY.
START IS-FILE
    INVALID KEY PERFORM RE-KEY-PROC.
READ IS-FILE
    AT END PERFORM FINISH-PROC.
```

The statements above are used in a program to sequentially access records from an indexed sequential file. All of the record keys between 50 and 100 are given below.

0053 0067 0075 0081 0090 0098

Match the results given below with the value in NUMERO which provides the result.

_____ 1. 0067
_____ 2. 0074
_____ 3. 0097

A. Control will be transferred to RE-KEY-PROC.
B. Control will be transferred to FINISH-PROC.
C. The record corresponding to NUMERO will be read.

41. In order to sequentially access records from an indexed sequential file at some record other than the first record in the file, the program must include several entries in a specific order. Arrange the following in the appropriate order.

a. Working-Storage Section in the Data Division.
b. START statement in the Procedure Division.
c. NOMINAL KEY clause in the Environment Division.
d. READ statement with the AT END option in the Procedure Division.
e. Statement to set NOMINAL KEY variable to the value of the record at which sequential access is to begin.

42. Which clauses are required for a File-Control paragraph that describes a sequentially accessed indexed sequential file that will be accessed beginning with the first record in the file?

43. Match the following with the appropriate options of the READ statement. (There may be more than one matching item.)

_____ 1. AT END
_____ 2. INTO
_____ 3. INVALID KEY

A. Allows execution of the READ statement to have the effects of READ and MOVE statements.
B. Is not permitted with the READ statement. Refers to a randomly accessed file.
C. Is required when the READ statement refers to a randomly accessed file.
D. Is used to specify action to take place when the end of a sequentially accessed file is reached.
E. Is optional when the READ statement refers to a randomly accessed file.

44. Match the statements below that could be used to update records in a file with the access modes for indexed sequential input/output files below. (There may be more than one matching item.)

______ 1. RANDOM		A. START
______ 2. SEQUENTIAL		B. READ with AT END option.
		C. READ with INVALID KEY option.
		D. REWRITE

45. Match the following. (There may be more than one matching item.)

______ 1. Indexed Sequential file Random access	A. ACCEPT clause required.
	B. May be opened as I-O.
______ 2. Indexed Sequential file Sequential access	C. May be opened as OUTPUT.
	D. May use READ with INVALID KEY option.
______ 3. Standard Sequential file	E. NOMINAL KEY clause required.
	F. REWRITE statement used to update records.
	G. WRITE statement used to update records.

46. The Johnson Corporation is converting its master tape file for sales to a master sequential disk file with the same organization.

The file-name for the tape is SALES-TAPE and the record-name is TAPE-RECORD. The record lengths are fixed at 100 characters and recorded in blocks of five with standard labels.

The disk file name is SALES-DISK and the record-name is DISK-RECORD with the same organization as the tape file. Write the program to create the disk file. Assume following device numbers.

(In all subsequent problems, if the device numbers are not mentioned, the following will be assumed.)

Computer	—IBM 370
Tape Unit	—Model 2400.
Disk Unit	—Model 2311.
Printer	—Model 1403.
Card Reader	—Model 2540.
Card Punch	—Model 2540.

47. The Acme Manufacturing Company wishes to create a sequential disk file from a set of records.

a. Write the necessary entries for the Environment and Data Divisions based on the following information:

Computer to be used—Use own computer if available

Input—File-name	FILE-IN	Record-name	CARD-IN.
Output—File-name	DISK-SEQ	Record-name	DISK-IN
		Record Size 80 characters.	

All records on the disk are in blocks of five and are of a fixed length.

Output Device—Disk Model 2311.

b. Write the necessary procedural statements to create the disk file by transferring the input data to the output file. Assume both records are of the same size. The program should branch to a routine called FULL-DISK when the disk is full and display a message on the console to that effect.

48. The Bryan Tool Corporation is converting its present master tape file to a disk with the same organization. You are asked to write a program based on the following:

a. Write the necessary entries for the Environment and Data Divisions based on the following:

Computer to be used—Use own computer if available

Input—File-name	TAPE-FILE	Record-name	TAPE-IN
		Record Size 100 characters.	

All records on the tape are in blocks of five and are of a fixed length. All label records are standard.

Output—File-name	DISK-SEQ	Record-name	DISK-IN
		Record Size 100 characters.	

b. Write the necessary procedural statements to create the disk file by transferring the input data to the output file. The program should branch to a routine called FULLDISK when the disk is full and display a message on the console to that effect.

49. Write a program to update the Bryan Tool Corporation sequential disk file (exercise 48) for new addresses. The following information is provided:

Disk Record Record name RECORD-IN has the following format:

Field	Record Positions
Date	1-6
Customer Number	7-12
Customer Name	13-27
Address—Street	28-42
—City and State	43-57
Balance	58-64
Credit Limit	65-71
Unused in this program	72-100

Changes Card File-name CHANGES Record-name CHANGE-CARD

Field	Record Positions
Date	1-6
Customer Number	7-12
Street	13-27
City and State	28-42
Unused	43-80

If a customer number is not in the disk file, the program should branch to a routine called NOT-IN-FILE, where the appropriate message together with the customer number will be displayed on the console.

50. Using the same information in exercise 48, the Bryan Tool Corporation wishes to create an indexed sequential file from its present master tape file. The Record Key will be the customer number.

If the record is not written for any reason, the program should branch to ERROR-ROUTINE where the proper message will be displayed.

When the program is complete, display INDEXED FILE FOR CUSTOMER ACCOUNTS CREATED on console.

51. An indexed file contains the following record:

DISK-RECORD

Field	Record Position	
Balance	1-6	
Date	7-12	
Cumulative Disbursements	13-20	
Cumulative Receipts	21-28	
Minimum Balance	29-36	
Class of Stock	37-40	
Stock Number	41-47	
Unit Price	48-53	XXX.XXX
Amount	54-61	XXXXXX.XX
Unit	62-65	
Description	66-95	
Unused	96-100	

Write the necessary entries for

a. The File-Control paragraph. The file name is DISK-FILE and the record name is DISK-RECORD. The disk device is a model 2311. The elementary variable stock-number contains the identifier for each record.
b. The File Section entry. The records in the file are blocked in groups of five.
c. The procedural statements to sequentially access all records and list them on a model 1403 printer in the same format as the input record leaving two spaces between each field. Appropriate headings for each field should be printed and totals of the balance and amount should be accumulated and printed at end of report.

52. Using the same record as exercise 51, the outside auditors wish to sample check our inventory as follows:

A listing of the accounts in the same print format as exercise 51 for the following groups of accounts.
Five accounts starting with stock-number 1456845, 5152467, and 8759415.

Write the necessary entries for

a. The File-Control paragraph for above.
b. The procedural statements to access the specified records and then print them.

53. Using the disk record described in exercise 51, you wish to add records to the indexed sequential disk file without having to recreate the file. Write the necessary entries for the File-Control paragraph and Procedure Division to accomplish this.
54. Using the same record as exercise 51, you wish to access randomly to print selected critical inventory items.

    The finder records will contain the following information:

    File-name FINDER-FILE Record-name FINDER-RECORD

Field	**Positions**
Date	1-6
Stock Number	7-13

    The input device is model 2540.

    Write the necessary entries for
    a. The File-Control paragraph.
    b. The procedural statements to access the records and print the selected records in the same format as exercise 51. If record is not found, display a message on the console to that effect and proceed to the next record.
55. Using the same record as exercise 51, write a program to update the records in the disk randomly.

    The format for the transaction record is as follows: (The record will be read on a model 2540 device.)

    File-name TRANSACTION-FILE Record-name TRANSACTION-CARD

Field	**Positions**	
Date	1-6	
Stock Number	7-13	
Transaction Code	14	
Receipts—1		
Disbursements—2		
Quantity	15-19	
Amount	20-26	xxxxx.xx
Unused	27-80	

    Write the necessary entries for the
    a. File-Control paragraph.
    b. The procedural entries for the following:
        1. *Receipts Record.*
           Add Quantity to Balance.
           Add Quantity to Cumulative Receipts.
           Add Amount to Amount in Disk Record.
        2. *Disbursement Record.*
           Subtract Quantity from Balance.
           Add Quantity to Cumulative Disbursements.
           Subtract Amount from Amount in Disk Record.
        3. If Stock Number not found in Disk Record, branch to ERROR-ROUTINE and display appropriate message and read another transaction record.
        4. Compute new Unit Price by dividing Balance into Amount.
        5. Print TRANSACTION REGISTER as follows: Same print format as exercise 51.
           List the old balance, then the transaction record, and finally the new balance.
        6. Write new record back on disk file.

Problem 1

Create a file of disk records. These disk file records are to be sequentially organized. The disk records shall be eighty positions long and contain the following fields. (Records are to be in blocks of 960 characters.)

Disk Positions	Output Field
1	Code M
2-7	Customer number
8-27	Customer name
28-47	Street address
48-67	City/State address
68-72	Zip code
73-80	Blank

The information used to create these disk records is to be input from 80-position records. The input fields are identical in name but rearranged as follows:

Positions	Input Field
1	Code M
2-21	Customer name
22-41	Street address
42-61	City/State address
62-66	Zip code
67-72	Customer number
73-80	Blank

Create your own names for input and output devices and any other information needed.

Output

There is no printed output; output is on the disk.

Problem 2

Create a file of indexed records.

The input file, CARDIN, contains the following fields:

Positions	Field
1-2	Code characters X6
3-6	Account number (key field)
7-25	Name
26-31	Beginning balance, two decimals

The output file records, MASTACCT are to contain these fields:

Positions	Field or Constant
1	Character B
2-5	Account number (key field)
6-11	Beginning balance
12-17	Creation date (UDATE)
31-49	Name

Record length is 80 characters.

Block length is 960 characters.

Create your own names for input and output devices and any other information needed.

Output

There is no printed output; output is on the disk.

Problem 3

Print the contents of the indexed file created in problem 2. As part of the report, print the total amount of all the "beginning balance" fields in the entire file.

The output on the printer is to look like this:

```
                REPORT OF MASTER ACCOUNTS
ACCOUNT               BALANCE             NAME
1002                  $100.00             MONEMATERS
2116                $2,160.00             CASHINFLOW
2249                $1,728.50             BANKCREDIT
2367                  $750.60             CASHIN

                    $4,739.10
```

Create your own names for input and output devices and any other information needed.

The Printer Spacing Chart is as follows:

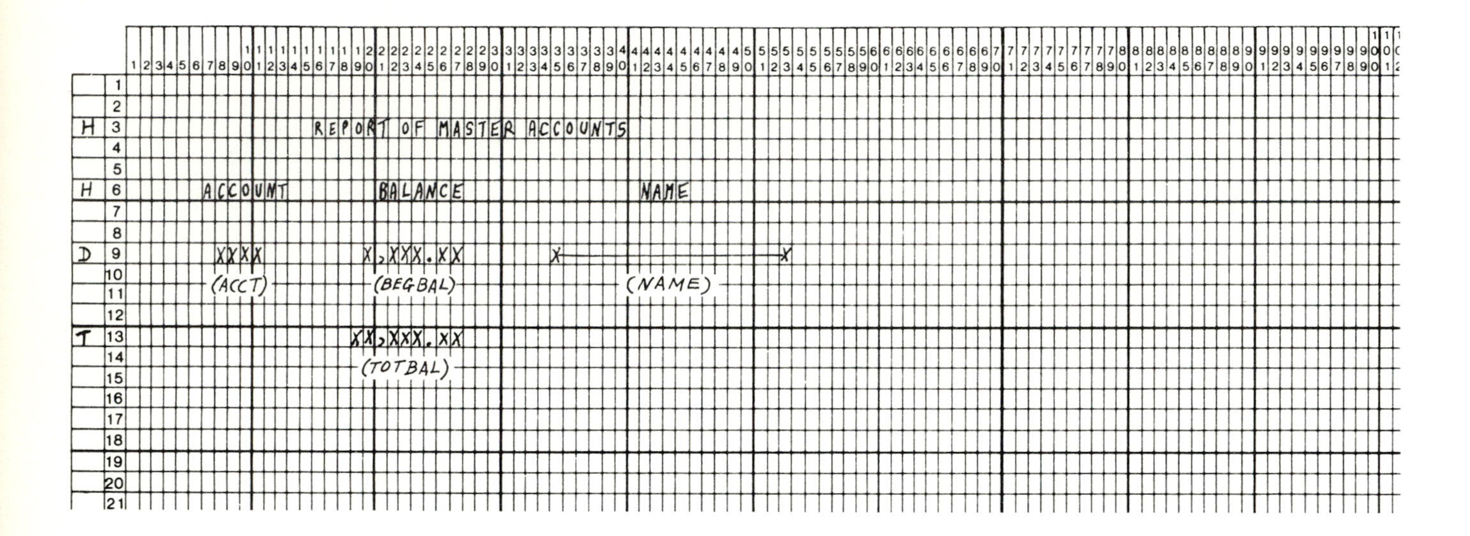

Output

Output is as follows:

```
              REPORT OF MASTER ACCOUNTS

ACCOUNT         BALANCE           NAME

 1002           $100.00         MONEMATERS

 2116         $2,160.00         CASHINFLOW

 2249         $1,728.50         BANKCREDIT

 2367           $750.60         CASHIN


              $4,739.10
```

Problem 4

NAMEFILE is an output file to be loaded with records read from CARDS in a file.

As each record is read from CARDS, the man number is used as the relative record number. The entire input record is written out on NAMEFILE in the relative record location corresponding to man number.

The input record contains man number in positions 1 to 5.

Code the necessary entries to create this relative file and load the records from CARDS. Use your own created names where necessary.

RECORD LENGTH	80 CHARACTERS
BLOCK LENGTH	960 CHARACTERS

Output

There is no printed output; output is on the disk.

14 Additional and Optional Features

Chapter Outline

Chapter Objectives

The learning objectives of this chapter are:

1. To introduce and explain some additional and optional features that can be useful in simplifying the coding process.
2. To describe and explain the use of the SPECIAL-NAMES paragraph of the Environment Division and how it can be used in COBOL programming. The various segments of the SPECIAL-NAMES paragraph can aid the programmer in solving problems relating to option switches and currency symbols other than the dollar sign, and can help to exchange the functions of the comma and period in editing currency amounts.
3. To explain the use of the RESERVE clause in the Input-Output Section in the File-Control paragraph of the Environment Division. To describe how the RESERVE clause may be used to provide additional input and output areas in addition to the buffers allocated by the compiler.
4. To describe and explain the use of the I-O-CONTROL paragraph of the Environment Division and some of the special techniques that may be used by the programmer to rerun programs at established points, the memory area that may be shared by different files, and the locations of files on a multiple file reel.
5. To explain the use of the RENAMES clause in the Data Division and how it may be used to permit alternate, possible overlapping and groupings of elementary items, and how it actually assigns a name to a segment of storage which can be addressed as a unique data item.
6. To explain and describe the operation of the LINAGE clause in the Data Division and how it may be used by the programmer as a means for specifying the depth of a logical page in terms of number of lines, as well as for providing top and bottom margins. In addition, a line number can be specified within the page body at which the footing area begins.
7. To explain the operation and functions of the CORRESPONDING options of the ADD, SUBTRACT, and MOVE statements in the Procedure Division and how these options may be used to process similar data items in different groups.
8. To explain the purpose and use of the Source Program Library Facility and how, through the use of the COPY statement, valuable programming time can be saved by copying prewritten library text into a source program for insertion in Environment, Data, or Procedure divisions at compile time.

Environment Division

Configuration Section—SPECIAL-NAMES Paragraph

The SPECIAL-NAMES paragraph enables the programmer to assign mnemonic-names or condition-names to the option switches used in the program. (The format for the SPECIAL-NAMES paragraph is shown in figure 14.1.) This paragraph also allows the programmer to assign a currency sign symbol other than $ and to exchange the function of the comma and period in editing currency amounts. If present, it must immediately follow the OBJECT-COMPUTER paragraph.

IMPLEMENTOR-NAME Clause

This clause is used to relate mnemonic-names to implementor-names, and/or to assign condition-names to the ON or OFF status of a software switch. Mnemonic-name can be any programmer-specified name as long as it conforms to the rules for formation of a data-name.

To illustrate the use of switches, consider that one program must determine whether or not a subsequent program is to produce a certain report. The report is to be produced only if switch 35 is set ON. The setting of the switch can be programmed as follows:

Program 1 (Setting switches):

SPECIAL-NAMES phrase:

```
SWITCH 35 IS REPORT-CONTROL

    ON STATUS IS . . .
    OFF STATUS IS . . .
```

Procedure Division statement:

```
IF . . .     DISPLAY 1 ON REPORT-CONTROL
ELSE         DISPLAY 0 ON REPORT-CONTROL.
```

The decision to produce the report or not can be programmed in one of the following two ways:

Program 2 (Sensing switches):

SPECIAL-NAMES phrase:

```
SWITCH 35 IS REPORT-CONTROL
    ON STATUS IS
    DO-REPORT.
```

Procedure Division statement:

```
IF DO-REPORT
PERFORM . . .
```

SPECIAL-NAMES phrase:

```
SWITCH 35 IS REPORT-CONTROL
    ON STATUS IS
    DO-REPORT.
```

Data Division syntax:

```
77  SWITCH-VALUE PIC 9 COMP-1.
```

Procedure Division statements:

```
ACCEPT SWITCH-VALUE FROM
    REPORT-CONTROL.
IF  SWITCH-VALUE = 1
    PERFORM . . .
```

Each implementor-name may be used only once in the SPECIAL-NAMES paragraph. If any implementor-name other than a switch name is used, its associated mnemonic-name may be used in ACCEPT and DISPLAY statements. ON STATUS and OFF STATUS apply only to switches.

```
SPECIAL-NAMES. [, implementor-name

    { IS mnemonic-name [, ON STATUS IS condition-name-1  [, OFF STATUS IS condition-name-2 ]]  }
    { IS mnemonic-name [, OFF STATUS IS condition-name-2  [, ON STATUS IS condition-name-1 ]]  }  ...
    { ON STATUS IS condition-name1 [, OFF STATUS IS condition-name-2 ]                          }
    { OFF STATUS IS condition-name-2 [, ON STATUS IS condition-name-1 ]                         }

    [, CURRENCY SIGN IS literal-9 ]

    [, DECIMAL-POINT IS COMMA ]] .
```

Figure 14.1 Format—SPECIAL NAMES paragraph.

The status of a switch (on or off) is specified by condition-names and is interrogated by testing the condition-names. The programmer can associate one condition-name with the ON STATUS, and the other with the OFF STATUS. Condition-names represent the equivalent of level-88 items. The mnemonic-names assigned to the switches themselves have no specific value to the programmer; it is the associated condition-name for the ON or OFF status of switches that allows the programmer to set the setting of any switch.

CURRENCY SIGN Clause

The literal that appears in the CURRENCY SIGN clause is used to represent the currency symbol in the PICTURE clause. The literal is limited to a single-character, nonnumeric literal and cannot be one of the characters that is assigned significance in the definition of the PICTURE clause character string.

Thus the character cannot be

1. Digits 0 through 9.
2. Alphabetic characters A, B, C, D, P, R, S, V, X or Z.
3. Special characters asterisk, plus, hyphen (minus), comma, period, left parenthesis, quotation mark, right parenthesis, space (blank), semicolon, or slash.

DECIMAL-POINT IS COMMA Clause

The DECIMAL-POINT IS COMMA clause states that in the current compilation, the functions of the comma and period are interchanged in the PICTURE clause character-string and in numeric literals.

The purpose of both the CURRENCY SIGN and DECIMAL-POINT IS COMMA clauses is to render COBOL more acceptable in countries where other conventions are observed.

Rules Governing the Use of the SPECIAL-NAMES Paragraph

1. In ACCEPT and DISPLAY statements, associated mnemonic-names may be used to identify the function-names identified with the input or output device.

   The following are examples of implementor-names.

IBM implementor-names

SYSIN
SYSOUT
SYSPUNCH
CONSOLE

Xerox implementor-names

Implementor-Names	*Meaning*
CONSOLE	Operator console
READER or CR	Card reader
PUNCH or CP	Card punch
PRINTER or LP	Line printer
PRINTER-F or LP-F	Line printer-formatted
SW0 through SW9	Ten software switches

2. When the mnemonic-name option is used in a WRITE statement with the ADVANCING option, the mnemonic-name must be defined as a function-name in the SPECIAL-NAMES paragraph. It is used to skip to channel 1–12 and to suppress spacing if desired. It may also be used for pocket selection for a card punch file.

   The following are examples of implementor-names and mnemonic-names.

   Switch-status names are established in the SPECIAL-NAMES paragraph of the Environment Division as illustrated below:

```
SPECIAL-NAMES.
    SW0      IS MONTH-END
    ON STATUS IS DO-MONTH-END.
```

   In the programmer's COBOL program, if the programmer wishes to know whether this is a month-end cycle, and wishes to execute a different portion of this program, he/she can merely ask in the Procedure Division code

```
IF DO-MONTH-END
    PERFORM MONTH-TOTAL-CALC
ELSE
    NEXT SENTENCE.
```

   The programmer can set up to ten run-time switches in this manner.

   The following are examples of implementor-names and their meanings.

Implementor-Name	*Action Taken*
CSP	Suppress spacing
C01–C12	Skip to channel 1 to 12 respectively
S01–S02	Pocket selection

   The following is an example of implementor-name and mnemonic-name.

```
SPECIAL-NAMES.
    C01 IS SKIP-TO-1.

PRT-HEADING.
    WRITE PRINTOUT FROM HDG-1
        AFTER ADVANCING SKIP-TO-1 LINES.
```

3. Mnemonic-names with characters enclosed in quotation marks are used in the CODE clause in the report description entry of the Report Writer Feature to identify output where more than one type of output is desired from a single input. (See Report Writer.)
4. Mnemonic-names may not be used in a source program except in the verb format that permits their usage.
5. The literal that appears in the CURRENCY SIGN IS clause is used in the PICTURE clause to represent the currency symbol. The literal is limited to a single character, must be nonnumeric, and must not be any of the following:
   a. Digits 0–9.
   b. Alphabetic characters A–Z and space.
   c. Special characters * – , . ; ( ) + ″ or ′.

   In the following example, the character M is used as the currency symbol in the PICTURE clause.

```
SPECIAL-NAMES.
    CURRENCY SIGN IS 'M'.
```

   If the clause CURRENCY SIGN IS literal is not present, then only the symbol $ can be used as the currency symbol in the PICTURE clause.

   The OPEN SWITCH is a function-name of this computer that enables the programmer to test the status of any one of the eight hardware option switches. The switch can set to an on or off status by the computer operator. The status of the switch can then be interrogated by testing the condition-name assigned to the switch. For example, the on status of switch 3 could be assigned a condition-name as follows:

```
SPECIAL-NAMES.
    OPTION SWITCH-3 ON STATUS IS PAPER-TAPE-INPUT.
```

The on status of option switch 3 can be tested by this conditional statement:

```
IF PAPER-TAPE-INPUT
    PERFORM PT-ROUTINE
ELSE
    PERFORM PC-ROUTINE.
```

This conditional statement goes to procedure with the procedure-name PT-ROUTINE if option switch 3 is on; it goes to PC-ROUTINE if option switch is off.

Suppose the programmer wants to perform one set of calculations if switch SW1 is on, and another set of calculations if SW1 is off. The paragraph entry might be,

```
SPECIAL-NAMES.
    SW1 IS BIT-ONE, ON STATUS IS BIT-ONE-ON, OFF
    STATUS IS BIT-ONE-OFF.
```

where SW1 is a function-name, BIT-ONE is a mnemonic-name, and BIT-ONE-ON and BIT-ONE-OFF are condition-names. Either or both of the condition-names can then be used in Procedure Division conditional test statements. One such statement might be,

```
IF BIT-ONE-ON, PERFORM PAR-3.
```

In this case, if switch SW1 is in an *on* status, control will be transferred to the paragraph named PAR-3. If switch SW1 is in an *off* status, control will fall through to the next statement after this one.

6. The clause DECIMAL-POINT IS COMMA means that the function of the comma and the period are exchanged in the PICTURE character string and in numeric literals. When the clause is used, the user must use a comma to represent a decimal point when required in numeric literals or in a PICTURE clause. The period is used for all functions ordinarily served by the comma.

Input-Output Section—FILE-CONTROL Paragraph

RESERVE Clause

The RESERVE clause is used to reserve additional input or output areas in addition to the buffers allocated by the compiler. (The format for the RESERVE clause is shown in figure 14.2.) The RESERVE clause allows the programmer to document the number of input or output or input/output areas in storage allocated to a specific file by the compiler. In American National Standard (ANS) COBOL, each file is assigned one storage area. The RESERVE clause indicates that storage areas are to be reserved for the file in addition to the usual single area.

The RESERVE clause allows the programmer to specify the number of input-output areas to be allocated for reading and writing physical records during program execution. If the RESERVE clause is specified and integer-1 is not zero (0), the number of input-output areas allocated is equal to the value of integer-1. If the RESERVE clause is omitted or if integer-1 is zero (0), the input-output control system will reserve the optimum number of input-output areas.

```
[ ; RESERVE integer-1  [ AREA  ] ]
                       [ AREAS ]
```

Figure 14.2 Format—RESERVE clause.

The following is an example of the RESERVE clause:

```
INPUT-OUTPUT SECTION.
FILE-CONTROL.
    SELECT CARD-FILE     ASSIGN TO READER
                         RESERVE 2 AREAS.
    SELECT PRINT-FILE    ASSIGN TO PRINTER
                         RESERVE 2 AREAS.
```

Two additional areas will be reserved for the input CARD-FILE and the output PRINT-FILE in addition to the buffers allocated by the compiler.

Rules Governing the Use of the RESERVE Clause

1. This clause specifies the number of buffers represented by the integer to be reserved for a standard sequential or an indexed sequential file that is accessed sequentially in addition to the one buffer that is reserved automatically.
2. If this clause is specified, the number of input-output areas allocated equals the value of integer-1.
3. If this clause is not specified, the number of input-output areas allocated is specified by the implementor.

I-O-CONTROL Paragraph

The I-O-CONTROL paragraph specifies some of the special techniques to be used in a program. The paragraph specifies the points at which rerun is to be established, the memory area that is to be shared by different files, and the location of files on a multiple file reel. (The format for the I-O-CONTROL paragraph is shown in figure 14.3.)

The I-O-CONTROL paragraph and its clauses are an optional part of the Environment Division.

RERUN Clause

When object programs process large volumes of data, it is desirable to indicate points at which the program may be restarted in case its running is interrupted by a power failure or some other unexpected event.

The restart facilities can be used to minimize the time lost in reprocessing a job that abnormally terminates. These facilities permit the automatic restarting of jobs that were abnormally terminated during execution.

A checkpoint is taken by periodically recording the contents of storage and registers during the execution of a program. The RERUN clause in the COBOL language facilitates the taking of checkpoint readings. Checkpoints are recorded onto a checkpoint data set.

```
I-O-CONTROL.

[ ; RERUN [ON {file-name-1        }] EVERY { { [END OF] {REEL} } OF file-name-2 } ] . . .
              {implementor-name   }         { {         {UNIT} }                  }
                                            { { integer-1 RECORDS }               }
                                            { integer-2 CLOCK-UNITS               }
                                            { condition-name                      }

[; SAME [RECORD] AREA FOR file-name-3 {, file-name-4} . . .] . . .

[; MULTIPLE FILE TAPE CONTAINS file-name-5 [POSITION integer-3]

     [, file-name-6 [POSITION integer-4]] . . .] . . .
```

Figure 14.3 Format—I-O-CONTROL paragraph.

Execution of a job can automatically be restarted at the beginning of a job step that abnormally terminated, or within the step.

The RERUN clause is used to establish any rerun or restart procedures. This clause specifies that checkpoint areas are to be written on the actual device at the time of the checkpoint, and that they can be read back into storage to restart the program from that point. A checkpoint record is the recording of the status of the problem program and main storage resources at desired intervals. The presence of this clause specifies that checkpoint records are to be taken. Checkpoint records are sequentially read and must be assigned to tape or mass storage devices.

The following are examples of rerun and checkpoint procedures—IBM.

To write single checkpoint records using tape:

```
//CHECKPT     DD     DSNAME=CHECK1,                          X
//                   VOLUME=SER=ND003,                       X
//                   UNIT=2400,DISP=(NEW,KEEP),              X
//                   LABEL=(,NL)
                     .
                     .
                     .
              ENVIRONMENT DIVISION.
                     .
                     .
                     .
                  RERUN ON UT-2400-S-CHECKPT EVERY
                  5000 RECORDS OF ACCT-FILE.
```

To write single checkpoint records using disk (note that more than one data set may share the same external-name):

```
//CHEK      DD     DSNAME=CHECK2,                           X
//                 VOLUME=(PRIVATE,RETAIN),                 X
//                    SER=DB030,                            X
//                 UNIT=2314,DISP=(NEW,KEEP),               X
//                 SPACE=(TRK,300)
                   .
                   .
                   .
              ENVIRONMENT DIVISION.
                   .
                   .
                   .
                  RERUN ON UT-2314-S-CHEK EVERY
                  20000 RECORDS OF PAYCODE.
                  RERUN ON UT-2314-S-CHEK EVERY
                  30000 RECORD OF IN-FILE.
```

To write multiple contiguous checkpoint records (on tape):

```
//CHEKPT      DD     DSNAME=CHECK3,                          X
//                   VOLUME=SER=111111,                      X
//                   UNIT=2400,DISP=(MOD,PASS),              X
//                   LABEL=(,NL)
                     .
                     .
                     .
              ENVIRONMENT DIVISION.
                     .
                     .
                     .
                  RERUN ON UT-2400-S-CHEKPT EVERY
                  10000 RECORDS OF PAY-FILE.
```

The RERUN clause specifies points in a program where checkpoints are to be recorded. This clause has no functional effect on the program and is therefore treated as a comment.

Rules Governing the Use of the RERUN Clause

1. This clause specifies when and where the rerun information is recorded. Rerun information is recorded in the following ways:
   a. If file-name-1 is specified, the rerun information is written on each reel or unit of an output file and the implementor specifies where, on the reel or file, the rerun information is to be recorded.
   b. If implementor-name is specified, the rerun information is written as a separate file on a device specified by the implementor.
2. There are seven forms of the RERUN clause, based on the several conditions under which rerun points can be established. The implementor must provide at least one of the specified forms of the RERUN clause
   a. When either the END OF REEL or END OF UNIT clause is used without the ON clause. In this case, the rerun information is written on file-name-2, which must be an output file.
   b. When either the END OF REEL or END OF UNIT clause is used and file-name-1 is specified in the ON clause. In this case, the rerun information is written on file-name-1, which must be an output file. In addition, normal reel or unit closing functions for file-name-2 are performed. File-name-2 may either be an input or an output file.
   c. When either the END OF REEL or END OF UNIT clause is used and implementor-name is specified in the ON clause. In this case, the rerun information is written on a separate rerun unit defined by the implementor. File-name-2 may be either an input or output file.
   d. When the integer-1 RECORDS clause is used. In this case, the rerun information is written on the device specified by implementor-name, which must be specified in the ON clause, whenever integer-1 records of file-name-2 have been processed. File-name-2 may be either an input or output file with any organization or access.
   e. When the integer-2 CLOCK-UNITS clause is used. In this case, the rerun information is written on the device specified by implementor-name, which must be specified in the ON clause, whenever an interval of time, calculated by an internal clock, has elapsed.
   f. When the condition-name clause is used and implementor-name is specified in the ON clause. In this case, the rerun information is written on the device specified by implementor-name whenever a switch assumes a particular status as specified by condition-name. In this case, the associated switch must be defined in the SPECIAL-NAMES paragraph of the Configuration Section of the Environment Division. The implementor specifies when the switch status is interrogated.
   g. When the condition-name clause is used and file-name-1 is specified in the ON clause. In this case, the rerun information is written on file-name-1, which must be an output file, whenever a switch assumes a particular status as specified by the condition-name. In this case, as in the previous paragraph, the associated switch must be defined in the SPECIAL-NAMES paragraph of the Configuration Section of the Environment Division. The implementor specifies when the switch status is interrogated.

SAME AREA Clause

The SAME AREA clause specifies that two or more files that do not represent sort or merge files are to use the same memory area during processing. The area being shared includes all storage areas assigned to the files specified; therefore, it is not valid to have more than one of the files open at the same time.

File-name-3 and file-name-4 are the names of the files. These files need not have the same organization or access. More than one SAME AREA clause may be included in the program; however, a specific file-name must not appear in more than one SAME AREA clause.

SAME RECORD AREA Clause

This clause specifies that two or more files are to use the same memory area for processing the current logical record. All of the files may be open at the same time. A logical record in the SAME RECORD AREA is considered as a logical record of each opened output file whose file-name appears in this SAME RECORD AREA clause, and of the most recently read input file whose file-name appears in this SAME RECORD AREA clause. This is equivalent to an implicit redefinition of the area, i.e., records are aligned on the leftmost character position.

If one or more file-names of a SAME AREA clause appear in a SAME RECORD AREA clause, all of the file-names in that SAME AREA clause must appear in the SAME RECORD AREA clause. However, additional file-names not appearing in that SAME AREA clause may also appear in that SAME RECORD AREA clause. The rule that only one of the files mentioned in a SAME AREA clause can be open at any given time takes precedence over the rule that all files mentioned in a SAME RECORD AREA clause can be open at any given time.

The files referenced in the SAME AREA or SAME RECORD AREA clauses need not have the same organization or access.

MULTIPLE FILE TAPE Clause

The MULTIPLE FILE TAPE clause documents the sharing of the same physical reel of magnetic tape by two or more files. Regardless of the number of files in a single reel, only those used in the object program need be specified.

In the following example, a reel may contain ABCFILE, GOODFILE, BESTFILE, and XYZFILE in the order listed. If only ABCFILE and BESTFILE will be used in a program, the MULTIPLE FILE TAPE clause would have the following forms:

```
I-O-CONTROL.
    MULTIPLE FILE TAPE CONTAINS ABCFILE POSITION 1, BESTFILE POSITION 3.
```

If all four files were used in a program, the MULTIPLE FILE TAPE clause would have this format:

```
I-O-CONTROL.
    MULTIPLE FILE TAPE CONTAINS ABCFILE, GOODFILE, BESTFILE, XYZFILE.
```

The MULTIPLE FILE clause is required when more than one file shares the same physical reel of tape. If all file-names have been listed in consecutive order, the POSITION clause need not be given. The following is an example of the MULTIPLE FILE TAPE clause:

```
.
.
INPUT-OUTPUT SECTION.
FILE-CONTROL.
    SELECT FILE-C      ASSIGN TO T1.
    SELECT FILE-A      ASSIGN TO T2.
    SELECT FILE-B      ASSIGN TO T3.
I-O-CONTROL.
    SAME AREA FOR FILE-B FILE-C FILE-A.
    MULTIPLE FILE TAPE CONTAINS FILE-A FILE-B FILE-C.
```

Although the files may be referenced in other phrases differently than they appear on the tape, they must be listed in their exact consecutive order in the MULTIPLE FILE phrase when the POSITION is omitted.

If any file in the sequence is not listed, the position relative to the beginning of the tape must be given. The following is an example of the MULTIPLE FILE TAPE clause—POSITION option:

```
MULTIPLE FILE TAPE CONTAINS FILE-C POSITION 3
    FILE-E POSITION 5     FILE-F POSITION 6
    FILE-Z POSITION 26.
```

Not more than one file on the same tape reel may be open at one time.

Data Division

RENAMES Clause

The RENAMES clause permits alternate, possibly overlapping, groupings of elementary items. The RENAMES data description entry consists of level number 66, the data-name, and the RENAMES clause. One or more RENAMES clauses may be written for a record. (The format for the RENAMES

66 data-name-1; RENAMES data-name-2 [{THROUGH / THRU} data-name-3]

Level-number 66, data-name-1 and the semicolon are shown in the format to improve clarity. Level-number and data-name-1 are not part of the RENAMES clause.

Figure 14.4 Format—RENAMES clause.

clause is shown in figure 14.4.) All RENAMES entries associated with a given logical record must immediately follow the last data description entry of that record.

Rules Governing the Use of the RENAMES Clause

1. Level number 66 must be used with the RENAMES clause.
2. All logical entries associated with a given logical record must immediately follow its last data description entry.
3. Data-name-2 and data-name-3 must be the name of elementary items or groupings of elementary items in the associated logical record, and cannot be the same data-name.
4. Data-name-3 cannot be subordinate to data-name-2.
5. A level 66 entry cannot rename another level 66 entry, nor can it rename a level 77, 88, or 01 entry.
6. Data-name-1 cannot be used as a qualifier and can be qualified only by level 01 or FD entries.
7. Data-name-2 and data-name-3 may be qualified.
8. An OCCURS clause may not appear in data-name-2, data-name-3, or any item subordinate to it.
9. Data-name-2 must precede data-name-3 in the record description entry.

The following is an example of the RENAMES clause.

```
01  CORRECTED-RECORD.
    05  GROUP-A.
        10  FIELD-1A.
            15  ITEM-1A     PICTURE XX.
            15  ITEM-2A     PICTURE XXX.
            15  ITEM-3A     PICTURE XX.
        10  FIELD-2A.
            15  ITEM-4A     PICTURE XX.
            15  ITEM-5A     PICTURE XX.
            15  ITEM-6A     PICTURE XX.
    05  GROUP-B REDEFINES GROUP-A.
        10  FIELD-1B.
            15  ITEM-1B     PICTURE XXXX.
            15  ITEM-2B     PICTURE XXX.
        10  FIELD-2B.
            15  ITEM-3B     PICTURE XXX.
            15  ITEM-4B     PICTURE XXX.
66  NEW-REC RENAMES ITEM-2A THRU ITEM-3B.

01  OUT-REC.
    05  FIELD-X.
        10  SUMMARY-GROUPX.
            15  FILE-1 PICTURE X.
            15  FILE-2 PICTURE X.
            15  FILE-3 PICTURE X.
    05  FIELD-Y.
        10  SUMMARY-GROUPY.
            15  FILE-1 PICTURE X.
            15  FILE-2 PICTURE X.
            15  FILE-3 PICTURE X.
    05  FIELD-Z.
        10  SUMMARY-GROUPZ.
```

```
                        15  FILE-1 PICTURE X.
                        15  FILE-2 PICTURE X.
                        15  FILE-3 PICTURE X.
                    66  SUM-X RENAMES FIELD-X.
                    66  SUM-XY RENAMES FIELD-X THRU FIELD-Y.
                    66  SUM-XYZ RENAMES FIELD-X THRU FIELD-Z.
```

Another example of the RENAMES clause follows:

```
01  MASTERRECORD
    05  ACCOUNT-NUMBER      PICTURE X(6).
    05  CHARGES-1           PICTURE 999V99.
    05  CHARGES-2           PICTURE 999V99.
    05  CHARGES-3           PICTURE 999V99.
    05  TOTAL-CHARGES       PICTURE 9999V99.
    05  TOTAL-PAYMENTS      PICTURE 9999V99.
    05  PAYMENTS-1          PICTURE 9999V99.
    05  PAYMENTS-2          PICTURE 9999V99.
    66  CHARGE-INFO RENAMES CHARGES-1 THRU TOTAL-CHARGES.
    66  TOTAL-INFO RENAMES TOTAL-CHARGES THRU TOTAL-PAYMENTS.
    66  PAYMENT-INFO RENAMES TOTAL-PAYMENTS THRU PAYMENTS-2.
    66  PAST-DUE      RENAMES PAYMENTS-2.
```

Consider the following data description:

```
01  PAY-MASTER.
    05  EMP-REC.
        10  EMP-NO              PICTURE 9(6).
        10  EMP-NAME            PICTURE X(25).
        10  DEPT                PICTURE 9(4).
        10  MAIL-STOP           PICTURE X(4).
        10  EXTENSION           PICTURE 9(4).
    05  PAY-REC.
        10  PAY-RATE            PICTURE 9V999.
        10  GROSS               PICTURE 9(4).
        10  GROSS-YTD           PICTURE 9(5)V99.
    66  TELEPHONE-DIRECTORY RENAMES
        EMP-NAME THRU EXTENSION.
```

By using the RENAMES clause, the programmer has created a new "pseudo record," which appears below:

```
05  TELEPHONE-DIRECTORY.
    10  EMP-NAME        PICTURE X(25).
    10  DEPT            PICTURE 9(4).
    10  MAIL-STOP       PICTURE X(4).
    10  EXTENSION       PICTURE 9(4).
```

The RENAMES clause actually assigns a name to a segment of storage, which can be addressed as a unique data item.

LINAGE Clause

The LINAGE clause provides a means for specifying the depth of a logical page in terms of the number of lines. It also provides for specifying the size of the top and bottom margins on the logical page, and the line number, within the page body, at which the footing area begins. (The format for the LINAGE clause is shown in figure 14.5.)

Rules Governing the Use of the LINAGE Clause

1. This clause provides a means for specifying the size of a logical page in terms of the number of lines. The logical page size is the sum of the values referenced by each phrase except the FOOTING phrase. If the LINES AT TOP or LINES AT BOTTOM phrases are not specified, the values for

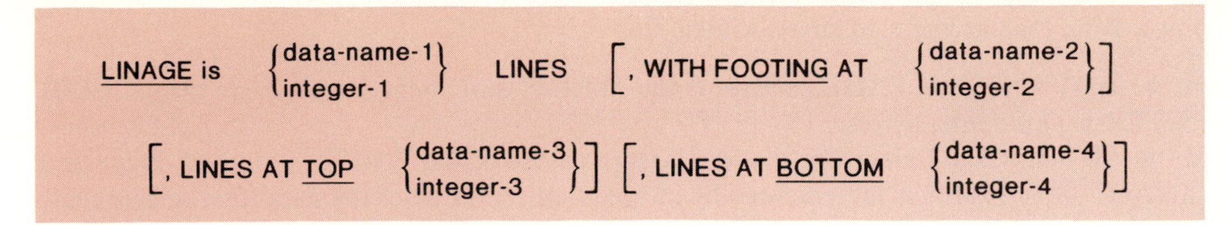

Figure 14.5 Format—LINAGE clause.

these functions are zero. If the FOOTING phrase is not specified, the assumed value is equal to integer-1, or the contents of the data referenced by data-name-1, whichever is specified.

There is not necessarily any relationship between the size of the logical page and the size of a physical page.

2. The value of integer-1 or the data item referenced by data-name-1 specifies the number of lines that can be written and/or spaced on the logical page. The value must be greater than zero. That part of the logical page in which these lines can be written and/or spaced is called the *page body*.
3. The value of integer-3 or the data item referenced by data-name-3 specifies the number of lines that comprise the top margin on the logical page. The value may be zero.
4. The value of integer-4 or the data item referenced by data-name-4 specifies the number of lines that comprise the bottom margin on the logical page. The value may be zero.
5. The value of integer-2 or the data item referenced by data-name-2 specifies the line number within the page body at which the footing area begins. The value must be greater than zero and not greater than the value of integer-1 or the data item referenced by data-name-1.

   The footing area comprises the area of the logical page between the line represented by the value of integer-2 or the data item referenced by data-name-2 and the line represented by the value of integer-1 or the data item referenced by data-name-1, inclusive.
6. The value of integer-1, integer-2, integer-3 and integer-4 is specified, and will be used at the time the file is opened by the execution of an OPEN statement with the OUTPUT phrase, to specify the number of lines that comprise each of the indicated sections of a logical page. The value of integer-2, if specified, will be used at that time to define the footing area. These values are used for all logical pages written for the file during a given execution of the program.
7. The values of the data items referenced by data-name-1, data-name-3, and data-name-4, if specified, will be used as follows:
   a. The values of the data items, at the time an OPEN statement with the OUTPUT phrase is executed for the file, will be used to specify the number of lines that are to comprise each of the indicated sections for the first logical page.
   b. The values of the data items, at the time a WRITE statement with the ADVANCING PAGE phrase is executed or a page overflow condition occurs, will be used to specify the number of lines that are to comprise each of the indicated sections for the next logical page.
8. The value of the data item referenced by data-name-2, if specified at the time an OPEN statement with the OUTPUT phrase is executed for the file, will be used to define the footing area for the first logical page. At the time a WRITE statement with the ADVANCING PAGE phrase is executed or a page overflow condition occurs, it will be used to define the footing area for the next logical page.

LINAGE-COUNTER

A LINAGE-COUNTER is generated by the presence of a LINAGE clause. The reserved word LINAGE-COUNTER is a name for a special register generated by the presence of a LINAGE clause in a file description entry. The implicit description is that of an unsigned integer whose size is equal to the size of integer-1 or the data item referenced by data-name-1 in the LINAGE clause. The value in the LINAGE-COUNTER at any given time represents the line number at which the device is positioned within the current page body.

The LINAGE-COUNTER may be referenced, but not modified, by a Procedure Division statement. Since more than one LINAGE-COUNTER may exist in a program, the programmer must qualify LINAGE-COUNTER by file-name when necessary.

The LINAGE-COUNTER is automatically modified, according to the following rules, during the execution of a WRITE statement to an associated file:

1. When the ADVANCING PAGE phrase of the WRITE statement is specified, the LINAGE-COUNTER is automatically reset to one (1).
2. When the ADVANCING identifier-2 or the integer phrase of the WRITE statement is specified, the LINAGE-COUNTER is incremented by integer-2 or the value of the data item referenced by identifier-2.
3. When the ADVANCING phrase of the WRITE statement is not specified, the LINAGE-COUNTER is incremented by the value of one (1).
4. The value of the LINAGE-COUNTER is automatically reset to one (1) when the device is repositioned to the first line that can be written on for each of the succeeding logical pages.
5. The value of the LINAGE-COUNTER is set to one (1) when an OPEN statement is executed for the associated file.
6. Data-name-1, data-name-2, data-name-3, and data-name-4 must reference elementary unsigned numeric data items.

The value of integers must be as follows:

Integer-1 must be greater than zero.

Integer-2 must not be greater than integer-1.

Integer-3 and integer-4 may be zero.

LINAGE-COUNTER Summary

The LINAGE COUNTER counter is automatically set to one (1) when a file with an OUTPUT phrase is opened, or the ADVANCING PAGE option of the WRITE statement is executed. This counter is automatically incremented by the number of lines implied in the WRITE statement. When the counter is equal to the value of FOOTING, overflow occurs. The counter may not be modified in the program.

For example, assume that the LINAGE clause is to be specified for a form of eleven inches in length, leaving a one-inch margin at the top and bottom of the form, with the totals to appear two lines before the last line of the form.

11 inches 6 lines to an inch = 66 lines		
Inch at top	6 lines	LINES AT TOP
Inch at bottom	6 lines	LINES AT BOTTOM
Total	12 lines	
Body	54 lines	LINAGE
Total—Logical Page	66 lines	
Footing	52	FOOTING

The entry in the File Section of the Data Division would appear as follows:

```
FD  PRINT-FILE
    LABEL RECORDS ARE OMITTED.
    LINAGE IS                    54 LINES
    WITH FOOTING AT              52
    LINES AT TOP                 6
    LINES AT BOTTOM              6.
```

The LINAGE clause can prove to be an invaluable asset to a programmer, who is completely relieved of the responsibility for accounting for all lines on a page through a line counter and signalling when an overflow should occur.

BLANK WHEN ZERO Clause

The BLANK WHEN ZERO clause is used when an item is to be filled with spaces when the value of the item is zero. The clause may be specified only for an item whose PICTURE is numeric at the

BLANK WHEN ZERO

Figure 14.6 Format—BLANK WHEN ZERO clause.

elementary level. The clause may not be specified for level 66 or 88 items. (The format for the BLANK WHEN ZERO clause is shown in figure 14.6.)

If data moved to data-name is:	and the PICTURE clause is:	then, the contents of data-name is:
00000	99,999 BLANK WHEN ZERO	
00020	99,999 BLANK WHEN ZERO	00,020
00000	99,999	00,000
00000	ZZ,ZZZ	
00000	$99,999 BLANK WHEN ZERO	
00000	$ZZ,ZZZ BLANK WHEN ZERO	
000000	$9,999.99 BLANK WHEN ZERO	
000000	$Z,ZZZ.ZZ BLANK WHEN ZERO	
000000	$*,*** ** BLANK WHEN ZERO	**********

JUSTIFIED Clause

The JUSTIFIED clause is used to override the normal positioning of alphabetic or alphanumeric data when it is moved to a larger area. (The format for the JUSTIFIED clause is shown in figure 14.7.) If

Figure 14.7 Format—JUSTIFIED clause.

an item is moved to a location that is larger than itself, it may be necessary to specify the position that the data is to occupy in the new area. In the absence of the JUSTIFIED clause, normal justification will be performed on the movement of data as follows:

Numeric data will be *right justified* after decimal alignment, and any unused positions at the right or left will be filled with zeros. The rightmost character will be placed in the rightmost positions in the new area (if no decimals are involved), and zeros will be supplied to the unused positions at the left.

Alphabetic and *Alphanumeric* data will be *left justified* after the move, and any unused character positions at the right will be filled with blanks. If the sending field is larger than the receiving area, excess characters at the right will be truncated.

If the programmer *wishes to alter* the normal justification of alphabetic or alphanumeric items, it can be done with the JUSTIFIED clause. If the JUSTIFIED clause is used, it will affect the positioning of the receiving area, as follows:

1. If the data being sent is longer than the receiving area, the leftmost characters will be truncated.
2. If the data being sent is shorter than the receiving area, the unused positions at the left will be filled with blanks.

The JUSTIFIED clause may be used only at the elementary level, and must not be specified for level-66 or level-88 data items. (Figure 14.8 shows an example of the JUSTIFIED clause.)

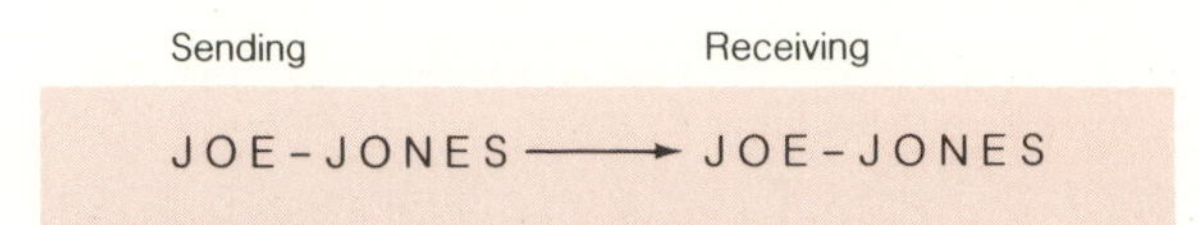

Figure 14.8 JUSTIFIED clause—example.

SYNCHRONIZED Clause

Some computer memories are organized in such a way that there are natural addressing boundaries in the computer memory (e.g., word boundaries, halfword boundaries, doubleword boundaries, byte boundaries). The manner in which data is stored is determined by the object program and need not respect these natural boundaries.

However, certain uses of data (e.g., in arithmetic operations or in subscripting) may be facilitated if the data is stored so as to be aligned on these natural boundaries. Specifically, additional machine operations in the object program may be required for the accessing and storage of data if portions of two or more data items appear between adjacent natural boundaries, or if certain natural boundaries divide a single data item into two branches.

Data items that are aligned in natural boundaries in such a way as to avoid additional machine operations are defined as being synchronized. A synchronized item is assumed to be introduced and carried in that form; conversion to synchronized form occurs only during the execution of a procedure (other than READ or WRITE) that stored data in the item.

Synchronization can be accomplished in two ways:

1. By use of the SYNCHRONIZED clause.
2. By recognizing the appropriate natural boundaries and organizing the data suitable without the use of the SYNCHRONIZED clause.

The SYNCHRONIZED clause is used to specify alignment of an elementary item in the natural boundaries of the computer memory. (The format of the SYNCHRONIZED clause is shown in figure 14.9.) This clause specifies that the COBOL processor, in creating the internal format for this item, must arrange the item in contiguous units of memory in such a way that no other data item appears in any of the memory units between the right and left natural boundaries delimiting these data items. If the size of the item is such that it does not itself utilize all of the storage area between the delimiting natural boundaries, the unused storage positions (or portion thereof) may not be used for any other data item.

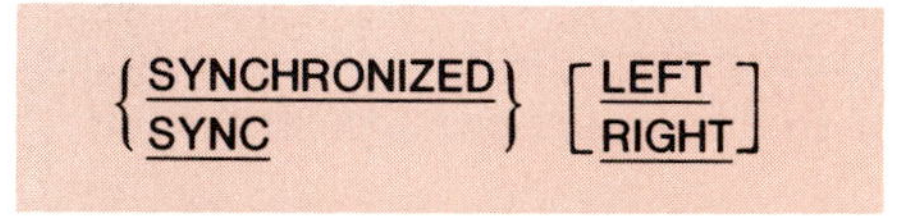

Figure 14.9 Format—SYNCHRONIZED clause.

Rules Governing the Use of the SYNCHRONIZED Clause

1. SYNCHRONIZED not followed by RIGHT or LEFT specifies that an elementary item is to be positioned between the natural boundaries in such a way as to effect utilization of the elementary data items. The specific positions are determined by the implementor.
2. If SYNCHRONIZED LEFT is specified, the leftmost character will occupy the leftmost position in the contiguous memory area. The right-hand positions of the area will be unoccupied.
3. If SYNCHRONIZED RIGHT is specified, the rightmost character will occupy the right-hand position in the contiguous memory area with the leftmost positions of the area unoccupied.
4. Whenever a SYNCHRONIZED item is referenced in the source program, the original size of the

items, as shown in the PICTURE clause, is used in determining any action that depends on size, such as justification, truncation, or overflow.

5. In the data description for an item, the sign appears in the normal operational size position regardless of whether the item is SYNCHRONIZED LEFT or SYNCHRONIZED RIGHT.
6. This clause is hardware-dependent and, in addition to the rules just stated, the implementor must specify how elementary items associated with this clause are handled. The user should consult individual reference manuals for particular computers for further information relative to this clause.

The following is an example of the use of the SYNCHRONIZED clause.

```
01  WORK-RECORD.
    05  WORK-CODE                    PICTURE X.
    05  COMP-TABLE OCCURS 10 TIMES.
        10  COMP-TYPE                PICTURE X.
        10  IA-Slack-Bytes           PICTURE XX. Inserted by compiler]
        10  COMP-PAY                 PICTURE S9(4)V99 USAGE COMP SYNC.
        10  COMP-HRS                 PICTURE S9(3)     USAGE COMP SYNC.
        10  COMP-NAME                PICTURE X(5).
        10  IE-Slack-Bytes           PICTURE XX. Inserted by compiler]
```

Procedure Division

CORRESPONDING Options

The CORRESPONDING options can be useful when it is necessary to process similar data items in different groups. The programmer can perform computations on elementary items or the movement of data items from one group to another just by specifying the group to which they belong. This option can be useful with the ADD, SUBTRACT, and MOVE statements, whereby elementary data items within groups are added to, subtracted from, or moved to elementary items with the same name in another group. This can save coding and prove as an invaluable aid to the programmer with the problem of working with similar data items in various groups that require addition, subtraction or movement.

The inherent problem in using the CORRESPONDING option in these statements is that the data names must be identical in both groups, and if these data items are used in other Procedure Division statements, the data names must be qualified for uniqueness. This may offset the advantage of using the similar names for the CORRESPONDING option. The programmer should carefully analyze the feasibility of using the CORRESPONDING option with the necessity of having to qualify data items in other Procedure Division statements.

ADD CORRESPONDING Statement

When the ADD CORRESPONDING statement is specified, the data items in the group item referenced preceding the word TO are added to and stored in the corresponding data items in the group item referenced following the word TO. (The format for the ADD CORRESPONDING statement is shown in figure 14.10.)

Rules Governing the Use of the ADD CORRESPONDING Statement

1. Elementary data items within identifier-1 are added to and stored in the corresponding data items in identifier-2. Identifier-1 and identifier-2 must be group items.

```
ADD  {CORRESPONDING}  identifier-1 TO identifier-2  [ ROUNDED ]
     {CORR         }

  [ ; ON SIZE ERROR imperative-statement ]
```

Figure 14.10 Format—ADD CORRESPONDING statement.

2. When the SIZE ERROR option is specified for CORRESPONDING option items, the test is made after the completion of all add operations. If any of the additions produce a size error condition, the resultant field for that addition remains unchanged, and the imperative statement specified in the SIZE ERROR option is executed.

   The following is an example of the ADD CORRESPONDING statement:

```
01  PRODUCT-RECORD.
    05  PRODUCT-1                PICTURE 9(5)V99.
    05  PRODUCT-2                PICTURE 9(5)V99.
    05  PRODUCT-3                PICTURE 9(5)V99.
    05  PRODUCT-4                PICTURE 9(5)V99.
    05  PRODUCT-5                PICTURE 9(5)V99.

WORKING-STORAGE SECTION.
01  PRODUCT-TOTAL-RECORD.
    05  PRODUCT-1-TOTAL          PICTURE 9(7)V99    VALUE ZEROS.
    05  PRODUCT-2-TOTAL          PICTURE 9(7)V99    VALUE ZEROS.
    05  PRODUCT-3-TOTAL          PICTURE 9(7)V99    VALUE ZEROS.
    05  PRODUCT-4-TOTAL          PICTURE 9(7)V99    VALUE ZEROS.
    05  PRODUCT-5-TOTAL          PICTURE 9(7)V99    VALUE ZEROS.

PROCEDURE DIVISION

    ADD CORRESPONDING PRODUCT-RECORD TO PRODUCT-TOTAL-RECORD
        ON SIZE ERROR DISPLAY 'PRODUCT TOTAL AREA TOO SMALL'
            UPON CONSOLE.
```

In the example just given, corresponding products will be added to the PRODUCT-TOTAL-RECORD. If any total area is too small to contain the sum, a message will be displayed on the console.

SUBTRACT CORRESPONDING Statement

When the SUBTRACT CORRESPONDING statement is specified, data items in the group referenced preceding the word FROM are subtracted from and stored in the corresponding data items in the group item referenced following the word FROM. (The format for the SUBTRACT CORRESPONDING statement is shown in figure 14.11.)

Rules Governing the Use of the SUBTRACT CORRESPONDING Statement

1. Elementary data items within identifier-1 are subtracted from corresponding data items in identifier-2, and the differences are stored in the corresponding identifier-2 data items.
2. When the CORRESPONDING option is used in conjunction with the SIZE ERROR option and a size error condition arises, the result for SUBTRACT is analogous to that of ADD.

   The following is an example of the SUBTRACT CORRESPONDING statement:

```
01  SALES-RECORD.
    05  PRODUCT-1-SALES    PICTURE 9(5).
    05  PRODUCT-2-SALES    PICTURE 9(5).
    05  PRODUCT-3-SALES    PICTURE 9(5).
    05  PRODUCT-4-SALES    PICTURE 9(5).
    05  PRODUCT-5-SALES    PICTURE 9(5).

01  INVENTORY-RECORD.
    05  PRODUCT-1-INV      PICTURE 9(5).
    05  PRODUCT-2-INV      PICTURE 9(5).
    05  PRODUCT-3-INV      PICTURE 9(5).
    05  PRODUCT-4-INV      PICTURE 9(5).
    05  PRODUCT-5-INV      PICTURE 9(5).

PROCEDURE DIVISION.

    SUBTRACT CORRESPONDING SALES-RECORD FROM INVENTORY-RECORD.
```

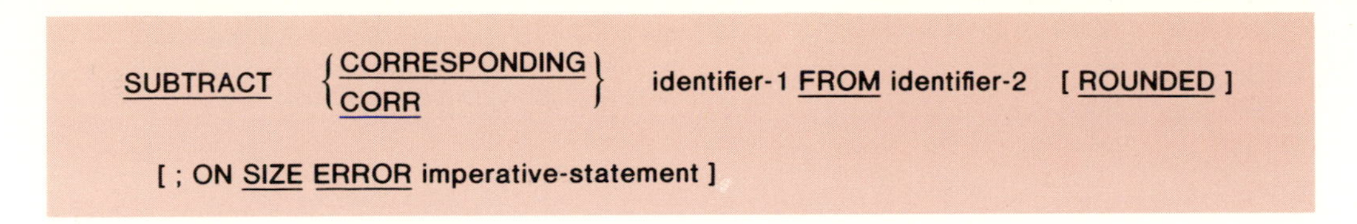

Figure 14.11 Format—SUBTRACT CORRESPONDING statement.

In the example just given, corresponding product sales will be subtracted from the INVENTORY-RECORD.

MOVE CORRESPONDING Statement

When the MOVE CORRESPONDING statement is specified, selected items subordinate to identifier-1 are moved to corresponding items subordinate to identifier-2. The selected data items are moved individually; editing, format conversion, fill, or truncation take place as appropriate, data item by data item. (The format for the MOVE CORRESPONDING statement is shown in figure 14.12.)

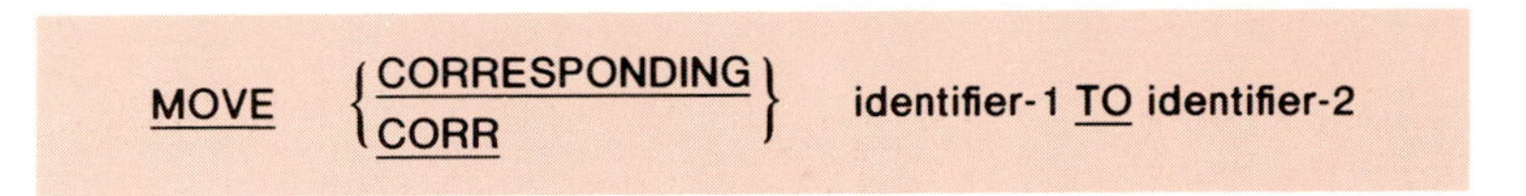

Figure 14.12 Format—MOVE CORRESPONDING statement.

In a MOVE CORRESPONDING statement, identifier-1 and identifier-2 must be group items. For each possible pair of data items subordinate to identifier-1 and identifier-2, a correspondence exists if the following rules are satisfied:

Rules Governing the Use of the MOVE CORRESPONDING Statement

1. Only corresponding data items having the same name and qualification as identifier-1 and identifier-2 are moved. (See figure 14.13.)
2. At least one of the items of the pair of matching items must be an elementary item.
3. The effect of a MOVE CORRESPONDING statement is equivalent to a series of simple MOVE statements.
4. Identifier-1 and identifier-2 must be group items.

```
MOVE CORRESPONDING INVENTORY-POSTING TO
    INVENTORY-RECORD.
```

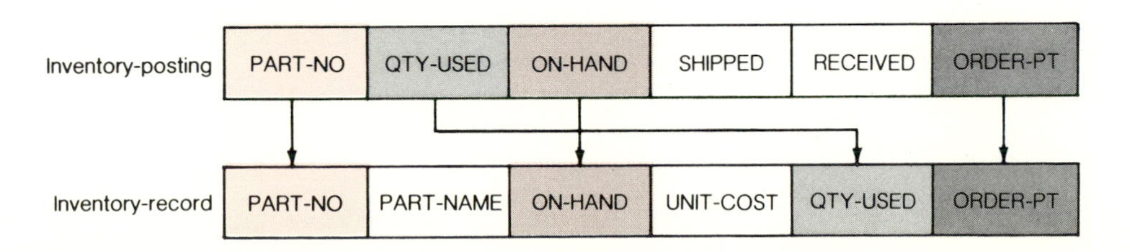

Figure 14.13 Data movement affected by MOVE CORRESPONDING statements.

MOVE CORRESPONDING CHANGE TO MASTER.

This Procedural sentence will place the data in the CHANGE record into the corresponding fields of the MASTER record as shown below:

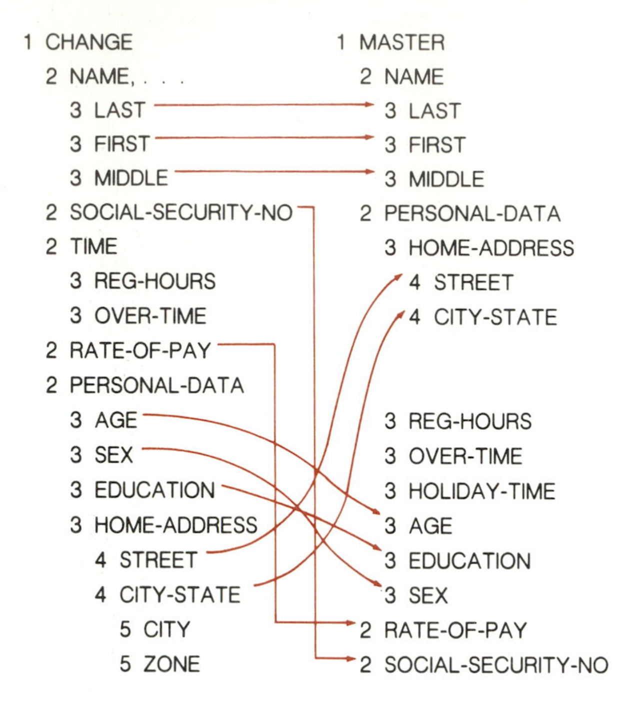

In this illustration, only data items with identical data-names will be moved.

Figure 14.14 MOVE CORRESPONDING statement—example.

5. An item subordinate to identifier-1 or identifier-2 is not considered corresponding if:
   a. It is an item identified by the word FILLER and any items subordinate to it, or
   b. An item identified by a REDEFINES, OCCURS, RENAMES, or USAGE IS INDEX clause and any items subordinate to it.
6. Either identifier may have REDEFINES or OCCURS clauses in its description or may be subordinate to a data item described with these clauses. If either identifier is described with an OCCURS clause, the items must be subscripted; each data item that corresponds will also have to be subscripted by the programmer.
7. Data-names with the level number 66, 77, or 88 (RENAMES clause, independent item clause, or condition-names) cannot be referenced by identifier-1 or identifier-2.
8. Each matched source item is moved in conformity with the description of the receiving item.

(See figure 14.14.)

For each corresponding pair of items, the effect of a MOVE CORRESPONDING statement is the same as if a separate MOVE statement had been written instead. For example, consider the statement MOVE CORRESPONDING ABLE TO BAKER (I), where the respective data descriptions are as follows:

```
10  ABLE . . .                        05  BAKER; OCCURS 10 TIMES . . .
    15  P . . .                           10  P
    15  Q . . .                           10  Q
    15  R REDEFINES Q . . .               10  R
        20  S . . .                           15  S
```

The effect is the same as if the following statements had been written:

```
MOVE P OF ABLE    TO P OF BAKER (I)
MOVE Q OF ABLE    TO Q OF BAKER (I)
```

Note that R is not moved (because neither R is elementary). S is not moved because S of ABLE is subordinate to a group item within the ABLE that is described with the REDEFINES clause.

A correspondence never exists for FILLER items.

Since a MOVE CORRESPONDING statement expands into a series of separate moves, the CORRESPONDING option should never be used for record-to-record or group-to-group moves where the sending and receiving groups have the same descriptions; a simple MOVE statement (without the CORRESPONDING option) produces more efficient object program coding. The COBOL compiler may abort when the MOVE CORRESPONDING statement is used in a program with a very large number of Data Division entries or with very long record descriptions in which "correspondences" would exist. The abort may be avoided in subsequent compilations by increasing the memory limits. If the memory limits cannot be increased, the abort can be circumvented by replacing the MOVE CORRESPONDING statement with elementary MOVE statements to accomplish the desired functions. The programmer must be aware of all the problems involved in modifying a program that contains corresponding data items. Additional qualification may be necessary.

Examples of MOVE CORRESPONDING statements that reflect the hierarchy of level structure follow:

```
01  A
    05  Z
    05  B
    05  C
        10  D
        10  E
            15  F
            15  G
                20  P
                20  Q
    05  H
        10  I
        10  J
    05  K
    05  L

01  X
    05  Y
        10  L
            15  M
            15  N
        10  C
            15  D
            15  E
                20  F
                20  G
                    25  U
                    25  V
        10  H
        10  B
    05  Z
    05  Q
```

MOVE CORRESPONDING A TO X is equivalent to

MOVE Z OF A TO Z OF X

MOVE CORRESPONDING A TO Y is equivalent to

```
MOVE  B  OF  A  TO  B  OF  Y  (elementary)
MOVE  D  OF  A  TO  D  OF  Y  (elementary)
MOVE  F  OF  A  TO  F  OF  Y  (elementary)
MOVE  L  OF  A  TO  L  OF  Y  (group)
MOVE  H  OF  A  TO  H  OF  Y  (group)
```

MOVE CORRESPONDING Y TO A is equivalent, with sending and receiving items reversed, to

MOVE CORRESPONDING A TO Y

ALTER procedure-name-1 TO [PROCEED TO] procedure-name-2

[procedure-name-3 TO [PROCEED TO] procedure-name-4] . . .

Figure 14.15 Format—ALTER statement.

ALTER Statement

Although good structured programming techniques forbid the use of the ALTER statement, it is introduced at this time as it is part of ANS COBOL and should be presented for those who choose to use it.

The ALTER statement is used to modify the effect of the unconditional GO TO statement elsewhere in the program, thus changing the sequence of operations to be performed. (The format for the ALTER statement is shown in figure 14.15.)

Rules Governing the Use of the ALTER Statement

1. A GO TO statement to be altered must be written as a single paragraph consisting solely of the unconditional GO TO statement preceded by a paragraph-name. (The format and an example for the GO TO statement—Format-3—are shown in figure 14.16.)
2. The ALTER statement replaces the procedure-name specified in the GO TO statement (if any) by the procedure-name specified in the ALTER statement.

The following is an example of an ALTER statement.

```
PARAGRAPH-1.
    GO TO BYPASS-PARAGRAPH.
PARAGRAPH-1A.

BYPASS-PARAGRAPH.
    ALTER PARAGRAPH-1 TO PROCEED TO PARAGRAPH-2.

PARAGRAPH-2.
```

Before the ALTER statement is executed, when control reaches PARAGRAPH-1, the GO TO statement transfers control to BYPASS- PARAGRAPH. After execution of the ALTER statement, however, when control reaches PARAGRAPH-1, the GO TO statement transfers control to PARAGRAPH-2.

Many experienced COBOL programmers suggest that use of the ALTER statement in connection with the GO TO statement is so difficult to document adequately, and so likely to lead to serious difficulties in program test and maintenance, that the ALTER statement should never be used under any circumstances. To accomplish the same purpose, there are alternative ways which are easier to understand and simpler to maintain.

Format 3

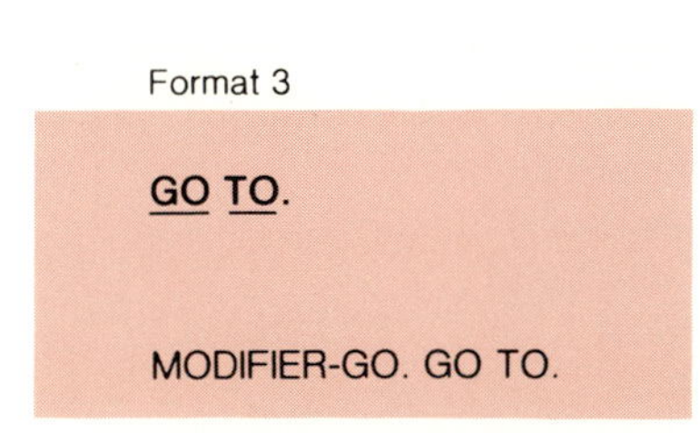
GO TO.

MODIFIER-GO. GO TO.

Figure 14.16 Format—GO TO statement—format 3 and example.

Source Program Library Facility

The writing of the data definitions for a master file that is accessed by many COBOL programs is a frequently encountered data processing problem. In order to avoid the repetitious writing of such data definitions, the data definitions are placed into the direct access file as recompilation master. Then when a source program needs the definition of the master file, just one COPY statement is written in place of the many data description entries.

The library module provides a capability for specifying text that is to be copied from a library. The COBOL library contains text that is available to a source program at compile time. Prewritten source program entries can be included in a source program at compile time. An installation can thus use standard file descriptions, record descriptions, or procedures without recoding them. The effect of the compilation of the library text is the same as if the text were actually written as part of the source program.

The COBOL library contains text that is available to a source program at compile time. These entries and procedures are contained in these programmer-created libraries; they are included in a source program by means of a COPY statement.

COBOL library text is placed in the COBOL library as a function independent of the COBOL program and according to implementor-defined techniques. An entry in the COBOL library may contain source program texts for the Environment, Data, or Procedure Divisions, or any combination of the three.

COPY Statement

The COPY statement permits the inclusion of prewritten Environment, Data, and Procedure Division entries at compile time in the source program. (See the format for the COPY statement in figure 14.17.)

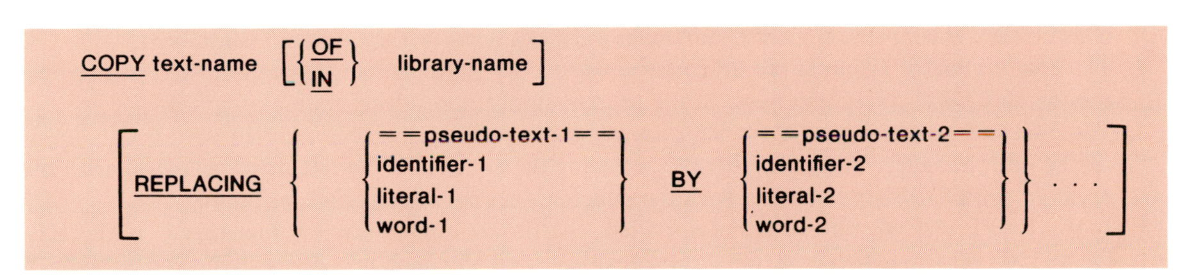

Figure 14.17 Format—COPY statement.

Text-name is the name of a member of a partitioned data set contained in the programmer's library; it identifies the library subroutine to the control program. A partitioned data set is one that is divided into sequentially organized members made up of one or more records. Each member has a unique name. Text-name must follow the rules for the formation of a program-name. The first eight characters are used to identify the name.

Partitioned organization is used mainly for the storage of sequential data such as programs, subroutines, compilers, and tables. The main advantage of a partitioned file is that it is possible for the programmer to retrieve specific members by specifying only the name of the data set wanted.

The COPY statement is used to insert library text into the source program, where it will be treated by the compiler as part of the source program.

When a COPY statement is processed, the library text associated with text-name is copied into the source program. It replaces the COPY statement, beginning with the reserved word COPY and ending with the period punctuation character.

For example, suppose the library entry named WORKREC consists of the following Data Division record:

```
05  AAA.
    10  HOURS-WORKED     PICTURE 99.
    10  SICK-LEAVE       PICTURE 99.
    10  HOLIDAY-PAY      PICTURE 99.
```

The programmer may copy this record into the source program with the following two statements:

```
01  WORK-RECORD       COPY WORKREC.
01  NEXT-RECORD.
```

After the record has been copied, the compiled source program will appear as if it had been written as:

```
01  WORK-RECORD.
    05  AAA.
        10  HOURS-WORKED       PICTURE 99.
        10  SICK-LEAVE         PICTURE 99.
        10  HOLIDAY-PAY        PICTURE 99.
01  NEXT-RECORD.
```

Library text is copied into the source program unchanged, and thus must conform to the rules for COBOL reference format. Library text may contain debugging lines, but may not contain any other COPY statements. If a COPY statement is specified on a debugging line, then the text that is the result of processing the COPY statement will appear as though it were specified on debugging lines with the following exception: comment lines in library text will appear as comment lines in the resultant source program.

Rules Governing the Use of the COPY Statement

1. Within one COBOL library, each text-name must be unique.
2. The COPY statement must be preceded by a space and terminated by the separator period.
3. A COPY statement may occur in the source program anywhere a character-string or a separator might occur, except that a COPY statement must not occur within a COPY statement.
4. The compilation of a source program containing COPY statements is logically equivalent to processing all COPY statements prior to the processing of the resulting source program.
5. The effect of processing a COPY statement is that the library text associated with text-name is copied into the source program, logically replacing the entire COPY statement, beginning with the reserved word COPY and ending with the punctuation character period, inclusive.
6. If the REPLACING phrase is not specified, the library text is copied unchanged. If the REPLACING phrase is specified, the library text is copied and each properly matched occurrence of pseudo-text-1, identifier-1, word-1, and literal-1 in the library text is replaced by the corresponding pseudo-text-2, identifier-2, word-2, or literal-2.
7. Comment lines appearing in the library text are copied into the source program unchanged.
8. Debugging lines are permitted within library text.
9. The text produced as a result of the complete processing of a COPY statement must not contain a COPY statement.
10. The syntactive correctness of the library text cannot be independently determined. The syntactive correctness of the entire COBOL source program cannot be determined until all COPY statements have been completely processed.
11. Library text must conform to the rules for COBOL reference format.
    If the library entry PAYLIB consists of the following Data Division record:

```
01  A.
    05  B     PICTURE S99.
    05  C     PICTURE S9(5)V99.
    05  D     PICTURE S9999 OCCURS 0 TO 52 TIMES
              DEPENDING ON B OF A.
```

the programmer can use the COPY statement in the Data Division of the program as follows:

```
01  PAYROLL COPY PAYLIB.
```

In this program, the library entry is then copied and appears in the source listing as follows:

```
01  PAYROLL.
    05  B     PICTURE S99.
    05  C     PICTURE S9(5)V99.
    05  D     PICTURE S9999 OCCURS 0 TO 52 TIMES
              DEPENDING ON B OF A.
```

Note that the data-name A has not been changed in the DEPENDING ON option. The REPLACING option can be used to change some (or all) of the names within the library entry to names the programmer wishes to reference in the program. An example of this follows:

```
01  PAYROLL COPY PAYLIB REPLACING A BY PAYROLL
    B BY PAY-CODE C BY GROSSPAY.
```

In this program the library entry is copied and appears in the source listing as follows:

```
01  PAYROLL.
    05  PAY-CODE    PICTURE S99.
    05  GROSSPAY    PICTURE S9(5)V99.
    05  D           PICTURE S9999 OCCURS 0 TO 52 TIMES
                    DEPENDING ON PAY-CODE OF PAYROLL.
```

The entry as it appears in the library remains unchanged.

Summary

There are additional features and optional features in the Environment, Data and Procedure divisions that can be an invaluable aid to the programmer.

In the Environment Division, the SPECIAL-NAMES paragraph enables the programmer to assign mnemonic-names or condition-names to the option switches and system tags used in the program. This paragraph also allows the programmer to assign a currency symbol other than the dollar sign and to exchange the function of the comma and period in editing currency amounts.

The implementor-name clause is used in the SPECIAL-NAMES paragraph to relate mnemonic-names to implementor-names, and/or to assign condition-names to the ON or OFF status of a software switch. The status of a switch (ON or OFF) is specified by the condition-name and is interrogated by testing the condition-name.

The literal that appears in the CURRENCY SIGN clause is used to represent the currency symbol in the PICTURE clause.

The DECIMAL-POINT IS COMMA clause states that in the current compilation the functions of the comma and period are interchanged in the PICTURE clause character-string and in numeric literals.

The purpose of both the CURRENCY SIGN and DECIMAL-POINT IS COMMA clauses is to render COBOL more acceptable in countries where other conventions are observed.

The RESERVE clause in the Input-Output Section, File-Control paragraph allows the programmer to specify the number of input-output areas to be allocated for reading and writing physical records during program execution. If the RESERVE clause is omitted or zero areas specified, the input-output control system will reserve the optimum number of input-output areas.

The I-O-CONTROL paragraph specifies some of the special techniques to be used in a program. It specifies the point at which rerun is to be established, the memory area which is to be shared by different files, and the location of files on a multiple file reel. The I-O-CONTROL paragraph and its clauses are an optional part of the Environment Division.

The RERUN clause is used to establish any rerun or restart procedures. This clause specifies that checkpoint areas are to be written on the actual device at the time of the checkpoint, and that they can be read into storage to restart the program from that point. This clause also specifies points in a program where checkpoints are to be recorded and has no functional effect on the program and is treated as a comment.

The SAME AREA clause specifies that two or more files will occupy the same area during processing. The area being shared includes all storage areas assigned to the files specified; therefore, it is not valid to have more than one of the files open at the same time.

The SAME RECORD AREA clause specifies that two or more files are to use the same memory area for processing of the current logical record. All of the files may be open at the same time.

The files referenced in the SAME AREA or SAME RECORD AREA clauses need not have the same organization or access.

The MULTIPLE FILE TAPE clause documents the sharing of the same physical reel of magnetic tape by two or more files. Regardless of the number of files on a single reel, only those used in the object program need be specified. This clause is required when more than one file shares the same physical reel of tape. If all file-names have been listed in consecutive order, the POSITION clause need not be given. If any file in the sequence is not listed, the position relative to the beginning of the tape must be given.

The RENAMES clause in the Data Division permits alternate, possibly overlapping, groupings of elementary items. The RENAMES description entry consists of level number 66, the data name, and the RENAMES clause. All RENAMES entries associated with a given logical record must immediately follow the last data description entry of that record. The RENAMES clause actually assigns a name to a segment of storage which can be addressed as a unique data item.

The LINAGE clause in the Data Division provides a means for specifying the depth of a logical page in terms of number of lines. It also provides for specifying the size of the top and bottom margins for the logical page and the line number, within the page body, at which the footing begins.

A LINAGE-COUNTER is generated by the presence of a LINAGE clause. The value in the LINAGE-COUNTER at any given time represents the line number at which the device is positioned within the current page body. The counter may be referenced but not modified by Procedure Division statements. The counter is automatically set to one (1) when a file with an OUTPUT phrase is opened, or the ADVANCING PAGE option of the WRITE statement is executed. The counter is automatically incremented by the number of lines implied in the WRITE statement. When the counter is equal to FOOTING, overflow occurs.

The LINAGE clause can prove to be an invaluable asset to a programmer who is completely relieved of the responsibility for accounting for all lines on a page through a line counter and signalling when an overflow should occur.

The BLANK WHEN ZERO clause permits the blanking of an item when its value is zero.

The JUSTIFIED clause is used to override the normal position of an alphabetic or alphanumeric data item when it is moved to a larger area.

The SYNCHRONIZED clause specifies the alignment of elementary data items on the natural boundaries of computer memory.

The CORRESPONDING options in the Procedure Division can be useful when it is necessary to process similar data items in different groups. The programmer can perform computations on elementary items or movement of data items from one group to another just by specifying the group to which they belong. This option is available with ADD, SUBTRACT and MOVE statements. This can save coding and prove an invaluable aid to the programmer with the problem of working with similar items in various groups that require addition, subtraction, or movement. The inherent problem is that the data name must be identical in both groups, and if these data items are used in other Procedure Division statements, qualification is necessary.

The ADD CORRESPONDING statement specifies that the data items in the group referenced preceding the word TO are added to and stored in the corresponding data items in the group item referenced following the word TO.

The SUBTRACT CORRESPONDING statement specifies that the data items in the group preceding the word FROM are subtracted from and stored in the corresponding data items in the group referenced following the word FROM.

The MOVE CORRESPONDING statement specifies that selected items subordinate to identifier-1 are moved to corresponding items subordinate to identifier-2. The selected data items are moved individually. Editing, format conversion, fill or truncation take place as appropriate, data item by data item. The use of the CORRESPONDING option can create problems when program modification is necessary.

Although the use of the ALTER statement is not recommended, it may be used to modify the effect of the unconditional GO TO statement elsewhere in the program, thus changing the sequence of operations to be performed.

The Source Program Library Module provides a capability for specifying text that is to be copied from the library. The library module may contain prewritten source entries in either the Environment, Data, or Procedure division that can be included in a source program at compile time.

The COPY statement permits the inclusion of prewritten Environment, Data, and Procedure division entries at compile time in the program where it will be treated by the compiler as part of the source program. When a COPY statement is processed, the library text associated with the text name is copied into the source program. It replaces the COPY statement, beginning with the reserved word COPY and ending with the period punctuation character. The library text must conform to the rules for COBOL reference format.

Questions for Review

1. What is the purpose of the Special-Names paragraph?
2. Briefly explain the use of the implementor-name clause in the Special-Names paragraph.
3. How are the CURRENCY SIGN and DECIMAL POINT IS COMMA clauses used?
4. What are the main uses of the RESERVE clause?
5. What is the main purpose of the I-O-CONTROL paragraph?
6. What are the main uses of the RERUN clause?
7. What do the SAME AREA and SAME RECORD AREA clauses specify?
8. What are the main purposes of the MULTIPLE FILE TAPE clause?
9. What is the main use of the RENAMES clause? Explain the operation of the RENAMES clause.
10. What is the main use of the LINAGE clause? Explain the operation of the LINAGE clause.
11. Give the main uses of the BLANK WHEN ZERO, JUSTIFIED, and SYNCHRONIZED clauses.
12. What is the main purpose of the CORRESPONDING option and how can it be useful to the programmer?
13. Explain the use of the CORRESPONDING option with the ADD, SUBTRACT, and MOVE statements.
14. How may the ALTER statement be used to modify unconditional GO TO statements? Why is it not recommended for structured programming?
15. What is a source program library facility and how may it be used?
16. What purposes does the COPY statement serve?

Matching Questions

A. *Match each item with its proper description.*

_____ 1. MOVE CORRESPONDING	A. Inclusion of prewritten library entries.
_____ 2. Special-Names	B. Selected items are moved individually to corresponding data items.
_____ 3. I-O-CONTROL	C. User-specified name.
_____ 4. CORRESPONDING	D. Modifies effect of unconditional GO TO statement.
_____ 5. Linage-Counter	E. Equates option switches with user-specified mnemonic-names.
_____ 6. Mnemonic-name	F. Specifying the group to which items belong.
_____ 7. SUBTRACT CORRESPONDING	G. Data items in group item referenced preceding the word TO are added to and stored in corresponding data items.
_____ 8. Implementor-name	H. Specifies some of the special techniques to be used in a program
_____ 9. COPY	I. Special register generated by the presence of a LINAGE clause
_____10. ADD CORRESPONDING	J. Data items in group referenced preceding the word FROM are subtracted from and stored in corresponding data items.
_____11. ALTER	K. Name of device or action.

B. *Match each clause with its proper description.*

______ 1. RENAMES	A. Additional input/output areas.
______ 2. BLANK WHEN ZERO	B. Functions of period are interchanged in PICTURE clause in current compilation.
______ 3. CURRENCY SIGN	C. Fills items with spaces when the value is zero.
______ 4. RERUN	D. Two or more files use the same memory area during processing.
______ 5. SAME RECORD AREA	E. Two or more files share the same physical reel of tape.
______ 6. JUSTIFIED	F. Used to represent currency symbol in PICTURE clause.
______ 7. DECIMAL POINT IS COMMA	G. Provides a means for specifying the depth of a logical page in terms of lines.
______ 8. MULTIPLE FILE TAPE	H. Checkpoint record.
______ 9. RESERVE	I. A segment of storage which can be addressed as a unique data item.
______10. LINAGE	J. Overrides the normal positioning of alphabetic or alphanumeric data when it is moved to a larger area.
______11. SAME AREA	K. Two or more files are to use the same memory area for processing of the current logical record.
______12. SYNCHRONIZED	L. Data items aligned on natural boundaries.

Exercises

Multiple Choice: Indicate the best *answer (questions 1–21).*

1. The Special-Names paragraph enables the programmer
   a. to assign condition-names to mnemonic-names.
   b. to assign mnemonic-names to option switches.
   c. to exchange the function of the period and the dollar sign.
   d. to assign a currency symbol to commas.
2. The RESERVE clause allows the programmer to
   a. reserve additional input/output areas for reading of physical records.
   b. document the number of input/output areas in storage allocated to specific files by the compiler.
   c. name additional input/output areas for the writing of physical records.
   d. All of the above.
3. The I-O-CONTROL paragraph specifies
   a. some of the special techniques to be used by the program.
   b. the location of files on a disk.
   c. points where memory areas are to be shared by different files.
   d. All of the above.
4. The RERUN clause is used to
   a. specify when and where the rerun information is recorded.
   b. establish restart procedures.
   c. specify checkpoint areas.
   d. All of the above.
5. The clause that specifies that two or more files are to use the same memory area for processing the current logical record is the
   a. SAME AREA clause.
   b. SAME RECORD AREA clause.
   c. SAME FILE clause.
   d. REDEFINES clause.
6. In the MULTIPLE FILE TAPE clause,
   a. all file-names must be listed in order.
   b. more than one file on the same tape reel may be opened at one time.
   c. more than one file shares the same reel of tape.
   d. the POSITION option must be used regardless of whether the tape file-names are listed in sequence.
7. The RENAMES clause
   a. must be a level 77 entry.
   b. permits alternate possible overlapping and groupings of elementary items.
   c. may be written for only one record.
   d. All of the above.

8. The LINAGE clause
   a. specifies the depth of a logical page.
   b. specifies the number of positions in a line.
   c. may specify a negative value.
   d. must establish a relationship between the size of a logical record and the size of a physical record.
9. The Linage-Counter
   a. is generated by the presence of a LINAGE clause.
   b. may be referred to.
   c. is set to one when an OPEN statement is executed for the associated file.
   d. All of the above.
10. The clause that is used to align items on a natural boundary is the
    a. SYNCHRONIZED clause.
    b. JUSTIFIED clause.
    c. REDEFINES clause.
    d. RENAMES clause.
11. The CORRESPONDING option can be used to
    a. process similar data items in different groups.
    b. perform computation on elementary items from one group to another.
    c. move data items from one group to another.
    d. All of the above.
12. In the execution of the MOVE CORRESPONDING statement,
    a. the data items need not have the same data-names to be moved from one group to another.
    b. both matching data items may be group items.
    c. no editing takes place in the receiving field.
    d. None of the above.
13. The COPY statement
    a. does not permit the inclusion of prewritten Procedure Division entries.
    b. need not be preceded by a space or terminated by a period.
    c. must refer to a unique text-name in the library.
    d. All of the above.
14. The ALTER statement
    a. is recommended for use in structured programming.
    b. modifies the effect of a conditional GO TO statement.
    c. replaces the procedure-name specified in the GO TO statement.
    d. does not change the sequence of operations to be performed.
15.
```
01  INPUT-RECORD.
    05  NAME
    05  HOME-ADDRESS
    05  FILLER
01  OUTPUT-RECORD.
    05  NAME
    05  FILLER
    05  HOME-ADDRESS
    05  FILLER
MOVE CORRESPONDING INPUT-RECORD TO OUTPUT-RECORD.
```
    The MOVE statement just given will cause
    a. the value of NAME and HOME-ADDRESS of INPUT-RECORD to be moved to NAME and HOME-ADDRESS of OUTPUT-RECORD.
    b. the movement of FILLER in the INPUT-RECORD to FILLER of the OUTPUT-RECORD.
    c. the movement of FILLER to only the FILLER item following HOME-ADDRESS in the OUTPUT-RECORD.
    d. the movement of FILLER to only the FILLER item preceding HOME-ADDRESS in the OUTPUT-RECORD.
16.
```
01  I-RECORD.
    05  EMPLOYEE-NUMBER
    05  EMPLOYEE-NAME
    05  RATE
01  W-RECORD.
    05  EMPLOYEE-NUMBER
    05  FILLER
    05  EMPLOYEE-NAME
    05  FILLER
    05  RATE
    05  FILLER
```

Statements that would move the corresponding items of I-RECORD to W-RECORD are

A. MOVE EMPLOYEE-NUMBER OF I-RECORD TO EMPLOYEE-NUMBER OF W-RECORD.
   MOVE EMPLOYEE-NAME OF I-RECORD TO EMPLOYEE-NAME OF W-RECORD.
   MOVE RATE OF I-RECORD TO RATE OF W-RECORD.

B. MOVE CORRESPONDING I-RECORD TO W-RECORD.

1. A is correct.
2. B is correct.
3. Both A and B are correct.
4. Neither A nor B are correct.

17. In order to move values from one record to another with the MOVE CORRESPONDING statement,
    a. the record names specified in the MOVE statement must be the same.
    b. variable names in levels other than 01 within the record names specified in the statement must be the same.
    c. there must be no FILLER items in the receiving record.
    d. None of the above.

18. FIX.
        GO TO IN-STATE.

    Which of the ALTER statements below could alter the GO TO statement above so that execution of paragraph FIX would cause control to be transferred to OUT-STATE?
    a. ALTER FIX TO PROCEED TO OUT-STATE.
    b. ALTER IN-STATE TO PROCEED TO OUT-STATE.
    c. ALTER FIX TO PROCEED TO IN-STATE.
    d. None of the above.

19. Which of the following paragraphs could be changed by the ALTER statement?
    a. PARA-1.
           GO TO Q R S
               DEPENDING ON LETTER.
    b. PARA-2.
           GO TO Q.
    c. PARA-3.
           WRITE ERROR-RECORDS.
           GO TO Q.
    d. PARA-4.
           ALTER Q TO PROCEED TO S.

20. The logical page depth is the sum of
    a. the top and bottom margins.
    b. page body and footing areas.
    c. top margin, page body, bottom margin and footing area.
    d. top margin, page body, and bottom margin.

21. The programmer can determine the current line number by
    a. setting up and maintaining a counter called Linage-Counter.
    b. interrogating Linage-Counter.
    c. incrementing Linage-Counter for each advance of the printer carriage.
    d. All of the above.

22. WRITE record-name
        AFTER ADVANCING mnemonic-name

    The option above may be used to skip to a new page when the mnemonic-name is ________________ or defined in the ________________ paragraph of the Configuration Section of the Environment Division.

23. Which of the following COPY statements is correct? (There may be more than one correct answer.)
    a. SELECT COPY FILEA.
    b. FD CARD-FILE
           COPY FILEB
    c. COPY CARD-RECORD
    d. SAVE-NUMBER COPY LIBRARY-1
    e. BEGIN. COPY OPEN-PARAGRAPH

24. DATA DIVISION.
    FILE SECTION.
    FD PRINT-FILE
        LABEL RECORDS ARE OMITTED
        LINAGE IS 60 LINES
        WITH FOOTING AT 50
        LINES AT TOP 3
        LINES AT BOTTOM 3.

The above FD entry describes a page with a logical page depth of ________ lines, a page body of ________ lines, and a top margin of ________ lines, and a bottom margin of ________ lines.

25. Write a FD entry for a printer file that has a logical page depth of forty lines, a top margin of four lines, a bottom margin of five lines, and a footing area beginning on the twenty-fifth line of the page body.

26. Write a statement to test the Linage-Counter for forty-four lines. If the statement is true, the program is to proceed to PRINT-FINAL, otherwise it is to continue processing.

27.
```
01  INPUT-RECORD.
    05  PART.
        10  NUMBER                  PICTURE 9(5).
        10  DESCRIPTION             PICTURE X(20).
    05  BALANCE.
        10  ON-HAND                 PICTURE 999.
        10  ON-ORDER.
            15  VENDOR-1            PICTURE 999.
            15  VENDOR-2            PICTURE 999.
        10  AVAILABLE               PICTURE 999.
            .
            .
            .
01  OUTPUT-RECORD.
    05  PART                        PICTURE X(25).
    05  BALANCE.
        10  AVAILABLE               PICTURE 999.
        10  ON-HAND                 PICTURE 999.
        10  ON-ORDER.
            15  VENDOR-1            PICTURE 999.
            15  VENDOR-2            PICTURE 999.
            .
            .
            .
    MOVE CORRESPONDING INPUT-RECORD TO OUTPUT-RECORD.
            .
            .
            .
```

Given the above MOVE statement and record description entries, which data items would be moved?

28. You are using an IBM model 370 computer. Write the necessary entries to accomplish the following:
    a. The Configuration Section of the Environment Division entries including the Special-Names paragraph to define the mnemonic-name FIRST-LINE as the first line of the form.
    b. The procedural WRITE statement with the AFTER ADVANCING option to write the output record RECORD-OUT after skipping to the first line of the next form after reaching the last line of the previous page.

29. A file called CUSTOMER-FILE with a record name of CUSTOMER-RECORD is to be outputted on a printer in the same format as the input record.
    Using the following hardware devices:

    Device
    Card Reader
    Printer

    Write the necessary entries to accomplish the following:
    a. Configuration Section including the Special-Names paragraph.
    b. File-Control paragraph including the RESERVE clause.
    c. The procedural entries to move the input record to the output including end-of-page procedures.

30.
```
01  CORRECTED-RECORD.
    05  GROUP-A.
        10  FIELD-1A.
            15  ITEM-1A     PICTURE XX.
            15  ITEM-2A     PICTURE XXX.
            15  ITEM-3A     PICTURE XX.
        10  FIELD-2A.
            15  ITEM-4A     PICTURE XX.
            15  ITEM-5A     PICTURE XX.
            15  ITEM-6A     PICTURE XX.
```

```
05  GROUP-B REDEFINES GROUP-A.
    10  FIELD-1B.
        15  ITEM-1B     PICTURE XXXX.
        15  ITEM-2B     PICTURE XXX.
    10  FIELD-2B.
        15  ITEM-3B     PICTURE XXX.
        15  ITEM-4B     PICTURE XXX.
```

In the above record it is necessary to group items ITEM-3A to ITEM-4B for future processing.
Write the necessary entry to accomplish the above using the RENAMES clause.

31. Since we are using the same Configuration Section in all of our programs, write the Configuration Section using the following library names:

Paragraph	**Library Name**
Source-Computer	S-COMPUTER
Object-Computer	O-COMPUTER
Special-Names	S-NAMES

32. In many of our programs, we are using the same headings and detail line formats. We have decided to put these formats in our library as follows:

Library Name	**Function**
SALES-HEAD	Name of report.
COLUMN-HEAD	Column headings of report.
DETAIL-RECORD	Detail line of the report.

Write the necessary entries for
 Working-Storage items for headings and detail using the COPY clause for the report heading (HDG-1), column headings (HDG-2), and detail line (DETAIL-LINE).

33. Most of our programs contain the same routine for printing headings on the first page and each overflow. We have decided to put these procedures in our library. Assume that the library name of the routine is HEADING-ROUTINE, and write the Procedure Division entry using the COPY clause for the paragraph called PRT-HDG.

Problem 1

Job Definition

A detail printed report titled STOCK INVENTORY REPORT is to be produced from a file of records arranged in ascending numerical order (by the material number field). Totals of on-hand cost are to be printed for each type of material.

Input

The input file consists of two different 80-position records.

Date Record

Positions	*Field Designation*
1-6	Date
79	Code: Letter O

Stock Record

Positions	*Field Designation*
7-9	Material number
12-16	Stock number
19-23	Unit cost (XXX.XX)
26-29	Item description
70-73	Quantity on hand
79	Code: Letter M

Processing

1. The individual detail records are to be printed with proper headings and editing.
2. The total lines are to be printed with proper headings and editing.

3. For each detail record, the quantity on hand is to be multiplied by the unit cost to calculate the on-hand cost. This cost shall be rounded to the nearest whole dollar.
4. A total on-hand cost is to be calculated for each group of records with the same material number.
5. A date card precedes the file of cards.

Output

A Stock Inventory Report is to be printed as follows:

```
                                    STOCK INVENTORY REPORT

                                          10/13/80

 MATERIAL    STOCK                  DESCRIPTION                  UNIT     QUANTITY       ON HAND
   NO.        NO.                                                COST      ON HAND         COST

    25       96543         CARBORUNDUM WHEELS                   10.25       4646          47,622

          THE TOTAL ON HAND COST IN DOLLARS FOR MATERIAL NUMBER  25 IS                   $47,622  **

   111       00986         STAINLESS SET SCREWS NSP               .42       5986           2,514

   111       01598         STAINLESS RODS                        8.59        934           8,023

   111       09346         HI GRADE CARBON                       4.82         52             251

   111       11632         CARBON STEEL                          5.96       1598           9,524

   111       11723         STAINLESS PINS                        9.17         52             477

   111       11725         STAINLESS TUBING                      1.15        915           1,052

   111       11899         STAINLESS FITTINGS                   15.67       1792          28,081

   111       55292         STEEL SHANK 4X9X1                      .14       4138             579

   111       62549         HEX STOCK TITANIUM                  100.48         89           8,943

   111       65342         TITANIUM BARS                        95.89         85           8,151

   111       72359         STEEL PLATE                          11.86         98           1,162

   111       81192         FLAT ROLLED STEEL SHEETS             15.92       1139          18,133

   111       81536         STEEL FLANGE                          4.80       1985           9,528

          THE TOTAL ON HAND COST IN DOLLARS FOR MATERIAL NUMBER 111 IS                   $96,418  **

   123       45678         ALLIGATOR PUMPS                     965.43       9999       9,653,335

          THE TOTAL ON HAND COST IN DOLLARS FOR MATERIAL NUMBER 123 IS                $9,653,335  **
```

The Printer Spacing Chart shows how the report is formatted.

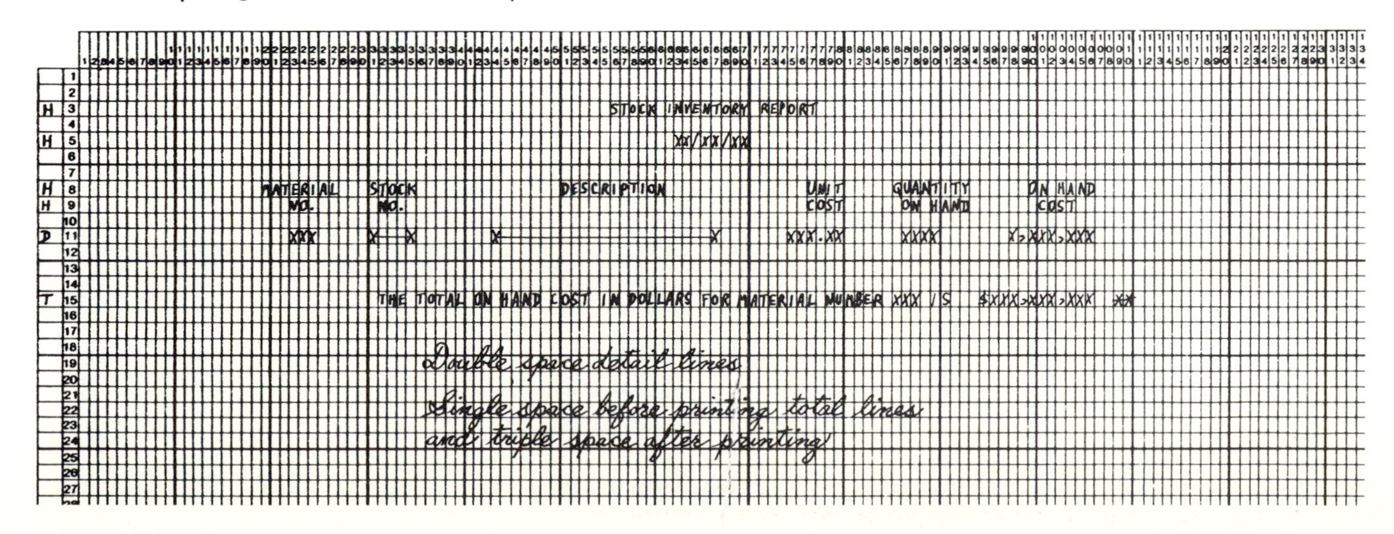

Problem 2

Write a COBOL program to process the following:

Input

Field	Positions	Format
Customer name	1-15	
Customer address	16-50	
Account number	51-55	
Previous balance	56-62	XXXXX.XX
Month sales	63-69	XXXXX.XX
Payments	70-76	XXXXX.XX
Not used	77-80	

Computations to Be Performed:

1. Service charge [rounded] = .015 × (Previous balance − Payments).
2. Amount due = (Previous balance − Payments) + Service charge + Month sales.

Output

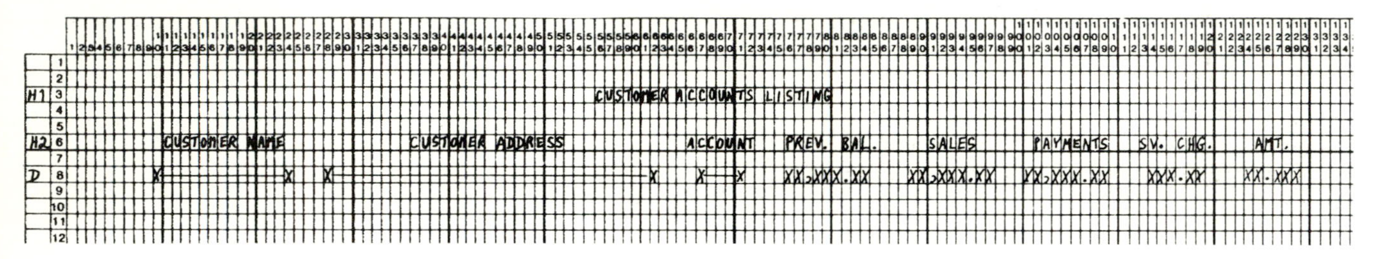

Problem 3: Insurance Premiums

Job Definition

In an insurance premium run, the monthly rates are determined by the risk class. Here we have a situation where there is a wide and irregular gap between the argument assigned to one table value and the argument assigned to the table value that follows. The premium rate assigned should be the one that matches or the next higher one in the table.

Input

The input file consists of an 80-position record with the following information:

Positions	Field Designation
1-10	Policy number
11-31	Name
32-37	Date
40-43	Risk class

Processing

A table is to be set up as follows:

Risk Class	Premium Rate
210	17.50
273	15.50
370	12.30
420	11.95
465	14.60
481	15.25
900	19.45
950	20.01
988	18.10
1030	8.55
1245	14.03
1366	19.99
1505	20.33
1666	12.22
1899	10.00

Search the table for the appropriate (or near appropriate) risk class and record the premium rate. If the appropriate risk class cannot be found, the premium rate for the next higher risk class must be used.

Output

The Insurance Premiums report is to be printed as follows:

INSURANCE PREMIUMS

PØLICY NUMBER	NAME	DATE	RISK CLASS	PREMIUM RATE
563 781 532	RØBERT PALMER	11/25/80	536	19.45
5310 562 386	SUZIE WILSØN	11/25/80	1302	19.99
7802 619 435	DAVE LEWIS	11/25/80	1723	10.00
4530 789 138	BØNNIE SELWØØD	11/25/80	798	19.45
400 831 562	GISELLE AGERGAARD	11/25/80	639	19.45
7319 530 088	DEAN HØPKINS	11/25/80	1699	10.00
7066 238 009	DØUG KEANS	11/25/80	901	20.01
4308 615 398	MIKE CØØPER	11/25/80	1086	14.03
8023 614 528	RØNNIE SALTZER	11/25/80	1630	12.22

The Printer Spacing Chart shows how the report is formatted.

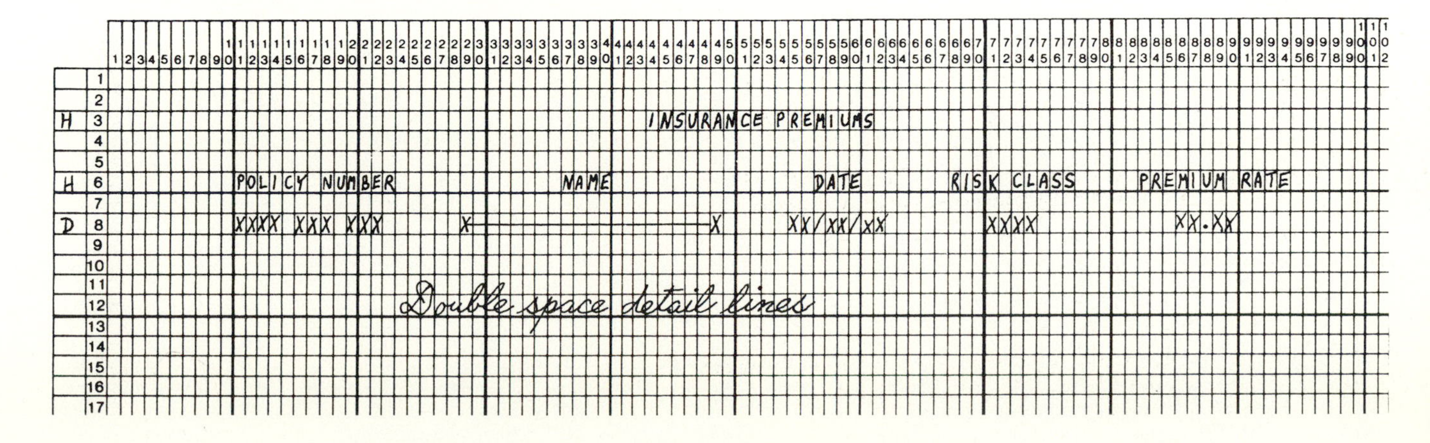

Problem 4: Daily Sales Report

Job Definition

The Duron Corp. maintains various departments in its stores. A Daily Sales Report is to be printed showing the daily sales of each department as well as a weekly total of each days' sales. A total of daily sales as well as the weekly total is to be shown for each store as well as an overall total for all days and totals.

Input

The input file consists of an 80-position record punched with the following information:

Positions	Field Designation
1-3	Store number
4-7	Department number
10-39	Daily sales (each day's sales consists of six positions with the format XXXX.XX).

Processing

1. Three arrays are to be set up for the department, store, and overall totals.
2. Each array should be added to the other array to arrive at daily totals for stores as well as overall totals.
3. Each array should be crossfooted to arrive at weekly totals.

Output

The Daily Sales Report is to be printed as follows:

DAILY SALES REPORT

STORE NO.	DEPARTMENT NO.	FIRST SALE	SECOND SALE	THIRD SALE	FORTH SALE	FIFTH SALE	TOTALS
123	1359	12.93	291.60	2,000.15	250.37	100.35	2,655.40
	1530	130.00	2.50	3,040.00	25.19	234.10	3,431.79
	2400	24.59	29.47	120.44	28.30	166.25	369.05
	3594	2.48	15.90	23.00	1.22	35.20	77.80
		170.00	339.47	5,183.59	305.08	535.90	6,534.04
530	1280	25.10	200.50	12.57	111.20	259.10	608.47
	3320	1,002.45	19.45	279.33	1,234.50	3,000.20	5,535.93
	7730	24.60	10.24	39.13	294.33	29.30	397.60
	9250	22.00	10.50	239.10	2.00	3.91	277.51
		1,074.15	240.69	570.13	1,642.03	3,292.51	6,819.51
	TOTALS	1,244.15	580.16	5,753.72	1,947.11	3,828.41	13,353.55

15

Declaratives and Linkage Sections

Chapter Outline

Chapter Objectives

The learning objectives of this chapter are:

1. To explain the use of declaratives in a COBOL program and how and where they are coded in the Procedure Division.
2. To describe how the execution of a program may be continued in spite of input-output errors.
3. To describe the format of the USE statement and how it is used in input-output error processing and report writer declaratives.
4. To explain the function of the Linkage Section in the Data Division and how it is used in interprogram communication.
5. To describe the operation of calling and called programs and the coding necessary in both the Data and Procedure divisions to accomplish the transfer of control between programs.
6. To explain the coding and function of the CALL and EXIT PROGRAM statements in interprogram communication.
7. To describe the operation of the ENTER statement and how it may be used in COBOL programming.

```
PROCEDURE DIVISION.

[DECLARATIVES.

{section-name SECTION  [segment-number].  declarative-sentence

[paragraph-name.  [sentence]  . . . ] . . . } . . .

END DECLARATIVES.]
```

Figure 15.1 Format—Declaratives section.

Declaratives Section

The Declaratives Section is written in the Procedure Division to specify any special circumstance under which a procedure is to be executed in the object program. Although the COBOL compiler provides error recovery routines in the case of input/output errors, the programmer may wish to specify additional procedures to supplement those supplied by the compiler. The programmer may wish to identify items that are to be monitored by the associated debugging sections. The Report Writer feature may also use declarative sections.

Since these procedures can be executed only at a time when an error occurs in the reading or writing of records, or during the execution of the debugging procedure, or before a report group is to be produced, they cannot appear in the regular sequence of procedural statements. The procedures are invoked nonsynchronously; that is, they are not executed as part of the sequential coding written by the programmer, but rather when a condition occurs which cannot normally be tested by the program.

The Declaratives Section is written as a subdivision at the beginning of the Procedure Division prior to the execution of the first procedure. A Declaratives Section consists of a group of declaratives procedures. Although the declaratives sections are located at the beginning of the Procedure Division, execution of the object program actually starts with the first procedure following the termination of the Declaratives Section. If declaratives are specified in a program, then the entire Procedure Division must be divided into sections.

The declaratives section subdivision of the Procedure Division must begin with the key word DECLARATIVES at the A margin followed by a period and a space. The declaratives section is terminated by the key words END DECLARATIVES followed by a period and a space. (The format for the Declaratives Section is shown in figure 15.1.) Both DECLARATIVES and END DECLARATIVES must appear on a line by themselves with no other coding permissible on the same line. Every declaratives section is terminated by the occurrence of another declaratives section or the words END DECLARATIVES.

A declaratives section consists of the following:

1. A section-name followed by the key word SECTION written at the A margin.
2. A USE statement must follow the section header on the same line after an intervening space or spaces, and terminated by a period.

Declaratives are instructions to the COBOL compiler and the run-time system to perform special operations during object programs execution before or after other standard operations. All declaratives sections must be grouped at the beginning of the Procedure Division preceded by the key word DECLARATIVES and followed by the key words END DECLARATIVES. On the line following the declaratives line, a section-name must be specified followed by a USE statement. The section continues until another section-name is executed or until the END DECLARATIVES line is used.

USE Statement

A USE statement must identify the conditions under which each Declarative section is executed. Procedures specified in a USE statement must be self-contained. One Declarative Section may be accessed from another Declarative Section via the PERFORM statement, but it is illegal to branch outside the declaratives area, or to the declaratives from outside.

The USE statement specifies procedures to be performed for each type of declarative that is added to the standard procedures provided by the compiler. The USE statement is itself never executed; rather, it defines the conditions calling for the execution of the USE procedures. The remainder of the section must consist of one or more procedural paragraphs that specify the procedures to be performed.

The following types of procedures are associated with the USE statement:

1. To specify procedures for input-output error handling that are in addition to the standard procedures provided by the input-output control system.
2. To specify procedures that are executed just before a report group named in the Report Section of the Data Division is produced. (This USE statement is discussed in the Report Writer section).
3. To specify items monitored by the associated Debugging section.

The exit from the declaratives section is inserted following the last statement of the section. All logical program paths within the section must lead to the exit point.

Input-Output Error-Processing Declaratives

These declaratives are used to specify procedures to be followed if an input/output error occurs during the processing. This option provides the users with input/output correction procedures in addition to those specified by the input-output control system.

When a USE statement is present, it must immediately follow a section header in the declaratives section, and must be followed by a period and a space. The remainder of the section must consist of one or more procedural programs that define the procedures to be used. USE is not an executable statement; rather, it defines the conditions under which the associated procedure, the declaratives section itself, is to be executed.

The USE statement specifies procedures for input-output error handling that are in addition to the standard procedures provided by the input-output control system. (The format for the input-output processing declarative is shown in figure 15.2.)

Rules Governing the Use of the Input-Output Error-Processing Declaratives

1. A USE statement, when present, must immediately follow a section header in the declaratives section and must be followed by a period followed by a space. The remainder of the section must consist of zero and one or more procedural paragraphs that define the procedures to be used.
2. The USE statement itself is never executed; it merely defines the conditions calling for the execution of the USE procedures.
3. The designated procedures are executed by the input-output system after completing the standard input-output routine, or upon recognition of the INVALID KEY or AT END condition, when the INVALID KEY or AT END phrase has not been specified in the input-output statement.
4. After execution of a USE procedure, control is returned to the invoking routine.
5. Within a USE procedure, there must not be any reference to any nondeclarative procedures. Conversely, in the nondeclarative portion there must be no reference to procedure-names that appear in the declarative portion, except the PERFORM statements which may refer to a USE statement or to the procedures associated with such a USE statement.

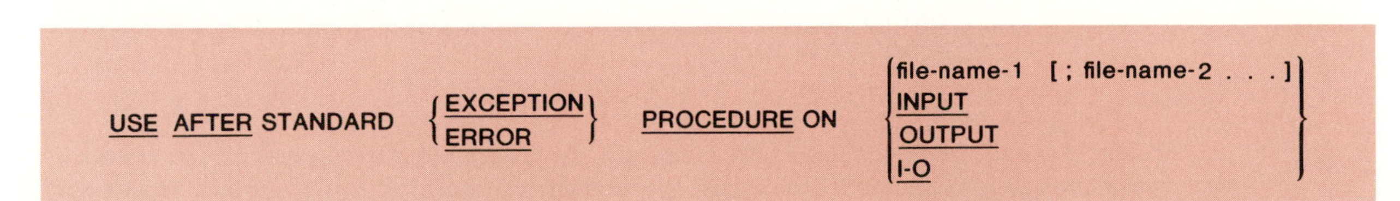
USE AFTER STANDARD
EXCEPTION
ERROR
PROCEDURE ON
file-name-1 [; file-name-2 . . .]
INPUT
OUTPUT
I-O

Figure 15.2 Format—input-output processing declarative.

6. Within a USE procedure, there must not be the execution of any statement that would cause the execution of a USE procedure that had previously been invoked and had not yet returned control to the invoking routine.
7. The same file-name can appear in a different specific arrangement of the format. Appearance of a file-name in a USE statement must not cause the simultaneous request for execution of more than one USE procedure.
8. The words ERROR and EXCEPTION are synonymous and may be used interchangeably.
9. The files implicitly or explicitly referenced in a USE statement need not all have the same organization or access.

If the INVALID KEY or AT END phrase for the READ, WRITE, START, REWRITE, or DELETE statements are specified and the condition occurs, control is transferred to the statement following the INVALID KEY or AT END phrase and the USE statement is not executed.

If the INVALID KEY or AT END phrase is not specified, a USE statement must be specified, either explicitly or implicitly, for those files, and then that procedure is executed.

INVALID KEY or AT END phrases must be specified for READ, WRITE, START, REWRITE, or DELETE verbs if no applicable USE procedures are specified for the files.

An exit form of this type of declarative can be affected by executing the last statement in the section (normal return) or by means of a GO TO statement to an exit statement. This is the normal return from an error declarative to the statement following the input/output statement that causes the error. The EXIT statement must be contained in the last paragraph of the section.

Continued Processing of a File

The continued processing of a file is permitted under the following conditions.

1. An error-processing procedure exists in the Declaratives section.
2. The detection of the error results in an automatic transfer to the error-processing procedure which permits the programmer to examine the error condition before it enters the process.

   The following is an example of the input-output processing declarative:

```
PROCEDURE DIVISION.
DECLARATIVES.
ERROR-PROCESSING SECTION.
    USE AFTER STANDARD ERROR PROCEDURE ON DISK-FILE.
        .
        .
DISK-ERROR-ROUTINE.
    IF MORE-ROOM
        PERFORM CREATE-DISK-FILE
        MOVE 'YES' TO DATA-FLAG
    ELSE
        PERFORM PRINT-ERROR-MESSAGE
        MOVE 'NO' TO DATA-FLAG.

CREATE-DISK-FILE
    .
    .
PRINT-ERROR-MESSAGE.
    .
    .
END DECLARATIVES

MAIN-ROUTINE SECTION.
CREATE-MASTER-DISK-FILE
    .
    .
```

3. At the conclusion of the processing of the error, it is the programmer's responsibility to update the parameters normally returned by the input/output control system. The programmer needs to reference the particular computer system reference manual for the necessary parameters that need updating.

USE BEFORE REPORTING identifier.

Figure 15.3 Format—USE BEFORE REPORTING declarative.

Report Writer-USE Statement

The USE statement specifies Procedure Division statements that are executed before a report group named in the Report Section of the Data Division is produced. (The format for the USE BEFORE REPORTING declarative is shown in figure 15.3.)

Rules Governing the Use of the Report Writer Declaratives

1. A USE statement, when present, must immediately follow a section header in the declaratives section and must be followed by a period followed by a space. The remainder of the section must consist of zero and one or more procedural paragraphs that identify the procedures to be used.
2. The identifier represents a report group. The identifier must not appear in a paragraph with a USE BEFORE REPORTING procedure.
   The GENERATE, INITIATE, or TERMINATE statements must not appear in a paragraph within a USE BEFORE REPORTING procedure.
   The USE BEFORE REPORTING procedure must not alter the value of any control data item.
3. The USE statement itself is never executed; it merely defines the conditions calling for the execution of the USE procedures.
4. The designated procedures are executed by the Report Writer just before the named report is produced, regardless of page or control breaks associated with the report group. The report group may be of any type except DETAIL.
5. There must not be any reference to any nondeclarative procedure. Conversely, in the nondeclarative portion, there must be no reference of procedure-names that appear in the declarative portion, except that PERFORM statements may refer to a USE declarative or to procedures associated with the USE declarative.
   The following is an example of the USE BEFORE REPORTING declarative:

```
PROCEDURE DIVISION.
DECLARATIVES.
PAGE-HEAD-RTN SECTION.
    USE BEFORE REPORTING PAGE-HEAD.
PAGE-HEAD-RTN-TEST.
    IF MONTH = SAVED-MONTH
        MOVE '(CONTINUED)' TO CONTINUED
    ELSE
        MOVE SPACES TO CONTINUED
        MOVE MONTH TO SAVED-MONTH.
END DECLARATIVES.

MAIN-ROUTINE.
    .
    .
    .
```

Note: When the programmer wishes to suppress the printing of a specified report group, the statement SUPPRESS may be used in the USE BEFORE REPORTING declarative section. When this statement is encountered, only the specified report group is not printed. The statement must be written for each report group whose printing is to be suppressed.

Additional information relative to the use of this declarative and examples of the use of this declarative can be found in the Report Writer Section.

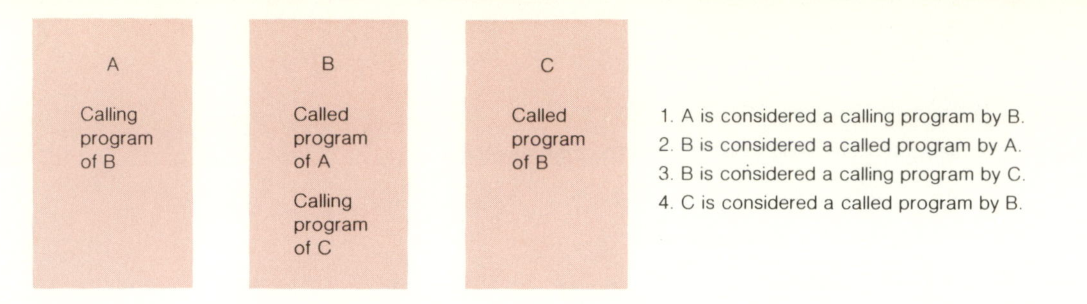
A
Calling program of B
B
Called program of A
Calling program of C
C
Called program of B
1. A is considered a calling program by B.
2. B is considered a called program by A.
3. B is considered a calling program by C.
4. C is considered a called program by B.

Figure 15.4 Calling and called programs—example.

Linkage Section

The Linkage Section describes data that is common between programs that communicate with each other within a single run unit. Program interaction requires that both programs have access to the same data items. The Linkage Section is used for describing data that is available through the calling program that is to be referred to in both the *calling* and *called* programs. A program that refers to another program is a *calling* program. A program that is referred to is a *called* program. (See figure 15.4.)

The Linkage Section is written in the Data Division to describe data from another program. The data item description entries and record description entries in the Linkage Section provide names and descriptions, but storage within the program is not reserved, inasmuch as the data area exists elsewhere. Any data description clause may be used to describe items in the Linkage Section, with one exception: the VALUE clause may not be specified for other than level-88 items.

This section must begin with the section header LINKAGE SECTION, followed by a period. (The general format for the LINKAGE Section is shown in figure 15.5.)

Concept of Interprogram Communication

Complex data processing systems problems are frequently solved by the use of separately compiled but logically coordinated programs, which at execution time form logical and physical subdivisions of a single run unit. This approach lends itself to dividing a large problem into smaller, more manageable portions that can be programmed and debugged independently. At execution time, control is transferred from program to program by the use of CALL and EXIT PROGRAM statements.

In COBOL terminology, a program is either a source program or an object program, depending on content; a source program is a syntactically correct set of COBOL statements, and an object program is the set of instructions, constants, and other machine-oriented data resulting from the operation of a compiler on a source program. A run unit is the total machine language necessary to solve a data processing problem; it includes one or more object programs as defined above, and it may include machine language from sources other than a COBOL compiler.

Either the 77-level-description entry or the 01-record-description entry is used to provide names and descriptions of the data from another program. No storage area is reserved for this data in the called program because the storage area already exists in the calling program. Any of the data description clauses may be used to describe items in the Linkage Section, except for the VALUE clause which

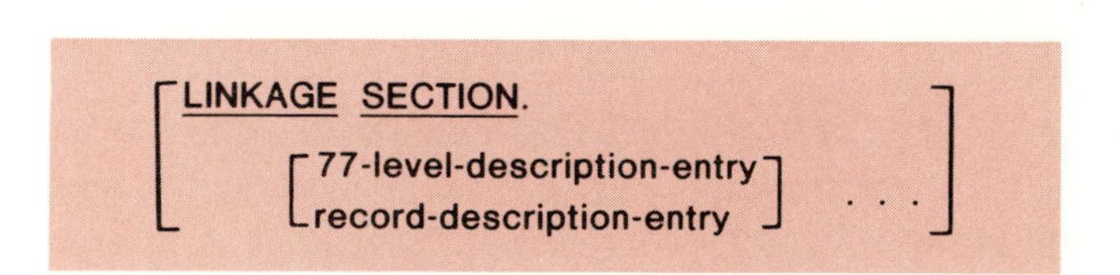
LINKAGE SECTION.
77-level-description-entry
record-description-entry
. . .

Figure 15.5 Format—Linkage section.

may not be used with other than level-88 items. In this section, the VALUE clause is meaningful only in condition-name entries.

Data Division in Interprogram Communication

The Linkage Section in a program is meaningful only if the object program is to function under control of a CALL statement in the calling program containing a USING phrase.

As stated earlier, the Linkage Section is used for describing data that is available through the calling program but is to be referred to in both the calling and called programs. No space is allocated in the program for data items referenced by data-names in the Linkage Section of that program. Procedure Division references to these data items are resolved at object time by equating the reference in the called program to the location used in the calling program. In the case of index-names, no such correspondence is established. Index-names in the called and calling programs always refer to separate indexes.

Data items defined in the Linkage Section of the called program may be referenced within the Procedure Division of the called program only if they are specified as operands of the USING phrase of the Procedure Division header or are subordinate to such operands, and if the object program is under the control of a CALL statement that specifies a USING phrase.

The structure of the Linkage Section is the same as that previously described for the Working-Storage section, beginning with a section header and followed by data description entries for noncontiguous data items or record description entries, or both. Each Linkage Section record-name and noncontiguous item name must be unique within a called program since it cannot be qualified.

Of those items defined in the Linkage Section, only data-name-1, data-name-2, . . . in the USING phrase of the Procedure Division header, data items subordinate to these data-names, and condition-names, and/or index-names associated with such data-names and/or subordinate data items may be referenced in the Procedure Division of the called program.

When one program invokes another program, the index-names in the called program and calling program always refer to separate indexes; therefore, each index-name must be initialized in the called program before referencing its associated array described in the Linkage Section. A calling program cannot SET a value in a called program's index-name, but a called program can use subscript values from a calling program if the subscripts are defined in the called program's Linkage Section.

Noncontiguous Linkage Storage

Items in the Linkage Section that bear no hierarchic relationship to one another do not have to be grouped into records. Instead they are classified and defined as noncontiguous elementary items. Each of these data items can be defined in a separate data description entry that begins with the special level number 77 or 01. The following clauses are required in each such data description entry:

Level number 77 or 01

Data-name

The PICTURE clause.

Other data description clauses, except the VALUE clause, are optional and may be used to complete the description of the item if necessary.

Linkage Records

Data elements in the Linkage Section that bear a definite hierarchic relationship to one another must be grouped into records according to the rules for formation of record descriptions. Any clause except the VALUE clause, used in an input or output record description can be used in the Linkage Section.

Subprogram Linkage Statements

Subprogram linkage statements are special statements that permit communication between object programs; these are the CALL and EXIT PROGRAM statements.

When the statement of a problem is subdivided into more than one program, the constituent programs must be able to communicate with each other. This communication may take two forms: transfer of control and reference to common data.

Calling and Called Programs

Transfer of Control

The CALL statement provides the means whereby control can be passed from one program to another within a run unit. A program that is activated by a CALL statement may itself contain CALL statements. However, results are unpredictable in those cases wherein circularity of control is initiated; that is, where program A calls program B, then program B calls program A or another program that calls program A.

When control is passed to a called program, execution proceeds in the normal way from the procedure statement to the procedure statement beginning with the first nondeclarative statement. If control reaches a STOP RUN statement, this signals the logical end of the run unit. If control reaches an EXIT PROGRAM statement, this signals the logical end of the called program *only,* and control then reverts to the point immediately following the CALL statement in the calling program. Stated briefly, the EXIT PROGRAM statement terminates only the program in which it occurs, whereas the STOP RUN statement terminates the entire run unit.

If the called program is not COBOL, then the termination of the run unit or the return to the calling program must be programmed in accordance with the language of the called program.

A COBOL program can refer to and pass control to other COBOL programs, or to programs written in other languages. A program in another language can refer to and pass control to a COBOL program. Control is returned from the called program to the first instruction following the calling sequence in the calling program.

A called program can also be a calling program, that is, a called program can in turn call another program.

Interprogram Data Storage

Program interaction requires that both programs have access to the same data items. In the calling program the common data items are described along with all other data items in the File Section or Working-Storage Section. At execution time, memory is allocated for the entire Data Division. In the called program, common data items are described in the Linkage Section. At execution time, memory space is not allocated for this section. Communication between the called program and the common data items stored in the calling program is affected through USING clauses contained in both programs. The USING clause in the calling program is contained in the CALL statement and the operands are a list of common data-names described in the Data Division. The USING clause in the called program follows the Procedure Division header, and the operands are a list of common data-names described in the Linkage Section. The data-names specified by the USING clause of the CALL statement indicate those data items available to a calling program that may be referred to by using the called program. The sequence of appearance of the data-names in the USING clause of the CALL statement and the USING clause in the Procedure Division is significant. Corresponding data-names refer to a single set of data that is available to the calling program. The correspondence is positional, and not by name. While the called program is being executed, every reference to an operand whose data-name appears in the called program's USING clause is treated as if it were a reference to the corresponding operand in the USING clause of the active CALL statement.

Thus, the interprogram communication feature allows a program to communicate with one or more other programs. This communication is provided by (1) the ability to transfer control from one program to another within a run unit, and (2) the capability of both programs to have access to the same data.

Specifying Linkage

Whenever a program calls another program, linkage must be established between the two. The calling program must state the entry point of the called program and must specify any arguments to be passed. The called program must have an entry point and must be able to accept the arguments. Further, the called program must establish the linkage for the return of control to the calling program.

Procedure Division in Interprogram Communication

Procedure Division Header

The Procedure Division in the called program is identified and must begin with the following header:

PROCEDURE DIVISION [USING data-name-1 [,data-name-2] . . .].

The USING phrase is present only if the object program is to function under the control of a CALL statement, and the CALL statement in the calling program contains a USING phrase.

Each of the operands in the USING phrase of the Procedure Division header must be defined as a data item in the Linkage Section of the program in which this header occurs, and it must have a 01- or 77-level number.

Within a called program, Linkage Section data items are processed according to their data descriptions given in the called program.

When the USING phrase is present, the object program operates as if data-name-1 of the Procedure Division header in the called program and data-name-1 in the USING phrase of the CALL statement in the calling program refer to a single set of data that is equally available to the called and calling programs. Their descriptions must define an equal number of character positions; however, they need not have the same name. Similarly, there is an equivalent relationship between data-name-2, . . . , in the USING phrase of the called program and data-name-2, . . , in the USING phrase of the CALL statement in the calling program. A data-name must not appear more than once in the USING phrase in the Procedure Division header of the called program; however, a given data-name may appear more than once in the same USING phrase of a CALL statement.

CALL Statement

A calling program must contain a CALL statement at the point where another program is to be called. The CALL statement allows communication between the COBOL object program and one or more subprograms or other language subprograms.

The CALL statement causes control to be transferred from one object program to another, within the run unit. The format of this statement is:

CALL literal-1 [USING data-name-1 [,data-name-2] . .].

Literal-1 is the name of the program being called (that is, the called program), and must be a nonnumeric literal. The program in which the CALL statement appears is known as the *calling program.* The literal must conform to the formation rules for a program-name, with only the first six characters used to identify the program. Literal-1 must be the same as the program-name specified in the PROGRAM-ID paragraph of the called program.

The following is an example of the CALL statement—USING option:

```
CALL 'SALESCALC'
    USING    UNITS-SOLD
             UNIT-PRICE
             SALES-AMT.
```

SALESCALC is the PROGRAM-ID of the called program. The USING phrase indicates that the fields UNITS-SOLD, UNIT-PRICE and SALES-AMT are to be used by the called program. These fields will be defined in either the File Section or the Working-Storage Section of the calling program.

Execution of a CALL statement causes control to pass to the called program, which is in its initial state the first time it is called within a run unit. On subsequent entries into the called program, the state is as it was upon the last exit from the program. This includes all possible changed data fields, the status and positioning of all files, and all alterable switch settings. Thus the programmer must be responsible for reinitializing the following:

1. GO TO statements that have been altered.
2. Data items.
3. PERFORM statements. If the programmer specifies a branch out of range of a PERFORM statement and then specifies an exit from the program, the range of the PERFORM statement is still in effect upon a subsequent entry.

Called programs may contain CALL statements. However, a called program must not contain a CALL statement that directly or indirectly calls the calling program.

USING Option
The USING option makes data items from a calling program available to a called program. When the USING option is specified in the CALL statement, it must be specified in the Procedure Division header of the called program, and vice versa. The format of the USING option is the same in the calling and called programs as shown here:
Format Within Calling Program

CALL literal-1 (USING data-name-1 [,data-name-2] . .)

Format Within Called Program

PROCEDURE DIVISION [USING data-name-1 [, data-name-2] . .]

The number of data-names specified in each of these USING options must be identical. The data-names specified by the USING option indicate the calling program data items that may be referred to in the called program. Each of the data-names must have been defined as a data item in the File Section, Working-Storage Section, or Linkage Section, and must have a level-number of 01 or 77.

The following is an example of the Linkage Section and division header:

Using the elements of the previous example, the Linkage Section in the called program is set up as follows:

```
LINKAGE SECTION.
01  SALES-FIELDS.
    05  UNITS-L       PICTURE 9(5).
    05  PRICE-L       PICTURE 999V99.
    05  SALES-L       PICTURE 9(8)V99.
```

The Procedure Division header should appear as follows:

```
PROCEDURE DIVISION     USING SALES-FIELDS.
```

The order of appearance of the data-names in the USING option of the CALL statement and the USING option in the Procedure Division header is critical. Corresponding data-names refer to a single set of data that is available to the called and calling program. The correspondence is positional, not by name. In the case of index-names, no such correspondence is established. Index-names in the called and calling programs always refer to separate indexes.

Rules Governing the Use of the CALL Statement

1. The CALL statement appears in the calling program. It may not appear in the called program.
2. *Literal-1* is a nonnumeric literal and is the name of the program being called.
   a. *Literal-1* must conform to rules for the formation of a program-name.
   b. The first six characters of the literal are used to make correspondence between the calling and called programs.
   c. If the called program is to be entered at the beginning of the Procedure Division, the literal must specify the program-name in the PROGRAM-ID paragraph of the called program.
3. If there is a USING clause in the CALL statement that invoked it, the called program must have a USING clause as part of its Procedure Division header.
4. The *data-names* specified in the USING option of the CALL statement indicates those data items available to a calling program that may be referred to in the called program. When the called subprogram is a COBOL program, each of the USING options of the calling program must be identified as a data item in the File Section, Working-Storage Section, or Linkage Section. If the called subprogram is in a language other than COBOL, the operands may either be a file-name or a procedure-name.
5. Names in the USING lists (that of CALL in the main program and that of the Procedure Division header of the called program) are paired on one-for-one correspondence, even though there is no necessary relationship between the actual names for the paired items; but the data-names must be equivalent.
6. The USING option is used only if there is a USING option in the called entry point at the beginning of the Procedure Division of the called program. The number of operands in the USING option of the CALL statement should be the same as the number of operands in the USING option of the Procedure Division header.

A *called program* is a program that is the object of a CALL statement, while a *calling program* is a program that executes a CALL to another program.

EXIT PROGRAM Statement

The EXIT PROGRAM statement marks the logical end of a called program. This statement is used in the same manner and performs as other EXIT statements with the exception that the EXIT PROGRAM statement will end the execution of called programs, while other EXIT statements are only NO-OPs.

Rules Governing the Use of the EXIT PROGRAM Statement

1. The statement must be preceded by a paragraph-name and must be the only statement in a paragraph.
2. If the control reaches an EXIT PROGRAM statement while the operation is under the control of a CALL statement, control returns to the point in the calling program immediately following the CALL statement.
3. If control reaches an EXIT PROGRAM statement and no CALL statement is active, control passes through the exit point to the first sentence of the next paragraph.

Operation of Calling and Called Programs

The execution of a CALL statement causes control to pass to the called program. The first time a called program is entered, its state is that of fresh copy of the program. Each subsequent time a called program is entered, the state is as it was upon the last exit from that program. The reinitiation of items in the called program is the responsibility of the programmer.

When a called program has a USING option in its Procedure Division header and linkage was affected by a CALL statement where literal-1 is the name of the called program, execution of the called program begins with the first instruction in the Procedure Division after the Declaratives Section.

When the USING option is present, the called program operates as though each occurrence of data-name-1, data-name-2, etc. in the Procedure Division had been replaced by the corresponding data-names for the USING option of the CALL statement of the calling program; that is, corresponding data-names refer to a single set of data which is available to the calling program. The correspondence is positional, not by name.

When control reaches the EXIT PROGRAM statement in the called program, control returns to the point in the calling program immediately following the CALL statement.

The following is the first of three examples comparing calling and called programs:

Calling Program

```
IDENTIFICATION DIVISION.
PROGRAM-ID. CALLPROG.
    .
    .
DATA DIVISION.
    .
    .
WORKING-STORAGE SECTION.
01  RECORD-1.
    05  SALARY     PICTURE S9(5)V99.
    05  RATE       PICTURE S9V99.
    05  HOURS      PICTURE S99V9.
        .
        .
PROCEDURE DIVISION.
        .
        .
    CALL "SUBPROG" USING RECORD-1.
        .
        .
        .
    STOP RUN.
```

Called Program

```
IDENTIFICATION DIVISION.
PROGRAM-ID. SUBPROG.
    .
    .
DATA DIVISION.
    .
    .
LINKAGE SECTION.
01  PAYREC.
    05  PAY           PICTURE S9(5)V99.
    05  HOURLY-RATE   PICTURE S9V99.
    05  HOURS         PICTURE S99V9.
    .
    .
PROCEDURE DIVISION USING PAYREC.
    .
    .
PAYREC-EXIT.
    EXIT PROGRAM.
```

Here is the second example comparing calling and called programs:

Calling Program

```
IDENTIFICATION DIVISION.
PROGRAM-ID. LK-IF-MOVE.

ENVIRONMENT DIVISION.
CONFIGURATION SECTION.
SOURCE-COMPUTER.
OBJECT-COMPUTER.

DATA DIVISION.
FILE SECTION.
WORKING-STORAGE SECTION.
01  R1.
    05  RG.
        10  RG1    PICTURE X(8).
        10  RG2    PICTURE X(8).
        10  RG3    PICTURE X(7).
    05  RN.
        10  RN1    PICTURE S9(2)    USAGE COMP.
        10  RN2    PICTURE S9(6)    USAGE COMP.
        10  RN3    PICTURE S(9)V99  USAGE COMP.
        10  RN4    PICTURE S9(3)    USAGE COMP-3.
        10  RN5    PICTURE 9(4).
        10  RN6    PICTURE XBXXBBXX.
        10  RN7    PICTURE $$99.99.

PROCEDURE DIVISION.
START-HERE.
    CALL 'LD-IF-MOVE' USING R1.
    STOP RUN.
```

Called Program

```
IDENTIFICATION DIVISION.
PROGRAM-ID. LD-IF-MOVE.

ENVIRONMENT DIVISION.
CONFIGURATION SECTION.
SOURCE-COMPUTER.
OBJECT-COMPUTER.

DATA DIVISION.
LINKAGE SECTION.
01  L1.
    05  LG.
        10  LG1    PICTURE X(8).
        10  LG2    PICTURE X(8).
        10  LG3    PICTURE X(7).
    05  LN.
        10  LN1    PICTURE S9(2)      USAGE COMP.
        10  LN2    PICTURE S9(6)      USAGE COMP.
        10  LN3    PICTURE S9(9)V99   USAGE COMP.
        10  LN4    PICTURE S9(3)      USAGE COMP-3.
        10  LN5    PICTURE 9(4).
        10  LN6    PICTURE XBXXBBXX.
        10  LN7    PICTURE $$99.99.

PROCEDURE DIVISION USING L1.
START-HERE.
    MOVE 789.56 TO LN7.
    DISPLAY LN7.
LINK-EXIT.
    EXIT PROGRAM.
```

And here is the third example comparing calling and called programs:

Calling Program

```
IDENTIFICATION DIVISION.
PROGRAM-ID. SUMMARY1.

ENVIRONMENT DIVISION.
CONFIGURATION SECTION.
SOURCE-COMPUTER.
OBJECT-COMPUTER.
INPUT-OUTPUT SECTION.
FILE-CONTROL.
SELECT CHARGEFILE      ASSIGN TO NCR-TYPE-32.
SELECT TOTALFILE       ASSIGN TO NCR-TYPE-32.

DATA DIVISION.
FILE SECTION.
FD CHARGEFILE     LABEL RECORD IS STANDARD
                  DATA RECORD IS CHANGERECORD.
01  CHANGERECORD.
    05  CNO                 PICTURE X(5).
    05  CSTORE-A            PICTURE S999V99.
    05  CSTORE-B            PICTURE S999V99.
    05  CSTORE-C            PICTURE S999V99.
    05  PAYMENT             PICTURE 9999V99.
    05  BALANCE             PICTURE S9999V99.
    05  RATE                PICTURE 99V99.
FD  TOTALFILE               LABEL RECORD IS STANDARD
                            DATA RECORD IS TOTALRECORD.
```

Called Program

```
IDENTIFICATION DIVISION.
PROGRAM-ID.      SUMMARY2.

ENVIRONMENT DIVISION.
CONFIGURATION SECTION.
SOURCE-COMPUTER.
OBJECT-COMPUTER.

DATA DIVISION.
LINKAGE SECTION.
77  W-BALANCE               PICTURE 9999V99.
01  CHARGE-RECORD.
    05  FILLER              PICTURE X(5).
    05  STORE-A             PICTURE S999V99.
    05  STORE-B             PICTURE S999V99.
    05  STORE-C             PICTURE S999V99.
    05  PAYMENT             PICTURE 9999V99.
    05  FILLER              PICTURE X(11).
01  TOTAL.
    05  FILLER              PICTURE X(5).
    05  TOTAL-CHARGES       PICTURE S9999V99.
    05  TOTAL-BALANCE       PICTURE S9999V99.
    05  FILLER              PICTURE X(7).

PROCEDURE DIVISION USING TOTAL,
                         W-BALANCE,
                         CHARGE-RECORD.
```

```
01  TOTALRECORD.
    05  TNO                      PICTURE X(5).
    05  TOTAL-CHARGES            PICTURE S9999V99.
    05  TOTAL-BALANCE            PICTURE S9999V99.
    05  TOTAL-BAL-DUE            PICTURE S9999V99.
WORKING-STORAGE SECTION.
77  WORK-TOTAL                   PICTURE S9999V99.
77  WORK-BALANCE                 PICTURE 9999V99.
01  FLAGS.
    05  MORE-DATA-FLAG           PICTURE XXX      VALUE 'YES'.
        88  MORE-DATA                             VALUE 'YES'.
        88  NO-MORE-DATA                          VALUE 'NO'.

PROCEDURE DIVISION.
MAIN-ROUTINE.
    OPEN    INPUT    CHARGEFILE
            OUTPUT   TOTALFILE.
    READ CHARGEFILE
        AT END MOVE 'NO' TO MORE-DATA-FLAG.
    PERFORM PROCESS-ROUTINE
        UNTIL NO-MORE-DATA.
    CLOSE   CHARGEFILE
            TOTALFILE.
    STOP RUN.

PROCESS-ROUTINE.
    MOVE CNO TO TNO.

    CALL 'SUMMARY2' USING TOTALRECORD,
                          WORK-BALANCE,
                          CHANGERECORD.

    WRITE TOTALRECORD.
    READ CHARGEFILE
        AT END MOVE 'NO' TO MORE-DATA-FLAG.

BEGIN.
    ADD STORE-A, STORE-B, STORE-C GIVING TOTAL-
    CHARGES.
    SUBTRACT PAYMENT FROM W-BALANCE.
    ADD  TOTAL-CHARGES, W-BALANCE
         GIVING TOTAL-BALANCE.
END-MODULE.
    EXIT PROGRAM.
```

ENTER Statement

The ENTER statement provides a means of allowing the use of more than one language in the same program. (The format for the ENTER statement is shown in figure 15.6.)

Rules Governing the Use of the ENTER Statement

1. The other language statements are executed in the object program as if they had been compiled in the object program following the ENTER statement.
2. The implementor will specify, for their compilers, all details on how the other language(s) are to be written.
3. If the statements in the entered language cannot be written in-line, a routine-name is given to identify the portion of the other language coding to be executed at this point in the procedure sequence. If the other language statements can be written in-line, routine-name is not used.
4. The language-name may refer to any programming language specified to be entered through COBOL by the implementor. Language-name is specified by the implementor.

```
ENTER language-name [ routine-name ] .
```

Figure 15.6 Format—ENTER statement.

5. A routine-name is a COBOL word and it may be referred to only in an ENTER sentence.
6. The sentence ENTER COBOL must follow the last other-language statement in order to indicate to the compiler where a return to COBOL source language takes place.

The following is an example of the ENTER statement:

```
MOVE TCODE TO WCODE.
ENTER NEAT/3.
     INSERTMFILE,WORKI
     ENTER COBOL
PERFORM READ-FILEIN.
```

In the example just given, the NEAT/3 instructions must be written in the NEAT/3 programming format. The ENTER COBOL statement must be written on a separate line immediately following the last NEAT/3 instruction in order to tell the compiler when to return to COBOL.

Summary

The Declaratives Section is written in the Procedure Division to specify any special circumstance under which a procedure is to be executed in the object program. Special declaratives are provided for supplying additional procedures for input-output errors, monitoring items in the debugging section and altering certain procedures in the report writer.

The Declaratives Section is written as a subdivision at the beginning of the Procedure Division beginning with the key word DECLARATIVES and ending with the key words END DECLARATIVES. Although the declaratives are written at the beginning of the Procedure Division, the procedures are invoked nonsynchronously as a result of a condition that occurred which cannot be normally tested by the program.

The USE statement must identify the condition under which each declarative section is executed. Procedures specified in a USE statement must be self contained. It is illegal to branch outside the declaratives area or to the declaratives from the outside, but one Declarative Section may be accessed from another Declarative Section via the PERFORM statement. The USE statement specifies procedures to be performed for each type of declarative that is an addition to the standard procedures provided by the compiler. The statement itself, although never executed, defines the conditions calling for the execution of the USE procedure. The types of procedures associated with the USE statement are (1) procedures for input-output error handling that are in addition to the standard procedures provided by the input-output control system, (2) procedures that are executed just before a report group named in the Report Section of the Data Division is produced, and (3) items monitored by the associated debugging section.

The USE statement must immediately follow a section header in the Declaratives Section. After execution of a USE statement, control is returned to the invoking routine.

The continued processing of a file is permitted if (1) an error-processing procedure exists in the declaratives section, (2) the detection of the error results in an automatic transfer to the error-processing procedure which permits the programmer to examine the error condition before it enters the process, and (3) at the conclusion of the processing of the error, it is the programmer's responsibility to update the parameters normally returned by the input-output control system.

The Report Writer USE statement specifies Procedure Division statements that are executed before a report group named in the Report Section of the Data Division is produced.

The Linkage Section describes data that is common between programs that communicate with each other within a single run unit. Program interaction requires that both programs, the calling program that calls another program, and the called program that is being referred to, have access to the same data items. The Linkage Section is written in the Data Division of the called program to describe the data being communicated, including data-name and description, but storage within the called program is not reserved, inasmuch as the data area exists elsewhere.

Control is transferred from program to program through the use of CALL and EXIT PROGRAM statements. The Linkage Section is meaningful here only if the object program is to function under control of a CALL statement in the calling program containing a USING phrase. The data items in the Linkage Section of the called program may be referenced within the Procedure Division of the called program only if they are specified as operands of the USING phrase. The Procedure Division header operands or all subordinate items to such operands in the called program is under control of the CALL statement with the USING phrase.

Items that bear no hierarchic relationship to one another do not have to be grouped into records; instead they are written in the same manner as other level-77 items. Items that have a definite hierarchic relationship must be grouped into records.

CALL and EXIT PROGRAM are special statements that permit communication between object programs. The CALL statement provides the means whereby control can be passed from one program to another within a run unit. The EXIT PROGRAM statement supplies the logical end of the called program and control is returned from the called program to the first instruction following the calling sequence in the calling program. A called program can itself in turn call another program.

At program execution time, common data items are described in the Linkage Section. The USING phrase in the called program follows the Procedure Division header. The sequence of appearance of data-names in the USING phrase of the CALL statement and the USING phrase in the Procedure Division is significant. Corresponding data-names refer to a single set of data that is equally available to the called and calling programs. Their descriptions must define an equal number of character positions; however, they need not have the same name.

A calling program must contain a CALL statement at the point where another program is to be called. The statement causes control to be transferred from one object program to one or more subprograms or other language subprograms.

The USING option makes data items from a calling program available to a called program if specified in the CALL statement. The USING option must be specified in the Procedure Division header of the called program. The number of data-names specified in each of the USING phrases must be identical.

The EXIT PROGRAM statement marks the logical end of a called program and is used in the same manner and performs the same as other exit statements, except that it terminates execution of the called program while other exits act as only NO-OPs. Control returns to the point in the calling program immediately following the CALL statement. If control reaches an EXIT PROGRAM statement and no CALL statement is active, control passes through the exit point to the first sentence of the next paragraph.

The ENTER statement provides a means of allowing the use of more than one language in the same program. Implementors will specify for their compilers all details on how the other language(s) are to be written. The sentence ENTER COBOL must follow the last other-language statement in order to indicate to the compiler where a return to COBOL source language takes place.

Illustrative Program: Subroutines—Depreciation Schedule

Job Definition

Print a depreciation schedule based on the straight-line method, the double declining-balance method, and sum-of-years-digits method of various assets. The depreciation schedule is to be printed for each asset, giving the annual depreciation and the current book value under each method until the limit is reached.

Input

A record for each asset contains the following information:

Positions	Field Designation
1–8	Serial number
9–28	Name of asset
29–36	Cost (XXX,XXX.XX)
37–42	Scrap value (X,XXX.XX)
43–44	Estimated life (years)
45–46	Limit (years)
47–80	Not used

Processing

The depreciation schedule is to be prepared using subroutines for each method based on the following information.

Assume the following facts are examples for determination of depreciation by applying the three depreciation methods and other factors detailed below: COST 6000, SCRAP VALUE 1680, ESTIMATED LIFE 8 years.

1. *Straight-Line Method:* The factor used in computing the annual depreciation is COST − SCRAP VALUE ÷ ESTIMATED LIFE = Annual Depreciation. The annual depreciation remains the same for the life of the asset.
   For example:

	COST	6000
−	SCRAP VALUE	1680
	DEPRECIATION	4320
÷	ESTIMATED LIFE	8 yr.
	Annual Depreciation	540

2. *Double Declining-Balance Method:* In this method, the rate of depreciation is determined by dividing 100% by the estimated life. This rate is then doubled and applied to the original cost, resulting in the first year's depreciation. Each succeeding year's depreciation is determined by subtracting the accumulated depreciation from the original cost and then applying the rate to the (declined) balance. *Scrap Value is not considered.*
   For example, ESTIMATED LIFE 8 is divided into 100% = 12.5%. This rate is doubled (25%).

1st year		COST	6000
	×	Rate	25%
		Annual Depreciation	1500
2nd year		COST	6000
	−	Accumulated Depr.	1500
		Declined Balance	4500
	×	Rate	25%
		Annual Depreciation	1125

The process is repeated for the life of the asset.

3. *Sum-of-Years-Digits Method:* The following steps are used in determining the annual depreciation by this method.
   a. Add the digits of the number of years in the ESTIMATED LIFE.
   b. The first year's depreciation is obtained by using a fraction.

   The numerator is the number of the year (in reverse sequence, highest number first) and the denominator is the sum of the digits. The sum of the digits may be determined by using the following formula.

   $\frac{N(N+1)}{2}$ in which N = the number of years in the ESTIMATED LIFE.

   For example,

	COST	6000
−	SCRAP VALUE	1680
	Depreciation	4320

   Formula for determining sum-of-the-years-digits.

$\frac{N(N+1)}{2}$	$\frac{8 \times (8+1)}{2} = 36$
1st year's depreciation	$8/36 \times 4320 = 960$
2nd year's depreciation	$7/36 \times 4320 = 840$
3rd year's depreciation	$6/36 \times 4320 = 720$
.	
.	
.	
8th year's depreciation	$1/36 \times 4320 = 120$

4. *Book Value:* The book value of the asset at the end of the period for each method is obtained by subtracting the accumulated depreciation from the original cost.
5. *Limit:* Each year's depreciation is to be printed until the limit is reached.
6. Each asset is to be printed on a separate sheet.

Output

A Depreciation Schedule report is to be printed as follows:

```
                                   DEPRECIATION SCHEDULE

         SERIAL NUMBER     657809                                       NAME   LOADER

COST     6,000.00               SCRAP  1,680.00              YEARS     8                   LIMIT    5

****************************************************************************************************
```

YEAR	DEPRECIATION ST-LINE	DEC-BAL	SUM-DIGITS	BOOK VALUE ST-LINE	DEC-BAL	SUM-DIGITS
1	540.00	1,500.00	960.00	5,460.00	4,500.00	5,040.00
2	540.00	1,125.00	840.00	4,920.00	3,375.00	4,200.00
3	540.00	843.75	720.00	4,380.00	2,531.25	3,480.00
4	540.00	632.81	600.00	3,840.00	1,898.44	2,880.00
5	540.00	474.61	480.00	3,300.00	1,423.83	2,400.00

```
                                    DEPRECIATION SCHEDULE

               SERIAL NUMBER    7840795                                      NAME  DISPLAY CASES

     COST      3,200.00              SCRAP      500.00               YEARS    10                  LIMIT    3

*****************************************************************************************************

                       DEPRECIATION                                                 BOOK VALUE

     YEAR            ST-LINE         DEC-BAL     SUM-DIGITS              ST-LINE          DEC-BAL        SUM-DIGITS


       1             270.00          640.00         490.91              2,930.00         2,560.00        2,709.09
       2             270.00          512.00         441.82              2,660.00         2,048.00        2,267.27
       3             270.00          409.60         392.73              2,390.00         1,638.40        1,874.54

                                    DEPRECIATION SCHEDULE

               SERIAL NUMBER   14756438                                      NAME  FURNITURE

     COST      6,050.00              SCRAP      500.00               YEARS    12                  LIMIT    5

*****************************************************************************************************

                       DEPRECIATION                                                 BOOK VALUE

     YEAR            ST-LINE         DEC-BAL     SUM-DIGITS              ST-LINE          DEC-BAL        SUM-DIGITS


       1             462.50        1,004.30         853.85              5,587.50         5,045.70        5,196.15
       2             462.50          837.58         782.69              5,125.00         4,208.12        4,413.46
       3             462.50          698.54         711.54              4,662.50         3,509.58        3,701.92
       4             462.50          582.59         640.38              4,200.00         2,926.99        3,061.54
       5             462.50          485.88         569.23              3,737.50         2,441.11        2,492.31

                                    DEPRECIATION SCHEDULE

               SERIAL NUMBER   38926042                                      NAME  AUTO

     COST     16,010.00              SCRAP    1,500.00               YEARS     7                  LIMIT    7

*****************************************************************************************************

                       DEPRECIATION                                                 BOOK VALUE

     YEAR            ST-LINE         DEC-BAL     SUM-DIGITS              ST-LINE          DEC-BAL        SUM-DIGITS


       1           2,072.85        4,546.84       3,627.50             13,937.15        11,463.16       12,382.50
       2           2,072.85        3,255.53       3,109.29             11,864.30         8,207.63        9,273.21
       3           2,072.85        2,330.96       2,591.07              9,791.45         5,876.67        6,682.14
       4           2,072.85        1,668.97       2,072.86              7,718.60         4,207.70        4,609.28
       5           2,072.85        1,194.98       1,554.64              5,645.75         3,012.72        3,054.64
       6           2,072.85          855.61       1,036.43              3,572.90         2,157.11        2,018.21
       7           2,072.85          612.61         518.21              1,500.05         1,544.50        1,500.00
```

Printer Spacing Chart

This Printer Spacing Chart shows how the report is formatted.

```
H  3                                            DEPRECIATION SCHEDULE
H  5             SERIAL NUMBER  X-----------X                                 NAME  X-----------------------X
                                  (SERIAL)                                               (ASSET)
H  7      COST   XXX,XXX.XX                SCRAP  X,XXX.XX          YEARS   XX                 LIMIT   XX
                   (COST)                          (SCRAPV)                (ESTLFE)                  (LIMIT)
  10 X-----------------------------------------------------------------------------------------------------X
H 15                        DEPRECIATION                                  BOOK VALUE
H 17       YEAR      ST-LINE         DEC-BAL    SUM-DIGITS          ST-LINE          DEC-BAL      SUM-DIGITS
D 21        XX     XXX,XXX.XX      XXX,XXX.XX   XXX,XXX.XX        XXX,XXX.XX       XXX,XXX.XX    XXX,XXX.XX
          (YEAR)   (ASTLOP)        (ADECBL)     (ASUMDPI)          (STLBKV)         (DBBKVL)      (SOBKVL)
```

Hierarchy Chart

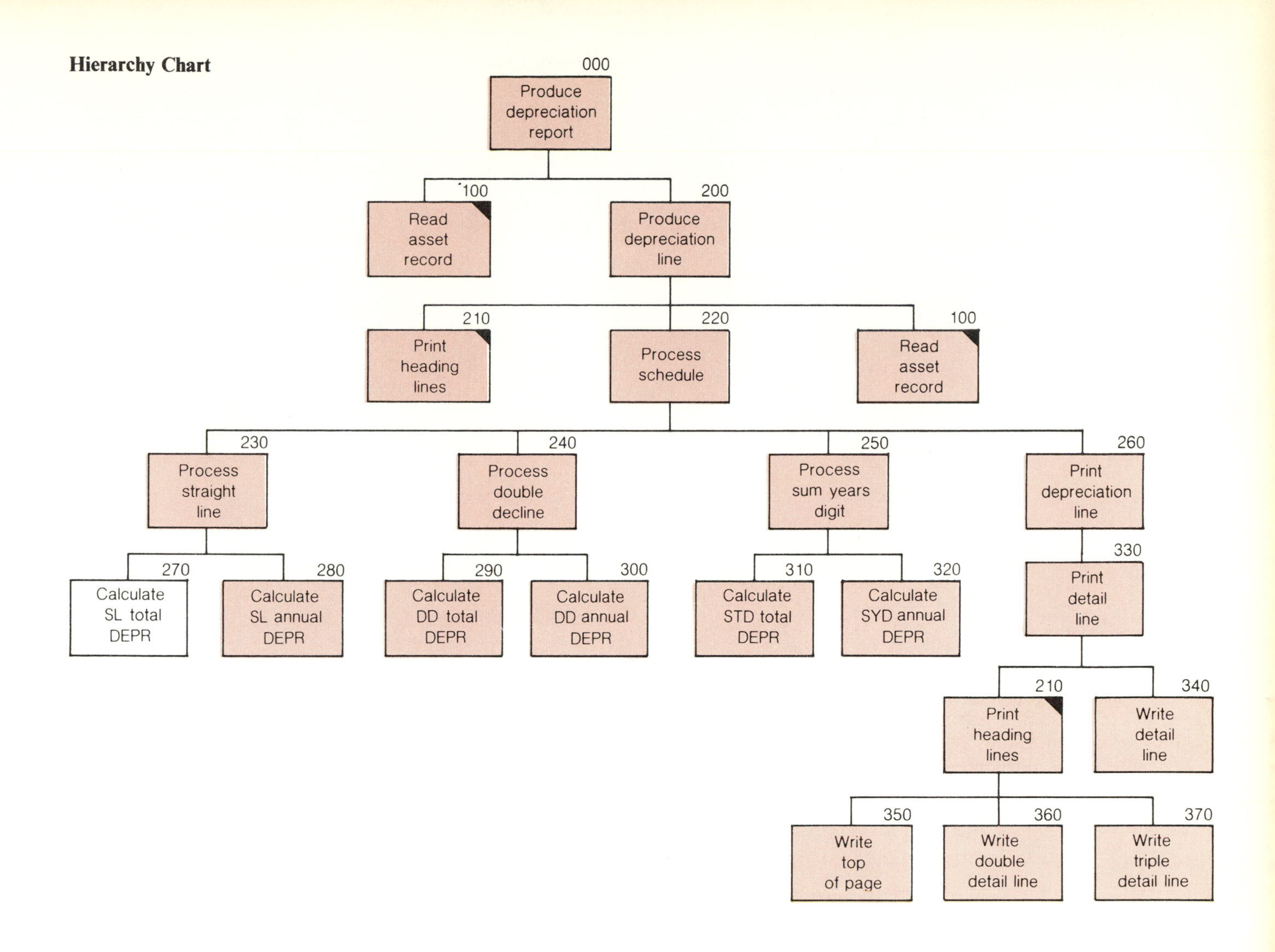

IPO Charts

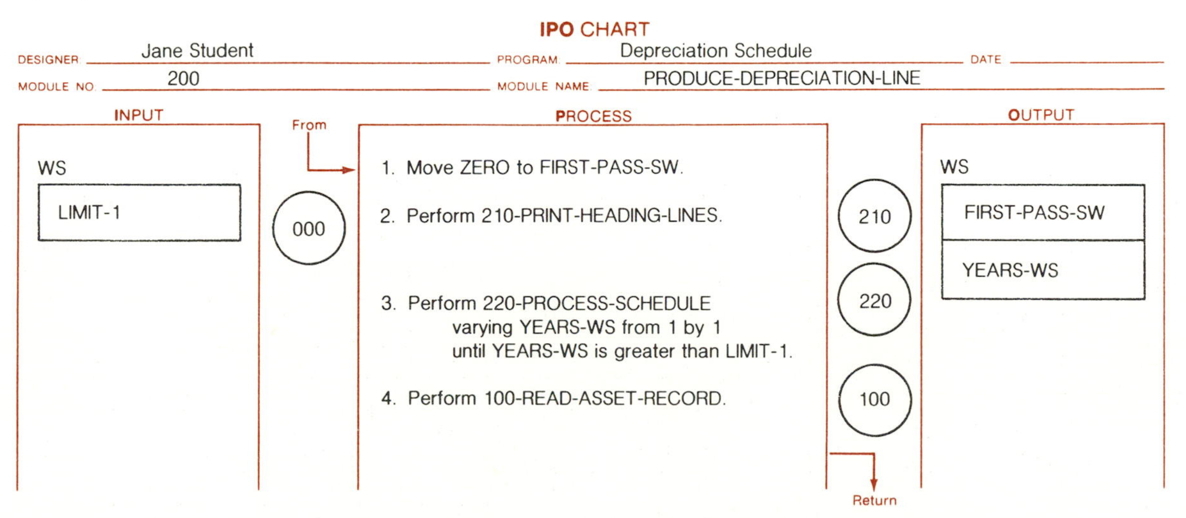

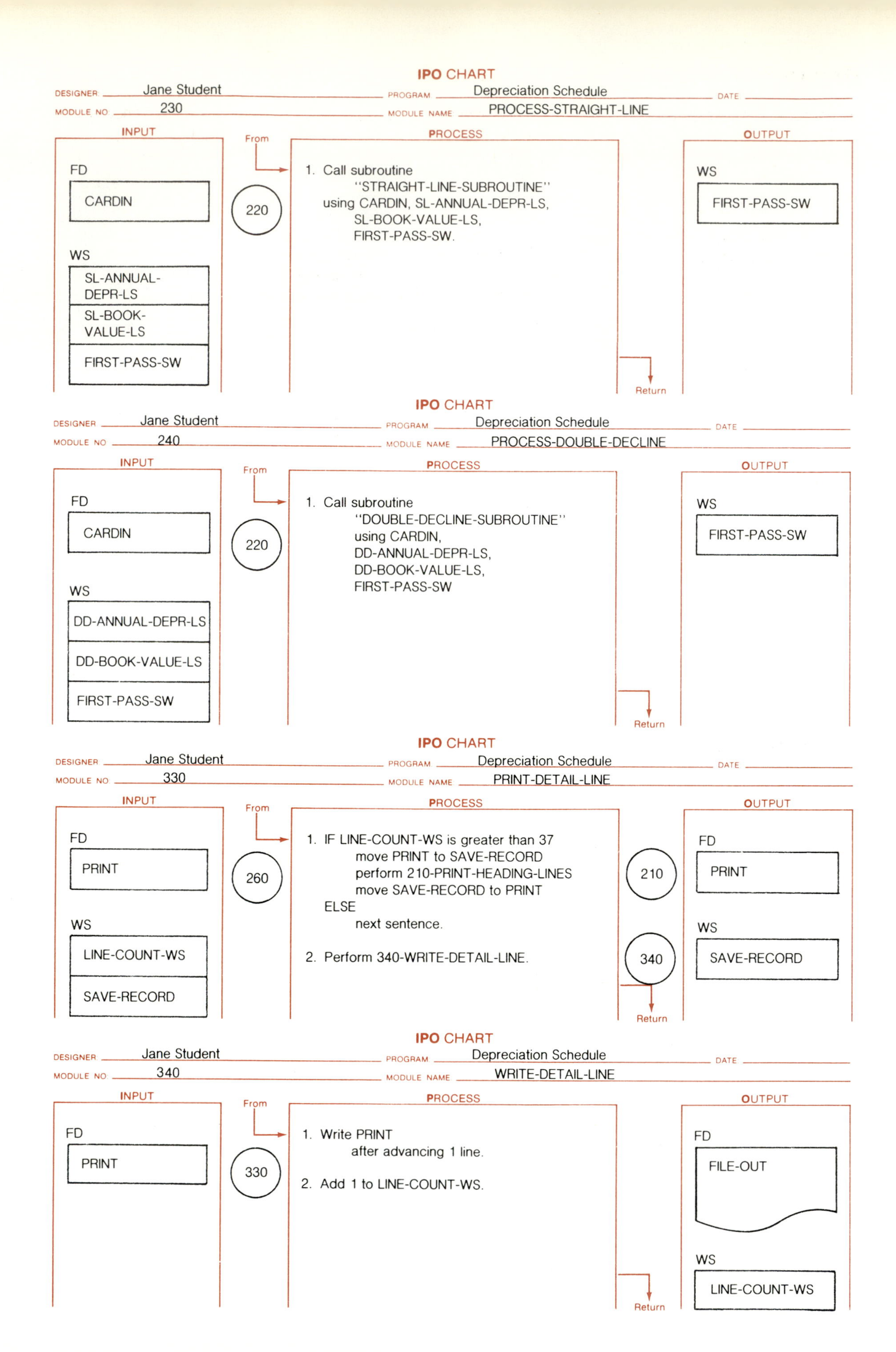

IPO CHART
DESIGNER Jane Student
PROGRAM Depreciation Schedule
DATE
MODULE NO 230
MODULE NAME PROCESS-STRAIGHT-LINE
INPUT
FD
CARDIN
WS
SL-ANNUAL-DEPR-LS
SL-BOOK-VALUE-LS
FIRST-PASS-SW
From
220
PROCESS
1. Call subroutine
"STRAIGHT-LINE-SUBROUTINE"
using CARDIN, SL-ANNUAL-DEPR-LS,
SL-BOOK-VALUE-LS,
FIRST-PASS-SW.
Return
OUTPUT
WS
FIRST-PASS-SW
IPO CHART
DESIGNER Jane Student
PROGRAM Depreciation Schedule
DATE
MODULE NO 240
MODULE NAME PROCESS-DOUBLE-DECLINE
INPUT
FD
CARDIN
WS
DD-ANNUAL-DEPR-LS
DD-BOOK-VALUE-LS
FIRST-PASS-SW
From
220
PROCESS
1. Call subroutine
"DOUBLE-DECLINE-SUBROUTINE"
using CARDIN,
DD-ANNUAL-DEPR-LS,
DD-BOOK-VALUE-LS,
FIRST-PASS-SW
Return
OUTPUT
WS
FIRST-PASS-SW
IPO CHART
DESIGNER Jane Student
PROGRAM Depreciation Schedule
DATE
MODULE NO 330
MODULE NAME PRINT-DETAIL-LINE
INPUT
FD
PRINT
WS
LINE-COUNT-WS
SAVE-RECORD
From
260
PROCESS
1. IF LINE-COUNT-WS is greater than 37
move PRINT to SAVE-RECORD
perform 210-PRINT-HEADING-LINES
move SAVE-RECORD to PRINT
ELSE
next sentence.
2. Perform 340-WRITE-DETAIL-LINE.
210
340
Return
OUTPUT
FD
PRINT
WS
SAVE-RECORD
IPO CHART
DESIGNER Jane Student
PROGRAM Depreciation Schedule
DATE
MODULE NO 340
MODULE NAME WRITE-DETAIL-LINE
INPUT
FD
PRINT
From
330
PROCESS
1. Write PRINT
after advancing 1 line.
2. Add 1 to LINE-COUNT-WS.
Return
OUTPUT
FD
FILE-OUT
WS
LINE-COUNT-WS

Flowcharts

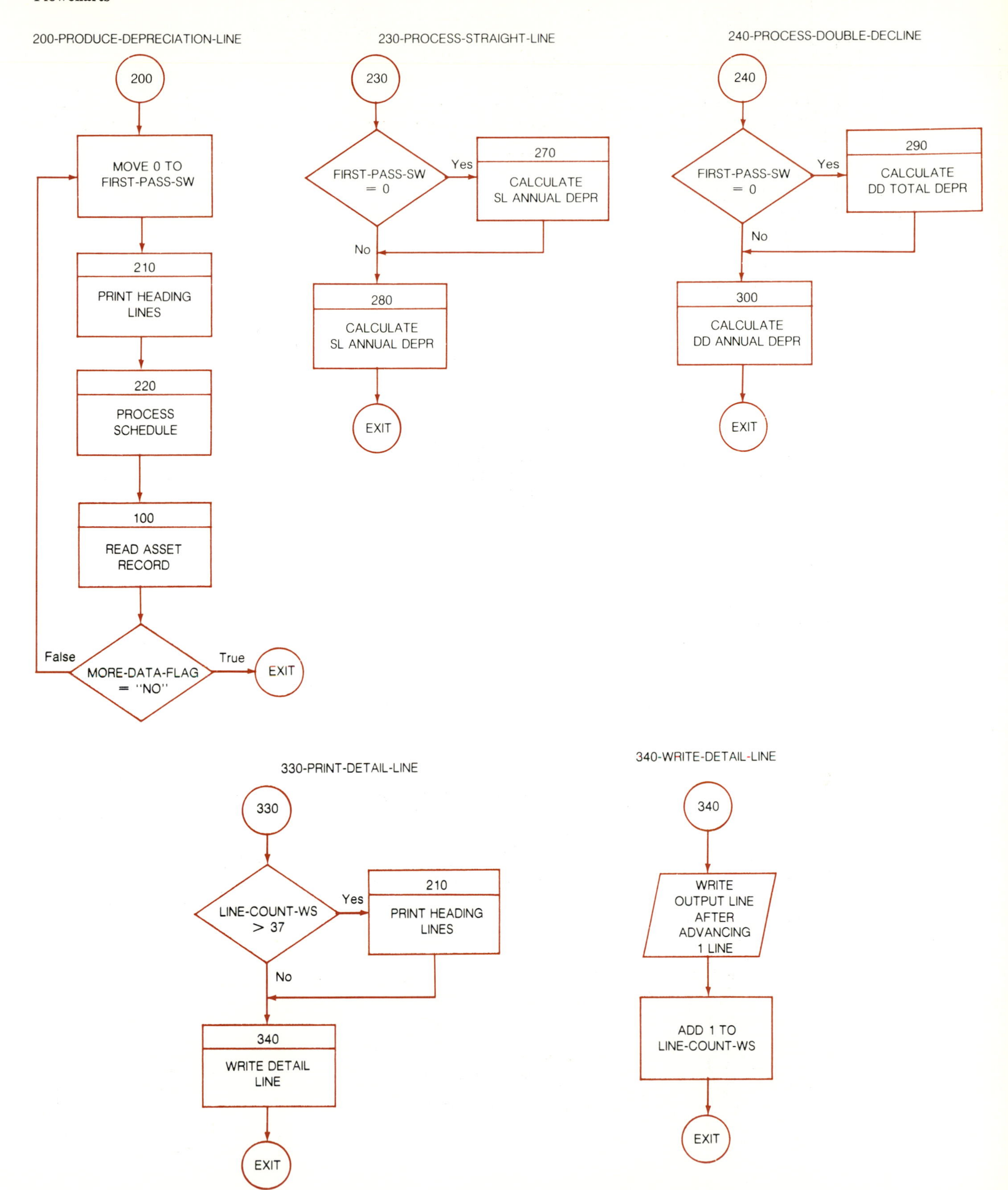
200-PRODUCE-DEPRECIATION-LINE
200
MOVE 0 TO FIRST-PASS-SW
210
PRINT HEADING LINES
220
PROCESS SCHEDULE
100
READ ASSET RECORD
MORE-DATA-FLAG = "NO"
False
True
EXIT
230-PROCESS-STRAIGHT-LINE
230
FIRST-PASS-SW = 0
Yes
270
CALCULATE SL ANNUAL DEPR
No
280
CALCULATE SL ANNUAL DEPR
EXIT
240-PROCESS-DOUBLE-DECLINE
240
FIRST-PASS-SW = 0
Yes
290
CALCULATE DD TOTAL DEPR
No
300
CALCULATE DD ANNUAL DEPR
EXIT
330-PRINT-DETAIL-LINE
330
LINE-COUNT-WS > 37
Yes
210
PRINT HEADING LINES
No
340
WRITE DETAIL LINE
EXIT
340-WRITE-DETAIL-LINE
340
WRITE OUTPUT LINE AFTER ADVANCING 1 LINE
ADD 1 TO LINE-COUNT-WS
EXIT

```
 1        *************************
 2
 3         IDENTIFICATION DIVISION.
 4
 5        *************************
 6
 7         PROGRAM-ID.
 8             DEPRECIATION-SCHEDULE.
 9         AUTHOR.
10             STEVE GONSOSKI.
11         INSTALLATION.
12             WEST LOS ANGELES COLLEGE.
13         DATE-WRITTEN.
14             3 FEBRUARY 1982.
15
16        *PURPOSE.
17        *    THIS PROGRAM PRINTS A DEPRECIATION SCHEDULE BASED ON THE
18        *    STRAIGHT-LINE METHOD, THE DOUBLE DECLINING-BALANCE METHOD,
19        *    AND SUM-OF-YEARS-DIGITS METHOD OF VARIOUS ASSETS.
20
21        *************************
22
23         ENVIRONMENT DIVISION.
24
25        *************************
26
27         CONFIGURATION SECTION.
28
29         SOURCE-COMPUTER.
30             LEVEL-66-ASCII.
31         OBJECT-COMPUTER.
32             LEVEL-66-ASCII.
33
34         INPUT-OUTPUT SECTION.
35
36         FILE-CONTROL.
37             SELECT FILE-IN      ASSIGN TO C1-CARD-READER.
38             SELECT FILE-OUT     ASSIGN TO P1-PRINTER.
39
40        *************************
41
42         DATA DIVISION.
43
44        *************************
45
46         FILE SECTION.
47
48         FD  FILE-IN
49             CODE-SET IS GBCD
50             LABEL RECORDS ARE OMITTED
51             DATA RECORD IS CARDIN.
52
53         01  CARDIN.
54
55             03  SERIAL-NUMBER        PICTURE 9(8).
56             03  NAME-OF-ASSET        PICTURE X(20).
57             03  COST                 PICTURE 9(6)V99.
58             03  SCRAP-VALUE          PICTURE 9(4)V99.
59             03  ESTIMATED-LIFE       PICTURE 9(2).
60             03  LIMIT-I              PICTURE 9(2).
61             03  FILLER               PICTURE X(34).
62
63         FD  FILE-OUT
64             CODE-SET IS GBCD
65             LABEL RECORDS ARE OMITTED
66             DATA RECORD IS PRINT.
67
68         01  PRINT                    PICTURE X(132).
69
70         WORKING-STORAGE SECTION.
71
72         77  LINE-COUNT-WS            PICTURE 9(2)     VALUE 38.
73         77  YEARS-WS                 PICTURE 9(2)     VALUE ZERO.
74         77  FIRST-PASS-SW            PICTURE 9(1)     VALUE ZERO.
75         77  SL-ANNUAL-DEPR-LS        PICTURE 9(6)V99.
76         77  SL-BOOK-VALUE-LS         PICTURE 9(6)V99.
77         77  DD-ANNUAL-DEPR-LS        PICTURE 9(6)V99.
78         77  DD-BOOK-VALUE-LS         PICTURE 9(6)V99.
79         77  SYD-ANNUAL-DEPR-LS       PICTURE 9(6)V99.
80         77  SYD-BOOK-VALUE-LS        PICTURE 9(6)V99.
81
82         01  FLAGS.
83
84             03  MORE-DATA-FLAG       PICTURE X(3)     VALUE "YES".
85                 88  MORE-DATA                         VALUE "YES".
86                 88  NO-MORE-DATA                      VALUE "NO ".
87
88         01  SAVE-RECORD              PICTURE X(132).
89
90         01  HDG-1.
91
92             03  FILLER               PICTURE X(52)    VALUE SPACES.
93             03  FILLER               PICTURE X(12)    VALUE "DEPRECIATION".
94             03  FILLER               PICTURE X(9)     VALUE " SCHEDULE".
95             03  FILLER               PICTURE X(59)    VALUE SPACES.
96
```

```
 97        01  HDG-2.
 98
 99            03  FILLER                 PICTURE X(19)    VALUE SPACES.
100            03  FILLER                 PICTURE X(7)     VALUE "SERIAL  ".
101            03  FILLER                 PICTURE X(8)     VALUE "NUMBER   ".
102            03  SERIAL-NUMBER-O        PICTURE Z(8).
103            03  FILLER                 PICTURE X(43)    VALUE SPACES.
104            03  FILLER                 PICTURE X(6)     VALUE "NAME   ".
105            03  NAME-OF-ASSET-O        PICTURE X(20).
106            03  FILLER                 PICTURE X(21)    VALUE SPACES.
107
108        01  HDG-3.
109
110            03  FILLER                 PICTURE X(10)    VALUE SPACES.
111            03  FILLER                 PICTURE X(6)     VALUE "COST   ".
112            03  COST-O                 PICTURE ZZZ,ZZZ.99.
113            03  FILLER                 PICTURE X(14)    VALUE SPACES.
114            03  FILLER                 PICTURE X(7)     VALUE "SCRAP   ".
115            03  SCRAP-VALUE-O          PICTURE Z,ZZZ.99.
116            03  FILLER                 PICTURE X(15)    VALUE SPACES.
117            03  FILLER                 PICTURE X(8)     VALUE "YEARS    ".
118            03  ESTIMATED-LIFE-O       PICTURE Z9.
119            03  FILLER                 PICTURE X(20)    VALUE SPACES.
120            03  FILLER                 PICTURE X(7)     VALUE "LIMIT   ".
121            03  LIMIT-I-O              PICTURE Z9.
122            03  FILLER                 PICTURE X(23)    VALUE SPACES.
123
124        01  HDG-4.
125
126            03  ASTRIKS                PICTURE X(132).
127
128        01  HDG-5.
129
130            03  FILLER                 PICTURE X(28)    VALUE SPACES.
131            03  FILLER                 PICTURE X(12)    VALUE "DEPRECIATION".
132            03  FILLER                 PICTURE X(48)    VALUE SPACES.
133            03  FILLER                 PICTURE X(10)    VALUE "BOOK VALUE".
134            03  FILLER                 PICTURE X(34)    VALUE SPACES.
135
136        01  HDG-6.
137
138            03  FILLER                 PICTURE X(11)    VALUE SPACES.
139            03  FILLER                 PICTURE X(4)     VALUE "YEAR".
140            03  FILLER                 PICTURE X(8)     VALUE SPACES.
141            03  FILLER                 PICTURE X(7)     VALUE "ST-LINE".
142            03  FILLER                 PICTURE X(7)     VALUE SPACES.
143            03  FILLER                 PICTURE X(7)     VALUE "DEC-BAL".
144            03  FILLER                 PICTURE X(3)     VALUE SPACES.
145            03  FILLER                 PICTURE X(10)    VALUE "SUM-DIGITS".
146            03  FILLER                 PICTURE X(16)    VALUE SPACES.
147            03  FILLER                 PICTURE X(7)     VALUE "ST-LINE".
148            03  FILLER                 PICTURE X(9)     VALUE SPACES.
149            03  FILLER                 PICTURE X(7)     VALUE "DEC-BAL".
150            03  FILLER                 PICTURE X(6)     VALUE SPACES.
151            03  FILLER                 PICTURE X(10)    VALUE "SUM-DIGITS".
152            03  FILLER                 PICTURE X(20)    VALUE SPACES.
153
154        01  DETAIL-LINE.
155
156            03  FILLER                 PICTURE X(12)    VALUE SPACES.
157            03  YEAR-O                 PICTURE Z9.
158            03  FILLER                 PICTURE X(6)     VALUE SPACES.
159            03  SL-ANNUAL-DEPR-O       PICTURE ZZZ,ZZZ.99.
160            03  FILLER                 PICTURE X(4)     VALUE SPACES.
161            03  DD-ANNUAL-DEPR-O       PICTURE ZZZ,ZZZ.99.
162            03  FILLER                 PICTURE X(4)     VALUE SPACES.
163            03  SYD-ANNUAL-DEPR-O      PICTURE ZZZ,ZZZ.99.
164            03  FILLER                 PICTURE X(13)    VALUE SPACES.
165            03  SL-BOOK-VALUE-O        PICTURE ZZZ,ZZZ.99.
166            03  FILLER                 PICTURE X(5)     VALUE SPACES.
167            03  DD-BOOK-VALUE-O        PICTURE ZZZ,ZZZ.99.
168            03  FILLER                 PICTURE X(5)     VALUE SPACES.
169            03  SYD-BOOK-VALUE-O       PICTURE ZZZ,ZZZ.99.
170            03  FILLER                 PICTURE X(21)    VALUE SPACES.
171
172       *************************
173
174        PROCEDURE DIVISION.
175
176       *************************
177
178        000-PRODUCE-DEPRECIATION-REPRT.
179
180            OPEN    INPUT   FILE-IN
181                    OUTPUT  FILE-OUT.
182            PERFORM 100-READ-ASSET-RECORD.
183            PERFORM 200-PRODUCE-DEPRECIATION-LINE
184                    UNTIL NO-MORE-DATA.
185            CLOSE   FILE-IN
186                    FILE-OUT.
187            STOP RUN.
188
189        100-READ-ASSET-RECORD.
190
191            READ FILE-IN
192                AT END MOVE "NO " TO MORE-DATA-FLAG.
193
194        200-PRODUCE-DEPRECIATION-LINE.
195
196            MOVE 0 TO FIRST-PASS-SW.
197            PERFORM 210-PRINT-HEADING-LINES.
```

```
198            PERFORM 220-PROCESS-SCHEDULE
199                VARYING YEARS-WS FROM 1 BY 1
200                UNTIL YEARS-WS IS > LIMIT-I.
201            PERFORM 100-READ-ASSET-RECORD.
202
203        210-PRINT-HEADING-LINES.
204
205            MOVE HDG-1 TO PRINT.
206            PERFORM 350-WRITE-TOP-OF-PAGE.
207            MOVE SERIAL-NUMBER TO SERIAL-NUMBER-O.
208            MOVE NAME-OF-ASSET TO NAME-OF-ASSET-O.
209            MOVE HDG-2 TO PRINT.
210            PERFORM 360-WRITE-DOUBLE-DETAIL-LINE.
211            MOVE COST TO COST-O.
212            MOVE SCRAP-VALUE TO SCRAP-VALUE-O.
213            MOVE ESTIMATED-LIFE TO ESTIMATED-LIFE-O.
214            MOVE LIMIT-I TO LIMIT-I-O.
215            MOVE HDG-3 TO PRINT.
216            PERFORM 360-WRITE-DOUBLE-DETAIL-LINE.
217            MOVE ALL "*" TO ASTRIKS.
218            MOVE HDG-4 TO PRINT.
219            PERFORM 370-WRITE-TRIPLE-DETAIL-LINE.
220            MOVE HDG-5 TO PRINT.
221            PERFORM 370-WRITE-TRIPLE-DETAIL-LINE.
222            MOVE HDG-6 TO PRINT.
223            PERFORM 360-WRITE-DOUBLE-DETAIL-LINE.
224            MOVE SPACES TO PRINT.
225            PERFORM 370-WRITE-TRIPLE-DETAIL-LINE.
226
227        220-PROCESS-SCHEDULE.
228
229            PERFORM 230-PROCESS-STRAIGHT-LINE.
230            PERFORM 240-PROCESS-DOUBLE-DECLINE.
231            PERFORM 250-PROCESS-SUM-YEARS-DIGIT.
232            PERFORM 260-PRINT-DEPRECIATION-LINE.
233            MOVE 1 TO FIRST-PASS-SW.
234
235        230-PROCESS-STRAIGHT-LINE.
236
237            CALL "STRAIGHT-LINE-SUBROUTINE"
238                USING CARDIN, SL-ANNUAL-DEPR-LS, SL-BOOK-VALUE-LS,
239                FIRST-PASS-SW.
240
241        240-PROCESS-DOUBLE-DECLINE.
242
243            CALL "DOUBLE-DECLINE-SUBROUTINE"
244                USING CARDIN, DD-ANNUAL-DEPR-LS, DD-BOOK-VALUE-LS,
245                FIRST-PASS-SW.
246
247        250-PROCESS-SUM-YEARS-DIGIT.
248
249            CALL "SUM-YEARS-DIGITS-SUBROUTINE"
250                USING CARDIN, SYD-ANNUAL-DEPR-LS, SYD-BOOK-VALUE-LS,
251                FIRST-PASS-SW.
252
253        260-PRINT-DEPRECIATION-LINE.
254
255            MOVE YEARS-WS              TO YEAR-O.
256            MOVE SL-ANNUAL-DEPR-LS     TO SL-ANNUAL-DEPR-O.
257            MOVE DD-ANNUAL-DEPR-LS     TO DD-ANNUAL-DEPR-O.
258            MOVE SYD-ANNUAL-DEPR-LS    TO SYD-ANNUAL-DEPR-O.
259            MOVE SL-BOOK-VALUE-LS      TO SL-BOOK-VALUE-O.
260            MOVE DD-BOOK-VALUE-LS      TO DD-BOOK-VALUE-O.
261            MOVE SYD-BOOK-VALUE-LS     TO SYD-BOOK-VALUE-O.
262            MOVE DETAIL-LINE           TO PRINT.
263            PERFORM 330-PRINT-DETAIL-LINE.
264
265        330-PRINT-DETAIL-LINE.
266
267            IF LINE-COUNT-WS IS > 37
268                MOVE PRINT TO SAVE-RECORD
269                PERFORM 210-PRINT-HEADING-LINES
270                MOVE SAVE-RECORD TO PRINT
271            ELSE
272                NEXT SENTENCE.
273            PERFORM 340-WRITE-DETAIL-LINE.
274
275        340-WRITE-DETAIL-LINE.
276
277            WRITE PRINT
278                AFTER ADVANCING 1 LINE.
279            ADD 1 TO LINE-COUNT-WS.
280
281        350-WRITE-TOP-OF-PAGE.
282
283            WRITE PRINT
284                AFTER ADVANCING PAGE.
285            MOVE 1 TO LINE-COUNT-WS.
286
287        360-WRITE-DOUBLE-DETAIL-LINE.
288
289            WRITE PRINT
290                AFTER ADVANCING 2 LINES.
291            ADD 2 TO LINE-COUNT-WS.
292
293        370-WRITE-TRIPLE-DETAIL-LINE.
294
295            WRITE PRINT
296                AFTER ADVANCING 3 LINES.
297            ADD 3 TO LINE-COUNT-WS.

THERE WERE 297 SOURCE INPUT LINES.
THERE WERE NO DIAGNOSTICS.
```

Subprogram

```
 1       *************************
 2
 3        IDENTIFICATION DIVISION.
 4
 5       *************************
 6
 7        PROGRAM-ID.
 8            STRAIGHT-LINE-SUBROUTINE.
 9        AUTHOR.
10            STEVE GONSOSKI.
11        INSTALLATION.
12            WEST LOS ANGELES COLLEGE.
13        DATE-WRITTEN.
14            3 FEBRUARY 1982.
15
16       *PURPOSE.
17       *    THIS SUBROUTINE CALCULATES THE DEPRECIATION OF AN ASSET
18       *    USING THE STRAIGHT-LINE METHOD.
19
20       *************************
21
22        ENVIRONMENT DIVISION.
23
24       *************************
25
26        CONFIGURATION SECTION.
27
28        SOURCE-COMPUTER.
29            LEVEL-66-ASCII.
30        OBJECT-COMPUTER.
31            LEVEL-66-ASCII.
32
33       *************************
34
35        DATA DIVISION.
36
37       *************************
38
39        WORKING-STORAGE SECTION.
40
41        77  DEPRECIATION-VALUE-WS    PICTURE 9(6)V99.
42
43        LINKAGE SECTION.
44
45        77  SL-ANNUAL-DEPR-LS        PICTURE 9(6)V99.
46        77  SL-BOOK-VALUE-LS         PICTURE 9(6)V99.
47        77  FIRST-PASS-SW            PICTURE 9(1).
48
49        01  CARDIN.
50
51            03  SERIAL-NUMBER        PICTURE 9(8).
52            03  NAME-OF-ASSET        PICTURE X(20).
53            03  COST                 PICTURE 9(6)V99.
54            03  SCRAP-VALUE          PICTURE 9(4)V99.
55            03  ESTIMATED-LIFE       PICTURE 9(2).
56            03  LIMIT-I              PICTURE 9(2).
57            03  FILLER               PICTURE X(34).
58
59        PROCEDURE DIVISION USING CARDIN,
60                                 SL-ANNUAL-DEPR-LS,
61                                 SL-BOOK-VALUE-LS,
62                                 FIRST-PASS-SW.
63
64        230-PROCESS-STRAIGHT-LINE.
65
66            IF FIRST-PASS-SW IS = 0
67                PERFORM 270-CALCULATE-SL-TOTAL-DEPR
68            ELSE
69                NEXT SENTENCE.
70            PERFORM 280-CALCULATE-SL-ANNUAL-DEPR.
71        END-OF-SUBROUTINE.
72            EXIT PROGRAM.
73
74        270-CALCULATE-SL-TOTAL-DEPR.
75
76            SUBTRACT COST FROM SCRAP-VALUE
77                GIVING DEPRECIATION-VALUE-WS.
78            DIVIDE DEPRECIATION-VALUE-WS BY ESTIMATED-LIFE
79                GIVING SL-ANNUAL-DEPR-LS.
80            MOVE COST TO SL-BOOK-VALUE-LS.
81
82        280-CALCULATE-SL-ANNUAL-DEPR.
83
84            SUBTRACT SL-ANNUAL-DEPR-LS FROM SL-BOOK-VALUE-LS.

THERE WERE 84 SOURCE INPUT LINES.
THERE WERE NO DIAGNOSTICS.
```

Subprogram

```
  1        **************************
  2
  3         IDENTIFICATION DIVISION.
  4
  5        **************************
  6
  7         PROGRAM-ID.
  8             DOUBLE-DECLINE-SUBROUTINE.
  9         AUTHOR.
 10             STEVE GONSOSKI.
 11         INSTALLATION.
 12             WEST LOS ANGELES COLLEGE.
 13         DATE-WRITTEN.
 14             3 FEBRUARY 1982.
 15
 16        *PURPOSE.
 17        *    THIS SUBROUTINE CALCULATES THE DEPRECIATION OF AN ASSET
 18        *    USING THE DOUBLE DECLINING-BALANCE METHOD.
 19
 20        **************************
 21
 22         ENVIRONMENT DIVISION.
 23
 24        **************************
 25
 26         CONFIGURATION SECTION.
 27
 28         SOURCE-COMPUTER.
 29             LEVEL-66-ASCII.
 30         OBJECT-COMPUTER.
 31             LEVEL-66-ASCII.
 32
 33        **************************
 34
 35         DATA DIVISION.
 36
 37        **************************
 38
 39         WORKING-STORAGE SECTION.
 40
 41         77  DEPRECIATION-RATE-WS     PICTURE 9V999.
 42         77  PERCENTAGE-LIFE-WS       PICTURE 9V999.
 43
 44         LINKAGE SECTION.
 45
 46         77  DD-BOOK-VALUE-LS         PICTURE 9(6)V99.
 47         77  DD-ANNUAL-DEPR-LS        PICTURE 9(6)V99.
 48         77  FIRST-PASS-SW            PICTURE 9(1).
 49
 50         01  CARDIN.
 51
 52             03  SERIAL-NUMBER        PICTURE 9(8).
 53             03  NAME-OF-ASSET        PICTURE X(20).
 54             03  COST                 PICTURE 9(6)V99.
 55             03  SCRAP-VALUE          PICTURE 9(4)V99.
 56             03  ESTIMATED-LIFE       PICTURE 9(2).
 57             03  LIMIT-I              PICTURE 9(2).
 58             03  FILLER               PICTURE X(34).
 59
 60         PROCEDURE DIVISION USING CARDIN,
 61                                  DD-ANNUAL-DEPR-LS,
 62                                  DD-BOOK-VALUE-LS,
 63                                  FIRST-PASS-SW.
 64
 65         240-PROCESS-DOUBLE-DECLINE.
 66
 67             IF FIRST-PASS-SW IS = 0
 68                 PERFORM 290-CALCULATE-DD-TOTAL-DEPR
 69             ELSE
 70                 NEXT SENTENCE.
 71             PERFORM 300-CALCULATE-DD-ANNUAL-DEPR.
 72         END-OF-SUBROUTINE.
 73             EXIT PROGRAM.
 74
 75         290-CALCULATE-DD-TOTAL-DEPR.
 76
 77             DIVIDE 1.0 BY ESTIMATED-LIFE
 78                 GIVING PERCENTAGE-LIFE-WS.
 79             MULTIPLY PERCENTAGE-LIFE-WS BY 2.0
 80                 GIVING DEPRECIATION-RATE-WS.
 81             MOVE COST TO DD-BOOK-VALUE-LS.
 82
 83         300-CALCULATE-DD-ANNUAL-DEPR.
 84
 85             MULTIPLY DD-BOOK-VALUE-LS BY DEPRECIATION-RATE-WS
 86                 GIVING DD-ANNUAL-DEPR-LS.
 87             SUBTRACT DD-ANNUAL-DEPR-LS FROM DD-BOOK-VALUE-LS.

THERE WERE 87 SOURCE INPUT LINES.
THERE WERE NO DIAGNOSTICS.
```

```
 1   *************************
 2
 3    IDENTIFICATION DIVISION.
 4
 5   *************************
 6
 7    PROGRAM-ID.
 8        SUM-YEARS-DIGITS-SUBROUTINE.
 9    AUTHOR.
10        STEVE GONSOSKI.
11    INSTALLATION.
12        WEST LOS ANGELES COLLEGE.
13    DATE-WRITTEN.
14        3 FEBRUARY 1982.
15
16   *PURPOSE.
17   *    THIS SUBROUTINE CALCULATES THE DEPRECIATION OF AN ASSET
18   *    USING THE SUM-OF-YEARS-DIGITS METHOD.
19
20   *************************
21
22    ENVIRONMENT DIVISION.
23
24   *************************
25
26    CONFIGURATION SECTION.
27
28    SOURCE-COMPUTER.
29        LEVEL-66-ASCII.
30    OBJECT-COMPUTER.
31        LEVEL-66-ASCII.
32
33   *************************
34
35    DATA DIVISION.
36
37   *************************
38
39    WORKING-STORAGE SECTION.
40
41    77  SUM-YEARS-WS              PICTURE 9(2)V99.
42    77  YEARS-DEPR-WS             PICTURE 9(2)V99.
43    77  DEPRECIATION-VALUE-WS     PICTURE 9(6)V99.
44
45    LINKAGE SECTION.
46
47    77  SYD-ANNUAL-DEPR-LS        PICTURE 9(6)V99.
48    77  SYD-BOOK-VALUE-LS         PICTURE 9(6)V99.
49    77  FIRST-PASS-SW             PICTURE 9(1).
50
51    01  CARDIN.
52
53        03  SERIAL-NUMBER         PICTURE 9(8).
54        03  NAME-OF-ASSET         PICTURE X(20).
55        03  COST                  PICTURE 9(6)V99.
56        03  SCRAP-VALUE           PICTURE 9(4)V99.
57        03  ESTIMATED-LIFE        PICTURE 9(2).
58        03  LIMIT-I               PICTURE 9(2).
59        03  FILLER                PICTURE X(34).
60
61    PROCEDURE DIVISION USING CARDIN,
62                             SYD-ANNUAL-DEPR-LS,
63                             SYD-BOOK-VALUE-LS,
64                             FIRST-PASS-SW.
65
66    250-PROCESS-SUM-YEARS-DIGIT.
67
68        IF FIRST-PASS-SW IS = 0
69            PERFORM 310-CALCULATE-SYD-TOTAL-DEPR
70        ELSE
71            NEXT SENTENCE.
72        PERFORM 320-CALCULATE-SYD-ANNUAL-DEPR.
73    END-OF-SUBROUTINE.
74        EXIT PROGRAM.
75
76    310-CALCULATE-SYD-TOTAL-DEPR.
77
78        COMPUTE SUM-YEARS-WS ROUNDED =
79            (ESTIMATED-LIFE * (ESTIMATED-LIFE + 1)) / 2.
80        SUBTRACT SCRAP-VALUE FROM COST
81            GIVING DEPRECIATION-VALUE-WS.
82        MOVE COST TO SYD-BOOK-VALUE-LS.
83        MOVE ESTIMATED-LIFE TO YEARS-DEPR-WS.
84
85    320-CALCULATE-SYD-ANNUAL-DEPR.
86
87        COMPUTE SYD-ANNUAL-DEPR-LS ROUNDED =
88            (YEARS-DEPR-WS / SUM-YEARS-WS)
89            * DEPRECIATION-VALUE-WS.
90        SUBTRACT 1 FROM YEARS-DEPR-WS.
91        SUBTRACT SYD-ANNUAL-DEPR-LS FROM SYD-BOOK-VALUE-LS.

THERE WERE 91 SOURCE INPUT LINES.
THERE WERE NO DIAGNOSTICS.
```

DEPRECIATION SCHEDULE

SERIAL NUMBER 657809 NAME LOADER

COST 6,000.00 SCRAP 1,680.00 YEARS 8 LIMIT 5

****

	DEPRECIATION			BOOK VALUE		
YEAR	ST-LINE	DEC-BAL	SUM-DIGITS	ST-LINE	DEC-BAL	SUM-DIGITS
1	540.00	1,500.00	960.00	5,460.00	4,500.00	5,040.00
2	540.00	1,125.00	840.00	4,920.00	3,375.00	4,200.00
3	540.00	843.75	720.00	4,380.00	2,531.25	3,480.00
4	540.00	632.81	600.00	3,840.00	1,898.44	2,880.00
5	540.00	474.61	480.00	3,300.00	1,423.83	2,400.00

DEPRECIATION SCHEDULE

SERIAL NUMBER 7840795 NAME DISPLAY CASES

COST 3,200.00 SCRAP 500.00 YEARS 10 LIMIT 3

****

	DEPRECIATION			BOOK VALUE		
YEAR	ST-LINE	DEC-BAL	SUM-DIGITS	ST-LINE	DEC-BAL	SUM-DIGITS
1	270.00	640.00	490.91	2,930.00	2,560.00	2,709.09
2	270.00	512.00	441.82	2,660.00	2,048.00	2,267.27
3	270.00	409.60	392.73	2,390.00	1,638.40	1,874.54

DEPRECIATION SCHEDULE

SERIAL NUMBER 14756438 NAME FURNITURE

COST 6,050.00 SCRAP 500.00 YEARS 12 LIMIT 5

****

	DEPRECIATION			BOOK VALUE		
YEAR	ST-LINE	DEC-BAL	SUM-DIGITS	ST-LINE	DEC-BAL	SUM-DIGITS
1	462.50	1,004.30	853.85	5,587.50	5,045.70	5,196.15
2	462.50	837.58	782.69	5,125.00	4,208.12	4,413.46
3	462.50	698.54	711.54	4,662.50	3,509.58	3,701.92
4	462.50	582.59	640.38	4,200.00	2,926.99	3,061.54
5	462.50	485.88	569.23	3,737.50	2,441.11	2,492.31

DEPRECIATION SCHEDULE

SERIAL NUMBER 38926042 NAME AUTO

COST 16,010.00 SCRAP 1,500.00 YEARS 7 LIMIT 7

****

	DEPRECIATION			BOOK VALUE		
YEAR	ST-LINE	DEC-BAL	SUM-DIGITS	ST-LINE	DEC-BAL	SUM-DIGITS
1	2,072.85	4,546.84	3,627.50	13,937.15	11,463.16	12,382.50
2	2,072.85	3,255.53	3,109.29	11,864.30	8,207.63	9,273.21
3	2,072.85	2,330.96	2,591.07	9,791.45	5,876.67	6,682.14
4	2,072.85	1,668.97	2,072.86	7,718.60	4,207.70	4,609.28
5	2,072.85	1,194.98	1,554.64	5,645.75	3,012.72	3,054.64
6	2,072.85	855.61	1,036.43	3,572.90	2,157.11	2,018.21
7	2,072.85	612.61	518.21	1,500.05	1,544.50	1,500.00

Questions for Review

1. Why is a Declaratives Section written? Give the steps necessary to write this section.
2. What is the primary function of the USE statement? List the types of procedures associated with the USE statement.
3. Where does the USE statement appear in the Declaratives Section?
4. Under what conditions is the continued processing of a file containing errors permitted?
5. What does the Report Writer declarative specify?
6. What is the function of the Linkage Section?
7. Briefly describe the concept of interprogram communication.
8. How is control transferred from program to program?
9. Explain the operation of the CALL and EXIT PROGRAM statements.
10. Briefly describe the interprogram communication.
11. What is the purpose of the CALL statement and USING option?
12. How is linkage satisfied in calling and called programs?
13. Explain the operation of the EXIT PROGRAM statement in a called program.
14. Explain the operation of calling and called programs.
15. What is the function of the ENTER statement and how is it used?

Matching Questions

Match each item with its proper description.

_____ 1. USING
_____ 2. Declaratives
_____ 3. Called program
_____ 4. USE BEFORE REPORTING
_____ 5. ENTER
_____ 6. Linkage Section
_____ 7. USE AFTER STANDARD EXCEPTION
_____ 8. EXIT PROGRAM
_____ 9. USE
_____10. Calling program
_____11. CALL

A. Specifies procedures for input-output handling that are in addition to the standard procedures by the input-output control system.
B. Identifies conditions under which each Declaratives section is executed.
C. Describes data that is common between programs that communicate with each other in a single run.
D. Marks the logical end of a called program.
E. A program that refers to another program.
F. Provides a means of allowing the use of more than one language in the same program.
G. Specifies any special circumstance under which a procedure is to be executed in the object program.
H. Specifies procedures that are executed just before a report group is produced.
I. Permits communication between the COBOL object program and one or more subprograms or other language subprograms.
J. Makes data items from a calling program available to a called program.
K. A program that is referred to.

Exercises

Multiple Choice: Indicate the best *answer (questions 1-13).*

1. The Declaratives Section
   a. specifies any special circumstance under which a procedure is to be executed.
   b. is written as a subdivision at the beginning of the Procedure Division.
   c. must contain a USE statement.
   d. All of the above.

2. A USE statement
   a. need not be self-contained.
   b. specifies specific procedures to be performed in the declarative.
   c. is executed as any other statement.
   d. All of the above.
3. The following type(s) of procedures are associated with the USE statement.
   a. Input-Output Error Handling Procedure.
   b. Report Writer Procedure.
   c. Items monitored by the associated Debugging Section.
   d. All of the above.
4. The Input-Output Error-Processing declaratives
   a. are never executed unless standard input-output error procedures do not exist.
   b. may not be specified when the INVALID KEY or AT END phrase has not been specified in the input/output statement.
   c. have control returned to the invoking routine after execution.
   d. All of the above.
5. The continued processing of a file is permitted
   a. if an error-processing procedure exists in the declaratives section.
   b. if the system updates all the parameters normally returned by the input/output control system.
   c. if the programmer writes instructions to transfer control·to the error-processing procedures.
   d. None of the above.
6. The Report Writer declarative
   a. is executed after a report group named is produced.
   b. may contain GENERATE, INITIATE, or TERMINATE statements.
   c. must not alter the value of any control data item.
   d. may refer to any nondeclarative procedure.
7. The Linkage Section
   a. describes data that is common between programs that communicate with each other.
   b. describes data that may be referred to only by the calling program.
   c. may contain VALUE clauses for all items.
   d. may contain only noncontiguous data items.
8. The following clauses are required in the Linkage Section data description entry.
   a. Level number 01 or 77.
   b. Data-name.
   c. PICTURE clause.
   d. All of the above.
9. The USING phrase
   a. is present only if the object program contains a Linkage Section.
   b. is present only if the object program is under control of a CALL statement.
   c. appears in the Data Division.
   d. refers to multiple sets of data.
10. The CALL statement
    a. appears in the called program.
    b. permits communication between the object program and one or more subprograms.
    c. contains a literal designating an entry point in the called program.
    d. is written in the USING phrase.
11. The USING option
    a. makes data items from a calling program available to a called program.
    b. may not appear in the called program.
    c. must not enter data-names that are defined in the File Section.
    d. does not need corresponding data-names in sequence.
12. The EXIT PROGRAM statement
    a. marks the logical end of a called program.
    b. causes control to return to the point in the calling program immediately following the CALL statement.
    c. if not under control of a CALL statement, allows control to pass through the exit point to the first sentence of the next paragraph.
    d. All of the above.
13. The ENTER statement
    a. is used to control subroutines.
    b. must contain the statement ENTER COBOL at the beginning of the procedure entered.
    c. allows the use of more than one language in the same program.
    d. specifies the details on how the other languages may be written.

14. The following is the Procedure Division of a program to create a file organization.

```
PROCEDURE DIVISION.
DECLARATIVES.
ERROR-PROCEDURE SECTION. USE AFTER STANDARD ERROR PROCEDURE
    ON D-FILE.
ERROR-ROUTINE.
    DISPLAY ERROR-CONDITION
        UPON CONSOLE.
    IF ERROR = 1
        PERFORM SYNONYM-ROUTINE
    ELSE
        DISPLAY "OTHER STANDARD ERROR " REC-ID
        PERFORM EOJ-PROC.

SYNONYM-ROUTINE.
    IF CC = 84
        AND HD = 9
        DISPLAY "OVERFLOW AREA FULL"
        PERFORM EOJ-PROC
    ELSE
        NEXT SENTENCE.

    IF CC = 84
        ADD 1 TO HD
        PERFORM ADJUST-HD
    ELSE
        NEXT SENTENCE.

    IF HH = 9
        PERFORM END-CYLINDER
    ELSE
        ADD 1 TO HH
        PERFORM WRITES.
END-CYLINDER.
    MOVE 84 TO CC.
    MOVE HD TO HH.
    PERFORM WRITES
ADJUST-HD.
    MOVE HD TO HH.
    PERFORM WRITES.
END DECLARATIVES.

FILE-CREATION SECTION.
INIT.
    OPEN      INPUT     C-FILE
              OUTPUT    D-FILE.
    PERFORM READS
        UNTIL NO-MORE-DATA.
    PERFORM EOJ-PROC.

READS.
    READ C-FILE
        AT END MOVE "NO" TO MORE-DATA-FLAG.
    MOVE CORRESPONDING C-REC TO D-REC.
    MOVE PART-NUM OF C-REC TO REC-ID-SAVE.
    DIVIDE REC-ID-SAVE BY 829 GIVING QUOTIENT
        REMAINDER TRACK-1.
    ADD 10 TO TRACK-1.
    MOVE CYL TO CC.
    MOVE HEAD TO HH.
    DISPLAY TRACK-ID, C-REC, CC, HH.
    WRITE D-REC
        INVALID KEY PERFORM INVALID-KEY-PROC.
```

```
WRITES.
    DISPLAY TRACK-ID, C-REC, CC, HH.
    WRITE D-REC
        INVALID KEY PERFORM INVALID-KEY-PROC.
    PERFORM READS.

INVALID-KEY-PROC.
    DISPLAY "INVALID KEY " REC-ID.
    PERFORM EOJ-PROC.

EOJ-PROC.
    CLOSE   C-FILE
            D-FILE.
    STOP RUN.
```

Explain the function of the Error-Processing declarative.

15.

```
PROCEDURE DIVISION.
DECLARATIVES.
RW-1 SECTION. USE BEFORE REPORTING PAGE-HED.
RW-2.
    IF CBL-CTR IS LESS THAN 1
        OR-EQUAL TO 1
        .
        .
        .
        MOVE 'BEGIN' TO BEGIN-FIELD-FOR-PH
        MOVE SPACES TO CONTINUED-FIELD-FOR-PH
    ELSE
        MOVE SPACES TO BEGIN-FIELD-FOR-PH
        MOVE 'CONTINUED' TO CONTINUED-FIELD-FOR-PH.
RW-3 SECTION. USE BEFORE REPORTING PAGE-FOOT.
RW-4.
    IF CBL-CTR IS GREATER THAN 1
        .
        .
        .
        MOVE MONTHNAME (MONTH) TO MONTH-FOR-PF
        MOVE 'CONTINUED ON NEXT PAGE' TO TEXT-FOR-PF
    ELSE
        MOVE SPACES TO MONTH-FOR-PF
        MOVE SPACES TO TEXT-FOR-PF.
END DECLARATIVES.
```

What is the main purpose of each of the above declarative procedural statements? Give examples to illustrate your point.

16. Write the Report Writer declarative to suppress printing for a footing group (FOOT-LINE) for job numbers 15-20.

17. Given the following information:

```
IDENTIFICATION DIVISION.
PROGRAM-ID. INVENTORY.
    .
    .
    .
DATA DIVISION.
    .
    .
    .
WORKING-STORAGE SECTION.
01  TRANS-RECORD.
    05  QUANTITY      PICTURE 9(5).
    05  UNIT-PRICE    PICTURE 999V99.
    05  AMOUNT        PICTURE 9(5)V99.
    .
    .
    .
```

```
PROCEDURE DIVISION.
       .
       .
       .
       CALL 'INVENTORYSUB' USING TRANS-RECORD.
       .
       .
       .
IDENTIFICATION DIVISION.
PROGRAM-ID. INVENTORYSUB.
       .
       .
       .
DATA DIVISION.
       .
       .
       .
LINKAGE SECTION.
01   INVREC.
     05  UNITS         PICTURE 9(5).
     05  PRICE         PICTURE 999V99.
     05  INV-BALANCE   PICTURE 9(5)V99.
       .
       .
       .
PROCEDURE DIVISION USING INVREC.
       .
       .
       .  EXIT PROGRAM.
```

In the aforementioned programs, explain the processing routines between the calling and called programs. How are the different values handled in each program by what data name? What happens with each execution of the CALL statement?

Problem 1

Job Definition

From the Sales Activity records, a report is to be written listing the following information: salesperson's number, salesperson's name, gross sales, sales returns, net sales, and average sales.

Input

Positions	Field	
1-5	Salesperson number	
6-25	Salesperson name	
26-32	Monday's sales	xxxxx.xx
33-39	Tuesday's sales	xxxxx.xx
40-46	Wednesday's sales	xxxxx.xx
47-53	Thursday's sales	xxxxx.xx
54-60	Friday's sales	xxxxx.xx
61-67	Not used	
68-74	Sales returns	xxxxx.xx
75-80	Not used	

Processing

Gross Sales = Monday, Tuesday, Wednesday, Thursday, Friday sales.
Net Sales = Gross Sales − Sales Returns
Average Sales = Net Sales divided by the number of days worked. (Use subroutine to calculate average sales)

If sales for a particular day are equal to zero, do not count that day in the calculation for the calculation for the average sale. For example, if sales are indicated for Monday through Thursday only, the average sale is determined by dividing net sales by 4.

Output

The Printer Spacing Chart shows how the report is formatted:

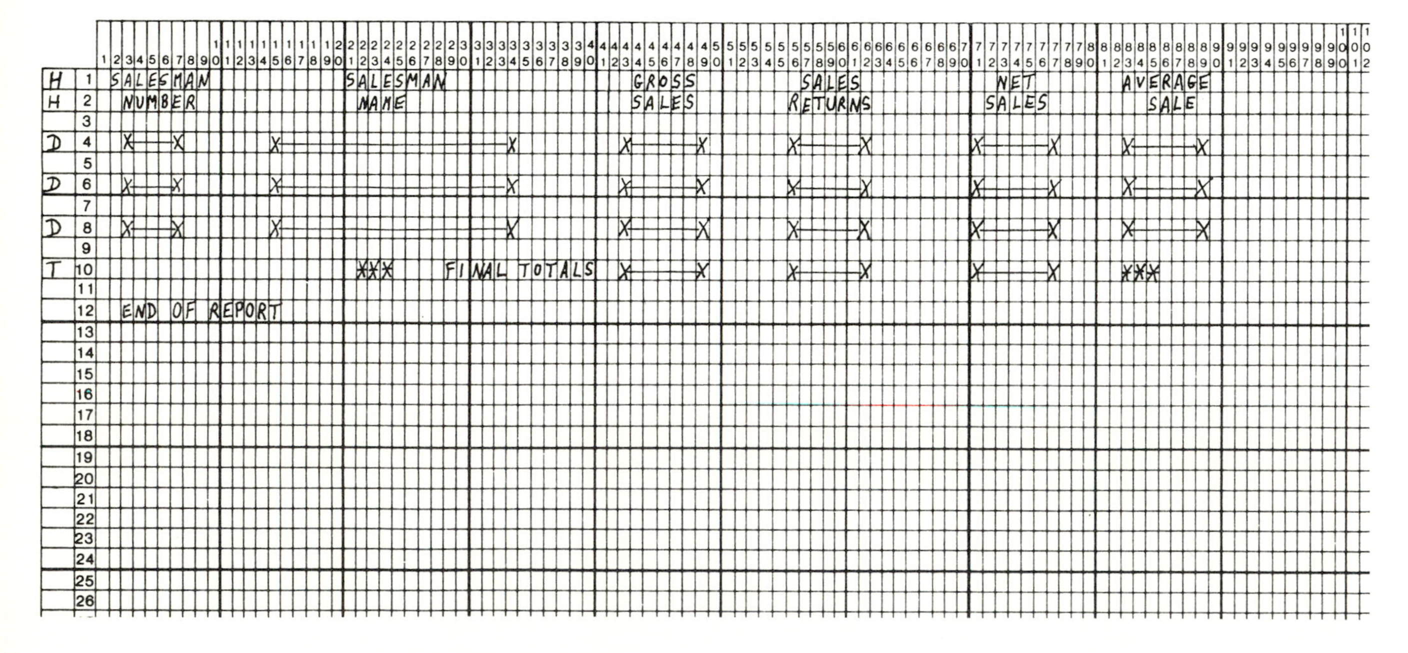

Problem 2

Job Definition

The area in which all salespersons work is divided into three districts: A, B, and C. Some salespersons work in only one district. Others may work in parts of two or more districts.

For each salesperson, the input file contains a record as shown below. The amounts in the district fields show total weekly sales made by that salesperson in each district. If the salesperson did not work a district or made no sales in that district, the field contains blanks.

Input

Field	Positions	Format
Salesperson name	1-25	
District A	26-32	xxxxx.xx
District B	33-39	xxxxx.xx
District C	40-46	xxxxx.xx
Not used	47-80	

Computations to Be Performed

The report must contain the commission earned in each district by each salesperson. In addition, total commissions must be accumulated for each salesperson and for each district. The percentage of commission is as follows (use subroutines to calculate commissions):

3 percent of gross sales .01 to 1,000.00 dollars
Plus 2 percent of the gross sales 1,000.01 to 5,000.00 dollars
Plus 1 percent of the gross sales over 5,000 dollars.

The desired report shows two things:

1. Total commission earned by each salesperson, by district, and total for all districts.
2. Total commissions paid for each district to all salespersons.

Output

COMMISSION REPORT

SALESMAN	DIST A	DIST B	DIST C	TOTAL
WHO DON IT	1.50	.60	.30	2.40
GEORGE DID IT	1.80	2.10	2.40	6.30
ETHIL DID NOT DO IT	3.00	6.00	9.00	18.00
BUT BERT IS CAPABLE	24.00	21.00	18.00	63.00
HOW ABOUT HUBERT	15.00	12.00	6.00	33.00
CARL SOMETIMES	27.00	34.00	120.00	181.00
MAX CAN	40.00	135.00	25.50	200.50
LOU KNOWS IT	165.00	15.96	60.50	241.46
JOHN & MARY		70.00	130.00	200.00
HOPE & CROSBY			42.00	42.00
SNOOPY				.00
MARSHALL		112.00	130.00	242.00
SAM & LOU	78.00	25.80		103.80
	355.30 *	434.46 *	543.70 *	1,333.46**

Problem 3

Job Definition

The program reads the data for each state and stores in particular arrays. The array information is transferred to the subroutine, which calculates averages and sends the results back to the main program. A different array information is sent to the subroutine for each calculation.

Input

Field	Positions
State Name	1-16
Not Used	17-20
Population	21-28
Not Used	29-30
Size	31-36
Not Used	37-40
Road Mileage	41-46
Not Used	47-80

Computations to be Performed

The program is to be designed to calculate the average of a set of numbers. The calculation is to be written into a subroutine. The main program reads in the specific data, calls the subroutine, and prints out the results. The data—population, size, and road mileage of each of the fifty states—is to be read in, and the average population, size and road mileage calculated.

The Printer Spacing Chart shows how the report is formatted:

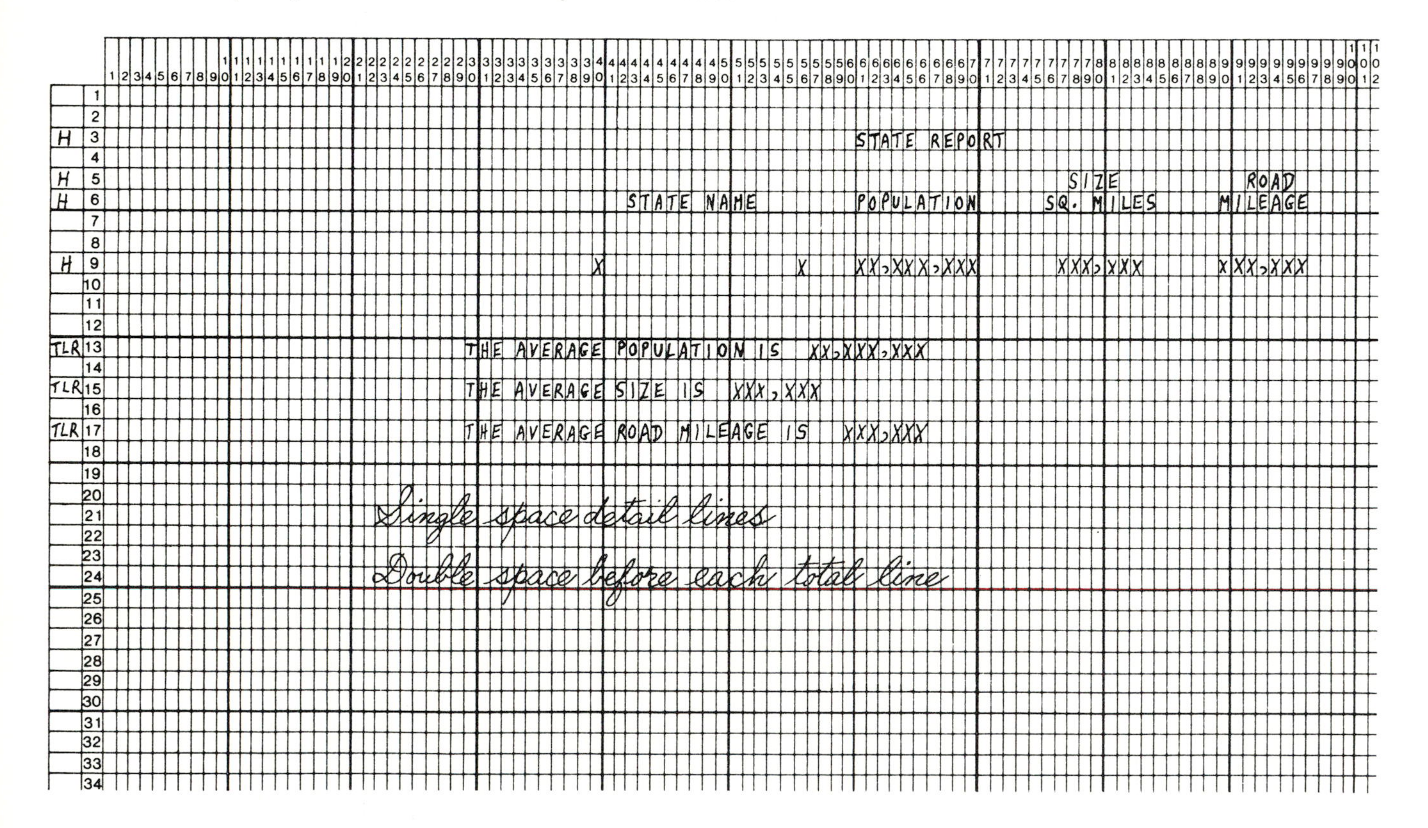

Output

STATE REPORT

STATE NAME	POPULATION	SIZE SQ. MILES	ROAD MILEAGE
ALABAMA	3,444,185	51,609	85,845
ALASKA	302,173	586,412	9,043
ARIZONA	1,772,482	113,909	51,415
ARKANSAS	1,923,295	53,104	78,088
CALIFORNIA	19,953,134	158,693	169,564
COLORADO	2,207,259	104,247	83,586
CONNECTICUT	3,032,217	5,009	13,734
DELAWARE	548,104	2,057	5,150
FLORIDA	6,789,443	58,560	98,129
GEORGIA	4,589,575	58,876	100,335
HAWAII	769,913	6,450	3,666
IDAHO	713,008	83,557	55,910
ILLINOIS	11,113,976	56,540	130,494
INDIANA	5,193,689	36,291	91,111
IOWA	2,825,041	56,290	112,944
KANSAS	2,249,071	82,264	134,770
KENTUCKY	3,219,311	40,395	69,791
LOUISIANA	3,643,180	48,523	54,124
MAINE	993,663	33,215	21,499
MARYLAND	3,922,399	10,577	26,859
MASSACHUSETTS	5,689,170	8,257	29,811
MICHIGAN	8,875,083	58,216	118,310
MINNESOTA	3,805,069	84,068	128,235
MISSISSIPPI	2,216,912	47,716	66,686
MISSOURI	4,677,399	69,686	114,966
MONTANA	694,409	147,138	77,932
NEBRASKA	1,483,791	77,227	98,017
NEVADA	488,738	110,540	49,659
NEW HAMPSHIRE	737,681	9,304	15,024
NEW JERSEY	7,168,164	7,836	32,422
NEW MEXICO	1,016,000	121,666	70,307
NEW YORK	18,241,266	49,576	107,776
NORTH CAROLINA	5,082,059	52,586	87,922
NORTH DAKOTA	617,761	70,665	106,247
OHIO	10,652,017	41,222	109,965
OKLAHOMA	2,559,253	69,919	108,509
OREGON	2,091,385	96,981	101,397
PENNSYLVANIA	11,793,909	45,333	114,497
RHODE ISLAND	949,723	1,214	5,540
SOUTH CAROLINA	2,509,518	31,055	60,295
SOUTH DAKOTA	666,257	77,047	82,720
TENNESSEE	3,924,164	42,244	80,656
TEXAS	11,196,780	267,338	251,489
UTAH	1,059,273	84,916	47,653
VERMONT	444,732	9,609	13,924
VIRGINIA	4,648,494	40,817	62,351
WASHINGTON	3,409,169	68,192	81,202
WEST VIRGINIA	1,744,237	24,181	36,323
WISCONSIN	4,417,933	56,154	104,290
WYOMING	332,416	97,914	40,602

THE AVERAGE POPULATION IS 4,047,956

THE AVERAGE SIZE IS 72,303

THE AVERAGE ROAD MILEAGE IS 76,015

Problem 4

Each day the current charges are merged with the accumulated charges in preparation for a monthly billing. The fields to be merged are telephone numbers with area codes.

The format for both files is as follows:

Positions	Field
1-3	Area code
5-11	Telephone number
15-35	Name
36-42	Charges
43-80	Other information

The input files are to be merged. Both input files are on magnetic tape and the output is to be put back on output tape. In addition, a report is to be printed as follows:

```
AREA CODE   TELEPHONE NUMBER   NAME   CHARGES
```

COBOL Programming Techniques

Chapter Outline

Chapter Objectives

The learning objectives of this chapter are:

1. To describe the use of various programming techniques that can increase the efficiency of the Environment, Data, and Procedure divisions.
2. To describe methods for improving the appearance of COBOL programs.
3. To explain the purpose of hierarchy charts and how they may be used for coding Procedure Division entries.
4. To describe the program technique of Conventional-Linkage and how it is applied to COBOL programming.

Introduction

A programmer can increase the efficiency of a COBOL program by having some knowledge of the processor on which the program is to run. Some machines, for example, must use library subroutines for the subdivision of large numbers while others do not. Some have limited accuracy in expressions used in conditional statements. The following discussion describes how a program may be made more efficient on a computer.

The writing of a computer program is a difficult and tedious task. Many times a program rapidly solves a particular problem without any regard for the efficiency of the program itself. In the writing of COBOL programs, this problem is more serious than in other programming languages. As mentioned earlier, COBOL does not produce a program as efficient as one written in the basic language of the particular computer. Thus it is imperative that the programmer adopt techniques that will increase the efficiency of the COBOL program.

Prior to the writing of the program, the problem should be properly defined, and the appropriate flowcharts, source document formats, and output formats should be available to the programmer.

Programming Standards

COBOL has extensive features that provide the capability for a source program to be highly self-documenting. These features make it possible for a new programmer to assume maintenance of a COBOL program with little or no prior training by the originating programmer. Yet, if an established set of standards is not placed in use, programmers will establish their own. Nonstandard methods of programming, even in COBOL, complicate program maintenance for a newly-assigned programmer.

One of the major objectives in the original design of COBOL was to have a COBOL program that would be essentially self-documenting. Features of the language tending to promote this objective are the compartmentalizing of the program into divisions, the requirement of an Identification Division, many of the options in the Environment Division, and particularly the use of a subset of the English language in the Procedure Division.

However, in the haste of getting a program "on the air," programmers may choose ways of writing their programs, which would make the later debugging and maintenance of those programs unnecessarily difficult. Generally speaking, the same program could have been written without any further expenditure of energy in a somewhat different manner that would aid rather than hinder future maintenance. The key to easier debugging and maintenance is the establishment, understanding, and enforcement of a small set of programming standards.

Source Program Appearance

In order to improve the "English" readability of a COBOL source program, many things can be done to present an organized appearance. For example, COBOL completely ignores input records that are entirely blank. Blank records can thus be interspersed within a program to set off paragraphs, sections, etc. A record containing all asterisks (*) in positions 7 through 72 can be used to divide portions of the source program into unmistakable divisions.

Source Program Conventions

Potentially difficult-to-read statements should be simplified. Improved readability and program logic understanding can result if the following suggestions are followed.

Use Indentation to Indicate Subordination

Those rules of COBOL that require certain items to begin in Area A do not dictate that these items necessarily begin in position 8. Such items might appear beginning in position 8, or in position 9, 10, or 11. Furthermore, language elements that must appear in Area B may begin in position 12 or any column thereafter. Therefore, the programmer is free to do the following:

1. Begin any division header in position 8, preceded by four blank lines, and followed by two blank lines.
2. Begin any section header in position 9, followed by one blank line.

3. Begin in position 10, preceded by one blank line, any of the following:
   a. a paragraph name in the Identification, Environment, or Procedure Division, or
   b. an FD, 01-, or 77-level entry in the Data Division, with not more than one of these items in a single record.
4. In the Data Division, begin any entry which is immediately subordinate to another entry two spaces to the right of that entry.
5. In the Procedure Division, whenever a statement is conditioned by an "IF," indent the conditioned statement to at least position 16, in order to show its dependency on the condition expressed in the "IF."
6. In the Procedure Division, start each new sentence in position 12. Start each subsequent line of the same sentence in position 16. (Avoid compound sentences.)

Start Each New Statement on a New Line

This suggestion is a requirement for division headers, section headers, and Data Division entries. It is not required, but it is suggested for the following items:

Paragraph-names. Place each paragraph-name on a line by itself. Any later rewrite or rearrangement of the contents of the paragraph need not disturb the line containing the paragraph-name.

Procedure Division verbs. Write each statement beginning with a verb on a line by itself. Any change made to that statement requires the modification of only that source program part.

Conditional statements. Write an IF condition on a line by itself, indenting all statements conditioned by it.

Make Each Nonconditioned Procedure Division Statement a Separate Sentence

It is not possible to make each Procedure Division statement a separate sentence where a series of Procedure Division statements are conditioned by an IF. In this case, of course, only the last such statement is followed by a period. Reading is made easier by the fact that any statements not followed by a period are conditioned statements.

Indent When Using GO TO with the DEPENDING ON Option

When using the DEPENDING ON option, indent each paragraph-name from the GO TO, putting each on a separate line. As an example:

```
12
GO TO
    PXXXX-ERROR-RTE
    PXXXX-VALID-TRX-RTE
    PXXXX-EXCEPTION-PRLS
    DEPENDING ON PROCESS-SWITCH.
```

Do Not Split Words or Numeric Literals Between Lines

A nonnumeric literal may be as much as 120 characters in length. It might, therefore, be split between two or maybe even three lines. On the other hand, a COBOL key word, a programmer-defined name, or a numeric literal need never be split between two lines. If a word cannot fit on a given line, begin it on the next line. Although the capability of splitting words and numeric literals exists in COBOL, the use of this option by the programmer tends to make the program difficult to read and maintain.

Use Blank Lines Freely to Improve Readability

A blank line may appear anywhere in a COBOL program, except between two lines across which a word or literal is split. In other words, a blank may appear anywhere except immediately preceding a continuation line. The use of blank lines makes any search for the major routines within a program much easier.

Use Section Names to Improve Understanding of Program Organization

The use of section identification is recommended. By giving two or more successive paragraphs a section-name, those paragraphs can be executed by a PERFORM statement that need not have a THRU clause.

Section identification in a large source module also provides clearer organization of COBOL routines.

The first record of each section should contain comments. If positions 7–9 and 70–72 of the source records containing comments are keyed with asterisks, such records are not only treated as comments due to the asterisk in position 7, but they also stand out clearly on the source listings, thus highlighting the comments contained between positions 11 through 68.

Section-names should be descriptive of the routine they identify. Consider the following example (remember, if sections are used then the entire Procedure Division must consist of sections):

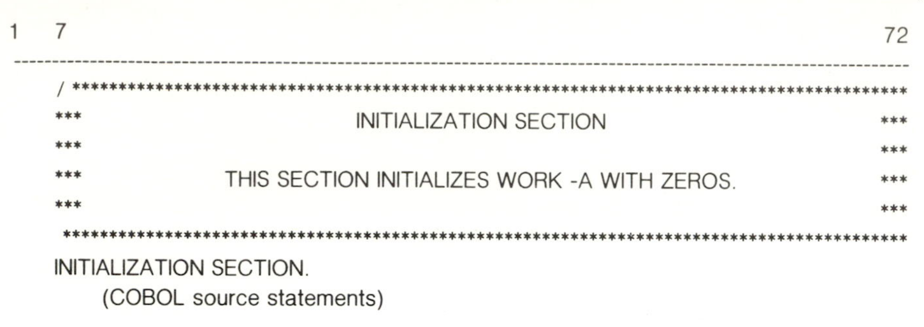

```
1     7                                                                              72
-----------------------------------------------------------------------------------------
      /******************************************************************************
      ***                     INITIALIZATION SECTION                                ***
      ***                                                                           ***
      ***          THIS SECTION INITIALIZES WORK -A WITH ZEROS.                     ***
      ***                                                                           ***
       ******************************************************************************
      INITIALIZATION SECTION.
          (COBOL source statements)
```

Use Comments Freely for Explanation

When a program is being written, by no means should all that is in the programmer's mind concerning the program being solved actually go into the code. In order to facilitate understanding of a program's logic, it may be necessary to include many comments along with pure program text. For example, blocks of comments should precede major sections of the program. Comments records identified by an * symbol in position 7 can provide such text comprehension. Such a line appears on the compilation listing, but serves no purpose other than comments.

Comment statements identified by a / symbol in position 7 provide for page ejection in the source listing prior to printing the comment. This type of comment should be used to clearly identify major sections of the program. *Note:* Many compilers do not support ejection on a source program listing. To cause ejection on these computers, the word "eject" is keyed on a record before the line that is to appear at the top of the new page is required.

While COBOL is designed to be as self-documenting as possible, often the logic used by a programmer may not be apparent to someone else reading the program. In such a case (and such cases are far more prevalent than programmers realize), comment lines are invaluable. For added readability, it is recommended that whenever a comment line is to be included, three comment lines be included as well. The first and third lines are blank, except for the asterisk in position 7, and the desired comments. Actually, the blank lines need not contain an asterisk in position 7.

The writing of the COBOL program can be simplified and the efficiency increased if the following programming techniques are applied. All four divisions of the COBOL program must be completed before the source program can be compiled and executed.

Identification Division

This is the simplest division of the four. It contains the information that identifies the program and is intended to provide information to the reader of the program. The name of the program must be stated, and other information about the program should be mentioned.

Entries That Should Appear in the Identification Division

1. The name of the division-Division Header.
2. The name of the program.
3. The name of the programmer.
4. When the program was written.
5. Comments that will explain the data processing job from which the program was written. An asterisk (*) in position 7 of the coding form will indicate that the purpose paragraph is treated as a comment.
6. Any other optional information.

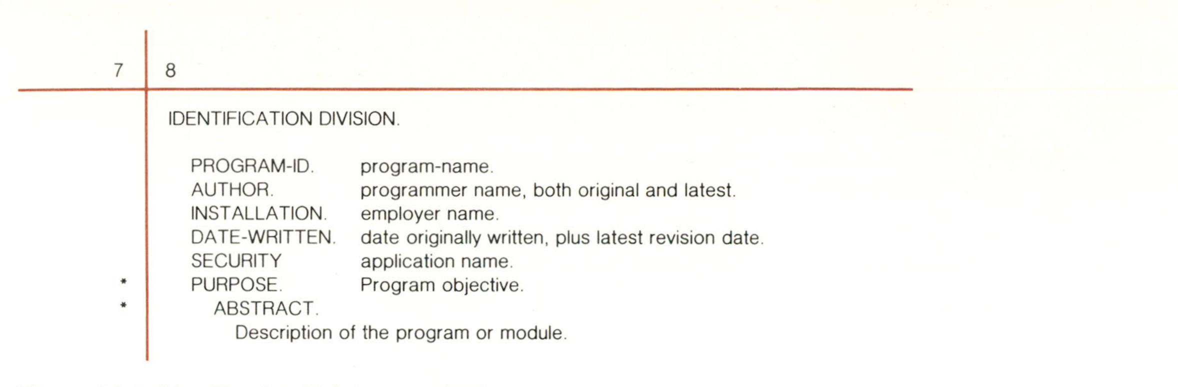

```
 7 | 8
   | IDENTIFICATION DIVISION.
   |
   |    PROGRAM-ID.      program-name.
   |    AUTHOR.          programmer name, both original and latest.
   |    INSTALLATION.    employer name.
   |    DATE-WRITTEN.    date originally written, plus latest revision date.
   |    SECURITY         application name.
 * |    PURPOSE.         Program objective.
 * |      ABSTRACT.
   |        Description of the program or module.
```

Figure 16.1 Identification Division standards.

Only items (1) and (2) are necessary for the proper execution of the program. Additional information is desirable for adequate documentation of the program. The reader of the program would be interested in the purpose statement where the intent of the program is mentioned. The pertinent information contained therein should provide the reader with a better understanding of the program. Therefore, it is essential that the purpose statement be as thorough as possible.

The standards shown in figure 16.1 are suggested for uniformity within an installation (see also figure 16.2).

Environment Division

The Environment Division is the only division in the COBOL program that is machine oriented. This division contains information about the equipment to be used when the object program is compiled and executed. Most importantly, it links the devices of the computer system and the data files to be processed.

In this division of a COBOL program, somewhat more flexibility is required from program to program than is required for the Identification Division. And yet, since the hardware does not change between programs, and there are certain common usages of input/output devices for a given programmer, the coding can be simplified, and reading of source program listings can be expedited, when standard entries are devised.

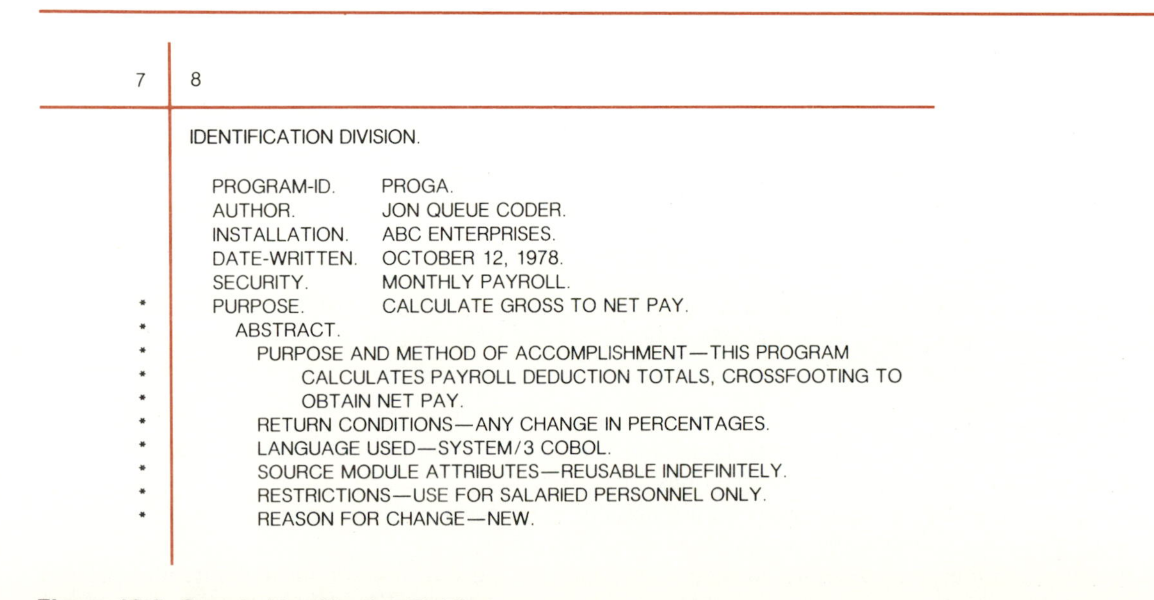

```
 7 | 8
   | IDENTIFICATION DIVISION.
   |
   |    PROGRAM-ID.      PROGA.
   |    AUTHOR.          JON QUEUE CODER.
   |    INSTALLATION.    ABC ENTERPRISES.
   |    DATE-WRITTEN.    OCTOBER 12, 1978.
   |    SECURITY.        MONTHLY PAYROLL.
 * |    PURPOSE.         CALCULATE GROSS TO NET PAY.
 * |      ABSTRACT.
 * |        PURPOSE AND METHOD OF ACCOMPLISHMENT—THIS PROGRAM
 * |            CALCULATES PAYROLL DEDUCTION TOTALS, CROSSFOOTING TO
 * |            OBTAIN NET PAY.
 * |        RETURN CONDITIONS—ANY CHANGE IN PERCENTAGES.
 * |        LANGUAGE USED—SYSTEM/3 COBOL.
 * |        SOURCE MODULE ATTRIBUTES—REUSABLE INDEFINITELY.
 * |        RESTRICTIONS—USE FOR SALARIED PERSONNEL ONLY.
 * |        REASON FOR CHANGE—NEW.
```

Figure 16.2 Sample Identification Division.

Advantages of Using the COBOL Source Library for Environment Division Coding

With the exception of the File-Control paragraph, the entire contents of the Environment Division can be standardized for all programs. For this reason it is strongly suggested that once such standard text has been devised, it be cataloged into the COBOL source language library. File-Control entries for each file to be processed can then be hand-coded, or these too can be included from separate source modules cataloged for each standard file.

Entries That Should Appear in the Environment Division

1. The name of the division-Division Header.
2. Configuration Section—Source and Object Computer Paragraphs.
3. Input-Output Section—File-Control and I-O-Control Paragraphs.

The external device names in the File-Control paragraph should be checked with the system programmers, as the names will vary with each data processing unit.

All files mentioned in the File-Control paragraph should be properly defined in the Data Division, and opened and closed in the Procedure Division.

Any special input/output techniques should be defined in the I-O-Control paragraph.

The following is the general structure of the Environment Division:

```
ENVIRONMENT DIVISION.
CONFIGURATION SECTION.
SOURCE-COMPUTER paragraph
OJBECT-COMPUTER paragraph
[SPECIAL-NAMES paragraph]
[INPUT-OUTPUT SECTION.
FILE-CONTROL paragraph
[I-O-CONTROL paragraph]]
```

Data Division

The Data Division describes the information to be processed by the object program. Each file mentioned in the Environment Division must be described in the Data Division. In addition, each data item within these files must be described. All data items that comprise the Working-Storage Section, such as constants and work areas, must also be described.

The Data Division should be written using copies of the source document formats and output formats as guides for writing the file description entries and the record description entries.

Unlike the Procedure Division, which is substantially different for each program, the Data Division offers opportunities for work-saving standardization. Most computer installations operate with a fixed number of permanent files. The Data Division entries for each such file can be cataloged into the COBOL source language library on disk, and included in each program requiring use of such file(s) via the COPY statement. By this means, data-names assigned to each field can be standardized in all programs. Updates to file or record descriptions can, in many cases, be accomplished by merely recataloging the revised source module, and recompiling all programs making use of this module.

Include Common Data Definitions with COPY Statements

The use of the COPY statement in data definition is a very powerful COBOL feature. The COPY statement permits the programmer to include in the source program at compile time prewritten Data Division or Environment Division entries from the COBOL source language library. There are many advantages in this approach:

1. Standardization of data definitions is achieved—every programmer uses identical file definitions with common data-names and descriptions.
2. Programming effort is minimized because only file description needs to be written, keyed, and debugged only once, and is then available to all programmers.
3. Program maintenance becomes considerably easier. Because of a common understanding of the data definitions, the programmer can pick up and more quickly modify programs written by others.

In addition, a modification of the data definition itself can be made once and recataloged. Then all affected modules can be recompiled, thus minimizing direct programming effort.

4. The library member becomes a final authority on file contents. Consider including comments for each field (group and elementary) containing a complete description of the usage, coding values or valid ranges, variance and inter-field relationships, etc. References may be made to other documents when further clarification is needed.

Use Meaningful Names

One feature of COBOL is its capability for using programmer-defined names of as long as thirty characters. While a name this long would seldom be used, the freedom to have names longer than five or eight or ten characters provides the programmer with the opportunity of defining names that are meaningful with regard to the items referenced by those names. Failure to take advantage of this capability may prevent a reader of the program from understanding the data involved. Experience shows that most successful naming conventions result in names of eight to twelve characters. Obviously this can vary depending upon the nature of the item to be named, but the general principle still applies.

All programs of any complexity eventually require maintenance. This task usually falls to the original programmer, if still available. While at the time of the original writing all programs may be clear, after the passage of time usually little is recalled. As a result, a programmer should consider the use of self-documenting names and frequent comment lines describing what is being accomplished. It is this very technique that enables one to perform successful maintenance with a minimum of effort.

In reading programs written by experienced, competent programmers, the repetitious use of the hyphen symbol in COBOL names is apparent. Not only does the hyphen enable construction of compound names, but it also ensures against the inadvertent use of a COBOL-reserved word.

Using the data-names below,

```
UNITS-I
UNITS-W
UNITS-O
```

the programmer can quickly locate the data-names in the Data Division when debugging the program. Also, since most entries require the transfer of input to output, the programmer can easily identify the items in the MOVE statement. (See suffixes in this chapter.)

One technique that reduces differences in program code produced by more than one programmer is the establishment of a list of standard abbreviations. For example, once standardized within an installation, the abbreviation "MAST" for "MASTER" is clear to everyone.

Entries That Should Appear in the Data Division

1. The name of the division-Division Header.
2. File Section-file description entries and record description entries.
3. Working-Storage Section record description entries for constants and work areas.
4. Linkage Section: record description entries used for subprograms.
5. Report Section: report description and report group description entries.

The following is the general structure of the Data Division:

```
DATA DIVISION.
FILE SECTION.
{file description entry
{record description entry} . . . } . . .
WORKING-STORAGE SECTION.
[data item description entry] . . .
[record description entry] . . .
LINKAGE SECTION.
[data item description entry] . . .
[record description entry] . . .
REPORT SECTION.
{report description entry
{report group description entry} . . . } . . .
```

File Section
The File Section specifies the characteristics of the file.

File Description Entries

The *FD* File-name is required and must agree with the name specified in the Environment Division.

The *RECORDING MODE* clause is optional but should be included if records are fixed, variable or undefined; otherwise, the compiler will generate an algorithm that does not always give V. This clause is required if the computer has more than one recording mode.

The *BLOCK CONTAINS* clause must be included when the records are blocked (e.g., when records are blocked on a tape). If the records vary in size, the character option should be used instead of records to specify the total number of characters in each block. When there is only one record per block, the clause may be omitted.

The *RECORD CONTAINS* clause is used when variable size records are used. This clause should specify the length of the shortest and longest record in the file. This clause may be omitted since the compiler determines the record size from the record description entries.

A good programming practice is to include this clause in every FD entry for the following reasons:

1. The compiler will check the agreement of the record count in the record description entry with the RECORD CONTAINS clause in the FD entry. This will assure that no data fields were erroneously omitted in the record description entry. Otherwise, the compiler will assume the count in the record description entries as being the correct record length.
2. It provides the programmer and reader of the program with the size of the record without the necessity of counting all field lengths stated in the record description entries.

The *LABEL RECORDS* clause must be included in every record description entry even if the files are located in cards or to be output on the printer where the OMITTED option is used.

The *DATA RECORDS* clause is optional, and each record-name should be written in sequence as it appears in the record description entries. The record description entries must appear immediately after the file description entries.

Record Description Entries
Level Numbers are required for each entry. Each level is a given number, always beginning with 01 for the data record itself. Each succeeding level is given a larger number to indicate a further breakdown of the data item. These larger numbers need not be in sequence. This may provide the programmer with more flexibility in assigning numbers. All level numbers may be written at the A margin, although only the 01- and 77-level numbers are required to be written at the A margin. However, indentation should be used, as this improves the readability of the program.

Data-names may be unique or otherwise qualified. The highest qualifier must be a unique name. In the File Section, the highest qualifier is the file-name; it is thus possible for two records to have the same name.

Names of independent items in the Working-Storage Section must be unique since they cannot be qualified. The highest qualifier in the Working-Storage Section is the record-name.

A recommended programming procedure for writing Data Division entries for output files for the printer is to describe the formats of the output file in the Working-Storage Section and to use the WRITE verb with the FROM option referring to the Working-Storage item. This technique provides the programmer with

1. The ability to define the heading with appropriate VALUE clauses (forbidden in the File Section) in the Working-Storage Section. Areas in the output can be blanked where necessary.
2. The ability to use one record description entry to define the output for headings, detail, and total lines.

The PICTURE clause tells the number of characters to be stored and what type they will be. PICTURE clauses are only found in the descriptions of elementary items. PICTURE and USAGE clauses must be compatible. For example, an alphabetic item cannot have a USAGE clause of COMPUTATIONAL.

The VALUE clause is used to assign initial values. The VALUE clause is not permitted in the File Section, except with level-88 (condition-name) entries. The VALUE clause must agree with its picture. For example,

```
77  DISCOUNT                  PICTURE SV99     VALUE +.02.
77  TOTAL-IDENTIFICATION      PICTURE A(15)
    VALUE 'PAY THIS AMOUNT'.
```

The USAGE clause is allowed at both the group and elementary levels. If this clause is omitted, the item's usage is assumed to be DISPLAY.

If a data item field is to be used in a series of arithmetic operations, it would be advisable to define an area in the Working-Storage Section in the computational mode. The data item would be moved from the input area to the Working-Storage area. After all the computations are completed, the computed item would be returned to the output area in the display mode.

This technique would save processing time of the compiler in changing the item from display to computational mode and back to display mode for each arithmetic operation.

Suffixes

To make it easier for programmers to locate items in a program listing, especially during the debugging stage, it is good practice to attach a suffix to a data-name to indicate where the item is to be found in the program. For example, if a QUANTITY item appears in the input and output records as well as in a Working-Storage area, the data-names may be assigned as follows:

Input item —QUANTITY-IN
Output item —QUANTITY-OUT
Working-Storage item—QUANTITY-WS

The same principle may be applied to all items that appear in these areas in order to make it simpler for the reader to know which fields are logically part of the same record or area.

MOVE CORRESPONDING Statement

The MOVE CORRESPONDING statement can save time in writing many MOVE statements of identical items, usually from an input to an output record. However, this technique involves qualification whenever an identical item is involved in a Procedure Division statement. To eliminate excessive qualifying, a REDEFINES or RENAMES statement may be used with the corresponding items. For example,

```
01  PAY-RECORD-IN.
    05  SAME-NAMES. (**)
        10  SAME-LAST-NAME       PICTURE . . .
        10  SAME-FIRST-NAME      PICTURE . . .
        10  SAME-PAYROLL         PICTURE . . .
    05  DIFF-NAMES REDEFINES SAME-NAMES.
        10  DIFF-LAST-NAME       PICTURE . . .
        10  DIFF-FIRST-NAME      PICTURE . . .
        10  DIFF-PAYROLL         PICTURE . . .
        .
        .
        .
01  PAY-RECORD-OUT.
    05  SAME-NAMES. (**)
        10  SAME-PAYROLL         PICTURE . . .
        10  FILLER               PICTURE . . .
        10  SAME-FIRST-NAME      PICTURE . . .
        10  FILLER               PICTURE . . .
        10  SAME-LAST-NAME       PICTURE . . .
        .
        .
        .
```

```
PROCEDURE DIVISION.
          .
          .
          .
     IF DIFF-PAYROLL IS EQUAL TO PAYROLL-WS
          AND DIFF-LAST-NAME IS NOT EQUAL TO LAST-NAME
          MOVE CORRESPONDING PAY-RECORD-IN TO PAY-RECORD-OUT
     ELSE
          NEXT SENTENCE.
          .
          .
          .
```

Note: Fields marked with a double asterisk (**) in the above must have exactly the same names for their subordinate fields in order to be considered corresponding. The same names must not be the redefining ones, or they will not be regarded as corresponding.

Level Numbers

The programmer should use widely incremented level numbers (i.e., 01, 05, 10, 15, etc., instead of 01, 02, 03, 04, etc.) in order to allow room for future insertions of group levels. For readability, indent level numbers. Use level-88 numbers for codes. Then, if the codes must be changed, the Procedure Division coding for these tests need not be changed.

Increasing the Efficiency of Data Division Entries

When writing Data Division statements, the COBOL programmer need not be concerned with data problems such as decimal alignment and mixed format; the compiler generates extra instructions to perform the necessary adjustments. Entries in the Data Division can significantly affect the amount of storage required by a program.

Conserving Storage-Decimal Alignment

Procedure Division operations are most efficient when the decimal positions of the data items involved are aligned. If they are not, the compiler generates instructions to align the decimal positions before any operations involving the data items can be executed.

In a typical source program, the frequency of the most common verbs written in the Procedure Division of a COBOL program averaged over a number of programs is:

MOVEs—50%
GO TO —20%
IF —15%

Miscellaneous (arithmetic, calculations, input/output)—15%.

An example of a pair of fields follows:

```
77  SENDFIELD     PICTURE 99V9       COMPUTATIONAL-3 (Sending Field).
77  RECFIELD      PICTURE 999V99     COMPUTATIONAL-3 (Receiving Field).
        MOVE SENDFIELD      TO RECFIELD.
```

Because the receiving field is one decimal position larger than the sending field, decimal alignment must be performed. Each time the move is executed, 2,250 bytes of storage are used. Adding one additional decimal position in the data sending or receiving field is small in cost compared to the savings possible in the Procedure Division.

Unequal-Length Fields

An intermediate operation may be required when handling fields of unequal length. For example, zeros may have to be inserted in numeric fields and blanks in the alphabetic or alphanumeric fields in order to pad out to the proper length. The compiler will have to generate instructions to perform these insertions.

To avoid these operations, the number of digits should be equal. Any increase in data fields is more than compensated for by the savings in the generated object program. For example,

```
SENDFIELD    PICTURE S999.
RECFIELD     PICTURE S99999.
Change SENDFIELD to PICTURE S99999.
```

Mixed Data Formats

When fields are used together in move, arithmetic, or relational statements, they should be in the same format whenever possible. Conversions require additional storage and longer execution time. Operations involving mixed data formats require one of the items to be converted to a matching data format before the operation is executed.

For maximum efficiency, avoid mixed data formats or use a one-time conversion; that is, move the data to a work area, thus converting it to the matching format. By referencing the work area in procedural statements, the data is converted only once instead of for each operation.

The following examples show what must logically be done before indicated operations can be performed when working with mixed data fields.

Display to Computational-3

To execute a move. No additional code is required (if proper alignment exists) because one instruction can both move and convert the data.

To execute a compare. Before a compare is executed, display data must be converted to the computational-3 format.

To perform arithmetic calculations. Before arithmetics are performed, display data is converted to the computational-3 format.

Computational-3 to Display

To execute a move. Before a move is executed, computational-3 data is converted to the display data format.

To execute a compare. Before a compare is executed, display data is converted to the computational-3 data format.

To perform arithmetic calculations. Before arithmetic calculations are performed, Display data is converted to the Computational-3 data format. The result is generated in a Computational-3 work area, which is then converted and moved to the Display result field.

Display to Display

To perform arithmetic calculations. Before arithmetic calculations are performed, display data is converted to the computational-3 format. The result is generated in the computational-3 work area, which is then converted and moved to the Display result field.

Sign Control

The absence or presence of a plus or minus sign in the description of an arithmetic field can affect the efficiency of a program. For numeric fields specified, the compiler attempts to insure that a positive is present so that the values are treated as absolute.

The use of unsigned numeric fields increases the possibility of error (an unintentional negative sign could cause invalid results) and requires additional generated code to control the sign. The use of unsigned fields should be limited to fields treated as absolute values.

For example, if data is defined as

```
A    PICTURE 999.
B    PICTURE S999.
C    PICTURE S999.
```

the following moves are made: MOVE B TO A. MOVE B TO C. Moving B to A causes four more bytes to be used than moving B to C because A is an absolute value.

Conditional Statements

Computing arithmetic values separately and then comparing them may produce more accurate results than including arithmetic statements in conditional statements. The final result of an expression included in a conditional statement is *limited to an accuracy of six decimal places*. The following example shows how separating computations from conditional statements can improve the accuracy.

The data is defined as

```
77  A    PICTURE S9V9999       COMPUTATIONAL-3.
77  B    PICTURE S9V9999       COMPUTATIONAL-3.
77  C    PICTURE S999V9(8)     COMPUTATIONAL-3.
```

and the following conditional statement is written.

```
IF A * B = C
     PERFORM EQUAL-X
ELSE
     NEXT SENTENCE.
```

the final result will be 99V9(6). Although the receiving field for the final result (C) specifies eight decimal positions, the final result actually obtained in the example specifies six decimal places. For increased accuracy, define the final result field as desired, perform the computation, and then make the desired comparison as follows:

```
77  X     PICTURE S999V9(8)     COMPUTATIONAL-3.
    COMPUTE X = A * B.
    IF X = C
         PERFORM EQUAL-X
    ELSE
         NEXT SENTENCE.
```

Summary: Basic Principles of Effective Coding

1. Match decimal places in related fields (decimal point alignment).
2. Match integer places in related fields (unequal length fields).
3. Do not mix usage of data formats (mixed data formats).
4. Include an S (sign) in all numeric pictures (sign control).
5. Keep arithmetic expressions out of conditionals (conditional statements).

Reusing Data Areas

The main storage area can be used more efficiently by writing different data descriptions for the same data area. For example, the coding that follows shows how the same area can be used as a work area for records of several input files that are not processed concurrently.

```
WORKING-STORAGE SECTION.
01  WORK-AREA-FILE-1     (Largest record description for FILE-1.)
01  WORK-AREA-FILE-2     REDEFINES WORK-AREA-FILE-1.
                         (Largest record description for FILE-2.)
```

Alternate Groupings and Descriptions

Program data can be described more efficiently by providing alternate groupings or data descriptions for the same data. As an example of alternate groupings, suppose a program makes references to both a field and its subfields, where it could be more efficient to describe the subfields with different usages.

This can be done with the REDEFINES clause, as follows:

```
01  PAYROLL-RECORD.
    05  EMPLOYEE-RECORD         PICTURE X(28).
    05  EMPLOYEE-FIELD REDEFINES EMPLOYEE-RECORD.
        10  NAME                PICTURE X(24).
        10  NUMBER              PICTURE S9(4)     USAGE COMP-3.
    05  DATE-RECORD             PICTURE X(10).
```

The following example of different data descriptions specified for the same data, illustrates how a table can be utilized.

```
05  VALUE-A.
    10  A1      PICTURE S9(9)     USAGE COMP-3      VALUE ZEROS.
    10  A2      PICTURE S9(9)     USAGE COMP-3      VALUE IS 1.
05  TABLE-A REDEFINES VALUE-A PICTURE S9(9)     USAGE COMP-3
        OCCURS 100 TIMES.
```

Data Formats in the Computer

The following examples illustrate how various COBOL data formats appear in the computer in Extended Binary Coded-Decimal Interchange Code (EBCDIC) format.

Numeric DISPLAY (Extended Decimal)
The value of an item is −1234.

```
PICTURE 9(4)     USAGE DISPLAY.     F1 F2 F3 F4
                                             Byte
PICTURE S9(4)    USAGE DISPLAY.     F1 F2 F3 D4
```

Hexadecimal F is treated arithmetically as a plus in the low-order byte. The hexadecimal D represents a negative sign.

COMPUTATIONAL-3 (Internal Decimal)
The value of an item is +1234.

```
PICTURE 9(4)     USAGE COMP-3.     01 23 4F
                                         Byte
PICTURE S9(4)    USAGE COMP-3.     01 23 4C
```

Hexadecimal F is treated arithmetically as plus.
Hexadecimal C represents a positive sign.

COMPUTATIONAL (Binary)
The value of an item is 1234.

```
PICTURE S9(4)     COMPUTATIONAL  0 000 0000 0000 0100  1101 0010
                                 Sign                  Byte
```

A 0 bit in the sign position means the number is positive. Negative numbers appear in the 2s complement form with a 1 in the sign position.

Redundant Coding

To avoid redundant coding of usage designations, use computational at the group level (this does not affect the object program).

For example, instead of

```
05  RECORD-A.
    10  A     PICTURE 99V9     USAGE COMP-3.
    10  B     PICTURE 99V9     USAGE COMP-3.
    10  C     PICTURE 99V9     USAGE COMP-3.
```

write

```
05  RECORD-A                   USAGE COMP-3.
    10  A     PICTURE 99V9.
    10  B     PICTURE 99V9.
    10  C     PICTURE 99V9.
```

Working-Storage Section

Grouping Related Data Items Together

Group similar fields together under their own group level, such as SWITCHES, SUBSCRIPTS, FLAGS, ACCUMULATORS, COUNTERS, etc. Literals that are used often and might change in the future should be defined in the Working-Storage Section. Rather than searching through an entire module to modify a frequently-used literal, only one modification should then be necessary in working storage. When this is done, however, consider using a name other than the spelled-out value of the literal. It may be extremely upsetting to find, after a long debugging session, that the item in working storage named FIVE has a value of 4. Conversely, literals that are used infrequently and that are not likely to change, should be used directly in Procedure Division statements, thereby increasing readability and causing more efficient code to be generated.

Procedure Division

The Procedure Division should be written using the Data Division as a guide to files, data-names, constants, and work areas used. The programmer actually writes the COBOL program using the program flowchart as a guide for the procedure entries.

The Procedure Division specifies the action, such as input/output, data movement, and arithmetic operations that are required to process the data. A series of English-like statements are written in the sequence of the program flowchart.

Entries That Appear in the Procedure Division

1. The name of the division-Division Header.
2. Any optional sections.
3. A series of procedural paragraphs specifying the actions to be performed.

Increasing the Efficiency of Procedure Division Entries

A program can be made more efficient in the Procedure Division with some of the techniques described in the following example of the general structure of the Procedure Division.

```
PROCEDURE DIVISION [USING data-name-1 [data-name-21] . . .].
[ [DECLARATIVES.
{section-name SECTION. USE Sentence.
{paragraph-name. {sentence} . . .} . . .} . . .
END DECLARATIVES.]
{section-name SECTION [priority].]
{paragraph-name. {sentence} . . .} . . .} . . .
```

Modular Programming

Modular programming, as its name implies, means the arrangement of a program into separate modules of logic.

Modularizing involves organizing the Procedure Division into at least three functional levels: a main-line routine, processing subroutines, and input/output subroutines. When the Procedure Division is modularized, programs are easier to maintain and document.

Advantages of modular programming are many; disadvantages are few. For example, a module that is written and tested but once can be used in more than one program by cataloging it in the source language library from which it can be retrieved by the COBOL COPY statement. Examples of such common modules are: program initialization, error routines, determination of control breaks, calculation routines, and end-of-job routines.

Maintenance is easier for programs written using the modular approach. A common module needs to be recompiled just once. The savings in terms of quicker and easier recompilations can be particularly significant at the testing stage, or in an installation with programs subject to a high degree of modification.

Modular programming can result in a big savings in main storage usage because of the ability to overlay, which thus provides room for maintenance changes as they occur in major programs.

A common modular approach might include one section that would make decisions and direct control among the various sections. As an example, every file maintenance program reads details and masters, and performs different logic for master high, low, or equal conditions. PERFORM or GO TO statements can then be executed for the processing sections required for each condition.

In summary, the advantages of modular programming include the following:

1. Use of prewritten, common modules
2. Better organized logic
3. Modular testing
4. Easier maintenance
5. Ability to overlay

Main-Line Routine

The main-line routine should be short and simple, and should contain all of the major logical decisions of the program. This routine controls the order in which second-level routines are executed. All second-level routines should be invoked from the main-line routine by PERFORM or CALL statements.

Processing Routines

Processing routines should be subdivided into as many functional levels as necessary, depending on the complexity of the program. These routines should have a single entry point as the first statement of the routine and a single exit, which should be the EXIT statement. It is easier to add code when the last statement is an EXIT statement.

Make All Conditional Expressions as Simple as Logically Possible

COBOL provides the facility for complex conditional expressions. It is normally not necessary to use this facility, however, and may lead to confusion on the part of less experienced programmers trying to read the program. There is an appropriate time for very complex conditional expressions, and that is when the only alternative to the complex expression is a series of individual simple IF conditions spread all over the program. In this case, the difficulty of jumping from one IF to another is greater than the difficulty in comprehending the complex expression. In all other cases, however, including the case where a number of different branches are to be made, the use of simple IF clauses is preferable.

Use of GO TO with the DEPENDING ON Option Instead of a Series of IF Statements

When the possible values of a control variable are the numeric values 1, 2, 3, etc., the GO TO statement with the DEPENDING ON option is more efficient in regard to storage space and execution time than a long series of relational expressions and IF statements.

The following is an example of the GO TO statement with the DEPENDING ON option:

```
           .
           .
           .
           PERFORM PROCESS-ROUTINE THRU PROCESS-ROUTINE-EXIT.
           .
           .
           .
       PROCESS-ROUTINE.
           GO TO HAIRCUT,
                   SHAMPOO,
                   STYLIST,
                   PERM-WAVE,
               DEPENDING ON TYPE-OPERATOR.
           PERFORM TYPE-ERROR-ROUTINE.
           GO TO PROCESS-ROUTINE-EXIT.
```

```
* SUBROUTINES.
*       THE FOLLOWING ARE SUBROUTINES CALLED UPON BY THE
*       PROCESS-ROUTINE.

 HAIRCUT.
     MULTIPLY CUSTOMERS-I BY 8.00
          GIVING GROSS-W.
     GO TO PROCESS-ROUTINE-EXIT.

 SHAMPOO.
     MULTIPLY CUSTOMERS-I BY 10.00
          GIVING GROSS-W.
     GO TO PROCESS-ROUTINE-EXIT.

 STYLIST.
     MULTIPLY CUSTOMERS-I BY 15.00
          GIVING GROSS-W.
     GO TO PROCESS-ROUTINE-EXIT.

 PERM-WAVE.
     MULTIPLY CUSTOMERS-I BY 16.00
          GIVING GROSS-W.

 PROCESS-ROUTINE-EXIT.
     EXIT.
```

Notice the extensive use of GO TO statements in the program just given. This is not a violation of structured programming techniques, as the top-down concept is preserved. The use of GO TOs improve the efficiency of the program as follows:

In the PROCESS-ROUTINE paragraph, if there is an error in the coding, the TYPE-ERROR-ROUTINE procedure will be performed, but return will be to the next statement after PERFORM, therefore the program has to be directed to the PROCESS-ROUTINE-EXIT.

In the subroutine paragraphs, the use of GO TOs prevents the program from erroneously going through all paragraphs of the subroutines.

Replace Blanks in Numeric Fields with Zeros

On certain COBOL compilers, if an input field read from a record and defined as numeric contains all blanks, a data exception would occur if an arithmetic operation is attempted.

A method of avoiding the situation is to perform a numeric class test on all numeric fields from a record. For example:

```
IF FIELD-NAME IS NOT NUMERIC
     PERFORM ERROR-ROUTINE.
ELSE
     NEXT SENTENCE
```

OPEN, READ, CLOSE, and WRITE Statements

The OPEN and READ statements must reference a file assigned in the Environment Division and described in the Data Division. The CLOSE statement must be written for each file opened. The WRITE statement references a record-name.

The Sequence of Divisions Is Important

Since the last statement in the Procedure Division denotes the end of the source program, it is imperative that all divisions remain in the proper sequence. If the sequence is disturbed, many diagnostic errors will be generated unnecessarily.

Intermediate Results

The compiler treats arithmetic statements as a succession of operations and sets up intermediate result fields to contain the results of these operations. The compiler can process complicated statements, but not always with the same efficiency of storage utilization as the source program. Because truncation may occur during compilation, unexpected intermediate results may occur.

Binary Data

If an operation involving binary operands requires an intermediate result greater than 18 digits, the compiler converts the operands to internal decimal before performing the operation. If the result field is binary, the result will be converted from internal decimal to binary.

If an intermediate result will not be greater than nine digits, the operation is performed most efficiently as binary data fields.

COBOL Library Subroutines

If a decimal multiplication operation requires an intermediate result greater than 30 digits, a COBOL library subroutine is used to perform the multiplication. The result of this multiplication is truncated to 30 digits.

A COBOL library subroutine is used to perform division if (1) the divisor is equal to or greater than 15 digits, (2) the length of the divisor plus the length of the dividend is greater than 16 bytes, or (3) the scaled dividend is greater than 30 digits (a scaled dividend is a number that has been multiplied by a power of 10 in order to obtain the desired number of decimal places in the quotient).

Intermediate Result Greater than 30 Digits

When the number of digits in a decimal intermediate result field is greater than 30, the field is truncated to 30. A warning message will be generated at compilation time, but the program flow will not be interrupted at execution time. This truncation may cause the result to be invalid.

ON SIZE ERROR Option

The ON SIZE ERROR option applies only to the final tabulated results, not to intermediate result fields.

A method of avoiding unexpected intermediate results is to make critical computations by assigning maximum (or minimum) values to all fields and analyzing the results by testing the critical computations for results expected.

Because of concealed intermediate results, the final result is not always obvious.

The necessity for computing the worst case (or best case) results can be eliminated by keeping statements simple. This can be accomplished by splitting up the statement and controlling the intermediate results to be sure unexpected final results are not obtained.

For example,

```
COMPUTE B = (A + 3) / C + 27.600.
```

First define adequate intermediate result fields, i.e.,

```
05  INTERMEDIATE-RESULT-A      PICTURE S9(6)V999.
05  INTERMEDIATE-RESULT-B      PICTURE S9(6)V999.
```

Then split up the expression as follows:

```
ADD A, 3 GIVING INTERMEDIATE-RESULT-A.
```

Then write:

```
DIVIDE C INTO INTERMEDIATE-RESULT-A GIVING INTERMEDIATE-RESULT-B.
```

Then compute the final results by writing:

```
ADD INTERMEDIATE-RESULT-B, 27.600 GIVING B.
```

Arithmetic Fields

Initialize arithmetic fields before using them in computation. Failure to do so may result in invalid results, or the job might terminate abnormally.

Comparison Fields

Numeric comparisons are usually done in the computational-3 format; therefore, computational-3 is the most efficient data format.

Because the compiler inserts slack bytes which can contain meaningless data, group comparisons should not be attempted when slack bytes are within the group unless the programmer knows the contents of the slack bytes.

OPEN and CLOSE Statements

Each opening or closing of a file requires the use of main storage that is directly proportional to the number of files being opened. Opening or closing more than one file with the same statement is faster than using a separate statement for each file. Separate statements, however, require less storage area.

For example, one statement OPEN INPUT FILE-A, FILE-B, FILE-C. rather than

```
OPEN INPUT FILE-A.
OPEN INPUT FILE-B.
OPEN INPUT FILE-C.
```

ACCEPT Verb

The ACCEPT verb does not provide for the recognition of the last card being read from the input device. When COBOL detects the end-of-file record, it drops through to the next statement. Because no indication of this is given by COBOL, the end-of-file detection requires special treatment. Thus the programmers must provide their own end record (some record other than the end-of-file record), which can be used to detect an end-of-file condition.

Paragraph-Names

Paragraph-names use storage when the PERFORM verb is used in the program. Use of paragraph-names for comments requires more storage than the use of a comment or a blank record. Use comment and/or a blank record for identifying inline procedures where paragraph-names are not required.

For example, avoid writing the following:

```
MOVE A TO B.
PERFORM JOES-ROUTINE.

JOES-ROUTINE. COMPUTE A = D + C * F.
```

The following coding is recommended:

```
      MOVE A TO B.
      PERFORM ROUTINE.
* THIS IS JOES-ROUTINE.
  ROUTINE.
      COMPUTE A = D + C * F.
```

COMPUTE Statement

The use of the COMPUTE statement generates more efficient coding than does the use of individual arithmetic statements because the compiler can keep track of internal work areas and does not have to store the results of intermediate calculations. It is the programmer's responsibility, however, to insure that the data is defined with the level of significance required in the answer.

IF Statement

Nested and computed IF statements should be avoided, as the logic is difficult to debug. Performing an IF operation for a variable greater than 256 bytes in length requires the generation of more instructions than are required for that of an IF operation of a variable of 256 bytes or less.

MOVE Statement

Performing a MOVE operation for an item greater than 256 bytes in length requires the generation of more instructions than is required for that of a MOVE statement for a variable of 256 bytes or less.

When a MOVE statement with the CORRESPONDING option is executed, data items are considered CORRESPONDING only if their respective data-names are the same, including all implied qualifications, up to but not including the data-names used in the MOVE statement itself.

PERFORM Verb

PERFORM is a useful verb if the programmer adheres to the following rules:

1. Always execute the last statement of a series of routines being operated on by a PERFORM statement. When branching out of the routine, make sure control will eventually return to the last statement of the routine. This statement should be an EXIT statement. Although no code is generated, the EXIT statement allows a programmer to immediately recognize the extent of a series of routines within the range of a PERFORM statement.
2. Always either PERFORM routine-name THRU routine name-exit, or PERFORM a section name. A PERFORM paragraph-name can cause trouble for the programmer trying to maintain the program. For example, if a paragraph must be broken into two paragraphs, the programmer must examine every statement to determine whether or not this paragraph is within the range of the PERFORM statement. Then all statements referencing the paragraph-name must be changed to PERFORM THRU statements. The EXIT statement can be used to provide a common exit point for both paragraphs.

READ INTO and WRITE FROM Options

Use READ INTO and WRITE FROM to do all the processing in the Working-Storage Section. This is suggested for two reasons.

1. Debugging is much simpler. Working-Storage areas are easier to locate in a dump and so are buffer areas. And, if files are blocked, it is much easier to determine which record in a block was being processed when the abnormal termination occurred.
2. Trying to access a record area after the AT END condition has occurred (for example, AT END MOVE HIGH-VALUE TO INPUT-RECORD) can cause problems if the record area is only in the File Section.

Note: The programmer should be aware that additional time is used to execute the move operations involved in each READ INTO or WRITE FROM instruction.

Hierarchy Chart

The hierarchy chart involves a treelike structure similar to any organization chart. (See figure 16.3.) It is composed of functions or actions. Each function on the hierarchy is represented as a box and can be described within that box as a verb (action) and an object (data affected). The verb-object format names as well as defines the functions.

The top box on the hierarchy describes the entire piece of software in terms of a single function. Each level below is a subset of the function above it. This hierarchy of functions is created by a technique known as *functional decomposition,* whereby a function is exploded into increasingly lower levels of detail until all functions have been defined. Determining the main function of the software is not a trivial exercise. It requires a great deal of insight, creativity, and experience on the part of the designer.

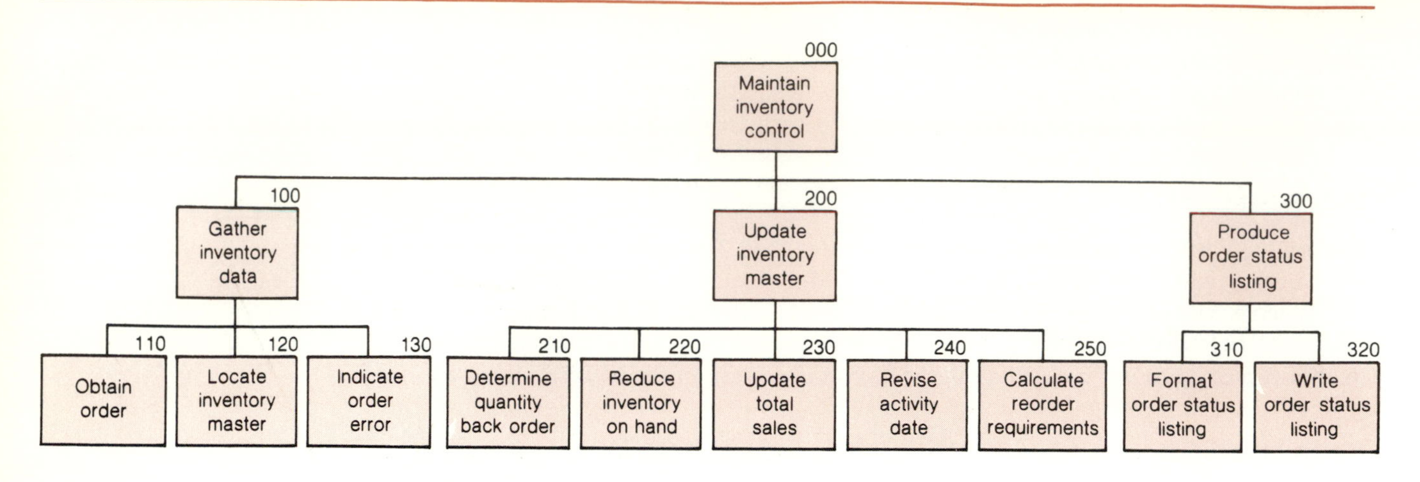

Figure 16.3 Hierarchy chart—example.

Top level functions on a hierarchy contain the control logic. They determine when and in what order lower level functions are to be invoked. They consist primarily of CALLs, PERFORMs, DO-WHILEs, and IFTHENELSE statements. Lower level functions are the workers; here sequential coding statements are found to predominate. (See figure 16.4.)

Conventional-Linkage for COBOL Programming

The following are *only* suggestions, inasmuch as no conventional linkage for structured COBOL programming has as yet been developed.

Conventional-Linkage combines both program design and a program structure. As a program design, Conventional-Linkage is top-down in nature. This concept requires that high-level coding be developed first (and possibly tested if physically modular; perhaps tested if logically modular and large). Then in the pattern of top-down design, each successively lower level is developed and coded. As a top-down program design, Conventional-Linkage requires that interfaces between COBOL sections and units of coding be predefined.

As a program structure (structured programming), proper use of Conventional-Linkage requires a rigid placement of program functions, and also the concepts of a single entry and exit point in a unit of code. For example, the code never branches out of set boundaries.

Conventional-Linkage helps to accomplish the aforementioned in the following manner:

1. Functions are almost always located in the same section, so that one is not concerned with where these operations will be performed. This allows the programmer to devote time to the logic of the application.
2. Sections are independent in that the programmer is not concerned with what is happening in other sections.
3. Basic interfaces are predefined, as, for example,
   a. how data is passed from section to section, and
   b. how sections communicate (switches, keys, etc.).

Conventional-Linkage consists of seven COBOL sections:

1. CONVENTIONAL-LINKAGE SECTION.
   Contains the control logic for the program.
2. INITIALIZATION SECTION.
   Executes housekeeping functions and runs through one processing cycle to get the program started.
3. READ-RECORD SECTION.
   Reads a record from appropriate input file designated by FILE-PRIORITY-SELECT SECTION.

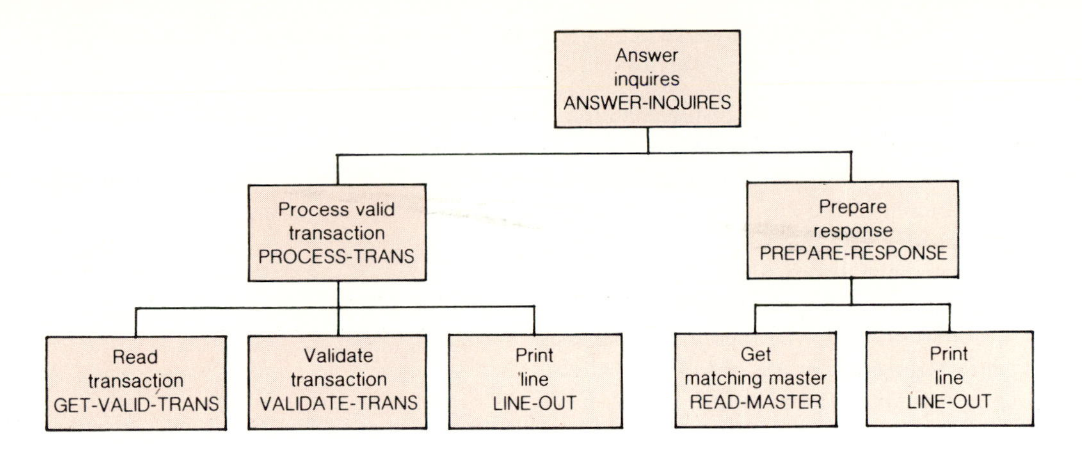

Figure 16.4 Hierarchy chart—example.

4. FILE-PRIORITY-SELECT SECTION.
   Selects next record to be processed based on lowest keys and highest priority.
5. CONTROL-BREAK-TEST SECTION.
   Determines whether a control break has occurred and takes the appropriate actions.
6. DETAIL-PROCESSING SECTION.
   Processes records, even to the extent of moving a record or parts of a record to a work area.
7. SUBROUTINES SECTION.

All units of coding may be used in more than one place or in different sections including the termination routines.

Conventional-Linkage

The control coding for the execution of a program using all sections that are performed as sections appears in the following manner:

```
CONVENTIONAL-LINKAGE SECTION.
    PERFORM INITIALIZATION.
    PERFORM DRIVER-SECTION.
DRIVER-SECTION.
    PERFORM READ-RECORD.
    PERFORM FILE-PRIORITY-SELECT.
    PERFORM CONTROL-BREAK-TEST.
    PERFORM DETAIL-PROCESSING.
```

Initialization Section

1. *Housekeeping.*
   a. Edit control data such as date, run code and so on. Control data may come from any source.
   b. Clear working storage fields that cannot be initialized with VALUE clauses.
   c. Open files.
2. *Prime Program.*
   a. Perform READ-RECORD once for each file.
   b. Perform FILE-PRIORITY-SELECT once.
   c. Move CONTROL-BREAK-HOLD to FIELD-CONTROL. CONTROL-BREAK-HOLD and FIELD-CONTROL are group items that contain keys tested in the CONTROL-BREAK-TEST SECTION.
   d. Perform DETAIL-PROCESSING.

Read-Record Section

The file to be read in is controlled by a switch (FILE-POINT) set in the FILE-PRIORITY-SELECT SECTION.

1. Read appropriate file (for end-of-file condition, see no. 6).
2. Sequence-check the file.
3. Move record key to field used by FILE-PRIORITY-SELECT. This may involve some modification or rearrangement of the key, and this field should have a name suggesting both file and function, such as FPS-SELECT-KEY.
4. Increment memory record count.
5. Edit if appropriate. This edit should be limited to testing of record type.
6. End of file. Each AT END imperative does the following:
   a. Turns on the appropriate end-of-file switch.
   b. Moves high-values to appropriate FPS-KEY.
   c. Performs TERMINATION-ROUTINE.
   d. Branches to READ-RECORD-EXIT.

File-Priority-Select Section

1. Selects record of lowest key. The FPS-SELECT-KEYs passed by READ-RECORD SECTION are used for comparison. When there are equal FPS-SELECT-KEYs, the program logic must determine which record has priority.
2. Moves the lowest FPS-SELECT-KEY (or priority FPS-SELECT-KEY) to CONTROL-BREAK-HOLD. CONTROL-BREAK-HOLD is never updated anywhere else in the program. CONTROL-BREAK-HOLD is used by the CONTROL-BREAK-TEST SECTION.
3. Set a switch (FILE-POINT). The values of this switch will:
   a. Tell READ-RECORD SECTION which file will be read next.
   b. Tell DETAIL-PROCESSING SECTION which record from which file is to be processed next.
   c. Be available to CONTROL-BREAK-TEST SECTION if needed.

   Note: After INITIALIZATION SECTION has been performed, FILE-POINT will never be updated anywhere other than in the FILE-PRIORITY-SELECT SECTION.

Control-Break-Test Section

1. This section determines whether a control break has occurred, and at what level. A control break is the result of an unequal compare between CONTROL-BREAK-HOLD and FIELD-CONTROL, which are control keys containing one or more segments. If the control key has only one segment, e.g., INVOICE-NO, no further compares are necessary. If the control key is made of multiple segments, such as WAREHOUSE-CENTER-NO, PRODUCER-NO, and INVOICE-NO, and an unequal comparison occurs, segments are tested individually starting with the highest level, e.g., WAREHOUSE-CENTER-NO, to discover the highest level of change.
2. In addition to testing for control breaks, this section contains two types of break functions, they are,
   a. the EOF-BREAK, which wraps up the last group processed. This may include:
      1. producing records,
      2. balancing,
      3. producing total records,
      4. checking results of processing and taking appropriate action,
      5. rolling counters,
      6. and anything else dictated by the logic of the program.
   b. the GROUP-BREAK, which does whatever is necessary to allow the next group to be processed. This may include,
      1. clearing counters and indexes,
      2. clearing work areas,
      3. resetting switches, and
      4. updating FIELD-CONTROL by moving in CONTROL-BREAK-HOLD.

3. EOF-BREAKs are performed from low level to high level. All EOF-BREAKs are completed before any GROUP-BREAKs are performed.
4. GROUP-BREAKs are performed from high level to low level.
5. The following example of CONTROL-BREAK-TEST coding is based on the following Working-Storage Section definitions of the control fields:

```
01  CONTROL-BREAK-HOLD.
    05  CB-WAREHOUSE-CENTER-NO      PICTURE X.
    05  CB-PRODUCER-NO              PICTURE X(6).
    05  CB-INVOICE-NO               PICTURE X(8).
01  FIELD-CONTROL.
    05  FC-WAREHOUSE-CENTER-NO      PICTURE X.
    05  FC-PRODUCER-NO              PICTURE X(6).
    05  FC-INVOICE-NO               PICTURE X(8).
FIELD-BREAK SECTION.
    IF CONTROL-BREAK = FIELD-CONTROL
        PERFORM CONTROL-BREAK-TEST-EXIT
    ELSE
        NEXT SENTENCE.
TEST-FOR-PC-BREAK.
    IF CB-WAREHOUSE-CENTER-NO = FC-WAREHOUSE-CENTER-NO
        NEXT SENTENCE
    ELSE
        PERFORM INVOICE-EOF-BREAK THRU
            INVOICE-GROUP-BREAK-EXIT
        PERFORM CONTROL-BREAK-TEST-EXIT.
TEST-FOR-PRODUCER-BREAK.
    IF CB-PRODUCER-NO = FC-PRODUCER-NO
        NEXT SENTENCE
    ELSE
        PERFORM INVOICE-EOF-BREAK THRU
            PRODUCER-EOF-BREAK-EXIT
        PERFORM PRODUCER-GROUP-BREAK THRU
            INVOICE-GROUP-BREAK-EXIT
        PERFORM FIELD-BREAK-EXIT.
INVOICE-BREAK.
        PERFORM INVOICE-EOF-BREAK THRU
            INVOICE-EOF-BREAK-EXIT
        PERFORM INVOICE-GROUP-BREAK THRU
            INVOICE-GROUP-BREAK-EXIT
        PERFORM FIELD-BREAK-EXIT.
INVOICE-EOF-BREAK.
        .
        .
        .
INVOICE-EOF-BREAK-EXIT.
        EXIT.
PRODUCER-EOF-BREAK.
        .
        .
        .
PRODUCER-EOF-BREAK-EXIT.
        EXIT.
WAREHOUSE-CENTER-EOF-BREAK.
        .
        .
        .
WAREHOUSE-CENTER-EOF-BREAK-EXIT.
        EXIT.
```

```
WAREHOUSE-CENTER-GROUP-BREAK.
          .
          .
          .
WAREHOUSE-CENTER-GROUP-BREAK-EXIT.
          EXIT.
PRODUCER-GROUP-BREAK.
          .
          .
          .
PRODUCER-GROUP-BREAK-EXIT.
          EXIT.
INVOICE-GROUP-BREAK.
          .
          .
          .
INVOICE-GROUP-BREAK-EXIT.
          EXIT.
          .
          .
          .
CONTROL-BREAK-TEST-EXIT.
          EXIT.
FIELD-BREAK-EXIT.
          EXIT.
```

Note: Normally, it is recommended in structured programming that the THRU option of the PERFORM verb *not* be used. However, there are situations wherein the use of the THRU option will save coding time, as in the example just given, and in CASE program structures.

Detail-Processing Section

1. This section uses FILE-POINT to select the processing routine to be applied. This section assumes different routines for different record types.
2. Processing routines may do some or all of the following:
   a. Set switches (for use in CONTROL-BREAK-TEST, etc.),
   b. Establish tables,
   c. Add to counters,
   d. Produce output, and
   e. Any other data manipulation required.

Subroutines Section

1. The routines residing in this section are generally those units of coding that are used in more than one place in a section or in
   a. Write routines,
   b. Headings,
   c. Conversions,
   d. Accumulations (formulas),
   e. Other arithmetic or mathematical functions,
   f. Balancing,
   g. Other, and
   h. TERMINATION-ROUTINE.
2. TERMINATION-ROUTINE is performed from READ-RECORD SECTION only.
   a. Tests for end-of-job condition. If not end-of-job, branches to exit.
   b. At end-of-job, wraps up the program by performing all levels of EOF-BREAKS.
   c. Closes files.
   d. Writes messages.
   e. Displays counts and balances.

Conclusion

Some of the programming techniques that will aid programmers in preparing efficient programs h
been explained and illustrated. Additional programming techniques will be found in the reference m
uals of the computer manufacturers. It is hoped that as programmers become more proficient in CO
programming, they will develop their own programming techniques.

Summary

COBOL does not produce a program as efficient as one written in the basic language of the particular computer. Therefore, it is imperative that the programmer adopt techniques that will increase the efficiency of the COBOL program.

To improve the appearance of a source program, a blank statement or statements with asterisks can be inserted to divide portions of the source program into unmistakable divisions.

Indentation of entries should be used to indicate the subordination of items within groups.

Paragraph-names should be on a line by themselves. Every Procedure Division verb should start a new line. An IF condition statement should be on a line by itself with all the statements conditioned by it on a separate line.

Indentation should be used when using the GO TO verb with the DEPENDING ON option. Each paragraph-name referenced should be indented on a separate line.

Numeric literals and words should never be split between lines.

The use of section identification by giving two or more successive paragraphs a section-name avoids the use of the THRU option of the PERFORM statement and provides a clearer organization of COBOL routines. However, once one section is specified, the entire Procedure Division must be coded in sections.

Comment lines (*symbol in position 7) should be used freely throughout the program to explain the purpose and uses of the entry.

In addition to the required entries of division name and program identification, the Identification Division optionally should include the name of the programmer, when the program was written and comment statement(s) to explain the purpose of the program.

The Environment Division is the only division in COBOL that is machine oriented. Since the hardware does not change between programs, with the exception of the FILE-CONTROL paragraph, the entire contents of the Environment Division can be standardized for all programs. Once such standard text has been devised, it can be catalogued into the COBOL source program library and copied whenever needed.

All files mentioned in the FILE-CONTROL paragraph should be checked with system programmers, as names vary in each data processing unit. The file should be properly defined in the Data Division and opened and closed in the Procedure Division.

The Data Division should be written using copies of the source document format as guides for writing the file description and record description entries. If a fixed number of permanent files is used, the Data Division entries can be catalogued into a COBOL source language library and included in each program requiring use of such file(s) via the COPY statement.

The use of the COPY statement has many benefits such as standardization of data definitions. The programming effort is minimized as file designations need be written and debugged only once and are available to all programs. This statement simplifies program maintenance and causes the library module to become the final authority on file contents.

Self-documenting data-names, which aid in the debugging and maintenance of the program, should be used.

The RECORD CONTAINS clause should be included as the compiler checks the agreement of the record count in the record description entry with the RECORD CONTAINS clause. This will assure that no data field has been omitted in the record description entry and at the same time provides the reader of the program with the record size without the necessity of counting individual entries.

Names of independent items in the Working-Storage Section must be unique as they cannot be qualified.

A recommended programming procedure is to describe all print lines, including heading, detail and footing, and use the WRITE verb with the FROM option referring to the Working-Storage items. This provides the programmer with the ability to define heading with the appropriate VALUE clauses (forbidden in the File Section), and the ability to use one record description entry to define heading, detail and total lines.

The PICTURE clause must be compatible with USAGE and VALUE clauses where they are used.

A suffix should be attached to a data-name to indicate where the item is found in the Data Division. This will aid in the debugging of the program.

The MOVE CORRESPONDING statement can save time in writing many MOVE statements of identical items usually from input to output. However, this technique involves qualification whenever an identical item is involved in Procedure Division statements, and can complicate maintenance of the program.

Procedure Division operations are most efficient when the decimal points and the data items involved are aligned.

An intermediate operation may be required when handling fields of unequal lengths. To avoid these operations, the number of digits should be equal.

When fields are used together in move, arithmetic, or relational statements, they should be in the same format whenever possible. Conversion requires additional storage and longer execution time.

The absence or presence of a plus or minus sign in the description of an arithmetic field can affect the efficiency of a program.

Computing arithmetic values separately and then comparing them may produce more accurate results than including arithmetic statements in conditional statements.

Main storage can be used more efficiently by writing different data descriptions for the same data area.

Program data can be described more efficiently by providing alternate groupings or data descriptions for the same data.

To avoid redundant coding of usage designations, use computational at the group level.

Group similar fields in the Working-Storage Section under their own group level such as switches, flags, counters, etc. Literals that are used often and might change in the future should be defined in working storage. In this manner, a change in the working-storage item will reflect in all Procedure Division statements where that item is being used. Conversely, literals that are used infrequently and that are not likely to change, should be used directly in Procedure Division statements, thereby increasing readability and causing more efficient code to be generated.

The Procedure Division should be written using the Data Division as a guide to file, data-name, constants and work areas used. The program flowchart is used as a guide for writing Procedure Division entries.

Modular programming is the arrangement of a program into separate modules of logic. Modularizing involves organizing the Procedure Division into at least three functional levels: a main-line routine, processing routines and input/output subroutines. When the Procedure Division is modularized, programs are easier to maintain and document.

The main-line routine should be short and simple and contain all of the major logical decisions in logical order in which the second level routines are executed.

Processing routines should be subdivided into as many functional levels as necessary, depending on the complexity of the program. These routines should have a single entry and exit point.

All conditional expressions should be as simple as logically possible.

The use of the GO TO with the DEPENDING ON option instead of a series of IF statements is more efficient in regard to storage space and execution time.

A method of avoiding a data exception in input data is to make sure that all numeric fields have leading zeros and no blanks.

When the number of digits in an intermediate result is greater than 30, a warning message will be generated at compilation time, but the program will continue executing.

The ON SIZE ERROR applies only to the final tabulated result, and not to any intermediate results.

Arithmetic fields should be initialized with zeros before using.

COMPUTATIONAL-3 is the most efficient data format of numeric operations.

Opening and closing of more than one file with the same statement is faster than using separate statements for each file.

The ACCEPT verb does not provide for the recognition of the last record being read from a file.

The use of paragraph-names for comments requires more storage than the use of a comment and blank record.

The COMPUTE statement generates more efficient coding than does the use of individual arithmetic statements.

Nested and computed IF statements should be avoided, as the logic is difficult to debug.

When a MOVE CORRESPONDING statement is executed, data items are considered corresponding only if their respective data-names are the same including all qualifications.

The last statement of a series of statements operated on by a PERFORM statement must be executed. An EXIT verb should be used when branching out of the routine occurs, so that control of the program can return to the last statement of the routine.

READ INTO and WRITE FROM statements should be used to do all processing in Working-Storage Section items, as debugging is simpler, and trying to access a record area after the AT END option has occurred causes problems if the record area is only in the File Section.

The hierarchy chart involves a treelike structure similar to an organization chart. It is composed of functions and actions. Each function in the hierarchy is represented as a box and can be described within that box. The top box in the hierarchy describes the entire piece of software in terms of a single function. Each level listed is a subset of the function above it. The hierarchy of functions is created by a technique known as functional decomposition, whereby a function is exploded into increasing lower levels of detail until all functions are defined.

Top-level functions in a hierarchy contain the control logic. They determine when and in what order lower-level functions are to be invoked.

Conventional-Linkage combines both program design and program structure. As a program design, Conventional-Linkage is top-down in structure. This concept requires that high-level coding be developed first. This is the pattern of top-down design, wherein each successive lower level is developed and coded. As a top-down program design, Conventional-Linkage requires that interfaces between COBOL sections and units of coding be predefined.

As a programming structure (structured programming), proper use of Conventional-Linkage requires the rigid placement of program functions and also the concept of a single entry and exit point in a unit of code.

In Conventional-Linkage, functions are almost always located in the same section, so that one is not concerned with where these operations will be performed. This allows the programmer to devote time to the logic of the application. Sections are independent in that the programmer is not concerned with what is happening in other sections.

Conventional-Linkage consists of seven COBOL sections:

1. The Conventional-Linkage Section contains the control logic for the program.
2. The Initialization Section executes housekeeping functions and runs through one processing cycle to get the program started.
3. The Read-Record Section reads a record from the appropriate input file designated by the File-Priority-Select Section.
4. The File-Priority-Select Section selects the next record to be processed based on the lowest keys and highest priority.
5. The Control-Break-Test Section determines whether a control break has occurred and takes the appropriate action.
6. The Detail-Processing Section processes records, even to the extent of moving a record or parts of a record to a work area.
7. The Subroutines Section involves all units of coding that may be used in more than one place or in different sections, including the termination routines.

Questions for Review

1. Why is it important to write an efficient COBOL program?
2. What are some suggestions for improving readability and program logic understanding?
3. What additional information should be included in the Identification Division?
4. What precautions should be taken in writing the Environment Division?
5. How can the use of the COPY statement improve COBOL programs? What are some of its advantages?
6. Why should the RECORD CONTAINS clause be included in every File Description entry in the Data Division?
7. Why is it desirable to write Data Division entries for output files in the Working-Storage Section?
8. When should the computational mode be used with data items?
9. Why is it important to attach suffixes to data-names?
10. What problems are involved in using the MOVE CORRESPONDING statement?
11. How could storage be conserved in decimal alignment?
12. How does the compiler align fields of unequal length?
13. What problem is caused by operations involving data items of mixed data format? How can this be overcome?
14. Explain the operation of sign control in the efficiency of a program.
15. Why is it possible to have an inaccurate answer as a result of an arithmetic statement in a conditional statement?
16. List the basic principles of effective coding.
17. How may alternate groupings provide a more efficient program?
18. How may the grouping of related data items together in the Working-Storage Section improve the efficiency of the program?
19. How may the Procedure Division be written more efficiently?
20. What is the danger of intermediate results in arithmetic statements?
21. What is the problem involved in using an ACCEPT statement?
22. Why is it more efficient to use a COMPUTE statement rather than a series of arithmetic statements?
23. Why is PERFORM a useful verb?
24. Why are the READ INTO and WRITE FROM options suggested for input/output operations?
25. What is a hierarchy chart?
26. What is Conventional-Linkage and what are its main advantages?
27. What are the seven sections of Conventional-Linkage and what is the main function of each?

Matching Questions

Match each item with its proper description.

_____ 1. Hierarchy chart	A. Applies only to the final tabulated results.
_____ 2. Mixed data format	B. Indicates where the item is to be found in a program.
_____ 3. ON SIZE ERROR	C. Machine oriented.
_____ 4. COPY	D. Requires one item to be converted to matched format.
_____ 5. Environment Division	E. Limited to 30 digits.
_____ 6. Suffix	F. Arranges program into separate modules of logic.
_____ 7. Conventional-Linkage	G. Includes prewritten Data and Environment Division entries.
_____ 8. Modular program	H. Treelike structure similar to an organization chart composed of functions and actions.
_____ 9. Alternate descriptions	I. Redefines fields.
_____10. Intermediate results	J. Combines both program design and program structure.

Exercises

Multiple Choice: Indicate the best *answer (questions 1-31).*

1. One of the problems in writing COBOL programs is that
   a. COBOL is not self-documenting.
   b. COBOL is difficult to understand.
   c. COBOL does not produce a program as efficient as one written in the basic language of a particular computer.
   d. All of the above.
2. COBOL source programs can be improved by
   a. using blank lines to separate paragraphs, sections, etc.
   b. the programmer having some knowledge of the processor on which the program will be run.
   c. using indentation to indicate subordination.
   d. All of the above.
3. To improve the readability of COBOL programs,
   a. as many statements as possible should be put on one line.
   b. split words or numeric literals between lines to conserve space.
   c. never use section names, as they only create confusion.
   d. None of the above.
4. In addition to the required Identification Division entries, the following are recommended:
   a. When the program was written.
   b. Comments that will explain the purpose of the program.
   c. The name of the programmer.
   d. All of the above.
5. The Environment Division
   a. is the only division in the COBOL program that is not machine oriented.
   b. contains information about the equipment to be used when the object program is compiled and executed.
   c. cannot be standardized, except for the File-Control paragraph, since the information is different for each program execution.
   d. may not contain any special input/output techniques.
6. The Data Division
   a. describes information to be processed by a source program.
   b. contains files that have to be mentioned in the Identification Division.
   c. contains constants and work areas which must be described in the Working-Storage Section.
   d. All of the above.
7. The section that may appear in the Data Division is the
   a. Input-Output Section.
   b. Declaratives Section.
   c. Linkage Section.
   d. All of the above.
8. The following clauses are required in the File Description entry.
   a. RECORDING MODE.
   b. BLOCK CONTAINS.
   c. RECORD CONTAINS.
   d. LABEL RECORDS.
9. The following are required in Record Description entries.
   a. Level numbers.
   b. Nonunique data-names.
   c. PICTURE clauses.
   d. VALUE clauses.
10. To make it easier for the programmer to locate items in a program listing,
    a. prefixes should be attached to data-names.
    b. suffixes should be attached to data-names.
    c. similar data-names should be used in all sections.
    d. data-names should consist of alphabetic and numeric characters.
11. The MOVE CORRESPONDING statement can save time
    a. when unlike items are to be moved.
    b. because qualification of data items is not necessary.
    c. in writing many MOVE statements for identical items, usually from input to output.
    d. All of the above.

12. Procedure Division operations are most efficient when
    a. decimal positions of the data items are aligned.
    b. equal length fields are used.
    c. mixed data formats are avoided.
    d. All of the above.
13. The final result of an expression in a conditional statement is limited to
    a. 9 decimal places.
    b. 6 decimal places.
    c. 8 decimal places.
    d. None of the above.
14. Program data can be described more efficiently by
    a. writing the same data description for different data areas.
    b. describing different subfields with the same usage.
    c. providing alternate groupings for the same data.
    d. All of the above.
15. COBOL data formats for numeric items include
    a. DISPLAY.
    b. COMPUTATIONAL.
    c. COMPUTATIONAL-3.
    d. All of the above.
16. To increase efficiency, the Working-Storage Section should contain
    a. literals that are used infrequently.
    b. literals that are not likely to change.
    c. similar items grouped together under their own group level.
    d. All of the above.
17. Modular programming
    a. is the arrangement of a program into separate modules of logic.
    b. makes programs difficult to maintain.
    c. makes programs difficult to document.
    d. All of the above.
18. Some of the advantages of modular programming include
    a. the use of prewritten common modules.
    b. better organized logic.
    c. modular testing.
    d. All of the above.
19. The main-line routine of a COBOL program should
    a. be long and complex.
    b. contain all the major logical decisions of the program.
    c. contain all main-line routines invoked by second-level routines.
    d. All of the above.
20. The following suggestions should lead to more efficient Procedure Divisions.
    a. Processing routines should be divided into many functional levels.
    b. All conditional expressions should be as simple as logically possible.
    c. Use the GO TO statement with the DEPENDING ON option instead of a series of IF statements.
    d. All of the above.
21. The following refer to input/output statements.
    a. The OPEN and READ statements must reference a file assigned in the Data Division.
    b. The CLOSE statement must be written for each file opened.
    c. The WRITE statement references a file-name.
    d. The READ statement references a record-name.
22. The end of a source program is indicated by the
    a. last statement in the Procedure Division.
    b. STOP RUN statement.
    c. delimiter.
    d. presence of a job control statement.
23. Intermediate results in the COBOL source library subroutines are limited to
    a. 18 digits.
    b. 16 decimal digits.
    c. 30 digits.
    d. 22 digits.

24. The ON SIZE ERROR option applies
    a. to intermediate results.
    b. to only multiplication and division results.
    c. to only data items in the COMPUTE statement.
    d. to final tabulated results.
25. The following are more efficient Procedure Division suggestions.
    a. Opening and closing more than one file with the same statement is faster than using separate statements.
    b. Use of paragraph-names for comments requires less storage than comments on individual records.
    c. Numeric comparisons done in the DISPLAY format are more efficient than COMPUTATIONAL-3 format comparisons.
    d. All of the above.
26. Additional efficient Procedure Division coding suggestions are:
    a. The use of individual arithmetic statements is more efficient than the use of the COMPUTE statement.
    b. Nested IF statements are easy to debug.
    c. The EXIT statement can be used to provide a common exit for a series of paragraphs.
    d. The use of READ INTO and WRITE FROM options makes debugging more complex.
27. The hierarchy chart
    a. involves a treelike structure.
    b. is similar to an organization chart.
    c. is composed of functions and actions.
    d. All of the above.
28. Conventional-Linkage
    a. is bottom-top in nature.
    b. requires high-level coding be developed first.
    c. does not require that interfaces between COBOL sections and units of coding be predefined.
    d. All of the above.
29. The housekeeping functions in Conventional-Linkage are provided in the following section:
    a. INITIALIZATION.
    b. HOUSEKEEPING.
    c. DETAIL-PROCESSING.
    d. READ-RECORD.
30. In Conventional-Linkage, the DETAIL-PROCESSING Section performs the following:
    a. Establishes tables.
    b. Adds to counters.
    c. Produces output.
    d. All of the above.
31. The SUBROUTINES Section of Conventional-Linkage may contain the following routines:
    a. Sequence-checking the file.
    b. Edit data.
    c. Rolling counters.
    d. Heading.
32. Write the minimum number of entries required for the following information for the Identification and Environment Divisions.
    a. The program is to be written to process data on an IBM 370 computer.
    b. The input/output devices to be used are the model 2540 card reader and a 2400 tape unit.
33. In the following COMPUTE statement, the accuracy of the result (X) will be affected by the number of integers and decimal digits returned in the various intermediate results.
    Rewrite the statement into a series of COMPUTE statements to assure that this does not occur. Assume all fields will not exceed eight digits.

    COMPUTE X = A + (B / C) + ((D ** E) * F) − G.

34. A program references a field and its subfields, each with different usages. The fields are as follows:

Field	Record Positions
Name	1-24
Number	25-28

    The number will be involved in numerous calculations. The entire field by itself will be used for display purposes.
    Write the record description entries for the aforementioned, so that both the name and number fields can be referenced as well as each subfield with different USAGE clauses.
35. Set up Working-Storage areas for WORK-AREA-FILE1 and WORK-AREA-FILE2 so that the same area can be used as a work area for records of several input files that can not be processed concurrently.

36. A program contains the following instructions:

```
77  FLD-A     PICTURE S9(5)V9999.
77  FLD-B     PICTURE S99V99.
.
.
.
PROCEDURE DIVISION.
.
.
.
    ADD FLD-A TO FLD-B.
```

What Picture clause must be changed to make the program correct and more efficient? Write the correct Picture clause for the item to be changed.

37. Which of the following PERFORM statements is incorrect?

A.
```
x PERFORM a THRU m
a
f
m
j
d PERFORM f THRU j
```

B.
```
x PERFORM a THRU m
a
d PERFORM f THRU j
f
m
j
```

C.
```
x PERFORM a THRU m
a
d PERFORM f THRU j
f
j
m
```

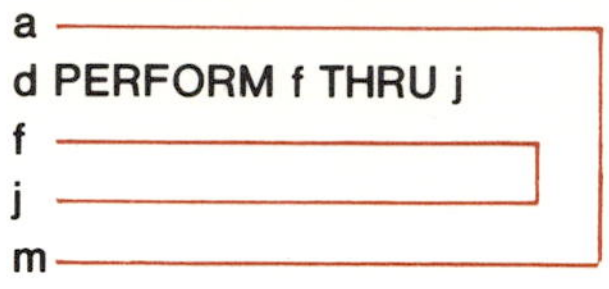

D.
```
x PERFORM a THRU m
a
d PERFORM f THRU j
h
m
f
j
```

Aged-Trial-Balance Report

Input

There are two record types in the customer file: master record and invoice record. The formats for the two records are as follows:

Field	Positions	Format
Master Record		
Code	1	Letter M
Customer Number	2-6	
Customer Name	7-30	
Not Used	31-36	
Credit Limit	37-43	XXXXX.XX
Not Used	44-80	
Invoice Record		
Code	1	Letter I
Customer Number	2-6	
Not Used	7-30	
Invoice Number	31-36	
Not Used	37-43	
Invoice Date	44-49	MM/DD/YY
Invoice Amount	50-56	XXXXX.XX
Not Used	57-80	

Computations to Be Performed

An Aged-Trial-Balance Report will be prepared according to the output format. The following calculations will be performed:

1. *Total Charges.* All invoice amounts will be added here.
2. *Current Charges.* Only the current month invoice amounts will be added here.
3. *Overdue Accounts. 30 Days*—Only the previous month invoice amounts will be added here.
4. *Overdue Accounts. 60 Days*—Only the second previous month invoice amounts will be added here.
5. *Overdue Accounts. 90 Days and Over*—All past due accounts beyond 60 days will be added here.
6. *Final Totals. Total Charges, Current Charges, Overdue Accounts 30, 60, and 90 Days and Over.*
7. *Percentages for Current Charges, and Overdue Accounts for each column.* Each total divided by Total Charges.

Output

AGED TRIAL BALANCE REPORT

CUSTOMER NUMBER	CUSTOMER NAME	CREDIT LIMIT	TOTAL CHARGES	CURRENT CHARGES	O V E R D U E A C C O U N T S 30 Days	60 days	90 DAYS AND OVER
10867	ALLEN & CO.	15,000.00	7,296.35	6,919.77	376.58	.00	.00
16535	ANDERSON AUTO SUPPLY	2,500.00	1,665,49	1,665.49	.00	.00	.00
17849	ANDREWS AND SONS INC.	750.00	146.64	.00	.00	146.64	.00
18978	ARGONAUT ENGINEERING	2,000.00	3,458.41	2,444.30	611.54	312.13	90.44
24743	BERKLEY PAPER CO.	6,300.00	5,289.00	1,185.50	2,652.45	1,400.05	51.00
25271	BEST DISTRIBUTING CO.	1,000.00	765.44	3.25	.00	.00	762.19
	TOTALS		18,621.33	12,218.31	3,640.57	1,858.82	903.63
	PERCENTAGES		100 %	65 %	20 %	10 %	5 %

Problem 2

Invoice

Input

There are two record types in the customer file: Name/Address record, and Transaction record.

Field	Positions	Format
Name/Address Record		
Code	1	Letter N
Account Number	2-6	
Name	7-26	
Address No. 1	27-44	
Address No. 2	45-62	
Address No. 3	63-80	
Transaction Record		
Code	1	Letter T
Account Number	2-6	
Not Used	7-8	
Item Number	9-14	
Description	15-29	
Quantity	30-34	
Unit Price	35-39	XXX.XX
Not Used	40-80	

Computations to Be Performed

Multiply quantity by unit price giving amount (round to two decimal places). An invoice like the one shown in the output format is to be prepared. The input file is so organized that all transaction records for a customer follow the customer's name/address record.

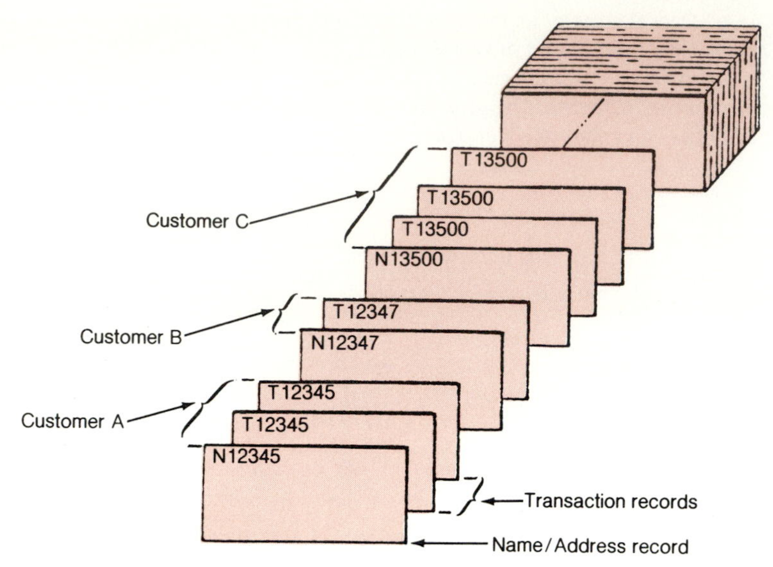

There will always be one name/address record for each customer, but there may be one or more transaction records per customer.

Output

```
                                I N V O I C E
ACCOUNT NUMBER   68252

NAME             JUDGE STORES INC.

ADDRESS          210 SO MAIN ST
                 LOS ANGELES
                 CALIF  90006

SHIPPING INSTRUCTIONS BY AIR

ITEM NUMBER      DESCRIPTION        QUANTITY     UNIT PRICE     AMOUNT

167242           GAS STOVES             4           150.75       603.00

267415           TABLES                 7           150.55     1,053.85

672637           TOP CHAIRS            15            10.00       150.00

786424           BLACK DECKS            5           110.00       550.00

                                              INVOICE TOTAL 2,356.85
```

Problem 3

Sales Problem

Input

Field		Positions	Format
Salesperson Number		1-6	
Area 1 Sales	TERR1	7-12	XXXX.XX
Area 2 Sales		13-18	XXXX.XX
Area 3 Sales		19-24	XXXX.XX
Area 4 Sales	TERR2	25-30	XXXX.XX
Area 5 Sales		31-36	XXXX.XX
Area 6 Sales		37-42	XXXX.XX
Area 7 Sales	TERR3	43-48	XXXX.XX
Area 8 Sales		49-54	XXXX.XX
Area 9 Sales		55-60	XXXX.XX

Computations to Be Performed

Write a subroutine to add the three area sales into their respective territories 1, 2, and 3. The subroutine is to also accumulate a line total for each salesperson as well as final totals for Territory 1, Territory 2, Territory 3, and Total fields.

There is one record for each salesperson. Salesperson operate in nine areas divided into three territories. The report is to show the total sales in each territory, a total of all sales for each salesperson, and a final total of all fields.

Output

```
                              S A L E S   R E P O R T

SALESMAN          TERRITORY          TERRITORY          TERRITORY          TOTAL
 NUMBER               1                  2                  3

123456             1,312.00          14,916.84          13,534.86        29,763.70
267024             4,646.55          10,131.90          12,751.59        27,530.04
360572             9,334.41           6,387.46          13,169.74        28,891.61

                  15,292.96          31,436.20          39,456.19        86,185.35  *
```

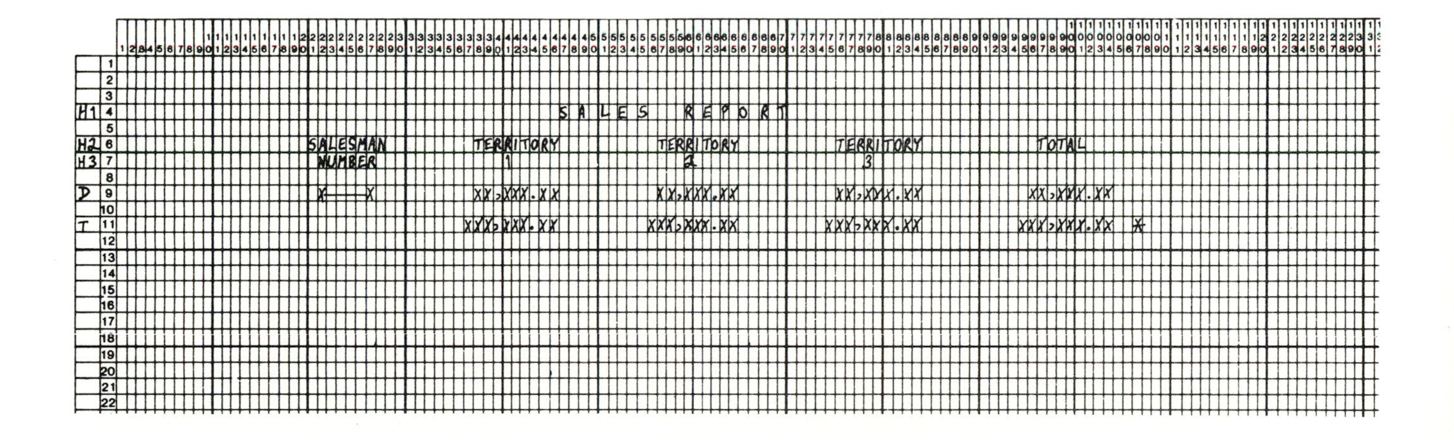

Problem 4

Write a program using the following information:

Input

Master

Field	*Positions*	*Format*
Code-M	1	
Balance	2-7	
Date	8-12	mm/dd/y
Cumulative Disb.	13-18	
Cumulative Receipts	19-24	
Not Used	25-32	
Stock Number	33-38	
Not Used	39-48	
Minimum Balance	49-54	
Unit	55-56	
Description	57-80	

Receipts

Field	Positions
Code-R	1
Quantity Received	2-7
Date	8-12
Not Used	13-32
Stock Number	33-38
Not Used	39-80

Disbursements

Field	Positions
Code-D	1
Quantity Disbursed	2-7
Date	8-12
Not Used	13-32
Stock Number	33-38
Not Used	39-80

Computations to Be Performed

1. Compute on-hand quantity = master balance + quantity received − quantity disbursed.
2. Compute accumulated disbursements = master cumulative disbursements + quantity disbursed.
3. Compute accumulated receipts = master cumulative receipts + quantity received.
4. If computed quantity on hand is less than minimum balance, print message ITEM BELOW MINIMUM.
5. Print report per output format.

Output

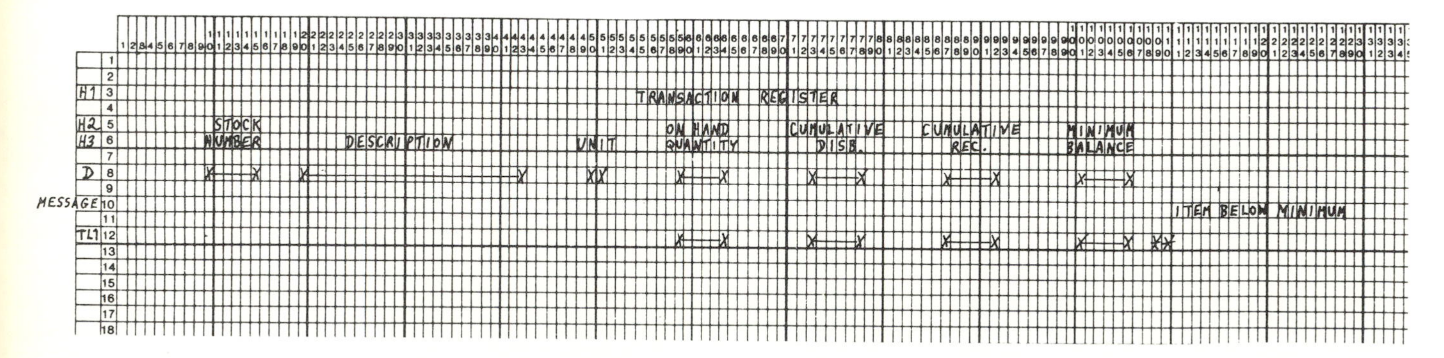

Appendixes

Appendix A

Debugging COBOL Programs

Compiler Diagnostics

Diagnostic messages are generated by the compiler and listed on the systems printer when errors are found in the source program (fig. A.1). A complete listing of diagnostic messages will be found in the programmer's guide reference manual for that particular computer.

Debugging Diagnostic Messages

1. Approach each diagnostic message in sequence as it appears in the compilation source listing. It is possible to get compound diagnostic messages as frequently as an earlier diagnostic message indicates the reason for a later diagnostic message. For example, a missing quotation mark for a nonnumeric literal could involve the inclusion of some clauses not intended for that particular literal. This could cause an apparently valid clause to be diagnosed as invalid because it is not complete, or because it is in conflict with something that preceded it.
2. Check for missing or superfluous punctuation or errors of this type.
3. Frequently, a seemingly meaningless message is clarified when the invalid syntax or format of the clause or statement in question is referenced.

Diagnostic Messages

The diagnostic messages associated with the compilation are always listed. The format of the diagnostic message for IBM computers is:

1. *Compiler Generated Number.* This is the number of a line in the source program related to the error.
2. *Message Identifier.* Message identification for the system.
3. *The Severity Level.* There are four severity levels as follows:

   (W) WARNING. This level indicates that an error was made in the source program. However, it is not serious enough to interfere with the execution of the program.

   (C) CONDITIONAL. This level indicates that an error was made, but the compiler usually makes a corrective assumption. The statement containing the error is retained. Execution can be attempted.

   (E) ERROR. This level indicates that a serious error was made. Usually the compiler makes no corrective assumption. The statement or operand containing the error is dropped. Compilation is completed, but execution of the program should not be attempted.

   (D) DISASTER. This level indicates that a serious error was made. Compilation is not completed and results are unpredictable.
4. *Message Text.* The text identifies the condition that causes the error and indicates the actions taken by the compiler.

Execution Output

The output generated by the program execution (in addition to data written on output files) may include

1. Data displayed on the console or on the printer.
2. Messages to the operator.
3. System informative messages.
4. System diagnostic messages.
5. A system dump.

A dump and system diagnostic messages are generated automatically during the program execution only if the program contains errors that cause the abnormal termination of the program.

```
CARD   ERROR MESSAGE

       ILA1100I-W   1 SEQUENCE ERROR IN SOURCE PROGRAM.
7      ILA1095I-W   WORD 'SECTION' OR 'DIVISION' MISSING. ASSUMED PRESENT.
12     ILA1132I-E   INVALID SYSTEM-NAME. SKIPPING TO NEXT CLAUSE.
13     ILA1132I-E   INVALID SYSTEM-NAME. SKIPPING TO NEXT CLAUSE.
19     ILA1056I-E   FILE-NAME NOT DEFINED IN A SELECT. DESCRIPTION IGNORED.
35     ILA1056I-E   FILE-NAME NOT DEFINED IN A SELECT. DESCRIPTION IGNORED.
46     ILA1077I-C   ALPHANUMERIC LIT CONTINUES IN A-MARGIN. ASSUME B-MARGIN.
50     ILA1077I-C   ALPHANUMERIC LIT CONTINUES IN A-MARGIN. ASSUME B-MARGIN.
51     ILA1077I-C   ALPHANUMERIC LIT CONTINUES IN A-MARGIN. ASSUME B-MARGIN.
55     ILA1077I-C   ALPHANUMERIC LIT CONTINUES IN A-MARGIN. ASSUME B-MARGIN.
56     ILA1077I-C   ALPHANUMERIC LIT CONTINUES IN A-MARGIN. ASSUME B-MARGIN.
56     ILA1076I-C   ALPHANUMERIC LIT EXCEEDS 120 CHARACTERS. TRUNCATED TO 120.
79     ILA1037I-E   * INVALID IN DATA DESCRIPTION. SKIPPING TO NEXT CLAUSE.
79     ILA2039I-C   PICTURE CONFIGURATION ILLEGAL. PICTURE CHANGED TO 9 UNLESS USAGE IS 'DISPLAY-ST',
                    THEN L(6)BDZ9BDZ9.
83     ILA1037I-E   ** INVALID IN DATA DESCRIPTION. SKIPPING TO NEXT CLAUSE.
83     ILA2039I-C   PICTURE CONFIGURATION ILLEGAL. PICTURE CHANGED TO 9 UNLESS USAGE IS 'DISPLAY-ST',
                    THEN L(6)BDZ9BDZ9.
85     ILA3001I-E   FILE-IN NOT DEFINED. DELETING TILL LEGAL ELEMENT FOUND.
85     ILA3001I-E   FILE-OUT NOT DEFINED. DELETING TILL LEGAL ELEMENT FOUND.
85     ILA4002I-E   OPEN STATEMENT INCOMPLETE. STATEMENT DISCARDED.
86     ILA4050I-E   SYNTAX REQUIRES RECORD-NAME . FOUND DNM=1-337 . STATEMENT DISCARDED.
86     ILA3001I-E   HDG-1 NOT DEFINED.
86     ILA3001I-E   0 NOT DEFINED.
87     ILA4050I-E   SYNTAX REQUIRES RECORD-NAME . FOUND DNM=1-337 . STATEMENT DISCARDED.
88     ILA4050I-E   SYNTAX REQUIRES RECORD-NAME . FOUND DNM=1-337 . STATEMENT DISCARDED.
89     ILA3001I-E   FILE-IN NOT DEFINED. STATEMENT DISCARDED.
94     ILA3001I-E   TOT-DED-WS NOT DEFINED. SUBSTITUTING TALLY .
105    ILA4091I-E   SYNTAX REQUIRES OPERAND. FOUND END OF PAGE . TEST DISCARDED.
106    ILA4050I-E   SYNTAX REQUIRES RECORD-NAME . FOUND DNM=1-337 . STATEMENT DISCARDED.
108    ILA5011I-W   HIGH ORDER TRUNCATION MIGHT OCCUR.
110    ILA4050I-E   SYNTAX REQUIRES RECORD-NAME . FOUND DNM=1-337 . STATEMENT DISCARDED.
114    ILA1077I-C   ALPHANUMERIC LIT CONTINUES IN A-MARGIN. ASSUME B-MARGIN.
113    ILA4003I-E   EXPECTING NEW STATEMENT. FOUND TO . DELETING TILL NEXT VERB OR PROCEDURE NAME.
116    ILA5011I-W   HIGH ORDER TRUNCATION MIGHT OCCUR.
117    ILA4050I-E   SYNTAX REQUIRES RECORD-NAME . FOUND DNM=1-337 . STATEMENT DISCARDED.
117    ILA3001I-E   FINAL-TOTAL NOT DEFINED.
```

Figure A.1 Diagnostic messages.

C110A	STOP literal
Explanation:	The programmer has issued a STOP literal statement in the American National Standard COBOL source program.
System Action:	Awaits operator response.
Programmer response:	Not applicable.
Operator response:	Operator should respond with end-of-block, or with any character in order to proceed with the program.

C111A	AWAITING REPLY
Explanation:	This message is issued in connection with the American National Standard COBOL ACCEPT statement.
System action:	Awaits operator reponse.
Programmer response:	Not applicable.
Operator response:	The operator should reply as specified by the programmer.

Figure A.2 Object time messages—console.

Operator Messages

The COBOL phase may issue operator messages. In the message, XX denotes a system-iterated two-character numeric file that is used to identify the program issuing the message. (See figure A.2.)

STOP Statement

The following message is generated by the STOP statement with the *literal* option:

XX C110A STOP 'literal'

This message is issued at the programmer's discretion to indicate possible alternative action to be taken by the operator.

The operator responds according to the instructions given both by the message and a on-the-job request form supplied by the programmer. If the job is to be resumed, the programmer presses the end-of-block key on the console.

ACCEPT Statement

The following message is generated by an ACCEPT statement with the FROM CONSOLE option:

```
XX C111A "AWAITING REPLY"
```

This message is issued by the object program when operator intervention is required.

The operator responds by entering the reply and by pressing the end-of-block key on the console. (The contents of the text field should be supplied by the programmer on-the-job request form.)

System Output

Informative and diagnostic messages may appear in the listing during the execution of the object program.

Each of these messages contains an identification code in the first column of the message to indicate the portion of the operating system that generated the message.

Dump

If a serious error occurs during the execution of the problem program, the programmer can request a printout of storage through the use of the DUMP option in the job-control cards. The job would be abnormally terminated, any remaining steps bypassed, and a program phase dump is generated. The programmer can use the dump to check out the program. In cases where a serious error occurs in other than the problem program (for example, in the control program), a dump is not produced. (*Note:* The program phase dump can be suppressed if the NODUMP option of the job-control statement has been specified.)

How to Use a Dump

When a job is abnormally terminated due to a serious error in the problem program, a message is written on the system output device, which indicates the following:

1. Type of interrupt (for example, program check).
2. The hexadecimal (IBM) address of the instruction that caused the interrupt.
3. Condition code.

The hexadecimal address of the instruction that caused the dump is subtracted from the load address of the module (which can be obtained from the map of main storage generated by the Linkage Editor) to obtain the relative instruction address as shown in the Procedure Division map. If the interrupt occurred within the COBOL program, the programmer can use the error address to locate the specific statement that caused a dump to be generated. Examination of the statement and fields associated with it may produce information as to the specific nature of the error.

A sample dump caused by a data exception follows. Invalid data (for example, data that did not correspond to its usage) was placed in the numeric field B as a result of redefinition. Letters identify the text corresponding to the letter in the program listing.

```
 CBL QUOTE,SEQ
00001    000010 IDENTIFICATION DIVISION.
00002    000020 PROGRAM-ID. TESTRUN.
00003    000030     AUTHOR. PROGRAMMER NAME.
00004    000040     INSTALLATION. NEW YORK PROGRAMMING CENTER.
00005    000050     DATE-WRITTEN.  FEBRUARY 4, 1971
00006    000060 DATE-COMPILED. 04/24/71
00007    000070    REMARKS. THIS PROGRAM HAS BEEN WRITTEN AS A SAMPLE PROGRAM FOR
00008    000080     COBOL USERS. IT CREATES AN OUTPUT FILE AND READS IT BACK AS
00009    000090     INPUT.
00010    000100
00011    000110 ENVIRONMENT DIVISION.
00012    000120 CONFIGURATION SECTION.
00013    000130 SOURCE-COMPUTER. IBM-360-H50.
00014    000140 OBJECT-COMPUTER. IBM-360-H50.
00015    000150 INPUT-OUTPUT SECTION.
00016    000160 FILE-CONTROL.
00017    000170     SELECT FILE-1 ASSIGN TO SYS008-UT-2400-S.
00018    000180     SELECT FILE-2 ASSIGN TO SYS008-UT-2400-S.
00019    000190
00020    000200 DATA DIVISION.                                    }
00021    000210 FILE SECTION.                                     }
00022    000220 FD  FILE-1                                        }
00023    000230     LABEL RECORDS ARE OMITTED                     }
00024    000240     BLOCK CONTAINS 5 RECORDS                      }
00025    000250     RECORDING MODE IS F                           }
00026    000255     RECORD CONTAINS 20 CHARACTERS                 }
00027    000260     DATA RECORD IS RECORD-1.                      }
00028    000270 01  RECORD-1.                                     }  (E)
00029    000280     05 FIELD-A PIC X(20).                         }
00030    000290 FD  FILE-2                                        }
00031    000300     LABEL RECORDS ARE OMITTED                     }
00032    000310     BLOCK CONTAINS 5 RECORDS                      }
00033    000320     RECORD CONTAINS 20 CHARACTERS                 }
00034    000330     RECORDING MODE IS F                           }
00035    000340     DATA RECORD IS RECORD-2.                      }
00036    000350 01  RECORD-2.                                     }
00037    000360     05 FIELD-A PIC X(20).                         }

00038    000370 WORKING-STORAGE SECTION.
00039    000380 01  FILLER.
00040    000390     02 COUNT PIC S99 COMP SYNC.
00041    000400     02 ALPHABET PIC X(26) VALUE IS "ABCDEFGHIJKLMNOPQRSTUVWXYZ".
00042    000410     02 ALPHA REDEFINES ALPHABET PIC X OCCURS 26 TIMES.
00043    000420     02 NUMBR PIC S99 COMP SYNC.
00044    000430     02 DEPENDENTS PIC X(26) VALUE "01234012340123401234012340".
00045    000440     02 DEPEND REDEFINES DEPENDENTS PIC X OCCURS 26 TIMES.
00046    000450 01  WORK-RECORD.
00047    000460     05 NAME-FIELD PIC X.
00048    000470     05 FILLER PIC X.
00049    000480     05 RECORD-NO PIC 9999.
00050    000490     05 FILLER PIC X VALUE IS SPACE.
00051    000500     05 LOCATION PIC AAA VALUE IS "NYC".
00052    000510     05 FILLER PIC X VALUE IS SPACE.
00053    000520     05 NO-OF-DEPENDENTS PIC XX.
00054    000530     05 FILLER PIC X(7) VALUE IS SPACES.
00055    000534 01  RECORDA.
00056    000535     02 A PICTURE S9(4) VALUE 1234.
00057    000536     02 B REDEFINES A PICTURE S9(7) COMPUTATIONAL-3.
00058    000540
00059    000550 PROCEDURE DIVISION.
00060    000560 BEGIN. READY TRACE.
00061    000570     NOTE THAT THE FOLLOWING OPENS THE OUTPUT FILE TO BE CREATED
00062    000580     AND INITIALIZES COUNTERS.
00063    000590 STEP-1. OPEN OUTPUT FILE-1. MOVE ZERO TO COUNT, NUMBR.
00064    000600     NOTE THAT THE FOLLOWING CREATES INTERNALLY THE RECORDS TO BE
00065    000610     CONTAINED IN THE FILE, WRITES THEM ON TAPE, AND DISPLAYS
00066    000620     THEM ON THE CONSOLE.
00067    000630 STEP-2. ADD 1 TO COUNT, NUMBR. MOVE ALPHA (COUNT) TO
00068    000640     NAME-FIELD.
00069    000645         COMPUTE B = B + 1. <--(D)
00070    000650     MOVE DEPEND (COUNT) TO NO-OF-DEPENDENTS.
00071    000660     MOVE NUMBR TO RECORD-NO.
00072    000670 STEP-3. DISPLAY WORK-RECORD UPON CONSOLE. WRITE RECORD-1 FROM
00073    000680     WORK-RECORD.
00074    000690 STEP-4. PERFORM STEP-2 THRU STEP-3 UNTIL COUNT IS EQUAL TO 26.
00075    000700     NOTE THAT THE FOLLOWING CLOSES THE OUTPUT FILE AND REOPENS
00076    000710     IT AS INPUT.
00077    000720 STEP-5. CLOSE FILE-1. OPEN INPUT FILE-2.
00078    000730     NOTE THAT THE FOLLOWING READS BACK THE FILE AND SINGLES
00079    000740     OUT EMPLOYEES WITH NO DEPENDENTS.
00080    000750 STEP-6. READ FILE-2 RECORD INTO WORK-RECORD AT END GO TO STEP-8.
00081    000760 STEP-7. IF NO-OF-DEPENDENTS IS EQUAL TO "0" MOVE "Z" TO
00082    000770     NO-OF-DEPENDENTS. EXHIBIT NAMED WORK-RECORD. GO TO STEP-6.
00083    000780 STEP-8. CLOSE FILE-2.
00084    000790     STOP RUN.
```

INTRNL NAME	LVL	SOURCE NAME	BASE	DISPL	INTRNL NAME	DEFINITION	USAGE	R	O	Q	M
DNM=1-148	FD	FILE-1	DTF=01		DNM=1-148		DTFMT				F
DNM=1-178	01	RECORD-1	BL=1	000	DNM=1-178	DS OCL20	GROUP				
DNM=1-199	02	FIELD-A	BL=1	000	DNM=1-199	DS 20C	DISP				
DNM=1-216	FD	FILE-2	DTF=02		DNM=1-216		DTFMT				F
DNM=1-246	01	RECORD-2	BL=2	000	DNM=1-246	DS OCL20	GROUP				
DNM=1-267	02	FIELD-A	BL=2	000	DNM=1-267	DS 20C	DISP				
DNM=1-287	01	FILLER	BL=3	000	DNM=1-287	DS OCL56	GROUP				
DNM=1-306	02	COUNT	BL=3	000	DNM=1-306	DS 1H	COMP				
DNM=1-321	02	ALPHABET	BL=3	002	DNM=1-321	DS 26C	DISP				
DNM=1-339	02	ALPHA	BL=3	002	DNM=1-339	DS 1C	DISP	R	O		
DNM=1-357	02	NUMBR	BL=3	01C	DNM=1-357	DS 1H	COMP				
DNM=1-372	02	DEPENDENTS	BL=3	01E	DNM=1-372	DS 26C	DISP				
DNM=1-392	02	DEPEND	BL=3	01E	DNM=1-392	DS 1C	DISP	R	O		
DNM=1-408	01	WORK-RECORD	BL=3	038	DNM=1-408	DS OCL20	GROUP				
DNM=1-432	02	NAME-FIELD	BL=3	038	DNM=1-432	DS 1C	DISP				
DNM=1-452	02	FILLER	BL=3	039	DNM=1-452	DS 1C	DISP				
DNM=1-471	02	RECORD-NO	BL=3	03A	DNM=1-471	DS 4C	DISP-NM				
DNM=1-490	02	FILLER	BL=3	03E	DNM=1-490	DS 1C	DISP				
DNM=2-000	02	LOCATION	BL=3	03F	DNM=2-000	DS 3C	DISP				
DNM=2-018	02	FILLER	BL=3	042	DNM=2-018	DS 1C	DISP				
DNM=2-037	02	NO-OF-DEPENDENTS	BL=3	043	DNM=2-037	DS 2C	DISP				
DNM=2-063	02	FILLER	BL=3	045	DNM=2-063	DS 7C	DISP				
DNM=2-082	01	RECORDA	BL=3	050	DNM=2-082	DS OCL4	GROUP				
DNM=2-102	02	A	BL=3	050	DNM=2-102	DS 4C	DISP-NM				
DNM=2-113	02	B ← J	BL=3	050	DNM=2-113	DS 4P	COMP-3	R			

MEMORY MAP

TGT	003E8
SAVE AREA	003E8
SWITCH	00430
TALLY	00434
SORT SAVE	00438
ENTRY-SAVE	0043C
SORT CORE SIZE	00440
NSTD-REELS	00444
SORT RET	00446
WORKING CELLS	00448
SORT FILE SIZE	00578
SORT MODE SIZE	0057C
PGT-VN TBL	00580
TGT-VN TBL	00584
SORTAB ADDRESS	00588
LENGTH OF VN TBL	0058C
LNGTH OF SORTAB	0058E
PGM ID	00590
A(INIT1)	00598
UPSI SWITCHES	0059C
OVERFLOW CELLS	005A4
BL CELLS ← N	005A4
DTFADR CELLS	005B0 ← F
TEMP STORAGE	005B8
TEMP STORAGE-2	005C0
TEMP STORAGE-3	005C0
TEMP STORAGE-4	005C0
BLL CELLS	005C0
VLC CELLS	005C4
SBL CELLS	005C4
INDEX CELLS	005C4
SUBADR CELLS	005C4
ONCTL CELLS	005CC
PFMCTL CELLS	005CC
PFMSAV CELLS	005CC
VN CELLS	005D0
SAVE AREA =2	005D4
XSASW CELLS	005D4
XSA CELLS	005D4
PARAM CELLS	005D4
RPTSAV AREA	005D8
CHECKPT CTR	005D8
IOPTR CELLS	005D8

```
              REGISTER ASSIGNMENT

                REG 6    BL =3  <--(K)
                REG 7    BL =1
                REG 8    BL =2

67     0006FC  41 40 6 002                 LA    4,002(0,6)             DNM=1-339
       000700  48 20 6 000                 LH    2,000(0,6)             DNM=1-306
       000704  4C 20 C 03A                 MH    2,03A(0,12)            LIT+2
       000708  1A 42                       AR    4,2
       00070A  5B 40 C 038                 S     4,038(0,12)            LIT+0
       00070E  50 40 D 1DC                 ST    4,1DC(0,13)            SBS=1
       000712  58 E0 D 1DC                 L     14,1DC(0,13)           SBS=1
       000716  D2 00 6 038 E 000           MVC   038(1,6),000(14)       DNM=1-432     DNM=1-339
69     00071C  FA 30 6 050 C 03C    (C)--> AP    050(4,6),03C(1,12)     DNM=2-113     LIT+4
70     000722  41 40 6 01E                 LA    4,01E(0,6)             DNM=1-392
       000726  48 20 6 000                 LH    2,000(0,6)             DNM=1-306
       00072A  4C 20 C 03A                 MH    2,03A(0,12)            LIT+2
       00072E  1A 42                       AR    4,2
       000730  5B 40 C 038                 S     4,038(0,12)            LIT+0
       000734  50 40 D 1E0                 ST    4,1E0(0,13)            SBS=2
       000738  58 E0 D 1E0                 L     14,1E0(0,13)           SBS=2
       00073C  D2 00 6 043 E 000           MVC   043(1,6),000(14)       DNM=2-37      DNM=1-392
       000742  92 40 6 044                 MVI   044(6),X'40'           DNM=2-37+1

// EXEC LNKEDT


         PHASE     XFR-AD   LOCORE   HICORE   DSK-AD    ESD TYPE   LABEL      LOADED   REL-FR

      PHASE***     0032A0   0032A0   004ADB   53 01 2   CSECT      TESTRUN    0032A0   0032A0  <--(B)

                                                        CSECT      IJFFBZZN   003C50   003C50
                                                        *  ENTRY   IJFFZZZN   003C50
                                                        *  ENTRY   IJFFBZZZ   003C50
                                                        *  ENTRY   IJFFZZZZ   003C50

                                                        CSECT      ILBDSAE0   0049F0   0049F0
                                                           ENTRY   ILBDSAE1   004A06

                                                        CSECT      ILBDMNS0   0049E8   0049E8

                                                        CSECT      ILBDDSP0   0041B8   0041B8
                                                        *  ENTRY   ILBDDSP1   004708
                                                        *  ENTRY   ILBDDSP2   0047A0
                                                        *  ENTRY   ILBDDSP3   004958

                                                        CSECT      ILBDIML0   004990   004990

                                                        CSECT      IJJCPD1    003FC0   003FC0
                                                           ENTRY   IJJCPD1N   003FC0
                                                        *  ENTRY   IJJCPD3    003FC0


// ASSGN SYS008,X'182'
// EXEC


0S03I PROGRAM CHECK INTERRUPTION - HEX LOCATION 0039BC - CONDITION CODE 0 - DATA EXCEPTION
0S00I JOB DTACHK    CANCELED
```

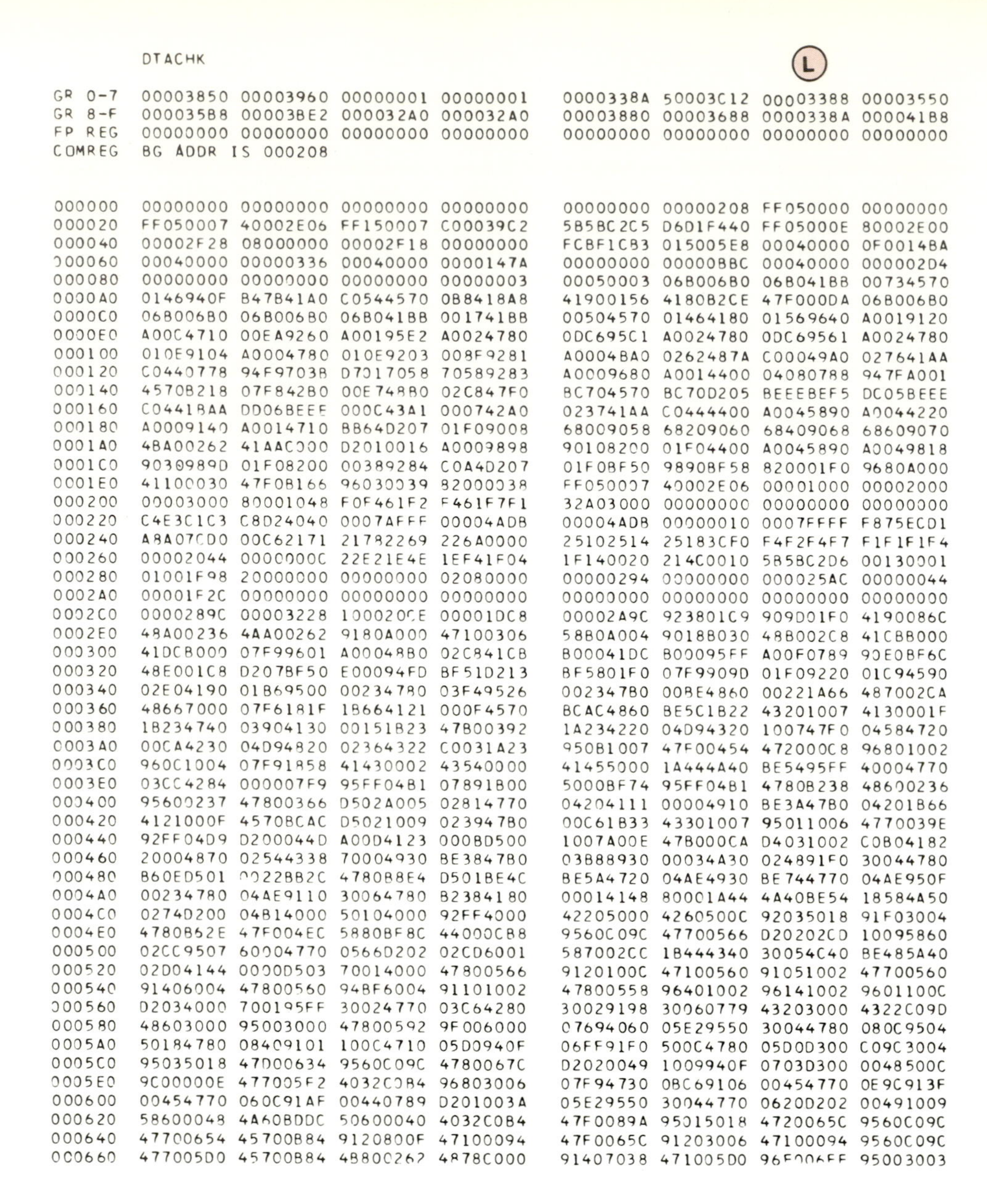

```
         DTACHK                                                                     (L)

GR 0-7   00003850 00003960 00000001 00000001    0000338A 50003C12 00003388 00003550
GR 8-F   000035B8 00003BE2 000032A0 000032A0    00003880 00003688 0000338A 000041B8
FP REG   00000000 00000000 00000000 00000000    00000000 00000000 00000000 00000000
COMREG   BG ADDR IS 000208

000000   00000000 00000000 00000000 00000000    00000000 00000208 FF050000 00000000
000020   FF050007 40002E06 FF150007 C00039C2    5B5BC2C5 D6D1F440 FF05000E 80002E00
000040   00002F28 08000000 00002F18 00000000    FCBF1CB3 015005E8 00040000 0F0014BA
000060   00040000 00000336 00040000 0000147A    00000000 00000BBC 00040000 000002D4
000080   00000000 00000000 00000000 00000003    00050003 06B006B0 06B041BB 00734570
0000A0   0146940F B47B41A0 C0544570 0B8418A8    41900156 4180B2CE 47F000DA 06B006B0
0000C0   06B006B0 06B006B0 06B041BB 001741BB    00504570 01464180 01569640 A0019120
0000E0   A00C4710 00EA9260 A00195E2 A0024780    0DC695C1 A0024780 0DC69561 A0024780
000100   010E9104 A0004780 010E9203 008F9281    A0004BA0 0262487A C00049A0 027641AA
000120   C0440778 94F9703B D7017058 70589283    A0009680 A0014400 04080788 947FA001
000140   4570B218 07F842B0 00E748B0 02C847F0    BC704570 BC70D205 BEEEBEF5 DC05BEEE
000160   C0441BAA DD06BEEE 000C43A1 000742A0    023741AA C0444400 A0045890 A0044220
000180   A0009140 A0014710 BB64D207 01F09008    68009058 68209060 68409068 68609070
0001A0   4BA00262 41AAC000 D2010016 A0009898    90108200 01F04400 A0045890 A0049818
0001C0   9030989D 01F08200 00389284 C0A4D207    01F0BF50 9890BF58 820001F0 9680A000
0001E0   41100030 47F0B166 96030039 82000038    FF050007 40002E06 00001000 00002000
000200   00003000 80001048 F0F461F2 F461F7F1    32A03000 00000000 00000000 00000000
000220   C4E3C1C3 C8D24040 0007AFFF 00004ADB    00004ADB 00000010 0007FFFF F875ECD1
000240   A8A07CD0 00C62171 21782269 226A0000    25102514 25183CF0 F4F2F4F7 F1F1F1F4
000260   00002044 0000000C 22E21E4E 1EF41F04    1F140020 214C0010 5B5BC2D6 00130001
000280   01001F98 20000000 00000000 02080000    00000294 00000000 000025AC 00000044
0002A0   00001F2C 00000000 00000000 00000000    00000000 00000000 00000000 00000000
0002C0   0000289C 00003228 100020CE 00001DC8    00002A9C 923801C9 909D01F0 4190086C
0002E0   48A00236 4AA00262 9180A000 47100306    58B0A004 9018B030 48B002C8 41CBB000
000300   41DCB000 07F99601 A00048B0 02C841CB    B00041DC B00095FF A00F0789 90E0BF6C
000320   48E001C8 D207BF50 E00094FD BF51D213    BF5801F0 07F9909D 01F09220 01C94590
000340   02E04190 01B69500 00234780 03F49526    002347B0 00BE4860 00221A66 487002CA
000360   48667000 07F6181F 1B664121 000F4570    BCAC4860 BE5C1B22 43201007 4130001F
000380   1B234740 03904130 00151B23 47B00392    1A234220 04D94320 100747F0 04584720
0003A0   00CA4230 04D94820 02364322 C0031A23    950B1007 47F00454 472000C8 96801002
0003C0   960C1004 07F91858 41430002 43540000    41455000 1A444A40 BE5495FF 40004770
0003E0   03CC4284 000007F9 95FF04B1 07891B00    5000BF74 95FF04B1 4780B238 48600236
000400   95600237 47800366 D502A005 02814770    04204111 00004910 BE3A47B0 04201B66
000420   4121000F 4570BCAC D5021009 023947B0    00C61B33 43301007 95011006 4770039E
000440   92FF04D9 D200044D A00D4123 000BD500    1007A00E 47B000CA D4031002 C0B04182
000460   20004870 02544338 70004930 BE3847B0    03B88930 00034A30 024891F0 30044780
000480   B60ED501 0022BB2C 4780B8E4 D501BE4C    BE5A4720 04AE4930 BE744770 04AE950F
0004A0   00234780 04AE9110 30064780 B2384180    00014148 80001A44 4A40BE54 18584A50
0004C0   0274D200 04B14000 50104000 92FF4000    42205000 4260500C 92035018 91F03004
0004E0   4780B62E 47F004EC 5880BF8C 44000CB8    9560C09C 47700566 D20202CD 10095860
000500   02CC9507 60004770 0566D202 02CD6001    587002CC 1B444340 30054C40 BE485A40
000520   02D04144 0000D503 70014000 47800566    9120100C 47100560 91051002 47700560
000540   91406004 47800560 94BF6004 91101002    47800558 96401002 96141002 9601100C
000560   D2034000 700195FF 30024770 03C64280    30029198 30060779 43203000 4322C09D
000580   48603000 95003000 47800592 9F006000    07694060 05E29550 30044780 080C9504
0005A0   50184780 08409101 100C4710 05D0940F    06FF91F0 500C4780 05D0D300 C09C3004
0005C0   95035018 47D00634 9560C09C 4780067C    D2020049 1009940F 0703D300 0048500C
0005E0   9C00000E 477005F2 4032C0B4 96803006    07F94730 0BC69106 00454770 0E9C913F
000600   00454770 060C91AF 00440789 D201003A    05E29550 30044770 0620D202 00491009
000620   58600048 4A60BDDC 50600040 4032C0B4    47F0089A 95015018 4720065C 9560C09C
000640   47700654 45700B84 9120800F 47100094    47F0065C 91203006 47100094 9560C09C
000660   477005D0 45700B84 4B800262 4878C000    91407038 471005D0 96F006FF 95003003
```

This sample dump (concluding on the facing page) was caused by a data exception. Invalid data (i.e., data which did not correspond to its usage) was placed in the numeric field B as a result of redefinition. The following notes illustrate the method of finding the specific statement in the program which caused the dump. Letters identifying the text correspond to letter in the program listing.

Ⓐ The program interrupt occurred at HEX LOCATION 0039BC. This is indicated in the SYSLST message printed just before the dump.

Ⓑ The linkage editor map indicates that the program was loaded into address 0032A0. This is determined by examining the load point of the control section TESTRUN. TESTRUN is the name assigned to the program module by the source coding:
PROGRAM-ID. TESTRUN.

Ⓒ The specific instruction which caused the dump is located by subtracting the load address from the interrupt address (i.e., subtracting 32A0 from 39BC). The result, 71C, is the relative interrupt address and can be found in the object code listing. In this case the instruction in question is an AP (add decimal).

DTACHK

```
0032E0   00005218 00000208 00000000 00000000   00000000 00000000 00000000 00000000
003300   00000000 --SAME--
003320   00000000 58C0F0C6 58E0C000 58D0F0CA   9500E000 4770F0A2 9610D048 92FFE000
003340   47F0F0AC 98CEF03A 90ECD00C 185D989F   F0BA9110 D0480719 07FF0700 00003BE2
003360   000032A0 000032A0 00003880 00003688   000038EC 00003BC8 C3D6C2C6 F0F0F0F1
003380   E3C5E2E3 D9E4D540 0001C1C2 C3C4C5C6   C7C8C9D1 D2D3D4D5 D6D7D8D9 E2E3E4E5
0033A0   E6E7E8E9 0001F0F1 F2F3F4F0 F1F2F3F4   F0F1F2F3 F4F0F1F2 F3F4F0F1 F2F3F4F0
0033C0   C1000000 000040D5 E8C34000 00404040   40404040 00000000 F1F2F3C4 00004C40
0033E0   01010014 00000000 00000000 00000000   0E000000 04000000 00009200 00000108
003400   00003430 00000000 10003C50 1160E2E8   E2F0F0F8 40400162 10000000 04000000
003420   00000000 86BCF018 41E0E001 58201044   010034E8 20000064 00003550 00003550
003440   00000014 000035B3 00640063 00000000   00000000 000049F0 01010014 00000000
003460   00000000 00000000 00000000 04000000   00008200 00000108 000034A8 00000000
003480   10003C50 1168E2E8 E2F0F0F8 40400272   00000000 20000000 00000000 86BCF018
0034A0   41E0E001 58201044 020035B8 00000064   00003620 00000000 00000014 00000000
0034C0   00640063 00000000 00004A06 000049F0   00000000 00000000 00000000 00000000
0034E0   00000000 00000000 00004770 30129261   10004110 100107F3 D20467CE 6017D201
003500   67D56274 C6C3D6C2 D6D3F8F0 F8F0F1F0   F1F2F1F1 F2F0F2F2 F2F1F3F0 F4F0F5F0
003520   F5F1F6F0 F6F1F7F0 0100DDA8 10006670   20006148 40005DC8 70004C40 41110004
003540   41110004 41110004 58110000 58F10010   45EF0018 41105342 07FB0000 000032B0
003560   000035A4 000035F8 00003DB4 000039B0   00003944 00004096 00003D0A 000032B0
003580   000062B8 00004478 00004C94 00005704   00005A4C 00005B68 0000373E 000035E4
0035A0   000036A6 060C40FF C4B2DE09 D2106276   D207601C D212F363 603B6276 96F06041
0035C0   4110601C 5840C65C 41200008 05301B24   47403018 95401000 47703012 92611000
0035E0   41101001 00000000 6276C494 58F0C340   077F9240 6820D206 00004218 000042B0
003600   00004348 000043E0 000062B8 00004478   00004510 000045F8 D500627C 00000000
003620   000001FF 00003800 00003982 00003968   00003F2C 00003BAA 00003BAA 00003C40
003640   000037DA 00003BAA 00003E6C 00003D60   00004090 00003DBE 00003B20 00003BC6
003660   00003BC6 00000203 02030001 04050104   00000203 00000105 00000404 00000104
003680   04040202 01030000 00202020 20210000   1C404040 40404000 00200000 00006148
0036A0   00180014 0F0F0000 000C1C0C 00000000   58F0C010 000036F4 10000006 0C000822
0036C0   00000000 00040D00 01E40267 00000003   7000004B 00000000 00000000 000038EC
0036E0   00000000 00000000 000033F8 00003550   000032A0 000033F8 50003C12 02AA1000
003700   00100C00 09EE0000 FFFFD201 6030C49A   4810C4A6 06104C10 C48C5010 D24C4810
003720   C4A60610 4C10C48C 5010D264 414062AE   5A40D24C F871D208 4000D205 00000000
003740   50D05362 41D053F6 5430536A 98675366   18809506 800041E0 568E58F0 52BA078F
003760   43680000 8C600004 89600002 8870001B   58B6536E 91508000 477054D0 91A08000
003780   47E054D0 00003958 00003550 01005366   70003934 000041B8 00003850 00003960
0037A0   00003550 000032A0 000033F8 50003C12   00003388 00003550 000035B8 00003BE2
0037C0   000032A0 000032A0 00003880 00003960   000041B8 00003850 00004708 00003550
0037E0   00015540 00003958 58F10010 45EF0008   180747F0 568ED703 532E532E 47F0568E
003800   49A053E2 58C053E6 078C91FF 53D14780   566845B0 55F445B0 00000000 00000000
003820   E000590C 478056AC 91FF53D0 47105686   41B05686 47F055F4 000032A0 91FF53D1
003840   47105598 00003550 000035B8 00003388   000033F8 00003470 00000000 0000001C
003860   00000000 0000338A 42F90000 88F00008   00003A3E 17671776 1767D201 60045366
003880   000049E8 000041B8 00004990 00003950   00003A3E 00003AE0 00003B2C 00003B88
0038A0   00003A5E 00003A72 00003B26 00003B58   00003A3E 504088AE 00000001 1C00001A
0038C0   5B5BC2D6 D7C5D540 5B5BC2C3 D3D6E2C5   5B5BC2C6 C3D4E4D3 F0E90000 C0000000
0038E0   E6D6D9D2 60D9C5C3 D6D9C420 58F0C004   051F0001 4004F6F0 404040AA 9640D048
003900   58F0C004 051F0001 4004F6F3 40404010   4110C040 5800D1C8 184005F0 5000F008
003920   4500F00C 000033F8 0A024100 D1C858F0   C00805EF 5810D1C8 96101020 5020D1BC
003940   5870D1BC D2016000 C038D201 601CC038   58F0C004 051F0001 4004F6F7 404040F1
003960   4830C03A 4A306000 4E30D1D0 D705D1D0   D1D0940F D1D64F30 D1D04030 60004830
003980   C03A4A30 601C4E30 D1D0D705 D1D0D1D0   940FD1D6 4F30D1D0 4030601C 41406002
0039A0   48206000 4C20C03A 1A425B40 C0385040   D1DC58E0 D1DCD200 6038E000 FA306050
0039C0   C03C4140 601E4820 60004C20 C03A1A42   5B40C038 5040D1E0 58E0D1E0 D2006043
0039E0   E0009240 60444830 601C4E30 D1D0F331   603AD1D6 96F0603D 58F0C004 051F0001
003A00   4004F7F2 4040404F 58F0C004 051F0002   00000014 0D0001C4 0038FFFF D2137000
```

(D) The left-hand column of the object code listing gives the compiler-generated line number associated with the instruction. It is line 69. As seen in the source listing, line 69 contains the COMPUTE statement.

(E) The DTF for FILE-1 precedes the DTF for FILE-2.

(F) DTFADR CELLS begin at relative location 5B0.

(G) Since the relocation factor is 32A0, the DTRADR CELLS begin at location 3850 in the dump.

(H) The DTF for FILE-1 begins at location 33F8, and the DTF for FILE-2 begins at location 3470.

Since the problem program in the previous illustration was interrupted because of a data exception, the programmer should locate the contents of field B at the time of the interruption. This can be done as follows:

(J) Locate data-name B in the glossary. It appears under the column headed SOURCE-NAME. Source-Name B has been assigned to base locator 3 (i.e., BL = 3) with a displacement of 050. The sum of the value of base locator 3 and the displacement value 50 is the address of data-name B.

Ⓚ The Register Assignment table lists the registers assigned to each base locator. Register 6 has been assigned to BL = 3.

Ⓛ The contents of the 16 general registers at the time of the interrupt are displayed at the beginning of the dump. Register 6 contains the address 00003388.

Ⓜ The location of data-name B can now be determined by adding the contents of register 6 and the displacement value 50. The result, 33D8, is the address of the leftmost byte of the 4-byte field B.

Note: Field B contains F1F2F3C4. This is external decimal representation and does not correspond to the USAGE COMPUTATIONAL-3 defined in the source listing.

Ⓝ The location assigned to a given data-name may also be found by using the BL CELLS pointer in the TGT Memory Map. The listing indicates that the BL cells begin at location 3844 (add 5A4 to the load point address, 32A0, of the object module). The first four bytes are the first BL cell, the second four bytes are the second BL cell, etc. Note that the third BL cell contains the value 3388. This is the same value as that contained in register 6.

Note: Some program errors may destroy the contents of the general registers or the BL cells. In such cases, alternate methods of locating the DTF's are useful.

Errors That Can Cause the Dump

A dump may be caused by one of many different types of errors. Several of these errors may occur at the COBOL language level, while others can occur at job-control levels.

The following are examples of COBOL language errors that can cause a dump.

1. *A* GO TO *statement with no procedure-name following it.* This statement may have been improperly initialized with an ALTER statement, and the execution of this statement will cause an invalid branch to be taken with unpredictable results.
2. *Moves of arithmetic calculations that have not been properly initialized.* For example, neglecting to initialize the object of an OCCURS clause with the DEPENDING ON option, and referencing data fields prior to the first READ statement may cause a program interrupt or dump.
3. Invalid data placed in a numeric field as a result of redefinition.
4. Input/output errors that are nonrecoverable.
5. An input file contains invalid data, such as blanks or partially blank numeric fields or data incorrectly specified by its data description.

The compiler does not generate a test to check the sign position for a valid configuration before the item is used as an operand. The programmer must test for valid data by means of the class test and by using either the EXAMINE or INSPECT statement to convert it to valid data.

For example, if the high-order positions of a numeric data field contain blanks and are to be involved in a calculation requiring a numeric PICTURE, the blank positions could be transformed to zero through the use of the INSPECT or EXAMINE verbs, thus creating a valid numeric field.

Locating Data in a Dump

The location assigned to a given data-name may be found by using the BL number and displacement given for that entry in the glossary and then locating the appropriate BL in the TGT. The hexadecimal sum of the glossary displacement and the contents of the cell should give the relative address of the desired area. This can be converted to an absolute address.

Programmers using the COBOL compiler have several methods available for testing their programs, debugging them, and revising them for increased efficiency in operation.

The COBOL debugging language can be used by itself or in conjunction with other COBOL statements. A dump can also be used for program checkout.

Appendix B

Job-Control Language

Introduction

Job-Control Language (JCL) cards establish the communication link between the COBOL programmer and the control system of the computer. The control system consists of a number of processing programs and a control program. The processing programs include the COBOL compiler, service programs, and any user-written programs.

The control program supervises the execution and loading of the processing programs; controls the location, storage, and retrieval of data; and schedules the jobs for continuous processing of problem programs.

Executing a COBOL Program

The basic operations to be performed to execute a COBOL program are:

A. *Compilation.* The process of translating a COBOL source program into a series of instructions comprehensible to the computer. In computer terminology, the input (source program) to the compiler is called the *source module.* The output (compiled source program) from the compiler is called the *object module.*

B. *Linkage Editing.* The Linkage Editor is a source program that prepares object modules for execution. It can also be used to combine two or more separately compiled object modules into a format suitable for execution as a single program. The executed output of the Linkage Editor is called a *load module.* The Linkage Editor may also combine previously edited load modules with or without one or more object modules to form one load module.

C. *Loading.* The loader is a service program the processes COBOL object and load modules, resolves any references to subprograms, and executes the loaded module. All these functions are specified in one step.

D. *Execution.* Actual execution is under the supervision of the control program, which obtains a load module, loads it into main storage, and initiates execution of the machine language instructions contained in the load module.

(See figure B.1.)

The JOB statement identifies the beginning of a job and the job to be performed. It may be also used by the installation's accounting routines.

The EXEC statement describes the job step and calls for execution.

Sample DOS JCL Statements

All JCL statements start in position 1. Slashes in positions 1 and 2 (//) identify the statements to the system as JCL, and not program or data sets. If commas are indicated in certain JCL statements, they must be included.

1. // JOB EXAMPLE
2. // OPTION LINK
3. // EXEC COBOL
4. Source Program Records
5. /*
6. // EXEC LNKEDT
7. // EXEC
8. Data Records
9. /*
10. /&

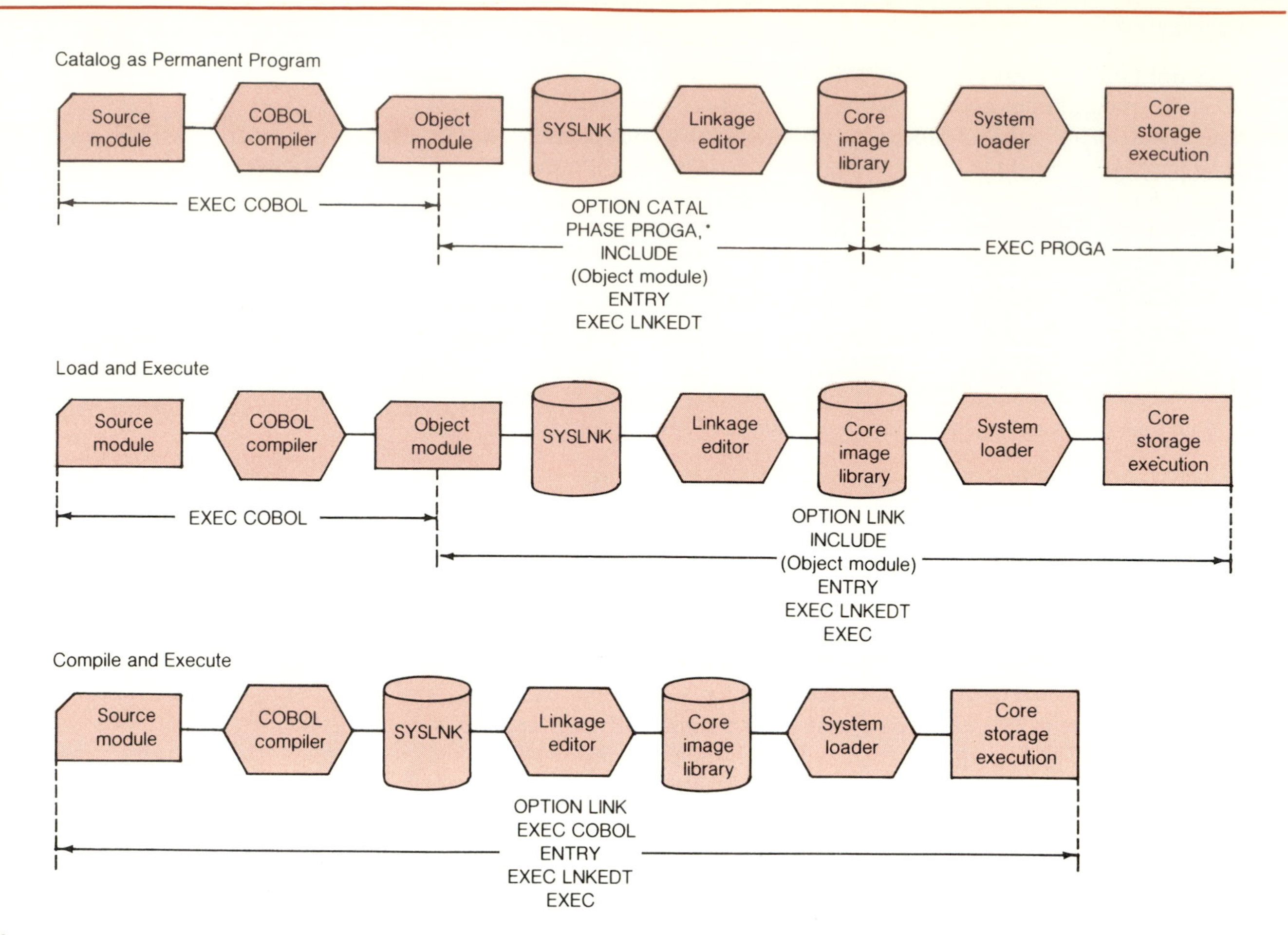

Figure B.1 Compilation and execution of a COBOL program.

Statement 1. The JOB statement identifies the program to the system. There must be at least one space before and after JOB. Name is limited to eight characters or less and need not be related to the PROGRAM-ID.

Statement 2. The OPTION LINK statement may vary from installation to installation depending upon the needs and requirements of the programmer. For example, if the program is unexecutable, a dump of storage may be requested.

Statement 3. The EXEC COBOL statement is calling for the execution of the COBOL compiler to translate the source program into the object module.

Statement 4. The source program records would be inserted here.

Statement 5. The delimiter (/*) signals the end of the source program and separates the source program from subsequent JCL statements.

Statement 6. The execution of the EXEC LNKEDT statement will prepare the object module for execution.

Statement 7. The EXEC statement controls the execution of the machine language instructions contained in the load module upon the data set that follows.

Statement 8. The data set containing the source data to be processed.

Statement 9. Another delimiter (/*) to signal the end of the data set and separate the data from any subsequent JCL statements.

Statement 10. The delimiter (/&) to signal the end of the program and to separate the executed program from any subsequent programs. An important note to remember is that the omission of this delimiter may cause two programs to be executed together.

It is beyond the scope of this text to discuss the various combinations of JCL statements permissible. There are many options available to the programmer, and they should be carefully studied in the pro-

```
// JOB PROG1
•
•
// EXEC COBOL
    (source deck - main program)
/*
•
•
// EXEC COBOL
    (source deck - first subprogram)
/*
•
•
// EXEC COBOL
    (source deck - second subprogram)
/*
•
•
// EXEC LNKEDT
•
•
// EXEC
```

Figure B.2 Sample structure of job deck for compiling, link editing and executing a main program and two subprograms.

grammer's guide reference manual for the particular computer. The foregoing is a typical example of the use and purpose of JCL statements. Following is a brief explanation of the format, use, and purpose of IBM DOS JCL statements.

Job-Control Language (DOS) (IBM)

Job-control statements prepare the system for the execution of COBOL programs. (See figure B.2.)

Job-Control Statements

Job-control statements are designed in an eighty-position format. The statements are entered into cards in essentially free form, but certain rules must be observed, as follows:

1. *Name.* Two slashes (//) identify the statement as a job-control statement. These characters must be in positions one and two of the statement followed immediately by at least one blank.
   Exception: The end-of-job statement delimiter contains /& in positions one and two; the end-of-data statement delimiter contains /* in positions one and two, and the comments statement contains an * in position one and a blank in position two.
2. *Operation.* This identifies the operation to be performed. It can be up to eight characters long with at least one blank following its last character.
3. *Operand.* This may be a blank or may contain one or more entries separated by commas. The last term must be followed by a blank, unless its last character is in position 71.
4. *Comments.* Optional programmer comments must be separated from the operand by at least one space.

   All JCL statements are read by the systems input device specified by the symbolic name SYSRDR.

Comment Statements

Comment statements (statements preceded by an asterisk in position one and followed by a blank) may be placed anywhere in the job stream. They may contain any character and are usually for communication with the operator; accordingly, they are written on the console printer as well as on the printer.

JOB Definition

A *JOB* is a specified unit of work to be performed under the control of the operating system. A typical JOB might be the processing of a COBOL program: compiling the source program, editing the module to form a phase, and then executing the phase.

A *JOBSTEP* is exactly what the name implies—one step in the processing of the JOB. Thus, in the JOB just mentioned, one job step is the compilation of the source statements, another is the line editing of the module, and the other is the execution of the phase. A compilation requires the execution of the COBOL compiler, an editing process implies the execution of the Linkage Editor, and, finally, the execution phase is the execution of the problem program itself.

Compilation of JOBSTEPs

The compilation of a COBOL program may necessitate more than one job step (more than one execution of the COBOL compiler). In some cases, a COBOL program consists of a main program and one or more subprograms. To compile such a program, a separate job step must be specified for the main program and for each of the subprograms. Thus the COBOL compiler is executed once for the main program and once for each subprogram. Each execution produces a module. The separate modules can then be combined into one phase by a single job step—the execution of the Linkage Eidtor.

There are four Job-Control statements that are used for job definition: the JOB statement, the EXEC statement, end-of-data statement (/*), and the end-of-job statement (/&). These are optional Job-Control statements that may be used to specify specific JCL functions.

The JOB statement defines the start of a job. One JOB statement is required for every job; it must be the first statement in the job deck. The programmer must name his or her job on the JOB statement.

The *EXEC* statement requests the execution of a program. Therefore, one EXEC statement is required for each job step within a job stream. The EXEC statement identifies the program that is to be executed (for example, the COBOL compiler, the Linkage Editor).

The end-of-data statement, also referred to as the slash asterisk (/*) statement, defines the end of a programmer's input data. The slash asterisk statement immediately follows the input data. For example, COBOL source statements would be placed immediately after the EXEC statement for the COBOL compiler; a /* statement would follow the last COBOL source statement. If input data is kept separate, the /* immediately follows each set of input data.

The end-of-job statement, also referred to as the slash ampersand (/&) statement, defines the end of the job. A /& statement must appear as the last statement in the job stream. If this statement is omitted, the preceding job would be combined with this job. (See figure B.3.)

Compilation

Compilation is the execution of the COBOL compiler. The programmer requests compilation by placing in the job stream an EXEC statement specifying the name of the COBOL compiler. Input to the compiler is a set of COBOL source statements consisting of either a main program or a subprogram.

Output from the COBOL compiler is dependent upon the options specified. This output may include a listing of source statements exactly as they appear in the input stream. Separate data and/or Procedure Division maps, a symbolic cross-reference list, and diagnostic messages can also be produced. The format of the compiler output is described and illustrated in the "Debugging" section.

The programmer can override any of the compiler options specified when the system was generated or can include some not specified by specifying the OPTION control statement in the compiler job step.

Editing

Editing is the execution of the Linkage Editor. The programmer requests editing by placing in the job stream an EXEC statement that contains the name LNKEDT, the name of the Linkage Editor.

Output from the Linkage Editor consists of one or more phases. A phase may be an entire program or it may be part of an overlay structure (multiple phases).

A phase produced by the Linkage Editor can be executed immediately after it is produced (that is, in the job step immediately following the Linkage Editor), or it can be executed later, either in a subsequent job step of the same job or a subsequent job step.

Statement	Function
// ASSGN	Input/output assignments.
// CLOSE	Closes a logical unit assigned to magnetic tape.
// DATE	Provides a date for the Communication Region.
// DLAB	Disk file label information.
// DLBL	Disk file label information.
// EXEC	Execute program.
// EXTENT	Disk file extent.
// JOB	Beginning of control information for a job.
// LBLTYP	Reserves storage for label information.
// LISTIO	Lists input/output assignments.
// MTC	Controls operations on magnetic tape.
// OPTION	Specifies one or more job control options.
// PAUSE	Creates a pause for operator intervention.
// RESET	Resets input/output assignments to standard assignments.
// RSTRT	Restarts a checkpointed program.
// TLBL	Tape label information.
// TPLAB	Tape label information.
// UPSI	Sets user-program switches.
// VOL	Disk/tape label information.
// XTENT	Disk file extent.
/*	End-of-data-file or end-of-job-step.
/&	End-of-job.
*	Comments.

Figure B.3 JOB CONTROL statement.

Phase Execution

Phase execution is the execution of the problem program; for example, the program written by the COBOL programmer. If the program is an overlay structure (multiple phases), the execution job step actually involves the execution of all phases in the program.

The programmer requests the execution of a phase by placing in the job stream an EXEC statement that specifies the name of the phase. However, if the phase to be executed was produced in the immediately preceding job step, it is not necessary to specify its name in the EXEC statement.

Sequence of JOB-CONTROL Statements

The job stream for a specific job always begins with a JOB statement and ends with a /& (end-of-job) statement. A specific job consists of one or more job steps. The beginning of a job step is indicated by the appearance of an EXEC statement. When an EXEC statement is encountered, it initiates the execution of the job step, which includes all preceding control statements up to but not including a previous EXEC statement.

ASSGN Statements

The ASSGN control statement assigns a logical input/output unit to a physical device. An ASSGN control statement must be present in the job stream for each data file assigned to an external storage device in the COBOL program where these assignments differ from those established at system generation time. Data files are assigned to programmer logical units in COBOL by means of the source language ASSGN clause. The ASSGN control statement may also be used to change a system standard assignment for the duration of the job. Device assignments made by the ASSGN statement are considered temporary until another ASSGN statement appears.

Job-Control Language (OS) (IBM)

The job-control language statements for an operating system (OS) are similar to those for a DOS system. The functions and rules for recording these statements are the same. There are some additional statements required in an OS system.

The types of JOB-CONTROL statements used to compile, linkage edit, and execute programs in an OS environment are:

Statement	**Function**
JOB	Indicates the beginning of a new job and describes that job.
EXEC	Indicates a job step and describes that job step; indicates the load module or catalogued procedure to be executed.
DD	Describes data sets, and controls device and volume assignment.
delimiter	Separates data sets in the input stream from control statements; it must follow each data set that appears in the input stream, e.g., after a COBOL source module records.
comment	Contains miscellaneous remarks and notes written by the programmer; it may appear anywhere in the job stream after the JOB statement.

The general format of control statements are as indicated in figure B.4.

Statement	Columns			Fields
	1	2	3	4
Job	/	/	name	JOB operand[1] comments[1]
Execute	/	/	name[1]	EXEC operand comments[1]
Data Definition	/	/	name[1]	DD operand comments[1]
Procedure	/	/	name[1]	PROC operand comments[1]
Command	/	/		operation(command) operand comments[1]
Delimiter	/	*		comments[1]
Null	/	/		
Comment	/	/	*	comments
Pend	/	/	name[1]	PEND

[1]Optional.

Figure B.4 General format of JOB-CONTROL statement.

The nine job-control language statements used to describe a job to the system are:

1. Job (JOB) statement.
2. Execute (EXEC) statement.
3. Data definition (DD) statement.
4. Delimiter statement.
5. Null statement.
6. Procedure (PROC) statement.
7. Procedure end (PEND) statement.
8. Comment statement.
9. Command statement.

A JOB-CONTROL statement consists of one or more 80-byte records. Most jobs are submitted to the operating system for execution in the form of 80-position records direct access devices. The operating system is able to distinguish a JOB-CONTROL statement from data included in the input stream. In positions 1 and 2 of all the statements except the delimiter statement, a // is coded. For the delimiter statement, a /* is coded in positions 1 and 2; this notifies the operating system that the statement is a delimiter statement. For a comment statement, a //* is coded in positions 1, 2, and 3 respectively. An example of a job-control procedure follows.

```
//JOB1          JOB
//STEP1         EXEC     PGM=IKFCBL00, PARM=DECK
//SYSUT1        DD       DSNAME=&&UT1, UNIT=SYSDA,SPACE=(TRK, (40) )
//SYSUT2        DD       DSNAME=&&UT2, UNIT=SYSSQ,SPACE=(TRK, (40) )
//SYSUT3        DD       DSNAME=&&UT3, UNIT=SYSSQ,SPACE=(TRK, (40) )
//SYSUT4        DD       DSNAME=&&UT4, UNIT=SYSSQ,SPACE=(TRK, (40) )
//SYSPRINT      DD       SYSOUT=A
//SYSPUNCH      DD       SYSOUT=B
//SYSIN         DD       *
    (source records)
/*
```

Parameters coded on these JCL statements help the job scheduler to regulate the execution of jobs and job steps, retrieve and dispose of data, allocate I/O resources, and communicate with the operator.

JOB Statement

The JOB statement indicates to the system at what point a job begins (fig. B.5). The name of the job is coded and is used to identify messages to the operator and to identify the program output. Additional information, such as accounting information, conditions for early termination of a job, job priority, and maximum amount of time, can be specified in parameters on the record (fig. B.6).

EXEC Statement

The EXEC statement marks the beginning of a job step and the end of the preceding step (fig. B.7). The program to be executed is identified or the catalogued procedure or in-stream procedure is called. A catalogued procedure is a set of job-control language statements that has been assigned a name and placed in a partitioned data set known as the procedure library.

The EXEC statement may also be used to provide job-step accounting information, to give conditions for bypassing or executing a job step, etc.

DD Statement

A DD statement identifies a data set and describes its attributes (fig. B.8). There must be a DD statement for each data set used or created in a job step. The DD statements are placed after the EXEC statement for the step. The DD statement provides such information as the name of the data set, the

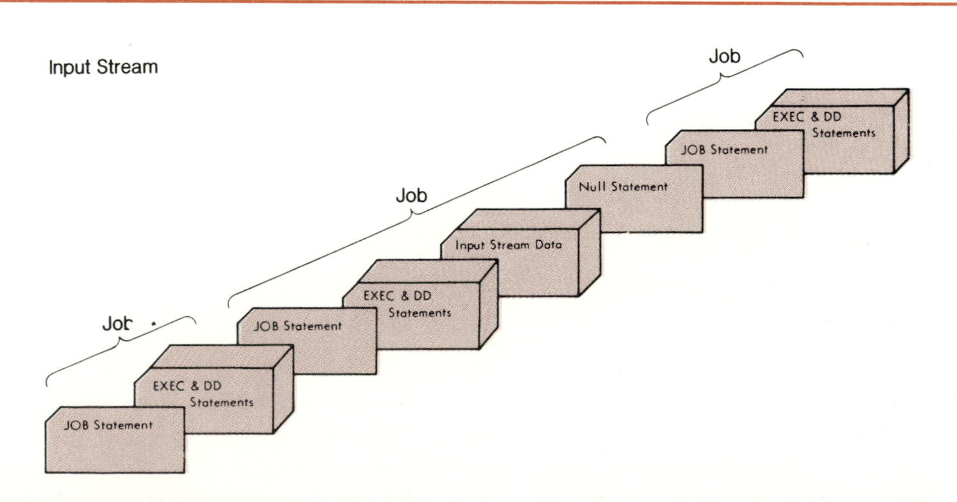

Figure B.5 Defining job boundaries.

```
Name        Operation  Operand

                       Positional Parameters

//jobname   JOB        [([account-number] [,accounting-information])1 2 3]

                       [,programmer-name]4 5

                       Keyword Parameters

                       [MSGLEVEL=(x,y)]6
                       [TIME=(minutes,seconds)]
                       [CLASS=jobclass]
                       [COND=((code,operator) [,(code,operator)]...7)8]
                       [PRTY=job priority]
                       [MSGCLASS=classname]
                       [REGION=(nnnnnxK[,nnnnnyK])]
                       [ROLL=(x,y)]
                       [TYPRUN=HOLD]
                       [RD=request]
                                   {*                    }
                       [RESTART=(  {stepname             }  [,checkid])]
                                   {stepname.procstepname}
```

[1]If the information specified (account-number and/or accounting-information) contains blanks, parentheses, or equal signs, the information must be delimited by single quotation marks instead of parentheses.
[2]If only account-number is specified, the delimiting parentheses may be omitted.
[3]The maximum number of characters allowed between the delimiting quotation marks is 142.
[4]If programmer-name contains any special characters other than the period, it must be enclosed within single quotation marks.
[5]The maximum number of characters allowed for programmer-name is 20.
[6]x = 0, 1, or 2 is the JCL message.
y = 0 or 1 is the allocation message level.
Note that the value 1 may be used in place of (1,1).
[7]The maximum number of repetitions allowed is 7.
[8]If only one test is specified, the outer pair of parentheses may be omitted.

Figure B.6 JOB statement.

```
               Oper-
Name           ation  Operand

                      Positional Parameters

//[stepname]1  EXEC    PGM=progname
                       PGM=*.stepname.ddname
                       PROC=procname
                       procname
                       PGM=*.stepname.procstep.ddname

                      Keyword Parameters

                      [{ACCT2         }                             3 4 5]
                      [{ACCT.procstep } = (accounting-information)       ]

                      [{COND2         }                                                6 7]
                      [{COND.procstep } = ((code,operator[,stepname[.procstep]])...)     ]

                      [{PARM2         }                       3 8 9]
                      [{PARM.procstep } = (option[,option]...)     ]

                      [{TIME          }                     ]
                      [{TIME.procstep } = (minutes,seconds) ]

                      [{REGION          }                     ]
                      [{REGION.procstep } = nnnnnxK[,nnnnnyK] ]

                      [{ROLL          }        ]
                      [{ROLL.procstep } = (x,y)]

                      [{RD          }          ]
                      [{RD.procstep } = request]

                      [{DPRTY          }                    ]
                      [{DPRTY.procstep } = (value 1, value 2)]
```

[1]Stepname is required when information from this control statement is referred to in a later job step.
[2]If this format is selected, it may be repeated in the EXEC statement once for each step in the cataloged procedure.
[3]If the information specified contains any special characters except hyphens, it must be delimited by single quotation marks instead of parentheses.
[4]If accounting-information contains any special characters except hyphens, it must be delimited by single quotation marks.
[5]The maximum number of characters allowed between the delimiting quotation marks or parentheses is 142.
[6]The maximum number of repetitions allowed is 7.
[7]If only one test is specified, the outer pair of parentheses may be omitted.
[8]If the only special character contained in the value is a comma, the value may be enclosed in quotation marks.
[9]The maximum number of characters allowed between the delimiting quotation marks or parentheses is 100.

Figure B.7 EXEC statement.

name of the volume on which it resides, the type of I/O device that holds the data set, the format of the records in the data set, whether the data set is old or new, the size of newly created data sets, and the method that will be used to create or access a data set. The name of the DD statement provides a symbolic link between the data set (on data file) named in the program and the actual name and location of the corresponding data set. This symbolic link allows one to relate the data set in the program to different data sets on different occasions.

```
Name                          | Operation | Operand
------------------------------+-----------+---------------------------------
    ( ddname          )  1    | DD        |
//  (                 )       |           |  (see below and next page)
    ( procstep.ddname )       |           |
```

```
Operand2
--------------------------------------------------------------------------------
        Positional Parameters

  [ *     ]  3
  [ DATA  ]
  [ DUMMY ]

        Keyword Parameters 4 5

[DDNAME=ddname]

[                   ( dsname                           ) ]  11
[                   ( dsname(element)                  ) ]
[  ( DSNAME )       ( *.ddname                         ) ]
[  (        )   =   ( *.stepname.ddname                ) ]
[  ( DSN    )       ( *.stepname.procstep.ddname       ) ]
[                   ( &&name                           ) ]
[                   ( &&name(element)                  ) ]

[        [ dsname                     ]                          ]  6
[  DCB=( [ *.ddname                   ]  [,subparameter-list])   ]
[        [ *.stepname.ddname          ]                          ]
[        [ *.stepname.procstep.ddname ]                          ]

  [ SEP=(subparameter list)7 ]  10
  [ AFF=ddname               ]

      Positional Subparameters   Keyword Subparameters

  [ UNIT=(name[,[n/P][,DEFER]][,SEP=(list of up to 8 ddnames)])8 ]  10 12
  [ UNIT=(AFF=ddname)                                            ]
                              Positional Subparameters

          ( TRK                   )
SPACE=(   ( CYL                   )   ,(primary-quantity[,secondary-quantity],
          ( average-record-length )

                                                   [ ,MXLG   ]
           [directory- or index-quantity])[,RLSE]  [ ,ALX    ]  [, ROUND])
                                                   [ ,CONTIG ]

SPACE=(ABSTR,(quantity,beginning-address[,directory- or index-quantity]))

              ( CYL                   )
SPLIT=(n,     (                       )  ,(primary-quantity[,secondary-quantity]))
              ( average-record-length )

              ( TRK                   )
SUBALLOC=(    ( CYL                   )  ,(primary-quantity[,secondary-quantity]
              ( average-record-length )

                                  ( ddname                   )
       [,directory-quantity]),    ( stepname.ddname          )  )
                                  ( stepname.procstep.ddname )

       Positional Subparameters

  ( VOLUME )
  (        ) =([PRIVATE],[RETAIN],[volume sequence number],[volume count])
  ( VOL    )

       Keyword Subparameters

  [ ,SER=(volume-serial-number[volume-serial-number]9...)   ]
  [                                                         ]
  [          ( dsname                     )                 ]
  [  ,REF=   ( *.ddname                   )                 ]
  [          ( *.stepname.ddname          )                 ]
  [          ( *.stepname.procstep.ddname )                 ]

  [                                       ( NL  )                              ]
  [ LABEL=([data-set-sequence-number],    ( SL  )  [ ,EXPDT=yyddd ] [,PASSWORD]) ]
  [                                       ( NSL )  [ ,RETPD=xxxx  ]            ]
  [                                       ( SUL )                              ]

  [         [ NEW ]  [ ,DELETE  ]  [ ,DELETE  ]     ]
  [         [ OLD ]  [ ,KEEP    ]  [ ,KEEP    ]     ]
  [ DISP=(  [ SHR ]  [ ,PASS    ]  [ ,CATLG   ]  )  ]
  [         [ MOD ]  [ ,CATLG   ]  [ ,UNCATLG ]     ]
  [                  [ ,UNCATLG ]                   ]
  [                                                 ]
  [ SYSOUT=classname                                ]
  [ SYSOUT=(x[,program-name][,form-no.])            ]
```

[1]The name field must be blank when concatenating data sets.

[2]All parameters are optional to allow a programmer flexibility in the use of the DD statement; however, a DD statement with a blank operand field is meaningless.

[3]If the positional parameter is specified, keyword parameters other than DCB cannot be specified.

[4]If subparameter-list consists of only one subparameter and no leading comma (indicating the omission of a positional subparameter) is required, the delimiting parentheses may be omitted.

[5]If subparameter-list is omitted, the entire parameter must be omitted.

[6]See "User-Defined Files" for the applicable subparameters.

[7]See the publication IBM System/360 Operating System: Job Control Language Reference.

[8]If only name is specified, the delimiting parentheses may be omitted.

[9]If only one volume-serial-number is specified, the delimiting parentheses may be omitted.

[10]The SEP and AFF parameters should not be confused with the SEP and AFF subparameters of the UNIT parameter.

[11]The value specified may contain special characters if the value is enclosed in apostrophes. If the only special character used is the hyphen, the value need not be enclosed in apostrophes. If DSNAME is a qualified name, it may contain periods without being enclosed in apostrophes.

[12]The unit address may contain a slash, and the unit type number may contain a hyphen, without being enclosed in apostrophes, e.g., UNIT=293/5,UNIT=2400-2.

Figure B.8 The DD statement.

DELIMITER and NULL Statements

The DELIMITER statement (or /* statement) and the NULL statement (or // statement) are markers in an input stream. The DELIMITER statement is used to separate data placed in the input stream from any JCL statement that may follow the data. The NULL statement can be used to mark the end of the JCL statements and data for a job.

PROC and PEND Statements

The PROC statement may appear as the first JCL statement in a catalogued or in-stream procedure. For catalogued procedures or in-stream procedures, the PROC statement is used to assign default values to parameters defined in a procedure. An in-stream procedure is a set of job-control language statements that appear in the input stream. The PROC statement is used to mark the beginning of an in-stream procedure. The PEND statement is used to mark the end of an in-stream procedure. In the example shown here, JOB2 is the name of the job, STEPA is the name of the single job step. The EXEC statement calls the cataloged procedure containing STEP1 to execute the job step (PROC = CATPROC).

```
//JOB2              JOB
//STEPA             EXEC    PROC=CATPROC
//STEP1.SYSIN       DD      *
    (source records)
/*
```

COMMENT Statement

The COMMENT statement can be inserted either before or after any JCL statement that follows the JOB statement, and can contain any information that would be helpful to one interested in the program.

COMMAND Statement

The COMMAND statement is used to enter commands through the input stream. Commands can activate and deactivate system input and output units, request printouts and displays, and perform a number of other operator functions.

Job-Control Fields

The name contains from one through eight alphanumeric characters, the first of which must be alphabetic. The name begins in position 3. It is followed by one or more blanks. The name is used as follows:

Name Field

1. To identify the control statement to the operating system
2. To enable other control statements in the job to refer to information contained in the named statement
3. To relate DD statements to files named in a COBOL source program

Operation Field

The operation field is preceded and followed by one or more blanks. It may contain one of the following operation codes:

```
JOB
EXEC
DD
PROC
PEND
```

If the statement is a delimiter statement, there is no operation field and comments may start after one blank.

Operand Field

The operand field is preceded and followed by one or more blanks and may continue through position 71 and onto one or more continuation records. It contains the parameters and subparameters that give required and optional information to the operating system. Parameters and subparameters are separated by commas. A blank in the operand field causes the system to treat the remaining data on the record as a comment. There are two types of parameters: positional and keyword.

Positional Parameters

Positional parameters are the first parameters in the operand field, and they must appear in the specified sequence. If a positional parameter is omitted and other positional parameters follow, the omission must be indicated by a comma. If other positional parameters do not follow, no comma is needed.

Keyword Parameters

A keyword parameter may be placed anywhere in the operand field following the positional parameters. A keyword parameter consists of a keyword, followed by an equal sign, followed by a single value or a list of parameters. If there is a subparameter list, it must be enclosed in parentheses or single quotation marks; the subparameters in the list must be separated by commas. Keyword parameters may appear in any sequence.

Comments Field

Optional comments must be separated from the last parameter (or the /* in a delimiter statement) by one or more blanks and may appear in the remaining positions up to and including position 71. An optional comment may be continued onto one or more continuation records. Comments can contain blanks. (See figure B.9.)

Figure B.9 Example of JOB CONTROL statement.

Job Control Statements

Card no.	Function
①	JOB CARD. Provides the following information to the system: A. NAME of job—IW016003 B. Accounting Information—(0160.33. . . . 198), the necessary information relative to the job being executed with the required positional parameter commas. C. Name of programmer—HENESYDR D. Job Priority—CLASS=I
②	EXEC card. Identifies program to be executed and the PROC statement is used to reference the catalogued procedure "COBOL."
③	DD card. Identifies data set with an asterisk (*) as following.
④	DD card. Calls for a dump of core storage in case of abnormal termination of the program during execution.
⑤	DD card. Identifies output is to be printed for data set.
⑥	DD card. Identifies data set with an asterisk (*) as following.
⑦	Null card. Marks the end of the JCL statements and data for job.

The delimiter statement cards (/*) will appear as markers behind the input stream after card 3 and card 6.

All statements without the characters // or /* in columns 1 and 2 are generated by the system.

Figure B.9 *(continued)*

```
①  //IW016003 JOB (0160,33,...,198),HENESYDR,CLASS=I                          JOB  456
②  //STEP1 EXEC PROC=COBOL                                                    2
           XXCOB      EXEC PGM=IKFCBL00,REGION=192K,                                        00000010
           XX         PARM='LOA,NOCLI,NODMA,NOPMA,SUP,NOXRE,CSY',TIME=2                     00000020
    ***                                                                       00000030
    ***                                                                       00000040
    ***       THIS PROCEDURE PROVIDED FOR THE USE OF LACCD STUDENTS.          00000050
    ***       IF MORE INFORMATION IS NEEDED, CONTACT:                         00000060
    ***                                                                       00000070
    ***          SOFTWARE GROUP - LACCD DATA PROCESSING DIV.                  00000080
    ***          2140 W. OLYMPIC BLVD, LOS ANGELES, CA 90006                  00000090
    ***                                                                       00000100
           XXSTEPLIB    DD DSNAME=SYS2.COBV4LIB,DISP=SHR                                    00000110
           XXSYSLIB     DD DSN=INST.COPYLIB,DISP=SHR                                        00000120
           XXSYSPRINT   DD SYSOUT=A                                                         00000130
           XXSYSPUNCH   DD DUMMY                                                            00000140
           XXSYSUT1     DD UNIT=SYSDA,SPACE=(460,(700,100))                                 00000150
           XXSYSUT2     DD UNIT=SYSDA,SPACE=(460,(700,100))                                 00000160
           XXSYSUT3     DD UNIT=SYSDA,SPACE=(460,(700,100))                                 00000170
           XXSYSUT4     DD UNIT=SYSDA,SPACE=(460,(700,100))                                 00000180
           XXSYSLIN     DD DSNAME=&LOADSET,DISP=(MOD,PASS),                                 00000190
           XX              UNIT=SYSDA,SPACE=(80,(720,100))                                  00000200
③  //SYSIN DD *                                                               3
     ALLOC. FOR IW016003 COB       STEP1
     155    ALLOCATED TO STEPLIB
     155    ALLOCATED TO SYSLIB
     D34    ALLOCATED TO SYSPRINT
     151    ALLOCATED TO SYSUT1
     150    ALLOCATED TO SYSUT2
     15A    ALLOCATED TO SYSUT3
     151    ALLOCATED TO SYSUT4
     150    ALLOCATED TO SYSLIN
     D11    ALLOCATED TO SYSIN
     - STEP WAS EXECUTED - COND CODE 0000
       SYS2.COBV4LIB                                   KEPT
       VOL SER NOS= SPOOL1.
       INST.COPYLIB                                    KEPT
       VOL SER NOS= SPOOL1.
       SYS77202.T223946.RV000.IW016003.R0004920        DELETED
       VOL SER NOS= CC3306.
       SYS77202.T223946.RV000.IW016003.R0004921        DELETED
       VOL SER NOS= CC3300.
       SYS77202.T223946.RV000.IW016003.R0004922        DELETED
       VOL SER NOS= CC3304.
       SYS77202.T223946.RV000.IW016003.R0004923        DELETED
       VOL SER NOS= CC3306.
       SYS77202.T223946.RV000.IW016003.LOADSET         PASSED
       VOL SER NOS= CC3300.


     STEP /COB      / START 77202.2239
     STEP /COB      / STOP  77202.2240 CPU    0MIN 03.52SEC STOR VIRT 100K
           XXGO         EXEC PGM=LOADER,                                                    00000210
           XX           PARM='NOMAP,NOPRINT,NOXREF,LET',                                    00000220
           XX           COND=(5,LT,COB),REGION=192K,TIME=1                                  00000230
           XXSYSLIN DD DSNAME=*.COB.SYSLIN,DISP=(OLD,DELETE)                                00000240
           XXSYSPRINT   DD SYSOUT=A                                                         00000250
           XXSYSOUT     DD SYSOUT=A           FOR "DISPLAY" VERBS                           00000260
           XXSYSLOUT    DD SYSOUT=A                                                         00000270
           XXSYSLIB     DD DSNAME=SYS2.COBV4SUB,DISP=SHR                                    00000280
           XXSORTLIB    DD DSNAME=SYS1.SORTLIB,DISP=SHR                                     00000290
    ***                                                                       00000300
    ***  UPDATED 11-14-76 TO REFLECT CHANGES IN STEPLIB AND/OR SYSLIB        00000310
    ***  UPDATED 11-19-76 TO ADD AN "INSTRUCTIONAL" COPYLIB                  00000320
    ***  UPDATED 02-17-77 TO ADD A SORTLIB DD CARD                           00000330
    ***  UPDATED 04-17-77 TO ADD A "SYSOUT" DD CARD                          00000340
    ***  UPDATED 04-26-77 TO INCLUDE A TIME LIMIT ON STEP "COB"              00000341
    ***  HOWARD DEAN                                                          00000350
④  //GO.SYSUDUMP DD SYSOUT=A                                                 178
⑤  //GO.PRINT DD SYSOUT=A                                                    180
⑥  //GO.CARDIN DD *                                                          181
⑦  //
     ALLOC. FOR IW016003 GO        STEP1
     150    ALLOCATED TO SYSLIN
     D34    ALLOCATED TO SYSPRINT
     D35    ALLOCATED TO SYSOUT
     D36    ALLOCATED TO SYSLOUT
     155    ALLOCATED TO SYSLIB
     155    ALLOCATED TO SORTLIB
     D37    ALLOCATED TO SYSUDUMP
     D38    ALLOCATED TO PRINT
     D11    ALLOCATED TO CARDIN
IEC020I 001-5,IW016003,GO,CARDIN,D11
IEC020I GET OR READ ISSUED AFTER END-OF-FILE
IEW1991 ERROR - USER PROGRAM HAS ABNORMALLY TERMINATED
COMPLETION CODE - SYSTEM=001  USER=0000
       SYS77202.T223946.RV000.IW016003.LOADSET         DELETED
       VOL SER NOS= CC3300.
       SYS2.COBV4SUB                                   KEPT
       VOL SER NOS= SPOOL1.
       SYS1.SORTLIB                                    KEPT
       VOL SER NOS= SPOOL1.
     STEP /GO       / START 77202.2240
     STEP /GO       / STOP  77202.2240 CPU    0MIN 02.41SEC STOR VIRT 144K
      JOB /IW016003/ START 77202.2239
      JOB /IW016003/ STOP  77202.2240 CPU    0MIN 05.93SEC
```

Appendix C

COBOL Differences— ANSI 1968 and ANSI 1974 COBOL

While much of the ANSI 1968 COBOL was included in the ANSI 1974 version, some of the material was changed, some deleted, and some new material was added. All of this is thoroughly detailed in publication X23. 1974 of the American National Standards Institute.

The major changes are listed as follows, according to modules.

Nucleus

1. The REMARKS paragraph and the NOTE statement have been deleted in favor of a generalized comment facility. An asterisk (*) in character position 7 (continuation column) now identifies any line as a comment line. A further refinement has been added. A slash (/) in character position 7 causes the line to be treated as a comment and causes a page ejection.
2. The EXAMINE statement has been deleted in favor of the more general and powerful INSPECT statement. The INSPECT statement provides the facility to count (Format-1), replace (Format-2), or count and replace (Format-3) occurrences of single characters or groups of characters in a data item.
3. Level-77 items need no longer precede Level-01 items in the Working-Storage Section.
4. The punctuation rules with regard to spaces have been relaxed. For example, spaces may now optionally precede the comma, period, or semicolon, and may optionally precede or follow a left parenthesis.
5. Two contiguous quotation marks may be used within a nonnumeric literal to represent a single occurrence of the character quotation mark.
6. A SIGN clause has been added that permits the specification of the position that the sign is to occupy in a signed numeric item (either leading or trailing) and/or that it is to occupy a separate character position.
7. The ACCEPT statement has been expanded to provide access to internal DATE, DAY, and TIME.
8. The GIVING identifier series has been added to the arithmetic statements; identifier series has been added to the COMPUTE statement; and INTO identifier series has been added to the DIVIDE statement.
9. The STRING statement has been added. This statement provides for the juxtapositioning within a single data item of the partial or complete contents of two or more data items. A companion statement, the UNSTRING statement, has also been added. This statement causes contiguous data within a single data item to be separated and placed in multiple receiving fields.
10. Certain ambiguities in abbreviated combined conditions with regard to NOT and the use of parentheses have been eliminated. Where any portion of an abbreviated combined condition is enclosed in parentheses, all subjects and operators required for the expansion of that portion must be included within the same set of parentheses.
11. The PROGRAM COLLATING SEQUENCE clause has been added to permit specification of the collating sequence used in nonnumeric comparisons. Native, ASCII, implementor-defined and user-defined collating sequences may be specified. This makes possible the processing of ASCII files without changing source program logic.

Table Handling

1. The left parenthesis enclosing subscripts need not be preceded by a space. Commas are not required between subscripts or indices. Literals and index-names may be mixed in a table reference.

2. A data description entry that contains an OCCURS DEPENDING ON clause may be followed, within that record description, only by data description entries that are subordinate to it. Thus, the "fixed" portion of a record must entirely precede any "variable" portion. The effect of the OCCURS DEPENDING ON clause was clarified to state explicitly that internal operations involving tables described with this clause reference only the portion of the table that is "active" (i.e., the actual size as defined by the current value of the operand of the DEPENDING ON phrase is used).
3. An index may be set up or down by a negative value.
4. The subject of the condition in the WHEN phrase of the SEARCH ALL statement must be a data item named in the KEY phrase of the referenced table; the object of this condition may not be such a data item. ANSI 1968 specified that either the subject or the object could be a data item named in the KEY phrase.

Sequential I-O

1. The FILE-LIMITS clause, the MULTIPLE REEL/UNIT clause, and the integer implementor-name phrase of the file control entry were deleted because it was felt that these functions could be handled better outside of the COBOL program.
2. The SEEK statement was deleted because it was felt to be redundant (it is implied by the READ, WRITE, etc.) and thus ineffective.
3. The OPEN REVERSED statement now positions a file at its end. The OPEN EXTEND statement was added to permit the addition of records at the end of an existing sequential file.
4. The USE AFTER STANDARD ERROR was changed to read USE AFTER STANDARD ERROR/EXCEPTION; the function was expanded to permit invocation of the associated on both error (e.g., boundary violation) or exception (i.e., AT END) conditions.
5. The AT END phrase of the READ statement was made optional; it must appear, however, if no applicable USE procedure appears.
6. The INVALID KEY phrase of the WRITE statement was deleted since there is no user-defined key for sequential files. Error and/or exception conditions can be monitored through appropriate USE statements.
7. The FILE STATUS clause was added to permit the system to convey information to the program concerning the status of I/O operations. Codes for "error," AT END, etc., have been defined.
8. The REWRITE statement has been added to permit the explicit updating of records on a sequential file.
9. The LINAGE clause was added to permit programmer definitions of logical page size and of the size of the top and bottom margins on the logical page.
10. The PAGE phrase was added to the WRITE statement to permit presentation of a line before or after advance to the top of the next logical page.
11. The facility to define, initialize, and access user-defined labels has been deleted.
12. The CODE-SET clause has been added to provide for the conversion of sequential nonmass storage files encoded in ASCII or implementor-specified codes from/to the native character code.

The Random Access module of the ANSI 1968 COBOL has been replaced by two new modules, the Relative I-O and Indexed I-O modules.

Relative I-O

Among the major features of the Relative I-O module are:

1. An ORGANIZATION IS RELATIVE clause.
2. A RELATIVE KEY clause.
3. An ACCESS MODE clause which specifies random, sequential, or dynamic access. Dynamic access permits the file to be accessed both randomly and sequentially.
4. FILE STATUS and USE AFTER STANDARD ERROR/EXCEPTION clauses as outlined in the Sequential I-O module. Here also the USE procedure may be used in place of the AT END and INVALID KEY phrases of the READ, WRITE, etc.
5. In addition to OPEN, CLOSE, READ, and WRITE, the DELETE, REWRITE, and START verbs are provided. The READ NEXT statement provides for the intermixing sequential with ran-

dom accesses of the file (when access mode is dynamic). The START statement provides the facility to position the file such that the next sequential READ statement will reference a specified record.

Indexed I-O

Among the major features of the Indexed I-O module are:

1. An ORGANIZATION IS INDEXED clause.
2. An ACCESS MODE clause with characteristics similar to that of the Relative I-O module.
3. FILE STATUS and USE procedures, as in the Relative I-O module.
4. The RECORD KEY specifies the data item that serves as the unique identifier for each record. The data item is known as the prime record key. The ALTERNATE KEY clause specifies additional (alternate) keys for the file. All insertion, updating, or deletion of records is done on the basis of the prime record key. Retrieval, however, may be on the basis of either prime or alternate record keys, thus providing more than one access path through the file.
5. As in the Relative I-O module, the new verbs DELETE, START, and REWRITE are available. READ NEXT and READ . . . KEY IS . . . are also available; the latter provides the means of specifying the key upon which retrieval is to be based (prime or alternate). The START statement also provides the means of specifying whether the prime area or alternate key is to be used for positioning the file.

Sort-Merge

The major change to the Sort module of the previous standard has been the addition of a MERGE statement to permit the combination of two or more identically ordered files. The MERGE statement parallels the SORT statement in format, except that no input procedure is provided. The COLLATING SEQUENCE phrase has been added to permit the overriding of the programming collating sequence when executing a SORT or MERGE statement.

Report Writer

The Report Writer module was completely rewritten in order to remove existing ambiguities and to provide a stronger and more useful facility. Care was taken in the rewrite not to imply that reports had to be presented on a printer (rather than on a type of graphic device).

Segmentation

1. There is no logical difference between fixed and fixed overlayable segments. (ANSI 1968 COBOL placed certain restrictions on the range of PERFORM's involving fixed overlayable segments.)
2. A PERFORM statement in a nonindependent segment may have only one of the following within its range: (1) nonindependent segments, or (2) sections wholly contained in a single independent segment, except that (2) reads "Sections wholly contained in the same independent segment." Where a SORT or MERGE statement appears in a segmented program, any associated input/output procedures are subject to the same constraints that apply to the range of a PERFORM (e.g., where the SORT is in a nonindependent segment, the associated input/output procedures must be either wholly contained in nonindependent segments or wholly contained in a single independent segment).

Library

The major changes introduced are:

1. The COPY statement may appear anywhere in the program that a COBOL word or separator may appear (ANSI 1968 COBOL permitted the COPY statement to appear only in certain specified places).
2. More than one library can be available.
3. All occurrences of a given literal, identifier, word or group of words in the library text can be replaced. (ANSI 1968 COBOL did not permit replacement of groups of words.)
4. The matching and replacement process has been significantly clarified.

Debug

The new Debug module provides a means by which the programmer can specify a debugging algorithm, including the conditions under which data items or procedures are to be monitored during program execution. The major features of this module are:

1. A USE FOR DEBUGGING statement, which permits full or selective procedure and data-name monitoring; control is passed to the procedure when the specified condition arises. Associated with the execution of each debugging section (i.e., the declarative procedure associated with the USE FOR DEBUGGING statement) is the special register DEBUG-ITEM. This is updated by the system each time a debugging section is executed with such information as the name (with occurrence numbers if it should be the name of a table element) that caused the execution, the line number upon which the name appears, etc. The USE FOR DEBUGGING statements and their associated declarative procedures are treated as comment lines if the WITH DEBUGGING MODE clause does not appear in the program. An object time switch is also provided, outside the COBOL program, through which the USE FOR DEBUGGING procedure can be "turned off" without the need to recompile the program.
2. Debugging Lines. Any line with a "D" in the continuation area is a debugging line and will be compiled and executed only if the WITH DEBUGGING MODE clause appears in the program. Where this compile item switch does not appear in the program, these lines are treated as comment lines. The setting of the object time switch has no effect on the execution of the debugging lines. Through the debugging line facility, the programmer has at his or her disposal the full power of the COBOL language for debugging purposes.

Inter-Program Communication

The new Inter-Program Communication module provides a facility by which a program can communicate with one or more other programs. This communication is made possible by: (a) enabling control to be transferred from one program to another within a run unit, and (b) enabling both programs to have access to the same data items. The major features of this module are:

1. The CALL statement causes control to be transferred from one object program to another. The CALL statement can be "static" (i.e., the name of the called program is known at compile time) or dynamic (i.e., the name of the called program is not known until program execution time). The USING phrase of the CALL statement names the data to be shared with the called program: a USING phrase in the Procedure Division header of the called program specifies the name by which this shared data is to be known in the called program. The ON OVERFLOW phrase of the CALL statement will cause control to be transferred to an associated imperative statement if there is not enough memory available at execution time to permit the loading of the called program.
2. The CANCEL statement releases the areas occupied by called programs that are no longer required to be in memory.
3. The EXIT PROGRAM statement marks the logical end of a called program and causes control to be returned to the calling program (i.e., the program in which the CALL statement appears).
4. The Linkage Section appears in a program that is to operate under the control of a CALL statement. It is used in the called program to describe data that is to be made available from the calling program through the CALL USING facility just described.

Communication

The new Communication module provides the ability to access, process, and create messages or portions thereof. It provides the ability to communicate through a Message Control System with local and remote communication devices. The major features of this module are:

1. The communication description entry (CD) specifies the interface area between the Message Control System (MCS) and a COBOL program. The CD specifies the input message queue structure, the symbolic names of destination for output messages, and such things as message data, message time, and text length.

2. The ENABLE and DISABLE statements notify the MCS to permit or inhibit the transfer of data between specified output queues and destinations for output, or between sources and input queues for input.
3. The RECEIVE statement makes available to the COBOL program a message, a portion thereof, and pertinent information about the message, from a queue maintained by the Message Control System.
4. The SEND statement causes a message or portion of a message to be released to one or more output queues maintained by the MCS.
5. The ACCEPT MESSAGE COUNT statement causes the number of messages in a queue to be made available.
6. The FOR INITIAL INPUT clause of the CD entry permits the MCS to schedule a program for execution upon receipt of a message for that program.

Substantive Changes That Are Included in the ANSI 1974 COBOL

The following is a summary of the substantive changes that are included in the ANSI 1974 COBOL. The codes used are as follows:

Module Affected

The mnemonic names that are used in these codes are the following:

Mnemonic Name	Meaning
NUC	Nucleus
TBL	Table Handling
SEQ	Sequential I-O
REL	Relative I-O
INX	Indexed I-O
SRT	Sort-Merge
RPW	Report Writer
SEG	Segmentation
LIB	Library
DEB	Debug
IPC	Inter-Program Communication
COM	Communication

Remarks

The code reflected under the remarks column is as follows:

1. Indicates the change will not impact existing programs. For example, a new verb or an additional capability for an old verb.
2. Indicates the change could impact existing programs and some reprogramming may be needed. For example, where the semantics or syntax of an existing verb are changed.
3. Indicates that the change impacts an area that was implementor-defined in the original standard. As such it may or may not affect existing programs.

Additions to the reserved word list that will impact existing programs are not included in the list.

Language elements associated with the Report Writer modules are not assigned codes because the report writer specifications were completely rewritten, and comparison with the previous standard is therefore not meaningful.

Substantive Change	Module Affected	Remarks
1. Space may immediately precede or may immediately follow a parenthesis (except in a PICTURE character-string).	NUC	(1) Relaxes punctuation rules.
2. Period, comma, or semicolon may be preceded by a space.	NUC TBL	(1) Relaxes punctuation rules.
3. Semicolon and comma are interchangeable.	NUC	(1)
4. An asterisk (*) in the continuation area (seventh character position) causes the line to be treated as a comment by the compiler. The comment line may appear in any division.	NUC	(1) New feature; replaces the NOTE statement and REMARKS paragraph.
5. A stroke (slash, '/' virgule) in the continuation area (seventh character position) of a line causes page ejection of the compilation listing. (The line is treated as comment.)	NUC	(1)
6. A phrase or clause (as well as sentence or entry) may be continued by starting subsequent lines in area B.	NUC	(1)
7. Two contiguous quotation marks may be used to represent a single quotation mark character in a nonnumeric literal.	NUC	(1) New feature.
8. Last line in a program may be a comment line.	NUC	(1)
9. Mnemonic-name must have at least one alphabetic character.	NUC	(3) ANSI 1968 had no such restriction.
10. Number of qualifiers permitted is implementor-defined, but must be at least five.	NUC	(2) ANSI 1968 specified no such lower limit.
11. Complete set of qualifiers for a name may not be the same as the partial list of qualifiers for another name.	NUC	(2)
12. REMARKS paragraph is deleted.	NUC	(2) Function was replaced by the comment line.
13. Continuation of Identification Division comment-entries must not have a hyphen in the continuation indicator area.	NUC	(2)
14. PROGRAM COLLATING SEQUENCE clause specifies that the collating sequence associated with alphabet-name is used in nonumeric comparisons.	NUC	(1) New feature.
15. SPECIAL-NAMES paragraph: 'L', '/', and '=' may not be specified in the CURRENCY SIGN clause.	NUC	(2) This restriction did not exist in ANSI 1968.
16. Alphabet-name clause relates a user-defined name to a specified collating sequence or character code set (ANSI, native, or implementor-specifier).	NUC	(1) New feature.
17. Alphabet-name clause: the literal phrase specifies a user-defined collating sequence.	NUC	(1) New feature.
18. Condition-name may be given the status of an implementor-defined switch. Switches are implementor-defined and may be either software or hardware switches.	NUC	(1) ANSI 1968 specified hardware switches only.
19. All items which are immediately subordinate to group item must have the same level number.	NUC	(2)
20. Level-77 items need not precede level-01 items in the Working-Storage Section.	NUC	(1) New feature.

Substantive Change	Module Affected	Remarks
21. Level numbers 02–49 may appear anywhere to the right of margin A. (Margin A is defined as being between character positions 7 and 8.)	NUC	(1)
22. Object of a REDEFINES clause can be subordinate to an item described with an OCCURS clause, but must not be referred to in the REDEFINES clause with a subscript or an index.	NUC	(1) New feature.
23. REDEFINES: No entry with lower level-number can appear between the redefined and redefining items.	NUC	(2) ANSI 1968 had no such restriction.
24. Multiple redefinition of same storage area permitted.	NUC	(3)
25. An asterisk used as a zero suppression symbol in a PICTURE clause and the BLANK WHEN ZERO clause may not appear in the same entry.	NUC	(2)
26. Alphabetic PICTURE character-string may contain the character B.	NUC	(1) New feature.
27. The number of digit positions that can be described by a numeric PICTURE character-string cannot exceed 18.	NUC	(2) ANSI 1968 had no such rule.
28. Stroke (/) permitted as an editing character.	NUC	(1) New feature.
29. PICTURE character-string is limited to 30 characters.	NUC	(3) ANSI 1968 defines limit as 30 symbols where one symbol could have been two characters.
30. SIGN clause allows the specification of the sign position.	NUC	(1) New feature.
31. A signed numeric literal cannot be used in a VALUE clause unless it is associated with a signed PICTURE character-string.	NUC	(2)
32. If the item is numeric edited, the literal in the VALUE clause must be nonnumeric.	NUC	(2)
33. In the Procedure Division, a section may contain zero or more paragraphs and a paragraph may contain zero or more sentences.	NUC	(1) New feature.
34. The unary + is permitted in arithmetic expressions.	NUC	(1) New feature.
35. The TO is not required in the EQUAL TO of a relation condition.	NUC	(1) ANSI 1968 required the word TO.
36. In relation and sign conditions, arithmetic expressions must contain at least one reference to a variable.	NUC	(2)
37. Comparison of nonnumeric operands; if one of the operands is described as numeric, it is treated as though it were moved to an alphanumeric item of the same size, and the contents of this alphanumeric item were then compared to the nonnumeric operand.	NUC	(3)
38. Abbreviated combined relation condition: When a portion is enclosed in parentheses, all subjects and operators required for the expansion of that	NUC	(2) No such restriction appeared in ANSI 1968.

Substantive Change	Module Affected	Remarks
portion must be included within the same set of parentheses.		
39. Abbreviated combined relation condition: If NOT is immediately followed by a relational operator, it is interpreted as part of the relational operator.	NUC	(2) In ANSI 1968, NOT was a logical operator in such cases.
40. Class condition: The numeric test cannot be used with a group item composed of elementary items described as signed.	NUC	(3)
41. In an arithmetic operation, the composite of operands must not contain more than 18 decimal digits.	NUC	(2) ANSI 1968 specified limits only for ADD and SUBTRACT.
42. ACCEPT identifier FROM DATE/DAY/TIME allows the programmer to access the date, day, and time.	NUC	(1) New feature.
43. ADD statement: the GIVING identifier series.	NUC	(1) New feature.
44. COMPUTE statement: the identifier series.	NUC	(1) New feature.
45. DISPLAY statement: If the operand is a numeric literal, it must be an unsigned integer.	NUC	(2)
46. DIVIDE statement: the INTO identifier series and the GIVING identifier series.	NUC	(2)
47. DIVIDE statement: the remainder item can be numeric edited.	NUC	(1) New feature.
48. GO TO statement; the word TO is not required.	NUC	(1) ANSI 1968 requires the word TO.
49. EXAMINE statement and the special register TALLY were deleted.	NUC	(2) Function was replaced by the INSPECT statement.
50. INSPECT statement provides ability to count or replace occurrences of single characters or groups of characters.	NUC	(1) New feature.
51. MOVE statement: A scaled integer item (i.e., the rightmost character of the PICTURE character is a P) may be moved to an alphanumeric or an alphanumeric edited item.	NUC	(1) New feature.
52. MULTIPLY statement: the BY identifier series and the GIVING identifier series.	NUC	(1) New feature.
53. PERFORM statement: Format-4 (PERFORM . . . VARYING, not using index-names) identifiers need not be described as integers.	NUC	(1) New feature.
54. PERFORM statement: Changing the FROM variable during execution can affect the number of times the procedures are executed in a Format-4 PERFORM if more than one AFTER phrase is specified.	NUC	(2)
55. PERFORM statement: There is no logical difference to the user between fixed and fixed overlayable segments.	NUC	(1) ANSI 1968 did not permit fixed overlayable segments to be treated the same as a fixed segment.
56. A PERFORM statement in a nonindependent segment can have in its range only one of the following: a. Nonindependent segment (fixed/fixed overlayable).	NUC SEG	(3)

Substantive Change	Module Affected	Remarks
b. Section and/or paragraphs wholly contained in a single independent segment.		
57. A PERFORM statement in an independent segment can have in its range only one of the following: a. Nonindependent segments (fixed/fixed overlayable). b. Sections and/or paragraphs wholly contained in the same independent segment as that PERFORM statement.	NUC SEG	(3)
58. PERFORM statement: Control is passed only once for each execution of a format-2 PERFORM statement (i.e., an independent segment referred to by such a PERFORM is made available in its initial state only once for each execution of that PERFORM statement).	NUC SEG	(3)
59. STOP statement: If the operand is a numeric literal, it must be an unsigned integer.	NUC	(2)
60. STRING statement provides for the juxtaposition of the partial or complete contents of two or more data items into a single data item.	NUC	(1) New feature.
61. STRING statement: Delimiter identifiers need not be fixed length items.	NUC	(1)
62. SUBTRACT statement: the GIVING identifier series.	NUC	(1) New feature.
63. UNSTRING statement permits contiguous data in sending field to be separated and placed into multiple receiving fields.	NUC	(1) New feature.
64. Commas are not required between subscripts or index-names.	TBL	(1)
65. Literal subscripts may be mixed with index-names when referencing a table item.	TBL	(1) New feature.
66. The DEPENDING phrase is now required in the format-2 of the OCCURS clause.	TBL	(2) ANSI 1968 has no restriction.
67. Integer-1 cannot be zero in format-2 of the OCCURS clause.	TBL	(2)
68. A data description entry with an OCCURS DEPENDING clause may be followed within that record, only by entries subordinate to it (i.e., only the last part of the record may have a variable number of occurrences).	TBL	(2) This rule did not appear in the ANSI 1968.
69. When a group item, having subordinate to it an entry that specifies format-2 of the OCCURS clause, is referenced, only part of the table area that is defined by the value of the operand of the DEPENDING phrase will be used in the operation (i.e., the actual size of a variable length item is used, not the maximum size).	TBL	(2)
70. If SYNCHRONIZED is specified for an item containing an OCCURS clause, any implicit FILLER generated for items in the same table are generated for each occurrence of those items.	TBL	(3)

Substantive Change	Module Affected	Remarks
71. The results of SEARCH ALL operation are predictable only when the data in the table is ordered as described by the ASCENDING/DESCENDING KEY clause associated with identifier-1.	TBL	(3)
72. The subject of the condition in the WHEN phrase of the SEARCH ALL statement must be a data item named in the KEY phrase of the table; the object of this condition may not be a data item named in the KEY phrase.	TBL	(2) ANSI 1968 specified that either the subject or object could be a data item named in the KEY phrase.
73. SEARCH . . . VARYING identifier-2: If identifier-2 is an index-data item, it is incremented as the associated index is incremented.	TBL	(3) In ANSI 1968, the data item is incremented by the same amount as occurrence number, i.e., by one.
74. In Format-2 of the SET statement, the literal may be negative.	TBL	(1) New feature.
75. File control entry: The ASSIGN TO implementor-name-1 OR implementor-name-n clause for the GIVING file of SORT statement was deleted.	SRT	(2)
76. MERGE statement.	SRT	(1) New feature.
77. RELEASE . . . FROM identifier is placed in Level 1 of Sort-Merge module.	SRT	(1) Was a Level-2 feature.
78. RETURN . . . INTO identifier is placed in Level 1 of Sort-Merge module.	SRT	(1) Was a Level-2 feature.
79. SORT statement: the USING file-name series.	SRT	(1) ANSI 1968 allowed only one file name.
80. SORT statement: semicolon deleted from format.	SRT	(2)
81. SORT statement: COLLATING SEQUENCE phrase provides the ability to override the program collating sequence.	SRT	(1) New feature.
82. No more than one file-name from a multiple file reel can appear in a SORT statement.	SRT	(2)
83. Where a SORT or MERGE statement appears in a segmented program, then any associated input/output procedures are subject to the same constraints that apply to the range of a PERFORM.	SRT	(2) No such restriction in ANSI 1968.
84. Segment-numbers are permitted in declaratives.	SEG	(1)
85. PAGE-COUNTER and LINE-COUNTER are described as unsigned integers that must handle values from 0 to 999999.	RPW	
86. The value in LINE-COUNTER must not be changed by the user.	RPW	
87. LINE-COUNTER, PAGE-COUNTER, and sum counter must not be used as subscripts in the Report Section.	RPW	
88. PAGE-COUNTER is always generated.	RPW	
89. PAGE-COUNTER does not need to be qualified in the Report Section.	RPW	
90. LINE-COUNTER is always generated.	RPW	

Substantive Change	Module Affected	Remarks
91. LINE-COUNTER does not need to be qualified in the Report Section.	RPW	
92. The words LINE and LINES are optional in the PAGE clause.	RPW	
93. The DATA RECORDS clause and the REPORT clause are mutually exclusive.	RPW	
94. A report may not be sent to more than one file.	RPW	
95. RESET is no longer a clause; it is a phrase under the SUM clause.	RPW	
96. Multiple SUM clauses may be specified in an item; multiple UPON phrases may be specified.	RPW	
97. Up to three hierarchical levels are permitted in a report group description.	RPW	
98. A report group level-01 entry cannot be elementary.	RPW	
99. An entry that contains a LINE NUMBER clause must not have a subordinate entry that also contains a LINE NUMBER clause.	RPW	
100. An entry that contains a COLUMN NUMBER clause but no LINE NUMBER clause must be subordinate to an entry that contains a LINE NUMBER clause.	RPW	
101. An entry that contains a VALUE clause must also have a COLUMN NUMBER clause.	RPW	
102. In the CODE clause, a mnemonic-name has been replaced by a literal (a two-character nonnumeric literal placed in the first two character positions of the logical record).	RPW	
103. If the CODE clause is specified for any report in a file, it must be specified for all reports in the same file.	RPW	
104. Control data items may not be subscripted or indexed.	RPW	
105. Each data-name in the CONTROL clause must identify a different data item.	RPW	
106. The GROUP INDICATE clause may only appear in a DETAIL report group entry that defines a printable item (contains a COLUMN and PICTURE clause).	RPW	
107. LINE clause integers must not exceed three significant digits in length.	RPW	
108. The NEXT PAGE phrase of the LINE clause is no longer legal in RH, PH, and PF groups.	RPW	
109. A relative LINE NUMBER clause can no longer be the first LINE NUMBER clause in a PAGE FOOTING group.	RPW	
110. A NEXT GROUP clause without a LINE clause is no longer legal.	RPW	
111. Integer-2 in the NEXT GROUP clause must not exceed three significant digits in length.	RPW	
112. If the PAGE clause is omitted, only a relative NEXT GROUP clause may be specified.	RPW	

Substantive Change	Module Affected	Remarks
113. The NEXT PAGE phrase of the NEXT GROUP clause must not be specified in a PAGE FOOTING report group.	RPW	
114. The NEXT GROUP clause must not be specified in a REPORT FOOTING report group.	RPW	
115. The phrases of the PAGE clause may be written in any order.	RPW	
116. In the PAGE clause, the maximum size of the integer is three significant digits.	RPW	
117. It is no longer possible to sum upon an item in another report.	RPW	
118. Source-sum correlation is not required. (Operands of a SUM clause need not be operands of a SOURCE clause in DETAIL groups.)	RPW	
119. TYPE clause data-names may not be subscripted or indexed.	RPW	
120. PAGE HEADING and PAGE FOOTING report groups may be specified only if a PAGE clause is specified in the corresponding report group description entry.	RPW	
121. In CONTROL FOOTING, PAGE HEADING, PAGE FOOTING, and REPORT FOOTING report groups, SOURCE clauses and USE statements may not reference: a. Group data items containing control data items. b. Data items subordinate to a control data item. c. A redefinition or renaming of any part of a control data item. In PAGE HEADING and PAGE FOOTING report groups, SOURCE clauses and USE statements must not reference control data-name.	RPW	
122. In summary reporting, only one detail group is allowed.	RPW	
123. The description of a report must include at least one body group.	RPW	
124. Report files must be opened with either the OPEN INPUT or OPEN EXTEND statement.	RPW	
125. A file described with a REPORT clause cannot be referenced by any input-output statement except the OPEN or CLOSE statement.	RPW	
126. The SUPPRESS statement.	RPW	
127. If no GENERATE statements have been executed for a report during the interval between the execution of an INITIATE statement and a TERMINATE statement for that report, the TERMINATE statement does not cause the Report Writer Control System to perform any of the related processing.	RPW	
128. A USE procedure may refer to a DETAIL group.	RPW	
129. FILE STATUS clause: data-name is updated by the system at the completion of each input-output operation.	SEQ REL INX	(1) New feature.

Substantive Change	Module Affected	Remarks
130. ACCESS MODE IS DYNAMIC clause: provides ability to access a file sequentially or randomly in the same program.	REL INX	(1) New feature.
131. ALTERNATE RECORD KEY clause: allows specification of multiple keys, any of which can be used to access an indexed file.	INX	(1) New feature.
132. ACTUAL KEY clause deleted.		(2)
133. RELATIVE KEY clause added for relative organization.	REL	(1) New feature.
134. RECORD KEY clause added for indexed organization.	INX	(1) New feature.
135. FILE-LIMITS clause deleted.		(2)
136. PROCESSING MODE clause deleted.		(2)
137. FILE-CONTROL paragraph: except for the ASSIGN clause, the order of clauses following file-name is optional.	SEQ REL INX	(1)
138. ORGANIZATION IS RELATIVE clause.	REL	(2) New feature.
139. ORGANIZATION IS SEQUENTIAL clause.	SEQ	(2) New feature.
140. ORGANIZATION IS INDEXED clause.	INX	(2) New feature.
141. MULTIPLE REEL/UNIT clause deleted.		(2)
142. RESERVE . . . ALTERNATIVE AREAS deleted.		(2)
143. RESERVE integer AREAS allows the user to specify the exact number of areas to be used.	SEQ REL INX	(1) New feature.
144. The file description entry for file-name must be equivalent to that used when this file was created.	SEQ REL INX	(3) No such rule in ANSI 1968.
145. The data-name option of the LABEL RECORDS clause was deleted.	SEQ REL INX	(2) ANSI 1968 provided for user-defined label records.
146. Data-name in the VALUE OF clause must be an implementor-name.	SEQ	(2) ANSI 1968 provided for user-defined field in label records.
147. LINAGE clause permits programmer definition of logical page size.	SEQ	(1) New feature.
148. CLOSE . . . FOR REMOVAL statement.	SEQ	(1) New feature.
149. DELETE statement.	REL INX	(1) New feature.
150. OPEN REVERSED positions file at its end.	SEQ	(2)
151. OPEN INPUT or OPEN I-O makes a record available to the program.	SEQ REL INX	(1) New feature.
152. OPEN EXTEND statement: adds records to an existing file.	SEQ	(1) New feature.
153. The OPEN and CLOSE statements with the NO REWIND phrase apply to all devices that claim support for this function.	SEQ	(1) ANSI 1968 restricted the application of this phrase.
154. The OPEN REVERSED statement applies to all devices that claim support for this function.	SEQ	(1) ANSI 1968 restricted the application of this phrase.
155. READ statement: AT END phrase required only if no applicable USE AFTER ERROR/ EXCEPTION procedure specified.	SEQ REL INX	(1) New feature.

Substantive Change	Module Affected	Remarks
156. READ statement: INVALID KEY phrase required only if no applicable USE AFTER ERROR/EXCEPTION procedure specified.	REL INX	(1) New feature.
157. READ statement: INTO phrase placed in Level 1.	SEQ REL INX	(1) Level 2 feature in ANSI 1968.
158. READ . . . NEXT statement: use to retrieve the next logical record from a file when the access mode is dynamic.	REL	(1) New feature.
159. REWRITE statement.	SEQ REL INX	(1) New feature.
160. SEEK statement was deleted.		(2)
161. START statement: provides for logical positioning within a relative or indexed file for sequential retrieval of records.	REL	(1) New feature.
162. USE statement: the label processing options are deleted.	SEQ REL	(2) ANSI 1968 provided for the processing of user-defined labels.
163. USE . . . ERROR/EXCEPTION statement.	SEQ REL INX	(1) New feature.
164. Recursive invocation of USE procedures prohibited.	SEQ REL INX	(2)
165. WRITE statement: INVALID KEY phrase deleted.	SEQ	(2)
166. WRITE statement: INVALID KEY phrase required only if no applicable USE AFTER ERROR/EXCEPTION procedure specified.	REL INX	(1)
167. WRITE statement: FROM phrase placed in Level 1.	SEQ REL INX	(1) Level 1 feature in ANSI 1968.
168. WRITE statement: BEFORE/AFTER PAGE phrase provides ability to skip to top of a page.	SEQ	(1)
169. WRITE statement: END-OF-PAGE phrase.	SEQ	(1) New feature.
170. Debugging line: defined by a 'D' in the continuation column.	DEB	(1) New feature.
171. WITH DEBUGGING MODE clause: a compile time switch; in addition an object time switch can be used to suppress coding at object time.	DEB	(1) New feature.
172. USE FOR DEBUGGING statement.	DEB	(1) New feature.
173. DEBUG-ITEM.	DEB	(1) New feature.
174. Linkage Section.	IPC	(1) New feature.
175. Procedure Division header: the USING phrase.	IPC	(1) New feature.
176. CALL identifier statement.	IPC	(1) New feature.
177. CALL identifier ON OVERFLOW statement.	IPC	(1) New feature.
178. CANCEL statement.	IPC	(1) New feature.
179. EXIT PROGRAM statement.	IPC	(1) New feature.
180. COPY statement may appear anywhere a COBOL word may appear.	LIB	(1) New feature.
181. Identifier, COBOL word, or a group of COBOL words may be replaced.	LIB	(1) New feature.

Substantive Change	Module Affected	Remarks
182. Multiple libraries are permitted.	LIB	(1) New feature.
183. Library-name is a user-defined word.	LIB	(1) New feature.
184. Communication description entry (CD).	COM	(1) New feature.
185. ACCEPT cd-name MESSAGE COUNT statement.	COM	(1) New feature.
186. ENABLE statement.	COM	(1) New feature.
187. DISABLE statement.	COM	(1) New feature.
188. RECEIVE statement.	COM	(1) New feature.
189. SEND statement.	COM	(1) New feature.

Elements Deleted from ANSI 1968 COBOL

The following elements were deleted from the ANSI 1968 standards.

REMARKS Paragraph. The REMARKS paragraph of the Identification Division was deleted and the function replaced by the asterisk (*) comment line.

EXAMINE Statement. The EXAMINE statement and the special register TALLY were deleted in favor of the new and more powerful INSPECT statement.

NOTE Statement. The NOTE statement was deleted and the function replaced by the asterisk (*) comment line.

FILE-LIMITS Clause. This clause was deleted from the file control entry because the function could be handled better outside the COBOL program.

SEEK Statement. This statement was redundant; it is implied by the READ, WRITE, etc.

MULTIPLE REEL/UNIT Clause. This clause was deleted from the file control entry because the function could be handled better outside the COBOL program.

ACTUAL KEY Clause. This clause was replaced by the RELATIVE KEY clause.

RESERVE integer ALTERNATE AREAS Clause. This clause was replaced by the RESERVE integer AREAS Clause.

OR implementor-name. This clause was deleted from the file control entry because the function could be handled better outside the COBOL program.

Integer implementor-name. This clause was deleted from the file control entry because the function could be handled better outside the COBOL program.

PROCESSING MODE IS SEQUENTIAL Clause. This clause was deleted from the file control entry as not being needed in a synchronous environment.

USE . . . LABEL Statement. An extensive revision to label processing is currently under way to remove ambiguities and provide for the processing of ANSI standard labels. This work was not completed in time for inclusion in the 1974 revision. In order not to hinder the introduction of this new facility, it was decided to define only a minimum label processing capability in the revised standard.

LABEL RECORDS IS data-name Clause. An extensive revision to label processing is currently under way to remove ambiguities and provide for the processing of ANSI standard labels. This work was not completed in time to be included in the 1974 revision. In order not to hinder the introduction of this new facility, it was decided to define only a minimum label processing capability in the revised standard.

Appendix D
Ten Problems

Problem 1: Shampoo Payroll Problem

In a beauty salon, operators are paid by the amount and type of work they do. The shampoo operators receive $4.00 per customer, the hair cutters receive $5.50 per customer, the hair setters receive $6.00 per customer, the stylists receive $8.00 per customer, and the permanent wave operators receive $10.00 per customer.

Given: Operator's name .
Type of operator
Number of customers

Input

Field	Positions
Name	1–25
Type	26
Customers	27–29
Blanks	30–80

Formula

Gross-pay = Rate × Customers

Printed Output

```
TYPE OF OPERATOR.
     SHAMPOO - - - - - - 1
     HAIR CUTTERS- - - - 2
     HAIR SETTERS- - - - 3
     STYLISTS- - - - - - 4
     PERMANENT WAVE- - - 5
```

NAME OF OPERATOR	TYPE	NO. OF CUSTOMERS	GROSS PAY
SUSAN CALDWELL	1	100	$400.00
BETTY JANE CLANCY	1	120	$480.00
RUTH ANN CORBETT	1	150	$600.00
MARGARET CUSHING	2	100	$550.00
ROSEMARY DUPUIS	2	75	$412.50
MAURICE ERICKSON	2	50	$275.00
LILLIAN FELLING	3	100	$600.00
NANCY HAMILTON	3	125	$750.00
BARBARA HICKMAN	3	110	$660.00
JOSEPHINE HOUSTON	4	50	$400.00
LORETTA JOHNSON	4	40	$320.00
ELAINE LEONARD	4	30	$240.00
LORRAINE CLARK	5	30	$300.00
TERRY MCDONNELL	5	35	$350.00
MARY ANN PALMER	5	40	$400.00

Problem 2: Department Store Problem

Given: Customer's name
Customer's address
Customer's account no.
Last month's balance
Payments made
Purchases made

Input

Field	Positions	
Customer's name	1–15	
Customer's address	16–50	
Account no.	51–55	
Last balance	56–60	(xxx.xx)
Month sales	61–65	(xxx.xx)
Payments	66–70	(xxx.xx)
Blanks	71–80	

Formulas to be Used

Service charge = $.015 \times$ (last-balance − payments) (rounded)
Amount due = (last-balance − payments) + service charge + month's sales

Output

NAME OF CUSTOMER	ADDRESS OF CUSTOMER	ACCOUNT	PREVIOUS BALANCE	SALES	PAYMENT	SERVICE CHARGE	AMT DUE
DAVID ANDERSON	18745 MOBILE ST., RESEDA, CALIF.	62986	$100.00	$100.00	$20.00	$1.20	$181.20
BETTY L. BREWER	10321 LUNDY DR., INGLEWOOD, CLAIF.	61477	$350.25	$50.00	$30.00	$4.80	$375.05
ARTHUR BROWN	12145 MADISON ST., L.A. , CALIF.	38940	$450.00	$30.00	$50.00	$6.00	$436.00
THOMAS CASSIDY	3726 HOPE AV., LYNWOOD, CALIF.	62180	$121.50	$40.00	$20.00	$1.52	$143.02
BOB CHAMBERS	3840 HOPE AVE., LYNWOOD, CALIF.	58920	$320.00	$15.50	$35.00	$4.28	$304.78
JACK T. CROSS	6421 BELMAR ST., RESEDA, CALIF.	43313	$105.80	$150.85	$20.00	$1.29	$237.94
KENT B. DAVID	11621 PENN DRIVE. ENCINO, CALIF.	84082	$320.75	$75.75	$35.00	$4.29	$365.79
SAMUAL FELLOW	10732 LINDLEY AV., ENCINO, CALIF.	41750	$290.60	$40.37	$30.00	$3.91	$304.88
MICHAEL FISHER	6345 TAMPA AV., TARZANA, CALIF.	30040	$444.35	$65.25	$50.00	$5.92	$465.52
GLADYS BUTTONS	3701 BALBOA AV., VAN NUYS, CALIF.	67542	$375.00	$89.30	$45.00	$4.95	$424.25
PATRICK HANEY	4218 VICTORY ST., L.A., CALIF.	72111	$450.10	$27.95	$55.00	$5.93	$428.98
LYNN HUBBARD	13245 VENTURA BLVD., RESEDA, CALIF.	64375	$195.75	$36.45	$25.00	$2.56	$209.76
MARVID JACOBS	13118 VENTURA BLVD., RESEDA, CALIF.	62550	$225.95	$44.95	$30.00	$2.94	$243.84
LINDA JOHNSON	20715 VAN NUYS ST., ENCINO, CALIF.	58214	$300.00	$87.50	$35.00	$3.98	$356.48
HOWARD DEYES	2181 SHERMAN WAY, RESEDA, CALIF.	46615	$279.80	$101.75	$30.00	$3.75	$355.30

Problem 3: Bank Balance Problem

Given: in an 80-position record
- Account number
- Type code: either a 1 digit or a 2 digit:
  - 1 indicates a checking account
  - 2 indicates a savings account
- Deposits
- Withdrawals
- Last Balance

Input

Field	Positions	
Account no.	1–5	
Type code	6	
Blanks	7–8	
Deposits	9–16	(xxxxxx.xx)
Withdrawals	17–24	(xxxxxx.xx)
Last balance	25–32	(xxxxxx.xx)
Blanks	33–80	

Formulas

New balance = Last-balance − withdrawals + deposits.
Interest = .015 × (last-balance − withdrawals + deposits)
New balance = Last-balance − withdrawals + deposits + interest.

Output

ACCOUNT NO.	NEW BALANCE	TYPE
61788	$7,358.75	2
61003	$10,500.00	1
58440	$102,685.50	1
57905	$38,650.00	1
60756	$37,500.00	1
61880	$25,590.00	1
59425	$50,500.50	1
63740	$45,000.00	1
65500	$10,500.00	1
62711	$5,250.00	1
60912	$10,500.00	1
57280	$7,500.00	1
59014	$20,000.00	1
60545	$50,500.00	1

Problem 4: "Africa" Payroll Problem

Each month a payroll is to be processed in the following manner:

A file contains a master record for every employee in the company. Each record contains the employee's name, number, the regular and overtime hours worked during the month, the wages earned so far that year, the rate of pay, and the number of dependents. The payroll is processed in the usual manner: computation of GROSS-PAY, FICA, WH-TAX and NET PAY. The results are used to print checks and to create new master records. These new records contain the new YTD-GROSS and zeros in the hours field (both regular and overtime). These new records create the CARD-OUT file, which will be used next month as CARD-IN. (During the month the regular hours and overtime hours are added in by another program.)

This company has a subsidiary in Africa, and whose employees, though U.S. citizens, are not required to pay income tax. These employees have A's in front of their numbers, other employee numbers will have spaces in these character positions.

This program, which will use a table of income tax exemptions according to the number of dependents, looks like this:

DEPENDENTS										
0	1	2	3	4	5	6	7	8	9	10
$0	$56	$112	$168	$224	$280	$336	$392	$448	$504	$560

Subtract from GROSS-PAY to find the taxable amount. The formulas that the program will use are:

In America:
Gross-Pay = (regular hours × rate) + (overtime hours × 1½ rate).
FICA = .067 × gross pay (if YTD is less than $32,400).
WH-TAX = .18 × (gross-pay − sub) sub is exemption according to dependents chart.
NET PAY = gross-pay − FICA − WH-TAX.

In Africa:
Gross-Pay = (regular hours × rate) + (overtime hours × 1½ rate).
WH-TAX = 0.
FICA = .067 × gross pay (if YTD is less than $32,400).
NET PAY = gross-pay − FICA.

Card-In

Field	Positions	
Employee Name	1–29	
Employee Number	30–39	
Blank	40	
Rate	41–43	(x.xx)
Dependents	44–45	
Regular Hours	46–50	(xxx.xx)
Overtime Hours	51–55	(xxx.xx)
YTD-Gross	56–62	(xxxxx.xx)
Blanks	63–80	

Print-Out

	Field	Print Positions
CHECK-LINE-1		
	Name	1–29
	Blanks	30–106
	Date	107–114
	Blanks	115–120
CHECK-LINE-2		
	Blanks	1–90
	Net-Pay	91–96
	Blanks	97–120
CHECK-LINE-3		
	Employee Name	1–29
	Blanks	30–31
	Employee Number	32–41
	Blanks	42–43
	Gross Pay	44–50
	Blanks	51–52
	Wh-Tax	53–57
	Blanks	58–59
	FICA	60–64
	Blanks	65–66
	Net-Pay	67–73
	Blanks	74–120

Output

```
GUY T. GOODWIN                                                                                  11 MAY 82
                                                                                     $430.22
GUY T. GOODWIN                     000020981    $504.40  $40.39  $33.79    $430.22

PAUL A. EVANS                                                                                   11 MAY 82
                                                                                     $336.42
PAUL A. EVANS                      000053487    $420.00  $55.44  $28.14    $336.42

MILTON C. MORGAN                                                                                11 MAY 82
                                                                                     $451.04
MILTON C. MORGAN                   000064378    $464.00  $12.96   $0.00    $451.04

JOHN L. REED                                                                                    11 MAY 82
                                                                                     $448.76
JOHN L. REED                       000052887    $569.19  $82.29  $38.14    $448.76

JAMES F. KING                                                                                   11 MAY 82
                                                                                     $990.16
JAMES F. KING                     A000053219  $1,061.27   $0.00  $71.11    $990.15

MARY DIXON                                                                                      11 MAY 82
                                                                                     $430.73
MARY DIXON                        A000062986    $461.66   $0.00  $30.93    $430.73
```

Problem 5: Sales and Commission Problem

Commissions are paid to salespersons based upon the number of units that are sold. The unit commission varies with the product sold and the total commission is based upon the number of units sold of each particular product.

Sales are determined by the number of units sold times the individual product selling price.

Given:

Product	Commission Rate	Selling Price
1	$ .10	$ 16.00
2	$ .20	$ 30.00
3	$ .30	$ 43.00
4	$ .40	$ 60.00
5	$ .50	$ 75.00

Input Record File

Field	Positions
Territory Number	1–2
Salesperson Number	3–5
Date	6–11
Name	12–30
Units Sold	31–35
Product Number	36

PROBLEM

1. Prepare a table of commissions for the five different products so that the product number itself will serve as a subscript.

2. Prepare a table of selling prices for the five different products so that the product number itself will serve as a subscript.

3. Write a COBOL program that will read both the commission table and price table and then process each data record to calculate the commission for each salesperson. At the same time, prepare a report of the number of units sold and the amount of sales for each product by salesperson, by territory and an overall total of sales.

Output

MONTHLY SALES AND COMMISSION REPORT
MAY 1982

PAGE 1

TERRITORY NUMBER	SALESMAN NUMBER	DATE	NAME	PRODUCT NUMBER	UNITS SOLD	TOTAL SALES	COMMISSION
10	111	010382	JONES HENRY	1	100	$1,600.00	$160.00
10	111	011082	JONES HENRY	2	2,301	$69,030.00	$13,806.00
10	111	011782	JONES HENRY	3	60	$2,580.00	$774.00
10	111	012482	JONES HENRY	4	20,502	$1,230,120.00	$492,048.00
					22,963	$1,303,330.00	$506,788.00
10	222	010382	SMITH ROBERT	1	55	$880.00	$88.00
10	222	010382	SMITH ROBERT	2	70	$2,100.00	$420.00
10	222	010382	SMITH ROBERT	3	800	$34,400.00	$10,320.00
					925	$37,380.00	$10,828.00
			TERRITORY-TOTAL		23,888	$1,340,710.00	$517,616.00
16	431	012482	HODGES JAMES	5	10	$750.00	$375.00
16	431	011782	HODGES JAMES	4	623	$37,380.00	$14,952.00
16	431	010382	HODGES JAMES	3	40,500	$1,741,500.00	$522,450.00
					41,133	$1,779,630.00	$537,777.00
			TERRITORY-TOTAL		41,133	$1,779,630.00	$537,777.00
24	350	012482	JOHNSTON HOWARD	5	8	$600.00	$300.00
24	350	011082	JOHNSTON HOWARD	3	42	$1,806.00	$541.80
					50	$2,406.00	$841.80
24	565	012482	MONTGOMERY DAN	4	820	$49,200.00	$19,680.00
24	565	011782	MONTGOMERY DAN	1	7,255	$116,080.00	$11,608.00
24	565	011782	MONTGOMERY DAN	1	7,255	$116,080.00	$11,608.00
					15,330	$281,360.00	$42,896.00
			TERRITORY-TOTAL		15,380	$283,766.00	$43,737.80
			TOTAL SALES			$3,404,106.00	**

END OF REPORT

Problem 6: Payroll Register Problem

Write a COBOL program that will calculate and print the Payroll Register as indicated.

Input Record File

Field	Positions	
Month	1–3	
Day	4–5	
Year	6–7	
Department	14–16	
Serial	17–21	
Gross Earnings	57–61	(xxx.xx)
Insurance	62–65	(xx.xx)
Withholding Tax	69–72	(xx.xx)
Not Used	73–75	
Miscellaneous Deductions	76–79	(xx.xx)
Code (letter E)	80	

Calculations

1. FICA = Gross Earnings × .067 (round to two decimal positions).
2. State UCI = Gross Earnings × .01 (round to two decimal positions).
3. Net Earnings = Gross Earnings − Insurance − FICA − Withholding Tax − State UCI − Miscellaneous Deductions.
4. If the Net Earnings are zero or negative, branch to an error routine.
5. The Department Earnings value is the sum of the Net Earnings for each employee.
6. Calculate totals for all columns by department as well as an overall total for the entire payroll.

Printed Output

Heading Line	Field	Print Positions
1	WEEKLY PAYROLL REGISTER	40–62
2	WEEK ENDING	40–51
3	EMPLOYEE NO.	3–14
	GROSS	22–26
	WITHHOLDING	51–61
	STATE	65–69
4	DEPT.	2–6
	SERIAL	10–15
	EARNINGS	20–27
	INSURANCE	31–39
	FICA	44–47
	TAX	56–58
	UCI	66–68
	MISC. DEDNS.	74–85
	NET EARNINGS	89–100

Detail	Print Positions	
Department	3–5	
Serial	10–14	
Gross Earnings	21–26	(xxxx.xx)
Insurance	33–37	(xxx.xx)
FICA	43–47	(xxx.xx)
Withholding Tax	54–58	(xxx.xx)
State UCI	65–68	(xx.xx)
Miscellaneous Deductions	76–80	(xxx.xx)
Net Earnings	92–97	(xxxx.xx)

Output

WEEKLY PAYROLL REGISTER
WEEK ENDING MAY 11 1982

PAGE 1

EMPLOYEE NO. DEPT.	SERIAL	GROSS EARNINGS	INSURANCE	FICA	WITHHOLDING TAX	STATE UCI	MISC. DEDNS.	NET EARNINGS
200	10670	$202.00	$3.10	$13.53	$28.00	$2.02	$0.00	$155.35
200	10695	$203.00	$3.10	$13.60	$28.00	$2.03	$5.00	$151.27
200	10700	$204.00	$3.10	$13.67	$28.00	$2.04	$0.00	$157.19
200	10703	$205.00	$3.10	$13.74	$28.00	$2.05	$0.00	$158.11
200	10725	$207.00	$3.10	$13.87	$28.00	$2.07	$10.00	$149.96
200	10730	$208.00	$3.10	$13.94	$28.00	$2.08	$0.00	$160.88
200	10742	$209.00	$3.10	$14.00	$28.00	$2.09	$0.00	$161.81
200	10800	$210.00	$3.10	$14.07	$28.00	$2.10	$0.00	$162.73
200	10890	$211.00	$3.10	$14.14	$28.00	$2.11	$1.00	$162.65
DEPT. TOTALS:		$1859.00	$27.90	$124.56	$252.00	$18.59	$16.00	$1419.95
300	10904	$212.00	$3.10	$14.20	$28.00	$2.12	$0.00	$164.58
300	10905	$213.00	$3.10	$14.27	$28.00	$2.13	$0.00	$165.50
300	10906	$214.00	$3.10	$14.34	$28.00	$2.14	$0.00	$166.42
300	10907	$215.00	$3.10	$14.41	$28.00	$2.15	$0.00	$167.34
DEPT. TOTALS:		$854.00	$12.40	$57.22	$112.00	$8.54	$0.00	$663.84
400	11215	$218.00	$3.10	$14.61	$28.00	$2.18	$0.00	$170.11
400	11225	$219.00	$3.10	$14.67	$28.00	$2.19	$0.00	$171.04
400	11240	$220.00	$3.15	$14.74	$28.50	$2.20	$0.00	$171.41
400	11250	$221.00	$3.15	$14.81	$28.50	$2.21	$0.00	$172.33
DEPT. TOTALS:		$878.00	$12.50	$58.83	$113.00	$8.78	$0.00	$684.89
600	12330	$225.00	$3.15	$15.08	$28.50	$2.25	$2.00	$174.02
600	12340	$226.00	$3.15	$15.14	$28.50	$2.26	$2.00	$174.95
600	12350	$227.00	$3.15	$15.21	$28.50	$2.27	$0.00	$177.87
600	12366	$228.00	$3.15	$15.28	$28.50	$2.28	$0.00	$178.79
600	12400	$229.00	$3.15	$15.34	$28.50	$2.29	$0.00	$179.72
DEPT. TOTALS:		$1135.00	$15.75	$76.05	$142.50	$11.35	$4.00	$885.35
T O T A L S :		$4726.00	$68.55	$316.66	$619.50	$47.26	$20.00	$3654.03

Problem 7: Updated Payroll Problem

Write a COBOL program to update a master TAPE FILE with a current file. Both input files are in Social Security Number sequence. Input TAPE records are in blocks of ten 52-character records. The updated output TAPE blocks will be the same size.

The exception list shall be double spaced.

Input Tape Record

Field	Positions	
Employee Name	6–25	
Social Security Number	26–34	
Old Year-to-Date Gross Earnings	35–41	(xxxxx.xx)
Old Year-to-Date Withholding Tax	42–47	(xxxx.xx)
Old Year-to-Date FICA	48–53	(xxxx.xx)

Input Record

Field	Positions	
Department Number	1–2	
Clock Number	3–5	
Social Security Number	26–34	
Current Gross Earnings	62–68	(xxxxx.xx)
Current Withholding Tax	69–74	(xxxx.xx)
Current FICA	75–79	(xxx.xx)
Code (Digit 1)	80	

Operations To Be Performed

1. New Year-to-Date Gross = Old Year-to-Date Gross plus Current Gross.
2. New Year-to-Date Withholding Tax = Old Year-to-Date Withholding Tax plus Current Withholding Tax.
3. New Year-to-Date FICA = Old Year-to-date FICA plus Current FICA.
4. If New Year-to-Date FICA record exceeds $2,170.80, print Department Number, Clock Number, Employee Name, Social Security Number, New Year-to-Date Gross, New Year-to-Date Withholding Tax and Excess FICA amount.

Output Updated Tape Record

Field	Positions
Department Number	1–2
Clock Number	3–5
Employee Name	6–25
Social Security Number	26–34
New Year-to-Date Gross Earnings	35–41
New Year-to-Date Withholding Tax	42–47
New Year-to-Date FICA	48–52

Output Printed Record

Field	Print Positions	
Department Number	4–5	
Clock Number	9–11	
Employee Name	15–34	
Social Security Number	37–47	(xxx-xx-xxxx)
New Year-to-Date Gross Earnings	52–60	(xx,xxx.xx)
New Year-to-Date Withholding Tax	64–71	(x,xxx.xx)
Excess FICA	77–81	(xx.xx)

XX EMPLOYEES OVER $2,170.80

Output

```
                          EXCEPTION LIST
                      EMPLOYEES OVER $2,170.80                        PAGE  1

DEPT CLOCK      EMPLOYEE NAME        SOC SEC NO.     YTD GROSS   YTD W/H     EX FICA

 75   925    FOX WILLIAM            130-09-5294    38,695.45 4,829.43        421.80

 37   857    PHILLIPS ROBERT        364-20-8841    33,234.70 2,925.63         55.93

 15   375    JACKSON KENNETH        543-01-2232    32,782.50 2,685.04         25.63

 42   902    SAWYER DAVID           556-32-0201    35,822.82 3,290.43        229.33

 21   472    YOUNG SAMUEL           557-16-7782    34,377.86 3,621.30        132.52

 87   524    HEPNER ELMER           559-10-9299    36,007.25 3,972.90        241.69

 63   708    HORNE ALBERT           578-20-1141    32,921.84 1,874.87         34.96

                      7 EMPLOYEES OVER $2,170.80
```

Problem 8: Sales Problem (Report Writer Feature)

Report

```
               D A I L Y      S A L E S      R E G I S T E R

WEEK OF 05-01-82                                         PAGE  01

          ENTRY     CUSTOMER    SALESMAN          SALE
           DAY       NUMBER      NUMBER          AMOUNT

            01       08257        071          $ 1,189.80
            01       11234        079             168.06
            01       29031        079              63.00
            01       79992        095              87.74

                          DAY 01 SALES      $    1,508.60 *

            17       02965        037          $12,716.92
            17       09002        001             842.17
            17       01179        002           7,071.12
            17       13605        001           3,092.72
            17       27654        009             217.90

                          DAY 17 SALES      $   23,940.83 *

            18       00390        092          $    27.00
            18       05006        056             897.32
            18       12125        181             371.98

                          DAY 18 SALES      $    1,296.30 *

            19       00298        100          $ 2,020.60
            19       00106        024           1,494.73

                          DAY 19 SALES      $    3,515.33 *

            21       00256        003          $    79.53
            21       00652        008              95.18
            21       18569        090             421.15
            21       20106        132          $   706.42
            21       00321        005             590.10

                          DAY 21 SALES      $    1,892.38 *


NUMBER OF SALES   19       TOTAL SALES $      32,153.44 **

END OF REPORT
```

Input Record File

Field	Positions	
Month	1–2	
Day	3–4	
Year	5–6	
Salesman Number	7–9	
Customer Number	10–14	
Sales Amount	51–57	(xxxxx.xx)

Calculations

Compute the total sales values for each day of the week.
Compute the total sales values for the week.
Compute the number of sales for the week.

Problem 9: Commission Problem (Report Writer Feature)

Report

```
                         S A L E S   C O M M I S S I O N   R E P O R T

                                FOR THE MONTH OF JANUARY 1982

SALESMAN       CUSTOMER       INVOICE          NET AMOUNT       RATE        COMMISSION

  4490          115121         25460        $  1,250.00          10       $    125.00
  4490           78345         25198           8,255.12           8            660.41
  4490           72914         44483             690.70          14             96.70
                   SALESMAN 4490 TOTAL   $    10,195.82  *              $      882.11    *

  2513           14983         14152        $    110.20           6       $      6.61
  2513          712129         13444          10,986.00          12          1,318.32
  2513           11110         12136           9,850.40           8            788.03
                   SALESMAN 2513 TOTAL   $    20,946.60  *              $    2,112.96    *


          TOTAL FOR MONTH OF JANUARY  $       31,142.42 **          $        2,995.07 **
```

Input Record File

Field	Positions	
Code (digit 5)	1	
Invoice Number	2–6	
Customer Number	13–19	
Net Amount	35–42	(xxxxxx.xx)
Salesman Number	43–46	
Commission Rate	54–55	(.xx)
Commission Amount	56–62	(xxxxx.xx)

Calculations

Find the total sales and total commissions for each salesman for the month.
Find the total sales and total commissions for the entire force for the month.

Problem 10: Hospital Problem (Sort Feature)

There is a record for the number of patients in each hospital of the United States.

A report is prepared indicating the various patient totals for cities and counties within each state of the United States. An overall total is indicated for the entire United States.

Input Record File

Field	Positions
Date	1–6
State	7–8
County	9–11
City	12–14
Hospital Number	15–18
Number of Patients	70–75

Operations Required

1. Sort data records in the following sequence; major-State, intermediate-County and minor-City.
2. Prepare listing per output record format.

Output Record Format

```
                                   HOSPITAL PATIENT REPORT
                                    JANUARY 31 1982                PAGE      1

      STATE         COUNTY           CITY                        NUMBER OF PATIENTS

        AL            SEC              NBI                                1,500
                                     COUNTY TOTAL                         1,500
                                     STATE  TOTAL                         1,500
        AR            ABA              CAL                                3,716
                                     COUNTY TOTAL                         3,716
                                     STATE  TOTAL                         3,716
        CA            LA               LA                                   800
                                     COUNTY TOTAL                           800
        CA            ORA              LB                                   400
                                     COUNTY TOTAL                           400
        CA            SCL              SUN                                2,667
                                     COUNTY TOTAL                         2,667
        CA            SD               SD                                 1,626
                                     COUNTY TOTAL                         1,626
        CA            VEN              PH                                   963
                                     COUNTY TOTAL                           963
                                     STATE  TOTAL                         6,456
        DE            BAN              OXA                                1,329
        DE            BAN              RON                                1,965
                                     COUNTY TOTAL                         3,294
                                     STATE  TOTAL                         3,294
        FL            SD               AR                                 3,048
                                     COUNTY TOTAL                         3,048
                                     STATE  TOTAL                         3,048
        IL            COK              CHI                                1,200
                                     COUNTY TOTAL                         1,200
                                     STATE  TOTAL                         1,200
        MD            BAL              BAL                                  200
        MD            BAL              PIK                                4,004
                                     COUNTY TOTAL                         4,204
        MD            CAR              TOW                                  924
                                     COUNTY TOTAL                           924
        MD            PGS              SSP                                2,400
                                     COUNTY TOTAL                         2,400
                                     STATE  TOTAL                         7,528
        NE            CEN              FON                                3,468
                                     COUNTY TOTAL                         3,468
                                     STATE  TOTAL                         3,468
        WA            KNG              STL                                4,068
                                     COUNTY TOTAL                         4,068
                                     STATE  TOTAL                         4,068
        WN            DUN              RED                                3,936
                                     COUNTY TOTAL                         3,936
                                     STATE  TOTAL                         3,936
```

This appendix contains input data for end-of-chapter problems and the ten problems in appendix D.

Appendix E

Input Data

Input Data for the End-of-Chapter Problems

Chapter 6

```
•••••••••1•••••••••2•••••••••3•••••••••4•••••••••5•••••••••6•••••••••7•••••••••8
CHAPTER 6 PRØBLEM 1
INPUT DATA

A B C D E F G H I J K L M N Ø P Q R S T U V W X Y Z
 A B C D E F G H I J K L M N Ø P Q R S T U V W X Y Z
0 1 2 3 4 5 6 7 8 9 0 1 2 3 4 5 6 7 8 9
 0 1 2 3 4 5 6 7 8 9 0 1 2 3 4 5 6 7 8 9
```

```
•••••••••1•••••••••2•••••••••3•••••••••4•••••••••5•••••••••6•••••••••7•••••••••8
CHAPTER 6 PRØBLEM ?
INPUT DATA

68832SWARØBERT JØNES
68322SWAJACK SMITH
68832STHHENRY KAHN
68324STHMARGARET KAISER
68325STHJUSTIN KRAMER
```

```
•••••••••1•••••••••2•••••••••3•••••••••4•••••••••5•••••••••6•••••••••7•••••••••8
CHAPTER 6 PRØBLEM 3
INPUT DATA

01039864STEVEN LEWIS          A
02002491DAVID MAIN            A
03964111MICHAEL MELTØN        C
04049923JEAN MYERS            D
05123941HARØLD ØWENS          F
```

```
•••••••••1•••••••••2•••••••••3•••••••••4•••••••••5•••••••••6•••••••••7•••••••••8
CHAPTER 6 PRØBLEM 4
INPUT DATA

502126934RØN PATTERSØN        31367062734
419638319THØMAS PATRICK       12346024692
214906184MARIA PEREZ          06243012486
436704125LEE RICHARDSØN       00979001958
383807581JØHN SANDERS         04774009548
```

Chapter 7

```
•••••••••1•••••••••2•••••••••3•••••••••4•••••••••5•••••••••6•••••••••7•••••••••8
CHAPTER 7 PRØBLEM 1
INPUT DATA

55047214400040T
55047392355000T
55097419400105T
74094193150000T
74153284205000T
99151272400150T
99151438350000T
99305277400000T
```

```
•••••••••1•••••••••2•••••••••3•••••••••4•••••••••5•••••••••6•••••••••7•••••••••8
CHAPTER 7 PRØBLEM 2
INPUT DATA

008000120000050
```

```
•••••••••1•••••••••2•••••••••3•••••••••4•••••••••5•••••••••6•••••••••7•••••••••8
CHAPTER 7 PRØBLEM 3
INPUT DATA

S1552JØHN HØFFMAN     00523816404198OPAUL FRIEDMAN                        0305004
S1631RICHARD KING     01437471909238OBARBARA SMITH                        0305004
S1679LARRY HAM    N   05655425701108OCARL JEFFERSØN                       1946832
S1741PAULA LØNDØNRD   10326190608208OHERBERT HØWARD                       0003419
S1832ED GRIFFEN       23947239309308ORØN MARTINEZ                         0072734
```

```
•••••••••1•••••••••2•••••••••3•••••••••4•••••••••5•••••••••6•••••••••7•••••••••8
CHAPTER 7 PRØBLEM 4
INPUT DATA

123456442500008254250O
333255333750008252982 5
```

Chapter 8

```
.........1.........2.........3.........4.........5.........6.........7.........8
CHAPTER 8 PRØBLEM 1
INPUT DATA

2452310053042                            0040000
2461340500001                            0400000
2514160046667                            0050000
3198740250025                            0050000
4319425050050                            7500000
```

```
.........1.........2.........3.........4.........5.........6.........7.........8
CHAPTER 8 PRØBLEM 2
INPUT DATA

LØAD DEPARTMENT2680JØHN CALDWELL       X
LØAD DEPARTMENT2701BILL BRØØKS         X
LØAD DEPARTMENT2712DAVID HAMILTØN      X
PAYRØLL SECTIØN4014CHUNG LEE           X
PAYRØLL SECTIØN4115LARRY HØØPER        X
PAYRØLL SECTIØN4216BENNY BRØWN         X
PAYRØLL SECTIØN4229JØHN JØHNSØN        X
PAYRØLL SECTIØN4238RØBERT CARLSØN      X
```

```
.........1.........2.........3.........4.........5.........6.........7.........8
CHAPTER 8 PRØBLEM 3
INPUT DATA

    1002791245     196280                                                      S
    1002791245     253105                                                      S
    1002791245     082271                                                      S
    1002789449     348849                                                      S
    1002789449     090152                                                      S
    1050691248     039923                                                      S
    1050691248     105398                                                      S
    1050691248     002075                                                      S
    1050729231     653114                                                      S
    1050729231     860152                                                      S
    1050729231     085320                                                      S
    1050729231     008501                                                      S
    1050798882     514980                                                      S
```

```
.........1.........2.........3.........4.........5.........6.........7.........8
CHAPTER 8 PRØBLEM 4
INPUT DATA

00050    05000     1
00750    02500     4
01500    01530     3
12000    00963     2
00500    00120     2
```

Chapter 9

```
.........1.........2.........3.........4.........5.........6.........7.........8
CHAPTER 9 PRØBLEM 1
INPUT DATA

J0281683121250
J0281594783000
J0283671291550
J0283700494230
J0285713124030
```

```
.........1.........2.........3.........4.........5.........6.........7.........8
CHAPTER 9 PRØBLEM 2
INPUT DATA

05018951562128436900003500020000                    5000350025002500
05001600564491212941625941621941                    9211974897480000
06202300581010334877775877005897                    8199941994190928
06204400495149142094820814800824                    0519024302430519
07502196591804330870503750012500                    8300750035005000
```

```
.........1.........2.........3.........4.........5.........6.........7.........8
CHAPTER 9 PRØBLEM 3
INPUT DATA

HENRY HINES              009505006000000000000
JACK SMITH               012000000555550750000
WALTER REID              004600000000000042000
JANE DØE                 015926001720360917940
CHARLES BRØWN            079291182629106512300
```

```
•••••••••1•••••••••2•••••••••3•••••••••4•••••••••5•••••••••6•••••••••7•••••••••8
CHAPTER 9 PRØBLEM 4
INPUT DATA

12345000651004751HAMMER-BALL PEEN  EA2468100246ACME HDWE CØ•, INC
24762000013246953BØILER-STEAM      EA2468100246ACMECØ•, INC
47672000011189752WASHING MACHINE   EA2468100246ACMECØ•, INC
67302000821004875NAILS-STEEL WIRE  LB2468100246ACME HDWE CØ•, INC
15762000671000752LAG SCREWS        DZ2468212481E•C• MØRGAN CØ•
38576000076001065CLIPS-FILE        GR2468212481E•C• MØRGAN CØ•
69251000052006521PAINT             GL2468212481E•C• MØRGAN CØ•
07603001105000151NUTS HEX 1/8      DZ2468328762WILLIAMS TØØL CØ•
07603001105000151NUTS HEX 1/8      DZ2468328762WILLIAMS TØØL CØ•
39827000037264721GRADERS           EA2468328762WILLIAMS TØØL CØ•
```

Chapter 10

```
•••••••••1•••••••••2•••••••••3•••••••••4•••••••••5•••••••••6•••••••••7•••••••••8
CHAPTER 10 PRØBLEM 1
TABLE DATA

BAKER T V  5120374FRANKS R E 8963201GRANT W E  4520312GUNTHER K L7860021
JØHNSØN T P2363210JØNSØN M   8865632JUGGERS Z T2770296JUSTICE P  3062981
KRAMER F R 2345123LEARNER R T4432423

INPUT DATA

JØHNSØN T P
JØNSØN M
JUGGERS Z T
JUSTICE P
KAMBERLEIN W
```

```
•••••••••1•••••••••2•••••••••3•••••••••4•••••••••5•••••••••6•••••••••7•••••••••8
CHAPTER 10 PRØBLEM 2
TABLE DATA

10625251063625106454010654501066550106747510686501069450107050010718OO

INPUT DATA

1062                               155
1065                               325
1066                               400
1068                               050
1071                               250
```

```
•••••••••1•••••••••2•••••••••3•••••••••4•••••••••5•••••••••6•••••••••7•••••••••8
CHAPTER 10 PRØBLEM 3
TABLE DATA

019001
039002
059003
079004
099005

INPUT DATA

10865    00149
12850    00052
22560    00168
25647    00121
26841    00185
36875    12356
47250    25863
78250    65324
```

```
•••••••••1•••••••••2•••••••••3•••••••••4•••••••••5•••••••••6•••••••••7•••••••••8
CHAPTER 10 PRØBLEM 4
TABLE DATA

SØAP      375BLEACH    238DETERGENT567CLEANSER 319PØWDER   276
525425725500455

INPUT DATA
011582100250
011582500030
011582300100
011582400020
011582200015
011582300157
011582100079
011582400524
011582500017
```

Chapter 11

```
•••••••••1•••••••••2•••••••••3•••••••••4•••••••••5•••••••••6•••••••••7•••••••••8
CHAPTER 11 PRØBLEM 1
INPUT DATA

555096862WILLIAM T. FØSTER    10816 SLAUSØN AVE.      LØS ANGELES CA90527
420689235STEVE M. JENSØN      3740 ØVERLAND ST.       LØS ANGELES CA92065
559236156GISELLE L. AGERGAARD 3716 TILDEN AVE.        CULVER CITY CA90230
750236579TØM J. WILLIAMS      10816 KINGSLAND ST.     BEVERLY HILLCA95762
755453376KATHY S. BUTLER      4562 W. 57TH ST.        LØS ANGELES CA96476
554364794CATHY M. GØRING      7584 WILSHIRE BL.       LØS ANGELES CA92367
558147691RØBERT L. PALMER     2496 CULVER BL.         CULVER CITY CA92030
```

```
•••••••••1•••••••••2•••••••••3•••••••••4•••••••••5•••••••••6•••••••••7•••••••••8
CHAPTER 11 PRØBLEM 2
INPUT DATA

0156709750HØLMES      AR
0156905510FRANKLIN    BR
0157007350STEVENS     SR
0157112775WALKER      FS
0157303300JØNES       DR
0157401875RØSEMAN     BS
0157704750MØRGAN      RS
0157810815GREEN       HR
0157907400MITCHELL    TR
0158202900PENNEY      JR
0158314450MØRGAN      RS
0158808250CRAIG       AR
0159011975WARD        MR
```

```
•••••••••1•••••••••2•••••••••3•••••••••4•••••••••5•••••••••6•••••••••7•••••••••8
CHAPTER 11 PRØBLEM 3
INPUT DATA

050189515621284369000035000200000                    5000350025002500
050016005644912129416259416219 41                    9211974897480000
062023005810103348777758770058 97                    8199941994190928
062044004951491420948208148008 24                    0519024302430519
075021965918043308705037500125 00                    8300750035005000
```

```
•••••••••1•••••••••2•••••••••3•••••••••4•••••••••5•••••••••6•••••••••7•••••••••8
CHAPTER 11 PRØBLEM 4
INPUT DATA

00015                         15245728   01675      EASTAINLESS FITTINGS
00567                         16116374   02604      DZSTEEL SHANK 4X9X1
01234                         22815362   00275      EASTEEL FLANGE
01598                         30409017   00596      FTCARBØN STEEL
13253                         32639705   00067      DZSTAINLESS SET SCREWS
01139                         42723598   01594      SHFLAT RØLLED STEEL SHEETS
00005                         59673111   45247      EAHEX STØCK TITANIUM
03695                         62207147   00352      GRSTAINLESS PINS
00152                         73984634   09589      EACARBØN STEEL
```

Chapter 12

```
•••••••••1•••••••••2•••••••••3•••••••••4•••••••••5•••••••••6•••••••••7•••••••••8
CHAPTER 12 PRØBLEM 1
INPUT DATA

    HVARTSØN     123 WØØD LANE  DE MØINES, CAL.
    G BELEBØR    784 GRAND DRIVESEMMDALE, VA.
    HMBREIGHT    NEW SPRING BLVDHEER, MD.
    ACCALIPHANDERSTRETCH BLVD   MITTAK, ALA.
    D DIERR      1 MADISØN RØAD HEARØLD, N.M.
```

```
•••••••••1•••••••••2•••••••••3•••••••••4•••••••••5•••••••••6•••••••••7•••••••••8
CHAPTER 12 PRØBLEM 2
INPUT DATA

1231720040101012 3CTBARTØNSKI    135724681250080023601250000009485
12317200401013654FAJACKSØN       142120300000030012501040015006191
12317200402041576MTRICHTER       102021501875100015701560104010215
12317200402047652SØTRACTØN       065000000000015010301520000003350
12317201003074680CJSYLVESTER     102020503750000024001840080011860
12317201003078543SRPØLIKØFF      120012100000000015001920104006870
12317201004202075FRLICHTMAN      105013500000080025001870000007570
12317201004206782NTGARVER        115512460800100015751950010007826
12317211516501464LR JØHNSØN      113015700000000018702405105008025
12317211516505765CRBØNHAM        105010052000100024702870000010395
12317211528703736DPHARRIGAN      102011000000070018601575010006355
12317211528709425SJGILLESPIE     107009000500000020401860000006370
```

```
•••••••••1•••••••••2•••••••••3•••••••••4•••••••••5•••••••••6•••••••••7•••••••••8
CHAPTER 12 PRØBLEM 3
INPUT DATA

12317201231525402162125140490 2
123172012315254021621251404902
123172947817050201401705020140
123172947817050201401705020140
12317292613204&264021005015680
12317292613204&264021005015680
123172647137504142153611915600
123172647137504142153611915600
123172971237624146151621736022
123172971237624146151621736022
1231722468147 2J232150000008494
1231722468147 2J232150000008494
12317281211472017216350700313M
12317281211472017216350700313M
123172114144272167144091220074
123172114144272167144091220074
```

```
•••••••••1•••••••••2•••••••••3•••••••••4•••••••••5•••••••••6•••••••••7•••••••••8
CHAPTER 12 PRØBLEM 4
INPUT DATA

0011234560    SMITH              JW     401              0028515
0011892750    JØNES              RA     396              0015216
0018929016    MAUS               JB     625              0018255
0020238648    GØLDMAN            H      317              0010025
0020333367    WØLFE              DJ     095              0002660
```

Chapter 13

```
•••••••••1•••••••••2•••••••••3•••••••••4•••••••••5•••••••••6•••••••••7•••••••••8
CHAPTER 13 PRØBLEM 1
INPUT DATA

MJØHNSØN MFG. CØ.    27 S. MAIN ST.      BILØXI, MISS.        14765249501
MCØLE PRØDUCTS INC.  395 HARRISØN ST.    DES MØINES, IØWA     52675350795
MSAVØY ELECTRØNICS   1425 S. PACIFIC ST. SAN FRANCISCØ, CALI 96317409523
MHENSHEY PUBLISHING  399 RIVERDALE AVE.  BRØØKLYN, N. Y.      12795510704
MSTAR FURNISHINGS CØ.4070 BLANKE AVE.    CHICAGØ, ILL.        30562935247
```

```
•••••••••1•••••••••2•••••••••3•••••••••4•••••••••5•••••••••6•••••••••7•••••••••8
CHAPTER 13 PRØBLEM 2
INPUT DATA

X61002MØNEMATERS         010000
X62116CASHINFLØW         216000
X62249BANKCREDIT         172850
X62367CASHIN             075060
```

```
•••••••••1•••••••••2•••••••••3•••••••••4•••••••••5•••••••••6•••••••••7•••••••••8
CHAPTER 13 PRØBLEM 3
NØ INPUT DATA
```

```
•••••••••1•••••••••2•••••••••3•••••••••4•••••••••5•••••••••6•••••••••7•••••••••8
CHAPTER 13 PRØBLEM 4
INPUT DATA

13762JØHN R. MASØN    325 CULVER ST.    JAMESTØWN, N. Y.
34965SIMØN T. FRASER  695 AMES AVE.     TERRE HAUTE, INDIANA
43925MØRTØN M. WØLFF  3751 LØCKE RD.    DAVIE, FLØRIDA
53862HELEN Z. MALKIND 6947 TØRRENT AVE. HEMSTEAD, N. Y.
97541JØSEPH ANDREWS   14162 S. LAKE ST. ST. ALBANS, VERMØNT
```

Chapter 14

```
•••••••••1•••••••••2•••••••••3•••••••••4•••••••••5•••••••••6•••••••••7•••••••••8
CHAPTER 14 PRØBLEM 1
INPUT DATA

DATE CARD

101382                                                                       Ø

STØCK CARD RECØRD

      025  96543  01025  CARBØRUNDUM WHEELS                       4646     M
      111  00986  00042  STAINLESS SET SCREWS NSP                 5986     M
      111  01598  00859  STAINLESS RØDS                           0934     M
      111  09346  00482  HI GRADE CARBØN                          0052     M
      111  11632  00596  CARBØN STEEL                             1598     M
      111  11723  00917  STAINLESS PINS                           0052     M
      111  11725  00115  STAINLESS TUBING                         0915     M
      111  11899  01567  STAINLESS FITTINGS                       1792     M
      111  55292  00014  STEEL SHANK 4X9X1                        4138     M
      111  62549  10048  HEX STØCK TITANIUM                       0089     M
      111  65342  09589  TITANIUM BARS                            0085     M
      111  72359  01186  STEEL PLATE                              0098     M
      111  81192  01592  FLAT RØLLED STEEL SHEETS                 1139     M
      111  81536  00480  STEEL FLANGE                             1985     M
      123  45678  96543  ALLIGATØR PUMPS                          9999     M
```

```
.........1.........2.........3.........4.........5.........6.........7.........8
CHAPTER 14 PROBLEM 2
INPUT DATA

DAVID ANDERSON 18745 MOBILE ST., RESEDA, CALIF.   62986012358302583690368521
BETTY L. BREWER10321 LUNDY DR., INGLEWOOD, CALIF. 61477258325008632591532675
ARTHUR BROWN   12145 MADISON ST., L.A. 45, CALIF. 38940108535714735791852752
THOMAS CASSIDY 3726 HOPE AV., LYNWOOD, CALIF.     62180023698520536082536985
BOB CHAMBERS   3840 HOPE AV., LYNWOOD, CALIF.     58920609752025836975283572
JACK T. CROSS  6421 BELMAR ST., RESEDA, CALIF.    43313035725903591420638527
KENT B. DAVIS  11621 PENN DRIVE, ENCINO, CALIF.   84082253697110369852369782
SAMUAL FELLOWS 10732 LINDLEY AV., ENCINO, CALIF.  41750052368115836521032158
MICHAEL FISHER 6345 TAMPA AV., TARZANA, CALIF.    30040203215230528324406852
GLADYS BUTTONS 3701 BALBOA AV., VAN NUYS, CALIF.  67542203215836985214985202
```

```
.........1.........2.........3.........4.........5.........6.........7.........8
CHAPTER 14 PROBLEM 3
TABLE DATA

02101750
02731550
03701230
04201195
04651460
04811525
09001945
09502001
09881810
10300855
12451403
13661999
15052033
16661222
18991000

INPUT DATA

0563781532ROBERT PALMER          112580  0536
5310562386SUZIE WILSON           112580  1302
7802619435DAVE LEWIS             112580  1723
4530789138BONNIE SELWOOD         112580  0798
0400831562GISELLE AGERGAARD      112580  0639
7319530088DEAN HOPKINS           112580  1699
7066238009DOUG KEANS             112580  0901
4308615398MIKE COOPER            112580  1086
8023614528RONNIE SALTZER         112580  1630
```

```
.........1.........2.........3.........4.........5.........6.........7.........8
CHAPTER 14 PROBLEM 4
INPUT DATA

1231359  0012930291602000150250370100035
1231530  0130000002503040000025190234410
1232400  0024590029470120440028300166625
1233594  0002480015900023000001220035200
5301280  0025100200500012570111200259100
5303320  1002450019450279331234503000200
5307730  0024600010240039130294330029300
5309250  0022000010500239100002000003910
```

Chapter 15

```
.........1.........2.........3.........4.........5.........6.........7.........8
CHAPTER 15 PROBLEM 1
INPUT DATA

08064LANSBERG PHILLIP R  070295736296430800252000000001209502       0130275
14926SULLIVAN JOHN L     152609307432400040805358001000700000       0084352
27569PARIS HENRY Q       072367500000001829732003046901876224       0100112
38497JENKINS JOSEPH P    242392105396030575909012976432409771       0364372
48504MARSHALL KATE L     162843107531531310003001702504826604       0296394
69002GARCIA MARIO S      092462115731510071465046275709750422       0220041
72675MILLER HIRAM T      000000024627131095207159375010015662       1247529
86592SPAULDING MARCIA T  147639411279260763841097027509856778       0973942
90020MCFADDEN SIMON R    004739261499932653749320204018462551       0820279
```

```
.........1.........2.........3.........4.........5.........6.........7.........8
CHAPTER 15 PROBLEM 2
INPUT DATA

WHO DON IT                 00050000000200000001000
GEORGE DID IT              00060000000700000008000
ETHIL DID NOT DO IT        00100000002000000030000
BUT BERT IS CAPABLE        00800000007000000060000
HOW ABOUT HUBERT           00500000004000000020000
CARL SOMETIMES             00900000012000000600000
MAX CAN                    01500000075000000085000
LOU KNOWS IT               10500000005320000252500
JOHN & MARY                00000000030000000700000
HOPE & CROSBY              00000000000000000160000
SNOOPY                     00000000000000000000000
MARSHALL                   00000000052000000700000
SAM & LOU                  03400000008600000000000
```

```
.........1.........2.........3.........4.........5.........6.........7.........8
CHAPTER 15 PROBLEM 3
INPUT DATA

ALABAMA              03444165  051609    085845
ALASKA               00302173  586412    009043
ARIZONA              01772482  113909    051415
ARKANSAS             01923295  053104    078088
CALIFORNIA           19953134  158693    169564
COLORADO             02207259  104247    083586
CONNECTICUT          03032217  005009    013734
DELAWARE             00548104  002057    005150
FLORIDA              06789443  058560    098129
GEORGIA              04589575  058876    100335
HAWAII               00769913  006450    003666
IDAHO                00713008  083557    055910
ILLINOIS             11113976  056540    130494
INDIANA              05193689  036291    091111
IOWA                 02825041  056290    112944
KANSAS               02249071  082264    134770
KEONTUCKY            03219311  040395    069791
LOUISIANA            03643180  048523    054124
MAINE                00993663  033215    021499
MARYLAND             03922399  010577    026859
MASSACHUSETTS        05689170  008257    029811
MICHIGAN             08875083  058216    118310
MINNESOTA            03805069  084068    128235
MISSISSIPPI          02216912  047716    066686
MISSOURI             04677399  069686    114966
MONTANA              00694409  147138    077932
NEBRASKA             01483791  077227    098017
NEVADA               00488738  110540    049659
NEW HAMPSHIRE        00737681  009304    015024
NEW JERSEY           07168164  007836    032422
NEW MEXICO           01016000  121666    070307
NEW YORK             18241266  049576    107776
NORTH CAROLINA       05082059  052586    087922
NORTH DAKOTA         00617761  070665    106247
OHIO                 10652017  041222    109965
OKLAHOMA             02559253  069919    108509
OREGON               02091385  096981    101397
PENNSYLVANIA         11793909  045333    114497
RHODE ISLAND         00949723  001214    005540
SOUTH CAROLINA       02509516  031055    060295
SOUTH DAKOTA         00666257  077047    082720
TENNESSEE            03924164  042244    080656

TEXAS                11196730  267338    251489
OUTAH                01059273  084916    047653
VERMONT              00444732  009609    013924
VIRGINIA             04648494  040817    062351
WASHINGTON           03409169  068192    081202
WEST VIRGINIA        01744237  024181    036323
WISCONSIN            04417933  056154    104290
WYOMING              00332416  097914    040602
```

```
.........1.........2.........3.........4.........5.........6.........7.........8
CHAPTER 15 PROBLEM 4
INPUT DATA

ACCUMULATED CHARGES FILE

213 2363210    ADAMS KEN J            0147621
213 3474836    JOHNSON FRANK M        1094162
213 4597094    BABETT HELEN T         0007937
213 8570702    HARRIS MARTIN R        0023142
714 4950674    WILSON HAROLD M        0100462
714 6400600    HAMILTON GEORGE L      2963284
714 9887250    LEWIS DORIS P          0064296
805 3559249    ABRAMS JOHN L          0124736
805 7763467    DELGADO PHILLIP T      0170021
805 5570241    SIMONS SUSAN T         0095034

CURRENT CHARGES FILE

213 4597094    BABETT HELEN T         0014256
213 3474836    JOHNSON FRANK M        0004826
714 4950674    WILSON HAROLD M        0049025
714 6400600    HAMILTON GEORGE L      0120162
714 9887250    LEWIS DORIS P          0003694
805 5570241    SIMON SUSAN T          0009437
805 7763467    DELGADO PHILLIP T      0014293
```

```
•••••••••1•••••••••2•••••••••3•••••••••4•••••••••5•••••••••6•••••••••7•••••••••8
CHAPTER 16 PRØBLEM 1
INPUT DATA

M10867ALLEN & CØ.                    1500000
I10867                        246817       0615760691977
I10867                        384350       0507760037658
M16535ANDERSØN AUTØ SUPPLY           0250000
I16535                        148643       0626760166549
M17849ANDREWS AND SØNS INC.          0075000
I17849                        008564       0415760014664
M18978ARGØNAUT ENGINEERING           0200000
I18978                        146541       0120760009044
I18978                        285978       0401760031213
I18978                        692468       0529760061154
I18978                        705694       0604760244430
M24743BERKLEY PAPER CØ.              0630000
I24743                        001751       0215760005100
I24743                        249702       0410760140005
I24743                        367498       0527760265245
I24743                        876530       0602760118550
M25271BEST DISTRIBUTING CØ.          0100000
I25271                        010100 Z     0130760076219
I25271                        824692 Z     0605760000325
```

```
•••••••••1•••••••••2•••••••••3•••••••••4•••••••••5•••••••••6•••••••••7•••••••••8
CAHPTER 16 PRØBLEM 2
INPUT DATA

N68252JUDGE STØRES INC.   210 SØ MAIN ST    LØS ANGELES       CALIF   90006
T68252  167242GAS STØVES     0000415075
T68252  267415TABLES         00007150550
T68252  672637TØP CHAIRS     0001501000
T68252  786424BLACK DECKS    0000511000
N09621SMITH MANUFACTURING 136920 9TH ST NE  BERNALILLØ        NEW MEXICØ 56120
T09621  439167SHEARS         0010002765
T09621  629408GASKET CØRK    0300000115
T09621  102139SPRIDGET WHITE 0005075000
```

```
•••••••••1•••••••••2•••••••••3•••••••••4•••••••••5•••••••••6•••••••••7•••••••••8
CHAPTER 16 PRØBLEM 3
INPUT DATA

123456046714026871057615875624246802369258482604568421302461
267024135790204162124703024757382713605720205794359264710101
360572157932468295307214138429475627024690234706514378567890
```

```
•••••••••1•••••••••2•••••••••3•••••••••4•••••••••5•••••••••6•••••••••7•••••••••8
CHAPTER 16 PRØBLEM 4
INPUT DATA

MASTER RECØRDS

M014925052820147620123 91     142658           007500EATERMINAL CLIP
M1157260602227642124765 4      283549           105000DZFUSE 15A
M0973750430210256307690 2      370294           072000EATERMINAL BAR
M3294950701229439126409 8      423025           300000EABREAKER 30A
M4073020629252402732620 6      501020           200000EABREAKER 60A
M0175360728203642902742 9      645555           010000EAMØTØR 1/2 HP 60 CYC
M2489950228248430752907 1      730946           200000EACH 18 BREAKER 15A
M0248010920210121509857 6      894345           025500DZTWIN SØCKET 1500
M7420511030289430570590 0      942024           500000GRSØCKT ADAPT BRN

RECEIPTS RECØRDS

R00525011042                  142658
R00200011152                  370294
R10100011062                  423025
R00500511252                  645555
R20070011192                  730946
R02500011202                  942024

DISBURSEMENTS RECØRDS

D00400011052                  142658
D00525511152                  283549
D09963411072                  423025
D01050011262                  501020
D25035411082                  730946
D07506211302                  942024
```

Input Data for the Ten Problems in Appendix D

```
.........1.........2.........3.........4.........5.........6.........7.........8
PROBLEM 1
INPUT DATA

SUSAN CALDWELL            1100
BETTY JANE CLANCY         1120
RUTH ANN CORBETT          1150
MARGARET CUSHING          2100
ROSEMARY DUPUIS           2075
MAURICE ERICKSON          2050
LILLIAN FELLING           3100
NANCY HAMILTON            3125
BARBARA HICKMAN           3110
JOSEPHINE HOUSTON         4050
LORETTA JOHNSON           4040
ELAINE LEONARD            4030
LORRAINE CLARK            5030
TERRY MCDONNELL           5035
MARY ANN PALMER           5040
```

```
.........1.........2.........3.........4.........5.........6.........7.........8
PROBLEM 2
INPUT DATA

DAVID ANDERSON 18745 MOBILE ST., RESEDA, CALIF.   629861000010000 2000
BETTY L. BREWER10321 LUNDY DR., INGLEWOOD, CLAIF. 6147735025 5000 3000
ARTHUR BROWN   12145 MADISON ST., L.A. , CALIF.   3894045000 3000 5000
THOMAS CASSIDY 3726 HOPE AV., LYNWOOD, CALIF.     6218012150 4000 2000
BOB CHAMBERS   3840 HOPE AVE., LYNWOOD, CALIF.    5892032000 1550 3500
JACK T. CROSS  6421 BELMAR ST., RESEDA, CALIF.    433131058015085 2000
KENT B. DAVID  11621 PENN DRIVE. ENCINO, CALIF.   8408232075 7575 3500
SAMUAL FELLOW  10732 LINDLEY AV., ENCINO, CALIF.  4175029060 4037 3000
MICHAEL FISHER 6345 TAMPA AV., TARZANA, CALIF.    3004044435 6525 5000
GLADYS BUTTONS 3701 BALBOA AV., VAN NUYS, CALIF.  6754237500 8930 4500
PATRICK HANEY  4218 VICTORY ST., L.A., CALIF.     7211145010 2795 5500
LYNN HUBBARD   13245 VENTURA BLVD., RESEDA, CALIF.6437519575 3645 2500
MARVID JACOBS  13118 VENTURA BLVD., RESEDA, CALIF.6255022595 4495 3000
LINDA JOHNSON  20715 VAN NUYS ST., ENCINO, CALIF. 5821430000 8750 3500
HOWARD DEYES   2181 SHERMAN WAY, RESEDA, CALIF.   466152798010175 3000
```

```
.........1.........2.........3.........4.........5.........6.........7.........8
PROBLEM 3
INPUT DATA

610031  0320000003650000001500000
584401  0500000000413100005681650
579051  0006500000405265007852650
607561  0125000002500000005000000
618801  0005200000000900002516000
594251  0010000000007890005028950
637401  0025000000001000004260000
655001  0000000000009000001140000
627111  0002750000000755000505050
609121  0012750000072250001645000
572801  0002500000100000000825000
590141  0001000001000000002990000
605451  0006750000000500004987500
617882  0095000000040000000175000
```

```
.........1.........2.........3.........4.........5.........6.........7.........8
PROBLEM 4
INPUT DATA

GUY T. GOODWIN                 000020981 287051600001050195007O
PAUL A. EVANS                  000053487 300021400000000002751640
MILTON C. MORGAN               000064378 290071600000000003645092
JOHN L. REED                   000052887 350021600000175052256 4
JAMES F. KING                 A000053219 312041600012010104709O
MARY DIXON                    A000062986 275031600000525077502O
```

```
.........1.........2.........3.........4.........5.........6.........7.........8
PROBLEM 5
INPUT DATA

1020304050      16003000430060007500
10111010382JONES HENRY           001001
10111011082JONES HENRY           023012
10111011782JONES HENRY           000603
10111012482JONES HENRY           205024
10222010382SMITH ROBERT          000551
10222010382SMITH ROBERT          000702
10222010382SMITH ROBERT          008003
16431012482HODGES JAMES          000105
16431011782HODGES JAMES          006234
16431010382HODGES JAMES          405003
24350012482JOHNSTON HOWARD       000085
24350011082JOHNSTON HOWARD       000423
24565012482MONTGOMERY DAN        008204
24565011782MONTGOMERY DAN        072551
```

```
.........1.........2.........3.........4.........5.........6.........7.........8
PROBLEM 6
INPUT DATA

              20010670                                  202000310    28000800000
              20010695                                  203000310    28000800500
              20010700                                  204000310    28000800000
              20010703                                  205000310    28000800000
              20010725                                  207000310    28000801000
              20010730                                  208000310    28000800000
              20010742                                  209000310    28000800000
              20010800                                  210000310    28000800000
              20010890                                  211000310    28000800100
              30010904                                  212000310    28000800000
              30010905                                  213000310    28000800000
              30010906                                  214000310    28000800000
              30010907                                  215000310    28000800000
              40011215                                  218000310    28000800000
              40011225                                  219000310    28000800000
              40011240                                  220000315    28500850000
              40011250                                  221000315    28500850000
              60012330                                  225000315    28500850200
              60012340                                  226000315    28500850200
              60012350                                  227000315    28500850000
              60012366                                  228000315    28500850000
              60012400                                  229000315    28500850000
```

```
.........1.........2.........3.........4.........5.........6.........7.........8
PROBLEM 7
INPUT TAPE RECORD

     FOX WILLIAM           1300952943113030321960208573
     PHILLIPS ROBERT       3642088412748980195040184182
     JOHNSON BEN           5430122222000000037000134000
     WASHINGTON GEORGE     5430122232006700037660134449
     MONTGOMERY ALEX       5430122242013367038330134896
     SMITH JOSEPH          5430122251020033039000068342
     BROWN WALLACE         5430122261526700039670102289
     DUNIGAN HENRY         5430122272033350040330136234
     JONES WILLIAM         5430122281566667043660104967
     DELANEY JERRY         5430122291086667045660072807
     HALLECK FRANCES       5430122301400033037000093802
     REID PATRICIA         5430122312007000037060134469
     JACKSON KENNETH       5430122322718833179001182162
     ALEXANDER CHARLES     5430122331493333046332100053
     HALL GEORGE           5430122400566672056000037967
     SIPLE CHARLES         5430122411037526056330069514
     GOODMAN HENRY         5430122420576737060333038641
     CAMM FRED J           5430122431203333063331080623
     DENTON TERRENCE       5430122480675262076660045243
     GOODSALL PHILLIP      5430122490766750086660051372
     SAWYER DAVID          5563202012921521219360195742
     YOUNG SAMUEL          5571677822823188241420189154
     HEPNER ELMER          5591092992933817264860196566
     HORNE ALBERT          5782011412728123124990182784
```

```
.........1.........2.........3.........4.........5.........6.........7.........8
PROBLEM 7
INPUT RECORD

75925                      130095294                   0756515160983506871
37857                      364208841                   0574490097523384911
42252                      543012222                   0200000018500134001
21663                      543012223                   0203350018833136241
21074                      543012224                   0206683019167138481
37185                      543012225                   0210017019500140711
15296                      543012226                   0213350019835142941
42307                      543012227                   0216675020167145171
37418                      543012228                   0233333021833156331
63529                      543012229                   0243333022833163031
42630                      543012230                   0200017018500134011
63741                      543012231                   0200350018533134231
15375                      543012232                   0559417089503374811
63852                      543012233                   0246667023167165271
37963                      543012240                   0283333028000189831
21074                      543012241                   0283333028167189831
15185                      543012242                   0301667030167202121
75296                      543012246                   0316667031667212171
87307                      543012248                   0383333038333256831
15141                      543012249                   0433333043333290331
42902                      556320201                   0660761109683442711
21472                      557167782                   0614598120710411781
87524                      559109299                   0666908132430446831
63708                      578201141                   0564061062497377921
```

```
.........1.........2.........3.........4.........5.........6.........7.........8
PROBLEM 8
INPUT DATA

05018207108257                                     0118980
05018207911234                                     0016806
05018207929031                                     0006300
05018209579992                                     0008774
05178203702965                                     1271692
05178200109002                                     0084217
05178200201179                                     0707112
05178200113605                                     0309272
05178200927654                                     0021790
05188209200390                                     0002700
05188205605006                                     0089732
05188218112125                                     0037198
05198210000298                                     0202060
05198202400106                                     0149473
05218200300256                                     0007953
05218200800652                                     0009518
05218209018569                                     0042115
05218213220106                                     0070642
05218200500321                                     0059010
```

```
.........1.........2.........3.........4.........5.........6.........7.........8
PROBLEM 9
INPUT DATA

525460      0115121                001250004490     100012500
525198      0078345                008255124490     080066041
544483      0072914                000690704490     140009670
514152      0014983                000110202513     060000661
513444      0712129                010986002513     120131832
512136      0011110                009850402513     080078803
```

```
.........1.........2.........3.........4.........5.........6.........7.........8
PROBLEM 10
INPUT DATA

010682ARABACALRBHE                                                   000929
010682CABRALB SFCC                                                   000100
010682MDCARTBWDTUG                                                   000231
010682CAVENPH UBIE                                                   000321
010682DEBANBXAFTUM                                                   000443
010682CASD SD SJDG                                                   000542
010682MDBALPIKJHBP                                                   001001
010682MDBALBALMDGN                                                   000100
010682CALA LA VLZJ                                                   000200
010682WAKNGSTLMWAG                                                   000300
010682MDPGSSSPPZCK                                                   000600
010682WAKNGSTLVAMN                                                   000717
010682FLSD AR QEIN                                                   000762
010682NECENFBNRXBJ                                                   000867
010682WNDUNREDKPYI                                                   000984
010682CALA LA VLZJ                                                   000200
010682WAKNGSTLMWAG                                                   000300
010682ILCBKCHINFXL                                                   000400
010682ALSECNBIBBAY                                                   000500
010682MDBALBALMDGN                                                   000100
010682MDPGSSSPPZCK                                                   000600
010682WAKNGSTLVAMN                                                   000717
010682CASCLSUNAQKD                                                   000889
010682ARABACALRBHE                                                   000929
010682CABRALB SFCC                                                   000100
010682MDCARTBWDTUG                                                   000231
010682MDBALPIKJHBP                                                   001001
010682DEBANRBNHWHL                                                   000655
010682FLSD AR QEIN                                                   000762
010682NECENFBNRXBJ                                                   000867
010682WNDUNREDKPYI                                                   000984
010682CALA LA VLZJ                                                   000200
010682WAKNGSTLMWAG                                                   000300
010682ILCBKCHINFXL                                                   000400
010682ALSECNBIBBAY                                                   000500
010682MDPGSSSPPZCK                                                   000600
010682WAKNGSTLVAMN                                                   000717
010682CASCLSUNAQKD                                                   000889
010682ARABACALRBHE                                                   000929
010682CABRALB SFCC                                                   000100
010682MDCARTBWDTUG                                                   000231
010682CAVENPH UBIE                                                   000321
010682DEBANBXAFTUM                                                   000443
010682CASD SD SJDG                                                   000542
010682MDBALPIKJHBP                                                   001001
010682DEBANRBNHWHL                                                   000655
010682FLSD AR QEIN                                                   000762
010682NECENFBNRXBJ                                                   000867
010682WNDUNREDKPYI                                                   000984
010682CALA LA VLZJ                                                   000200
010682WAKNGSTLMWAG                                                   000300
010682ILCBKCHINFXL                                                   000400
010682ALSECNBIBBAY                                                   000500
010682MDPGSSSPPZCK                                                   000600
010682WAKNGSTLVAMN                                                   000717
010682CASCLSUNAQKD                                                   000889
010682ARABACALRBHE                                                   000929
010682CABRALB SFCC                                                   000100
010682MDCARTBWDTUG                                                   000231
010682CAVENPH UBIE                                                   000321
010682DEBANBXAFTUM                                                   000443
010682CASD SD SJDG                                                   000542
010682MDBALPIKJHBP                                                   001001
010682DEBANRBNHWHL                                                   000655
010682FLSD AR QEIN                                                   000762
010682NECENFBNRXBJ                                                   000867
010682WNDUNREDKPYI                                                   000984
```

Appendix F

Useful Reference Information*

Picture and Edit Characters

Picture character	*Data type*	*Specification*	*Additional explanation*
X	alphanumeric	The associated position in the value will contain any character from the COBOL character set.	
A	alphabetic	The associated position in the value will contain an alphabetic character or a space.	
9	numeric or numeric edited	The associated position in the value will contain any digit.	
V	numeric	The decimal point in the value will be assumed to be at the location of the V. The V does not represent a character position.	
	numeric edited	The associated position in the value will contain a point or a space.	A space will occur if the entire data item is suppressed.
$	numeric edited	a. (simple insertion) The associated position in the value will contain a dollar sign. b. (floating insertion) The associated position in the value will contain a dollar sign, a digit, or a space.	The leftmost $ in a floating string does not represent a digit position. If the string of $ is specified only to the left of a decimal point, the rightmost $ in the picture corresponding to a position that precedes the leading nonzero digit in the value will be printed. A string of $ that extends to the right of a decimal point will have the same effect as a string to the left of the point unless the value is zero; in this case blanks will appear. All positions corresponding to $ positions to the right of the printed $ will contain digits; all to the left will contain blanks.
	numeric edited	The associated position in the value will contain a comma, space, or dollar sign.	A comma included in a floating string is considered part of the floating string. A space or dollar sign could appear in the position in the value corresponding to the comma.
S	numeric	A sign (+ or −) will be part of the value of the data item. The S does not represent a character position.	

*From IBM Publication #SR29–0286–2.

Rules for Forming User-supplied Words

User-supplied name	*Use*	*Number of charac-ters*	*Type of characters*	*Restrictions*
file name	name a file	1 to 30	at least one must be alphabetic; no spaces; cannot begin or end with hyphen.	name must be unique
record name	name a record			name must be unique or qualifiable
data name	name a data item			
condition name	name a value of a data item			name must be unique
paragraph name	name a paragraph	1 to 30	no alphabetic character required; no spaces; cannot begin or end with hyphen.	name must be unique
program name	name a program			first eight characters must be unique
library name	name a library entry			

Valid Procedure Division Entries

Term	*Definition*	
	Content	*Terminating symbol(s)*
statement	a basic valid combination of words and symbols used in the Procedure Division	a space, a comma followed by a space, or a period followed by a space (If a period is used, the statement is also a sentence.)
sentence	a sequence of one or more statements	a period followed by a space
paragraph	a sequence of one or more sentences, the first one being preceded by a paragraph name	another paragraph name or the end of the Procedure Division
section	a sequence of one or more successive paragraphs, the first one being preceded by a section header and the word SECTION.	another section name or the end of the Procedure Division

Relational and Logical Operators

Type of operation	*Operator (operation symbol)*	*Operation*
Relational	IS GREATER THAN (>)	is greater than
	IS LESS THAN (<)	is less than
	IS EQUAL TO (=)	is equal to
Logical	OR	logical inclusive OR (either or both are true)
	AND	logical conjunction (both are true)
	NOT	logical negation

Level Numbers

Number	*In area*	*Purpose*
01	A	Record description entries
02 through 49	A or B	Subdivisions of level-01 entries. May be group or elementary items.
66	A or B	RENAMES clause entries.
77	A	Independent data item (not a subdivision; not subdivided)
88	A or B	Condition name (must use the VALUE clause)

Types of Valid Comparisons

First operand \ Second operand		Group	Elementary: Alphanumeric	Elementary: Alphabetic	Elementary: Numeric	Elementary: Literal
Group		C	C	C	C	C
Elementary	*Alphanumeric*	C	C	C	C	C
	Alphabetic	C	C	C	I	C
	Numeric	C	C	I	N	C
	Literal	C	C	C	C	I

C — Compared logically (one character at a time, according to collating sequence shown in figure 1)
N — Compared algebraically (numeric values are compared)
I — Invalid comparison

Example

IF TOTAL GREATER THAN MAXIMUM PERFORM MESSAGE.

first operand (TOTAL) — operator (GREATER THAN) — second operand (MAXIMUM)

condition

Explanation
To use this chart find the data type (determined by the picture) of the first operand in the column headed First Operand. Then find the data type of the second operand across the top of the figure opposite Second Operand. Extend imaginary lines into the figure from the data types of the first and second operands. In the block where these two lines intersect is a letter that tells you how the values are compared.

USAGE Options

USAGE	*Type of numeric variable*	*Bytes required*	*Required for*	*Used to*
DISPLAY	External decimal	1 per digit	nonnumeric data. data in printer, tape, or card files. data to be accepted or displayed.	
COMP-3	Packed decimal	1 per 2 digits plus 1 byte for the low-order digit and sign		avoid conversions in arithmetic operations. store numeric data in less space than DISPLAY.

Arithmetic Operations

	Conversion	*Alignment*
Computation fields	Any external decimal (DISPLAY) value is converted to packed decimal.	Decimal points are aligned, padded with zeros if necessary.
Result field	The result is converted if the result variable is not packed decimal.	Decimal point of result is aligned with decimal point of result variable; truncation and/or padding may occur.

Summary of Arithmetic Statements and Their Options

Arithmetic statements / *Allowable options*	*GIVING variable-name*	*variable-name ROUNDED*	*SIZE ERROR statement*	*REMAINDER variable-name*
ADD {identifier-1 / literal-1} [identifier-2 / literal-2] . . . TO identifier-m	X *	X	X	
SUBTRACT {identifier-1 / literal-1} [identifier-2 / literal-2] . . . FROM identifier-m	X	X	X	
MULTIPLY {identifier-1 / literal-1} BY identifier-2	X	X	X	
DIVIDE {identifier-1 / literal-1} INTO identifier-2	X	X	X	X
DIVIDE {identifier-1 / literal-1} BY {identifier-2 / literal-2}	X	X	X	X
COMPUTE identifier-1 [identifier-2] = arithmetic-expression		X	X	

*The reserved word TO is omitted when the GIVING option is specified.

Arithmetic Operators

Hierarchy of evaluation	*Operator*	*Meaning*	*Example*	
			Arithmetic expression	*COBOL expression*
1	+	unary plus sign	+2	+2
	−	unary minus sign	−2	−2
2	**	exponentiation	3^2	3 ** 2
3	*	multiplication	3 × 2	3 * 2
	/	division	3 ÷ 2	3 / 2
4	+	addition	3 + 2	3 + 2
	−	subtraction	3 − 2	3 − 2

1. Parentheses modify the order of evaluation; operations enclosed in parentheses are performed first, beginning with the innermost pair of parentheses.
2. When 2 operators of same level in hierarchy appear in the same expression, the operations are performed from left to right.
3. Every operator must be preceded by and followed by a space, except for unary signs.

Specific Conversions in Certain Moves

MOVE sending-variable TO receiving-variable

Sending variable \ *Receiving variable*	*Alphanumeric* (Alphanumeric move)	*External decimal* (Numeric move)	*Packed decimal* (Numeric move)	*Edited* (Edit move)
Alphanumeric	No conversion	Whole numbers only	Whole numbers only	Whole numbers only
External decimal	Whole numbers only; no conversion	No conversion	Converted to packed decimal	Value is edited
Packed decimal	Whole numbers only; no conversion	Converted to external decimal	No conversion	Converted to external decimal and value is edited
Edited	No conversion	Invalid move	Invalid move	Invalid move

How to use chart
Find the data type of the sending variable in the leftmost column. Then find the data type of the receiving variable in the row across the top of the chart. Extend imaginary lines into the chart from the two data types. These two lines will intersect in a block that tells what conversion takes place.

Types of Moves

Sending variable \ *Receiving variable*	*Group*	*Alphabetic*	*Alphanumeric*	*External decimal*	*Packed decimal*	*Edit*
Group	A	A	A	AU	AU	I
Alphabetic	A	A	A	I	I	I
Alphanumeric	A	A	A	N*	N*	E*
External decimal	AU	I	A*	N	N	E
Packed decimal	AU	I	A*	N	N	E
Edit	A	I	A	I	I	I

A Alphanumeric move
E Edit move
AU Alphanumeric move (value of receiving field is unpredictable)
N Numeric move
* Integers only
I Invalid

How to use chart
Find the data type of the sending variable in the leftmost column. Then find the data type of the receiving variable in the row across the top of the chart. Extend imaginary lines into the chart from the two data types. These two lines will intersect in a block that indicates the type of move.

Effects of Types of Moves

Type of move	*Receiving item*	*Compiler action during move*	*Alignment*	*Padding if necessary*	*Truncation if necessary*
alphanumeric	group	none	at left of value	on right with spaces	on right
	alphabetic or alphanumeric	any necessary conversion	at left of value	on right with spaces	on right
numeric	external decimal or packed decimal	any necessary conversion	at decimal point	on left and right with zeros	on left and right
edit	edited	editing and any necessary conversion	at decimal point	on left and right with zeros (unless suppressed)	on left and right

Additional Edit Characters

Picture character	*Data type*	*Specification*	*Additional explanation*
Z	numeric edited	A leading zero in the associated position will be suppressed and replaced with a blank.	No Z may be to the right of a 9.
*	numeric edited	A leading zero in the associated position will be suppressed and replaced with an asterisk.	No * may be to the right of a 9.
B	numeric edited	The associated position in the value will contain a blank, $, or *.	When a B is included in a string of $'s, *'s, or Z's, a digit, $, or * could appear in its position.
0	numeric edited	The associated position in the value will contain a zero, blank, $, or *.	When a 0 is included in a string of $'s, *'s, or Z's, a zero, $, or * could appear in its position.
CR DB	numeric edited	A negative indicator will be inserted into the value when value is negative; two blanks will appear otherwise.	These symbols must be at right end of picture.
−	numeric edited	A minus sign is inserted into the value when value is negative; one blank will appear otherwise.	A − may be at either end of picture; may be "floated" with the same rules as $.
+	numeric edited	The appropriate sign (+ or −) will be inserted into the value.	A + may be at either end of picture; may be "floated" with the same rules as $.

Rules for User-Supplied Portions of the OCCURS Clause

Format

OCCURS integer-2 TIMES [DEPENDING ON data-name-2] [INDEXED BY index-name-1]

Rules

1. Integer-2 must specify the maximum table size for any execution of the program.
2. Data-name-2 must have positive integer values.
3. Data-name-2 may be qualified but not subscripted.
4. Index-name-1 is defined by its appearance in INDEXED BY option.
5. Values of index-name-1 are relative addresses that correspond to occurrence numbers of the table.

Summary of Data Description Clauses

Level number {data-name-1 / FILLER} appropriate clauses

Clause Format	Purpose	Restrictions
REDEFINES data-name-2	Allows alternate descriptions of data to apply to the same storage area.	1. Must immediately follow data-name-1. 2. Level number of data-name-1 must be the same as that of data-name-2. 3. Level number must not be 66 or 88. 4. Data-name-1 must be in working storage. 5. Data-name-2 cannot contain an OCCURS clause and cannot be subordinated to a name for which an OCCURS clause is specified. (A name subordinate to data-name-2 may contain an OCCURS clause without the DEPENDING ON option.)
OCCURS integer-2 TIMES [DEPENDING ON data-name-1] [INDEXED BY index-name-1]	Specifies table size.	May not be specified for a level-01 or -77 name.
PICTURE IS character-string	Gives form of numeric variable or editing requirements of elementary data items.	Only characters in "Picture and Edit Characters" (on page 816) and "Additional Edit Characters" (on page 821) may be used in the character string.
USAGE IS {DISPLAY / COMP-3}	Specifies the manner in which data is to be stored.	May be written at the group or elementary level.
VALUE IS literal	1. Defines the initial value of an item in working storage. 2. Gives the value associated with condition name.	1. In the Working-Storage section may assign a value or a condition name. 2. In the File section may be used for condition names only.

Examples and Effects of the REDEFINES Clause

Used to redefine	Example	Effect
picture	01 A PIC 99. 01 B REDEFINES A PIC XX.	The value of A originally defined as numeric data is treated as alphanumeric data if referred to by B, without moving the value to B.
subdivision	01 SALARY-RECORD. 05 SOCIAL-SECURITY PIC 9(9). 05 SALARY PIC 999V99. 05 MONTH PIC 99. 01 WAGE-RECORD REDEFINES SALARY-RECORD. 05 SOCIAL-SECURITY PIC 9(9). 05 WAGE PIC 9V99. 05 HOURS PIC 99. 05 WEEK PIC 99.	Two kinds of records with different subdivision can be processed differently without moving one record or a portion of the record to another variable. Five digits for monthly salary in a salaried employee record can be referred to by SALARY or three digits for hourly pay and two for hours in an hourly paid employee record can be referred to by WAGE and HOURS, respectively, without moving the values to WAGE and HOURS.

Note: The REDEFINES clause must not be used at the 01 level in the File Section. The above entries appear in the Working-Storage Section.

Guide to File Description Entries

Device type	*labels**	*blocking***
Card	N	N
Printer	N	N
Tape	O	O
Disk sequential	R	O
Disk indexed	R	O

N—not permitted
O—optional
R—required
* LABEL RECORDS clause is always required.
** BLOCK CONTAINS clause is only required when blocking is used.

Guide to Environment Division Entries

FILE-CONTROL clauses / file access and organization	*SELECT*	*ASSIGN*	*RECORD KEY*	*NOMINAL KEY*	*ACCESS*	*RESERVE*
Sequential access Standard sequential	R	R	N	N	O	P
Sequential access Indexed sequential	R	R	R	S	O	N
Random access Indexed sequential	R	R	R	R	R	N

P = required for printer files with END-OF-PAGE option; optional otherwise
R = required
O = optional (ACCESS IS SEQUENTIAL will be assumed if omitted)
N = not permitted
S = only if access will not begin at first record in file

Guide to File Maintenance for Indexed Files

Desired effect	*Access*	*Required statements*	*Required options*	*Optional options*
update a record in the file	SEQUENTIAL	*START	INVALID KEY	—
		READ	AT END	INTO
		REWRITE		FROM
	RANDOM	READ	INVALID KEY	INTO
		REWRITE		FROM
add a record to the file	RANDOM	WRITE	INVALID KEY	FROM

* Only if access is to begin at some record other than the first record in the file

Appendix G

Reserved Word List

The following is a list of ANSI-reserved words used in COBOL.

ACCEPT
ACCESS
ADD
ADVANCING
AFTER
ALL
ALPHABETIC
ALSO
ALTER
ALTERNATE
AND
ARE
AREA
AREAS
ASCENDING
ASSIGN
AT
AUTHOR

BEFORE
BLANK
BLOCK
BOTTOM
BY

CALL
CANCEL
CD
CF
CH
CHARACTER
CHARACTERS
CLOCK-UNITS
CLOSE
COBOL
CODE
CODE-SET
COLLATING
COLUMN
COMMA
COMMUNICATION
COMP
COMPUTATIONAL
COMPUTE
CONFIGURATION
CONTAINS
CONTROL
CONTROLS
COPY
CORR
CORRESPONDING
COUNT
CURRENCY

DATA
DATE
DATE-COMPILED
DATE-WRITTEN
DAY
DE
DEBUG-CONTENTS
DEBUG-ITEM
DEBUG-LINE
DEBUG-NAME
DEBUG-SUB-1
DEBUG-SUB-2
DEBUG-SUB-3
DEBUGGING
DECIMAL-POINT
DECLARATIVES
DELETE
DELIMITED
DELIMITER
DEPENDING
DESCENDING
DESTINATION
DETAIL
DISABLE
DISPLAY
DIVIDE
DIVISION
DOWN
DUPLICATES
DYNAMIC

EGI
ELSE
EMI
ENABLE
END
END-OF-PAGE
ENTER
ENVIRONMENT
EOP
EQUAL
ERROR
ESI
EVERY
EXCEPTION
EXIT
EXTEND

FD
FILE
FILE-CONTROL
FILLER
FINAL
FIRST
FOOTING
FOR
FROM

GENERATE
GIVING
GO
GREATER
GROUP

HEADING
HIGH-VALUE
HIGH-VALUES

I-O
I-O-CONTROL
IDENTIFICATION
IF
IN
INDEX
INDEXED
INDICATE
INITIAL
INITIATE
INPUT
INPUT-OUTPUT
INSPECT
INSTALLATION
INTO
INVALID
IS

JUST
JUSTIFIED

KEY

LABEL
LAST
LEADING
LEFT
LENGTH
LESS
LIMIT
LIMITS
LINAGE
LINAGE-COUNTER
LINE
LINE-COUNTER
LINES
LINKAGE
LOCK
LOW-VALUE
LOW-VALUES

MEMORY
MERGE
MESSAGE
MODE
MODULES
MOVE
MULTIPLE
MULTIPLY

NATIVE
NEGATIVE
NEXT
NO
NOT
NUMBER
NUMERIC

OBJECT-COMPUTER
OCCURS
OF
OFF
OMITTED
ON
OPEN
OPTIONAL
OR
ORGANIZATION
OUTPUT
OVERFLOW

PAGE
PAGE-COUNTER
PERFORM
PF
PH
PIC
PICTURE
PLUS
POINTER
POSITION
POSITIVE
PRINTING
PROCEDURE
PROCEDURES
PROCEED
PROGRAM
PROGRAM-ID

QUEUE
QUOTE
QUOTES

RANDOM
RD
READ
RECEIVE
RECORD
RECORDS
REDEFINES
REEL
REFERENCES
RELATIVE
RELEASE
REMAINDER
REMOVAL
RENAMES
REPLACING
REPORT
REPORTING
REPORTS
RERUN
RESERVE
RESET
RETURN
REVERSED
REWIND
REWRITE
RF
RH
RIGHT
ROUNDED
RUN

SAME
SD
SEARCH
SECTION
SECURITY
SEGMENT
SEGMENT-LIMIT
SELECT
SEND
SENTENCE
SEPARATE
SEQUENCE
SEQUENTIAL
SET
SIGN
SIZE
SORT
SORT-MERGE
SOURCE
SOURCE-COMPUTER
SPACE
SPACES
SPECIAL-NAMES
STANDARD
STANDARD-1
START
STATUS
STOP
STRING
SUB-QUEUE-1
SUB-QUEUE-2
SUB-QUEUE-3
SUBTRACT
SUM
SUPPRESS
SYMBOLIC
SYNC
SYNCHRONIZED

TABLE
TALLYING
TAPE
TERMINAL
TERMINATE
TEXT
THAN
THROUGH
THRU
TIME
TIMES
TO
TOP
TRAILING
TYPE

UNIT
UNSTRING
UNTIL
UP
UPON
USAGE
USE
USING

VALUE
VALUES
VARYING

WHEN
WITH
WORDS
WORKING-STORAGE
WRITE

ZERO
ZEROES
ZEROS

+
−
*
/
**
>
<
=

Appendix H

COBOL Reference Format Notation

The following are the general COBOL reference formats used in this text.

Identification Division

General Identification Division Format

```
IDENTIFICATION DIVISION.
PROGRAM-ID. program-name.
[AUTHOR. [comment-entry] ...]
[INSTALLATION. [comment-entry] ...]
[DATE-WRITTEN. [comment-entry] ...]
[DATE-COMPILED. [comment-entry] ...]
[SECURITY. [comment-entry] ...]
```

Environment Division

General Environment Division Formats

```
ENVIRONMENT DIVISION.
CONFIGURATION SECTION.
SOURCE-COMPUTER. computer-name [WITH DEBUGGING MODE].
OBJECT-COMPUTER. computer-name

    [                         { WORDS      } ]
    [ , MEMORY SIZE integer   { CHARACTERS } ]
    [                         { MODULES    } ]

    [, PROGRAM COLLATING SEQUENCE IS alphabet-name]

    [, SEGMENT-LIMIT IS segment-number] .

[SPECIAL-NAMES. [, implementor-name

    { IS mnemonic-name [, ON STATUS IS condition-name-1 [, OFF STATUS IS condition-name-2]] }
    { IS mnemonic-name [, OFF STATUS IS condition-name-2 [, ON STATUS IS condition-name-1]] }
    { ON STATUS IS condition-name-1 [, OFF STATUS IS condition-name-2]                      }
    { OFF STATUS IS condition-name-2 [, ON STATUS IS condition-name-1]                      }

    [                     { STANDARD-1                                                  } ]
    [                     { NATIVE                                                      } ]
    [ , alphabet-name IS  { implementor-name                                            } ]
    [                     {           [ { THROUGH } literal-2                 ]         } ]
    [                     { literal-1 [ { THRU    }                           ]         } ] ...
    [                     {           [ ALSO literal-3 [, ALSO literal-4] ... ]         } ]
    [                     {             [ { THROUGH } literal-6                 ]       } ]
    [                     {   literal-5 [ { THRU    }                           ]  ...  } ]
    [                     {             [ ALSO literal-7 [, ALSO literal-8] ... ]       } ]

    [, CURRENCY SIGN IS literal-9]

    [, DECIMAL-POINT IS COMMA] .]

INPUT-OUTPUT SECTION.
FILE-CONTROL.
    {file-control-entry} ...
```

```
I-O-CONTROL.
     [; RERUN [ON {file-name-1        }]
                  {implementor-name   }

              { {[END OF] {REEL} }                    }
              { {         {UNIT} } OF file-name-2     }
        EVERY { {integer-1 RECORDS }                  } ] ...
              { integer-2 CLOCK-UNITS                 }
              { condition-name                        }

     [        [RECORD     ]                                    ]
     [; SAME  [SORT       ] AREA FOR file-name-3 {, file-name-4} ... ].
     [        [SORT-MERGE ]                                    ]
     [; MULTIPLE FILE TAPE CONTAINS file-name-5 [POSITION integer-3]
           [, file-name-6 [POSITION integer-4]] ...] ... .]]
```

General File Control Entry Formats

Format-1

```
SELECT [OPTIONAL] file-name
     ASSIGN TO implementor-name-1 [, implementor-name-2] ...
     [; RESERVE integer-1  [AREA ]]
                           [AREAS]
     [; ORGANIZATION IS SEQUENTIAL]
     [; ACCESS MODE IS SEQUENTIAL]
     [; FILE STATUS IS data-name-1].
```

Format-2

```
SELECT file-name
     ASSIGN TO implementor-name-1 [, implementor-name-2] ...
     [; RESERVE integer-1 [AREA ]]
                          [AREAS]
      ; ORGANIZATION IS RELATIVE
     [                   { SEQUENTIAL [, RELATIVE KEY IS data-name-1] } ]
     [; ACCESS MODE IS   { {RANDOM }                                  } ]
     [                   { {DYNAMIC} , RELATIVE KEY IS data-name-1    } ]
[; FILE STATUS IS data-name-2].
```

Format-3

```
SELECT file-name
     ASSIGN TO implementor-name-1 [, implementor-name-2] ...
     [; RESERVE integer-1  AREA  ]
                           AREAS ]
      ; ORGANIZATION IS INDEXED
     [                   SEQUENTIAL ]
     [; ACCESS MODE IS   RANDOM     ]
     [                   DYNAMIC    ]
      ; RECORD KEY IS data-name-1
     [; ALTERNATE RECORD KEY IS data-name-2 [WITH DUPLICATES]] ...
     [; FILE STATUS IS data-name-3].
```

Format-4

```
SELECT file-name ASSIGN TO implementor-name-1 [, implementor-name-2] ...
```

General Data Division Formats

```
DATA DIVISION.
[FILE SECTION.
[FD file-name

   [; BLOCK CONTAINS [integer-1 TO] integer-2 {RECORDS   }]
                                              {CHARACTERS}
   [; RECORD CONTAINS [integer-3 TO] integer-4 CHARACTERS]
    ; LABEL {RECORD IS  } {STANDARD}
            {RECORDS ARE} {OMITTED }
   [; VALUE OF implementor-name-1 IS {data-name-1}
                                      {literal-1  }
        [, implementor-name-2 IS {data-name-2}] ... ]
                                  {literal-2  }
   [; DATA {RECORD IS  } data-name-3 [, data-name-4] ...]
           {RECORDS ARE}
   [; LINAGE IS {data-name-5} LINES [, WITH FOOTING AT {data-name-6}]
                {integer-5  }                            {integer-6  }
        [, LINES AT TOP {data-name-7}] [, LINES AT BOTTOM {data-name-8}]]
                        {integer-7  }                      {integer-8  }
   [; CODE-SET IS alphabet-name]

   [; {REPORT IS  } report-name-1 [, report-name-2] ...].
      {REPORTS ARE}
  [record-description-entry] ...] ...

  [SD file-name
   [; RECORD CONTAINS [integer-1 TO] integer-2 CHARACTERS]
   [; DATA {RECORD IS  } data-name-1 [, data-name-2] ...]
           {RECORDS ARE}
   {record-description-entry}...] ...]

[WORKING-STORAGE SECTION.
[77-level-description-entry] ...]
[record-description-entry  ]

[LINKAGE SECTION.
[77-level-description-entry] ...]
[record-description-entry  ]

[REPORT SECTION.
[RD report-name
   [; CODE literal-1]
  [; {CONTROL IS  } {data-name-1 [, data-name-2] ...              }]
     {CONTROLS ARE} {FINAL [, data-name-1 [, data-name-2] ...]    }
  [; PAGE {LIMIT IS  } integer-1 [LINE ] [, HEADING integer-2]
          {LIMITS ARE}           [LINES]
     [, FIRST DETAIL integer-3] [, LAST DETAIL integer-4]
     [, FOOTING integer-5]].
{report-group-description-entry} ...] ...]
```

General Data Description Entry Formats

Format-1

```
level-number { data-name-1 }
             { FILLER      }
    [; REDEFINES data-name-2]
    [; { PICTURE } IS character-string]
       { PIC     }
    [; [USAGE IS] { COMPUTATIONAL }]
                  { COMP          }
                  { DISPLAY       }
                  { INDEX         }
    [; [SIGN IS] { LEADING  } [SEPARATE CHARACTER]]
                 { TRAILING }
    [; OCCURS { integer-1 TO integer-2 TIMES DEPENDING ON data-name-3 }
              { integer-2 TIMES                                        }
      [{ ASCENDING  } KEY IS data-name-4 [, data-name-5] . . . ] . . .
       { DESCENDING }
      [INDEXED BY index-name-1 [, index-name-2] . . .]]
    [; { SYNCHRONIZED } [ LEFT  ]]
       { SYNC         } [ RIGHT ]
    [; { JUSTIFIED } RIGHT]
       { JUST      }
    [; BLANK WHEN ZERO]
    [; VALUE IS literal] .
```

Format-2

```
66  data-name-1; RENAMES data-name-2 [ { THROUGH } data-name-3 ] .
                                       { THRU    }
```

Format-3

```
88  condition-name; { VALUE IS   } literal-1 [ { THROUGH } literal-2 ]
                    { VALUES ARE }             { THRU    }
    [, literal-3 [ { THROUGH } literal-4 ]] . . . .
                   { THRU    }
```

General Report Group Description Entry Formats

Format-1

```
01    [data-name-1]

      [ ; LINE NUMBER IS { integer-1 [ON NEXT PAGE] } ]
                         { PLUS integer-2           }

      [                   { integer-3       } ]
      [ ; NEXT GROUP IS   { PLUS integer-4  } ]
      [                   { NEXT PAGE       } ]

                    ( { REPORT HEADING }                                )
                    ( { RH             }                                )
                    ( { PAGE HEADING }                                  )
                    ( { PH           }                                  )
                    ( { CONTROL HEADING }  { data-name-2 }              )
                    ( { CH              }  { FINAL       }              )
      ; TYPE IS     ( { DETAIL }                                        )
                    ( { DE     }                                        )
                    ( { CONTROL FOOTING }  { data-name-3 }              )
                    ( { CF              }  { FINAL       }              )
                    ( { PAGE FOOTING }                                  )
                    ( { PF           }                                  )
                    ( { REPORT FOOTING }                                )
                    ( { RF             }                                )

      [; [USAGE IS] DISPLAY] .
```

Format-2

```
level-number [data-name-1]

      [ ; LINE NUMBER IS { integer-1 [ON NEXT PAGE] } ]
                         { PLUS integer-2           }

      [; [USAGE IS] DISPLAY] .
```

Format-3

```
level-number [data-name-1]

    [; BLANK WHEN ZERO]

    [; GROUP INDICATE]

    [ ; { JUSTIFIED } RIGHT ]
        { JUST      }

    [ ; LINE NUMBER IS { integer-1 [ON NEXT PAGE] } ]
                       { PLUS integer-2           }

    [; COLUMN NUMBER IS integer-3]

    ; { PICTURE } IS character-string
      { PIC     }

    ( ; SOURCE IS identifier-1                                      )
    ( ; VALUE IS literal                                            )
    ( {; SUM identifier-2 [, identifier-3] . . .                    )
    (     [UPON data-name-2 [, data-name-3] . . .] } . . .          )
    (   [ RESET ON { data-name-4 } ]                                )
    (   [          { FINAL       } ]                                )

    [; [USAGE IS] DISPLAY] .
```

Procedure Division

General Procedure Division Formats

Format-1

PROCEDURE DIVISION [USING data-name-1 [, data-name-2] . . .] .

[DECLARATIVES.

{section-name SECTION [segment-number] . declarative-sentence

[paragraph-name.[sentence] . . .] . . .} . . .

END DECLARATIVES.]

{section-name SECTION [segment-number]

[paragraph-name. [sentence] . . .] . . .} . . .

Format-2

PROCEDURE DIVISION [USING data-name-1 [, data-name-2] . . .] .

[paragraph-name. [sentence] . . .] . . .} . . .

COBOL Verbs

General Verb Formats

```
ACCEPT identifier [FROM mnemonic-name]

                           { DATE }
ACCEPT identifier FROM     { DAY  }
                           { TIME }

ACCEPT cd-name MESSAGE COUNT

    { identifier-1 }   { , identifier-2 }
ADD {              }   {                } . . . TO identifier-m [ROUNDED]
    { literal-1    }   { , literal-2    }

    [, identifier-n [ROUNDED]] . . . [; ON SIZE ERROR imperative-statement]

    { identifier-1 }   { identifier-2 }  [ , identifier-3 ]
ADD {              } , {              }  [                ] . . .
    { literal-1    }   { literal-2    }  [ , literal-3    ]

    GIVING identifier-m [ROUNDED] [, identifier-n [ROUNDED]] . . .
    [; ON SIZE ERROR imperative-statement]

    { CORRESPONDING }
ADD {               }  identifier-1 TO identifier-2 [ROUNDED]
    { CORR          }

    [; ON SIZE ERROR imperative-statement]

ALTER procedure-name-1 TO [PROCEED TO] procedure-name-2
    [, procedure-name-3 TO [PROCEED TO] procedure-name-4] . . .

     { identifier-1 }
CALL {              }  [USING data-name-1 [, data-name-2] . . .]
     { literal-1    }

    [; ON OVERFLOW imperative-statement]

       { identifier-1 }  [ , identifier-2 ]
CANCEL {              }  [                ] . . .
       { literal-1    }  [ , literal-2    ]

                     [ { REEL }  [ WITH NO REWIND ] ]
                     [ { UNIT }  [ FOR REMOVAL    ] ]
CLOSE file-name-1    [                              ]
                     [        { NO REWIND }         ]
                     [ WITH   { LOCK      }         ]

   [                  [ { REEL }  [ WITH NO REWIND ] ] ]
   [                  [ { UNIT }  [ FOR REMOVAL    ] ] ]
   [  , file-name-2   [                              ] ] . . .
   [                  [        { NO REWIND }         ] ]
   [                  [ WITH   { LOCK      }         ] ]

CLOSE file-name-1 [WITH LOCK] [, file-name-2 [WITH LOCK]] .

COMPUTE identifier-1 [ROUNDED] [, identifier-2 [ROUNDED]] . . .
      = arithmetic-expression [; ON SIZE ERROR imperative-statement]

DELETE file-name RECORD [; INVALID KEY imperative-statement]

        { INPUT  [TERMINAL] }                   { identifier-1 }
DISABLE {                   }  cd-name WITH KEY {              }
        { OUTPUT            }                   { literal-1    }

        { identifier-1 }  [ , identifier-2 ]
DISPLAY {              }  [                ] . . . [UPON mnemonic-name]
        { literal-1    }  [ , literal-2    ]

       { identifier-1 }
DIVIDE {              }  INTO identifier-2 [ROUNDED]
       { literal-1    }

    [, identifier-3 [ROUNDED]] . . . [; ON SIZE ERROR imperative-statement]

       { identifier-1 }        { identifier-2 }
DIVIDE {              }  INTO  {              }  GIVING identifier-3 [ROUNDED]
       { literal-1    }        { literal-2    }

    [, identifier-4 [ROUNDED]] . . . [; ON SIZE ERROR imperative-statement]

       { identifier-1 }      { identifier-2 }
DIVIDE {              }  BY  {              }  GIVING identifier-3 [ROUNDED]
       { literal-1    }      { literal-2    }

    [, identifier-4 [ROUNDED]] . . . [; ON SIZE ERROR imperative-statement]

       { identifier-1 }        { identifier-2 }
DIVIDE {              }  INTO  {              }  GIVING identifier-3 [ROUNDED]
       { literal-1    }        { literal-2    }

    REMAINDER identifier-4 [; ON SIZE ERROR imperative-statement]
```

```
DIVIDE {identifier-1} BY {identifier-2} GIVING identifier-3 [ROUNDED]
       {literal-1   }    {literal-2   }
    REMAINDER identifier-4 [; ON SIZE ERROR imperative-statement]

ENABLE {INPUT  [TERMINAL]} cd-name WITH KEY {identifier-1}
       {OUTPUT           }                  {literal-1   }

ENTER language-name [routine-name] .

EXIT [PROGRAM] .

GENERATE {data-name  }
         {report-name}

GO TO [procedure-name-1]

GO TO procedure-name-1 [, procedure-name-2] . . . , procedure-name-n
    DEPENDING ON identifier

IF condition; {statement-1  } {; ELSE statement-2    }
              {NEXT SENTENCE} {; ELSE NEXT SENTENCE  }

INITIATE report-name-1 [, report-name-2] . . .

INSPECT identifier-1 TALLYING

{, identifier-2 FOR {, {{ALL    } {identifier-3}} [{BEFORE} INITIAL {identifier-4}] } . . . } . . .
                    {  {{LEADING} {literal-1   }} [{AFTER }         {literal-2   }] }
                    {  { CHARACTERS            }                                  }

INSPECT identifier-1 REPLACING

{ CHARACTERS BY identifier-6   BEFORE  INITIAL  identifier-7                                    }
{               literal-4      AFTER            literal-5                                       }
{ { {ALL    } {, {identifier-5} BY {identifier-6} [{BEFORE} INITIAL {identifier-7}] } . . . } . . . }
{ { {LEADING} {  {literal-3   }    {literal-4   } [{AFTER }         {literal-5   }] }         }
{ { {FIRST  } {                                                                    }         }

INSPECT identifier-1 TALLYING

{, identifier-2 FOR {, {{ALL    } {identifier-3}} [{BEFORE} INITIAL {identifier-4}] } . . . } . . .
                    {  {{LEADING} {literal-1   }} [{AFTER }         {literal-2   }] }
                    {  { CHARACTERS            }                                  }

    REPLACING

{ CHARACTERS BY {identifier-6} [{BEFORE} INITIAL {identifier-7}]                                  }
{               {literal-4   } [{AFTER }         {literal-5   }]                                  }
{ { , {ALL    } {, {identifier-5} BY {identifier-6} [{BEFORE} INITIAL {identifier-7}] } . . . } . . . }
{ {   {LEADING} {  {literal-3   }    {literal-4   } [{AFTER }         {literal-5   }] }         }
{ {   {FIRST  } {                                                                    }         }

MERGE file-name-1 ON {ASCENDING } KEY data-name-1 [, data-name-2] . . .
                     {DESCENDING}
                  [ ON {ASCENDING } KEY data-name-3 [, data-name-4] . . . ] . . .
                  [    {DESCENDING}                                       ]
    [COLLATING SEQUENCE IS alphabet-name]
    USING file-name-2, file-name-3 [, file-name-4] . . .
    {OUTPUT PROCEDURE IS section-name-1 [{THROUGH} section-name-2]}
    {                                   [{THRU   }               ]}
    {GIVING file-name-5                                           }

MOVE {identifier-1} TO identifier-2 [, identifier-3] . . .
     {literal     }

MOVE {CORRESPONDING} identifier-1 TO identifier-2
     {CORR         }
```

```
MULTIPLY {identifier-1} BY identifier-2 [ROUNDED]
         {literal-1   }

    [, identifier-3 [ROUNDED]] . . . [; ON SIZE ERROR imperative-statement]

MULTIPLY {identifier-1} BY {identifier-2} GIVING identifier-3 [ROUNDED]
         {literal-1   }    {literal-2   }

    [, identifier-4 [ROUNDED]] . . . [; ON SIZE ERROR imperative-statement]

     { INPUT file-name-1 [REVERSED        ] [, file-name-2 [REVERSED        ]] . . . }
     {                   [WITH NO REWIND  ]                [WITH NO REWIND  ]        }
OPEN { OUTPUT file-name-3 [WITH NO REWIND] [, file-name-4 [WITH NO REWIND]] . . .    } . . .
     { I-O file-name-5 [, file-name-6] . . .                                          }
     { EXTEND file-name-7 [, file-name-8] . . .                                       }

     { INPUT file-name-1 [, file-name-2] . . .  }
OPEN { OUTPUT file-name-3 [, file-name-4] . . . } . . .
     { I-O file-name-5 [, file-name-6] . . .    }

PERFORM procedure-name-1 [{THROUGH} procedure-name-2]
                         [{THRU   }                 ]

PERFORM procedure-name-1 [{THROUGH} procedure-name-2] {identifier-1} TIMES
                         [{THRU   }                 ] {integer-1   }

PERFORM procedure-name-1 [{THROUGH} procedure-name-2] UNTIL condition-1
                         [{THRU   }                 ]

PERFORM procedure-name-1 [{THROUGH} procedure-name-2]
                         [{THRU   }                 ]

    VARYING {identifier-2 } FROM {indentifier-3}
            {index-name-1 }      {index-name-2 }
                                 {literal-1    }

         BY {identifier-4} UNTIL condition-1
            {literal-3   }

   [ AFTER {identifier-5} FROM {identifier-6}
   [       {index-name-3}      {index-name-4}
   [                           {literal-3   }

         BY {identifier-7} UNTIL condition-2
            {literal-4   }

     AFTER {identifier-8} FROM {identifier-9}
           {index-name-5}      {index-name-6}
                               {literal-5   }

         BY {identifier-10} UNTIL condition-3 ]]
            {literal-6    }

READ file-name RECORD [INTO identifier] [; AT END imperative-statement]

READ file-name [NEXT] RECORD [INTO identifier]
    [; AT END imperative-statement]

READ file-name RECORD [INTO identifier] [; INVALID KEY imperative-statement]

READ file-name RECORD [INTO identifier]
    [; KEY IS data-name]
    [; INVALID KEY imperative-statement]

RECEIVE cd-name {MESSAGE} INTO identifier-1 [; NO DATA imperative-statement]
                {SEGMENT}

RELEASE record-name [FROM identifier]

RETURN file-name RECORD [INTO identifier] ; AT END imperative-statement

REWRITE record-name [FROM identifier]

REWRITE record-name [FROM identifier] [; INVALID KEY imperative-statement]
```

```
SEARCH identifier-1 [ VARYING { identifier-2   } ] [; AT END imperative-statement-1]
                              { index-name-1   }

    ; WHEN condition-1 { imperative-statement-2 }
                       { NEXT SENTENCE          }

  [ ; WHEN condition-2 { imperative-statement-3 } ] ...
                       { NEXT SENTENCE          }

SEARCH ALL identifier-1 [; AT END imperative-statement-1]

    ; WHEN { data-name-1 { IS EQUAL TO } { identifier-3            } }
           {             { IS =        } { literal-1               } }
           {                             { arithmetic-expression-1 } }
           { condition-name-1                                        }

         [ AND { data-name-2 { IS EQUAL TO } { identifier-4            } } ] ...
         [     {             { IS =        } { literal-2               } } ]
         [     {                             { arithmetic-expression-2 } } ]
         [     { condition-name-2                                        } ]

         { imperative-statement-2 }
         { NEXT SENTENCE          }

SEND cd-name FROM identifier-1

                                   { WITH identifier-2 }
SEND cd-name [FROM identifier-1]   { WITH ESI          }
                                   { WITH EMI          }
                                   { WITH EGI          }

  [                          { { identifier-3 } [ LINE  ] } ]
  [ { BEFORE } ADVANCING     { { integer      } [ LINES ] } ]
  [ { AFTER  }               {                            } ]
  [                          { { mnemonic-name }          } ]
  [                          { { PAGE          }          } ]

SET { identifier-1 [, identifier-2] ...    } TO { identifier-3 }
    { index-name-1 [, index-name-2] ...    }    { index-name-3 }
                                                { integer-1    }

SET index-name-4 [, index-name-5] ... { UP BY   } { identifier-4 }
                                      { DOWN BY } { integer-2    }

SORT file-name-1 ON { ASCENDING  } KEY data-name-1 [, data-name-2] ...
                    { DESCENDING }

                 [ ON { ASCENDING  } KEY data-name-3 [, data-name-4] ... ] ...
                      { DESCENDING }

    [COLLATING SEQUENCE IS alphabet-name]

    { INPUT PROCEDURE IS section-name-1 [ { THROUGH } section-name-2 ] }
    {                                     { THRU    }                  }
    { USING file-name-2 [, file-name-3] ...                            }

    { OUTPUT PROCEDURE IS section-name-3 [ { THROUGH } section-name-4 ] }
    {                                      { THRU    }                  }
    { GIVING file-name-4                                                }

                       [     { IS EQUAL TO        }           ]
                       [     { IS =               }           ]
START file-name        [ KEY { IS GREATER THAN    } data-name ]
                       [     { IS >               }           ]
                       [     { IS NOT LESS THAN   }           ]
                       [     { IS NOT <           }           ]

    [; INVALID KEY imperative-statement]

STOP { RUN     }
     { literal }
```

```
                                                        {identifier-3}
STRING {identifier-1} [, identifier-2] . . . DELIMITED BY {literal-3   }
       {literal-1   } [, literal-2   ]                    {SIZE        }

   [  {identifier-4} [, identifier-5]                    {identifier-6} ]
   [ ,{literal-4   } [, literal-5   ] . . . DELIMITED BY {literal-6   } ]
   [                                                     {SIZE        } ]

   INTO identifier-7 [WITH POINTER identifier-8]
   [; ON OVERFLOW imperative-statement]

SUBTRACT {identifier-1} [, identifier-2] . . . FROM identifier-m [ROUNDED]
         {literal-1   } [, literal-2   ]

   [, identifier-n [ROUNDED]] . . . [; ON SIZE ERROR imperative-statement]

SUBTRACT {identifier-1} [, identifier-2] . . . FROM {identifier-m}
         {literal-1   } [, literal-2   ]            {literal-m   }

   GIVING identifier-n [ROUNDED] [, identifier-o [ROUNDED]] . . .
   [; ON SIZE ERROR imperative-statement]

SUBTRACT {CORRESPONDING} identifier-1 FROM identifier-2 [ROUNDED]
         {CORR         }

   [; ON SIZE ERROR imperative-statement]

SUPPRESS PRINTING

TERMINATE report-name-1 [, report-name-2]

UNSTRING identifier-1

   [ DELIMITED BY [ALL] {identifier-2} [, OR [ALL] {identifier-3} ] . . . ]
   [                    {literal-1   } [           {literal-2   } ]       ]

    INTO identifier-4 [, DELIMITER IN identifier-5] [, COUNT IN identifier-6]
       [, identifier-7 [, DELIMITER IN identifier-8] [, COUNT IN identifier-9]] . . .
   [WITH POINTER identifier-10] [TALLYING IN identifier-11]
   [; ON OVERFLOW imperative-statement]

                                                      {file-name-1 [, file-name-2] . . .}
                                                      {INPUT                            }
USE AFTER STANDARD {EXCEPTION} PROCEDURE ON          {OUTPUT                           }.
                   {ERROR    }                        {I-O                              }
                                                      {EXTEND                           }

                                                      {file-name-1 [, file-name-2] . . .}
USE AFTER STANDARD {EXCEPTION} PROCEDURE ON          {INPUT                            }.
                   {ERROR    }                        {OUTPUT                           }
                                                      {I-O                              }

USE BEFORE REPORTING identifier.

                     {cd-name-1                          }
                     {[ALL REFERENCES OF] identifier-1   }
USE FOR DEBUGGING ON {file-name-1                        }
                     {procedure-name-1                   }
                     {ALL PROCEDURES                     }

   [ cd-name-2                          ]
   [ [ALL REFERENCES OF] identifier-2   ]
   [ file-name-2                        ] . . . .
   [ procedure-name-2                   ]
   [ ALL PROCEDURES                     ]

WRITE record-name [FROM identifier-1]

   [                          {identifier-2} [LINE ] ]
   [ {BEFORE}  ADVANCING      {integer     } [LINES] ]
   [ {AFTER }                 {mnemonic-name}        ]
   [                          {PAGE         }        ]

   [; AT {END-OF-PAGE} imperative-statement]
   [     {EOP        }                     ]

WRITE record-name [FROM identifier] [; INVALID KEY imperative-statement]
```

Conditions

General Condition Formats

Relation Condition

```
                              ⎧ IS [NOT] GREATER THAN ⎫
⎧ identifier-1           ⎫    ⎪ IS [NOT] LESS THAN    ⎪   ⎧ identifier-2           ⎫
⎨ literal-1              ⎬    ⎨ IS [NOT] EQUAL TO     ⎬   ⎨ literal-2              ⎬
⎪ arthimetic-expression-1⎪    ⎪ IS [NOT] >            ⎪   ⎪ arithmetic-expression-2⎪
⎩ index-name-1           ⎭    ⎪ IS [NOT] <            ⎪   ⎩ index-name-2           ⎭
                              ⎩ IS [NOT] =            ⎭
```

Class Condition

```
identifier IS [NOT] { NUMERIC    }
                    { ALPHABETIC }
```

Sign Condition

```
                                  ⎧ POSITIVE ⎫
arithmetic-expression is [NOT]    ⎨ NEGATIVE ⎬
                                  ⎩ ZERO     ⎭
```

Condition-Name Condition

condition-name

Switch-Status Condition

condition-name

Negated Simple Condition

NOT simple-condition

Combined Condition

```
condition  { { AND } condition } ...
           { { OR  }           }
```

Abbreviated Combined Relation Condition

```
relation-condition  { { AND } [NOT] [relational-operator] object } ...
                    { { OR  }                                    }
```

Miscellaneous

Miscellaneous Formats

Qualification

```
{data-name-1   }  [{OF} data-name-2] . . .
{condition-name}  [{IN}            ]

paragraph-name    [{OF} section-name]
                  [{IN}             ]

text-name [{OF} library-name]
          [{IN}             ]
```

Subscripting

```
{data-name     }  (subscript-1 [, subscript-2 [, subscript-3] ])
{condition-name}
```

Indexing

```
{data-name     }  ( {index-name-1 [{±} literal-2]}
{condition-name}    {literal-1                   }

   [, {index-name-2 [{±} literal-4]}  [, {index-name-3 [{±} literal-6]}]])
   [  {literal-3                   }  [  {literal-5                   }]]
```

Identifier: Format-1

```
data-name-1 [{OF} data-name-2] . . . [(subscript-1 [, subscript-2
            [{IN}            ]

   [, subscript-3] ])]
```

Identifier: Format-2

```
data-name-1 [{OF} data-name-2] . . . [( {index-name-1 [{±} literal-2]}
            [{IN}            ]       [  {literal-1                   }

   [, {index-name-2 [{±} literal-4]}  [, {index-name-3 [{±} literal-6]}]])]
   [  {literal-3                   }  [  {literal-5                   }]]
```

COPY Statement

General Copy Statement Format

```
COPY text-name [{OF} library-name]
               [{IN}             ]

   [             { {==pseudo-text-1==}      {==pseudo-text-2==} }      ]
   [ REPLACING   {,{identifier-1     } BY   {identifier-2     } } . . .]
   [             { {literal-1        }      {literal-2        } }      ]
   [             { {word-1           }      {word             } }      ]
```

Glossary

These definitions are intended to be either reference material or introductory material to be reviewed prior to reading the detailed language specifications for an element of COBOL. For this reason, these definitions are, in most instances, brief and do not include syntactical rules.

These definitions are in accordance with their meaning as used in describing COBOL and may not have the same meaning for other languages.

Abbreviated Combined Relation Condition
The combined condition that results from the explicit omission of a common subject or a common subject and common relational operator in a consecutive sequence of relational operator in a consecutive sequence of relation conditions.

ACCESS
The manner in which files are referenced by the computer. Access can be sequential (records are referred to one after another in the order in which they appear on the file), or it can be random (the individual records can be referred to in a nonsequential manner).

ACCESS MODE
The manner in which records are to be operated upon within a file.

Actual Decimal Point
The physical representation, using either of the decimal point characters (. or ,), of the decimal point position in a data item. When specified, it will appear in a printed report, and it requires an actual space in storage.

Alphabetic Character
A character which is one of the twenty-six characters of the alphabet, or a space. In COBOL, the term does *not* include any other characters.

Alphabet-Name
A user-defined word, in the SPECIAL-NAMES paragraph of the Environment Division, which assigns a name to a specific character set and/or collating sequence.

Alphanumeric Character
Any character in the computer's character set.

Alphanumeric Edited Character
A character within an alphanumeric character string which contains at least one B or 0.

Alternate Record Key
A key, other than the prime record key, whose contents identify a record within an indexed file.

Arithmetic Expression
A statement containing any combination of data-names, numeric literals, and figurative constants, joined together by one or more arithmetic operators in such a way that the statement as a whole can be reduced to a single numeric value.

An arithmetic expression can be an identifier or a numeric elementary item, a numeric literal, such identifiers and literals separated by arithmetic operators, two arithmetic expressions separated by an arithmetic operator, or an arithmetic expression enclosed in parentheses.

Arithmetic Operator
A symbol (single character or two-character set) or COBOL verb which directs the system to perform an arithmetic operation. A single character or a fixed two-character combination that belongs to the following set:

Character	*Meaning*
+	addition
−	subtraction
*	multiplication
/	division
**	exponentiation

Ascending Key
A key upon the values of which data is ordered starting with the lowest value of key up to the highest value in accordance with the rules for comparing data items.

Assumed Decimal Point
A decimal point position that does not involve the existence of an actual character in a data item. It does not occupy an actual space in storage, but is used by the compiler to properly align a value for calculation.

The assumed decimal point has logical meaning but no physical representation.

At-End Condition
A condition caused

1. during the execution of a READ statement for a sequentially accessed file.

2. during the execution of a RETURN statement, when no next

logical record exists for the associated sort or merge file.

3. during the execution of a SEARCH statement, when the search operation terminates without satisfying the condition specified in any of the associated WHEN phrases.

BLOCK
A physical unit of data that is normally composed of one or more logical records. For mass storage files, a block may contain a portion of a logical record. The size of a block has no direct relationship to the size of the file within which the block is contained or to the size of the logical record(s) that is either contained within the block or that overlaps the block.

In COBOL, a group of characters or records that is treated as an entity when moved into or out of the computer. The term is synonymous with the term physical record.

Body Group
Generic term for a report group of TYPE DETAIL, CONTROL HEADING or CONTROL FOOTING.

Buffer
A portion of main storage into which data is read or from which it is written.

Byte
A sequence of eight adjacent binary bits.

Called Program
A program which is the object of a CALL statement combined at object time with the calling program to produce a run unit.

Calling Program
A program which executes a CALL to another program.

Channel
A device that directs the flow of information between the computer main storage and the input/output devices.

Character
One of a set of indivisible symbols that can be arranged in sequences to express information. These symbols include the letters A through Z, the decimal digits 0 through 9, punctuation symbols, and any other symbols which would be accepted by the data-processing system.

The character is the basic indivisible unit of the language.

Character Position
A character position is the amount of physical storage required to store a single standard data format character described as usage is DISPLAY. Further characteristics of the physical storage are defined by the implementor.

Character Set
All the valid COBOL characters. The complete set of fifty-one characters.

Character-String
A sequence of contiguous characters that form a COBOL word, a literal, a PICTURE character-string, or a comment-entry.

Checkpoint
A reference point in a program at which information about the contents of main storage can be recorded so that, if necessary, the program can be restarted at an intermediate point.

Class Condition
A statement that the content of an item is wholly alphabetic or wholly numeric. The statement may be true or false.

Clause
A clause is an ordered set of consecutive COBOL character-strings whose purpose is to specify an attribute of an entry.

COBOL Character Set
The complete COBOL character set consists of the following fifty-one characters:

Character	*Meaning*
0,1,...,9	digit
A,B,...,Z	letter
	space (blank)
+	plus sign
−	minus sign (hyphen)
*	asterisk
/	stroke (virgule, slash)
=	equal sign
$	currency sign
,	comma (decimal point)
;	semicolon
.	period (decimal point)
"	quotation mark
(	left parenthesis
)	right parenthesis
>	greater than symbol
<	less than symbol

COBOL Word
(See word)

Collating Sequence
The arrangement of all valid characters in the order of their relative precedence. The collating sequence of a computer is part of the computer design; each acceptable character has a predetermined place in the sequence. The sequence in which the characters that are acceptable in a computer are ordered for purposes of sorting, merging, and comparing.

Column
A character position within a print line. The columns are numbered from 1, by 1, starting at the leftmost character position of the print line and extending to the rightmost position of the print line.

COLUMN Clause
A COBOL clause used to identify a specific position within a report line.

Combined Condition
A condition that is the result of connecting two or more conditions with the 'AND' or the 'OR' logical operator.

Comment-Entry
An entry in the Identification Division that may be any combination of characters from the computer character set.

Comment Line
A source program line represented by an asterisk in the indicator area of the line and any characters from the computer's character set in areas A and B of that line. The comment line serves only for documentation in a program. A special form of comment line represented by stroke (/) in the indicator area of the line, and any characters from the computer's character set in areas A and B of that line causes page ejection prior to printing the comment.

Compiler
A program which translates a source program into a machine-language-object program.

Compiler-Directing Statement
A statement, beginning with a compiler-directing verb, which causes the compiler to take specific action during compilation.

Compile Time
The time at which a COBOL source program is translated, by a COBOL compiler, to a COBOL object program.

Complex Condition
A condition in which one or more logical operators act upon one or more conditions. (See Negated Simple Condition, Combined Condition, Negated Combined Condition.)

Compound Condition
A statement that tests two or more relational expressions. It may be true or false.

Computer-Name
A system-name that identifies the computer upon which the program is to be compiled or run.

Condition
One of a set of specified values a data item can assume.

A simple conditional expression; relation condition, class condition, condition-name condition, sign condition, NOT condition.

A status of a program at execution time for which a truth value can be determined. Where the term 'condition' (condition-1, condition-2, . . .) appears in these language specifications in or in reference to 'condition' (condition-1, condition-2, . . .) of a general format, it is a conditional expression consisting of either a simple condition optionally parenthesized, or a combined condition consisting of the syntactical correct combination of simple conditions, logical operators, and parentheses, for which a truth value can be determined.

Conditional Expression
A simple condition or a complex condition specified in an IF, PERFORM, or SEARCH statement. (See Simple Condition and Complex Condition.)

Conditional Statement
A syntactically correct statement, which is made up of data-names and/or figurative constants and/or logical operators, and is so constructed that it tests a truth value. The subsequent action of the object program is dependent on this truth value.

Conditional Variable
A data item that can assume more than one value; the value(s) it assumes has a condition-name assigned to it.

Condition-Name
A user-defined word assigned to a specific value, set of values, or range of values within the complete set of values that a conditional variable may possess; or the user-defined word assigned to a status of an implementor-defined switch or device.

Condition-Name Condition
The proposition, for which a truth value can be determined, that the value of a conditional variable is a member of the set of values attributed to a condition-name associated with the conditional variable.

Configuration Section
A section of the Environment Division that describes the overall specifications of the source and object computers.

Connective
A reserved word or a punctuation character that is used to: associate a data-name, paragraph-name, condition-name, or text-name with its qualifier; link two or more operands written in a series; and form conditions (logical connectives). (See Logical Operator.)

CONSOLE
A COBOL mnemonic-name associated with the console typewriter.

Contiguous Items
Items that are described by consecutive entries in the Data Division, and that bear a definite relationship to each other.

Control Break
A change in the value of a data item that is referenced in the CONTROL clause. More generally, a change in the value of a data item that is used to control the hierarchical structure of a report.

Control Break Level
The relative position within a control hierarchy at which the most major control break occurred.

Control Data Item
A data item, a change in whose contents may produce a control break. A data item that is tested each time a report line is to be printed. If the value of the data item has changed, a control break occurs and special actions are performed before the line is printed.

Control Data-Name
A data-name that appears in a CONTROL clause and refers to a control data item.

CONTROL FOOTING
A report group that is presented at the end of the control group of which it is a member.

Control Group
A set of body groups that is presented for a given value of a control data item or of FINAL. Each control group may begin with a CONTROL HEADING and end with a CONTROL FOOTING, and contain DETAIL report groups.

CONTROL HEADING
A report group that is presented at the beginning of the control group of which it is a member.

Control Hierarchy
A designated sequence of report subdivisions defined by the positional order of FINAL and the data-names within a CONTROL clause.

Counter
A data item used for storing numbers or number representations in a manner that permits these numbers to be increased or decreased by the value of another number, or to be changed or reset to zero or to an arbitrary positive or negative value.

Currency Sign
The character "$" of the COBOL character set.

Currency Symbol
The character defined by the CURRENCY SIGN clause in the SPECIAL-NAMES paragraph. If no CURRENCY SIGN clause is present in a COBOL source program, then the currency symbol is identical to the currency sign.

Current Record
The record that is available in the record area associated with the file.

Current Record Pointer
A conceptual entity that is used in the selection of the next record.

Data Clause
A clause that appears in a data description entry in the Data Division and provides information describing a particular attribute of a data item.

Data Description Entry
An entry in the Data Division that is used to describe the characteristics of a data item. It consists of a level number,

followed by an optional data-name, followed by data clauses that fully describe the format the data will take. An elementary data description entry (or item) cannot logically be subdivided further. A group data description entry (or item) is made up of a number of related groups and/or elementary items.

Data Division
One of the four main component parts of a COBOL program. The Data Division describes the files to be used in the program and the records contained within the files. It also describes any internal Working-Storage or Linkage Section records that will be needed.

Data Item
A character or a set of contiguous characters (excluding in either case literals) defined as a unit of data by the COBOL program.

Data-Name
A name assigned by the programmer to a data item in a COBOL program. It must contain at least one alphabetic character. The data-name names a data item described in a data description entry in the Data Division. When used in the general formats, 'data-name' represents a word that can neither be subscripted, indexed, nor qualified unless specifically permitted by the rules for that format.

Debugging Line
A debugging line is any line with a 'D' in the indicator area of that line.

Debugging Section
A debugging section is a section that contains a USE FOR DEBUGGING statement.

Declaratives
A set of one or more special purpose sections, written at the beginning of the Procedure Division, the first of which is preceded by the key word DECLARATIVES and the last of which is followed by the key words END DECLARATIVES. A declarative is composed of a section header, followed by a USE compiler-directing sentence, followed by a set of zeros, and one or more associated paragraphs.

Declarative-Sentence
A compiler-directing sentence consisting of a single USE statement terminated by the separator period.

Delimiter
A character or a sequence of contiguous characters that identifies the end of a string of characters and separates that string of characters from the following string of characters. A delimiter is not a part of the string of characters that it delimits.

Descending Key
A key upon the values of which data is ordered starting with the highest value of key down to the lowest value of key, in accordance with the rules for comparing data items.

Digit
Any of the numerals from 0 through 9. In COBOL, the term is not used in reference to any other symbol.

Digit Position
A digit position is the amount of physical storage required to store a single digit. This amount may vary depending on the usage of the data item describing the digit position. Further characteristics of the physical storage are defined by the implementor.

Division
A set of zeros and one or more sections of paragraphs, called the division body, which are formed and combined in accordance with a specific set of rules. There are four divisions in a COBOL program: (1) the Identification Division, which names the program; (2) the Environment Division, which indicates the machine equipment and equipment features to be used in the program; (3) the Data Division, which defines the nature and characteristics of the data to be processed; (4) and the Procedure Division, which consists of statements directing the processing of data in a specified manner at execution time.

Division Header
A combination of words followed by a period and a space that indicates that beginning of a division. The four division headers are:
IDENTIFICATION DIVISION.
ENVIRONMENT DIVISION.
DATA DIVISION.
PROCEDURE DIVISION. [USING data-name-1 [data-name-2 . . . .].

Division-Name
The name of one of the four divisions of a COBOL program.

Dynamic Access
An access mode in which specific logical records can be obtained from or placed into a mass storage file in a non-sequential manner (see Random Access) and obtained from a file in a sequential manner (see Sequential Access) during the scope of the same OPEN statement.

EBCDIC Character
Any one of the symbols included in the eight-bit EBCDIC (Extended Binary-Coded-Decimal Interchange Code) set. All fifty-one COBOL characters are included.

Editing Character
A single character or fixed two-character combination used to create proper formats for output reports, and belonging to the following set:

Character	*Meaning*
B	space
0	zero
+	plus
−	minus
CR	credit
DB	debit
Z	zero suppress
*	check protect
$	currency sign
,	comma (decimal point)
.	period (decimal point)
/	stroke (virgule, slash)

Elementary Item
A data item that is described as not being further logically subdivided.

End of Procedure Division
The physical position in a COBOL source program after which no further procedures appear.

Entry
Any descriptive set of consecutive clauses terminated by a period and written in the Identification Division, Environment Division, or Data Division of a COBOL source program.

Environment Clause
A clause that appears as part of an Environment Division entry.

Environment Division
One of the four main component parts of a COBOL program. The Environment Division describes the computers upon which the source program is compiled and those on which the object program is executed, and

provides a linkage between the logical concept of files and their records, and the physical aspects of the devices on which files are stored.

Execution Time
The time at which an object program actually performs the instructions coded in the Procedure Division, using the actual data provided.

Exponent
A number indicating how many times another number (the base) is to be repeated as a factor. Positive numbers denote multiplication, negative exponents denote division, fractional exponents denote a root of a quantity. In COBOL, exponentiation is indicated by the symbol ** followed by the exponent.

Extend Mode
The state of a file after execution of an OPEN statement, with the EXTEND phrase specified for that file and before the execution of a CLOSE statement for that file.

Figurative Constant
A compiler-generated value referenced through the use of certain reserved words. The reserved word represents a numeric value, a character, or a string of repeated values or characters. The word can be written in a COBOL program to represent the values or characters without being defined in the Data Division.

File
A collection of records.

File Clause
A clause that appears as part of any of the following Data Division entries:

File description (FD)
Sort-merge file description (SD)
Communication description (CD)

FILE-CONTROL
The name and header of an Environment Division paragraph in which the data files for a given source program are named and assigned to specific input/output devices.

File Description Entry
An entry in the File Section of the Data Division that provides information about the identification and physical structure of a file. This entry is composed of the level indicator FD, followed by a file-name, and then followed by a set of file clauses as required.

File-Name
A user-defined word that names a file described in a file description entry or a sort-merge file description entry within the File Section of the Data Division. The file-name is assigned to a set of input data or output data and must include at least one alphabetic character.

File Organization
The permanent logical file structure established at the time a file is created.

File Section
A section of the Data Division that contains descriptions of all externally stored data (or files) used in a program. Such information is given in one or more file description entries.

Floating-Point Literal
A numeric literal whose value is expressed in floating-point notation; that is, as a decimal number followed by an exponent which indicates the actual placement of the decimal point.

Format
A specific arrangement of a set of data.

Function-Name
A name, specified by the computer manufacturer, which identifies system logical units, printer and card-punch control characters, and report codes. When a function-name is associated with a mnemonic-name in the Environment Division, the mnemonic-name may be substituted in any format in which such substitutions are valid.

Group Item
A named contiguous set of elementary or group items. The data item is made up of a series of logically related elementary items. It can be part of a record or a complete record.

Header Label
A record that identifies the beginning of a physical file or a volume.

High-Order End
The leftmost character in a string of characters.

Identification Division
One of the four main component parts of a COBOL program. The Identification Division identifies the source program and the object program and, in addition, may include such documentation as the author's name, the installation where written, date written, purpose of the program, etc.

Identifier
A data-name, followed as required by the syntactically correct combination of qualifiers, subscripts, and indices necessary to make a unique reference to a data item.

Imperative Statement
A statement that begins with an imperative verb and specifies an unconditional action to be taken. An imperative statement may consist of a sequence of imperative statements.

Implementor-Name
A system-name that refers to a particular feature available on that implementor's computing system.

Index
A computer storage position or register, the contents of which represent the identification of a particular element in a table.

Index Data Item
A data item in which the value associated with an index-name can be stored in a form specified by the implementor.

Indexed Data-Name
An identifier that is composed of a data-name, followed by one or more index-names enclosed in parentheses.

Indexed File
A file with indexed organization.

Indexed Organization
The permanent logical file structure in which each record is identified by the value of one or more keys within that record.

Index-Name
A user-defined word that names an index associated with a specific table.

Input File
A file that is opened in the input mode.

Input Mode
The state of a file after execution of an OPEN statement (with the INPUT phrase specified) for that file and before the execution of a CLOSE statement for that file.

Input-Output File
A file that is opened in the I-O mode.

Input-Output Section
The section of the Environment Division that names the files and the external media required by an object program and which provides information required for the transmission and handling of data during execution of the object program.

INPUT PROCEDURE
A set of statements that is executed each time a record is released to the sort file. Input procedures are optional; whether they are used or not depends upon the logic of the program.

Integer
A numeric literal or a numeric data item that does not include any character positions to the right of the assumed decimal point. Where the term "integer" appears in general formats, the integer must not be a numeric data item, and must not be signed, nor zero, unless explicitly allowed by the rules of that format.

Invalid Key Condition
A condition, at object time, caused when a specific value of the key associated with an indexed or relative file is determined to be invalid.

I-O-CONTROL
The name of an Environment Division paragraph in which object program requirements for specific input-output techniques, rerun points, sharing of same areas by several data files, and multiple file storage on a single input-output device are specified.

I-O-Mode
The state of a file after execution of an OPEN statement (with the I-O phrase specified) for that file and before the execution of a CLOSE statement for that file.

Key
A data item which identifies the location of a record, or a set of data items which serve to identify the ordering of data.

Key of Reference
The key, either prime or alternative, currently being used to access records within an indexed file.

Key Word
A reserved word whose employment is essential to the meaning and structure of a COBOL statement. In this text, key words are indicated in the formats of statements by underscoring. Key words are included in the reserved word list.

Language-Name
A system-name that specifies a particular programming language.

Level Indicator
Two alphabetic characters that identify a specific type of file or a position in a hierarchy.

Level-Number
A user-defined word that indicates the position of a data item in the hierarchical structure of a logical record or which indicates special properties of a data description entry. A level-number is expressed as a one- or two-digit number. Level-numbers in the range 1 through 49 indicate the position of a data item in the hierarchical structure of a logical record. Level-numbers in the range 1 through 9 may be written either as a single digit or as a zero followed by a significant digit. Level numbers 66, 77, and 88 identify special properties of a data description entry.

Library-Name
A user-defined word that names a COBOL library that is to be used by the compiler for a given source program compilation.

Library-Text
A sequence of character strings and/or separators in a COBOL library.

Line
(See Report Line)

Line Number
An integer that denotes the vertical position of a report line on a page.

Linkage Section
The section in the Data Division of the called program that describes data items available from the calling program. These data items may be referred to by both the calling and called programs.

Literal
A character-string whose value is implied by the ordered set of characters comprising the string. The numeric literal 7 expresses the value 7, and the nonnumeric literal 'CHARACTERS' expresses the value CHARACTERS.

Logical Operator
One of the reserved words AND, OR, or NOT. In the formation of a condition, both or either of AND and OR can be used as logical connectives. NOT can be used for logical negation. The three logical operators and their meanings are:

OR (logical connective—either or both)
AND (logical connective—both)
NOT (logical negation)

Logical Record
The most inclusive data item, and identified by a level 01 entry. It consists of one or more related data items.

Low-Order End
The rightmost character of a string of characters.

Main Program
The highest level COBOL program involved in a step. (Programs written in other languages that follow COBOL linkage conventions are considered COBOL programs in this sense.)

Mantissa
The decimal part of a logarithm. Therefore, the part of a floating-point number that is expressed as decimal fraction.

Mass Storage
A storage medium on which data may be organized and maintained in both a sequential and a nonsequential manner.

Mass Storage Control System (MSCS)
An input-output control system that directs or controls the processing of mass storage files.

Mass Storage File
A collection of records that is assigned to a mass storage medium.

Merge File
A collection of records to be merged by a MERGE statement. The merge file is created and can be used only by the merge function.

Mnemonic-Name
A user-defined word associated with a specific implementor-name in the Environment Division. It may then be written in place of the implementor-name in any format wherein such substitution is valid.

MODE
The manner in which records of a file are accessed or processed.

MSCS
(See Mass Storage Control System)

Name
A word composed of not more than thirty characters, and which defines a COBOL operand.

Negated Combined Condition
The 'NOT' logical operator immediately followed by a parenthesized combined condition.

Negated Simple Condition
The 'NOT' logical operator immediately followed by a simple condition.

Next Executable Sentence
The next sentence to which control will be transferred after execution of the current statement is complete.

Next Executable Statement
The next statement to which control will be transferred after execution of the current statement is complete.

Next Record
The record which logically follows the current record of a file.

Noncontiguous Items
Elementary data items, in the Working-Storage and Linkage Sections, which bear no hierarchic relationship to other data items.

Nonnumeric Item
A data item whose description permits its contents to be composed of any combination of characters taken from the computer's character set. Certain categories of nonnumeric items may be formed from restricted character sets.

Nonnumeric Literal
A character-string enclosed by quotation marks. The string of characters may include any character in the computer's character set. To represent a single quotation mark character within a nonnumeric literal, two contiguous quotation marks must be used.

Numeric Character
A character that belongs to one of the set of digits 0 through 9.

Numeric Edited Item
A numeric character which is in such a form that it may be used in a printed output. It may consist of decimal digits 0 through 9, the decimal point, commas, the dollar sign, etc., as the programmer wishes.

Numeric Item
A data item whose description restricts its contents to a value represented by characters chosen from the digits 0 through 9; if signed, the item may also contain a +, −, or other representation of an operational sign.

Numeric Literal
A literal composed of one or more numeric characters that also may contain either a decimal point, or an algebraic sign, or both. The decimal point must not be the rightmost character. The algebraic sign, if present, must be the leftmost character.

OBJECT-COMPUTER
The name of an Environment Division paragraph in which the computer environment, within which the object program is executed, is described.

Object of Entry
A set of operands and reserved words, within a Data Division entry, which immediately follows the subject of the entry.

Object Program
A set or group of executable machine-language instructions and other material designed to interact with data to provide problem solutions. In this context, an object program is generally the machine language result of the operation of a COBOL compiler on a source program. Where there is no danger of ambiguity, the word "program" alone may be used in place of the phrase "object program."

Object Time
The time at which an object program is executed.

Open Mode
The state of a file after execution of an OPEN statement for that file and before the execution of a CLOSE statement for that file. The particular mode is specified in the OPEN statement as either INPUT, OUTPUT, I-O, or EXTEND.

Operand
Whereas the general definition of operand is "that component which is operated upon," for the purposes of this text, any lowercase word (or words) that appears in a statement or entry format may be considered to be an operand and, as such, is an implied reference to the data indicated by the operand.

Operational Sign
An algebraic sign, associated with a numeric data item or a numeric literal, to indicate whether its value is positive or negative.

Optional Word
A reserved word that is included in a specific format only to improve the readability of the language, and whose presence is optional to the user when the format in which the word appears is used in a source program.

Output File
A file that is opened in either the output mode or extend mode.

Output Mode
The state of a file after execution of an OPEN statement, with the OUTPUT or EXTEND phrase specified for that file and before the execution of a CLOSE statement for that file.

OUTPUT PROCEDURE
A set of statements to which control is given during execution of a SORT statement after the sort function is completed, or during execution of a MERGE statement after the merge function has selected the next record in merged order. Output procedures are optional; whether they are used or not depends upon the logic of the program.

Overflow Condition
In string manipulation, a condition that occurs when the sending area(s) contains untransferred characters after the receiving area(s) has been filled.

Overlay
The technique of repeatedly using the same areas of internal storage during different stages in processing a problem.

Page
A vertical division of a report representing a physical separation of report data; the separation being based on internal reporting requirements and/or external characteristics of the reporting medium.

Page Body
That part of the logical page in which lines can be written and/or spaced.

PAGE FOOTING
A report group that is presented at the end of a report page as determined by the Report Writer Control System.

PAGE HEADING
A report group that is presented at the beginning of a report page as determined by the Report Writer Control System.

Paragraph
A set of one or more COBOL sentences making up a logical processing entity. In the Procedure Division, a paragraph-name followed by a period and a space and by zero, one, or more sentences. In the Identification Division, a paragraph header followed by zero, one, or more entries.

Paragraph Header
A reserved word, followed by a period and a space, which indicates the beginning of a paragraph in the Identification and Environment Divisions. The permissible paragraph headers are:

In the Identification Division:
PROGRAM-ID.
AUTHOR.
INSTALLATION.
DATE-WRITTEN.
DATE-COMPILED.
SECURITY.
In the Environment Division:
SOURCE-COMPUTER.
OBJECT-COMPUTER.
SPECIAL-NAMES.
FILE-CONTROL.
I-O-CONTROL.

Paragraph-Name
A user-defined word that identifies and begins a paragraph in the Procedure Division.

Parameter
A variable that is given a specific value for a specific purpose or process. In COBOL, parameters are often used to pass data values between calling and called programs.

Phrase
A phrase is an ordered set of one or more consecutive COBOL character-strings that form a portion of either a COBOL procedural statement or a COBOL clause.

Physical Record
(See Block.)

Prime Record Key
A key whose contents uniquely identify a record within an indexed file.

Printable Group
A report group that contains at least one print line.

Printable Item
A data item, the extent and contents of which are specified by an elementary report entry. This elementary report entry contains a COLUMN NUMBER clause, a PICTURE clause, and a SOURCE, SUM, or VALUE clause.

Procedure
A paragraph or group of logically successive paragraphs, or a section or group of logically successive sections within the Procedure Division.

Procedure Division
One of the four main components of a COBOL program. The Procedure Division contains instructions for solving a problem. The Procedure Division may contain imperative statements, conditional statements, paragraphs, procedures, and sections.

Procedure-Name
A user-defined word that is used to name a paragraph or section in the Procedure Division. It consists of a paragraph-name (which may be qualified) or a section-name.

Program-Name
A user-defined word in the Identification Division that identifies a COBOL source program.

Pseudo-Text
A sequence of character-strings and/or separators bounded by, but not including, pseudo-text delimiters.

Pseudo-Text Delimiter
Two contiguous equal sign (=) characters used to delimit pseudo-text.

Punctuation Character
A character that belongs to the following:

Character	*Meaning*
,	comma
;	semicolon
.	period
"	quotation mark
(	left parenthesis
)	right parenthesis
	space
=	equal sign

Qualified Data-Name
An identifier that is composed of a data-name followed by one or more sets of either of the connectives OF and IN followed by a data-name qualifier.

Qualifier
A data-name that is used in a reference together with another data name at a lower level in the same hierarchy. A section-name that is used in a reference together with a paragraph-name specified in that section. A library-name that is used in a reference together with a text-name associated with that library.

Random Access
An access mode in which specific logical records are obtained from or placed into a mass storage file in a nonsequential manner. An access mode in which the program-specified value of a key data item identifies the logical record that is obtained from, deleted from or placed into a relative or indexed file.

Record
(See Logical Record.)

Record Area
A storage area allocated for the purpose of processing the record described in a record description entry in the File Section.

Record Description
(See Record Description Entry)

Record Description Entry
The total set of data description entries associated with a particular record.

Record Key
A key, either the prime record key or an alternate record key, whose contents identify a record within an indexed file.

Record-Name
A user-defined word that names a record described in a record description entry in the Data Division.

Reel
A module of external storage associated with a tape device.

Reference Format
A format that provides a standard method for describing COBOL source programs.

Relation
(See Relational Operator)

Relational Operator
A reserved word, a relation character, a group of consecutive reserved words, or

a group of consecutive reserved words and relation characters used in the construction of a relation condition. The permissible operators and their meanings are:

Relational Operator	*Meaning*
IS [NOT] GREATER THAN IS [NOT] >	Greater than or not greater than
IS [NOT] LESS THAN IS [NOT] <	Less than or not less than
IS [NOT] EQUAL TO IS [NOT] =	Equal to or not equal to

Relation Character
A character that expresses a relationship between two operands. The following are COBOL relation characters:

Character	*Meaning*
>	greater than
<	less than
=	equal to

Relation Condition
The proposition, for which a truth value can be determined, that the value of an arithmetic expression or data item has a specific relationship to the value of another arithmetic expression or data item. (See Relational Operator.)

Relative File
A file with relative organization.

Relative Key
A key whose contents identify a logical record in a relative file.

Relative Organization
The permanent logical file structure in which each record is uniquely identified by an integer value greater than zero, which specifies the record's ordinal position in the file.

Report
A presentation of a set of processed data described in a Report File.

Report Clause
A clause, in the Report Section of the Data Division, that appears in a report description entry or a report group description entry.

Report Description Entry
An entry in the Report Section of the Data Division that names and describes the format of a report to be produced. It is composed of the level indicator RD, followed by a report name, followed by a set of report clauses as required.

Report File
An output file whose file description entry contains a REPORT clause. The contents of a report file consist of records that are written under control of the Report Writer Control System. The file consists of a collection of records that can be used to print a report in the desired format.

REPORT FOOTING
A report group that is presented only at the end of a report.

Report Group
In the Report Section of the Data Division, an 01 level-number entry and its subordinate entries. A set of related data that makes up a logical entity in a report.

Report Group Description Entry
An entry in the Report Section of the Data Division that is composed of the level-number 01, the optional data-name, a TYPE clause, and an optional set of report clauses.

REPORT HEADING
A report group that is presented only at the beginning of a report.

Report Line
A division of a page representing one row of horizontal character positions. Each character position of a report line is aligned vertically beneath the corresponding character position of the report line above it. Report lines are numbered from 1, by 1, starting at the top of the page.

Report-Name
A user-defined word that names a report described in a report description entry within the Report Section of the Data Division.

Report Section
The section of the Data Division that contains one or more report description entries and their associated report group description entries.

Report Writer Control System (RWCS)
An object time control system, provided by the implementor, that accomplishes the construction of reports.

Report Writer Logical Record
A record that consists of the Report Writer print line and associated control information necessary for its selection and vertical positioning.

Reserved Word
A COBOL word specified in the list of words that may be used in COBOL source programs, but which must not appear in the programs as user-defined words or system-names.

Routine
A set of statements in a program that causes the computer to perform an operation or series of related operations.

Routine-Name
A user-defined word that identifies a procedure written in a language other than COBOL.

Run Unit
A set of one or more object programs which function, at object time, as a unit to provide problem solutions. The compiler considers a run unit to be the highest level calling program plus all called subprograms.

RWCS
(See Report Writer Control System)

Section
A set of zero, one, or more paragraphs or entries, called a section body, the first of which is preceded by a section header. Each section consists of the section header and the related section body. A logically related sequence of one or more paragraphs. A section must always be named.

Section Header
A combination of words followed by a period and a space that indicates the beginning of a section in the Environment, Data and Procedure Divisions.

In the Environment and Data Divisions, a section header is composed of reserved words followed by a period and a space. The permissible section headers are:

In the Environment Division:
CONFIGURATION SECTION.
INPUT-OUTPUT SECTION.
In the Data Division:
FILE SECTION.
WORKING-STORAGE SECTION.
LINKAGE SECTION.

COMMUNICATION SECTION.
REPORT SECTION.
In the Procedure Division, a section header is composed of a section-name, followed by the reserved word SECTION, followed by a segment-number (optional), followed by a period and a space.

Section-Name
A user-defined word that names a section in the Procedure Division.

Segment-Number
A user-defined word which classifies sections in the Procedure Division for purposes of segmentation. Segment-numbers may contain only the characters 0, 1, ... , 9. A segment-number may be expressed either as a one- or two-digit number.

Sentence
A sequence of one or more statements, the last of which is terminated by a period followed by a space.

Separator
A punctuation character used to delimit character-strings.

Sequential Access
An access mode in which logical records are obtained from or placed into a file in a consecutive predecessor-to-successor logical record sequence determined by the order of the records in the file. The order of the records is established by the programmer when creating the file.

Sequential File
A file with sequential organization.

Sequential Organization
The permanent logical file structure in which a record is identified by a predecessor-successor relationship when the record is placed into the file.

Sequential Processing
The processing of logical records in the order in which records are accessed.

77-Level-Description-Entry
A data description entry that describes a noncontiguous data item with the level-number 77.

Sign Condition
The proposition, for which a truth value can be determined, that the algebraic value of a data item or an arithmetic expression is either less than, greater than, or equal to zero.

Simple Condition
An expression that can have two values, and causes the object program to select between alternate paths of control, depending on the value found. Any single condition can be chosen from the set:

relation condition
class condition
condition-name condition
switch-status condition
sign condition
(simple condition)

Sort File
A collection of records to be sorted by a SORT statement. The sort file is created and can be used by the sort function only.

Sort-Merge File Description Entry
An entry in the File Section of the Data Division that is composed of the level indicator SD, followed by a file-name, and then followed by a set of file clauses as required.

SOURCE-COMPUTER
The name of an Environment Division paragraph in which the computer environment, within which the source program is compiled, is described.

Source Item
An identifier designated by a SOURCE clause that provides the value of a printable item.

Source Program
Although it is recognized that a source program may be represented by other forms and symbols, in this text it always refers to a syntactically correct set of COBOL statements beginning with an Identification Division and ending with the end of the Procedure Division. In the contexts where there is no danger of ambiguity, the word "program" alone may be used in place of the phrase "source program."

Special Character
A character that is neither numeric nor alphabetic. A character that belongs to the following set:

Character	*Meaning*
+	plus sign
−	minus sign
*	asterisk
/	stroke (virgule, slash)
=	equal sign
$	currency sign
,	comma (decimal point)
;	semicolon
.	period (decimal point)
"	quotation mark
(	left parenthesis
)	right parenthesis
>	greater than symbol
<	less than symbol

Special-Character Word
A reserved word which is an arithmetic operator or a relation character.

SPECIAL-NAMES
The name of an Environment Division paragraph in which implementor-names are related to user-specified mnemonic-names.

Special Registers
Compiler-generated storage areas whose primary use is to store information produced in conjunction with the user of specific COBOL features.

Standard Data Format
The concept used in describing the characteristics of data in a COBOL Data Division under which the characteristics or properties of the data are expressed in a form oriented to the appearance of the data on a printed page of infinite length and breadth, rather than a form oriented to the manner in which the data is stored internally in the computer, or on a particular external medium.

Statement
A syntactically valid combination of words and symbols written in the Procedure Division and beginning with a verb. A statement combines COBOL reserved words and user-defined operands.

Subject of Entry
An operand or reserved word that appears immediately following the level indicator or the level-number in a Data Division entry. It serves to reference the entry.

Subprogram
(See Called Program.)

Subscript
An integer whose value identifies a particular element in a table.

Subscripted Data-Name
An identifier that is composed of a data-name followed by one or more subscripts enclosed in parentheses.

Sum Counter
A signed numeric data item established by a SUM clause in the Report Section of the Data Division. The sum counter is used by the Report Writer Control System to contain the results of designated summing operations that take place during the production of a report.

Switch-Status Condition
The proposition, for which a truth value can be determined, that an implementor-defined switch, capable of being set to an on or off status, has been set to a specified status.

System-Name
A COBOL word that is used to communicate with the operating environment.

Table
A set of logically consecutive items of data that are defined in the Data Division by means of the OCCURS clause. A collection and arrangement of data in a fixed form for ready reference. Such a collection follows some logical order, expressing particular values (functions) corresponding to other values (arguments) by which they are referenced.

Table Element
A data item belongs to the set of repeated items comprising a table. An argument together with its corresponding function(s) make up a table element.

Test-Condition
A statement that, taken as a whole, may be either true or false, depending on the circumstances existing at the time the expression is evaluated.

Text-Name
A user-defined word that identifies library text.

Text-Word
Any character-string or separator, except space, in a COBOL library or in pseudo-text.

Truth Value
The representation of the result of the evaluation of a condition in terms of one of two values, true or false.

Unary Operator
A plus (+) or a minus (−) sign, which precedes a variable or a left parenthesis in an arithmetic expression and which has the effect of multiplying the expression of +1 or −1 respectively.

Unit
A module of mass storage the dimensions of which are determined by each implementor.

User-Defined Word
A COBOL word that must be supplied by the user to satisfy the format of a clause or statement.

Variable
A data item whose value may be changed by execution of the object program. A variable used in an arithmetic expression must be a numeric elementary item.

Verb

A word that expresses an action to be taken by a COBOL compiler or object program.

Volume
A module of external storage. For tape devices, it is a reel; for mass storage devices, it is a unit.

Word
A character-string of not more than thirty characters, which forms a user-defined word, a system-name, or a reserved word. The string may be chosen from the following: the letters A through Z, the digits 0 through 9, and the hyphen (-). The hyphen may not appear as either the first or last character.

Working-Storage Section
The section of the Data Division that describes working storage data items, composed of either noncontiguous items or of working-storage records or of both.

Index